Food Chemistry

Springer

Berlin
Heidelberg
New York
Barcelona
Hong Kong
London
Milan
Paris
Singapore
Tokyo

H.-D. Belitz · W. Grosch

Food Chemistry

Translation from the Fourth German Edition
by M. M. Burghagen, D. Hadziyev, P. Hessel, S. Jordan
and C. Sprinz

Second Edition with 460 Figures and 531 Tables

 Springer

Professor Dr.-Ing. H.-D. Belitz †

Professor Dr.-Ing. W. Grosch
Institut für Lebensmittelchemie der Technischen Universität München
and Deutsche Forschungsanstalt für Lebensmittelchemie, München
Lichtenbergstraße 4
D-85748 Garching, FRG

Translators:

First edition
Professor Dr. D. Hadziyev

Second edition:
Peter Hessel (chapters 0 and 1)
Christiane Sprinz (chapter 2)
Dr. Sabine Jordan (chapter 3)
Dr. Margaret Burghagen (chapters 4–23)

ISBN 3-540-64692-2 2nd Ed. (soft cover) Springer-Verlag Berlin Heidelberg New York
ISBN 3-540-64704-X 2nd Ed. (hard cover) Springer-Verlag Berlin Heidelberg New York
ISBN 3-540-15043-9 1st Ed. Springer-Verlag Berlin Heidelberg New York

Library of Congress Cataloging-in-Publication Data

Belitz, H.-D, (Hans-Dieter)
[Lehrbuch der Lebensmittelchemie. English]
Food chemistry / H.-D. Belitz, W. Grosch. – 2nd ed. / translation from the fourth German edition
by M. Burghagen ... [et al.]
Includes bibliographical references and index.
ISBN 3-540-64704-X (hard). – ISBN 3-540-64692-2 (soft)
1. Food-Analysis. I. Grosch, W. (Werner) II. Title.
TX545.B3513 1999   664'.07–dc21   99-10777

Springer-Verlag ist ein Unternehmen der Fachverlagsgruppe BertelsmannSpringer.
© Springer-Verlag Berlin Heidelberg 1987 and 1999
Printed in Germany

Cover design: de'blik, Berlin
Typesetting: Fotosatz-Service Köhler GmbH, 97084 Würzburg
Production: ProduServ GmbH, Verlagsservice, Berlin
SPIN: 10768171   52/3012 – 5 4 3 2 1 – Printed on acid-free paper

In Memoriam
Joseph Schormüller (1903–1974)
Professor of Food Chemistry at the
Technische Universität Berlin

Preface to the Second English Edition

The second edition of "Food Chemistry" is a translation of the fourth German edition of this textbook. The text has been corrected only in a few places, e.g., most of the production data are presented for the year 1996.

The preparation of this edition was greatly delayed due to the deaths of Professor Dr. H.-D. Belitz in March 1993 and of Professor Dr. D. Hadziyev, who translated the first edition, in July 1995. H.-D. Belitz worked on the preparation of the second edition.

Dr. Margaret Burghagen translated most of the extensive changes incorporated into this new edition and revised the entire text. I am greatly indebted to her for her excellent work. It was a pleasure to work with her.

I gratefully acknowledge the help of my colleagues who made valuable criticisms and contributed to the improvement of the text. I particularly thank Dr. M.C. Kühn, Holland.

I would also like to thank Mrs. R. Jauker for assistance in completing the manuscript and for proofreading and my son B. Grosch for assistance in preparing the index.

Garching, January 1999   W. Grosch

Preface to the First English Edition

The two German editions of the "Lehrbuch für Lebensmittelchemie" were so well accepted not only as a university textbook, but also as a first comprehensive source of information for people in science, industry, official food control and administration, that the publishing house, Springer-Verlag (Heidelberg), decided to edit an English version.

The first English edition is actually the second German edition which was revised for this purpose.

We are specially thankful to our colleague Prof. Dr. D. Hadziyev for the translation of the book.

Garching, December 1986   H.-D. Belitz, W. Grosch

Preface to the Fourth German Edition

For the fourth edition, all the chapters have been carefully revised and updated. Furthermore, one chapter and various sections have been added and others have been reworked. Some of the changes that have been made are:

- The relationships between the water content and reactivity of food have been presented in terms of the kinetics of phase transitions.
- The formation of mutagenic compounds from amino acids by thermal reactions has been discussed in greater detail.
- Technologically important properties of proteins, such as the formation of foams, gels and emulsions, have been summarized.
- The chapter on enzymes contains revised sections on the relationships between structure and catalytic activity and on the temperature dependency of the reaction rate.
- In the chapter on carbohydrates, the *Maillard* reaction and the section on starch have been discussed in greater detail and a section on halodeoxy sugars has been added.
- The chapter on aroma substances and the sections on the aroma substances of individual foods have been completely revised.
- The chapter on food additives has been modified to include new sweeteners (suosan, guanidines, alitame, sucralose) and a more detailed presentation of emulsifiers.
- Dioxins have been included in the chapter on food contamination.
- The chapters on individual foods contain more detailed presentations of analysis (meat, fats, fruit), various constituents (cereal proteins, legume proteins, proteinase inhibitors, phenolic compounds) and of some technical processes (micelle and gel formation in the case of milk and baking process and ageing of baked products).
- A short chapter on drinking water, mineral and table water has been added.
- The references included in each chapter have been updated.

We would like to thank all our readers who have helped us with their constructive criticism in the preparation of this manuscript. For the completion and proofreading of the manuscript, we are indebted to Mrs. R. Berger, Mrs. Ch. Hoffmann, Mrs. I. Hofmeier, Mrs. G. Nominacher-Ullrich and Mrs. K. Wüst. We are very grateful to Springer Verlag for their consideration of our wishes and for the pleasant cooperation.

Garching, June 1992   H.-D. Belitz, W. Grosch

Preface to the First German Edition

The very rapid development of food chemistry and technology over the last two decades, which is due to a remarkable increase in the analytical and manufacturing possibilities, makes the complete lack of a comprehensive, teaching or reference text particularly noticeable. It is hoped that this textbook of food chemistry will help to fill this gap. In writing this volume we were able to draw on our experience from the lectures which we have given, covering various scientific subjects, over the past fifteen years at the Technical University of Munich.

Since a separate treatment of the important food constituents (proteins, lipids, carbohydrates, flavor compounds, etc.,) and of the important food groups (milk, meat, eggs, cereals, fruits, vegetables, etc.,) has proved successful in our lectures, the subject matter is also organized in the same way in this book.

Compounds which are found only in particular foods are discussed where they play a distinctive role while food additives and contaminants are treated in their own chapters. The physical and chemical properties of the important constituents of foods are discussed in detail where these form the basis for understanding either the reactions which occur, or can be expected to occur, during the production, processing, storage and handling of foods or the methods used in analyzing them. An attempt has also been made to clarify the relationship between the structure and properties at the level of individual food constituents and at the level of the whole food system.

The book focuses on the chemistry of foodstuffs and does not consider national or international food regulations. We have also omitted a broader discussion of aspects related to the nutritional value, the processing and the toxicology of foods. All of these are an essential part of the training of a food chemist but, because of the extent of the subject matter and the consequent specialization, must today be the subject of separate books. Nevertheless, for all important foods we have included brief discussions of manufacturing processes and their parameters since these are closely related to the chemical reactions occurring in foods.

Commodity and production data of importance to food chemists are mainly given in tabular form. Each chapter includes some references which are not intended to form an exhaustive list. No preference or judgement should be inferred from the choice of references; they are given simply to encourage further reading. Additional literature of a more general nature is given at the end of the book.

This book is primarily aimed both at students of food and general chemistry but also at those students of other disciplines who are required or choose to study food chemistry as a supplementary subject. We also hope that this comprehensive text will prove useful to both food chemists and chemists who have completed their formal education.

We thank sincerely Mrs. A. Mödl (food chemist), Mrs. R. Berger, Mrs. I. Hofmeier, Mrs. E. Hortig, Mrs. F. Lynen and Mrs. K. Wüst for their help during the preparation of the manuscript and its proofreading. We are very grateful to Springer Verlag for their consideration of our wishes and for the agreeable cooperation.

Garching, July 1982   H.-D. Belitz, W. Grosch

Table of Contents

5   Aroma Substances 319

Introduction

Foods are materials which, in their naturally occurring, processed or cooked forms, are consumed by humans as nourishment and for enjoyment.

The terms "nourishment" and "enjoyment" introduce two important properties of foods: the nutritional value and the hedonic value. The former is relatively easy to quantify since all the important nutrients are known and their effects are defined. Furthermore, there are only a limited number of nutrients. Defining the hedonic value of a food is more difficult because such a definition must take into account all those properties of a food, such as visual appeal, smell, taste and texture, which interact with the senses. These properties can be influenced by a large number of compounds which in part have not even been identified. Besides their nutritional and hedonic values, foods are increasingly being judged according to properties which determine their handling. Thus, the term "convenience foods". An obvious additional requirement of a food is that it be free from toxic materials.

Food chemistry is involved not only in elucidating the composition of the raw materials and end-products, but also with the changes which occur in food during its production, processing, storage and cooking. The highly complex nature of food results in a multitude of desired and undesired reactions which are controlled by a variety of parameters. To gain a meaningful insight into these reactions, it is necessary to break up the food into model systems. Thus, starting from compositional analyses (detection, isolation and structural characterization of food constituents), the reactions of a single constituent or of a simple mixture can be followed. Subsequently, an investigation of a food in which an individual reaction dominates can be made. Inherently, such a study starts with a given compound and is thus not restricted to any one food or group of foods. Such general studies of reactions involving food constituents are supplemented by special investigations which focus on chemical processes in individual foods. Research of this kind, is from the very beginning, closely associated with economic and technological aspects and contributes, by understanding the basics of the chemical processes occurring in foods, both to resolving specific technical problems and to process optimization.

A comprehensive evaluation of foods requires that analytical techniques keep pace with the available technology. As a result a major objective in food chemistry is concerned with the application and continual development of analytical methods. This aspect is particularly important when following possible contamination of foods with substances which may involve a health risk. Thus, there are close links with environmental problems.

Food chemistry research is aimed at establishing objective standards by which the criteria mentioned above – nutritional value, hedonic value, absence of toxic compounds and convenience – can be evaluated. These are a prerequisite for the industrial production of high quality food in bulk amounts.

This brief outline thus indicates that food chemistry, unlike other branches of chemistry which are concerned either with particular classes of compounds or with particular methods, is a subject which, both in terms of the actual chemistry and the methods involved, has a very broad field to cover.

0 Water

0.1 Foreword

Water (moisture) is the predominant constituent in many foods (Table 0.1). As a medium water supports chemical reactions, and it is a direct reactant in hydrolytic processes. Therefore, removal of water from food or binding it by increasing the concentration of common salt or sugar retards many reactions and inhibits the growth of microorganisms, thus improving the shelf lives of a number of foods. Through physical interaction with proteins, polysaccharides, lipids and salts, water contributes significantly to the texture of food.

Table 0.1. Moisture content of some foods

Food	Moisture content (weight-%)	Food	Moisture content (weight-%)
Meat	65–75	Cereal flour	12–14
Milk	87	Coffee beans,	
Fruits,		roasted	5
vegetables	70–90	Milk powder	4
Bread	35	Edible oil	0
Honey	20		
Butter,			
margarine	16–18		

The function of water is better understood when its structure and its state in a food system are clarified. Special aspects of binding of water by individual food constituents (cf. 1.4.3.3, 3.5.2 and 4.4.3) and meat (cf. 12.5) are discussed in the indicated sections.

0.2 Structure

0.2.1 Water Molecule

The six valence electrons of oxygen in a water molecule are hybridized to four sp³ orbitals

that are elongated to the corners of a somewhat deformed, imaginary tetrahedron (Fig. 0.1). Two hybrid orbitals form O–H covalent bonds with a bond angle of 105° for H–O–H, whereas the other 2 orbitals hold the nonbonding electron pairs (n-electrons). The O–H covalent bonds, due to the highly electronegative oxygen, have a partial (40%) ionic character.

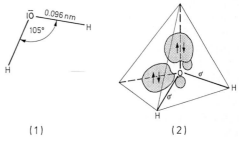

Fig. 0.1. Water. **1** Molecular geometry, **2** orbital model

Each water molecule is tetrahedrally coordinated with four other water molecules through hydrogen bonds. The two unshared electron pairs (n-electrons or sp³ orbitals) of oxygen act as H-bond acceptor sites and the H–O bonding orbitals act as hydrogen bond donor sites (Fig. 0.2). The dissociation energy of this hydrogen bond is about 25 kJ mole^{-1}.

The simultaneous presence of two acceptor sites and two donor sites in water permits asso-

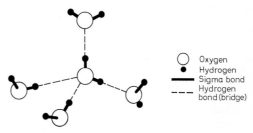

Fig. 0.2. Tetrahedral coordination of water molecules

ciation in a three-dimensional network stabilized by H-bridges. This structure which explains the special physical properties of water is unusual for other small molecules. For example, alcohols and compounds with iso-electric dipoles similar to those of water, such as HF or NH_3, form only linear or two-dimensional associations.

The above mentioned polarization of H−O bonds is transferred via hydrogen bonds and extends over several bonds. Therefore, the dipole moment of a complex consisting of increasing numbers of water molecules (multi-molecular dipole) is higher as more molecules become associated and is certainly much higher than the dipole moment of a single molecule. Thus, the dielectric constant of water is high and surpasses the value, which can be calculated on the basis of the dipole moment of a single molecule. Proton transport takes place along the H-bridges. It is actually the jump of a proton from one water molecule to a neighboring water molecule. Regardless of whether the proton is derived from dissociation of water or originates from an acid, it will sink into the unshared electron pair orbitals of water:

$$(0.1)$$

In this way a hydrated $H_3O^\oplus$ ion is formed with an exceptionally strong hydrogen bond (dissociation energy about $100 \ kJ \ mol^{-1}$). A similar mechanism is valid in transport of $OH^\ominus$ ions, which also occurs along the hydrogen bridges:

$$(0.2)$$

Since the transition of a proton from one oxygen to the next occurs extremely rapidly ($\nu > 10^{12} \ s^{-1}$), proton mobility surpasses the mobilities of all other ions by a factor of 4−5, except for the stepwise movement of $OH^\ominus$ within the structure; its rate of exchange is only 40% less than that of a proton.

H-bridges in ice extend to a larger sphere than in water (see the following section). The mobility of protons in ice is higher than in water by a factor of 100.

0.2.2 Liquid Water and Ice

The arrangements of water molecules in "liquid water" and in ice are still under intensive investigation. The outlined hypotheses agree with existing data and are generally accepted.

Due to the pronounced tendency of water molecules to associate through H-bridges, liquid water and ice are highly structured. They differ in the distance between molecules, coordination number and time-range order (duration of stability). Stable ice-I is formed at 0°C and 1 atm pressure. It is one of nine known crystalline polymorphic structures, each of which is stable in a certain temperature and pressure range. The coordination number in ice-I is four, the O−H···O (nearst neighbor) distance is 0.276 nm (0°C) and the H-atom between neighboring oxygens is 0.101 nm from the oxygen to which it is bound covalently and 0.175 nm from the oxygen to which it is bound by a hydrogen bridge. Five water molecules, forming a tetrahedron, are loosely packed and kept together mostly through H-bridges.

When ice melts and the resultant water is heated (Table 0.2), both the coordination number and the distance between the nearest neighbors increase. These changes have opposite influences on the density. An increase in coordination number (i.e. the number of water molecules arranged in an orderly fashion around each water molecule) increases the density, whereas an increase in distance between nearest neighbors decreases the density. The

Table 0.2. Coordination number and distance between two water molecules

	Coordination number	O−H···O Distance
Ice (0 °C)	4	0.276 nm
Water (1.5 °C)	4.4	0.290 nm
Water (83 °C)	4.9	0.305 nm

effect of increasing coordination number is predominant during a temperature increase from 0 to 4 °C. As a consequence, water has an unusual property: its density in the liquid state at 0 °C (0.9998 g cm^{-3}) is higher than in the solid state (ice-I, $\varrho = 0.9168$ g cm^{-3}). Water is a structured liquid with a short time-range order. The water molecules, through H-bridges, form short-lived polygonal structures which are rapidly cleaved and then reestablished giving a dynamic equilibrium. Such fluctuations explain the lower viscosity of water, which otherwise could not be explained if H-bridges were rigid.

The hydrogen-bound water structure is changed by solubilization of salts or molecules with polar and/or hydrophobic groups. In salt solutions the n-electrons occupy the free orbitals of the cations, forming "aqua complexes". Other water molecules then coordinate through H-bridges, forming a hydration shell around the cation and disrupting the natural structure of water.

Hydration shells are formed by anions through ion-dipole interaction and by polar groups through dipole-dipole interaction or H-bridges, again contributing to the disruption of the structured state of water.

Aliphatic groups which can fix the water molecules by dispersion forces are no less disruptive. A minimum of free enthalpy will be attained when an ice-like water structure is arranged around a hydrophobic group (tetrahedral-four-coordination). Such ice-like hydration shells around aliphatic groups contribute, for example, to stabilization of a protein, helping the protein to acquire its most thermodynamically favorable conformation in water.

The highly structured, three-dimensional hydrogen bonding state of ice and water is reflected in many of their unusual properties.

Table 0.3. Some physical constants of water, methanol and dimethyl ether

	F_p (°C)	K_p (°C)
H_2O	0.0	100.0
CH_3OH	−98	64.7
CH_3OCH_3	−138	−23

Additional energy is required to break the structured state. This accounts for water having substantially higher melting and boiling points and heats of fusion and vaporization than methanol or dimethyl ether (cf. Table 0.3). Methanol has only one hydrogen donor site, while dimethyl ether has none but does have a hydrogen bond acceptor site; neither is sufficient to form a structured network as found in water.

0.3 Effect on Storage Life

Drying and/or storage at low temperatures are among the oldest methods for the preservation of food with high water contents. Modern food technology tries to optimize these methods. A product should be dried and/or frozen only long enough to ensure wholesome quality for a certain period of time.

Naturally, drying and/or freezing must be optimized for each product individually. It is therefore necessary to know the effect of water on storage life before suitable conditions can be selected.

0.3.1 Water Activity

In 1952, W.J. Scott came to the conclusion that the storage quality of food does not depend on the water content, but on water activity (a_w), which is defined as follows:

$$a_w = P/P_0 = ERH/100 \qquad (0.3)$$

P = partial vapor pressure of food moisture at temperature T

P_0 = saturation vapor pressure of pure water at T

ERH = equilibrium relative humidity at T.

The relationship between water content and water activity is indicated by the sorption isotherm of a food (Fig. 0.3).

At a low water content ($<50\%$), even minor changes in this parameter lead to major changes in water activity. For that reason, the sorption isotherm of a food with lower water content is shown with an expanded ordinate in Fig. 0.3b, as compared with Fig. 0.3a.

Figure 0.3b shows that the desorption isotherm, indicating the course of a drying pro-

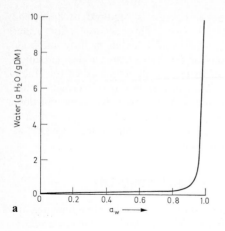

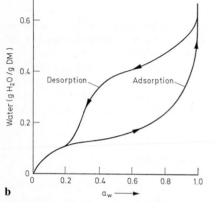

Fig. 0.3. Moisture sorption isotherm (according to *T. P. Labuza et al.*, 1970). **a** Food with high moisture content; **b** Food with low moisture content (DM: Dry matter)

cess, lies slightly above the adsorption isotherm pertaining to the storage of moisture-sensitive food. As a rule, the position of the hysteresis loop changes when adsorption and desorption are repeated with the same sample. The effect of water activity on processes that can influence food quality is presented in Fig. 0.4. Decreased water activity retards the growth of microorganisms, slows enzyme catalyzed reactions (particularly involving hydrolases; cf. 2.2.2.1) and, lastly, retards non-enzymatic browning. In contrast, the rate of lipid autoxidation increases in dried food systems (cf. 3.7.2.1.4).

Foods with a_w values between 0.6 and 0.9 (examples in Table 0.4) are known as "intermediate moisture foods" (IMF). These foods are largely protected against microbial spoilage.

Table 0.4. Water activity of some food

Food	a_w	Food	a_w
Leberwurst	0.96	Marmalades	0.82−0.94
Salami	0.82−0.85	Honey	0.75
Dried fruits	0.72−0.80		

One of the options for decreasing water activity, and thus improving the shelf life of food, is to use additives with high water binding capacities (humectants). Table 0.5 shows that, in addition to common salt, glycerol, sorbitol and sucrose have potential as humectants. How-

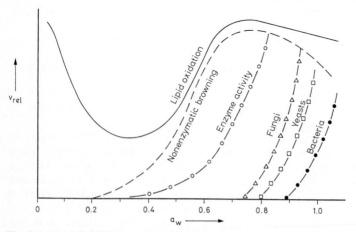

Fig. 0.4. Food shelf life (storage stability) as a function of water activity (according to *T. P. Labuza*, 1971)

Table 0.5. Moisture content of some food or food ingredients at a water activity of 0.8

	Moisture content (%)		Moisture content (%)
Peas	16	Glycerol	108
Casein	19	Sorbitol	67
Starch		Saccharose	56
(potato)	20	Sodium chloride	332

ever, they are also sweeteners and would be objectionable from a consumer standpoint in many foods in the concentrations required to regulate water activity.

0.3.2 Water Activity as an Indicator

Water activity is only of limited use as an indicator for the storage life of foods with a low water content, since water activity indicates a state that applies only to ideal, i.e. very dilute solutions that are at a thermodynamic equilibrium. However, foods with a low water content are non-ideal systems whose metastable (fresh) state should be preserved for as long as possible. During storage, such foods do not change thermodynamically, but according to kinetic principles. A new concept based on phase transition, which takes into account the change in physical properties of foods during contact between water and hydrophilic ingredients, is better suited to the prediction of storage life. This will be briefly discussed in the following sections (0.3.3−0.3.5).

0.3.3 Phase Transition of Foods Containing Water

The physical state of metastable foods depends on their composition, on temperature and on storage time. For example, depending on the temperature, the phases could be glassy, rubbery or highly viscous. The kinetics of phase transitions can be measured by means of differential scanning calorimetry (DSC), producing a thermogram that shows temperature T_g as the characteristic value for the transition from glassy to rubbery (plastic). Foods be-

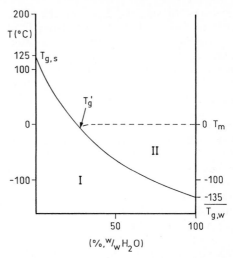

Fig. 0.5. State diagram, showing the approximate T_g temperatures as a function of mass fraction, for a gelatinized starch-water system (according to *Van den Berg*, 1986).
States: I = glassy; II = rubbery;
$T_{g,s}$ and $T_{g,w}$= phase transition temperatures of dehydrated starch and water; T_m = melting point (ice).

come plastic when their hydrophilic components are hydrated. Thus the water content affects the temperature T_g, for example in the case of gelatinized starch (Fig. 0.5).
Table 0.6 shows the T_g of some mono- and oligosaccharides and the difference between melting points T_m.
During the cooling of an aqueous solution below the freezing point, part of the water

Table 0.6. Phase transition temperature T_g and melting point T_m of mono- and oligosaccharides

Compound	T_g	[°C]	T_m
Glycerol	− 93		18
Xylose	9.5		153
Ribose	− 10		87
Xylitol	− 18.5		94
Glucose	31		158
Fructose	100		124
Galactose	110		170
Mannose	30		139.5
Sorbitol	− 2		111
Sucrose	52		192
Maltose	43		129
Maltotriose	76		133.5

crystallizes, causing the dissolved substance to become enriched in the remaining fluid phase (unfrozen water). In the thermogram, temperature T_g' appears, at which the glassy phase of the concentrated solution turns into a rubber-like state. The position of T_g' ($-5\,°C$) on the T_g curve is shown by the example of gelatinized starch (Fig. 0.5); the quantity of unfrozen water W_g' at this temperature is 27% by weight. Table 0.7 lists the temperatures T_g' for aqueous solutions (20% by weight) of carbohydrates and proteins. In the case of oligosaccharides composed of three glucose molecules, maltotriose has the lowest T_g' value in comparison with panose and isomaltotriose. The reason is probably that in aqueous solution, the effective chain length of linear oligosaccharides is greater than that of branched compounds of the same molecular weight.

Table 0.7. T_g' and W_g' of aqueous solutions (20% by weight) of carbohydrates and proteins[a]

Substance	T_g'	W_g'
Glycerol	−65	0.85
Xylose	−48	0.45
Ribose	−47	0.49
Ribitol	−47	0.82
Glucose	−43	0.41
Fructose	−42	0.96
Galactose	−41.5	0.77
Sorbitol	−43.5	0.23
Sucrose	−32	0.56
Lactose	−28	0.69
Trehalose	−29.5	0.20
Raffinose	−26.5	0.70
Maltotriose	−23.5	0.45
Panose	−28	0.59
Isomaltotriose	−30.5	0.50
Potato starch (DE 10)	−8	
Potato starch (DE 2)	−5	
Hydroxyethylcellulose	−6.5	
Tapioca (DE 5)	−6	
Waxy corn (DE 0.5)	−4	
Gelatin	−13.5	0.46
Collagen, soluble	−15	0.71
Bovine serum albumin	−13	0.44
α-Casein	−12.5	0.61
Sodium caseinate	−10	0.64
Gluten	−5 to −10	0.07 to 0.41

[a] Phase transition temperature T_g' (°C) and water content W_g' (g per g of substance) of maximum freeze-concentrated glassy structure.

In the case of homologous series of oligo- and polysaccharides, T_g and T_g' increase with the molecular weight up to a certain limit (Fig. 0.6).

Table 0.8 lists the phase transition temperatures T_g' of some fruits and vegetables.

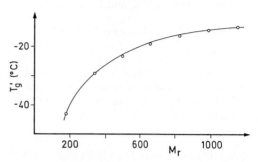

Fig. 0.6. Phase transition temperatures T_g' (aqueous solution, 20% by weight) of the homologous series glucose to maltoheptaose as a function of molecular weight M_r

Table 0.8. Phase transition temperature T_g' of some fruits and vegetables

Fruit/vegetable	T_g' (°C)
Strawberries	−33 to −41
Peaches	−36.5
Bananas	−35
Apples	−42
Tomatoes	−41.5
Peas (blanched, frozen)	−25
Carrots	−25.5
Broccoli, stalks	−26.5
Broccoli, flower buds	−11.5
Spinach (blanched, frozen)	−17
Potatoes	−11

0.3.4 WLF Equation

The viscosity of a food is extremely high at temperature T_g or T_g' (about 10^{13} Pa.s). As the temperature rises, the viscosity decreases, which means that processes leading to a drop in quality will accelerate. In the temperature range of T_g to about ($T_g + 100\,°C$), the change in viscosity does not follow the equation of *Arrhenius* (cf. 2.5.4.2), but a relationship formulated by *M. L. Williams, R. F. Landel* and *J. D. Ferry* (the WLF equation):

$$\log \frac{\eta}{\varrho T} \bigg/ \frac{\eta_g}{\varrho_g T_g} = -\frac{C_1(T-T_g)}{C_2+(T-T_g)}$$

Viscosity (η) and density (ϱ) at temperature T; viscosity (η_g) and density (ϱ_g) at phase transition temperature T_g; C_1 and C_2: constants.

According to the WLF equation, the rate of which in our example water crystallizes in ice cream at temperatures slightly above T'_g rises exponentially (Fig. 0.7). If the *Arrhenius* equation were to be valid, crystallization would accelerate linearly at a considerably slower rate after exceeding T'_g (Fig. 0.7).

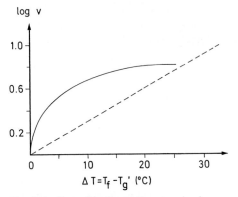

Fig. 0.7. Crystallization of water in ice cream (according to *Levine* and *Slade*, 1990).
v: crystallization velocity;
T_f: temperature in the freezer compartment;
T'_g: phase transition temperature.
The *Arrhenius* kinetics (– – – –) is shown for comparison.

0.3.5 Conclusion

In summary, we find that the rate of a food's chemical and enzymatic reactions as well as that of its physical processes becomes almost zero when the food is stored at the phase transition temperature of T_g or T'_g. Measures to improve storage life by increasing T_g or T'_g can include the extraction of water through drying and/or an immobilization of water by

Table 0.9. Unwanted chemical, enzymatic and physical processes in the production and storage of foods, depending on phase transition temperature T_g or T'_g and delayed by the addition of starch partial hydrolysates (lower DE value)

Process
1. Agglomeration and lumping of foods in the amorphous state
2. Recrystallization
3. Enzymatic reaction
4. Collapse of structure in case of freeze-dried products
5. Non-enzymatic browning

means of freezing, or by adding polysaccharides. Table 0.9 shows examples of how the drop in quality of certain foods can be considerably delayed when T_g or T'_g are increased by the addition of polysaccharides and approximated to the storage temperature.

0.4 Literature

Fennema, O. R., Water and ice. In: Principles of food science, Part I (Ed.: Fennema, O.R.), p. 13. New York: Marcel Dekker, Inc., 1976

Franks, F., Water, ice and solutions of simple molecules. In: Water relations of foods (Ed.: Duckworth, R. B.), p. 3. London: Academic Press, 1975

Hardman, T.M. (Ed.), Water and food quality. London: Elsevier Applied Science, 1989

Heiss, R., Haltbarkeit und Sorptionsverhalten wasserarmer Lebensmittel. Berlin: Springer-Verlag, 1968

Karel, M., Water activity and food preservation. In: Principles of food science, Part II (Editors: Karel, M., Fennema, O.R., Lund, D.B.), p. 237. New York: Marcel Dekker, Inc., 1975

Labuza, T.P., Kinetics of lipidoxidation in foods. Crit. Rev. Food Technol. 2, 355 (1971)

Polesello, A., Gianciacomo, R., Application of near infrared spectrophotometry to the nondestructive analysis of food: a review of experimental results. Crit. Rev. Food Sci. Nutr., 18, 203 (1982/83)

Slade, W., Levine, H., Beyond water activity – Recent advances based on an alternative approach to the assessment of food quality and safety. Crit. Rev. Food Sci. Nutr. 30, 115 (1991)

1 Amino Acids, Peptides, Proteins

1.1 Foreword

Amino acids, peptides and proteins are important constituents of food. They supply the required building blocks for protein biosynthesis. In addition, they directly contribute to the flavor of food and are precursors for aroma compounds and colors formed during thermal or enzymatic reactions in production, processing and storage of food. Other food constituents, e.g. carbohydrates, take part in such reactions. Proteins also contribute significantly to the physical properties of food through their ability to build or stabilize gels, foams, emulsions and fibrillar structures.

Table 1.0 provides information about the most important protein sources and their share in worldwide protein production. In addition to plants and animals, protein producers include algae (*Chlorella, Scenedesmus, Spirulina* spp.), yeasts and bacteria (single-cell proteins [SCP]). Among the C sources we use are glucose, molasses, starch, sulfite liquor, waste water, the higher n-alkanes, and methanol. Yeast of the genus *Candida* grow on paraffins, for example, and supply about 0.75 t of protein per t of carbohydrate. Bacteria of the species *Pseudomonas* in aqueous methanol produce about 0.3 t of protein per t of alcohol. Because of the high nucleic acid content of yeasts and bacteria (6–17% of dry weight), it is necessary to isolate protein from the cell mass. The future importance of single-cell proteins depends on price and on the technological properties.

In other raw materials, too, protein enrichment occurs for various reasons: protein concentration in the raw material may be too low for certain purposes, the sensory characteristics of the material (color, taste) may not be acceptable, or undesirable constituents may be present. Some products rich in protein also result from other processes, e.g., in oil and starch production. Enrichment results from the extraction of the constituents (protein concentrate) or from extraction and subsequent separation of protein from the solution, usually through thermal coagulation or isoelectric precipitation (protein isolate). Protein concentrates and protein isolates serve to enhance the nutritional value and to achieve the enhancement of the above mentioned physical properties of foods. They are added, sometimes after modification (cf. 1.4.6.1), to traditional foods, such as meat and cereal products, but they are also used in the production of novel

Table 1.0. Protein production (World, 1978/79)

Protein source	Protein quantity (million t/a)	Yield (kg/ha)	Price (US$/kg)
Grain	140	200– 700	1
Oilseeds	40	500–1200	0.8
Legumes[a]	8.6	200–1000	1
Vegetables[b]	8.3		7
Meat	18	50– 200	17
Fish	13		11
Milk	15	50– 400	12
Eggs	3		10

[a] Without oilseeds.   [b] Roots and tubers.

food items such as meat, fish and milk substitutes. Raw materials in which protein enrichment takes place include:

- Legumes such as soybeans (cf. 16.3.1.2.1) and broad beans;
- Wheat and corn, which provide gluten as a by-product of starch production;
- Potatoes; from the natural sap left over after starch production, proteins can be isolated by thermal coagulation;
- Eggs, which are processed into different whole egg, egg white and egg yolk products (cf. 11.4);
- Milk, which supplies casein (cf. 10.2.9) and whey protein (cf. 10.2.10);
- Fish, which supplies protein concentrates after fat extraction (cf. 13.1.6.13 and 1.4.6.3.2);
- Blood from slaughter animals, which is processed into blood meal, blood plasma concentrate (cf. 12.6.1.10) and globin isolate.
- Green plants grown for animal fodder, such as alfalfa, which are processed into leaf protein concentrates through the thermal coagulation of cell sap proteins.

1.2 Amino Acids

1.2.1 General Remarks

There are about 20 amino acids in a protein hydrolysate. With a few exceptions, their general structure is:

$$R—CH—COOH \atop |{}NH_2 \qquad (1.0)$$

In the simplest case, R=H (aminoacetic acid or glycine). In other amino acids, R is an aliphatic, aromatic or heterocyclic residue and may incorporate other functional groups. Table 1.1 shows the most important "building blocks" of proteins. There are about 200 amino acids found in nature (Fig. 1.1). Some of the more uncommon ones, which occur mostly in plants in free form, are covered in Chapter 17 on vegetables.

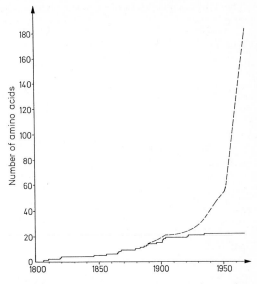

Fig. 1.1. Discovery of naturally occurring amino acids (according to *Meister*, 1965). – – – Amino acids, total; — protein constituents

1.2.2 Classification, Discovery and Occurrence

1.2.2.1 Classification

There are a number of ways of classifiying amino acids. Since their side chains are the deciding factors for intra- and intermolecular interactions in proteins, and hence, for protein properties, amino acids can be classified as:

- Amino acids with nonpolar, uncharged side chains: e.g., glycine, alanine, valine, leucine, isoleucine, proline, phenylalanine, tryptophan and methionine.
- Amino acids with uncharged, polar side chains: e.g., serine, threonine, cysteine, tyrosine, asparagine and glutamine.
- Amino acids with charged side chains: e.g., aspartic acid, glutamic acid, histidine, lysine and arginine.

Based on their nutritional/physiological roles, amino acids can be differentiated as:

- Essential amino acids:
  Valine, leucine, isoleucine, phenylalanine, tryptophan, methionine, threonine, histidine (essential for infants), lysine and arginine ("semi-essential").

Table 1.1. Amino acids (protein building blocks) with their corresponding three ond one letter symbols

Structure	Name	Structure	Name	Structure	Name
COOH, $H_2N—CH_2$	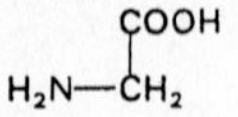 Glycine (Gly. G)	COOH, $H_2N—CH$, CH_2, CH_2, S, CH_3	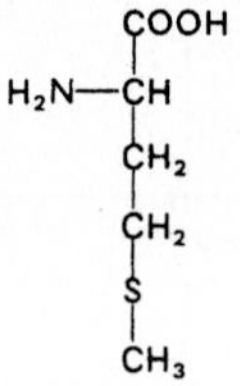 L–Methionine (Met. M)	COOH, $H_2N—CH$, CH_2, COOH	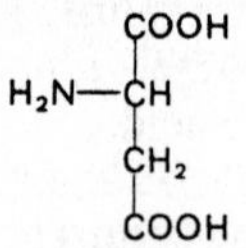 L–Aspartic acid (Asp. D)
COOH, $H_2N—CH$, CH_3	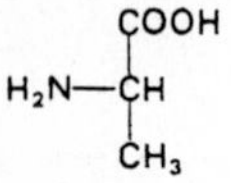 L–Alanine (Ala. A)	COOH, $H_2N—CH$, CH_2OH	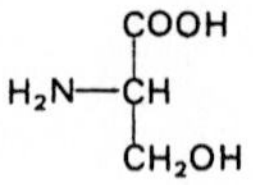 L–Serine (Ser. S)	COOH, $H_2N—CH$, CH_2, CH_2, COOH	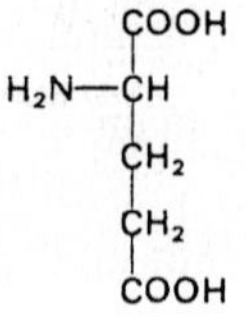 L–Glutamic acid (Glu. E)
COOH, $H_2N—CH$, CH, H_3C CH_3	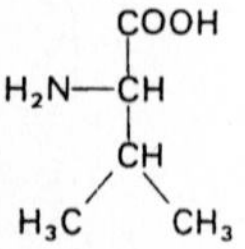 L–Valine (Val. V)	COOH, $H_2N—CH$, HC—OH, CH_3	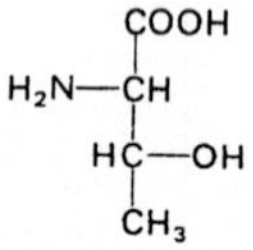 L–Threonine (Thr. T)	COOH, $H_2N—CH$, CH_2, CH_2, CH_2, CH_2NH_2	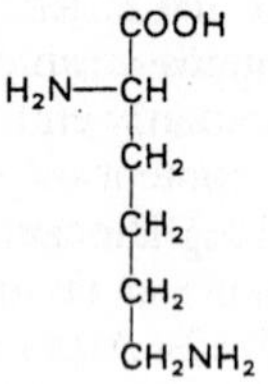 L–Lysine (Lys. K)
COOH, $H_2N—CH$, CH_2, CH, H_3C CH_3	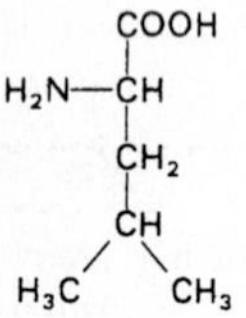 L–Leucine (Leu. L)	COOH, $H_2N—CH$, CH_2SH	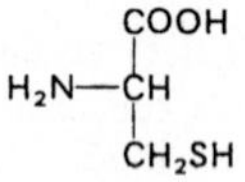 L–Cysteine (Cys. C)		
COOH, $H_2N—CH$, $H_3C—CH$, CH_2, CH_3	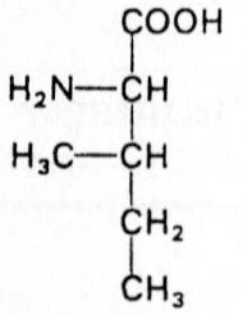 L–Isoleucine (Ile. I)	COOH, HN, OH	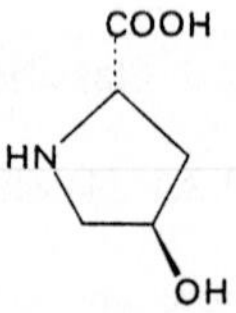 L–4–Hydroxy-proline	COOH, $H_2N—CH$, CH_2, CH_2, HO—CH, CH_2NH_2	L–5–Hydroxy-lysine
COOH, HN	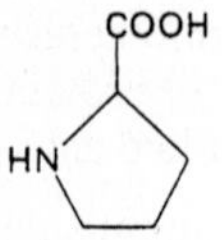 L–Proline (Pro. P)	COOH, $H_2N—CH$, CH_2, (benzene ring), OH	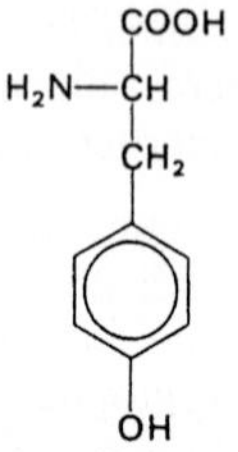 L–Tyrosine (Tyr. Y)	COOH, $H_2N—CH$, CH_2, (imidazole ring), HN—N	L–Histidine (His. H)
COOH, $H_2N—CH$, CH_2, (benzene ring)	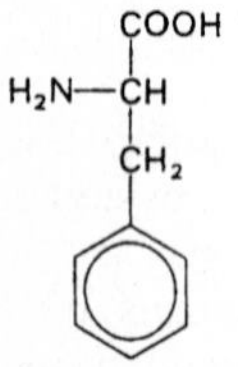 L–Phenylalanine (Phe. F)	COOH, $H_2N—CH$, CH_2, $CONH_2$	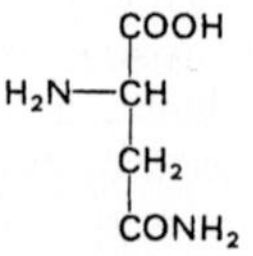 L–Asparagine[a] (Asn. N)		
COOH, $H_2N—CH$, CH_2, (indole ring), NH	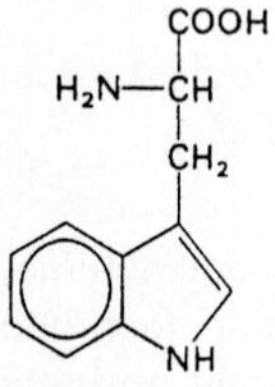 L–Tryptophan (Trp. W)	COOH, $H_2N—CH$, CH_2, CH_2, $CONH_2$	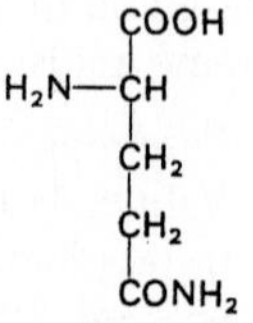 L–Glutamine[a] (Gln. Q)	COOH, $H_2N—CH$, CH_2, CH_2, CH_2, NH, C, HN NH_2	L–Arginine (Arg. R)

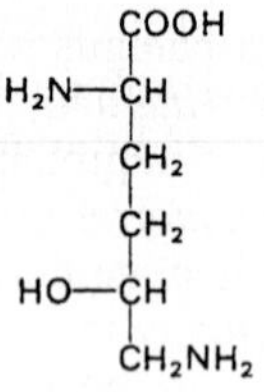

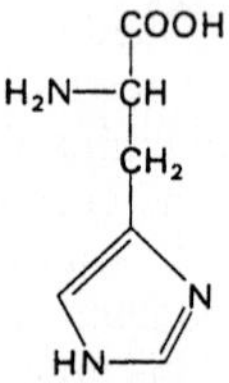

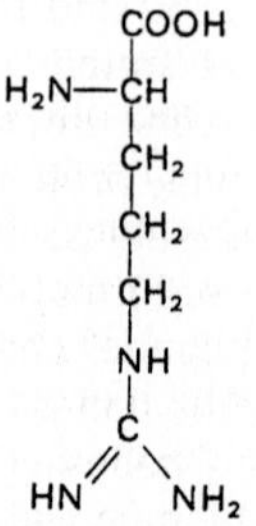

[a] When no distinction exists between the acid and its amide then the symbols (Asx, B) and (Glx, Z) are valid.

- Nonessential amino acids:
  Glycine, alanine, proline, serine, cysteine, tyrosine, asparagine, glutamine, aspartic acid and glutamic acid.

1.2.2.2 Discovery and Occurrence

Alanine was isolated from silk fibroin by *Th. Weyl* in 1888. It is present in most proteins and is particularly enriched in silk fibroin (35%). Gelatin and zein contain about 9% alanine, while its content in other proteins is 2–7%. Alanine is considered nonessential for humans.

Arginine was first isolated from lupin seedlings by *E. Schulze* and *E. Steiger* in 1886. It is present in all proteins at an average level of 3–6%, but is particularly enriched in protamines. The arginine content of peanut protein is relatively high (11%). Biochemically, arginine is of great importance as an intermediary product in urea synthesis. Arginine is a semi-essential amino acid for humans. It appears to be required under certain metabolic conditions.

Asparagine from asparagus was the first amino acid isolated by *Vauguelin* and *Robiquet* in 1806. Its occurrence in proteins (edestin) was confirmed by *Damodaran* in 1932. In glycoproteins the carbohydrate component may be bound N-glycosidically to the protein moiety through the amide group of asparagine (cf. 11.2.3.1.1 and 11.2.3.1.3).

Aspartic Acid was isolated from legumes by *H. Ritthausen* in 1868. It occurs in all animal proteins, primarily in albumins at a concentration of 6–10%. Alfalfa and corn proteins are rich in aspartic acid (14.9% and 12.3%, respectively) while its content in wheat is low (3.8%). Aspartic acid is nonessential.

Cystine was isolated from bladder calculi by *W.H. Wolaston* in 1810 and from horns by *L. Moerner* in 1899. Its content is high in keratins (9%). Cystine is very important since the peptide chains of many proteins are connected by two cysteine residues, i.e. by disulfide bonds. A certain conformation may be fixed within a single peptide chain by disulfide bonds. Most proteins contain 1–2% cystine. Although it is itself nonessential, cystine can partly replace methionine which is an essential amino acid.

Glutamine was first isolated from sugar beet juice by *Schulze* and *Bosshard* in 1883. Its occurrence in protein (edestin) was confirmed by *Damodaran* in 1932. Glutamine is readily converted into pyrrolidone carboxylic acid, which is stable between pH 2.2 and 4.0, but is readily cleaved to glutamic acid at other pH's:

$$CH_2\!-\!CH_2\!-\!CONH_2$$
$$CH\!-\!NH_2 \longrightarrow \quad HOOC\diagup\!\!\!\bigcirc\!\!\!=\!O + NH_3$$
$$COOH \tag{1.1}$$

Glutamic Acid was first isolated from wheat gluten by *H. Ritthausen* in 1866. It is abundant in most proteins, but is particularly high in milk proteins (21.7%), wheat (31.4%), corn (18.4%) and soya (18.5%). Molasses also contains relatively high amounts of glutamic acid. Monosodium glutamate is used in numerous food products as a flavor enhancer.

Glycine is found in high amounts in structural protein. Collagen contains 25–30% glycine. It was first isolated from gelatin by *H. Braconnot* in 1820. Glycine is a nonessential amino acid although it does act as a precursor of many compounds formed by various biosynthetic mechanisms.

Histidine was first isolated in 1896 independently by *A. Kossel* and by *S. G. Hedin* from protamines occurring in fish. Most proteins contain 2–3% histidine. Blood proteins contain about 6%. Histidine is essential in infant nutrition.

5-Hydroxylysine was isolated by *van Slyke et al.* (1921) and *Schryver et al.* (1925). It occurs in collagen. The carbohydrate component of glycoproteins may be bound O-glycosidically to the hydroxyl group of the amino acid (cf. 12.3.2.3.1).

4-Hydroxyproline was first obtained from gelatin by *E. Fischer* in 1902. Since it is abundant in collagen (12.4%), the determination of hydroxyproline is used to detect the presence of connective tissue in comminuted meat products. Hydroxyproline is a nonessential amino acid.

Isoleucine was first isolated from fibrin by *P. Ehrlich* in 1904. It is an essential amino acid.

Meat and cereal proteins contain 4–5% isoleucine; egg and milk proteins, 6–7%.

Leucine was isolated from wool and from muscle tissue by *H. Braconnot* in 1820. It is an essential amino acid and its content in most proteins is 7–10%. Cereal proteins contain variable amounts (corn 12.7%, wheat 6.9%). During alcoholic fermentation, fusel oil is formed from leucine and isoleucine.

Lysine was isolated from casein by *E. Drechsel* in 1889. It makes up 7–9% of meat, egg and milk proteins. The content of this essential amino acid is 2–4% lower in cereal proteins in which prolamin is predominant. Crab and fish proteins are the richest sources (10–11%). Along with threonine and methionine, lysine is a limiting factor in the biological value of many proteins, mostly those of plant origin. The processing of foods results in losses of lysine since its ε-amino group is very reactive (cf. *Maillard* reaction).

Methionine was first isolated from casein by *J.H. Mueller* in 1922. Animal proteins contain 2–4% and plant proteins contain 1–2% methionine. Methionine is an essential amino acid and in many biochemical processes its main role is as a methyl-donor. It is very sensitive to oxygen and heat treatment. Thus, losses occur in many food processing operations such as drying, kiln-drying, puffing, roasting or treatment with oxidizing agents. In the bleaching of flour with NCl_3 (nitrogen trichloride), methionine is converted to the toxic methionine sulfoximide:

$$H_3C-\underset{\underset{NH}{\|}}{\overset{\overset{O}{\|}}{S}}-CH_2-CH_2-\underset{\underset{NH_2}{|}}{CH}-COOH \qquad (1.2)$$

Phenylalanine was isolated from lupins by *E. Schulze* in 1881. It occurs in almost all proteins (averaging 4–5%) and is essential for humans. It is converted *in vivo* into tyrosine, so phenylalanine can replace tyrosine nutritionally.

Proline was discovered in casein and egg albumen by *E. Fischer* in 1901. It is present in numerous proteins at 4–7% and is abundant in wheat proteins (10.3%), gelatin (12.8%) and casein (12.3%). Proline is nonessential.

Serine was first isolated from sericin by *E. Cramer* in 1865. Most proteins contain about 4–8% serine. In phosphoproteins (casein, phosvitin) serine, like threonine, is a carrier of phosphoric acid in the form of O-phosphoserine. The carbohydrate component of glycoproteins may be bound O-glycosidically through the hydroxyl group of serine and/or threonine [cf. 10.1.2.1.1 (κ-casein) and 13.1.4.2.4].

Threonine was discovered by *W.C. Rose* in 1935. It is an essential amino acid, present at 4.5–5% in meat, milk and eggs and 2.7–4.7% in cereals. Threonine is often the limiting amino acid in proteins of lower biological quality. The "bouillon" flavor of protein hydrolysates originates partly from a lactone derived from threonine (cf. 5.3.1.3).

Tryptophan was first isolated from casein hydrolysates, prepared by hydrolysis using pancreatic enzymes, by *F.G. Hopkins* in 1902. It occurs in animal proteins in relatively low amounts (1–2%) and in even lower amounts in cereal proteins (about 1%). Tryptophan is exceptionally abundant in lysozyme (7.8%). It is completely destroyed during acidic hydrolysis of protein. Biologically, tryptophan is an important essential amino acid, primarily as a precursor in the biosynthesis of nicotinic acid.

Tyrosine was first obtained from casein by *J. Liebig* in 1846. Like phenylalanine, it is found in almost all proteins at levels of 2–6%. Silk fibroin can have as much as 10% tyrosine. It is converted through dihydroxyphenylalanine by enzymatic oxidation into brown-black colored melanins.

Valine was first isolated by *P. Schutzenberger* in 1879. It is an essential amino acid and is present in meat and cereal proteins (5–7%) and in egg and milk proteins (7–8%). Elastin contains notably high concentrations of valine (15.6%).

1.2.3 Physical Properties

1.2.3.1 Dissociation

In aqueous solution amino acids are present, depending on pH, as cations, zwitterions or anions:

R—CH—COOH $\xrightleftharpoons[+\,H^{\oplus}]{-\,H^{\oplus}}$ R—CH—COO$^{\ominus}$
| |
NH$_3^{\oplus}$ NH$_3^{\oplus}$

$\xrightleftharpoons[+\,H^{\oplus}]{-\,H^{\oplus}}$ R—CH—COO$^{\ominus}$ (1.3)
                   |
                   NH$_2$

With the cation denoted as $^+$A, the dipolar zwitterion as $^+$A$^-$ and the anion as A$^-$, the dissociation constant can be expressed as:

$$\frac{[^{\oplus}A^{\ominus}][H^{\oplus}]}{[^{\oplus}A]} = K_1 \qquad \frac{[A^{\ominus}][H^{\oplus}]}{[^{\oplus}A^{\ominus}]} = K_2 \quad (1.4)$$

At a pH where only dipolar ions exist, i.e. the isoelectric point, pI, $[^+A] = [A^-]$:

$$[^{\oplus}A] = \frac{[^{\oplus}A^{\ominus}][H^{\oplus}]}{K_1} = [A^{\ominus}] = \frac{[^{\oplus}A^{\ominus}]K_2}{[H^{\oplus}]}$$

$[H^{\oplus}] = (K_1 \cdot K_2)^{0,5}$

pI $= 0{,}5\,(pK_1 + pK_2)$ (1.5)

The dissociation constants of amino acids can be determined, for example, by titration of the acid. Figure 1.2 shows titration curves for glycine, histidine and aspartic acid. Table 1.2 lists the dissociation constants for some amino acids. In amino acids the acidity of the carboxyl group is higher and the basicity of the amino group lower than in the corresponding

carboxylic acids and amines (cf. pK values for propionic acid, 2-propylamine and alanine). As illustrated by the comparison of pK values of 2-aminopropionic acid (alanine) and 3-aminopropionic acid (β-alanine), the pK is influenced by the distance between the two functional groups.

The reasons for this are probably as follows: in the case of the cation → zwitterion transition, the inductive effect of the ammonium group; in the case of the zwitterion → anion transition, the stabilization of the zwitterion through hydration caused by dipole repulsion (lower than in relation to the anion).

$$\quad (1.6)$$

($\oplus$—$\ominus$, zwitterion; $+\!\!\rightarrow$ water dipole)

Table 1.2. Amino acids: dissociation constants and isoelectric points at 25°C

Amino acid	pK$_1$	pK$_2$	pK$_3$	pK$_4$	pI
Alanine	2.34	9.69			6.0
Arginine	2.18	9.09	12.60		10.8
Asparagine	2.02	8.80			5.4
Aspartic acid	1.88	3.65	9.60		2.8
Cysteine	1.71	8.35	10.66		5.0
Cystine	1.04	2.10	8.02	8.71	5.1
Glutamine	2.17	9.13			5.7
Glutamic acid	2.19	4.25	9.67		3.2
Glycine	2.34	9.60			6.0
Histidine	1.80	5.99	9.07		7.5
4-Hydroxyproline	1.82	9.65			5.7
Isoleucine	2.36	9.68			6.0
Leucine	2.36	9.60			6.0
Lysine	2.20	8.90	10.28		9.6
Methionine	2.28	9.21			5.7
Phenylalanine	1.83	9.13			5.5
Proline	1.99	10.60			6.3
Serine	2.21	9.15			5.7
Threonine	2.15	9.12			5.6
Tryptophan	2.38	9.39			5.9
Tyrosine	2.20	9.11	10.07		5.7
Valine	2.32	9.62			6.0
Propionic acid	4.87				
2-Propylamine	10.63				
β-Alanine	3.55	10.24			6.9
γ-Aminobutyric acid	4.03	10.56			7.3

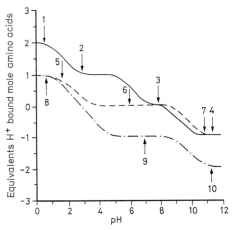

Fig. 1.2. Calculated titration curves for glycine (---), histidine (—) and aspartic acid (-·-·-). Numerals on curves are related to charge of amino acids in respective pH range: 1 $^{++}$His$^-$, 2 $^{++}$His$^-$, 3 $^+$His$^-$, 4 His$^-$, 5 $^+$Gly, 6 $^+$Gly$^-$, 7 Gly$^-$, 8 $^+$Asp. 9 $^+$Asp^{--}, 10 Asp^{--}

1.2.3.2 Configuration and Optical Activity

Amino acids, except for glycine, have at least one chiral center and, hence, are optically active. All amino acids found in proteins have the same configuration on the α-C-atom: they are considered L-amino acids or (S)-amino acids* in the *Cahn-Ingold-Prelog* system (with L-cysteine an exception; it is in the (R)-series). D-amino acids (or (R)-amino acids) also occur in nature, for example, in a number of peptides of microbial origin:

$$(1.7)$$

L-Amino acid
(S)-Amino acid

D-Amino acid
(R)-Amino acid

Isoleucine, threonine and 4-hydroxyproline have two asymmetric C-atoms, thus each has four isomers:

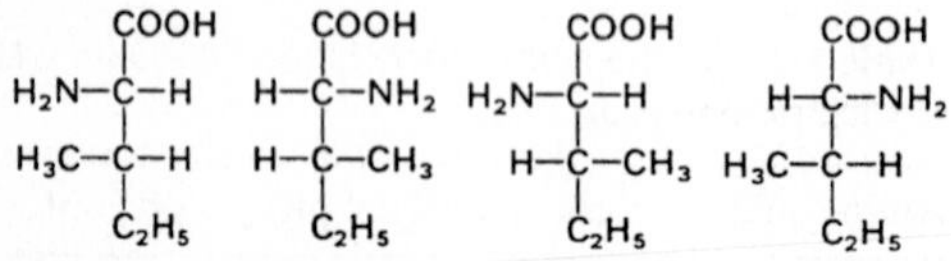

L-Isoleucine
(2S:3S)-
Isoleucine
(Common
in proteins)

D-Isoleucine
(2R:3R)-
Isoleucine

L-allo-
Isoleucine
(2S:3R)-
Isoleucine

D-allo-
Isoleucine
(2R:3S)-
Isoleucine

$$(1.8)$$

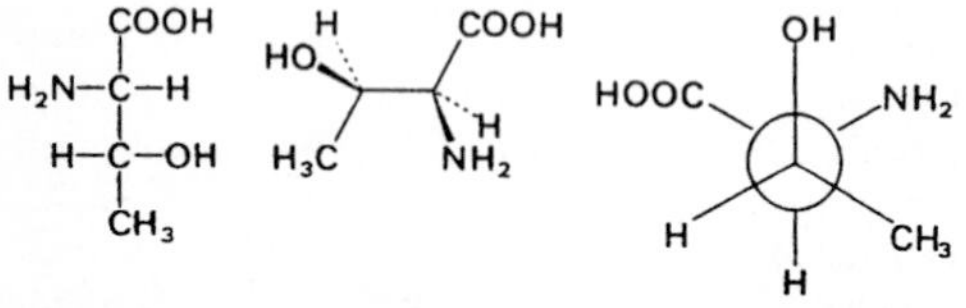

(Fischer-
projection)

(dotted line-
wedge)

(Newman-
projection)

L-Threonine, (2S:3R)-Threonine (Common in proteins)

$$(1.9)$$

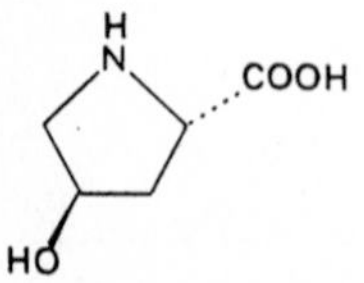

L-4-Hydroxyproline, (2S:4R)-Hydroxyproline
(Common in proteins)

$$(1.10)$$

---

* As with carbohydrates, D,L-nomenclature is pre-ferred with amino acids.

The specific rotation of amino acids in aqueous solution is strongly influenced by pH. It passes through a minimum in the neutral pH range and rises after addition of acids or bases (Table 1.3).

Table 1.3. Amino acids: specific rotation ($[\alpha]_D^t$)

Amino acid	Solvent system	Temperature (°C)	$[\alpha]_D$
L-Alanine	0.97 M HCl	15	+ 14.7°
	water	22	+ 2.7°
	3 M NaOH	20	+ 3.0°
L-Cystine	1.02 M HCl	24	− 214.4°
L-Glutamic acid	6.0 M HCl	22.4	+ 31.2°
	water	18	+ 11.5°
	1 M NaOH	18	+ 10.96°
L-Histidine	6.0 M HCl	22.7	+ 13.0°
	water	25.0	− 39.01°
	0.5 M NaOH	20	− 10.9°
L-Leucine	6.0 M HCl	25.9	+ 15.1°
	water	24.7	− 10.8°
	3.0 M NaOH	20	+ 7.6°

There are various possible methods of separating racemates, which generally occur in amino acid synthesis (cf. 1.2.5), or of detecting D-amino acids in foods, which can be formed during food processing (cf. 1.4.4.11). Selective crystallization of an over-saturated solution of racemate after seeding with an enantiomer is used, as is the fractioned crystallization of diastereomeric salts or other derivatives, such as (S)-phenylethylammonium salts of N-acetylamino acids. With enzymatic methods, asymmetric synthesis is used, e.g., of acyl-amino acid anilides from acylamino acids and aniline through papain:

$$\text{D,L—R—CO—NH—CHR}^1\text{—COOH} \xrightarrow[\text{Papain}]{\text{Aniline}}$$

L—R—CO—NH—CHR^1—CO—NH—C$_6$H$_5$

+ D—R—CO—NH—CHR^1—COOH   (1.11)

or asymmetric hydrolysis, e.g., of amino acid esters through esterases, amino acid amides through amidases or N-acylamino acids through aminoacylases:

$$D,L—H_2N—CHR—COOR^1 \xrightarrow{\text{Esterase}}$$

$$L—H_2NCHR—COOH + D—H_2N—CHR—COOR^1$$

$$D,L—H_2N—CHR—CONHR^1 \xrightarrow{\text{Amidase}}$$

$$L—H_2N—CHR—COOH + D—H_2N—CHR—CONHR^1$$

$$D,L—R—CO—NH—CHR^1—COOH \xrightarrow{\text{Acylase}}$$

$$L—H_2N—CHR^1—COOH$$

$$+ D—R—CO—NH—CHR^1—COOH \qquad (1.12)$$

Another possibility is the chromatographic separation of suitable chiral derivatives, e.g. of N-trifluoroacetylamino acid-2-(R,S)-butyl-ester on achiral phases or suitable achiral derivatives on chiral phases (Fig. 1.3).

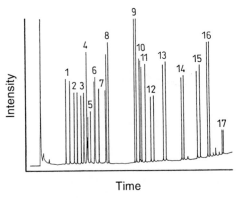

Fig. 1.3. Gas chromatogram of N-pentafluoropro-panoyl-DL-amino acid isopropylesters on Chirasil-Val (N-propionyl-L-valine-tert-butylamide-poly-siloxane) (1: D-, L-Ala, 2: D-, L-Val, 3. D-, L-Thr, 4: Gly, 5: D-, L-Ile, 6: D-, L-Pro, 7: D-, L-Leu, 8: D-, L-Ser, 9: D-, L-Cys, 10: D-, L-Asp, 11: D-, L-Met, 12: D-, L-Phe, 13: D-, L-Glu, 14: D-, L-Tyr, 15: D-, L-Orn, 16: D-, L-Lys, 17: D-, L-Trp; according to *Frank* et al., 1977)

1.2.3.3 Solubility

The solubilities of amino acids in water are highly variable. Besides the extremely soluble proline, hydroxyproline, glycine and alanine are also quite soluble. Other amino acids (cf.

Table 1.4. Solubility of amino acids in water (g/100 g H_2O)

Amino acid	Temperature (°C)				
	0	25	50	75	100
L-Alanine	12.73	16.51	21.79	28.51	37.30
L-Aspartic acid	0.209	0.500	1.199	2.875	6.893
L-Cystine	0.005	0.011	0.024	0.052	0.114
L-Glutamic acid	0.341	0.843	2.186	5.532	14.00
Glycine	14.18	24.99	39.10	54.39	67.17
L-Histidine	–	4.29	–	–	–
L-Hydroxy-proline	28.86	36.11	45.18	51.67	–
L-Isoleucine	3.791	4.117	4.818	6.076	8.255
L-Leucine	2.270	2.19	2.66	3.823	5.638
D,L-Methionine	1.818	3.381	6.070	10.52	17.60
L-Phenylalanine	1.983	2.965	4.431	6.624	9.900
L-Proline	127.4	162.3	206.7	239.0	–
D,L-Serine	2.204	5.023	10.34	19.21	32.24
L-Tryptophan	0.823	1.136	1.706	2.795	4.987
L-Tyrosine	0.020	0.045	0.105	0.244	0.565
L-Valine	8.34	8.85	9.62	10.24	–

Table 1.4) are significantly less soluble, with cystine and tyrosine having particularly low solubilities. Addition of acids or bases improves the solubility through salt formation. The presence of other amino acids, in general, also brings about an increase in solubility. Thus, the extent of solubility of amino acids in a protein hydrolysate is different than that observed for the individual components.

The solubility in organic solvents is not very good because of the polar characteristics of the amino acids. All amino acids are insoluble in ether. Only cysteine and proline are relatively soluble in ethanol (1.5 g/100 g at 19 °C). Methionine, arginine, leucine (0.0217 g/100 g; 25 °C), glutamic acid (0.00035 g/100 g; 25 °C), phenylalanine, hydroxy-proline, histidine and tryptophan are sparingly soluble in ethanol. The solubility of isoleucine in hot ethanol is relatively high (0.09 g/100 g at 20 °C; 0.13 g/100 g at 78–80 °C).

1.2.3.4 UV-Absorption

Aromatic amino acids such as phenylalanine, tyrosine and tryptophan absorb in the UV-range of the spectrum with absorption maxima at 200–230 nm and 250–290 nm (Fig. 1.4). Dissociation of the phenolic HO-group of tyrosine shifts the absorption curve by about

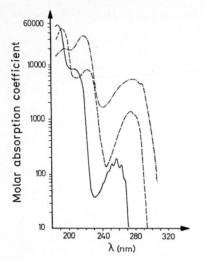

Fig. 1.4. Ultraviolet absorption spectra of some amino acids. (according to *Luebke, Schroeder*, and *Kloss*, 1975). ‑‑‑‑‑ Trp. ‑‑‑ Tyr. ⸺ Phe

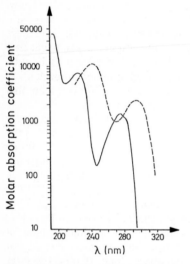

Fig. 1.5. Ultraviolet absorption spectrum of tyrosine as affected by pH. (according to *Luebke, Schroeder* and *Kloss*, 1975) ⸺ 0.1 mol/l HCl, ‑‑‑ 0.1 mol/l NaOH

20 nm towards longer wavelengths (Fig. 1.5). Absorption readings at 280 nm are used for the determination of proteins and peptides. Histidine, cysteine and methionine absorb between 200 and 210 nm.

1.2.4 Chemical Reactions

Amino acids show the usual reactions of both carboxylic acids and amines. Reaction specificity is due to the presence of both carboxyl and amino groups and, occasionally, of other functional groups. Reactions occurring at 100–220 °C, such as in cooking, frying and baking, are particularly relevant to food chemistry.

1.2.4.1 Esterification of Carboxyl Groups

Amino acids are readily esterified by acid-catalyzed reactions. An ethyl ester hydrochloride is obtained in ethanol in the presence of HCl:

$$R-\underset{\overset{|}{NH_3^\oplus Cl^\ominus}}{CH}-COOH + R'-OH \xrightarrow{H^\oplus} R-\underset{\overset{|}{NH_3^\oplus Cl^\ominus}}{CH}-COOR' + H_2O$$

$$(1.13)$$

The free ester is released from its salt by the action of alkali. A mixture of free esters can then be separated by distillation without decomposition. Fractional distillation of esters is the basis of a method introduced by *Emil Fischer* for the separation of amino acids:

$$R-\underset{\overset{|}{NH_3^\oplus X^\ominus}}{CH}-COOR' \xrightarrow{B} R-\underset{\overset{|}{NH_2}}{CH}-COOR' + BH^\oplus X^\ominus$$

$$(1.14)$$

Free amino acid esters have a tendency to form cyclic dipeptides or open-chain polypeptides:

$$(1.15)$$

$$(1.16)$$

tert-butyl esters, which are readily split by acids, or benzyl esters, which are readily cleaved by HBr/glacial acetic acid or catalytic hydrogenation, are used as protective groups in peptide synthesis.

1.2.4.2 Reactions of Amino Groups

1.2.4.2.1 Acylation

Activated acid derivatives, e.g. acid halogenides or anhydrides, are used as acylating agents:

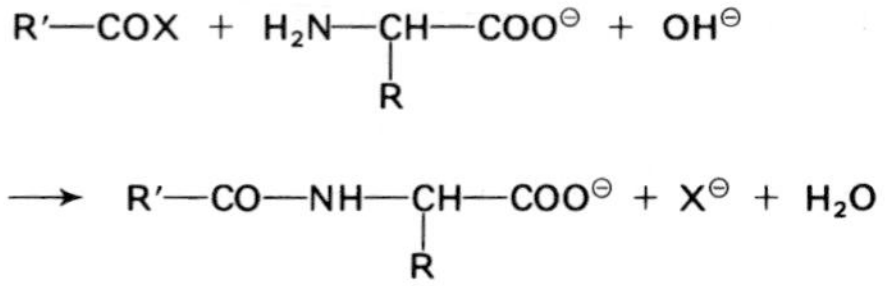

$$R'—COX + H_2N—\underset{\underset{R}{|}}{C}H—COO^\ominus + OH^\ominus$$

$$\longrightarrow R'—CO—NH—\underset{\underset{R}{|}}{C}H—COO^\ominus + X^\ominus + H_2O$$

$$(1.17)$$

N-acetyl amino acids are being considered as ingredients in chemically-restricted diets and for fortifying plant proteins to increase their biological value. Addition of free amino acids to food which must be heat treated is not problem free. For example, methionine in the presence of a reducing sugar can form methional by a *Strecker* degradation mechanism, imparting an off-flavor to food. Other essential amino acids, e.g., lysine or threonine, can lose their biological value through similar reactions. Feeding tests with rats have shown that N-acetyl-L-methionine and N-acetyl-L-threonine have nutritional values equal to those of the free amino acids (this is true also for humans with acetylated methionine). The growth rate of rats is also increased significantly by the α- or ε-acetyl or α,ε-diacetyl derivatives of lysine.

Some readily cleavable acyl residues are of importance as temporary protective groups in peptide synthesis.

The trifluoroacetyl residue is readily removed by mild base-catalyzed hydrolysis:

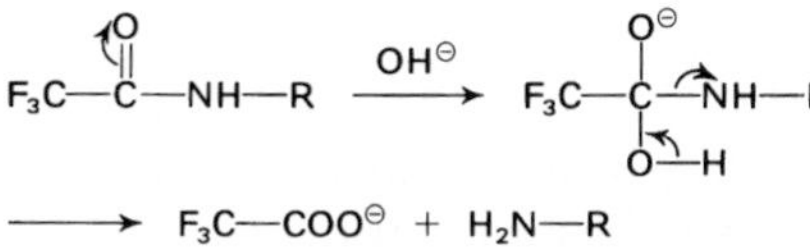

$$F_3C—\overset{\overset{O}{||}}{C}—NH—R \xrightarrow{OH^\ominus} F_3C—\underset{\underset{O-H}{|}}{C}^\ominus NH—R$$

$$\longrightarrow F_3C—COO^\ominus + H_2N—R \qquad (1.18)$$

The phthalyl residue can be readily cleaved by hydrazinolysis:

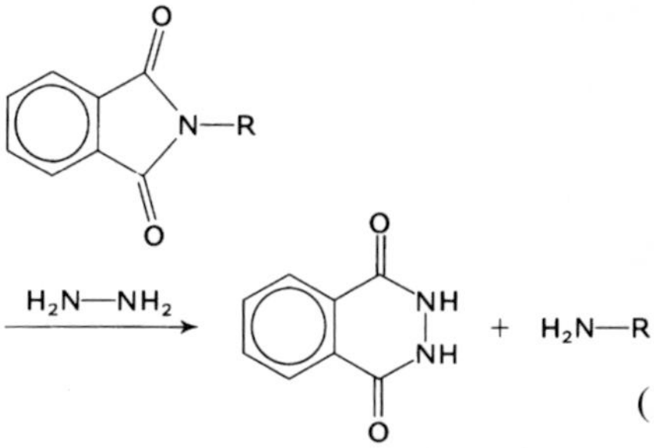

$$(1.19)$$

The benzyloxycarbonyl group can be readily removed by catalytic hydrogenation or by hydrolysis with HBr/glacial acetic acid:

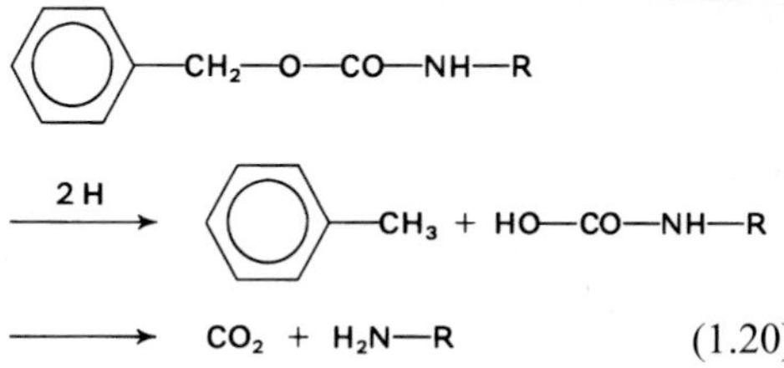

$$\longrightarrow CO_2 + H_2N—R \qquad (1.20)$$

$$\longrightarrow \text{—CH}_2Br + CO_2 + H_2N—R$$

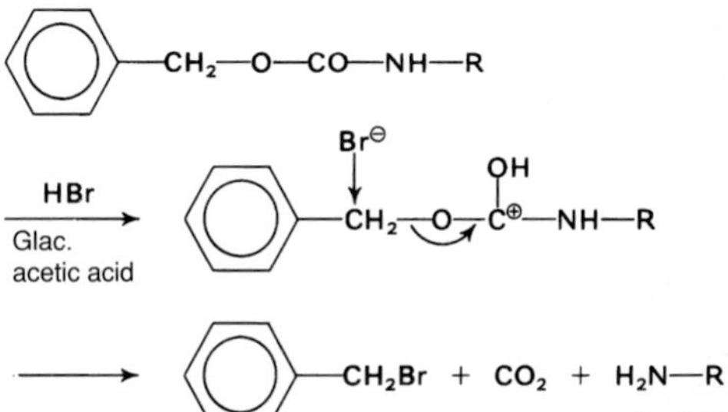

$$(1.21)$$

The *tert*-alkoxycarbonyl residues, e.g., the *tert*-butyloxycarbonyl groups, are cleaved under acid-catalyzed conditions:

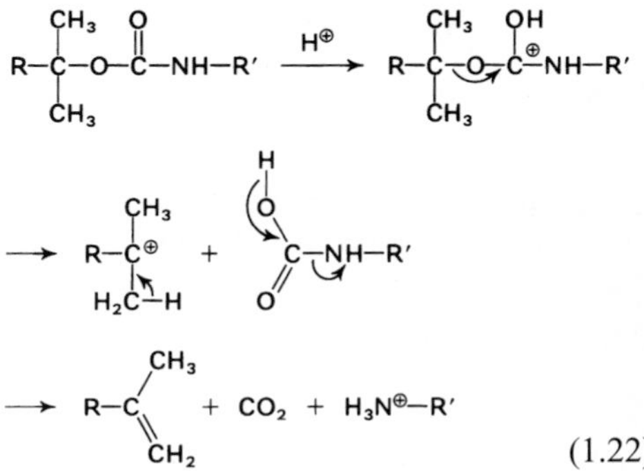

$$(1.22)$$

N-acyl derivatives of amino acids are transformed into oxazolinones (azlactones) by elimination of water:

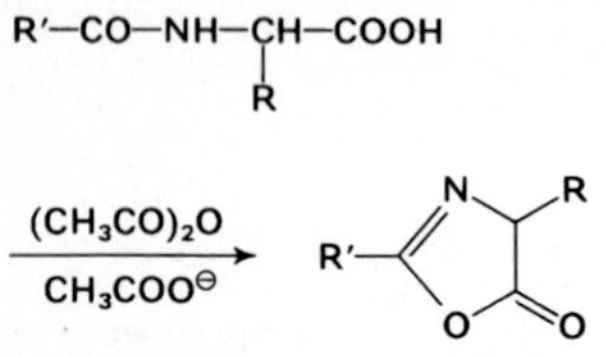

R′—CO—NH—CH—COOH
                |
                R

$$\xrightarrow[\text{CH}_3\text{COO}^\ominus]{(\text{CH}_3\text{CO})_2\text{O}}$$

(1.23)

These are highly reactive intermediary products which form a mesomerically stabilized anion. The anion can then react, for example, with aldehydes. This reaction is utilized in amino acid synthesis with glycine azlactone as a starting compound:

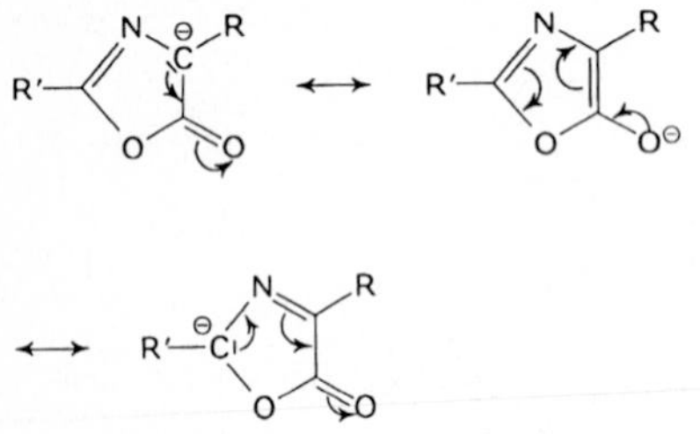

(1.24)

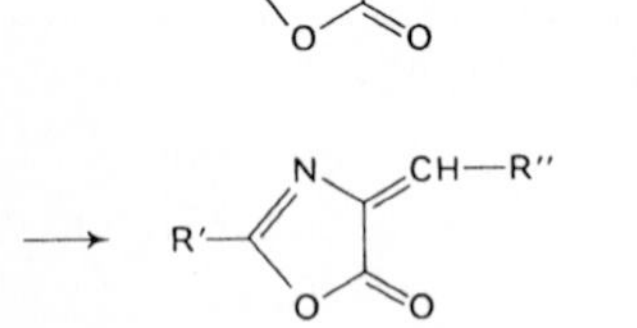

(1.25)

Acylation of amino acids with 5-dimethylaminonaphthalene-1-sulfonyl chloride (dansyl chloride, DANS-Cl) is of great analytical importance:

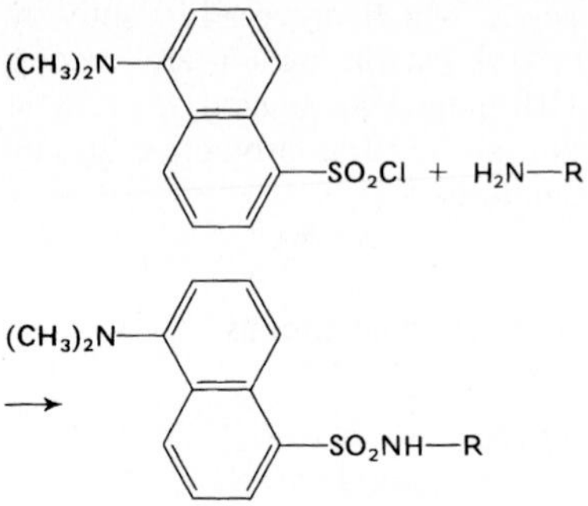

(1.26)

The aryl sulfonyl derivatives are very stable against acidic hydrolysis. Therefore, they are suitable for the determination of free N-terminal amino groups or free ε-amino groups of peptides or proteins. Dansyl derivatives which fluoresce in UV-light have a detection limit in the nanomole range, which is lower than that of 2,4-dinitrophenyl derivatives by a factor of 100.

Dimethylaminoazobenzenesulfonylchloride (DABS-Cl) and 9-fluoroenylmethylchloroformate (FMOC) also lead to sensitively detectable amino derivatives that are suitable for separation and detection by column chromatography (cf. 1.4.1.1):

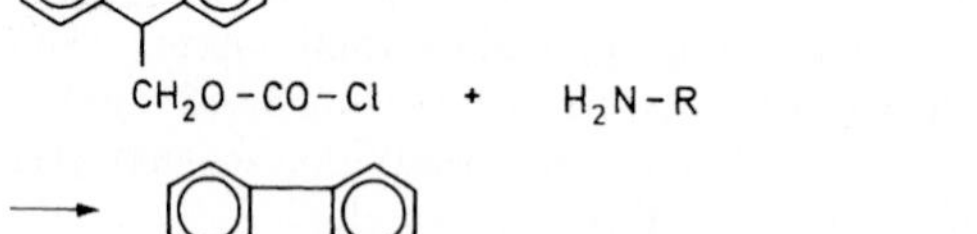

(1.27)

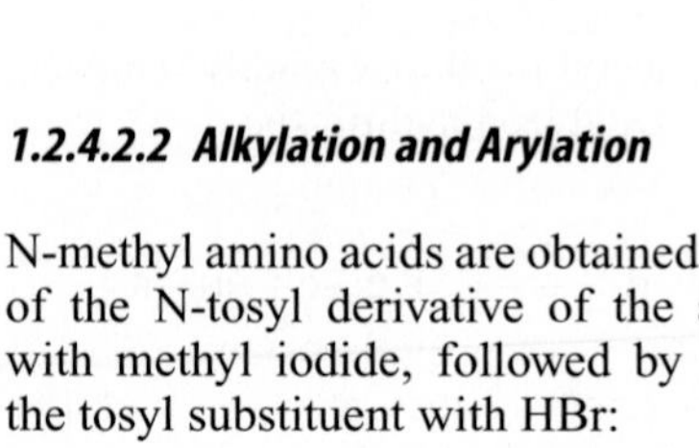

(1.28)

1.2.4.2.2 Alkylation and Arylation

N-methyl amino acids are obtained by reaction of the N-tosyl derivative of the amino acid with methyl iodide, followed by removal of the tosyl substituent with HBr:

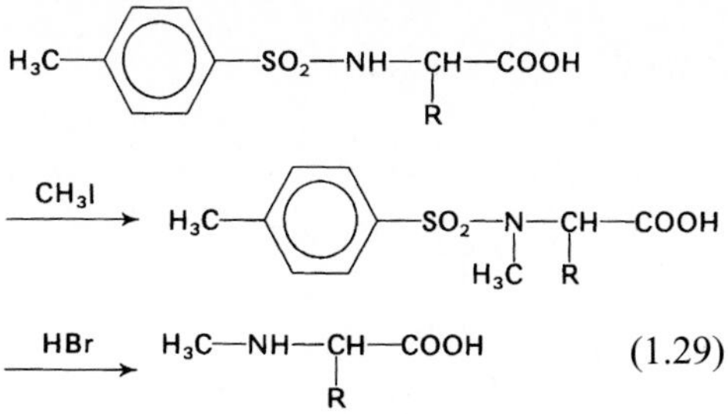

$$H_3C-\!\!\bigcirc\!\!-SO_2-NH-CH-COOH \quad (R)$$

$$\xrightarrow{CH_3I} \quad H_3C-\!\!\bigcirc\!\!-SO_2-N-CH-COOH \quad (CH_3, R)$$

$$\xrightarrow{HBr} \quad H_3C-NH-CH-COOH \quad (R) \qquad (1.29)$$

The N-methyl compound can also be formed by methylating with HCHO/HCOOH the benzylidene derivative of the amino acid, formed initially by reaction of the amino acid with benzaldehyde. The benzyl group is then eliminated by hydrogenolysis:

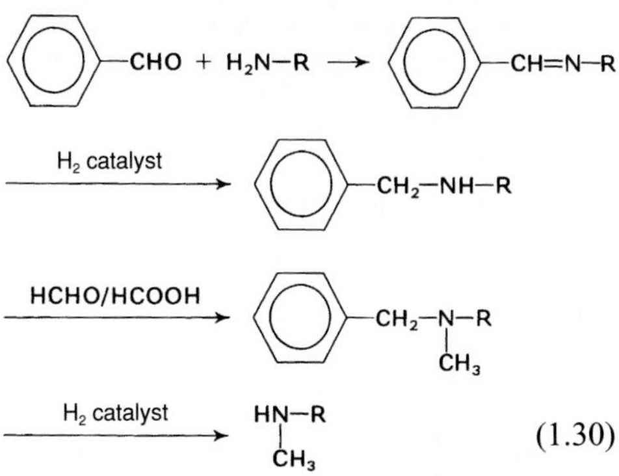

$$\bigcirc\!\!-CHO + H_2N-R \rightarrow \bigcirc\!\!-CH=N-R$$

$$\xrightarrow{H_2\ catalyst} \bigcirc\!\!-CH_2-NH-R$$

$$\xrightarrow{HCHO/HCOOH} \bigcirc\!\!-CH_2-N-R \quad (CH_3)$$

$$\xrightarrow{H_2\ catalyst} \quad HN-R \quad (CH_3) \qquad (1.30)$$

Dimethyl amino acids are obtained by reaction with formaldehyde, followed by reduction with sodium borohydride:

$$2\,HCHO + H_2N-R \xrightarrow[pH\ 9,\ 0\,°C]{NaBH_4} (CH_3)_2N-R \qquad (1.31)$$

The corresponding reactions with proteins are being considered as a means of protecting the ε-amino groups and, thus, of avoiding their destruction in food through the *Maillard* reaction (cf. 1.4.6.2.2).

Direct reaction of amino acids with methylating agents, e. g. methyl iodide or dimethyl sulfate, proceeds through monomethyl and dimethyl compounds to trimethyl derivatives (or generally to N-trialkyl derivatives) denoted as betaines:

$$H_2N-CH-COOH \xrightarrow{CH_3I} (CH_3)_3N^{\oplus}-CH-COO^{\ominus} \quad (R) \qquad (1.32)$$

As shown in Table 1.5, betaines are widespread in both the animal and plant kingdoms.

Table 1.5. Occurrence of trimethyl amino acids $(CH_3)_3N^+\!-CHR-COO^-$ (betaines)

Amino acid	Betaine	Occurrence
β-Alanine	Homobetaine	Meat extract
γ-Amino-butyric acid	Actinine	Mollusk (shell-fish)
Glycine	Betaine	Sugar beet, other samples of animal and plant origin
Histidine	Hercynine	Mushrooms
β-Hydroxy-γ-amino-butyric acid	Carnitine	Mammals muscle tissue, yeast, wheat germ, fish, liver, whey, mollusk (shell-fish)
4-Hydroxy-proline	Betonicine	Jack beans
Proline	Stachydrine	Stachys, orange leaves, lemon peel, alfalfa, Aspergillus oryzae

Derivatization of amino acids by reaction with 1-fluoro-2,4-dinitrobenzene (FDNB) yields N-2,4-dinitrophenyl amino acids (DNP-amino acids), which are yellow compounds and crystallize readily. The reaction is important for labeling N-terminal amino acid residues and free ε-amino groups present in peptides and proteins; the DNP-amino acids are stable under conditions of acidic hydrolysis (cf. Reaction 1.33).

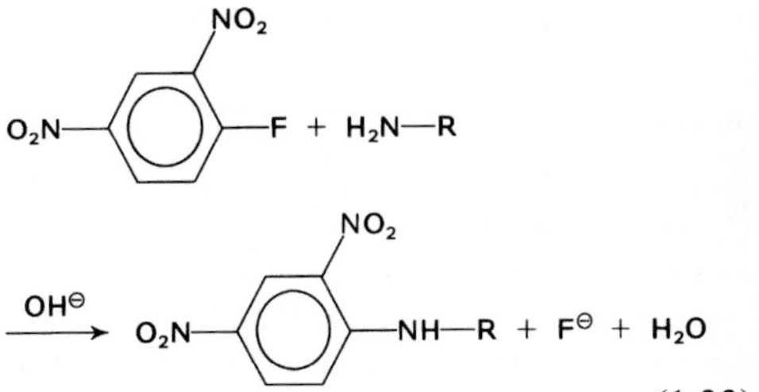

$$O_2N-\!\!\bigcirc\!\!(NO_2)-F + H_2N-R$$

$$\xrightarrow{OH^{\ominus}} O_2N-\!\!\bigcirc\!\!(NO_2)-NH-R + F^{\ominus} + H_2O \qquad (1.33)$$

Another arylation reagent is 7-fluoro-4-nitrobenzo-2-oxa-1,3-diazol (NBD-F), which is also used as a chlorine compound (NBD-Cl) and which leads to derivatives that are suited for an amino acid analysis through HPLC separation:

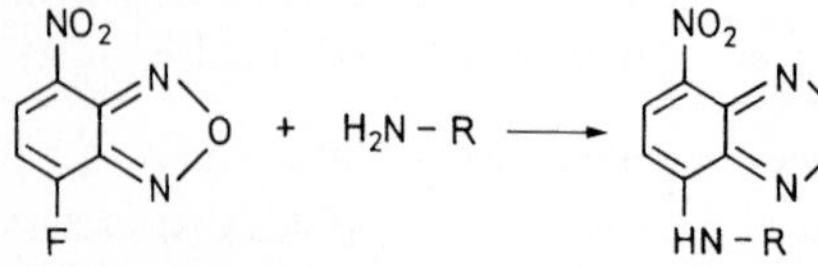
(1.34)

Reaction of amino acids with triphenylmethyl chloride (tritylchloride) yields N-trityl derivatives, which are alkali stable. However, the derivative is cleaved in the presence of acid, giving a stable triphenylmethyl cation and free amino acid:

$$(C_6H_5)_3C{-}Cl + H_2N{-}R \xrightarrow{OH^\ominus} (C_6H_5)_3C{-}NH{-}R \xrightarrow{H^\oplus} (C_6H_5)_3C^\oplus + H_2N{-}R \quad (1.35)$$

The reaction with trinitrobenzene sulfonic acid is also of analytical importance. It yields a yellow-colored derivative that can be used for the spectrophotometric determination of protein:

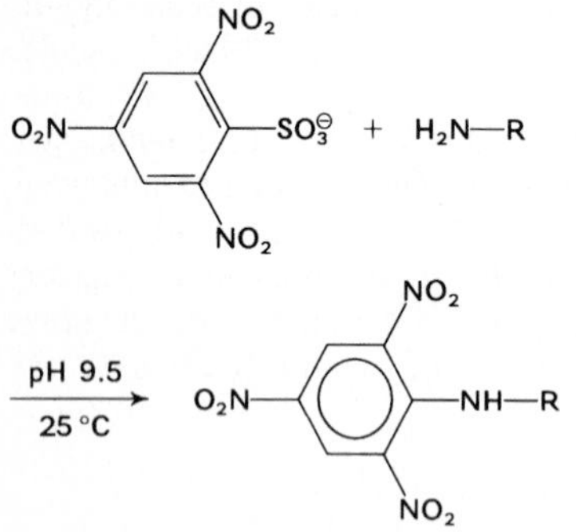
(1.36)

The reaction is a nucleophilic aromatic substitution proceeding through an intermediary addition product (*Meisenheimer* complex). It occurs under mild conditions only when the benzene ring structure is stabilized by electron-withdrawing substituents on the ring (cf. Reaction 1.37).

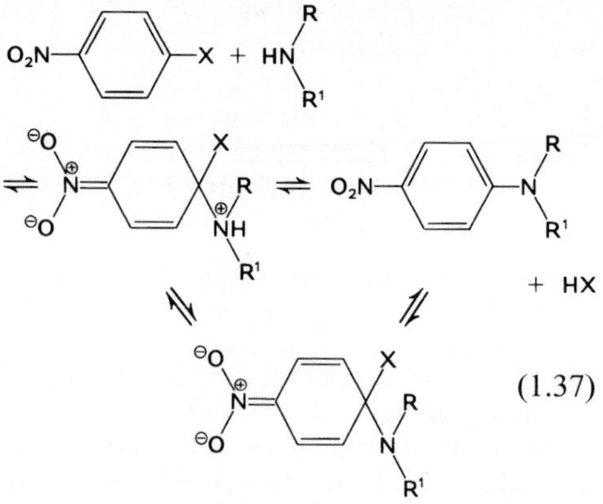
(1.37)

The formation of the *Meisenheimer* complex has been verified by isolating the addition product from the reaction of 2,4,6-trinitroanisole with potassium ethoxide (cf. Reaction 1.38).

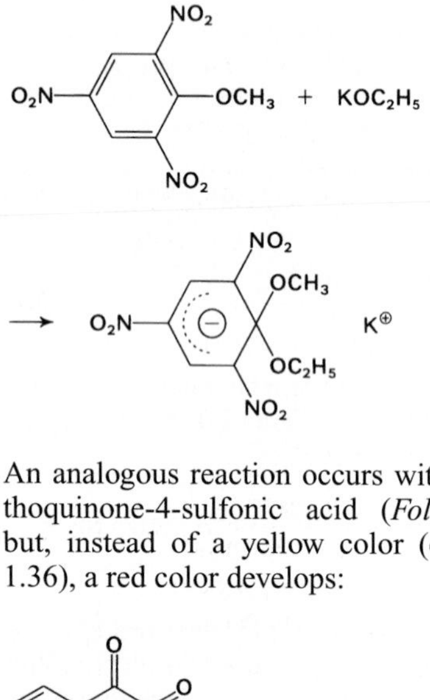
(1.38)

An analogous reaction occurs with 1,2-naphthoquinone-4-sulfonic acid (*Folin* reagent) but, instead of a yellow color (cf. Formula 1.36), a red color develops:

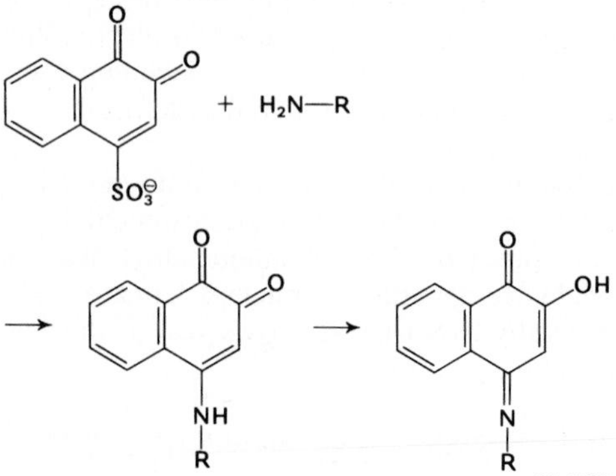
(1.39)

1.2.4.2.3 Carbamoyl and Thiocarbamoyl Derivatives

Amino acids react with isocyanates to yield carbamoyl derivatives which are cyclized into 2,4-dioxoimidazolidines (hydantoins) by boiling in an acidic medium:

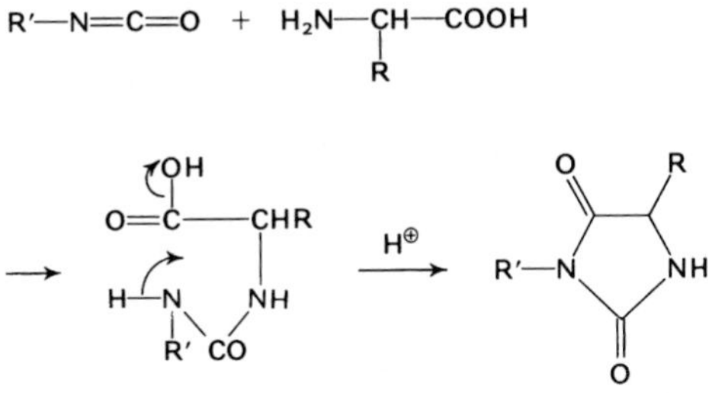

$$(1.40)$$

A corresponding reaction with phenylisothiocyanate can degrade a peptide in a stepwise fashion (*Edman* degradation). The reaction is of great importance for revealing the amino acid sequence in a peptide chain. The phenylthiocarbamoyl derivative (PTC-peptide) formed in the first step (coupling) is cleaved non-hydrolytically in the second step (cleavage) with anhydrous trifluoroacetic acid into anilinothiazolinone as derivative of the N-terminal amino acid and the remaining peptide which is shortened by the latter. Because of its instability, the thiazolinone is not suited for an identification of the N-terminal amino acid and is therefore – after separation from the remaining peptide, in the third step (conversion) – converted in aqueous HCl via the phenylthiocarbamoylamino acid into phenylthiohydantoin, while the remaining peptide is fed into a new cycle.

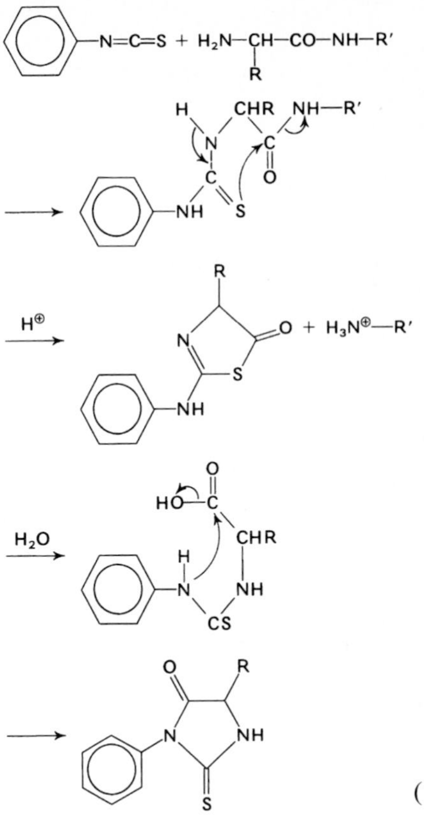

$$(1.41)$$

1.2.4.2.4 Reactions with Carbonyl Compounds

Amino acids react with carbonyl compounds, forming azomethines. If the carbonyl compound has an electron-withdrawing group, e.g., a second carbonyl group, transamination and decarboxylation occur:

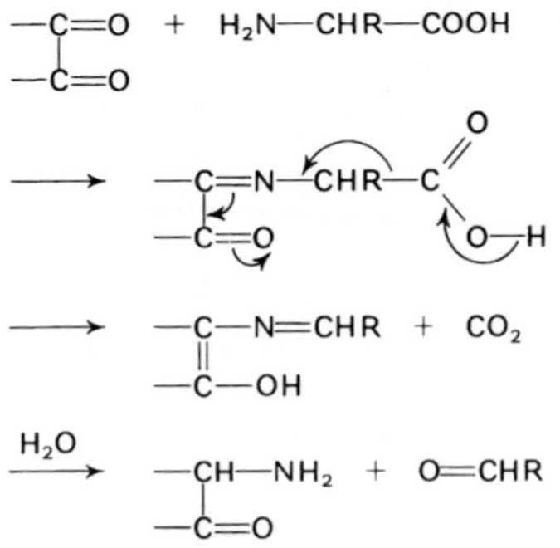

$$(1.42)$$

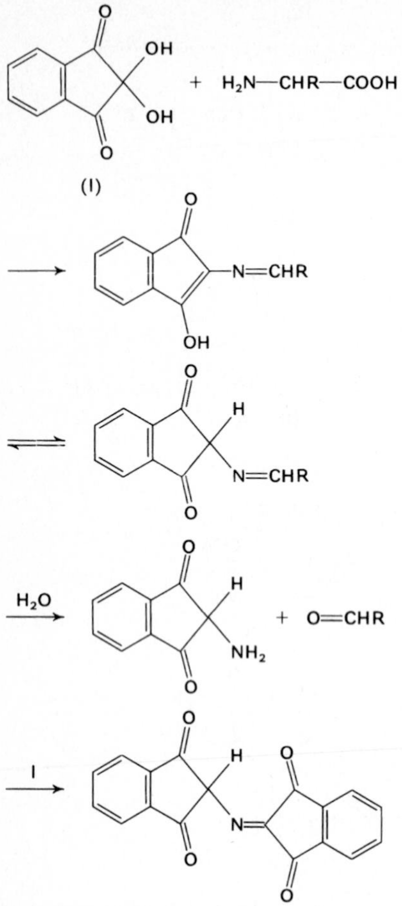

(I)

(1.43)

(I)

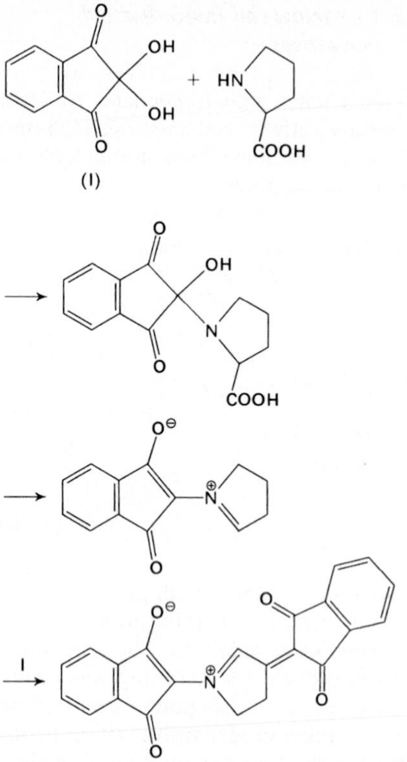

(1.44)

The reaction of amino acids with o-phthaldi-aldehyde (OPA) and mercaptoethanol leads to fluorescent isoindole derivatives ($\lambda_{ex} = 330$ nm, $\lambda_{em} = 455$ nm)

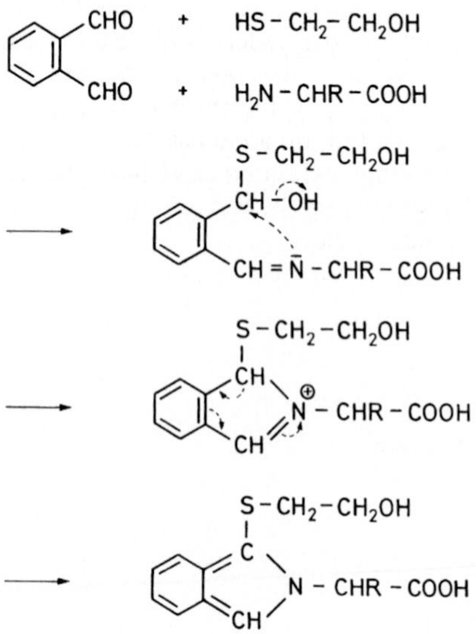

The reaction is known as the *Strecker* degra-dation and plays a role in food since food can be an abundant source of dicarbonyl com-pounds generated by the *Maillard* reaction (cf. 4.2.4.4.7). The aldehydes which are form-ed from amino acids (*Strecker* aldehydes) are aroma compounds.

The ninhydrin reaction is a special case of the *Strecker* degradation. It is an important reaction for the quantitative determination of amino acids using spectrophotometry (cf. Reaction 1.43). The detection limit lies at 1−0.5 nmol.

The resultant blue-violet color has an absorp-tion maximum at 570 nm, except for reaction with proline, which yields a yellow-colored compound with $\lambda_{max} = 440$ nm:

(1.45)

The derivatives can be used for amino acid analysis via HPLC separation. The detection limit lies at 1 pmol. Mercaptoethanol can be replaced by other thiols, such as N,N-dimethyl-2-mercaptoethylamine. If a chiral thiol is used, the separation of enantiomeric amino acids is possible. One disadvantage of the method is that proline and hydroxyproline are not covered.

Fluorescamine reacts with primary amines and amino acids – at room temperature under alkaline conditions – to form fluorescent pyrrolidones (λ_{ex} = 390 nm, λ_{em} = 474 nm). The detection limit lies at 50–100 pmol:

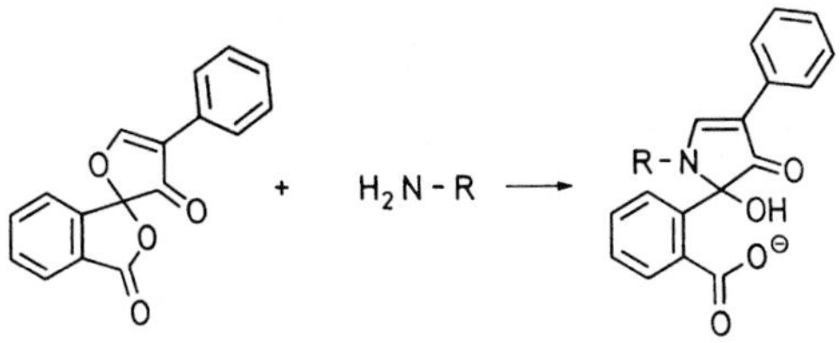

$$(1.46)$$

The excess reagent is very quickly hydrolyzed into water-soluble and non-fluorescent compounds.

1.2.4.3 Reactions Involving Other Functional Groups

The most interesting of these reactions are those in which α-amino and α-carboxyl groups are blocked, that is, reactions occurring with peptides and proteins. These reactions will be covered in detail in sections dealing with modification of proteins (cf. 1.4.4 and 1.4.6.2). A number of reactions of importance to free amino acids will be covered in the following sections.

1.2.4.3.1 Lysine

A selective reaction may be performed with either of the amino groups in lysine. Selective acylation of the ε-amino group is possible using the lysine-Cu²⁺ complex as a reactant:

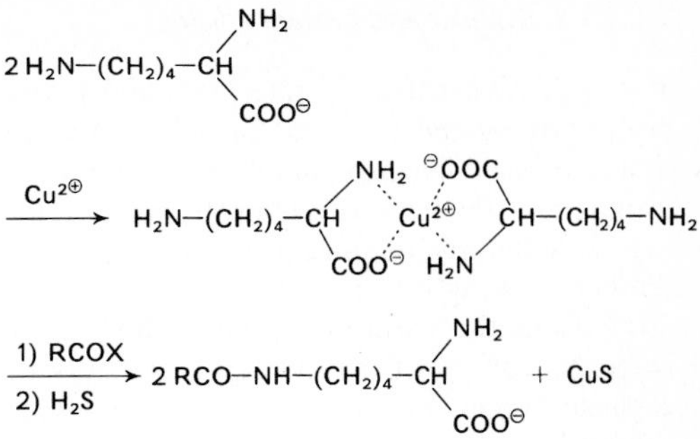

$$(1.47)$$

Selective reaction with the α-amino group is possible using a benzylidene derivative:

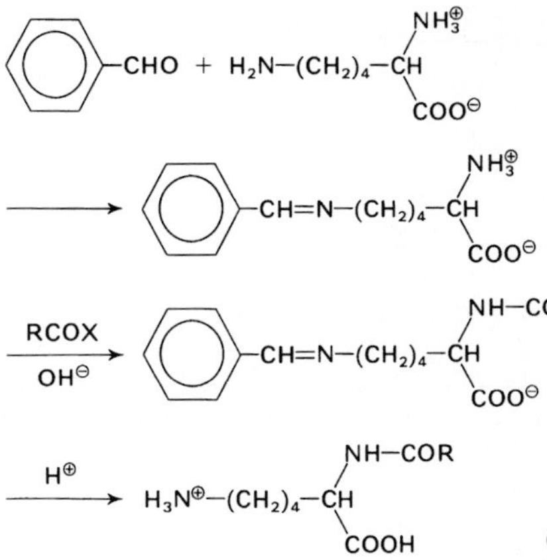

$$(1.48)$$

ε-N-benzylidene-L-lysine and ε-N-salicylidene-L-lysine are as effective as free lysine in growth feeding tests with rats. Browning reactions of these derivatives are strongly retarded, hence they are of interest for lysine fortification of food.

1.2.4.3.2 Arginine

In the presence of α-naphthol and hypobromite, the guanidyl group of arginine gives a red compound with the following structure:

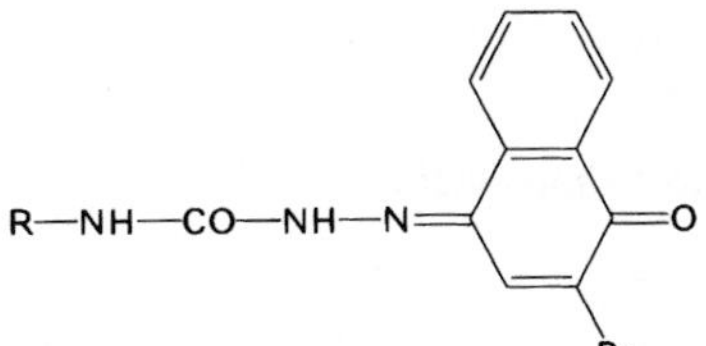

$$(1.49)$$

1.2.4.3.3 Aspartic and Glutamic Acids

The higher esterification rate of β- and γ-carboxyl groups can be used for selective reactions. On the other hand the β- and γ-carboxyl groups are more rapidly hydrolyzed in acid-catalyzed hydrolysis since protonation is facilitated by having the ammonium group further away from the carboxyl group. Alkali-catalyzed hydrolysis of methyl or ethyl esters of aspartic or glutamic acids bound to peptides can result in the formation of isopeptides.

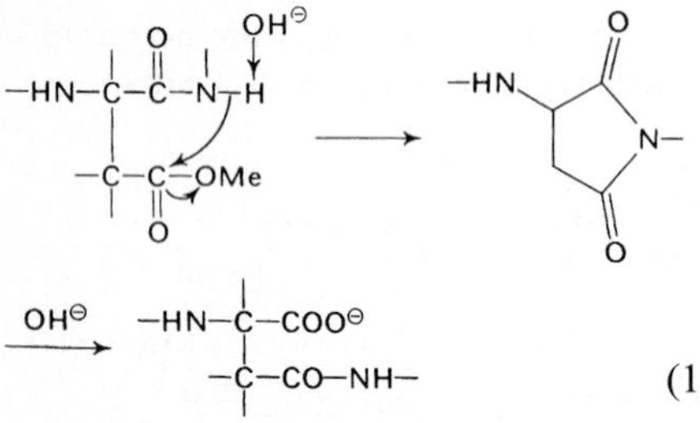

$$-HN-\overset{O}{\underset{}{C}}-\overset{OH^\ominus}{\underset{}{C}}-N\!-\!H \quad\longrightarrow\quad -HN$$

$$\overset{OH^\ominus}{\longrightarrow} \quad -HN-\overset{|}{\underset{|}{C}}-COO^\ominus \qquad\qquad (1.50)$$
$$-\overset{|}{\underset{|}{C}}-CO-NH-$$

1.2.4.3.4 Serine and Threonine

Acidic or alkaline hydrolysis of protein can yield α-keto acids through β-elimination of a water molecule:

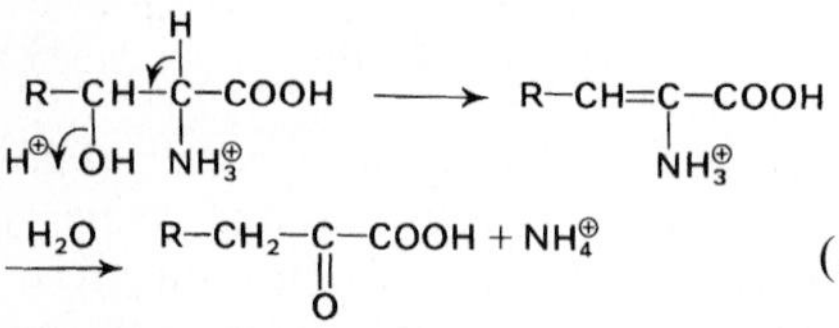

$$R-\overset{H}{\underset{H^\oplus OH}{CH}}\overset{}{\underset{NH_3^\oplus}{C}}-COOH \longrightarrow R-CH=\overset{}{\underset{NH_3^\oplus}{C}}-COOH$$

$$\overset{H_2O}{\longrightarrow} R-CH_2-\overset{}{\underset{O}{C}}-COOH + NH_4^\oplus \qquad (1.51)$$

In this way, α-ketobutyric acid formed from threonine can yield another amino acid, α-aminobutyric acid, via a transamination reaction. Reaction 1.51 is responsible for losses of hydroxy amino acids during protein hydrolysis.

Reliable estimates of the occurrence of these amino acids are obtained by hydrolyzing protein for varying lengths of time and extrapolating the results to zero time.

1.2.4.3.5 Cysteine and Cystine

Cysteine is readily converted to the corresponding disulfide, cystine, even under mild oxidative conditions, such as treatment with I_2 or potassium hexacyanoferrate(III). Reduction of cystine to cysteine is possible using sodium borohydride or thiol reagents (mercaptoethanol, dithiothreitol):

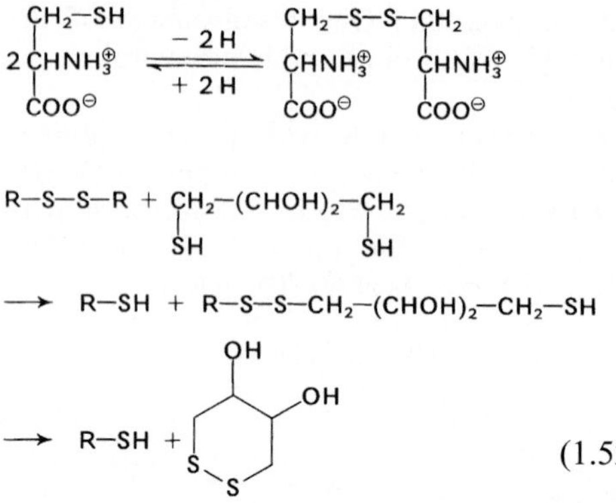

$$2\;\overset{CH_2-SH}{\underset{COO^\ominus}{\overset{|}{CHNH_3^\oplus}}}\;\underset{+2H}{\overset{-2H}{\rightleftharpoons}}\;\overset{CH_2-S-S-CH_2}{\underset{COO^\ominus\qquad COO^\ominus}{\overset{|}{CHNH_3^\oplus}\quad\overset{|}{CHNH_3^\oplus}}}$$

$$R-S-S-R + \overset{}{\underset{SH}{CH_2}}-(CHOH)_2-\overset{}{\underset{SH}{CH_2}}$$

$$\longrightarrow\; R-SH + R-S-S-CH_2-(CHOH)_2-CH_2-SH$$

$$\longrightarrow\; R-SH + \qquad\qquad\qquad\qquad (1.52)$$

The equilibrium constants for the reduction of cystine at pH 7 and 25 °C with mercaptoethanol or dithiothreitol are 1 and 10^4, respectively.

Stronger oxidation of cysteine, e.g., with performic acid, yields the corresponding sulfonic acid, cysteic acid:

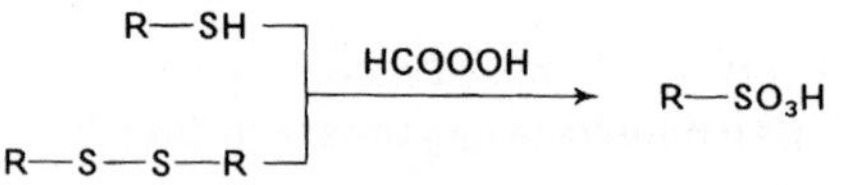

$$\left.\begin{array}{l} R-SH \\ \\ R-S-S-R \end{array}\right\}\;\xrightarrow{HCOOOH}\; R-SO_3H \qquad (1.53)$$

Reaction of cysteine with alkylating agents yields thioethers. Iodoacetic acid, iodoacetamide, dimethylaminoazobenzene iodoacetamide, ethylenimine and vinylpyridine are the most commonly used alkylating agents:

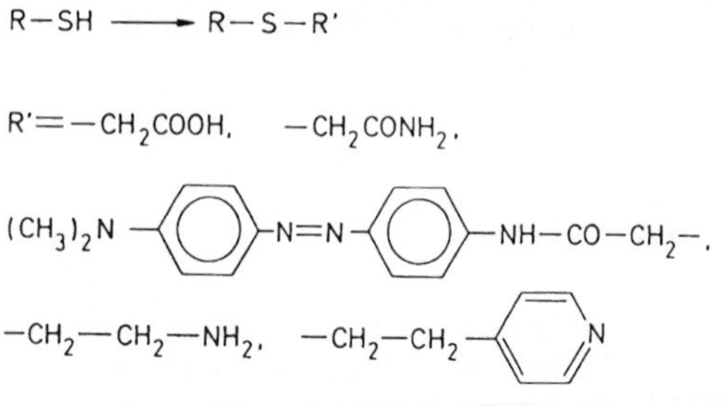

$$R-SH \longrightarrow R-S-R'$$

$$R'=-CH_2COOH, \quad -CH_2CONH_2,$$

$$(CH_3)_2N-\!\!\!\!\bigcirc\!\!\!\!-N=N-\!\!\!\!\bigcirc\!\!\!\!-NH-CO-CH_2-,$$

$$-CH_2-CH_2-NH_2, \quad -CH_2-CH_2-\!\!\!\!\bigcirc$$

$$(1.54)$$

1.2.4.3.6 Methionine

Methionine is readily oxidized to the sulfoxide and then to the sulfone. This reaction can result in losses of this essential amino acid during food processing:

$$R-S-CH_3 \rightarrow R-\overset{\displaystyle O}{\underset{}{S}}-CH_3 \rightarrow R-\overset{\displaystyle O}{\underset{\displaystyle O}{S}}-CH_3$$

(1.55)

1.2.4.3.7 Tyrosine

Tyrosine reacts, like histidine, with diazotized sulfanilic acid (*Pauly* reagent). The coupled-reaction product is a red azo compound:

(1.56)

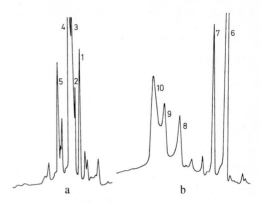

Fig. 1.6. Reaction products from cysteine heated in tributyrin (reflux at 200–220°C for 2h) (according to *Severin* and *Ledl*, 1975). Gas chromatographic separation of a light (a) and a less volatile (b) fraction. Separation conditions: 20% carbowax on kieselgel, column lenght 5 m, diameter 1 cm; column temperature 70° (a), 80° (b), and after 15 min programming with 5°C/min till 220°C. 1 2-Methylthiazoline, 2 2-methylthiazolidine, 3 2-propylthiazole, 4 2-propylthiazoline, 5 butyric acid, 6 N-ethylbutyric acid amide, 7 butyric acid amide, 8, 9, 10 butyric acid mono- and diglycerides, 11 2,5-dimethyldioxopiperazine, 12 N,N′-dibutyrylcystamine

1.2.4.4 Reactions of Amino Acids at Higher Temperatures

Reactions at elevated temperatures are important during the preparation of food. Frying, roasting, boiling and baking develop the typical aromas of many foods in which amino acids participate as precursors. Studies with food and model systems have shown that the characteristic odorants are formed via the *Maillard* reaction and that they are subsequent products, in particular of cysteine, methionine, ornithine and proline (cf. 5.3.1, 12.9).

The following model reaction, in which cysteine was kept in tributyrin for 2 h at 200–220 °C, provides a concept of subsequent products which can be formed when amino acids are heated in the presence of a triglyceride.
Figure 1.6 presents the separation of the reaction products by gas chromatography. The following reaction pathways are assumed to occur (the numerals in Reaction 1.57 refer to Fig. 1.6):

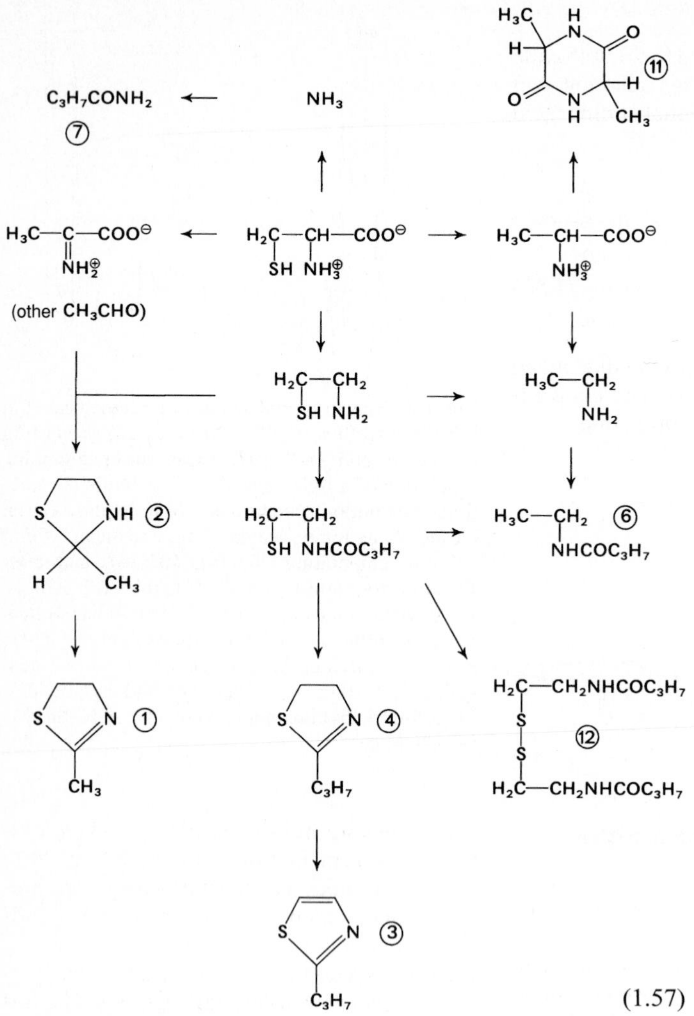

The reaction scheme (1.57) is shown here.

$$C_3H_7CONH_2 \quad \longleftarrow \quad NH_3$$
$$\textcircled{7}$$

$$H_3C-\underset{\underset{NH_2^{\oplus}}{\parallel}}{C}-COO^{\ominus} \quad \longleftarrow \quad H_2C-\underset{\underset{SH \; NH_3^{\oplus}}{|}}{CH}-COO^{\ominus} \quad \longrightarrow \quad H_3C-\underset{\underset{NH_3^{\oplus}}{|}}{CH}-COO^{\ominus}$$

(other CH_3CHO)

$$H_2C-CH_2$$
$$\underset{SH \quad NH_2}{|} \quad \longrightarrow$$

$$H_3C-CH_2$$
$$\underset{NH_2}{|}$$

$$\underset{H}{\overset{S}{\diagup}}\underset{CH_3}{\diagdown} NH \;\textcircled{2}$$

$$H_2C-CH_2 \atop SH \; NHCOC_3H_7 \quad \longrightarrow \quad H_3C-CH_2 \;\textcircled{6} \atop NHCOC_3H_7$$

$$\underset{CH_3}{\diagup S \diagdown N} \;\textcircled{1}$$

$$\underset{C_3H_7}{\diagup S \diagdown N} \;\textcircled{4}$$

$$H_2C-CH_2NHCOC_3H_7 \atop S \atop S \atop H_2C-CH_2NHCOC_3H_7} \;\textcircled{12}$$

$$\underset{C_3H_7}{\diagup S \diagdown N} \;\textcircled{3}$$

$$(1.57)$$

Fat acts as an acylating agent at high temperatures. For example, 2-propylthiazoline is obtained as the main reaction product from the acylation of cysteamine (generated by the decarboxylation of cysteine), followed by cyclization. Dehydrogenation to the corresponding thiazole can occur in the presence of oxygen or free radicals. The formation of N-ethyl-butyramide in relatively high yield is explained by a radical-type elimination of sulfur during pyrolysis of cystine. The occurrence of alanine and ethylamine in the pyrolysis reaction products is explained in a similar fashion.

In the late 1970s it was shown that charred surface portions of barbecued fish and meat as well as the smoke condensates captured in barbecuing have a highly mutagenic effect in microbial tests (*Salmonella typhimurium* tester strain TA 98). In model tests it could be demonstrated that pyrolyzates of amino acids and proteins are responsible for that effect. Table 1.6 lists the mutagenic compounds isolated from amino acid pyrolyzates. They are pyridoindoles, pyridoimidazoles and tetraazafluoroanthenes.

Table 1.6. Mutagenic compounds from pyrolysates of amino acids and proteins

Mutagenic compound	Short form	Pyrolized compound	Structure
3-Amino-1,4-dimethyl-5H-pyrido[4,3-b]indole	Trp-P-1	Tryptophan	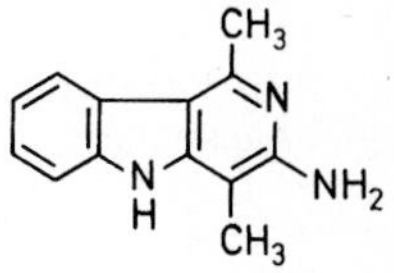
3-Amino-1-methyl-5H-pyrido[4,3-b]indole	Trp-P-2	Tryptophan	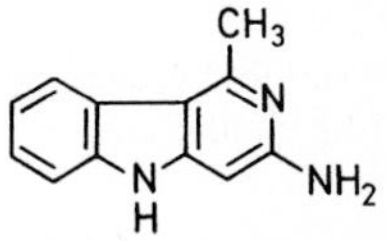
2-Amino-6-methyldipyrido[1,2-a:3′,2′-d]imidazole	Glu-P-1	Glutamic acid	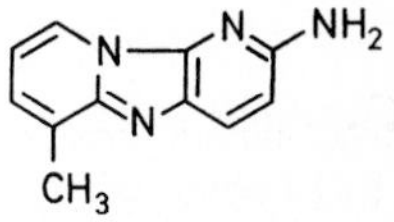
2-Aminodipyrido[1,2-a:3′,2′-d]imidazole	Glu-P-2	Glutamic acid	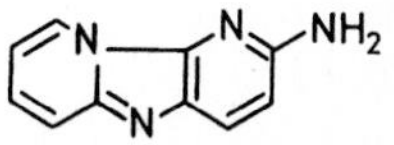
3,4-Cyclopentenopyrido[3,2-a]carbazole	Lys-P-1	Lysine	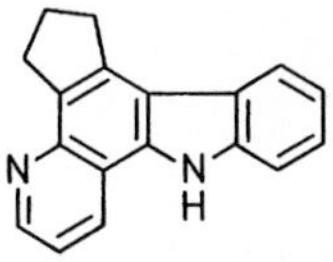
4-Amino-6-methyl-1H-2,5,10,10b-tetraazafluoroanthene	Orn-P-1	Ornithine	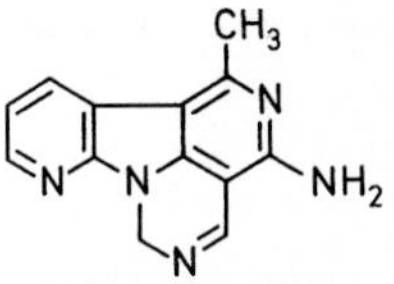
2-Amino-5-phenylpyridine	Phe-P-1	Phenylalanine	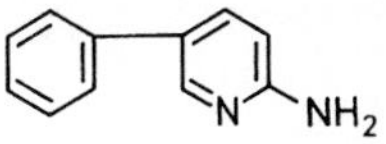
2-Amino-9H-pyrido[2,3-b]indole	AαC	Soya globulin	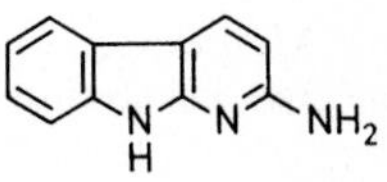
2-Amino-3-methyl-9H-pyrido[2,3-b]indole	MeAαC	Soya globulin	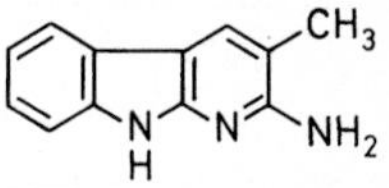

At the same time, it was found that mutagenic compounds of amino acids and proteins can also be formed at lower temperatures. The compounds listed in Table 1.7 were obtained from meat extract, deep-fried meat, grilled fish and heated model mixtures on the basis of creatine, an amino acid (glycine, alanine, threonine) and glucose. For the most part they were imidazoquinolines and imidazoquinoxalines.

Table 1.7. Mutagenic compounds from various heated foods and from model systems

Mutagenic compound	Short form	Food Model system[a]	Structure
2-Amino-3-methylimidazo-[4,5-f]quinoline	IQ	1,2,3	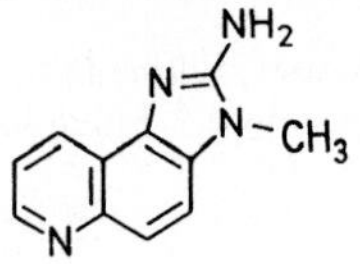
2-Amino-3,4-methylimidazo-[4,5-f]quinoline	MeIQ	3	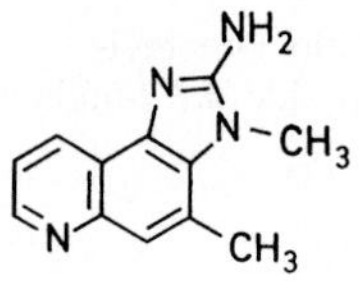
2-Amino-3,8-dimethylimidazo-[4,5-f]quinoxaline	MeIQx	2,3	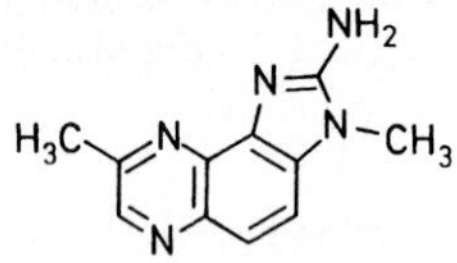
2-Amino-3,4,8-trimethyl-imidazo-[4,5-f]quinoxaline	4,8-Di MeIQx	2,3,5,6	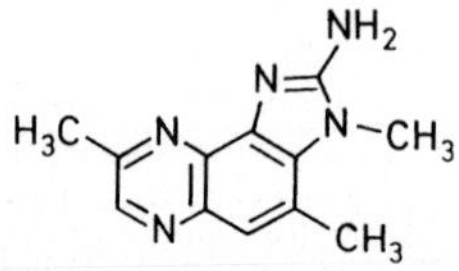
2-Amino-3,7,8-trimethyl-imidazo-[4,5-f]quinoxaline	7,8-Di MeIQx	4	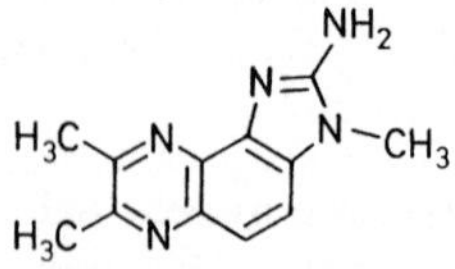
2-Amino-1-methyl-6-phenyl-imidazo[4,5-b]pyridine	PhIP	2	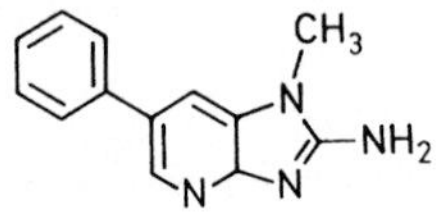

[a] 1: Meat extract; 2: Deep-fried meat; 3: Grilled fish; 4: Model mixture of creatinine, glycine, glucose; 5: as 4, but alanine; 6: as 4, but threonine

It is assumed that they are formed from creatinine, subsequent products of the *Maillard* reaction (pyridines, pyrazines, cf. 4.2.4.4) and amino acids in the following manner:

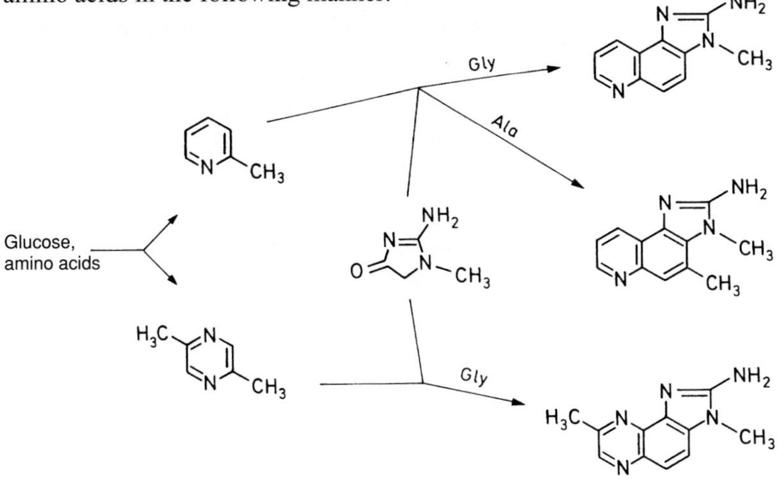

(1.58)

Toxicity is based on the heteroaromatic amino function. The compounds listed in Table 1.6 can be deaminated by nitrite in weakly acid solution and can thus be inactivated, while the compounds listed in Table 1.7 do not react in the imidazole ring because of their guanidine structure.

The β-carbolines norharmane (I, R=H) and harmane (I, R=CH$_3$) are well known as components of tobacco smoke. They are formed by a reaction of tryptophan and formaldehyde or acetaldehyde:

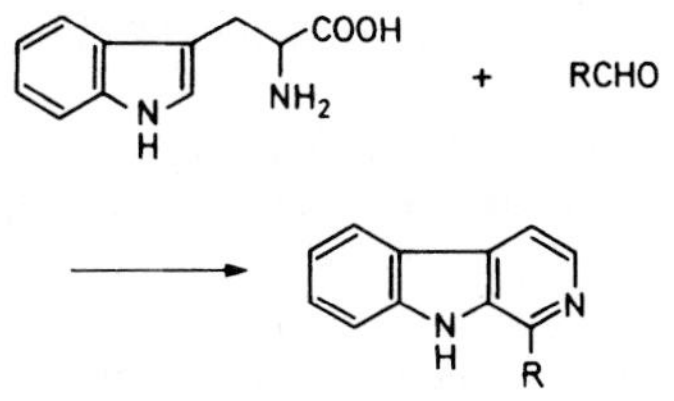

(1.59)

Tetrahydro-β-carboline-3-carboxylic acid (II) and (1S, 3S)-(III) and (1R, 3S)-methyltetrahydro-β-carboline-3-carboxylic acid (IV) were detected in beer (II: 2–11 mg/L, III + IV: 0.3–4 mg/L) and wine (II: 0.8–1.7 mg/L, III + IV: 1.3–9.1 mg/L). The ratio of diastereomers III and IV (Formula 1.60) was always near 2:1.

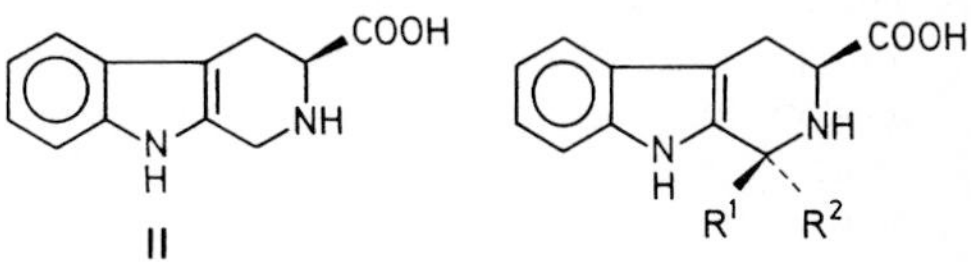

III   R^1 = CH$_3$, R^2 = H
IV   R^1 = H,   R^2 = CH$_3$

(1.60)

The compounds are pharmacologically active.

1.2.5 Synthetic Amino Acids Utilized for Increasing the Biological Value of Food (Food Fortification)

The daily requirements of humans for essential amino acids and their occurrence in some important food proteins are presented in Table 1.8. The biological value of a protein (g protein formed in the body/100 g food protein) is determined by the absolute content of essential amino acids, by the relative proportions of essential amino acids, by their ratios to nonessential amino acids and by factors such as digestibility and availability. The most important (more or less expensive) *in vivo* and *in vitro* methods for determining the biological valence are based on the following principles:

Table 1.8. Adult requirement for essential amino acids and their occurrence in various food

Amino acid	1	2	3	4	5	6	7	8	9
Isoleucine	10–11	3.5	4.0	4.6	3.9	3.6	3.4	5.0	3.5
Leucine	11–14	4.2	5.3	7.1	4.3	5.1	6.5	8.2	5.4
Lysine	9–12	3.5	3.7	4.9	3.6	4.4	2.0	3.6	5.4
Methionine + Cystine	11–14	4.2	3.2	2.6	1.9	2.1	3.8	3.4	1.9
Methionine		2.0	1.9	1.9	1.2	0.9	1.4	2.2	0.8
Phenylalanine + Tyrosine	13–14	4.5	6.1	7.2	5.8	5.5	6.7	8.9	6.0
Phenylalanine		2.4	3.5	3.5	3.1	3.3	4.6	4.7	2.5
Threonine	6– 7	2.2	2.9	3.3	2.9	2.7	2.5	3.7	3.8
Tryptophan	3	1.0	1.0	1.0	1.0	1.0	1.0	1.0	1.0
Valine	11–14	4.2	4.3	5.6	3.6	3.3	3.8	6.4	4.1
Tryptophan[a]			1.7	1.4	1.4	1.5	1.1	1.0	1.3

1: Daily requirement in mg/kg body weight.
2–8: Relative value related to Trp = 1 (pattern).
2: Daily requirements, 3: eggs, 4: bovine milk, 5: potato, 6: soya, 7: wheat flour, 8: rice, and 9: *Torula*-yeast.

[a] Tryptophan (%) in raw protein.

- Replacement of endogenous protein after protein depletion.
  The test determines the amount of endogenous protein that can be replaced by 100 g of food protein. The test person is given a nonprotein diet and thus reduced to the absolute N minimum. Subsequently, the protein to be examined is administered, and the N balance is measured. The biological valence (BV) follows from

$$BV = \frac{\text{Urea-N (non-protein diet)} + \text{N balance}}{\text{N intake}} \times 100$$

"Net protein utilization" (NPU) is based on the same principle and is determined in animal experiments. A group of rats is fed a non-protein diet (Gr 1), while the second group is fed the protein to be examined (Gr 2). After some time, the animals are killed, and their protein content is analyzed. The biological valence follows from

$$NPU = \frac{\text{Protein content Gr 2} - \text{protein content Gr 1}}{\text{Protein intake}} \times 100$$

- Utilization of protein for growth.
  The growth value (protein efficiency ratio = PER) of laboratory animals is calculated according to the following formula:

$$PER = \frac{\text{Weight gain (g)}}{\text{Available protein (g)}}$$

- Maintenance of the N balance.
- Plasma concentration of amino acids.
- Calculation from the amino acid composition.
- Determination by enzymatic cleavage *in vitro*.

Table 1.9 lists data about the biological valence of some food proteins, determined according to different methods.

Table 1.9. Biological valence of some food proteins determined according to different methods[a]

Protein from	Biological valence			Limiting amino acid
	BV	NPU	PER	
Chicken egg	94	93	3.9	
Cow's milk	84	81	3.1	Met
Fish	76	80	3.5	Thr
Beef	74	67	2.3	Met
Potatoes	73	60	2.6	Met
Soybeans	73	61	2.3	Met
Rice	64	57	2.2	Lys, Tyr
Beans	58	38	1.5	Met
Wheat flour (white)	52	57	0.6	Lys, Thr

[a] The methods are explained in the text.

The highest biological value observed is for a blend of 35 % egg and 65 % potato proteins. The biological value of a protein is generally limited by:

- Lysine: deficient in proteins of cereals and other plants
- Methionine: deficient in proteins of bovine milk and meat
- Threonine: deficient in wheat and rye
- Tryptophan: deficient in casein, corn and rice.

Since food is not available in sufficient quantity or quality in many parts of the world, increasing its biological value by addition of essential amino acids is gaining in importance. Illuminating examples are rice fortification with L-lysine and L-threonine, supplementation of bread with L-lysine and fortification of soya and peanut protein with methionine. Table 1.10 lists data about the increase in biological valence of some food proteins through the addition of amino acids. Synthetic amino acids

Table 1.10. Increasing the biological valence (PER[a]) of some food proteins through the addition of amino acids

Protein from	without	Addition (%)			
		0.2 Lys	0.4 Lys	0.4 Lys 0.2 Thr	0.4 Lys 0.07 Thr	0.4 Lys 0.07 Thr 0.2 Thr
Casein (Reference)	2.50				
Wheat flour	0.65	1.56	1.63	2.67	
Corn	0.85		1.08		2.50	2.59

[a] The method is explained in the text.

are used also for chemically defined diets which can be completely absorbed and utilized for nutritional purposes in space travel, in pre- and post-operative states, and during therapy for maldigestion and malabsorption syndromes.

The fortification of animal feed with amino acids (0.05−0.2%) is of great significance.

These demands have resulted in increased production of amino acids. Table 1.11 gives

Table 1.11. World production of amino acids, 1982

Amino acid	t/year	Process[a]				Mostly utilized as
		1	2	3	4	
L-Ala	130		+		+	Flavoring compound
D,L-Ala	700	+				Flavoring compound
L-Arg	500			+	+	Infusion Therapeutics
L-Asp	250		+		+	Therapeutics Flavoring compound
L-Asn	50				+	Therpeutics
L-CySH	700				+	Baking additive Antioxidant
L-Glu	270.000			+		Flavoring compound, flavor enhancer
L-Gln	500			+		Therapeutics
Gly	6.000	+				Sweetener
L-His	200			+	+	Therapeutics
L-Ile	150			+	+	Infusion
L-Leu	150			+	+	Infusion
L-Lys	32.000			+	+	Feed ingredient
L-Met	150		+			Therapeutics
D,L-Met	110.000	+				Feed ingredient
L-Phe	150			+	+	Infusion
L-Pro	100			+	+	Infusion
L-Ser	50			+	+	Cosmetics
L-Thr	160			+	+	Food additive
L-Trp	200			+	+	Infusion
L-Tyr	100				+	Infusion
L-Val	150		+		+	Infusion

[a] 1: Chemical synthesis, 2: protein hydrolysis, 3: microbiological procedure, 4: isolation from raw materials.

data for world production in 1982. The production of L-glutamic acid, used to a great extent as a flavor enhancer, is exceptional. Production of methionine and lysine is also significant.

Four main processes are distinguished in the production of amino acids: chemical synthesis, isolation from protein hydrolysates, enzymatic and microbiological methods of production, which is currently the most important. The following sections will further elucidate the important industrial processes for a number of amino acids.

1.2.5.1 Glutamic Acid

Acrylnitrile is catalytically formylated with CO/H$_2$ and the resultant aldehyde is transformed through a *Strecker* reaction into glutamic acid dinitrile which yields D,L-glutamic acid after alkaline hydrolysis. Separation of the racemate is achieved by preferential crystallization of the L-form from an oversaturated solution after seeding with L-glutamic acid:

$$H_2C{=}CH{-}CN \xrightarrow[\;[Co(CO)_4]_2\;]{CO/H_2} OHC{-}CH_2{-}CH_2{-}CN$$

$$\xrightarrow{HCN/NH_3} \underset{\underset{NH_2}{|}}{NC{-}CH}{-}CH_2{-}CH_2{-}CN$$

$$\xrightarrow{OH^{\ominus}} \text{D,L-Glu} \qquad (1.61)$$

A fermentation procedure with various selected strains of microorganisms (*Brevibacterium flavum*, *Brev. roseum*, *Brev. saccharolyticum*) provides L-glutamic acid in yields of 50 g/l of fermentation liquid:

$$CH_3COONH_4 \text{ (20 g/l)} \xrightarrow[pH\ 7.5]{MO} \text{L-Glu (50 g/l)} \qquad (1.62)$$

1.2.5.2 Aspartic Acid

Aspartic acid is obtained in 90% yield from fumaric acid by using the aspartase enzyme:

$$\text{Fumaric acid} \xrightarrow[NH_3]{Aspartase} \text{L-Asp} \qquad (1.63)$$

1.2.5.3 Lysine

A synthetic procedure starts with caprolactam, which possesses all the required structural features, except for the α-amino group which is introduced in several steps:

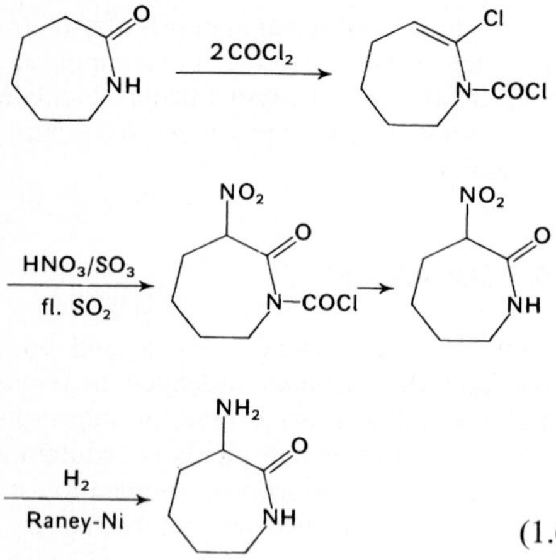

(1.64)

Separation of isomers is done at the α-amino caprolactam (Acl) step through the sparingly soluble salt of the L-component with L-pyrrolidone carboxylic acid (Pyg):

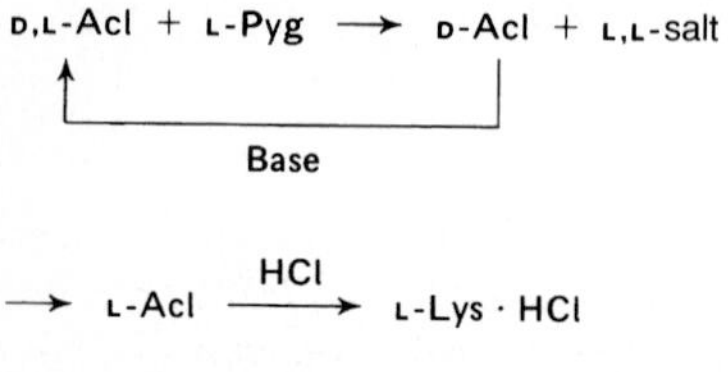

D,L-Acl + L-Pyg $\longrightarrow$ D-Acl + L,L-salt

Base

$\longrightarrow$ L-Acl $\xrightarrow{\text{HCl}}$ L-Lys · HCl   (1.65)

More elegant is selective hydrolysis of the L-enantiomer by an L-α-amino-ε-caprolactamase which occurs in several yeasts, for example in *Cryptococcus laurentii*. The racemization of the remaining D-isomers is possible with a racemase of *Achromobacter obae*. The process can be performed as a one-step reaction: the racemic aminocaprolactam is incubated with intact cells of *C. laurentii* and *A. obae*, producing almost 100% L-lysine.

In another procedure, acrylnitrile and ethanal react to yield cyanobutyraldehyde which is then transformed by a *Bucherer* reaction into cyanopropylhydantoin. Catalytic hydrogena-

tion of the nitrile group, followed by alkaline hydrolysis yields D,L-lysine. The isomers can be separated through the sparingly soluble L-lysine sulfanilic acid salt:

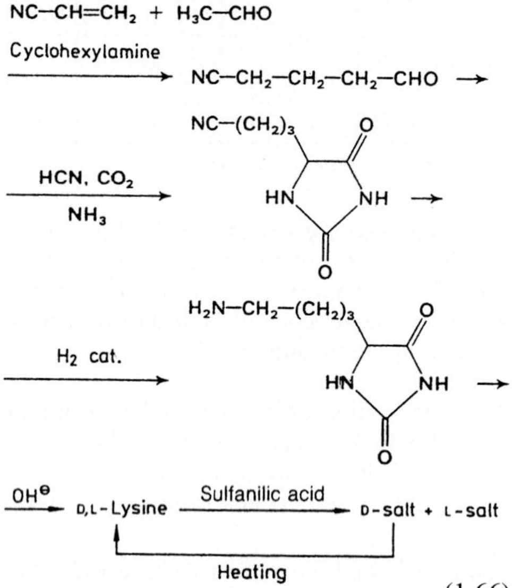

NC—CH=CH₂ + H₃C—CHO

Cyclohexylamine

$\xrightarrow{\hspace{2cm}}$ NC—CH₂—CH₂—CH₂—CHO $\longrightarrow$

OH⊖ D,L-Lysine $\xrightarrow{\text{Sulfanilic acid}}$ D-salt + L-salt

Heating

(1.66)

Fermentation with a pure culture of *Brevibacterium lactofermentum* or *Micrococcus glutamicus* produces L-lysine directly:

CH₃COONH₄ $\xrightarrow[\text{pH 7-8.5}]{\text{MO}}$ L-Lys (40–90 g/l)

(<15 g/l)

(1.67)

1.2.5.4 Methionine

Interaction of methanethiol with acrolein produces an aldehyde which is then converted to the corresponding hydantoin through a *Bucherer* reaction. The product is hydrolyzed by alkaline catalysis. Separation of the resultant racemate is usually not carried out since the D-form of methionine is utilized by humans via transamination:

CH₃SH + H₂C=CH—CHO $\longrightarrow$ H₃C—S—CH₂CH₂—CHO

$\xrightarrow[\text{2) OH⊖}]{\text{1) HCN, CO}_2\text{, NH}_3}$ D,L-Met

(1.68)

1.2.5.5 Phenylalanine

Benzaldehyde is condensed with hydantoin, then hydrogenation using a chiral catalyst gives a product which is about 90% L-phenylalanine:

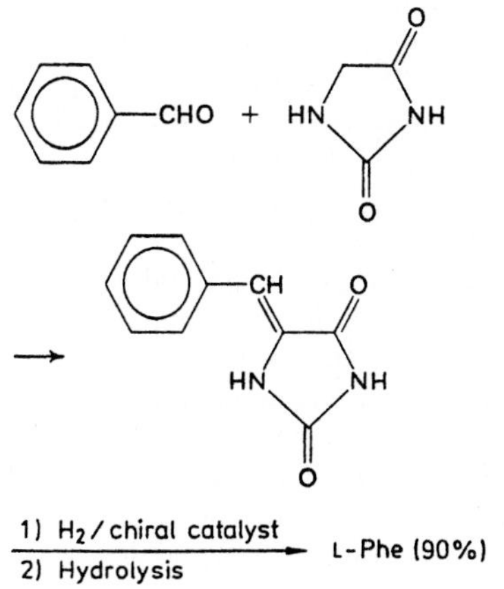

$$\begin{array}{l}\text{1) } H_2 / \text{chiral catalyst} \\ \text{2) Hydrolysis}\end{array} \longrightarrow \text{L-Phe (90\%)} \qquad (1.69)$$

1.2.5.6 Threonine

Interaction of a copper complex of glycine with ethanal yields the *threo* and *erythro* isomers in the ratio of 2:1. They are separated on the basis of their differences in solubility:

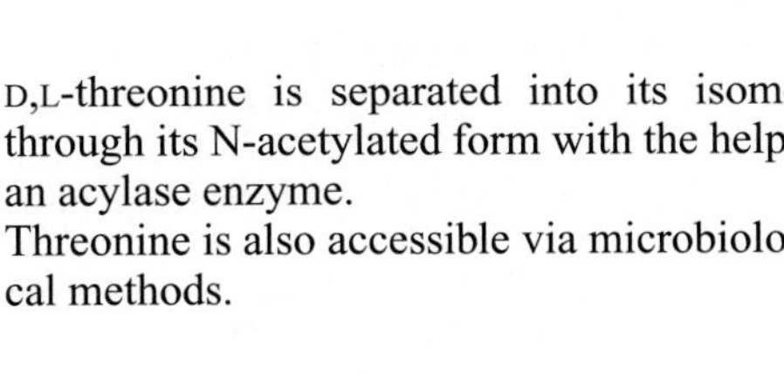

$$\qquad (1.70)$$

D,L-threonine is separated into its isomers through its N-acetylated form with the help of an acylase enzyme.
Threonine is also accessible via microbiological methods.

1.2.5.7 Tryptophan

Tryptophan is obtained industrially by a variation of the *Fischer* indole synthesis. Addition of hydrogen cyanide to acrolein gives 3-cya-nopropanal which is converted to hydantoin through a *Bucherer* reaction. The nitrile group is then reduced to an aldehyde group. Reaction with phenylhydrazine produces an indole derivative. Lastly, hydantoin is saponified with alkali:

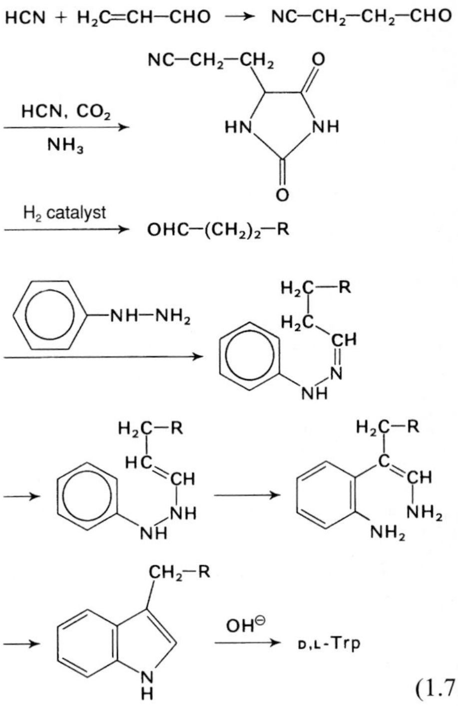

$$HCN + H_2C{=}CH{-}CHO \longrightarrow NC{-}CH_2{-}CH_2{-}CHO$$

$$\xrightarrow{\ OH^{\ominus}\ } \text{D,L-Trp} \qquad (1.71)$$

L-Tryptophan is also produced through enzymatic synthesis from indole and serine with the help of tryptophan synthase:

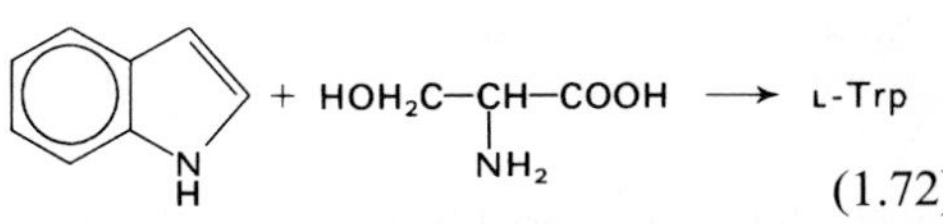

$$(1.72)$$

1.2.6 Sensory Properties

Free amino acids can contribute to the flavor of protein-rich foods in which hydrolytic processes occur (e.g. meat, fish or cheese).
Table 1.12 provides data on taste quality and taste intensity of amino acids. Taste quality is influenced by the molecular configuration:

Table 1.12. Taste of amino acids in aqueous solution at pH 6−7

sw − sweet, bi − bitter, neu − neutral

Amino acid	Taste			
	L-Compound		D-Compound	
	Quality	Intensity[a]	Quality	Intensity[a]
Alanine	sw	12−18	sw	12−18
Arginine	bi		neu	
Asparagine	neu		sw	3−6
Aspartic acid	neu		neu	
Cystine	neu		neu	
Glutamine	neu		neu	
Glutamic acid	meat broth like		neu	
Glycine[b]	sw	25−35		
Histidine	bi	45−50	sw	2−4
Isoleucine	bi	10−12	sw	8−12
Leucine	bi	11−13	sw	2−5
Lysine	sw		sw	
	bi	80−90		
Methionine	sulphurous		sulphurous	
			sw	4−7
Phenylalanine	bi	5−7	sw	1−3
Proline	sw	25−40	neu	
	bi	25−27		
Serine	sw	25−35	sw	30−40
Threonine	sw	35−45	sw	40−50
Tryptophan	bi	4−6	sw	0.2−0.4
Tyrosine	bi	4−6	sw	1−3

1-Aminocycloalkane-1-carboxylic acid[b]

Cyclobutane derivative	sw	20−30	
Cyclopentane derivative	sw	3−6	
	bi	95−100	
Cyclohexane derivative	sw	1−3	
	bi	45−50	
Cyclooctane derivative	sw	2−4	
	bi	2−5	
Caffeine	bi	1−1.2	
Saccharose	sw	10−12	

[a] Recognition threshold value (mmol/l).
[b] Compounds not optically active.

sweet amino acids are primarily found among members of the D-series, whereas bitter amino acids are generally within the L-series. Consequently amino acids with a cyclic side chain (1-aminocycloalkane-1-carboxylic acids) are sweet and bitter.

The taste intensity of a compound is reflected in its recognition threshold value. The recognition threshold value is the lowest concentration needed to recognize the compound reliably, as assessed by a taste panel. Table 1.12 shows that the taste intensity of amino acids is dependent on the hydrophobicity of the side chain.

L-Tryptophan and L-tyrosine are the most bitter amino acids with a threshold value of $c_{t\,bitter} = 4−6$ mmol/l. D-Tryptophan, with $c_{t\,sweet} = 0.2−0.4$ mmol/l, is the sweetest amino acid. A comparison of these threshold values with those of caffeine ($c_{t\,bi} = 1−1.2$ mmole/l) and sucrose ($c_{t\,sw} = 10−12$ mmol/l) shows that caffeine is about 5 times as bitter as L-tryptophan and that D-tryptophan is about 37 times as sweet as sucrose.

L-Glutamic acid has an exceptional position. In higher concentrations it has an imitation meat broth flavor, while in lower concentrations, it enhances the characteristic flavor of a given food (flavor enhancer, cf. 8.6.1). L-Methionine has a sulfur-like flavor.

The bitter taste of the L-amino acids can interfere with the utilization of these acids, e.g. in chemically defined diets.

1.3 Peptides

1.3.1 General Remarks, Nomenclature

Peptides are formed by binding amino acids together through an amide linkage.
On the other hand peptide hydrolysis results in free amino acids:

$$2\,H_2N\!-\!\underset{\underset{R}{|}}{CH}\!-\!COOH$$

$$\underset{+\,H_2O}{\overset{-\,H_2O}{\rightleftharpoons}}\quad H_2N\!-\!\underset{\underset{R}{|}}{CH}\!-\!CO\!-\!NH\!-\!\underset{\underset{R}{|}}{CH}\!-\!COOH \qquad (1.73)$$

Functional groups not involved in the peptide synthesis reaction should be blocked. The protecting or blocking groups must be removed after synthesis under conditions which retain the stability of the newly formed peptide bonds:

$$X\!-\!NH\!-\!\underset{\underset{R^1}{|}}{CH}\!-\!COOH + H_2N\!-\!\underset{\underset{R^2}{|}}{CH}\!-\!COY$$

$$\overset{-\,H_2O}{\longrightarrow}\quad X\!-\!NH\!-\!\underset{\underset{R^1}{|}}{CH}\!-\!CO\!-\!NH\!-\!\underset{\underset{R^2}{|}}{CH}\!-\!COY$$

$$\overset{-\,X,\,-\,Y}{\longrightarrow}\quad H_2N\!-\!\underset{\underset{R^1}{|}}{CH}\!-\!CO\!-\!NH\!-\!\underset{\underset{R^2}{|}}{CH}\!-\!COOH \qquad (1.74)$$

Peptides are denoted by the number of amino acid residues as di-, tri-, tetrapeptides, etc., and the term "oligopeptides" is used for those with 10 or less amino acid residues. Higher molecular weight peptides are called polypeptides. The transition of "polypeptide" to "protein" is rather undefined, but the limit is commonly assumed to be at a molecular weight of about 10 kdal, i.e. about 100 amino acid residues are needed in the chain for it to be called a protein.

Peptides are interpreted as acylated amino acids:

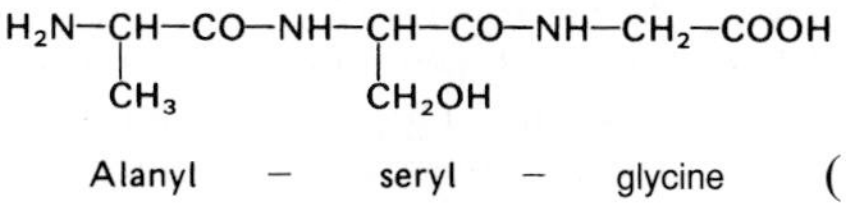

$$H_2N-\underset{\underset{CH_3}{|}}{CH}-CO-NH-\underset{\underset{CH_2OH}{|}}{CH}-CO-NH-CH_2-COOH$$

Alanyl — seryl — glycine   (1.75)

The first three letters of the amino acids are used as symbols to simplify designation of peptides (cf. Table 1.1). Thus, the peptide shown above can also be given as:

Ala—Ser—Gly   (1.76)

One-letter symbols (cf. Table 1.1) are used for amino acid sequences of long peptide chains.

D-Amino acids are denoted by the prefix D-. In compounds in which a functional group of the side chain is involved, the bond is indicated by a perpendicular line. The tripeptide glutathione (γ-glutamyl-cysteinyl-glycine) is given as an illustration along with its corresponding disulfide, oxidized glutathione:

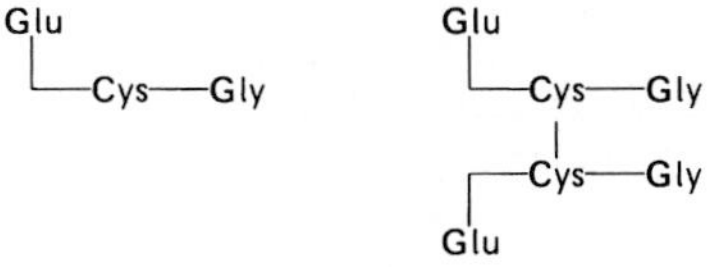

Glu
 └─Cys──Gly

Glu
 └─Cys──Gly
        |
    ┌─Cys──Gly
    Glu   (1.77)

By convention, the amino acid residue with the free amino group is always placed on the left. The amino acids of the chain ends are denoted as N-terminal and C-terminal amino acid residues. The peptide linkage direction in cyclic peptides is indicated by an arrow, i.e. $-CO \rightarrow NH-$.

1.3.2 Physical Properties

1.3.2.1 Dissociation

The pK values and isoelectric points for some peptides are listed in Table 1.13. The acidity of the free carboxyl groups and the basicity of the free amino groups are lower in peptides than in the corresponding free amino acids. The amino acid sequence also has an influence (e.g., Gly-Asp/Asp-Gly).

Table 1.13. Dissociation constants and isoelectric points of various peptides (25°C)

Peptide	pK_1	pK_2	pK_3	pK_4	pK_5	pI
Gly-Gly	3.12	8.17				5.65
Gly-Gly-Gly	3.26	7.91				5.59
Ala-Ala	3.30	8.14				5.72
Gly-Asp	2.81	4.45	8.60			3.63
Asp-Gly	2.10	4.53	9.07			3.31
Asp-Asp	2.70	3.40	4.70	8.26		3.04
Lys-Ala	3.22	7.62	10.70			9.16
Ala-Lys-Ala	3.15	7.65	10.30			8.98
Lys-Lys	3.01	7.53	10.05	11.01		10.53
Lys-Lys-Lys	3.08	7.34	9.80	10.54	11.32	10.93
Lys-Glu	2.93	4.47	7.75	10.50		6.10
His-His	2.25	5.60	6.80	7.80		7.30

1.3.3 Sensory Properties

While the taste quality of amino acids does depend on configuration, peptides, except for the sweet dipeptide esters of aspartic acid (see below), are neutral or bitter in taste with no relationship to configuration (Table 1.14). As with amino acids, the taste intensity is influenced by the hydrophobicity of the side chains (Table 1.15). The taste intensity does not appear to be dependent on amino acid sequence (Table 1.14).

Bitter tasting peptides can occur in food after proteolytic reactions. For example, the bitter taste of cheese is a consequence of faulty ripening. Therefore, the wide use of proteolytic enzymes to achieve well-defined modifications of food proteins, without producing a bitter taste, causes some problems. Removal of the bitter taste of a partially hydrolyzed protein is outlined in the section dealing with proteins modified with enzymes (cf. 1.4.6.3.2).

Table 1.14. Taste-threshold values of various peptides: effect of configuration and amino acid sequence (tested in aqueous solution at pH 6–7); bi – bitter

Peptide[a]	Taste	
	Quality	Intensity[b]
Gly-Leu	bi	19–23
Gly-D-Leu	bi	20–23
Gly-Phe	bi	15–17
Gly-D-Phe	bi	15–17
Leu-Leu	bi	4– 5
Leu-D-Leu	bi	5– 6
D-Leu-D-Leu	bi	5– 6
Ala-Leu	bi	18–22
Leu-Ala	bi	18–21
Gly-Leu	bi	19–23
Leu-Gly	bi	18–21
Ala-Val	bi	60–80
Val-Ala	bi	65–75
Phe-Gly	bi	16–18
Gly-Phe	bi	15–17
Phe-Gly-Phe-Gly	bi	1.0–1.5
Phe-Gly-Gly-Phe	bi	1.0–1.5

[a]   L-Configuration if not otherwise designated.
[b]   Recognition threshold value in mmol/l.

The sweet taste of aspartic acid dipeptide esters (I) was discovered by chance in 1969 for α-L-aspartyl-L-phenylalanine methyl ester ("Aspartame", "NutraSweet"). The corresponding peptide ester of L-aminomalonic acid (II) is also sweet.

A comparison of structures I, II and III reveals a relationship between sweet dipeptides and

$$(1.78)$$

sweet D-amino acids. The required configuration of the carboxyl and amino groups and the side chain substituent, R, is found only in peptide types I and II.

Since the discovery of the sweetness of compounds of type I, there has been a systematic study of the structural prerequisites for a sweet taste.

The presence of L-aspartic acid was shown to be essential, as was the peptide linkage through the α-carboxyl group.

R^1 may be an H or CH_3 group[*], while the R^2 and R^3 groups are variable within a certain range. Several examples are presented in Table 1.16. The sweet taste intensity passes through a maximum with increasing length and volume of the R^2 residue (e.g., COO-fenchyl ester is $22-23 \times 10^3$ times sweeter than sucrose). The size of the R^3 substituent is limited to a narrow range. Obviously, the R^2 substituent has the greatest influence on taste intensity.

---

* Data are not yet available for compounds with $R^1 > CH_3$.

Table 1.15. Bitter taste of dipeptide A–B: dependence of recognition threshold value (mmol/l) on side chain hydrophobicity (0: sweet or neutral taste)

A / B		Asp	Glu	Asn	Gln	Ser	Thr	Gly	Ala	Lys	Pro	Val	Leu	Ile	Phe	Tyr	Trp
		0	0	0	0	0	0	0	0	85	26	21	12	11	6	5	5
Gly	0[a]	–	–	–	–	–	–	0	0	–	45	75	21	20	16	17	13
Ala	0	–	–	–	–	–	–	0	0	–	–	70	20	–	–	–	–
Pro	26	–	–	–	–	–	–	–	–	–	–	–	6	–	–	–	–
Val	21	–	–	–	–	–	–	65	70	–	–	20	10	–	–	–	–
Leu	12	–	–	–	–	–	–	20	20	–	–	–	4.5	–	–	3.5	0.4
Ile	11	43	43	33	33	33	33	21	21	23	4	9	5.5	5.5	–	–	0.9
Phe	6	–	–	–	–	–	–	17	–	–	2	–	1.4	–	0.8	0.8	–
Tyr	5	–	–	–	–	–	–	–	–	–	–	–	4	–	–	–	–
Trp	5	–	28	–	–	–	–	–	–	–	–	–	–	–	–	–	–

[a]   Threshold of the amino acid (cf. Table 1.12).

The following examples show that R^2 should be relatively large and R^3 relatively small: L-Asp-L-Phe-OMe (aspartame, $R^2 - CH_2C_6H_5$, $R^3 = COOMe$) is almost as sweet ($f_{sac,g}$ (1) = 180) as L-Asp-D-Ala-OPr ($f_{sac,g}$ (0.6) = 170), while L-Asp-D-Phe-OMe has a bitter taste.

In the case of acylation of the free amino group of aspartic acid, the taste characteristics depend on the introduced group. Thus, D-Ala-L-Asp-L-Phe-OMe is sweet ($f_{sac,g}$ (0.6) = 170), while L-Ala-L-Asp-L-Phe-OMe is not. It should be noted that superaspartame is extremely sweet (cf. 8.8.17.2).

The intensity of the salty taste of Orn-β-Ala depends on the pH (Table 1.18).

Some peptides exhibit a salty taste, e.g. ornithyl-β-alanine hydrochloride (Table 1.17) and may be used as substitutes for sodium chloride.

Table 1.16. Taste of dipeptide esters of aspartic acid[a] and of amino malonic acid[b]

R^2	R^3	Taste[c]
COOCH₃	H	8
n-C₃H₆	COOCH₃	4
n-C₄H₇	COOCH₃	45
n-C₄H₉	COOC₂H₅	5
n-C₆H₁₃	CH₃	10
n-C₇H₁₅	CH₃	neutral
COOCH(CH₃)₂	nC₃H₇	17
COOCH(CH₃)₂	n-C₄H₉	neutral
COOCH₃	CH₂C₆H₅	bitter
CH(CH₃)C₂H₅	COOCH₃	bitter
CH₂CH(CH₃)₂	COOCH₃	bitter
CH₂C₆H₅	COOCH₃	140
COO-2-methyl-cyclohexyl	COOCH₃	5– 7,000
COO-fenchyl	COOCH₃	22–33,000

[a] Formula 1.78 I, $R^1 = H$.
[b] Formula 1.78 II, $R^1 = H$.
[c] For sweet compounds the factor $f_{sac,g}$ is given, related to the threshold value of a 10% saccharose solution (cf. 8.8.1.1).

Table 1.17. Peptides with a salty taste

Peptide[a]	Taste	
	Threshold (mmol/l)	Quality[b]
Orn-βAla.HCl	1.25	3
Orn-γAbu.HCl	1.40	3
Orn-Tau.HCl	3.68	4
Lys-Tau.HCl	5.18	4
NaCl	3.12	3

[a] Abbreviations: Orn, ornithine; β-Ala, β-alanine, γ-Abu, γ-aminobutyric acid; Tau, taurine.
[b] The quality of the salty taste was evaluated by rating it from 0 to 5 on a scale in comparison with a 6.4 mmol/L NaCl solution (rated 3); 4 is slightly better, 5 clearly better than the control solution.

Table 1.18. Effect of HCl on the salty taste of Orn-β-Ala[a]

Equivalents HCl	pH	Taste	
		salty[b]	sour[c]
0	8.9	0	
0.79	7.0	0	
0.97	6.0	1	
1.00	5.5	2	
1.10	4.7	3	+/–
1.20	4.3	3.5	+
1.30	4.2	3	++

[a] Peptide solution: 30 mmol/L.
[b] The values 1, 3 and 5 correspond in intensity to 0.5%, 0.25% and 0.1% NaCl solutions respectively.
[c] Very weak (+) and slightly sour (++).

1.3.4 Individual Peptides

Peptides are widespread in nature. They are often involved in specific biological activities (peptide hormones, peptide toxins, peptide antibiotics). A number of peptides of interest to food chemists are outlined in the following sections.

1.3.4.1 Glutathione

Glutathione (γ-L-glutamyl-L-cysteinyl-glycine) is widespread in animals, plants and microorganisms. A noteworthy feature is the binding of glutamic acid through its γ-carboxyl group. The peptide is the coenzyme of glyoxalase.

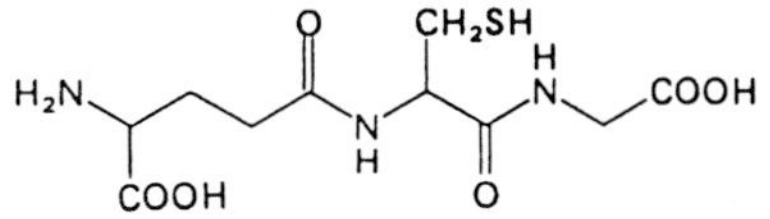

(1.79)

It is involved in active transport of amino acids and, due to its ready oxidation, is also involved in many redox-type reactions. It influences the rheological properties of wheat flour dough through thiol-disulfide interchange with wheat gluten. High concentrations of reduced glutathione in flour bring about reduction of protein disulfide bonds and a corresponding decrease in molecular weight of some of the protein constituents of dough gluten (cf. 15.4.1.4.1).

1.3.4.2 Carnosine, Anserine and Balenine

These peptides are noteworthy since they contain a β-amino acid, β-alanine, bound to L-histidine or 1-methyl- or 3-methyl-L-histidine, and are present in meat extract and in muscle of vertebrates (cf. Formula 1.80).

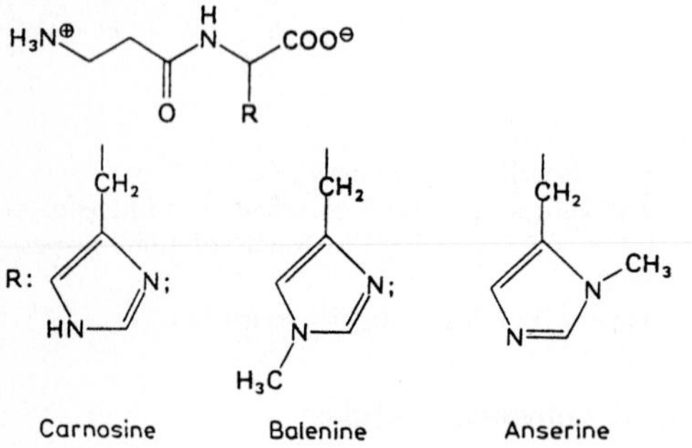

Data on the amounts of these peptides present in meat are given in Table 1.19. Carnosine is predominant in beef muscle tissue, while anserine is predominant in chicken meat. Balenine is a characteristic constituent of whale muscle, although it appears that sperm whales do not have this dipeptide. The amounts found in commercial sperm whale meat extract are probably due to the presence of meat from other whale species. These peptides are used analytically to identify the meat extract. Their physiological roles are not clear. Their buffering capacity in the pH range of 6−8 may be of some importance. They may also be involved in revitalizing exhausted muscle, i.e. in the muscle regaining its excitability and ability to contract. Carnosine may act as a neurotransmitter for nerves involved in odor perception.

Carnosine   Balenine   Anserine

(1.80)

1.3.4.3 Nisin

This peptide is formed by several strains of *Streptococcus lactis* (Langfield-N-group). It contains a number of unusual amino acids, namely dehydroalanine, dehydro-β-methylalanine, lanthionine, β-methyl-lanthionine, and therefore also five thioether bridges (cf. Formula 1.81).

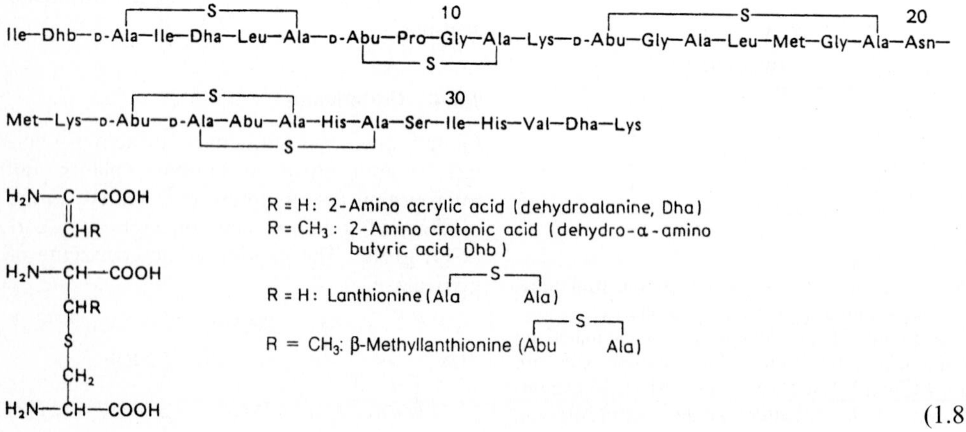

(1.81)

Table 1.19. Occurrence of carnosine, anserine and balenine (%) in meat[a]

Meat	Carnosine	Anserine	Balenine	Σ[b]
Beef muscle tissue	0.15−0.35	0.01−0.05		0.2−0.4
Beef meat extract	3.1 −5.7	0.4 −1.0		4.4−6.2
Chicken meat[c]	0.01−0.1	0.05−0.25		
Chicken meat extract	0.7 −1.2	2.5 −3.5		
Whale meat				ca. 0.3
Whale meat extract a[d]	3.1 −5.9	0.2 −0.6	13.5−23.0	16−30
Whale meat extract b[e]	2.5 −4.5	1.2 −3.0	0 − 5.2	3.5−12

[a] The results are expressed as % of the moist tissue weight, or of commercially available extracts containing 20% moisture.
[b] β-Alanine peptide sum.
[c] Lean and deboned chiken meat.
[d] Commercial extract mixture of various whales.
[e] Commercial extract mixture, with sperm whale prevailing.

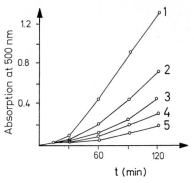

Fig. 1.7. Browning of some lysine derivatives (0.1 M lysine or lysine derivative, 0.1 M glucose in 0.1 M phosphate buffer pH 6.5 at 100 °C in sealed tubes. (according to *Finot* et al., 1978.) 1 Lys, 2 Ala-Lys, 3 Gly−Lys, 4 Glu−Lys, 5 Lys

The peptide subtilin is related to nisin. Nisin is active against Gram-positive microorganisms (lactic acid bacteria, *Streptococci*, *Bacilli*, *Clostridia* and other anaerobic spore-forming microorganisms). Nisin begins to act against the cytoplasmic membrane as soon as the spore has germinated. Hence, its action is more pronounced against spores than against vegetative cells. Nisin is permitted as a preservative in several countries. It is used to suppress anaerobes in cheese and cheese products, especially in hard cheese and processed cheese to inhibit butyric acid fermentation. The use of nisin in the canning of vegetables allows mild sterilization conditions.

1.3.4.4 Lysine Peptides

A number of peptides, such as:

Gly−Lys, Ala−Lys, Glu−Lys, Lys, Gly⌋

Lys, Lys, Gly−Lys
Glu⌋ Glu Gly⌋ (1.82)

have been shown to be as good as lysine in rat growth feeding tests. These peptides substantially retard the browning reaction with glucose (Fig. 1.7), hence they are suitable for lysine

fortification of sugar-containing foods which must be heat treated.

1.3.4.5 Other Peptides

Other peptides occur commonly and in variable levels in protein rich food as degradation products of proteolytic processes.

1.4 Proteins

Like peptides, proteins are formed from amino acids through amide linkages. Covalently bound hetero constituents can also be incorporated into proteins. For example, phosphoproteins such as milk casein (cf. 10.1.2.1.1) or phosvitin of egg yolk (cf. 11.2.4.1.2) contain phosphoric acid esters of serine and threonine residues.

Glycoproteins, such as ϰ-casein (cf. 10.1.2.1.1), various components of egg white (cf. 11.2.3.1) and egg yolk (cf. 11.2.4.1.2), collagen from connective tissue (cf. 12.3.2.3.1) and serum proteins of some species of fish (cf. 13.1.4.2.4), contain one or more monosaccharide or oligosaccharide units bound O-glycosidically to serine, threonine or δ-hydroxylysine or N-glycosidically to asparagine:

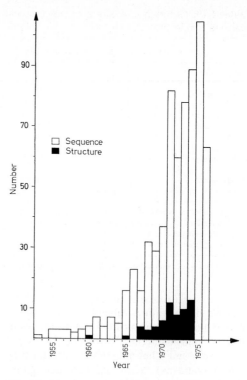

$$(1.83)$$

R : H, CH$_3$; R^1 : H, Sugar residue; Ac : Acetyl

The structure of a protein is dependent on the amino acid sequence (the primary structure) which determines the molecular conformation (secondary and tertiary structures). Proteins sometimes occur as molecular aggregates which are arranged in an orderly geometric fashion (quaternary structure). The histograms in Fig. 1.8 provide data on the proteins with known amino acid sequences (1953–1976) and conformations (1960–1974).

Fig. 1.8. Number of elucidated amino acid sequences (1953–1976) and protein structures (years 1960–1974)

1.4.1 Amino Acid Sequence

1.4.1.1 Amino Acid Composition, Subunits

Sequence analysis can only be conducted on a pure protein. First, the amino acid composition is determined after acidic hydrolysis. The procedure (separation on a single cation-exchange resin column and color development with ninhydrin reagent or fluorescamine) has been standardized and automated (amino acid analyzers). Figure 1.9 shows a typical amino acid chromatogram.

As an alternative to these established methods, the derivatization of amino acids with the subsequent separation and detection of derivatives is possible (pre-column derivatization). Various derivatization reagents can be selected, such as:

- 9-Fluorenylmethylchloroformate (FMOC, cf. 1.2.4.2.1)
- Phenylisothiocyanate (PITC, cf. 1.2.4.2.3)
- Dimethylaminoazobenzenesulfonylchloride (DABS-Cl, cf. 1.2.4.2.1)
- Dimethylaminonaphthalenesulfonylchloride (DANS-Cl, cf. 1.2.4.2.1)
- 7-Fluoro-4-nitrobenzo-2-oxa-1,3-diazole (NBDF, cf. 1.2.4.2.1)
- 7-Chloro-4-nitrobenzo-2-oxa-1,3-diazole (NBDCl, cf. 1.2.4.2.1)
- o-Phthaldialdehyde (OPA, cf. 1.2.4.2.4)

It is also necessary to know the molecular weight of the protein. This is determined by gel column chromatography, ultracentrifugation or SDS-PAG electrophoresis. Furthermore, it is necessary to determine whether the

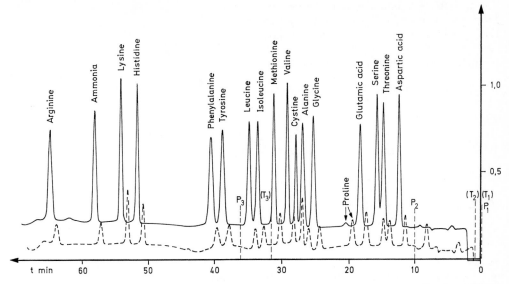

Fig. 1.9. Amino acid chromatogram. Separation of a mixture of amino acids (10 nmol/amino acid) by an amino acid analyzer. Applied is a single ion exchange column: Durrum DC-4A, 295 × 4 mm, buffers $P_1/P_2/P_3$: 0.2 N Na-citrate pH 3.20/0.2 N Na-citrate pH 4.25/1.2 N Na-citrate and NaCl of pH 6.45. Temperatures $T_1/T_2/T_3$: 48/56/80 °C. Flow rate: 25 ml/h; absorbance reading after color development with ninhydrin at 570/440 nm: —/-----

protein is a single molecule or consists of a number of identical or different polypeptide chains (subunits) associated through disulfide bonds or noncovalent forces. Dissociation into subunits can be accomplished by a change in pH, by chemical modification of the protein, such as by succinylation, or with denaturing agents (urea, guanidine hydrochloride, sodium dodecyl sulfate). Disulfide bonds, which are also found in proteins which consist of only one peptide chain, can be cleaved by oxidation of cystine to cysteic acid or by reduction to cysteine with subsequent alkylation of the thiol group (cf. 1.2.4.3.5) to prevent reoxidation. Separation of subunits is achieved by chromatographic or electrophoretic methods.

1.4.1.2 Terminal Groups

N-terminal amino acids can be determined by treating a protein with 1-fluoro-2,4-dinitro-benzene (*Sanger's* reagent; cf. 1.2.4.2.2) or 5-dimethylaminonaphthalene-1-sulfonyl chloride (dansyl chloride; cf. 1.2.4.2.1). Another possibility is the reaction with cyanate, follow-

ed by elimination of the N-terminal amino acid in the form of hydantoin, and separation and recovery of the amino acid by cleavage of the hydantoin (cf. 1.2.4.2.3). The N-terminal amino acid (and the amino acid sequence close to the N-terminal) is accessible by hydrolysis with aminopeptidase, in which case it should be remembered that the hydrolysis rate is dependent on amino acid side chains and that proline residues are not cleaved. A special procedure is required when the N-terminal residue is acylated (N-formyl- or N-acetyl amino acids, or pyroglutamic acid).
Determination of C-terminal amino acids is possible via the hydrazinolysis procedure recommended by *Akabori*:

$$H_2N-CH-CO-(HN-CH-CO-)HN-CH-COOH$$
$$\qquad R_1 \qquad\qquad R_{2-n} \qquad\qquad R_m$$

$$\xrightarrow[100\,°C]{H_2N-NH_2} H_2N-CH-CO-NH-NH_2$$
$$\qquad\qquad\qquad\qquad R_{1-n}$$

$$+\ H_2N-CH-COOH$$
$$\qquad\qquad R_m$$

$$(1.84)$$

The C-terminal amino acid is then separated from the amino acid hydrazides, e.g., by a cation exchange resin, and identified. It is possible to mark the C-terminal amino acid through selective titration via oxazolinone:

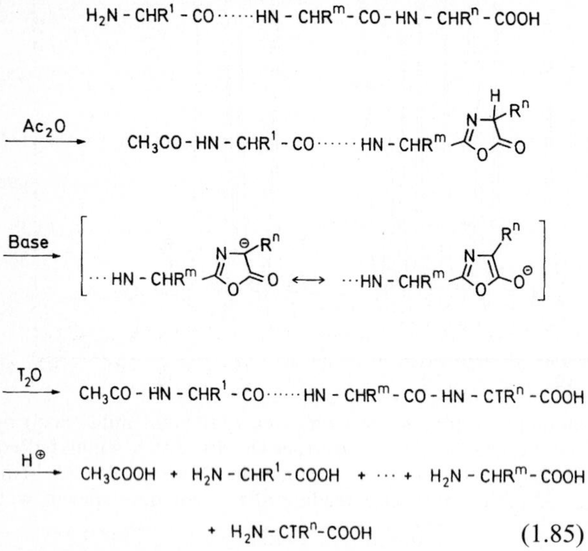

$$H_2N - CHR^1 - CO \cdots HN - CHR^m - CO - HN - CHR^n - COOH$$

$$\xrightarrow{Ac_2O} \quad CH_3CO - HN - CHR^1 - CO \cdots HN - CHR^m \diagdown$$

$$\xrightarrow{Base} \quad \left[\cdots HN - CHR^m \diagdown \longleftrightarrow \cdots HN - CHR^m \diagdown \right]$$

$$\xrightarrow{T_2O} \quad CH_3CO - HN - CHR^1 - CO \cdots HN - CHR^m - CO - HN - CTR^n - COOH$$

$$\xrightarrow{H^\oplus} \quad CH_3COOH + H_2N - CHR^1 - COOH + \cdots + H_2N - CHR^m - COOH$$

$$+ \ H_2N - CTR^n - COOH \qquad (1.85)$$

The C-terminal amino acids can be removed enzymatically by carboxypeptidase A which preferentially cleaves amino acids with aromatic and large aliphatic side chains, carboxypeptidase B which preferentially cleaves lysine, arginine and amino acids with neutral side chains or carboxypeptidase C which cleaves with less specificity but cleaves proline.

1.4.1.3 Partial Hydrolysis

Longer peptide chains are usually fragmented. The fragments are then separated and analyzed individually for amino acid sequences. Selective enzymatic cleavage of peptide bonds is accomplished primarily with trypsin, which cleaves exclusively Lys-X- and Arg-X-bonds, and chymotrypsin, which cleaves peptide bonds with less specificity (Tyr-X, Phe-X, Trp-X and Leu-X). The enzymatic attack can be influenced by modification of the protein. For example, acylation of the ε-amino group of lysine limits tryptic hydrolysis to Arg-X (cf. 1.4.4.1.3 and 1.4.4.1.4), whereas substitution of the SH-group of a cysteine residue with an

aminoethyl group introduces a new cleavage position for trypsin into the molecule ("pseudolysine residue"):

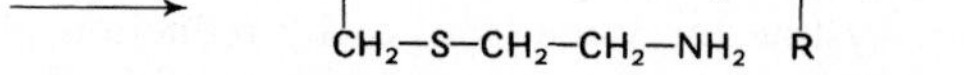

$$-NH-CH-CO-NH-CH-CO-$$
$$\quad\quad | \quad\quad\quad\quad\quad\quad |$$
$$\quad\quad CH_2-SH \quad\quad\quad R$$

$$H_2C-CH_2 \quad -NH-CH-CO-NH-CH-CO-$$
$$\quad \diagdown \quad / \quad\quad\quad\quad | \quad\quad\quad\quad\quad\quad |$$
$$\quad\quad NH \quad\quad\quad\quad\quad\quad\quad\quad\quad\quad\quad\quad R$$
$$\xrightarrow{\quad\quad} \quad\quad\quad\quad CH_2-S-CH_2-CH_2-NH_2$$

$$\xrightarrow{Trypsin} \quad -NH-CH-COOH \quad + \ H_2N-CH-CO-$$
$$\quad\quad\quad\quad\quad | \quad\quad\quad\quad\quad\quad\quad\quad\quad\quad |$$
$$\quad\quad\quad\quad CH_2-S-CH_2-CH_2-NH_2 \quad R$$
$$(1.86)$$

Also suited for the specific enzymatic hydrolysis of peptide chains is the endoproteinase Glu-C from *Staphylococcus aureus* V8. It cleaves GLU-X bonds (ammonium carbonate buffer pH 7.8 or ammonium acetate buffer pH 4.0) as well as Glu-X plus Asp-X bonds (phosphate buffer pH 7.8).

The most important chemical method for selective cleavage uses cyanogen bromide (BrCN) to attack Met-X-linkages (Reaction 1.87).

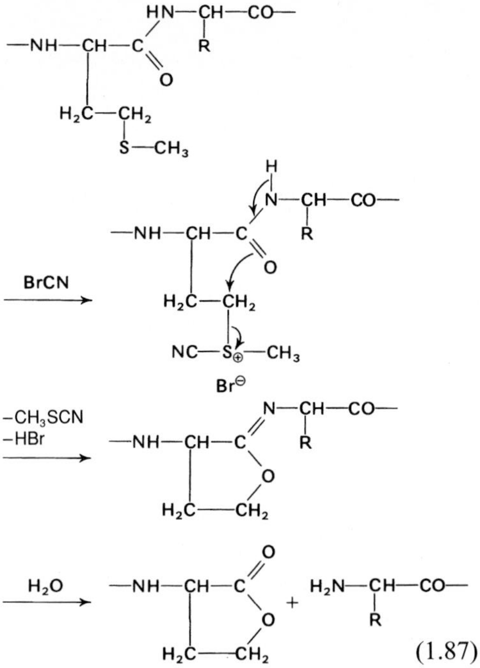

Hydrolysis of proteins with dilute acids preferentially cleaves aspartyl-X-bonds.

Separation of peptide fragments is achieved by gel and ion-exchange column chromatography using a volatile buffer as eluent (pyridine, morpholine acetate) which can be removed by freeze-drying of the fractions collected. The separation of peptides and proteins by reversed-phase HPLC has gained great importance, using volatile buffers mixed with organic, water-soluble solvents as the mobile phase.

The fragmentation of the protein is performed by different enzymic and/or chemical techniques, at least by two enzymes of different specifity. The arrangement of the obtained peptides in the same order as they occur in the intact protein is accomplished with the aid of overlapping sequences. The principle of this method is illustrated for subtilisin BPN′ as an example in Fig. 1.10.

1.4.1.4 Sequence Analysis

The *Edman* degradation is by far the most important method in sequence analysis. It involves stepwise degradation of peptides with phenylisothiocyanate (cf. 1.2.4.2.3) or suitable derivatives, e.g. dimethylaminoazobenzene isothiocyanate (DABITC). The resultant phenylthiohydantoin is either identified directly or the amino acid is recovered. The stepwise reactions are performed in solution or on peptide bound to a carrier, i.e. to a solid phase. Both approaches have been automated ("sequencer"). Carriers used include resins containing amino groups (e.g. amino polystyrene) or glass beads treated with amino alkylsiloxane:

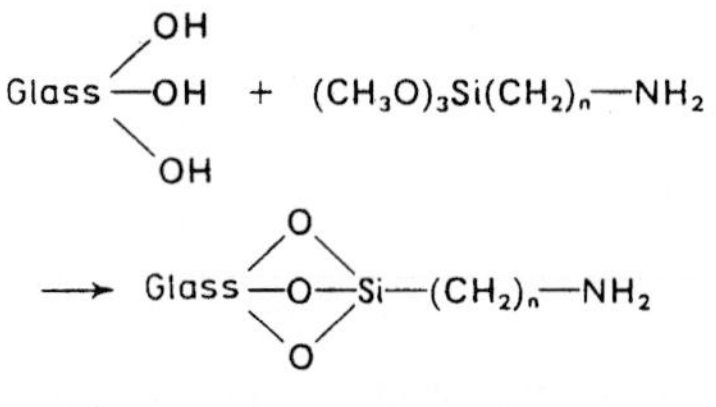

$$\text{Glass}\begin{array}{c}\text{OH}\\ \text{—OH}\\ \text{OH}\end{array} + (CH_3O)_3Si(CH_2)_n\text{—}NH_2$$

$$\longrightarrow \text{Glass}\begin{array}{c}O\\ \text{—O—Si—}(CH_2)_n\text{—}NH_2\\ O\end{array} \qquad (1.89)$$

The peptides are then attached to the carrier by carboxyl groups (activation with carbodiimide or carbonyl diimidazole, as in peptide synthesis) or by amino groups. For example, a pep-

Hydrolysis of proteins with strong acids reveals a difference in the rates of hydrolysis of peptide bonds depending on the adjacent amino acid side chain. Bonds involving amino groups of serine and threonine are particularly susceptible to hydrolysis. This effect is due to N → O-acyl migration via the oxazoline and subsequent hydrolysis of the ester bond:

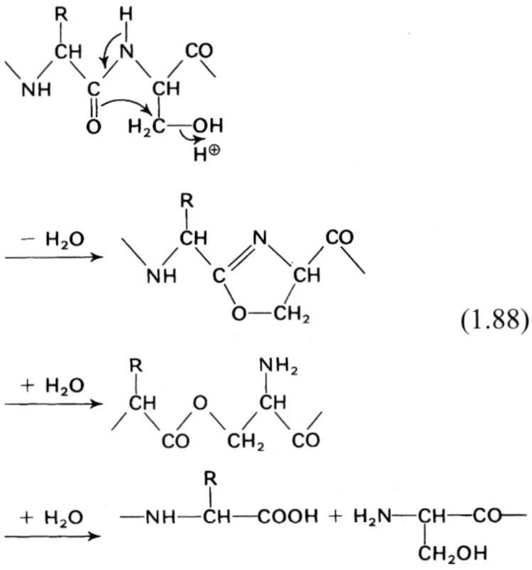

(1.87)

(1.88)

```
       C             C             C      T             C             C             T
NH2-Ala-Gln┴Ser-Val-Pro-Tyr┼Gly┬Val-Ser-Gln┼Ile┬Lys┴Ala-Pro-Ala-Leu┬His┴Ser-Gln┬Gly-Tyr┴Thr-Gly-Ser┬Asn┬Val-Lys┬
                        10                    Lys                     20                              27
       P   P             P   P                P   P             P                           P   P
```

```
                                  T
Val┬Ala-Val┬ Ile-Asp-Ser-Gly-Ile-Asp-Ser-Ser-His-Pro-Asn-Leu┴Lys┴Val-Ala-Gly-Gly-Ala-Ser-Met-Val-Pro-Ser┬Glu-Thr-
28 │   30 │                      40                         Lys                        50                   55
   P       P                                               P                                       P
```

```
       C             C             C                         C     C                         C
Pro-Asn-Phe┼Gln-Asp- Asp-Asn┴Ser-His-Gly-Thr-His┴Val-Ala-Gly-Thr-Val-Ala┬Ala┬Leu┴Asn-Asn┴Ser-Ile-Gly-Val-Leu┼Gly-
56         │ 60                      70                      P   P               80             │ 83
           P                                                                                  P
```

```
                         C       T   C                                     C     C
Val-Ala-Pro-Ser-Ser-Ala-Leu┬Tyr┴Ala-Val-Lys┴Val-Leu┼Gly-Asn-Ala-Gly-Ser-Gly-Gln-Tyr┴Ser-Trp┼Ile-Ile-Asn-Gly┬Ile-
84                      90                  100                              P           110 │ 111
                          P                                                                  P
```

```
      C       C             C                               C                              C   T
Gln┬Trp┴Ala┬Ile-Ala┬Asn-Asn-Met┬Asp-Val-Ile-Asn-Met┼Ser-Leu-Gly-Gly-Pro-Ser-Gly-Ser-Ala-Ala-Leu┴Lys┴Ala┬Ala-Val-
112│        │       │120             130                                         P             P       139
   P   P   P        P   P                            P
```

```
      T
Asp┬Lys┴Ala-Val-Ala-Ser-Gly┬Val┬Val-Val┬Val┬Ala┬Ala-Ala-Gly-Asn-Gln-Gly-Ser-Thr-Gly-Ser-Ser-Ser-Thr┬Val-Gly-Tyr-
140│                        150                  160                                        167
   P                        P   P       P   P   P                                           P
```

```
      T                                                            C   T             C
Pro-Gly┬Lys┴Tyr-Pro-Ser-Val┬Ile-Ala┬Val-Gly┬Ala-Val- Asp-Ser-Ser-Asn-Gln┼Arg┴Ala-Ser┬Phe┼Ser-Ser-Val-Gly-Pro-Glu-
168│170                     180                                        P           │190                195
   P                        P   P       P   P                                      P   P
```

```
                                  C                         T   C             C                    C
Leu┬Asp-Val-Met-Ala-Pro-Gly-Val-Ser┼Ile-Gln┴Ser-Thr-Leu-Pro-Gly-Asn-Lys┬Tyr┴Gly-Ala-Tyr┼Asn-Gly-Thr-Ser-Met┴Ala-
196│        200                      210                            P         220                223
   P                                 P
```

```
      C                         C     C       T             C     C             T             C
Ser-Pro-His┼Val-Ala-Gly-Ala┬Ala┬Ala┬Leu┴Ile┬Leu┴Ser-Lys┴His-Pro-Asn┬Trp┴Thr-Asn┬Thr-Gln-Val-Arg┴Ser-Ser-Leu┼Gln┬
224        P                 P   P   P   P           240                        P                   250 251 P   P
```

```
         C   T                   C   C       T       C
Asn-Thr┬Thr-Thr┴Lys┴Leu┬Gly-Asp-Ser┬Phe┬Tyr┴Tyr┴Gly-Lys┴Gly-Leu┴Ile-Asn-Val-Gln┬Ala┬Ala-Ala-GlnCOOH
252     P             P            260 │   │       P                       270 │   │        275
                                       P   P                                   P   P
```

Fig. 1.10. Subtilisin BPN′; peptide bonds hydrolyzed by trypsin (T), chymotrypsin (C), and pepsin (P)

tide segment from the hydrolysis of protein by trypsin has lysine as its C-terminal amino acid. It is attached to the carrier with p-phenylene-diisothiocyanate through the α- and ϵ-amino groups. Mild acidic treatment of the carrier under conditions of the *Edman* degradation splits the first peptide bond. The *Edman* procedure is then performed on the shortened peptide through second, third and subsequent repetitive reactions:

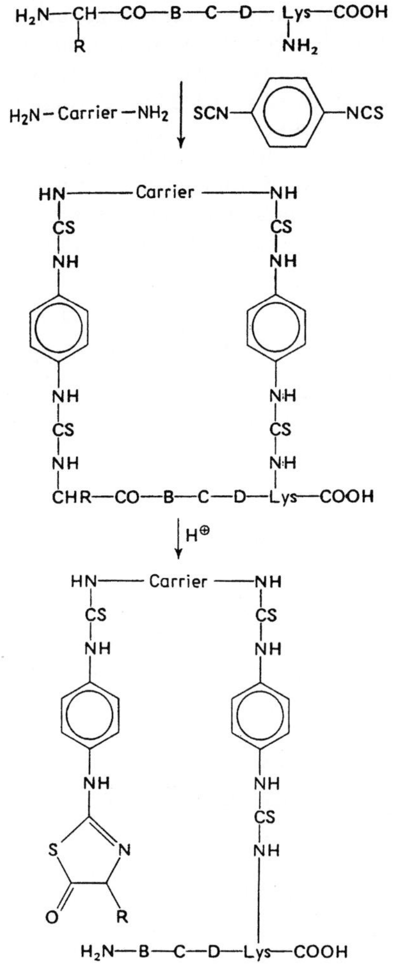

$$(1.90)$$

Microvariants allow working in the picomole range. In the reaction chamber, the protein is fixed on a glass-fiber disc, and the coupling and cleaving reagents are added and removed in a carrier gas stream (vapour-phase sequentiation).

Methods other than the *Edman* degradation can provide additional information. These include determination of terminal residues with amino- and carboxypeptidases, as already discussed, or mass spectrometric analysis of suitable volatile peptide derivatives.

1.4.1.5 Derivation of Amino Acid Sequence from the Nucleotide Sequence of the Coding Gene

The number of proteins whose coding gene is localized and isolated in the genome increases constantly. In these cases, the analysis of the nucleotide sequence of the gene is much simpler than the analysis of the amino acid sequence of the expressed protein. A considerable portion of the amino acid sequences known today has therefore already been derived from the nucleotide sequences in question.

For the sequencing of deoxyribonucleic acid (DNA), two methods have proven particularly successful: base-specific cleavage of terminally labeled molecules, and primed enzymatic synthesis.

In the first method, the single- or double-stranded DNA is fragmented by phosphatases which cleave very specifically at certain bases, the so-called restriction nucleases. The use of several enzymes of different specificity provides overlapping fragments from which the original total sequence can be reconstructed. The fragments are labeled radioactively at one end, usually by transferring the γ-phosphate residue of γ-^{32}P-ATP to the 5' hydroxy group with the aid of a polynucleotide kinase:

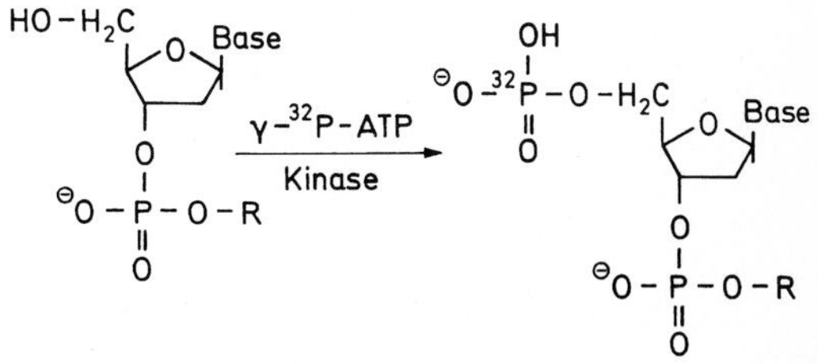

$$(1.91)$$

The DNA fragments are separated and cleaved in four parallel batches each by one of four specific chemical reactions at the guanine (G), guanine + adenine (G + A), cytosine (C) or cytosine + thymine bases (C + T).

For cleavage at guanine, dimethylsulfate is used to methylate N-7 of the base (cf. Formula 1.92). The attack of a hydroxyl ion at C-8 results in the opening of the imidazole ring. Piperidine then replaces the opened methyl guanidine at the 2-deoxyribose, forming an aldimine. This compound decomposes with the elimination of both phosphate groups in the 3'- and 5'-position. This reaction can be extended to adenine.

For cleavage at cytosine and thymine, the pyrimidine ring is opened with hydrazine. The reaction for thymine is presented in Formula 1.93. An urea derivative substituted with methyl pyrazolidone is formed first. Elimination of the pyrazolidone ring and the replacement of the urea remaining on the sugar by piperidine produces the same aldimine as shown in Formula 1.92. Chain cleavage occurs again with elimination of both phosphate groups. The reaction can be restricted to cytosine.

The fragments obtained are subsequently separated according to their molecular size by polyacrylamide gel (PAG) electrophoresis. The sequence can be read directly from an autoradiogram, as shown in Fig. 1.11.

Of the above mentioned synthetic sequencing methods, usually the chain-terminating version is used. The starting material is a hybrid from the single-stranded DNA to be sequenced and a short complementary oligonucleotide, the so-called primer, which is extended by means of DNA polymerase and the four ^{32}P-marked 2'-deoxyribonucleoside triphosphates. Four parallel batches receive, in addition to the four 2'-deoxynucleotides, one of the four 2', 3'-dideoxyribonucleoside triphosphates, whose incorporation causes chain termination. The synthetics, which all have the

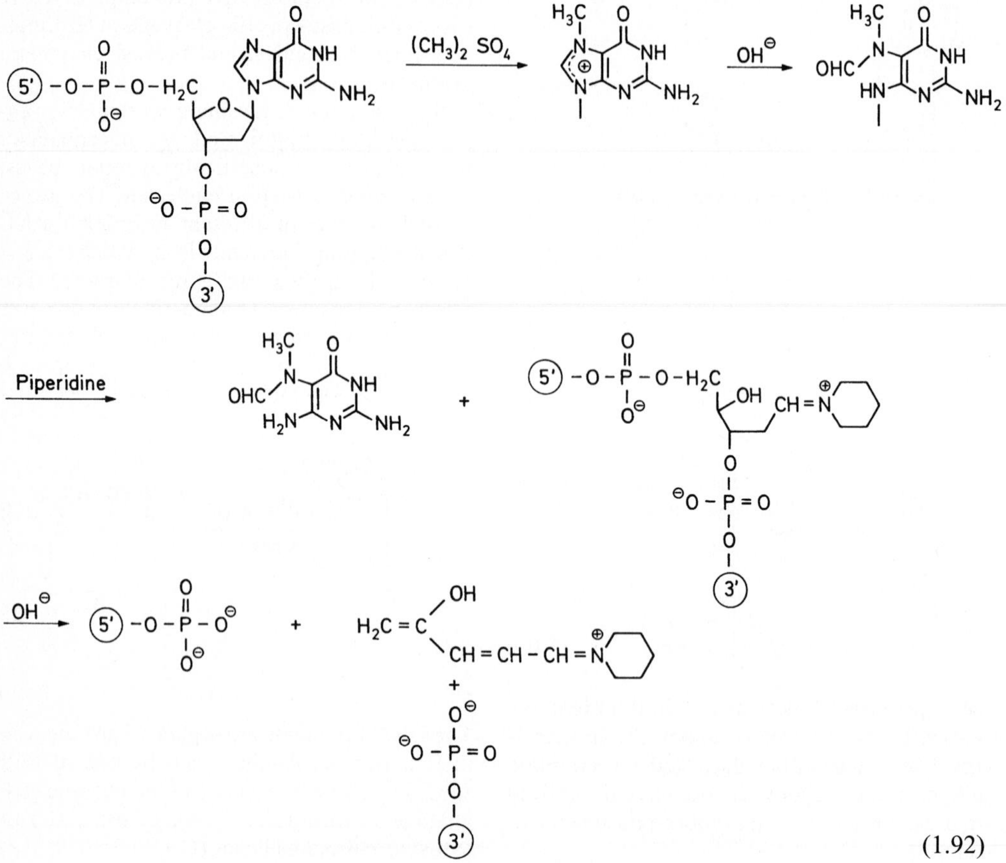

(1.92)

H_2N-NH_2

Piperidine

(1.93)

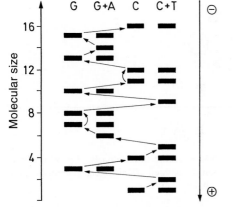

Fig. 1.11. Schematic representation of the sequence analysis of the DNA fragment ^{32}pGpCpTpGpCpT-pApGpGpTpGpCpCpGpApGpC by specific chemical cleavage and PAG electrophoresis of the fragments. (The fragments obtained by cleavage at guanine, guanine + adenine, cytosine, and cytosine + thymine are applied side by side in the positions G, G + A, C and C + T. The sequence is read according to increasing molecular size.)

same 5′ end, are separated electrophoretically by molecular size. The sequence can be read directly from the autoradiogram, in analogy to the cleavage method (Fig. 1.11).

1.4.2 Conformation

Information about conformation is available through X-ray crystallographic analysis of protein crystals and by measuring the distance (≤30 nm) between selected protons of the peptide chain (NH_i–NH_{i+1}, NH_{i+1}–$C_\alpha H_i$, NH_{i+1}–$C_\beta H_i$, $C_\alpha H_i$–$C_\alpha H_{i+1}$, $C_\alpha H_i$–$C_\beta H$) by means of H-NMR spectroscopy in solution. This assumes that, in many cases, the conformation of the protein in crystalline form is similar to that of the protein in solution. In 1960 *Kendrew et al.* succeeded in elucidating the structure of myoglobin (17.8 kdal) with a resolution of 0.2 nm. As an example the calculated electron density distributions of 2,5-dioxopiperazine based on various degrees of resolution is presented in Fig. 1.12. Individual atoms are well revealed at 0.11 nm. Such a resolution has not been achieved with proteins. Reliable localization of the C_α-atom of the peptide chain requires a resolution of less than 0.3 nm.

1.4.2.1 Extended Peptide Chains

X-ray structural analysis and other physical measurements of a fully extended peptide chain reveal the lengths and angles of bonds (see the "ball and stick" representation in Fig. 1.13). The peptide bond has partial (40%) double bond character with π electrons shared between the C′–O and C′–N bonds. The resonance energy is about 83.6 kJ/mole:

$$\text{(1.94)}$$

Normally the bond has a trans-configuration, i.e. the oxygen of the carbonyl group and the hydrogen of the NH group are in the trans-position; a cis-configuration which has 8 kJmol^{-1} more energy occurs only in exceptional cases (e.g. in small cyclic peptides or in proteins before proline residues).

Thus in ribonuclease A, two X-Pro bonds have trans-conformation (Pro-42 and Pro-117), and two have cis-conformation (Pro-93 and Pro-114). The equilibrium between the two isomers is catalyzed by specific enzymes (peptidyl-prolyl-cis/trans-isomerases). This accel-

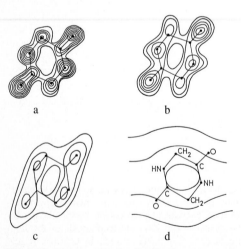

Fig. 1.12. Electron density distribution patterns for 2,5-dioxopiperazine with varying resolution extent. **a** 0.11 nm, **b** 0.15 nm, **c** 0.20 nm, **d** 0.60 nm (after *Perutz*, 1962)

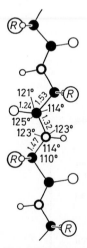

Fig. 1.13. Structure of an elongated peptide chain. ● Carbon, ○ oxygen, O nitrogen, ○ hydrogen and ® side chain

erates the folding of a peptide chain (cf. 1.4.2.3.2), which in terms of the biosynthesis occurs initially in all-trans-conformation. Six atoms of the peptide bonds, C_i^α, C_i', O_i, N_{i+1}, C_{i+1}^α and H_{i+1}, lie in one plane (cf. Fig. 1.14). For a trans-peptide bond, ω_i is 180°. The position of two neighboring planes is determined by the numerical value of the angles ψ_i (rotational bond between a carbonyl carbon and an α-carbon) and ϕ_i (rotational bond between an amide-N and an α-carbon). For an extended peptide chain, $\psi_i = 180°$ and $\phi_i = 180°$. The position of side chains can also be described by a series of angles χ_i^{1-n}.

1.4.2.2 Secondary Structure (Regular Structural Elements)

The primary structure gives the sequence of amino acids in a protein chain while the secondary structure reveals the arrangement of the chain in space. The peptide chains are not in an extended or unfolded form (ψ_i, $\phi_i \neq 180°$). It can be shown with models that ψ_i and ϕ_i, at a permissible minimum distance between nonbonding atoms (Table 1.20), can assume only particular angles. Figure 1.15 presents the permissible ranges for amino acids other than glycine (R $\neq$ H). The range is broader for glycine (R = H). Figure 1.16 demonstrates that most of 13 different proteins with a total of

Table 1.20. Minimal distances for nonbonded atoms (Å)

	C	N	O	H
C	3.20[a] (3.00)[b]	2.90 (2.80)	2.80 (2.70)	2.40 (2.20)
N		2.70 (2.60)	2.70 (2.60)	2.40 (2.20)
O			2.70 (2.60)	2.40 (2.20)
H				2.00 (1.90)

[a] Normal values
[b] Extreme values

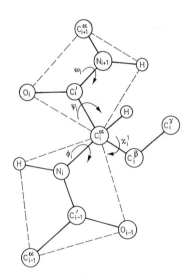

Fig. 1.14. Definitions for torsion angles in a peptide chain

$\omega_i = 0°$ for $C_i^\alpha - C_i'/N_{i+1} - C_{i+1}^\alpha \rightarrow$ cis,
$\psi_i = 0°$ for $C_i^\alpha - N_i/C_i' - O_i \rightarrow$ trans,
$\varphi_i = 0°$ for $C_i^\alpha - C_i'/N_i - H \rightarrow$ trans,
$\chi_i = 0°$ for $C_i^\alpha - N_i C_i^\beta - C_i \rightarrow$ cis;

The angles are positive when the rotation is clockwise and viewed from the N-terminal side of a bond or (for X) from the atom closer to the main chain respectively. (according to *Schulz* and *Schirmer*, 1979)

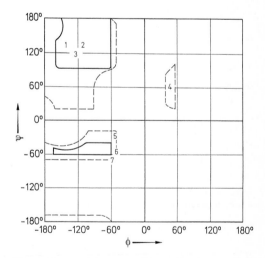

Fig. 1.15. ϕ,ψ-Diagram (*Ramachandran* plot). Allowed conformations for amino acids with a C^β-atom obtained by using normal (–) and lower limit (---) contact distances for non-bonded atoms, from Table 1.20. β-Sheet structures: antiparallel (1); parallel (2), twisted (3). Helices: α-, left-handed (4), 3_{10} (5), α, right-handed (6), π (7)

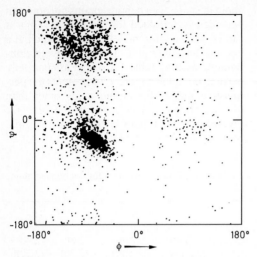

180°

ψ

0°

-180°
-180°   0°   180°

ϕ ———▶

Fig. 1.16. ϕ, ψ-Diagram for observed values of 13 different proteins containing a total of 2,500 amino acids. (according to *Schulz* and *Schirmer*, 1979)

about 2,500 amino acid residues have been shown empirically to have values of ψ,ϕ-pairs within the permissible range. When a multitude of equal ψ,ϕ-pairs occurs consecutively in a peptide chain, the chain acquires regular repeating structural elements. The types of structural elements are compiled in Table 1.21.

1.4.2.2.1 *β-Sheet*

Three regular structural elements (pleated-sheet structures) have values in the range of $\phi = -120°$ and $\psi = +120°$. The peptide chain is always lightly folded on the C_α atom (cf. Fig. 1.17), thus the R side chains extend perpendicularly to the extension axis of the chain, i.e. the side chains change their projections alternately from $+z$ to $-z$. Such a pleated structure is stabilized when more chains are present.

Table 1.21. Regular structural elements (secondary structures) in polypeptides

Structure	Φ (°)	ψ (°)	n^a	d^b (Å)	r^c (Å)	Comments
β-Pleated sheet, parallel	−119	+113	2.0	3.2	1.1	Occurs occasionally in neighbouring chain sectors of globular proteins
β-Pleated sheet, antiparallel	−139	+135	2.0	3.4	0.9	Common in proteins and synthetic polypeptides
3_{10}-Helix	−49	−26	3.0	2.0	1.9	Observed at the ends of α-helixes
α-Helix, left-handed coiling	−57	−47	3.6	1.5	2.3	Common in globular proteins, as α "coiled coil" in fibrous proteins
α-Helix, right-handed coiling	+57	+47	3.6	1.5	2.3	Poly-D-amino acids poly-(β-benzyl)-L-aspartate
π-Helix	−57	−70	4.4	1.15	2.8	Hypothetical
Polyglycine II	−80	+150	3.0	3.1		Similar to antiparallel β-pleated-sheet formation
Polyglycine II, left-handed coiling	+80	−150	3.0	3.1		Synthetic polyglycine is a mixture of right- and left-handed helices; in some silk fibroins, the left-handed helix occurs
Poly-L-proline I	−83	+158	3.3	1.9		Synthetic poly-L-proline, only cis-peptide bonds
Poly-L-proline II	−78	+149	3.0	3.1		As left-handed polyglycine II, as triple helix in collagen

[a] Amino acid residues per turn.
[b] The rise along the axis direction, per residue.
[c] The radius of the helix.

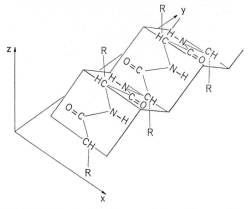

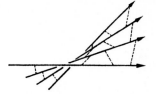

Fig. 1.19. Diagrammatic presentation of a twisted sheet structure of parallel peptide chains (according to *Schulz* and *Schirmer*, 1979)

Fig. 1.17. A pleated sheet structure of a peptide chain

Subsequently, adjacent chains interact along the x-axis by hydrogen bonding, thus providing the cross-linking required for stability.

When adjacent chains run in the same direction, the peptide chains are parallel. This provides a stabilized, planar, parallel sheet structure. When the chains run in opposite directions, a planar, antiparallel sheet structure is stabilized (Fig. 1.18). The lower free energy, twisted sheet structures, in which the main axes of the neighboring chains are arranged at an angle of 25° (Fig. 1.19), are more common than planar sheet structures.

The β structures can also be regarded as special helix with a continuation of 2 residues per turn. With proline, the formation of a β structure is not possible.

1.4.2.2.2 Helical Structures

There are three regular structural elements in the range of $\phi = -60°$ and $\psi = -60°$ (cf. Fig. 1.15) in which the peptide chains are coiled like a threaded screw. These structures are stabilized by intrachain hydrogen bridges which extend almost parallel to the chain axis, cross-linking the CO and NH groups, i.e., the CO group of amino acid residue i with the NH group of residue i + 3 (3_{10}-helix), 1 + 4 (α-helix) or i + 5 (π-helix).

The most common structure is the α-helix and for polypeptides from L-amino acids, exclusively the right-handed α-helix (Fig. 1.20). The left-handed α-helix is energetically unfavourable for L-amino acids, since the side chains here are in close contact with the backbone. No α-helix is possible with proline. The 3_{10}-helix was observed only at the ends of α-helices but not as an independent regular structure. The π-helix is hypothetical. Two helical conformations are known of polyproline (I and II). Polyproline I contains only cis-peptide bonds and is right-handed, while polyproline II contains trans-peptide bonds and is left-handed. The stability of the two conformations depends on the solvent and other factors. In water, polyproline II predominates. Polyglycine can also occur in two conforma-

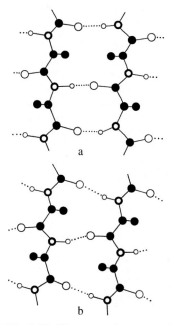

Fig. 1.18. Diagrammatic presentation of antiparallel (**a**) and parallel (**b**) peptide chain arrangements

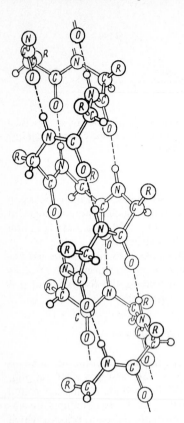

Fig. 1.20. Right-handed α-helix

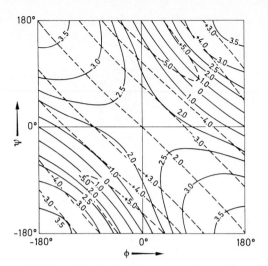

Fig. 1.21. φ,ψ-Diagram with marked helix parameters n (- - -) and d (—). (according to *Schulz* and *Schirmer*, 1979)

1.4.2.2.3 Reverse Turns

An important conformational feature of globular proteins are the reverse turns β-turns and β-bends. They occur at "hairpin" corners, where the peptide chain changes direction abruptly. Such corners involve four amino acid residues often including proline and glycine. Several types of turns are known; of greatest importance are type I (42% of 421 examined turns), type II (15%) and type III (18%); see Fig. 1.22.

In type I, all amino acid residues are allowed, with the exception of proline in position 3. In type II, glycine is required in position 3. In type III, which corresponds to a 3_{10}-helix, all amino acids are allowed. The sequences of the β-bends of lysozyme are listed in Table 1.22 as an example.

tions. Polyglycine I is a β-structure, while polyglcine II corresponds largely to the polyproline II-helix. A helix is characterized by the angles φ and ψ, or by the parameters derived from these angles: n, the number of amino acid residues per turn; d, the rise along the main axis per amino acid residue; and r, the radius of the helix. Thus, the equation for the pitch, p, is p = n · d. The parameters n and d are presented within a φ, ψ plot in Fig. 1.21.

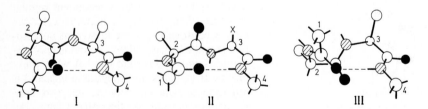

Fig. 1.22. Turns of the peptide chains (β-turns), types I–III. ○ = carbon, ⊘ = nitrogen, ● = oxygen. The α-C atoms of the amino acid residues are marked 1–4. X = no side chain allowed

Table 1.22. β-Turns in the peptide chain of egg white lysozyme

Residue Number	Sequence			
20– 23	Y	R	G	Y
36– 39	S	N	F	N
39– 42	N	T	Q	A
47– 50	T	D	G	S
54– 57	G	I	L	E
60– 63	S	R	W	W
66– 69	D	G	R	T
69– 72	T	P	G	S
74– 77	N	L	C	N
85– 88	S	S	D	I
100–103	S	D	G	D
103–106	D	G	M	N

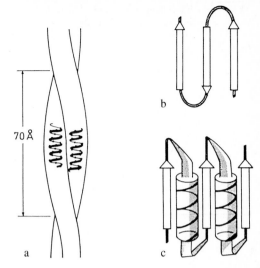

Fig. 1.23. Superhelix secondary structure (according to *Schulz* and *Schirmer*, 1979). **a** coiled-coil α-helix, **b** β-meander, **c** βαβαβ-structure

1.4.2.2.4 Super-Secondary Structures

Analysis of known protein structures has demonstrated that regular elements can exist in combined forms. Examples are the coiled-coil α-helix (Fig. 1.23, a), chain segments with antiparallel β-structures (β-meander structure; Fig. 1.23, b) and combinations of α-helix and β-structure (e. g., βαβαβ; Fig. 1.23 c).

1.4.2.3 Tertiary and Quaternary Structures

Proteins can be divided into two large groups on the basis of conformation: (a) fibrillar (fibrous) or scleroproteins, and (b) folded or globular proteins.

1.4.2.3.1 Fibrous Proteins

The entire peptide chain is packed or arranged within a single regular structure for a variety of fibrous proteins. Examples are wool keratin (α-helix), silk fibroin (β-sheet structure) and collagen (a triple helix). Stabilization of these structures is achieved by intermolecular bonding (electrostatic interaction and disulfide linkages, but primarily hydrogen bonds and hydrophobic interactions).

1.4.2.3.2 Globular Proteins

Regular structural elements are mixed with randomly extended chain segments (randomly coiled structures) in globular proteins. The proportion of regular structural elements is highly variable: 20–30% in casein, 45% in lysozyme and 75% in myoglobin (Table 1.23). Five structural subgroups are known in this group of proteins: (1) α-helices occur only; (2) β-structures occur only; (3) α-helical and β-structural portions occur in separate segments on the peptide chain; (4) α-helix and β-structures alternate along the peptide chain; and (5) α-helix and β-structures do not exist.

The process of peptide chain folding is not yet fully understood. It begins spontaneously, probably arising from one center or from several centers of high stability in larger proteins. The tendency to form regular structural elements shows a very different development in the various amino acid residues. Table 1.24 lists data which were derived from the analysis of globular proteins of known conformation. The data indicate, for example, that Met, Glu, Leu and Ala are strongly helix-forming. Gly and Pro on the other hand show a strong helix-breaking tendency. Val, Ile and Leu promote the formation of pleated-sheet structures, while Asp, Glu and Pro prevent them. Pro and Gly are important building blocks of turns. By means of such data it is possible to forecast the expected conformations for a given amino acid sequence.

Table 1.23. Proportion of "regular structural elements" present in various globular proteins

Protein	α-Helix	β-Structure	n_G	n	%
Myoglobin	3– 16[a]			14	
	20– 34			15	
	35– 41			7	
	50– 56			7	
	58– 77			20	
	85– 93			9	
	99–116			18	
	123–145			23	
			151	173	75
Lysozyme	5– 15			11	
	24– 34			11	
		41–54		14	
	80– 85			6	
	88– 96			9	
	97–101			5	
	109–125			7	
			129	63	49
α_{S1}-Casein			199		ca 30
β-Casein			209		ca 20

[a] Position number of the amino acid residue in the sequence.

n_G: Total number of amino acid residues.
n: Amino acid residues within the regular structure.
%:   Percentage of the amino acid residues present in regular structure.

Table 1.24. Normalized frequencies[a] of amino acid residues in the regular structural elements of globular proteins

Amino acid	α-Helix (P_α)	Pleated sheet (P_β)	β-Turn (P_t)
Ala	1.29	0.90	0.78
Cys	1.11	0.74	0.80
Leu	1.30	1.02	0.59
Met	1.47	0.97	0.39
Glu	1.44	0.75	1.00
Gln	1.27	0.80	0.97
His	1.22	1.08	0.69
Lys	1.23	0.77	0.96
Val	0.91	1.49	0.47
Ile	0.97	1.45	0.51
Phe	1.07	1.32	0.58
Tyr	0.72	1.25	1.05
Trp	0.99	1.14	0.75
Thr	0.82	1.21	1.03
Gly	0.56	0.92	1.64
Ser	0.82	0.95	1.33
Asp	1.04	0.72	1.41
Asn	0.90	0.76	1.28
Pro	0.52	0.64	1.91
Arg	0.96	0.99	0.88

[a] Shown is the fraction of an amino acid in a regular structural element, related to the fraction of all amino acids of the same structural element. P = 1 means random distribution; P > 1 means enrichment, P < 1 means depletion. The data are based on an analysis of 66 protein structures.

Folding of the peptide chain packs it densely by formation of a large number of intermolecular noncovalent bonds. Data on the nature of the bonds involved are provided in Table 1.25.

The H-bonds formed between main chains, main and side chains and side-side chains are of particular importance for folding. The portion of polar groups involved in H-bond build-up in proteins of Mr > 8.9 kdal appears to be fairly constant at about 50%.

The hydrophobic interaction of the nonpolar regions of the peptide chains also plays an important role in protein folding. These interactions are responsible for the fact that nonpolar groups are folded to a great extent towards the interior of the protein globule. The surface areas accessible to water molecules have been calculated for both unfolded and native folded forms for a number of monomeric proteins with known conformations. The proportion of the accessible surface in the stretched state, which tends to be buried in the interior of the globule as a result of folding, is a simple linear function of the molecular weight (M). The gain in free energy for the folded surface is 10 kJ nm^{-2}. Therefore, the total hydrophobic contribution to free energy due to folding is:

$$\Delta G_{HP} = 88\,M + 79 \cdot 10^{-5}\,M^2\ [\text{J} \cdot \text{mol}^{-1}] \qquad (1.95)$$

This relation is valid for a range of $6.108 \leq M \leq 34.409$, but appears to be also valid for larger molecules since they often consist of

Table 1.25. Bond-types in proteins

Type	Examples	Bond strength (kJ/mole)
Covalent bonds	$-S-S-$	ca. -230
Electrostatic bonds	$-COO \cdots H_3N^+-$ $>C=O \ O=C<$	-21 $+1.3$
Hydrogen bonds	$-O-H \cdots O<$ $>N-H \cdots O=C<$	-16.7 -12.5
Hydrophobic bonds	$-CH \Big\langle {CH_3 \ H_3C \atop CH_3 \ H_3C} \Big\rangle CH-$	0.01^b
	$-Ala \cdots Ala-$	-3
	$-Val \cdots Val-$	-8
	$-Leu \cdots Leu-$	-9
	$-Phe \cdots Phe-$	-13
	$-Trp \cdots Trp-$	-19

[a] For $\varepsilon = 4$.
[b] Per $Å^2$-surface area.

several loose associations of independent globular portions called structural domains (Fig. 1.24).
Proteins with disulfide bonds fold at a significantly slower rate than those without disulfide bonds. Folding is not limited by the reaction rate of disulfide formation. Therefore the folding process of disulfide-containing proteins seems to proceed in a different way. The re-

Fig. 1.24. Globular protein with two-domain structure (according to *Schulz* and *Schirmer*, 1979)

verse process, protein unfolding, is very much slowed down by the presence of disulfide bridges which generally impart great stability to globular proteins. This stability is particularly effective against denaturation. An example is the *Bowman-Birk* inhibitor from soybean (Fig. 1.25) which inhibits the activity of trypsin and chymotrypsin. Its tertiary structure is stabilized by seven disulfide bridges. The reactive sites of inhibition are Lys[16]-Ser[17] and Leu[43]-Ser[44], i.e. both sites are located in relatively small rings, each of which consists of nine amino acid residues held in ring form by a disulfide bridge. The thermal stability of this inhibitor is high.
As examples of the folding of globular proteins, Fig. 1.26 shows schematically the course of the peptide chains in the β-chain of hemoglobin, in triosephosphate isomerase and carboxypeptidase. Other protein conformations are shown in the following figures:

- Fig. 8.7 (cf. 8.8.4): Thaumatin and monellin (two-dimensional)

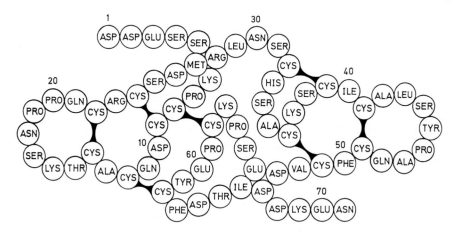

Fig. 1.25. *Bowman-Birk* inhibitor from soybean (after *Ikenaka* et al., 1974)

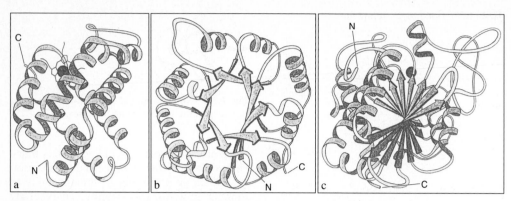

Fig. 1.26. Tertiary structures (schematic: spiral: α-helix, arrow: pleated sheet) of the β-chain of hemoglobin (**a**), of triosephosphate isomerase (**b**) and carboxypeptidase (**c**). (according to *Walton*, 1981)

– Fig. 8.8 (cf. 8.8.5): Thaumatin and monellin
  (three-dimensional)
– Fig. 11.3 (cf. 11.2.3.1.4): Lysozyme (three-
  dimensional)

1.4.2.3.3 Quaternary Structures

In addition to the free energy gain by folding of a single peptide chain, association of more than one peptide chain (subunit) can provide further gains in free energy. For example, hemoglobin (4 associated peptide chains) $\Delta G^0 = -46\,kJ\,mole^{-1}$ and the trypsin-trypsin inhibitor complex (association of 2 peptide chains) $\Delta G^0 = -75.2\,kJ\,mole^{-1}$. In principle such associations correspond to the folding of a larger peptide chain with several structural domains without covalently binding the subunits. Table 1.26 lists some proteins which partially exhibit quaternary structures.

1.4.2.4 Denaturation

The term denaturation denotes a reversible or irreversible change of native conformation (tertiary structure) without cleavage of covalent bonds (except for disulfide bridges). Denaturation is possible with any treatment that cleaves hydrogen bridges, ionic or hydrophobic bonds. This can be accomplished by: changing the temperature, adjusting the pH, increasing the interface area, or adding organic solvents, salts, urea, guanidine hyrochloride or detergents such as sodium dodecyl sulfate. Denaturation is generally reversible when the

Table 1.26. Examples of globular proteins

Name	Origin	Molecular weight (Kdal)	Number of subunits
Lysozyme	Chicken egg	14.6	1
Papain	*Papaya latex*	20.7	1
α-Chymotrypsin	Pancreas (beef)	23	1
Trypsin	Pancreas (beef)	23.8	1
Pectinesterase	Tomato	27.5	
Chymosin	Stomach (calf)	31	
β-Lactoglobulin	Milk	35	2
Pepsin A	Stomach (swine)	35	1
Peroxidase	Horseradish	40	1
Hemoglobin	Blood	64.5	4
Avidin	Chicken egg	68.3	4
Alcohol-dehydrogenase	Liver (horse)	80	2
	Yeast	150	4
Hexokinase	Yeast	104	2
Lactate dehydrogenase	Heart (swine)	135	4
Glucose oxidase	*P. notatum*	152	
Pyruvate kinase	Yeast	161	8
	A. niger	186	
β-Amylase	Sweet potato	215	4
Catalase	Liver (beef)	232	4
	M. lysodeikticus	232	
Adenosine triphosphatase	Heart (beef)	284	6
Urease	Jack beans	483	6
Glutamine synthetase	*E. coli*	592	12
Arginine decarboxylase	*E. coli*	820	10

peptide chain is stabilized in its unfolded state by the denaturing agent and the native conformation can be reestablished after removal of the agent. Irreversible denaturation occurs when the unfolded peptide chain is stabilized by interaction with other chains (as occurs for instance with egg proteins during boiling). During unfolding reactive groups, such as thiol groups, that were covered or blocked, may be exposed. Their participation in the formation of disulfide bonds may also cause an irreversible denaturation.

An aggregation of the peptide chains caused by the folding of globular proteins is connected with reduced solubility or swellability. Thus the part of wheat gluten that is soluble in acetic acid diminishes as heat stress increases (Fig. 1.27). As a result of the reduced rising capacity of gluten caused by the pre-treatment, the volume of bread made of recombined flours is smaller (Fig. 1.28).

In the case of fibrous proteins, denaturation, through destruction of the highly ordered structure, generally leads to increased solubility or rising capacity. One example is the thermally caused collagen-to-gelatin conversion, which occurs when meat is cooked (cf. 12.3.2.3.1).

The thermal denaturation of the whey proteins β-lactoglobulin and α-lactalbumin has been well-studied. The data in Table 1.27 based on reaction kinetics and the *Arrhenius* diagram (Fig. 1.29) indicate that the activation energy of the overall reaction in the range of 80–90 °C changes. The higher E_a values at lower temperatures must be attributed to folding, which is the partial reaction that determines the reaction rate at temperatures < 90 °C. At higher temperatures (> 95 °C), the aggregation to which the lower activation energy corresponds predominates.

The values in Table 1.27 determined for activation entropy also support the above mentioned attribution. In the temperature range of 70–90C, $\Delta S^{\#}$ is always positive, which indi-

Table 1.27. Denaturation of β-lactoglobulins A and B (β-LG-A, β-LG-B) and of α-lactalbumin (a-LA)

Protein	n	ϑ (°C)	E_a (kJ mol^{-1})	ln (k$_o$) (s^{-1})	ΔS^+ (kJ mol^{-1} K^{-1})
β-LG-A	1.5	70– 90	265.21	84.16	0.445
		95–150	54.07	14.41	− 0.136
β-LG-B	1.5	70– 90	279.96	89.43	0.487
		95–150	47.75	12.66	− 0.150
α-LA	1.0	70– 80	268.56	84.92	0.452
		85–150	69.01	16.95	− 0.115

n: reaction order, δ: temperature, E_a: activation energy, k_o: reaction rate constant, ΔS^+: activation entropy.

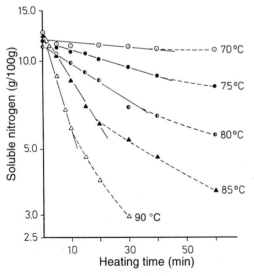

Fig. 1.27. Solubility of moist gluten (wheat) in diluted acetic acid after various forms of thermal stress (according to *Pence* et al., 1953)

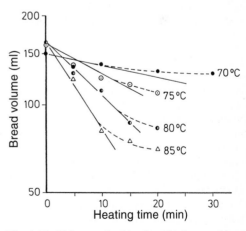

Fig. 1.28. Volume of white bread of recombined flours using thermally treated liquid gluten (wheat) (according to *Pence* et al., 1953)

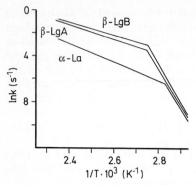

Fig. 1.29. *Arrhenius* diagram for the denaturation of the whey proteins β-lactoglobulin A, β-lactoglobulin B and α-lactalbumin B (according to *Kessler*, 1988)

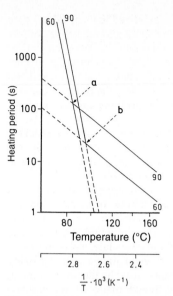

Fig. 1.30. Lines of equal denaturation degrees of β-lactoglobulin B. [The steeper lines correspond to the folding (60%, 90%), the flatter lines to aggregation (60%, 90%); at point a, 60% are folded and 90% can be aggregated, corresponding to an overall reaction of 60%; at point b, 90% are folded and 60% can be aggregated, corresponding to an overall reaction of 60%; according to *Kessler*, 1988]

cates a state of greater disorder than should be expected with the predominance of the folding reaction. On the other hand, the negative $\Delta S^{\#}$ values at 95–105°C indicate a state of greater order than should be expected considering that aggregation predominates in this temperature range. Detailed studies of the kind described above allow optimal control of thermal processes. In the case of milk processing, the data have made it possible, for example, to avoid the separation of whey proteins in heating equipment and to optimize the properties of yogurt gels (cf. 10.1.3.3 and 10.2.1.2).

Figure 1.30 shows the denaturation of β-LG in a diagram that combines the heating period with the temperature (cf. 2.5.4.3) in the form of straight lines of equal denaturation degrees. This allows us to read directly the time/temperature combinations required for a certain desired effect. At 85°C/136 s for example, only 60% of the β-LG-B are folded, so that only 60% can aggregate, although 90% would be potentially able to aggregate: at this temperature, the folding determines the overall reaction, as shown above. Conversely, 90% of the protein is potentially folded at 95°C/21 s while only 60% can be aggregated. At this temperature, aggregation determines the overall reaction.

Denaturation of biologically active proteins is usually associated with loss of activity. The fact that denatured proteins are more readily digested by proteolytic enzymes is also of interest.

1.4.3 Physical Properties

1.4.3.1 Dissociation

Proteins, like amino acids, are amphoteric. Depending on pH, they can exist as polyvalent cations, anions or zwitter ions. Proteins differ in their α-carboxyl and α-amino groups – since these groups are linked together by peptide bonds, the uptake or release of protons is limited to free terminal groups. Therefore, most of the dissociable functional groups are derived from side chains. Table 1.28 lists pK values of some protein groups. In contrast to free amino acids, these values fluctuate greatly for proteins since the dissociation is influenced by neighboring groups in the macromolecule. For example, in lysozyme the γ-carboxyl group of Glu[35] has a pK of 6–6.5, while the pK of the β-carboxyl group of Asp[66] is 1.5–2, of Asp[52] is 3–4.6 and of Asp[101] is 4.2–4.7.

The total charge of a protein, which is the absolute sum of all positive and negative charges, is differentiated from the so-called

Table 1.28. pK values of protein side chains

Group	pK (25°C)	Group	pK (25°C)
α-Carboxyl-	3– 4	Imidazolium-	4– 8
β,γ-Carboxyl-	3– 5	Hydroxy-	
α-Ammonium-	7– 8	(aromatic)	9–12
ε-Ammonium-	9–11	Thiol	8–11
Guanidinium-	12–13		

net charge which, depending on the pH, may be positive, zero or negative. By definition the net charge is zero and the total charge is maximal at the isoelectric point. Lowering or raising the pH tends to increase the net charge towards its maximum, while the total charge always becomes less than at the isolectric point.

Since proteins interact not only with protons but also with other ions, there is a further differentiation between an isoionic and an isoelectric point. The isoionic point is defined as the pH of a protein solution at infinite dilution, with no other ions present except for H^+ and HO^-. Such a protein solution can be acquired by extensive dialysis (or, better, electrodialysis) against water. The isoionic point is constant for a given substance while the isoelectric point is variable depending on the ions present and their concentration. In the presence of salts, i.e. when binding of anions is stronger than that of cations, the isoelectric point is lower than the isoionic point. The reverse is true when cationic binding is dominant. Figure 1.31 shows the shift in pH of an isoionic serum albumin solution after addition of various salts. The shift in pH is consistently positive, i.e. the protein binds more anions than cations.

The titration curve of β-lactoglobulin at various ionic strengths (Fig. 1.32) shows that the isoelectric point of this protein, at pH 5.18, is independent of the salts present. The titration curves are, however, steeper with increasing ionic strength, which indicates greater suppression of the electrostatic interaction between protein molecules.

At its isoelectric point a protein is the least soluble and the most likely to precipitate ("isoelectric precipitation") and is at its maximal crystallization capacity. The viscosity of solubilized proteins and the swelling power of insoluble proteins are at a minimum at the isoelectric point.

When the amino acid composition of a protein is known, the isoelectric point can be estimated according to the following formula:

$$pI = -10 \log Q_{pI} + 7.0 \qquad (1.96)$$

where QpI is the sum of deviations of the isoelectric points of all participating amino acids from the neutral point:

$$Q_{pI} = \frac{4.2 \cdot n\,Asp + 3.8\,m\,Glu}{3.8\,q\,Arg + 2.6\,r\,Lys + 0.5\,s\,His} \qquad (1.96)$$

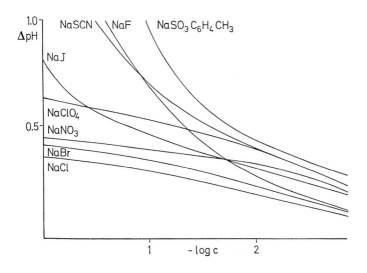

Fig. 1.31. pH-shift of isoionic serum albumin solutions by added salts. (according to *Edsall* and *Wymann*, 1958)

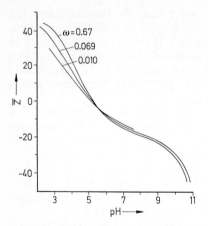

Fig. 1.32. Titration curves for β-lactoglobulin at various ionic strengths ω. (according to *Edsall* and *Wyman*, 1958)

The formula fails when acid or alkaline groups occur in masked form.

1.4.3.2 Optical Activity

The optical activity of proteins is due not only to asymmetry of amino acids but also to the chirality resulting from the arrangement of the peptide chain. Information on the conformation of proteins can be obtained from a record-ing of the optical rotatory dispersion (ORD) or the circular dichroism (CD), especially in the range of peptide bond absorption wavelengths (190−200 nm). The *Cotton* effect occurs in this range and reveals quantitative information on secondary structure. An α-helix or a β-structure gives a negative *Cotton* effect, with absorption maxima at 199 and 205 nm, while a randomly coiled conformation shifts the maximum to shorter wavelengths, i.e. results in a positive *Cotton* effect (Fig. 1.33).

1.4.3.3 Solubility, Hydration and Swelling Power

Protein solubility is variable and is influenced by the number of polar and apolar groups and their arrangement along the molecule. Gen-erally, proteins are soluble only in strongly polar solvents such as water, glycerol, form-amide, dimethylformamide or formic acid. In a less polar solvent such as ethanol, proteins are rarely noticeably soluble (e.g. prolamines). The solubility in water is dependent on pH and on salt concentration. Figure 1.34 shows these relationships for β-lactoglobulin.

At low ionic strengths, the solubility rises with increase in ionic strength and the solubility minimum (isoelectric point) is shifted from pH 5.4 to pH 5.2. This shift is due to preferen-tial binding of anions to the protein.

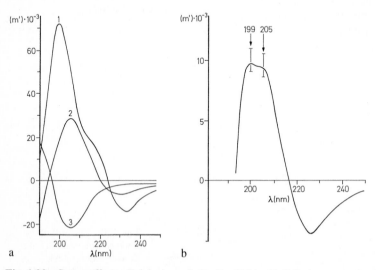

Fig. 1.33. *Cotton* effect. **a** Polylysine α-helix (1, pH 11−11.5) β-sheet structure (2, pH 11−11.3 and heated above 50 °C) and random coiled (3, pH 5−7). **b** Ribonuclease with 20% α-helix, 40% β-sheet structure and 40% random coiled region. (according to *Luebke, Schroeder*, and *Kloss*, 1975)

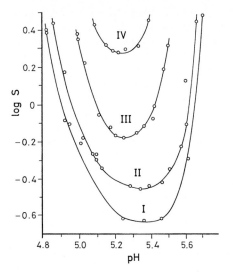

Fig. 1.34. β-Lactoglobulin solubility as affected by pH and ionic strength I. 0.001, II. 0.005, III. 0.01, IV. 0.02

If a protein has enough exposed hydrophobic groups at the isoelectric point, it aggregates due to the lack of electrostatic repulsion via intermolecular hydrophobic bonds, and (isoelectric) precipitation will occur. If on the other hand, intermolecular hydrophobic interactions are only poorly developed, a protein will remain in solution even at the isoelectric point, due to hydration and steric repulsion.

As a rule, neutral salts have a two-fold effect on protein solubility. At low concentrations they increase the solubility ("salting in" effect) by suppressing the electrostatic protein-protein interaction (binding forces).

The log of the solubility (S) is proportional to the ionic strength (μ) at low concentrations (cf. Fig. 1.34.):

$$\log S = k \cdot \mu. \tag{1.98a}$$

Protein solubility is decreased ("salting out" effect) at higher salt concentrations due to the ion hydration tendency of the salts. The following relationship applies (S_0: solubility at $\mu = 0$; K: salting out constant):

$$\log S = \log S_0 - K \cdot \mu \tag{1.98b}$$

Cations and anions in the presence of the same counter ion can be arranged in the following orders (*Hofmeister* series) based on their salting out effects:

$$K^+ > Rb^+ > Na^+ > Cs^+ > Li^+ > NH_4^+;$$
$$SO_4^{2-} > citrate^{2-} > tartrate^{2-} > acetate^-$$
$$> Cl^- > NO_3^- > Br^- > J^- > CNS^-.$$
$$\tag{1.99}$$

Multivalent anions are more effective than monovalent anions, while divalent cations are less effective than monovalent cations.

Since proteins are polar substances, they are hydrated in water. The degree of hydration (g water of hydration/g protein) is variable. It is 0.22 for ovalbumin (in ammonium sulfate), 0.06 for edestin (in ammonium sulfate), 0.8 for β-lactoglobulin and 0.3 for hemoglobin. Approximately 300 water molecules are sufficient to cover the surface of lysozyme (about 6000 Å²), that is one water molecule per 20 Å².

The swelling of insoluble proteins corresponds to the hydration of soluble proteins in that insertion of water between the peptide chains results in an increase in volume and other changes in the physical properties of the protein. For example, the diameter of myofibrils (cf. 12.2.1) increases to 2.5 times the original value during rinsing with 1.0 mol/L NaCl, which corresponds to a six-fold volume increase (cf. 12.5). The amount of water taken up by swelling can amount to a multiple of the protein dry weight. For example, muscle tissue contains 3.5–3.6 g water per g protein dry matter.

The water retention capacity of protein can be estimated with the following formula:

$$a = f_c + 0.4 f_p + 0.2 f_n \tag{1.100}$$

(a: g water/g protein; f_c, f_p, f_n: fraction of charged, polar, neutral amino acid residues).

1.4.3.4 Foam Formation and Foam Stabilization

In several foods, proteins function as foam-forming and foam-stabilizing components, for example in baked goods, sweets, desserts and beer. This varies from one protein to

another. Serum albumin foams very well, while egg albumin does not. Protein mixtures such as egg white can be particularly well suited (cf. 11.4.2.2). In that case, the globulins facilitate foam formation. Ovomucin stabilizes the foam, egg albumin and conalbumin allow its fixation through thermal coagulation.

Foams are dispersions of gases in liquids. Proteins stabilize by forming flexible, cohesive films around the gas bubbles. During impact, the protein is adsorbed at the interface via hydrophobic areas; this is followed by partial unfolding (surface denaturation). The reduction of surface tension caused by protein adsorption facilitates the formation of new interfaces and further gas bubbles. The partially unfolded proteins associate while forming stabilizing films.

The more quickly a protein molecule diffuses into interfaces and the more easily it is denatured there, the more it is able to foam. These values in turn depend on the molecular mass, the surface hydrophobicity, and the stability of the conformation.

Foams collapse because large gas bubbles grow at the expense of smaller bubbles (disproportionation). The protein films counteract this disproportionation. That is why the stability of a foam depends on the strength of the protein film and its permeability for gases. Film strength depends on the adsorbed amount of protein and the ability of the adsorbed molecules to associate. Surface denaturation generally releases additional amino acid side chains which can enter into intermolecular interactions. The stronger the cross-linkage, the more stable the film. Since the smallest possible net charge promotes association, the pH of the system should lie in the range of the isoelectric points of the proteins that participate in film formation.

In summary, the ideal foam-forming and foam-stabilizing protein is characterized by a low molecular weight, high surface hydrophobicity, good solubility, a small net charge in terms of the pH of the food, and easy denaturability.

Foams are destroyed by lipids and organic solvents such as higher alcohols, which due to their hydrophobicity displace proteins from the gas bubble surface without being able to form stable films themselves. Even a low concentration of egg yolk, for example, prevents the bursting of egg white. This is attributed to a disturbance of protein association by the lecithins.

The foam-forming and foam-stabilizing characteristics of proteins can be improved by chemical and physical modification. Thus a partial enzymatic hydrolysis leads to smaller, more quickly diffusing molecules, better solubility, and the release of hydrophobic groups. Disadvantages are the generally lower film stability and the loss of thermal coagulability. The characteristics can also be improved by introducing charged or neutral groups (cf. 1.4.6.2) and by partial thermal denaturation (e.g. of whey proteins). Recently, the addition of strongly alkaline proteins (e. g. clupeines) is being tested, which apparently increases the association of protein in the films and allows the foaming of fatty systems.

1.4.3.5 Gel Formation

Gels are disperse systems of at least two components in which the disperse phase in the dispersant forms a cohesive network. They are characterized by the lack of fluidity and elastic deformability. Gels are placed between solutions, in which repulsive forces between molecules and the disperse phase predominate, and precipitates, where strong intermolecular interactions predominate. We differentiate between two types of gel, the *polymeric networks* and the *aggregated dispersions*, although intermediate forms are found as well.

Examples of *polymeric networks* are the gels formed by gelatin (cf. 12.3.2.3.1) and polysaccharides such as agarose (cf. 4.4.4.1.2) and carrageenan (4.4.4.3.2). Formation of a three-dimensional network takes place through the aggregation of unordered fibrous molecules via partly ordered structures, e. g. while double helices are formed (cf. 4.4.4.3.2, Fig. 4.14, Fig. 12.21). Characteristic for gels of this type is the low polymer concentration (~1%) as well as transparency and fine texture. Gel formation is caused by setting a certain pH, by adding certain ions, or by heating/cooling. Since aggregation takes place mostly via intermolecular hydrogen bonds which easily break when heated, polymeric networks are thermo-reversible, i.e. the gels are formed when a solution cools, and they melt again when it is heated.

Examples of *aggregated dispersions* are the gels formed by globular proteins after heating and denaturation. The thermal unfolding of the protein leads to the release of amino acid side chains which may enter into intermolecular interactions. The subsequent association occurs while small spherical aggregates form which combine into linear strands whose interaction establishes the gel network. Before gel can be formed in the unordered type of aggregation, a relatively high protein concentration (5–10%) is necessary. The aggregation rate should also be slower than the unfolding rate, since otherwise coarse and fairly unstructured gels are formed, such as in the area of the isoelectric point. The degree of denaturation necessary to start aggregation seems to depend on the protein. Since partial denaturation releases primarily hydrophobic groups, intermolecular hydrophobic bonds generally predominate, which results in the thermoplastic (thermo-irreversible) character of this gel type, in contrast to the thermoreversible gel type stabilized by hydrogen bonds. Thermoplastic gels do not liquefy when heated, but they can soften or shrink. In addition to hydrophobic bonds, disulfide bonds formed from released thiol groups can also contribute to cross-linkage, as can intermolecular ionic bonds between proteins with different isoelectric points in heterogeneous systems (e. g. egg white).

Gel formation can be improved by adding salt. The moderate increase in ionic strength increases interaction between charged macromolecules or molecule aggregates through charge shielding without precipitation occurring. An example is the heat coagulation of soybean curd (tofu, cf. 16.3.1.2.3) which is promoted by calcium ions.

1.4.3.6 Emulsifying Effect

Emulsions are disperse systems of one or more immiscible liquids. They are stabilized by emulsifiers – compounds which form interface films and thus prevent the disperse phases from flowing together (cf. 8.15). Due to their amphipathic nature, proteins can stabilize emulsions such as milk (cf. 10.1.2.3). The suitability of a protein as an emulsifier depends on the rate at which it diffuses into the interface, on its adsorbability there and on the deformability of its conformation under the influence of interfacial tension (surface denaturation). The diffusion rate depends on the temperature and the molecular weight, which in turn can be influenced by the pH and the ionic strength. The adsorbability depends on the exposure of hydrophilic and hydrophobic groups and thus on the amino acid profile, as well as on the pH, the ion strength and the temperature. The conformative stability depends in the amino acid composition, the molecular weight and the intramolecular disulfide bonds.

Therefore, a protein with ideal qualities as an emulsifier for an oil-in-water emulsion would have a relatively low molecular weight, a balanced amino acid composition in terms of charged, polar and nonpolar residues, good water solubility, well-developed surface hydrophobicity, and a relatively stable conformation.

1.4.4 Chemical Reactions

The chemical modification of protein is of importance for a number of reasons. It provides derivatives suitable for sequence analysis, identifies the reactive groups in catalytically active sites of an enzyme, enables the binding of protein to a carrier (protein immobilization) and provides changes in protein properties which are important in food processing. In contrast to free amino acids and except for the relatively small number of functional groups on the terminal amino acids, only the functional groups on protein side chains are available for chemical reactions.

1.4.4.1 Lysine Residue

Reactions involving the lysine residue can be divided into several groups: (a) reactions leading to a positively charged derivative; (b) reactions eliminating the positive charge; (c) derivatizations introducing a negative charge; and (d) reversible reactions. The latter are of particular importance.

1.4.4.1.1 Reactions Which Retain the Positive Charge

Alkylation of the free amino group of lysine with aldehydes and ketones is possible, with a simultaneous reduction step:

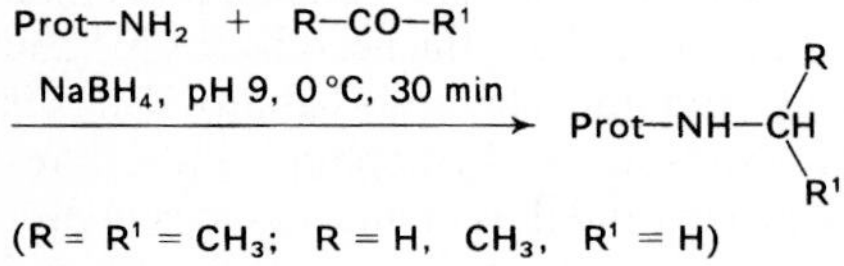

$$\text{Prot—NH}_2 \; + \; \text{R—CO—R}^1$$

$$\xrightarrow{\text{NaBH}_4,\ \text{pH 9, 0 °C, 30 min}} \quad \text{Prot—NH—CH}\genfrac{}{}{0pt}{}{R}{R^1}$$

$$(\text{R} = \text{R}^1 = \text{CH}_3; \quad \text{R} = \text{H},\ \text{CH}_3, \quad \text{R}^1 = \text{H}) \tag{1.101}$$

A dimethyl derivative $[\text{Prot—N(CH}_3)_2]$ can be obtained with formaldehyde $(\text{R}=\text{R}_1=\text{H})$ (cf. 1.2.4.2.2).

Guanidination can be accomplished by using O-methylisourea as a reactant. α-Amino groups react at a much slower rate than ε-amino groups:

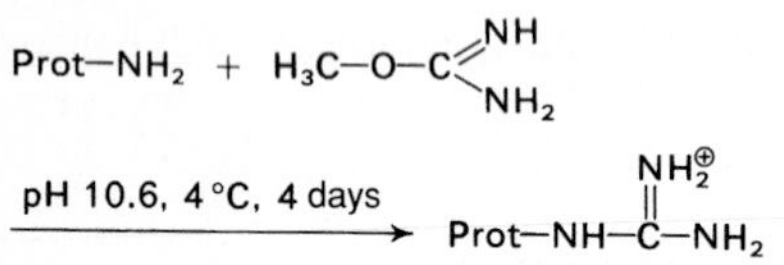

$$\text{Prot—NH}_2 \; + \; \text{H}_3\text{C—O—C}\genfrac{}{}{0pt}{}{NH}{NH_2}$$

$$\xrightarrow{\text{pH 10.6, 4 °C, 4 days}} \quad \text{Prot—NH—}\overset{\overset{\displaystyle NH_2^{\oplus}}{\|}}{C}\text{—NH}_2 \tag{1.102}$$

This reaction is used analytically to assess the amount of biologically available ε-amino groups and for measuring protein digestibility.

Derivatization with imido esters is also possible. The reactant is readily accessible from the corresponding nitriles:

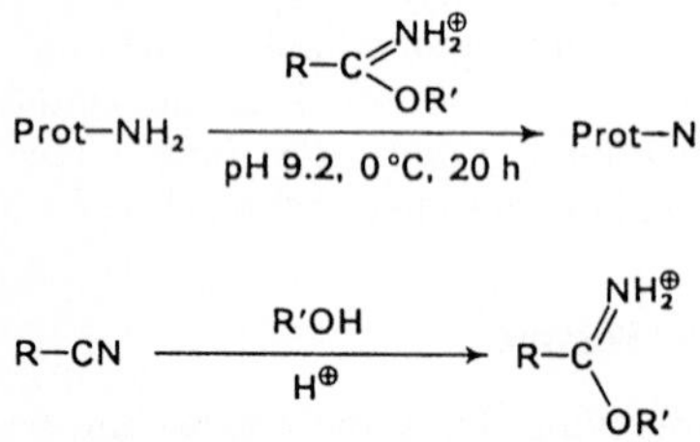

$$\text{Prot—NH}_2 \; \xrightarrow[\text{pH 9.2, 0 °C, 20 h}]{\text{R—C}\genfrac{}{}{0pt}{}{NH_2^{\oplus}}{OR'}} \; \text{Prot—NH—}C\genfrac{}{}{0pt}{}{NH_2^{\oplus}}{R}$$

$$\text{R—CN} \; \xrightarrow[\text{H}^{\oplus}]{\text{R'OH}} \; \text{R—C}\genfrac{}{}{0pt}{}{NH_2^{\oplus}}{OR'} \tag{1.103}$$

Proteins can be cross-linked with the use of a bifunctional imido ester (cf. 1.4.4.10).

Treatment of the amino acid residue with amino acid carboxyanhydrides yields a polycondensation reaction product:

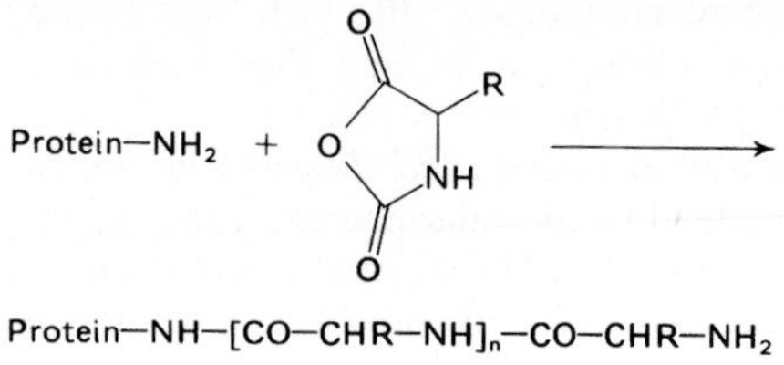

$$\text{Protein—NH}_2 \; + \; \text{[oxazolidinedione, R]} \longrightarrow$$

$$\text{Protein—NH—[CO—CHR—NH]}_n\text{—CO—CHR—NH}_2 \tag{1.104}$$

The value n depends on reaction conditions. The carboxyanhydrides are readily accessible through interaction of the amino acid with phosgene:

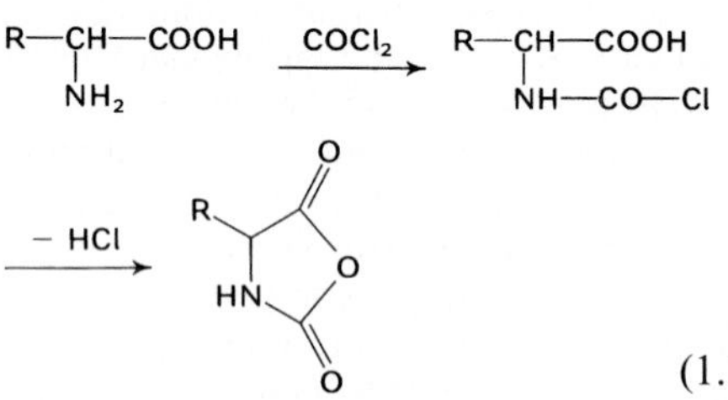

$$\text{R—CH—COOH} \; \xrightarrow{\text{COCl}_2} \; \text{R—CH—COOH}$$
$$\qquad\quad | \qquad\qquad\qquad\quad\ |$$
$$\qquad\ \text{NH}_2 \qquad\qquad\qquad \text{NH—CO—Cl}$$

$$\xrightarrow{-\text{HCl}} \quad \text{[oxazolidinedione, R]} \tag{1.105}$$

1.4.4.1.2 Reactions Resulting in a Loss of Positive Charge

Acetic anhydride reacts with lysine, cysteine, histidine, serine, threonine and tyrosine residues. Subsequent treatment of the protein with hydroxylamine (1 M, 2 h, pH 9, 0 °C) leaves only the acetylated amino groups intact:

$$\text{Prot—NH}_2 \; \xrightarrow[\text{pH 7–9.5, 0 °C}]{(\text{CH}_3\text{CO})_2\text{O}} \; \text{Prot—NH—CO—CH}_3 \tag{1.106}$$

Carbamoylation with cyanate attacks α- and ε-amino groups as well as cysteine and tyrosine residues. However, their derivatization is reversible under alkaline conditions:

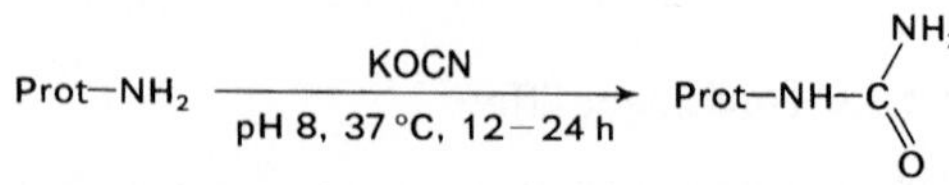

$$\text{Prot—NH}_2 \; \xrightarrow[\text{pH 8, 37 °C, 12–24 h}]{\text{KOCN}} \; \text{Prot—NH—}C\genfrac{}{}{0pt}{}{NH_2}{O} \tag{1.107}$$

Arylation with 1-fluoro-2,4-dinitrobenzene (*Sanger's* reagent; FDNB) and trinitrobenzene sulfonic acid was outlined in Section 1.2.4.2.2.

FDNB also reacts with cysteine, histidine and tyrosine.

4-Fluoro-3-nitrobenzene sulfonic acid, a reactant which has good solubility in water, is also of interest for derivatization of proteins:

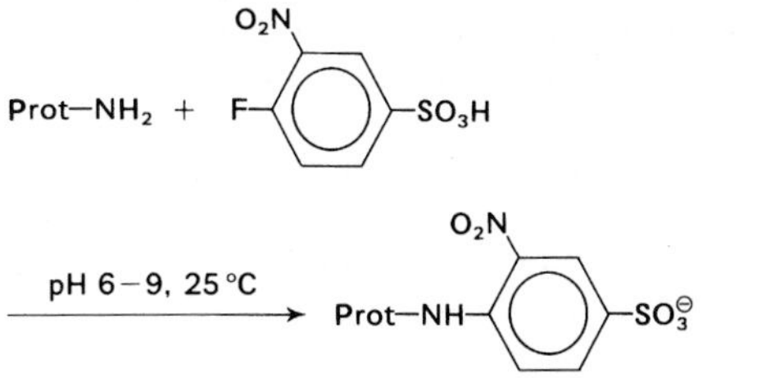

$$\text{Prot—NH}_2 + F\text{—}\underset{O_2N}{\bigcirc}\text{—SO}_3H$$

$$\xrightarrow{\text{pH } 6-9,\ 25\,°C}\ \text{Prot—NH—}\underset{O_2N}{\bigcirc}\text{—SO}_3^\ominus \tag{1.108}$$

Deamination can be accomplished with nitrous acid:

$$\text{Prot—NH}_2 \xrightarrow[\text{pH } 4.35,\ 0\,°C]{\text{HNO}_2}\ \text{Prot—OH} + N_2 \tag{1.109}$$

This reaction involves α- and ε-amino groups as well as tryptophan, tyrosine, cysteine and methionine residues.

1.4.4.1.3 Reactions Resulting in a Negative Charge

Acylation with dicarboxylic acid anhydrides, e.g. succinic acid anhydride, introduces a carboxyl group into the protein:

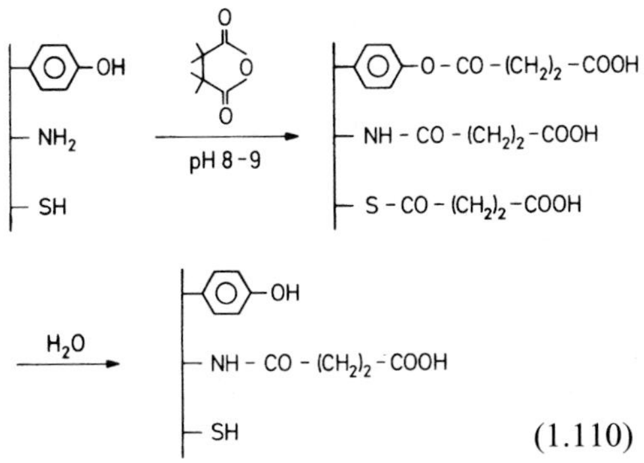

$$\tag{1.110}$$

Introduction of a fluorescent acid group is possible by interaction of the protein with pyridoxal phosphate followed by reduction of the intermediary *Schiff* base:

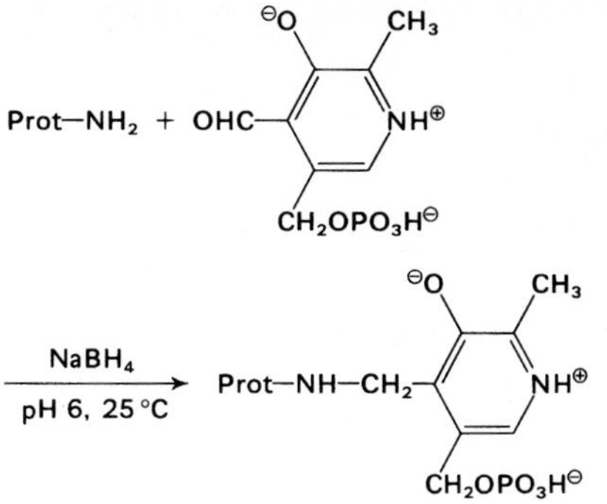

$$\text{Prot—NH}_2 + \text{OHC}\cdots \xrightarrow[\text{pH } 6,\ 25\,°C]{\text{NaBH}_4}\ \text{Prot—NH—CH}_2\cdots \tag{1.111}$$

1.4.4.1.4 Reversible Reactions

N-Maleyl derivatives of proteins are obtained at alkaline pH by reaction with maleic acid anhydride. The acylated product is cleaved at pH < 5, regenerating the protein:

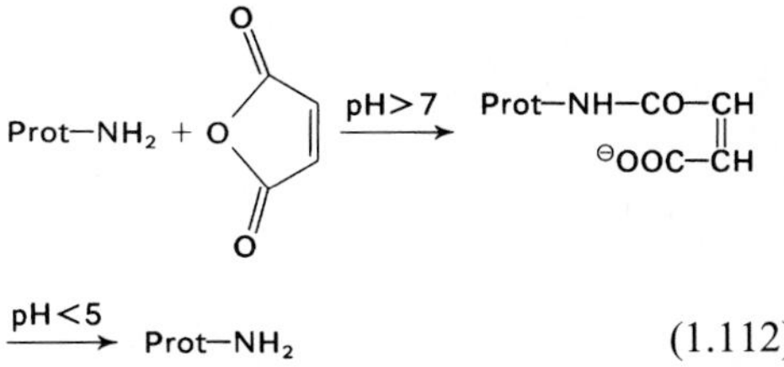

$$\xrightarrow{\text{pH}<5}\ \text{Prot—NH}_2 \tag{1.112}$$

The half-life (τ) of ε-N-maleyl lysine is 11 h at pH 3.5 and 37°C. More rapid cleavage is observed with the 2-methyl-maleyl derivative (τ < 3 min at pH 3.5 and 20°C) and the 2,2,3,3-tetrafluoro-succinyl derivative (τ very low at pH 9.5 and 0°C). Cysteine binds maleic anhydride through an addition reaction. The S-succinyl derivative is quite stable. This side reaction is, however, avoided when protein derivatization is done with exo-cis-3,6-end-oxohexahydrophthalic acid anhydride:

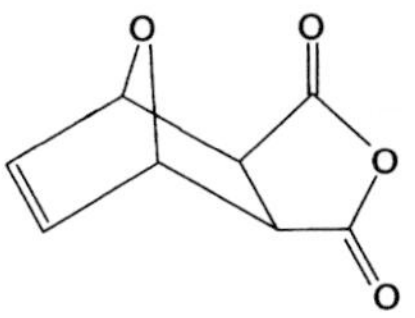

$$\tag{1.113}$$

For ε-N-acylated lysine, τ = 4−5 h at pH 3 and 25°C.

Acetoacetyl derivatives are obtained with diketene:

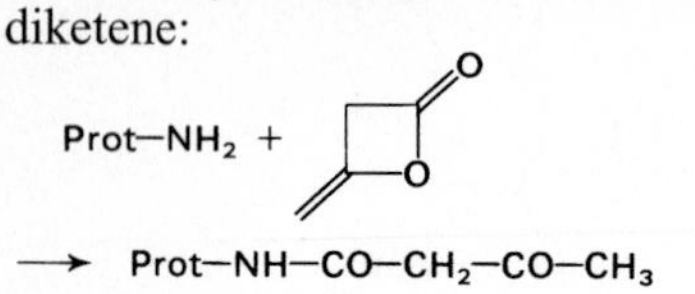

Prot—NH$_2$ +

$$\longrightarrow \text{Prot—NH—CO—CH}_2\text{—CO—CH}_3 \quad (1.114)$$

This type of reaction also occurs with cysteine and tyrosine residues. The acyl group is readily split from tyrosine at pH 9.5. Complete release of protein from its derivatized form is possible by treatment with phenylhydrazine or hydroxylamine at pH 7.

1.4.4.2 Arginine Residue

The arginine residue of proteins reacts with α- or β-dicarbonyl compounds to form cyclic derivatives:

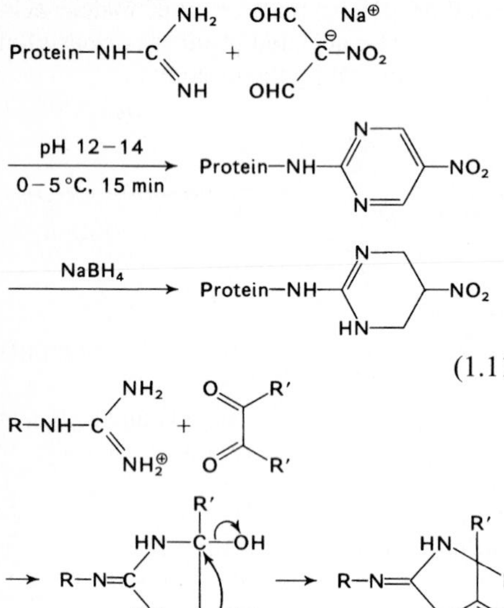

$$(1.115)$$

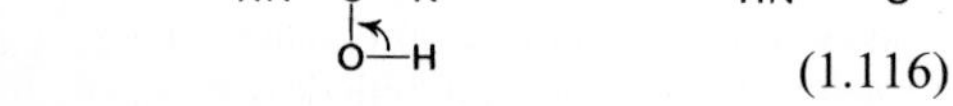

$$(1.116)$$

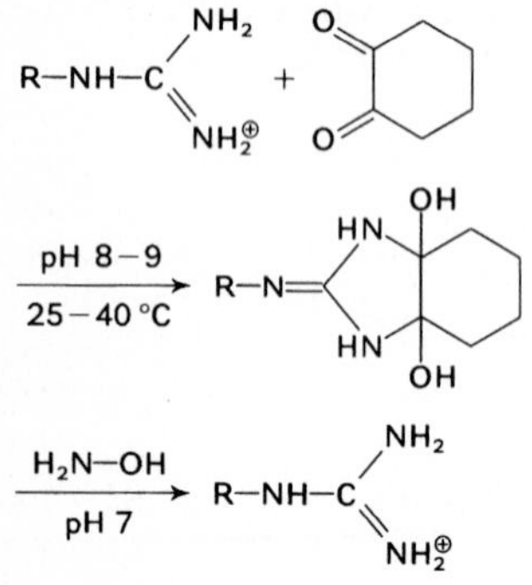

$$(1.117)$$

The nitropyrimidine derivative absorbs at 335 nm. The arginyl bond of this derivative is not cleaved by trypsin but it is cleaved in its tetrahydro form, obtained by reduction with NaBH$_4$ (cf. Reaction 1.115). In the reaction with benzil, an iminoimidazolidone derivative is obtained after a benzilic acid rearrangement (cf. Reaction 1.116).
Reaction of the arginine residue with 1,2-cyclohexanedione is highly selective and proceeds under mild conditions. Regeneration of the arginine residue is again possible with hydroxylamine (cf. Reaction 1.117).

1.4.4.3 Glutamic and Aspartic Acid Residues

These amino acid residues are usually esterified with methanolic HCl. There can be side reactions, such as methanolysis of amide derivatives or N,O-acyl migration in serine or threonine residues:

$$\text{Protein—COOH} \xrightarrow[0\,°C]{\text{CH}_3\text{OH/HCl}} \text{Protein—COOCH}_3 \quad (1.118)$$

Diazoacetamide reacts with a carboxyl group and also with the cysteine residue:

$$\text{Protein—COOH} + \text{N}_2\text{CH}_2\text{—CONH}_2$$
$$\longrightarrow \text{R—COOCH}_2\text{CONH}_2 \quad (1.119)$$

Amino acid esters or other similar nucleophilic compounds can be attached to a carboxyl group of a protein with the help of a carbodiimide:

$$\text{Protein—COOH} + \text{H}_2\text{N—CH}_2\text{—COOCH}_3$$
$$\xrightarrow{\text{R—N=C=N—R}} \text{Protein—CO—NH—CH}_2\text{—COOCH}_3 \quad (1.120)$$

Amidation is also possible by activating the carboxyl group with an isooxazolium salt (*Woodward* reagent) to an enolester and its conversion with an amine.

Protein – COOH   +

$$\text{(isoxazolium structure with } R^2, R^3, R^1, O^{\oplus}, N, H\text{)}$$

$$\xrightarrow{\quad} \text{Protein - CO - O -} \overset{R^1}{\underset{|}{C}} = \overset{R^2}{\underset{|}{C}} - \overset{R^3}{\underset{|}{C}} = N - OH$$

$$\xrightarrow{R-NH_2}\quad \text{Protein - CO - NH - R}$$

$$+\quad O = \overset{R^1}{\underset{|}{C}} - \overset{R^2}{\underset{|}{C}}H - \overset{R^3}{\underset{|}{C}} = N - OH \qquad (1.121)$$

1.4.4.4 Cystine Residue (cf. also Section 1.2.4.3.5)

Cleavage of cystine is possible by a nucleophilic attack:

$$\text{Protein–S–S–Protein} + Y^{\ominus}$$

$$\xrightarrow{\quad} \text{Protein–S–Y + Protein–S}^{\ominus} \qquad (1.122)$$

The nucleophilic reactivity of the reagents decreases in the series: hydride > arsenite and phosphite > alkanethiol > aminoalkanethiol > thiophenol and cyanide > sulfite > OH⁻ > p-nitrophenol > thiosulfate > thiocyanate. Cleavage with sodium borohydride and with thiols was covered in Section 1.2.4.3.5. Complete cleavage with sulfite requires that oxidative agents (e.g. Cu^{2+}) be present and that the pH be higher than 7:

$$\text{RSSR} + SO_3^{2\ominus} \xrightarrow{\quad} \text{RSSO}_3^{\ominus} + \text{RS}^{\ominus}$$

$$2\,\text{RS}^{\ominus} \xrightarrow{Cu^{2\oplus}} \text{RSSR} \qquad (1.123)$$

The resultant S-sulfo derivative is quite stable in neutral and acidic media and is fairly soluble in water. The S-sulfo group can be eliminated with an excess of thiol reagent.

Cleavage of cystine residues with cyanides (nitriles) is of interest since the thiocyanate formed in the reaction is cyclized into a 3-acyl-2-iminothiazolidine under cleavage of the N-acyl bond:

$$\text{R–CO–NH–CH–CO–NH–R'}$$
$$\underset{R''–S–S}{\overset{|}{C}H_2}$$

$$\xrightarrow{CN^{\ominus}} \quad \text{R–CO–NH–CH–CO–NH–R'}$$
$$\underset{NC–S}{\overset{|}{C}H_2} \quad + R''–S^{\ominus}$$

$$\xrightarrow{\quad} \text{R–CO–N——CH–CO–NH–R'}$$
$$HN=\overset{|}{C}\underset{S}{\overset{}{}}\overset{|}{C}H_2$$

$$\xrightarrow{\quad} \text{R–COOH} + \text{HN——CH–CO–NH–R'}$$
$$\qquad\qquad HN=\overset{}{C}\underset{S}{}\overset{}{C}H_2 \qquad (1.124)$$

This reaction can be utilized for the selective cleavage of peptide chains. Initially, all the disulfide bridges are reduced with dithiothreitol, and then are converted to mixed disulfides through reaction with 5,5'-dithio-bis-(2-nitrobenzoic acid). These mixed disulfides are then cleaved by cyanide at pH 7.

Electrophilic cleavage occurs with Ag^+ and Hg^+ or Hg^{2+} as follows:

$$2\,Ag^{\oplus} + 2\,RSSR \xrightarrow{\quad} 2\,RSAg + 2\,RS^{\oplus}$$

$$2\,RS^{\oplus} + 2\,OH^{\ominus} \xrightarrow{\quad} 2\,RSOH \xrightarrow{\quad} RSO_2H + RSH$$

$$RSH + Ag^{\oplus} \xrightarrow{\quad} RSAg + H^{\oplus}$$

_____

$$3\,Ag^{\oplus} + 2\,RSSR + 2\,OH^{\ominus} \xrightarrow{\quad} 3\,RSAg + RSO_2H + H^{\oplus} \qquad (1.125)$$

Electrophilic cleavage with H^+ is possible only in strong acids (e.g. 10 mol/L HCl). The sulfenium cation which is formed can catalyze a disulfide exchange reaction:

$$RSSR + H^{\oplus} \xrightarrow{\quad} RSH + RS^{\oplus}$$

$$RS^{\oplus} + R'SSR' \xrightarrow{\quad} RSSR' + R'S^{\oplus} \qquad (1.126)$$

In neutral and alkaline solutions a disulfide exchange reaction is catalyzed by the thiolate anion:

$$RSSR + OH^{\ominus} \rightleftharpoons R–SOH + RS^{\ominus}$$

$$R'SSR' + RS^{\ominus} \rightleftharpoons R'SSR + R'S^{\ominus} \qquad (1.127)$$

1.4.4.5 Cysteine Residue (cf. also Section 1.2.4.3.5)

A number of alkylating agents yield derivatives which are stable under the conditions for acidic hydrolysis of proteins. The reaction with ethylene imine giving an S-aminoethyl derivative and, hence, an additional linkage position in the protein for hydrolysis by trypsin, was mentioned in Section 1.4.1.3. Iodoacetic acid, depending on the pH, can react with cysteine, methionine, lysine and histidine residues:

$$\text{Protein–SH} \xrightarrow{ICH_2COOH} \text{Protein–S–}CH_2\text{–COOH} \quad (1.128)$$

The introduction of methyl groups is possible with methyl iodide or methyl isourea, and the introduction of methylthio groups with methylthiosulfonylmethane:

$$\text{Protein-SH} \xrightarrow{CH_3I} \text{Protein–S–}CH_3 \quad (1.129)$$

$$\text{Protein–SH} + CH_3\text{–O–C}\begin{matrix} \nearrow NH_2 \\ \searrow NH_2^{\oplus} \end{matrix} \longrightarrow \text{Protein–S–}CH_3 + O{=}C\begin{matrix} \nearrow NH_2 \\ \searrow NH_2 \end{matrix} \quad (1.130)$$

$$\text{Protein–SH} + CH_3\text{–S–}SO_2\text{–}CH_3 \longrightarrow \text{Protein–S–S–}CH_3 + CH_3SO_3H \quad (1.131)$$

Maleic acid anhydride and methyl-p-nitrobenzene sulfonate are also alkylating agents:

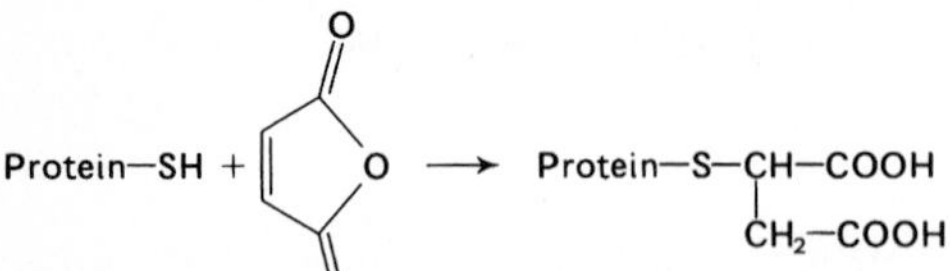

(1.132)

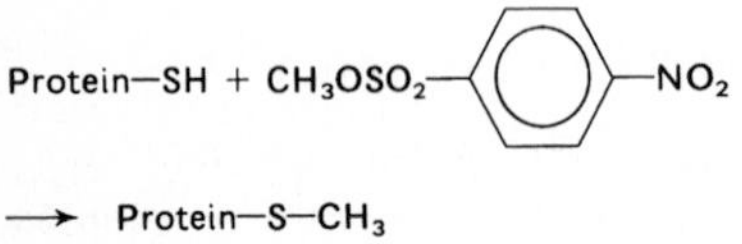

(1.133)

A number of reagents make it possible to measure the thiol group content spectrophotometrically. The molar absorption coefficient, ε, for the derivative of azobenzene-2-sulfenylbromide, ε_{353}, is 16,700 M^{-1} cm^{-1} at pH 1:

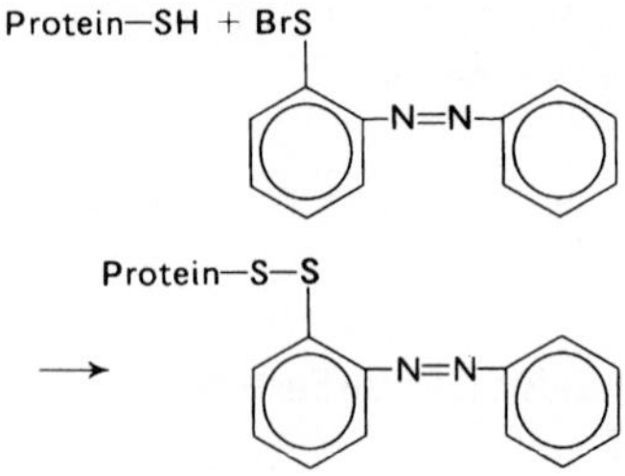

(1.134)

5,5′-Dithiobis-(2-nitrobenzoic acid) has a somewhat lower ε_{412} of 13,600 at pH 8 for its product, a thionitrobenzoate anion:

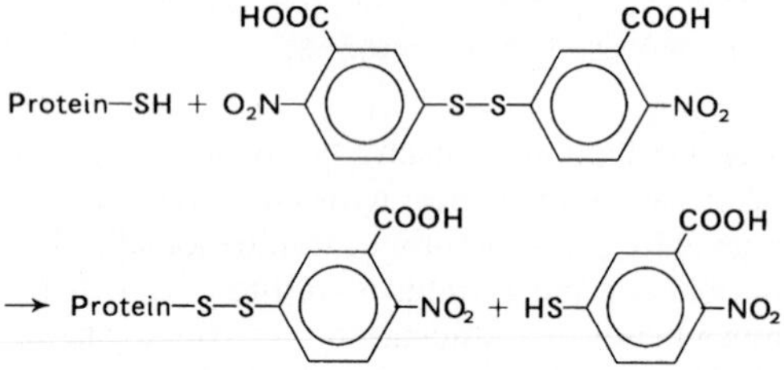

(1.135)

The derivative of p-mercuribenzoate has an ε_{250} of 7,500 at pH 7, while the derivative of N-ethylmaleic imide has an ε_{300} of 620 at pH 7:

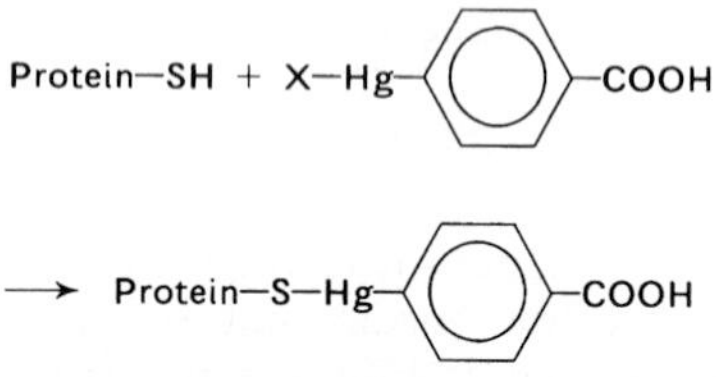

(1.136)

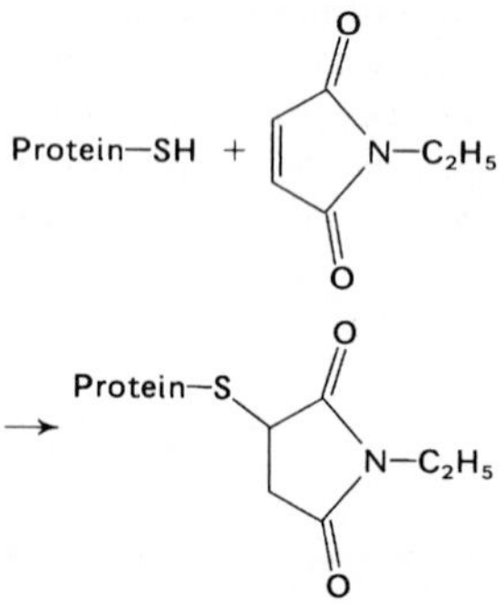

(1.137)

Especially suitable for the specific isolation of cysteine-containing peptides of great sensitivity is N-dimethylaminoazobenzenemaleic acid imide (DABMA).

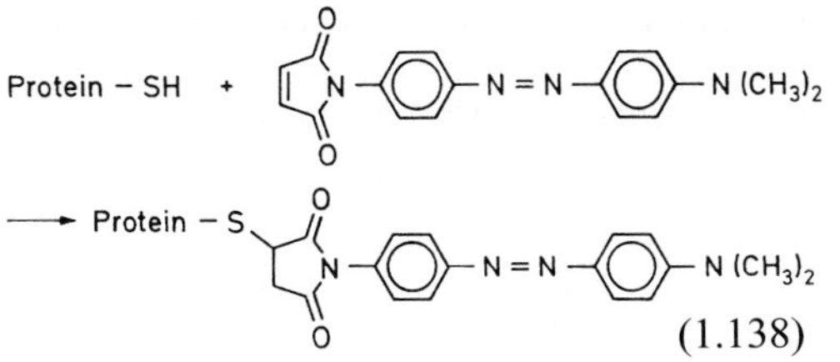

(for example, 1-chloro-3-tosylamido-7-aminoheptan-2-one inactivates trypsin and 1-chloro-3-tosylamido-4-phenylbutan-2-one inactivates chymotrypsin) by N-alkylation of the histidine residue:

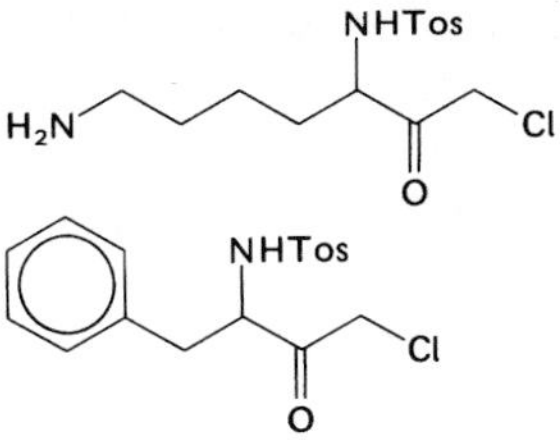

$$(1.138)$$

$$(1.140)$$

1.4.4.6 Methionine Residue

Methionine residues are oxidized to sulfoxides with hydrogen peroxide. The sulfoxide can be reduced, regenerating methionine, using an excess of thiol reagent (cf. 1.2.4.3.6). α-Halogen carboxylic acids, β-propiolactone and alkyl halogenides convert methionine into sulfonium derivatives, from which methionine can be regenerated in an alkaline medium with an excess of thiol reagent:

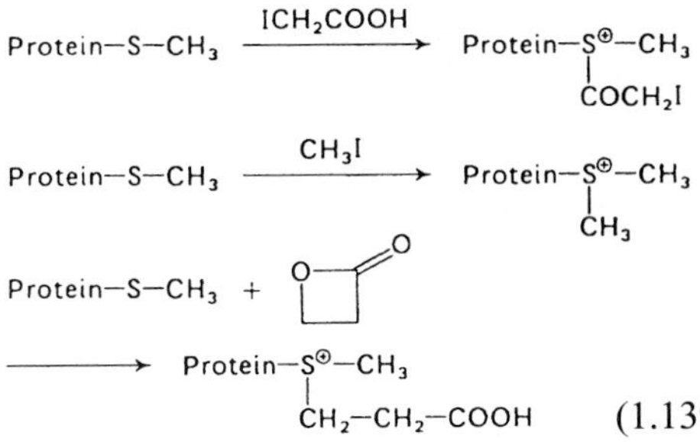

1.4.4.8 Tryptophan Residue

N-Bromosuccinimide oxidizes the tryptophan side chain and also tyrosine, histidine and cysteine:

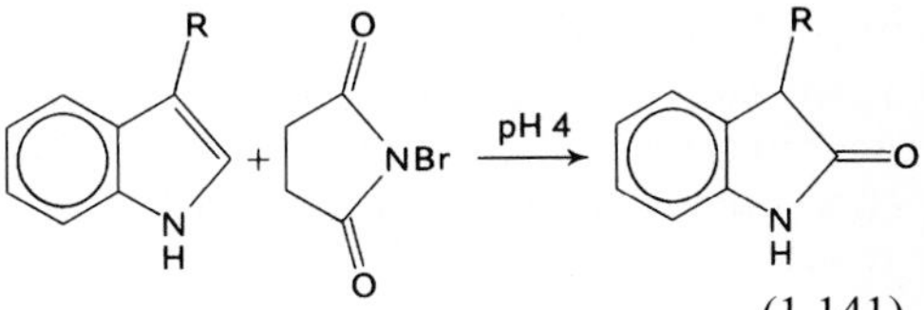

$$(1.141)$$

The reaction is used for the selective cleavage of peptide chains and the spectrophotometric determination of tryptophan.

$$(1.139)$$

1.4.4.9 Tyrosine Residue

Selective acylation of tyrosine can occur with acetylimidazole as a reagent:

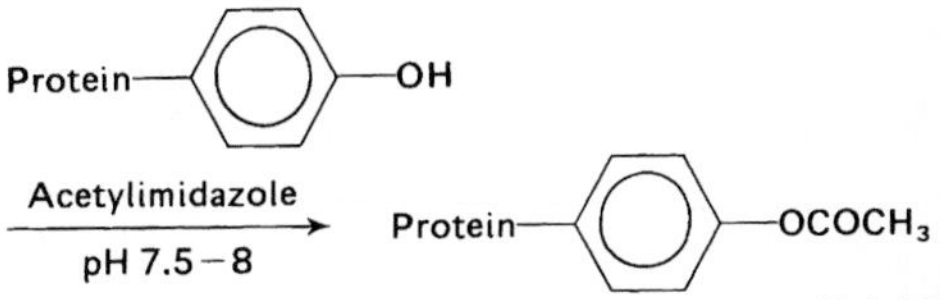

Reaction with cyanogen bromide (BrCN), which splits the peptide bond on the carboxyl side of the methionine molecule, was outlined in Section 1.4.1.3.

$$(1.142)$$

1.4.4.7 Histidine Residue

Selective modification of histidine residues present on active sites of serine proteinases is possible. Substrate analogues such as halogenated methyl ketones inactivate such enzymes

Diazotized arsanilic acid reacts with tyrosine and with histidine, lysine, tryptophan and arginine:

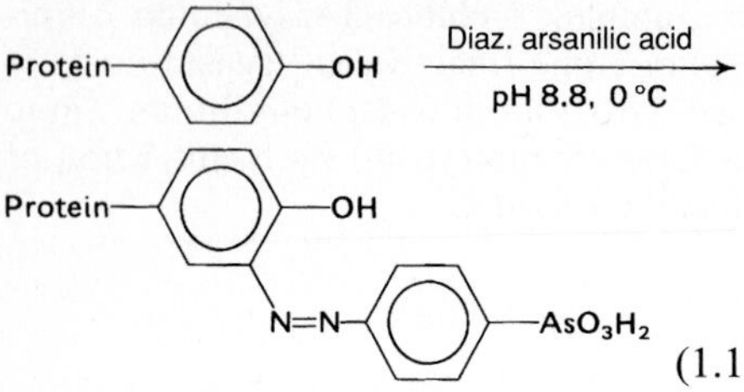

$$(1.143)$$

Tetranitromethane introduces a nitro group into the ortho position:

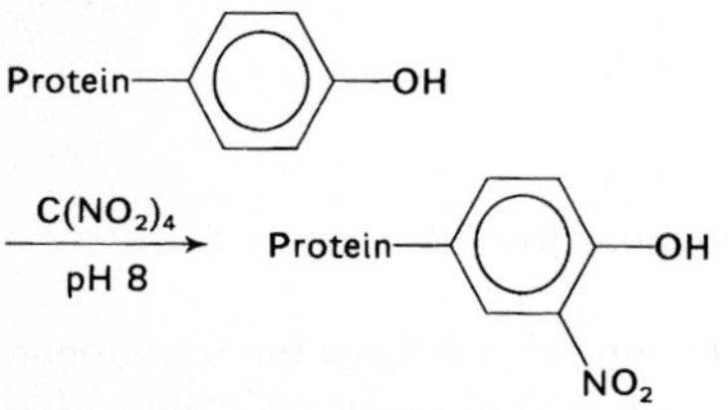

$$(1.144)$$

1.4.4.10 Bifunctional Reagents

Bifunctional reagents enable intra- and intermolecular cross-linking of proteins. Examples are bifunctional imidoester, fluoronitrobenzene, isocyanate derivatives and maleic acid imides:

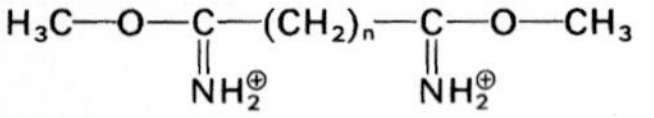

$$(1.145)$$

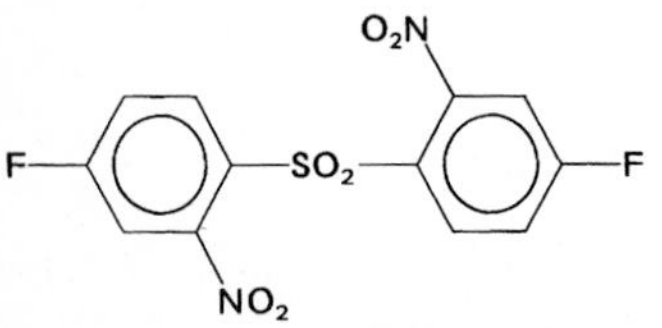

$$(1.146)$$

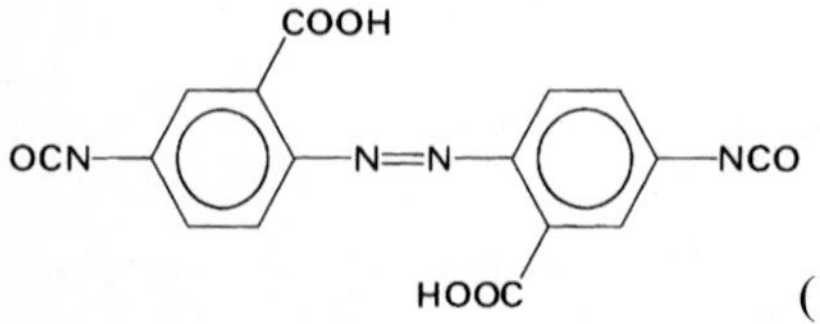

$$(1.147)$$

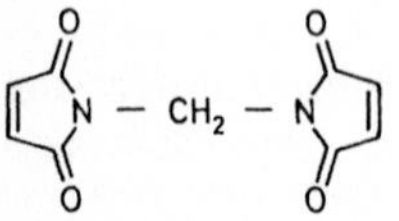

$$(1.148)$$

1.4.4.11 Reactions Involved in Food Processing

The nature and extent of the chemical changes induced in proteins by food processing depend on a number of parameters, for example, composition of the food and processing conditions, such as temperature, pH or the presence of oxygen. As a consequence of these reactions, the biological value of proteins may be decreased:

- Destruction of essential amino acids
- Conversion of essential amino acids into derivatives which are not metabolizable
- Decrease in the digestibility of protein as a result of intra- or interchain cross-linking.

Formation of toxic degradation products is also possible. The nutritional/physiological and toxicological assessment of changes induced by processing of food is a subject of some controversy and opposing opinions.

The *Maillard* reaction of the ε-amino group of lysine prevails in the presence of reducing sugars, for example, lactose or glucose, which yield protein-bound ε-N-deoxylactulosyl-1-lysine or ε-N-deoxyfructosyl-1-lysine, respectively. Lysine is not biologically available in these forms. Acidic hydrolysis of such primary reaction products yields lysine as well as the degradation products furosine and pyridosine in a constant ratio (cf. 4.2.4.4):

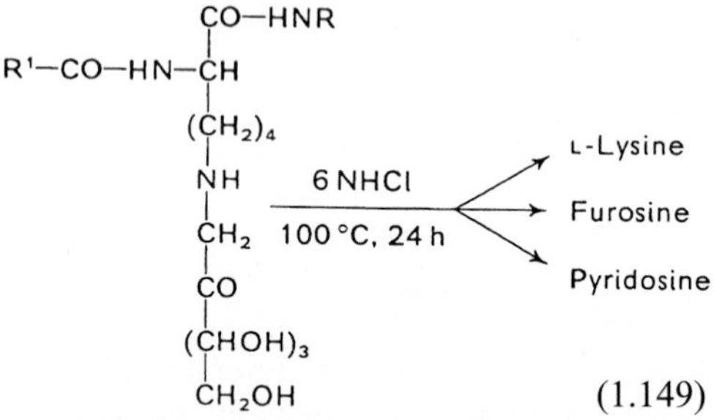

$$(1.149)$$

A nonreducing sugar (e.g. sucrose) can also cause a loss of lysine when conditions for sugar hydrolysis are favorable.

Losses of available lysine, cystine, serine, threonine, arginine and some other amino acids occur at higher pH values. Hydrolysates of alkali-treated proteins often contain some unusual compounds, such as ornithine, β-aminoalanine, lysinoalanine, ornithinoalanine,

lanthionine, methyllanthionine and D-alloisoleucine, as well as other D-amino acids.
The formation of these compounds is based on
the following reactions: 1,2-elimination in the
case of hydroxy amino acids and thio amino
acids results in 2-amino-acrylic acid (dehydroalanine) and 2-aminocrotonic acid (dehydro-
aminobutyric acid), respectively:

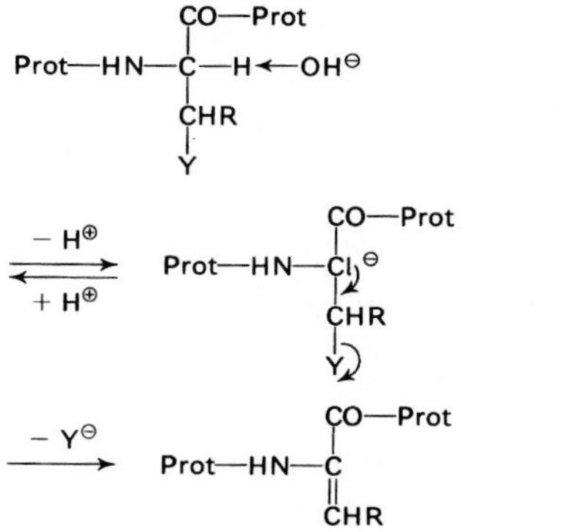

$$
\begin{array}{c}
\text{CO—Prot} \\
| \\
\text{Prot—HN—C—H} \leftarrow \text{OH}^{\ominus} \\
| \\
\text{CHR} \\
| \\
Y
\end{array}
$$

$$
\xrightarrow[+\,H^{\oplus}]{-\,H^{\oplus}}
\begin{array}{c}
\text{CO—Prot} \\
| \\
\text{Prot—HN—C}|^{\ominus} \\
| \\
\text{CHR} \\
| \\
Y
\end{array}
$$

$$
\xrightarrow{-\,Y^{\ominus}}
\begin{array}{c}
\text{CO—Prot} \\
| \\
\text{Prot—HN—C} \\
\| \\
\text{CHR}
\end{array}
\qquad (1.150)
$$

R = H, CH$_3$; Y = OH, OPO$_3$H$_2$, SH, SR^1, SSR^1

In the case of cystine, the eliminated thiol-
cysteine can form a second dehydroalanine
residue:

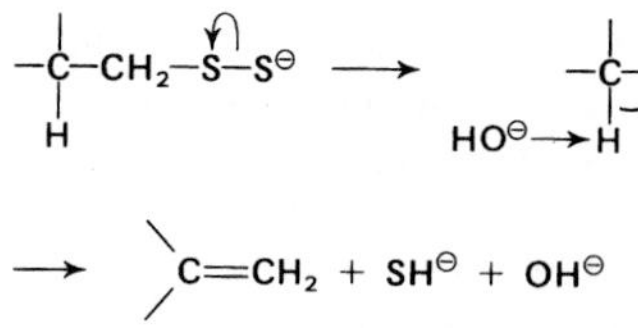

$$
\begin{array}{c}
| \\
-\text{C—CH}_2\text{—S—S}^{\ominus} \\
| \\
H
\end{array}
\longrightarrow
\begin{array}{c}
| \\
-\text{C—CH}_2\text{—S}^{\ominus} + \text{S}^{\text{o}} \\
| \\
HO^{\ominus} \rightarrow H
\end{array}
$$

$$
\longrightarrow \quad \ce{>C=CH2} + \text{SH}^{\ominus} + \text{OH}^{\ominus} \qquad (1.151)
$$

Alternatively, cleavage of the cystine disulfide
bond can occur by nucleophilic attack on sulfur, yielding a dehydroalanine residue via
thiol and sulfinate intermediates:

$$
\text{R—S—S—R} + 2\,\text{OH}^{\ominus} \longrightarrow
$$
$$
\text{R—S—O}^{\ominus} + \text{R—S}^{\ominus} + \text{H}_2\text{O} \qquad (1.152)
$$

$$
\text{R—S—O}^{\ominus} + \text{R—S—S—R}
$$
$$
\longrightarrow \quad
\begin{array}{c}
\text{R—S—S—R} + \text{R—S}^{\ominus} \\
\| \\
O
\end{array}
\qquad (1.153)
$$

$$
\begin{array}{c}
\text{R—S—S—R} + 2\,\text{OH}^{\ominus} \\
\| \\
O
\end{array}
$$
$$
\longrightarrow \quad \text{R—SO}_2^{\ominus} + \text{R—S}^{\ominus} + \text{H}_2\text{O}
\qquad (1.154)
$$

Intra- and interchain cross-linking of proteins
can occur in dehydroalanine reactions involv-

ing additions of amines and thiols. Ammonia
may also react via an addition reaction:

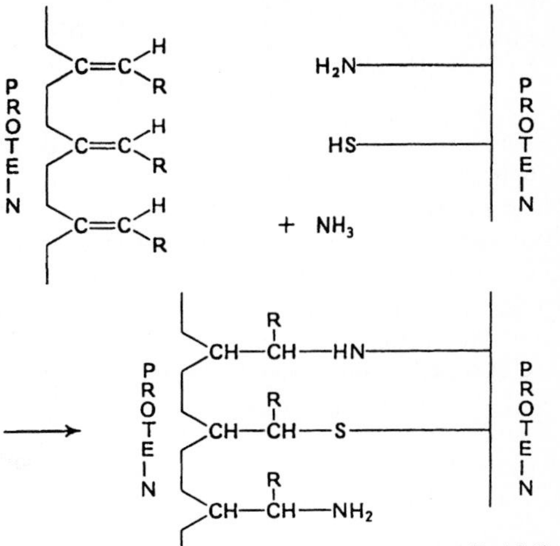

$$
(1.155)
$$

Acidic hydrolysis of such a cross-linked protein yields the unusual amino acids listed in
Table 1.29. Ornithine is formed during cleavage of arginine (Reaction 1.56).

Table 1.29. Formation of unusual amino acids by
alkali treatment of protein

Name	Formula	
3-N^6-Lysinoalanine (R=H) 3-N^6-Lysino-3-methyl-alanine (R=CH$_3$)	COOH | CHNH$_2$ | CHR	COOH | CHNH$_2$ | NH—(CH$_2$)$_4$
3-N^5-Ornithinoalanine (R=H) 3-N^5-Ornithino-3-methylalanine (R=CH$_3$)	COOH | CHNH$_2$ | CHR	COOH | CHNH$_2$ | NH—(CH$_2$)$_3$
Lanthionine (R=H) 3-Methyllanthionine (R=CH$_3$)	COOH | CHNH$_2$ | CHR	COOH | CHNH$_2$ | S——CH$_2$
3-Aminoalanine (R=H) 2,3-Diamino butyric acid (R=CH$_3$)	COOH | CHNH$_2$ | CHRNH$_2$	

$$\begin{array}{c}\text{COOH}\\|\\\text{CHNH}_2\\|\\(\text{CH}_2)_3\\|\\\text{NH}-\text{C}\\\quad\backslash\\\quad\quad\text{NH}_2\end{array}\xrightarrow{\text{OH}^\ominus}\begin{array}{c}\text{COOH}\\|\\\text{CHNH}_2\\|\\(\text{CH}_2)_3\\|\\\text{NH}_2\end{array}+\quad\text{O}=\text{C}\begin{array}{c}\text{NH}_2\\\\\text{NH}_2\end{array}$$

(1.156)

Formation of D-amino acids occurs through abstraction of a proton via a C2-carbanion. The reaction with L-isoleucine is particularly interesting. L-Isoleucine is isomerized to D-alloisoleucine which, unlike other D-amino acids, is a diastereoisomer and so has a retention time different from L-isoleucine, making its determination possible directly from an amino acid chromatogram:

$$\begin{array}{c}\text{CO}-\\|\\-\text{HN}-\text{C}-\text{H}\\|\\\text{H}_3\text{C}-\text{C}-\text{H}\\|\\\text{C}_2\text{H}_5\end{array}\underset{+\text{H}^\oplus}{\overset{-\text{H}^\oplus}{\rightleftarrows}}\begin{array}{c}\text{CO}-\\|\\-\text{HN}-\text{C}\mid^\ominus\\|\\\text{H}_3\text{C}-\text{C}-\text{H}\\|\\\text{C}_2\text{H}_5\end{array}$$

L-Isoleucine

$$\underset{-\text{H}^\oplus}{\overset{+\text{H}^\oplus}{\rightleftarrows}}\begin{array}{c}\text{CO}-\\|\\\text{H}-\text{C}-\text{NH}-\\|\\\text{H}_3\text{C}-\text{C}-\text{H}\\|\\\text{C}_2\text{H}_5\end{array}$$

D-allo-Isoleucine   (1.157)

Heating proteins in a dry state at neutral pH results in the formation of isopeptide bonds between the ε-amino groups of lysine residues and the β- or γ-carboxamide groups of asparagine and glutamine residues:

$$\begin{array}{l}\text{P}\\\text{R}\\\text{O}\\\text{T}\\\text{E}\\\text{I}\\\text{N}\end{array}\begin{array}{l}-\text{NH}_2\quad\text{H}_2\text{NOC}-\text{CH}_2-\\\\-\text{NH}_2\quad\text{H}_2\text{NOC}-(\text{CH}_2)_2-\end{array}\begin{array}{l}\text{P}\\\text{R}\\\text{O}\\\text{T}\\\text{E}\\\text{I}\\\text{N}\end{array}$$

$$\xrightarrow{-\text{NH}_3}\begin{array}{l}\text{P}\\\text{R}\\\text{O}\\\text{T}\\\text{E}\\\text{I}\\\text{N}\end{array}\begin{array}{l}-\text{NH}-\text{OC}-\text{H}_2\text{C}-\\\\-\text{NH}-\text{OC}-(\text{H}_2\text{C})_2-\end{array}\begin{array}{l}\text{P}\\\text{R}\\\text{O}\\\text{T}\\\text{E}\\\text{I}\\\text{N}\end{array}$$

(1.158)

These isopeptide bonds are cleaved during acidic hydrolysis of protein and, therefore, do not contribute to the occurrence of unusual amino acids. A more intensive heat treatment of proteins in the presence of water leads to a more extensive degradation.

Oxidative changes in proteins primarily involve methionine, which relatively readily forms methionine sulfoxide:

$$\begin{array}{c}\text{CO}-\\|\\-\text{HN}-\text{C}-\text{H}\\|\\(\text{CH}_2)_2\\|\\\text{S}\\|\\\text{CH}_3\end{array}\longrightarrow\begin{array}{c}\text{CO}-\\|\\-\text{HN}-\text{C}-\text{H}\\|\\(\text{CH}_2)_2\\|\\\text{S}=\text{O}\\|\\\text{CH}_3\end{array}$$

(1.159)

The formation of methionine sulfoxide was observed in connection with lipid peroxidation, phenol oxidation and light exposure in the presence of oxygen and sensitizers such as riboflavin.

After in vivo reduction to methionine, protein-bound methionine sulfoxide is apparently biologically available.

Figure 1.35 shows the effect of alkaline treatment of a protein isolate of sunflower seeds. Serine, threonine, arginine and isoleucine concentrations are markedly decreased with increasing concentrations of NaOH. New amino acids (ornithine and alloisoleucine) are formed. Initially, lysine concentration decreases, but increases at higher concentrations of alkali. Lysinoalanine behaves in the opposite manner. The extent of formation of D-amino acids as a result of alkaline treatment of proteins is shown in Table 1.30.

Table 1.30. Formation of D-amino acids by alkali treatment of proteins[a] (1% solution in 0.1 N NaOH, pH ~ 12.5, temperature 65°C)

Protein	Heating time (h)	D-Asp (%)	D-Ala	D-Val	D-Leu	D-Pro	D-Glu	D-Phe
Casein	0	2.2	2.3	2.1	2.3	3.2	1.8	2.8
	1	21.8	4.2	2.7	5.0	3.0	10.0	16.0
	3	30.2	13.3	6.1	7.0	5.3	17.4	22.2
	8	32.8	19.4	7.3	13.6	3.9	25.9	30.5
Wheat gluten	0	3.3	2.0	2.1	1.8	3.2	2.1	2.3
	3	29.0	13.5	3.9	5.6	3.2	25.9	23.3
Promine D (soya protein)	0	2.3	2.3	2.6	3.3	3.2	1.8	2.3
	3	30.1	15.8	6.6	8.0	5.8	18.8	24.9
Lactalbumin	0	3.1	2.2	2.9	2.7	3.1	2.9	2.3
	3	22.7	9.2	4.8	5.8	3.6	12.2	16.5

[a] Results in % correspond to D- + L-amino acids = 100%.

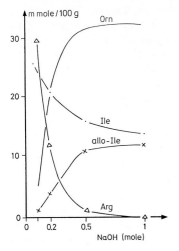

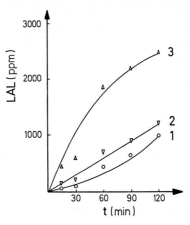

Fig. 1.36. Formation of lysinoalanine (LAL) by heating casein (5% solution at 100°C) (according to *Sternberg* and *Kim*, 1977) 1 pH 5.0, 2 pH 7.0, 3 pH 8.0

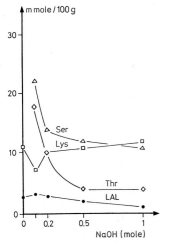

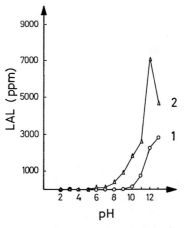

Fig. 1.35. Amino acid contents of a sunflower seed protein isolate heated in sodium hydroxide solutions at 80°C for 16 h. (according to *Mauron*, 1975)

Fig. 1.37. Lysinoalanine (LAL) formation from wheat gluten (2) and corn gluten (1). Protein contents of the glutens: 70%; heated as 6.6% suspension at 100°C for 4 h. (according to *Sternberg* and *Kim*, 1977)

Data presented in Figs. 1.36 and 1.37 clearly show that the formation of lysinoalanine is influenced not only by pH but also by the protein source. An extensive reaction occurs in casein even at pH 5.0 due to the presence of phosphorylated serine residues, while noticeable reactions occur in gluten from wheat or in zein from corn only in the pH range of 8–11. Figure 1.38 illustrates the dependence of the reaction on protein concentration.

Table 1.31 lists the contents of lysinoalanine in food products processed industrially or prepared under the "usual household conditions".

The contents are obviously affected by the food type and by the processing conditions.

In the radiation of food, o-hydroxyphenylalanine called o-tyrosine is formed through the reaction of phenylalanine with OH-radicals. In hydrolysates, the compound can be detected with the help of HPLC (fluorescence detection or electrochemical detection). It is under discussion as an indicator for food radiation. The amount formed depends on the irradiated

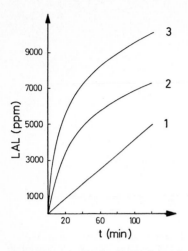

Fig. 1.38. Lysinoalanine (LAL) formation as influenced by casein concentration. (1): 5%, (2): 15%, and (3) 20% all at pH 12.8. (according to *Sternberg* and *Kim*, 1977)

Table 1.31. Lysinoalanine content of various foods

Food	Origin/Treatment		Lysinoalanine (mg/kg protein)
Frankfurter	CP[a]	Raw	0
		Cooked	50
		Roasted in oven	170
Chicken drums	CP	Raw	0
		Roasted in oven	110
		Roasted in micro wave oven	200
Egg white, fluid	CP		15
Egg white		Boiled	
		(3 min)	140
		(10 min)	270
		(30 min)	370
		Baked	
		(10 min/150 °C)	350
		(30 min/150 °C)	1,100
Dried egg white	CP		160–1,820[b]
Condensed milk, sweetened	CP		360–540
Condensed milk, unsweetened	CP		590–860
Milk product for infants	CP		150–640
Infant food	CP		<55–150
Soya protein isolate	CP		0–370
Hydrolyzed vegetable protein	CP		40–500
Cocoa powder	CP		130–190
Na-caseinate	CP		45–560
Na-caseinate	CP		430–6,900
Ca-caseinate	CP		250–4,320

[a] Commercial product.
[b] Variation range for different brand name products.

dose and on the temperature. In samples of chicken and pork, fish and shrimps, <0.1 mg/kg (non-radiated controls), 0.5–0.8 mg/kg (5 kGy, −18 °C) and 0.8–1.2 mg/kg (5 kGy, 20 °C) were found.

1.4.5 Enzyme-Catalyzed Reactions

1.4.5.1 Foreword

A great number and variety of enzyme-catalyzed reactions are known with protein as a substrate. These include hydrolytic reactions (cleavage of peptide bonds or other linkages, e.g., the ester linkage in a phosphoprotein), transfer reactions (phosphorylation, transfer of sugar residues and methyl groups) and redox reactions (thiol oxidation, disulfide reduction, amino group oxidation or incorporation of hydroxyl groups). Table 1.32 is a compilation of some examples.

Some of these reactions are covered in Section 1.4.6.3 or in the sections related to individual foodstuffs. Only enzymes that are involved in hydrolysis of peptide bonds (proteolytic enzymes, peptidases) will be covered in the following sections.

1.4.5.2 Proteolytic Enzymes

Processes involving proteolysis play a role in the production of many foods. Proteolysis can occur as a result of proteinases in the food itself, e.g. autolytic reactions in meat, or due to microbial proteinases, e.g., the addition of pure cultures of selected microorganisms during the production of cheese.

This large group of enzymes is divided up as shown in Table 1.33. The two subgroups formed are: peptidases (exopeptidases) that cleave amino acids or dipeptides stepwise from the terminal ends of proteins, and proteinases (endopeptidases) that hydrolyze the linkages within the peptide chain, not attacking the terminal peptide bonds. Further division is possible, for example, by taking into account the presence of a given amino acid residue in the active site of the enzyme. The most important types of proteolytic enzymes are presented in the following sections.

Table 1.32. Enzymatic reactions affecting proteins

Hydrolysis
 – Endopeptidases
 – Exopeptidases

Proteolytic induced aggregation
 – Collagen biosynthesis
 – Blood coagulation
 – Plastein reaction

Cross-linking
 – Disulfide bonds
    Protein disulfide isomerase
    Protein disulfide reductase (NAD(P)H)
    Protein disulfide reductase (glutathione)
    Sulfhydryloxidase
    Lipoxygenase
    Peroxidase
 – Aldol-, aldimine condensation and subsequent
    reactions (connective tissue)
    Lysyloxidase

Phosphorylation, dephosphorylation
 – Protein kinase
 – Phosphoprotein phosphatase

Hydroxylation
 – Proline hydroxylase
 – Lysine hydroxylase

Glycosylation
 – Glycoprotein-β-galactosyltransferase

Methylation and demethylation
 – Protein(arginine)-methyl-transferase
 – Protein(lysine)-methyl-transferase
 – Protein-O-methyl-transferase

Acetylation, deacetylation
 – ε-N-Acetyl-lysine

1.4.5.2.1 Serine Peptidases

Enzymes of this group, in which activity is confined to the pH range of 7–11, are denoted as alkaline proteinases. Typical representatives from animal sources are trypsin, chymotrypsin, elastase, plasmin and thrombin. Serine proteinases are produced by a great number of bacteria and fungi, e.g. *Bacillus cereus, B. firmus, B. licheniformis, B. megaterium, B. subtilis, Serratia marcescens, Streptomyces fradiae, S. griseus, Trititrachium album, Aspergillus flavus, A. oryzae* and *A. sojae*.
These enzymes have in common the presence of a serine and a histidine residue in their active sites (cf. 2.4.2.5). Cleavage of protein occurs through an acyl-enzyme (cf. Formula

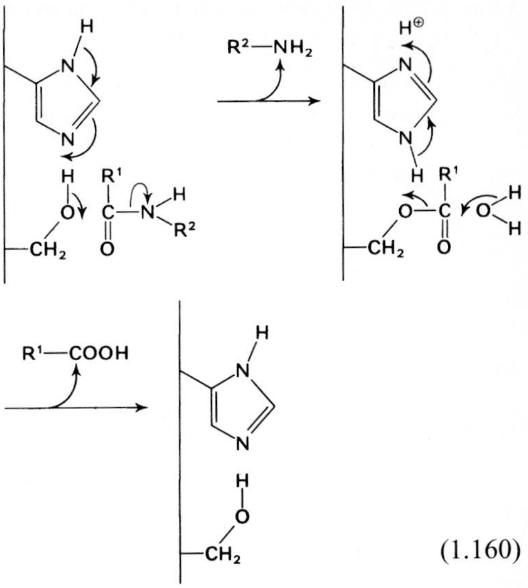

$$(1.160)$$

1.160 and Fig. 2.17) formed as an intermediate.
Inactivation of these enzymes is possible with reagents such as diisopropylfluorophosphate (DIFP) or phenylmethanesulfonylfluoride (PMSF). These reagents irreversibly acylate the serine residue in the active site of the enzymes:

$$E-CH_2OH + FY \longrightarrow E-CH_2OY + HF$$

$$(Y: -PO(i\,C_3H_7O)_2, -SO_2-CH_2C_6H_5) \quad (1.161)$$

Irreversible inhibition can also occur in the presence of halogenated methyl ketones which alkylate the active histidine residue (cf. 2.4.1.1), or as a result of the action of proteinase inhibitors, which are also proteins, by interaction with the enzyme to form inactive complexes. These natural inhibitors are found in the organs of animals and plants (pancreas, colostrum, egg white, potato tuber and seeds of many legumes; cf. 16.2.3). The specificity of serine proteinases varies greatly (cf. Table 1.34). Trypsin exclusively cleaves linkages of amino acid residues with a basic side chain (lysyl or arginyl bonds) and chymotrypsin preferentially cleaves bonds of amino acid residues which have aromatic side chains (phenylalanyl, tyrosyl or tryptophanyl bonds). Enzymes of microbial origin often are less specific.

Table 1.33. Classification of proteolytic enzymes (peptidases)

EC-No.[a]	Enzyme group	Comments	Examples
	Exopeptidases	Cleave proteins/peptides stepwise from N- or C-terminals	
3.4.11.	Aminopeptidases	Cleave amino acids from N-terminal	Various aminopeptidases
3.4.13.	Dipeptidases	Cleave dipeptides	Various dipeptidases (carnosinase, anserinase)
3.4.14.	Dipeptidyl- and tripeptidyl-peptidases	Cleave dipeptides from N-terminal	Cathepsin C
3.4.15.	Peptidyl-dipeptidases	Cleave dipeptides from C-terminal	Carboxycathepsin,
3.4.16.	Serine carboxypeptidases	Cleave amino acids from C-terminal, serine in the active site	Carboxypeptidase C, cathepsin A
3.4.17.	Metalocarboxypeptidases	Cleave amino acids from C-terminal, Zn^{2+} or Co^{2+} in the active site	Carboxypeptidases A and B
3.4.18.	Cysteine carboxypeptidases	Cleave amino acids from C-terminal, cysteine in the active site	Lysosomal carboxy-peptidase B
	Endopeptidases	Cleave protein/peptide bonds other than terminal ones	
3.4.21.	Serine peptidases	Serine in the active site	Chymotrypsins A, B and C, α- and β-trypsin, microbial alkaline proteinases
3.4.22.	Cysteine peptidases	Cysteine in the active site	Papain, ficin, bromelain, cathepsin B
3.4.23.	Aspartic peptidases	Aspartic acid (2 residues) in the active site	Pepsin, cathepsin D, rennin (chymosin)
3.4.24.	Metalopeptidases	Metal ions in the active site	Collagenase, microbial neutral proteinases

[a] cf. 2.2.6

1.4.5.2.2 Cysteine Peptidases

Typical representatives of this group of enzymes are: papain (from the sap of a tropical, melonlike fruit tree, *Carica papaya*), bromelain (from the sap and stem of pineapples, *Ananas comosus*), ficin (from *Ficus latex* and other *Ficus spp.*) and a Streptococcus proteinase. The range of activity of these enzymes is very wide and, depending on the substrate, is pH 4.5–10, with a maximum at pH 6–7.5 The mechanism of enzyme activity appears to be similar to that of serine proteinases. A cysteine residue is present in the active site. A thioester is formed as a covalent intermediary product. The enzymes are highly sensitive to oxidizing agents. Therefore, as a rule they are used in the presence of a reducing agent (e.g., cysteine) and a chelating agent (e.g., EDTA). Inactivation of the enzymes is possible with oxidative agents, metal ions or alkylating reagents (cf. 1.2.4.3.5 and 1.4.4.5). In general these enzymes are not very specific (cf. Table 1.34).

1.4.5.2.3 Metalo Peptidases

This group includes exopeptidases, carboxypeptidases A and B, aminopeptidases, dipeptidases, prolidase and prolinase, and proteinases from bacteria and fungi, such as *Bacillus cereus*, *B. megaterium*, *B. subtilits*, *B. thermoproteolyticus* (thermolysin), *Streptomyces griseus* (pronase; it also contains carboxy- and aminopeptidases) and *Aspergillus oryzae*.
Most of these enzymes contain one mole of Zn^{2+} per mole of protein, but prolidase and prolinase contain one mole of Mn^{2+}. The metal ion acts as a *Lewis* acid in carboxypeptidase A, establishing contact with the carbonyl group of the peptide bond which is to be cleaved.

Table 1.34. Specificity of proteolytic enzymes [based on cleavage of oxidized B chain of bovine insulin; strong cleavage: ↓, weak cleavage (↓)]

No.[a]	Phe	Val	Asn	Gln	His	Leu	Cys	Gly	Ser	His	Leu	Val	Glu	Ala	Leu	Tyr	Leu	Val	Cys	Gly	Glu	Arg	Gly	Phe	Phe	Tyr	Thr	Pro	Lys	Ala

Serine peptidases: rows 1–6
Cysteine peptidases: rows 9–11
Metalopeptidases: rows 12–14
Aspartic peptidases: rows 15–19

[a] Enzymes.

1) Trypsin (bovine)
2) Chymotrypsin A (bovine)
3) Chymotrypsin C (porcine)
4) Aspergillopeptidase C (*Aspergillus oryzae*)
5) Proteinase from *Streptomyces griseus* (trypsin-like)
6) Subtilisin BPN′

7) Proteinase from *Aspergillus oryzae*
8) Proteinase from *Aspergillus flavus*
9) Papain (*Papaya carica*)
10) Ficin III (*Ficus glabrata*)
11) Chymopapain (*Charica papaya*)

12) Proteinase II from *Aspergillus oryzae*
13) Proteinase from *Bacillus subtilis*
14) Thermolysin (*Bacillus thermoproteolyticus Rokko*)
15) Pepsin A (porcine)

16) Rennin (calf)
17) Proteinase from *Candida albicans*
18) Proteinase from *Mucor miehei*
19) Proteinase from *Rhizopus chinensis*

Figure 1.39 shows the arrangement of other participating residues in the active site, as revealed by X-ray structural analysis of the enzyme-substrate complex.

The enzymes are active in the pH 6–9 range; their specificity is generally low (cf. Table 1.34).

Inhibition of these enzymes is achieved with chelating agents (e.g. EDTA) or sodium dodecyl sulfate.

1.4.5.2.4 Aspartic Peptidases

Typical representatives of this group are enzymes of animal origin, such as pepsin and rennin (called Lab-enzyme in Europe), active in the pH range of 2–4, and cathepsin D, which has a pH optimum between 3 and 5 depending on the sbstrate and on the source of the enzyme. At pH 6–7 rennin cleaves a bond of κ-casein with great specificity, thus causing curdling of milk (cf. 10.1.2.1.1).

Aspartic proteinases of microbial origin can be classified as pepsin-like or rennin-like enzymes. The latter are able to coagulate milk. The pepsin-like enzymes are produced, for example, by *Aspergillus awamori, A. niger, A. oryzae, Penicillium spp.* and *Trametes sangui-*

nea. The rennin-like enzymes are produced, for example, by *Aspergillus usamii* and *Mucor spp.*, such as *M. pusillus*.

There are two carboxyl groups, one in undissociated form, in the active site of aspartic proteinases. The mechanism postulated for cleavage of peptide bonds is illustrated in Reaction 1.162. The nucleophilic attack of a water molecule on the carbonyl carbon atom of the peptide bond is catalyzed by the side chains of

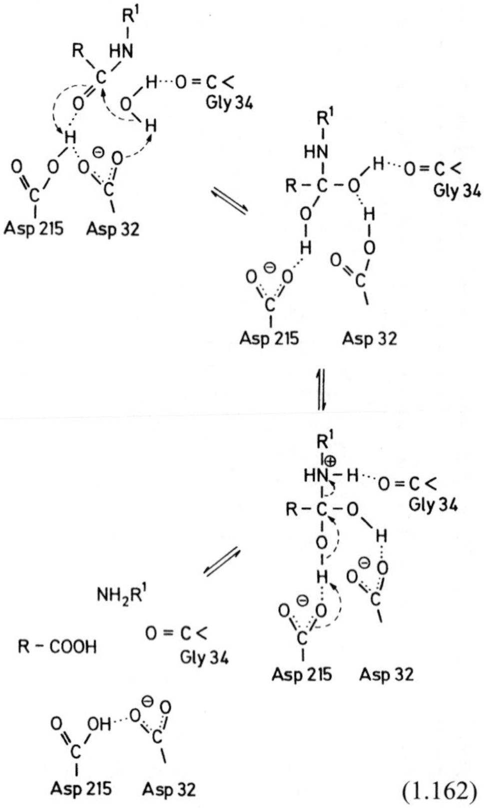

(1.162)

Asp-32 (basic catalyst) and Asp-215 (acid catalyst). The numbering of the amino acid residues in the active site applies to the aspartic proteinase from *Rhizopus chinensis*.

Inhibition of these enzymes is achieved with various diazoacetylamino acid esters, which apparently react with carboxyl groups on the active site, and with pepstatin. The latter is isolated from various *Streptomycetes* as a peptide mixture with the general formula (R: isovale-

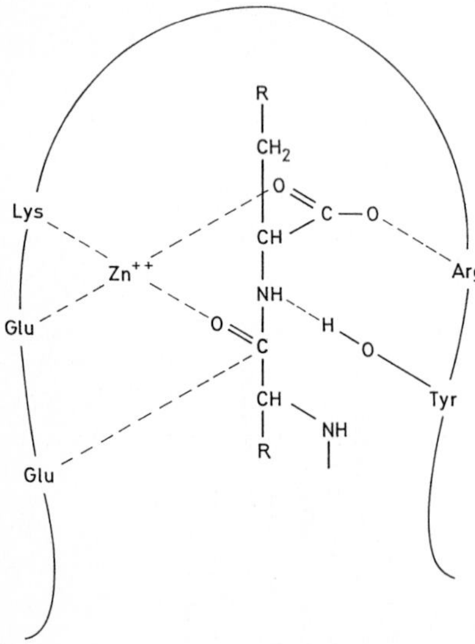

Fig. 1.39. Carboxypeptidase A active site. (according to *Lowe* and *Ingraham*, 1974)

ric or n-caproic acid; AHMHA: 4-amino-3-hydroxy-6-methyl heptanoic acid):

R—Val—Val—AHMHA—Ala—AHMAH   (1.163)

The specifity of aspartic proteinases is given in Table 1.34.

1.4.6 Chemical and Enzymatic Reactions of Interest to Food Processing

1.4.6.1 Foreword

Standardization of food properties to meet nutritional/physiological and toxicological demands and requirements of food processing operations is a perennial endeavor. Food production is similar to a standard industrial fabrication process: on the one hand is the food commodity with all its required properties, on the other hand are the components of the product, each of which supplies a distinct part of the required properties. Such considerations have prompted investigations into the relationships in food between macroscopic physical and chemical properties and the structure and reactions at the molecular level. Reliable understanding of such relationships is a fundamental prerequisite for the design and operation of a process, either to optimize the process or to modify the food components to meet the desired properties of the product.

Modification of proteins is still a long way from being a common method in food processing, but it is increasingly being recognized as essential, for two main reasons:

Firstly, proteins fulfill multipurpose functions in food. Some of these functions can be served better by modified than by native proteins.

Secondly, persistent nutritional problems the world over necessiate the utilization of new raw materials.

Modifying reactions can ensure that such new raw materials (e. g., proteins of plant or microbial origin) meet stringent standards of food safety, palatability and acceptable biological value. A review will be given here of several protein modifications that are being used or are being considered for use. They involve chemical or enzymatic methods or a combina-

Table 1.35. Properties of protein in food

Properties with	
nutritional/physiological relevance	processing relevance
Amino acid composition Availability of amino acids	Solubility, dispersibility Ability to coagulate Water binding/holding capacity Gel formation Dough formation, extensibility, elasticity Viscosity, adhesion, cohesion Whippability Foam stabilization Emulsifying ability Emulsion stabilization

tion of both. Examples have been selected to emphasize existing trends. Table 1.35 presents some protein properties which are of interest to food processing. These properties are related to the amino acid composition and sequence and the conformation of proteins. Modification of the properties of proteins is possible by changing the amino acid composition or the size of the molecule, or by removing or inserting hetero constituents. Such changes can be accomplished by chemical and/or enzymatic reactions.

From a food processing point of view, the aims of modification of proteins are:

- Blocking the reactions involved in deterioration of food (e. g., the *Maillard* reaction)
- Improving some physical properties of proteins (e. g., texture, foam stability, whippability, solubility)
- Improving the nutritional value (increasing the extent of digestibility, inactivation of toxic or other undesirable constituents, introducing essential ingredients such as some amino acids).

1.4.6.2 Chemical Modification

Table 1.36 presents a selection of chemical reactions of proteins that are pertinent to and of current importance in food processing.

Table 1.36. Chemical reactions of proteins significant in food

Reactive group	Reaction	Product
–NH₂	Acylation	–NH–CO–R
–NH₂	Reductive alkylation with HCHO	–N(CH₃)₂
–CONH₂	Hydrolysis	–COOH
–COOH	Esterification	–COOR
–OH	Esterification	–O–CO–R
–SH	Oxidation	–S–S–
–S–S–	Reduction	–SH
–CO–NH–	Hydrolysis	–COOH + H₂N–

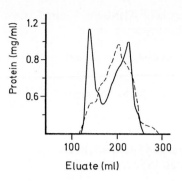

Fig. 1.41. Gel column chromatography of an acetic acid (0.2 mol/L) wheat protein extract. Column: Sephadex G-100 (— before and – – – after succinylation). (according to *Grant*, 1973)

1.4.6.2.1 Acylation

Treatment with succinic anhydride (cf. 1.4.4.1.3) generally improves the solubility of protein.

For example, succinylated wheat gluten is quite soluble at pH 5 (cf. Fig. 1.40). This effect is related to disaggregation of high molecular weight gluten fractions (cf. Fig. 1.41). In the case of succinylated casein it is obvious that the modification shifts the isoelectric point of the protein (and thereby the solubility minimum) to a lower pH (cf. Fig. 1.42). Succinylation of leaf proteins improves the solubility as well as the flavor and emulsifying properties. Succinylated yeast protein has not only an increased solubility in the pH range of 4–6, but is more heat stable above pH 5. It has better emulsifying properties, surpassing many other proteins (Table 1.37), and has increased whippability.

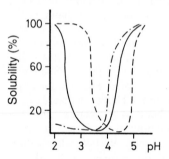

Fig. 1.42. Solubilities of native (– – –) and succinylated casein (— 50% and –·–·– 76%) as a function of pH. (according to *Schwenke* et al., 1977)

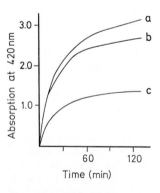

Fig. 1.43. Hydrolysis of a reductively methylated casein by bovine α-chymotrypsin. Modification extents: a 0%, b 33%, and c 52%. (according to *Galembeck* et al., 1977)

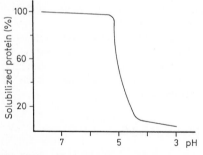

Fig. 1.40. Solubility of succinylated wheat protein as a function of pH (0.5% solution in water). (according to *Grant*, 1973)

Table 1.37. Emulsifying property of various proteins[a]

Protein	Emulsifying Activity Index ($m^2 \times g^{-1}$)	
	pH 6.5	pH 8.0
Yeast protein (88%) succinylated	322	341
Yeast protein (62%) succinylated	262	332
Sodium dodecyl sulfate (0.1%)	251	212
Bovine serum albumin	–	197
Sodium caseinate	149	166
β-Lactoglobulin	–	153
Whey protein powder A	119	142
Yeast protein (24%) succinylated	110	204
Whey protein powder B	102	101
Soya protein isolate A	41	92
Hemoglobin	–	75
Soya protein isolate B	26	66
Yeast protein (unmodified)	8	59
Lysozyme	–	50
Egg albumin	–	49

[a] Protein concentration: 0.5% in phosphate buffer of pH 6.5.

Gluten $\xrightarrow{\text{reduction}}$ reduced gluten
(cohesive, elastic)

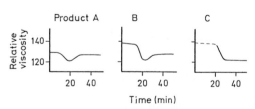

A: readily soluble, soft, adhesive, non-elastic
B: cohesive, elastic
C: sparingly soluble, strong, cohesive and non-elastic

Fig. 1.44. Properties of modified wheat gluten. (after *Lasztity*, 1975)

Fig. 1.45. Viscosity curves during reduction of different wheat glutens. For sample designation see Fig. 1.44. (according to *Lasztity*, 1975)

Introduction of aminoacyl groups into protein can be achieved by reactions involving amino acid carboxy anhydrides (Fig. 1.46), amino acids and carbodiimides (Fig. 1.47) or by BOC-amino acid hydroxysuccinimides with subsequent removal of the aminoprotecting group (BOC):

$$H_2N-CH-COOH \; : \; Ala, Trp, Gly, Met \qquad (1.164)$$

Feeding tests with casein with attached methionine, as produced by the above method, have demonstrated a satisfactory availability of methionine (Table 1.38). Such covalent attachment of essential amino acids to a protein may avoid the problems associated with food supplementation with free amino acids: losses in processing, development of undesired aroma due to methional, etc.

Gliadin (1.4% Ala)

Edestin (3.8% Ala)

$+ \; H_3C-CH$

pH 6.8, 48 h → Polyalanyl-gliadin (23.6% Ala)
pH 6.8, 0.005 M SDS 72 h → Polyalanyl-edestin (6.2% Ala)

Fig. 1.46. Reaction of proteins with D,L-alanine carboxy anhydride. (according to *Sela* et al., 1962 and *St. Angelo* et al., 1966)

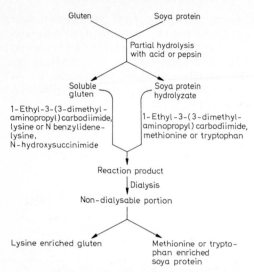

Fig. 1.47. Covalent binding of lysine to gluten (according to *Li-Chan* et al., 1979) and of methionine or tryptophan to soya protein (according to *Voutsinas* and *Nakai*, 1979), by applying a carbodiimide procedure

Table 1.38. Feeding trial (rats) with modified casein: free amino acid concentration in plasma and PER value

Diet	µmole/100 ml plasma				
	Lys	Thr	Ser	Gly	Met
Casein	101	19	34	32	5
Met-casein[a]	96	17	33	27	39

	PER[b]
Casein (10%)	2.46
Casein (10%) + Met (0.2%)	3.15
Casein (5%) + Met-casein[a] (5%)	2.92

[a] Covalent binding of methionine to ε-NH$_2$ groups of casein.
[b] Protein Efficiency Ratio (cf. 1.2.5).

Table 1.39, using β-casein as an example, shows to what extent the association of a protein is affected by its acylation with fatty acids of various chain lengths.

1.4.6.2.2 Alkylation

Modification of protein by reductive methylation of amino groups with formaldehyde/NaBH$_4$ retards *Maillard* reactions. The resul-

Table 1.39. Association of acylated β-casein A

Protein	SD[a]	Mono-mer	Poly-mer	$S_{20,w}^{0}$	$S_{20,w}^{1\%}$
	(%)	(%)	(%)	(S·10^{13})	(S·10^{13})
β-Casein					
A (I)	–	11	89	12.6	6.3
Acetyl-I	96	41	59	4.8	4.7
Propionyl-I	97	24	76	10.5	5.4
n-Butyryl-I	80	8	92	8.9	8.3
n-Hexanoyl-I	85	0	100	7.6	11.6
n-Octanoyl-I	89	0	100	6.6	7.0
n-Decanoyl-I	83	0	100	5.0	6.5

[a] Substitution degree.

tant methyl derivative, depending on the degree of substitution, is less accessible to proteolysis (Fig. 1.43). Hence, its value from a nutritional/physiological point of view is under investigation.

1.4.6.2.3 Redox Reactions Involving Cysteine and Cystine

Disulfide bonds have a strong influence on the properties of proteins. Wheat gluten can be modified by reduction of its disulfide bonds to sulfhydryl groups and subsequent reoxidation of these groups under various conditions (Fig. 1.44). Reoxidation of a diluted suspension in the presence of urea results in a weak, soluble, adhesive product (gluten A), whereas reoxidation of a concentrated suspension in the presence of a higher concentration of urea yields an insoluble, stiff, cohesive product (gluten C). Additional viscosity data have shown that the disulfide bridges in gluten A are mostly intramolecular while those in gluten C are predominantly intermolecular (Fig. 1.45).

1.4.6.3 Enzymatic Modification

Of the great number of enzymatic reactions with protein as a substrate (cf. 1.4.5), only a small number have so far been found to be suitable for use in food processing.

1.4.6.3.1 Dephosphorylation

Figure 1.48 uses β-casein as an example to show that the solubility of a phosphoprotein in

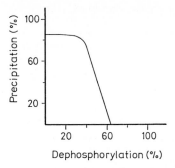

Fig. 1.48. Solubility of β-casein, partially dephosphorylated by phosphoprotein phosphatase: Precipitation: pH 7.1: 2.5 mg/ml protein: 10 mmol/L CaCl₂: 35 °C; 1 h. (according to *Yoshikawa* et al., 1974)

the presence of calcium ions is greatly improved by partial enzymatic dephosphorylation.

1.4.6.3.2 Plastein Reaction

The plastein reaction enables peptide fragments of a hydrolysate to join enzymatically through peptide bonds, forming a larger polypeptide of about 3 kdal:

$$R—CO—NH—R^1 + E—OH$$
$$\rightleftharpoons R—CO—O—E + H_2N—R^1$$
$$\xrightarrow{H_2O} R—COOH + E—OH$$
$$\xrightarrow{H_2N—R^2} R—CO—NH—R^2 + E—OH$$

$$(1.165)$$

The reaction rate is affected by, among other things, the nature of the amino acid residues. Hydrophobic amino acid residues are preferably linked together (Fig. 1.49). Incorporation of amino acid esters into protein is affected by the alkyl chain length of the ester. Short-chain alkyl esters have a low rate of incorporation, while the long-chain alkyl esters have a higher rate of incorporation. This is especially important for the incorporation of amino acids with a short side chain, such as alanine (cf. Table 1.40).

The plastein reaction can help to improve the biological value of a protein. Figure 1.50 shows the plastein enrichment of zein with tryptophan, threonine and lysine. The amino

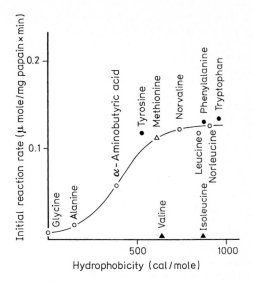

Fig. 1.49. Plastein reaction with papain: incorporation rates of amino acid esters as function of side chain hydrophobicity. (according to *Arai* et al., 1978)

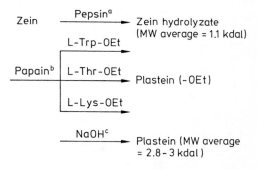

Fig. 1.50. Zein enrichment with Trp, Thr, and Lys by a plastein reaction. (according to *Aso* et al., 1974)
[a] 1% Substrate, E/S = 1/50, pH 1.6 at 37 °C for 72 h
[b] 50% substrate, hydrolyzate/AS-OEt = 10/1, E/S = 3/100 at 37 °C for 48 h
[c] 0.1 mol/L in 50% ethanol at 25 °C for 5 h

acid composition of such a zein-plastein product is given in Table 1.41.

Enrichment of a protein with selected amino acids can be achieved with the corresponding amino acid esters or, equally well, by using suitable partial hydrolysates of another protein.

Figure 1.51 presents the example of soya protein enrichment with sulfur-containing amino

Table 1.40. Plastein reaction catalyzed by papain: rate of incorporation of amino acid esters[a]

Aminoacyl residue	OEt	OnBu	OnHex	OnOct
L-Ala	0.016	0.054	0.133	0.135
D-Ala	0.0	–	0.0	–
α-Methylala	0.0	–	0.0	–
L-Val	0.005	–	0.077	–
L-Norval	0.122	–	0.155	–
L-Leu	0.119	–	0.140	–
L-Norleu	0.125	–	0.149	–
L-Ile	0.005	–	0.048	–

[a] μmole $\times$ mg papain^{-1} $\times$ min^{-1}.

Table 1.41. Amino acid composition of various plasteins (weight-%)

	1	2	3	4	5	6
Arg	1.56	1.33	1.07	1.06	1.35	1.74
His	1.07	0.95	0.81	0.75	0.81	1.06
Ile	4.39	6.39	6.58	5.49	6.23	5.67
Leu	20.18	23.70	23.05	23.75	25.28	23.49
Lys	0.20	0.20	0.24	2.14	3.24	0.19
Phe	6.63	7.26	6.82	7.34	7.22	6.98
Thr	2.40	2.18	9.23	2.36	2.46	2.13
Trp	0.38	9.71	0.25	0.40	0.42	0.33
Val	3.62	5.23	5.77	5.53	6.18	6.20
Met	1.58	1.87	1.67	1.89	2.06	2.04
Cys	1.00	0.58	0.88	0.81	0.78	0.92
Ala	7.56	7.51	8.05	7.97	7.93	8.77
Asp	4.61	3.38	3.42	3.71	3.60	3.91
Glu	21.70	12.48	14.03	14.77	12.95	13.02
Gly	1.48	1.15	1.23	1.29	1.27	1.52
Pro	10.93	8.42	9.10	9.73	9.14	9.37
Ser	4.42	3.40	3.89	3.93	3.74	4.28
Tyr	4.73	5.35	4.97	5.00	6.08	5.54

1) Zein hydrolyzate; 2) Trp-plastein; 3) Thr-plastein; 4) Lys-plastein; 5) Ac-Lys-plastein; 6) Control without addition of amino acid ethyl esters.

Table 1.42. PER-values for various proteins and plasteins

Protein	PER value (rats)
Casein	2.40
Soya protein (I)	1.20
Plastein SW[a] + I (1:2)	2.86
Plastein-Met[b] + I (1:3)	3.38

[a] From hydrolyzate I and wool keratin hydrolyzate.
[b] From hydrolyzate I and Met-OEt.
PER (cf. 1.2.5).

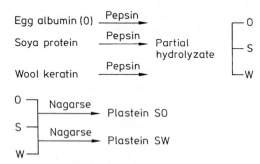

Fig. 1.51. Protein enrichment with sulfur amino acids applying plastein reaction. (according to *Yamashita* et al., 1971)

The plastein reaction also makes it possible to improve the solubility of a protein, for example, by increasing the content of glutamic acid (Fig. 1.53). A soya protein with 25% glutamic acid yields a plastein with 42% glutamic acid.

Soya protein has a pronounced solubility minimum in the pH range of 3–6. The minimum is much less pronounced in the case of the unmodified plastein, whereas the glutamic acid-enriched soya plastein has a satisfactory solubility over the whole pH range (Fig. 1.54) and is also resistant to thermal coagulation (Fig. 1.55).

Proteins with an increased content of glutamic acid show an interesting sensory effect: partial hydrolysis of modified plastein does not result in a bitter taste, rather it generates a pronounced "meat broth" flavor (Table 1.43).

Elimination of the bitter taste from a protein hydrolysate is also possible without incorporation of hydrophilic amino acids. Bitter-tasting peptides, such as Leu-Phe, which are released

acids through "adulteration" with the partial hydrolysate of wool keratin. The PER (protein efficiency ratio) values of such plastein products are significantly improved, as is seen in Table 1.42.

Figure 1.52 shows that the production of plastein with an amino acid profile very close to that recommended by FAO/WHO can be achieved from very diverse proteins.

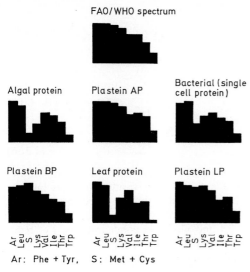

Fig. 1.52. Amino acid patterns of some proteins and their corresponding plasteins. (according to *Arai* et al., 1978)

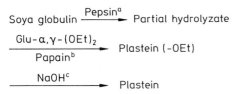

Fig. 1.53. Soy globulin enrichment with glutamic acid by a plastein reaction. (according to *Yamashita* et al., 1975)

[a] pH 1.6
[b] Partial hydrolyzate/Glu-α-γ-(OEt)$_2$ = 2: 1, substrate concentration: 52.5 %, E/S = 1/50, pH 5.5 at 37 °C for 24 h; sample contains 20% acetone
[c] 0.2 mol/L, at 25 °C for 2 h

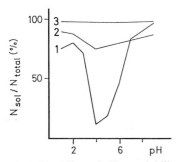

Fig. 1.54. Effect of pH on solubility of soy protein and modified products (1 g/100 ml water). 1 Soy protein, 24.1% Glu; 2 Plastein 24.8% Glu; 3 Glu-plastein with 41.9% Glu. (according to *Yamashita* et al., 1975)

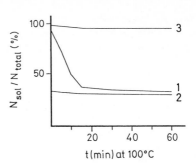

Fig. 1.55. Solubility of soy protein and modified products (800 mg/10 ml water) as a function of heating time at 100 °C. 1 Soy protein 24.1% Glu; 2 Plastein 24.8% Glu; 3 Glu-plastein, 41.9% Glu. (according to *Yamashita* et al., 1975)

Table 1.43. Taste of glutamic acid enriched plasteins

Enzyme	pH	Substrate[a]	Hydrolysis[b]	Taste[c] bitter	Taste[c] meat broth type
Pepsin	1.5	G	67	1	1.3
		P	73	4.5	1.0
α-Chymotrypsin	8.0	G	48	1	1.0
		P	72	4.5	1.0
Molsin	3.0	G	66	1.0	5.0
		P	74	1.3	1.3
Pronase	8.0	G	66	1.0	4.3
		P	82	1.3	1.2

[a] G: Glu-plastein, P: plastein; 1 g/100 ml.
[b] N_{sol} (10% TCA)/N_{total} (%).
[c] 1: no taste, 5: very strong taste.

by partial hydrolysis of protein, react preferentially in the subsequent plastein reaction and are incorporated into higher molecular weight peptides with a neutral taste.

The versatility of the plastein reaction is also demonstrated by examples wherein undesired amino acids are removed from a protein. A phenylalanine-free diet, which can be prepared by mixing amino acids, is recommended for certain metabolic defects. However, the use of a phenylalanine-free higher molecular weight peptide is more advantageous with respect to sensory and osmotic propertis. Such peptides can be prepared from protein by the plastein reaction. First, the protein is partially hydrolyzed with pepsin. Treatment with pronase under suitable conditions then preferentially releases amino acids with long hydrophobic side chains. The remaining peptides are separated by gel chromatography and then subjected to the plastein reaction in the presence of added tyrosine and tryptophan (Fig. 1.56). This yields a plastein that is practically phenylalanine-free and has a predetermined ratio of other amino acids, including tyrosine (Table 1.44).

The plastein reaction can also be carried out as a one-step process (Fig. 1.57), thus putting these reactions to economic, industrial-scale use.

1.4.6.3.3 Associations Involving Cross-Linking

Cross-linking between protein molecules is achieved with peroxidase. The cross-linking occurs between tyrosine residues when a protein is incubated with peroxidase/H_2O_2 (cf. Reaction 1.166).

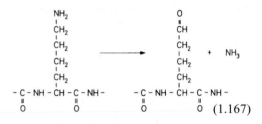

(1.166)

Incubation of protein with peroxidase/H_2O_2/catechol also results in cross-linking. The reactions in this case are the oxidative deamination of lysine residues, followed by aldol and aldimine condensations, i.e. reactions analogous to those catalyzed by lysyl oxidase in connective tissue:

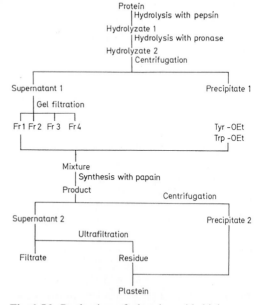

(1.167)

Table 1.44. Amino acid composition (weight-%) of plasteins with high tyrosine and low phenylalanine contents from fish protein concentrate (FPC) and soya protein isolate (SPI)

Amino acid	FPC	FPC-Plastein	SPI	SPI-Plastein
Arg	7.05	4.22	7.45	4.21
His	2.31	1.76	2.66	1.41
Ile	5.44	2.81	5.20	3.83
Leu	8.79	3.69	6.73	2.43
Lys	10.68	10.11	5.81	3.83
Thr	4.94	4.20	3.58	4.39
Trp	1.01	2.98	1.34	2.80
Val	5.88	3.81	4.97	3.24
Met	2.80	1.90	1.25	0.94
Cys	0.91	1.41	1.78	1.82
Phe	4.30	0.05	4.29	0.23
Tyr	3.94	7.82	3.34	7.96
Ala	6.27	4.82	4.08	2.56
Asp	11.13	13.67	11.51	18.00
Glu	17.14	27.17	16.94	33.56
Gly	4.42	3.94	4.88	3.89
Pro	3.80	4.25	6.27	2.11
Ser	4.59	3.58	5.45	4.67

Fig. 1.56. Production of plasteins with high tyrosine and low phenylalanine contents. (according to *Yamashita* et al., 1976)

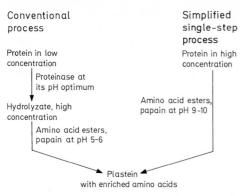

Conventional process		Simplified single-step process
Protein in low concentration		Protein in high concentration
↓ Proteinase at its pH optimum		
Hydrolyzate, high concentration	Amino acid esters, papain at pH 9-10	
↓ Amino acid esters, papain at pH 5-6		
→ Plastein ← with enriched amino acids		

Fig. 1.57. An outline for two- and single-step plastein reactions. (according to *Yamashita* et al., 1979)

Table 1.45. Content of dityrosine in some proteins after their oxidation with horseradish peroxidase/H_2O_2 (pH 9.5, 37°C, 24 h. Substrate/enzyme = 20:1)

Protein	Tyrosine content prior to oxidation (g/100 g protein)	Tyrosine decrease (%)	Dityrosine content (g/100 g protein)
Casein	6.3	21.8	1.37
Soyamine[a]	3.8	11.5	0.44
Bovine serum albumin	4.56	30.7	1.40
Gliadin	3.2	5.4	0.17

[a] Protein preparation from soybean.

Table 1.45 presents some of the proteins modified by peroxidase/H_2O_2 treatment and includes their ditryrosine contents.

1.4.7 Texturized Proteins

1.4.7.1 Foreword

The protein produced for nutrition in the world is currently about 20% from animal sources and 80% from plant sources. The plant proteins are primarily from cereals (57%) and oilseed meal (16%). Some nonconventional sources of protein (single cell proteins, leaves) have also acquired some importance.

Proteins are responsible for the distinct physical structure of a number of foods, e.g. the fibrous structure of muscle tissue (meat, fish), the porous structure of bread and the gel structure of some dairy and soya products.

Many plant proteins have a globular structure and, although available in large amounts, are used to only a limited extent in food processing. In an attempt to broaden the use of such proteins, a number of processes were developed in the mid-1950's which confer a fiber-like structure to globular proteins. Suitable processes give products with cooking strength and a meat-like structure. They are marketed as meat extenders and meat analogues and can be used whenever a lumpy structure is desired.

1.4.7.2 Starting Material

The following protein sources are suitable for the production of texturized products: soya; casein; wheat gluten; oilseed meals such as from cottonseed, groundnut, sesame, sunflower, safflower or rapeseed; zein (corn protein); yeast; whey; blood plasma; or packing plant offal such as lungs or stomach tissue.

The required protein content of the starting material varies and depends on the process used for texturization. The starting material is often a mixture such as soya with lactalbumin, or protein plus acidic polysaccharide (alginate, carrageenan or pectin).

The suitability of proteins for texturization varies, but the molecular weight should be in the range of 10–50 kdal. Proteins of less than 10 kdal are weak fiber builders, while those higher than 50 kdal are disadvantageous due to their high viscosity and tendency to gel in the alkaline pH range. The proportion of amino acid residues with polar side chains should be high in order to enhance intermolecular binding of chains. Bulky side chains obstruct such interactions, so that the amounts of amino acids with these structures should be low.

1.4.7.3 Texturization

The globular protein is unfolded during texturization by breaking the intramolecular binding forces. The resultant extended protein chains are stabilized through interaction with neighboring chains. In practice, texturization is achieved in one of two ways:

- The starting protein is solubilized and the resultant viscous solution is extruded through a spinning nozzle into a coagulating bath (spin process).
- The starting protein is moistened slightly and then, at high temperature and pressure, is extruded with shear force through the orifices of a die (extrusion process).

1.4.7.3.1 Spin Process

The starting material (protein content >90%, e.g. a soya protein isolate) is suspended in water and solubilized by the addition of alkali. The 20% solution is then aged at pH 11 with constant stirring. The viscosity rises during this time as the protein unfolds. The solution is then pressed through the orifices of a die (5,000–15,000 orifices, each with a diameter of 0.01–0.08 mm) into a coagulating bath at pH 2–3. This bath contains an acid (citric, acetic, phosphoric, lactic or hydrochloric) and, usually, 10% NaCl. Spinning solutions of protein and acidic polysaccharide mixtures also contain earth alkali salts. The protein fibers are extended further (to about 2- to 4-times the original length) in a "winding up" step and are bundled into thicker fibers with diameters of 10–20 mm. The molecular interactions are enhanced during stretching of the fiber, thus increasing the mechanical strength of the fiber bundles.

The adherent solvent is then removed by pressing the fibers between rollers, then placing them in a neutralizing bath ($NaHCO_3 + NaCl$) of pH 5.5–6 and, occasionally, also in a hardening bath (conc. NaCl).

The fiber bundles may be combined into larger aggregates with diameters of 7–10 cm.

Additional treatment involves passage of the bundles through a bath containing a binder and other additives (a protein which coagulates when heated, such as egg protein; modified starch or other polysaccharides; aroma compounds; lipids). This treatment produces bundles with improved thermal stability and aroma. A typical bath for fibers which are to be processed into a meat analogue might consist of 51% water, 15% ovalbumin, 10% wheat gluten, 8% soya flour, 7% onion powder, 2% protein hydrolysate, 1% NaCl, 0.15% monosodium glutamate and 0.5% pigments.

Finally, the soaked fiber bundles are heated and sliced.

1.4.7.3.2 Extrusion Process

The moisture content of the starting material (protein content about 50%, e.g., soya flour) is adjusted to 30–40% and additives (NaCl, buffers, aroma compounds, pigments) are incorporated. Aroma compounds are added in fat as a carrier, when necessary, after the extrusion step to compensate for aroma losses. The protein mixture is fed into the extruder (a thermostatically controlled cylinder or conical body which contains a polished, rotating screw with a gradually decreasing pitch) which is heated to 120–180 °C and develops a pressure of 30–40 bar. Under these conditions the mixture is transformed into a plastic, viscous state in which solids are dispersed in the molten protein. Hydration of the protein takes place after partial unfolding of the globular molecules and stretching and rearrangement of the protein strands along the direction of mass transfer.

The process is affected by the rotation rate and shape of the screw and by the heat transfer and viscosity of the extruded material and its residence time in the extruder.

As the molten material exits from the extruder, the water vaporizes, leaving behind vacuoles in the ramified protein strands.

The extrusion process is more economical than the spin process. However, it yields fiber-like particles rather than well-defined fibers. A great number and variety of extruders are now in operation. As with other food processes, there is a trend toward developing and utilizing high-temperature/short-time extrusion cooking.

1.5 Literature

Aeschbach, R., Amado, R., Neukom, H.: Formation of dityrosine cross-links in proteins by oxidation of tyrosine residues. Biochim. Biophys. Acta *439*, 292 (1976)

Akimoto, H., Kawai, A., Nomura, H., Nagao, M., Kawachi, T., Sugimura, T.: Synthesis of potent mutagens in tryptophan pyrolysates. Chem. Lett. *1977*, 1061

Arai, S., Yamashita, M., Fujimaki, M.: Nutritional improvement of food proteins by means of the plastein reaction and its novel modification. Adv. Exp. Med. Biol. *105*, 663 (1978)

Ariyoshi, Y.: The structure-taste relationships of aspartyl dipeptide esters. Agric. Biol. Chem. *40*, 983 (1976)

Aso, K., Yamashita, M., Arai, S., Fujimaki, M.: Tryptophan-, threonine-, and lysine-enriched plasteins from zein. Agric. Biol. Chem. *38*, 679 (1974)

Aso, K., Yamashita, M., Arai, S., Suzuki, J., Fujimaki, M.: Specificity for incorporation of α-amino acid esters during the plastein reaction by papain. J. Agric. Food Chem. *25*, 1138 (1977)

Belitz, H.-D., Wieser, H.: Zur Konfigurationsabhängigkeit des süßen oder bitteren Geschmacks von Aminosäuren und Peptiden. Z. Lebensm. Unters. Forsch. *160*, 251 (1976)

Bodanszky, M.: Peptide Chemistry. Springer-Verlag: Berlin, 1988

Boggs, R.W.: Bioavailability of acetylated derivatives of methionine, threonine and lysine. Adv. Exp. Med. Biol. *105*, 571 (1978)

Bosin, T.R., Krogh, S., Mais, D.: Identification and quantitation of 1,2,3,4-Tetrahydro-β-carboline-3-carboxylic acid and 1-methyl-1,2,3,4-tetrahydro-β-carboline-3-carboxylic acid in beer and wine. J. Agric. Food Chem. *34*, 843 (1986)

Bott, R.R., Davies, D.R.: Pepstatin binding to *Rhizopus chinensis* aspartyl proteinase. In: Peptides: Structure and function (Eds.: Hruby, V.J., Rich, D.H.), p. 531. Pierce Chemical Co.: Rockford, Ill. 1983

Brussel, L.B.P., Peer, H.G., van der Heijden, A.: Structure-taste relationship of some sweet-tasting dipeptide esters. Z. Lebensm. Unters. Forsch. *159*, 337 (1975)

Chen, C., Pearson, A.M., Gray, J.I.: Meat Mutagens. Adv. Food Nutr. Res. *34*, 387, (1990)

Cherry, J.P. (Ed.): Protein functionality in foods. ACS Symposium Series 147, American Chemical Society: Washington, D.C. 1981

Creighton, T.E.: Proteins: structures and molecular properties. W.H. Freeman and Co.: New York. 1983

Croft, L.R.: Introduction to protein sequence analysis, 2nd edn., John Wiley and Sons, Inc.: Chichester. 1980

Edsall, J.T., Wyman, J.: Biophysical chemistry. Vol. I, Academic Press: New York. 1958

Einsele, A.: Biomass from higher n-alkanes. In: Biotechnology (Eds.: Rehm, H.-J., Reed, G.), Vol. 3, p. 43, Verlag Chemie: Weinheim. 1983

Evans, M.T.A., Irons, L., Petty, J.H.P.: Physico-chemical properties of some acyl derivatives of β-Casein. Biochim. Biophys. Acta *243*, 259 (1971)

Faust, U., Präve. P.: Biomass from methane and methanol. In: Biotechnology (Eds.: Rehm, H.-J., Reed, G.), Vol. 3, p. 83, Verlag Chemie: Weinheim. 1983

Finot, P.-A., Mottu, F., Bujard, E., Mauron, J.: N-substituted lysines as sources of lysine in nutrition. Adv. Exp. Med. Biol. *105*, 549 (1978)

Frank, H., Nicholson, G.J., Bayer, E.: Rapid gas chromatographic separation of amino acid enantiomers with a novel chiral stationary phase. J. Chromatogr. Sci. *15*, 174 (1977)

Fujino, M., Wakimasu, M., Tanaka, K., Aoki, H., Nakajima, N.: L-aspartyl-aminomalonic acid diesters. Naturwissenschaften *60*, 351 (1973)

Galembeck, F., Ryan, D.S., Whitaker, J.R., Feeney, R.E.: Reactions of proteins with formaldehyde in the presence and absence of sodium borohydride. J. Agric. Food Chem. *25*, 238 (1977)

Glazer, A.N.: The chemical modification of proteins by group-specific and site-specific reagents. In: The proteins (Eds.: Neurath, H., Hill, R.L., Boeder, C.-L.), 3rd edn., Vol. II, p. 1, Academic Press: New York-London. 1976

Grant, D.R.: The modification of wheat flour proteins with succinic anhydride. Cereal Chem. *50* 417 (1973)

Gross, E., Morell, J.L.: Structure of nisin. J. Am. Chem. Soc. *93*, 4634 (1971)

Haagsma, N., Slump, P.: Evaluation of lysinoalanine determinations in food proteins. Z. Lebensm. Unters. Forsch. *167*, 238 (1978)

Hudson, B.J.F. (Ed.): Developments in food proteins-1. Applied Science Publ.: London. 1982

Ikenaka, T., Odani, S., Koide, T.: Chemical structure and inhibitory activities of soybean proteinase inhibitors. Bayer-Symposium V "Proteinase inhibitors", p. 325, Springer-Verlag: Berlin-Heidelberg. 1974.

Kasai, H., Yamaizumi, Z., Shiomi, T., Yokoyama, S., Miyazawa, T., Wakabayashi, K., Nagao, M., Sugimura, T., Nishimura, S.: Structure of a potent mutagen isolated from fried beef. Chem. Lett. *1981*, 485

Kasai, H., Yamaizumi, Z., Wakabayashi, K., Nagao, M., Sugimura, T., Yokoyama, S., Miyazawa, T., Spingarn, N.E., Weisberger, J.H., Nishimura, S.: Potent novel mutagens produced by broiling fish under normal conditions. Proc. Jpn. Acad. Ser. B 56, 278 (1980)

Kessler, H.G.: Lebensmittel- und Bioverfahrenstechnik; Molkereitechnologie. 3rd edn., Verlag A. Kessler: Freising, 1988

Kinsella, J.E., Shetty, K.J.: Yeast proteins: Recovery and functional properties. Adv. Exp. Med. Biol. *105*, 797 (1978)

Kinsella, J.E.: Functional properties of proteins in foods: A survey. Crit. Rev. Food Sci. Nutr. *7*, 219 (1976)

Kinsella, J.E.: Texturized proteins: Fabrication, flavoring, and nutrition. Crit. Rev. Food Sci. Nutr. *10*, 147 (1978)

Kostka, V. (Ed.): Aspartic proteinases and their inhibitors. Walter de Gruyter: Berlin. 1985

Lásztity, R.: Rheologische Eigenschaften von Weizenkleber und ihre Beziehungen zu molekularen Parametern. Nahrung *19*, 749 (1975)

Li-Chan, E., Helbig, N., Holbek, E., Chau, S., Nakai, S.: Covalent attachment of lysine to wheat gluten for nutritional improvement. J. Agric. Food Chem. *27*, 877 (1979)

Lottspeich, F., Henschen, A., Hupe, K.-P. (Eds.): High performance liquid chromatography in protein and peptide chemistry. Walter de Gruyter: Berlin. 1981

Lowe, J.N., Ingraham, L.L.: An introduction to biochemical reaction mechanisms. Prentice Hall, Inc.: Englewood Cliffs, N.J. 1974

Lübke, K., Schröder, E., Kloss, G.: Chemie und Biochemie der Aminosäuren, Peptide und Proteine. Georg Thieme Verlag: Stuttgart. 1975

Masters, P.M., Friedman, M.: Racemization of amino acids in alkali-treated food proteins. J. Agric. Food Chem. *27*, 507 (1979)

Mauron, J.: Ernährungsphysiologische Beurteilung bearbeiteter Eiweißstoffe. Dtsch. Lebensm. Rundsch. *71*, 27 (1975)

Mazur, R.H., Goldkamp, A.H., James, P.A., Schlatter, J.M.: Structure-taste relationships of aspartic acid amides. J. Med. Chem. *13*, 1217 (1970)

Mazur, R.H., Reuter, J.A., Swiatek, K.A., Schlatter, J.M.: Synthetic sweetener. 3. Aspartyl dipeptide esters from L- and D-alkylglycines. J. Med. Chem. *16*, 1284 (1973)

Meister, A.: Biochemistry of the amino acid. 2nd edn., Vol. I, Academic Press: New York-London. 1965

Morrissay, P.A., Mulvihill, D.M., O'Neill, E.M.: Functional properties of muscle proteins. In: Development in Food Proteins – 5; (Ed.: Hudson, B.J.F.), p. 195, Elsevier Applied Science: London, 1987

Nagao, M., Yahagi, T., Kawachi, T., Seino, Y., Honda, M., Matsukura, N., Sugimura, T., Wakabayashi, K., Tsuji, K., Kosuge, T.: Mutagens in foods, and especially pyrolysis products of protein. Dev. Toxicol. Environ. Sci. 2nd (Prog. Genet. Toxicol.), p. 259 (1977)

Needleman, S.B. (Ed.): Protein sequence determination. Springer-Verlag: Berlin-Heidelberg. 1970

Needleman, S.B. (Ed.): Advanced methods in protein sequence determination. Springer-Verlag: Berlin-Heidelberg. 1977

Oura, E.: Biomass from carbohydrates. In: Biotechnology (Eds.: Rehm, H.-J., Reed, G.), Vol. 3, p. 3, Verlag Chemie: Weinheim, 1983

Pence, J.W., Mohammad, A., Mecham, D.K.: Heat denaturation of gluten. Cereal Chem. *30*, 115 (1953)

Perutz, M.F.: Proteins and nucleic acids. Elsevier Publ. Co.: Amsterdam. 1962

Poindexter, E.H., Jr., Carpenter, R.D.: Isolation of harmane and norharmane from cigarette smoke. Chem. Ind. *1962*, 176

Puigserver, A.J., Sen, L.C., Clifford, A.J., Feeney, R.E., Whitaker, J.R.: A method for improving the nutritional value of food proteins: Covalent attachment of amino acids. Adv. Exp. Med. Biol. *105*, 587 (1978)

Repley, J.A., Careri, G.: Protein hydration and function. Adv. Protein Chem. *41*, 38 (1991)

Richmond, A.: Phototrophic microalgae. In: Biotechnology (Eds.: Rehm, H.-J., Reed, G.), Vol. 3, p. 109, Verlag Chemie: Weinheim. 1983

Schulz, G.E., Schirmer, R.H.: Principle of protein structure. Springer-Verlag: Berlin-Heidelberg. 1979

Schmitz, M.: Möglichkeiten und Grenzen der Homoarginin-Markierungsmethode zur Messung der Proteinverdaulichkeit beim Schwein. Dissertation, Universität Kiel. 1988

Schwenke, K.D.: Beeinflussung funktioneller Eigenschaften von Proteinen durch chemische Modifizierung. Nahrung *22*, 101 (1978)

Seki, T., Kawasaki, Y., Tamura, M., Tada, M., Okai, H.: Further study on the salty peptide ornithyl-β-alanine. Some effects of pH and additive ions on saltiness. J. Agric. Food Chem. *38*, 25 (1990)

Sela, M., Lupu, N., Yaron, A., Berger, A.: Water-soluble polypeptidyl gliadins. Biochem. Biophys. Acta *62*, 594 (1962)

Severin, Th., Ledl, F.: Thermische Zersetzung von Cystein in Tributyrin. Chem. Mikrobiol. Technol. Lebensm. *1*, 135 (1972)

Sheppard, R.C. (Ed.): Amino-acids, peptides, and proteins, Vol. 1 ff, The Chemical Society, Berlington House, London. 1970

Shinoda, I., Tada, M., Okai, H.: A new salty peptide, ornithyl-β-alanine hydrochloride. Pept. Chem., 21st. Proceeding of the Symposium on Peptide Chemistry 1983, p. 43 (1984)

Soda, K., Tanaka, H., Esaki, N.: Amino acids. In: Biotechnology (Eds.: Rehm, H.-J. Reed, G.),

Vol. 3, p. 479, Verlag Chemie: Weinheim, 1983

St. Angelo, A.J., Conkerton, E.J., Dechary, J.M., Altschul, A.M.: Modification of edestin with N-carboxyl-D,L-alanine anhydride. Biochim. Biophys. Acta *121*, 181 (1966)

Sternberg, M., Kim, C.Y.: Lysinoalanine formation in protein food ingredients. Adv. Exp. Med. Biol. *86B*, 73 (1977)

Sugimura, T., Kawachi, T., Nagao, M., Yahagi, T., Seino, Y., Okamoto, T., Shudo, K., Kosuge, T., Tsuji, K. et al.: Mutagenic principle(s) in tryptophan and phenylalanine pyrolysis products. Proc. Jpn. Acad. *33*, 58 (1977)

Sulser, H.: Die Extraktstoffe des Fleisches. Wissenschaftliche Verlagsgesellschaft mbH: Stuttgart. 1978

Traub, W., Piez, K.A.: The chemistry and structure of collagen. Adv. Protein Chem. *25*, 267 (1971)

Treleano, R., Belitz, H.-D., Jugel, H., Wieser, H.: Beziehungen zwischen Struktur und Geschmack bei Aminosäuren mit cyclischen Seitenketten. Z. Lebensm. Unters. Forsch. *167*, 320 (1978)

Tschesche, H. (Ed.): Modern Methods in Protein Chemistry, Vol. 2, Walter de Gruyter: Berlin, 1985

Voutsinas, L.P., Nakai, S.: Covalent binding of methionine and tryptophan to soy protein. J. Food Sci. *44*, 1205 (1979).

Wakabayashi, K., Tsuji, K., Kosuge, T., Takeda, K., Yamaguchi, K., Shudo, K., Iitaka, Y., Okamoto, T., Yahagi, T. et al.: Isolation and structure determination of a mutagenic substance in L-lysine pyrolysate. Proc. Jpn. Acad. Ser. B*54*, 569 (1978)

Walton, A.G.: Polypeptides and protein structure, Elsevier North Holland, Inc., New York-Oxford. 1981

Watanabe, M., Arai, S.: The plastein reaction and its applications. In: Developments in Food Proteins − 6; (Ed.: Hudson, B.J.F.), p. 179, Elsevier Applied Science: London. 1988

Whitaker, J.R., Fujimaki, M. (Eds.): Chemical deterioration of proteins, ACS Symposium Series 123, American Chemical Society: Washington D.C. 1980

Wieser, H., Belitz, H.-D.: Zusammenhänge zwischen Struktur und Bittergeschmack bei Aminosäuren und Peptiden. I. Aminosäuren und verwandte Verbindungen. Z. Lebensm. Unters. Forsch. *159*, 65 (1975)

Wieser, H., Belitz, H.-D.: Zusammenhänge zwischen Struktur und Bittergeschmack bei Aminosäuren und Peptiden. II. Peptide und Peptidderivate. Z. Lebensm. Unters. Forsch. *160*, 383 (1976)

Wieser, H., Jugel, H., Belitz, H.-D.: Zusammenhänge zwischen Struktur und Süßgeschmack bei Aminosäuren. Z. Lebensm. Unters. Forsch. *164*, 277 (1977)

Wittmann-Liebold, B., Salnikow, J., Erdmann, V.A. (Eds.): Advanced Methods in Protein Microsequence Analysis. Springer-Verlag: Berlin. 1986

Yamashita, M., Arai, S., Fujimaki, M.: A low-phenylalanine, high-tyrosine plastein as an acceptable dietetic food. J. Food Sci. *41*, 1029 (1976)

Yamashita, M., Arai, S., Amano, Y., Fujimaki, M.: A novel one-step process for enzymatic incorporation of amino acids into proteins: Application to soy protein and flour for enhancing their methionine levels. Agric. Biol. Chem. *43*, 1065 (1979)

Yamashita, M., Arai, S., Tsai, S.-J., Fujimaki, M.: Plastein reaction as a method for enhancing the sulfur-containing amino acid level of soybean protein. J. Agric. Food Chem. *19*, 1151 (1971)

Yamashita, M., Arai, S., Kokubo, S., Aso, K., Fujimaki, M.: Synthesis and characterization of a glutamic acid enriched plastein with greater solubility. J. Agric. Food Chem. *23*, 27 (1975)

Yoshikawa, M., Tamaki, M., Sugimoto, E., Chiba, H.: Effect of dephosphorylation on the self-association and the precipitation of β-Casein. Agric. Biol. Chem. *38*, 2051 (1974)

2 Enzymes

2.1 Foreword

Enzymes are proteins with powerful catalytic activity. They are synthesized by biological cells and, in all organisms, they are involved in chemical reactions related to metabolism. Therefore, enzyme-catalyzed reactions also proceed in many foods and thus enhance or deteriorate food quality. Relevant to this phenomenon are the ripening of fruits and vegetables, the aging of meat and dairy products, and the processing steps involved in the making of dough from wheat or rye flours and the production of alcoholic beverages by fermentation technology.

Enzyme inactivation or changes in the distribution patterns of enzymes in subcellular particles of a tissue can occur during storage or thermal treatment of food. Since such changes are readily detected by analytical means, enzymes often serve as suitable indicators for revealing such treatment of food. Examples are the detection of pasteurization of milk, beer or honey, and differentiation between fresh and deep frozen meat or fish.

Enzyme properties are of interest to the food chemist since enzymes are available in increasing numbers for enzymatic food analysis or for utilization in industrial food processing. Examples of both aspects of their use are provided in this chapter in section 2.6.4 on food analysis and in section 2.7, which covers food processing.

Details of enzymes which play a role in food science are restricted in this chapter to only those enzyme properties which are able to provide an insight into the build-up or functionality of enzymes or can contribute to the understanding of enzyme utilization in food analysis or food processing and storage.

2.2 General Remarks, Isolation and Nomenclature

2.2.1 Catalysis

Let us consider the catalysis of an exergonic reaction:

$$A \underset{k_{-1}}{\overset{k_1}{\rightleftharpoons}} P \qquad (2.1)$$

with a most frequently occurring case in which the reaction does not proceed spontaneously. Reactant A is metastable, since the activation energy, E_A, required to reach the activated transition state in which chemical bonds are formed or cleaved in order to yield product P, is exceptionally high (Fig. 2.1).

The reaction is accelerated by the addition of a suitable catalyst. It transforms reactant A into intermediary products (EA and EP in Fig. 2.1), the transition states of which are at a lower energy level than the transition state of a non-catalyzed reaction ($A^{\neq}$ in Fig. 2.1). The molecules of the species A contain enough energy

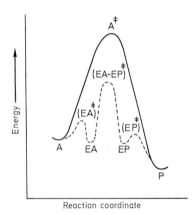

Fig. 2.1. Energy profile of an exergonic reaction A → P; — without and − − − with catalyst E

Table 2.1. Examples of catalyst activity

Reaction	Catalyst	Activation energy $(kJ \cdot mol^{-1})$	$k_{rel}(25\,^\circ C)$
1. $H_2O_2 \rightarrow H_2O + 1/2\ O_2$	Absent	75	1.0
	$I^{\ominus}$	56.5	$\sim 2.1 \cdot 10^3$
	Catalase	26.8	$\sim 3.5 \cdot 10^8$
2. Casein + n $H_2O \rightarrow$	$H^{\oplus}$	86	1.0
(n + 1) Peptides	Trypsin	50	$\sim 2.1 \cdot 10^6$
3. Ethylbutyrate	$H^{\oplus}$	55	1.0
+ $H_2O \rightarrow$ butyric acid	Lipase	17.6	$\sim 4.2 \cdot 10^6$
+ ethanol			
4. Saccharose + $H_2O \rightarrow$	$H^{\oplus}$	107	1.0
Glucose + Fructose	Invertase	46	$\sim 5.6 \cdot 10^{10}$
5. Linoleic acid	Absent	150–270	1.0
+ $O_2 \rightarrow$ Linoleic acid	Cu^{2+}	30–50	$\sim 10^2$
hydroperoxide	Lipoxygenase	16.7	$\sim 10^7$

to combine with the catalyst and, thus, to attain the "activated state" and to form or break the covalent bond that is necessary to give the intermediary product which is then released as product P along with free, unchanged catalyst. The reaction rate constants, k_{+1} and k_{-1}, are therefore increased in the presence of a catalyst. However, the equilibirum constant of the reaction, i.e. the ratio $k_{1+}/k_{-1} = K$, is not altered.

Activation energy levels for several reactions and the corresponding decreases of these energy levels in the presence of chemical or enzymatic catalysts are provided in Table 2.1. Changes in their reaction rates are also given. In contrast to reactions 1 and 5 (Table 2.1) which proceed at measurable rates even in the absence of catalysts, hydrolysis reactions 2, 3 and 4 occur only in the presence of protons as catalysts. However, all reaction rates observed in the case of inorganic catalysts are increased by a factor of at least several orders of magnitude in the presence of suitable enzymes. Because of the powerful activity of enzymes, their presence at levels of 10^{-8} to 10^{-6} mol/l is sufficient for *in vitro* experiments. However, the enzyme concentrations found in living cells are often substantially higher.

2.2.2 Specificity

In addition to an enzyme's ability to substantially increase reaction rates, there is a unique enzyme property related to its high specificity for both the compound to be converted (substrate specificity) and for the type of reaction to be catalysed (reaction specificity).

The activities of allosteric enzymes (cf. 2.5.1.3) are affected by specific regulators or effectors. Thus, the activities of such enzymes show an additional regulatory specificity.

2.2.2.1 Substrate Specificity

The substrate specificity of enzymes shows the following differences. The occurrence of a distinct functional group in the substrate is the only prerequisite for a few enzymes, such as some hydrolases. This is exemplified by non-specific lipases (cf. Table 3.21) or peptidases (cf. 1.4.5.2.1) which generally act on an ester or peptide covalent bond.

More restricted specificity is found in other enzymes, the activities of which require that the substrate molecule contains a distinct structural feature in addition to the reactive functional group. Examples are the proteinases trypsin and chymotrypsin which cleave only ester or peptide bonds with the carbonyl group derived from lysyl or arginyl (trypsin) or tyrosyl, phenylalanyl or tryptophanyl residues (chymotrypsin). Many enzymes activate only one single substrate or preferentially catalyze the conversion of one substrate while other substrates are converted into products with a lower reaction rate (cf. examples in Table 2.2 and 3.24). In the latter cases a reliable assessment of specificity is possible only when the enzyme is available in purified form, i.e. all

Table 2.2. Substrate specificity of a legume α-glucosidase

Substrate	Relative activity (%)	Substrate	Relative activity (%)
Maltose	100	Cellobiose	0
Isomaltose	4.0	Saccharose	0
Maltotriose	41.5	Phenyl-α-glucoside	3.1
Panose	3.5	Phenyl-α-maltoside	29.7
Amylose	30.9		
Amylopectin	4.4		

other accompanying enzymes, as impurities, are completely removed.

An enzyme's substrate specificity for stereoisomers is remarkable. When a chiral center is present in the substrate in addition to the group to be activated, only one enantiomer will be converted to the product. Another example is the specificity for diastereoisomers, e.g. for cis-trans geometric isomers.

Enzymes with high substrate specificity are of special interest for enzymatic food analysis. They can be used for the selective analysis of individual food constituents, thus avoiding the time consuming separation techniques required for chemical analyses, which can result in losses.

2.2.2.2 Reaction Specificity

The substrate is specifically activated by the enzyme so that, among the several thermodynamically permissible reactions, only one occurs. This is illustrated by the following example: L(+)-lactic acid is recognized as a substrate by four enzymes, as shown in Fig. 2.2, although only lactate-2-monooxygenase decarboxylates the acid oxidatively to acetic acid. Lactate dehydrogenase and lactate-malate transhydrogenase form a common reaction product, pyruvate, but by different reaction pathways (Fig. 2.2). This may suggest that reaction specificity should be ascribed to the different cosubstrates, such as NAD^+ or oxalacetate. But this is not the case since a change in cosubstrates stops the reaction. Obviously, the enzyme's reaction specificity as well as the substrate specificity are predetermined by the structure and chemical properties of the protein moiety of the enzyme.

Of the four enzymes considered, only the lactate racemase reacts with either of the enantiomers of lactic acid, yielding a racemic mixture.

Therefore, enzyme reaction specificity rather than substrate specificity is considered as a basis for enzyme classification and nomenclature (cf. 2.2.6).

2.2.3 Structure

Enzymes are globular proteins with greatly differing particle sizes (cf. Table 1.26). As outlined in section 1.4.2, the protein structure is determined by its amino acid sequences and by its conformation, both secondary and tertiary, derived from this sequence. Larger enzyme molecules often consist of two or more peptide chains (subunits or protomers, cf. Table 1.26) arranged into a specified quaternary structure (cf. 1.4.2.3). Section 2.4.1 will show that the three dimensional shape of the enzyme molecule is actually responsible for its specificity and its effective role as a catalyst. On the other hand, the protein nature of the enzyme restricts its activity to a relatively narrow pH range (for pH optima, cf. 2.5.3) and heat treatment leads readily to loss of activity by denaturation (cf. 1.4.2.4 and 2.5.4.4).

Some enzymes are complexes consisting of a protein moiety bound firmly to a nonprotein component which is involved in catalysis, e.g. a "prosthetic" group (cf. 2.3.2). The activities of other enzymes require the presence of a cosubstrate which is reversibly bound to the protein moiety (cf. 2.3.1).

2.2.4 Isolation and Purification

Most of the enzyme properties are clearly and reliably revealed only with purified enzymes. As noted under enzyme isolation, prerequisites for the isolation of a pure enzyme are selected protein chemical separation methods carried out at $0-4\,^\circ C$ since enzymes are often not stable at higher temperatures.

Tissue Disintegration and Extraction. Disintegration and homogenization of biological tissue requires special precautions: procedures should be designed to rupture the majority of

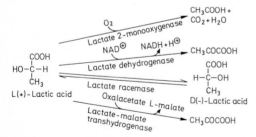

Fig. 2.2. Examples of reaction specificity of some enzymes

the cells in order to release their contents so that they become accessible for extraction. The tissue is usually homogenized in the presence of an extraction buffer which often contains an ingredient to protect the enzymes from oxidation and from traces of heavy metal ions. Particular difficulty is encountered during the isolation of enzymes which are bound tenaciously to membranes which are not readily solubilized. Extraction in the presence of tensides may help to isolate such enzymes. As a rule, large amounts of tissue have to be homogenized because the enzyme content in proportion to the total protein isolated is low and is usually further diminished by the additional purification of the crude enzyme isolate (cf. example in Table 2.3).

Enzyme Purification. Removal of protein impurities, usually by a stepwise process, is essentially the main approach in enzyme purification. As a first step, fractional precipitation, e.g. by ammonium sulfate saturation, is often used or the extracted proteins are fractionated by molecular weight e.g., column gel chromatography. The fractions containing the desired enzyme activity are collected and purified further, e.g., by ion-exchange chromatography. Additional options are also available, such as various forms of preparative

electrophoresis, e.g. disc gel electrophoresis or isoelectric focusing. The purification procedure can be substantially shortened by using affinity column chromatography. In this case, the column is packed with a stationary phase to which is attached the substrate or a specific inhibitor of the enzyme. The enzyme is then selectively and reversibly bound and, thus, in contrast to the other inert proteins, its elution is delayed.

Control of Purity. Previously, the complete removal of protein impurities was confirmed by crystallization of the enzyme. This "proof" of purity can be tedious and is open to criticism. Today, electrophoretic methods of high separation efficiency or HPLC are primarily used.

The behavior of the enzyme during chromatographic separation is an additional proof of purity. A purified enzyme is characterized by a symmetrical elution peak in which the positions of the protein absorbance and enzyme activity coincide and the specific activity (expressed as units per amount of protein) remains unchanged during repeated elutions.

During a purification procedure, the enzyme activities are recorded as shown in Table 2.3. They provide data which show the extent of purification achieved after each separation

Table 2.3. Isolation of a glucosidase from beans (*Phaseolus vidissimus*)

No. Isolation step	Protein (mg)	α-Glucosidase			
		Activity (μcat)	Specific activity (μcat/mg)	Enrichment (-fold)	Yield (%)
1. Extraction with 0.01 mol/L acetate buffer of pH 5.3					
2. Saturation to 90% with ammonium sulfate followed by solubilization in buffer of step 1	44,200	3,840	0.087	1	100
3. Precipitation with polyethylene glycol (20%). Precipitate is then solubilized in 0.025 mol/L Tris-HCl buffer of pH 7.4	7,610	3,590	0.47	5.4	93
4. Chromatography on DEAE-cellulose column, an anion exchanger	1,980	1,650	0.83	9.5	43
5. Chromatography on SP-Sephadex C-50, a cation exchanger	130	845	6.5	75	22
6. Preparative isoelectric focusing	30	565	18.8	216	15

step and show the enzyme yield. Such a compilation of data readily reveals the undesired separation steps associated with loss of activity and suggests modifications or adoption of other steps.

2.2.5 Multiple Forms of Enzymes

Chromatographic or electrophoretic separations of an enzyme can occasionally result in separation of the enzyme into *"isoenzymes"*, i.e. forms of the enzyme which catalyze the same reaction although they differ in their protein structure. The occurrence of multiple enzyme forms can be the result of the following:

a) Different compartments of the cell produce genetically independent enzymes with the same substrate and reaction specificity, but which differ in their primary structure. An example is glutamate-oxalacetate transaminase occurring in mitochondria and also in muscle tissue sarcoplasm. This is the indicator enzyme used to differentiate fresh from frozen meat (cf. 12.10.1.2).
b) Protomers associate to form polymers of differing size. An example is the glutamate dehydrogenase occurring in tissue as an equilibrium mixture of molecular weights $M_r = 2.5 \cdot 10^5 - 10^6$.
c) Different protomers combine in various amounts to form the enzyme. For example, lactate dehydrogenase is structured from a larger number of subunits with the reaction specificity given in Fig. 2.2. It consists of five forms (A_4, A_3B, A_2B_2, AB_3 and B_4), all derived from two protomers, A and B.

2.2.6 Nomenclature

The Nomenclature Commitee of the "International Union of Biochemistry and Molecular Biology" (IUBMB) adopted rules last amended in 1992 for the systematic classification and designation of enzymes based on reaction specificity. All enzymes are classified into six major classes according to the nature of the chemical reaction catalyzed:

1. Oxidoreductases.
2. Transferases.
3. Hydrolases.
4. Lyases (cleave C–C, C–O, C–N, and other groups by elimination, leaving double bonds, or conversely adding groups to double bonds).
5. Isomerases (involved in the catalysis of isomerizations within one molecule).
6. Ligases (involved in the biosynthesis of a compound with the simultaneous hydrolysis of a pyrophosphate bond in ATP or a similar triphosphate).

Each class is then subdivided into subclasses which more specifically denote the type of reaction, e.g. by naming the electron donor of an oxidation-reduction reaction or by naming the functional group carried over by a transferase or cleaved by a hydrolase enzyme.

Each subclass is further divided into sub-subclasses. For example, sub-subclasses of oxidoreductases are denoted by naming the acceptor which accepts the electron from its respective donor.

Each enzyme is classified by adopting this system. An example will be analyzed. The enzyme ascorbic acid oxidase catalyzes the following reaction:

$$\text{L-Ascorbic acid} \quad + \quad \tfrac{1}{2}O_2$$
$$\rightleftharpoons \quad \text{L-Dehydroascorbic acid} \quad + \quad H_2O \qquad (2.2)$$

Hence, its systematic name is L-ascorbate: oxygen oxidoreductase, and its systematic number is E.C. 1.10.3.3 (cf. Formula 2.3). The systematic names are often quite long. Therefore, the short, trivial names along with the systematic numbers are often convenient for enzyme designation. Since enzymes of different biological origin often differ in their properties, the source and, when known, the subcellular fraction used for isolation are specified in addition to the name of the enzyme preparation; for example, "ascorbate oxidase (E.C. 1.10.3.3) from cucumber". When known, the subcellular fraction of origin (cytoplasmic, mitochondrial or peroxisomal) is also specified.

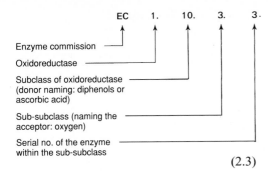

Enzyme commission

Oxidoreductase

Subclass of oxidoreductase
(donor naming: diphenols or
ascorbic acid)

Sub-subclass (naming the
acceptor: oxygen)

Serial no. of the enzyme
within the sub-subclass

$$EC \quad 1. \quad 10. \quad 3. \quad 3. \tag{2.3}$$

A number of enzymes of interest to food chemistry are described in Table 2.4. The number of the section in which an enzyme is dealt with is given in the last column.

2.2.7 Activity Units

The catalytic activity of enzymes is exhibited only under specific conditions, such as pH, ionic strength, buffer type, presence of cofactors and suitable temperature. Therefore, the rate of substrate conversion or product formation can be measured in a test system designed to follow the enzyme activity. The International System of Units (SI) designation is mol s^{-1} and its recommended designation is the "katal" (kat*). Decimal units are formed in the usual way, e.g.:

$$\mu kat = 10^{-6} \, kat = \mu mol \cdot s^{-1} \tag{2.4}$$

Concentration of enzymatic activity is given as $\mu kat \, l^{-1}$. The following activity units are derived from this:

a) The *specific catalytic activity*, i.e. the activity of the enzyme preparation in relation to the protein concentration.
b) The *molar catalytic activity*. This can be determined when the pure enzyme with a known molecular weight is available. It is expressed as "katal per mol of enzyme" (kat mol^{-1}). When the enzyme has only one active site or center per molecule, the molar catalytic activity equals the "exchange

---

* The old definition in the literature may also be used: 1 enzyme unit (U) $\triangleq 1 \, \mu mol \, min^{-1}$ (1 U $\triangleq$ 16.67 $\cdot 10^{-9}$ kat).

number", which is defined as the number of substrate molecules converted per unit time by each active site of the enzyme molecule.

2.3 Enzyme Cofactors

Rigorous analysis has demonstrated that numerous enzymes are not pure proteins. In addition to protein, they contain metal ions and/or low molecular weight nonprotein organic molecules. These nonprotein hetero constituents are denoted as cofactors which are indispensable for enzyme activity.

According to the systematics (Fig. 2.3), an apoenzyme is the inactive protein without a cofactor. Metal ions and coenzymes participating in enzymatic activity belong to the cofactors which are subdivided into prosthetic groups and cosubstrates. The prosthetic group is bound firmly to the enzyme. It can not be removed by, e.g. dialysis, and during enzyme catalysis it remains attached to the enzyme molecule. Often, two substrates are converted by such enzymes, one substrate followed by the other, returning the prosthetic group to its original state. On the other hand, during metabolism, the cosubstrate reacts with at least two enzymes. It transfers the hydrogen or the functional group to another enzyme and, hence, is denoted as a "transport metabolite" or as an "intermediary substrate". It is distinguished from a true substrate by being regenerated in a

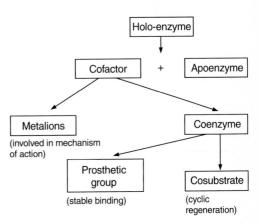

Fig. 2.3. Systematics of cofactor-containing enzymes (according to *A. Schellenberger*, 1989)

Table 2.4. Systematic classification of some enzymes of importance to food chemistry

Class/subclass		Enzyme	EC-Number	In text found under
1.	*Oxidoreductases*			
1.1	CH−OH as donor			
1.1.1	With NAD⊕ or NADP⊕ as acceptor	Alcohol dehydrogenase	1.1.1.1	2.6.1
		Butanediol dehydrogenase	1.1.1.4	2.7.2.1.5
		L-Iditol 2-dehydrogenase	1.1.1.14	2.6.1
		L-Lactate dehydrogenase	1.1.1.27	2.6.1
		Malate dehydrogenase	1.1.1.37	2.6.1
		Galactose 1-dehydrogenase	1.1.1.48	2.6.1
		Glucose-6-phosphate 1-dehydrogenase	1.1.1.49	2.6.1
1.1.3	With oxygen as acceptor	Glucose oxidase	1.1.3.4	2.6.1 and 2.7.2.1.1
		Xanthine oxidase	1.1.3.22	2.3.3.2
1.2	Aldehyde group as donor			
1.2.1	With NAD⊕ or NADP⊕ as acceptor	Aldehyde dehydrogenase	1.2.1.3	2.7.2.1.4
1.8	S-Compound as donor			
1.8.5	With quinone or related compound as acceptor	Glutathione dehydrogenase (ascorbate)	1.8.5.1	15.2.2.7
1.10	Diphenol or dienol as donor			
1.10.3	With oxygen as acceptor	Ascorbate oxidase	1.10.3.3	2.2.6
1.11	Hydroperoxide as acceptor	Catalase	1.11.1.6	2.7.2.1.2
		Peroxidase	1.11.1.7	2.3.2.2 and 2.5.4.4
1.13	Acting on single donors			
1.13.11	Incorporation of molecular oxygen	Lipoxygenase	1.13.11.12	2.5.4.4 and 3.7.2.2
1.14	Acting on paired donors			
1.14.18	Incorporation of one oxygen atom	Monophenol monooxygenase (Polyphenol oxidase)	1.14.18.1	2.3.3.2
2.	*Transferases*			
2.7	Transfer of phosphate			
2.7.1	HO-group as acceptor	Hexokinase	2.7.1.1	2.6.1
		Glycerol kinase	2.7.1.30	2.6.1
		Pyruvate kinase	2.7.1.40	2.6.1
2.7.3	N-group as acceptor	Creatine kinase	2.7.3.2	2.6.1
3.	*Hydrolases*			
3.1	Cleavage of ester bonds			
3.1.1	Carboxylester hydrolases	Carboxylesterase	3.1.1.1	3.7.1.1
		Triacylglycerol lipase	3.1.1.3	2.5.4.4 and 3.7.1.1
		Phospholipase A_2	3.1.1.4	3.7.1.2
		Acetylcholinesterase	3.1.1.7	2.4.2.5
		Pectinesterase	3.1.1.11	4.4.5.2
		Phospholipase A_1	3.1.1.32	3.7.1.2

Table 2.4 (continued)

Class/subclass		Enzyme	EC-Number	In text found under
3.1.3	Phosphoric monoester hydrolases	Alkaline phosphatase	3.1.3.1	2.5.4.4
3.1.4	Phosphoric diester hydrolases	Phospholipase C	3.1.4.3	3.7.1.2
		Phospholipase D	3.1.4.4	3.7.1.2
3.2	Hydrolyzing O-glycosyl compounds			
3.2.1	Glycosidases	α-Amylase	3.2.1.1	4.4.5.1.1
		β-Amylase	3.2.1.2	4.4.5.1.2
		Glucan-1,4-α-D-glucosidase (Glucoamylase)	3.2.1.3	4.4.5.1.3
		Cellulase	3.2.1.4	4.4.5.3
		Polygalacturonase	3.2.1.15	2.5.4.4 and 4.4.5.2
		Lysozyme	3.2.1.17	2.7.2.2.11 and 11.2.3.1.4
		α-D-Glucosidase (Maltase)	3.2.1.20	2.6.1
		β-D-Glucosidase	3.2.1.21	2.6.1
		α-D-Galactosidase	3.2.1.22	
		β-D-Galactosidase (Lactase)	3.2.1.23	2.7.2.2.7
		β-Fructofuranosidase (Invertase, saccharase)	3.2.1.26	2.7.2.2.8
		1,3-β-D-Xylanase	3.2.1.32	2.7.2.2.10
		α-L-Rhamnosidase	3.2.1.40	2.7.2.2.9
		Pullulanase	3.2.1.41	4.4.5.1.4
		Exopolygalacturonase	3.2.1.67	4.4.5.2
3.2.3	Hydrolysing S-glycosyl compounds	Thioglucosidase (Myrosinase)	3.2.3.1	2.7.2.2.12
3.4	Peptidases[a]			
3.4.21	Serine endopeptidases[a]	Microbial serine proteinases e.g. Subtilisin	3.4.21.62	1.4.5.2.1
3.4.22	Cysteine endopeptidases[a]	Papain	3.4.22.2	1.4.5.2.2
		Ficin	3.4.22.3	1.4.5.2.2
		Bromelain	3.4.22.33	1.4.5.2.2
3.4.23	Aspartic acid endopeptidases[a]	Chymosin (Rennin)	3.4.23.4	1.4.5.2.4
3.4.24	Metalloendopeptidases[a]	Thermolysin	3.4.24.27	1.4.5.2.3
3.5	Acting on C−N bonds, other than peptide bonds			
3.5.2	In cyclic amides	Creatininase	3.5.2.10	2.6.1
4.	*Lyases*			
4.2	C−O-Lyases			
4.2.2	Acting on polysaccharides	Pectate lyase	4.2.2.2	4.4.5.2
		Exopolygalacturonate lyase	4.2.2.9	4.4.5.2
		Pectin lyase	4.2.2.10	4.4.5.2
5.	*Isomerases*			
5.3	Intramolecular oxidoreductases			
5.3.1	Interconverting aldoses and ketoses	Xylose isomerase	5.3.1.5	2.7.2.3
		Glucose-6-phosphate isomerase	5.3.1.9	2.6.1

[a] cf. Table 1.33.

subsequent reaction. Therefore the concentration of the intermediary substrates can be very low. In food analysis higher amounts of cosubstrates are often used without regeneration. Only those cofactors with enzymatic activities of importance in enzymatic analysis of food and/or in food processing will be presented. Some cofactors are related to water-soluble vitamins (cf. 6.3). The metal ions are dealt with separately in section 2.3.3.

2.3.1 Cosubstrates

2.3.1.1 Nicotinamide Adenine Dinucleotide

Transhydrogenases (e.g. lactate dehydrogenase, alcohol dehydrogenase) dehydrogenate or hydrogenate their substrates with the help of a pyridine cosubstrate (Fig. 2.4); its nicotinamide residue accepts or donates a hydride ion (H^-) at position 4:

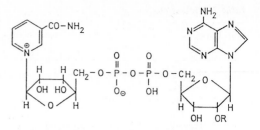

Fig. 2.4. Nicotinamide adenine dinucleotide (NAD) and nicotinamide adenine dinucleotide phosphate (NADP); R= H : NAD; R = PO$_3$H$_2$: NADP

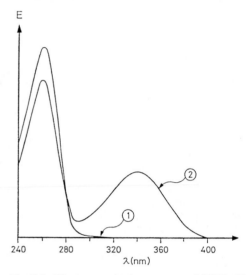

Fig. 2.5. Electron excitation spectra of NAD (1) and NADH (2)

The reaction proceeds stereospecifically (cf. 2.4.1.2.1); ribose phosphate and the $-CONH_2$ group force that pyridine ring of the cosubstrate to become planar on the enzyme surface. The role of Zn^{2+} ions in this catalysis is outlined in section 2.3.3.1. The transhydrogenases differ according to the site on the pyridine ring involved in or accessible to H-transfer. For example, alcohol and lactate dehydrogenases transfer the pro-R-hydrogen from the A* side, whereas glutamate or glucose dehydrogenases transfer the pro-S-hydrogen from the B* side.

The oxidized and reduced forms of the pyridine cosubstrate are readily distinguished by absorbance readings at 340 nm (Fig. 2.5). Therefore, whenever possible, enzymatic reactions which are difficult to measure directly are coupled with an NAD(P)-dependent indicator reaction (cf. 2.6.1.1) for food analysis.

2.3.1.2 Adenosine Triphosphate

The nucleotide adenosine triphosphate (ATP) is an energy-rich compound. Various groups are cleaved and transferred to defined substrates during metabolism in the presence of ATP.

_____
* Until the absolute configuration of the chiral center is determined, the two sides of the pyridine ring are denoted as A and B.

One possibility, the transfer of orthophosphates by kinases, is utilized in the enzymatic analysis of food (cf. Table 2.16).

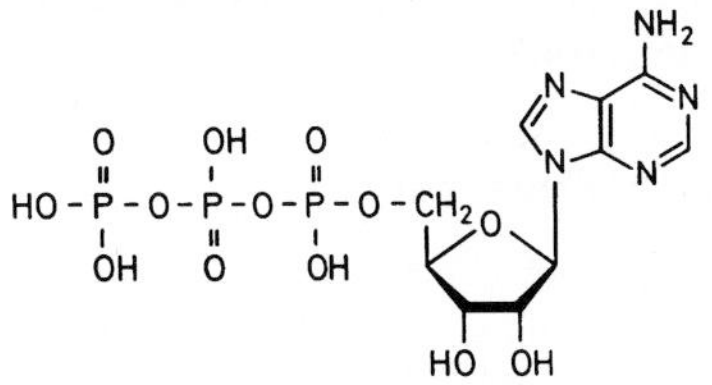

$$(2.6)$$

$$\text{ATP} + \text{H}_2\text{O} \longrightarrow \text{ADP} + \text{H}_3\text{PO}_4$$
$$(\Delta G^0 \text{ at pH 7} = -50 \text{ kJ mol}^{-1}) \qquad (2.7)$$

2.3.2 Prosthetic Groups

2.3.2.1 Flavins

Riboflavin (7,8-dimethyl-10-ribityl-isoalloxazine), known as vitamin B_2 (cf. 6.3.2), is the building block of flavin mononucleotide (FMN) and flavin adenine dinucleotide (FAD). Both act as prosthetic groups for electron transfer reactions in a number of enzymes.

Due to the much wider redox potential of the flavin enzymes, riboflavin is involved in the transfer of either one or two electrons. This is different from nicotinamides which participate in double electron transfer only. Values between $+0.19$ V (stronger oxidizing effect than $NAD^{\oplus}$) and -0.49 V (stronger reducing effect than NADH) have been reported.

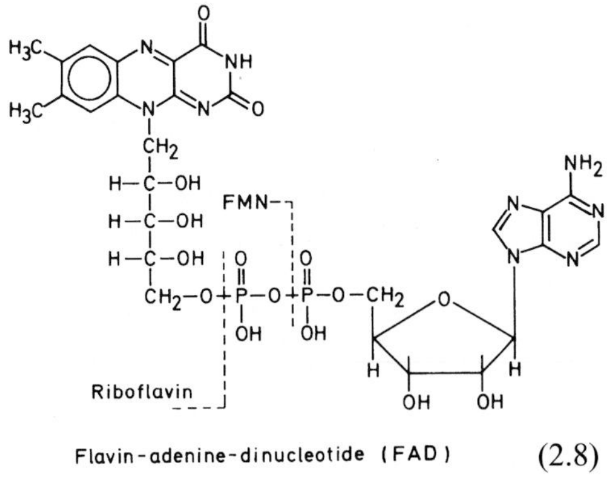

Flavin-adenine-dinucleotide (FAD) $\qquad (2.8)$

An example for a flavin enzyme is glucose oxidase, an enzyme often used in food processing to trap residual oxygen (cf. 2.7.2.1.1).

The enzyme isolated and purified from *Aspergillus niger* is a dimer ($M_r = 168,000$) with two noncovalently bound FAD molecules. In contrast to xanthine oxidase (cf. 2.3.3.2), for example, this enzyme has no heavy metal ion. During oxidation of a substrate, such as the oxidation of β-D-glucose to δ-D-gluconolactone, the flavoquinone is reduced by two single electron transfers:

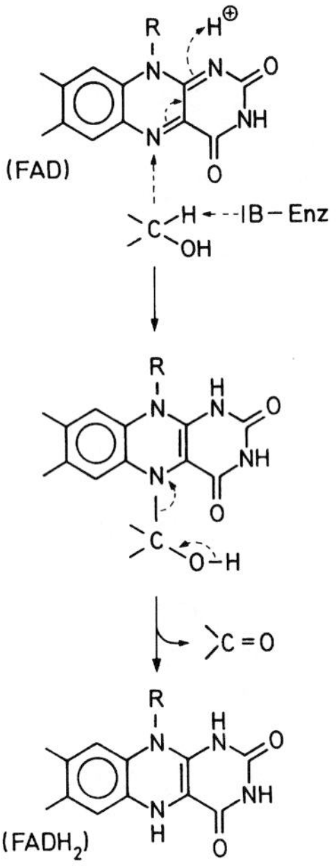

$$(2.9)$$

Like glucose oxidase, many flavin enzymes transfer the electrons to molecular oxygen, forming H_2O_2 and flavoquinone. The following intermediary products appear in this reaction:

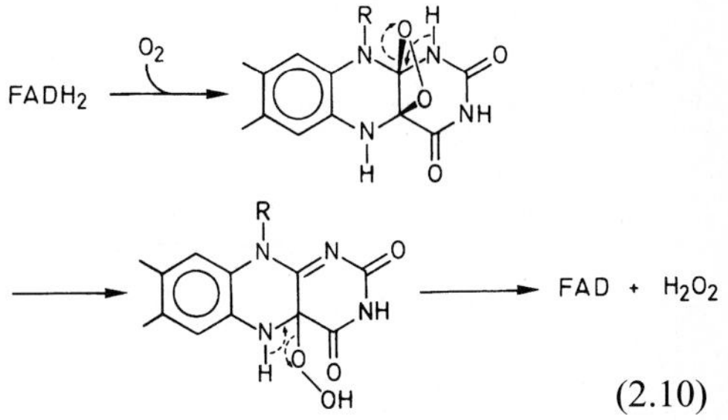

$$(2.10)$$

2.3.2.2 Hemin

Peroxidases from food of plant origin and several catalases contain ferri-protoporphyrin IX (hemin, cf. Formula 2.11) as their prosthetic group and as the chromophore responsible for the brown color of the enzymes:

$$(2.11)$$

In catalytic reactions there is a change in the electron excitation spectra of the peroxidases (Fig. 2.6a) which is caused by a valence change of the iron ion (Fig. 2.6b). Intermediary compounds I (green) and II (pale red) are formed during this change by reaction with H_2O_2 and reducing agent AH. The reaction cycle is completed by another single electron transfer.

Some verdoperoxidases, which are green in color (as suggested by their name) and found in various foods of animal origin, e.g. milk, contain an unidentified Fe-protoporphyrin as their prosthetic group.

2.3.2.3 Pyridoxal Phosphate

Pyridoxal phosphate (Formula 2.12) and pyridoxamine (Formula 2.13), derived from it, are designated as vitamin B_6 (cf. 6.3.3) and are essential ingredients of food:

$$(2.12)$$

$$(2.13)$$

Coupled to the enzyme as a prosthetic group through a lysyl residue, pyridoxal phosphate is involved in conversion reactions of amino acids. In the first step of catalysis, the amino group of the amino acid substrate displaces the 6-amino group of lysine from the aldimine linkage (cf. Reaction 2.14).

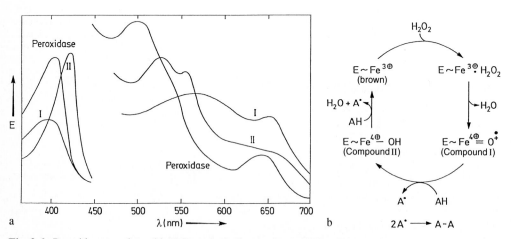

Fig. 2.6. Peroxidase reaction with H_2O_2 and a hydrogen donor (AH). **a** Electron excitation spectra of peroxidase and intermediates I and II; **b** mechanism of catalysis

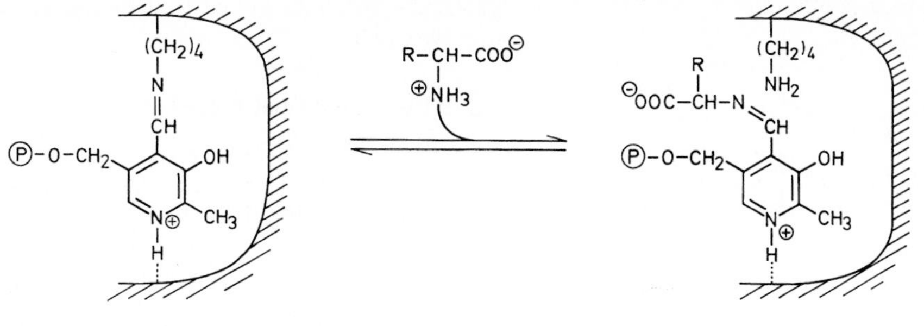

$$ \tag{2.14} $$

The positively charged pyridine ring then exerts an electron shift towards the α-C-atom of the amino acid substrate; the shift being supported by the release of one substituent of the α-C-atom. In Fig. 2.7 is shown how the ionization of the proton attached to the α-C-atom leads to *transamination* of the amino acid with formation of an α-keto acid. The reaction may also proceed through a *decarboxylation* (Fig. 2.7) and yield an amine. Which of these two pathway options will prevail is decided by the structure of the protein moiety of the enzyme.

2.3.3 Metal Ions

Metal ions are indispensable cofactors and stabilizers of the conformation of many enzymes. They are especially effective as cofactors with enzymes converting small molecules. They influence the substrate binding and participate in catalytic reactions in the form of a *Lewis* acid or play the role of an electron carrier. Only the most important ions will be discussed.

2.3.3.1 Magnesium, Calcium and Zinc

Mg^{2+} ions activate some enzymes which hydrolyze phosphoric acid ester bonds (e.g. phosphatases; cf. Table 2.4) or transfer phosphate residues from ATP to a suitable acceptor (e.g. kinases; cf. Table 2.4). In both cases, Mg^{2+} ions act as an electrophilic *Lewis* acid, polarize the P–O-linkage of the phosphate residue of the substrate or cosubstrate and, thus, facilitate a nucleophilic attack (water with hydrolases; ROH in the case of kinases). An example is the hexokinase enzyme (cf.

Table 2.16) which, in glycolysis, is involved in catalyzing the phosphorylation of glucose to glucose-6-phosphate with ATP as cosubstrate. The effect of a Mg^{2+} ion within the enzyme-

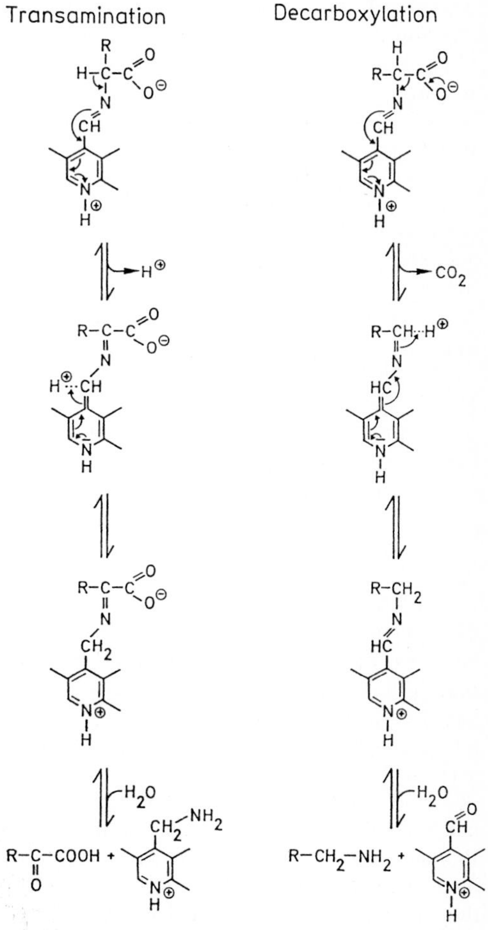

Transamination   Decarboxylation

Fig. 2.7. The role of pyridoxal phosphate in transamination and decarboxylation of amino acids

substrate complex is obvious from the following formulation:

(2.15)

Ca^{2+} ions are weaker *Lewis* acids than Mg^{2+} ions. Therefore, the replacement of Mg^{2+} by Ca^{2+} may result in an inhibition of the kinase enzymes. Enhancement of the activity of other enzymes by Ca^{2+} is based on the ability of the ion to interact with the negatively charged sites of amino acid residues and, thus, to bring about stabilization of the enzyme conformation (e. g. α-amylase; cf. 4.4.5.1.1). The activation of the enzyme may be also caused by the involvement of the Ca^{2+} ion in substrate binding (e.g. lipase; cf. 3.7.1.1).

The Zn^{2+} ion, among the series of transition metals, is a cofactor which is not involved in redox reactions under physiological conditions. As a *Lewis* acid similar in strength to Mg^{2+}, Zn^{2+} participates in similar reactions. Hence, substituting the Zn^{2+} ion for the Mg^{2+} ion in some enzymes is possible without loss of enzyme activity.

Both metal ions can function as stabilizers of enzyme conformation and their direct participation in catalysis is readily revealed in the case of alcohol dehydrogenase. This enzyme isolated from horse liver consists of two identical polypeptide chains, each with one active site. Two of the four Zn^{2+} ions in the enzyme readily dissociate. Although this dissociation has no effect on the quaternary structure, the enzyme activity is lost. As described under section 2.3.1.1, both of these Zn^{2+} ions are involved in the formation of the active site. In catalysis they polarize the substrate's C–O linkage and, thus, facilitate the transfer of hydride ions from or to the cosubstrate. Unlike the dissociable ions, removal of the two residual Zn^{2+} ions is possible only under drastic conditions, namely disruption of the enzyme's quaternary structure which is maintained by these two ions.

2.3.3.2 Iron, Copper and Molybdenum

The redox system of Fe^{3+}/Fe^{2+} covers a wide range of potentials (Table 2.5) depending on the attached ligands. Therefore, the system is exceptionally suitable for bridging large potential differences in a stepwise electron transport system. Such an example is encountered in the transfer of electrons by the cytochromes as members of the respiratory chain (cf. textbook of biochemistry) or in the biosynthesis of unsaturated fatty acids (cf. 3.2.4), and by some individual enzymes.

Iron-containing enzymes are attributed either to the heme (examples in 2.3.2.2) or to the nonheme Fe-containing proteins. The latter case is exemplified by lipoxygenase, for which the mechanism of activity is illustrated in section 3.7.2.2, or by xanthine oxidase.

Xanthine oxidase from milk ($M_r = 275,000$) reacts with many electron donors and acceptors. However, this enzyme is most active with substrates such as xanthine or hypoxanthine as electron donors and molecular oxygen as the electron acceptor. The enzyme is assumed to have two active sites per molecule, with each having 1 FAD moiety, 4 Fe-atoms and 1 Mo-atom. During the oxidation of xanthine to uric acid:

(2.16)

oxygen is reduced by two one-electron steps to H_2O_2 by an electron transfer system in which the following valence changes occur:

(2.17)

Table 2.5. Redox potentials of Fe^{3+}/Fe^{2+} complex compounds at pH 7 (25 °C) as affected by the ligand

Redox-System	E_0' (Volt)
$[Fe^{III}(o\text{-phen}^a)_3]^{3+}/[Fe^{II}(o\text{-phen})_3]^{2+}$	+ 1.10
$[Fe^{III}(OH_2)_6]^{3+}/[Fe^{II}(OH_2)_6]^{2+}$	+ 0.77
$[Fe^{III}(CN)_6]^{3-}/[Fe^{II}(CN)_6]^{4-}$	+ 0.36
Cytochrome a (Fe^{3+})/Cytochrome a (Fe^{2+})	+ 0.29
Cytochrome c (Fe^{3+})/Cytochrome c (Fe^{2+})	+ 0.26
Hemoglobin (Fe^{3+})/Hemoglobin (Fe^{2+})	+ 0.17
Cytochrome b (Fe^{3+})/Cytochrome b (Fe^{2+})	+ 0.04
Myoglobin (Fe^{3+})/Myoglobin (Fe^{2+})	0.00
$(Fe^{III}EDTA)^{1-}/(Fe^{II}EDTA)^{2-}$	− 0.12
$(Fe^{III}(oxin^b)_3)/(Fe^{II}(oxin)_3)^{1-}$	− 0.20
Ferredoxin (Fe^{3+})/Ferredoxin (Fe^{2+})	− 0.40

[a] o-phen: o-Phenanthroline.
[b] oxin: 8-Hydroxyquinoline.

Under certain conditions the enzyme releases a portion of the oxygen when only one electron transfer has been completed. This yields $O_2^{\ominus}$, the superoxide radical anion, with one unpaired electron. This ion can initiate lipid peroxidation by a chain reaction (cf. 3.7.2.1.4); hence, participation of xanthine oxidase in the generation of an "oxidation" flavor in milk is postulated.

Polyphenol oxidases and ascorbic acid oxidase, which occur in food, are known to have a Cu^{2+}/Cu^{1+} redox system as a prosthetic group. Polyphenol oxidases play an important role in the quality of food of plant origin because they cause the "*enzymatic browning*" for example in potatoes, apples and mushrooms. Tyrosinases, catecholases, phenolases or cresolases are enzymes that react with oxygen and a large range of mono and diphenols.

Polyphenol oxidase catalyzes two reactions: first the hydroxylation of a monophenol to o-diphenol (EC 1.14.18.1, monophenol monooxygenase) followed by an oxidation to o-quinone (EC 1.10.3.1, o-diphenol: oxygen oxidoreductase). Both activities are also known as cresolase and catecholase activity.

At its active site, polyphenol oxidase contains two $Cu^{1\oplus}$ ions with two histidine residues each in the ligand field. In an "ordered mechanism" (cf. 2.5.1.2.1) the enzyme first binds oxygen and later monophenol with participation of the intermediates shown in Fig. 2.8. The Cu ions change their valency ($Cu^{1\oplus} \rightarrow Cu^{2\oplus}$). The newly formed complex ([] in Fig. 2.8) has a strongly polarized O−O=bonding, resulting in a hydroxylation to o-diphenol. The cycle closes with the oxidation of o-diphenol to o-quinone.

2.4 Theory of Enzyme Catalysis

It has been illustrated with several examples (Table 2.1) that enzymes are substantially better catalysts than are protons or other ionic species used in nonenzymatic reactions. Enzymes invariably surpass all chemical catalysts in relation to substrate and reaction specificities.

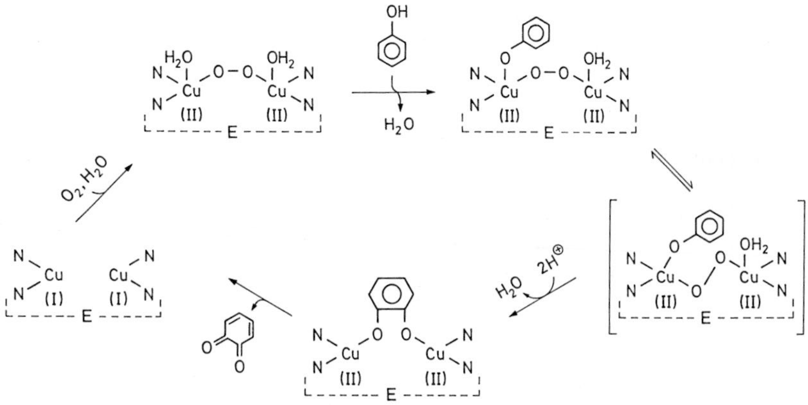

Fig. 2.8. Mechanism of polyphenol oxidase activity

Theories have been developed to explain the exceptional efficiency of enzyme activity. They are based on findings which provide only indirect insight into enzyme catalysis. Examples are the identification of an enzyme's functional groups involved in catalysis, elucidation of their arrangement within the tertiary structure of the enzyme, and the detection of conformational changes induced by substrate binding. Complementary studies involve low molecular weight model substrates, the reactions of which shed light on the active sites or groups of the enzyme and their coordinated interaction with other factors affecting enzymatic catalysis.

2.4.1 Active Site

An enzyme molecule is, when compared to its substrate, often larger in size by a factor of several orders of magnitude. For example, glucose oxidase ($M_r = 1.5 \cdot 10^5$) and glucose ($M_r = 180$). This strongly suggests that in catalysis only a small locus of an active site has direct contact with the substrate. Specific parts of the protein structure participate in the catalytic process from the substrate binding to the product release from the so-called *active site*. These parts are amino acid residues which bind substrate and, if required, cofactors and assist in conversion of substrate to product.

Investigations of the structure and function of the active site are conducted to identify the amino acid residues participating in catalysis, their steric arrangement and mobility, the surrounding micro-environment and the catalysis mechanism.

2.4.1.1 Active Site Localization

Several methods are generally used for the identification of amino acid residues present at the active site since data are often equivocal. Once obtained, the data must still be interpreted with a great deal of caution and insight.

The influence of pH on the activity assay (cf. 2.5.3) provides the first direct answer as to whether dissociable amino acid side chains, in charged or uncharged form, assist in catalysis. The data readily obtained from this assay must again be interpreted cautiously since neighboring charged groups, hydrogen bonds or the hydrophobic environment of the active site can affect the extent of dissociation of the amino acid residues and, thus, can shift their pK values (cf. 1.4.3.1).

Selective labeling of side chains which form the active site is also possible by chemical modification. When an enzyme is incubated with reagents such as iodoacetic acid (cf. 1.2.4.3.5) or dinitrofluorobenzene (cf. 1.2.4.2.2), resulting in a decrease of activity, and subsequent analysis of the modified enzyme shows that only one of the several available functional groups is bound to reagent (e.g. one of several −SH groups), then this group is most probably part of the active site. Selective labeling data when an inhibiting substrate analogue is used are more convincing. Because of its similarity to the chemical structure of the substrate, the analogue will be bound covalently to the enzyme but not converted into product. We will consider the following examples:

N-tosyl-L-phenylalanine ethyl ester (Formula 2.18) is a suitable substrate for the proteinase chymotrypsin which hydrolyzes ester bonds. When the ethoxy group is replaced by a chloromethyl group, an inhibitor whose structure is similar to the substrate is formed (N-tosyl-L-phenylalanine chloromethylketone, TPCK).

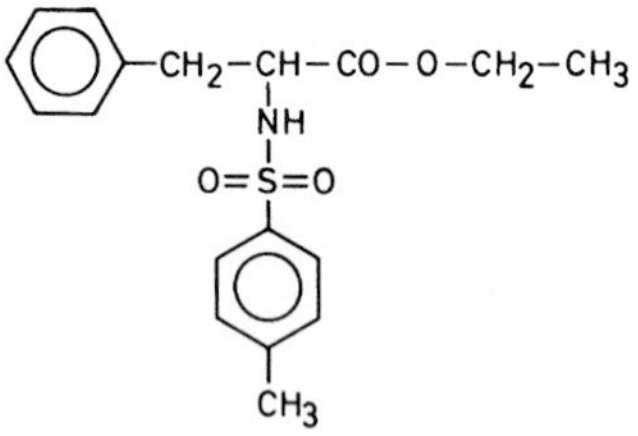

(2.18)

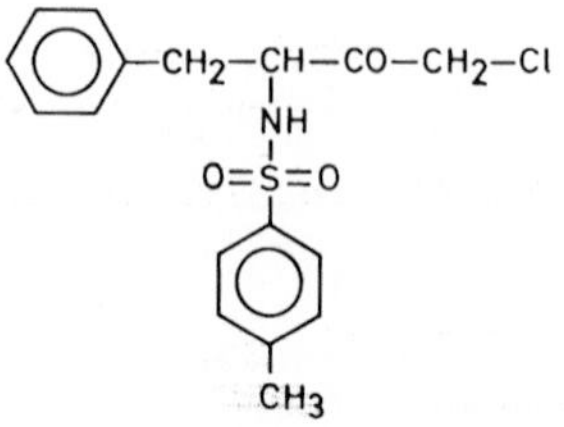

(2.19)

Thus, the substrate analogue binds specifically and irreversibly to the active site of chymotrypsin. Analysis of the enzyme inhibitor complex reveals that, of the two histidine residues present in chymotrypsin, only His[57] is alkylated at one of its ring nitrogens. Hence, the modified His residue is part of the active site (cf. mechanism of chymotrypsin catalysis, Fig. 2.17). TPCK binds highly specifically, thus the proteinase trypsin is not inhibited. The corresponding inhibiting substrate analogue, which binds exclusively to trypsin, is N-tosyl-L-lysine chloromethylketone (TLCK):

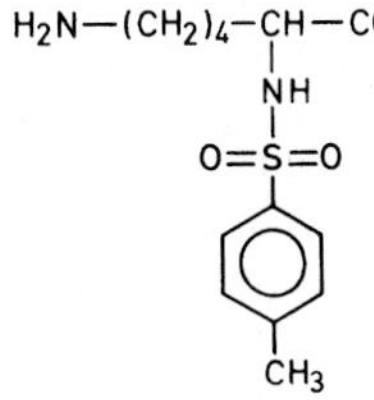

$$(2.20)$$

Reaction of diisopropylfluorophosphate (DIFP)

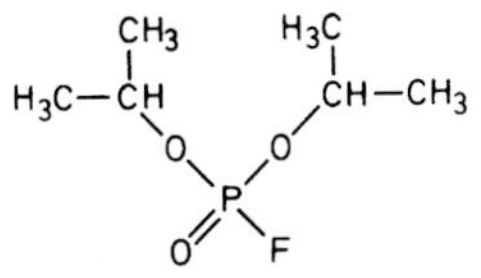

$$(2.21)$$

with a number of proteinases and esterases alkylates the unusually reactive serine residue at the active site. Thus, of the 28 serine residues present in chymotrypsin, only Ser[195] is alkylated, while the other 27 residues are unaffected by the reagent. It appears that the reactivity of Ser[195] is enhanced by its interaction with the neighboring His[57] (cf. mechanism of catalysis in Fig. 2.17). The participation of a carboxyl group at the active site in β-glucosidase catalysis has been confirmed with the help of conduritol B-epoxide, an inhibiting substrate analogue:

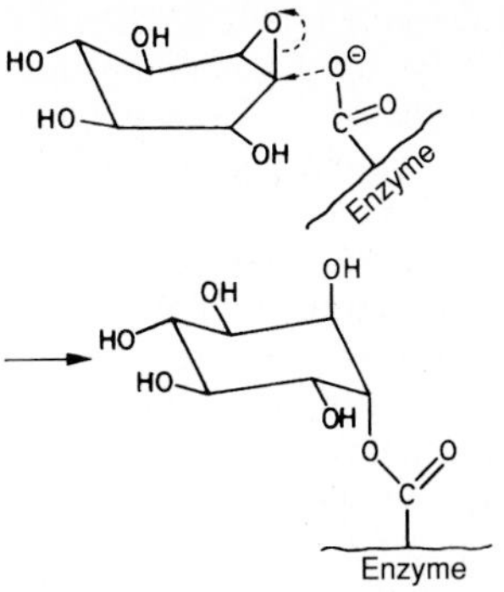

$$(2.22)$$

A lysine residue is involved in enzyme catalysis in a number of lyase enzymes and in enzymes in which pyridoxal phosphate is the cosubstrate. An intermediary *Schiff* base product is formed between an ε-amino group of the enzyme and the substrate or pyridoxal phosphate (cf. 2.3.2.3). The reaction site is then identified by reduction of the *Schiff* base with NaBH$_4$.

An example of a "lysine" lyase is the aldolase enzyme isolated from rabbit muscle. The intermediary product formed with dihydroxyacetone phosphate (cf. mechanism in Fig. 2.19) is detected as follows:

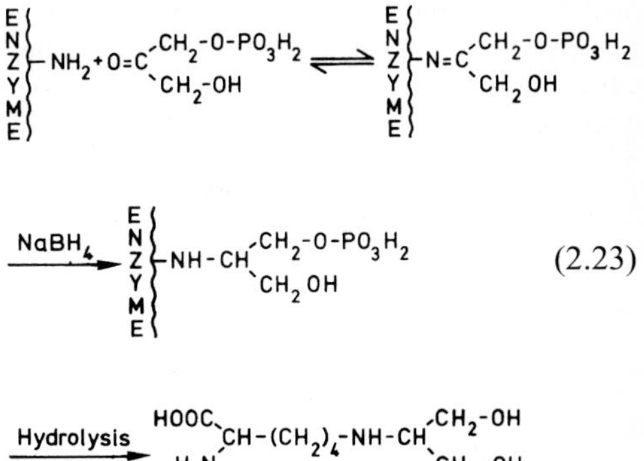

$$(2.23)$$

2.4.1.2 Substrate Binding

2.4.1.2.1 Stereospecificity

Enzymes react stereospecifically. Before being bound to the binding locus, the substrates are distinguished by their cis, trans-isomerism and also by their optical antipodes. The latter property was illustrated by the reactions of L(+)-lactic acid (Fig. 2.2). There are distinct recognition areas on the binding locus. Alcohol dehydrogenase will be used to demonstrate this. This enzyme removes two hydrogen atoms, one from the methylene group and the other from the hydroxyl group, to produce acetaldehyde. However, the enzyme recognizes the difference between the two methylene hydrogens since it always stereospecifically removes the same hydrogen atom. For example, yeast alcohol dehydrogenase always removes the pro-R-hydrogen from the C-1 position of a stereospecifically deuterated substrate and transfers it to the C-4 position of the nicotinamide ring of NAD:

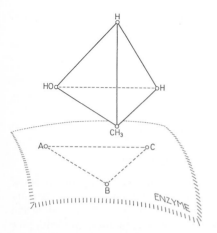

$$(2.24)$$

To explain the stereospecificity, it has been assumed that the enzyme must bind simultaneously to more than one point of the molecule. Thus, when two substituents (e.g. the methyl and hydroxyl groups of ethanol; Fig. 2.9) of the prochiral site are attached to the enzyme surface at positions A and B, the position of the third substituent is fixed. Therefore, the same substituent will always be bound to reactive position C, e.g. one of the two methylene hydrogens in ethanol. In other words, the two equal substituents in a symmetrical molecule are differentiated by asymmetric binding to the enzyme.

2.4.1.2.2 "Lock and Key" Hypothesis

To explain substrate specificity, *E. Fischer* proposed a hypothesis a century ago in which he depicted the substrate as being analogous to a key and the enzyme as its lock. According to this model, the active site has a geometry which is complementary only to its substrate (Fig. 2.10). In contrast, there are many possibilities for a "bad" substrate to be bound to the enzyme, but only one provides the properly positoned enzyme-substrate complex, as illustrated in Fig. 2.10, which is converted to the product.

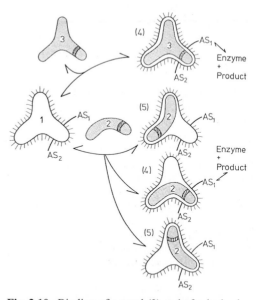

Fig. 2.10. Binding of a good (3) and of a bad substrate (2) by the active site (1) of the enzyme. (according to *W. P. Jencks*, 1969). (4) A productive enzyme-substrate complex; (5) a nonproductive enzyme substrate complex. As$_1$ and As$_2$: reactive amino acid residues of the enzyme involved in conversion of substrate to product

Fig. 2.9. A model for binding of a prochiral substrate (ethanol) by an enzyme

The proteinases chymotrypsin and trypsin are two enzymes for which secondary and tertiary structures have been elucidated by x-ray analysis and which have structures supporting the lock and key hypothesis to a certain extent. The binding site in chymotrypsin and trypsin is a three-dimensional hydrophobic pocket (Fig. 2.11). Bulky amino acid residues such as aromatic amino acids fit neatly into the pocket (chymotrypsin, Fig. 2.11 a), as do substrates with lysyl or arginyl residues (trypsin, Fig. 2.11 b). Instead of Ser[189], the trypsin peptide chain has Asp[189] which is present in the deep cleft in the form of a carboxylate anion and which attracts the positively charged lysyl or arginyl residues of the substrate. Thus, the substrate is stabilized and realigned by its peptide bond to face the enzyme's Ser[195] which participates in hydrolysis (transforming locus).

The peptide substrate is hydrolyzed by the enzyme elastase by the same mechanism as for chymotrypsin. However, here the pocket is closed to such an extent by the side chains of Val[216] and Thr[226] that only the methyl group of alanine can enter the cleft (Fig. 2.11 c). Therefore, elastase has specificity for alanyl peptide bonds or alanyl ester bonds.

2.4.1.2.3 Induced-fit Model

The conformation of a number of enzymes is changed by the binding of the substrate. An example is carboxypeptidase A, in which the Try[248] located in the active site moves approximately 12 Å towards the substrate, glycyl-L-phenylalanine, to establish contact. This and other observations support the dynamic induced-fit model proposed by *Koshland* (1964). Here, only the substrate has the power to induce a change in the tertiary structure to the active form of the enzyme. Thus, as the substrate molecule approaches the enzyme surface, the amino acid residues A and B change their positions to conform closely to the shape of the substrate (I, in Fig. 2.12). Groups A and B are then in the necessary position for reaction with the substrate.

Diagrams II and III (Fig. 2.12) illustrate the case when the added compound is not suitable as substrate. Although group C positioned the substrate correctly at its binding site, the shape of the compound prevents groups A and B from being aligned properly in their active positions and, thus, from generating the product.

In accordance with the mechanisms outlined above, one theory suitable for enzymes following the lock and key mechanism and the

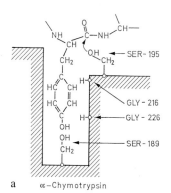

a   α–Chymotrypsin

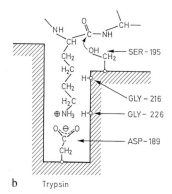

b   Trypsin

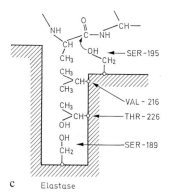

c   Elastase

Fig. 2.11. A hypothesis for substrate binding by α-chymotrypsin, trypsin and elastase enzymes (according to *D. Shotton*, 1971)

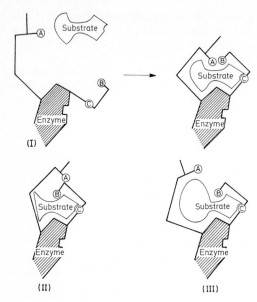

Fig. 2.12. A schematic presentation of "induced-fit model" for an active site of an enzyme (according to *D. E. Koshland*, 1964).
— Polypeptide chain of the enzyme with catalytically active residues of amino acids, A and B; the residue C binds the substrate

other theory for enzymes operating with the dynamic induced-fit model, the substrate specificity of any enzyme-catalyzed reaction can be explained satisfactorily.

In addition, the relationship between enzyme conformation and its catalytic activity thus outlined also accounts for the extreme sensitivity of the enzyme as catalyst. Even slight interferences imposed on their tertiary structure which affect the positioning of the functional groups result in loss of catalytic activity.

2.4.2 Reasons for Catalytic Activity

Even though the rates of enzymatically catalyzed reactions vary, they are very high compared to the effectiveness of chemical catalysts (examples in Table 2.2). The factors responsible for the high increase in reaction rate are outlined below. They are of different importance for the individual enzymes.

2.4.2.1 Steric Effects – Orientation Effects

The specificity of substrate binding contributes substantially to the rate of an enzyme-catalyzed reaction.

Binding to the active site of the enzyme concentrates the reaction partners in comparison with a dilute substrate solution. In addition, the reaction is now the favored one since binding places the substrate's susceptible reactive group in the proximity of the catalytically active group of the enzyme.

Therefore the contribution of substrate binding to the reaction rate is partially due to a change in the molecularity of the reaction. The intermolecular reaction of the two substrates is replaced by an intramolecular reaction of an enzyme-substrate complex. The consequences can be clarified by using model compounds which have all the reactive groups within their molecules and, thus, are subjected to an intramolecular reaction. Their reactivity can then be compared with that of the corresponding bimolecular system and the results expressed as a ratio of the reaction rates of the intramolecular (k_1) to the intermolecular (k_2) reactions. Based on their dimensions, they are denoted as "effective molarity". As an example, let us consider the cleavage of p-bromophenylacetate in the presence of acetate ions, yielding acetic acid anhydride:

$$Br\text{---}\langle\bigcirc\rangle\text{---}O\text{---}CO\text{---}CH_3 + CH_3COO^\ominus + Na^\oplus$$
$$\longrightarrow Br\text{---}\langle\bigcirc\rangle\text{---}O^\ominus Na^\oplus + (CH_3CO)_2O$$

(2.25)

Intramolecular hydrolysis is substantially faster than the intermolecular reaction (Table 2.6). The effective molarity sharply increases when the reactive carboxylate anion is in close proximity to the ester carbonyl group and, by its presence, retards the mobility of the carbonyl group. Thus, the effective molarity increases (Table 2.6) as the C−C bond mobility decreases. Two bonds can rotate in a glutaric acid ester, whereas only one can rotate in a succinic acid ester. The free rotation is effectively blocked in a bicyclic system. Hence, the reaction rate is sharply increased. Here, the rigid steric arrangement of the acetate ion and of the ester group provides a configuration that imitates that of a transition state.

Table 2.6. Relative reaction rate for the formation of acid anhydrides

Model		k_1/k_2 (mol/l)
I.	CH₃—COOR + CH₃—COO⊖	
II.	H₂C⟨ CH₂—COOR / CH₂—COO⊖	$9.5 \cdot 10^2$
III.	H₂C—COOR | H₂C—COO⊖	$2.2 \cdot 10^5$
IV.		$5 \cdot 10^7$

R: Br—⟨O⟩—

In contrast to the examples given in Table 2.6, examples should be mentioned in which substrates are not bound covalently by their enzymes. The following model will demonstrate that other interactions can also promote close positioning of the two reactants.

Hydrolysis of p-nitrophenyldecanoic acid ester is catalyzed by an alkylamine:

$$(2.26)$$

The reaction rate in the presence of decylamine is faster than that in the presence of ethylamine by a factor of 700. This implies that the reactive amino group has been oriented very close to the susceptible carbonyl group of the ester by the establishment of a maximal number of hydrophobic contacts. Correspondingly, there is a decline in the reaction rate as the alkyl amine group is lengthened further.

2.4.2.2 Structural Complementarity to Transition State

It is assumed that the active conformation of the enzyme matches the transition state of the reaction. This is supported by affinity studies which show that a compound with a structure analogous to the transition state of the reaction ("*transition state analogs*") is bound better than the substrate.

Hydroxamic acid, for example, is such a transition state analog which inhibits the reaction of triosephosphate isomerase (Fig. 2.13). Comparisons between the *Michaelis* constant and the inhibitor constant show that the inhibitor has a 30 times higher affinity to the active site than the substrate.

The active site is complementary to the transition state of the reaction to be catalyzed. This assumption is supported by a reversion of the concept. It has been possible to produce catalytically active monoclonal antibodies directed against transition state analogs. The antibodies accelerate the reaction approximating the transition state of the analog. However, their catalytic activity is weaker compared to enzymes because only the environment of the antibody which is complementary to the transition state causes the acceleration of the reaction.

Transition state analog inhibitors were used to show that in the binding the enzyme displaces the hydrate shell of the substrates. The reaction rate can be significantly increased by removing the hydrate shell between the participants.

Other important factors in catalytic reactions are the distortion of bonds and shifting of charges. The substrate's bonds will be strongly polarized by the enzyme, and thus highly reactive, through the precise positioning of an acid or base group or a metall ion (*Lewis* acid, cf. 2.3.3.1) (example see Formula 2.15). These hypotheses are supported by investigations using suitable transition state analog inhibitors.

2.4.2.3 Entropy Effect

An interpretation in thermodynamic terms takes into account that a loss of entropy occurs during catalysis due to the loss of freedom of rotation and translation of the reactants. This

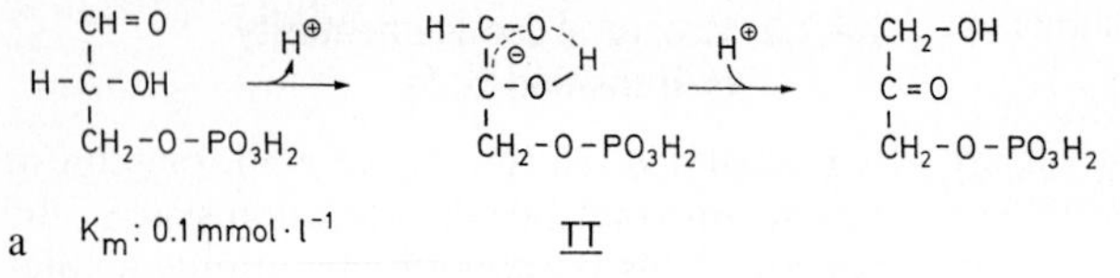

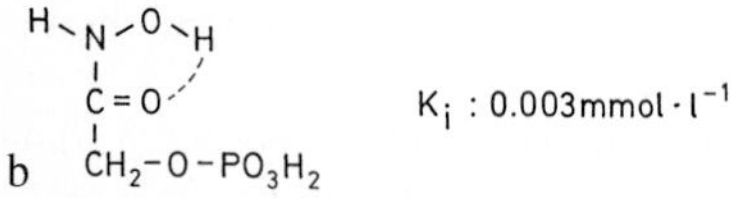

$$CH=O$$
$$|$$
$$H-C-OH \quad \xrightarrow{H^{\oplus}} \quad H-C\!\cdots\!O\!\cdots\!H / C-O \quad \xrightarrow{H^{\oplus}} \quad CH_2-OH$$
$$| \qquad\qquad\qquad | \qquad\qquad\qquad\qquad |$$
$$CH_2-O-PO_3H_2 \qquad\qquad CH_2-O-PO_3H_2 \qquad\qquad C=O$$
$$CH_2-O-PO_3H_2$$

a K_m: 0.1 mmol · l^{-1} TT

$$H\!\cdots\!N\!\!-\!\!O\!\cdots\!H$$
$$|$$
$$C=O$$
$$|$$
b $CH_2-O-PO_3H_2$ K_i : 0.003 mmol · l^{-1}

Fig. 2.13. Example of a transition state analog inhibitor **a** reaction of triosephosphate isomerase, TT: postulated transition state; **b** inhibitor

entropy effect is probably quite large in the case of the formation of an enzyme-substrate complex since the reactants are fairly rigidly positioned before the transition state is reached. Consequently, the conversion of the enzyme-substrate complex to the transition state is accompanied by little or no change of entropy. As an example, a reaction running at 27 °C with a decrease in entropy of 140 J K^{-1} mol^{-1} is considered. Calculations indicate that this decrease leads to a reduction in free activation energy by about 43 kJ. This value falls in the range of the amount by which the activation energy of a reaction is lowered by an enzyme (cf. Table 2.1) and which can have the effect of increasing the reaction rate by a factor of 10^8.

The catalysis by chymotrypsin, for example, shows how powerful the entropy effects can be. In section 2.4.2.5 we will see that this catalysis is a two-step event proceeding through an acylated enzyme intermediate. Here we will consider only the second step, deacylation, thereby distinguishing the following intermediates:

a) N-acetyl-L-tyrosyl-chymotrypsin
b) Acetyl-chymotrypsin.

In case a) deacylation is faster by a factor of 3540 since the carbonyl group is immobilized by insertion of the bulky N-acetyl-L-tyrosyl group into a hydrophobic pocket on the enzyme (Fig. 2.14a) at the correct distance from the attacking nucleophilic OH$^{\ominus}$ ion derived from water (cf. 2.4.2.5). In case b) the immobilization of the small acetyl group is not pos-

sible (Fig. 2.14b) so that the difference between the ground and transition states is very large. The closer the ground state is to the transition state, the more positive will be the entropy of the transition state, $\Delta S^{\#}$; a fact that as mentioned before can lead to a considerable increase in reaction rate. The thermodynamic data in Table 2.7 show that the difference in reaction rates depends, above all, on an entropy effect; the enthalpies of the transition states scarcely differ.

2.4.2.4 General Acid-Base Catalysis

When the reaction rate is affected by the concentration of hydronium ($H_3O^{\oplus}$) or OH$^{\ominus}$ ions from water, the reaction is considered to be specifically acid or base catalyzed. In the so-called general acid or base catalysis the reaction rate is affected by prototropic groups located on the side chains of the amino acid residues. These groups involve proton donors (denoted as general acids) and proton acceptors (general bases). Most of the amino acids located on the active site of the enzyme in-

Table 2.7. Thermodynamic data for transition states of two acyl-chymotrypsins

Acyl-enzyme	$\Delta G^{\#}$ (kJ · mol^{-1})	$\Delta H^{\#}$ (kJ · mol^{-1})	$\Delta S^{\#}$ (J·K^{-1}·mol^{-1})
N-Acetyl-L-tyrosyl	59.6	43.0	− 55.9
Acetyl	85.1	40.5	− 149.7

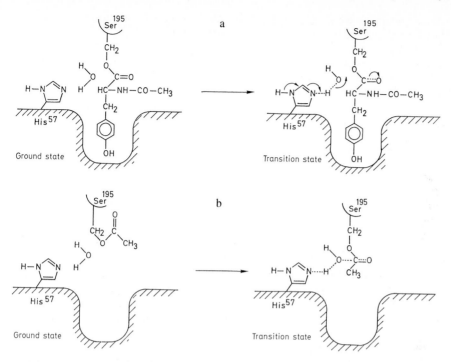

Fig. 2.14. Influence of the steric effect on deacylation of two acyl-chymotrypsins (according to *M.L. Bender* et al., 1964). **a** N-acetyl-L-tyrosyl-chymotrypsin, **b** acetyl-chymotrypsin

fluence the reaction rate by general acid-base catalysis.

As already mentioned, the amino acid residues in enzymes have prototropic groups which have the potential to act as a general acid or as a general base. Of these, the imidazole ring of histidine is of special interest since it can perform both functions simultaneously:

His[12] serves as a general base, removing the proton from a water molecule. This is followed by nucleophilic attack of the intermediary $OH^{\ominus}$ ions on the electrophilic phosphate group. This attack is supported by the concerted action of the general acid His[119].

$$H \overset{\oplus}{+} \underset{\underset{\text{General}}{\text{base}}}{\overset{R}{N}} \underset{\text{acid}}{\overset{R}{N-H}} \rightleftarrows H-N \overset{R}{\underset{\text{General}}{N}} N \cdot + H^{\oplus} \qquad (2.27)$$

The imidazole ring ($pK_2 = 6.1$) can cover the range of the pH optima of many enzymes.

Thus, two histidine residues are involved in the catalytic activity of ribonuclease, a phosphodiesterase. The enzyme hydrolyzes pyrimidine-2′,3′-cyclic phosphoric acids. As shown in Fig. 2.15, cytidine-2′,3′-cyclic phosphoric acid is positioned between two imidazole groups at the binding locus of the acitve site.

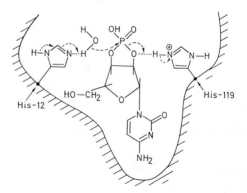

Fig. 2.15. Hydroylysis of cytidine-2′,3′-phosphate by ribonuclease. (reaction mechanism according to *D. Findlay*, 1962)

Another concerted general acid-base catalysis is illustrated by triose phosphate isomerase, an enzyme involved in glycolysis. Here, the concerted action involves the carboxylate anion of a glutamic acid residue as a general base with a general acid which has not yet been identified:

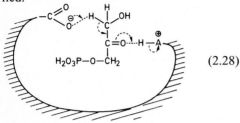

$$(2.28)$$

The endiol formed from dihydroxyacetone-3-phosphate in the presence of enzyme isomerizes into glyceraldehyde-3-phosphate.

These two examples show clearly the significant differences to chemical reactions in solutions. The enzyme driven acid-base catalysis takes place selectively at a certain locus of the active site. The local concentration of the amino acid residue acting as acid or base is fairly high due to the perfect position relative to the substrate. On the other hand, in chemical reactions in solutions all reactive groups of the substrate are nonspecifically attacked by the acid or base.

2.4.2.5 Covalent Catalysis

Studies aimed at identifying the active site of an enzyme (cf. 2.4.1.1) have shown that, during catalysis, a number of enzymes bind the substrate by covalent linkages. Such covalent linked enzyme-substrate complexes form the corresponding products much faster than compared to the reaction rate in a non-catalyzed reaction.

Examples of enzyme functional groups which are involved in covalent bonding and are responsible for the transient intermediates of an enzyme-substrate complex are compiled in Table 2.8. Nucleophilic catalysis is dominant (examples 1–6, Table 2.8), since amino acid residues are present in the active site of these enzymes, which only react with substrate by donating an electron pair (nucleophilic catalysis). Electrophilic reactions occur mostly by involvement of carbonyl groups (example 7, Table 2.8) or with the help of metal ions.

Table 2.8. Examples of covalently linked enzyme-substrate intermediates

Enzyme	Reactive functional group	Intermediate
1. Chymotrypsin	HO-(Serine)	Acylated enzyme
2. Papain	HS-(Cysteine)	Acylated enzyme
3. β-Amylase	HS-(Cysteine)	Maltosyl-enzyme
4. Aldolase	ε-H_2N-(Lysine)	*Schiff* base
5. Alkaline phosphatase	HO-(Serine)	Phosphoenzyme
6. Glucose-6-phosphatase	Imidazole-(Histidine)	Phosphoenzyme
7. Histidine decarboxylase	O=C< (Pyruvate)	*Schiff* base

A number of peptidase and esterase enzymes react covalently in substitution reactions by a two-step nucleophilic mechanism. In the first step, the enzyme is acylated; in the second step, it is deacylated. Chymotrypsin will be discussed as an example of this reaction mechanism. Its activity is dependent on His[57] and Ser[195], which are positioned in close proximity within the active site of the enzyme because of folding of the peptide chain (Fig. 2.16).

Because Asp[102] is located in hydrophobic surroundings, it can polarize the functional groups in close proximity to it. Thus, His[57] acts as a strong general base and abstracts a proton from the OH-group of the neighboring Ser[195] residue (step 'a', Fig. 2.17). The oxygen remaining on Ser[195] thus becomes a strong nucleophile and attacks the carbon of the carbonyl group of the peptide bond of the substrate. At this stage an amine (the first product) is released (step 'b', Fig. 2.17) and the transient covalently-bound acyl enzyme is formed. A deacylation step follows. The previous position of the amine is occupied by a water molecule. Again, His[57], through support from Asp[102], serves as a general base, abstracting the proton from water (step 'c', Fig. 2.17). This is followed by nucleophilic attack of the resultant $OH^\ominus$ ion on the carbon of the carbonyl group of the acyl enzyme (step 'd', Fig. 2.17), resulting in free enzyme and the second product of the enzymic conversion.

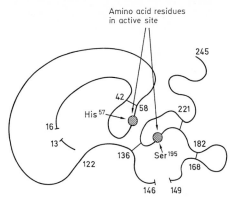

Amino acid residues
in active site

Fig. 2.16. Polypeptide chain conformation in the chymotrypsin molecule (according to *A.L. Lehninger*, 1977)

An exceptionally reactive serine residue has been identified in a great number of hydrolase enzymes, e.g., trypsin, subtilisin, elastase, acetylcholine esterase and some lipases. These enzymes appear to hydrolyze their substrates by a mechanism analogous to that of chymotrypsin. Hydrolases such as papain, ficin and bromelain, which are distributed in plants, have a cysteine residue instead of an "active" serine residue in their active sites. Thus, the transient intermediates are thioesters.

Enzymes involved in the cleavage of carbohydrates can also function by the above mechanism. Figure 2.18 shows that amylose hydrolysis by β-amylase occurs with the help of four functional groups in the active site. The enzyme-substrate complex is subjected to a nucleophilic attack by an SH-group on the carbon involved in the α-glycosidic bond. This transition step is facilitated by the carboxylate anion in the role of a general base and by the imidazole ring as an acid which donates a proton to glycosidic oxygen. In the second transition state the imidazole ring, as a general base in the presence of a water molecule, helps to release maltose from the maltosylenzyme intermediate.

Lysine is another amino acid residue actively involved in covalent enzyme catalysis (cf. 2.4.1.1). Many lyases react covalently with a substrate containing a carbonyl group. They catalyze, for example, aldol or retroaldol condensations important for the conversion and

cleavage of monosaccharides or for decarboxylation reactions of β-keto acids. As an example, the details of the reaction involved will be considered for aldolase (Fig. 2.19). The enzyme-substrate complex is first stabilized by electrostatic interaction between the phosphate residues of the substrate and the charged groups present on the enzyme. A covalent intermediate, a *Schiff* base, is then formed by nucleophilic attack of the ε-amino group of the "active" lysine on a carbonyl group of the substrate. The *Schiff* base cation facilitates the retroaldol cleavage of the substrate, whereas a negatively charged group on the enzyme (e.g. a thiolate or carboxylate anion) acts as a general base, i.e. binds the free proton. Thus, the first product, glyceraldehyde-3-phosphate, is released. An enamine

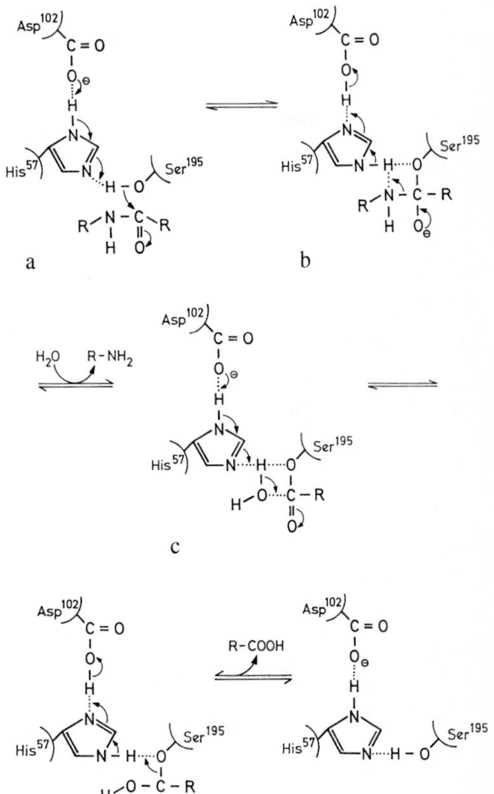

Fig. 2.17. Postulated reaction mechanism for chymotrypsin activity (according to *D.M. Blow* et al., 1969)

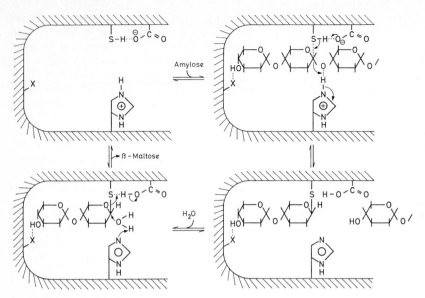

Fig. 2.18. Postulated mechanism for hydrolysis of amylose by β-amylase

rearrangement into a ketimine structure is followed by release of dihydroxyacetone phosphate.

This is the mechanism of catalysis by aldolases which occur in plant and animal tissues (lysine aldolases or class I aldolases). A second group of these enzymes often produced by microorganisms contains a metal ion (metallo-aldolases). This group is involved in accelerating retroaldol condensations through electrophilic reactions with carbonyl groups:

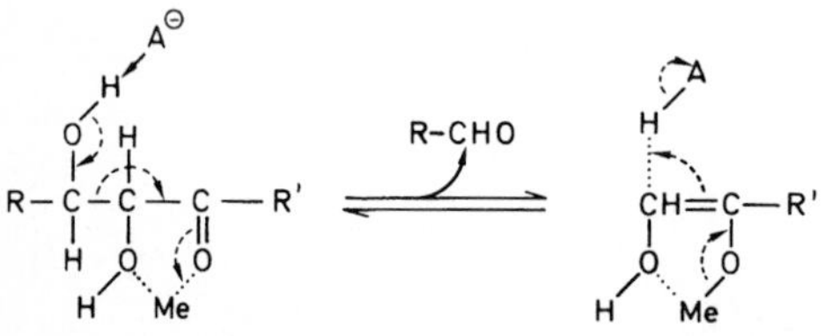

$$R-C-C-C-R' \rightleftharpoons CH=C-R'$$

Me : Metal ion, probably Zn²⊕   (2.29)

Other examples of electrophilic metal catalysis are given under section 2.3.3.1. Electrophilic reactions are also carried out by enzymes which have an α-keto acid (pyruvic acid or α-keto butyric acid) at the transforming locus of the active site. One example of such an enzyme is histidine decarboxylase in which the N-terminal amino acid residue is

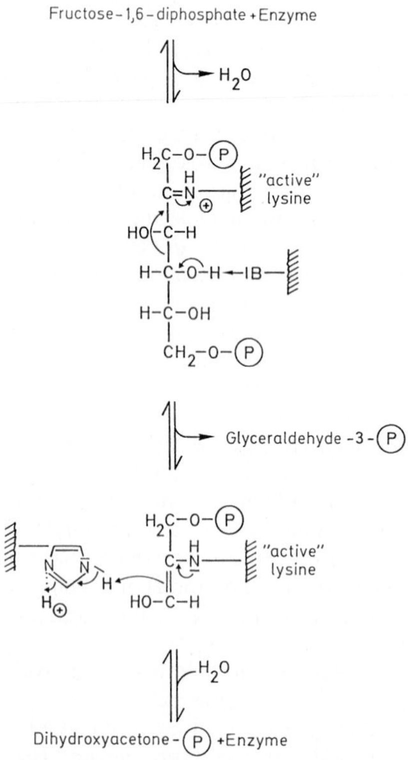

Fructose–1,6–diphosphate + Enzyme

Glyceraldehyde -3- (P)

Dihydroxyacetone - (P) + Enzyme

Fig. 2.19. Aldolase of rabbit muscle tissue. A model for its activity; P: PO₃H₂

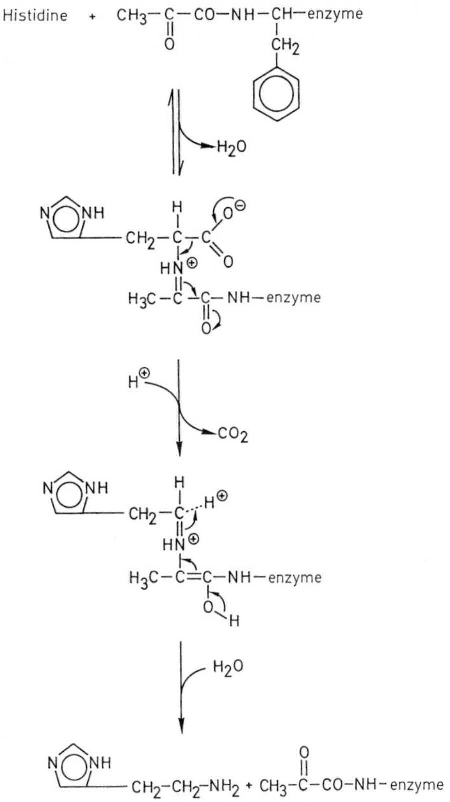

Fig. 2.20. A proposed mechanism for the reaction of histidine decarboxylase

bound to pyruvate. Histidine decarboxylation is initiated by the formation of a *Schiff* base by the reaction mechanism in Fig. 2.20.

2.4.3 Closing Remarks

The hypotheses discussed here allow some understanding of the fundamentals involved in the action of enzymes. However, the knowledge is far from the point where the individual or combined effects which regulate the rates of enzyme-catalyzed reactions can be calculated.

2.5 Kinetics of Enzyme-Catalyzed Reactions

Enzymes in food can be detected only indirectly by measuring their catalytic activity and, in this way, differentiated from other enzymes. This is the rationale for acquiring knowledge needed to analyze the parameters which influence or determine the rate of an enzyme-catalyzed reaction.

The reaction rate is dependent on the concentrations of the components involved in the reaction. Here we mean primarily the substrate and the enzyme. Also, the reaction can be influenced by the presence of activators and inhibitors. Finally, the pH, the ionic strength of the reaction medium, the dielectric constant of the solvent (usually water) and the temperature exert an effect.

2.5.1 Effect of Substrate Concentration

2.5.1.1 Single-Substrate Reactions

2.5.1.1.1 Michaelis-Menten Equation

Let us consider a single-substrate reaction. Enzyme E reacts with substrate A to form an intermediary enzyme-substrate complex, EA. The complex then forms the product P and releases the free enzyme:

$$E + A \underset{k_{-1}}{\overset{k_1}{\rightleftharpoons}} EA \overset{k_2}{\longrightarrow} E + P \qquad (2.30)$$

In order to determine the catalytic activity of the enzyme, the decrease in substrate concentration or the increase in product concentration as a function of time can be measured. The activity curve obtained (Fig. 2.21) has the following regions:

a) The maximum activity which occurs for a few msec until an equilibrium is reached between the rate of enzyme-substrate formation and rate of breakdown of this complex.
Measurements in this pre-steady state region which provide an insight into the reaction steps and mechanism of catalysis are difficult and time consuming. Hence,

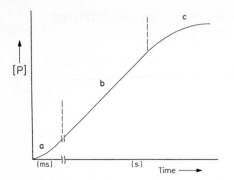

Fig. 2.21. Progress of an enzyme-catalyzed reaction

further analysis of the pre-steady state will be ignored.

b) The usual procedure is to measure the enzyme activity when a steady state has been reached. In the steady state the intermediary complex concentration remains constant while the concentration of the substrate and end product are changing. For this state, the following is valid:

$$\frac{dEA}{dt} = -\frac{dEA}{dt} \qquad (2.31)$$

c) The reaction rate continuously decreases in this region in spite of an excess of substrate. The decrease in the reaction rate can be considered to be a result of:
Enzyme denaturation which can readily occur, continuously decreasing the enzyme concentration in the reaction system, or the product formed increasingly inhibits enzyme activity or, after the concentration of the product increases, the reverse reaction takes place, converting the product back into the initial reactant.

Since such unpredictable effects should be avoided during analysis of enzyme activities, as a rule the initial reaction rate, v_0, is measured as soon as possible after the start of the reaction.

The basics of the kinetic properties of enzymes in the steady state were given by *G. E. Briggs* and *J. B. S. Haldane* (1925) and are supported by earlier mathematical models proposed by *L. Michaelis* and *M. L. Menten* (1913).

The following definitions and assumptions should be introduced in relation to the reaction in Equation 2.30:

$[E_0]$ = total enzyme concentration available at the start of the catalysis.
$[E]$ = concentration of free enzyme not bound to the enzyme-substrate complex, EA, i.e. $[E] = [E_0] - [EA]$.
$[A_0]$ = total substrate concentration available at the start of the reaction. Under these conditions, $[A_0] \gg [E_0]$. Since in catalysis only a small portion of A_0 reacts, the substrate concentration at any time, $[A]$, is approximately equal to $[A_0]$.

When the initial reaction rate, v_0, is considered, the concentration of the product, $[P]$, is 0. Thus, the reaction in Equation 2.30 takes the form:

$$\frac{dP}{dt} = v_0 = k_2 (EA) \qquad (2.32)$$

The concentration of enzyme-substrate complex, $[EA]$, is unknown and can not be determined experimentally for Equation 2.32. Hence, it is calculated as follows:
The rate of formation of EA, according to Equation 2.30, is:

$$\frac{dEA}{dt} = k_1 (E)(A_0) \qquad (2.33)$$

and the rate of EA breakdown is:

$$-\frac{dEA}{dt} = k_{-1} (EA) + k_2 (EA) \qquad (2.34)$$

Under steady-state conditions the rates of breakdown and formation of EA are equal (cf. Equation 2.31):

$$k_1 (E)(A_0) = (k_{-1} + k_2) (EA) \qquad (2.35)$$

Also, the concentration of free enzyme, $[E]$, can not be readily determined experimentally. Hence, free enzyme concentration from the above relationship ($[E] = [E_0] - [EA]$) is substituted in Equation 2.35:

$$k_1 [(E_0) - (EA)] (A_0) = (k_{-1} + k_2)(EA) \qquad (2.36)$$

Solving Equation 2.36 for the concentration of the enzyme-substrate complex, [EA], yields:

$$[EA] = \frac{(E_0)(A_0)}{\frac{k_{-1} + k_2}{k_1} + (A_0)}$$ (2.37)

The quotient of the rate constants in Equation 2.37 can be simplified by defining a new constant, K_m, called the *Michaelis* constant:

$$[EA] = \frac{(E_0)(A_0)}{K_m + (A_0)}$$ (2.38)

Substituting the value of [EA] from Equation 2.38 in Equation 2.32 gives the *Michaelis–Menten* equation for v_0 (initial reaction rate):

$$v_0 = \frac{k_2(E_0)(A_0)}{K_m + (A_0)}$$ (2.39)

Equation 2.39 contains a quantity, $[E_0]$, which can be determined only when the enzyme is present in purified form. In order to be able to make kinetic measurements using impure enzymes, *Michaelis* and *Menten* introduced an approximation for Equation 2.39 as follows.
In the presence of a large excess of substrate, $[A_0] \gg K_m$ in the denominator of Equation 2.39. Therefore, K_m can be neglected compared to $[A_0]$:

$$v_0 = \frac{k_2(E_0)(A_0)}{(A_0)} = V$$ (2.40)

Thus, a zero order reaction rate is obtained. It is characterized by a rate of substrate breakdown or product formation which is independent of substrate concentration, i.e. the reaction rate, V, is dependent only on enzyme concentration. This rate, V is denoted as the maximum velocity.
From Equation 2.40 it is obvious that the catalytic activity of the enzyme must be measured in the presence of a large excess of substrate. To eliminate the $[E_0]$ term, V is introduced into Equation 2.39 to yield:

$$v_0 = \frac{V(A_0)}{K_m + (A_0)}$$ (2.41)

If $[A_0] = K_m$, the following is derived from Equation 2.41:

$$v_0 = \frac{V}{2}$$ (2.42)

Thus, the *Michaelis* constant, K_m, is equal to the substrate concentration at which the reaction rate is half of its maximal value. K_m is independent of enzyme concentration. The lower the value of K_m, the higher the affinity of the enzyme for the substrate, i.e. the substrate will be bound more tightly by the enzyme and most probably will be more efficiently converted to product. Usually, the values of K_m are within the range of 10^{-2} to 10^{-5} mol $\cdot$ l^{-1}.
From the definition of K_m:

$$K_m = \frac{k_{-1} + k_2}{k_1}$$ (2.43)

it follows that K_m approaches the enzyme-substrate dissociation constant, K_s, only if $k_{+2} \ll k_{-1}$.

$$k_2 \ll k_{-1} \cap K_m \approx \frac{k_{-1}}{k_1} = K_s$$ (2.44)

Some values for the constants k_{+1}, k_{-1}, and k_0 are compiled in Table 2.9. In cases in which the catalysis proceeds over more steps than shown in Equation 2.30 the constant k_{+2} is replaced by k_0. The rate constant, k_{+1}, for the formation of the enzyme-substrate complex has values in the order of 10^6 to 10^8: in a few cases it approaches the maximum velocity ($\sim 10^9$ l $\cdot$ mol^{-1} s^{-1}), especially when small molecules of substrate readily diffuse through the solution to the active site of the enzyme. The values for k_{-1} are substantially lower in most cases, whereas k_0 values are in the range of 10^1 to 10^6 s^{-1}.

Table 2.9. Rate constants for some enzyme catalyzed reactions

Enzyme	Substrate	k_1 (l $\cdot$ mol^{-1} s^{-1})	k_{-1} (s^{-1})	k_0 (s^{-1})
Fumarase	Fumarate	$> 10^9$	$4.5 \cdot 10^4$	10^3
Acetylcholine esterase	Acetylcholine	10^9		10^3
Alcohol dehydrogenase (liver)	NAD	$5.3 \cdot 10^5$	74	
	NADH	$1.1 \cdot 10^7$	3.1	
	Ethanol	$> 1.2 \cdot 10^4$	> 74	10^3
Catalase	H_2O_2	$5 \cdot 10^6$		10^7
Peroxidase	H_2O_2	$9 \cdot 10^6$	< 1.4	10^6
Hexokinase	Glucose	$3.7 \cdot 10^6$	$1.5 \cdot 10^3$	10^3
Urease	Urea	$> 5 \cdot 10^6$		10^4

A further extreme case to be considered is if $[A_0] \ll K_m$, which occurs at about $[A_0] < 0.05$ K_m. In that case, $[A_0]$ in the denominator of Equation 2.39 can be neglected:

$$v_0 = \frac{k_2 (E_0)(A_0)}{K_m} \qquad (2.45)$$

and, considering that $k_2 [E_0] = V$, it follows that:

$$v_0 = \frac{V}{K_m} (A_0) \qquad (2.46)$$

In this case the *Michaelis–Menten* equation reflects a first-order reaction in which the rate of substrate breakdown depends on substrate concentration. In using a kinetic method for the determination of substrate concentration (cf. 2.6.1.3), the experimental conditions must be selected such that Equation 2.46 is valid.

2.5.1.1.2 Determination of K_m and V

In order to determine values of K_m and V, the catalytic activity of the enzyme preparation is measured as a function of substrate concentration. Very good results are obtained when $[A_0]$ is in the range of $0.1 K_m$ to $10 K_m$.

A graphical evaluation of the result is obtained by inserting the data into Equation 2.41. As can be seen from a plot of the data in Fig. 2.22, the equation corresponds to a rectangular hyperbola. This graphical approach yields correct values for K_m only when the maximum velocity, V, can be accurately determined.

For a more reliable extrapolation of V, Equation 2.41 is transformed into a straight-line equation. Most frequently, the *Lineweaver–Burk* plot is used which is the reciprocal form of Equation 2.41:

$$\frac{1}{v_0} = \frac{K_m}{V} \cdot \frac{1}{(A_0)} + \frac{1}{V} \qquad (2.47)$$

Figure 2.23 graphically depicts a plot of $1/v_0$ versus $1/[A_0]$. The values V and K_m are obtained from the intercepts of the ordinate ($1/V$) and of the abscissa ($- 1/K_m$), respectively. If the data do not fit a straight line, then the system deviates from the required steady-state kinetics; e.g., there is inhibition by excess substrate or the system is influenced by allosteric effects (cf. 2.5.1.3; allosteric enzymes do not obey *Michaelis–Menten* kinetics).

A disadvantage of the *Lineweaver–Burk* plot is the possibility of departure from a straight line since data taken in the region of saturating substrate concentrations or at low substrate concentrations can be slightly inflated. Thus, values taken from the straight line may be somewhat overestimated.

A procedure which yields a more uniform *distribution* of the data on the straight line is that proposed by *Hofstee* (the *Eadie–Hofstee* plot). In this procedure the *Michaelis-Menten* equation, 2.41, is algebraically rearranged into:

$$v_0 (A_0) + v_0 K_m = V \cdot (A_0) \quad (a)$$

$$v_0 + \frac{v_0}{(A_0)} \cdot K_m = V \quad (b)$$

$$v_0 = - K_m \frac{v_0}{(A_0)} + V \quad (c) \qquad (2.48)$$

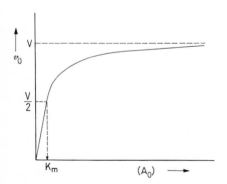

Fig. 2.22. Determination of *Michaelis* constant, K_m, according to equation (2.41)

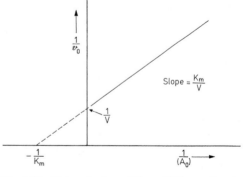

Fig. 2.23. Determination of K_m and V (according to *Lineweaver* and *Burk*)

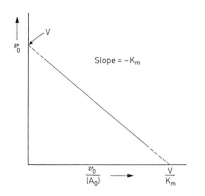

Fig. 2.24. Determination of K_m and V (according to *Hofstee*)

When Equation 2.48c is plotted using the substrate-reaction velocity data, a straight line with a negative slope is obtained (Fig. 2.24) where y is v_0 and x is $v_0/[A_0]$. The y and x intercepts correspond to V and V/K_m, respectively.

Single-substrate reactions, for which the kinetics outlined above (with some exceptions, cf. 2.5.1.3) are particularly pertinent, are those catalyzed by lyase enzymes and certain isomerases. Hydrolysis by hydrolase enzymes can also be considered a single-substrate reaction when the water content remains unchanged, i.e., when it is present in high concentration (55.6 mol/l). Thus, water, as a reactant, can be disregarded.

Characterization of an enzyme-substrate system by determining values for K_m and V is important in enzymatic food analysis (cf. 2.6.4) and for assessment of enzymatic reactions occurring in food (e.g. enzymatic browning of sliced potatoes, cf. 2.5.1.2.1) and for utilization of enzymes in food processing, e.g., aldehyde dehydrogenase (cf. 2.7.2.1.4).

2.5.1.2 Two-Substrate Reactions

For many enzymes, for examples, oxidoreductase and ligase-catalyzed reactions, two or more substrates or cosubstrates are involved.

2.5.1.2.1 *Order of Substrate Binding*

In the reaction of an enzyme with two substrates, the binding of the substrates can occur sequentially in a specific order. Thus, the binding mechanism can be divided into catalysis which proceeds through a ternary adsorption complex (enzyme + two substrates) or through a binary complex (enzyme + one substrate), i.e. when the enzyme binds only one of the two available substrates at a time.

A ternary enzyme-substrate complex can be formed in two ways. The substrates are bound to the enzyme in a random fashion ("random mechanism") or they are bound in a well-defined order ("ordered mechanism").

Let us consider the reaction

$$A + B \underset{}{\overset{E}{\rightleftharpoons}} P + Q \qquad (2.49)$$

If the enzyme reacts by a "random mechanism", substrates A and B form the ternary enzyme-substrate complex, EAB, in a random fashion and the P and Q products dissociate randomly from the ternary enzyme-product complex, EPQ:

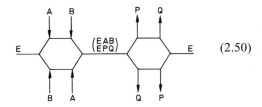

(2.50)

Creatine kinase from muscle (cf. 12.3.6) is an example of an enzyme which reacts by a random mechanism:

Creatine + ATP
$$\rightleftharpoons \text{Creatine-phosphate} + \text{ADP} \qquad (2.51)$$

In an "ordered mechanism" the binding during the catalyzed reaction according to equation 2.49 is as follows:

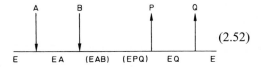

(2.52)

Alcohol dehydrogenase reacts by an "ordered mechanism", although the order of the binding of substrates NAD^+ and ethanol is decided by the ethanol concentration. NAD^+ is absorbed first at low concentrations (< 4 mmol/l):

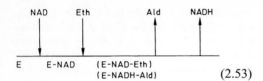

$$(2.53)$$

Eth : Ethanol
Ald : Acetaldehyde

When the ethanol concentration is increased to 7–8 mmol/l, ethanol is absorbed first, followed by the cosubstrate. The order of removal of products (acetaldehyde and NADH) is, however, not altered.

Polyphenol oxidase from potato tubers also reacts by an "ordered mechanism". Oxygen is absorbed first, followed by phenolic substrates. The main substrates are chlorogenic acid and tyrosine. Enzyme affinity for tyrosine is greater and the reaction velocity is higher than for chlorogenic acid. The ratio of chlorogenic acid to tyrosine affects enzymatic browning to such an extent that it is considered to be the major problem in potato processing. The deep-brown colored melanoidins are formed quickly from tyrosine but not from chlorogenic acid. In assessing the processing quality of potato cultivars, the differences in phenol oxidase activity and the content of ascorbic acid in the tubers should also be considered in relation to "enzymatic browning". Ascorbic acid retards formation of melanoidins by its ability to reduce o-quinone, the initial product of enzymatic oxidation (cf. 18.1.2.5.7).

In enzymatic reactions where functional group transfers are involved, as a rule only binary enzyme-substrate complexes are formed by the so-called "ping pong mechanism".

A substrate is adsorbed by enzyme, E, and reacts during alteration of the enzyme (a change in the oxidation state of the prosthetic group, a conformational change, or only a change in covalent binding of a functional group). The modified enzyme, which is denoted F, binds the second substrate and the second reaction occurs, which regenerates the initial enzyme, E, and releases the second product:

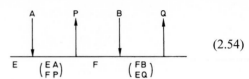

$$(2.54)$$

The glycolytic enzyme hexokinase reacts by a "ping pong mechanism":

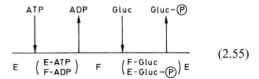

$$(2.55)$$

2.5.1.2.2 Rate Equations for a Two-Substrate Reaction

Here the reaction rate is distinguished by its dependence on two reactants, either two molecules of the same compound or two different compounds. The rate equations can be derived by the same procedures as used for single-substrate catalysis. Only the final forms of the equations will be considered.

When the catalysis proceeds through a ternary enzyme-substrate complex, EAB, the general equation is:

$$v_0 = \frac{V}{1 + \frac{K_a}{(A_0)} + \frac{K_b}{(B_0)} + \frac{K_{ia} \cdot K_b}{(A_0)(B_0)}} \qquad (2.56)$$

When compared to the rate equation for a single-substrate reaction (Equation 2.41), the difference becomes obvious when the equation for a single-substrate reaction is expressed in the following form:

$$v_0 = \frac{V}{1 + \frac{K_a}{(A_0)}} \qquad (2.57)$$

The constants K_a and K_b in Equation 2.56 are defined analogously to K_m, i.e. they yield the concentrations of A or B for $v_0 = V/2$ assuming that, at any given moment, the enzyme is saturated by the other substrate (B or A). Each of the constants, like K_m (cf. Equation 2.43), is composed of several rate constants. K_{ia} is the inhibitor constant for A.

When the binding of one substrate is not influenced by the other, each substrate occupies its own binding locus on the enzyme and the substrates form a ternary enzyme-substrate complex in a defined order ("ordered mechanism"), the following is valid:

$$K_{ia} \cdot K_b = K_a \cdot K_b \qquad (2.58)$$

or from Equation 2.56:

$$v_0 = \frac{V}{1 + \frac{K_a}{(A_0)} + \frac{K_b}{(B_0)} + \frac{K_a \cdot K_b}{(A_0)(B_0)}} \qquad (2.59)$$

However, when only a binary enzyme-substrate complex is formed, i.e. one substrate or one product is bound to the enzyme at a time by a "ping pong mechanism", the denominator term $K_{ia} \cdot K_b$ must be omitted since no ternary complex exists. Thus, Equation 2.56 is simplified to:

$$v_0 = \frac{V}{1 + \frac{K_a}{(A_0)} + \frac{K_b}{(B_0)}} \qquad (2.60)$$

For the determination of rate constants, the initial rate of catalysis is measured as a function of the concentration of substrate B (or A) for several concentrations of A (or B). Evaluation can be done using the *Lineweaver–Burk* plot. Reshaping Equation 2.56 for a "random mechanism" leads to:

$$\frac{1}{v_0} = \left[\frac{K_b}{V} + \frac{K_{ia} \cdot K_b}{(A_0)V} \right] \frac{1}{(B_0)} + \left[1 + \frac{K_a}{(A_0)} \right] \frac{1}{V}$$
$$(2.61)$$

First, $1/v_0$ is plotted against $1/[B_0]$. The corresponding slopes and ordinate intercepts are taken from the straight lines obtained at various values for $[A_0]$ (Fig. 2.25):

$$\text{Slope} = \frac{K_b}{V} + \frac{K_{ia} K_b}{V} \cdot \frac{1}{(A_0)}$$
$$\text{Ordinate intercept} = \frac{1}{V} + \frac{K_a}{V} \cdot \frac{1}{(A_0)} \qquad (2.62)$$

and are then plotted against $1/[A_0]$. In this way two straight lines are obtained (Fig. 2.26a and b), with slopes and ordinate intercepts which provide data for calculating constants K_a, K_b,

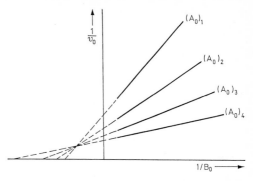

Fig. 2.25. Evaluation of a two-substrate reaction, proceeding through a ternary enzyme-substrate complex (according to *Lineweaver* and *Burk*). $[A_0]_4 > [A_0]_3 > [A_0]_2 > [A_0]_1$

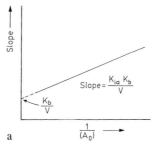

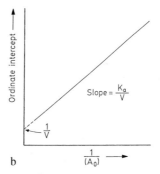

Fig. 2.26. Plotting slopes (a) and ordinate intercepts (b) from Fig. 2.25 versus $1/[A_0]$

K_{ia}, and the maximum velocity, V. If the catalysis proceeds through a "ping pong mechanism", then plotting $1/v_0$ versus $1/[B_0]$ yields a family of parallel lines (Fig. 2.27) which are then subjected to the calculations described above.

A comparison of Figs. 2.25 and 2.27 leads to the conclusion that the dependence of the ini-

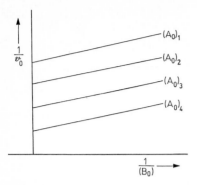

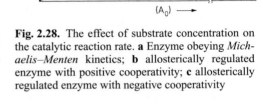

Fig. 2.27. Evaluation of a two-substrate reaction, proceeding through a binary enzyme-substrate complex (according to *Lineweaver* and *Burk*). $[A_0]_4 > [A_0]_3 > [A_0]_2 > [A_0]_1$

Fig. 2.28. The effect of substrate concentration on the catalytic reaction rate. **a** Enzyme obeying *Michaelis–Menten* kinetics; **b** allosterically regulated enzyme with positive cooperativity; **c** allosterically regulated enzyme with negative cooperativity

tial catalysis rate on substrate concentration allows the differentiation between a ternary and a binary enzyme-substrate complex. However, it is not possible to differentiate an "ordered" from a "random" reaction mechanism by this means.

2.5.1.3 Allosteric Enzymes

We are already acquainted with some enzymes consisting of several protomers (cf. Table 1.26). When the protomer activities are independent of each other in catalysis, the *Michaelis–Menten* kinetics, as outlined under sections 2.5.1.1 and 2.5.1.2, are valid. However, when the subunits cooperate, the enzymes deviate from these kinetics. This is particularly true in the case of positive cooperation when the enzyme is activated by the substrate. In this kind of plot, v_0 versus $[A_0]$ yields not a hyperbolic curve but a saturation curve with a sigmoidal shape (Fig. 2.28).

Thus, enzymes which do not obey the *Michaelis–Menten* model of kinetics are allosterically regulated. These enzymes have a site which reversibly binds the allosteric regulator (substrate, cosubstrate or low molecular weight compound) in addition to an active site with a binding and transforming locus. Allosteric enzymes are, as a rule, engaged at control sites of metabolism. An example is tetrameric phos-

phofructokinase, the key enzyme in glycolysis. In glycolysis and alcoholic fermentation it catalyzes the phosphorylation of fructose-6-phosphate to fructose-1,6-diphosphate. The enzyme is activated by its substrate in the presence of ATP. The prior binding of a substrate molecule which enhances the binding of each succeeding substrate molecule is called positive cooperation.

The two enzyme-catalyzed reactions, one which obeys *Michaelis–Menten* kinetics and the other which is regulated by allosteric effects, can be reliably distinguished experimentally by comparing the ratio of the substrate concentration needed to obtain the observed value of 0.9 V to that needed to obtain 0.1 V. This ratio, denoted as R_s, is a measure of the cooperativity of the interaction.

$$R_s = \frac{(A_0)_{0.9V}}{(A_0)_{0.1V}} \tag{2.63}$$

For all enzymes which obey *Michaelis–Menten* kinetics, $R_s = 81$ regardless of the value of K_m or V. The value of R_s is either lower or higher than 81 for allosteric enzymes. $R_s < 81$ is indicative of positive cooperation. Each substrate molecule, often called an effector, accelerates the binding of succeeding substrate molecules, thereby increasing the catalytic activity of the enzyme (case b in Fig. 2.28). When $R_s > 81$, the system shows negative cooperation. The effector (or allosteric inhibitor) decreases the binding of the next substrate molecule (case c in Fig. 2.28).

Various models have been developed in order to explain the allosteric effect. Only the symmetry model proposed by *J. Monod, J. Wyman* and *J.P. Changeux* (1965) will be described in its simplified form: specifically, when the substrate acts as a positive allosteric regulator or effector. Based on this model, the protomers of an allosteric enzyme exist in two conformations, one with a high affinity (R-form) and the other with a low affinity (T-form) for the substrate. These two forms are interconvertible. There is an interaction between protomers. Thus, binding of the allosteric regulator by one protomer induces a conformational change of all the subunits and greatly increases the activity of the enzyme.

Let us assume that the R- and T-forms of an enzyme consisting of four protomers are in an equilibrium which lies completely on the side of the T-form:

$$\underset{\text{(T-Form)}}{\text{⦾}} \ \underset{A}{\overset{A}{\rightleftharpoons}} \ \underset{\text{(R-Form)}}{\boxplus} \quad (2.64)$$

Addition of substrate, which here is synonymous to the allosteric effector, shifts the equilibrium from the low affinity T-form to the substantially more catalytically active R-form. Since one substrate molecule activates four catalytically active sites, the steep rise in enzyme activity after only a slight increase in substrate concentration is not unexpected. In this model it is important that the RT conformation is not permitted. All subunits must be in the same conformational state at one time to conserve the symmetry of the protomers. The equation given by *A.V. Hill* in 1913, derived from the sigmoidal absorption of oxygen by hemoglobin, is also suitable for a quantitative description of allosteric enzymes with sigmoidal behavior:

$$v_0 = \frac{V\,(A_0)^n}{K' + (A_0)^n} \quad (2.65)$$

The equation says that the catalytic rate increases by the nth power of the substrate

concentration when $[A_0]$ is small in comparison to K. The *Hill* coefficient, n, is a measure of the sigmoidal character of the curve and, therefore, of the extent of the enzyme's cooperativity. For n = 1 (Equation 2.65) the reaction rate is transformed into the *Michaelis–Menten* equation, i.e. in which no cooperativity factor exists.

In order to assess the experimental data, Equation 2.65 is rearranged into an equation of a straight line:

$$\log \frac{v_0}{V - v_0} = n \log (A_0) - \log K' \quad (2.66)$$

The slope of the straight line obtained by plotting the substrate concentration as log $[A_0]$ versus log $[v_0/(V-v_0)]$ is the *Hill* coefficient, n (Fig. 2.29). The constant K incorporates all the individual K_m values involved in all the steps of substrate binding and transformation. The value of K_m is obtained by using the substrate concentration, denoted as $[A_0]_{0.5v}$, at which $v_0 = 0.5$ V. Under these conditions, the following is derived from Equation 2.66:

$$\log \frac{0.5\,V}{0.5\,V} = 0 = n \cdot \log (A_0)_{0.5\,V} - \log K' \quad \text{(a)}$$
$$K' = (A_0)_{0.5\,V}^n \quad \text{(b)} \quad (2.67)$$

2.5.2 Effect of Inhibitors

The catalytic activity of an enzyme, in addition to substrate concentration, is affected by the type and concentration of inhibitors, i.e. compounds which decrease the rate of catalysis, and activators, which have the opposite

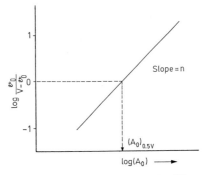

Fig. 2.29. Linear presentation of *Hill's* equation

effect. Metal ions and compounds which are active as prosthetic groups or which provide stabilization of the enzyme's conformation or of the enzyme-substrate complex (cf. 2.3.2 and 2.3.3) are activators. The effect of inhibitors will be discussed in more detail in this section.

Inhibitors are found among food constituents. Proteins which specifically inhibit the activity of certain peptidases (cf. 16.2.3), amylases or β-fructofuranosidase are examples. Furthermore, food contains substances which nonselectively inhibit a wide spectrum of enzymes. Phenolic constituents of food (cf. 18.1.2.5) and mustard oil (cf. 17.1.2.6.5) belong to this group. In addition, food might be contaminated with pesticides, heavy metal ions and other chemicals from a polluted environment (cf. Chapter 9) which can become inhibitors under some circumstances. These possibilities should be taken into account when enzymatic food analysis is performed.

Food is usually heat treated (cf. 2.5.4) to suppress undesired enzymatic reactions. As a rule, no inhibitors are used in food processing. An exception is the addition of, for example, SO_2 to inhibit the activity of phenolase (cf. 8.12.6).

Much data concerning the mechanism of action of enzyme inhibitors have been compiled in recent biochemical research. These data cover the elucidation of the effect of inhibitors on funtional groups of an enzyme, their effect on the active site and the clarification of the general mechanism involved in an enzyme-catalyzed reaction (cf. 2.4.1.1).

Based on kinetic considerations, inhibitors are divided into two groups: inhibitors bound *irreversibly* to enzyme and those bound *reversibly*.

2.5.2.1 Irreversible Inhibition

In an irreversible inhibition the inhibitor binds mostly covalently to the enzyme; the EI complex formed does not dissociate:

$$E + I \xrightarrow{k_1} EI \qquad (2.68)$$

The rate of inhibition depends on the reaction rate constant k_1 in Equation 2.68, the enzyme concentration, [E], and the inhibitor concentration, [I]. Thus, irreversible inhibition is a function of reaction time. The reaction cannot be reversed by diluting the reaction medium. These criteria serve to distinguish irreversible from reversible inhibition.

Examples of irreversible inhibition are the reactions of SH-groups of an enzyme with iodoacetic acid:

$$\text{Enz-SH} + \text{ICH}_2\text{COOH}$$
$$\longrightarrow \text{Enz-S-CH}_2\text{-COOH} + \text{HI} \qquad (2.69)$$

and other reactions with the inhibitors described in section 2.4.1.1.

2.5.2.2 Reversible Inhibition

Reversible inhibition is characterized by an equilibrium between enzyme and inhibitor:

$$E + I \rightleftharpoons EI \qquad (a)$$

$$\frac{(E) \cdot (I)}{(EI)} = K_i \qquad (b) \qquad (2.70)$$

The equilibrium constant or dissociation constant of the enzyme-inhibitor complex, K_i, also known as the inhibitor constant, is a measure of the extent of inhibition. The lower the value of K_i, the higher the affinity of the inhibitor for the enzyme.

Kinetically, three kinds of reversible inhibition can be distinguished: competitive, non-competitive and uncompetitive inhibition (examples in Table 2.10). Other possible cases, such as allosteric inhibition and partial competitive or partial non-competitive inhibition, are omitted in this treatise.

2.5.2.2.1 Competitive Inhibition

Here the inhibitor binds to the active site of the free enzyme, thus preventing the substrate from binding. Hence, there is competition between substrate and inhibitor:

$$E + I \rightleftharpoons EI \ (a) \qquad E + A \rightleftharpoons EA \ (b) \qquad (2.71)$$

According to the steady-state theory for a single-substrate reaction, we have:

$$v_0 = \frac{V(A_0)}{K_m\left(1 + \frac{(I)}{K_i}\right) + (A_0)} \qquad (2.72)$$

Table 2.10. Examples of reversible enzyme inhibition

Enzyme	EC-Number	Sustrate	Inhibitor	Inhibition type[a]	K_i(mmol/l)
Glucose dehydrogenase	1.1.1.47	Glucose/NAD	Glucose-6-phosphate	C	$4.4 \cdot 10^{-5}$
Glucose-6-phosphate dehydrogenase	1.1.1.49	Glucose-6-phosphate/NADP	Phosphate	C	$1 \cdot 10^{-1}$
Succinate dehydrogenase	1.3.99.1	Succinate	Fumarate	C	$1.9 \cdot 10^{-3}$
Creatine kinase	2.7.3.2	Creatine/ATP	ADP	NC	$2 \cdot 10^{-3}$
Glucokinase	2.7.1.2	Glucose/ATP	D-Mannose	C	$1.4 \cdot 10^{-2}$
			2-Deoxyglucose	C	$1.6 \cdot 10^{-2}$
			D-Galactose	C	$6.7 \cdot 10^{-1}$
Fructose-biphosphatase	3.1.3.11	D-Fructose-1,6-biphosphate	AMP	NC	$1.1 \cdot 10^{-4}$
α-Glucosidase	3.2.1.20	p-Nitrophenyl-α-D-glucopyranoside	Saccharose	C	$3.7 \cdot 10^{-2}$
			Turanose	C	$1.1 \cdot 10^{-2}$
Cytochrome c oxidase	1.9.3.1	Ferrocytochrome c	Azide	UC	

[a] C: competitive, NC: noncompetitive, and UC: uncompetitive.

In the presence of inhibitors, the *Michaelis* constant is apparently increased by the factor:

$$1 + \frac{(I)}{K_i} \qquad (2.73)$$

Such an effect can be useful in the case of enzymatic substrate determinations (cf. 2.6.1.3). When inhibitor activity is absent, i.e. [I] = 0, Equation 2.72 is transformed into the *Michaelis–Menten* equation (Equation 2.41). The *Lineweaver–Burk* plot (Fig. 2.30a) shows that the intercept 1/V with the ordinate is the same in the presence and in the absence of the inhibitor, i.e. the value of V is not affected although the slopes of the lines differ. This shows that the inhibitor can be fully dislodged by the substrate from the active site of the enzyme when the substrate is present in high concentration. In other words, inhibition can be overcome at high substrate concentrations (see application in Fig. 2.49). The inhibitor constant, K_i, can be calculated from the corresponding intercepts with the abscissa in Fig. 2.30a by calculating the value of K_m from the abscissa intercept when [I] = 0.

2.5.2.2.2 Non-Competitive Inhibition

The non-competitive inhibitor is not bound to the active site of the enzyme but to some other site. Therefore, the inhibitor can react equally with free enzyme or with enzyme-substrate complex. Thus, three processes occur in parallel:

$$E + A \rightleftharpoons EA \ (a) \quad E + I \rightleftharpoons EI \ (b)$$
$$EA + I \rightleftharpoons EAI \ (c) \qquad (2.74)$$

Postulating that EAI and EI are catalytically inactive and the dissociation constants K_i and K_{EAi} are numerically equal, the following equation is obtained by rearrangement of the equation for a single-substrate reaction into its reciprocal form:

$$\frac{1}{v_0} = \frac{K_m}{V}\left(1 + \frac{(I)}{K_i}\right)\frac{1}{(A_0)} + \frac{1}{V}\left(1 + \frac{(I)}{K_i}\right) \qquad (2.75)$$

The double-reciprocal plot (Fig. 2.30b) shows that, in the presence of a noncompetitive inhibitor; K_m is unchanged whereas the values of V are decreased such that V becomes V/(1 + [I]/K_i), i.e. non-competitive inhibition can not be overcome by high concentrations of substrate.

This also indicates that, in the presence of inhibitor, the amount of enzyme available for catalysis is decreased.

2.5.2.2.3 Uncompetitive Inhibition

In this case the inhibitor reacts only with enzyme-substrate complex:

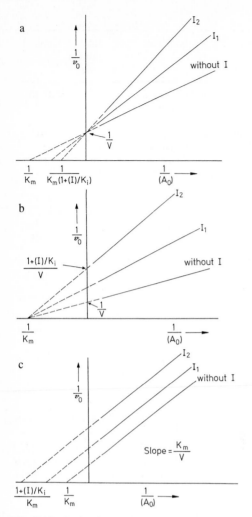

a

b

c

Fig. 2.30. Evaluation of inhibited enzyme-catalyzed reaction according to *Lineweaver* and *Burk*, $[I_1] < (I_2)$. **a** Competitive inhibition, **b** noncompetitive inhibition, **c** uncompetitive inhibition

$$E + A \rightleftharpoons EA \begin{array}{c} \nearrow E + P \\ \searrow \\ EAI \end{array} \qquad (2.76)$$

Rearranging Equation 2.76 into an equation for a straight line, the reaction rate becomes:

$$\frac{1}{v_0} = \frac{K_m}{V} \cdot \frac{1}{(A_0)} + \frac{1}{V} \left(1 + \frac{(I)}{K_i}\right) \qquad (2.77)$$

The double reciprocal plot (Fig. 2.30c) shows that in the presence of an uncompetitive inhibitor, both the maximum velocity, V, and K_m are changed but not the ratio of K_m/V. Hence the slopes of the lines are equal and in the presence of increasing amounts of inhibitor, the lines plotted are parallel. Uncompetitive inhibition is rarely found in single-substrate reactions. It occurs more often in two-substrate reactions.

In conclusion, it can be stated that the three types of reversible inhibition are kinetically distinguishable by plots of reaction rate versus substrate concentration using the procedure developed by *Lineweaver* and *Burk* (Fig. 2.30).

2.5.3 Effect of pH on Enzyme Activity

Each enzyme is catalytically active only in a narrow pH range and, as a rule, each has a pH optimum which is often between pH 5.5 and 7.5 (Table 2.11).

The optimum pH is affected by the type and ionic strength of the buffer used in the assay. The reasons for the sensitivity of the enzyme to changes in pH are two-fold:

a) sensitivity is associated with a change in protein structure leading to irreversible denaturation,

Table 2.11. pH Optima of various enzymes

Enzyme	Source	Substrate	pH Optimum
Pepsin	Stomach	Protein	2
Chymotrypsin	Pancreas	Protein	7.8
Papain	Tropical plants	Protein	7–8
Lipase	Microorganisms	Olive oil	5–8
α-Glucosidase (maltase)	Microorganisms	Maltose	6.6
β-Amylase	Malt	Starch	5.2
β-Fructofuranosidase (invertase)	Tomato	Saccharose	4.5
Pectin lyase	Microorganisms	Pectic acid	9.0–9.2
Xanthine oxidase	Milk	Xanthine	8.3
Lipoxygenase, type I[a]	Soybean	Linoleic acid	9.0
Lipoxygenase, type II[a]	Soybean	Linoleic acid	6.5

[a] See 3.7.2.2.

b) the catalytic activity depends on the quantity of electrostatic charges on the enzyme's active site generated by the prototropic groups of the enzyme (cf. 2.4.2.4).

In addition the ionization of dissociable substrates as affected by pH can be of importance to the reaction rate. However, such effects should be determined separately. Here, only the influences mentioned under b) will be considered with some simplifications.

An enzyme, E, its substrate, A, and the enzyme-substrate complex formed, EA, depending on pH, form the following equilibria:

$$E^{n+1} \underset{}{\overset{-H^{\oplus}}{\rightleftharpoons}} E^{n} \underset{}{\overset{-H^{\oplus}}{\rightleftharpoons}} E^{n-1}$$

$$E^{n+1}A \underset{}{\overset{-H^{\oplus}}{\rightleftharpoons}} E^{n}A \underset{}{\overset{-H^{\oplus}}{\rightleftharpoons}} E^{n-1}A \qquad (2.78)$$

Which of the charged states of E and EA are involved in catalysis can be determined by following the effect of pH on V and K_m.

a) Plotting K_m versus pH reveals the type of prototropic groups involved in substrate binding and/or maintaining the conformation of the enzyme. The results of such a plot, as a rule, resemble one of the four diagrams shown in Fig. 2.31.
Figure 2.31a: K_m is independent of pH in the range of $4-9$. This means that the forms

E^{n+1}, E^{n}, and E^{n-1}, i.e. enzyme forms which are neutral, positively or negatively charged on the active site, can bind substrate.
Figures 2.31b and c: K_m is dependent on one prototropic group, the pK value of which is below (Fig. 2.31b) or above (Fig. 2.31c) neutrality. In the former case, E^{n} and E^{n-1} are the active forms, while in the latter, E^{n+1} and E^{n} are the active enzyme forms in substrate binding.
Figure 2.31d: K_m is dependent on two prototropic groups; the active form in substrate binding is E^{n}.

b) The involvement of prototropic groups in the conversion of an enzyme-substrate complex into product occurs when the enzyme is saturated with substrate, i.e. when equation 2.40 which defines V is valid ($[A_0] \gg K_m$). Thus, a plot of V versus pH provides essentially the same four possibilities presented in Fig. 2.31, the difference being that, here, the prototropic groups of EA, which are involved in the conversion to product, are revealed.

In order to better understand the form of the enzyme involved in catalysis, a hypothetical enzyme-substrate system will be assayed and interpreted. We will start from the assumption that data are available for v_0 (initial velocity) as a function of substrate concentration at several pH's, e.g., for the *Lineweaver* and *Burk*. The values for K_m and V are obtained from the family of straight lines (Fig. 2.32) and plotted against

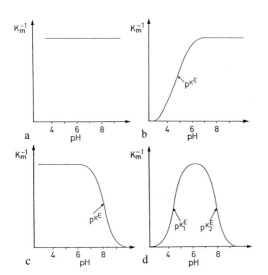

Fig. 2.31. The possible effects of pH on the *Michaelis* constant, K_m

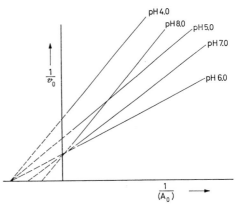

Fig. 2.32. Determination of V and K_m at different pH values

pH. The diagram of $K_m^{-1} = f(pH)$ depicted in Fig. 2.33a corresponds to Fig. 2.31c which implies that neutral (E^n) and positively charged (E^{n+1}) enzyme forms are active in binding the substrate.

Figure 2.33b: V is dependent on one prototropic group, the pK value of which is below neutrality. Therefore, of the two enzyme-substrate complexes, E^{n+1} A and E^n A, present in the equilibrium state, only the latter complex is involved in the conversion of A to the product.

In the example given above, the overall effect of pH on enzyme catalysis can be illustrated as follows:

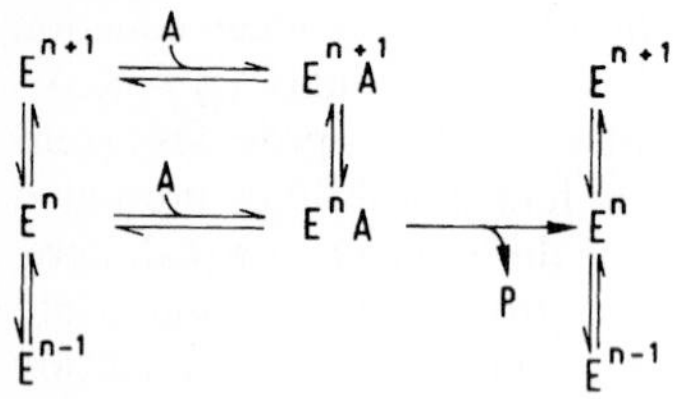

$$
\begin{array}{ccccc}
E^{n+1} & \underset{\longleftarrow}{\overset{A}{\rightleftharpoons}} & E^{n+1}A & & E^{n+1} \\
\updownarrow & & \updownarrow & & \updownarrow \\
E^n & \underset{\longleftarrow}{\overset{A}{\rightleftharpoons}} & E^nA & \longrightarrow & E^n \\
\updownarrow & & \updownarrow & \searrow & \updownarrow \\
E^{n-1} & & E^{n-1} & P & E^{n-1}
\end{array}
\qquad (2.79)
$$

This schematic presentation is also in agreement with the diagram of $V/K_m = f(pH)$ (Fig. 2.33c) which reveals that, overall, two prototropic groups are involved in the enzyme-catalyzed reaction.

An accurate determination of the pK values of prototropic groups involved in enzyme-cataly-

zed reactions is possible using other assays (cf. *J.R. Whitaker*, 1972). However, identification of these groups solely on the basis of pK values is not possible since the pK value is often strongly influenced by surrounding groups. Pertinent to this claim is our recollection that the pH of acetic acid in water is 4.75, whereas in 80% acetone it is about 7. Therefore, the enzyme activity data as related to pH have to be considered only as preliminary data which must be supported and verified by supplementary investigations.

2.5.4 Influence of Temperature

Thermal processes are important factors in the processing and storage of food because they allow the control of chemical, enzymatic and microbial changes. Undesired changes can be delayed or stopped by refrigerated storage. Heat treatment may either accelerate desirable chemical or enzymatic reactions or inhibit undesirable changes by inactivation of enzymes or microorganisms. Table 2.12 informs about quality deterioration caused by enzymes which can be eliminated e.g., by thermal inactivation.

Temperature and time are two parameters responsible for the effects of a thermal treatment. They should be selected carefully to make sure that all necessary changes, e.g., killing of pathogens, are guaranteed, but still all undesir-

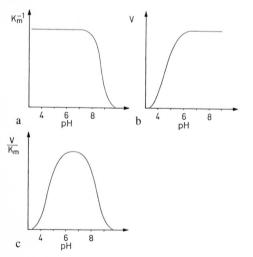

Fig. 2.33. Evaluation of K_m and V versus pH for a hypothetical case

Table 2.12. Thermal inactivation of enzymes to prevent deterioration of food quality

Food product	Enzyme	Quality loss
Potato products, apple products	Monophenol oxidase	Enzymatic browning
Semi-ripe peas	Lipoxygenase, peroxidase	Flavor defects; bleaching
Fish products	Proteinase, thiaminase	Texture (liquefaction), loss of vitamine B_1
Tomato purée	Polygalacturonase	Texture (liquefaction)
Apricot products	β-Glucosidase	Color defects
Oat flakes	Lipase, lipoxygenase	Flavor defects (bitter taste)
Broccoli Cauliflower	Cystathionine β-Lyase (cystine-lyase)	Off-flavor

ed changes such as degradation of vitamins are kept as low as possible.

2.5.4.1 Time Dependence of Effects

The reaction rates for different types of enzymatic reactions have been discussed in section 2.5.1. The inactivation of enzymes and the killing of microorganisms can be depicted as a reaction of 1st order:

$$c_t = c_0 e^{-kt} \qquad (2.80)$$

with c_0 and c_t = concentrations (activities, germ counts) at times 0 and t, and k = rate constant for the reaction. For c_t and t follows from equation 2.80:

$$\log c_t = -\frac{k}{2,3} \cdot t + \log c_0 \qquad (2.81)$$

$$t = \frac{2,3}{k} \log \frac{c_0}{c_t} \qquad (2.82)$$

$c_0/c_t = 10$ gives:

$$t = \frac{2,3}{k} = D \qquad (2.83)$$

The co-called "D-value" represents the time needed to reduce the initial concentration (activity, germ count) by one power of ten. It refers to a certain temperature which has to be stated in each case. For example: *Bacillus cereus* $D_{121\,°C}$ = 2.3 s, *Clostridium botulinum* $D_{121\,°C}$ = 12.25 s. For a heat treatment process, the D-value allows the easy determination of the holding time required to reduce the germ count to a certain level. If the germ count of *B. cereus* or *Cl. botulinum* in a certain food should be reduced by seven powers of ten, the required holding times are $2.3 \times 7 = 16.1$ s and $12.25 \times 7 = 85.8$ s.

2.5.4.2 Temperature Dependence of Effects

A relationship exists for the dependence of reaction rate on temperature. It is expressed by an equation of *Arrhenius*:

$$k = A \cdot e^{-E_a/RT} \qquad (2.84)$$

with k = rate constant for the reaction rate, E_a = activation energy, R = general gas constant

and A = Arrhenius factor. For the relationship between k and T, the *Arrhenius* equation is only an approximation. According to the theory of the transition state (cf. 2.2.1), A is transferred via the active state $A^{\neq}$ into P. A and $A^{\neq}$ are in equilibrium.

$$A \underset{k_{-1}}{\overset{k_1}{\rightleftharpoons}} A^{\neq} \longrightarrow P \qquad (2.85)$$

For the reaction rate follows:

$$k = M \cdot \frac{A}{A^{\neq}} = M \cdot \frac{k_1}{k_{-1}} = M \cdot K^{\neq} \qquad (2.86)$$

with

$$M = \frac{k_B \cdot T}{h} = \frac{R \cdot T}{N_A \cdot h} \qquad (2.87)$$

($K^{\neq}$ equilibrium constant, k_B *Boltzmann* constant, h: *Planck* constant, N_A: *Avogadro* number).

For the equilibrium constant follows:

$$K^{\neq} = e^{-\Delta G^{\neq}/RT} \qquad (2.88)$$

Resulting for the equilibrium constant in:

$$k = \frac{k_B \cdot T}{h} e^{-\Delta G^{\neq}/RT} \qquad (2.89)$$

and for the free activation enthalpy:

$$\Delta G^{\neq} = -RT \ln \frac{k \cdot h}{k_B \cdot T} \qquad (2.90)$$

If k is known for any temperature, $\Delta G^{\neq}$ can be calculated according to equation 2.90. Furthermore, the following is valid:

$$\Delta G^{\neq} = \Delta H^{\neq} - T\Delta S^{\neq} \qquad (2.91)$$

A combination with equation 2.90 results in:

$$-RT \ln \frac{k \cdot h}{k_B \cdot T} = \Delta H^{\neq} - T\Delta S^{\neq} \qquad (2.92)$$

and

$$\log \frac{k}{T} = -\log \frac{h}{k_B} - \frac{\Delta H^{\neq}}{2.3\,RT} + \frac{T\Delta S^{\neq}}{2.3\,R} \qquad (2.93)$$

It is possible to determine $\Delta H^{\neq}$ graphically based on the above equation if k is known for several temperatures and log k/T is plotted against 1/T. If $\Delta G^{\neq}$ and $\Delta H^{\neq}$ are known, $\Delta S^{\neq}$ can be calculated from equation 2.91.

The activation entropy is contained in the *Arrhenius* factor A as can be seen by comparing the empirical *Arrhenius* equation 2.84 with equation 2.89 which is based on the transition state hypothesis:

$$k = A \cdot e^{-E_a/RT} \qquad (2.94\,a)$$

$$k = \frac{k_B}{h} \cdot e^{-\Delta S^{\neq}/R} \cdot T \cdot e^{-\Delta H^{\neq}/RT} \qquad (2.94\,b)$$

Activation energy E_a and activation enthalpy $\Delta H^{\neq}$ are linked with each other as follows:

$$\frac{d \ln k}{dT} = \frac{E_a}{RT^2} \qquad (2.95)$$

$$\frac{d \ln k}{dT} = \frac{1}{T} + \frac{\Delta H^{\neq}}{RT^2} = \frac{RT + \Delta H^{\neq}}{RT^2} \qquad (2.96)$$

$$E_a = \Delta H^{\neq} + RT \qquad (2.97)$$

Using plots of log k against 1/T, the activation energy of the *Arrhenius* equation can be determined. For enzyme catalyzed reactions, E_a is 10–60, for chemical reactions this value is 50–150 and for the inactivation of enzymes, the unfolding of proteins, and the killing of microorganisms, 250–350 kJ/mol are required.

For enzymes which are able to convert more than one substrate or compound into product, the activation energy may be dependent on the substrate. One example is alcohol dehydrogenase, an important enzyme for aroma formation in semiripened peas (Table 2.13). In this case the activation energy for the reverse reaction is only slightly influenced by substrate.

Table 2.13. Alcohol dehydrogenase from pea seeds: activation energy of alcohol dehydrogenation and aldehyde reduction

Alcohol	E_a (kJ·mole^{-1})	Aldehyde	E_a (kJ·mole^{-1})
Ethanol	20		
n-Propanol	37	n-Propanal	20
2-Propenol	18		
n-Butanol	40	n-Butanal	21
n-Hexanol	37	n-Hexanal	18
2-trans-hexenol	15	2-trans-Hexenal	19
		2-trans-Heptenal	18

Under consideration of the temperature dependence of the rate constant k in equation 2.80, the implementation of the expression from *Arrhenius* equation 2.84 leads to:

$$c_t = c_o \cdot e^{-k_o \cdot t \cdot e^{-E_a/RT}} \qquad (2.98)$$

For a constant effect follows:

$$\frac{c_t}{c_o} = const. = e^{-k_o \cdot t \cdot e^{-E_a/RT}} \qquad (2.99)$$

and

$$\ln t = \frac{E_a}{RT} + const. \qquad (2.100)$$

When plotting ln t against 1/T, a family of parallel lines results for each of different activation energies E_a with each line from a family corresponding to a constant effect c_t/c_0 (cf. equation 2.99) (Fig. 2.34).

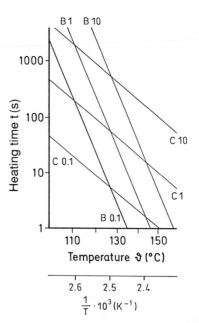

Fig. 2.34. Lines of equal microbiological and chemical effects for heat-treated milk (lines B10, B1, and B0.1 correspond to a reduction in thermophilic spores by 90, 9, and 1 power of ten compared to the initial load; lines C10, C1, and C0.1 correspond to a thiamine degradation of 30%, 3%, and 0.3%; according to *Kessler*, 1988)

For very narrow temperature ranges, sometimes a diagram representing log t against temperature δ (in °C) is favourable. It corresponds to:

$$\log\frac{t}{t_B} = -\frac{E_a}{2.3\,R\cdot T_B\cdot T}(\vartheta - \vartheta_B) = \frac{1}{z}(\vartheta - \vartheta_B)$$

$$(2.101)$$

with t_B as reference time and T_B or δ_B as reference temperature in K respectively °C. For log t/t_B the following is valid:

$$z = \frac{2.3\,R\cdot T_B\cdot T}{E_a}\qquad(2.102)$$

This z-value, used in practice, states the temperature increase in °C required to achieve a certain effect in only one tenth of the time usually needed at the reference temperature. However, due to the temperature dependence of the z-value (equation 2.101), linearity can be expected for a very narrow temperature range only. A plot according to equation 2.100 is therefore more favourable.

In the literature, the effect of thermal processes is often described by the Q_{10} value. It refers to the ratio between the rates of a reaction at temperatures $\delta + 10$ (°C) and δ (°C):

$$Q_{10} = \frac{k_{\vartheta + 10}}{k_\vartheta} = \frac{t_\vartheta}{t_{\vartheta + 10}}\qquad(2.103)$$

The combination of equations 2.101 and 2.103 shows the relationship between the Q_{10} value and z-value:

$$\frac{\log Q_{10}}{10} = \frac{E_a}{2.3\,RT^2} = \frac{1}{z}\qquad(2.104)$$

2.5.4.3 Temperature Optimum

Contrary to common chemical reactions, enzyme-catalyzed reactions as well as growth of microorganisms show a so-called temperature optimum, which is a temperature-dependent maximum resulting from the overlapping of two counter effects with significantly different activation energies (cf. 2.5.4.2):

- increase in reaction or growth rate
- increase in inactivation or killing rate

For starch hydrolysis by microbial α-amylase, the following activation energies, which lie

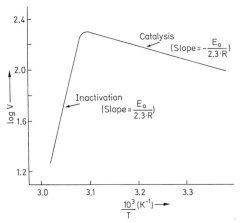

Fig. 2.35. Fungal α-amylase. Amylose hydrolysis versus temperature. *Arrhenius* diagram for assessing the activation energy of enzyme catalysis and enzyme inactivation; V = total reaction rate

between the limits stated in section 2.5.4.2, were derived from e.g. the *Arrhenius* diagram (Fig. 2.35):

- E_a (hydrolysis) = 20 kJ · mol^{-1}
- E_a (inactivation) = 295 kJ · mol^{-1}

As a consequence of the difference in activation energies, the rate of enzyme inactivation is substantially faster with increasing temperature than the rate of enzyme catalysis. Based on activation energies for the above example, the following relative rates are obtained (Table 2.14). Increasing δ from 0 to 60°C increases the hydrolysis rate by a factor of 5, while the

Table 2.14. α-Amylase activity as affected by temperature: relative rates of hydrolysis and enzyme inactivation

Temperature (°C)	Relative rate[a]	
	hydrolysis	inactivation
0	1.0	1.0
10	1.35	$1.0 \cdot 10^2$
20	1.8	$0.7 \cdot 10^4$
40	3.0	$1.8 \cdot 10^7$
60	4.8	$1.5 \cdot 10^{10}$

[a] Activation energies of 20 kJ · mole^{-1} for hydrolysis and 295 kJ · mole^{-1} for enzyme inactivation were used for calculation according to *J.R. Whitaker* (1972).

rate of inactivation is accelerated by more than 10 powers of ten.

The growth of microorganisms follows a similar temperature dependence and can also be depicted according to the *Arrhenius* equation (Fig. 2.36) by replacing the value k by the growth rate and assuming E_a is the reference value μ of the temperature for growth.

For maintaining food quality, detailed knowledge of the relationship between microbial growth rate and temperature is important for optimum production processes (heating, cooling, freezing).

The highly differing activation energies for killing microorganisms and for normal chemical reactions have triggered a trend in food technology towards the use of high-temperature short-time (HTST) processes in production. These are based on the findings that at higher temperatures the desired killing rate of microorganisms is higher than the occurrence of undesired chemical reactions.

2.5.4.4 Thermal Stability

The thermal stability of enzymes is quite variable. Some enzymes lose their catalytic activity at lower temperatures, while others are capable of withstanding – at least for a short period of time – a stronger thermal treatment. In a few cases enzyme stability is lower at low temperatures than in the medium temperature range.

Lipase and alkaline phosphatase in milk are thermolabile (Fig. 2.37), whereas acid phosphatase is relatively stable. Therefore, alkaline phosphatase is used to distinguish raw from pasteurized milk because its activity is easier to determine than that of lipase. Of all the enzymes in the potato tuber (Fig. 2.38), peroxidase is the last one to be thermally inactivated. Such inactivation patterns are often found among enzymes in vegetables. In such cases, peroxidase is a suitable indicator for controlling the total inactivation of all the enzymes e. g., in assessing the adequacy of a blanching process. However, newer developments aim to limit the enzyme inactivation to such enzymes responsible for quality deterioration during storage. For example semiripened pea seeds in which lipoxygenase is responsible for spoilage. However, lipoxygenase is more sensitive than peroxidase, thus a sufficient but gentle blanching requires the inactivation of lipoxygenase only. Inactivation of peroxidase is not necessary.

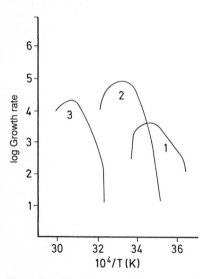

Fig. 2.36. Growth rate and temperature for 1) psychrophilic (*Vibrio AF-1*), 2) mesophilic (*E. coli K-12*) and 3) thermophilic (*Bacillus cereus*) microorganisms (according to *Herbert*, 1989)

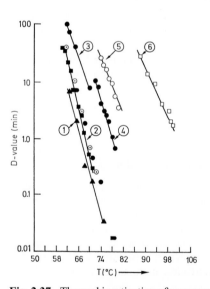

Fig. 2.37. Thermal inactivation of enzymes of milk. 1 Lipase (inactivation extent, 90%), 2 alkaline phosphatase (90%), 3 catalase (80%), 4 xanthine oxidase (90%), 5 peroxidase (90%), and 6 acid phosphatase (99%)

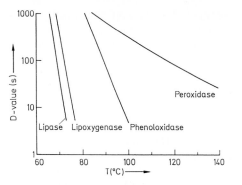

Fig. 2.38. Thermal inactivation (90%) of enzymes present in potato tuber

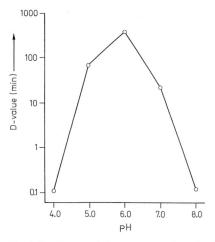

Fig. 2.39. Pea seed lipoxygenase. Inactivation extent at 65 °C as affected by pH

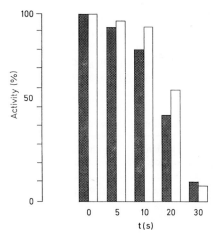

Fig. 2.40. Blanching of semiripened peas at 95 °C; lipoxygenase inactivation (according to *S. Svensson*, 1977). ■ Experimentally found, □ calculated

All the changes which occur in proteins outlined in section 1.4.2.4 also occur during the heating of enzymes. It the case of enzymes the consequences are even more readily observed since a slight conformational change at the active site can result in total loss of activity.

The inactivation or killing rates for enzymes and microorganisms depend on several factors. Most significant is the pH. Lipoxygenase isolated from pea seeds (Fig. 2.39) denatures most slowly at its isoelectric point (pH 5.9) as do many other enzymes.

Table 2.21 contains a list of technically useful proteinases and their thermal stability. However, these data were determined using isolated enzymes. They may not be transferrable to the same enzymes in food because in its natural environment an enzyme usually is much more stable. In additional studies, mostly related to heat transfer in food, some successful procedures to calculate the degree of enzyme inactivation based on thermal stabilty data of isolated enzymes have been developed. An example for the agreement between calculated and experimental results is presented in Fig. 2.40.

Peroxidase activity can partially reappear during storage of vegetables previously subjected to a blanching process to inactivate enzymes. The reason for this recurrence, which is also observed for alkaline phosphatase of milk, is not known yet.

Enzymes behave differently below the freezing point. Changes in activity depend on the type of enzyme and on a number of other factors which are partly contrary. The activity is positively influenced by increasing the concentration of enzyme and substrate due to formation of ice crystals. A positive or negative change might be caused by changes in pH. Viscosity increase of the medium results in negative changes because the diffusion of the substrate is restricted. In completely frozen food (T < phase transition temperature T'_g, cf. 0.3.3 and Table 0.8), a state reached only during deep-freezing, the catalytic activity stops temporarily. Relatively few enzymes are irreversibly destroyed by freezing.

2.5.5 Influence of Water

Up to a certain extent, enzymes need to be hydrated in order to develop activity. Hydration of e.g. lysozyme was determined by IR and NMR spectroscopy. As can be seen in Table 2.15, first the charged polar groups of the side chains hydrate, followed by the uncharged ones. Enzymatic activity starts at a water content of 0.2 g/g protein, which means even before a monomolecular layer of the polar groups with water has taken place. Increase in hydration resulting in a monomolecular layer of the whole available enzyme surface at 0.4 g/g protein raises the activity to a limiting value reached at a water content of 0.9 g/g protein. Here the diffusion of the substrate to the enzyme's active site seems to be completely guaranteed.

For preservation of food it is mandatory to inhibit enzymatic activity completely if the storage temperature is below the phase transition temperature T_g or T_g' (cf. 0.3.3). With help of a model system containing glucose oxidase, glucose and water as well as sucrose and maltodextrin (10 DE) for adjustment of T_g' values in the range of -9.5 to $-32\,°C$, it was found that glucose was enzymatically oxidized only in such samples that were stored for two months above the T_g' value and not in those kept at storage temperatures below T_g'.

2.6 Enzymatic Analysis

Enzymatic food analysis involves the determination of food constituents, which can be both substrates or inhibitors of enzymes, and the determination of enzyme activity in food.

2.6.1 Substrate Determination

2.6.1.1 Principles

Qualitative and quantitative analysis of food constituents using enzymes can be rapid, highly sensitive, selective and accurate (examples in Table 2.16). Prior purification and separation steps, as a rule, are not necessary in the enzymatic analysis of food.

In an enzymatic assay, spectrophotometric or electrochemical determination of the reactant or the product is the preferred approach. When this is not applicable, the determination is performed by a coupled enzyme assay. The coupled reaction includes an auxiliary reaction in which the food constituent is the reactant to be converted to product, and an indicator reaction which involves an indicator enzyme and its reactant or product, the formation or breakdown of which can be readily followed analytically. In most cases, the indicator reaction follows the auxiliary reaction:

$$
\begin{array}{ll}
A + B \xrightleftharpoons{\text{Auxiliary reaction}} P + Q & \\
P + C \xrightleftharpoons{\text{Indicator reaction}} R + S & (2.106)
\end{array}
$$

Reactant A is the food constituent which is being analyzed. C or R or S is measured. The equilibrium state of the coupled indicator reaction is concentration dependent. The reaction has to be adjusted in some way in order to remove, for example, P from the auxiliary reaction before an equilibrium is achieved. By

Table 2.15. Hydration of Lysozyme

$\dfrac{\text{g Water}}{\text{g Protein}}$	Hydration sequence	Molecular changes
0.0	Charged groups	Relocation of protons
	Uncharged, polar groups (formation of clusters)	New orientation of disulfide bonds
0.1	Saturation of COOH groups Saturation of polar groups in side chains	Change in conformation
0.2	Peptide-NH	Start of enzymatic activity
0.3	Peptide-CO Monomolecular hydration of polar groups Apolar side chains	
0.4	Complete enzyme hydration	

Table 2.16. Examples of enzymatic analysis of food constituents [a]

Constituent	Auxiliary reaction	Indicator reaction
Glucose	β-D-Glucose[b] + O_2 $\xrightarrow{\text{Glucose oxidase}}$ δ-D-Gluconolactone + H_2O_2(a_H)	o-Dianisidine + H_2O_2 $\xrightarrow{\text{Peroxidase}}$ Oxid. o-dianisidine (a_I)
	Glucose + ATP $\xrightleftharpoons{\text{Hexokinase}}$ Glucose-6P(b_H)	Glucose-6P + NADP$^\oplus$ $\xrightarrow{\text{Glucose-6P dehydrogenase}}$ Gluconate-6P + NADPH + H$^\oplus$ (b_I)
Fructose	Fructose + ATP $\xrightleftharpoons{\text{Hexokinase}}$ Fructose-6P	
	Fructose-6P $\xrightleftharpoons{\text{Glucosephosphate isomerase}}$ Glucose-6P	As glucose-6P (b_I)
Sorbitol	D-Sorbitol + NAD $\xrightleftharpoons{\text{Sorbitol dehydrogenase}}$ Fructose + NADH + H$^\oplus$	
Maltose	Maltose + H_2O $\xrightarrow{\alpha\text{-Glucosidase}}$ 2 Glucose	As glucose ($b_H + b_I$)
Starch	Starch + (n − 1) H_2O $\xrightarrow{\text{Amyloglucosidase}}$ n-Glucose	As glucose ($b_H + b_I$)
Galactose	β-D-Galactose + NAD$^\oplus$ $\xrightleftharpoons{\text{Galactose dehydrogenase}}$ D-Galactono-γ-lactone + NADH + H$^\oplus$	
Ethanol	Ethanol + NAD$^\oplus$ $\xrightleftharpoons{\text{Alcohol dehydrogenase}}$ Acetaldehyde + NADH + H$^\oplus$	
Glycerol	Glycerol + ATP $\xrightleftharpoons{\text{Glycerol kinase}}$ sn-Glycerol-3P + ATP	ADP + Phosphoenolpyruvate $\xrightleftharpoons{\text{Pyruvate kinase}}$ ATP + Pyruvate (c)
		Pyruvate + NADH + H$^\oplus$ $\xrightleftharpoons{\text{Lactate dehydrogenase}}$ Lactate + NAD$^\oplus$ (d)
Lactate	L-Lactate assay is achieved by a reversed reaction of d), and D-lactate assay with a dehydrogenase specific for D-enantiomer.	
Creatinine and Creatine	Creatinine + H_2O $\xrightleftharpoons{\text{Creatininase}}$ Creatine	
	Creatine + ATP $\xrightleftharpoons{\text{Creatine kinase}}$ Creatine-P + ADP; ADP is determined through c) and d)	
Individual amino acids	R–CH(NH$_2$)COOH $\xrightarrow{\text{Amino acid decarboxylase}^d}$ R–CH$_2$–NH$_2$ + CO$_2$	
L-Malate	L-Malate + NAD$^\oplus$ $\xrightleftharpoons{\text{Malate dehydrogenase}}$ Oxalacetate + NADH + H$^\oplus$	

[a] For saccharose and lactose see Fig. 2.41.
[b] The content of α-anomeric form is accessible through mutarotation.
[c] After hydrolysis this method is suitable for the assay of acylglycerols.
[d] Specific decarboxylases are availabe as exemplified by those for L-tryosine, L-lysine, L-glutamic acid, L-aspartic acid, or L-arginine

$$Glucose + ATP \xrightarrow{\ HK\ } Glucose-6P \qquad\qquad (a)$$

$$Glucose-6P + NADP^{\oplus} \xrightarrow{\ G\,6P-DH\ } 6\text{-Phosphogluconate} + NADPH + H^{\oplus}\ (b)$$

$$Lactose \xrightarrow{\ \beta-Ga\ } Glucose + Galactose \qquad\qquad (c)$$

$$Saccharose \xrightarrow{\ \beta-F\ } Glucose + Fructose \qquad\qquad (d)$$

(2.105)

using several sequential auxiliary reactions with one indicator reaction, it is possible to simultaneously determine several constituents in one assay. An example is the analysis of glucose, lactose and saccharose (cf. Reaction 2.105).
First, glucose is phosphorylated with ATP in an auxiliary reaction (a). The product, glucose-6-phosphate, is the substrate of the NADP-dependent indicator reaction (b). Addition of β-galactosidase starts the lactose analysis (c) in which the released glucose, after phosphorylation, is again measured through the indicator reaction [(b) of Reaction 2.105 and also Fig. 2.41]. Finally, after addition of β-fructosidase, saccharose is cleaved (d) and the released glucose is again measured through reactions (a) and (b) as illustrated in Fig. 2.41.

2.6.1.2 End-Point Method

This procedure is reliable when the reaction proceeds virtually to completion. If the sub-

strate is only partly consumed, the equilibrium is displaced in favor of the products by increasing the concentration of reactant or by removing one of the products of the reaction. If it is not possible to achieve this, a standard curve must be prepared. In contrast to kinetic methods (see below), the concentration of substrate which is to be analyzed in food must not be lower than the *Michaelis* constant of the enzyme catalyzing the auxiliary reaction. The reaction time is readily calculated when the reaction rate follows first-order kinetics for the greater part of the enzymatic reaction.
In a two-substrate reaction the enzyme is saturated with the second substrate. Since Equation 2.41 is valid under these conditions, the catalytic activity of the enzyme needed for the assay can be determined for both one- and two-substrate reactions. The examples shown in Table 2.17 suggest that enzymes with low K_m values are desirable in order to handle the substrate concentrations for the end-point method with greater flexibility.
Data for K_m and V are needed in order to calculate the reaction time required. A prerequi-

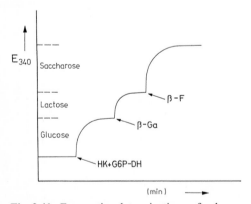

Fig. 2.41. Enzymatic determination of glucose, saccharose and lactose in one run. After adding cosubstrates, ATP and NADP, the enzymes are added in the order: hexokinase (HK), glucose-6-phosphate dehydrogenase (G6P-DH), β-galactosidase (β-Ga) and β-fructosidase (β-F)

Table 2.17. Enzyme concentrations used in the end-point method of enzymatic food analysis

Substrate	Enzyme	K_m (mol/l)	Enzyme concentration (µcat/l)
Glucose	Hexokinase	$1.0 \cdot 10^{-4}$ (30 °C)	1.67
Glycerol	Glycerol kinase	$5.0 \cdot 10^{-5}$ (25 °C)	0.83
Uric acid	Urate oxidase	$1.7 \cdot 10^{-5}$ (20 °C)	0.28
Fumaric acid	Fumarase	$1.7 \cdot 10^{-6}$ (21 °C)	0.03

site is a reaction in which the equilibrium state is displaced toward formation of product with a conversion efficiency of 99%.

2.6.1.3 Kinetic Method

Substrate concentration is obtained using a method based on kinetics by measuring the reaction rate. To reduce the time required per assay, the requirement for the quantitative conversion of substrate is abandoned. Since kinetic methods are less susceptible to interference than the endpoint method, they are advantageous for automated methods of enzymatic analysis.

The determination of substrate using kinetic methods is possible only as long as Equation 2.46 is valid. Hence, the following is required to perform the assay:

a) For a two-substrate reaction, the concentration of the second reactant must be so high that the rate of reaction depends only on the concentration of the substrate which is being analyzed.
b) Enzymes with high *Michaelis* constants are required; this enables relatively high substrate concentrations to be determined.
c) If enzymes with high *Michaelis* constants are not available, the apparent K_m is increased by using competitive inhibitors.

In order to explain requirement c), let us consider the example of the determination of glycerol as given in Table 2.16. This reaction allows the determination of only low concentrations of glycerol since the K_m values for participating enzymes are low: 6×10^{-5} mol/l to 3×10^{-4} mol/l.

In the reaction sequence the enzyme pyruvate kinase is competitively inhibited by ATP with respect to ADP. The expression $K_m(1+(I)/K_I)$ (cf. 2.5.2.2.1) may in these circumstances assume a value of 6×10^{-3} mol/l, for example. This corresponds to an apparent increase by a factor of 20 for the K_m of ADP (3×10^{-4} mol/l). The ratio $(S)/K_m(1+[I]/K_I)$ therefore becomes 1×10^{-3} to 3×10^{-2}. Under these conditions, the auxilary reaction (Table 2.16) with pyruvate kinase follows pseudo-first-order kinetics with respect to ADP over a wide range of concentrations and, as a result of the inhibition by ATP, it is also the rate-determining step of the overall reaction. It is then possible to kinetically determine higher concentrations of glycerol.

2.6.2 Determination of Enzyme Activity

In the foreword of this chapter it was emphasized that enzymes are suitable indicators for identifying heat-treated food. However, the determination of enzyme activity reaches far beyond this possibility: it is being used to an increasing extent for the evaluation of the quality of raw food and for optimizing the parameters of particular food processes. In addition, the activities of enzyme preparations have to be controlled prior to use in processing or in enzymatic food analysis.

The measure of the catalytic activity of an enzyme is the rate of the reaction catalyzed by the enzyme. The conditions of an enzyme activity assay are optimized with relation to: type and ionic strength of the buffer, pH, and concentrations of substrate, cosubstrate and activators used. The closely controlled assay conditions, including the temperature, are critical because, in contrast to substrate analysis, the reliability of the results in this case often can not be verified by using a weighed standard sample.

Temperature is a particularly important parameter which strongly influences the enzyme assay. Temperature fluctuations significantly affect the reaction rate (cf. 2.5.4); e.g., a 1°C increase in temperature results in about a 10% increase in activity. Whenever possible, the incubation temperature should by maintained at 25°C.

The substrate concentration in the assay is adjusted ideally so that Equation 2.40 is valid, i.e. $[A_0] \gg K_m$. Difficulties often arise while trying to achieve this condition: the substrate's solubility is limited; spectrophotometric readings become unreliable because of high light absorbance by the substrate; or the high concentration of the substrate inhibits enzyme activity. For such cases procedures exist to assess the optimum substrate concentration which will support a reliable activity assay.

2.6.3 Enzyme Immunoassay

Food compounds can be determined specifically and sensitively by immunological methods. These are based on the specific reaction of an antibody containing antiserum with the antigen, the substance to be determined. The antiserum is produced by immunization of rabbits for example. Because only compounds with a high molecular weight ($M_r >$ 5.000) display immunological activity, for low molecular compounds (haptens) covalent coupling to a protein is necessary. The antiserum produced with the "conjugate" contains antibodies with activities against the protein as well as the hapten.

Prior to the application, the antiserum is tested for its specificity against all proteins present in the food to be analyzed. As far as possible all unspecificities will be removed. For example, it is possible to treat an antiserum intended to be used for the determination of peanut protein with proteins from other nuts in such a way that it specifically reacts with peanut protein only. However, there are also cases in which the specificity could not be increased because of the close immunochemical relationship between the proteins. This happens, for example, with proteins from almonds, peach and apricot kernels.

The general principle of the competitive immunoassay is shown in Fig. 2.42. Excess amounts of marked and unmarked antigens compete for the antibodies present. The concentration of the unmarked antigen to be determined is the only variable if the concentration of the marked antigen and the antibody concentration are kept on a constant level during the examination. Following the principle of mass action, the unknown antigen concentration can be calculated indirectly based on the proportion of free marked antigen. Older methods still require the formation of a precipitate for the detection of an antibody-antigen reaction (cf. 12.10.2.3.2). Immunoassays are much faster and more sensitive.

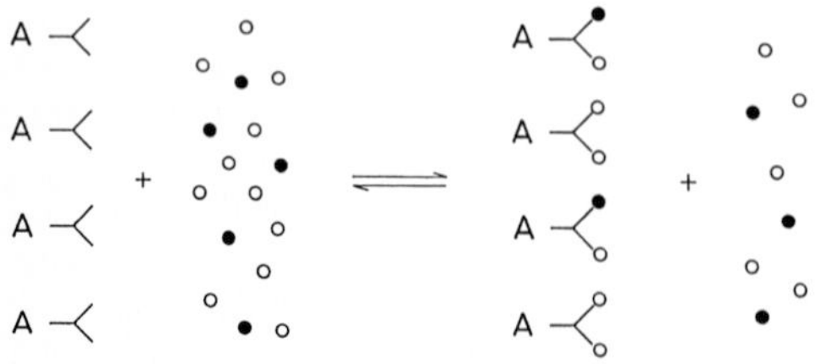

Radioisotopes (^{3}H, ^{14}C) and enzymes are used to mark antigens. Furthermore, fluorescent and luminescent dyes as well as stable radicals are important. Horseradish peroxidase, alkaline phosphatase from calf stomach, and β-D-galactosidase from *E. coli* are often used as indicator enzymes because they are available in high purity, are very stable and their activity can be determined sensitively and precisely. Enzymes are bound to antigens or haptens by covalent bonds, e.g., by reaction with glutaraldehyde or carbodiimide.

Enzyme immunoassays are increasingly used in food analysis (examples see Table 2.18). Laboratories employing these methods need no specific equipment contrary to use of radio immunological methods (RIA). Furthermore, for radio immunoassays free antigens have always to be separated from the ones bound to antibodies (heterogeneous immunoassay) while an enzyme immunoassay is suitable for homogeneous tests if the activity of the indicator enzyme is inhibited by the formation of an antigen-antibody-complex.

A specific heterogeneous method used mainly in food analysis is the ELISA *test* (*enzyme linked immunosorbent assay*). It is possible to

Table 2.18. Examples for application of enzyme immunoassay in food analysis

Detection and quantification
Type of meat
Soya protein in meat products
Myosin in muscle meat
Cereal proteins as well as papain in beer
Gliadins (absence of gluten in foods)
Veterinary drugs and fattening aids, e.g. penicillin in milk, natural or synthetic estrogens in meat
Toxins (aflatoxins, enterotoxins, ochratoxins) in food
Pesticides (atrazine, aldicarb, carbofuran)
Glycoalkaloids in potatoes

Fig. 2.42. Principle of an immunoassay. Marked antigens (●) and unmarked antigens (○) compete for the binding sites of the antibodies A

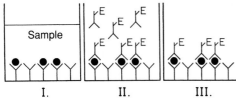

Fig. 2.43. Principle of non-competitive ELISA

Y Immobilized antibody,

● antigen,

Y^E enzyme-marked antibody

increase sensitivity and specificity of the non-competitive ELISA test by a second immuno-chemical reaction. It works as shown in Fig. 2.43. A plastic carrier holds the antibodies, e.g. against a toxin, by adsorption. When the sample is added, the toxin (antigen) reacts with the excess amount of antibodies (I in Fig. 2.43). The second antibody marked with an enzyme (e.g. alkaline phosphatase) and with specificity for the antigen forms a sandwich complex (II). Unbound enzyme-marked anti-bodies are washed out. The remaining enzyme activity is determined (III) and the antigen concentration in the sample can be calculated based on measured standards and a calibration curve.

2.7 Enzyme Utilization in the Food Industry

Enzyme-catalyzed reactions in food process-ing have been used unintentionally since ancient times. The enzymes are either an inte-gral part of the food or are obtained from microorganisms. Addition of enriched or puri-fied enzyme preparations of animal, plant or, especially, microbial origin is a recent prac-tice. Such intentionally used additives provide a number of advantages in food processing: exceptionally pronounced substrate specificity (cf. 2.2.2), high reaction rate under mild reaction conditions (temperature, pH), and a fast and continuous, readily controlled reac-tion process with generally modest operational costs and investment. Examples for the appli-cation of microbial enzymes in food process-ing are given in Table 2.19.

2.7.1 Technical Enzyme Preparations

2.7.1.1 Production

The methods used for industrial-scale enzyme isolation are outlined in principle under sec-tion 2.2.4. In contrast to the production of highly purified enzymes for analytical use, the production of enzymes for technical pur-poses is directed to removing the interfering activities which would be detrimental to pro-cessing and to staying within economically acceptable costs. Selective enzyme precipita-tion by changing the ionic strength and/or pH, adsorption on inorganic gels such as calcium phosphate gel or hydroxyl apatite, chromato-graphy on porous gel columns and ultrafil-tration through membranes are among the fractionation methods commonly used. Ion-exchange chromatography, affinity chromato-graphy (cf. 2.2.4) and preparative electropho-resis are relatively expensive and are seldom used. A few temperature-stable enzymes are heat treated to remove the other contaminating and undesired enzyme activities.

Commercial enzyme preparations are avail-able with defined catalytic activity. The activi-ty is usually adusted by the addition of suitable inert fillers such as salts or carbohydrates. The amount of active enzyme is relatively low, e.g., proteinase preparations contain 5–10% proteinase, whereas amylase preparations used for treamtent of flour contain only 0.1% pure fungal α-amylase.

2.7.1.2 Immobilized Enzymes

Enzymes in solution are usually used only once. The repeated use of enzymes fixed to a carrier is more economical. The use of en-zymes in a continous process, for example, immobilized enzymes used in the form of a stationary phase which fills a reaction column where the reaction can be controlled simply by adjustment of the flow rate, is the most advanced technique. Immobilized enzymes are produced by various methods (Fig. 2.44).

2.7.1.2.1 Bound Enzymes

An enzyme can be bound to a carrier by cova-lent chemical linkages, or in many cases, by physical forces such as adsorption, by charge

Table 2.19. Examples for the use of microbial enzymes in food processing

EC Number	Enzyme[a]	Biological Origin	Application[b]
Oxidoreductases			
1.1.1.39	Malate dehydrogenase (decarboxylating)	*Leuconostoc oenos*	10
1.1.3.4	Glucose oxidase	*Aspergillus niger*	7, 10, 16
1.11.1.6	Catalase	*Micrococcus lysodeicticus*	
		Aspergillus niger	1, 2, 7, 10, 16
Hydrolases			
3.1.1.1	Carboxylesterase	*Mucor miehei*	2, 3
3.1.1.3	Triacylglycerol lipase	*Aspergillus niger, A. oryzae, Candida lipolytica, Mucor javanicus, M. miehei, Rhizopus arrhizus, R. niveus*	2, 3
3.1.1.11	Pectinesterase	*Aspergillus niger*	9, 10, 17
3.1.1.20	Tannase	*Aspergillus niger, A. oryzae*	10
3.2.1.1	α-Amylase	*Bacillus licheniformis, B. subtilis, Aspergillus oryzae*	3, 8, 9, 10, 12 14, 15
		Aspergillus niger, Rhizopus delemar, R. oryzae	8, 9, 10, 12, 14, 15
3.2.1.2	β-Amylase	*Bacillus cereus, B. magatherium, B. subtilis*	8, 10
3.2.1.3	Glucan-1,4-α-D-glucosidase (glucoamylase)	*Aspergillus oryzae*	3, 9, 10, 12, 14 15, 18
		Aspergillus niger, Rhizopus arrhizus, R. delemar, R. niveus, R. oryzae, Trichoderma reesei	9, 10, 12, 14, 15, 18
3.2.1.4	Cellulase	*Aspergillus niger, A. oryzae, Rhizopus delemar, R. oryzae, Sporotrichum dimorphosporum, Thielavia terrestris, Trichoderma reesei*	9, 10, 18
3.2.1.6	Endo-1,3(4)-β-D-glucanase	*Bacillus circulans, B. subtilis, Aspergillus niger, A. oryzae, Penicillum emersonii, Rhizopus delemar, R. oryzae*	10
3.2.1.7	Inulinase	*Kluyveromyces fragilis*	12
3.2.1.11	Dextranase	*Klebsiella aerogenes, Penicillium funicolosum, P. lilacinum*	12
3.2.1.15	Polygalacturonase	*Aspergillus niger, Penicillium simplicissimum, Trichoderma reesei*	3, 9, 10, 17
		Aspergillus oryzae, Rhizopus oryzae	3, 9, 10
		Aspergillus niger	9, 10, 17
3.2.1.20	α-D-Glucosidase	*Aspergillus niger, A. oryzae, Rhizopus oryzae*	8
3.2.1.21	β-D-Glucosidase	*Aspergillus niger, Trichoderma reesei*	9
3.2.1.22	α-D-Galactosidase	*Aspergillus niger, Mortierella vinacea sp., Saccharomyces carlsbergensis*	12
3.2.1.23	β-D-Galactosidase	*Aspergillus niger, A. oryzae, Kluyveromyces fragilis, K. lactis*	1, 2, 4, 18
3.2.1.26	β-D-Fructofuranosidase	*Aspergillus niger, Saccharomyces carlsbergensis, S. cerevisiae*	14
3.2.1.32	Xylan endo-1,3-β-D-xylosidase	*Streptomyces sp., Aspergillus niger, Sporotrichum dimorphosporum*	8, 10, 13
3.2.1.41	α-Dextrin endo-1,6-α-glucosidase (pullunanase)	*Bacillus acidopullulyticus*	8, 10, 12, 14, 15
		Klebsiella aerogenes	8, 10, 12
3.2.1.55	α-L-Arabinofuranosidase	*Aspergillus niger*	9, 10, 17

Table 2.19 (continued)

EC Number	Enzyme[a]	Biological Origin	Application[b]
3.2.1.58	Glucan-1,3-β-D-glucosidase	*Trichoderma harzianum*	10
3.2.1.68	Isoamylase	*Bacillus cereus*	8, 10
3.2.1.78	Mannan endo-1,4-β-D-mannanase	*Bacillus subtilis, Aspergillus oryzae, Rhizopus delemar, R. oryzae, Sporotrichum dimorphosporum, Trichoderma reesei*	13
		Aspergillus niger	13, 17
3.4.21.62–67	Microbial serine proteinases[c]	*Bacillus licheniformis*	5, 6, 10, 11
3.4.23.18–30	Microbial carboxyl proteinases	*Aspergillus melleus, Endothia parasitica, Mucor miehei, M. pusillus*	2
		Aspergillus oryzae	2, 5, 6, 8, 9, 10, 11, 15, 18
3.4.24.25–40	Microbial metallo-proteinases	*Bacillus cereus, B. subtilis*	10, 15
Lyases			
4.2.2.10	Pectin lyase	*Aspergillus niger*	9, 10, 17
Isomerases			
5.3.1.5	Xylose isomerase[d]	*Actinoplanes missouriensis, Arthrobacter sp., Bacillus coagulans, Streptomyces albus, S. olivaceus, S. olivochromogenes, S. rubiginosus*	8, 9, 10, 12

[a] Principal activity.
[b] 1) Milk, 2) Cheese, 3) Fats and oils, 4) Ice cream, 5) Meat, 6) Fish, 7) Egg, 8) Cereal and starch, 9) Fruit and vegetables, 10) Beverages (soft drinks, beer, wine), 11) Soups and broths, 12) Sugar and honey, 13) Cacao, chocolate, coffee, tea, 14) Confectionery, 15) Bakery, 16) Salads, 17) Spices and flavors, 18) Diet food.
[c] Similar to Subtilisin.
[d] Some enzymes also convert D-glucose to D-fructose, cf. 2.7.2.3.

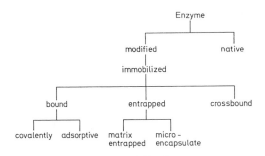

Fig. 2.44. Forms of immobilized enzymes

attraction, H-bond formation and/or hydrophobic interactions. The covalent attachment to a carrier, in this case an activated matrix, is usually achieved by methods employed in peptide and protein chemistry. First, the matrix is activated. In the next step, the enzyme is coup-led under mild conditions to the reactive site on the matrix, usually by reaction with a free amino group. This is illustrated by using cellulose as a matrix (Fig. 2.45). Another possibility is a process of copolymerization with suitable monomers. Generally, covalent attachment of the enzyme prevents leaching or "bleeding".

2.7.1.2.2 Enzyme Entrapment

An enzyme can be entrapped or enclosed in the cavities of a polymer network by polymerization of a monomer such as acrylamide or N,N′-methylene-bis-acrylamide in the presence of enzyme, and still remain accessible to substrate through the network of pores. Furthermore, suitable processes can bring about enzyme encapsulation in a semipermeable membrane (microencapsulation) or confinement in hollow fiber bundles.

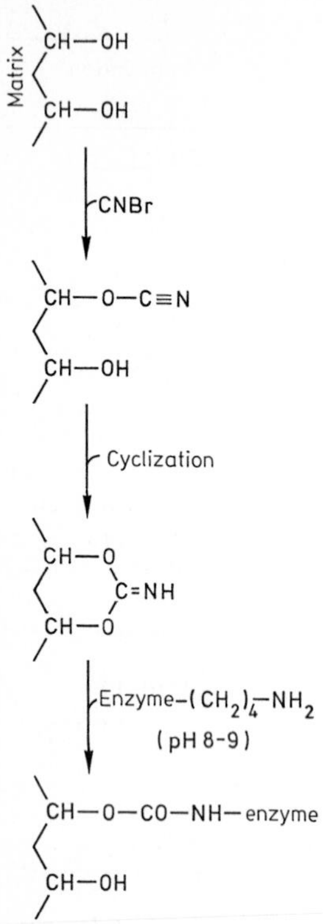

Matrix

CH—OH

CH—OH

↓ CNBr

CH—O—C≡N

CH—OH

↓ Cyclization

CH—O
‌   C=NH
CH—O

↓ Enzyme–(CH$_2$)$_4$–NH$_2$
   (pH 8-9)

CH—O—CO—NH—enzyme

CH—OH

Fig. 2.45. Enzyme immobilization by covalent binding to a cellulose matrix

2.7.1.2.3 Cross-Linked Enzymes

Derivatization of enzymes using a bifunctional reagent, e.g. glutaraldehyde, can result in cross-linking of the enzyme and, thus, formation of large, still catalytically active insoluble complexes. Such enzyme preparations are relatively unstable for handling and, therefore, are used mostly for analytical work.

2.7.1.2.4 Properties

The properties of an immobilized enzyme are often affected by the matrix and the methods used for immobilization.

Kinetics. As a rule, higher substrate concentrations are required for saturation of an entrapped enzyme than for a free, native enzyme. This is due to a decrease in the concentration gradient which takes place in the pores of the polymer network. Also, there is an increase in the "apparent" *Michaelis* constant for an enzyme bound covalently to a matrix carrying an electrostatic charge. This is also true when the substrate and the functional groups of the matrix carry the same charge. On the other hand, opposite charges bring about an increase of substrate affinity for the matrix. Consequently, this decreases the "apparent" K_m.

pH Optimum. Negatively charged groups on a carrier matrix shift the pH optimum of the covalently bound enzyme to the alkaline region, whereas positive charges shift the pH optimum towards lower pH values. The change in pH optimum of an immobilized enzyme can amount to one to two pH units in comparison to that of a free, native enzyme.

Thermal Inactivation. Unlike native enzymes, the immobilized forms are often more heat stable (cf. example for β-D-glucosidase, Fig. 2.46). Heat stability and pH optima changes induced by immobilization are of great interest in the industrial utilization of enzymes.

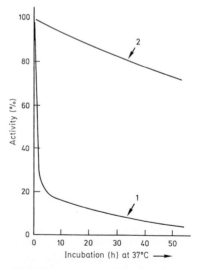

Fig. 2.46. Thermal stabilities of free and immobilized enzymes (according to *O.R. Zaborsky*, 1973). 1 β-D-glucosidase, free, 2 β-D-glucosidase, immobilized

2.7.2 Individual Enzymes

2.7.2.1 Oxidoreductases

Broader applications for the processing industry, besides the familiar use of glucose oxidase, are found primarily for catalase and lipoxygenase, among the many enzymes of this group. A number of oxidoreductases have been suggested or are in the experimental stage of utilization, particularly for aroma improvement (examples under 2.7.2.1.4 and 2.7.2.1.5).

2.7.2.1.1 Glucose Oxidase

The enzyme produced by fungi such as *Aspergillus niger* and *Penicillium notatum* catalyzes glucose oxidation by consuming oxygen from the air. Hence, it is used for the removal of either glucose or oxygen (Table 2.19). The H_2O_2 formed in the reaction is occasionally used as an oxidizing agent, but it is usually degraded by catalase.

Removal of glucose during the production of egg powder using glucose oxidase (cf. 11.4.3) prevents the *Maillard* reaction responsible for discoloration of the product and deterioration of its whippability. Similar use of glucose oxidase for some meat and protein products would enhance the golden-yellow color rather than the brown color of potato chips or French fries which is obtained in the presence of excess glucose.

Removal of oxygen from a sealed package system results in suppression of fat oxidation and oxidative degradation of natural pigments. For example, the color change of crabs and shrimp from pink to yellow is hindered by dipping them into a glucose oxidase/catalase solution. The shelf life of citrus fruit juices, beer and wine can be prolonged with such enzyme combinations since the oxidative reactions which lead to aroma deterioration are retarded.

2.7.2.1.2 Catalase

The enzyme isolated from liver or microorganisms is important as an auxiliary enzyme for the decomposition of H_2O_2:

$$2\,H_2O_2 \rightleftharpoons 2\,H_2O + O_2 \qquad (2.107)$$

Hydrogen peroxide is a by-product in the treatment of food with glucose oxidase. It is added to food in some specific canning procedures. An example is the pasteurization of milk with H_2O_2, which is important when the thermal process is shut down by technical problems. Milk thus stabilized is also suitable for cheesemaking since the sensitive casein system is spared from heat damage. The excess H_2O_2 is then eliminated by catalase.

2.7.2.1.3 Lipoxygenase

The properties of this enzyme are described under section 3.7.2.2 and its utilization in the bleaching of flour and the improvement of the rheological properties of dough is covered under section 15.4.1.4.3.

2.7.2.1.4 Aldehyde Dehydrogenase

During soya processing, volatile degradation compounds (hexanal, etc.) with a "bean-like" aroma defect are formed because of the enzymatic oxidation of unsaturated fatty acids. These defects can be eliminated by the enzymatic oxidation of the resultant aldehydes to carboxylic acids. Since the flavor threshold values of these acids are high, the acids generated do not interfere with the aroma improvement process.

$$\text{n-Hexanal} + \text{NAD}^{\oplus}$$
$$\rightleftharpoons \text{Caproic acid} + \text{NADH} + \text{H}^{\oplus} \qquad (2.108)$$

Of the various aldehyde dehydrogenases, the enzyme from beef liver mitochondria has a particularly high affinity for n-hexanal (Table 2.20). Hence its utilization in the production of soya milk is recommended.

2.7.2.1.5 Butanediol Dehydrogenase

Diacetyl formed during the fermentation of beer can be a cause of a flavor defect. The enzyme from *Aerobacter aerogenes*, for example, is able to correct this defect by reducing the diketone to the flavorless 2,3-butanediol:

$$\text{CH}_3\text{-CO-CO-CH}_3 + \text{NADH} + \text{H}^{\oplus}$$
$$\rightleftharpoons \underset{\substack{| \quad |\\ \text{OH} \ \text{OH}}}{\text{CH}_3\text{-CH-CH-CH}_3} + \text{NAD}^{\oplus} \qquad (2.109)$$

Table 2.20. *Michaelis* constants for aldehyde dehydrogenase (ALD) from various sources

Substrate	K_m (µmol/l)			
	ALD (bovine liver)			ALD
	Mitochondria	Cytosol	Microsomes	Yeast
Ethanal	0.05	440	1,500	30
n-Propanal	–	110	1,400	–
n-Butanal	0.1	< 1	–	–
n-Hexanal	0.075	< 1	< 1	6
n-Octanal	0.06	< 1	< 1	–
n-Decanal	0.05	–	–	–

Such a process is improved by the utilization of yeast cells which, in addition to the enzyme and NADH, contain a system able to regenerate the cosubstrate. In order to prevent contamination of beer with undesirable cell constituents, the yeast cells are encapsulated with gelatin.

2.7.2.2 Hydrolases

Most of the enzymes used in the food industry belong to the class of hydrolase enzymes (cf. Table 2.19).

2.7.2.2.1 Proteinases

The mixture of proteolytic enzymes used in the food industry contains primarily endopeptidases (specificity and classification under section 1.4.5.2). These enzymes are isolated from animal organs, higher plants or microorganisms, i.e. from their fermentation media (Table 2.21). Examples of their utilization are as follows. Proteinases are added to wheat flour in the production of some bakery products to modify rheological properties of dough and, thus, the firmness of the end-product. During such dough treatment, the firm or hard wheat gluten is partially hydrolyzed to a soft-type gluten (cf. 15.4.1.4.5).

In the dairy industry the formation of casein curd is achieved with chymosin or rennin (cf. Table 2.19) by a reaction mechanism described under section 10.1.2.1.1. Casein is also precipitated through the action of other proteinases by a mechanism which involves secondary proteolytic activity resulting in

Table 2.21. Proteinases utilized in food processing

Name	Source	pH optimum	Optimal stability pH range
A. Proteinases of animal origin			
Pancreatic proteinase[a]	Pancreas	9.0[b]	3–5
Pepsin	Gastric lining of swine or bovine	2	
Chymosin	Stomach lining of calves or genetically engineered microorganisms	6–7	5.5–6.0
B. Proteinases of plant origin			
Papain	Tropical melon tree (*Carica papaya*)	7–8	4.5–6.5
Bromelain	Pineapple (fruit and stalk)	7–8	
Ficin	Figs (*Ficus carica*)	7–8	
C. Bacterial proteinases			
Alkaline proteinases e.g. subtilisin	*Bacillus subtilis*	7–11	7.5–9.5
Neutral proteinases e.g. thermolysin	*Bacillus thermoproteolyticus*	6–9	6–8
Pronase	*Streptomyces griseus*		
D. Fungal proteinases			
Acid proteinase	*Aspergillus oryzae*	3.0–4.0[d]	5
Neutral proteinase	*Aspergillus oryzae*	5.5–7.5[d]	7.0
Alkaline proteinase	*Aspergillus oryzae*	6.0–9.5[d]	7–8
Proteinase	*Mucor pussillus*	3.5–4.5[d]	3–6
Proteinase	*Rhizopus chinensis*	5.0	3.8–6.5

[a] A mixture of trypsin, chymotrypsin, and various peptidases with amylase and lipase as accompanying enzymes.
[b] With casein as a substrate.
[c] A mixture of various proteinases including amino- and carboxypeptidases.
[d] With hemoglobin as as substrate.

diminished curd yields and lower curd strength. Rennin is essentially free of other undesirable proteinases and is, therefore, especially suitable for cheesemaking. However, there is a shortage of rennin since it has to be isolated from the stomach of a suckling calf. However, it is now possible to produce this enzyme using genetically engineered microorganism. Proteinases from *Mucor miehei*, *M. pusillus* and *Endothia parasitica* are a suitable replacement for rennin.

Plant proteinases (cf. Table 2.21) and also those of microorganisms are utilized for ripening and tenderizing meat. The practical problem to be solved is how to achieve uniform distribution of the enzymes in muscle tissue. An optional method appears to be injection of the proteinase into the blood stream immediately before slaughter, or rehydration of the freeze-dried meat in enzyme solutions.

Cold turbidity in beer is associated with protein sedimentation. This can be eliminated by hydrolysis of protein using plant proteinases (cf. Table 2.21). Utilization of papain was suggested by *Wallerstein* in 1911. Production of complete or partial protein hydrolysates by enzymatic methods is another example of an industrial use of proteinases. This is used in the liquefaction of fish proteins to make products with good flavors.

One of the concerns in the enzymatic hydrolysis of proteins is to avoid the release of bitter-tasting peptides and/or amino acids (cf. 1.2.6 and 1.3.3). Their occurrence in the majority of proteins treated (an exception is collagen) can not be ignored, especially when the extent of hydrolysis yields peptide fragments with a molecular weight below 6,000.

2.7.2.2.2 α- and β-Amylases

Amylases are either produced by bacteria or yeasts (Table 2.19) or they belong to the components of malt preparations. The high temperature-resistant bacterial amylases, particularly those of *Bac. licheniformis* (Fig. 2.47) are of interest for the hydrolysis of corn starch (gelatinization at 105–110 °C). The hydrolysis rate of these enzymes can be enhanced further by adding Ca^{2+} ions. α-Amylases added to the wort in the beer production process accelerate starch degradation. These enzymes are also used in the baking industry (cf. 15.4.1.4.8).

2.7.2.2.3 Glucan-1,4-α-D-Glucosidase (Glucoamylase)

Glucoamylase cleaves β-D-glucose units from the non-reducing end of an 1,4-α-D-glucan. The α-1,6-branching bond present in amylopectin is cleaved at a rate about 30 times slower than the α-1,4-linkages occurring in straight chains. The enzyme preparation is produced from bacterial and fungal cultures.

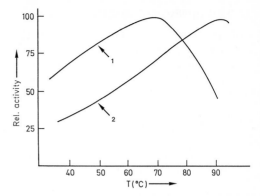

Fig. 2.47. The activity of α-amylase as influenced by temperature. 1 α-amylase from *Bacillus subtilis*, 2 from *Bacillus licheniformis*

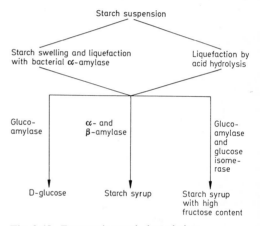

Fig. 2.48. Enzymatic starch degradation

The removal of transglucosidase enzymes which catalyze, for example, the transfer of glucose to maltose, thus lowerung the yield of glucose in the starch saccharification process, is important in the production of glucoamylase.

The starch saccharification process is illustrated in Fig. 2.48. In a purely enzymatic process (left side of the figure), the swelling and gelatinization and liquefaction of starch can occur in a single step using heat-stable bacterial α-amylase (cf. 2.7.2.2.2). The action of amylases yields starch syrup which is a mixture of glucose, maltose and dextrins (cf. 19.1.4.3.2).

2.7.2.2.4 Pullulanase (Isoamylase)

Pullulanase (cf. 4.4.5.1.4) is utilized in the brewing process and in starch hydrolysis. In combination with β-amylase, it is possible to produce a starch sirup with a high maltose content.

2.7.2.2.5 Endo-1,3(4)-β-D-Glucanase

In the brewing process, β-glucans from barley increase wort viscosity and impede filtration. Enzymatic endo-hydrolysis reduces viscosity.

2.7.2.2.6 α-D-Galactosidase

This and the following enzymes (up to and including section 2.7.2.2.9) attack the non-reducing ends of di-, oligo- and polysaccharides with release of the terminal monosaccharide. The substrate specificity is revealed by the name of the enzyme, e.g., α-D-galactosidase:

(2.110)

In the production of sucrose from sugar beets (cf. 19.1.4.1.2), the enzymatic preparation from *Mortiella vinacea* hydrolyzes raffinose and, thus, improves the yield of granular sugar in the crystallization step. Raffinose in amounts > 8% effectively prevents crystallization of sucrose.

Gas production (flatulence) in the stomach or intestines produced by legumes originates from the sugar stachyose (cf. 16.2.5). When this tetrasaccharide is cleaved by α-D-galactosidase, flatulency from this source is eliminated.

2.7.2.2.7 β-D-Galactosidase (Lactase)

This enzyme preparation from fungi (*Aspergillus niger)* or from yeast is of importance in the dairy industry. The preparation is used to hydrolyze lactose, the solubility of which is low, and, hence, interferes with the production of skim milk concentrate or ice cream. The immobilized form of the enzyme is also used successfully.

2.7.2.2.8 β-D-Fructofuranosidase (Invertase)

Enzyme preparations isolated from special yeast strains are used for saccharose (sucrose) inversion in the confectionery or candy industry. Invert sugar is more soluble and, because of the presence of free fructose, is sweeter than saccharose.

2.7.2.2.9 α-L-Rhamnosidase

Some citrus fruit juices and purées (especially those of grapes) contain naringin, a dihydrochalcone with a very bitter taste. Treatment of naringin with combined preparations of α-L-rhamnosidase and β-D-glucosidase yields the nonbitter aglycone compound naringenin (cf. 18.1.2.5.4).

2.7.2.2.10 Cellulases and Hemicellulases

The baking quality of rye flour and the shelf life of rye bread can be improved by partial hydrolysis of the rye pentosans. Technical pentosanase preparations are mixtures of β-glycosidases (1,3- and 1,4-β-D-xylanases, etc.). Solubilization of plant constituents by soaking in an enzyme preparation (maceration) is a mild and sparing process. Such preparations usually contain exo- and endo-cellulases, α- and β-mannosidases and pectolytic enzymes (cf. 2.7.2.2.13). Examples of the utilization are: production of fruit and vegetable purées (mashed products), disintegration of tea leaves, or production of dehydrated mashed potatoes. Some of these enzymes are used to prevent mechanical damage to cell walls during mashing and, thus, to prevent excessive leaching of gelatinized starch from the cells, which would make the purée too sticky.

Glycosidases (cellulases and amylases from *Aspergillus niger*) in combination with proteinases are recommended for removal of shells from shrimp. The shells are loosened and then washed off in a stream of water.

2.7.2.2.11 Lysozyme

The cell walls of gram-positive bacteria are formed from peptidoglycan (synonymous with murein). Peptidoglycan consists of repeating units of the disaccharide N-acetylglucosamine (NAG) and N-acetylmuramic acid (NAM) connected by β-1,4-glycosidic linkages, a tetrapeptide and a pentaglycine peptide bridge. The NAG and NAM residues in peptidoglycan alternate and form the linear polysaccharide chain.

Lysozyme (cf. 11.2.3.1.4) solubilizes peptidoglycan by cleaving the 1,4-β-linkage between NAG and NAM. In order to decrease the spore count in cheeses and, thus, prevent later undesired puffing by *Clostridia*, addition of lysozyme to cheeses has been suggested.

2.7.2.2.12 Thioglucosidase

Proteins from seeds of the mustard family (*Brassicaceae*), such as turnip, rapeseed or brown or black mustard, contain glucosinolates which can be enzymatically decomposed into pungent mustard oils (esters of isothiocyanic acid, R–N=C=S). The oils are usually isolated by steam distillation. The reactions of thioglycosidase and a few glucosinolates occurring in *Brassicaceae* are covered in section 17.1.2.6.5.

2.7.2.2.13 Pectolytic Enzymes

Pectolytic enzymes are described in section 4.4.5.2. Pectic acid which is liberated by pectin esterases flocculates in the presence of Ca^{2+} ions. This reaction is responsible for the undesired "cloud" flocculation in citrus juices. After thermal inactivation of the enzyme at about 90 °C, this reaction is not observable. However, such treatment brings about deterioration of the aroma of the juice. Investigations of the pectin esterase of orange peel have shown that the enzyme activity is affected by competitive inhibitors: oligogalacturonic acid and pectic acid (cf. Fig. 2.49). Thus, the increase in turbidity of citrus juice can be prevented by the addition of such compounds.

Pectinolytic enzymes are used for the clarification of fruit and vegetable juices. The mechanism of clarification is as follows: the core of the turbidity causing particles consists

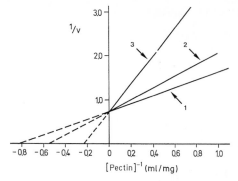

Fig. 2.49. Pectin esterase (orange) activity as affected by inhibitors (according to *F. Termote*, 1977). 1 Without inhibitor, 2 hepta- and octagalacturonic acids, 3 pectic acid

of carbohydrates and proteins (35%). The prototropic groups of these proteins have a positive charge at the pH of fruit juice (3.5). Negatively charged pectin molecules form the outer shell of the particle. Partial pectinolysis exposes the positive core. Aggregation of the polycations and the polyanions then follows, resulting in flocculation. Clarification of juice by gelatin (at pH 3.5 gelatin is positively charged) and the inhibition of clarification by alginates which are polyanions at pH 3.5 support this suggested model.

In addition, pectinolytic enzymes play an important role in food processing, increasing the yield of fruit and vegetable juices and the yield of oil from olive fruits.

2.7.2.2.14 Lipases

The mechanism of lipase activity is described under section 3.7.1.1. Lipase from microbial sources (e.g. *Candida lipolytica*) is utilized for enhancement of aromas in cheesemaking.

Limited hydrolysis of milk fat is also of interest in the production of chocolate milk. It enhances the "milk character" of the flavor. The utilization of lipase for this commodity is also possible.

Staling of bakery products is retarded by lipase, presumably through the release of mono- and diacylglycerols (cf. 15.4.4). The defatting of bones, which has to be carried out under mild conditions in the production of gelatin, is facilitated by using lipase-catalyzed hydrolysis.

150   2 Enzymes

2.7.2.2.15 Tannases

Tannases hydrolyze polyphenolic compounds (tannins):

$$\text{Digallate} \xrightarrow{\text{H}_2\text{O}} \text{2 Gallate} \qquad (2.111)$$

For example, preparations from *Aspergillus niger* prevent the development of turbidity in cold tea extracts.

2.7.2.3 Isomerases

Of this group of enzymes, glucose isomerse, which is used in the production of starch syrup with a high content of fructose (cf. 19.1.4.3.5), is very important. The enzyme used industrially is of microbial origin. Since its activity for xylose isomerization is higher than for glucose, the enzyme is classified under the name "xylose isomerase" (cf. Table 2.4).

2.8 Literature

Acker, L., Wiese, R.: Über das Verhalten der Lipase in wasserarmen Systemen. Z. Lebensm. Unters. Forsch. *150*, 205 (1972)

Baudner, S., Dreher, R.M.: Immunchemische Methoden in der Lebensmittelanalytik – Prinzip und Anwendung. Lebensmittelchemie *45*, 53 (1991)

Bender, M.L., Bergeron, R.J., Komiyama, M.: The bioorganic chemistry of enzymatic catalyis. John Wiley & Sons: New York. 1984

Bergmeyer, H.U., Bergmeyer, J., Graßl, M.: Methods of enzymatic analysis. 3rd edn., Vol. 1ff., Verlag Chemie: Weinheim. 1983ff

Bergmeyer, H.U., Gawehn, K.: Grundlagen der enzymatischen Analyse. Verlag Chemie: Weinheim. 1977

Betz, A.: Enzyme. Verlag Chemie: Weinheim. 1974

Birch, G.G., Blakebrough, N., Parker, K.J.: Enzyme and food processing. Applied Science Publ.: London. 1981

Blow, D.M., Birktoft, J.J., Harley, B.S.: Role of a buried acid group in the mechanism of action of chymotrypsin: Nature *221*, 337 (1969)

Eriksson, C.E.: Enzymic and non-enzymic lipid degradation in foods. In: Industrial aspects of biochemistry (Ed.: Spencer, B.), Federation of European Biochemical Societies, Vol. 30, p. 865, North Holland/American Elsevier: Amsterdam. 1974

Fennema, O.: Actvity of enzymes in partially frozen aqueous systems. In: Water relations of foods (Ed.: Duckworth, R.B.), p. 397, Academic Press: London-New York-San Francisco. 1975

Gray, C.J.: Enzyme-catalysed reactions. Van Nostrand Reinhold Comp.: London. 1971

Guibault, G.G.: Analytical uses of immobilized enzymes. Marcel Dekker, Inc.: New York. 1984

International Union of Biochemistry and Molecular Biology: Enzyme nomenclature 1992. Academic Press: New York-San Francisco-London. 1992

Kessler, H.G.: Lebensmittel- und Bioverfahrenstechnik, Molkereitechnologie, 3. Auflage, Verlag A. Kessler, Freising, 1988

Kilara, A., Shahani, K.A.: The use of immobilized enzymes in the food industry: a review. Crit. Rev. Food Sci. Nutr. *12*, 161 (1979)

Koshland, D.E.: Conformation changes at the active site during enzyme action. Fed. Proc. *23*, 719 (1964)

Koshland, D.E., Neet, K.E.: The catalytic and regulatory properties of proteins. Ann. Rev. Biochem. *37*, 359 (1968)

Lehninger, A.L.: Biochemie. Verlag Chemie: Weinheim. 1977

Levine, H., Slade, L.: A polymer physico-chemical approach to the study of commercial starch hydrolysis products (SHPs). Carbohydrate Polymers *6*, 213 (1986)

Matheis, G.: Polyphenoloxidase and enzymatic browning of potatoes *(Solanum)*. Chem. Microbiol. Technol. Lebensm. *12*, 86 (1989)

Morris, B.A., Clifford, M.N. (Eds.): Immunoassays in food analysis. Elsevier Applied Science Publ.: London. 1985

Morris, B.A., Clifford, M.N., Jackman, R. (Eds.): Immunoassays for veterinary and food analysis. Elsevier Applied Science, London, 1988

Page, M.I., Williams, A. (Eds.): Enzyme mechanisms. The Royal Society of Chemistry, London, 1987

Palmer, T.: Understanding enzymes. 2nd edn., Ellis Horwood Publ.: Chichester. 1985

Phipps, D.A.: Metals and metabolism. Clarendon Press: Oxford. 1978

Plückthun, A.: Wege zu neuen Enzymen: Protein Engineering und katalytische Antikörper. Chem. unserer Zeit *24*, 182 (1990)

Potthast, K., Hamm, R., Acker, L.: Enzymic reactions in low moisture foods. In: Water relations of foods (Ed.: Duckworth, R.B.), p. 365, Academic Press: London-New York-San Francisco. 1975

Reed, G.: Enzymes in food processing. 2nd edn., Academic Press: New York-London. 1975

Richardson, T.: Enzymes. In: Principles of food science, Part I (Ed.: Fennema, O.R.), Marcel Dekker, Inc.: New York-Basel. 1977

Schellenberger, A. (Ed.): Enzymkatalyse, Springer-Verlag, Berlin, 1989

Schwimmer, S.: Source book of food enzymology. AVI Publ. Co.: Westport, Conn. 1981

Scrimgeour, K.G.: Chemistry and control of enzyme reactions. Academic Press: London-New York. 1977

Segel, I.H.: Biochemical calculations. John Wiley and Sons, Inc.: New York. 1968

Shotton, D.: The molecular architecture of the serine proteinases. In: Proceedings of the international research conference on proteinase inhibitors (Eds.: Fritz, H., Tschesche, H.), p. 47, Walter de Gruyter: Berlin-New York. 1971

Suckling, C.J. (Eds.): Enzyme chemistry. Chapman and Hall: London. 1984

Svensson, S.: Inactivation of enzymes during thermal processing. In: Physical, chemical and biological changes in food caused by thermal processing (Eds.: Hoyem, T., Kvalle, O.), p. 202, Applied Science Publ.: London. 1977

Termote, F., Rombouts, F.M., Pilnik, W.: Stabilization of cloud in pectinesterase active orange juice by pectic acid hydrolysates. J. Food Biochem. *1*, 15 (1977)

Teuber, M.: Production of chymosin (EC 3.4.23) by microorganisms and its use for cheesemaking. Int. Dairy Federation, Ann. Sessions Copenhagen 1989, B-Doc 162

Uhlig, H.: Enzyme arbeiten für uns – Technische Enzyme und ihre Anwendung. Carl Hanser Verlag, München, 1991

Whitaker, J.R.: Principles of enzymology for the food sciences. Marcel Dekker, Inc.: New York. 1972

Whitaker J.R., Sonnet, P.E. (Eds.): Biocatalysis in Agricultural Biotechnology. ACS Symp. Ser. 389, American Chemical Society, Washington DC, 1989

Zaborsky, O.R.: Immoiblized enzymes. CRC-Press: Cleveland, Ohio. 1973

3 Lipids

3.1 Foreword

Lipids are formed from structural units with a pronounced hydrophobicity. This solubility characteristic, rather than a common structural feature, is unique for this class of compounds. Lipids are soluble in organic solvents but not in water. Water insolubility is the analytical property used as the basis for their facile separation from proteins and carbohydrates. Some lipids are surface-active since they are amphiphilic molecules (contain both hydrophilic and hydrophobic moieties). Hence, they are polar and thus distinctly different from neutral lipids. The two approaches generally accepted for lipid classification are presented in Table 3.1.

The majority of lipids are derivatives of fatty acids. In these so-called acyl lipids the fatty acids are present as esters and in some minor lipid groups in amide form (Table 3.1). The acyl residue influences strongly the hydrophobicity and the reactivity of the acyl lipids.

Some lipids act as building blocks in the formation of biological membranes which surround cells and subcellular particles. Such lipids occur in food, but usually at less than 2% (cf. Table 3.16). Nevertheless, even as minor food constituents they deserve particular attention, since their high reactivity may strongly influence the organoleptic quality of the food.

Primarily triacylglycerols (also called triglycerides) are deposited in some animal tissues and organs of some plants. Lipid content in such storage tissues can rise to 15–20% or higher and so serve as a commercial source for isolation of triacylglycerols. When this lipid is refined, it is available to the consumer as an edible oil or fat. The nutritive/physiological importance of lipids is based on their role as fuel molecules (37 kJ/g triacylglycerols) and as a source of essential fatty acids and vita-

mins. Apart from these roles, some other lipid properties are indispensable in food handling or processing. These include the pleasant creamy or oily mouthfeel, and the ability to solubilize many taste and aroma constituents of

Table 3.1. Lipid classification

A. Classification according to "acyl residue" characteristics

I. Simple lipids (not saponifiable)

Free fatty acids, isoprenoid lipids (steroids, carotenoids, monoterpenes), tocopherols

II. Acyl lipids (saponifiable)	Constituents
Mono-, di-, triacyl- glycerols	Fatty acid, glycerol
Phospholipids (phosphatides)	Fatty acid, glycerol or sphingosine, phosphoric acid, organic base
Glycolipids	Fatty acid, glycerol or sphingosine, mono-, di- or oligosaccharide
Diol lipids	Fatty acid, ethane, propane, or butane diol
Waxes	Fatty acid, fatty alcohol
Sterol esters	Fatty acid, sterol

B. Classification according to the characteristics "neutral-polar"

Neutral lipids	Polar (amphiphilic) lipids
Fatty acids (> C_{12})	Glycerophospholipid
Mono-, di-, triacyl- glycerols	Glyceroglycolipid
Sterols, sterol esters	Sphingophospholipid
Carotenoids	Sphingoglycolipid
Waxes	
Tocopherols[a]	

[a] Tocopherols and quinone lipids are often considered as "redox lipids".

food. These properties are of importance for food to achieve the desired texture, specific mouthfeel and aroma, and a satisfactory aroma retention. In addition, some foods are prepared by deep frying, i.e. by dipping the food into fat or oil heated to a relatively high temperature.

The lipid class of compounds also includes some important food aroma substances or precursors which are degraded to aroma compounds. Some lipid compounds are indispensable as food emulsifiers, while others are important as fat- or oil-soluble pigments or food colorants.

3.2 Fatty Acids

3.2.1 Nomenclature and Classification

Acyl lipid hydrolysis releases aliphatic carboxylic acids which differ in chemical structure. They can be divided into groups according to chain length, number, position and configuration of their double bonds, and the occurrence of additional functional groups along the chains. The fatty acid distribution pattern in food is another criterion for differentiation.

Table 3.2 compiles the major fatty acids which occur in food. Palmitic, oleic and linoleic acids frequently occur in higher amounts, while the other acids listed, though widely distributed, as a rule occur only in small amounts (major vs minor fatty acids). Percentage data

of acid distribution make it obvious that unsaturated fatty acids are the predominant form in nature.

Fatty acids are usually denoted in the literature by a "shorthand description", e.g. $18:2$ $(9, 12)$ for linoleic acid. Such an abbreviation shows the number of carbon atoms in the acid chain and the number, positions and configurations of the double bonds. All bonds are considered to be cis; whenever trans-bonds are present, an additional "tr" is shown. As will be outlined later in a detailed survey of lipid structure, the carbon skeleton of lipids should be shown as a *zigzag* line (Table 3.2).

3.2.1.1 Saturated Fatty Acids

Unbranched, straight-chain molecules with an even number of carbon atoms are dominant among the saturated fatty acids (Table 3.6). The short-chain, low molecular weight fatty acids ($<14:0$) are triglyceride constituents only in fat and oil of milk, coconut and palmseed. In the free form or esterified with low molecular weight alcohols, they occur in nature only in small amounts, particularly in food processed with the aid of microorganisms as well as in fruits, in which they are aroma substances.

Odor and taste threshold values of fatty acids are compiled in Table 3.3 for cream, butter and cocoa fat. The data for cream and coconut fat indicate lower odor than taste threshold values of C_4- and C_6- fatty acids, while it is the reverse for C_8- up to C_{14}- fatty acids.

Table 3.2. Structures of the major fatty acids

Abbreviated designation	Structure[a]	Common name	Proportion (%)[b]
$14:0$	∿∿∿∿COOH	Myristic acid	2
$16:0$	∿∿∿∿COOH	Palmitic acid	11
$18:0$	∿∿∿∿COOH	Stearic acid	4
$18:1 (9)$	∿∿∿∿COOH	Oleic acid	34
$18:2 (9, 12)$	∿∿∿∿COOH	Linoleic acid	34
$18:3 (9, 12, 15)$	∿∿∿∿COOH	Linolenic acid	5

[a] Numbering of carbon atoms starts with carboxyl group-C as number 1.
[b] A percentage estimate based on world production of edible oils.

Table 3.3. Aroma threshold values (odor and/or taste) of free fatty acids in different food items

Fatty acid	Aroma threshold (mg/kg) in				
	Cream		Sweet cream butter	Coconut fat	
	Odor	Taste		Odor	Taste[a]
4:0	50	60	40	35	160
6:0	85	105	15	25	50
8:0	200	120	455	> 1000	25
10:0	> 400	90	250	> 1000	15
12:0	> 400	130	200	> 1000	35
14:0	> 400	> 400	5000	> 1000	75
16:0	n.d.	n.d.	10000	n.d.	n.d.
18:0	n.d.	n.d.	15000	n.d.	n.d.

[a] Quality of taste: 4:0 rancid, 6:0 rancid, like goat, 8:0 musty, rancid, soapy, 10:0, 12:0 and 14:0 soapy
n.d.: not determined.

Table 3.4. Threshold values[a] of fatty acids depending on the pH-value of an aqueous solution

Fatty acids	Threshold (mg/kg) at pH		
	3.2	4.5	6.0
4:0	0.4	1.9	6.1
6:0	6.7	8.6	27.1
8:0	2.2	8.7	11.3
10:0	1.4	2.2	14.8

[a] Odor and taste.

The aroma threshold increases remarkably with higher pH-values (Table 3.4) since only the undissociated fatty acid molecule is aroma active. Additive effects can be observed in mixtures: examples No. 1 and 2 in Table 3.5 demonstrate that the addition of a mixture of C_4–C_{12} fatty acids to cream will produce a rancid soapy taste if the capryl, capric and lauryl acid contents rise form 30 to 40% of their threshold value concentration. A further increase of these fatty acids to about 50% of the threshold concentration, as in mixture No. 3, results in a musty rancid odor.

Some high molecular weight fatty acids (>18:0) are found in legumes (peanut butter). They can be used, like lower molecular weight homologues, for identification of the source of triglycerides (cf. 14.5.2.3). Fatty acids with odd numbers of carbon atoms, such as valeric (5:0) or enanthic (7:0) acids (Table 3.6) are present in food only in traces. Some of these short-chain homologues are important as food aroma constituents. Pentadecanoic and heptadecanoic acids are odd-numbered fatty acids present in milk and a number of plant oils. The common name "margaric acid" for 17:0 is an erroneous designation. M.E. Chevreul (1786–1889), who first discovered that fats are glycerol esters of fatty acids, coined the word "margarine" to denote a product from oleomargarine (a fraction of edible beef tallow), believing that the product contained a new fatty acid, 17:0. Only later was it clarified that such margarine or "17:0 acid" was a mixture of palmitic and stearic acids.

Branched-chain acids, such as iso- (with an isopropyl terminal group) or anteiso- (a secondary butyl terminal group) are rarely found in food. Pristanic and phytanic acids have been detected in milk fat (Table 3.6). They are isoprenoid acids obtained from the degradation of the phytol side chain of chlorophyll.

3.2.1.2 Unsaturated Fatty Acids

The unsaturated fatty acids, which dominate lipids, contain one, two or three allyl groups in their acyl residues (Table 3.7). Acids with isolated double bonds (a methylene group inserted between the two cis-double bonds) are usually denoted as isolene-type or nonconjugated fatty acids.

Table 3.5. Odor and taste of fatty acid mixtures in cream

No.	Fatty acid mixtures of					Odor	Taste
	4:0	6:0	8:0	10:0	12:0		
	Concentration in % of aroma threshold[a]						
1	28	17	29	31	30	n. O.	n. T.
2	28	17	40	42	37	n. O.	rancid, soapy
3	28	17	52	53	45	musty, rancid	rancid, soapy
4	48	29	29	31	30	musty, rancid	n. T.
5	48	29	40	42	37	musty, rancid	rancid, soapy

[a] The concentration of each fatty acid is based on the threshold values indicated in Table 3.3 for odor for 4:0 and 6:0 and for taste for 8:0–12:0.
n. O. = no difference in odor from that of cream.
n. T. = no difference in taste from that of cream.

Table 3.6. Saturated fatty acids

Abbreviated designation	Structure	Systematic name	Common name	Melting point (°C)
A. Even numbered straight chain fatty acids				
4:0	$CH_3(CH_2)_2COOH$	Butanoic acid	Butyric acid	− 7.9
6:0	$CH_3(CH_2)_4COOH$	Hexanoic acid	Caproic acid	− 3.9
8:0	$CH_3(CH_2)_6COOH$	Octanoic acid	Caprylic acid	16.3
10:0	$CH_3(CH_2)_8COOH$	Decanoic acid	Capric acid	31.3
12:0	$CH_3(CH_2)_{10}COOH$	Dodecanoic acid	Lauric acid	44.0
14:0	$CH_3(CH_2)_{12}COOH$	Tetradecanoic acid	Myristic acid	54.4
16:0	$CH_3(CH_2)_{14}COOH$	Hexadecanoic acid	Palmitic acid	62.9
18:0	$CH_3(CH_2)_{16}COOH$	Octadecanoic acid	Stearic acid	69.6
20:0	$CH_3(CH_2)_{18}COOH$	Eicosanoic acid	Arachidic acid	75.4
22:0	$CH_3(CH_2)_{20}COOH$	Docosanoic acid	Behenic acid	80.0
24:0	$CH_3(CH_2)_{22}COOH$	Tetracosanoic acid	Lignoceric acid	84.2
26:0	$CH_3(CH_2)_{24}COOH$	Hexacosanoic acid	Cerotic acid	87.7
B. Odd numbered straight chain fatty acids				
5:0	$CH_3(CH_2)_3COOH$	Pentanoic acid	Valeric acid	− 34.5
7:0	$CH_3(CH_2)_5COOH$	Heptanoic acid	Enanthic acid	− 7.5
9:0	$CH_3(CH_2)_7COOH$	Nonanoic acid	Pelargonic acid	12.4
15:0	$CH_3(CH_2)_{13}COOH$	Pentadecanoic acid		52.1
17:0	$CH_3(CH_2)_{15}COOH$	Heptadecanoic acid	Margaric acid	61.3
C. Branched chain fatty acids				
	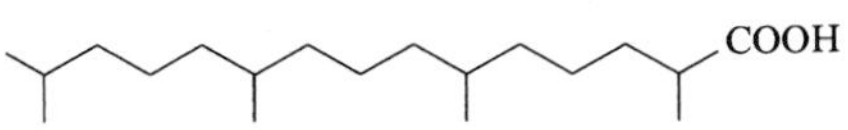 COOH	2,6,10,14-Tetra-methyl-penta-decanoic acid	Pristanic acid	
	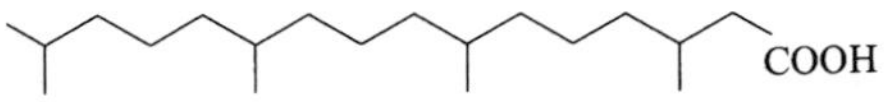 COOH	3,7,11,15-Tetra-methyl-hexa-decanoic acid	Phytanic acid	

Table 3.7. Unsaturated fatty acids

Abbreviated designation	Structure	Common name	Melting point (°C)
A. Fatty acids with nonconjugated cis double bonds			
	ω9-Family		
18:1 (9)	$CH_3-(CH_2)_7-CH=CH-CH_2-(CH_2)_6-COOH$	Oleic acid	13.4
22:1 (13)	$-(CH_2)_{10}-COOH$	Erucic acid	34.7
24:1 (15)	$-(CH_2)_{12}-COOH$	Nervonic acid	42.5
	ω6-Family		
18:2 (9, 12)	$CH_3-(CH_2)_4-(CH=CH-CH_2)_2-(CH_2)_6-COOH$	Linoleic acid	− 5.0
18:3 (6, 9, 12)	$-(CH=CH-CH_2)_3-(CH_2)_3-COOH$	γ-Linolenic acid	
20:4 (5, 8, 11, 14)	$-(CH=CH-CH_2)_4-(CH_2)_2-COOH$	Arachidonic acid	− 49.5
	ω3-Family		
18:3 (9, 12, 15)	$CH_3-CH_2-(CH=CH-CH_2)_3-(CH_2)_6-COOH$	α-Linolenic acid	− 11.0
20:5 (5, 8, 11, 14, 17)	$-(CH=CH-CH_2)_5-(CH_2)_2-COOH$	EPA[a]	
22:6 (4, 7, 10, 13, 16, 19)	$-(CH=CH-CH_2)_6-CH_2-COOH$	DHA[a]	
	Δ9-Family		
18:1 (9)	$CH_3-(CH_2)_7-CH=CH-CH_2-(CH_2)_6-COOH$	Oleic acid	13.4
16:1 (9)	$CH_3-(CH_2)_5-$	Palmitoleic acid	0.5
14:1 (9)	$CH_3-(CH_2)_3-$	Myristoleic acid	
B. Fatty acids with nonconjugated trans-double bonds			
18:1 (tr9)	$CH_3-(CH_2)_7-CH\overset{tr}{=}CH-(CH_2)_7-COOH$	Elaidic acid	46
18:2 (tr9, tr12)	$CH_3-(CH_2)_4-CH\overset{tr}{=}CH-CH_2-CH\overset{tr}{=}CH-(CH_2)_7-COOH$	Linolelaidic acid	28
C. Fatty acids with conjugated double bonds			
18:3 (9, tr11, tr13)	$CH_3-(CH_2)_3-CH\overset{tr}{=}CH-CH\overset{tr}{=}CH-CH\overset{c}{=}CH-(CH_2)_7-COOH$	α-Eleostearic acid	48
18:3 (tr9, tr11, tr13)	$CH_3-(CH_2)_3-CH\overset{tr}{=}CH-CH\overset{tr}{=}CH-CH\overset{tr}{=}CH-(CH_2)_7-COOH$	β-Eleostearic acid	71.5
18:4 (9, 11, 13, 15)[b]	$CH_3-CH_2-(CH=CH)_4-(CH_2)_7-COOH$	Parinaric acid	85

[a] EPA: Eicosapentanoic acid, DHA: Docosahexanoic acid.
[b] Geometry of the double bond was not determined.

The structural relationship that exists among the unsaturated, nonconjugated fatty acids derived from a common biosynthetic pathway is distinctly revealed when the double bond position is determined by counting from the methyl end of the chain (it should be emphasized that position designation using this method of counting requires the suffix "ω"). Acids with the same methyl ends are then combined into groups. Thus, three family groups exist: ω3 (linolenic type), ω6 (linoleic type) and ω9 (oleic acid type; Table 3.7). Using this classification, the common structural features abundantly found in C_{18} fatty acids (Table 3.2) are also found in less frequently occurring fatty acids. Thus, erucic acid (20:1), occurring only in the mustard family of seeds (*Brassicaceae*, cf. 14.3.2.2.5), belongs to the ω9 group, arachidonic acid (20:4), occurring in meat, liver, lard and lipids of chicken eggs, belongs to the ω6 group, while the $C_{20}-C_{22}$ fatty acids with 5 and 6 double bonds, occurring in fish lipids, belong to the ω3 group.

Linoleic acid can not be synthesized by the human body. This acid and other members of the ω6 family are considered as essential fatty acids required as building blocks for biologically active membranes. Whether α-linolenic acid, which belongs to the ω3 family and which is synthesized only by plants, plays a nutritional role as an essential fatty acid is disputed.

A formal relationship exists in some olefinic unsaturated fatty acids with regard to the position of the double bond when counted from the carboxyl end of the chain. Oleic, palmitoleic and myristoleic acids belong to such a Δ9 family (cf. Table 3.7); the latter two fatty acids are minor constituents in foods of animal or plant origin.

Unsaturated fatty acids with an unusual structure are those with one trans-double bond and/or conjugated double bonds (Table 3.7). Such trans-unsaturated acids are formed as artifacts in the industrial processing of oil or fat (heat treatment, oil hardening). However, several occur in nature. The trans-analogue of oleic acid is found in mutton tallow, while that of linoleic acid is found in *Chilopsis linearis* seeds. Conjugated fatty acids with diene, triene or tetraene systems occur frequently in several

Table 3.8. Taste of unsaturated fatty acids emulsified in water

Compound	Threshold (mmol/l)	Quality
Oleic acid	9–12	bitter, burning, pungent
Elaidic acid	22	slightly burning
Linoleic acid	4–6	bitter, burning, pungent
Linolelaidic acid	11–15	bitter, burning, scratchy
γ-Linolenic acid	3–6	bitter, burning, pungent
α-Linolenic acid	0.6–1.2	bitter, burning, pungent, like fresh walnut
Arachidonic acid	6–8	bitter, repugnant off-taste

seed oils, but do not play a role in human nutrition. Table 3.7 presents, as an example, two naturally occurring acids with conjugated triene systems which differ in the configuration of one double bond at position 9 (cis, trans).

Unsaturated fatty acids emulsified in water taste bitter with a relatively low threshold value for α-linolenic acid (Table 3.8). Thus an off-taste can be present due to fatty acids liberated, as indicated in Table 3.8, by the enzymatic hydrolysis of unsaturated triacyl glycerides which are tasteless in an aqueous emulsion.

3.2.1.3 Substituted Fatty Acids

Hydroxy Fatty Acids. Ricinoleic acid is the best known of the straight-chain hydroxy fatty acids. Its structure is 12-OH, 18:1 (9). It is an optically active acid with a D(+)-configuration:

$$(3.1)$$

Ricinoleic acid is the main acid of castor bean oil, comprising up to 90% of the total acids. Hence, it can serve as an indicator for the presence of this oil in edible oil blends.

D-2-Hydroxy saturated 16:0 to 25:0 fatty acids with both even and odd numbers of carbons in their chains occur in lipids in green leaves of a great number of vegetables.

γ- or ∂-Lactones are obtained from 4- and 5-hydroxycarboxylic acids (C_8 to C_{16}) by the elimination of water. ∂-Lactones have been found in milk fat and fruits. They are very active aroma components (cf. 5.3.2.3).

Oxo Fatty Acids. Natural oxo (or keto) acids are less common than hydroxy acids. About 1% of milk fat consists of saturated (C_{10}–C_{24}) oxo fatty acids, with an even number of carbon atoms, in which the carbonyl group is located on C-5 to C-13: One of 47 identified compounds of this substance class has the following structure:

$$CH_3-(CH_2)_4-CH=CH-CH_2-CH_2-\overset{\overset{O}{\|}}{C}-(CH_2)_7-COOH \tag{3.2}$$

Furan Fatty Acids. These occur in fish liver oil in a range of 1–6% and up to 25% in some freshwater fish. Furan fatty acids are also part of the minor constituents of some plant oils and butter (Table 3.9). They are also present in fruits (lemon, strawberry), vegetables (cabbage, potato) and mushrooms (champignons).
Two of these acids have the following formulas

$$\tag{3.3}$$

I: n = 8;   II: n = 10

Photooxidation (cf. 3.7.2.1.4) of these acids can deteriorate especially the quality of soybean oil.
Substituted fatty acids are also derived by autoxidation or enzymatic peroxidation of unsaturated fatty acids, which will be dealt with in more detail in 3.7.2.3 and 3.7.2.4.1.

Table 3.9. Furan fatty acids I and II in plant oils and butter

Oil	Concentration (mg/kg)	
	I[a]	II[a]
Soya oil	120–170	130–230
Wheat germ oil	100–130	105–150
Rapeseed oil	6– 16	7– 20
Corn oil	8– 11	9– 13
Butter	13–139	24–208

[a] I: 10,13-epoxy-11,12-dimethyloctadeca-10,12-dienoic acid.
II: 12,15-epoxy-13,14-dimethyleicosa-12,14-dienoic acid (Formula 3.3).

3.2.2 Physical Properties

3.2.2.1 Carboxyl Group

Carboxylic acids have a great tendency to form dimers which are stabilized by hydrogen bonds:

$$\tag{3.4}$$

The binding energy of the acid dimer dissolved in hexane is 38 kJ/mole. Also, the fatty acid molecules are arranged as dimers in the crystalline lattice (cf. Fig. 3.2).
The acidic character of the carboxyl group is based on proton dissociation and on the formation of the resonance-stabilized carboxylate anion:

$$\tag{3.5}$$

The pK_S values for the C_2–C_9 short-chain acid homologues range from 4.75–4.95. The pK_S of 7.9 for linoleic acid deviates considerably from this range. This unexpected and anomalous behavior, which has not yet been clarified, is clearly illustrated in the titration curves for propionic, caprylic and linoleic acids recorded under identical conditions (Fig. 3.1).

3.2.2.2 Crystalline Structure, Melting Points

Melting properties of fats depend on the arrangement of the acyl residues in the crystal lattice in addition to other factors attributed solely to the structure of triglycerides.
Calculations of the energy content of the carbon chain conformation have revealed that at room temperature 75% of the C–C bonds of a saturated fatty acid are present in a fully staggered *zigzag* or "trans" conformation and only 25% in the energetically slightly less favorable skew conformation.
The unsaturated fatty acids, because of their double bonds, are not free to rotate and hence have rigid kinks along the carbon chain skeleton. However, a molecule is less bent by a trans

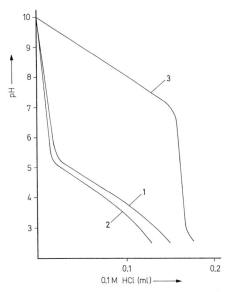

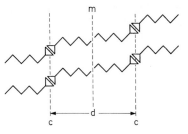

Fig. 3.2. Arrangement of caproic acid molecules in crystal (according to *J.M. Mead* et al., 1965). Results of a X-ray diffraction analysis reveal a strong diffraction in the plane of carboxyl groups (c) and a weak diffraction at the methyl terminals (m): d: identity period

Fig. 3.1. Fatty acid titration curves (according to *G.S. Bild* et al., 1977). Aqueous solutions (0.1 mol/l) of Na-salts of propionic (1), caprylic (2) and linoleic acids (3) were titrated with 0.1 mol/l HCl

than by a cis double bond. Thus, this cis-configuration in oleic acid causes a bending of about 40°:

$$\text{(structure)} \qquad (3.6)$$

The corresponding elaidic acid, with a trans-configuration, has a slightly shortened C-chain, but is still similar to the linear form of stearic acid:

$$\text{(structure)} \qquad (3.7)$$

The extent of molecular crumpling is also increased by an increase in the number of cis double bonds. Thus, the four cis double bonds in arachidonic acid increase the deviation from a straight line to 165°:

$$\text{(structure)} \qquad (3.8)$$

When fatty acids crystallize, the saturated acids are oriented as depicted by the simplified pattern in Fig. 3.2. The dimer molecular ar-

rangement is thereby retained. The principal reflections of the X-ray beam are from the planes (c) of high electron density in which the carboxyl groups are situated. The length of the fatty acid molecule can be determined from the "main reflection" site intervals (distance d in Fig. 3.2). For stearic acid (18:0), this distance is 2.45 nm.

The crystalline lattice is stabilized by hydrophobic interaction along the acyl residues. Correspondingly, the energy and therefore the temperature required to melt the crystal increase with an increased number of carbons in the chain.

Odd-numbered as well as unsaturated fatty acids can not be uniformly packed into a crystalline lattice as can the saturated and even-numbered acids. The odd-numbered acids are slightly interfered by their terminal methyl groups.

The consequence of less symmetry within the crystal is that the melting points of even-numbered acids (C_n) exceed the melting points of the next higher odd-numbered (C_{n+1}) fatty acids (cf. Table 3.6).

The molecular arrangement in the crystalline lattice of unsaturated fatty acids is not strongly influenced by trans double bonds, but is strongly influenced by cis double bonds. This difference, due to steric interference as mentioned above, is reflected in a decrease in melting points in the fatty acid series 18:0, 18:1 (tr9) and 18:1 (9). However, this ranking should be considered as reliable only when the double bond positions within the molecules are fairly comparable. Thus, when a cis double bond is at the end of the carbon chain, the deviation from the form of a straight extended acid is not as large as in oleic acid. Hence, the melting point of such an acid is higher. The melting point of cis-2-octadecenoic acid is in agreement with this rule; it even surpasses the 9-trans isomer of the same acid (Table 3.10).

The melting point decreases with an increasing number of isolated cis- double bonds (Table 3.10). This behavior can be explained by the changes in the geometry of the molecules, as can be seen when comparing the geometric structures of oleic and arachidonic acid.

3.2.2.3 Urea Adducts

When urea crystallizes, channels with a diameter of 0.8–1.2 nm are formed within its crystals and can accomodate long-chain hydrocarbons. The stability of such urea adducts of fatty acids parallels the geometry of the acid molecule. Any deviation from a straight-chain arrangement brings about weakening of the adduct. A tendency to form inclusion compounds decreases in the series $18:0 > 18:1 (9) > 18:2 (9, 12)$.

A substitution on the acyl chain prevents adduct formation. Thus, it is possible to separate branched or oxidized fatty acids or their methyl esters from the corresponding straight-chain compounds on the basis of the formation of urea adducts. This principle is used as a

method for preparative-scale enrichment and separation of branched or oxidized acids from a mixture of fatty acids.

3.2.2.4 Solubility

Long-chain fatty acids are practically insoluble in water; instead, they form a floating film on the water surface. The polar carboxyl groups in this film are oriented toward the water, while the hydrophobic tails protrude into the gaseous phase. The solubility of the acids increases with decreasing carbon number; butyric acid is completely soluble in water.

Ethyl ether is the best solvent for stearic acid and other saturated long-chain fatty acids since it is sufficiently polar to attract the carboxyl groups. A truly nonpolar solvent, such as petroleum ether, is not suitable for free fatty acids.

The solubility of fatty acids increases with an increase in the number of cis double bonds. This is illustrated in Fig. 3.3 with acetone as a solvent. The observed differences in solubility can be utilized for separation of saturated from unsaturated fatty acids. The mixture of acids is dissolved at room temperature and cooled stepwise to −80 °C. However, the separation efficiency of such a fractional crystallization is limited since, for example, stearic acid is substantially more soluble in acetone containing oleic acid than in pure acetone. This mutual effect on solubility has not been considered in Fig. 3.3.

3.2.2.5 UV-Absorption

All unsaturated fatty acids which contain an isolated cis double bond absorb UV light at a wavelength close to 190 nm. Thus, the acids can not be distinguished spectrophotometrically. Conjugated fatty acids absorb light at various wavelenths depending on the length of conjugation and configuration of the double bond system.

Figure 3.4 illustrates such behavior for several fatty acids. See 3.2.3.2.2 for the conversion of an isolene-type fatty acid into a conjugated fatty acid.

Table 3.10. The effect of number, configuration and double bond position on melting points of fatty acids

Fatty acid		Melting point (°C)
18:0	Stearic acid	69
18:1 (tr9)	Elaidic acid	46
18:1 (2)	cis-2-Octadecenoic acid	51
18:1 (9)	Oleic acid	13.4
18:2 (9, 12)	Linoleic acid	− 5
18:2 (tr9, tr12)	Linolelaidic acid	28
18:3 (9, 12, 15)	α-Linolenic acid	− 11
20:0	Arachidic acid	75.4
20:4 (5, 8, 11, 14)	Arachidonic acid	− 49.5

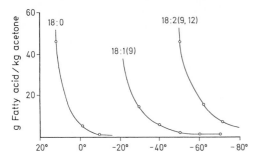

Fig. 3.3. Fatty acid solubility in acetone (according to *J. M. Mead* et al., 1965)

3.2.3 Chemical Properties

3.2.3.1 Methylation of Carboxyl Groups

The carboxyl group of a fatty acid must be depolarized by methylation in order to facilitate gas chromatographic separation or separation by fractional distillation. Reaction with diazomethane is preferred for analytical purposes. Diazomethane is formed by alkaline hydrolysis of N-nitroso-N-methyl-p-toluene sulfonamide.

The gaseous CH_2N_2 released by hydrolysis is swept by a stream of nitrogen into a receiver containing the fatty acid solution in ether-methanol (9:1 v/v). The reaction:

$$R-COOH + CH_2N_2 \longrightarrow R-COOCH_3 + N_2 \qquad (3.9)$$

proceeds under mild conditions without formation of by-products. Further possibilities for methylation include: esterification in the presence of excess methanol and a *Lewis* acid (BF_3) as a catalyst; or the reaction of a fatty acid silver salt with methyl iodide:

$$R-COOAg + CH_3I \longrightarrow R-COOCH_3 + AgI \qquad (3.10)$$

3.2.3.2 Reactions of Unsaturated Fatty Acids

A number of reactions which are known for olefinic hydrocarbons play an important role in the analysis and processing of lipids containing unsaturated fatty acids.

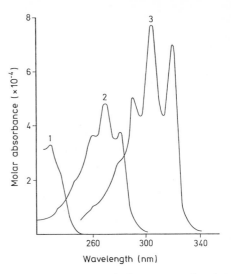

Fig. 3.4. Electron excitation spectra of conjugated fatty acids (according to *H. Pardun*, 1976). 1 9,11-isolinoleic acid, 2 α-elaeostearic acid, 3 parinaric acid

3.2.3.2.1 Halogen Addition Reactions

The number of double bonds present in an oil or fat can be determined through their iodine number (cf. 14.5.2.1). The fat or oil is treated with a halogen reagent which reacts only with the double bonds. Substitution reactions generating hydrogen halides must be avoided. IBr in an inert solvent, such as glacial acetic acid, is a suitable reagent:

$$(3.11)$$

The number of double bonds is calculated by titrating the unreacted IBr reagent with thiosulfate.

3.2.3.2.2 Transformation of Isolene-Type Fatty Acids to Conjugated Fatty Acids

Allyl systems are labile and are readily converted to a conjugated double bond system in the presence of a base (KOH or K-tert-butylate):

$$- CH = CH - C - CH = CH - \xrightarrow{\qquad} \qquad (3.12)$$

$$- CH = CH - CH = CH - CH_2 -$$

During this reaction, an equilibrium is established between the isolene and the conjugated forms of the fatty acid, the equilibrium state being dependent on the reaction conditions. This isomerization is used analytically since it provides a way to simultaneously determine linoleic, linolenic and arachidonic acids in a fatty acid mixture. The corresponding conjugated diene, triene and tetraene systems of these fatty acids have a maximum absorbance at distinct wavelengths (cf. Fig. 3.4). The assay conditions can be selected to isomerize only the naturally occurring cis double bonds and to ignore the trans fatty acids formed, for instance, during oil hardening (cf. 14.4.2).

3.2.3.2.3 Formation of a π-Complex with Ag$^+$ Ions

Unsaturated fatty acids or their triacylglycerols, as well as unsaturated aldehydes obtained through autoxidation of lipids (cf. 3.7.2.1.5), can be separated by "argentation chromatography". The separation is based on the number, position and configuration of the double bonds present. The separation mechanism involves interaction of the π-electrons of the double bond with Ag$^+$ions, forming a reversible π-complex of variable stability:

$$C = C + Ag^\oplus \rightleftharpoons \underset{Ag}{C - C} \qquad (3.13)$$

The complex stability increases with increasing number of double bonds. This means a fatty acid with two cis double bonds will not migrate as far as a fatty acid with one double bond on a thin-layer plate impregnated with a silver salt. The R_f values increase for the series 18:2 (9, 12) < 18:1 (9) < 18:0. Furthermore, fatty acids with isolated double bonds form a stronger Ag$^+$ complex than those with conjugated bonds. Also, the complex is stronger with a cis- than with a trans-configuration. The complex is also more stable, the further the

double bond is from the end of the chain. Finally, a separation of nonconjugated from conjugated fatty acids and of isomers that differ only in their double bond configuration is possible by argentation chromatography.

3.2.3.2.4 Hydrogenation

In the presence of a suitable catalyst, e. g. Ni, hydrogen can be added to the double bond of an acyl lipid. This heterogeneous catalytic hydrogenation occurs stereo selectively as a cis-addition. Catalyst-induced isomerization from an isolene-type fatty acid to a conjugated fatty acid occurs with fatty acids with several double bonds:

(1) Isomerization
(2) Hydrogenation

(3.14)

Since diene fatty acids form a more stable complex with a catalyst than do monoene fatty acids, the former are preferentially hydrogenated. Since nature is not an abundant source of the solid fats which are required in food processing, the partial and selective hydrogenation, just referred to, plays an important role in the industrial processing of fats and oils (cf. 14.4.2).

3.2.4 Biosynthesis of Unsaturated Fatty Acids

The biosynthetic precursors of unsaturated fatty acids are saturated fatty acids in an activated form (cf. a biochemistry textbook). These are aerobiocally and stereospecifically dehydrogenated by dehydrogenase action in plant as well as animal tissues. A flavoprotein and ferredoxin are involved in plants in the electron transport system which uses oxygen as a terminal electron acceptor (cf. Reaction 3.15).

To obtain polyunsaturated fatty acids, the double bonds are introduced by a stepwise process. A fundamental difference exists between mammals and plants. In the former, oleic acid synthesis is possible, and, also, addi-

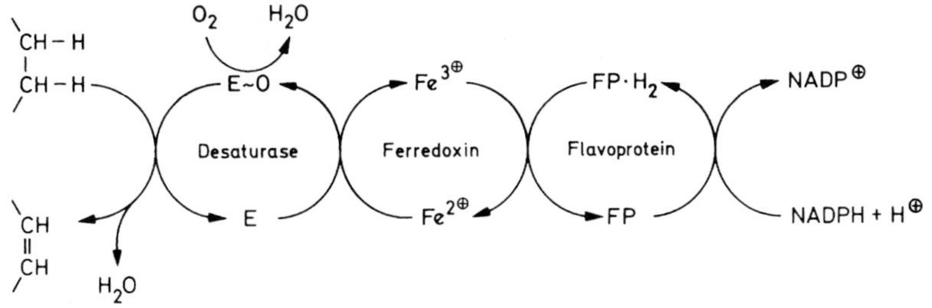

$$
\begin{array}{c}
\text{CH-H} \\
| \\
\text{CH-H} \\
/
\end{array}
\quad
\begin{array}{c}
O_2 \quad H_2O
\end{array}
\qquad
\begin{array}{c}
Fe^{3\oplus}
\end{array}
\qquad
\begin{array}{c}
FP\cdot H_2
\end{array}
\qquad
NADP^{\oplus}
$$

E-O — Desaturase — Ferredoxin — Flavoprotein

$$
\begin{array}{c}
\text{CH} \\
\| \\
\text{CH} \\
/ \quad H_2O
\end{array}
\qquad E \qquad Fe^{2\oplus} \qquad FP \qquad NADPH + H^{\oplus}
$$

(3.15)

tional double bonds can be inserted towards the carboxyl end of the fatty acid molecule. For example, γ-linolenic acid can be formed from the essential fatty acid linoleic acid and, also, arachidonic acid (Fig. 3.5) can be formed by chain elongation of γ-linolenic acid. In a diet deficient in linoleic acid, oleic acid is dehydrogenated to isolinoleic acid and its derivatives (Fig. 3.5), but none of these acquire the physiological function of an essential acid such as linoleic acid.

Plants can introduce double bonds into fatty acids in both directions: towards the terminal CH_3-group or towards the carboxyl end. Oleic acid (oleoyl-CoA ester or β-oleoyl-phosphatidylcholine) is thus dehydrogenated to linoleic and then to linolenic acid. In addition synthesis of the latter can be achieved by another pathway involving stepwise dehydrogenation of lauric acid with chain elongation reactions involving C_2 units (Fig. 3.5).

3.3 Acylglycerols

Acylglycerols (or acylglycerides) comprise the mono-, di- or triesters of glycerol with fatty acids (Table 3.1). They are designated as neutral lipids. Edible oils or fats consist nearly completely of triacylglycerols.

3.3.1 Triacylglycerols (TG)

3.3.1.1 Nomenclature, Classification

Glycerol, as a trihydroxylic alcohol, can form triesters with one, two or three different fatty acids. In the first case a triester is formed with three of the same acyl residues (e.g. tripalmitin; P_3). The mixed esters involve two or three different acyl residues, e.g., dipalmito-olein (P_2O) and palmito-oleo-linolein (POL). The rule of this shorthand designation is that the

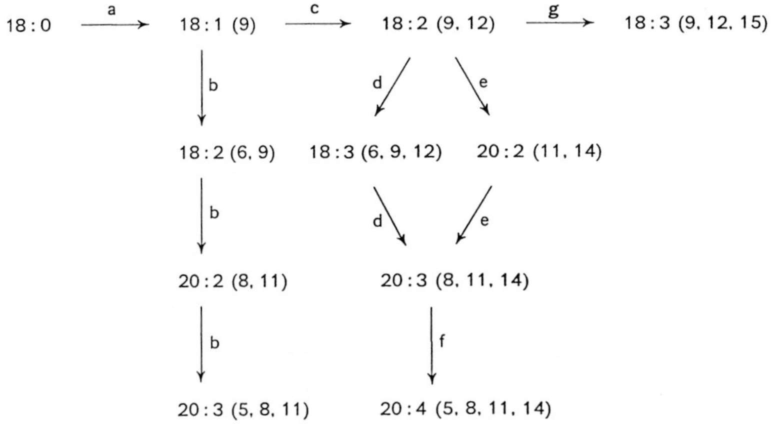

18:0 —a→ 18:1 (9) —c→ 18:2 (9, 12) —g→ 18:3 (9, 12, 15)

|b   d/ \e

18:2 (6, 9)   18:3 (6, 9, 12)   20:2 (11, 14)

|b   d\   /e

20:2 (8, 11)   20:3 (8, 11, 14)

|b |f

20:3 (5, 8, 11)   20:4 (5, 8, 11, 14)

Fig. 3.5. Biosynthesis of unsaturated fatty acids
Synthesis pathways: a, c, g in higher plants; a, c, g and a, c, d, f in algae; a, b and d, f (main pathway for arachidonic acid) or e, f in mammals

acid with the shorter chain or, in the case of an equal number of carbons in the chain, the chain with fewer double bonds, is mentioned first. The Z number gives the possible different triacylglycerols which can occur in a fat (oil), where n is the number of different fatty acids identified in that fat (oil):

$$Z = \frac{n^3 + n^2}{2} \qquad (3.16)$$

For n = 3, the possible number of triglycerols (Z) is 18. However, such a case where a fat (oil) contains only three fatty acids is rarely found in nature. One exception is Borneo tallow (cf. 14.3.2.2.3), which contains essentially only 16:0, 18:0 and 18:1 (9) fatty acids.

Naturally, the Z value also takes into account the number of possible positional isomers within a molecule, for example, by the combination of POS, PSO and SOP. When only positional isomers are considered and the rest disregarded, Z is reduced to Z':

$$Z' = \frac{n^3 + 3n^2 + 2n}{6} \qquad (3.17)$$

Thus, when n = 3, Z' = 10.
A chiral center exists in a triacylglycerol when the acyl residues in positions 1 and 3 are different:

$$\begin{array}{ll} & \text{CH}_2-\text{O}-\text{CO}-\text{R}_1 \\ & \quad | * \\ \text{R}_1-\text{CO}-\text{O}-\text{CH} & \qquad * \text{chiral} \\ & \quad | & \qquad \text{center} \qquad (3.18) \\ & \text{CH}_2-\text{O}-\text{CO}-\text{R}_2 \end{array}$$

In addition enantiomers may be produced by 1-monoglycerides, all 1,2-diglycerides and 1,3-diglycerides containing unlike substituents.
In the stereospecific numbering of acyl residues (prefix sn), the L-glycerol molecule is shown in the *Fischer* projection with the secondary HO-group pointing to the left. The top carbon is then denoted C-1. Actually, in a *Fischer* projection, the horizontal bonds denote bonds in front and the vertical bonds those behind the plane of the page:

$$\begin{array}{ll} \text{CH}_2-\text{OH} & \text{sn-1} \\ \quad | & \\ \text{HO}-\text{C}-\text{H} & \text{sn-2} \qquad (3.19) \\ \quad | & \\ \text{CH}_2-\text{OH} & \text{sn-3} \end{array}$$

For example, the nomenclature for a triacylglycerol which contains P, S and O:

sn-POS = sn-1-Palmito-2-oleo-3-stearin.

This assertion is only possible when a stereospecific analysis (cf. 3.3.1.4) provides information on the fatty acids at positions 1, 2 and 3. rac-POS = sn-POS and sn-SOP in the molar ratio 1:1, i.e. the fatty acid in position 2 is fixed while the other two acids are equally distributed at positions 1 and 3.

POS = mixture of sn-POS, sn-OPS, sn-SOP, sn-PSO, sn-OSP and sn-SPO

3.3.1.2 Melting Properties

TG melting properties are affected by fatty acid composition and their distribution within the glyceride molecule (Table 3.11).
Mono-, di- and triglycerides are polymorphic, i.e. they crystallize in different modifications, denoted as α, β' and β. These forms differ in their melting points (Table 3.11) and crystallographic properties.
During the cooling of melted acylglycerols, one of the three polymorphic forms is yielded. This depends also on the temperature gradient chosen. The α-form has the lowest melting point. This modification is transformed first into the β'-form upon heating and then into the β-form. The β-form is the most stable and, hence, also has the highest melting point (Table 3.11). These changes are typically monotropic, i.e. they proceed in the order of lower to higher stability.
Crystallization of triglycerides from a solvent system generally yields β-form crystals.
X-ray analysis as well as measurements by Raman spectroscopy revealed that saturated triglycerols in their crystalline state exist in a chair form (Fig. 3.6a): The "tuning fork" configuration for the β'-modification was not verified. The different properties of the three forms are based on the crystallization in different systems.
α-form: hexagonal system; the melting point is relatively low, since areas of the methyl ends are freely arranged as in liquid crystals.
β'-form: (Fig. 3.6b): orthorhombic system; the carbon chains are perpendicular to each other.

Table 3.11. Triacylglycerols and their polymorphic forms

Compound	Melting point (°C) of polymorphic form		
	α	β′	β
Tristearin	55	63.2	73.5
Tripalmitin	44.7	56.6	66.4
Trimyristin	32.8	45.0	58.5
Trilaurin	15.2	34	46.5
Triolein	− 32	− 12	4.5−5.7
1,2-Dipalmitoolein	18.5	29.8	34.8
1,3-Dipalmitoolein	20.8	33	37.3
1-Palmito-3-stearo-2-olein	18.2	33	39
1-Palmito-2-stearo-3-olein	26.3	40.2	
2-Palmito-1-stearo-3-olein	25.3	40.2	
1,2-Diacetopalmitin	20.5	21.6	42.3

β-form: (Fig. 3.6c): triclinic system; parallel arrangement of the carbon chains.

Unsaturated fatty acids interfere with the orderly packing of molecules in the crystalline lattice, thereby causing a decrease in the melting point of the crystals.

TG such as 1,3-diaceto-palmitin, i.e. a triglyceride with one long and two short-chain fatty acids, exists in the exceptionally stable α-form. Since films of such TG's can expand by 200 to 300 times their normal length, they are of interest for application as protective coating for fat-containing foods. In edible fats and oils, more than the three mentioned polymorphic forms can be present, e.g., 4−6 forms are being discussed for cocoa butter. In order to classify fats and oils, that form is used that is predominant after solidification.

3.3.1.3 Chemical Properties

Hydrolysis, methanolysis and interesterification are the most important chemical reactions for TG's.

Hydrolysis. The fat or oil is cleaved or saponified by treatment with alkali (e.g. alcoholic KOH):

$$R-CO-O-\begin{bmatrix} -O-CO-R \\ \\ -O-CO-R \end{bmatrix} + 3\,KOH \longrightarrow HO-\begin{bmatrix} -OH \\ \\ -OH \end{bmatrix} + 3\,RCOOK$$

(3.20)

After acidification and extraction, the free fatty acids are recovered as alkali salts (commonly called soaps). This procedure is of interest for analysis of fat or oil samples. Commercially, the free fatty acids are produced by cleaving

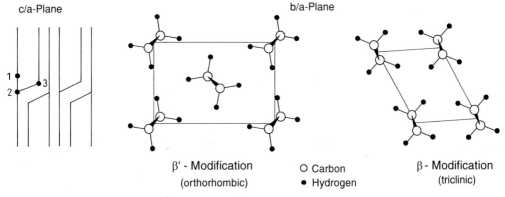

c/a-Plane   b/a-Plane

β′ - Modification   ○ Carbon   β - Modification
(orthorhombic)   ● Hydrogen   (triclinic)

Fig. 3.6. Arrangement of the β′- and β-form of saturated triacylglycerols in the crystalline lattice (Cartesian coordinates a, b, c)

Table 3.12. Crystallization patterns of edible fats or oils

β-Type	β'-Type	β-Type	β'-Type
Coconut oil	Cottonseed oil	Peanut butter	Tallow
Corn germ oil	Butter	Sunflower oil	Whale oil
Olive oil	Palm oil	Lard	
Palm seed oil	Rapeseed oil		

triglycerides with steam under elevated pressure and temperature and by increasing the reaction rate in the presence of an alkaline catalyst (ZnO, MgO or CaO) or an acidic catalyst (aromatic sulfonic acid).

Methanolysis. The fatty acids in TG are usually analyzed by gas liquid chromatography, not as free acids, but as methyl esters. The required transesterification is most often achieved by Na-methylate (sodium methoxide) in methanol and in the presence of 2,2-dimethoxypropane to bind the released glycerol. Thus, the reaction proceeds rapidly and quantitatively even at room temperature.

Interesterification. This reaction is of industrial importance (cf. 14.4.3) since it can change the physical properties of fats or oils or their mixtures without altering the chemical structure of the fatty acids. Both intra- and intermolecular acyl residue exchanges occur in the

reaction until an equilibrium is reached which depends on the structure and composition of the TG molecules. The usual catalyst for interesterification is Na-methylate.

The principle of the reaction will be elucidated by using a mixture of tristearin (SSS) and triolein (OOO) or stearodiolein (OSO). Two types of interesterification are recognized:

a) A single-phase interesterification where the acyl residues are randomly distributed:

$$
\begin{array}{c}
\underset{(50\%)}{\text{S S S}} + \underset{(50\%)}{\text{O O O}} \\
\downarrow {\scriptstyle (\text{NaOCH}_3)} \\
\underset{(12,5\%)}{\text{S S S}} \quad \underset{(12,5\%)}{\text{S O S}} \quad \underset{(25\%)}{\text{O S S}} \quad \underset{(25\%)}{\text{S O O}} \quad \underset{(12,5\%)}{\text{O S O}} \quad \underset{(12,5\%)}{\text{O O O}}
\end{array} \tag{3.22}
$$

b) A directed interesterification in which the reaction temperature is lowered until the higher melting and least soluble TG molecules in the mixture crystallize. These molecules cease to participate in further reactions, thus the equilibrium is continuously changed. Hence, a fat (oil) can be divided into high and low melting point fractions, e.g.:

$$
\begin{array}{c}
\text{O S O} \\
\downarrow {\scriptstyle (\text{NaOCH}_3)} \\
\underset{(33.3\%)}{\text{S S S}} \quad \underset{(66.7\%)}{\text{O O O}}
\end{array} \tag{3.23}
$$

3.3.1.4 Structural Determination

Apart from identifying a fat or oil from an unknown source (cf. 14.5.2), TG structural analysis is important for the clarification of the relationship existing between the chemical structure and the melting or crystallization properties, i.e. the consistency.

An introductory example: cocoa butter and beef tallow, the latter used during the past century for adulteration of cocoa butter, have very similar fatty acid compositions, especially when the two main saturated fatty acids, 16:0 and 18:0, are considered together (Table 3.13). In spite of their compositions, the two fats differ significantly in their melting properties. Cocoa butter is hard and brittle and melts in a narrow temperature range (28–36 °C).

$$\tag{3.21}$$

Table 3.13. Average fatty acid and triacylglycerol composition (weight-%) of cocoa butter, tallow and Borneo tallow (a cocoa butter substitute)

	Cocoa butter	Edible beef tallow	Borneo tallow[a]
16:0	25	36	20
18:0	37	25	42
20:0	1		1
18:1 (9)	34	37	36
18:2 (9,12)	3	2	1
SSS[b]	2	29	4
SUS	81	33	80
SSU	1	16	1
SUU	15	18	14
USU		2	
UUU	1	2	1

[a] cf. 14.3.2.2.3
[b] S: Saturated, and U: unsaturated fatty acids.

Edible beef tallow, on the other hand, melts at a higher temperature (approx. 45 °C) and over a wider range and has a substantially better plasticity. The melting property of cocoa butter is controlled by the presence of a different pattern of triglycerols: SSS, SUS and SSU (cf. Table 3.13). The chemical composition of Borneo tallow (Tenkawang fat) is so close to that of cocoa butter that the TG distribution patterns shown in Table 3.13 are practically indistinguishable. Also, the melting properties of the two fats are similar, consequently, Borneo tallow is currently used as an important substitute for cocoa butter. Analysis of the TG's present in fat (oil) could be a tedious task, when numerous TG compounds have to be separated. The composition of milk fat is particularly complex. It contains more than 150 types of TG molecules.

The separation by HPLC using reverse phases is the first step in TG analysis. It is afforded by the chain length and the degree of unsaturation of the TG's. As shown in Fig. 3.7 the oils from different plant sources yield characteristic patterns in which distinct TG's predominate.

TG's differing only in the positions of the acyl residues are not separated. However, in some cases it is possible to separate positional isomeric triglycerols after bromination of the

double bonds because triglycerols with a brominated acyl group in β-position are more polar compared to those in α-position.

The separation capacity of the HPLC does not suffice for mixtures of plant oils with complex triglycerol composition. Therefore it is advisable to perform a preseparation of the triglycerols according to their number of double bonds by "argentation chromatography" (cf. 3.2.3.2.3).

Various hypotheses have been advanced, supported by results of TG biosynthesis, to predict the TG composition of a fat or oil when all the fatty acids occurring in the sample are known. The values calculated with the aid of the *1,3-random-2-random* hypothesis agree well with values found experimentally for plant oil or fat. The hypothesis starts with two separated fatty acid pools. The acids in both pools are randomly distributed and used as such for TG biosynthesis. The primary HO-groups (positions 1 and 3 of glycerol) from the first pool are esterified, while the secondary HO-group is esterified in the second pool. The proportion of each TG is then determined (as mole %):

$$\beta\text{-XYZ (mol-\%)} = 2 \cdot \begin{bmatrix} \text{mol-\% X in} \\ \text{1,3-Position} \end{bmatrix} \cdot \begin{bmatrix} \text{mol-\% Y in} \\ \text{2-Position} \end{bmatrix} \begin{bmatrix} \text{mol-\% Z in} \\ \text{1,3-Position} \end{bmatrix} \cdot 10^{-4} \quad (3.24)$$

The data required in order to apply the formula are obtained as follows: after partial hydrolysis of fat (oil) with pancreatic lipase (cf. 3.7.1.1), the fatty acids bound at positions 1 and 3 are determined. The fatty acids in position 2 are calculated from the difference between the total acids and those acids in positions 1 and 3.

Table 3.14 illustrates the extent of agreement for the TG composition of sunflower oil obtained experimentally and by calculation using the 1,3-random-2-random hypothesis. However, both approaches disregard the differences between positions 1 and 3. In addition, the hypothesis is directed to plants, of which the fats and oils consist of only major fatty acids.

Stereospecific Analysis. Biochemically, the esterified primary OH-groups of glycerol are differentiated from each other; thus, the determination of fatty acids in positions 1, 2 and 3 is possible. The reaction sequence of one of the

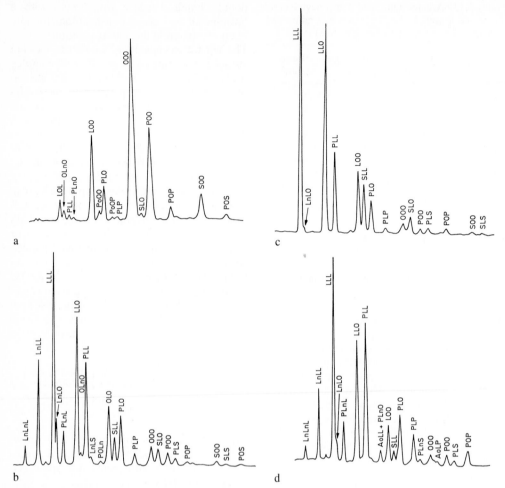

Fig. 3.7. Composition of triacylglycerols present in edible fats or oils as determined by HPLC.
a Olive oil, **b** soybean oil, **c** sunflower oil, **d** wheat germ oil.
Fatty acids: P palmitic, S stearic, O oleic, L linoleic, Ln linolenic, Ao eicosanoic

many procedures designed to carry out a stereospecific analysis is presented in Fig. 3.8. First, the TG (I) is hydrolyzed under controlled conditions to a diacylglycerol using pancreatic lipase (cf. 3.7.1.1). Phosphorylation with a diacylglycerol kinase follows. The enzyme reacts stereospecifically since it phosphorylates only the 1,2- or (S)- but not the 2,3-diglycerol. Subsequently, compound I will be hydrolyzed to a monoacylglycerol (III). The distribution of the acyl residues in positions 1, 2 and 3 is calculated from the results of the fatty acid analysis of compounds I, II and III.

Individual TG's or their mixtures can be analyzed with this procedure. Based on these results (some are presented in Table 3.15), general rules for fatty acid distribution in plant oils or fats can be deduced:
- The primary HO-groups in positions 1 and 3 of glycerol are preferentially esterified with saturated acids.
- Oleic and linolenic acids are equally distributed in all positions, with some exceptions, such as cocoa butter (cf. Table 3.15).
- The remaining free position, 2, is then filled with linoleic acid.

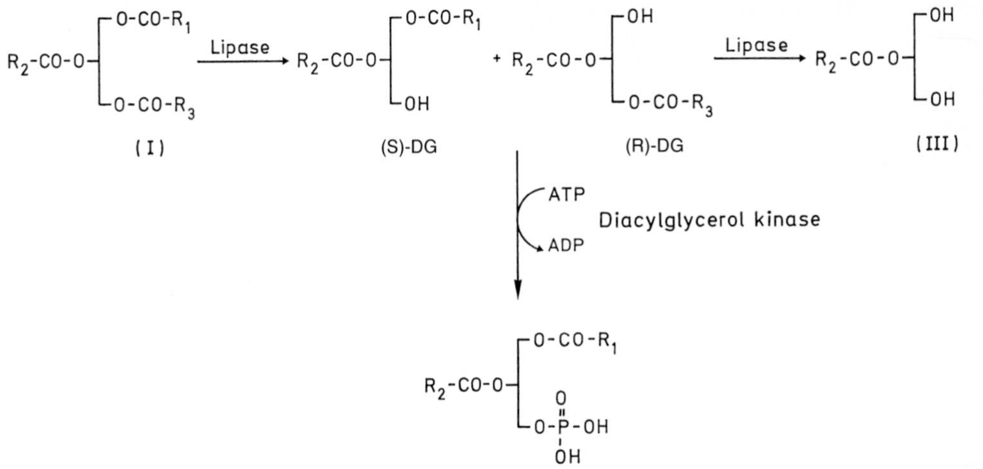

```
        ┌─O-CO-R₁                    ┌─O-CO-R₁          ┌─OH                     ┌─OH
R₂-CO-O─┤        Lipase   R₂-CO-O─┤          + R₂-CO-O─┤        Lipase  R₂-CO-O─┤
        └─O-CO-R₃  ───────>        └─OH              └─O-CO-R₃ ───────>         └─OH
          (I)                       (S)-DG              (R)-DG                   (III)
```

Diacylglycerol kinase — ATP → ADP

```
        ┌─O-CO-R₁
R₂-CO-O─┤   O
        │   ‖
        └─O-P-OH
            OH
         (II)
```

Fig. 3.8. Stereospecific analysis of triacylglycerols

Table 3.14. Triacylglycerol composition (mole-%) of a sunflower oil. A comparison of experimental values with calculated values based on a 1,3-random-2-random hypothesis

Triacyl-glycerol[a]	Found	Calcu-lated	Triacyl-glycerol[a]	Found	Calcu-lated
β-StOSt	0.3	0.5	β-OStL	0.5	0.2
β-StStO	0.2	trace	β-OOL	8.1	6.5
β-StOO	2.3	1.6	β-OLO	3.1	4.2
β-OStO	0.1	trace	β-StLL	13.2	14.0
β-StStL	0.3	0.2	β-LStL	1.3	0.3
β-StLSt	2.2	1.7	β-OLL	20.4	21.9
OOO	1.3	1.2	β-LOL	8.4	8.7
β-StOL	4.4	4.2	LLL	28.1	28.9
β-StLO	4.0	5.3	Others	0.9	0.9

St: Stearic, O: oleic, and L: linoleic acid.
[a] Prefix β: The middle fatty acid is esterified at the β- or sn-2-position, the other two acids are at the sn-1 or sn-3 positions.

Table 3.15. Results of stereospecific analysis of some fats and oils[a]

Fat/Oil	Posi-tion	16:0	18:0	18:1 (9)	18:2 (9,12)	18:3 (9,12,15)
Peanut	1	13.6	4.6	59.2	18.5	–
	2	1.6	0.3	58.5	38.6	–
	3	11.0	5.1	57.3	18.0	–
Soya	1	13.8	5.9	22.9	48.4	9.1
	2	0.9	0.3	21.5	69.7	7.1
	3	13.1	5.6	28.0	45.2	8.4
Cocoa	1	34.0	50.4	12.3	1.3	–
	2	1.7	2.1	87.4	8.6	–
	3	36.5	52.8	8.6	0.4	–
Liver (rat)	1	19.0	2.53	6.2	5.3	–
	2	1.2	0.2	13.0	18.0	–
	3	6.0	0.8	14.0	10.0	–

[a] In order to simplify the Table other fatty acids present in fat/oil are not listed.

Results compiled in Table 3.15 show that for oil or fat of plant origin, there is little difference in acyl residues between positions 1 and 3. Therefore, the 1,3-random-2-random hypothesis provides results that agree well with experimental findings.

The fatty acid pattern in animal fats is strongly influenced by the fatty acid composition of animal feed. A steady state is established only after 4–6 months of feeding with the same feed composition. The examples given in Table 3.15 show that positions 1 and 3 in triglycerides of animal origin show much greater variability than in fats or oils of plant origin. Therefore, any prediction of TG types in animal fat should be calculated from three separate fatty acid pools (1-random-2-random-3-random hypothesis).

The specific distribution of saturated fatty acids in the triglycerols of fats and oils of plant origin serves as an evidence of *ester oils*.

Ester oils are produced by esterification of glycerol with purified fatty acids obtained from olive oil residues. In this case the saturated acyl groups are equally distributed between all three positions of the glycerol molecule, whereas in olive oil saturated acyl groups are attached to position 1 and 3. As proof, the amount of 2-MG containing palmitic acid is determined after hydrolysis of the triglycerols with a lipase (pancreas). Values above 2% are indicative of an adulteration of the olive oil with an ester oil.

The positional specific distribution of palmitic acid is unfavorable for the use of fats and oils of plant origin in infant food, as this acid is liberated by lipolysis in the gastric tract. Palmitic acid then forms insoluble salts with Ca^{2+}-ions from the food, possibly resulting in severe bilious attacks. The fatty acids of human milk consist of up to 25% of palmitic acid; 70% are bound to the 2-position of the triglycerols. During lipolysis 2-monopalmitin is formed that is easily resorbed.

3.3.1.5 Biosynthesis

A TG molecule is synthesized in the fat cells of mammals and plants from L-glycerol-3-phosphate and fatty acid-CoA esters (Fig. 3.9). The L-glycerol-3-phosphate supply is provided by the reduction of dihydroxy acetone phosphate by NAD^+-dependent glycerol phosphate dehydrogenase. The dihydroxy acetone phosphate originates from glycolysis.

The lipid bodies (oleosomes, spherosomes) synthesized are surrounded by a membrane and are deposited in storage tissues.

The TG fatty acid composition within a plant species depends on the environment, especially the temperature. A general rule is that plants in cold climates produce a higher proportion of unsaturated fatty acids. Obviously, the mobility of TG's is thus retained. In the sunflower (cf. Fig. 3.10), this rule is highly pronounced; whereas in safflower, only a weak response to temperature variations is observed (Fig. 3.10).

3.3.2 Mono- and Diacylglycerols (MG, DG)

3.3.2.1 Occurrence, Production

The occurrence of MG and DG in edible oils or fats or in raw food is very low. However,

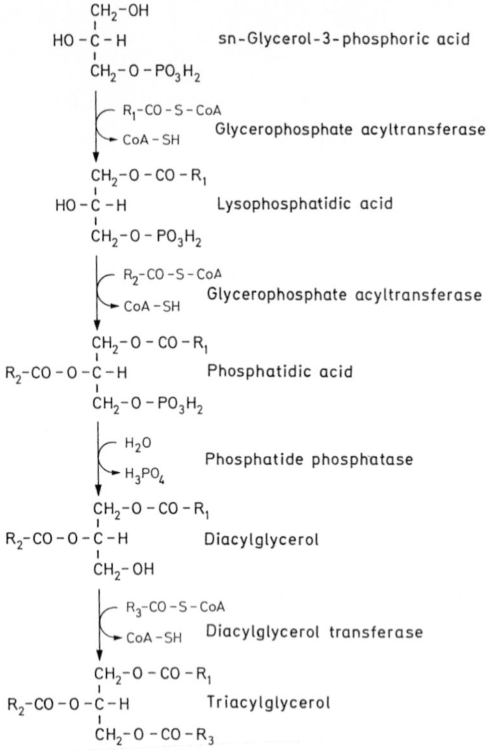

Fig. 3.9. Biosynthesis of triacylglycerols

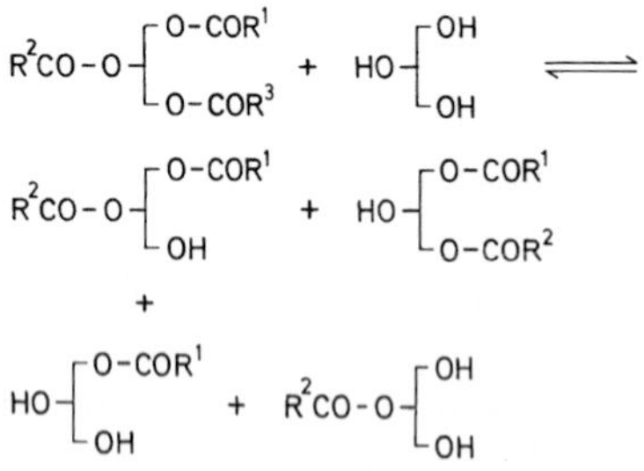

their levels may be increased by the action of hydrolases during food storage or processing. MG and DG are produced commercially by fat glycerolysis (200 °C, basic catalyst)

(3.25)

From the equilibrium (cf. Formula 3.25) that contains 40−60% MG, 45−35% DG and 15−5% TG, the MG are separated by distillation under high vacuum. The amount of 1-MG (90−95%) is predominant over the amount of 2-MG.

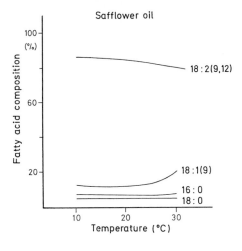

Safflower oil

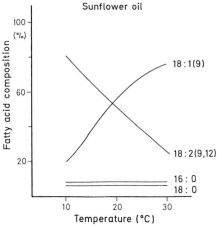

Sunflower oil

Fig. 3.10. The effect of climate (temperature) on the fatty acid composition of triacylglycerols

3.3.2.2 Physical Properties

MG and DG crystallize in different forms (polymorphism; cf. 3.3.1.2). The melting point of an ester of a given acid increases for the series 1,2-DG < TG < 2-MG < 1,3-DG < 1-MG:

	Melting Point (°C) β-form
Tripalmitin	65.5
1,3-Dipalmitin	72.5
1,2-Dipalmitin	64.0
1-Palmitin	77.0
2-Palmitin	68.5

MG and DG are surface-active agents. Their properties can be further modified by esterification with acetic, lactic, fumaric, tartaric or citric acids. These esters play a significant role as emulsifiers in food processing (cf. 8.15.3.1).

3.4 Phospho- and Glycolipids

3.4.1 Classes

Phospho- and glycolipids, together with proteins, are the building blocks of biological membranes. Hence, they invariably occur in all foods of animal and plant origin. Examples are compiled in Table 3.16. As surface-active compounds, phospho- and glycolipids contain hydrophobic moieties (acyl residue, N-acyl sphingosine) and hydrophilic portions (phosphoric acid, carbohydrate). Therefore, they are capable of forming orderly structures (micelles or planar layers) in aqueous media; the bilayer structures are found in all biological membranes. Examples for the composition of membrane lipids are listed in Table 3.17.

3.4.1.1 Phosphatidyl Derivatives

The following phosphoglycerides are derived from phosphatidic acid. Phosphatidyl choline or lecithin (phosphate group esterified with the OH-group of choline):

$$
\begin{array}{l}
{}^{(1)}\ CH_2-O-CO-R_1 \\
R_2-CO-O-CH\ ^{(2)} \\
{}^{(3)}\ CH_2-O-P-O-CH_2-CH_2-\overset{\oplus}{N}-CH_3 \\
\end{array}
\qquad (3.26)
$$

Table 3.16. Composition of lipids of various foods[a]

	Milk	Soya	Wheat	Apple
Total lipids	3.6	23.0	1.5	0.088
Triacylglycerols	94	88	41	5
Mono-, and diacylglycerols	1.5		1	
Sterols	< 1		1	15
Sterol esters			1	2
Phospholipids	1.5	10	20	47
Glycolipids		1.5	29	17
Sulfolipids				1
Others		0.54	7	15

[a] Total lipids as %, while lipid fractions are expressed as percent of the total lipids.

Table 3.17. Lipid composition of biomembranes

	Lipid content of membrane (%)	Lipid composition (%)			
		Neutral lipids	Glycero-glyolipids	Glycero-phospholipids	Sphingo-lipids
Chloroplast (spinach)	52	29	45	9	no data
Mitochondria (bovine heart)	26	8	no data	92	0
Endoplasmic reticulum (heart, swine)	25	32	no data	55	11
Myelin (mammalian brain)	78	25	no data	32	31

Phosphatidyl serine (phosphate group esterified with the HO-group of the amino acid serine):

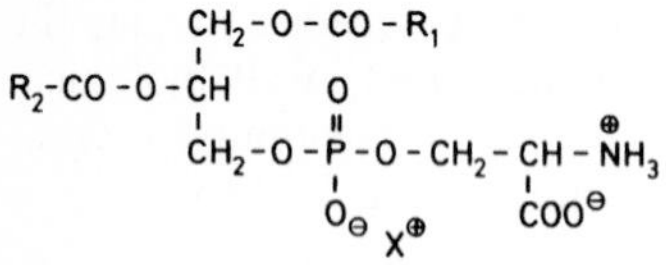

$$
\begin{array}{l}
CH_2-O-CO-R_1 \\
| \\
R_2-CO-O-CH \qquad O \\
\qquad | \qquad\quad \| \\
\qquad CH_2-O-P-O-CH_2-\overset{\oplus}{CH}-\overset{\oplus}{NH_3} \\
\qquad\qquad | \\
\qquad\qquad O_{\ominus}\ {}_{X^{\oplus}}\qquad COO^{\ominus}
\end{array}
\qquad (3.27)
$$

Phosphatidyl ethanolamine (phosphate group esterified with ethanolamine):

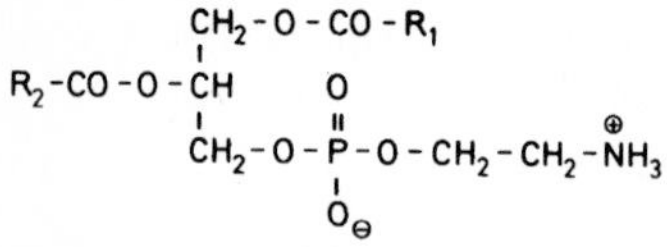

$$
\begin{array}{l}
CH_2-O-CO-R_1 \\
| \\
R_2-CO-O-CH \qquad O \\
\qquad | \qquad\quad \| \\
\qquad CH_2-O-P-O-CH_2-CH_2-\overset{\oplus}{NH_3} \\
\qquad\qquad | \\
\qquad\qquad O_{\ominus}
\end{array}
\qquad (3.28)
$$

Phosphatidyl inositol (phosphate group esterified with inositol):

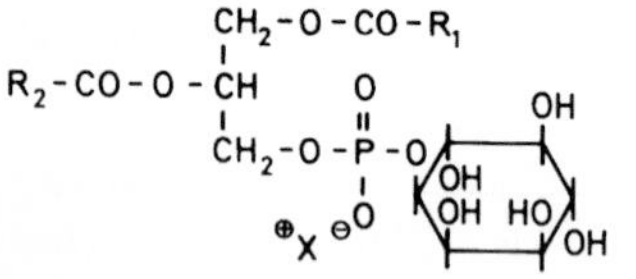

$$ (3.29) $$

A mixture of phosphatidyl serine and phosphatidyl ethanolamine was once referred to as cephalin.

Only one acyl residue is cleaved by hydrolysis (cf. 3.7.1.2.1) with phospholipase A. This yields the corresponding lyso-compounds from lecithin or phosphatidyl ethanolamine. Some of these lyso-derivatives occur in nature, e.g., in cereals.

Phosphatidyl glycerol is invariably found in green plants, particularly in chloroplasts:

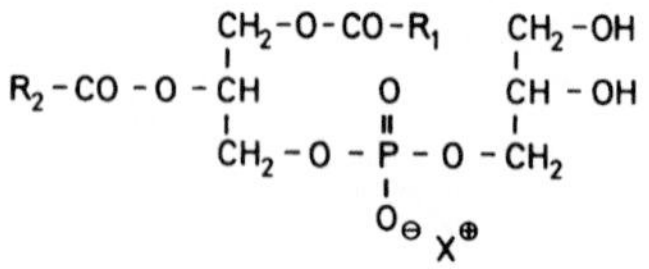

$$ (3.30) $$

L-α-Phosphatidyl-D-glycerol

Cardiolipin, first identified in beef heart, is also a minor constituent of green plant lipids. Its chemical structure is diphosphatidyl glycerol:

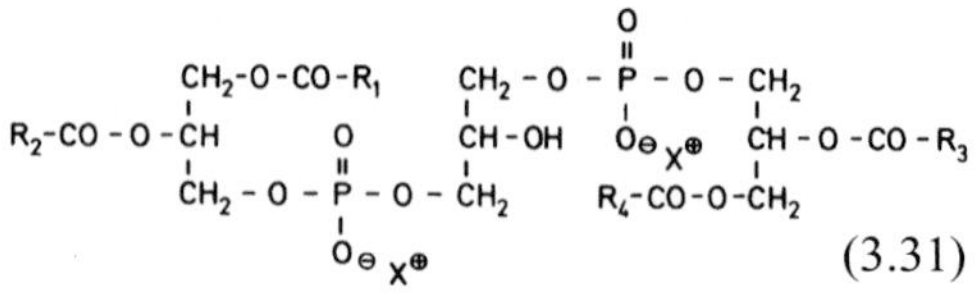

$$ (3.31) $$

Diphosphatidyl glycerol (cardiolipin)

The plasmalogens occupy a special place in the class of phospho-glycerides. They are phosphatides in which position 1 of glycerol is linked to a straight-chain aldehyde with 16 or 18 carbons. The linkage is an enol-ether type with a double bond in the cis-configuration. Plasmalogens of 1-O-(1-alkenyl)-2-O-acylglycerophospholipid type occur in small amounts in animal muscle tissue and also in milk fat. The enol-ether linkage, unlike the ether bonds of a 1-O-alkylglycerol (cf. 3.6.2), is readily hydrolyzed even by weak acids.

$$\begin{array}{l} CH_2 - O - CH = CH - R_1 \\ R_2 - CO - O - CH \quad\quad O \\ \quad\quad CH_2 - O - \overset{\overset{\displaystyle O}{\|}}{\underset{\underset{\displaystyle O_\ominus}{}}{P}} - O - CH_2 - CH_2 - \overset{\oplus}{N}H_3 \end{array}$$

(3.32)

Plasmalogen

Phospholipids are sensitive to autoxidation since they contain an abundance of linoleic acid. The other acid commonly present is palmitic acid.

Phospholipids are soluble in chloroform-methanol and poorly soluble in water-free acetone. The pK$_s$ value of the phosphate group is between 1 and 2. Phosphatidyl choline and phosphatidyl ethanolamine are zwitter-ions at pH 7.

Phospholipids can be hydrolyzed stepwise by alcoholic KOH. Under mild conditions, only the fatty acids are cleaved, whereas, with strong alkalies, the base moiety is released. The bonds between phosphoric acid and glycerol or phosphoric acid and inositol are stable to alkalies, but are readily hydrolyzed by acids. Phosphatidyl derivatives, together with triacylglycerols and sterols, occur in the lipid fraction of lipoproteins (cf. 3.5.1).

Lecithin. Lecithin plays a significant role as a surface-active agent in the production of emulsions. "Raw lecithin", especially that of soya and that isolated from egg yolk, is available for use on a commercial scale. "Raw lecithins" are complex mixtures of lipids with phosphatidyl cholines, ethanolamines and inositols as main components (Table 3.18).

The major phospholipids of raw soya lecithin are given in Table 3.18. The manufacturer

Table 3.18. Composition of glycerol phospholipids in soya "raw lecithin" and in the resulting fractions[a]

	Unfrac-tionated	Ethanol soluble fraction	Ethanol insoluble fraction
Phosphatidyl ethanolamine	13–17	16.3	13.3
Phosphatidyl choline	20–27	49	6.6
Phosphatidyl inositol	9	1	15.2

[a] Values in weight%.

often separates lecithin into ethanol-soluble and ethanol-insoluble fractions.

Pure lecithin is a W/O emulsifier with a HLB-value (cf. 8.15.2.3) of about 3. The HLB-value rises to 8–11 by hydrolysis of lecithin to lyso-phosphatidyl choline; an o/w emulsifier is formed. Since the commercial lecithins are complex mixtures of lipids their HBL-values lie within a wide range. The ethanol-insoluble fraction (Table 3.18) is suitable for stabilization of W/O emulsions and the ethanol-soluble fraction for O/W emulsions. To increase the HLB-value, "hydroxylated lecithins" are produced by hydroxylation of the unsaturated acyl groups with hydrogen peroxide in the presence of lactic-, citric- and tartaric acid.

3.4.1.2 Glyceroglycolipids

These lipids consist of 1,2-diacylglycerols and a mono-, di- or, less frequently, tri- or tetrasaccharide bound in position 3 of glycerol. Galactose is predominant as the sugar component among plant glycerolipids. The chloroplasts are particularly abundant in these lipids (Table 3.17).

$$\begin{array}{l} H_2C - O - CO - R^1 \\ R^2 - CO - O - CH \\ HO - CH_2 \quad O - CH_2 \end{array}$$

(3.33)

Monogalactosyl diacylglycerol (MGDG) (1,2-diacyl-3-β-D-galactopyranosyl-L-glycerol)

(3.34)

Digalactosyl diacylglycerol (DGDG) (1,2-diacyl-3-(α-D-galactopyranosyl-1,6-β-D-galactopyranosyl)-L-glycerol)

6-O-acyl-MGDG and 6-0-acyl-DGDG are minor components of plant lipids.

Sulfolipids are glyceroglycolipids which are highly soluble in water since they contain a sugar moiety esterified with sulfuric acid. The sugar moiety is 6-sulfochinovose. Sulfolipids occur in chloroplasts but are also detected in potato tubers:

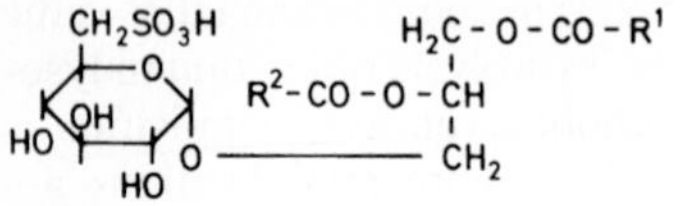

$$\text{(3.35)}$$

Sulfolipid
(1,2-diacyl-(6-sulfo-α-D-chinovosyl-1,3)-L-glycerol)

3.4.1.3 Sphingolipids

Sphingolipids contain sphingosine, an amino alcohol with a long unsaturated hydrocarbon chain (D-*erythro*-1,3-dihydroxy-2-amino-trans-4-octadecene) instead of glycerol:

$$\text{(3.36)}$$

Sphingolipids which occur in plants, e.g., wheat, contain phytosphingosines:

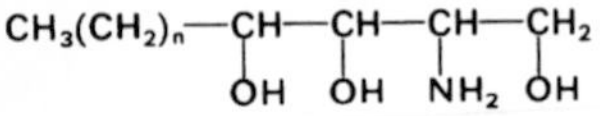

$$CH_3(CH_2)_n—CH—CH—CH—CH_2$$
$$\qquad\quad OH\ \ OH\ \ NH_2\ OH$$
$$\text{(3.37)}$$
n: 13, 14, 15, 16

The amino group in sphingolipids is linked to a fatty acid to form a carboxy amide, denoted as ceramide. The primary hydroxyl group is either esterified with phosphoric acid (sphingophospholipid: ceramide-phosphate-base) or bound glycosidically to a mono- di-, or oligo-saccharide (sphingoglycolipid: ceramide-phosphate-sugar$_n$). In the third group of sphingolipids the ceramide moiety is linked by a phosphate residue to the carbohydrate building blocks. These compounds are also referred to as phytoglycolipids.

Sphingophospholipids. Sphingomyelin is one example of a sphingophospholipid. It is the most abundant sphingolipid and is found in myelin, the fatty substance of the sheath around nerve fibers. The structure of sphingomyelin is:

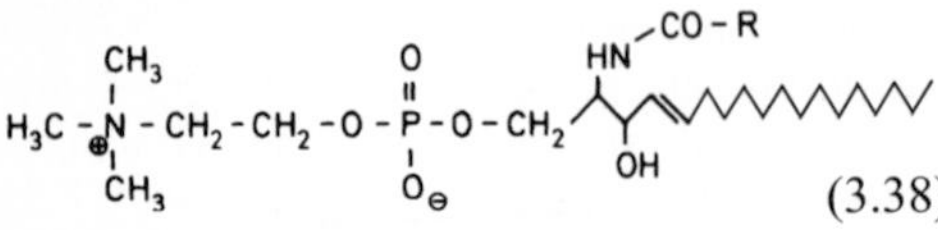

$$\text{(3.38)}$$

Sphingoglycolipids are found in tissue of animal origin, milk and in plants (especially cereals). Based on structural properties of the carbohydrate building blocks, one differentiates neutral and acid glycosphingolipids. The sulfatides and gangliosides also belong to this group.

Lactosylceramide in milk and the ceramide glycosides of wheat are examples of neutral glycosphingolipids that contain, next to glucose and mannose, also saturated (14:0–28:0) and monounsaturated (16:1–26:1) 2-hydroxy- or 2,3-dihydroxy fatty acids.

Formula 3.40a depicts a sphingoglycolipid of wheat.

Gangliosides contain sialic acid (N-acetyl-neuraminic acid; cf. Formula 3.40b). In the ganglioside fraction of milk monosialosyl-lactosyl-ceramide (cf. Formula 3.41) was identified.

Phytosphingolipids. These lipids also have a complex structure. Total hydrolysis yields phytosphingosine, inositol, phosphoric acid and various monosaccharides (galactose, arabinose, mannose, glucosamine, glucuronic acid).

Phytosphingolipids are isolated from soya and peanuts (cf. Formula 3.42).

3.4.2 Analysis

3.4.2.1 Extraction, Removal of Nonlipids

A solvent mixture of chloroform/methanol (2 + 1 v/v) is suitable for a quantitative extraction of lipids. Addition of a small amount of BHA (cf. 3.7.3.2.2) is recommended for the stabilization of lipids against autoxidation.

Nonlipid impurities were earlier removed by shaking the extracts with a special salt solution under rather demanding conditions. An improved procedure, by which emulsion formation is avoided, is based on column chromatography with dextran gels.

3.4.2.2 Separation and Identification of Classes of Components

Isolated and purified lipids can be separated into classes by thin layer chromatography

$$\text{D-Glc}p\text{-}\beta\text{-}(1\rightarrow4)\text{-D-Man}p\text{-}\beta\text{-}(1\rightarrow4)\text{-D-Man}p\text{-}\beta\text{-}(1\rightarrow4)\text{-D-Glc}p\text{-}\beta\text{-}(1\rightarrow1)\text{-O-CH}_2\text{-CH-CH-CH-(CH}_2)_{13}\text{-CH}_3$$

with substituent
$$\text{NH-CO-CH-(CH}_2)_{21}\text{-CH}_3$$
$$\overset{|}{\text{OH}}$$
$$\overset{|}{\text{OH}} \quad \overset{|}{\text{OH}}$$

(3.39)

$$\text{CH}_3\text{-(CH}_2)_{12}\text{-CH=CH-CH-CH-CH}_2\text{-O - Glucose-Galactose-N-Acetylgalactosamine-Galactose}$$
$$\overset{|}{\text{OH}} \quad \overset{|}{\text{NH}}$$
$$\overset{|}{\text{CO}}$$
$$\overset{|}{\text{R}}$$

N-Acetylneuraminic acid

(3.40a)

$$\text{HO - CH}_2\text{-CH-CH}$$
with OH, OH substituents and $\text{CH}_3\text{-CO-HN}$

COOH, OH, O, OH ring

(3.40b)

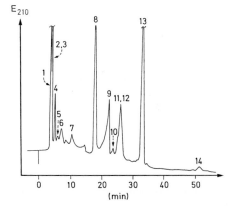

(3.41)

$$\text{CH}_3\text{-(CH}_2)_{13}\text{-CH-CH-CH-CH}_2\text{-O-}\overset{\overset{\displaystyle O}{\|}}{P}\text{-O - Inositol-Glucuronic acid-Glucosamine}$$
$$\overset{|}{\text{OH}} \quad \overset{|}{\text{OH}} \quad \overset{|}{\text{NH}} \qquad \overset{|}{O_\ominus} \text{ } X^\oplus$$
$$\overset{|}{\text{CO}}$$
$$\overset{|}{\text{R}_1}$$

Galactose
Arabinose
Mannose

(3.42)

using developing solvents of different polarity. Figure 3.12 shows an example of the separation of neutral and polar lipids. Identification of polar lipids is based on using spraying reagents which react on the plates with the polar moiety of the lipid molecules. For example, phosphoric acid is identified by the molybdenum-blue reaction, monosaccharides by orcinol-FeCl$_3$, choline by bismuth iodide (*Dragendorff* reagent), ethanolamine and serine by the ninhydrin reaction, and sphingosine by a chlorine-benzidine reagent.

When sufficient material is available, it is advisable to perform a preliminary separation of lipids by column chromatography on magnesium silicate (florisil), silicic acid, hydrophobic dextran gel or a cellulose-based ion-exchanger, such as DEAE-cellulose.

Also the HPLC-analysis of phospho- and glycolipids is of growing importance. Figure 3.11 for example demonstrates the separation of soya "raw lecithin".

Fig. 3.11. HPLC-analysis of soy "raw lethicin" (according to *N. Sotirhos* et al. 1986)

1 Triacylglycerols, 2 free fatty acids, 3 phosphatidyl glycerol, 4 cerebrosides, 5 phytosphingosine, 6 diphosphatidyl glycerol, 7 digalactosyldiacyl glycerol, 8 phosphatidyl ethanolamine, 9 phosphatidyl inositol, 10 lysophosphatidyl ethanolamine, 11 phosphatidic acid, 12 phosphatidyl serine, 13 phosphatidyl choline, 14 lysophosphatidyl choline

Developing solvent

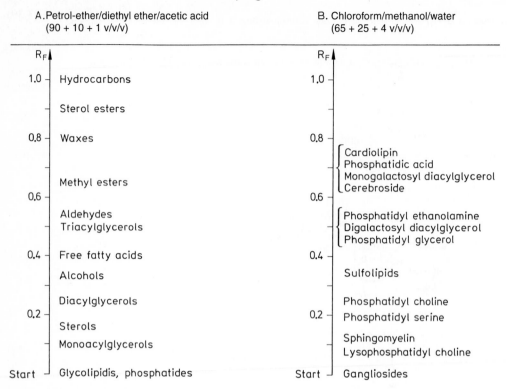

A. Petrol-ether/diethyl ether/acetic acid (90 + 10 + 1 v/v/v)

B. Chloroform/methanol/water (65 + 25 + 4 v/v/v)

Fig. 3.12. Separation of lipid classes by thin layer chromatography using silicagel as an adsorbent. R_f values in two solvent systems

3.4.2.3 Analysis of Lipid Components

Fatty acid composition is determined after methanolysis of the lipid. For positional analysis of acyl residues (positions 1 or 2 in glycerol), phosphatidyl derivatives are selectively hydrolyzed with phospholipases (cf. 3.7.1.2.1) and the fatty acids liberated are analyzed by gas chromatography.

The sphingosine base can also be determined by gas chromatography after trimethylsilyl derivatization. The length of the carbon skeleton, of interest for phytosphingosine, can be determined by analyzing the aldehydes released after the chain has been cleaved by periodate:

$$CH_3 - (CH_2)_n - \underset{\underset{OH}{|}}{CH} - \underset{\underset{OH}{|}}{CH} - \underset{\underset{NH_2}{|}}{CH} - CH_2 - OH$$

$$\downarrow IO_4^{\ominus}$$

$$CH_3 - (CH_2)_n - CHO + NH_3 + 3\, CH_2O$$

(3.43)

The monosaccharides in glycolipids can also be determined by gas chromatography. The lipids are hydrolyzed with trifluoroacetic acid and then derivatized to an acetylated glyconic acid nitrile. By using this sugar derivative, the chromatogram is simplified because of the absence of sugar anomers (cf. 4.2.4.6).

3.5 Lipoproteins, Membranes

3.5.1 Lipoproteins

3.5.1.1 Definition

Lipoproteins are aggregates, consisting of proteins, polar lipids and triacylglycerols, which are water soluble and can be separated into protein and lipid moieties by an extraction procedure using suitable solvents. This indicates that only noncovalent bonds are involved in the formation of lipoproteins. The aggregates are primarily stabilized by hydrophobic interactions between the apolar side chains of hydrophobic regions of the protein and the acyl residues of the lipid. In addition, there is a contribution to stability by ionic forces between charged amino acid residues and charges carried by the phosphatides. Hydrogen bonds, important for stabilization of the secondary structure of protein, play a small role in binding lipids since phosphatidyl derivatives have only a few sites available for such linkages. Hydrogen bonds can exist to a greater extent between proteins and glycolipids; however, such lipids have not yet been found as lipoprotein components, but rather as building blocks of biological membranes. An exception may be their occurrence in wheat flour, where they are responsible for gluten stability of dough. Here, the lipoprotein complex consists of prolamine and glutelin attached to glycolipids by hydrogen bonds and hydrophobic forces. Although the presence of covalent bonds between lipids and proteins cannot be completely excluded, experimental results do not support such an assumption.

3.5.1.2 Classification

Lipoproteins exist as globular particles in an aqueus medium. They are solubilized from biological sources by buffers with high ionic strength, by a change of pH or by detergents in the isolating medium. The latter, a more drastic approach, is usually used in the recovery of lipoproteins from membranes.
Lipoproteins are characterized by ultracentrifugation. Since lipids have a lower density (0.88−0.9 g/ml) than proteins (1.3−1.35 g/ml), the separation is possible because of

differences in the ratios of lipid to protein within a lipoprotein complex. The lipoproteins of blood plasma have been thoroughly studied. They are separated by a stepwise centrifugation in solutions of NaCl into three fractions with different densities (Fig. 3.13). The "very low density lipoproteins" (VLDL; density <1.006 g/ml), the "low density lipoproteins" (LDL; 1.063 g/ml) and the "high density lipoproteins" (HDL; 1.21 g/ml) float, and the sediment contains the plasma proteins. The VLDL fraction can be separated further by electrophoresis into chylomicrons (the lightest lipoprotein, density <1.000 g/ml) and pre-β-lipoprotein.
Lipoproteins in the LDL fraction from an electrophoretic run have a mobility close to that of blood plasma β-globulin. Therefore, the LDL fraction is denoted as β-lipoprotein. An analogous designation of α-lipoprotein is assigned to the HDL fraction.
Chylomicrons, the diameters of which range from 1,000−10,000 Å, are small droplets of

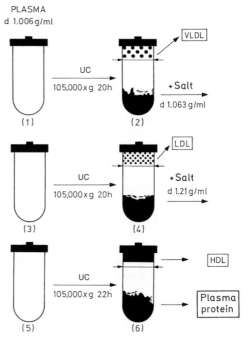

Fig. 3.13. Plasma protein fractionation by a preparative ultracentrifugation (UC) method (according to *D. Seidel*, 1971)

Table 3.19. Composition of typical lipoproteins

Source	Lipoprotein	Particle weight (kdal)	Protein (%)	Glycero-phospho-lipids (%)	Cholesterol		Triacyl-glycerols (%)
					free (%)	esterified (%)	
Human	Chylomicron	10^9-10^{10}	1-2	4	2.5-3	3-4	85-90
blood serum	Pre-β-lipoprotein	$5-100 \cdot 10^6$	8.3	19.2	7.4	11.1	54.2
	LDL (β-lipoprotein)	$2.3 \cdot 10^6$	22.7	27.9	8.5	28.8	10.5
	HDL (α-lipoprotein)	$1-4 \cdot 10^5$	58.1	24.7	2.9	9.2	5.9
Egg yolk	β-Lipovitellin	$4 \cdot 10^5$	78	12	0.9	0.1	9
(chicken)	LDL	$2-10 \cdot 10^6$	18	22	3.5	0.2	58
Bovine milk	LDL	$3.9 \cdot 10^6$	12.9	52	0	0	35.1

triacylglycerol stabilized in the aqueous medium by a membrane-like structure composed of protein, phosphatides and cholesterol. The role of chylomicrons in blood is to transport triacylglycerols to various organs, but preferentially from the intestines to adipose tissue and the liver. The milk fat globules (cf. 10.1.2.3) have a structure similar to that of chylomicrons. The composition of plasma lipoprotein is presented in Table 3.19.

Some diseases related to fat metabolism can be clinically diagnosed by the content and composition of the plasma lipoprotein fractions.

Two lipoproteins have been isolated from egg yolk (Table 3.19). An LDL similar to plasma lipoprotein occurs in a soluble form. β-Lipovitellin, the composition of which corresponds to an HDL (Table 3.19), has the properties of a membrane constituent.

Electron microscopy studies have revealed that the fat globules in milk have small particles attached to their membranes; these are detached by detergents and have been identified as LDL (cf. Table 3.19).

3.5.2 Involvement of Lipids in the Formation of Biological Membranes

Membranes that compartmentalize the cells and many subcellular particles are formed from two main building blocks: proteins and lipids (phospholipids and cholesterol). Differences in membrane structure and function are reflected by the compositional differences of membrane proteins and lipids (see examples in Table 3.17).

Studies of membrane structure are difficult since the methods for isolation and purification profoundly change the organization and functionality of the membrane.

Model membranes are readily formed. The major forces in such events are the hydrophobic interactions between the acyl tails of phospholipids, providing a bilayer arrangement. In addition, the amphipathic character of the lipid molecules makes membrane formation a spontaneous process. The acyl residues are sequestered and oriented in the nonpolar interior of the bilayer, whereas the polar hydrophilic head groups are oriented toward the outer aqueous phase.

Another arrangement in water that satisfies both the hydrophobic acyl tails and the hydrophilic polar groups is a globular micelle. Here, the hydrocarbon tails are sequestered inside, while the polar groups are on the surface of the sphere. There is no bilayer in this arrangement. The favored structure for most phospho- and glycolipids in water is a bimolecular arrangement, rather than a micelle. Two model systems can exist for such bimolecular arrangements. The first is a lipid vesicle, known as a liposome, the core of which is an aqueous compartment surrounded by a lipid bilayer, and the second is a planar, bilayer membrane. The latter, together with the micellar model, is presented in Fig. 3.14. Globular proteins, often including enzymes, are found in animal cell membranes and are well embedded or inserted into the bimolecular layer. Some of

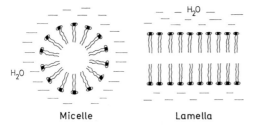

Fig. 3.14. Arrangement of polar acyl lipids in aqueous medium. ∞ Polar lipid tails; ≈ hydrophobic lipid tails

$$H_2C\text{---}O\text{---}CO\text{---}R$$
$$H_2C\text{---}O\text{---}CO\text{---}R$$

$$
\begin{array}{l}
H_3C \\
HC\text{---}O\text{---}CO\text{---}R \\
HC\text{---}O\text{---}CO\text{---}R \\
H_3C
\end{array}
\qquad (3.44)
$$

$$(3.45)$$

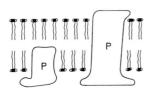

Fig. 3.15. Fluid mosaic model of a biological membrane. The protein (P) is not fixed but is mobile in the phospholipid phase

these so-called integral membrane proteins protrude through both sides of the membrane (fluid mosaic model, Fig. 3.15). Although integral proteins interact extensively with the hydrophobic acyl tails of membrane lipids, they are mobile within the lipid membrane.

3.6 Diol Lipids, Higher Alcohols, Waxes and Cutin

3.6.1 Diol Lipids

The diol lipids which occur in both plant and animal tissues are minor lipid constituents. The diol content is about 1% of the content of glycerol. Exceptions are sea stars, sea urchins and mollusks, the lipids of which in summer contain 25–40% diol lipids. This proportion decreases sharply in winter and spring. Neutral and polar lipids derived from ethylene glycol, propane-(1,2 and 1,3)-diol and butane-(1,3; 1,4- and 2,3)-diol have been identified in the diol lipid fraction. Several of those isolated from corn oil have the following structures:

In a glycodiol lipid one hydroxyl group of ethanediol is esterified with a fatty acid. Diol lipids with structures analogous to phosphatidyl choline or plasmalogen have also been identified.

3.6.2 Higher Alcohols and Derivatives

3.6.2.1 Waxes

Higher alcohols occur either free or bound in plant and animal tissues. Free higher alcohols are abundant in fish oil and include:

Cetyl alcohol	$C_{16}H_{33}OH$
Stearyl alcohol	$C_{18}H_{37}OH$
Oleyl alcohol	$C_{18}H_{35}OH$

Waxes are important derivatives of higher alcohols. They are higher alcohols esterified with long-chain fatty acids. Plant waxes are usually found on leaves or seeds. Thus, cabbage leaf wax consists of the primary alcohols C_{12} and C_{18}–C_{28} esterified with palmitic acid and other acids. The dominant components are stearyl and ceryl alcohol ($C_{26}H_{53}OH$). In addition to primary alcohols, esters of secondary alcohols, e.g., esters of nonacosane-15-ol, are present:

$$H_3C\text{---}(CH_2)_{13}\text{---}\underset{\underset{OH}{|}}{CH}\text{---}(CH_2)_{13}\text{---}CH_3 \qquad (3.46)$$

The role of waxes is to protect the surface of plant leaves, stems and seeds from dehydration and infections by microorganisms. Waxes are removed together with oils by solvent extraction of nondehulled seeds. Waxes are oil-soluble at elevated temperatures but crystallize at room temperature, causing undesired oil turbidity. Ceryl cerotate (ceryl alcohol esterified with cerotic acid, $C_{25}H_{51}COOH$)

$$H_3C\text{---}(CH_2)_{24}\text{---}CO\text{---}O\text{---}CH_2\text{---}(CH_2)_{24}\text{---}CH_3 \quad (3.47)$$

is removed from seed hulls during extraction of sunflower oil. Waxes are removed by an oil refining winterization step during the production of clear edible oil.

Waxes are present in fish oils, especially in sperm whale blubber and whale head oil, which contain a "reservoir" of spermaceti wax.

3.6.2.2 Alkoxy Lipids

The higher alcohols, $16:0$, $18:0$ and $18:1$ (9), form mono- and diethers with glycerol. Such alkoxy-lipids are widely distributed in small amounts in mammals and sea animals. Examples of confirmed structures are shown in Formula 3.48.

The elucidation of ether lipid structure is usually accomplished by cleavage by concentrated HI at elevated temperatures.

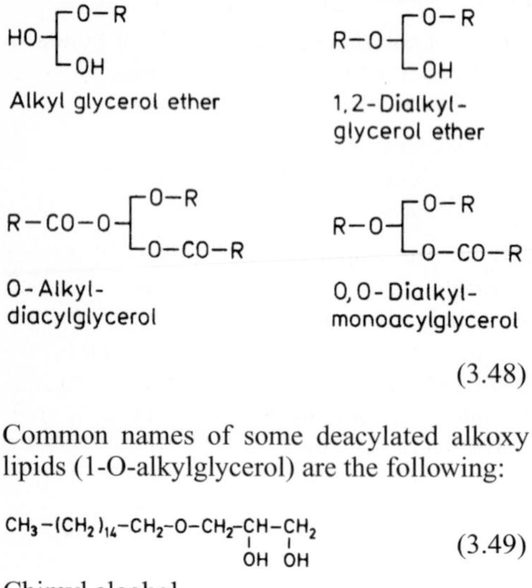

Alkyl glycerol ether

1,2-Dialkyl-glycerol ether

O-Alkyl-diacylglycerol

0,0-Dialkyl-monoacylglycerol

(3.48)

Common names of some deacylated alkoxy lipids (1-O-alkylglycerol) are the following:

$CH_3-(CH_2)_{14}-CH_2-O-CH_2-CH-CH_2$
            $\quad\quad\quad\quad OH \;\; OH$

(3.49)

Chimyl alcohol

$CH_3-(CH_2)_{16}-CH_2-O-CH_2-CH-CH_2$
            $\quad\quad\quad\quad OH \;\; OH$

(3.50)

Batyl alcohol

$CH_3-(CH_2)_7-CH=CH-(CH_2)_7-CH_2-O-CH_2-CH-CH_2$
                        $\quad\quad\quad\quad\quad\quad OH \;\; OH$

Selachyl alcohol   (3.51)

3.6.3 Cutin

Plant epidermal cells are protected by a suberized or waxy cuticle. An additional layer of

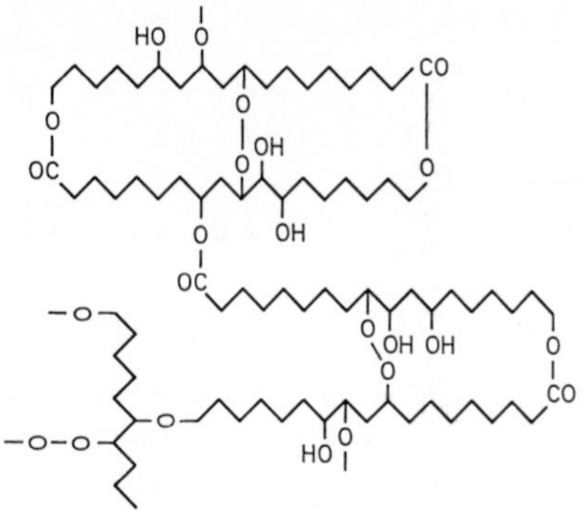

Fig. 3.16. A structural segment of cutin (according to C. *Hitchcock* and B.W. *Nichols*, 1971)

epicuticular waxes is deposited above the cuticle in many plants. The waxy cuticle consists of cutin. This is a complex, high molecular weight polyester which is readily solubilized in alkali. The structural units of the polymer are hydroxy fatty acids. The latter are similar in structure to the compounds given in 3.7.2.4.1. A segment of the postulated structure of cutin is presented in Fig. 3.16.

3.7 Changes in Acyl Lipids of Food

3.7.1 Enzymatic Hydrolysis

Hydrolases, which cleave acyl lipids, are present in food and microorganisms. The release of short-chain fatty acids ($<C_{14}$), e.g., in the hydrolysis of milk fat, has a direct effect on food aroma. Lipolysis is undesirable in fresh milk since the free C_4-C_{12} fatty acids (cf. Tables 3.3 to 3.5 for odor threshold values) are responsible for the rancid aroma defect. On the other hand, lipolysis occurring during the ripening of cheese is a desired and favorable process because the short-chain fatty acids are involved in the build-up of specific cheese aromas. Likewise, slight hydrolysis of milk fat is advantageous in the production of chocolate.

Linoleic and linolenic acid released by hydrolysis and present in emulsified form affect the flavor of food even at low concentrations. They cause a bitter-burning sensation (cf. Table 3.8). In addition, they decompose by autoxidation

Table 3.20. Lipid hydrolysis occurring during potato tuber homogenization

	μmoles/g[a]	
	Acyl lipids	Free fatty acids
Potato	2.34	0.70
Homogenate[b]	2.04	1.40
Homogenate[b] kept for 10 min at 0 °C	1.72	1.75
Homogenate kept for 10 min at 25 °C	0.54	2.90

[a] Potato tissue fresh weight.
[b] Sliced potatoes were homogenized for 30 sec at 0 °C.

(cf. 3.7.2.1) or enzymatic oxidation (cf. 3.7.2.2) into compounds with an intensive odor. In fruits and vegetables enzymatic oxidation in conjunction with lipolysis occur, as a rule, at a high reaction rate, especially when tissue is sliced or homogenized (an example for rapid lipolysis is shown in Table 3.20). Also, enzymatic hydrolysis of a small amount of the acyl lipids present can not be avoided during disintegration of oil seeds. Since the release of higher fatty acids promotes foaming, they are removed during oil refining (cf. 14.4.1).
Enzymes with lipolytic activity belong to the carboxyl-ester hydrolase group of enzymes (cf. Table 2.4).

3.7.1.1 Triacylglycerol Hydrolases (Lipases)

Lipases (cf Table 2.4) hydrolyze only emulsified acyl lipids; they are active on a water/lipid interface. Lipases differ from esterase enzymes since the latter cleave only water-soluble esters, such as triacetylglycerol.
Lipase activity is detected, for example, in milk, oilseeds (soybean, peanut), cereals (oats, wheat), fruits and vegetables and in the digestive tract of mammals. Many microorganisms release lipase-type enzymes into their culture media.
As to their specificity, fat-splitting enzymes which preferentially cleave primary HO-group esters are distinguished from those which indiscriminantly hydrolyze all three ester bonds of acyl glycerols (Table 3.21).

Table 3.21. Examples of lipase specificity

Hydrolyzed from a triacylglycerol	Lipase source
Acyl residues in positions 1 and 3	Pancreas, milk, *Pseudomonas fragi*, *Penicillium roqueforti*
Acyl residues in positions 1, 2 and 3	Oats, Castor bean, *Aspergillus flavus*
Oleic and linoleic acids in position 1, 2 and 3	*Geotrichum candidum*

The lipase secreted by the swine pancreas has been the most studied. Its molecular weight is 48 kdal. The enzyme cleaves the following types of acyl glycerols with a decreasing rate of hydrolysis: triacyl- >diacyl- ≫ monoacylglycerols. Table 3.21 shows that pancreatic lipase reacts with acyl residues at positions 1 and 3. The third acyl residue of a triacylglycerol is cleaved (cf. Reaction 3.52) only after acyl migration, which requires a longer incubation time.

$$(3.52)$$

The smaller the size of the oil droplet, the larger the oil/water interface and, therefore, the higher the lipase activity. This relationship should not be ignored when substrate emulsions are prepared for the assay of enzyme activities.
A model for pancreatic lipase has been suggested to account for the enzyme's activity on the oil/water interface (Fig. 3.17). The lipase's "hydrophobic head" is bound to the oil droplet by hydrophobic interactions, while the enzyme's active site aligns with and binds to the substrate molecule. The active site resembles that of serine proteinase. The splitting of the ester bond occurs with the involvement of Ser, His and Asp residues on the

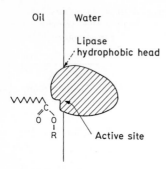

Fig. 3.17. A hypothetical model of pancreatic lipase fixation of an oil/water interphase (according to *H. Brockerhoff*, 1974)

enzyme by a mechanism analogous to that of chymotrypsin (cf. 2.4.2.5). The dissimilarity between pancreatic lipase and serine protein-ase is in the active site: lipase has a leucine residue within this site in order to establish hydrophobic contact with the lipid substrate and to align it with the activity center.

Lipase-catalyzed reactions are accelerated by Ca^{2+} ions since the liberated fatty acids are precipitated as insoluble Ca-salts.

The properties of milk lipase closely resemble those of pancreatic lipase.

Lipases of microbiological origin are often very heat stable. As can be seen from the example of a lipase of *Pseudomonas fluorescence* (Table 3.22), such lipases are not inactivated by pasteurization, ultra high temperature treatment, as well as drying procedures, e.g., the production of dry milk. These lipases can be the cause of decrease in quality of such products during storage.

A lipase of microbial origin with exceptional specificity (Table 3.21) has been detected. It hydrolyzes fatty acids only when they have a double bond in position 9. It is used to elucidate triacylglyceride structure. The use of lipases in food processing was outlined under 2.7.2.2.14.

Lipase activities in foods can be measured very sensitively with fluorochromic substrates, e.g., 4-methyl umbelliferyl fatty acid esters. Of course it is not possible to predict the storage stability of a food item with regard to lipolysis based only on such measurements.

The substrate specificity of the lipases, which can vary widely, is of essential importance for the aroma quality. Therefore, individual fatty acids can increase in different amounts even at the same lipase activity measured against a standard substrate. Since the odor and taste threshold values of the fatty acids differ greatly (cf. Tables 3.3–3.5), the effects of the lipases on the aroma are very variable. It is not directly possible to predict the point of time when rancid aroma notes will be present from the determination of the lipase activity. More precise information about the changes to be expected is obtained through storage experiments during which the fatty acids are quantitatively determined by gas chromatographic analysis. Table 3.23 shows the change in the concentrations of free fatty acids in sweet cream butter together with the resulting rancid aroma notes.

Table 3.22. Heat inactivation of a lipase of *Pseudomonas fluorescence* dissolved in skim milk

Temperature °C	D-value[a] (min)
100	23.5
120	7.3
140	2.0
160	0.7

[a] Time for 90% decrease in enzyme activity (cf. 2.5.4.1).

Table 3.23. Free fatty acids in butter (sweet cream) samples of different quality

Fatty acid	Butter				
	A (mg/kg)	B	C	D	E
4:0	0	5	38	78	119
6:0	0	4	28	25	46
8:0	8	22	51	51	86
10:0	38	58	104	136	229
12:0	78	59	142	137	231
14:0	193	152	283	170	477
Aroma[a]	2.3	2.8	3.0	4.6	5.4

[a] Classification: 2 not rancid, 3 slightly rancid, 4 rancid, 5 very rancid.

3.7.1.2 Polar-Lipid Hydrolases

These enzymes are denoted as phospholipases, lysophospholipases or glycolipid hydrolases, depending on the substrate.

3.7.1.2.1 Phospholipases

Phospholipase A_1. The enzyme is present together with phospholipase A_2 in many mammals and bacteria. It cleaves specifically the sn-1 ester bonds of diacylphosphatides (Formula 3.53).

Phospholipase A_2. Enzymes with sn-2 specificity isolated form snake and bee venoms. They are very stable, are activated by Ca^{2+}-ions and are amongst the smallest enzyme molecules (molecular weight about 14,000).

P-Lip.: Phospholipase   (3.53)

Phospholipase B. The existence of phospholipase B, which hydrolyzes in a single-step reaction both acyl groups in diacylphosphatides, is controversial. Other than the phospholipases A_1, A_2, C and D, the B-type could not be isolated in its pure form. A phospholipase B hydrolyzing also lysolecithin was enriched from germinating barley.

Phospholipase C. It hydrolyzes lecithin to a 1,2-diacylglyceride and phosphoryl choline. The enzyme is found in snake venom and in bacteria.

Phospholipase D. This enzyme cleaves the choline group in the presence of water or an alcohol, such as methanol, ethanol or glycerol, yielding free or esterified phosphatidic acid. For example:

Phosphatidylcholine + ROH

$\longrightarrow$ Phosphatidyl-OR + Choline   (3.54)

R: H, CH_3, CH_3CH_2, $CH_2(OH)$—CH(OH)—CH_2

Phospholipase D cannot cleave phosphatidyl inositol. The enzyme is present in cereals, such as rye and wheat, and in legumes. It was isolated and purified from peanuts.

Lysophospholipases. The enzymes, hydrolyzing only lysophosphatides, are abundant in animal tissue and bacteria. There are lysophospholipases that split preferentially 1-acylphosphatides while others prefer 2-acylphosphatides, and a third group doesn't differentiate at all between the two lysophosphatide types.

3.7.1.2.2 Glycolipid Hydrolases

Enzymes that cleave the acyl residues of mono- and digalactosyl-diacylglycerides are localized in green plants. A substrate specificity study for such a hydrolase from potato (Table 3.24) shows that plants also contain enzymes that are able to hydrolyze polar lipids in general. The potato enzyme preferentially cleaves the acyl residue from monoacylglycerols and lysolecithin, whereas triacylglycerols, such as triolein, are not affected.

3.7.2 Peroxidation of Unsaturated Acyl Lipids

Acyl lipid constituents, such as oleic, linoleic and linolenic acids, have one or more allyl groups within the fatty acid molecule (cf. Table 3.7) and thus are readily oxidized to hydroperoxides. The latter, after subsequent degradation reaction, yield a great number of other compounds. Therefore, under the usual conditions of food storage, unsaturated acyl lipids cannot be considered as stable food constituents.

Table 3.24. Purified potato acyl hydrolase: substrate specificity

Substrate	Relative activity (%)	Substrate	Relative activity (%)
Monolein	100	Lecithin	13
Diolein	21	Monogalactosyl-	
Triolein	0.2	diacylglycerol	31
Methyloleate	28	Digalactosyl-	
Lysolecithin	72	diacylglycerol	17

Autoxidation should be distinguished from *lipoxygenase catalysis* in the process denoted as *lipid peroxidation*. Both oxidations provide hydroperoxides, but the latter occurs only in the presence of the enzyme.

Lipid peroxidation provides numerous volatile and nonvolatile compounds. Since some of the volatiles are exceptionally odorous compounds, lipid peroxidation is detected even in food with unsaturated acyl lipids present as minor constituents, or in food in which only a small portion of lipid was subjected to oxidation.

Induced changes in food aroma are frequently assessed by consumers as objectionable, for example, as rancid, fishy, metallic or cardboardlike, or as an undefined old or stale flavor. On the other hand, the fact that some volatile compounds, at a level below their off-flavor threshold values, contribute to the pleasant aroma of many fruits and vegetables and to rounding-off the aroma of many fat- or oil-containing foods should not be neglected.

3.7.2.1 Autoxidation

Autoxidation is quite complex and involves a great number of interrelated reactions of intermediates. Hence, autoxidation of food is usually imitated by the study of a model system in which, for example, changes of one unsaturated fatty acid or one of its intermediary oxidation products are recorded in the presence of oxygen under controlled experimental conditions.

Model system studies have revealed that the rate of autoxidation is affected by fatty acid composition, degree of unsaturation, the presence and activity of pro- and antioxidants, partial pressure of oxygen, the nature of the surface being exposed to oxygen and the storage conditions (temperature, light exposure, moisture content, etc.) of fat/oil-containing food. Thus, the autoxidation rate can vary considerably.

An extreme case (cf. Fig. 3.18-1) demonstrates what has invariably been found in food: the initial oxidation products are detectable only after a certain elapsed storage time. When this *induction period*, which is typical for a given autoxidation process, has expired, a steep rise occurs in the reaction rate. The prooxidant concentration is high in some foods. In these

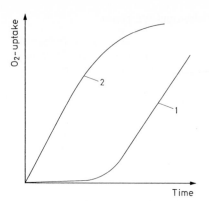

Fig. 3.18. Autoxidation of unsaturated acyl lipids. Prooxidant concentration: 1 low, 2 high

cases, illustrated in Fig. 3.18-2, the induction period may be nonexistent.

3.7.2.1.1 *Fundamental Steps of Autoxidation*

The length of the induction period and the rate of oxidation depend, among other things, on the fatty acid composition of the lipid (Table 3.25); the more allyl groups present, the shorter the induction period and the higher the oxidation rate.

Both phenomena, the induction period and the rise in reaction rate in the series, oleic, linoleic and linolenic acid can be explained as follows: Oxidation proceeds by a sequential free radical chain-reaction mechanism. Relatively stable radicals that can abstract H-atoms from the activated methylene groups in an olefinic compound are formed. On the basis of this assumption and, in addition, on the fact that the oxidation rate is exponential, *Farmer* et al. (1942) and *Bolland* (1949) proposed an aut-

Table 3.25. Induction period and relative rate of oxidation for fatty acids at 25°C

Fatty acid	Number of allyl groups	Induction period (h)	Oxidation rate (relative)
18:0	0		1
18:1 (9)	1	82	100
18:2 (9, 12)	2	19	1,200
18:3 (9, 12, 15)	3	1.34	2,500

Start: Formation of peroxy (RO$_2^{\cdot}$),
alkoxy (RO$^{\cdot}$) or alkyl (R$^{\cdot}$) radicals

Chain propagation:

(1) R$^{\cdot}$ + O$_2$ ⟶ RO$_2^{\cdot}$ k_1: 10^9 lmol^{-1} s^{-1}

(2) RO$_2^{\cdot}$ + RH ⟶ ROOH + R$^{\cdot}$ k_2: 10–60 lmol^{-1} s^{-1}

(3) RO$^{\cdot}$ + RH ⟶ ROH + R$^{\cdot}$

Chain branching:

(4) ROOH ⟶ RO$^{\cdot}$ + $^{\cdot}$OH

(5) 2 ROOH ⟶ RO$_2^{\cdot}$ + RO$^{\cdot}$ + H$_2$O

Chain termination:

(6) 2 R$^{\cdot}$ ⟶ ⎫
(7) R$^{\cdot}$ + RO$_2^{\cdot}$ ⟶ ⎬ Stable products
(8) 2 RO$_2^{\cdot}$ ⟶ ⎭

Fig. 3.19. Basic steps in the autoxidation of olefins

oxidation mechanism for olefinic compounds and, thus, also for unsaturated fatty acids. This mechanism has several fundamental steps. As shown in Fig. 3.19, the oxidation process is essentially a radical-induced chain reaction divided into initiation (start), propagation, branching and termination steps. Autoxidation is initiated by free radicals of frequently unknown origin.

Measured and calculated reaction rate constants for the different steps of the radical chain reaction show that due to the stability of the peroxy free radicals (ROO$^{\cdot}$), the whole process is limited by the conversion of these free radicals into monohydroperoxide molecules (ROOH). This reaction is achieved by abstraction of an H-atom from a fatty acid molecule [reaction step 2 (RS-2 in Fig. 3.19)]. The H-abstraction is the slowest and, hence, the rate limiting step in radical (R$^{\cdot}$) formation. Peroxidation of unsaturated fatty acids is accelerated autocatalytically by radicals generated from the degradation of hydroperoxides by a monomolecular reaction mechanism (RS-4 in Fig. 3.19). This reaction is promoted by heavy metal ions or heme(in)-containing molecules

(cf. 3.7.2.1.4). Also, degradation of hydroperoxides is considered as a starting point in discussions pertinent to volatile reaction products (cf. 3.7.2.1.5).

After a while, the hydroperoxide concentration reaches a level at which it begins to generate free radicals by a bimolecular degradation mechanism (RS-5 in Fig. 3.19). Reaction RS-5 is exothermic, unlike the endothermic monomolecular decomposition of hydroperoxides (RS-4 in Fig. 3.19) which needs approx. 150 kJ/mol. However, in most foods, RS-5 is of no relevance since fat (oil) oxidation makes a food unpalatable well before reaching the necessary hydroperoxide level for the RS-5 reaction step to occur. RS-4 and RS-5 (Fig. 3.19) are the branching reactions of the free radical chain.

At room temperature, a radical may inititiate the formation of 100 hydroperoxide molecules before chain termination occurs. In the presence of air (oxygen partial pressure >130 mbar), all alkyl radicals are transformed into peroxy radicals through the rapid radical chain reaction 1 (RS-1, Fig. 3.19). Therefore, chain termination occurs through collision of two peroxy radicals (RS-8, Fig. 3.19).

Termination reactions RS-6 and RS-7 in Fig. 3.19 play a role when, for example, the oxygen level is low, e.g. in the inner portion of a fatty food.

The hypothesis presented in Fig. 3.19 is valid only for the initiation phase of autoxidation. The process becomes less and less clear with increasing reaction time since, in addition to hydroperoxides, secondary products appear that partially autoxidize into tertiary products. The stage at which the process starts to become difficult to survey depends on the stability of the primary products. It is instructive here to compare the difference in the structures of monohydroperoxides derived from linoleic and linolenic acids.

3.7.2.1.2 Monohydroperoxides

The peroxy radical formed in RS-1 (Fig. 3.19) is slow reacting and therefore it selectively abstracts the most weakly bound H-atom from a fat molecule. It differs in this property from, for example, the substantially more reactive hydroxy (HO·) and alkoxy (RO·) radicals (cf. 3.7.2.1.4). RS-2 in Fig. 3.19 has a high reaction rate only when the energy for H-abstraction is clearly lower than the energy released in binding H to O during formation of hydroperoxide groups (about 376 kJ mol^{-1}).

Table 3.26 lists the energy inputs needed for H-abstraction from the carbon chain segments or groups occurring in fatty acids. The peroxy radical abstracts hydrogen more readily from a methylene group of a 1,4-pentadiene system than from a single allyl group. In the former case, the 1,4-diene radical that is generated is more effectively stabilized by resonance, i.e. electron delocalization over 5 C-atoms. Such considerations explain the difference in rates of autoxidation for unsaturated fatty acids and show why, at room temperature, the unsaturated fatty acids are attacked very selectively by peroxy radicals while the saturated acids are stable.

The general reaction steps shown in Fig. 3.19 are valid for all unsaturated fatty acids. In the case of oleic acid, H-atom abstraction occurs

Table 3.26. Energy requirement for a H-atom abstraction

	D_{R-H} (kJ/mole)
$\overset{\overset{\text{H}}{\mid}}{\text{CH}_2-}$	422
$\text{CH}_3-\overset{\overset{\text{H}}{\mid}}{\text{CH}}-$	410
$-\overset{\overset{\text{H}}{\mid}}{\text{CH}}-\text{CH}=\text{CH}-$	322
$-\text{CH}=\text{CH}-\overset{\overset{\text{H}}{\mid}}{\text{CH}}-\text{CH}=\text{CH}-$	272

on the methylene group adjacent to the double bond, i.e. positions 8 and 11 (Fig. 3.20). This would give rise to four hydroperoxides. In reality, they have all been isolated and identified as autoxidation products of oleic acid. The configuration of the newly formed double bond of

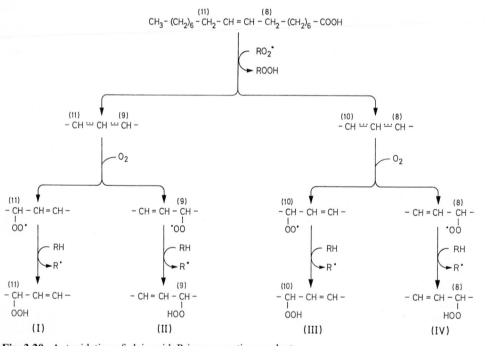

Fig. 3.20. Autoxidation of oleic acid. Primary reaction products:
I 11-Hydroperoxyoctadec-9-enoic acid; II 9-hydroperoxyoctadec-10-enoic acid, III 10-hydroperoxyoctadec-8-enoic acid, IV 8-hydroperoxyoctadec-9-enoic acid

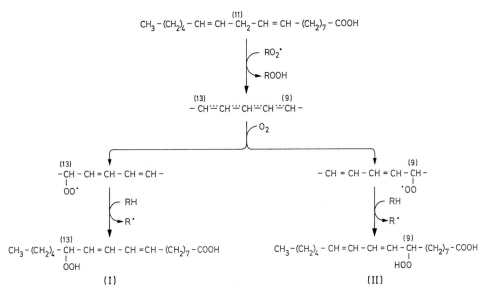

Fig. 3.21. Autoxidation of linoleic acid. Primary reaction products. I 13-Hydroperoxyoctadeca-9,11-dienoic acid, II 9-hydroperoxyoctadeca-10,12-dienoic acid

the hydroperoxides is affected by temperature. This configuration has 33% of cis and 67% of the more stable trans-configuration at room temperature.

Oxidation of the methylene group in position 11 of linoleic acid is activated especially by the two neighboring double bonds. Hence, this is the initial site for abstraction of an H-atom (Fig. 3.21). The pentadienyl radical generated is stabilized by formation of two hydroperoxides at positions 9 and 13, each retaining a conjugated diene system. These hydroperoxides have an UV maximum absorption at 235 nm and can be separated by high performance liquid chromatography as methyl esters, either directly or after reduction to hydroxydienes (Fig. 3.22).

The monoallylic groups in linoleic acid (positions 8 and 14 in the molecule), in addition to the bis-allylic group (position 11), also react to a small extent, giving rise to four hydroperoxides (8-, 10-, 12- and 14-OOH), each isomer having two isolated double bonds. The proportion of these minor monohydroperoxides is about 4% of the total (Table 3.27).

Autoxidation of linolenic acid yields four monohydroperoxides (Table 3.27). Formation of the monohydroperoxides is easily achieved

by H-abstraction from the bis-allylic groups in positions 11 and 14. The resultant two pentadiene radicals then stabilize analogously to linoleic acid oxidation (Fig. 3.21); each radical corresponds to two monohydroperoxides. However, the four isomers are not formed in

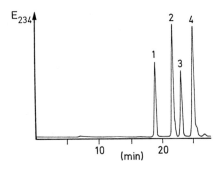

Fig. 3.22. Autoxidation of linoleic acid methyl ester. Analysis of primary products (according to reduction of the hydroperoxy group) by HPLC (after *H.W.S. Chan* and *G. Levett*, 1977). 1 13-Hydroxy-cis-9, trans-11-octadecadienoic acid methyl ester, 2 13-hydroxy-trans-9,trans-11-octadecadienoic acid methyl ester, 3 9-hydroxy-trans-10, cis-12-octadecadienoic acid methyl ester, 4 9-hydroxy-trans-10, trans-12-octadecadienoic acid methyl ester

Table 3.27. Monohydroperoxides formed by autoxidation ($^3O_2$) and photooxidation ($^1O_2$) of unsaturated fatty acids

Fatty acid	Monohydroperoxide			
	Position of		Proportion (%)	
	HOO- group	double bond	$^3O_2$	$^1O_2$
Oleic acid	8	9	27	
	9	10	23	48
	10	8	23	52
	11	9	27	
Linoleic acid	8	9, 12	1.5	
	9	10, 12	46.5	32
	10	8, 12	0.5	17
	12	9, 13	0.5	17
	13	9, 11	49.5	34
	14	9, 12	1.5	
Linolenic acid	9	10, 12, 15	31	23
	10	8, 12, 15		13
	12	9, 13, 15	11	12
	13	9, 11, 15	12	14
	15	9, 12, 16		13
	16	9, 12, 14	46	25

equimolar amounts; the 9- and 16-isomers predominate (Table 3.27). The configuration of the conjugated double bonds again depends on the reaction conditions. Cis-hydroperoxides are the main products at temperatures <40 °C. Competition exists between conversion of the peroxy radical to monohydroperoxide and reactions involving β-fragmentation and cyclization. Allyl peroxy radicals can undergo β-fragmentation which results, after a new oxygen molecule is attached, in a peroxy radical positional isomer, e.g. rearrangement of an oleic acid peroxy radical:

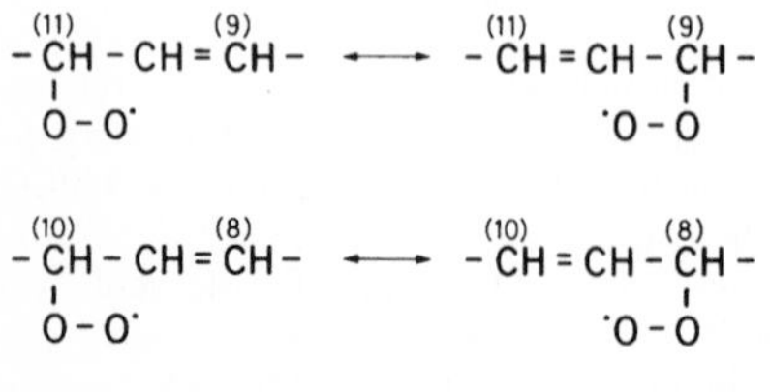

(3.55)

However, hydroperoxides can also be isomerized by such a reaction pathway. When they interact with free radicals (H-abstraction from -OOH group) or with heavy metal ions (cf. Reaction 3.64), they are again transformed into peroxy radicals. Thus, the 13-hydroperoxide of linoleic acid isomerizes into the 9-isomer and vice versa:

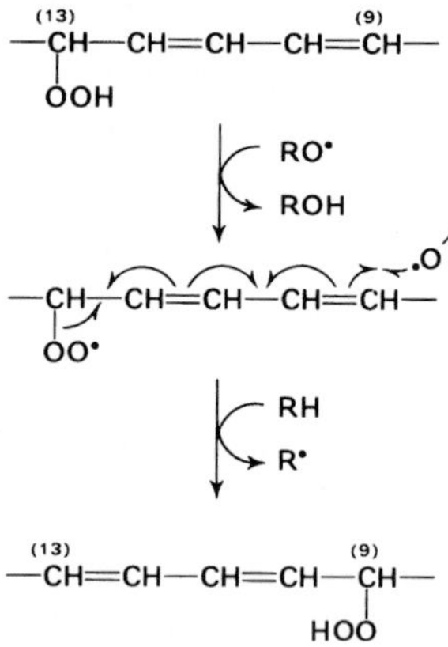

(3.56)

3.7.2.1.3 Hydroperoxide-Epidioxides

Peroxy radicals which contain isolated β, γ double bonds are prone to cyclization reactions in competition with reactions leading to monohydroperoxides. A hydroperoxide-epidioxide results through attachment of a second oxygen molecule and abstraction of a hydrogen atom:

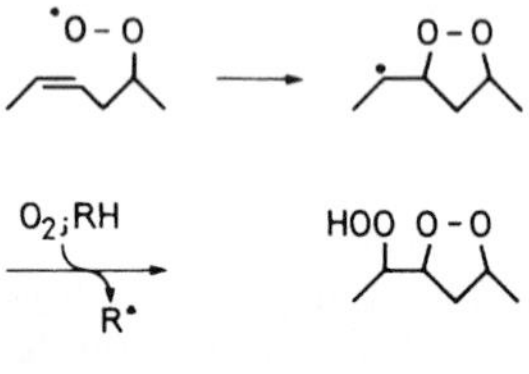

(3.57)

Peroxy radicals with isolated β, γ double bonds are formed as intermediary products after autoxidation and photooxidation (reaction with singlet O_2) of unsaturated fatty acids having two or more double bonds.

For this reason the 10- and 12-peroxy radicals obtained from linoleic acid readily form hydroperoxy-epidioxides. While such radicals are only minor products in autoxidation, in photooxidation they are generated as intermediary products in yields similar to the 9- and 13-peroxy radicals, which do not cyclize.

Ring formation by 10- and 12-peroxy radicals decreases formation of the corresponding monohydroperoxides (Table 3.27; reaction with $^1O_2$).

Among the peroxy radicals of linolenic acid which are formed by autoxidation, the isolated β, γ double bond system exists only for the 12- and 13-isomers, and not for the 9- and 16-isomers. Also, the tendency of the 12- and 13-peroxy radicals of linolenic acid to form hydroperoxy-epidioxides results in the formation of less monohydroperoxide of the corresponding isomers as opposed to the 9- and 16-isomers (Table 3.27).

Peroxy radicals interact rapidly with antioxidants which may be present to give monohydroperoxides (cf. 3.7.3.1). Thus, it is not only the chain reaction which is inhibited by antioxidants, but also β-fragmentation and peroxy radical cyclization.

$$(3.58)$$

R^1 : $CH_3 - (CH_2)_3$; R^2 : $(CH_2)_7 - COOH$

Fragmentation occurs when a hydroperoxide-epidioxide is heated, resulting in formation of aldehydes and aldehydic acids. For example, hydroperoxide-epidioxide fragments derived from the 12-peroxy radical of linoleic acid are formed as shown in Reaction 3.58.

Peroxy radicals formed from fatty acids with three or more double bonds can form bicycloendoperoxides with an epidioxide radical as intermediate. This is illustrated in Reaction 3.74.

3.7.2.1.4 Initiation of a Radical Chain Reaction

Since autoxidation of unsaturated acyl lipids frequently results in deterioration of food quality, an effort is made to at least decrease the rate of this deterioration process. However, pertinent measures are only possible when more knowledge is acquired about the reactions involved during the induction period of autoxidation and how they trigger the start of autoxidation.

In recent decades model system studies have revealed that two fundamentally different groups of reactions are involved in initiating autoxidation.

The first group is confined to the initiating reactions which overcome the energy barrier required for the reaction of molecular oxygen with an unsaturated fatty acid. Photosensitized oxidation (photooxidation) which provides the "first" hydroperoxides belongs to this group of reactions. These hydroperoxides are then converted further into radicals by the second group of reactions. Heavy metal ions and heme(in) proteins are involved in this second reaction group. Some enzymes which generate the superoxide radical anion can be placed in between these two delineated reaction groups since at least H_2O_2 is necessary as reactant for the formation of radicals.

Photooxidation. In order to understand photooxidation and to differentiate it from autoxidation, the electronic configuration of the molecular orbital energy levels for oxygen should be known. As presented in Fig. 3.23, the allowed energy levels correspond to $^3\Sigma^-$ g, $^1\Delta$g and $^1\Sigma^+$ g.

The notation for the molecular orbital of O_2 is $(\sigma\, 2s)^2\, (\sigma^*\, 2s)^2\, (\sigma\, 2p)^2\, (\pi\, 2p)^4\, (\pi^*\, 2p)^2$.

In the ground state, oxygen is a triplet ($^3O_2$). As seen from the above notation, the term $(\pi^*\, 2p)^2$ accounts for two unpaired electrons in the oxygen molecule. These are the two antibonding π orbitals available: $\pi^*\, 2p_y$ and $\pi^*\, 2p_z$. The two electrons occupy these orbitals alone. The net angular momentum of the unpaired electrons has three components, hence the term "triplet". When the electrons are paired, the angular momentum can not be split into components and this represents a singlet state. In the triplet state, oxygen reacts preferentially with radicals, i.e. molecules having one unpaired electron. In contrast, direct reactions of triplet-state oxygen with molecules which have all electrons paired, as in the case of fatty acids, are prevented by spin barriers. For this reason the activation energy of the reaction

$$RH + {}^3O_2 \rightarrow ROOH \qquad (3.59)$$

is so high (146–273 kJ/mole) that it does not occur without some assistance.

Oxygen goes from the ground state to the short-lived 1-singlet-state ($^1O_2$) by the uptake of 92 kJ/mole of energy (Fig. 3.23). The previously unpaired single electrons are now paired on the π^*2p_y antibonding orbital. The reactivity of this molecule resembles ethylenic or general olefinic π electron pair reactions, but it is more electrophilic. Hence, in the reaction with oleic acid, the 1-singlet-state oxygen attacks the 9–10 double bond, generating two monohydroperoxides, the 9- and 10-isomers (cf. Table 3.27). The second singlet-state of oxygen ($^1\Sigma^+$ g) has a much shorter life than the 1-singlet-state and plays no role in the oxidation of fats or oils.

For a long time it has been recognized that the stability of stored fat (oil) drops in the presence of light. Light triggers lipid autoxidation. Low amounts of some compounds participate as sensitizers.

According to *Schenk* and *Koch* (1960), there are two types of sensitizers. Type I sensitizers are those which, once activated by light (sen*), react directly with substrate, generating substrate radicals. These then trigger the autoxidation process. Type II sensitizers are those which activate the ground state of oxygen to the $^1O_2$ singlet state.

Type I and II photooxidation compete with each other. Which reaction will prevail depends on the structure of the sensitizer but also on the concentration and the structure of the substrate available for oxidation.

Table 3.27 shows that the composition of hydroperoxide isomers derived from an unsaturated acid by autoxidation ($^3O_2$) differs from that obtained in the reaction with $^1O_2$. The isomers can be separated by analysis of hydroperoxides using high performance liquid chromatography and, thus, one can distinguish Type I from Type II photooxidation. Such studies have revealed that sensitizers, such as chlorophylls a and b, pheophytins a and b and riboflavin, present in food, promote the Type II oxidation of oleic and linoleic acids.

As already stated, the Type II sensitizer, once activated, does not react with the substrate but with ground state triplet oxygen, transforming it with an input of energy into 1-singlet-state oxygen:

$$\text{Sen} \xrightarrow{\ h\nu\ } \text{Sen*}$$
$$\text{Sen*} + O_2 \longrightarrow \text{Sen} + {}^1O_2 \qquad (3.60)$$

The singlet $^1O_2$ formed now reacts directly with the unsaturated fatty acid by a mechanism of "cyclo-addition":

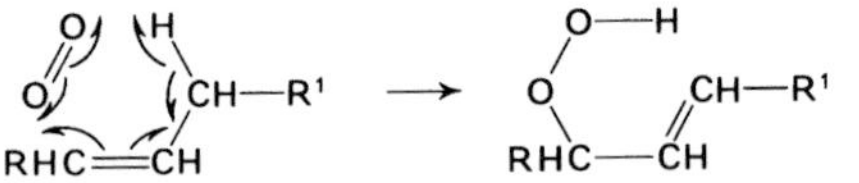

$$(3.61)$$

The fact that the number of hydroperoxides formed are double the number of isolated double bonds present in the fatty acid molecule is in agreement with the above reaction mechanism. The reaction is illustrated in Fig. 3.24 for the oxidation of linoleic acid. In addition to the two hydroperoxides with a conjugated diene system already mentioned (Fig. 3.21), two hydroperoxides are obtained with isolated double bonds.

Furan fatty acids (cf. 3.2.1.3) react much faster with $^1O_2$ than linoleic or linolenic acid. An endo peroxide is the main reaction product. Of much higher importance for the aroma of some fats and oils is the side reaction of branched furanoic fatty acids with $^1O_2$, depicted in Fig. 3.25. During this reaction the allyl group of C-9 to C-11 is oxidized to an 11-hydroperoxide. The following β-cleavage

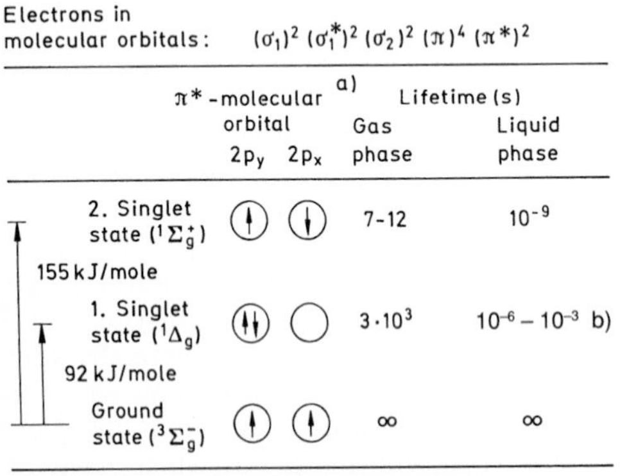

Electrons in
molecular orbitals: $(\sigma_1)^2\ (\sigma_1^*)^2\ (\sigma_2')^2\ (\pi)^4\ (\pi^*)^2$

π*-molecular orbital		Lifetime (s)[a]	
$2p_y$	$2p_x$	Gas phase	Liquid phase
2. Singlet state ($^1\Sigma_g^+$) ⊙ ⊙		7–12	10^{-9}
155 kJ/mole			
1. Singlet state ($^1\Delta_g$) ⊛ ○		$3 \cdot 10^3$	$10^{-6} – 10^{-3}$ [b]
92 kJ/mole			
Ground state ($^3\Sigma_g^-$) ⊙ ⊙		∞	∞

Fig. 3.23. Configuration of electrons in an oxygen molecule
[a] Electrons in $2p_x$ and $2p_y$ orbitals
[b] Dependent on solvent, e.g. 2 μs in water, 20 μs in D_2O and 7 μs in methanol

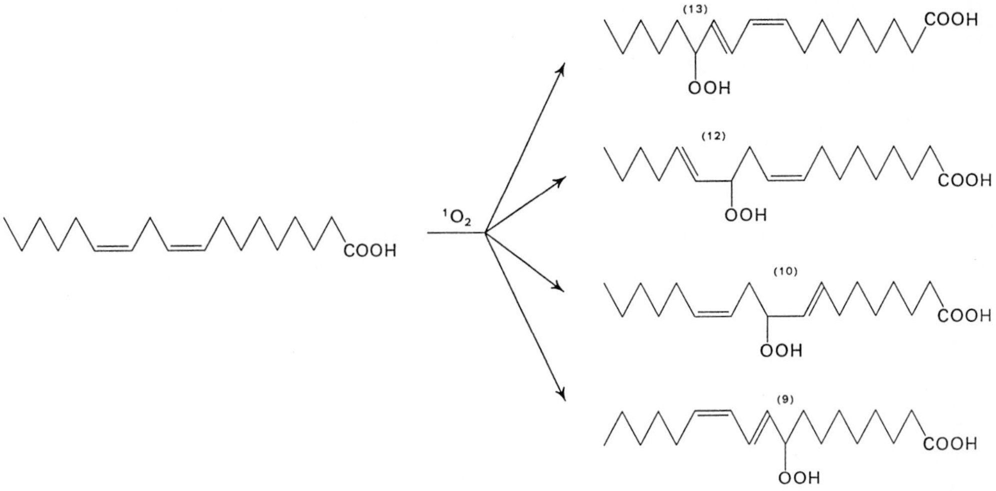

Fig. 3.24. Hydroperoxides derived from linoleic acid by type-2 photooxidation

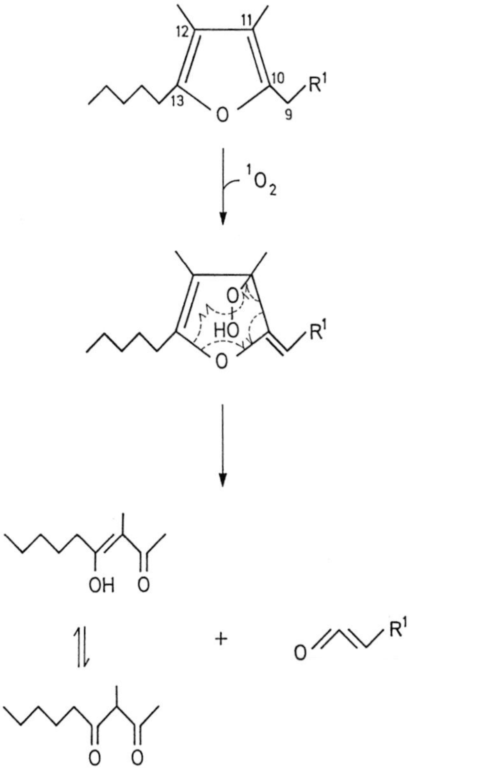

Fig. 3.25. Side reaction of a branched furan fatty acid with singlet oxygen (R^1: $(CH_2)_7COOH$)

of hydroperoxide results in the formation of a carbonyl group at C-11 and a hydroxyl radical which combines with C-13 after homolysis of the furan ring. The furan fatty acid thus is split into 3-methyl-2,4-nonanedione and a fragment of unknown structure. The dione is a very intensive aroma compound (Table 3.31) that contributes to the light induced off-flavor of soybean oil (cf. 14.3.2.2.5).

Formation of 1-singlet oxygen ($^1O_2$) is inhibited by carotenoids (car):

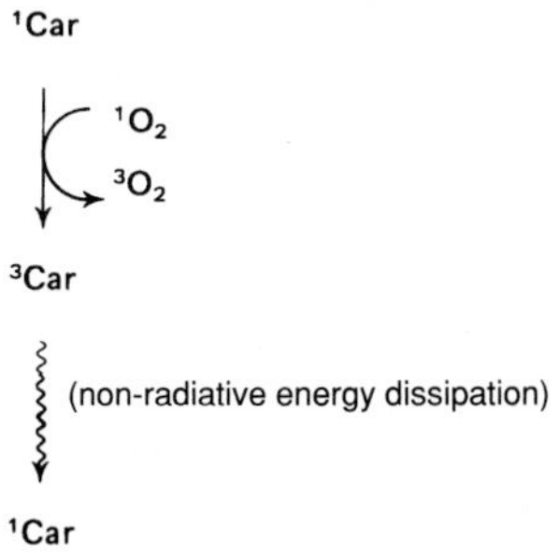

$$(3.62)$$

The quenching effect of carotenoids (transition of $^1O_2$ to $^3O_2$) is very fast (k = 3×10^{10} l mole^{-1}s^{-1}). They also prevent energy transfer from excited-state chlorophyll to $^3O_2$. Therefore, carotenoids are particularly suitable for protecting fat (oil)-containing food from Type II photooxidation.

Heavy Metal Ions. These ions are involved in the second group of initiation reactions, namely, in the decomposition of initially-formed hydroperoxides into radicals which then propel the radical chain reaction of the autooxidation process. Fats, oils and foods always contain traces of heavy metals, the complete removal of which in a refining step would be uneconomical. The metal ions, primarily Fe, Cu and Co, may originate from:

- Raw food. Traces of heavy metal ions are present in many enzymes and other metal-bound proteins. For example, during the crushing and solvent extraction of oilseeds, metal bonds dissociate and the free ions bind to fatty acids.
- From processing and handling equipment. Traces of heavy metals are solubilized during the processing of fat (oil). Such traces are inactive physiologically but active as prooxidants.
- From packaging material. Traces of heavy metals from metal foils or cans or from wrapping paper can contaminate food and diffuse into the fat or oil phase.

The concentration of heavy metal ions that results in fat (oil) shelf-life instability is dependent on the nature of the metal ion and the fatty acid composition of the fat (oil). Edible oils of the linoleic acid type, such as sunflower and corn germ oil, should contain less than 0.03 ppm Fe and 0.01 ppm Cu to maintain their stability. The concentration limit is 0.2 ppm for Cu and 2 ppm for Fe in fat with a high content of oleic and/or stearic acids, e.g. butter.

Heavy metal ions trigger the autoxidation of unsaturated acyl lipids only when they contain hydroperoxides. That is, the presence of a hydroperoxide group is a prerequisite for metal ion activity, which leads to decomposition of the hydroperoxide group into a free radical:

$$Me^{n\oplus} + ROOH \longrightarrow Me^{(n+1)\oplus} + RO^{\bullet} + OH^{\ominus}$$
$$(3.63)$$

$$Me^{(n+1)\oplus} + ROOH \longrightarrow RO_2^{\bullet} + H^{\oplus} + Me^{n\oplus}$$
$$(3.64)$$

Me: Heavy metal ion

Reaction rate constants for the decomposition of linoleic acid hydroperoxide are given in Table 3.28. As seen with iron, the lower oxidation state (Fe^{2+}) provides a ten-fold faster decomposition rate than the higher state (Fe^{3+}). Correspondingly, Reaction 3.63 proceeds much faster than Reaction 3.64 in which the reduced state of the metal ion is regenerated. The start of autoxidation then is triggered by radicals from generated hydroperoxides.

The decomposition rates for hydroperoxides emulsified in water depend on pH (Table 3.28). The optimal activity for Fe and Cu ions is in the pH range of 5.5–6.0. The presence of ascorbic acid, even in traces, accelerates the decomposition. Apparently, it sustains the reduced state of the metal ions.

The direct oxidation of an unsaturated fatty acid to an acyl radical by a heavy metal ion

$$RH + Me^{(n-1)\oplus} \longrightarrow R^{\bullet} + H^{\oplus} + Me^{n\oplus} \qquad (3.65)$$

proceeds, but at an exceptionally slow rate. It seems to be without significance for the initiation of autoxidation.

The autoxidation of acyl lipids is also influenced by the moisture content of food. The reaction rate is high for both dehydrated and water-containing food, but is minimal at a water activity (a_w) of 0.3 (Fig. 0.4). The following hypotheses are discussed to explain these differences: The high reaction rate in dehydrated food is due to metal ions with depleted hydration shells. In addition, ESR spectroscopic stu-

Table 3.28. Linoleic acid hydroperoxides[a]: decomposition by heavy metal or heme compounds at 23 °C. Relative reaction rates k_{rel} are given at two pH's[a]

Heavy metal ion[b]	k_{rel}		Heme compound[b]	k_{rel}	
	pH 7	pH 5.5		pH 7	pH 5.5
Fe^{3+}	1	10^2	Hematin	$4 \cdot 10^3$	$4 \cdot 10^4$
Fe^{2+}	14	10^3	Methemoglobin	$5 \cdot 10^3$	$7.6 \cdot 10^3$
Cu^{2+}	0.2	1.5	Cytochrome C	$2.6 \cdot 10^3$	$3.9 \cdot 10^3$
Co^{3+}	$6 \cdot 10^2$	1	Oxyhemoglobin	$1.2 \cdot 10^3$	
Mn^{2+}	0	0	Myoglobin	$1.1 \cdot 10^3$	
			Catalase	1	
			Peroxidase	1	

[a] Linoleic acid hydroperoxide is emulsified in a buffer.
[b] Reaction rate constant is related to reaction rate in presence of Fe^{3+} at pH 7 ($k_{rel} = 1$).

dies show that food drying promotes the formation of free radicals which might initiate lipid peroxidation. As the water content starts to increase, the rate of autoxidation decreases. It is assumed that this decrease in rate is due to hydration of ions and also of radicals. Above an a_w of 0.3, free water is present in food in addition to bound water. Free water appears to enhance the mobility of prooxidants, thus accounting for the renewed increase in autoxidation rate that is invariably observed at high moisture levels in food.

Heme(in) Compounds. Heme (Fe^{2+}) and hemin (Fe^{3+}) proteins are widely distributed in food. Lipid peroxidation in animal tissue is accelerated by hemoglobin, myoglobin and cytochrome C. These reactions are often responsible for rancidity or aroma defects occurring during storage of fish, poultry and cooked meat. In plant food the most important heme(in) proteins are peroxidase and catalase. Cytochrome P_{450} is a particularly powerful catalyst for lipid peroxidation, although it is not yet clear to what extent the compound affects food shelf life "in situ".
During heme catalysis, a Fe^{2+} protoporphyrin complex ($P-Fe^{2+}$), like in myoglobin, will be oxidized by air to $P-Fe^{3+}$ as indicated in Formula 3.66. The formed superoxide radical anion O_2^-, whose properties are discussed below, will further react yielding H_2O_2. Hydrogen peroxide will then oxidize $P-Fe^{3+}$ to the oxene species $P-Fe=O$. The reaction with H_2O_2 is accelerated by acid/base catalysis, facilitating the loss of the water mole-

cule; the hemin protein and one carboxylic group of the protoporphyrin system acts as proton acceptor and proton donor respectively.
Oxene is the active form of the hemin catalyst. It oxidizes two fatty acid hydroperoxide molecules to peroxy radicals that will then initiate lipid peroxidation.
In comparison with iron ions, some heme(in) compounds degrade the hydroperoxides more rapidly by several orders of magnitude (cf. Table 3.28). Therefore they are more effective as initiators of lipid peroxidation. Their activity is also negligibly influenced by a decrease in the pH-value.
However, the activity of a heme(in) protein for hydroperoxides is influenced by its steric accessibility to fatty acid hydroperoxides. Hydroperoxide binding to the Fe-porphyrin moiety of native catalase and peroxidase molecules is obviously not without interferences. The prosthetic group is free to promote hydroperoxide decomposition only after heat denaturation of the enzymes. Results from a model study involving peroxidase are given in Table 3.29. As the data show, heating the enzyme results in an increase of prooxidative activity by a factor of 10 and, as expected, in a concomitant drop of enzyme activity. Similar results are obtained in reaction systems containing catalase.
Suppression of peroxidase and catalase activity is of importance for the shelf life of heat-processed food. As long as the protein moiety has not been denatured, it is the lipoxygenase enzyme which is the most active for lipid per-

$$P-Fe^{2\oplus} + O_2 \longrightarrow P-Fe^{3\oplus} + O_2^{\ominus}$$

$$2O_2^{\ominus} + 2H^{\oplus} \longrightarrow H_2O_2 + O_2$$

$$P-Fe^{3\oplus} + H_2O_2 \longrightarrow P\overset{\oplus}{\underset{Fe^{3\oplus}\cdots O^{\ominus}-O-H}{\diagup}}^H \longrightarrow \underset{P-Fe^{4\oplus}-O^{\ominus}}{\overset{P-Fe^{3\oplus}=O}{\updownarrow}} + H_2O \qquad (3.66)$$

$$P-Fe^{4\oplus}-O^{\ominus} + ROOH \longrightarrow P-Fe^{4\oplus}\cdots^{\ominus}OH + ROO^{\bullet}$$

$$P-Fe^{4\oplus}\cdots O^{\ominus}-H + ROOH \longrightarrow P-Fe^{3\oplus} + ROO^{\bullet} + H_2O$$

Table 3.29. Rates of linoleic acid peroxidation in the presence of heat-treated horseradish peroxidase

Heat treatment 2 min at °C	Linoleic acid peroxidation (μmole O_2/min)	Enzyme activity (%)
25	8.7	100
53	10.5	100
90	47.5	80
120	79.5	50
140	96.0	14

oxidation (cf. 3.7.2.2). After lipoxygenase activity is destroyed by heat denaturation, its role is replaced by the heme(in) proteins. As already suggested, an assay of heme(in) protein enzyme activity does not necessarily reflect its prooxidant activity.

Activated Oxygen from Enzymatic Reactions. In enzymatic reactions oxygen can form three intermediates, which differ greatly in their activities and which are all ultimately reduced to water:

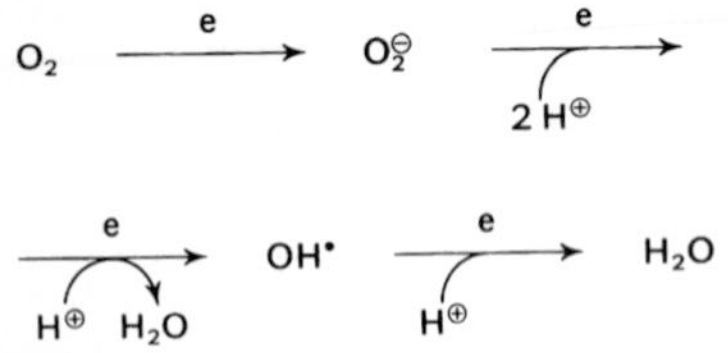

e: Electron

Oxygen takes up one electron to form a superoxide radical anion [O_2^-, $(\sigma\,2s)^2\,(\sigma^*\,2s)^2\,(\sigma\,2p)^2\,(\pi\,2p)^4\,(\pi^*\,2p)^3$; i.e. one electron-pair exists in the antibonding π molecular orbital]. This anion radical is a reducing agent with chemical properties dependent on pH, according to the equilibrium:

$$O_2^\ominus + H^\oplus \;\rightleftharpoons\; HO_2^\bullet \;(pK_s: 4.8) \qquad (3.68)$$

Based on its pK_s value under physiological conditions, this activated oxygen species occurs as an anion with its radical character suppressed. It acts as a nucleophilic reagent (e.g. it promotes phospholipid hydrolysis within the membranes) under such conditions, but is not directly able to abstract an H-atom and to initiate lipid peroxidation. The free

radical activity of the superoxide anion appears only in acidic media, wherein the perhydroxy radical form ($HO_2^\bullet$) prevails. The O_2^- has a finite stability since it slowly dismutates ($k = 0.35\ \text{l mol}^{-1}\text{s}^{-1}$):

$$2\,O_2^\ominus + 2\,H^\oplus \;\longrightarrow\; H_2O_2 + O_2 \qquad (3.69)$$

An enzyme with superoxide dismutase activity which significantly accelerates ($k = 2 \times 10^9$ l mol^{-1} s^{-1}) Reaction 3.69 occurs in numerous animal and plant tissues.

The superoxide radical anion, O_2^-, is generated by flavin enzymes, such as xanthine oxidase (cf. 2.3.3.2). The involvement of this enzyme in the development of milk oxidation flavor has been questioned for a long time.

Hydrogen peroxide, H_2O_2, is the second intermediate of oxygen reduction. In the absence of heavy metal ions, energy-rich radiation including UV light and elevated temperatures, H_2O_2 is a rather indolent and sluggish reaction agent. On the other hand, the hydroxy radical ($HO^\bullet$) derived from it is exceptionally active. During the abstraction of an H-atom,

$$R-H + HO^\bullet \;\longrightarrow\; R^\bullet + H_2O \qquad (3.70)$$

the energy input in the HO-bond formed is 497 kJ/mol, thus exceeding the dissociation energy for abstraction of hydrogen from each C–H bond by at least 75 kJ/mol (cf. Table 3.26). Therefore, the $HO^\bullet$ radical reacts nonselectively with all organic constituents of food. Consequently, it can directly initiate lipid peroxidation. However, in a complex system such as food, the following question is always pertinent: "Has the $HO^\bullet$ radical actually reached the unsaturated acyl lipid, or was it trapped prior to lipid oxidation by some other food ingredient?".

The reaction of the superoxide radical anion with hydrogen peroxide should be emphasized in relation to initiation of autoxidation. This is the so-called *Fenton* reaction in particular of an Fe-complex:

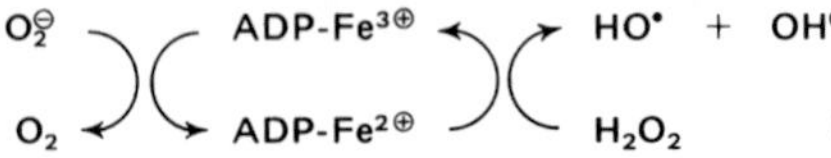

The Fe-complex (e.g. with ADP) occurs in food of plant and animal origin. The Fe^{2+} obtained by reduction with O_2^- can then reduce the H_2O_2 present and generate free HO· radicals.

3.7.2.1.5 Secondary Products

The primary products of autoxidation, the monohydroperoxides, are odorless and tasteless (such as linoleic acid hydroperoxides; cf. Table 3.35). Food quality is not affected until volatile compounds are formed. The latter are usually powerfully odorous compounds and, even in the very small amounts in which they occur, affect the odor and flavor of food.

From the numerous volatile secondary products of lipid peroxidation the following compounds will be discussed in detail

- odor-active carbonyl compounds
- malonic dialdehyde
- alkanes, alkenes

Odor-Active Monocarbonyl Compounds. Model expriments showed that the volatile fractions formed during the autoxidation of oleic, linoleic and linolenic acid contain mainly aldehydes and ketones (Table 3.30). Linoleic acid, a component of all lipids sensitive to

autoxidation, is a precursor of hexanal that is predominant in the volatile fraction. Therefore this substance, since it can easily be determined by headspace analysis, is used as an indicator for the characterization of off-flavors resulting from lipid peroxidation.

A comparison of the sensory properties (Table 3.31) shows that some carbonyl compounds, belonging to side components of the volatile fractions, may intensively contribute to an off-flavor due to their low threshold values. Food items containing linoleic acid, especially 2-cis-nonenal, trans-4,5-epoxy-2-trans-decenal and 1-octen-3-one, are very aroma active.

The rapid deterioration of food containing linolenic acid should not be ascribed solely to the preferential oxidation of this acid but also to the low odor threshold values of the carbonyl compounds formed, such as 3-cis-hexenal, 2-trans,6-cis-nonadienal and 1,cis-5-octadien-3-one (Table 3.31). Aldehydes with exceptionally strong aromas can be released in food by the autoxidation of some fatty acids, even if they are present in low amounts. An example is octadeca-cis-11,cis-15-dienoic acid (the precursor for 4-cis-heptenal), which occurs in

Table 3.30. Volatile compounds formed by autoxidation of unsaturated fatty acids (µg/g)[a]

Oleic acid		Linoleic acid		Linolenic acid	
Heptanal	50	Pentane[b]	+[c]	Propanal[b]	
Octanal	320	Pentanal	55	1-Penten-3-one	30
Nonanal	370	Hexanal	5,100	2tr-Butenal	10
Decanal	80	Heptanal	50	2tr-Pentenal	35
2tr-Decenal	70	2tr-Heptenal	450	2c-Pentenal	45
2tr-Undecenal	85	Octanal	45	2tr-Hexenal	10
		1-Octen-3-one	2	3tr-Hexenal	15
		1-Octen-3-hydroperoxide	+[c]	3c-Hexenal	90
		2c-Octenal	990	2tr-Heptenal	5
		2tr-Octenal	420	2tr,4c-Heptadienal	320
		3c-Nonenal	30	2tr,4tr-Heptadienal	70
		3tr-Nonenal	30	2c,5c-Octadienal	20
		2c-Nonenal	+[c]	3,5-Octadien-2-one	30
		2tr-Nonenal	30	2tr,6c-Nonadienal	10
		2c-Decenal	20	2,4,7-Decatrienal	85
		2tr,4tr-Nonadienal	30	1,5c-Octadien-3-one	+[c]
		2tr,4c-Decadienal	250	1,5c-Octadien-3-hydroperoxide	+[c]
		2tr,4tr-Decadienal	150		
		trans-4,5-Epoxy-2tr-decenal	+[c]		

[a] Each fatty acid in amount of 1 g was autoxidized at 20 °C by an uptake of 0.5 mole oxygen/mole fatty acid.
[b] Major compound of autoxidation.
[c] Detected, but not quantified.

Table 3.31. Sensory properties of aroma components resulting from lipid peroxidation

Compound	Flavor quality	Odor threshold (µg/kg) in oil		water nasal
		nasal	retronasal	
Aldehydes				
5:0	pungent, like bitter almonds	240	150	18
6:0	tallowy, green leafy	320	75	12
7:0	oily, fatty	3,200	50	5
8:0	oily, fatty, soapy	320	50	0.8
9:0	tallowy, soapy-fruity	13,500	260	5
10:0	orange peel like	6,700	850	5
5:1 (2tr)	pungent, apple	2,300	600	–
6:1 (2tr)	apple	10,000	400	50
6:1 (3c)	green, leafy	14	3	0.25
7:1 (2tr)	fatty, bitter almond	14,000	400	51
7:1 (4c)	cream, putty	2	1	0.8
8:1 (2c)	walnut	–	50	–
8:1 (2tr)	fatty, nutty	7,000	125	4
9:1 (2c)	fatty, green leafy	4.5	0.6	0.02
9:1 (2tr)	tallowy, cucumber	900	65	0.8
9:1 (3c)	cucumber	250	35	–
10:1 (2tr)	tallowy, orange	33,800	150	–
7:2 (2tr,4c)	frying odor, tallowy	4,000	50	–
7:2 (2tr,4tr)	fatty, oily	10,000	30	–
9:2 (2tr,4tr)	fatty, oily	2,500	460	–
9:2 (2tr,6c)	like cucumber	4	1.5	–
10:2 (2tr,4c)	frying odor	10	–	–
10:2 (2tr,4tr)	frying odor	180	40	0.2
10:3 (2tr,4c,7c)	cut beans	–	24	–
trans, 4,5-Epoxy-2tr-decenal	metallic	1.3	3	0.12
Ketones				
1-Penten-3-one	hot, fishy	0.7	3	1.3
1-Octen-3-one	like mushrooms, fishy	10	0.3	1
1,5c-Octadien-3-one	like geraniums, metallic	0.45	0.03	1.2×10^{-3}
3tr,5tr-Octadien-3-one	fatty, fruity	300	–	–
3tr,5c-Octadien-2-one	fatty, fruity	200	–	–
3-Methyl-2,4-nonanedione	like straw, fruity, like butter	23	1.5	0.03
Miscellaneous compounds				
1-Octen-3-hydroperoxide	metallic	240	–	–
2-Pentylfuran	like butter, like green beans	2,000	–	–

beef and mutton and often in butter (odor threshold in Table 3.31).

Also, the processing of oil and fat can provide an altered fatty acid profile. These can then provide new precursors for a new set of carbonyls. For example, 6-trans-nonenal, the precursor of which is octadeca-cis-9-trans-15-dienoic acid, is a product of the partial hydrogenation of linolenic acid. This aldehyde can

be formed during storage of partially hardened soya and linseed oils. The aldehyde, together with other compounds, is responsible for an off-flavor denoted as "hardened flavor".

Several reaction mechanisms have been suggested to explain the formation of volatile carbonyl compounds. The most probable mechanism is the β-scission of monohydroperoxides with formation of an intermediary short-lived alkoxy radical (Fig. 3.26). Such β-scission is catalyzed by heavy metal ions or heme(in) compounds (cf. 3.7.2.1.4).

There are two possibilities for β-scission of each hydroperoxide fatty acid (Fig. 3.26). Option "B", i.e. the cleavage of the C–C bond located further away from the double bond position, is the energetically preferred one since it leads to resonance-stabilized "oxo-ene" or "oxo-diene" compounds. Applying this β-scission mechanism ("B") to both major monohydroperoxide isomers of linoleic acid gives the following fragmentation products:

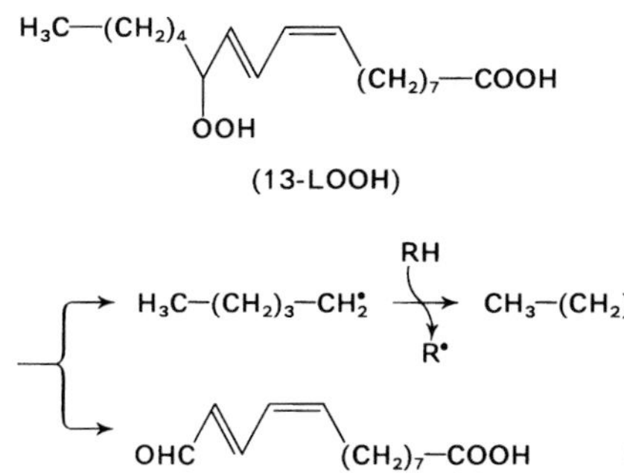

the C–C linkage adjacent to the double bond. The carbonium ion then splits into an oxo-acid and hexanal. The fact that linoleic acid 9-hydroperoxide gives rise to 2-nonenal is in agreement with this outline.

However, in the water-free fat or oil phase of food, the homolytic cleavage of hydroperoxides presented above is the predominant reaction mechanism. Since option "A" of the cleavage reaction is excluded (Fig. 3.26), some other reactions should be assumed to occur to account for formation of hexanal and

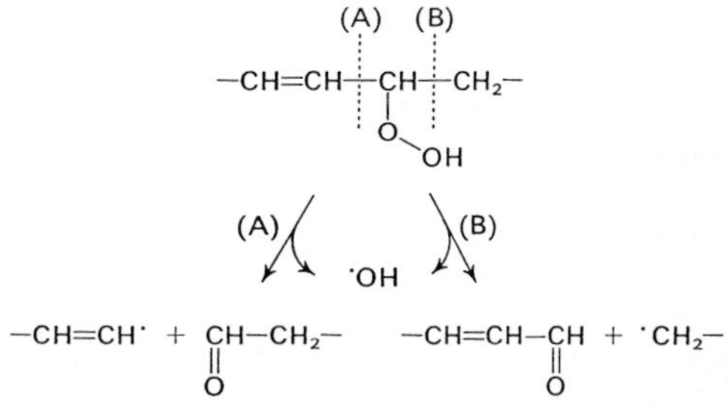

H₃C—(CH₂)₄ ⌇⌇⌇ (CH₂)₇—COOH
|
OOH

(13-LOOH)

RH
→ H₃C—(CH₂)₃—CH₂• →→ CH₃—(CH₂)₃—CH₃
        R•

→ OHC ⌇⌇⌇ (CH₂)₇—COOH   (3.72)

$$(A)\ (B)$$
$$-CH=CH \overset{|}{\underset{|}{}} CH \overset{|}{\underset{|}{}} CH_2-$$
$$\underset{OH}{O}$$

(A) ↙   ↘ (B)
·OH

$$-CH=CH\cdot + \underset{O}{\overset{\|}{C}}H-CH_2- \qquad -CH=CH-\underset{O}{\overset{\|}{C}}H + \cdot CH_2-$$

Fig. 3.26. β-Scission of monohydroperoxides (according to *H.T. Badings*, 1970)

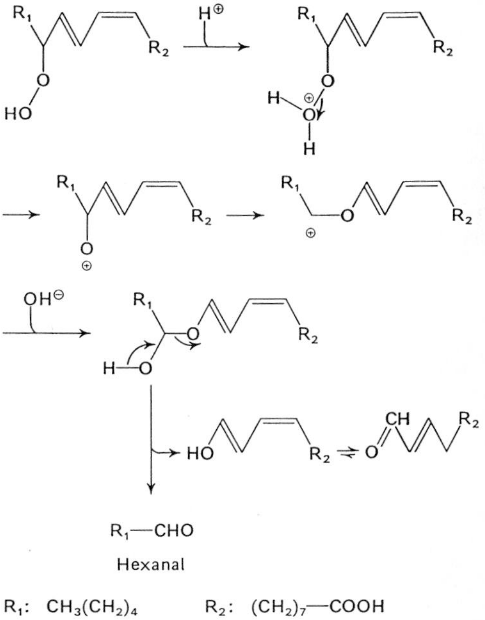

OOH
|
H₃C—(CH₂)₄ ⌇⌇⌇ (CH₂)₇—COOH

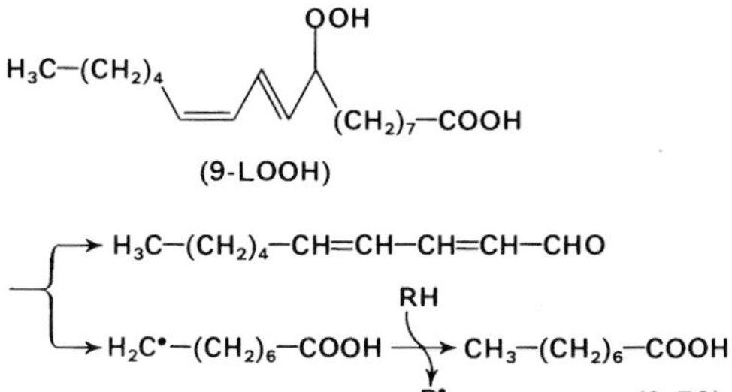

(9-LOOH)

→ H₃C—(CH₂)₄—CH=CH—CH=CH—CHO

RH
→ H₂C•—(CH₂)₆—COOH →→ CH₃—(CH₂)₆—COOH
        R•
                          (3.73)

From the volatile autoxidation products which contain the methyl end of the linoleic acid molecule, the formation of 2,4-decadienal and pentane can be explained by reaction 3.72.

The formation of hexanal among the main volatile compounds derived from linoleic acid (cf. Table 3.30) is still an open question. The preferential formation of hexanal in aqueous systems can be explained with an ionic mechanism. As shown in Fig. 3.27, the heterolytic cleavage is initiated by the protonation of the hydroperoxide group. After elimination of a water molecule, the oxo-cation formed is subjected to an insertion reaction exclusively on

R₁ —CHO

Hexanal

$R_1:\ CH_3(CH_2)_4 \qquad R_2:\ (CH_2)_7—COOH$

Fig. 3.27. Proton-catalyzed cleavage of linoleic acid 13-hydroperoxide (according to *G. Ohloff*, 1973)

other aldehydes from linoleic acid. The further oxidation reactions of monohydroperoxides and carbonyl compounds are among the possibilities.

The above assumption is supported by the finding that 2-alkenals and 2,4-alkadienals are oxidized substantially faster than the unsaturated fatty acids (Fig. 3.28). In addition, the autoxidation of 2,4-decadienal yields hexanal and other volatiles which coincide with those obtained from linoleic acid. Since saturated aldehydes oxidize slowly, as demonstrated by nonanal (Fig. 3.28), they will enrich the oxidation products and become predominant.

Also the delayed appearance of hexanal during the storage of linoleic acid containing fats and oils compared to pentane and 2,4-decadienal, supports the hypothesis that hexanal is not directly formed by a β-scission of the 13-hydroperoxide. It is mainly produced in a tertiary reaction, e.g., during the autoxidation of 2,4-decadienal.

Other studies to elucidate the multitude of aldehydes which arise suggest that the decomposition of minor hydroperoxides formed by autoxidation of linoleic acid (cf. Table 3.27) contribute to the profile of aldehydes. This suggestion is supported by pentanal, which originates from the 14-hydroperoxide.

The occurrence of 2,4-heptadienal (from the 12-hydroperoxide isomer) and of 2,4,7-decatrienal (from the 9-hydroperoxide isomer) as oxidation products is, thereby, readily explained by accepting the fragmentation mechanism outlined above (option "B" in Fig. 3.26) for the autoxidation of α-linolenic acid. The formation of other volatile carbonyls can then follow by autoxidation of these two aldehydes or from the further oxidation of labile monohydroperoxides.

Malonic Aldehyde. This dialdehyde is preferentially formed by autoxidation of fatty acids with three or more double bonds. The compound is odorless. In food it may be bound to proteins by a double condensation, cross-linking the proteins (cf. 3.7.2.4.3). Malonic aldehyde is formed from α-linolenic acid by a modified reaction pathway, as outlined under the formation of hydroperoxide-epidioxide (cf. 3.7.2.1.3). However, a bicyclic compound is formed here as an intermediary product that readily fragments to malonic aldehyde:

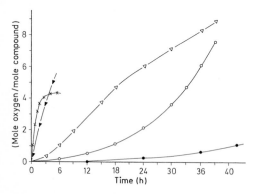

(3.74)

Alkanes, Alkenes. The main constituents of the volatile hydrocarbon fraction are ethane and pentane. Since these hydrocarbons are readily quantitated by gas chromatography using head-

Fig. 3.28. Reaction rate of an autoxidation process (according to *D.A. Lillard* and *E.A. Day*, 1964).
$-\triangledown-\triangledown-$ Linolenic acid methyl ester, $-\circ-\circ-$ linoleic acid methyl ester, $\times-\times-$ 2-nonenal, $\blacktriangledown-\blacktriangledown-\blacktriangledown$ 2,4-heptadienal, $-\bullet-\bullet-$ nonanal

space analysis, they can serve as suitable indicators for *in vivo* detection of lipid peroxidation. Pentane is probably formed from the 13-hydroperoxide of linoleic acid by the β-scission mechanism (cf. reaction 3.72). The corresponding pathway for 16-hydroperoxide of linolenic acid should then yield ethane.

3.7.2.2 Lipoxygenase: Occurrence and Properties

A lipoxygenase (linoleic acid oxygen oxidoreductase, EC 1.13.11.12) enzyme occurs in many plants and also in erythrocytes and leucocytes. It catalyzes the oxidation of some unsaturated fatty acids to their corresponding monohydroperoxides. These hydroperoxides have the same structure as those obtained by autoxidation. Unlike autoxidation, reactions catalyzed by lipoxygenase are characterized by all the features of enzyme catalysis: substrate specificity, peroxidation selectivity, occurrence of a pH optimum, susceptibility to heat treatment and a high reaction rate in the range of $0-20\,°C$. Also, the activation energy for linoleic acid peroxidation is rather low: 17 kJ/mol (as compared to the activation energy of a noncatalyzed reaction, see 3.7.2.1.4). Lipoxygenase oxidizes only fatty acids which contain a 1-cis,4-cis-pentadiene system. Therefore, the preferred substrates are linoleic and linolenic acids for the plant enzyme, and arachidonic acid for the animal enzyme; oleic acid is not oxidized.

Lipoxygenase is a metal-bound protein with an Fe-atom in its active center. The enzyme is activated by its product and during activation, Fe^{2+} is oxidized to Fe^{3+}. The catalyzed oxidation pathway is assumed to have the following reaction steps (cf. Fig. 3.29a): abstraction of a methylene H-atom from the substrate's 1,4-pentadiene system and oxidation of the H-atom to a proton. The pentadienyl radical bound to

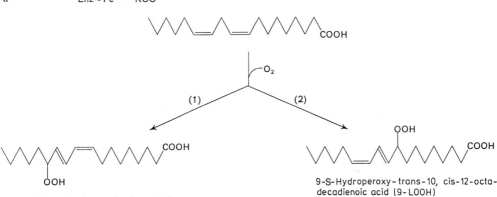

Fig. 3.29. Reactions of lipoxygenase – type I
a Proposed mechanism of reaction (according to *G.A. Veldink*, 1977); RH: linoleic acid; LOOH: linoleic acid hydroperoxide
b Specificity for linoleic acid oxidation. (1) Lipoxygenase from soybean (L-1; cf. Table 3.32); (2) lipoxygenase from tomato (cf. Table 3.32)

the enzyme is then rearranged into a conjugated diene system, followed by the uptake of oxygen. The peroxy radical formed is then reduced by the enzyme and, after attachment of a proton, the hydroperoxide formed is released.

With respect to the plant enzyme properties, two types of enzymes exist (although a transitional form may also exist). The enzyme that peroxidizes only free fatty acids with a high stereo- and regioselectivity is a type I lipoxygenase. It gives rise to an optically active hydroperoxide with a cis,trans-diene system (Fig. 3.29b). This type I enzyme forms preferentially either 9- or 13-hydroperoxides from free linoleic acid as the substrate (Table 3.32 and Fig. 3.29b).

Type II lipoxygenase, on the other hand, acts more like a catalyst of autoxidation with a much lower reaction specificity for linoleic acid. It produces both 9- and 13-hydroperoxides in equal amounts (as in the noncatalyzed autoxidation) and some other products, such as ketodiene fatty acids. Moreover, lipoxygenase type II also reacts with an esterified substrate. Thus, it does not require prior release of fatty acids by a lipase enzyme for activity in food.

The type II lipoxygenase can cooxidize carotenoids and chlorophyll and thus can degrade these pigments to colorless products. The property of the enzyme is utilized in flour "bleaching" (cf. 15.4.1.4.3). The involvement of the enzyme in cooxidation reactions can be explained by the possibility that the peroxy radicals are not fully converted to their hydroperoxides as in the reaction of the type I enzyme. Thus, a fraction of the free peroxy radicals are released by the enzyme. It can abstract an H-atom either from the unsaturated fatty acid present (pathway 2a in Fig. 3.30) or from a polyene (pathway 2b in Fig. 3.30).

The type II enzyme present in legumes produces a wide spectrum of volatile aldehydes from lipid substrates. These aldehydes, identical to those of a noncatalyzed autoxidation, can be further reduced to their alcohols, depending on the status of NADH-NAD+.

Table 3.32. Occurrence and properties of various lipoxygenases

Food	pH optimum	Peroxidation specificity[a]		
		9-LOOH (%)	13-LOOH (%)	Type
Soybean, L-1	9.0	5	95	I
Soybean, L-2	6.5	50	50	II
Peas, L-2	6.5	50	50	II
Peanut	6.0	0	100	I
Potato	5.5	95	5	I
Tomato	5.5	95	5	I
Wheat	6.0	90	10	I
Cucumber	5.5	75	25	
Apple	6.0	10	90	
Strawberry	6.5	23	77	
Gooseberry	6.5	45	55	II

[a] Against linoleic acid; 9- or 13-LOOH, cf. Fig. 3.29b.

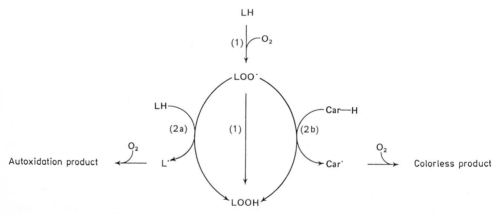

Fig. 3.30. Reactions of lipoxygenase type-II (according to F. Weber and W. Grosch, 1976). (1) Main catalysis pathway; (2a) and (2b) cooxidation pathways. LH: linoleic acid; Car-H: carotenoid; LOOH: linoleic acid hydroperoxide

3.7.2.3 Enzymatic Degradation of Hydroperoxides

Animals and plants degrade fatty acid hydroperoxides differently. In animal tissue, the enzyme glutathione peroxidase (cf. 7.3.2.8) catalyzes a reduction of the fatty acid hydroperoxides to the corresponding hydroxy acids, while in plants and mushrooms, the cleavage of the hydroperoxides by a lyase is predominant. This reaction is highly interesting with regard to food chemistry since the hydroperoxides, which are formed by lipoxygenase catalysis of linoleic and linolenic acid, are precursors of odorants. Those are important for fruits, vegetables and mushrooms, like the green-grassy or cucumberlike smelling aldehydes hexanal, 3-cis-hexenal ("leafy aldehyde"), 3-cis,6-cis nonadienal and the mushroomlike 1-octen-3(R)-ol (Table 3.33). The suggested mechanism is a β-cleavage of the hydroperoxide (Fig. 3.31).

The difference in volatile products in plants (aldehydes) and mushrooms (allyl alcohols) is due to the different substrate and reaction specificity of the hydroperoxide lyases. In the first case, in hydroperoxides with conjugated

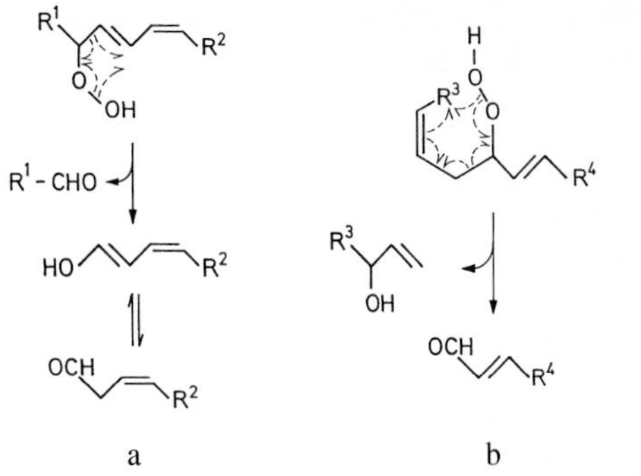

	R¹	R²
13 - LOOH	$CH_3(CH_2)_4$	$(CH_2)_7 COOH$
13 - LnOOH	$CH_3CH_2CH = CHCH_2$	$(CH_2)_7 COOH$
9 - LOOH	$HOOC(CH_2)_7$	$(CH_2)_4 CH_3$
9 - LnOOH	$HOOC(CH_2)_7$	$CH_2CH = CH CH_2CH_3$
	R³	R⁴
10 - LOOH	$CH_3(CH_2)_4$	$(CH_2)_6 COOH$
10 - LnOOH	$CH_3CH_2CH = CHCH_2$	$(CH_2)_6 COOH$

Fig. 3.31. Mechanism of the cleavage of hydroperoxides by lyases (according to *M. Wurzenberger* and *W. Grosch*, 1986)
a in plants, **b** in mushrooms

Table 3.33. Occurrence and properties of various hydroperoxide-lyases

Occurrence	Substrate	Products of the catalyses
Apple, tomato, cucumber, tea leaf (chloroplasts), soy beans, grape	13(S)-hydroperoxy-9-cis,11-trans-octadecadienoic acid (13-LOOH)	hexanal + 12-oxo-9-cis-dodecenoic acid
Apple, tomato, cucumber, tea leaf (chloroplasts), soy beans, grape	13(S)-hydroperoxy-9-cis,11-trans,15-cis-octadecatrienoic acid (13-LnOOH)	3-cis-hexenal + 12-oxo-9-cis-dodecenoic acid
Cucumber, pear	9(S)-hydroperoxy-10-trans,12-cis-octadecadienoic acid (9-LOOH)	3-cis-nonenal + 9-oxo-nonanoic acid
Cucumber, pear	9(S)-hydroperoxy-10-trans,12-cis, 15-cis-octadecatrienoic acid (9-LnOOH)	3-cis,6-cis-nonadienal+ 9-oxononanoic acid
Champignon	10(S)-hydroperoxy-8-trans,12-cis-octadecadienoic acid (10-LOOH)	1-octen-3(R)-ol +10-oxo-8-trans-decenoic acid
Champignon	10(S)-hydroperoxy-8-trans,12-cis,15-cis-octadecatrienoic acid (10-LnOOH)	1,5-cis-octadien-3(R)-ol + 10-oxo-8-trans-decenoic acid

diene systems (Fig. 3.31a), the bond between the C-atom bearing the HOO-group and the C-atom of the diene system is cleaved. In the second case (Fig. 3.31b), cleavage of hydroperoxides with isolated double bonds occurs in the opposite direction between the C-atom with the OOH-group and the C-atom with the adjacent methylene group. The 3-cis-aldehydes in plants formed by the splitting reaction can transform themselves into the respective 2-trans aldehydes. Isomerases that catalyze this reaction were identified in cucumbers, apples and tea chloroplasts.

The widespread presence of the C6- and C9-aldehydes in fruits and vegetables as well as the C8-alcohols in mushrooms (Table 3.33) permits the conclusion to be drawn that enzymatic-oxidative cleavage of linoleic and linolenic acid with the enzymes lipoxygenase, hydroperoxide-lyase and, if necessary, an aldehyde-isomerase generally contributes to the formation of aroma in these food items. This process is intensified when oxygen can permeate the cells freely by destruction of tissue (during the chopping of fruits and vegetables).

An allene oxide synthetase (AOS) that metabolizes fatty acid hydroperoxides was first detected in flaxseed. AOS has been also named hydroperoxide isomerase as it catalyzes the formation of α-ketols from monohydroperoxides (cf. Formula 3.75).

AOS is a cytochrome P450 which was found in abundance in many plants. In soya beans, activity was detected in immature seed coat and the pericarp.

AOS catalyzes a series of transformations via an allene oxide as intermediate which was identified as 12,13(S)-epoxy-9(Z),11-octadecadienoic acid when 13(S)-hydroperoxy-9(Z),11(E)-octadecadienoic acid was used as substrate. The allene oxide has a half-life of only 33 s at 0 °C.

R: CH_3—$(CH_2)_4$
R′: $(CH_2)_7$—COOH

(3.75)

Linoleic acid 9-hydroperoxide formed by lipoxygenase in potato is changed enzymatically by elimination of water into a fatty acid with a dienyl-ether structure:

(1′-trans, 3′-cis-Nonadienyloxa)-trans-8-nonenoic acid

(3.76)

In addition to lipoxygenase, lipoperoxidase activity has been observed in oats. The 9-hydroperoxide formed initially is reduced to 9-hydroxy-trans-10,cis-12-octadecadienoic acid.

Since hydroxy but not hydroperoxy acids taste bitter, this reaction should contribute to the bitter taste generated during the storage of oats (cf. 15.2.2.3).

3.7.2.4 Hydroperoxide-Protein Interactions

3.7.2.4.1 Products Formed from Hydroperoxides

Hydroperoxides formed enzymatically in food are usually degraded further. This degradation can also be of a nonenzymatic nature. In nonspecific reactions involving heavy metal ions, heme(in) compounds or proteins, hydroperoxides are transformed into oxo, expoxy, mono-, di- and trihydroxy acids (Table 3.34). Unlike hydroperoxides, i.e. the primary products of autoxidation, some of these derivatives are characterized as having a bitter taste (Table 3.35). Such compounds are detected in legumes and cereals. They may play a role in other foods rich in unsaturated fatty acids and proteins, such as fish and fish products.

In order to clarify the formation of the compounds presented in Table 3.34, the reaction sequences given in Fig. 3.32 have been assumed to occur. The start of the reaction is from the alkoxydiene radical generated from the 9- or 13-hydroperoxide by the catalytic action of heavy metal ions or heme(in) compounds (cf. 3.7.2.1.4). The alkoxydiene radical may disproportionate into a hydroxydiene and

Table 3.34. Products obtained by non-enzymic degradation of linoleic acid hydroperoxides

Product[a]	Hydroperoxide interaction with				
	Fe³⁺ cysteine	Hemo-globin	Soya homogenate	Pea homogenate	Wheat flour
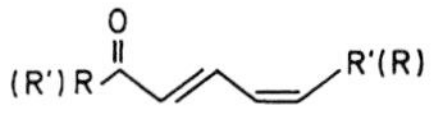 (R')R–C(=O)–CH=CH–R'(R) (oxo)	+	+	+	+	
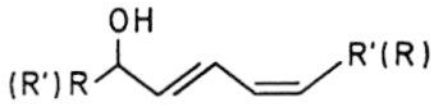 (R')R–CH(OH)–CH=CH–R'(R) (hydroxy)	+	+	+	+	+
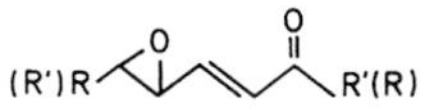 (R')R–(epoxy)–CH=CH–C(=O)–R'(R) (epoxy-oxo)	+		+		
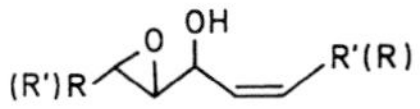 (R')R–(epoxy)–CH(OH)–CH=CH–R'(R) (epoxy-hydroxy)	+	+	+		
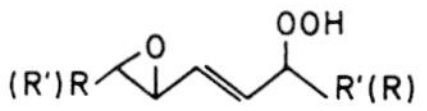 (R')R–(epoxy)–CH=CH–CH(OOH)–R'(R) (epoxy-hydroperoxy)	+		+		
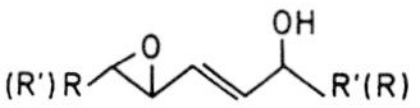 (R')R–(epoxy)–CH=CH–CH(OH)–R'(R) (epoxy-hydroxy)		+		+	+
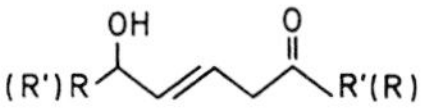 (R')R–CH(OH)–CH=CH–C(=O)–R'(R) (hydroxy-oxo)	+				
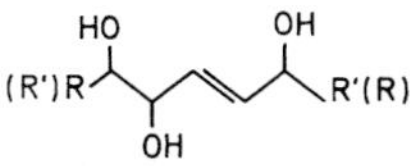 (R')R–CH(OH)–CH(OH)–CH=CH–CH(OH)–R'(R) (trihydroxy)	+		+	+	+

[a] As a rule a mixture of two isomers are formed with R: $CH_3(CH_2)_4$ and R': $(CH_2)_7COOH$.

Table 3.35. Taste of oxidized fatty acids

Compound	Threshold value for bitter taste (mmol/l)
13-Hydroperoxy-cis-9,trans-11-octa-decadienoic acid	not bitter[a]
9-Hydroperoxy-trans-10,cis-12-octa-decadienoic acid	not bitter[a]
13-Hydroxy-cis-9,trans-11-octa-decadienoic acid	7.6–8.5[a]
9-Hydroxy-trans-10,cis-12-octa-decadienoic acid	6.5–8.0[a]
9,12,13-Trihydroxy-trans-10-octa-decenoic acid 9,10,13-Trihydroxy-trans-11-octa-decenoic acid	0.6–0.9[b]

[a] A burning taste sensation.
[b] A blend of the two trihydroxy fatty acids was assessed.

an oxodiene fatty acid. Frequently this reaction is only of secondary importance since the alkoxydiene radical rearranges immediately to an epoxyallylic radical which is susceptible to a variety of radical combination reactions. Under aerobic conditions the epoxyallylic radical combines preferentially with molecular oxygen. The epoxyhydroperoxides formed are, in turn, subject to homolysis via an oxyradical. A disproportionation reaction leads to epoxyoxo and epoxyhydroxy compounds. Under anaerobic conditions the epoxyallylic radical combines with other radicals, e.g. hydroxy radicals (Fig. 3.32) or thiyl radicals (Fig. 3.33).

Of the epoxides produced, the allylic epoxides are known to be particularly susceptible to hydrolysis in the presence of protons. As shown in Fig. 3.32 trihydroxy fatty acids may result from the hydrolysis of an allylic epoxyhydroxy compound.

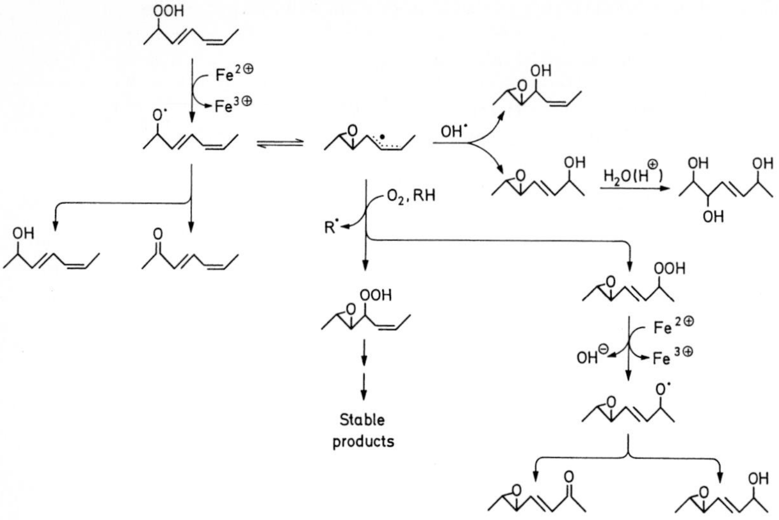

Fig. 3.32. Degradation of linoleic acid hydroperoxides to hydroxy-, epoxy- and oxo-fatty acids. The postulated reaction sequence explains the formation of identified products. Only segments of the structures are presented (according to *H.W. Gardner*, 1985)

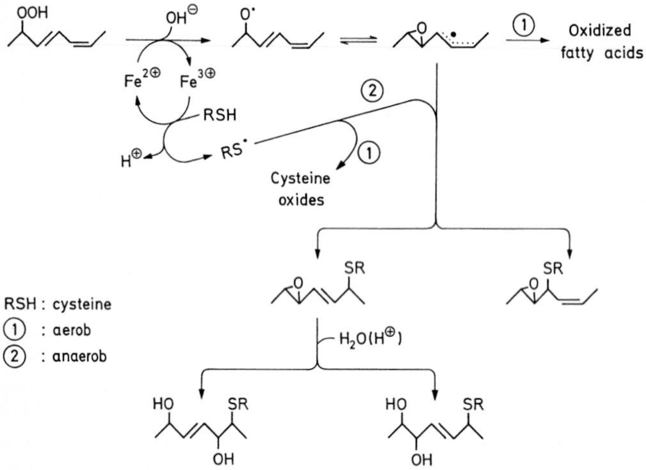

Fig. 3.33. Interaction of linoleic acid hydroperoxides with cysteine. A hypothesis to explain the reaction products obtained. Only segments of the structures are presented (according to *H.W. Gardner*, 1985)

3.7.2.4.2 Lipid-Protein Complexes

Studies related to the interaction of hydroperoxides with proteins have shown that, in the absence of oxygen, linoleic acid 13-hydroperoxide reacts with N-acetylcysteine, yielding an adduct of which one isomer is shown:

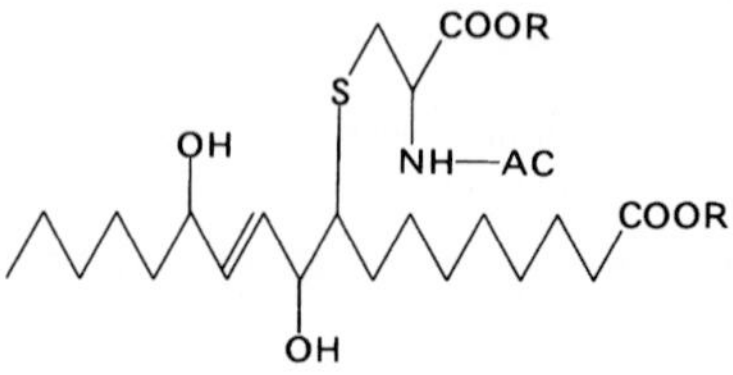

(3.77)

However, in the presence of oxygen, covalently bound amino acid-fatty acid adduct formation is significantly suppressed; instead, oxidized fatty acids are formed as listed in Table 3.34.

The difference in reaction products is explained in the reaction scheme shown in Fig. 3.33 which gives an insight into the different reaction pathways. The thiyl radical, derived from cysteine by abstraction of an H-atom, is added to the epoxyallyllic radical only in the absence of oxygen (pathway 2 in Fig. 3.33). In the presence of oxygen, oxidation of cysteine to cysteine oxide and of fatty acids to their more oxidized forms (Fig. 3.32) occur with a higher reaction rate than in the previous reaction.

As a consequence, a large portion of the oxidized lipid from protein-containing food stored in air does not have lipid-protein covalent bonds and, hence, is readily extracted with a lipid solvent such as chloroform/methanol (2:1).

3.7.2.4.3 Protein Changes

Some properties of proteins are changed when they react with hydroperoxides and their degradation products. This is reflected by changes in food texture, decreases in protein solubility (formation of cross-linked proteins), color (browning) and changes in nutritive value (loss of essential amino acids).

The radicals generated from hydroperoxides (cf. Fig. 3.33) can abstract H-atoms from protein (PH), preferentially from the amino acids Trp, Lys, Tyr, Arg, His, cysteine and cystine, wherein the phenolic HO-, S- or N-containing group reacts:

$$RO^{\bullet} + PH \longrightarrow P^{\bullet} + ROH \qquad (3.78)$$

$$2\,P^{\bullet} \longrightarrow P{-}P \qquad (3.79)$$

In Reaction 3.79, protein radicals combine with each other, resulting in the formation of a protein network. Malonicaldehyde is generated (cf. 3.7.2.1.5) under certain conditions during lipid peroxidation. As a bifunctional reagent, malonicaldehyde can crosslink proteins through a *Schiff* base reaction with the ε-NH$_2$ group of lysine:

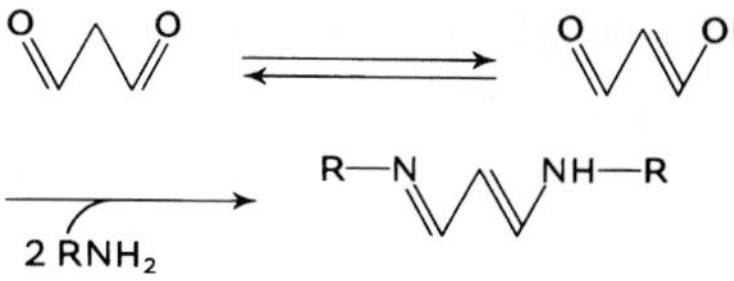

$$(3.80)$$

The *Schiff* base adduct is a conjugated fluorochrome that has distinct spectral properties (λ_{max} excitation ~ 350 nm; λ_{max} emission ~ 450 nm). Hence, it can be used for detecting lipid peroxidation and the reactions derived from it with the protein present.

Reactions resulting in the formation of a protein network like that oulined above also have practical implications, e.g., they are responsible for the decrease in solubility of fish protein during frozen storage.

Also, the monocarbonyl compounds derived from autoxidation of unsaturated fatty acids readily condense with protein-free NH$_2$ groups, forming *Schiff* bases that can provide brown polymers by repeated aldol condensations (Fig. 3.34). The brown polymers are often N-free since the amino compound can be readily eliminated by hydrolysis. When hydrolysis occurs in the early stages of aldol condensations (after the first or second condensation; cf. Fig. 3.34) and the released aldehyde, which has a powerful odor, does not reenter the reaction, the condensation process results not only in discoloration (browning) but also in a change in aroma.

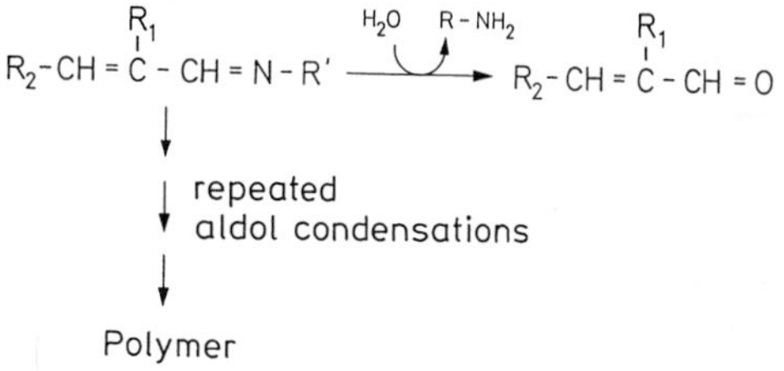

$$R_1{-}CH_2{-}CHO + H_2N{-}R'$$

$$\downarrow {-} H_2O$$

$$R_1{-}CH_2{-}CH = N{-}R'$$

$$\downarrow \; R_2{-}CHO$$
$$\downarrow {-} H_2O$$

$$R_2{-}CH = \overset{R_1}{\underset{}{C}}{-}CH = N{-}R' \xrightarrow{\; H_2O \quad R{-}NH_2 \;} R_2{-}CH = \overset{R_1}{\underset{}{C}}{-}CH = O$$

$$\downarrow$$

$$\overset{|}{\underset{\downarrow}{|}} \; \text{repeated aldol condensations}$$

$$\downarrow$$

Polymer

Fig. 3.34. Reaction of volatile aldehydes with protein amino groups

3.7.2.4.4 Decomposition of Amino Acids

Studies of model systems have revealed that protein cleavage and degradation of side chains, rather than formation of protein networks, are the preferred reactions when the water content of protein/lipid mixtures decreases. Several examples of the extent of losses of amino acids in a protein in the presence of an oxidized lipid are presented in Table 3.36. The strong dependence of this loss on the nature of the protein and reaction conditions is obvious. Degradation products obtained in model systems of pure amino acids and oxidized lipids are described in Table 3.37.

3.7.3 Inhibition of Lipid Peroxidation

Autoxidation of unsaturated acyl-lipids can be retarded by:

- Exclusion of oxygen. Possibilities are packaging under a vacuum or addition of glucose oxidase (cf. 2.7.2.1.1).
- Storage at low temperature in the dark. The autoxidation rate is thereby decreased substantially. However, in fruits and vegetables which contain the lipoxygenase enzyme, these precautions are not applicable. Food deterioration is prevented only after inactivation of the enzyme by a blanching process (cf. 2.6.4).
- Addition of antioxidants to food.

3.7.3.1 Antioxidant Activity

The peroxy and oxy free radicals formed during the propagation and branching steps of the autoxidation radical chain (cf. Fig. 3.19) are scavenged by antioxidants (AH; cf. Fig. 3.35).

Antioxidants containing a phenolic group play the major role in food. In reactions 1 and 2 in Fig. 3.35, they form radicals which are stabilized by an aromatic resonance system. In contrast to the acyl peroxy and oxy free radicals, they are not able to abstract a H-atom from an unsaturated fatty acid and therefore cannot initiate lipid peroxidation. The end-products formed in reactions 3 and 4 in Fig. 3.35 are relatively stable and in consequence the autoxidation radical chains are shortened.

Table 3.36. Amino acid losses occurring in protein reaction with peroxidized lipids

Reaction system		Reaction conditions		Amino acids lost
protein	lipid	time	T (°C)	(% loss)
Cyto-chrome C	Linolenic acid	5 h	37	His (59), Ser (55), Pro (53), Val (49), Arg (42), Met (38), Cys (35)[a]
Trypsin	Linoleic acid	40 min	37	Met (83), His (12)[a]
Lysozyme	Linoleic acid	8 days	37	Trp (56), His (42), Lys (17), Met (14), Arg (9)
Casein	Linoleic acid ethyl ester	4 days	60	Lys (50), Met (47), Ile (30), Phe (30), Arg (29), Asp (29), Gly (29), His (28), Thr (27), Ala (27), Tyr (27)[a, b]
Oval-bumin	Linoleic acid ethyl ester	24 h	55	Met (17), Ser (10), Lys (9), Ala (8), Leu (8)[a, b]

[a] Trp analysis was not performed.
[b] Cystine analysis was not performed.

Table 3.37. Amino acid products formed in reaction with peroxidized lipid

Reaction system		Compounds formed from amino acids
amino acid	lipid	
His	Methyl linoleate	Imidazolelactic acid, Imidazoleacetic acid
Cys	Ethyl arachidonate	Cystine, H$_2$S, cysteic acid, alanine, cystine-disulfoxide
Met	Methyl linoleate	Methionine-sulfoxide
Lys	Methyl linoleate	Diaminopentane, aspartic acid, glycine, alanine, α-aminoadipic acid, pipecolinic acid, 1,10-diamino-1,10-dicarboxydecane

$$RO_2^{\cdot} + AH \longrightarrow ROOH + A^{\cdot} \quad (1)$$
$$RO^{\cdot} + AH \longrightarrow ROH + A^{\cdot} \quad (2)$$
$$RO_2^{\cdot} + A^{\cdot} \longrightarrow ROOA \quad (3)$$
$$RO^{\cdot} + A^{\cdot} \longrightarrow ROA \quad (4)$$

Fig. 3.35. Activity of an antioxidant as a radical scavenger. AH: Antioxidant

The reaction scheme (Fig. 3.35) shows that one antioxidant molecule combines with two radicals. Therefore, the maximum achievable stoichiometric factor is n = 2. In practice, the value of n is between 1 and 2 for the antioxidants used. Antioxidants, in addition to their main role as radical scavengers, can also partially reduce hydroperoxides to hydroxy compounds.

3.7.3.2 Antioxidants in Food

3.7.3.2.1 Natural Antioxidants

The unsaturated lipids in living tissue are relatively stable. Plants and animals have the necessary complement of antioxidant and of enzymes, for instance, glutathione peroxidase and superoxide dismutase, to effectively prevent lipid oxidation.

During the isolation of oil from plants (cf. 3.8.3), tocopherols are also isolated. A sufficient level is retained in oil even after refining, thus, tocopherols secure the stability of the oil end-product. Soya oil, due to its relatively high level of furan fatty acids and linolenic acid (cf. 14.3.2.2.5), is an exception. The tocopherol content of animal fat is influenced by animal feed.

The antioxidant activity of tocopherols increases from $\alpha \rightarrow \delta$. It is the reverse of the vitamin E activity (cf. 6.2.3) and of the rate of reaction with peroxy radicals. Table 3.38 demonstrates that α-tocopherol reacts with peroxy radicals faster than the other tocopherols and the synthetic antioxidant BHT.

The higher efficiency of γ-tocopherol in comparison to α-tocopherol is based on the higher stability of γ-tocopherol and on different reaction products formed during the antioxidative reaction.

Table 3.38. Rate constants of tocopherols and BHT for reaction 2 in Fig. 3.35 at 30 °C

Antioxidant	$k\,(l \cdot mol^{-1} \cdot s^{-1}) \cdot 10^{-5}$
α-Tocopherol	23.5
β-Tocopherol	16.6
γ-Tocopherol	15.9
δ-Tocopherol	6.5
2,6-Di-tert-butyl-p-cresol (BHT)	0.1

After opening of the chroman ring system, α-tocopherol is converted into an alkyl radical which in turn oxidizes to a hydroxy-alkylquinone (I in Formula 3.81). α-Tocopherol is a faster scavenger for peroxy radicals formed during autoxidation than γ-tocopherol (Table 3.38), but α-tocopherol then generates an alkyl radical which, in contrast to the slow reacting chromanoxyl radical, can start autoxidation of unsaturated fatty acids. Therefore, the peroxidation rate of an unsaturated fatty acid increases with higher α-tocopherol concentrations after going through a minimum. This prooxidative effect is smaller in the case of γ-tocopherol because in contrast to α-tocopherol, no opening of the chroman ring takes place but formation of diphenylether and biphenyl dimers occurs. The supposed explanation for these reaction products is: The peroxy radical of a fatty acid abstracts a hydrogen atom from γ-tocopherol (Formula 3.82). A chromanoxyl radical (I) is formed, that can transform into a chromanyl radical (II). Recombination of (I) and (II) results in the diphenylether dimer (III) and recombination of two radicals (II) into the biphenyl dimer (IV). Unlike p-quinone from the reaction of α-tocopherol, the dimer structures (III) and (IV) possess one or two phenolic OH-groups that are also antioxidatively active.

$$(3.81)$$

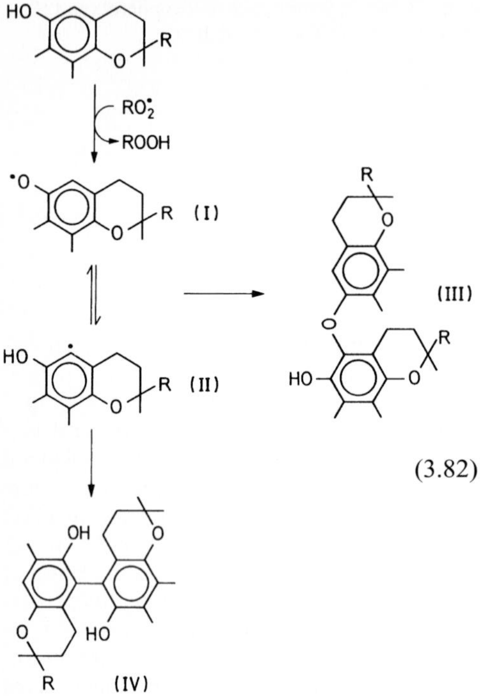

$$(3.82)$$

Ascorbic acid (cf. 6.3.9) is active as an antioxidant in aqueous media, but only at higher concentrations ($\sim 10^{-3}$ mol/l). A prooxidant activity is observed at lower levels (10^{-5} mol/l), especially in the presence of heavy metal ions. The effect of tocopherols is enhanced by the addition of fat soluble ascorbyl palmitate or ascorbic acid in combination with an emulsifier (e.g. lecithin) since the formed tocopherol radical from reaction 2 in Fig. 3.35 is rapidly reduced to α-tocopherol by vitamin C.

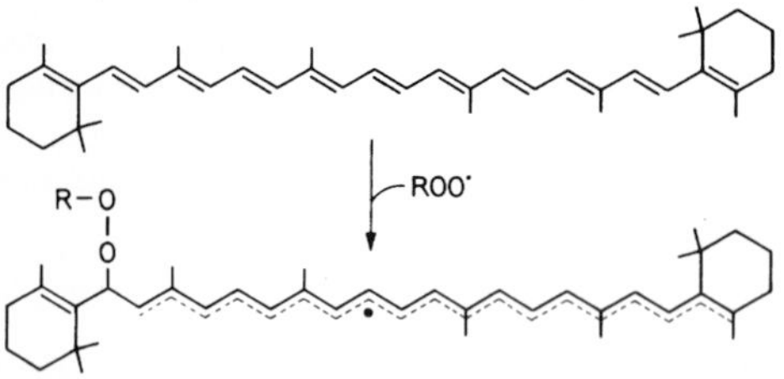

$$(3.83)$$

Carotinoids also can act as scavengers for alkyl radicals. Radicals stabilized by resonance are formed (Formula 3.83), unable to initiate lipid peroxidation. β-Carotenes are most active at a concentration of $5 \cdot 10^{-5}$ mol/l, while at higher concentrations the prooxida-

tive effect is predominant. Also the partial pressure of oxygen is critical, it should be below 150 mm Hg.

Flavanones (cf. 18.1.2.5.4) and flavonols (cf. 18.1.2.5.5) are phenolic compounds which are widely distributed in plant tissues, where they act as natural antioxidants. The protective effect of several herbs, spices (e.g. sage or rosemary) and tea extracts against fat (oil) oxidation is based on the presence of such natural antioxidants (cf. 21.2.5.1 and 22.1.1.4). Polyphenols in wood, such as lignin, undergo thermal cracking, resulting in volatile phenols, during the generation of smoke by burning wood or, even more so, sawdust. These phenols deposit on the food surface during smoking and then penetrate into the food, thus acting as antioxidants. It was also demonstrated that vanillin, in food items where its aroma is desired, plays an important role as an antioxidant. Finally, some of the *Maillard* reaction products, such as reductones (cf. 4.2.4.4), should be considered as naturally active antioxidants.

3.7.3.2.2 Synthetic Antioxidants

In order to be used as an antioxidant, a synthetic compound has to meet the following requirements: it should not be toxic; it has to be highly active at low concentrations (0.01–0.02%); it has to concentrate on the surface of the fat or oil phase. Therefore, strongly lipophilic antioxidants are particularly suitable (with low HLB values, e.g. BHA, BHT or tocopherols, dodecylgallate) for o/w emulsions. On the other hand, the more polar antioxidants, such as TBHQ and propyl gallate, are very active in fats and oils since they are enriched at the surface of fat and come in contact with air. Antioxidants should be stable under the usual food processing conditions. This stability is denoted as the "carry through" effect. Some of the synthetic antioxidants used worldwide are:

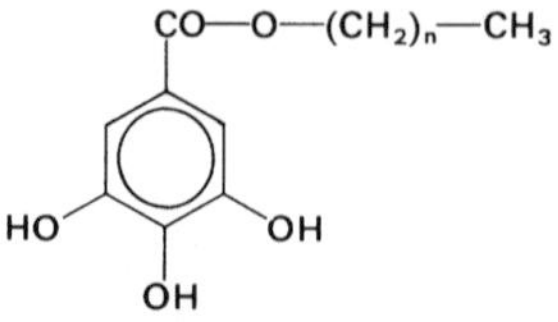

$$(3.84)$$

Propyl (n = 2); octyl (n = 7) and dodecyl (n = 11) gallate

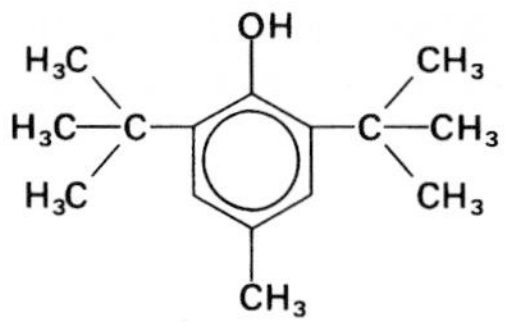

H_3C
H_3C—C
H_3C
OH
CH_3
C—CH_3
CH_3
CH_3 (3.85)

2,6-Di-tert-butyl-p-hydroxytoluene (BHT)

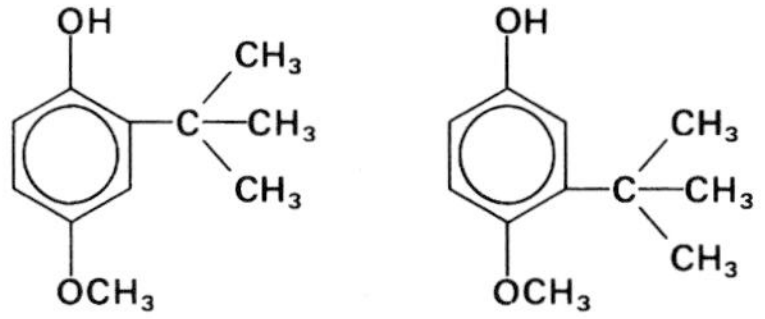

OH
C—CH_3
CH_3
OCH_3

OH
CH_3
C—CH_3
CH_3
OCH_3 (3.86)

tert-Butyl-4-hydroxyanisole (BHA)

Commercial BHA is a mixture of two isomers, 2- and 3-tert-butyl-4-hydroxyanisole

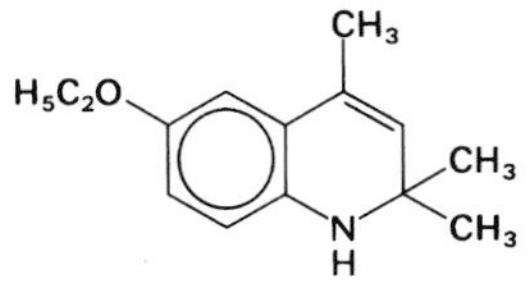

H_5C_2O
CH_3
CH_3
N
H
CH_3 (3.87)

6-Ethoxy-1,2-dihydro-2,2,4-trimethylquinoline (ethoxyquin)

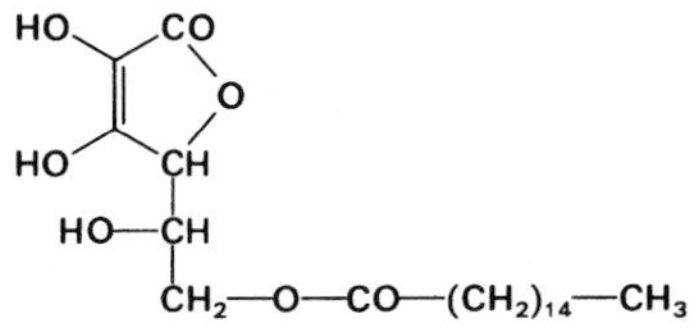

ESR spectroscopy has demonstrated that a large portion of ethoxyquin is present in oil as a free radical

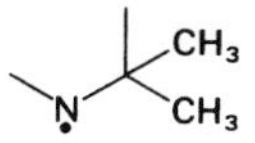

HO
HO
CO
O
CH
HO—CH
CH_2—O—CO—$(CH_2)_{14}$—CH_3 (3.88)

Ascorbyl palmitate

N
CH_3
CH_3 (3.89)

and stabilization by dimerization of the radical occurs. The radical, and not the dimer, is the active antioxidant.

tert-Butylhydroquinone (TBHQ) is a particularly powerful antioxidant used, for example, for stabilization of soya oil. The "carry through" properties are of importance in the use of BHA, TBHQ and BHT in food proces-
sing. All three antioxidants are steam distillable at higher temperatures. Utilization of antioxidants is often regulated by governments through controls on the use of food additives. In North America incorporation of antioxidants is permitted at a maximum level of 0.01% for any one antioxidant, and a maximum of 0.02% for any combination. The regulations related to permitted levels often vary from country to country.

The efficiency of an antioxidant can be evaluated by a comparative assay, making use of an "antioxidative factor" (AF):

$$AF = I_A/I_0 \qquad (3.90)$$

where I_A = oxidation induction period for a fat or oil (cf. 3.7.2.1.1) in the presence of an antioxidant and I_0 = oxidation induction period of a fat or oil without an antioxidant.

Hence, the efficiency of an antioxidant increases with an increase in the AF value. As illustrated by the data in Table 3.39, BHA in comparison with BHT shows a higher efficiency in a lard sample. This result is understandable since in BHT both tertiary butyl substituents sterically hinder the reaction with radicals to a certain extent (reaction 1 in Fig. 3.35). The effect on antioxidants depends not only on the origin of fat or oil but, also, on the processing steps used in the isolation and refining procedures. Hence, data in Table 3.39 serve only as an illustration.

BHA and BHT together at a given total concentration are more effective in extending shelf-life of a fat or oil than either antioxidant alone at the same level of use (Table 3.39).

To explain this, it is suggested that BHA, by participating in reaction 1 (Fig. 3.35), provides a phenoxy radical (I):

Table 3.39. Antioxidative factor (AF) values of some antioxidants (0.02%) in refined lard

Antioxidant	AF	Antioxidant	AF
d-α-Tocopherol	5	Octyl gallate	6
dl-γ-Tocopherol	12	Ascorbyl palmitate	4
BHA	9.5	BHA and	
BHT	6	BHT[a]	12

[a] Each compound is added in amount of 0.01%.

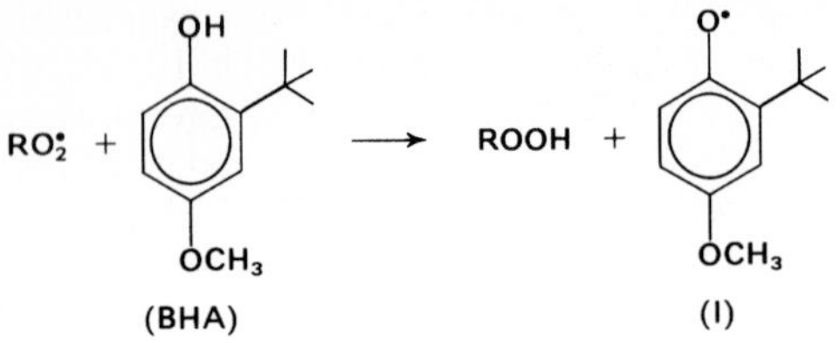

$$(3.91)$$

Table 3.40. Synergistic action of citric (C) and phosphoric acids (P) in combination with lauryl gallate (LG) on oxidation of fats and oils

Added to fat/oil	AF value after addition of				
	0.01% C	0.01% P	0.01% LG	0.01% LG + 0.01% C	0.01% LG + 0.01% P
0.2 ppm Cu	0.3	0.2	0.9	4.7	4.1
2 ppm Fe	0.6	0.5	0.1	5.7	0.2
2 ppm Ni	0.5	0.6	3.0	7.0	4.4

which is then regenerated into the original molecule by rapid interaction with BHT:

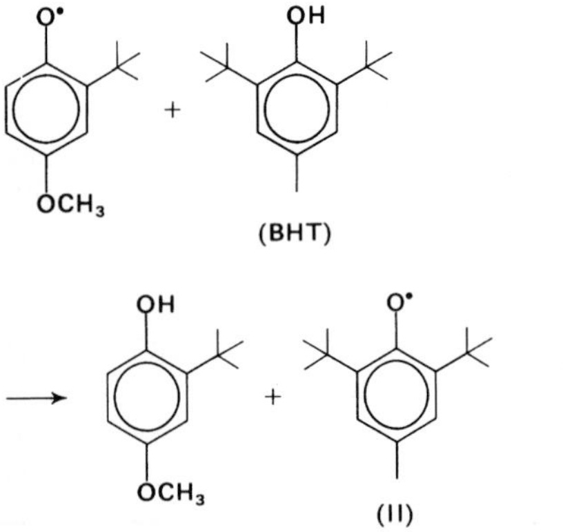

$$(3.92)$$

On the other hand, the phenoxy radical (II) derived from BHT can react further with an additional peroxy radical:

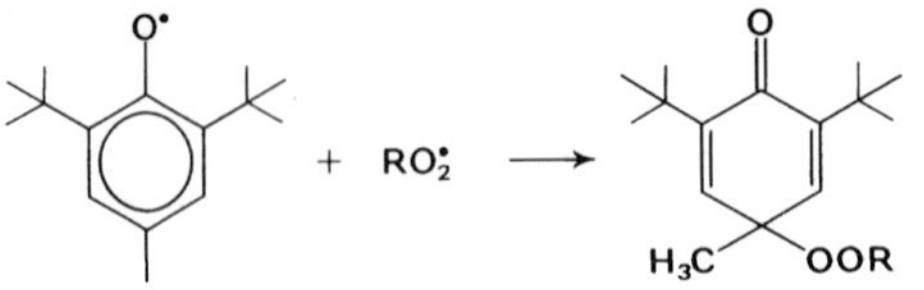

$$(3.93)$$

Propyl gallate (PG) increases the efficiency of BHA, but not that of BHT. Ascorbyl palmitate, which is by itself a rather weak antioxidant, substantially sustains the antioxidative activity of γ-d,l-tocopherol.

3.7.3.2.3 Synergists

Substances which enhance the activity of antioxidants are called synergists. The main examples are lecithin, amino acids, citric, phosphoric, citraconic and fumaric acids, i.e. compounds which complex heavy metal ions (chelating agents, sequesterants or scavengers of trace metals). Thus, initiation of heavy metalcatalyzed lipid autoxidation can be prevented (cf. 3.7.2.1.4). Results compiled in Table 3.40 demonstrate the synergistic activities of citric and phosphoric acids in combination with lauryl gallate. Whereas citric acid enhances the antioxidant effectiveness in the presence of all three metal ions, phosphoric acid is able to do so with copper and nickel, but not with iron. Also, use of citric acid is more advantageous since phosphoric acid promotes polymerization of fat or oil during deep frying.

The synergistic effect of phospholipids is different. Addition of dipalmitoylphosphatidylethanolamine (0.1–0.2 weight %) to lard enhances the antioxidative activity of α-tocopherol, BHA, BHT and propyl gallate, while phosphatidylcholine shows no activity.

The reaction of ascorbic acid with tocopherol radicals as described in 3.7.3.2 is a synergistic effect.

3.7.4 Fat or Oil Heating (Deep Frying)

Deep frying is one of the methods of food preparation used both in the home and in industry. Meat, fish, doughnuts, potato chips or french fries are dipped into fat (oil) heated to about 180 °C. After several minutes of frying, the food is sufficiently tender to be served.

The frying fat or oil changes substantially in its chemical and physical properties after prolonged use. Data for a partially hydrogenated soybean oil compiled in Table 3.41 indicate that heating of oil causes reactions involving double bonds. This will result in a decrease in iodine number. As can be deduced from changes in the composition of fatty acids (Table 3.41), in the case of soybean oil, linoleic and linolenic acid are the most affected. Peroxides formed at elevated temperatures fragment immediately with formation of hydroxy compounds thus increasing the hydroxyl number (Table 3.41). Therefore, determination of peroxide values to evaluate the quality of fat or oil in deep frying is not appropriate.

Unsaturated TG polymerize during heating thus increasing the viscosity of the fat. Di- and trimeric TG are formed. The increase of these components can be monitored by means of gel permeation chromatography (GPC) (Fig. 3.36).

Before or after methanolysis of the oil sample, GPC is a valuable first tool to analyze the great number of reaction products formed during deep frying. Monomeric methyl esters are further fractionated via the urea adducts, while the cyclic fatty acids enrich themselves in the supernatant. Dimeric methyl esters can be preseparated by RP-HPLC and further analyzed by GC/MS after silylation of the OH-groups.

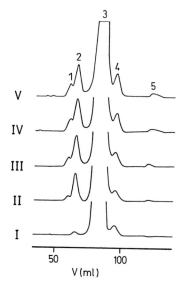

Fig. 3.36. Gel permeation chromatography of heated soybean oil (according to *J.A. Rojo* and *E.G. Perkins*, 1987). Oil samples (composition and heating conditions see Table 3.41) were analyzed immediately (I), as well as after 8 h (II), 24 h (III), 48 h (IV) and 80 h (V), *1* Trimeric TG, *2* Dimeric TG, *3* TG, *4* DG, *5* free fatty acids

Table 3.41. Characteristics of partially hydrogenated soybean oil before and after simulated deep fat frying[a]

Characteristics	Fresh oil	Heated oil
Iodine number	108.9	101.3
Saponification number	191.4	195.9
Free fatty acids[b]	0.03	0.59
Hydroxyl number	2.25	9.34
DG	1.18	2.73
Composition of fatty acids (weight %)		
14:0	0.06	0.06
16:0	9.90	9.82
18:0	4.53	4.45
18:1 (9)	45.3	42.9
18:2 (9, 12)	37.0	29.6
18:3 (9, 12, 15)	2.39	1.67
20:0	0.35	0.35
22:0	0.38	0.38
Other	0.50	0.67

[a] The oil was heated for 80 h (8 h/day) at 195 °C. Batches of moist cotton balls containing 75 % by weight of water were fried at 30-min intervals (17 frying operations/day) in order to simulate the deep frying process.
[b] Weight % calculated as oleic acid.

A great number of volatile and nonvolatile products are obtained during deep frying of oil or fat. The types of reactions involved in and responsible for changes in oil and fat structures are compiled in Table 3.42. Some of the reactions presented will be outlined in more detail.

3.7.4.1 Autoxidation of Saturated Acyl Lipids

The selectivity of autoxidation decreases above 60 °C since the hydroperoxides formed are subjected to homolysis giving hydroxy and alkoxy radicals (Reaction RS-4 in Fig. 3.19) which, due to their high reactivity, can abstract H-atoms even from saturated fatty acids.

Numerous compounds result from these reactions. For example, Table 3.43 lists a series of aldehydes and methyl ketones derived preferentially from tristearin. Both classes of compounds are also formed by thermal degrada-

Table 3.42. A review of reactions occurring in heat treated fats and oils

Fat/oil heating	Reaction	Products
1. Deep frying without food	Autoxidation Isomerization Polymerization	Volatile acids aldehydes esters alcohols Epoxides Branched chain fatty acids Dimeric fatty acids Mono- and bicyclic compounds Aromatic compounds Compounds with trans double bonds Hydrogen, CO_2.
2. Deep frying with food added	As under 1. and in addition hydrolysis	As under 1. and in addition free fatty acids, mono- and diacylglycerols and glycerol

Table 3.43. Volatile compounds formed from heat-treated tristearin[a]

Class of compound	Portion	C-number	Major compounds
Alcohols	2.7	4–14	n-Octanol n-Nonanol n-Decanol
γ-Lactones	4.1	4–14	γ-Butyrolactone γ-Pentalactone γ-Heptalactone
Alkanes	8.8	4–17	n-Heptadecane n-Nonane n-Decane
Acids	9.7	2–12	Caproic acid Valeric acid Butyric acid
Aldehydes	36.1	3–17	n-Hexanal n-Heptanal n-Octanal
Methyl ketones	38.4	3–17	2-Nonanone 2-Heptanone 2-Decanone

[a] Tristearin is heated in air at 192 °C.

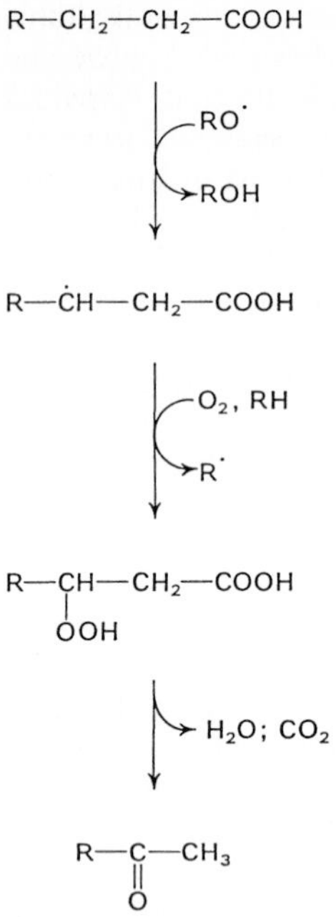

$$R—CH_2—CH_2—COOH$$

Fig. 3.37. Autoxidation of saturated fatty acids. Postulated reaction steps involved in formation of methyl ketones

tion of free fatty acids. These acids are formed by triglyceride hydrolysis or by the oxidation of aldehydes.

Methyl ketones are obtained by thermally induced β-oxidation followed by a decarboxylation reaction (Fig. 3.37). Aldehydes are obtained from the fragmentation of hydroperoxides by a β-scission mechanism (Fig. 3.38) occurring nonselectively at elevated temperatures (compare the difference with 3.7.2.1.5).

Unsaturated aldehydes with a double bond conjugated to the carbonyl group are easily degraded during the deep frying process (Formula 3.94). Addition of water results in the formation of a 3-hydroxyaldehyde that is split by retro aldol condensation catalyzed by heat. Examples of this mechanism are the degradation of 2-trans, 6-cis-nonadienal to 4-cis-heptenal and acetaldehyde, as well as the cleavage of 2,4-decadienal into 2-octenal and acetaldehyde.

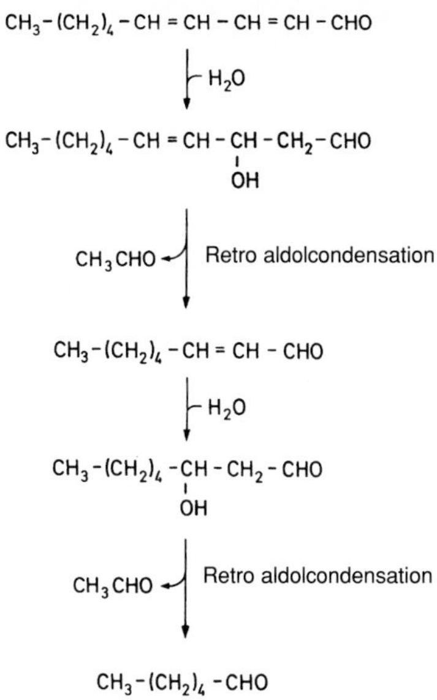

$$CH_3-(CH_2)_4-CH=CH-CH=CH-CHO$$

$$\downarrow\ H_2O$$

$$CH_3-(CH_2)_4-CH=CH-CH-CH_2-CHO$$
$$\underset{OH}{|}$$

$$CH_3CHO \leftarrow |\ \text{Retro aldolcondensation}$$

$$CH_3-(CH_2)_4-CH=CH-CHO \qquad\qquad (3.94)$$

$$\downarrow\ H_2O$$

$$CH_3-(CH_2)_4-CH-CH_2-CHO$$
$$\underset{OH}{|}$$

$$CH_3CHO \leftarrow |\ \text{Retro aldolcondensation}$$

$$CH_3-(CH_2)_4-CHO$$

Some volatiles are important odorous compounds, e.g., the mixture of 2-trans-4-trans-decadienal and 2-cis-4-trans-decadienal, contribute to the pleasant deep-fried flavor. Since these aldehydes are formed by thermal degradation of linoleic acid, fats or oils containing this acid provide a better aroma during deep frying than hydrogenated fats. However, when a fat is used for a prolonged period of time, the unpleasant aroma of volatile compounds, e.g. of trans-4,5-epoxy-trans-2-decenal,

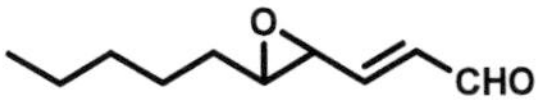

$$\text{(structure)}\qquad (3.95)$$

becomes noticeable. This is a sign of the advanced stage of fat or oil deterioration.

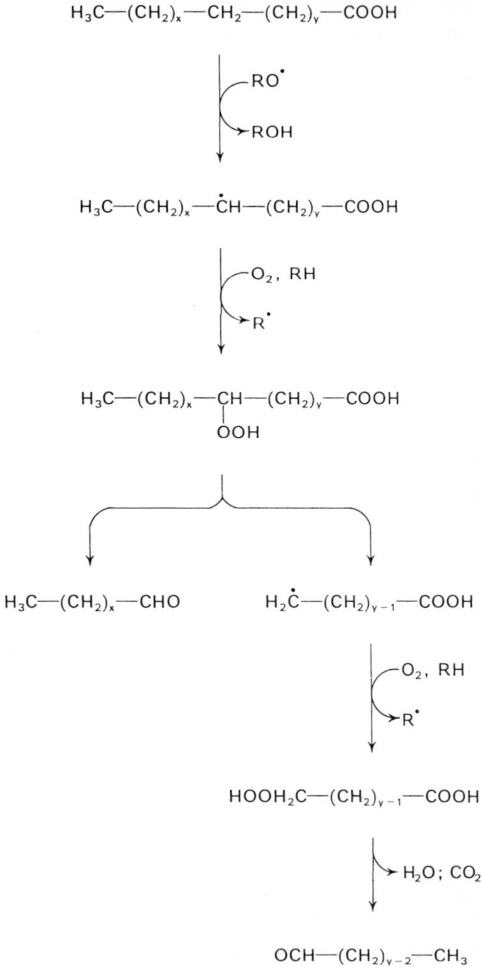

3.7.4.2 Polymerization

Under deep frying conditions, the isolenic fatty acids are isomerized into conjugated fatty acids which in turn interact by a 1,4-cycloaddition, yielding so-called *Diels-Alder* adducts (cf. Reaction 3.96).
The side chains of the resultant tetra-substituted cyclohexene derivatives are shortened by oxidation to oxo, hydroxy or carboxyl groups. In addition, the cyclohexene ring is readily dehydrogenated to an aromatic ring, hence compounds related to benzoic acid can be formed.

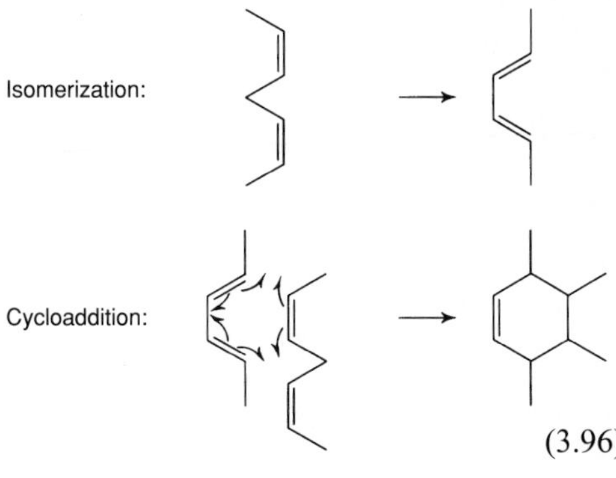

$$H_3C-(CH_2)_x-CH_2-(CH_2)_y-COOH$$

$$\downarrow\ \begin{matrix}RO^\bullet\\ ROH\end{matrix}$$

$$H_3C-(CH_2)_x-\overset{\bullet}{C}H-(CH_2)_y-COOH$$

$$\downarrow\ \begin{matrix}O_2,\ RH\\ R^\bullet\end{matrix}$$

$$H_3C-(CH_2)_x-\underset{OOH}{\overset{|}{C}H}-(CH_2)_y-COOH$$

$$H_3C-(CH_2)_x-CHO \qquad H_2\overset{\bullet}{C}-(CH_2)_{y-1}-COOH$$

$$\downarrow\ \begin{matrix}O_2,\ RH\\ R^\bullet\end{matrix}$$

$$HOOH_2C-(CH_2)_{y-1}-COOH$$

$$\downarrow\ H_2O;\ CO_2$$

$$OCH-(CH_2)_{y-2}-CH_3$$

Fig. 3.38. Autoxidation of saturated fatty acids. Hypothetical reactions involved in formation of volatile aldehydes

Isomerization:

Cycloaddition:

$$\qquad\qquad (3.96)$$

The fatty acid or triacylglycerol radicals formed by H-abstraction in the absence of oxygen can dimerize and then form a ring structure:

2 R—CH=CH—CH₂—CH=CH—R

Dimerization:

(3.97)

On the other hand, polymers with ether and peroxide linkages are formed in the presence of oxygen. They also may contain hydroxy, oxo or epoxy groups. The following structures, among others, have been identified:

(3.98)

Such compounds are undesirable in deep-fried oil or fat since they permanently diminish the flavoring characteristics of the oil or fat and, because of their HO-groups, behave like surface-active agents, i.e. they foam.

Disregarding the odor or taste deficiencies developed in a fat or oil heated for a prolonged period of time, the oil is considered spoiled when its petroleum ether-insoluble oxidized fatty acids reach a level $\geq 1\%$ (or $\geq 0.7\%$ at

Table 3.44. Relative stability of various fats and oils in deep-frying (RSDF)

Oil/fat	RSDF	Oil/fat	RSDF
Sunflower	1.0	Coconut	2.4
Rapeseed	1.0	Edible beef tallow	2.4
Soya	1.0	Soya oil,	
Peanut	1.2	hydrogenated	2.3
Palm	1.5	Peanut oil,	
Lard	2.0	hydrogenated	4.4
Butter fat	2.3		

the decreased smokepoint temperature of $\leq 170\,^{\circ}\mathrm{C}$). The fats or oils differ in their heat stability (Table 3.44). The stability is increased by hydrogenation of the double bonds.

3.7.5 Radiolysis

Alkyl and acyloxy radicals are formed during radiolysis of acyl lipids. These will further react to form volatile compounds. The formation of alkanes and alkenes, that lack one or two C-atoms, from the original acyl residue are of interest for the detection of an irradiation (Fig. 3.39).

The proposed indicators for the irradiation of meat are the hydrocarbons 14:1, 15:0, 16:1,

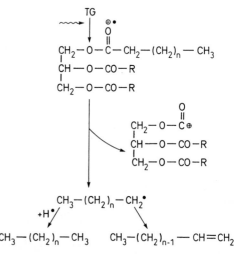

Fig. 3.39. Formation of alkanes and 1-alkenes during radiolysis of saturated triacyl glycerols

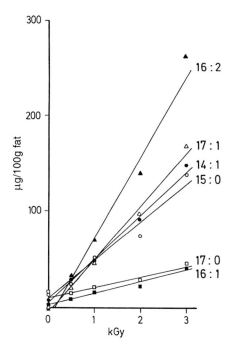

Fig. 3.40. Increase in the concentration of hydrocarbons relative to the radiation dose during irradiation of chicken meat (according to *W. W. Nawar* et al., 1990)

16:2, 17:0 and 17:1 which are formed during radiolysis of palmitic, oleic and stearic acid. It was demonstrated that their concentrations in fat increased depending on the radiation dose, e. g., in chicken meat (Fig. 3.40).

3.7.6 Microbial Degradation of Acyl Lipids to Methyl Ketones

Fatty acids of short and medium chain lengths present in milk fat, coconut and palm oils are degraded to methyl ketones by some fungi. A number of *Penicillium* and *Aspergillus* species, as well as several *Ascomycetes*, *Phycomycetes* and *fungi imperfecti* are able to do this.

The microorganisms first hydrolyze the triglycerides enzymatically (cf. 3.7.1) and then they degrade the free acids by a β-oxidation pathway (Fig. 3.41). The fatty acids $<C_{14}$ are transformed to methyl ketones, the C-

Table 3.45. Sensory properties of methyl ketones

Compound	Odor description	Odor threshold (ppb; in water)
2-Pentanone	Fruity, like bananas	2,300
2-Hexanone		930
2-Heptanone	Fragrant, herbaceous	650
2-Octanone	Flowery, refreshing	190
2-Nonanone	Flowery, fatty	190

skeletons of which have one C-atom less than those of the fatty acids. Apparently, the thiohydrolase activity of these fungi is higher than the β-ketothiolase activity. Hence, ester hydrolysis occurs instead of thioclastic cleavage of the thioester of a β-keto acid (see a textbook of biochemistry). The β-keto acid released is rapidly decarboxylated enzymatically; a portion of the methyl ketones is reduced to the corresponding secondary alcohols.

The odor threshold values for methyl ketones are substantially higher than those for aldehydes (cf. Tables 3.31 and 3.45). Nevertheless, they act as aroma constituents, particularly in flavors of mold-ripened cheese (cf. 10.2.8.3). However, methyl ketones in coconut or palm oil or in milk fat provide an undesirable, unpleasant odor denoted as "perfume rancidity".

3.8 Unsaponifiable Constituents*

Disregarding a few exceptions, fats and oils contain an average of 0.2–1.5% unsaponifiable compounds (Table 3.46). They are isolated from a soap solution (alkali salts of fatty acids) by extraction with an organic solvent.

---

* The free higher alcohols described under 3.6.2 and the deacylated alkoxy-lipids belong to this class.

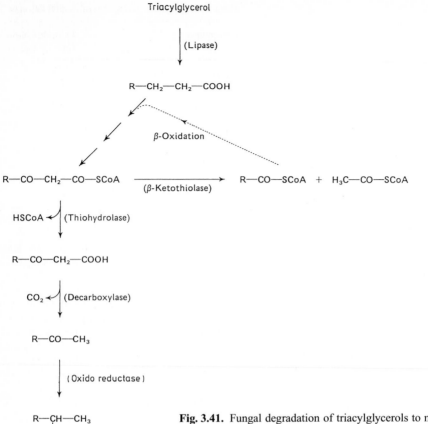

Triacylglycerol

(Lipase)

R—CH₂—CH₂—COOH

β-Oxidation

R—CO—CH₂—CO—SCoA ⟶ R—CO—SCoA + H₃C—CO—SCoA
(β-Ketothiolase)

HSCoA (Thiohydrolase)

R—CO—CH₂—COOH

CO₂ (Decarboxylase)

R—CO—CH₃

(Oxido reductase)

R—CH—CH₃
   OH

Fig. 3.41. Fungal degradation of triacylglycerols to methyl ketones (according to *J. E. Kinsella* and *D. H. Hwang*, 1976)

Table 3.46. Content of unsaponifiables in various fats and oils

Fat/oil	Unsaponifiable (weight-%)	Fat/oil	Unsaponifiable (weight-%)
Soya	0.6−1.2	Shea	3.6−10.0
Sunflower	0.3−1.2	Lard	0.1− 0.2
Cocoa	0.2−0.3	Shark (refined)	15 −17
Peanut	0.2−4.4		
Olive	0.4−1.1	Herring (refined)	0.7− 1.0
Palm	0.3−0.9		
Rapeseed	0.7−1.1		

The unsaponifiable matter contains hydrocarbons, steroids, tocopherols and carotenoids. In addition, contaminants or fat or oil additives, such as mineral oil, plasticizers or pesticide residues, can be found.

Each class of compounds in the unsaponifiable matter is represented by a number of components, the structures and properties of which have been thoroughly elucidated in the past decade or two, thus reflecting the advance in the analytical chemistry of fats and oils. Studies aimed at elucidating the constituents, and their structures, of unsaponifiable matter are motivated by a desire to find compounds which can serve as a reliable indicator for the identity of a fat or an oil.

3.8.1 Hydrocarbons

All edible oils contain hydrocarbons with an even or an odd C-number (C_{11} to C_{35}). Olive, rice and fish oils are particularly rich in this class of compounds. The main hydrocarbon constituent of olive oil (1−7 g/kg) and rice oil (3.3 g/kg) is a linear triterpene (C_{30}), squalene:

$$(3.99)$$

This compound is used as an analytical indicator for olive oil (cf. Table 14.25).

Squalene is present in a substantially higher concentration in fish liver oil. For example, shark liver oil has up to 30% squalene, and 7% pristane (2,6,10,14-tetramethylpentadecane) and some phytane (3,7,11,15-tetramethylhexadecane).

Apart from traces of polycyclic hydrocarbons (cf. 9.7), low levels of alkylbenzenes are detected. Such compounds occurring in olive oil are shown in Formula 3.100.

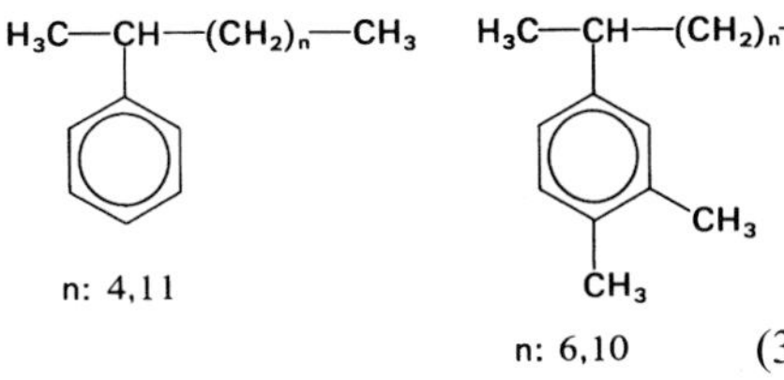

n: 4,11

n: 6,10 (3.100)

3.8.2 Steroids and Steroid Derivatives

3.8.2.1 Structure, Nomenclature

The steroid skeleton is made up of four condensed rings; A, B, C and D. The first three are in the chair conformation, whereas ring D is usually planar. While rings B and C, and C and D are fused in a trans-conformation, rings A and B can be fused in a trans- or in a cis-conformation. A characteristic of steroids is the presence of an alcoholic HO-group in position C-3.

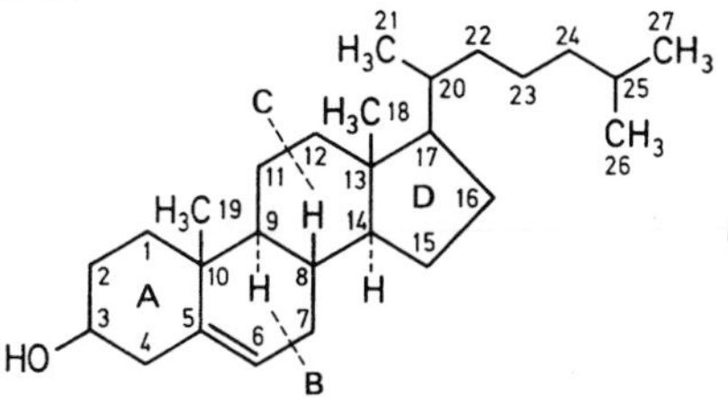

$$(3.101)$$

Conformational isomers introduced by fusing rings A and B in cholest-5-ene-3-β-ol (cholesterol; cf. Formula 3.101) are not possible since the C-5 position has a double bond.

By convention, the steric arrangement of substituents and H-atoms is related to the angular methyl group attached at C-10. When the plane containing the four rings is assumed to be the plane of this page, the substituent at C-10, by definition, is above the plane; all substituents below the plane are denoted by dashed or dotted lines. They are said to be α-oriented and have a trans-conformation. Substituents above the plane are termed β-oriented and are shown by solid line bonds and, in relation to the angular C-10 methyl group, are of cis-conformation.

In cholesterol (cf. Formula 3.101) the HO-group, the angular methyl group at C-13, the side chain on C-17 and the H-atom on C-8 are β-oriented (cis), whereas the H-atoms at C-9, C-14 and C-17 are α-oriented (trans). Steroids that are not methylated at position C-4 are denoted as desmethyl steroids.

3.8.2.2 Steroids of Animal Food

3.8.2.2.1 Cholesterol

Cholesterol (cf. Formula 3.101) is obtained biosynthetically from squalene (see a textbook of biochemistry). It is the main steroid of mammals and occurs in lipids in free form or esterified with saturated and unsaturated fatty acids. The content of cholesterol in some foods is illustrated by the data in Table 3.47.

Autoxidation of cholesterol, which is accelerated manyfold by 18:2 and 18:3 fatty acid peroxy radicals, proceeds through the inter-

Table 3.47. Cholesterol content of some food

Food	Amount (mg/100 g)
Calf brain	2,000
Egg yolk[a]	1,010
Pork kidney	410
Pork liver	340
Butter	240
Pork meat, lean	70
Beef, lean	60
Fish (Halibut; *Hypoglossus vulgaris*)	50

[a] Egg white is devoid of cholesterol.

mediary 3β-hydroxycholest-5-en-7α- and 7β-hydroperoxides, of which the 7β-epimer is more stable because of its quasi-equatorial conformation and, hence, is formed predominantly. Unlike autoxidation, the photosensitized oxidation (reaction with a singlet oxygen) of cholesterol yields 3β-hydroxycholest-6-en-5α-hydroperoxide. Among the many derivatives obtained by the further degradation of the hydroperoxides, cholest-5-en-3β,7α-diol, cholest-5-en-3β,7β-diol, 3β-hydroxycholest-5-en-7-one, 5,6β-epoxy-5β-cholestan-3β-ol and 5α-cholestan-3β,5,6β-triol have been identified as major products. These so-called "oxycholesterols" have been detected as side components in some food items (dried egg yolk, whole milk powder, butter oil and heated meat).

Cholesterol is the precursor in animal organisms for the biosynthesis of other steroids, such as sex hormones and bile acids. The latter, due to their interface activity, play a role in the emulsification and absorption of triacyl lipids in the intestine. Cholic acid is a bile acid with three α-oriented HO-groups:

$$(3.102)$$

Bile acids present in bile or intestines are almost always bound to glycine e.g., cholic acid as glycocholic acid, and to taurine e.g., taurocholic acid.

Products of cholesterol metabolism include C_{19}-sterols which produce the specific smell of boar in boar meat. Five aroma components (Table 3.48) were identified; 5α-androst-16-en-3α-ol (Formula 3.103) has also been detected in truffels (cf. 17.1.2.6.1).

$$(3.103)$$

Table 3.48. Odor-active C_{19}-steroids

Compound	Odor threshold (mg/kg; oil)
5α-Androst-16-en-3-one	0.6
5α-Androst-16-en-3α-ol	0.9
5α-Androst-16-en-3β-ol	1.2
4,16-Androstadien-3-one	7.8
5,16-Androstadien-3β-ol	8.9

3.8.2.2.2 Vitamin D

Cholecalciferol (vitamin D_3) is formed by photolysis of 7-dehydrocholesterol, a precursor in cholesterol biosynthesis. As shown in Fig. 3.42, UV radiation opens the B-ring. The precalciferol formed is then isomerized to vitamin D_3 by a rearrangement of the double bond which is influenced by temperature. Side-products, such as lumi- and tachisterol, have no vitamin D activity. Cholecalciferol is converted into the active hormone, 1,25-dihydroxy-cholecalciferol, by hydroxylation reactions in liver and kidney.

7-Dehydrocholesterol, the largest part of which is supplied by food intake and which accumulates in human skin, is transformed by UV light into vitamin D_3. The occurrence and the physiological significance of the D vitamins are covered in Section 6.22.

The main steroid of yeasts, ergosterol (ergosta-5,7,22-trien-3β-ol, i.e. two conjugated double bonds in ring B and a third in the side chain) is known as provitamin D_2.

3.8.2.3 Plant Steroids (Phytosterols)

3.8.2.3.1 Desmethylsterols

Cholesterol, long considered to be an indicator of the presence of animal fat, also occurs in small amounts in plants (Table 3.49). Campe-, stigma- and sitosterol, which are predominant in the sterol fraction of some plant oils, are structurally related to cholesterol; only the side chain on C-17 is changed. The following structural segments (only ring D and the side chain) show these differences (Formulas 3.104−3.107):

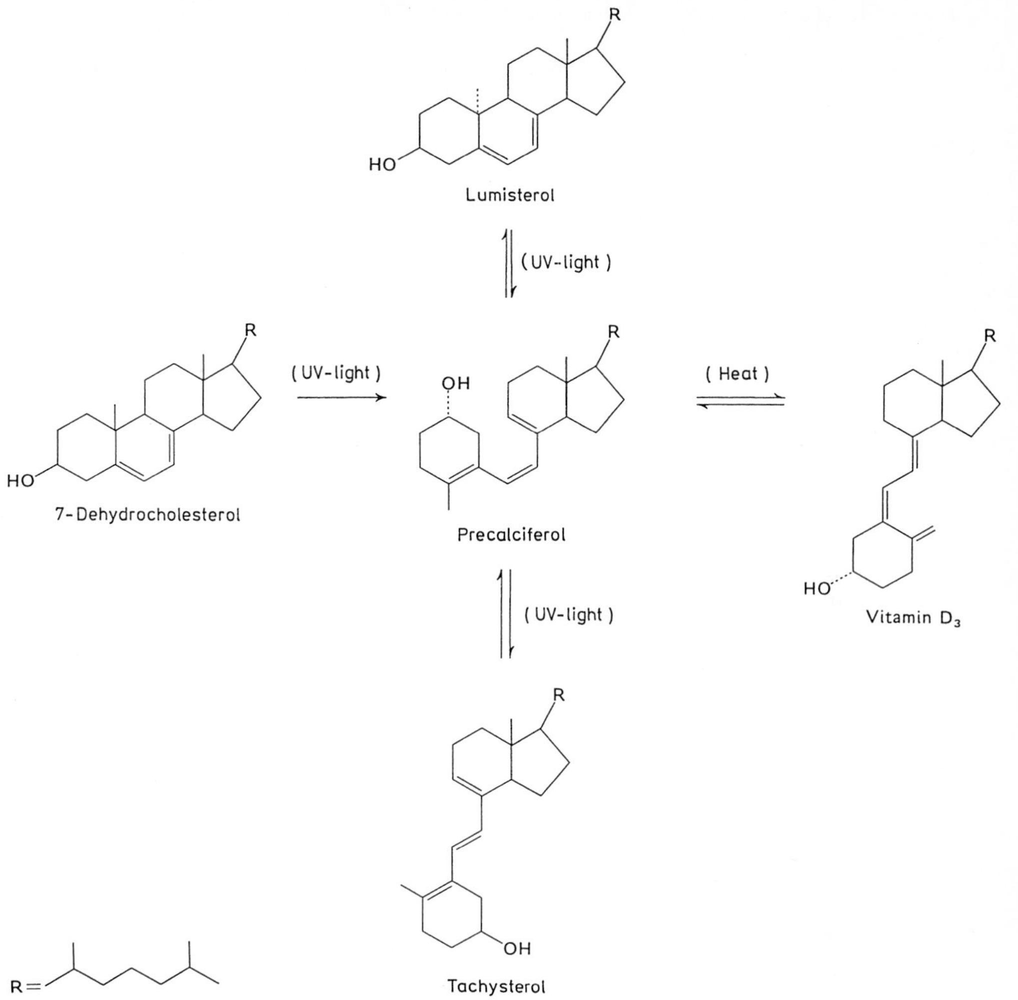

Lumisterol

(UV-light)

7-Dehydrocholesterol   (UV-light) →   Precalciferol   (Heat)   Vitamin D₃

(UV-light)

Tachysterol

R=

Fig. 3.42. Photochemical conversions of provitamin D₃

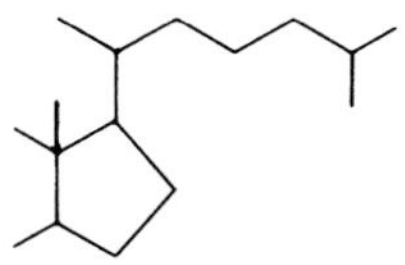

Cholesterol   (3.104)

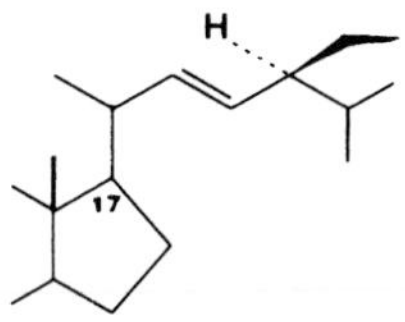

Stigmasterol (double bond at C-22)   (3.106)

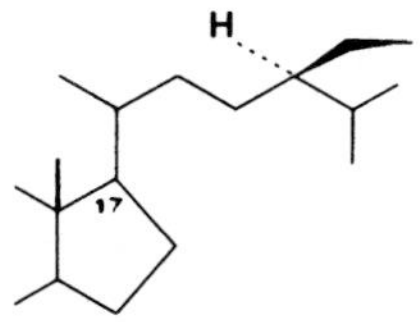

β-Sitosterol (24-α-ethyl-cholesterol)   (3.105)

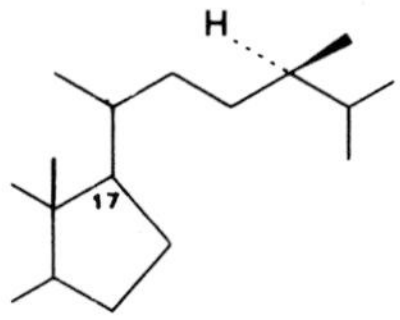

Campesterol (24-α-methyl-cholesterol)   (3.107)

Table 3.49. Average sterol composition of plant oils[a]

Component	Sun-flower	Peanut	Soya	Cotton-seed	Corn	Olive	Palm
Cholesterol	0.5	6.2	0.5	0.5	0.5	0.5	0.5
Brassicasterol	0.5	0.5	0.5	0.5	0.5	0.5	0.5
Campesterol	242	278	563	276	2655	19	88
Stigmasterol	236	145	564	17	499	0.5	42
β-Sitosterol	1961	1145	1317	3348	9187	732	252
Δ^5-Avenasterol	163	253	46	85	682	78	0.5
Δ^7-Stigmasterol	298	0.5	92	0.5	96	0.5	51
Δ^7-Avenasterol	99	34	63	18	102	30	0.5
24-Methylene-cycloartenol	204	0.5	53	0.5	425	580	0.5

[a] Values in mg/kg.

Δ^5-Avenasterol is a sitosterol derivative:

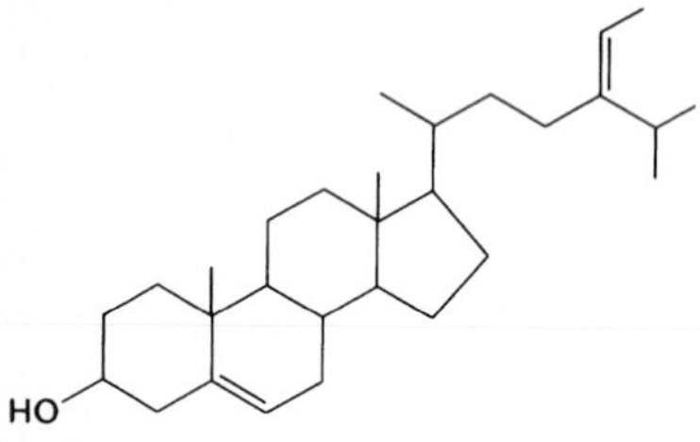

Δ^5-Avenasterol

(3.108)

In addition to Δ^5-sterols Δ^7-sterols occur in plant lipids; for example:

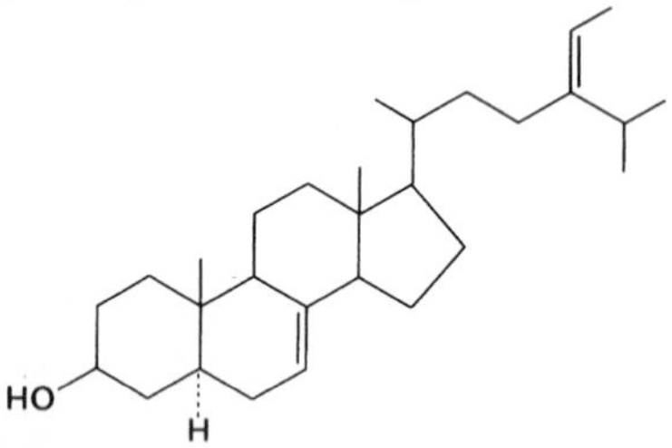

Δ^7-Avenasterol

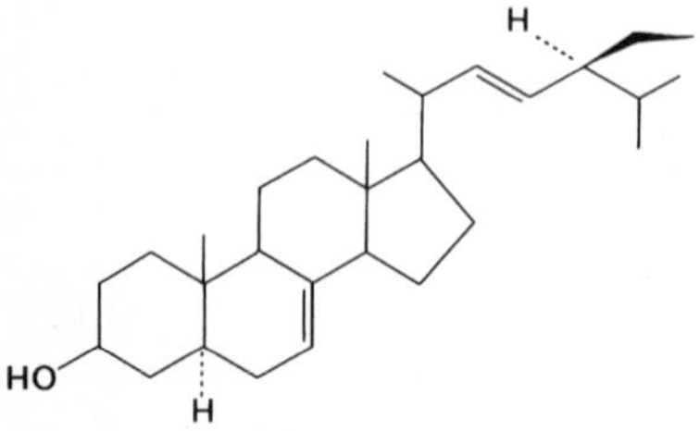

(3.109)

Δ^7-Stigmasterol

(3.110)

Plant lipids contain 0.15–0.9% sterols, with sitosterol as the main component (Table 3.49). In order to identify blends of fats (oils), the data on the predominant steroids are usually expressed as a quotient. For example, the ratio of stigmasterol/campesterol is determined in order to detect adulteration of cocoa butter. As seen from Table 3.50, this ratio is significantly lower in a number of cocoa butter substitutes than in pure cocoa butter. The phytosterol fraction (e.g. sito- and campesterol) has to be determined in order to detect the presence of plant fats in animal fats.

Table 3.50. Ratio of stigmasterol to campesterol in various fats

Fat	Sterols (g/kg)	% Stigmasterol[a] / % Campesterol
Cocoa	1.8	2.8 –3.5
Tenkawang	2.15	0.42–0.55
Coconut oil	0.75	1.47
Peanut oil, hydrogenated		0.72
Coberine[b]		0.31–0.60
Calvetta[b]		0.58–0.61

[a] The ratio is calculated from peak area after the unsaponifiables (sterol fraction) were separated by gas chromatography.
[b] Cocoa butter substitutes.

3.8.2.3.2 Methyl- and Dimethylsterols

Sterols with α-oriented C-4 methyl groups occur in oils of plant origin. The main compounds are:

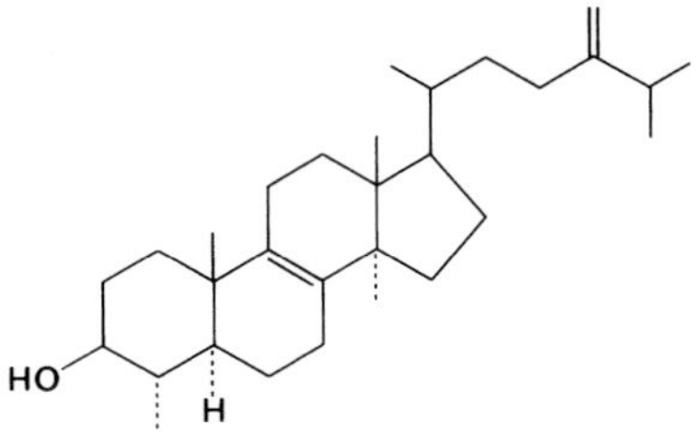

(3.111)

4α,14α-Dimethyl-24-methylene-5α-cholest-8-en-3β-ol (Obtusifoliol)

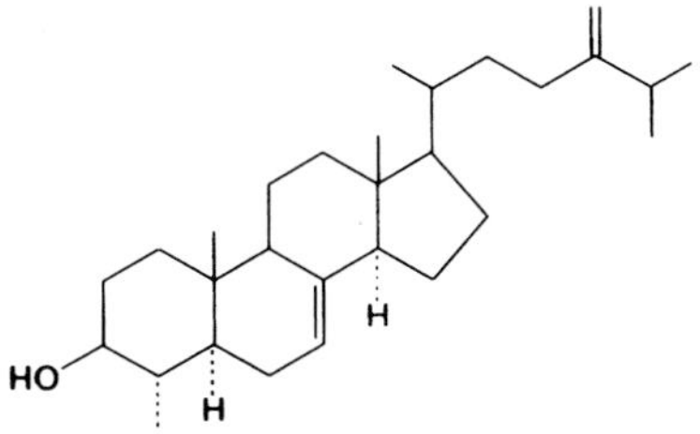

(3.112)

4α-Methyl-24-methylene-5α-cholest-7-en-3β-ol (Gramisterol)

Gas chromatographic-mass spectrometric studies have also revealed the presence of 4,4-dimethylsterols in the steroid fraction of many plant oils:

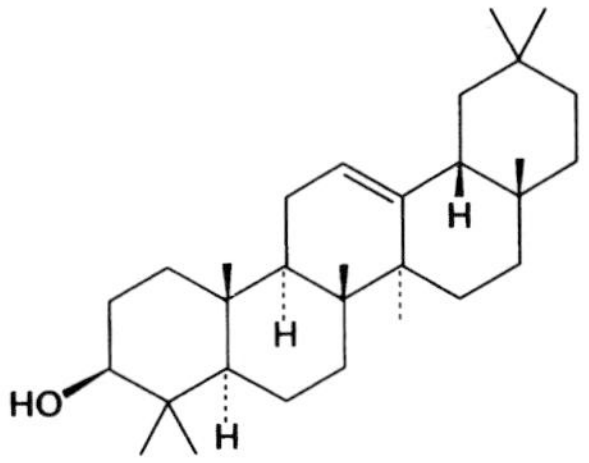

(3.113)

β-Amyrine

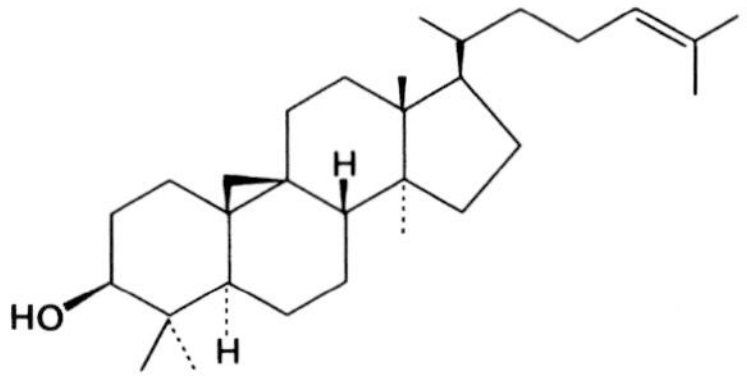

(3.114)

Cycloartenol

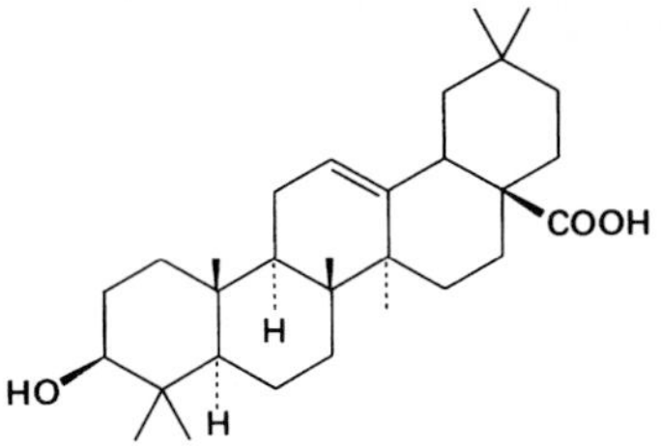

(3.115)

Oleanolic acid

Oleanolic acid has long been known as a constituent of olive oil. Methyl- and dimethylsterols are important in identifying fats and oils (cf. Fig. 3.44).

3.8.2.4 Analysis

Qualitative determination of sterols is conducted using the *Liebermann-Burchard* reaction, in which a mixture of glacial acetic and concentrated sulfuric acids reacts directly with the fat or oil or the unsaponifiable fraction. Several modifications of this basic assay have been developed which, depending on the steroid and the oxidizing agent used, result in the production of a green or red color. The reaction is more sensitive when the SO_3 oxidizing agent is replaced by the Fe^{3+} ion. The conversion of sterols into a chromophore is based on the reaction sequence given in Fig. 3.43. As shown, the assay is applicable only to sterols containing a double bond, such as in the B ring of cholesterol.

Sterols are separated as 3,5-dinitrobenzoic acid derivatives by thin layer chromatography and, after reaction with 1,3-diaminopropane, are determined quantitatively with high sensitivity in the form of a *Meisenheimer* adduct. Today, gas chromatographic analysis of silylated sterols is used more often. One application of this method is illustrated by the detection of 2% coberine in cocoa butter (Fig. 3.44). Coberine is a cocoa butter substitute made by blending palm oil and shea butter (the shea is an African tree with seeds that yield a thick white fat, shea butter).

The content of egg (more accurately, the yolk) in pasta products or cookies can be calculated after the cholesterol content has been determined, usually by gas chromatography or HPLC. Vitamin D determination requires specific

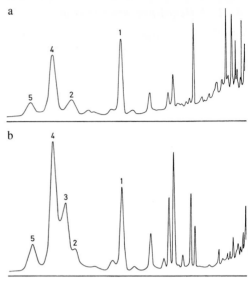

Pentaenyl cation (λ_M : 620 nm)

Fig. 3.43. Sterol detection according to *Lieberman-Burchard*. Reactions involved in color development

Fig. 3.44. Gas chromatographic separation of the triterpene alcohol fraction after silylation: cocoa butter **a** and cocoa butter + 2% coberine **b** (according to *A. Fincke*, 1976). Peak 1: cholesterol added as a marker compound; peak 2: β-amyrine; peak 3: butyrospermol and α-amyrine; peak 4: cycloartenol; peak 5: 24-methylenecycloartenol

procedures in which precautions are taken with regard to the compound's sensitivity to light. A chemical method uses thin layer chromatographic separation of unsaponifiables, elution of vitamin D from the plate and photometric reading of the color developed by antimony (III) chloride. An alternative method recommends the use of HPLC.

3.8.3 Tocopherols and Tocotrienols

3.8.3.1 Structure, Importance

The methyl derivatives of tocol [2-methyl-2(4′,8′,12′-trimethyltridecyl)chroman-6-ol] are denoted tocopherols. In addition the corresponding methyl derivatives of tocotrienol occur in food.

All four tocopherols and tocotrienols, with the chemical structures given in Fig. 3.45, are found primarily in cereals (especially wheat germ oil), nuts and rapeseed oils. These redox-type lipids are of nutritional/physiological and analytical interest. As antioxidants (cf. 3.7.3.2.1), they prolong the shelf lives of many foods containing fat or oil. The significance of tocopherols such as vitamin E was outlined in 6.2.3.

About 60–70% of the tocopherols in oilseeds are retained during the oil extraction and refining process (cf. 14.4.1 and Table 3.51). Some oils with very similar fatty acid compositions can be distinguished by their distinct tocopherol spectrum. To illustrate this, two examples are provided. The amount of β-tocopherol in wheat germ oil is quite high (Table 3.51), hence it serves as an indicator of that oil. The blending of soya oil with sunflower oil is detectable by an increase in the content of linolenic acid (cf. 14.5.2.3). However, it is possible to make a final conclusive decision about the presence and quantity of soya oil in sunflower oil only after an analysis of the composition of the tocopherols.

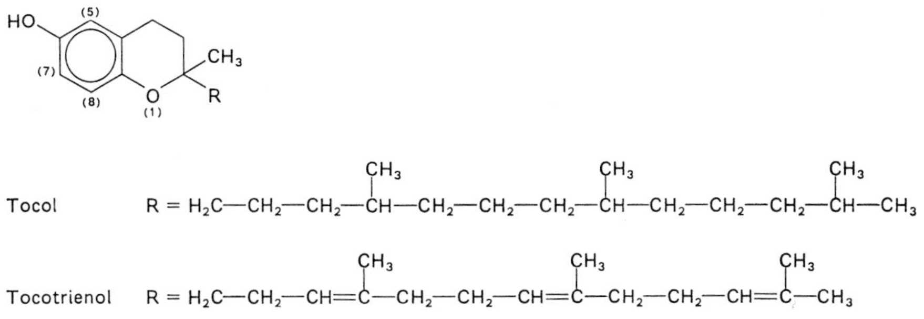

Tocol	R = H₂C—CH₂—CH₂—CH—CH₂—CH₂—CH₂—CH—CH₂—CH₂—CH₂—CH—CH₃		

(with CH₃ branches)

Tocotrienol   R = H₂C—CH₂—CH=C—CH₂—CH₂—CH=C—CH₂—CH₂—CH=C—CH₃ (with CH₃ branches)

Substitution	Tocopherols (T)	Tocotrienols (T-3)
5,7,8-Trimethyl	α-T	α-T-3
5,8-Dimethyl	β-T	β-T-3
7,8-Dimethyl	γ-T	γ-T-3
8-Methyl	δ-T	δ-T-3

Fig. 3.45. Tocopherols and tocotrienols present in food

Table 3.51. Tocopherols and Tocotrienols in plant oils[a]

Oil	α-T	α-T-3	β-T	β-T-3	γ-T	γ-T-3	δ-T	δ-T-3
Sunflower	56.4	< 0.02	2.45	0.2	0.4	0.02	0.09	
Peanut	14.1	< 0.02	0.4	0.4	13.1	0.03	0.92	
Soya	17.9	< 0.02	2.8	0.4	60.4	0.08	37.1	
Cottonseed	40.3	< 0.02	0.2	0.9	38.3	0.09	0.5	
Corn	27.2	5.4	0.2	1.1	56.6	6.2	2.5	
Olive	9.0	< 0.02	0.2	0.4	0.5	0.03	0.04	
Palm (raw)	20.6	39.2	< 0.1	2.5	< 0.1	42.6	2.6	10.1
Wheat germ	133.0	< 2.6	71.0	18.1	26.0		27.1	
Almond	20.7		0.3		0.9			
Apricot kernel	0.5				22.4		0.3	
Peach kernel	6.4		1.3		1.0			
Cocoa butter	0.3		< 0.1		5.3		< 0.1	
Palm oil, middle fraction	< 0.1		< 0.1		0.43		< 0.1	
Shea fat stearin	< 0.1		< 0.1		0.43		< 0.1	

[a] Average composition; indicated in mg/100 g.

The tocopherol pattern is also different in almond and apricot kernel oil (Table 3.51) whose fatty acid compositions are very similar. Therefore adulteration of marzipan with persipan can be detected by the analysis of the tocopherols.

3.8.3.2 Analysis

Isolation of tocopherols is accompanied by losses due to oxidation. Therefore, the edible oil is dissolved in acetone at 20–25 °C in the presence of ascorbyl palmitate as an antioxidant. The major portion of triacylglycerols is separated by crystallization at −80 °C. Tocopherols remaining in solution are then analyzed by thin layer or gas chromatography (after silylation of the phenolic HO-group) or by HPLC (cf. Fig. 3.46). UV spectrophotometry is also possible. However, the fluorometric method based on an older colorimetric procedure developed by *Emmerie* and *Engel* is even more sensitive. It involves reduction of the Fe (III) ion to Fe (II) by tocopherols and the reac-

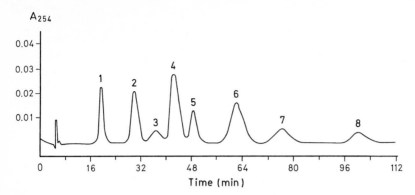

Fig. 3.46. Tocopherol and tocotrienol analysis by HPLC (according to *J. F. Cavins* and *G. E. Inglett*, 1974). 1 α-Tocopherol, 2 α-tocotrienol, 3 β-tocopherol, 4 γ-tocopherol, 5 β-tocotrienol, 6 γ-tocotrienol, 7 δ-tocopherol, and 8 δ-tocotrienol

tion of the reduced iron with 2,2′-bipyridyl to form an intensive red colored complex.

3.8.4 Carotenoids

Carotenoids are polyene hydrocarbons biosynthesized from eight isoprene units (tetraterpenes) and, correspondingly, have a 40-C skeleton.

Table 3.52. Carotenoids in various food

Food	Concentration (ppm)[a]	Food	Concentratioin (ppm)[a]
Carrots	54	Peaches	27
Spinach	26–76	Apples	0.9–5.4
Tomatoes	51	Peas	3–7
Apricots	35	Lemons	2–3

[a] On dry weight basis.

They provide the intensive yellow, orange or red color of a great number of foods of plant origin (Table 3.52). They are synthesized only by plants (see a textbook of biochemistry). However, they reach animal tissues via the feed (pasture, fodder) and can be modified and deposited there.

A well known example is the chicken egg yolk, which is colored by carotenoids. The carotenoids in green plants are masked by chlorophyll. When the latter is degraded, the presence of carotenoids is readily revealed (e.g. the green pepper becomes red after ripening).

3.8.4.1 Chemical Structure, Occurrence

Other carotenoids are derived by hydrogenation, dehydrogenation and/or cyclization of the basic structure of the C_{40}-carotenoids (cf.

Formula 3.116). The cyclization reaction can occur at one or both end groups. The differences in C_9-end groups are denoted by Greek letters (cf. Formula 3.117).

(3.117)

A semisystematic nomenclature used at times has two Greek letters as a prefix for the generic name "carotene", denoting the structure of both C_9- end groups (cf. Formulas III, IV, VI or X: cf. Formulas 3.120, 3.121, 3.122 and 3.128, respectively). Designations such as α-, β- or γ-carotene are common names.

Carotenoids are divided into two main classes: carotenes and xanthophylls. In contrast to carotenes, which are pure polyene hydrocarbons, xanthophylls contain oxygen in the form of hydroxy, epoxy or oxo groups. Some carotenoids of importance to food are presented in the following sections.

3.8.4.1.1 Carotenes

Acyclic or Aliphatic Carotenes

Carotenes I, II and III (cf. Formulas 3.118–120) are intermediary or precursor compounds which, in biosynthesis after repeated dehydrogenizations, provide lycopene (IV; see a textbook of biochemistry). Lycopene is the red color of the tomato (and also of wild rose hips). In yellow tomato cultivars, lycopene precursors are present together with β-carotene (Table 3.53).

Table 3.53. Carotenes (ppm) in some tomato cultivars

Cultivar	Phytoene (I)	Phytofluene (II)	β-Carotene (VII)	ξ-Carotene (III)	γ-Carotene (V)	Lycopene (IV)
Campbell	24.4	2.1	1.4	0	1.1	43.8
Ace Yellow	10.0	0.2	trace	0	0	0
High Beta	32.5	1.7	35.6	0	0	0
Jubilee	68.6	9.1	0	12.1	4.3	5.1

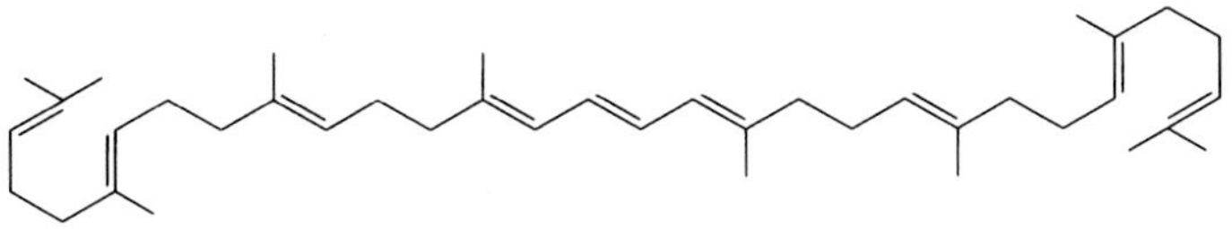

(3.118)

Phytoene (I)

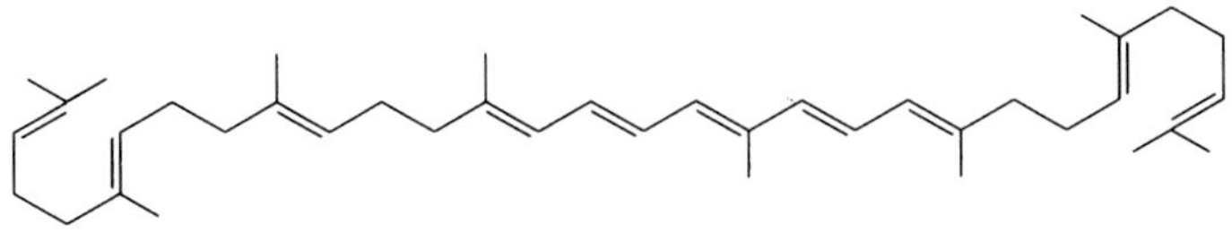

(3.119)

Phytofluene (II)

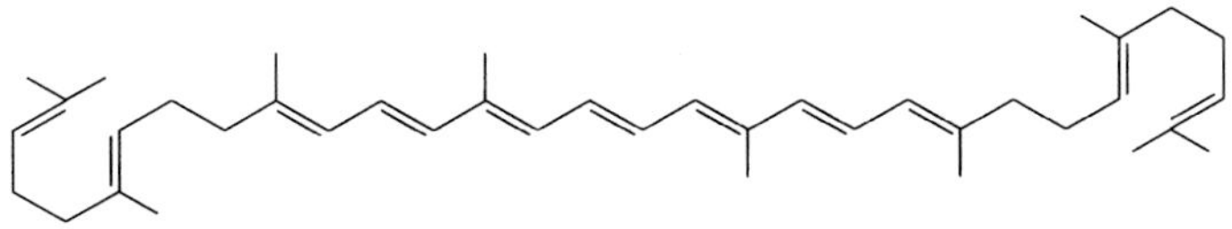

(3.120)

ξ-Carotene (7,8,7′,8′-tetrahydro-ψ,ψ-carotene) (III)

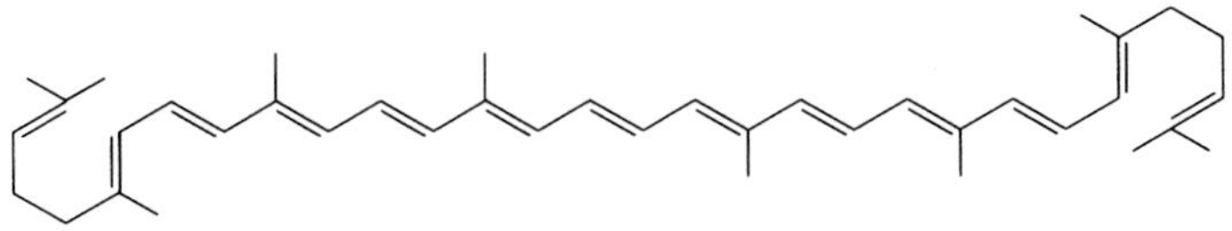

(3.121)

Lycopene (ψ,ψ-carotene) (IV)

Monocyclic Carotenes

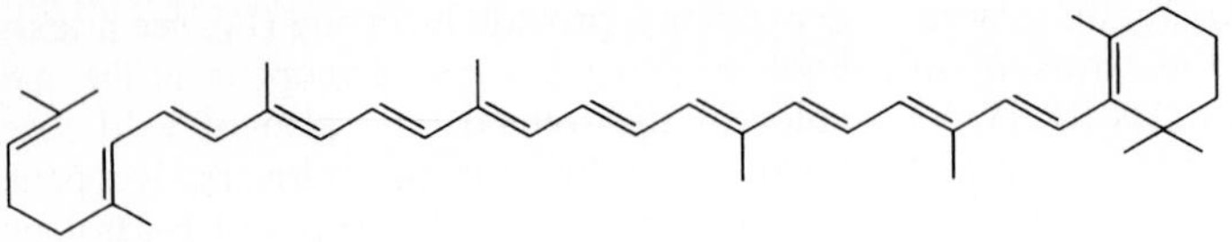

(3.122)

γ-Carotene (ψ,β-carotene) (V)

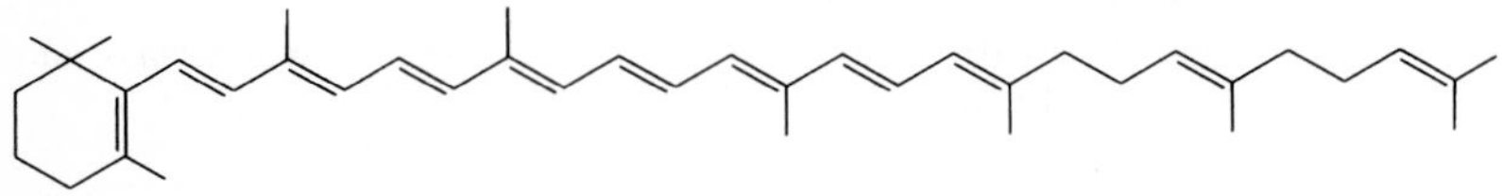

(3.123)

β-Zeacarotene (Va)

Bicyclic Carotenes

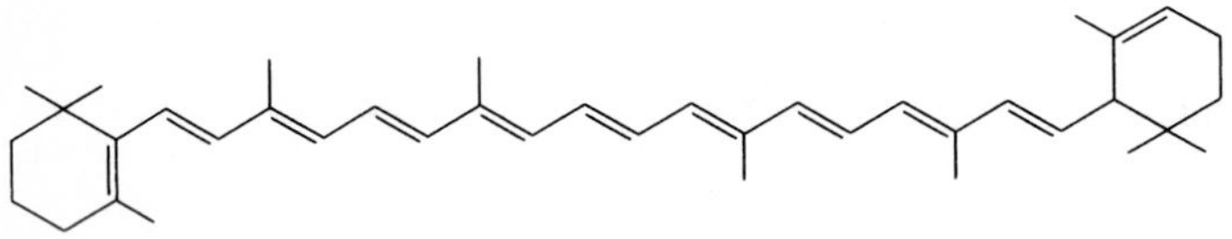

(3.124)

α-Carotene (β,ε-carotene) (VI)

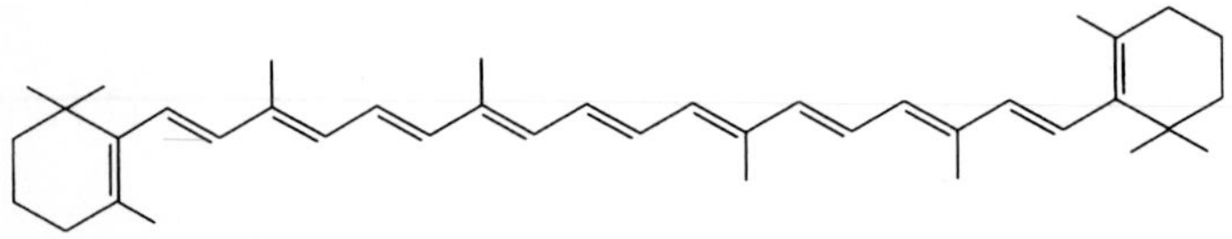

(3.125)

β-Carotene (β,β-carotene) (VII)

The importance of β-carotene as provitamin A is covered under 6.2.1.

3.8.4.1.2 Xanthophylls

Hydroxy Compounds

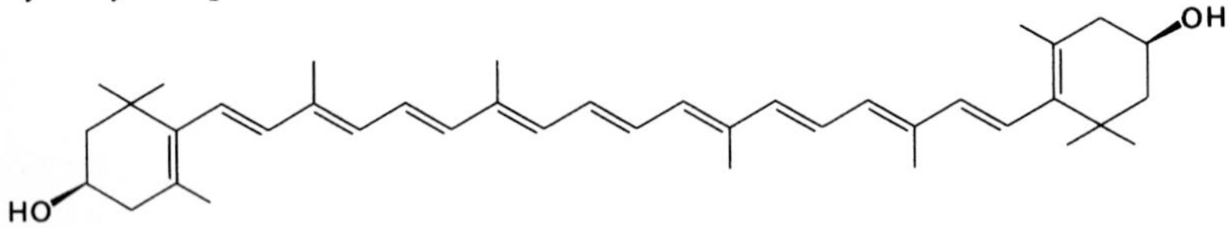

(3.126)

Zeaxanthin (β,β-carotene-3,3′-diol) (VIII)

This xanthophyll is present in corn (*Zea mays*).

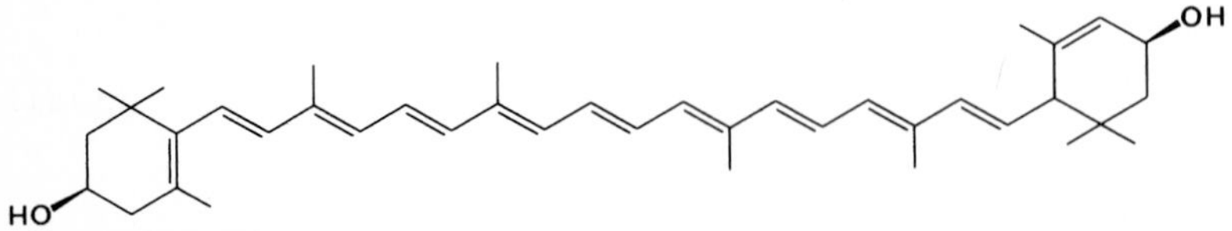

(3.127)

Lutein (β,ε-carotene-3,3′-diol (IX)

This xanthophyll occurs in green leaves and in egg yolk.

Keto Compounds

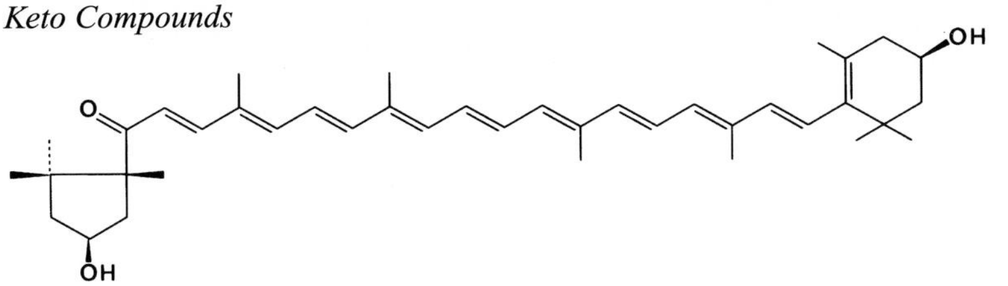

(3.128)

Capsanthin (3,3′-dihydroxy-β,ϰ-carotene-6′-one) (X)

This xanthophyll is the major carotene of paprika peppers.

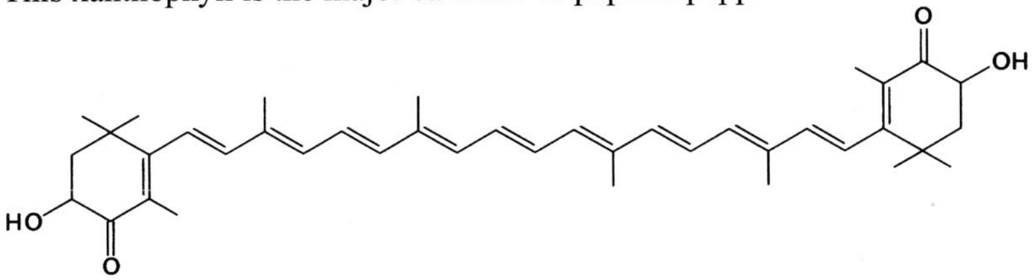

(3.129)

Astaxanthin (XI)

Astaxanthin is present in crab and lobster shells and, in combination with proteins, provides three blue hues (α-, β- and γ-crustacyanin) and one yellow pigment. During the cooking of crabs and lobsters, the red astaxanthin is released from a green carotenoid-protein complex. Astaxanthin usually occurs in lobster shell as an ester, e.g., dipalmitic ester.

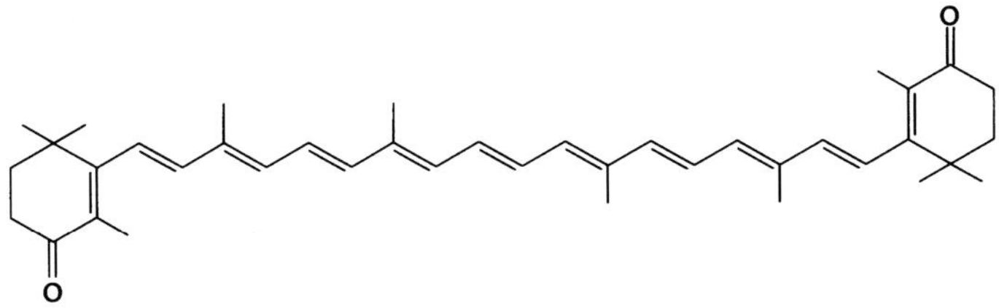

(3.130)

Canthaxanthin (XII)

This xanthophyll is used as a food colorant (cf. 3.8.4.5).

Epoxy Compounds

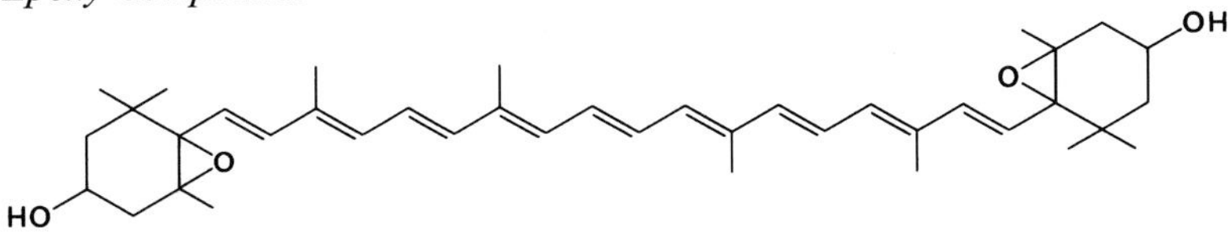

(3.131)

Violaxanthin (zeaxanthin-diepoxide) (XIII)

Violaxanthin is present in orange juice (cf. Table 3.54) and it also occurs in green leaves.

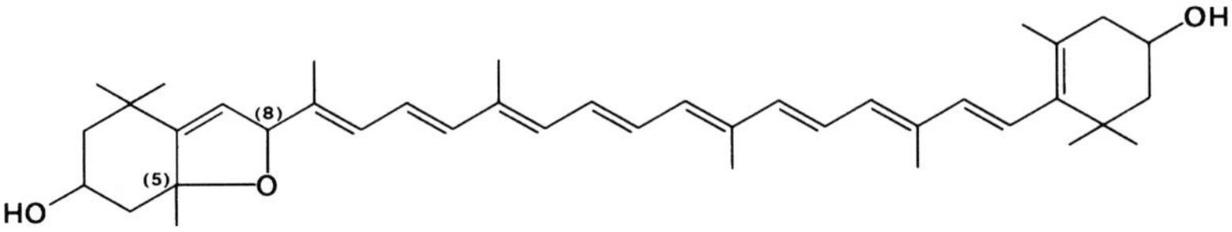

(3.132)

Mutatoxanthin (5,8-epoxy-5,8-dihydro-β,β-carotene-3,3′-diol) (XVI)

This epoxy carotenoid is present in oranges (cf. Table 3.54).

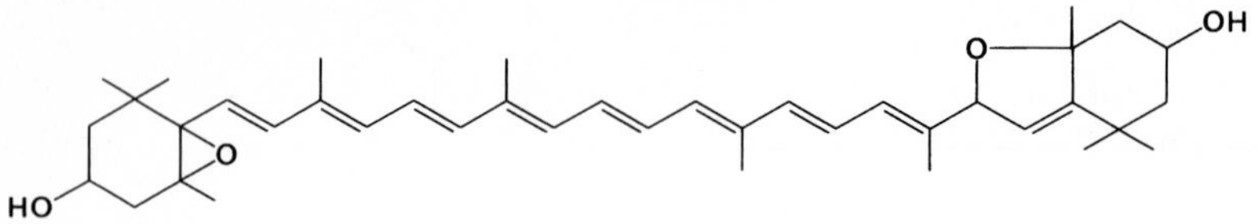

(3.133)

Luteoxanthin (XIV)

Luteoxanthin is the major carotenoid of oranges (cf. Table 3.54).

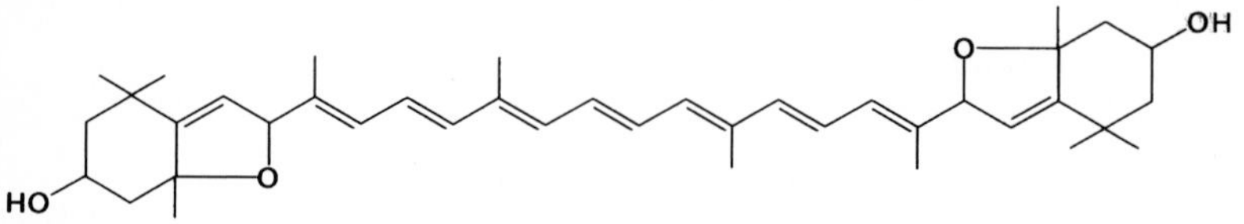

(3.134)

Auroxanthin (XV)

This carotenoid is a constituent of oranges (cf. Table 3.54).

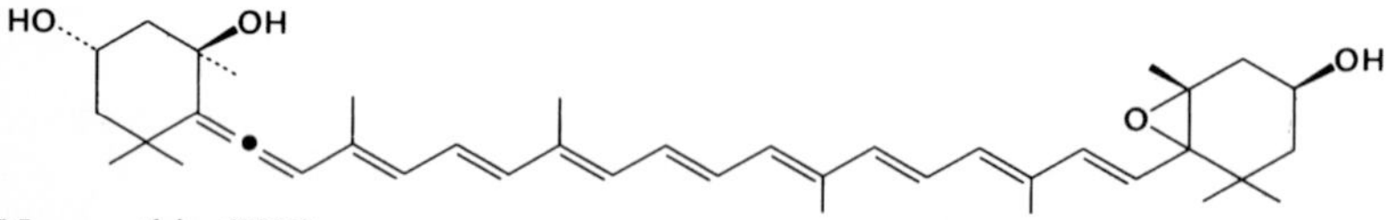

(3.135)

Neoxanthin (XX)

Dicarboxylic Acids and Esters

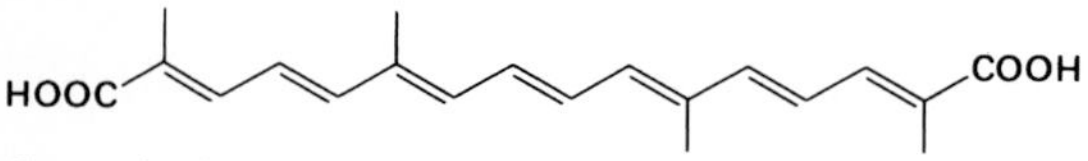

(3.136)

Crocetin (XVII)

This carboxylic acid carotenoid is the yellow pigment of saffron. It occurs in plants as a diester, i.e. glycoside with the disaccharide gentiobiose. The diester, called crocin, is therefore water-soluble.

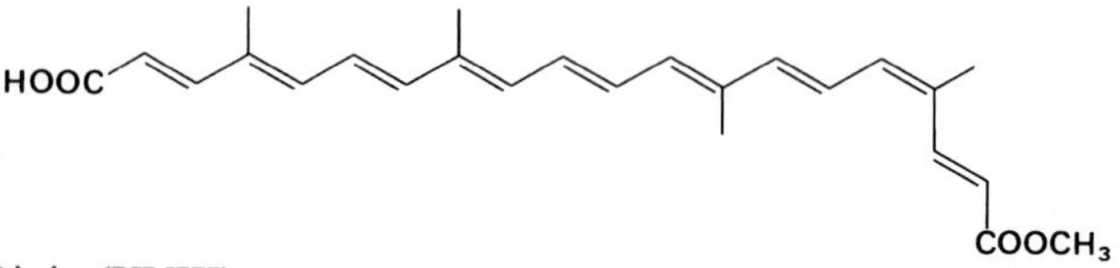

(3.137)

Bixin (XVIII)

Bixin is the main pigment of annato extract. Annato originates from the West Indies and the pigment is isolated from the seed pulp of the tropical bush *Bixa orellana*. Bixin is the monomethyl ester of norbixin, a dicarboxylic acid homologous to crocetin.

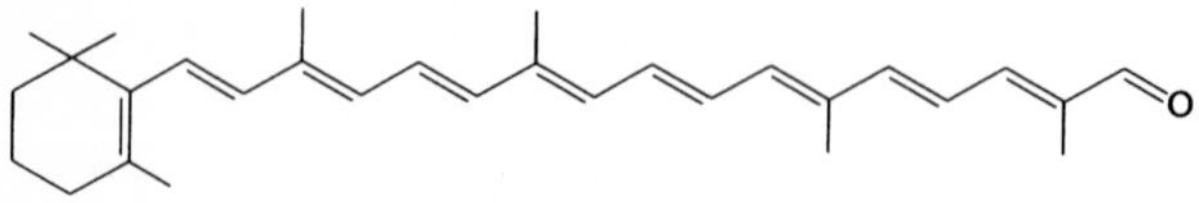

(3.138)

β-apo-8′-carotenal* (XIX)

---

* The prefix "apo" indicates a compound derived from a carotenoid by removing part of its structure.

Table 3.54. Major carotenoid components in orange juice

Carotenoid	As percent of total carotenoids
Phytoene (I)	13
ξ-Carotene (III)	5.4
Cryptoxanthin (3-Hydroxy-β-carotene)	5.3
Antheraxanthin (5,6-Epoxyzeaxanthin)	5.8
Mutatoxanthin (XVI)	6.2
Violaxanthin (XIII)	7.4
Luteoxanthin (XIV)	17.0
Auroxanthin (XV)	12.0

Table 3.55. Absorption wavelength maxima for some carotenoids

Compound	Conjugated double bonds	Wavelength, nm (petroleum ether)		
A. Effect of the number of conjugated double bonds				
Phytoene (I)	3	275	285	296
Phytofluene (II)	5	331	348	367
ξ-Carotene (III)	7	378	400	425
Neurosporene	9	416	440	470
Lycopene (IV)	11	446	472	505
B. Effect of the ring structure				
γ-Carotene (V)	11	431	462	495
β-Carotene (VII)	11	425[a]	451	483

[a] Maximum absorption wavelength is not unequivocal (cf. Fig. 3.47).

Carotenoids are, as a rule, present in plants as a complex mixture. For example, the orange has more than 50 well characterized compounds, of which only those that exceed 5% of the total carotenoids are presented in Table 3.54.

Hydroxy-carotinoids are often present as esters of fatty acids; e.g., orange juice contains 3-hydroxy-β-carotene (cryptoxanthin) esterified with lauric, myristic and palmitic acid. The quantitative analysis of this ester fraction is used as proof of an adulteration of orange juice with mandarin juice.

3.8.4.2 Physical Properties

Carotenoids are very soluble in apolar solvents, including edible fats and oils, but they are not soluble in water. Hence, they are denoted "lipochromes". Carotenoids are readily extracted from plant sources with petroleum ether, ether or benzene. Ethanol and acetone are also suitable solvents.

The color of carotenoids is the result of the presence of a conjugated double bond system in the molecules. The electron excitation spectra of such systems are of interest for elucidation of their structure and for qualitative and quantitative analyses.

Carotenoids show three distinct maxima in the visible spectrum, with wavelength positions dependent on the number of conjugated double bonds (Table 3.55). The fine structure of the spectrum is better distinguished in the case of acyclic lycopene (IV) than bicyclic β-carotene, since the latter is no longer a fully planar molecule. The methyl groups positioned on the rings interfere with those on the polyenic chain. Such steric effects prevent the total overlapping of π orbitals; consequently, a hypsochromic shift (a shift to a shorter wavelength) is observed for the major absorption bands (Fig. 3.47a).

Oxo groups in conjugation with the polyene system shift the major absorption bands to longer wavelengths (a bathochromic effect) with a simultaneous quenching of the fine structure of the spectrum (Fig. 3.47b). The hydroxyl groups in the molecule have no influence on the spectra.

A change of solvent system alters the position of absorption maxima. For example, replacing hexane with ethanol leads to a bathochromic shift.

Most of the carotenoids in nature and, thus, in food are of the trans-double bond configuration. When a mono-cis- or di-cis-compound occurs, the prefix "neo" is used. When one bond of all trans-double bonds is rearranged into this cis-configuration, there is a small shift in absorption maxima with a new minor "cis band" shoulder on the side of the shorter wavelength.

3.8.4.3 Chemical Properties

Carotenoids are highly sensitive to oxygen and light. When these factors are excluded, carot-

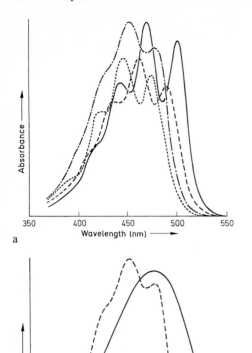

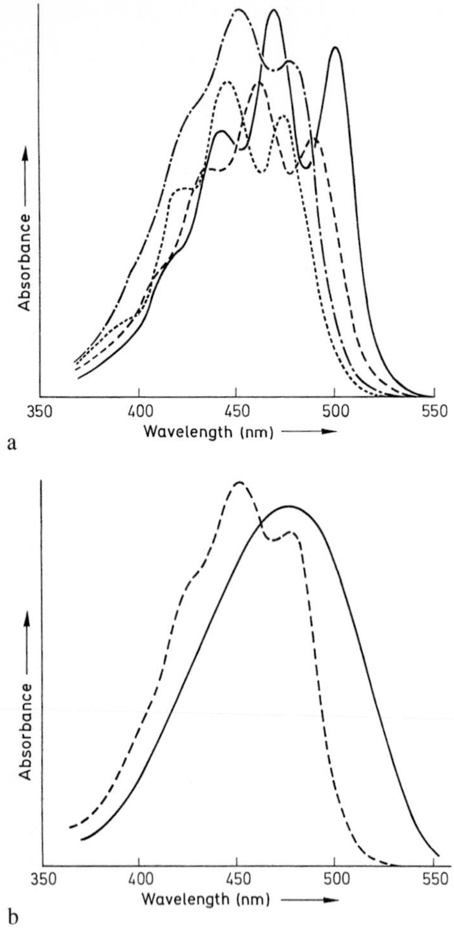

a

b

Fig. 3.47. Electron excitation spectra of carotenoids (according to *O. Isler*, 1971).
a — Lycopene (IV), - - - - γ-carotene (V), ······ α-carotene (VI), —·—·—β-carotene (VII); **b** Canthaxanthin (XII) before —— and after - - - - oxo groups reduction with NaBH₄

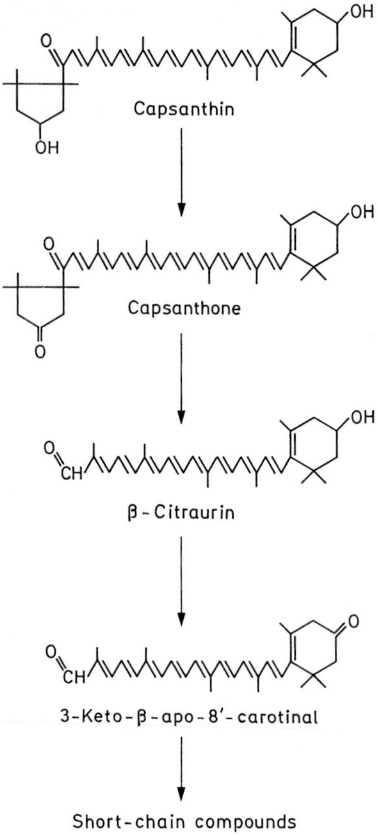

Fig. 3.48 structures:

Capsanthin

Capsanthone

β – Citraurin

3-Keto-β-apo-8′-carotinal

Short-chain compounds

Fig. 3.48. Oxidative degradation of capsanthin during storage of paprika (according to *T. Philip* and *F.J. Francis*, 1971)

The color change in paprika from red to brown, as an example, is due partly to a slow *Maillard* reaction, but primarily to oxidation of capsanthin (Fig. 3.48) and to some as yet unclear polymerization reactions.

3.8.4.4 Precursors of Aroma Compounds

Aroma compounds are formed during the oxidative degradation of carotenoids. Such compounds, their precursors and the foods in which they occur are listed in Table 3.56. The mentioned ionones and β-damascenone belong to the class of C_{13}-norisoprenoides. Other than β-ionone, α-ionone is a chiralic aroma compound whose R-enantiomer is present almost in optical purity in the food items listed in Table 3.56. α- and β-Damascone

enoids in food are stable even at high temperatures. Their degradation is, however, accelerated by intermediary radicals occurring in food due to lipid peroxidation (cf. 3.7.2). The cooxidation phenomena in the presence of lipoxygenase (cf. 3.7.2.2) are particularly visible.
Changes in extent of coloration often observed with dehydrated paprika and tomato products are related to oxidative degradation of carotenoids. Such discoloration is desirable in flours (flour bleaching; cf. 15.4.1.4.3).

(Formula 3.139), present in black tea are probably derived from α- and β-carotene. Chirospecific analysis (cf. 5.2.4) indicated that α-damascone occurs as racemate.

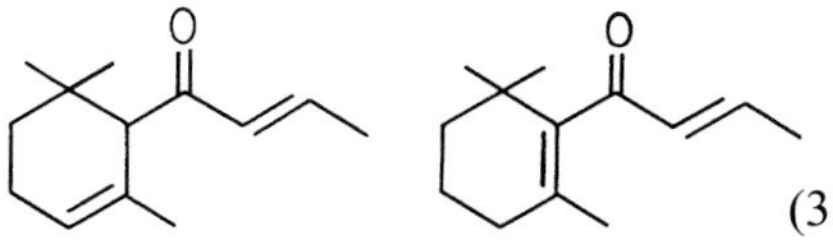

α-Damascone   β-Damascone   (3.139)

The odor thresholds of the R- and S-form (about 1 µg/kg, water) differ rarely.
Of all C_{13}-norisoprenoids, β-damascenone and β-ionone, smelling like honey and violets respectively, have the lowest odor threshold values (Table 3.56). Precursor of β-damascenone is neoxanthine, out of which the *Grasshopper ketone* (I in Formula 3.141) is formed by oxidative cleavage. The oxygen function migrates from the C-9 to the C-7 position by reduction of I to form an allentriol, elimination and attachment of HO-ions. In acid medium, 3-hydroxy-β-damascone and β-damascenone result from the intermediate (II).

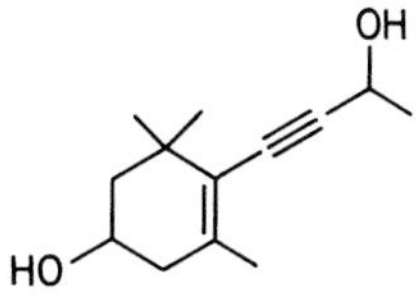

(3.140)

Besides the *Grasshopper ketone,* another enindiol (Formula 3.140) was identified in grape juices. When heated (pH 3), this enindiol yields 3-hydroxy-β-damascone as main and β-damascenone as minor product.
Hydroxylated C_{13}-norisoprenoids (i.e. *Grasshopper ketone*, 3-hydroxy-β-damascone) often occur in plants as glycosides, and can be liberated from these by enzymatic or acid hydrolysis and then transformed into aroma compounds. Therefore the aroma profile changes when fruits are heated during the production of juice or marmalade. An example is the formation of vitispirane (II in Formula 3.142) by hydrolysis of glycosidic bound 3-hydroxy-7,8-dihydro-β-ionol (I) in wine. The odor threshold of vitispirane is relatively high (800 µg/kg, wine) but is clearly exceeded in some port varieties.

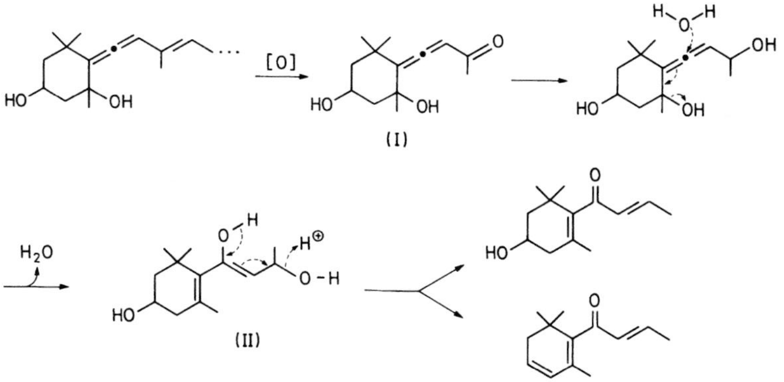

(I)

(II)

(3.141)

(pH3; 100°C)

R: Glycosyl residue

ROH

II

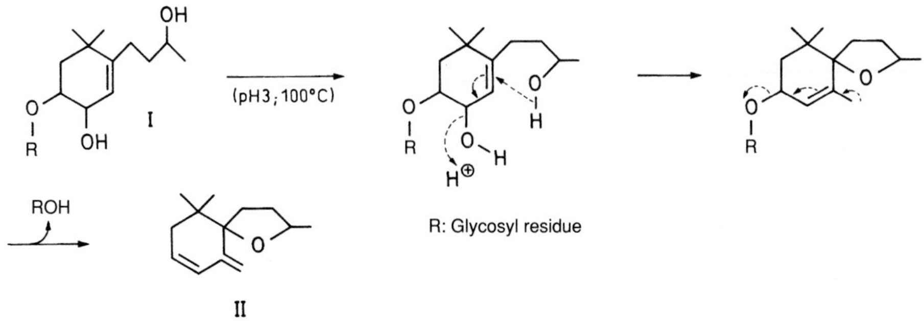

(3.142)

Table 3.56. Aroma compounds formed in oxidative degradation of carotenoids

Precursor[a]	Aroma compound	Odor threshold (µg/l, water)	Occurrence
Lycopene (I)	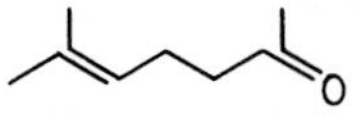 6-Methyl-5-hepten-2-one	50	Tomato
	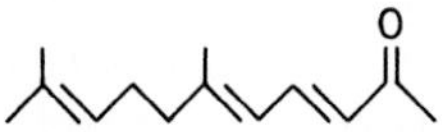 Pseudo ionone	800	Tomato
Dehydrolycopene	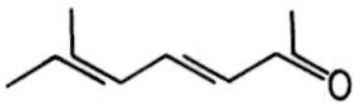 6-Methyl-3,5-heptadien-2-one	380	Tomato
α-Carotene (VI)	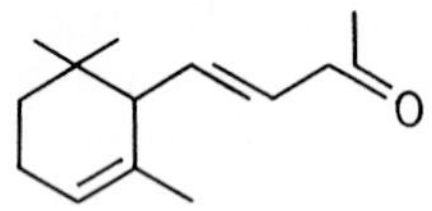 α-Ionone	R(+): 0.5–5 S(−): 20–40	Raspberry, black tea carrots, vanilla
β-Carotene (VII)	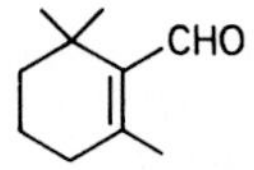 β-Cyclocitral	5	Tomato
	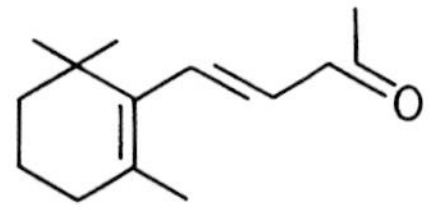 β-Ionone	0.007	Tomato, raspberry, blackberry, passion fruit, black tea
Neoxanthin (XX)	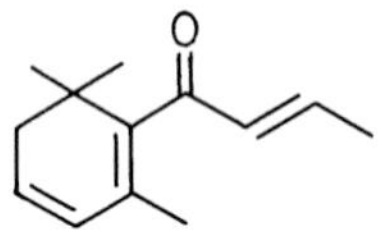 β-Damascenone	0.002	Tomato, coffee, black tea, wine, beer, honey, apple
	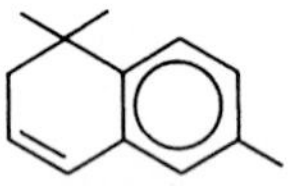 1,2-Dihydro-1,1,6-trimethylnaphthalene	2	Wine, peach, strawberry

[a] Roman numerals refer to the chemical structures presented in 3.8.4.1.

1,2-Dihydro-1,1,6-trimethyl naphthalene (Table 3.56) can be formed by a degradation of neoxanthin and other carotinoids during the storage of wine. It smells like kerosene (threshold 20 µg/kg, wine). It is thought that this odorant contributes considerably to the typical aroma of white wine that was stored for a long period in the bottle. The compound may cause an off-flavor in pasteurized passion fruit juice.

3.8.4.5 Use of Carotenoids in Food Processing

Carotenoids are utilized as food pigments to color margarine, ice creams, various cheese products, beverages, sauces, meat, and confectionery and bakery products. Plant extracts and/or individual compounds are used.

3.8.4.5.1 Plant Extracts

Annato is a yellow oil or aqueous alkaline extract of fruit pulp of *Raku* or *Orleans* shrubs or brushwood (*Bixa orellana*). The major pigments of annato are bixin (XVIII) and norbixin, both of which give dicarboxylic acids upon hydrolysis.

Oleoresin from paprika is a red, oil extract containing about 50 different pigments. The aqueous extract of saffron (more accurately, from the pistils of the flower *Crocus sativus*) contains crocin (XVII) as its main constituent. It is used for coloring beverages and bakery products.

Raw, unrefined palm oil contains 0.05–0.2% carotenoids with α- and β-carotenes, in a ratio of 2:3, as the main constituents. It is of particular use as a colorant for margarine.

3.8.4.5.2 Individual Compounds

β-Carotene (VII), canthaxanthin (XII), β-apo-8′-carotenal (XIX) and the carboxylic acid ethyl ester derived from the latter are synthesized for use as colorants for edible fats and oils. These carotenoids, in combination with surface-active agents, are available as microemulsions (cf. 8.15.1) for coloring foods with a high moisture content.

3.8.4.6 Analysis

The total lipids are first extracted from food with ispropanol/petroleum ether (3:1 v/v) or with acetone. Alkaline hydrolysis follows, removing the extracted acyl-lipids and the carotenoids from the unsaponifiable fraction. This is the usual procedure when alkali-stable carotenoids are analyzed. Although carotenoids are generally alkali stable, there are exceptions. When alkali-labile carotenoids are present, the acyl lipids are removed instead by a saponification method using column chromatography as the separation technique.

A preliminary separation of the lipids into classes of carotenoids is carried out when a complex mixture of carotenoids is present. For example, column chromatography is used with

Table 3.57. Separation of carotenoids into classes by column chromatography using neutral aluminum oxide (6% moisture) as an adsorbent

P: Petroleum ether, D: diethyl ether

Elution with	Carotenoids in effluent
100% P	Carotenes
5% D in P	Carotene-epoxides
20–59% D in P	Monohydroxy-carotenoids
100% D	Dihydroxy-carotenoids
5% Ethanol in D	Dihydroxy-epoxy-carotenoids

Al_2O_3 as an adsorbent (Table 3.57). Additional separation into classes or individual compounds is achieved by thin layer chromatography. Thin layers made of MgO or $ZnCO_3$ are suitable. These adsorbent layers permit separation of carotenoids into classes according to the number, position and configuration of double bonds.

Identification of carotenoids is based on chromatographic data and on electron excitation spectra (cf. 3.8.4.2), supplemented when necessary with tests specific to each group. For example, a hypsochromic effect after addition of $NaBH_4$ suggests the presence of oxo or aldehyde groups, whereas the same effect after addition of HCl suggests the presence of a 5,6-epoxy group. The latter "blue hue shift" is based on a rearrangement reaction:

(5,6-Epoxide)   (5,8-Epoxide)

(3.143)

Such rearrangements can also occur during chromatographic separations of carotenoids on silicic acid. Hence, this adsorbent is a potential source of artifacts.

Epoxy group rearrangement in the carotenoid molecule can also occur during storage of food with a low pH, such as orange juice.

Elucidation of the structure of carotenoids requires, in addition to VIS/UV spectrophotometry, supplemental data from mass spectrometry and IR spectroscopy. Carotenoids are determined photometrically with high sensitivity based on their high molar absorbancy coefficients. This is often used for simultaneous qualitative and quantitative analysis. New separation methods based on high performance liquid chromatography have also proved advantageous for the qualitative and quantitative analysis of carotenoids present as a highly complex mixture in food.

3.9 Literature

Andersson, R. E., Hedlund, C. B., Jonsson, U.: Thermal inactivation of a heat-resistant lipase produced by the psychotrophic bacterium *Pseudomonas fluoreszens*. J. Dairy Sci. *62*, 361 (1979)

Axelrod, B.: Lipoxygenases. In: Food related enzymes (Ed.: Whitaker, J.R.), p. 324, Advances in Chemistry Series 136, American Chemical Society: Washington D.C. 1974

Badings, H.T.: Cold storage defects in butter and their relation to the autoxidation of unsaturated fatty acids. Ned. Melk Zuiveltijdschr. *24*, 147 (1970)

Barnes, P.J.: Lipid composition of wheat germ and wheat germ oil. Fette Seifen Anstrichm. *84*, 256 (1982)

Bergelson, L. D.: Diol lipids. New types of naturally occurring lipid substances. Fette Seifen Anstrichm. *75*, 89 (1973)

Britton, G., Goodwin, T.W.: Biosynthesis of carotenoids. In: Methods in enzymology (Eds.: Colowick, S.P., Kaplan, N.O.), Vol. XVIII, part C, p. 654, Academic Press: New York. 1971

Chan, H.W.-S. (Ed.): Autoxidation of unsaturated lipids. Academic Press: London. 1987

Chang, S.S., Peterson, R.J., Ho, C.-T.: Chemistry of deep fat fried flavor. In: Lipids as a source of flavor (Ed.: Supran, M.K.), p. 18, ACS Symposium Series 75, American Chemical Society: Washington, D.C. 1978

Christie, W.W.: Lipid analysis. 2. Aufl. Pergamon Press: Oxford. 1982

Christie, W.W.: High-performance liquid chromatography and lipids. Pergamon Press: Oxford. 1987

Christopoulou, C.N., Perkins, E.G.: Isolation and characterization of dimers formed in used soybean oil. J. Am. Oil Chem. Soc. *66*, 1360 (1989)

Dennis, E.A.: Phospholipases. In: The enzymes (Ed.: Boyer, P.D.) 3rd edn., Vol. XVI, p. 307, Academic Press: New York. 1983

Eriksson, C.E.: Enzymic and non-enzymic lipid degradation in foods. In: Industrial aspects of biochemistry (Ed.: Spencer, B.), Federation of European Biochemical Societies, Vol. 30, p. 865, North Holland/American Elsevier: Amsterdam. 1974

Fedeli, E.: Lipids of olives. Prog. Chem. Fats Other Lipids *15*, 57 (1977)

Fiebig, H.-J.: HPLC-Trennung von Triglyceriden. Fette Seifen Anstrichm. *87*, 53 (1985)

Foote, C.S.: Photosensitized oxidation and singlet oxygen: Consequences in biological systems. In: Free radicals in biology (Ed.: Pryor, W.A.), Vol. II, p. 85, Academic Press: New York. 1976

Frankel, E.N.: Recent advances in lipid oxidation. J. Sci. Food Agric. *54*, 495 (1991)

Galliard, T., Mercer, E.I. (Eds.): Recent advances in the chemistry and biochemistry of plant lipids. Academic Press: London. 1975

Galliard, T., Phillips, D.R., Reynolds, J.: The formation of cis-3-nonenal, trans-2-nonenal and hexanal from linoleic acid hydroperoxide isomers by a hydroperoxide cleavage enzyme system in cucumber (*cucumis sativus*) fruits. Biochim. Biophys. Acta *441*, 181 (1976)

Gardner, H.W.: Lipid hydroperoxide reactivity with proteins and amino acids: A review. J. Agric. Food Chem. *27*, 220 (1979)

Garti, N., Sato, K.: Crystallization and polymorphism of fats and fatty acids. Marcel Dekker: New York. 1988

Gertz, C., Herrmann, K.: Zur Analytik der Tocopherole und Tocotrienole in Lebensmitteln. Z. Lebensm. Unters. Forsch. *174*, 390 (1982)

Glass, R.L.: Alcoholysis, saponification and the preparation of fatty acid methyl esters. Lipids *6*, 919 (1971)

Gottstein, T., Grosch, W.: Model study of different antioxidant properties of α- and γ-tocopherol in fats. Fat Sci. Technol. *92*, 139 (1990)

Grosch, W.: Lipid degradation products and flavours. In: Food flavours (Eds.: Morton, I.D., MacLeod, A. J.), Part A, p. 325, Elsevier Scientific Publ. Co.: Amsterdam. 1982

Grosch, W.: Reaction of hydroperoxides – Products of low molecular weight. In: Autoxidation of unsaturated lipids (Ed.: H.W.-S. Chan), p. 95, Academic Press, London, 1987

Gunstone, F.D., Norris, F.A.: Lipids in foods. Pergamon Press: Oxford. 1983

Gunstone, F.D., Harwood, J.L., Padley, F.B.: The lipid handbook. Chapman and Hall: London. 1986

Guth, H., Grosch, W.: Detection of furanoid fatty acids in soya-bean oil. – Cause for the light-induced off-flavour. Fat Sci. Technol. *93*, 249 (1991)

Guth, H., Grosch, W.: Deterioration of soya-bean oil: Quantification of primary flavour compounds using a stable isotope dilution assay. Lebensm. Wiss. u. Technol. *23*, 513 (1990)

Hamilton, R.J., Bhati, A.: Recent advances in chemistry and technology of fats and oils. Elsevier Applied Science: London. 1987

Haslbeck, F., Senser, F., Grosch, W.: Nachweis niedriger Lipase-Aktivitäten in Lebensmitteln. Z. Lebensm. Unters. Forsch. *181*, 271 (1985)

Hitchcock, C., Nickols, B.W.: Plant lipid biochemistry. Academic Press: London. 1971

Homberg, E., Bielefeld, B.: Sterine und Methylsterine in Kakaobutter und Kakaobutter-Ersatzfetten. Dtsch. Lebensm. Rundsch. *78*, 73 (1982)

Hudson, B.J.F. (Eds.): Food antioxidations. Elsevier Applied Science: London. 1990

Isler, O. (Ed.): Carotenoids. Birkhäuser Verlag: Basel. 1971

Jeong, T.M., Itoh, T., Tamura, T., Matsumoto, T.: Analysis of methylsterol fractions from twenty vegetable oils. Lipids *10*, 634 (1975)

Johnson, A.R., Davenport, J.B.: Biochemistry and methodology of lipids. John Wiley and Sons: New York. 1971

Kindl, H., Wöber, G.: Biochemie der Pflanzen. Springer-Verlag: Berlin. 1975

Kinsella, J.E., Hwang, D.H.: Enzyme *penicillium roqueforti* involved in the biosynthesis of cheese flavour. Crit. Rev. Sci. Nutr. *8*, 191 (1976)

Kinsella, J.E.: Food lipids and fatty acids: Importance in food quality, nutrition, and health. Food Technol. *42* (10), 124 (1988)

Kinsella, J.E.: Seafoods and fish oils in human health and disease. Marcel Dekker: New York. 1987

Korycka-Dahl, M.B., Richardson, T.: Active oxygen species and oxidation of food constituents. Crit. Rev. Food Sci. Nutr. *10*, 209 (1978)

Labuza, T.P.: Kinetics of lipid oxidation in foods. Crit. Rev. Food Technol. *2*, 355 (1971)

Lillard, D.A., Day, E.A.: Degradation of monocarbonyls from autoxidizing lipids. J. Am. Chem. Soc. *41*, 549 (1964)

Litchfield, C.: Analysis of triglycerides. Academic Press: New York. 1972

Mahadevan, V.: Fatty alcohols: Chemistry metabolism. Prog. Chem. Fats Other Lipids *15*, 255 (1977)

Mead, J.F., Howton, D.R., Nevenzel, J.: Fatty acids, long-chain alcohols and waxes. In: Comprehensive biochemistry (Eds.: Florkin, M., Stotz, E.H.), Vol. 6, p. 1, Elsevier Publ. Co.: Amsterdam. 1965

Meijboom, P.W., Jongenotter, G.A.: Flavor perceptibility of straight chain, unsaturated aldehydes as a function of double-bond position and geometry. J. Am. Oil Chem. Soc. 680 (1981)

Min. D.B., Smouse, T.H. (Eds.): Flavor chemistry of lipid foods. American Oil Chemists' Society: Champaign. 1989

Nawar, W.W., Bradley, S.J., Lamanno, S.S., Richardson, G.G., Whiteman, R.C.: Volatiles from frying fats: A comparative study. In: Lipids as a source of flavor (Ed.: Supran, K.), p. 42, ACS Symposium Series 75, American Chemical Society: Washington, D.C. 1978

Nawar, W.W.: Volatiles from food irradiation. Food Rev. Internat. *2*, 45 (1986)

Ohloff, G.: Recent developments in the field of naturally-occurring aroma components. Fortschr. Chem. Org. Naturst. *35*, 431 (1978)

Pardun, H.: Die Pflanzenlecithine. Verlag für chemische Industrie H. Ziolkowsky KG: Augsburg. 1988

Perkins, E.G. (Ed.): Analysis of lipids and lipoproteins. American Oil Chemists' Society: Champaign, Ill. 1975

Philip, T., Francis, F.J.: Oxidation of capsanthin. J. Food Sci. *36*, 96 (1971)

Podlaha, O., Töregard, B., Püschl, B.: TG-type composition of 28 cocoa butters and correlation between some of the TG-type components. Lebensm. Wiss. Technol. *17*, 77 (1984)

Porter, N.A., Lehman, L.S., Weber, B.A., Smith, K.J.: Unified mechanism for polyunsaturated fatty acid autoxidation. Competition of peroxy radical hydrogen atom abstraction, β-scission, and cyclization. J. Am. Chem. Soc. *103*, 6447 (1981)

Pryde, E.H. (Ed.): Fatty acids. American Oil Chemists' Society: Champaign, Ill. 1979

Pryor, W.A., Stanley, J.P., Blair, E.: Autoxidation of polyunsaturated fatty acids. II. A suggested mechanism for the formation of TBA-reactive materials from prostaglandin – like endoperoxides. Lipids *11*, 370 (1976)

Rojo, J.A., Perkins, E.G.: Cyclic fatty acid monomer formation in frying fats. I. Determination and structural study. J. Am. Oil Chem. Soc. *64*, 414 (1987)

Seidel, D.: Plasmalipoproteine – Funktion und Charakterisierung. In: Fettstoffwechselstörungen (Hrsg.: Schettler, G.), S. 24, Georg Thieme Verlag: Stuttgart. 1971

Sherwin, E.R.: Oxidation and antioxidants in fat and oil processing. J. Am. Oil Chem. Soc. *55*, 809 (1978)

Simic, M.G., Karel, M. (Eds.): Autoxidation in food and biological systems. Plenum Press: New York. 1980

Slower, H.L.: Tocopherols in foods and fats. Lipids *6*, 291 (1971)

Smith, L.L.: Cholesterol autoxidation. Plenum Press: New York. 1981

Sotirhos, N., Ho, C.-T., Chang, S.S.: High performance liquid chromatographic analysis of soybean phospholipids. Fette Seifen Anstrichm. *88*, 6 (1986)

Sven, D.J., Rackis, J.J.: Lipid – derived flavors of legume protein products. J. Am. Oil Chem. Soc. *54*, 468 (1977)

Szuhaj, B.F., List, G.R.: Lecithins. American Oil Chemists' Society: Champaign, 1985

Thiele, O.W.: Lipide, Isoprenoide mit Steroiden. Georg Thieme Verlag: Stuttgart. 1979

Van Niekerk, P.J., Burger, A.E.C.: The estimation of the composition of edible oil mixtures. J. Am. Oil Chem. Soc. *62*, 531 (1985)

Veldink, G.A., Vliegenthart, J.F.G., Boldingh, J.: Plant lipoxygenases. Prog. Chem. Fats Other Lipids *15*, 131 (1977)

Werkhoff, P., Bretschneider, W., Güntert, M., Hopp, R., Surburg, H.: Chirospecific analysis in flavor and essential oil chemistry. Part B. Direct enantiomer resolution of trans-α-ionone and trans-α-damascone by inclusion gas chromatography. Z. Lebensm. Unters. Forsch. *192*, 111 (1991)

Winterhalter, P., Schreier, P.: Natural precursors of thermally induced C_{13} norisoprenoids in quince. In: Thermal generation of aromas (Eds.: T.H. Parliment, R.J. McGorrin, C.-T. Ho) p. 320, ACS Symposium Series 409, American Chemical Society: Washington D.C. 1989

Wolfram, G.: ω-3- und ω-6-Fettsäuren – Biochemische Besonderheiten und biologische Wirkungen. Fat Sci. Technol. *91*,459 (1989)

Woo, A.H., Lindsay, R.C.: Statistical correlation of quantitative flavor intensity assessments and individual free fatty acid measurements for routine detection and prediction of hydrolytic rancidity off-flavours in butter. J. Food Sci. *48*, 1761 (1983)

4 Carbohydrates

4.1 Foreword

Carbohydrates are the most widely distributed and abundant organic compounds on earth. They have a central role in the metabolisms of animals and plants. Carbohydrate biosynthesis by green plants from carbon dioxide and water in the presence of light energy, i.e. photosynthesis, supports the existence of all other organisms.

Carbohydrates are a basic food, accounting for a large portion of total nutrient intake. Even the nondigestible carbohydrates, acting as ballast material, are of importance in a balanced daily nutrition. Other important functions in food are fulfilled by carbohydrates. They act for instance as sweetening, gel- or paste-forming and thickening agents, stabilizers and are also precursors for aroma and coloring substances generated within the food by a series of reactions during handling and processing.

The term carbohydrates goes back to times when it was thought that all compounds of this class were hydrates of carbon, on the basis of their empirical formula, e.g. glucose, $C_6H_{12}O_6$ ($6C + 6H_2O$). Meanwhile, many compounds were identified which deviated from this general formula, but retained common reactions and, hence, were also classed as carbohydrates. These are exemplified by deoxysugars, amino sugars and sugar carboxylic acids. Carbohydrates are commonly divided into monosaccharides, oligosaccharides and polysaccharides. Monosaccharides are polyhydroxy-aldehydes or -ketones, generally with an unbranched C-chain. Well known representatives are glucose, fructose and galactose. Oligosaccharides are carbohydrates which are obtained from monosaccharides by elimination of water, e.g. by the reaction:

$$\text{n } C_6H_{12}O_6 \xrightarrow{-(n-1) H_2O} C_{6n}H_{10n+2}O_{5n+1} \quad (4.1)$$

Well known representatives of disaccharides are saccharose (sucrose), maltose and lactose, and of trisaccharides, raffinose, and tetrasaccharides, stachyose.

In polysaccharides, consisting of n monosaccharides, the number n is rather large. Hence, the properties of these high molecular weight polymers differ greatly from other carbohydrates. Unlike mono- or oligosaccharides, polysaccharides are in many cases insoluble or at best not readily soluble in water. They do not have a sweet taste and are essentially inert. Well known representatives are starch, cellulose and pectin.

4.2 Monosaccharides

4.2.1 Structure and Nomenclature

4.2.1.1 Nomenclature

Monosaccharides are polyhydroxy-aldehydes (aldoses), formally considered to be derived from glyceraldehyde, or polyhydroxyketones (ketoses), derived from dihydroxyacetone by inserting CHOH units into the carbon chains. The resultant compounds in the series of aldoses are denoted by the total number of carbons as trioses, for the starting glyceraldehyde, and tetroses, pentoses, hexoses, etc. The ketose series begins with the simplest ketose, dihydroxyacetone, a triulose, followed by tetruloses, pentuloses, hexuloses, etc. The position of the keto group is designated by a numerical prefix, e.g. 2-pentulose, 3-hexulose.

When a monosaccharide carries a second carbonyl group, it is denoted as a −dialdose (2 aldehyde groups), -osulose (aldehyde and keto groups) or −diulose (2 keto groups). Substitution of an HO-group by an H-atom gives rise to a deoxy sugar, and by an H_2N-group to aminodeoxy compounds (cf. Formula 4.2).

All monosaccharides starting with tetroses and 2-pentuloses cyclize to five- and six-membered lactols respectively by intramolecular

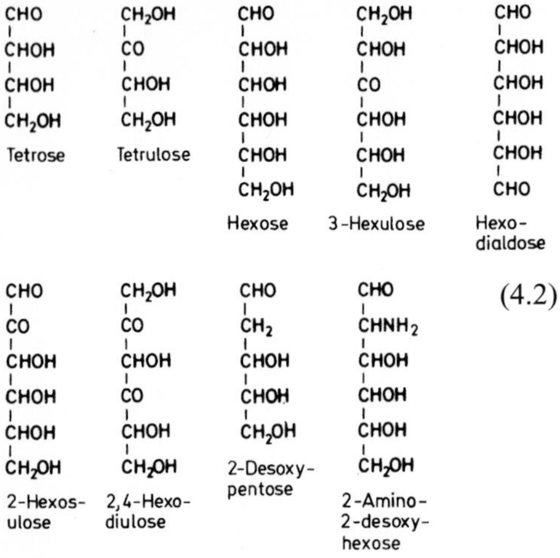

CHO	CH₂OH	CHO	CH₂OH	CHO
CHOH	CO	CHOH	CHOH	CHOH
CHOH	CHOH	CHOH	CO	CHOH
CH₂OH	CH₂OH	CHOH	CHOH	CHOH
Tetrose	Tetrulose	CHOH	CHOH	CHOH
		CH₂OH	CH₂OH	CHO
		Hexose	3-Hexulose	Hexo-dialdose

(4.2)

CHO	CH₂OH	CHO	CHO
CO	CO	CH₂	CHNH₂
CHOH	CHOH	CHOH	CHOH
CHOH	CO	CHOH	CHOH
CHOH	CHOH	CH₂OH	CHOH
CH₂OH	CH₂OH	2-Desoxy-pentose	CH₂OH
2-Hexos-ulose	2,4-Hexo-diulose		2-Amino-2-desoxy-hexose

hemiacetal formation. With the exception of erythrose, monosaccharides are crystallized in these cyclic forms and, even in solution, there is an equilibrium between the open chain carbonyl form and cyclic hemiacetals, with the latter predominating. The tendency to cyclize is pronounced in simple ω-hydroxyaldehydes, and is even more pronounced in monosaccharides, as shown by $\Delta G°$-values and equilibrium concentrations in 75% aqueous ethanol (cf. Formula 4.3).

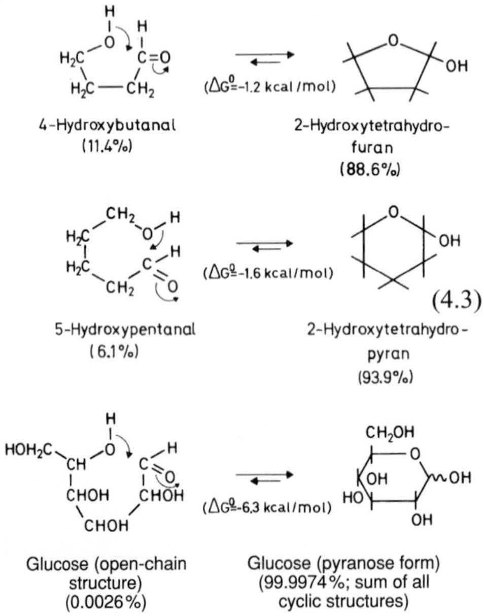

(4.3)

Lactols can be considered as tetrahydrofuran or tetrahydropyran derivatives, hence, they are also denoted as furanoses or pyranoses.

4.2.1.1 Configuration

Glyceraldehyde, a triose, has one chiral center, so it exists as an enantiomer pair, i.e. in D- and L-forms. It is possible by cyanhydrin synthesis to obtain from each enantiomer a pair of diastereomeric tetroses (with a new chiral center at carbon position 2):

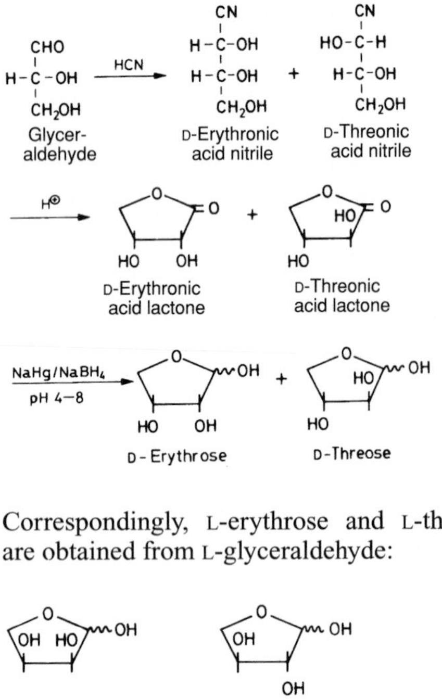

(4.4)

Correspondingly, L-erythrose and L-threose are obtained from L-glyceraldehyde:

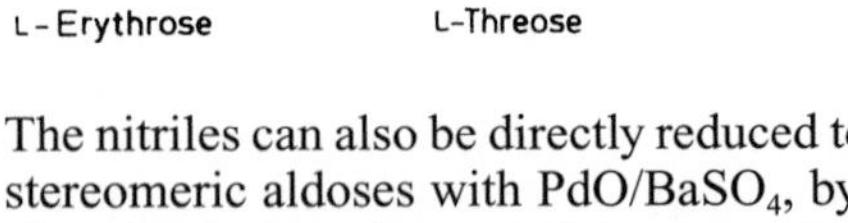

(4.5)

The nitriles can also be directly reduced to diastereomeric aldoses with PdO/BaSO₄, by passing the lactone intermediate stage. Another reaction for the formation of monosaccharides is the nitroalkane synthesis. The epimeric nitro compounds, obtained by the reaction of an aldose with nitromethane as anions, are separated and converted to the corresponding aldoses by an acinitroalkane cleavage (Nef-reaction):

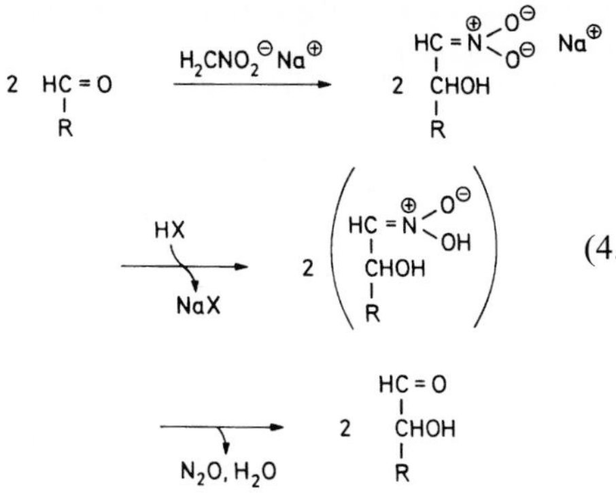

$$2 \ \underset{R}{\underset{|}{HC}} = O \quad \xrightarrow{H_2CNO_2^{\ominus} Na^{\oplus}} \quad 2 \ \underset{R}{\overset{HC \ = \ N \nearrow O^{\oplus}_{\searrow O^{\ominus}}}{\underset{|}{\overset{|}{CHOH}}}} \ Na^{\oplus}$$

$$\xrightarrow[NaX]{HX} \quad 2 \left(\underset{R}{\overset{HC \ = \ N \nearrow O^{\oplus}_{\searrow OH}}{\underset{|}{\overset{|}{CHOH}}}} \right) \qquad (4.6)$$

$$\xrightarrow[N_2O, \ H_2O]{} \quad 2 \ \underset{R}{\overset{HC \ = \ O}{\underset{|}{\overset{|}{CHOH}}}}$$

After repeated cyanhydrin reactions, four tetroses will provide a total of eight pentoses (each tetrose provides a pair of new diastereomers with one more chiral center), which can then yield sixteen stereoisomeric hexoses. The

compounds derived from D-glyceraldehyde are designated as D-aldoses and those from L-glyceraldehyde as L-aldoses.

An important degradation reaction of aldoses proceeds via the disulfone of the dithioacetal:

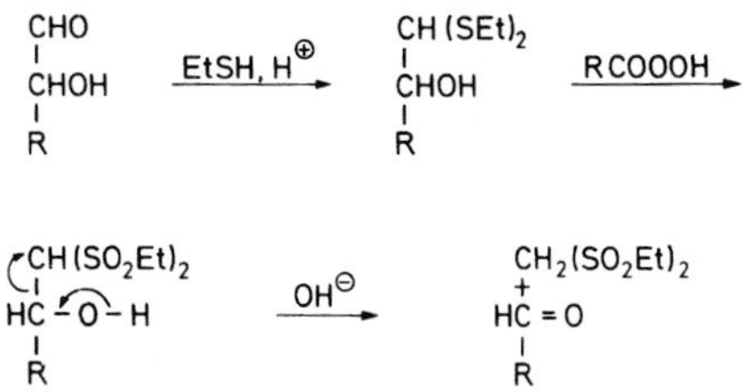

$$\underset{R}{\overset{CHO}{\underset{|}{\overset{|}{CHOH}}}} \quad \xrightarrow{EtSH, \ H^{\oplus}} \quad \underset{R}{\overset{CH(SEt)_2}{\underset{|}{\overset{|}{CHOH}}}} \quad \xrightarrow{RCOOOH}$$

$$(4.7)$$

$$\underset{R}{\overset{CH(SO_2Et)_2}{\underset{|}{\overset{|}{HC \ \overset{\frown}{} O^{-} H}}}} \quad \xrightarrow{OH^{\ominus}} \quad \underset{R}{\overset{CH_2(SO_2Et)_2}{+}{\underset{|}{\overset{|}{HC \ = \ O}}}}$$

Figure 4.1 shows the formulas and names for D-aldoses using simplified *Fischer* projections. The occurrence of aldoses of importance in food is compiled in Table 4.1. Epimers are

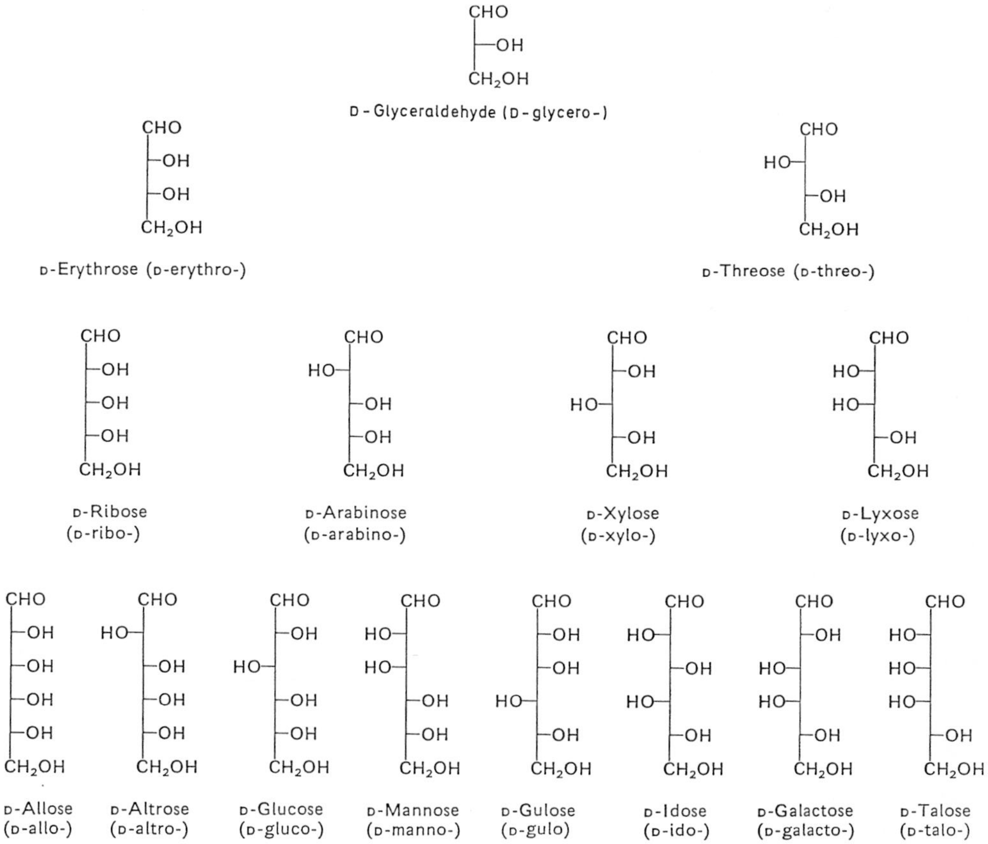

Fig. 4.1. D-Aldoses in *Fischer* projection

monosaccharides which differ in configuration at only one chiral C-atom. D-Glucose and D-mannose are 2-epimers. D-glucose and D-galactose are 4-epimers.

Table 4.1. Occurrence of aldoses

Name, structure	Occurrence
Pentoses	
D-Apiose (3-C-Hydroxy-methyl-D-glycero-tetrose)	Parsley, celery seed
L-Arabinose	Plant gums, hemicelluloses, pectins, glycosides
2-Deoxy-D-ribose	Deoxyribonucleic acid
D-Lyxose	Yeast-nucleic acid
2-O-Methyl-D-xylose	Hemicelluloses
D-Ribose	Ribonucleic acid
D-Xylose	Xylanes, hemicelluloses plant gums, glycosides
Hexoses	
L-Fucose (6-Deoxy-L-galactose)	Human milk, seaweed (algae), plant gums and mucilage
D-Galactose	Widespread in oligo- and poly-saccharides
D-Glucose	Widespread in plants and animals
D-Mannose	Widespread as polysaccharide building blocks
L-Rhamnose (6-Deoxy-L-mannose)	Plant gums and mucilage, glycosides

The enantiomers D- and L-tetrulose, by formally inserting additional CHOH-groups between the keto and existing CHOH-groups, form a series of D- and L-2-ketoses. Figure 4.2 gives D-2-ketoses in their simplified *Fischer* projections.

Data are provided in Table 4.2 on the occurrence of ketoses of interest in food.

For simplified presentation of oligo- and polysaccharide structures, abbreviations are used which, as with amino acids, are of three characters. Usually, the first three letters are from the name of the monosaccharide. Figure 4.1 gives the configuration prefix derived from the trivial names, representing a specified configuration applied in monosaccharide classification. Thus, systematic names for D-glucose and D-fructose are D-gluco-hexose and D-arabino-2-hexulose. Such nomenclature makes it possible to systematically denote all monosaccharides that contain more than four chiral centers. According to this procedure the portion of the molecule adjacent to the carbonyl group is given the maximal possible prefix, while the portion furthest from the carbonyl group is denoted first. In the case of ketoses, the two

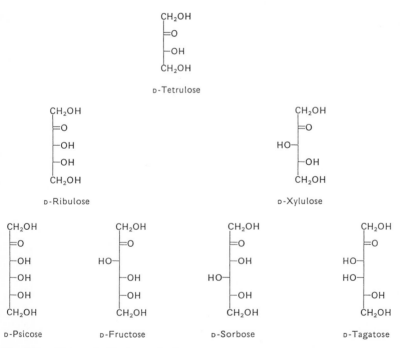

Fig. 4.2. D-Ketoses in *Fischer* projection

Table 4.2. Occurrence of ketoses

Name, structure	Occurrence
Hexulose	
D-Fructose	Present in plants and honey
D-Psicose	Found in residue of fermented molasses
Heptulose	
D-manno-2-Heptulose	Avocado fruit
Octulose	
D-glycero-D-manno-2-Octulose	Avocado fruit
Nonulose	
D-erythro-L-gluco-2-Nonulose	Avocado fruit

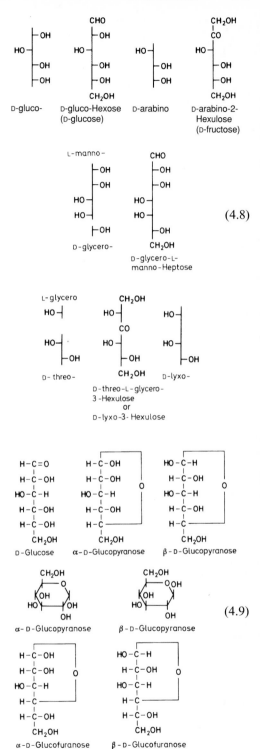

(4.8)

(4.9)

portions of the molecule separated by the keto group are given. In a combined prefix designation, as with aldoses, the portion which has the C-atom furthest from the keto group is mentioned first. However, when a monosaccharide does not have more than four chiral centers, a designation in the ketose series may omit the two units separated by the keto group. The examples in Formula 4.8 illustrate the rule.

Hemiacetal or lactol formation provides a new chiral center on C-1. Thus, there are two additional diastereomers for each pyranose and furanose. These isomers are called anomers and are denoted as α and β-forms. Formula 4.9 illustrates the two anomeric D-glucose molecules in both *Fischer* and *Haworth* projections.

Based on the cis-arrangement of the two adjacent HO-groups in positions C-1 and C-2, α-D-glucopyranose, unlike its β-anomer, increases the conductivity of boric acid. A borate complex is formed which is a stronger acid than boric acid itself (cf. Formula 4.10).

In *Haworth* projections the HO-group of α/β-anomers of the D-series usually occurs below/above the pyranose or furanose ring planes, while in the L-series the reverse is true (cf. Formula 4.11).

Each monosaccharide can exist in solution together with its open chain molecule in a total of five forms. Due to the strong tendency for

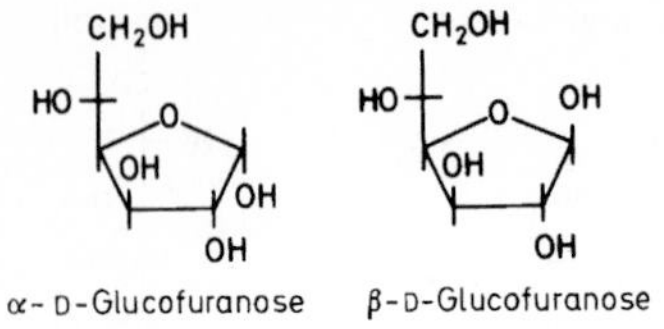

α-D-Glucofuranose   β-D-Glucofuranose

(4.9 cont.)

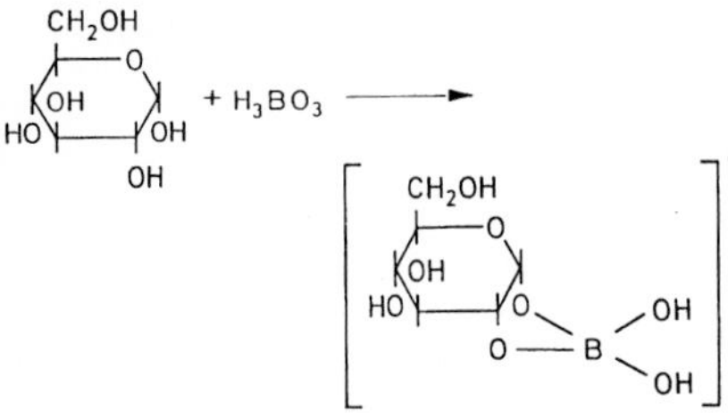

(4.10)

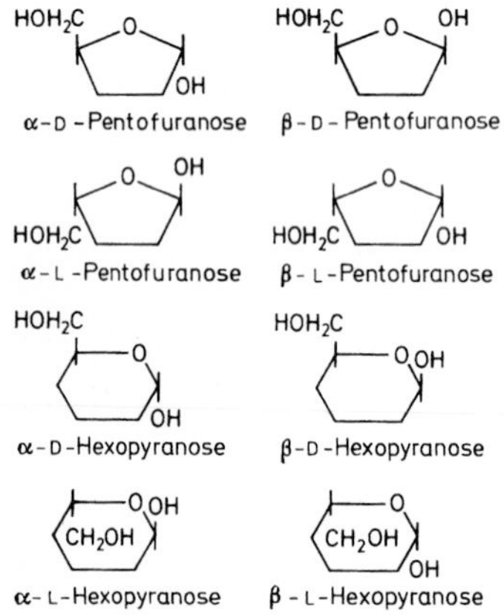

α-D-Pentofuranose   β-D-Pentofuranose

α-L-Pentofuranose   β-L-Pentofuranose

α-D-Hexopyranose   β-D-Hexopyranose

(4.11)

α-L-Hexopyranose   β-L-Hexopyranose

an open chain molecule to be in a cyclic form, the amount of the open chain form is negligible. The contribution of the different cyclic forms to the equilibrium state in a solution depends on the configuration pattern of the sugar. An aqueous D-glucose solution is nearly exclusively the two pyranoses, with 36 % α- and 64 % β-anomer, while the furanose ring form is less than 1 %. The equilibrium state varies greatly among sugars (cf. Table 4.6).

The transition into various forms occurs through the open-chain carbonyl compound. The acid- and base-catalyzed ring opening is the rate limiting step of the reaction:

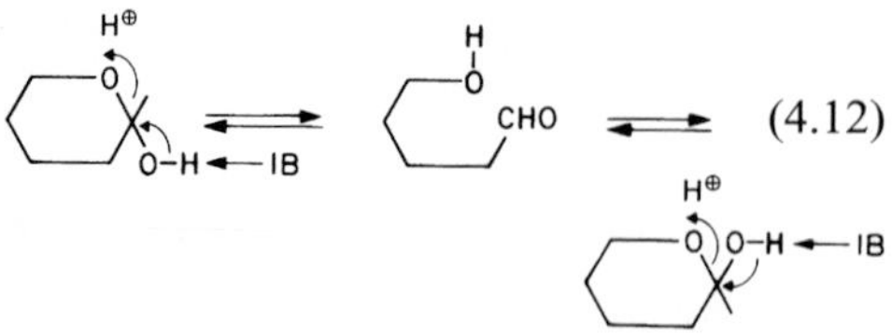

(4.12)

Table 4.3. Mutarotation rate of 2,3,4,6-tetramethyl-D-glucose (0.09 mol/l) in benzene

Catalyst	k (min^{-1})	k$_{rel}$
–	7.8×10^{-5}	1.0
Pyridine (0.1 mol/l)	3.7×10^{-4}	4.7
p-Cresol (0.1 mol/l)	4.2×10^{-4}	5.4
Pyridine + p-cresol (0.1 mol/l)	7.9×10^{-3}	101
2-Pyridone (0.1 mol/l)	1.8×10^{-1}	2,307
Benzoic acid (0.1 mol/l)	2.2	28,205

2,3,4,6-Tetramethyl-D-glucose reaches equilibrium in benzene rapidly through the concerted action of cresol and pyridine as acid-base catalysts (Table 4.3). A bifunctional reagent, 2-pyridone, can also be efficient as an acid-base catalyst in both polar and nonpolar solvents:

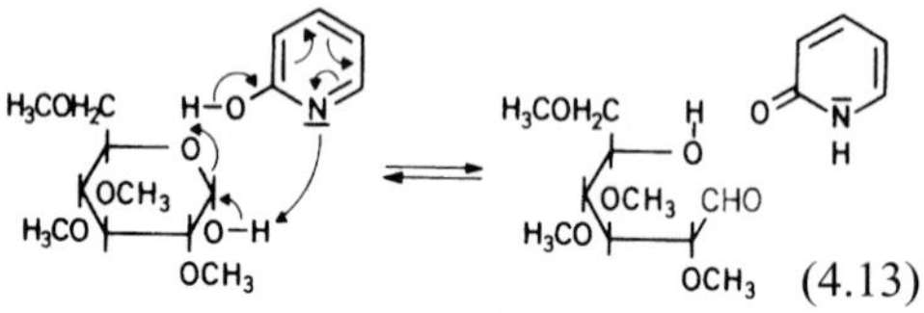

(4.13)

Water can also be a bifunctional catalyst:

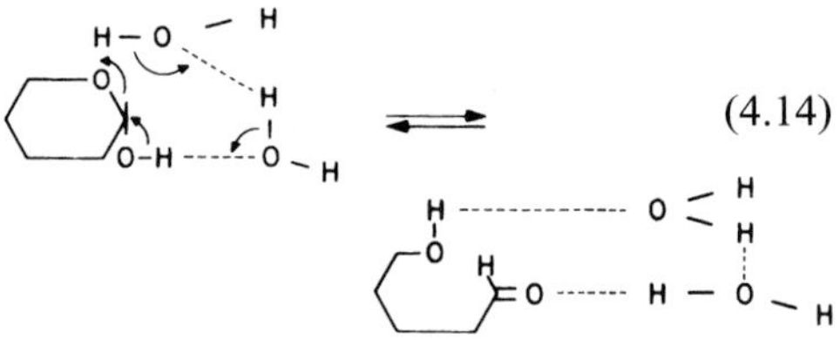

(4.14)

The reaction rate for the conversion of the α- and β-forms (so-called mutarotation) has a wide minimum in an aqueous medium and in a pH range of 2–7, as illustrated in section 10.1.2.2 with lactose, and the rate increases beyond this pH range.

4.2.1.3 Conformation

The preferred conformation for a pyranose is the so-called chair conformation and not the twisted-boat conformation, since the former has the highest thermodynamic stability. The two chair C-conformations are $^4C_1$ (the superscript corresponds to the number of the C-atom in the upper position of the chair and the subscript to that in the lower position; often designated as C1 or "O-outside") and $^1C_4$ (often designated as 1C, the mirror image of C1, and C-1 in upper and C-4 in lower positions, or simply the "O-inside" conformer). The $^4C_1$-conformation is preferred in the series of D-pyranoses, with most of the bulky groups, e.g., HO and, especially, CH_2OH, occupying the roomy equatorial positions. The interaction of the bulky groups is low in such a conformation, hence the conformational stability is high. This differs from the $^1C_4$-conformation, in which most of the bulky groups are crowded into axial positions, thus imparting a thermodynamic instability to the molecule. β-D-Glucopyranose in $^4C_1$-conformation is an exception. All substituents are arranged equatorially, while in $^1C_4$ all are axial. α-D-Glucopyranose in $^4C_1$-conformation has one axial group and is also thermodynamically more stable than $^1C_4$ (cf. Table 4.5):

β-D-Glucose ($^4C_1$)   α-D-Glucose ($^4C_1$)

β-D-Glucose ($^1C_4$)   α-D-Glucose ($^1C_4$)

(4.15)

The arrangement of substituents differs in α-D-idopyranose. Here, all the substituents are in axial positions in the $^4C_1$-conformation (axial HO-groups at 1, 2, 3, 4), except for the CH_2OH-group, which is equatorial. However, the $^1C_4$-conformation is thermodynamically more favorable despite the fact that the CH_2OH-group is axial (cf. Table 4.5):

α-D-Idose ($^4C_1$)   α-D-Idose ($^1C_4$)

(4.16)

A second exception (or rather an extreme case) is α-D-altropyranose. Both conformations (O-outside and O-inside) have practic-

ally the same stability in this sugar (cf. Table 4.5).

The free energy of the conformers in the pyranose series can be calculated from partial interaction energies (derived from empirical data). Only the 1,3-diaxial interactions (with exception of the interactions between H-atoms), 1,2-gauche or staggered (60°) interactions of two HO-groups and that between HO-groups and the CH_2OH-group will be considered. The partial interaction energies are compiled in Table 4.4 and conformational free energies, calculated from these data for various conformers, are presented in Table 4.5. In addition to the interaction energies an effect is considered which destabilizes the anomeric HO-group in equatorial position, while it stabilizes this group in axial position. This is called the *anomeric effect*. The effect is attributed to repulsion forces between the parallel dipoles resulting from the polarized bonds C5−O5 and C1−O1 (equatorial). The repulsion forces the anomeric HO-group to take up the more stable axial or α-position:

(4.17)

Table 4.4. Free energies of unfavorable interactions between substituents on the tetrahydropyran ring

Interaction	Energy kJ/mole[a]
$H_{ax} - O_{ax}$	1.88
$H_{ax} - C_{ax}$	3.76
$O_{ax} - O_{ax}$	6.27
$O_{ax} - C_{ax}$	10.45
$O_{eq} - O_{eq}/O_{ax} - O_{eq}$	1.46
$O_{eq} - C_{eq}/O_{ax} - C_{eq}$	1.88
Anomeric effect[b]	
for O_{eq}^{c2}	2.30
for O_{ax}^{c2}	4.18

[a] Aqueous solution, room temperature.
[b] To be considered only for an equatorial position of the anomeric HO-group.

Table 4.5. Calculated conformational free energies ($E_{conf.}$) for hexopyranoses

Hexopyranose	Conformation	$E_{conf.}$ (kJ/mole)
α-D-Glucose	$^4C_1$	10.03
	$^1C_4$	27.38
β-D-Glucose	$^4C_1$	8.57
	$^1C_4$	33.44
α-D-Mannose	$^4C_1$	10.45
β-D-Mannose	$^4C_1$	12.33
α-D-Galactose	$^4C_1$	11.91
β-D-Galactose	$^4C_1$	10.45
α-D-Idose	$^4C_1$	18.18
	$^1C_4$	16.09
β-D-Idose	$^4C_1$	16.93
	$^1C_4$	22.36
α-D-Altrose	$^4C_1$	15.26
	$^1C_4$	16.09

The other substituents influence the anomeric effect, particularly the HO-group in C-2 position. Here, due to an antiparallel dipole formation, the axial position enhances stabilization better than the equatorial position. Correspondingly, in an equilibrium state in water, D-mannose is 67% in its α-form, while α-D-glucose or α-D-galactose are only 36% and 32%, respectively (Table 4.6). The anomeric effect (dipole-dipole interaction) increases as the dielectric constant of the solvent system decreases e.g., when water is replaced by methanol.
Alkylation of the lactol HO-group also enhances the anomeric effect (Table 4.7).
Conformational isomers of furanose also occur since its ring is not planar. There are two basic conformations, the envelope (E) and the twist (T), which are the most stable; in solution

a mixture exists of conformers similar in energy (cf. Formula 4.18).

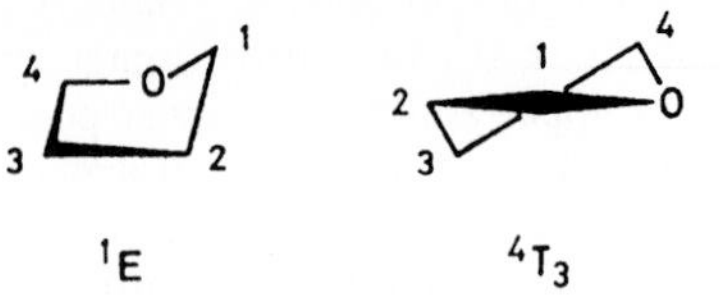

^1E $^4T_3$ (4.18)

An anomeric effect preferentially forces the anomeric HO-group into the axial position. This is especially the case when the HO-group attached to C-2 is also axial. When pyranose ring formation is prevented or blocked, as in 5-O-methyl-D-glucose, the twisted $^3T_2$-conformer becomes the dominant form:

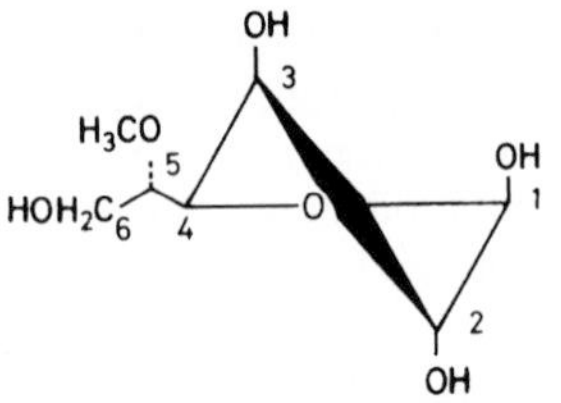

(4.19)

5-O-Methyl-β-D-glucose ($^3T_2$)

A pyranose is generally more stable than a furanose, hence, the former and not the latter conformation is predominant in most monosaccharides (Table 4.6).
The composition of isomers in aqueous solution, after equilibrium is reached, is compiled for a number of monosaccharides in Table 4.6. Evidence for such compositions is obtained by polarimetry, by oxidation with bromine, which occurs at a higher reaction rate with β- than α-pyranose and, above all, by nuclear magnetic resonance spectroscopy.

Table 4.6. Equilibrium composition[a] of aldoses and ketoses in aqueous solution

Compound	T (°C)	α-Pyranose	β-Pyranose	α-Furanose	β-Furanose
D-Glucose	20	36	64	–	–
D-Mannose	20	67	33	–	–
D-Galactose	20	32	64	1	3
D-Idose	60	31	37	16	16
D-Ribose	40	20	56	6	18
D-Xylose	20	35	65	–	–
D-Fructose	20	–	76	4	20

[a] Values in %.

Table 4.7. Methylglucoside isomers in methanol (1% HCl) at equilibrium state[a]

Compound	α-Pyrano-side	β-Pyrano-side	α-Furano-side	β-Furano-side
Methyl-D-glucoside	66	32.5	0.6	0.9
Methyl-D-mannoside	94	5.3	0.7	0
Methyl-D-galactoside	58	20	6	16
Methyl-D-xyloside	65	30	2	3

[a] Values in %.

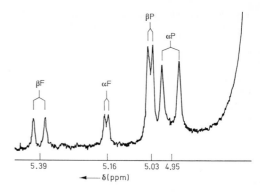

Fig. 4.3. ^{1}H-NMR spectrum of D-idose (according to *Angyal*, 1969)

In proton magnetic resonance (^{1}H-NMR) spectroscopy of sugars, the hydrogen atoms bound to oxygen, which complicate the spectrum, are replaced by derivatization (O-acetyl derivatives) or are exchanged for deuterium when the sugar is in D_2O solution. The chemical shift of the retained H-atoms bound covalently to carbon is distinct. Due to less shielding by the two oxygens in α position, protons on anomeric carbon atoms appear at a lower magnetic field than other protons, the chemical shift increasing in the order pyranoses < furanoses in the range of $\delta = 4.5-5.5$ (free monosaccharides). They are also the only protons that appear as doublets because they couple with H-2 only. Furthermore, an axial proton (β-form of D-series) appears at higher field than an equatorial proton (α-form of D-series). The sugar conformation is elucidated by coupling effects of neighboring protons: equatorial-equatorial, equatorial-axial (small coupling constants) or axial-axial (larger coupling constants).

Figure 4.3 shows the ^{1}H-NMR spectrum for D-idose, beginning with the pyranose (P) and followed by the furanose (F) ring conformations, in increasing δ-value order. The high coupling constant recorded at the lowest δ-value indicates the presence of a diaxial H-atom arrangement on C-1 and C-2, which is incompatible with either β-pyranose conformer but is compatible with the α-pyranose $^{1}C_4$-conformation. Hence, the second signal belongs to β-pyranose which, in both $^{1}C_4$- and

$^{4}C_1$-conformations, has a small coupling constant. In furanose the signal at lower field with higher coupling constant is due to the cis-arrangement of protons at C-1 and C-2 (β-form), while at higher field and with lower coupling constant the signal is due to trans-arrangement (α-form).

Elucidation of sugar conformation can also be achieved by ^{13}C-nuclear magnetic resonance spectroscopy. Like ^{1}H-NMR, the chemical shifts differ for different C-atoms and are affected by the spatial arrangement of ring substituents.

4.2.2 Physical Properties

4.2.2.1 Hygroscopicity and Solubility

The moisture uptake of sugars is variable and depends, for example, on the sugar structure, isomers present and sugar purity. Solubility decreases as the sugars cake together, as often happens in sugar powders or granulates. On the other hand, the retention of food moisture by concentrated sugar solutions, e.g., glucose syrup, is utilized in the baking industry.

The solubility of mono- and oligosaccharides in water is good. Anomers may differ substantially in their solubility, as exemplified by α- and β-lactose (cf. 10.1.2.2). Monosaccharides are soluble to a small extent in ethanol and are insoluble in organic solvents such as ether, chloroform or benzene.

4.2.2.2 Optical Rotation, Mutarotation

Specific rotation constants, designated as $[\alpha]$ for sodium D-line light at $20-25\,°C$, are listed in Table 4.8 for some important mono- and oligosaccharides. The specific rotation constant $[\alpha]_\lambda^t$ at a selected wavelength and temperature is calculated from the angle of rotation, α, by the equation:

$$[\alpha]_\lambda^t = \frac{100 \cdot \alpha}{l \cdot c} \tag{4.20}$$

where l is the polarimeter tube length in decimeters and c the number of grams of the optically active sugar in 100 ml of solution. The molecular rotation, $[M]$, is suitable for comparison of the rotational values of compounds with differing molecular weights:

$$[M]_\lambda^t = \frac{M[\alpha]_\lambda^t}{100} \tag{4.21}$$

where M represents the compound's molecular weight. Since the rotational value differs for anomers and also for pyranose and furanose conformations, the angle of rotation for a freshly prepared solution of an isomer changes until an equilibrium is established. This phenomenon is known as mutarotation. When an equilibrium exists only between two isomers, as with glucose (α- and β-pyranose forms), the reaction rate follows first order kinetics:

$$-\frac{dc_\alpha}{dt} = k \cdot c_\alpha - k' \cdot c_\beta \tag{4.22}$$

A simple mutarotation exists in this example, unlike complex mutarotations of other sugars, e.g., idose which, in addition to pyranose, is also largely in the furanose form. Hence, the order of its mutarotation kinetics is more complex.

Table 4.8. Specific rotation of various mono- and oligosaccharides

Compound	$[\alpha]_D^a$	Compound	$[\alpha]_D^a$
Monosaccharides		*Oligosaccharides (continued)*	
L-Arabinose	+ 105	Kestose	+ 28
α-	+ 55.4	Lactose	+ 53.6
β-	+ 190.6	β-	+ 34.2
D-Fructose	− 92	Maltose	+ 130
β-	− 133.5	α-	+ 173
D-Galactose	+ 80.2	β-	+ 112
α-	+ 150.7	Maltotriose	+ 160
β-	+ 52.8	Maltotetraose	+ 166
D-Glucose	+ 52.7	Maltopentaose	+ 178
α-	+ 112	Maltulose	+ 64
β-	+ 18.7	Manninotriose	+ 167
D-manno-2-		Melezitose	+ 88.2
Heptulose	+ 29.4	Melibiose	+ 143
D-Mannose	+ 14.5	β-	+ 123
α-	+ 29.3	Palatinose	+ 97.2
β-	− 17	Panose	+ 154
D-Rhamnose	− 7.0	Raffinose	+ 101
D-Ribose	− 23.7	Saccharose	+ 66.5
D-Xylose	+ 18.8	α-*Schardinger-*	
α-	+ 93.6	Dextrin	+ 151
		β-*Schardinger-*	
Oligosaccharides		Dextrin	+ 162
(including disaccharides)		γ-*Schardinger-*	
Cellobiose	+ 34.6	Dextrin	+ 180
β-	+ 14.2	Stachyose	+ 146
Gentianose	+ 33.4		
Gentiobiose	+ 10		
α-	+ 31		
β-	− 3		

[a] Temperature: $20-25\,°C$.

4.2.3 Sensory Properties

Mono- and oligosaccharides and their corresponding sugar alcohols, with a few exceptions, are sweet. β-D-Mannose has a sweet-bitter taste, and some oligosaccharides are bitter, e.g. gentiobiose.

The most important sweeteners are saccharose (sucrose), starch syrup (a mixture of glucose, maltose and malto-oligosaccharides) and glucose. Invert sugar, fructose-containing glucose syrups, fructose, lactose and sugar alcohols, such as sorbitol, mannitol and xylitol, are also of importance.

The sugars differ in quality of sweetness and taste intensity. Saccharose is distinguished from other sugars by its pleasant taste even at high concentrations. The taste intensity of oligosaccharides drops regularly as the chain length increases.

The taste intensity can be measured by determining the recognition threshold of the sugar (the lowest concentration at which sweetness is still perceived) or by comparison with a reference substance (isosweet concentrations). The threshold value is related to the affinity of sweet-taste chemoreceptors for the sweet substance and is of importance in elucidation of reltionships between the chemical structure of a compound and its taste. For practical purpose, the use of a reference substance is of greater importance: taste intensity is dependent on concentration and varies greatly among sweet compounds.

Saccharose is the reference substance usually chosen. Tables 4.9, 4.10 and 4.11 list some sugar sweetness threshold values and relative sweetness values. Only mean values are given with deviations omitted. The recognition threshold values for saccharose cited in the literature vary from 0.01 to 0.037 mol/l.

Taste quality and intensity are dependent not only on a compound's structure but on other taste reception parameters: temperature, pH and the presence of additional sweet or non-sweet compounds.

The temperature dependence of the taste intensity is especially pronounced in the case of D-fructose (Fig. 4.4). It is based on the varying intensity of sweetness of the different isomers: β-D-fructopyranose is the sweetest isomer, and its concentration decreases as the temperature increases (Fig. 4.5).

Table 4.9. Taste threshold values of sugars in water

Sugar	Recognition threshold		Detection threshold	
	mol/l	%	mol/l	%
Fructose	0.052	0.94	0.02	0.24
Glucose	0.090	1.63	0.065	1.17
Lactose	0.116	4.19	0.072	2.60
Maltose	0.080	2.89	0.038	1.36
Saccharose	0.024	0.81	0.011	0.36

Table 4.10. Relative sweetness of sugars and sugar alcohols to sucrose[a]

Sugar/sugar alcohol	Relative sweetness	Sugar/sugar alcohol	Relative sweetness
Saccharose	100	D-Mannitol	69
Galactitol	41	D-Mannose	59
D-Fructose	114	Raffinose	22
D-Galactose	63	D-Rhamnose	33
D-Glucose	69	D-Sorbitol	51
Invert sugar	95	Xylitol	102
Lactose	39	D-Xylose	67
Maltose	46		

[a] 10% aqueous solution.

Table 4.11. Concentration (%) of iso-sweet aqueous solutions of sugars and sugar alcohols

D-Fructose	D-Glucose	Lactose	Saccharose	D-Sorbitol	Xylitol
0.8	1.8	3.5	1.0		
1.7	3.6	6.5	2.0		
4.2	8.3	15.7	5.0	9.3	8.5
8.6	13.9	25.9	10.0	17.2	9.8
13.0	20.0	34.6	15.0		
16.1	27.8		20.0	28.2	18.5
	39.0		30.0		25.4
	48.2		40.0	48.8	

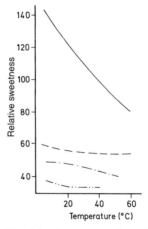

Fig. 4.4. Temperature dependence of the relative sweetness of some sugars (based on saccharose ≙ 100 at each temperature; —— D-fructose, – – – D-glucose, –·–·– D-galactose, –··–··– maltose) (according to *Shallenberger* and *Birch*, 1975)

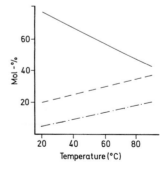

Fig. 4.5. Temperature dependence of the mutarotation equilibrium of D-fructose, (—— β-D-fructopyranose, – – – β-D-fructofuranose, –·–·– α-D-fructofuranose) (according to *Shallenberger* and *Birch*, 1975)

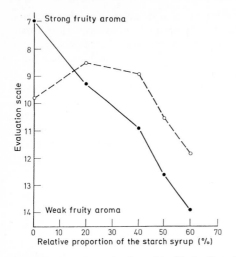

Fig. 4.6. Sensory evaluation of the "fruity flavor" of canned peaches at different ratios of saccharose/starch syrup (●—● 60°, ○—○ 50° Brix) (according to *Pangborn*, 1959)

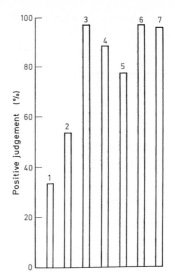

Fig. 4.7. Sensory evaluation of canned cherries prepared with different sweeteners 1, 2, 3: 60, 50, 40% saccharose, 4: 0.15% cyclamate, 5: 0.05% saccharin, 6: 10% saccharose +0.10% cyclamate, 7: 10% saccharose +0.02% saccharin (according to *Salunkhe*, 1963)

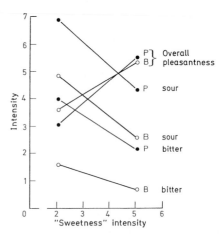

Fig. 4.8. Sensory evaluation of the categories "overall pleasantness", "sour" and "bitter" versus sweetness intensity. B bilberry (○—○) and P (●—●) cranberry juice (according to *Sydow*, 1974)

Furthermore, an interrelationship exists between the sugar content of a solution and the sensory assessment of the volatile aroma compounds present. Even the color of the solution might affect taste evaluation. Figures 4.6–4.9 clarify these phenomena, with fruit juice and canned fruits as selected food samples.

The overall conclusion is that the composition and concentration of a sweetening agent has to be adjusted for each food formulation to provide optimum sensory perception.

A prerequisite for a compound to be sweet is the presence in its structure of a proton donor/acceptor system (AH/B-system), which may be supplemented by a hyrophobic site X (cf. 8.8.1.1). This AH/B/X-system interacts with a complementary system of the taste receptor site located on the taste buds. Based on studies of the taste quality of sugar derivatives and deoxy sugars, the following AH/B/X-systems have been proposed for β-D-glucopyranose and β-D-fructopyranose respectively:

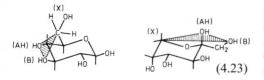

β-D-glucopyranose and β-L-glucopyranose are sweet. Molecular models show that the AH/B/X-system of both sugar components fits equally well with the complementary receptor system $AH_R/B_R/X_R$ (Formula 4.24):

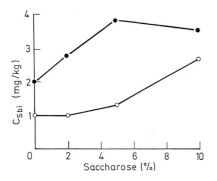

Fig. 4.9. Bitter taste threshold values of limonin (o—o) and naringin ($\times$ 10^{-1} •—•) in aqueous saccharose solution (according to *Guadagni*, 1973)

$$
\begin{array}{cc}
\text{CHO} & \text{CHO} \\
\text{H}-\text{C}-\text{OH} & \text{HO}-\text{C}-\text{H} \\
\text{HO}-\text{C}-\text{H} & \text{HO}-\text{C}-\text{H} \\
\text{H}-\text{C}-\text{OH} & \text{H}-\text{C}-\text{OH} \\
\text{H}-\text{C}-\text{OH} & \text{H}-\text{C}-\text{OH} \\
\text{CH}_2\text{OH} & \text{CH}_2\text{OH} \\
\text{D-Glucose} & \text{D-Mannose}
\end{array}
$$

(4.26)

$$
\begin{array}{ccc}
\text{CH}_2\text{OH} & \text{CH}_2\text{OH} & \text{CH}_2\text{OH} \\
\text{H}-\text{C}-\text{OH} & \text{CO} & \text{HO}-\text{C}-\text{H} \\
\text{HO}-\text{C}-\text{H} \leftarrow & \text{HO}-\text{C}-\text{H} \rightarrow & \text{HO}-\text{C}-\text{H} \\
\text{H}-\text{C}-\text{OH} & \text{H}-\text{C}-\text{OH} & \text{H}-\text{C}-\text{OH} \\
\text{H}-\text{C}-\text{OH} & \text{H}-\text{C}-\text{OH} & \text{H}-\text{C}-\text{OH} \\
\text{CH}_2\text{OH} & \text{CH}_2\text{OH} & \text{CH}_2\text{OH} \\
\text{D-Glucitol} & \text{D-Fructose} & \text{D-Mannitol} \\
\text{(sorbitol)} & &
\end{array}
$$

The alcohol name is derived from the sugar name by replacing the suffix -ose or -ulose with the suffix -itol. The sugar alcohols of importance in food processing are xylitol, the only one of the four pentitols (meso-ribitol, D,L-arabitol, mesoxylitol) used, and only D-glucitol (sorbitol) and D-mannitol of the ten stereoisomeric forms of hexitols [meso-allitol, meso-galactitol (dulcitol), D,L-glucitol (sorbitol), D,L-iditol, D,L-mannitol and D,L-altritol]. They are used as sugar substitutes in dietetic food formulations to decrease water activity in "intermediate moisture foods", as softeners, as crystallization inhibitors and for improving the rehydration characteristics of dehydrated food. Sorbitol is found in nature in many fruits, e. g., pears, apples and plums. Maltitol, the reduction product of the disaccharide maltose, is being considered for wider use in food formulations.

(4.24)

With asparagine enantiomers, the D-form is sweet, while the L-form is tasteless. Here, unlike D- and L-glucose, only the D-form interacts with the complementary $AH_R/B_R/X_R$-system:

(4.25)

D-Asparagine   L-Asparagine

These examples demonstrate that the AH/B/X-model, as a prerequisite for eliciting sweet taste responses, is correct in principle.

4.2.4 Chemical Reactions and Derivatives

4.2.4.1 Reduction to Sugar Alcohols

Monosaccharides can be reduced to the corresponding alcohols by $NaBH_4$, electrolytically or via catalytic hydrogenation. Two new alcohols are obtained from ketoses since a new chiral center is formed:

4.2.4.2 Oxidation to Aldonic, Dicarboxylic and Uronic Acids

Under mild conditions, e. g., with bromine water in buffered neutral or alkaline media, aldoses are oxidized to aldonic acids. Oxidation involves the lactol group exclusively. β-Pyranose is oxidized more rapidly than the α-form. Since the β-form is more acidic (cf. 4.2.1.3), it can be considered that the pyranose anion is the reactive form. The oxidation product is a δ-lactone which is in equilibrium with the γ-lactone and the free form of aldonic acid. The latter form prevails at pH > 3.

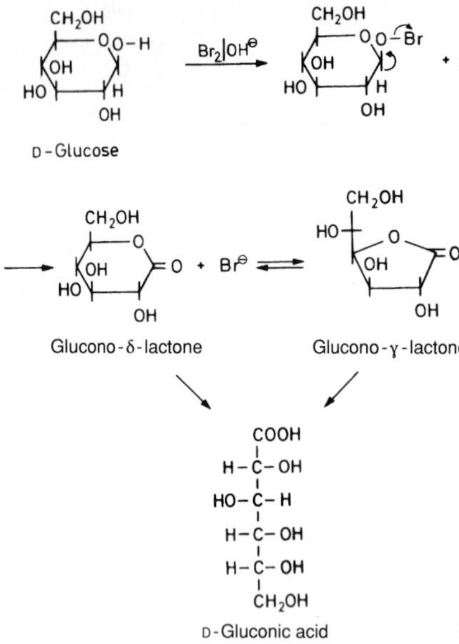

D-Glucose

Glucono-δ-lactone   Glucono-γ-lactone

$$\begin{array}{c} COOH \\ | \\ H-C-OH \\ | \\ HO-C-H \\ | \\ H-C-OH \\ | \\ H-C-OH \\ | \\ CH_2OH \end{array}$$

D-Gluconic acid   (4.27)

The transition of lactones from δ- to γ-form and vice versa probably proceeds through an intermediary bicyclic form.

The acid name is obtained by adding the suffix -onic acid (e.g. aldose → aldonic acid).

Glucono-δ-lactone is utilized in food when a slow acid release is required, as in baking powders, raw fermented sausages or dairy products.

Treatment of aldose with more vigorous oxidizing agents, such as nitric acid, brings about oxidation of the C-1 aldehyde group and the CH_2OH-group, resulting in formation of a dicarboxylic acid (nomenclature: stem name of the parent sugar + the suffix -aric acid, e.g. aldose → aldaric acid). Thus, galactaric acid (common or trivial name: mucic acid) is obtained from galactose:

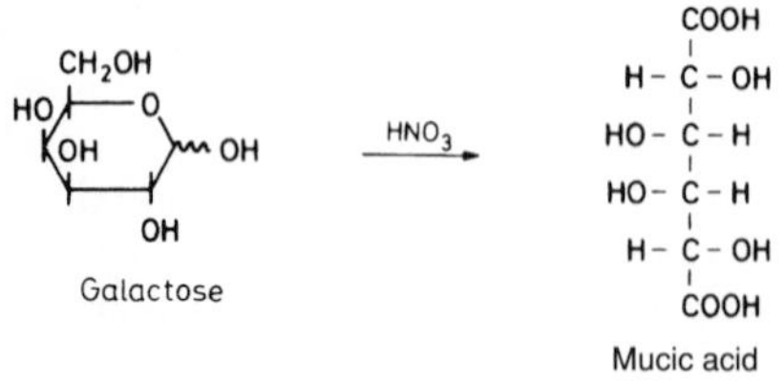

Galactose

$$\begin{array}{c} COOH \\ | \\ H-C-OH \\ | \\ HO-C-H \\ | \\ HO-C-H \\ | \\ H-C-OH \\ | \\ COOH \end{array}$$ (4.28)

Mucic acid

The dicarboxylic acid can, depending on its configuration, form mono- or dilactones.

Oxidation of the CH_2OH-group by retaining the carbonyl function at C-1, with the aim of obtaining uronic acids (aldehydocarboxylic acids), is possible only by protecting the carbonyl group during oxidation. A suitable way is to temporarily block the vicinal HO-groups by ketal formation which, after the oxidation at C-6 is completed, are deblocked under mild acidic conditions:

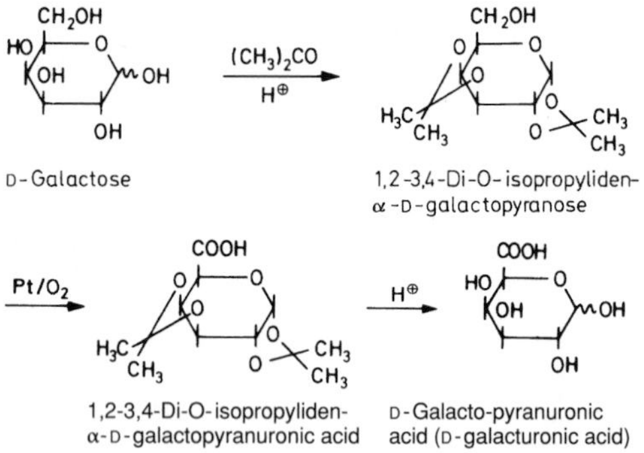

D-Galactose   1,2-3,4-Di-O-isopropyliden-
                               α-D-galactopyranose

1,2-3,4-Di-O-isopropyliden-   D-Galacto-pyranuronic
α-D-galactopyranuronic acid   acid (D-galacturonic acid)

(4.29)

An additional possibility for uronic acid synthesis is the reduction of monolactones of the corresponding aldaric acids:

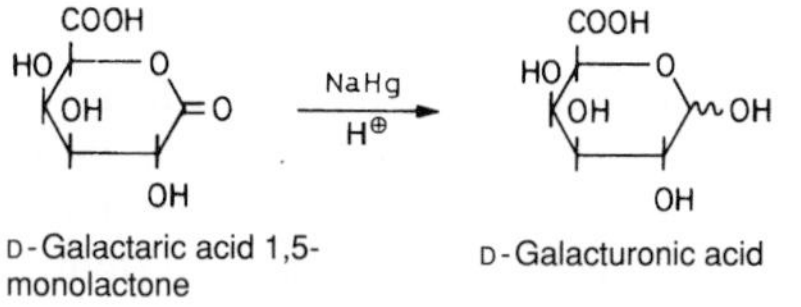

D-Galactaric acid 1,5-   D-Galacturonic acid
monolactone

(4.30)

Another industrially applied glucuronic acid synthesis involves first oxidation then hydrolysis of starch:

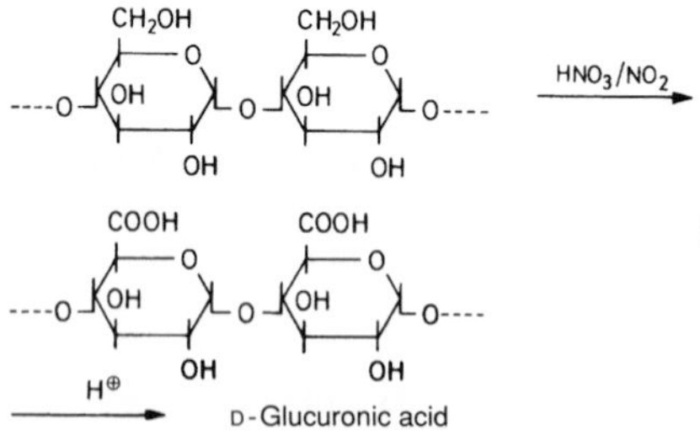

(4.31)

D-Glucuronic acid

Depending on their configuration, the uronic acids can form lactone rings in pyranose or furanose forms.

Uronic acid biosynthesis also occurs from the monosaccharide blocked at C-1 (cf. Reaction Sequence 4.32).

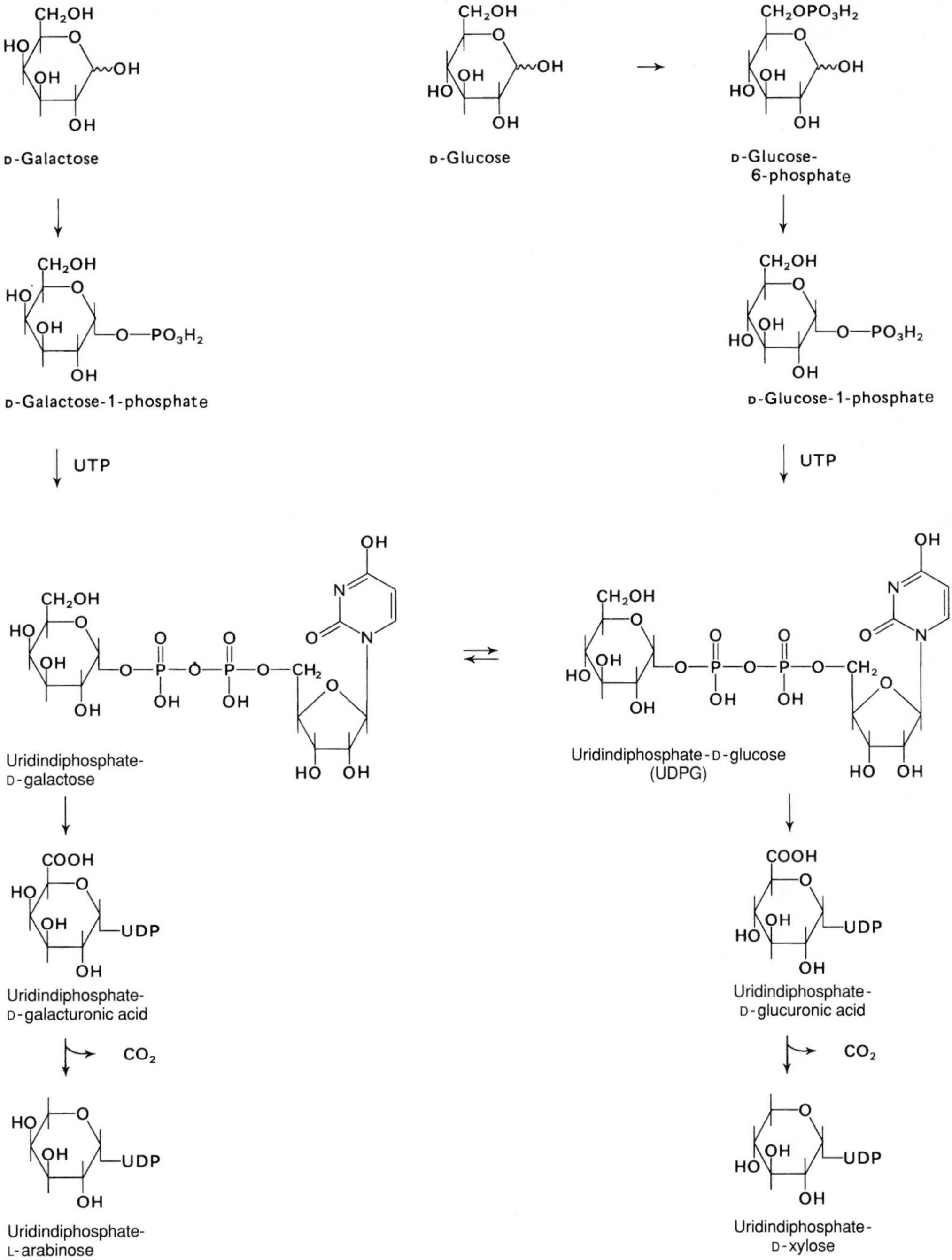

D-Galactose

D-Glucose → D-Glucose-6-phosphate

D-Galactose-1-phosphate

D-Glucose-1-phosphate

UTP

UTP

Uridindiphosphate-D-galactose

$\rightleftharpoons$

Uridindiphosphate-D-glucose (UDPG)

Uridindiphosphate-D-galacturonic acid

Uridindiphosphate-D-glucuronic acid

CO_2

CO_2

Uridindiphosphate-L-arabinose

Uridindiphosphate-D-xylose

(4.32)

A number of uronic acids occur fairly abundantly in nature. Some are constituents of polysaccharides of importance in food processing, such as gel-forming and thickening agents, e.g. pectin (D-galacturonic acid) and sea weed-derived alginic acid (D-mannuronic acid, L-guluronic acid).

4.2.4.3 Reactions in the Presence of Acids and Alkalis

Monosaccharides are fairly stable in a pH range of 3–7. However, at both ends of the pH limits, depending on conditions, extensive conversion can occur. Enolization with subsequent water elimination are the predominant reactions, retaining the C-chain length in an acidic medium. Treatment with alkali causes enolization and even fragmentation of the sugar molecule, followed by additional secondary reactions.

4.2.4.3.1 Reactions in Strongly Acidic Media

The reverse of glycoside hydrolysis (cf. 4.2.4.5) i.e. formation of glycosides, occurs in dilute mineral acids. All the possible disaccharides and higher oligosaccharides, but preferentially isomaltose and gentiobiose, are obtained from glucose:

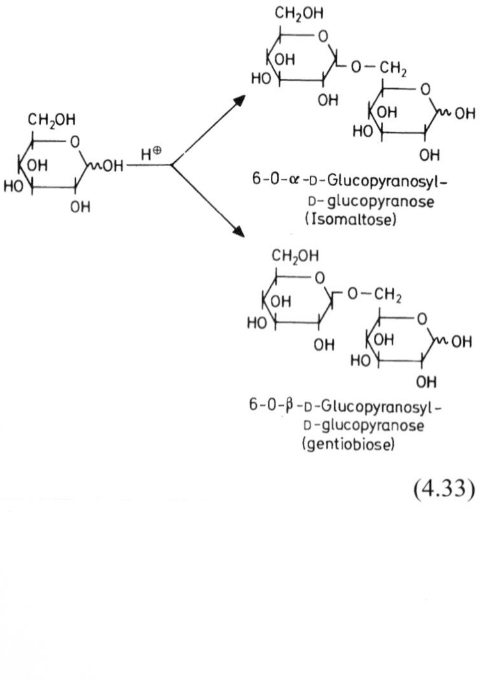

6-0-α-D-Glucopyranosyl-
D-glucopyranose
(Isomaltose)

6-0-β-D-Glucopyranosyl-
D-glucopyranose
(gentiobiose)

(4.33)

Such reversion-type reactions are also observed in acidic hydrolysis of starch.

In addition to the formation of intermolecular glycosides, intramolecular glycosidic bonds can be readily established when the sugar conformation is suitable. β-Idopyranose, which occurs in the $^1C_4$-conformation, is readily changed to 1,6-anhydroidopyranose, while the same reaction with β-D-glucopyranose ($^4C_1$-conformation) occurs only under drastic conditions, e.g., during pyrolysis of glucose, starch or cellulose. Heating glucose syrup above 100 °C can also form 1,6-anhydroglucopyranose, but only in traces:

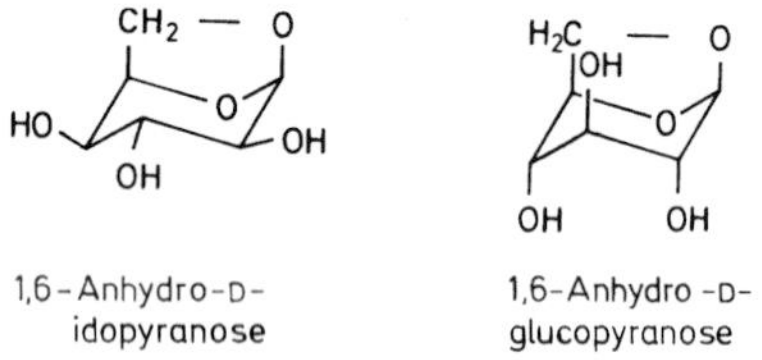

1,6–Anhydro–D–
idopyranose

1,6–Anhydro –D–
glucopyranose

(4.34)

Tricyclic compounds can be formed through intermolecular glycosidic bonds. For example, small amounts of di-D-fructopyranose-1,2′:2,1′-dianhydride are detected in molten saccharose:

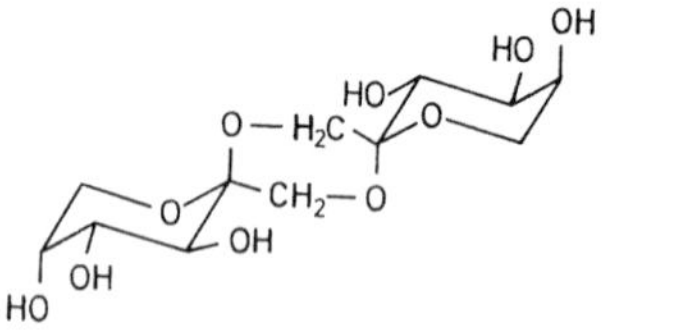

(4.35)

Heating monosaccharides in weakly acidic media, and more intensively at higher acid concentrations, leads, after a slow enolization step, to rapid proton-catalyzed β-eliminations of water which then, through several reaction steps, give furan derivatives. Dehydration of pentoses forms furfural and of hexoses, 5-hydroxymethyl furfural, with 3 molecules of water eliminated in each case. The reactivity of 2-ketohexoses is higher because the 1,2-enolization step occurs more easily than with an aldose. The sequence of reactions is shown with glucose and fructose taken as examples:

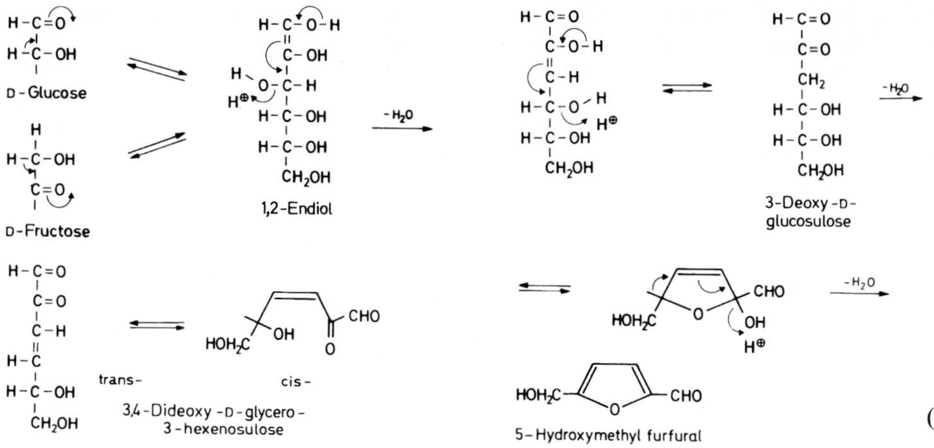
D-Glucose
D-Fructose
1,2-Endiol
3-Deoxy-D-glucosulose
trans-
cis-
3,4-Dideoxy-D-glycero-3-hexenosulose
5-Hydroxymethyl furfural

(4.36)

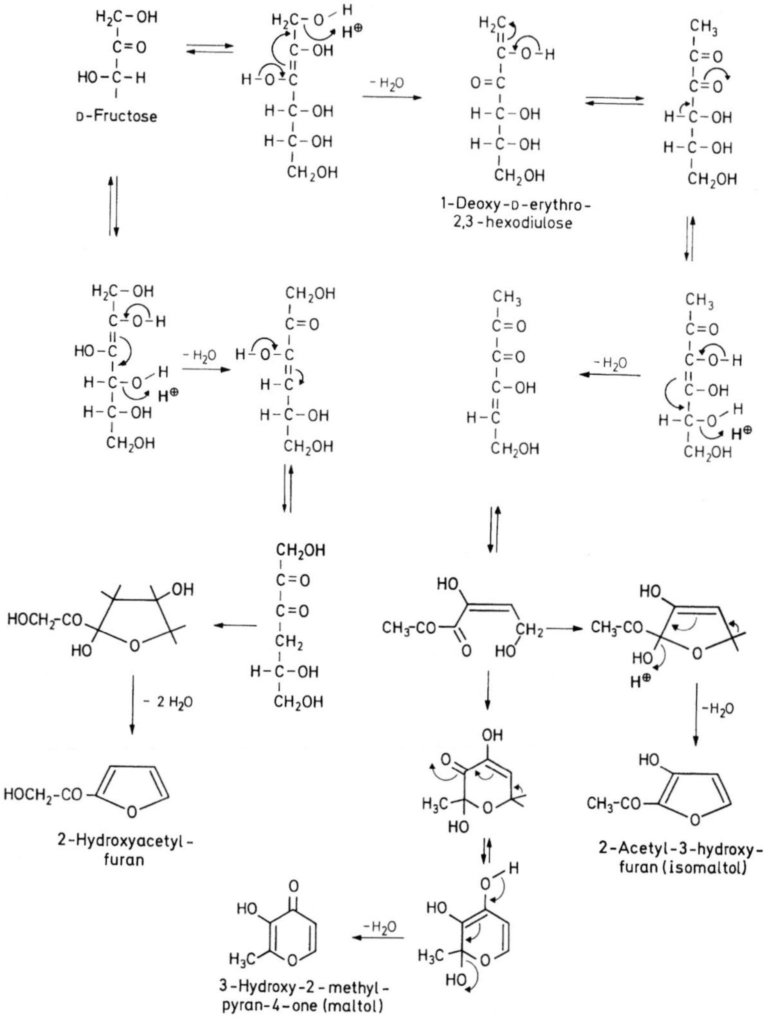
D-Fructose
1-Deoxy-D-erythro-2,3-hexodiulose
2-Hydroxyacetyl-furan
2-Acetyl-3-hydroxy-furan (isomaltol)
3-Hydroxy-2-methyl-pyran-4-one (maltol)

(4.37)

Fructose, in addition to the predominant 1,2-endiol, can form the 2,3-endiol (cf. Reaction 4.37), hence the spectrum of its degradation products is wider than with glucose. Elimination of an HO-group on C-1, and corresponding intermediary steps, form 2-acetyl-3-hydroxyfuran (isomaltol) or 3-hydroxy-2-methyl-pyran-4-one (maltol), whereas HO-group elimination from C-4 forms hydroxy-acetylfuran.

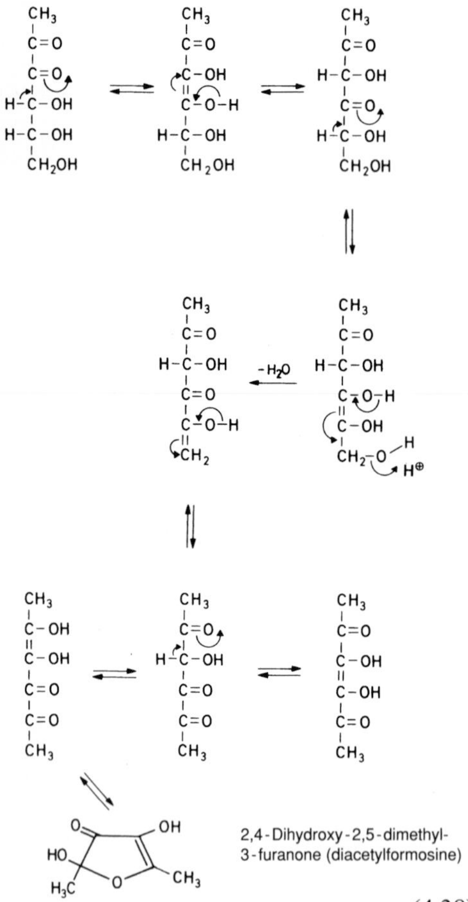

an intermediate (the latter, in cyclic form, is actually the end product) (cf. Reaction 4.38).

Compounds, such as diacetylformosine and several intermediary products of the reaction steps described above, which retain a carbonyl group in the vicinity of an endiol group are called reductones. Ascorbic acid belongs to this class of compounds, a class characterized by strong reducing power in acidic media, even at low temperatures.

Thus, Ag^+, Au^{3+} and Pt^{4+} are reduced to the metal state, Cu^{2+} to Cu^+, Fe^{3+} to Fe^{2+} and Br_2 and I_2 to their respective anions, Br^- and I^-. The reductone is oxidized to its dehydro compound in these reactions:

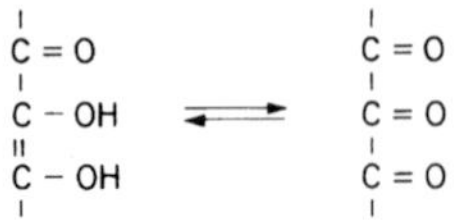

$$
\begin{array}{ccc}
\text{C}=\text{O} & & \text{C}=\text{O} \\
| & & | \\
\text{C}-\text{OH} & \rightleftharpoons & \text{C}=\text{O} \\
\| & & | \\
\text{C}-\text{OH} & & \text{C}=\text{O} \\
| & & |
\end{array}
\qquad (4.39)
$$

Reductones are stable at pH < 6, as resonance-stabilized monoanions, while at higher pH the dianion is unstable in the presence of oxygen:

$$
\begin{array}{ccc}
\text{C}=\text{O} & & \text{C}-\text{O}^{\ominus} \\
| & & \| \\
\text{C}-\text{OH} & \longleftrightarrow & \text{C}-\text{OH} \\
\| & & | \\
\text{C}-\text{O}^{\ominus} & & \text{C}=\text{O} \\
| & & |
\end{array}
\qquad (4.40)
$$

The reactions outlined above take place even under mild conditions (cf. 4.2.4.4) in the presence of amino compounds. Reductones formed in this way in food act as natural anti-oxidants.

2,4-Dihydroxy-2,5-dimethyl-3-furanone (diacetylformosine)

(4.38)

Formation of a number of other products can also be explained. Thus, 2,4-dihydroxy-2,5-dimethyl-3-furanone (diacetylformosine), a compound with a caramel-like odor, can be formed from 1-deoxy-2,3-hexodiulose through 1,6-di-deoxy-2,4,5-hexotriulose as

4.2.4.3.2 Reactions in Strongly Alkaline Solution

Aldoses and ketoses are readily enolized in the presence of alkali. Thus, in the presence of alkali, glucose, mannose and fructose are in equilibrium through the common 1,2-endiol. Also present is a small amount of psicose, which is derived from fructose by 2,3-enolization:

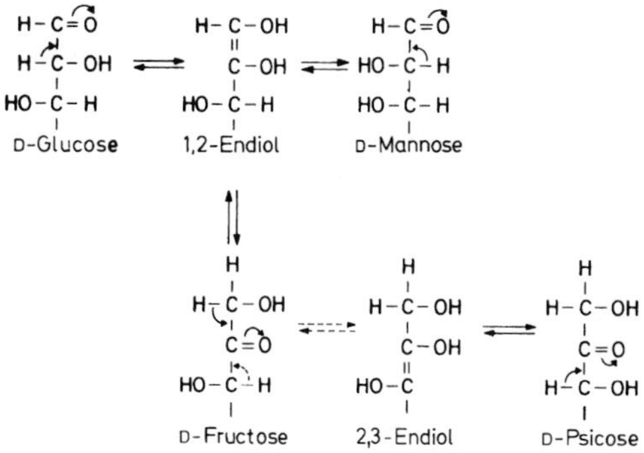

D-Glucose   1,2-Endiol   D-Mannose

D-Fructose   2,3-Endiol   D-Psicose

$$(4.41)$$

In this isomerization reaction, known as the *Lobry de Bruyn-van Ekenstein* rearrangement, one type of sugar can be transformed to another sugar in widely differing yields. The reaction is also applicable to disaccharides, transforming an aldose to a ketose. For example, in the presence of sodium aluminate as a catalyst, lactose (4-O-β-D-galactopyranosido-D-glucopyranose, I) is rearranged into lactulose (4-O-β-D-galactopyranosido-D-fructose):

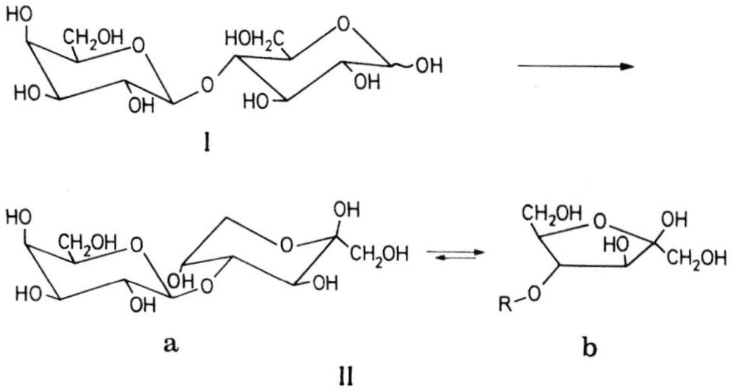

tion conditions, particularly the type of alkali present, other hydroxyacids are also formed due to enolization occurring along the molecule:

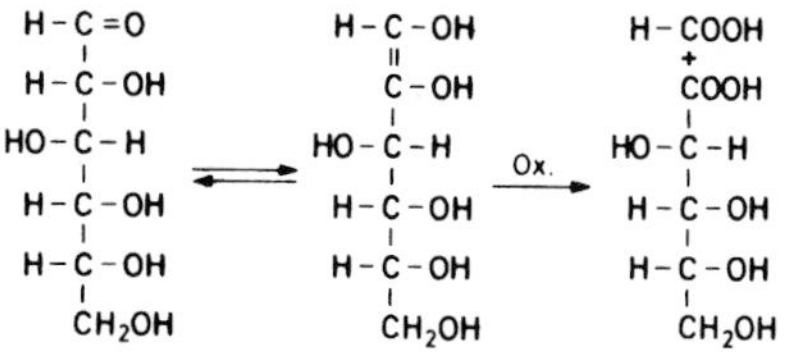

$$(4.43)$$

The nonstoichiometric sugar oxidation process in the presence of alkali is used for both qualitative and quantitative determination of reducing sugars (*Fehling's* reaction with alkaline cupric tartrate; *Nylander's* reaction with alkaline trivalent bismuth tartrate; or using *Benedict's* solution, in which cupric ion complexes with citrate ion).

Likewise, hydroxyaldehydes and hydroxyketones are formed by chain cleavage due to retroaldol reaction under nonoxidative conditions using dilute alkali at elevated tem-

I

$$(4.42)$$

a   b

II

In this disaccharide, fructose is present mainly as the pyranose (IIa) and, to a small extent, as the furanose (IIb).

Lactulose utilization in infant nutrition is under consideration since it acts as a bifidus factor and prevents obstipation.

Cleavage of the double bond of an endiol occurs in the presence of oxygen or other oxidizing agents, e.g., Cu^{2+}, resulting in two corresponding carboxylic acids. In such a reaction with glucose, the main products are D-arabinonic and formic acids. Depending on reac-

peratures or concentrated alkali even in the cold:

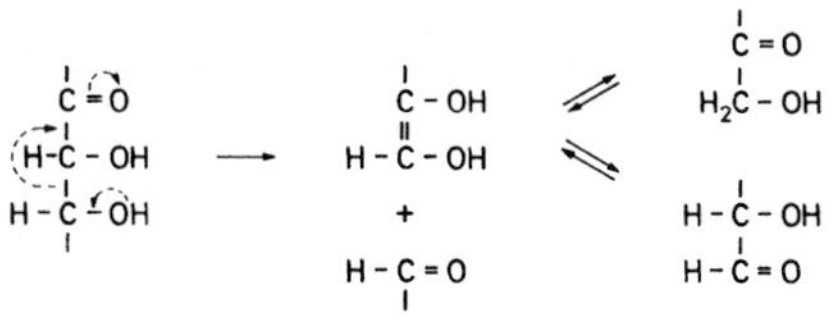

$$(4.44)$$

For example, the pair of substances 1-deoxytriose/-triulose and D-glyceric acid are produced in this way from 1-deoxy-D-erythro-2,3-hexodiulose (1-deoxyosone of glucose or fructose) via the β-dicarbonyl compound:

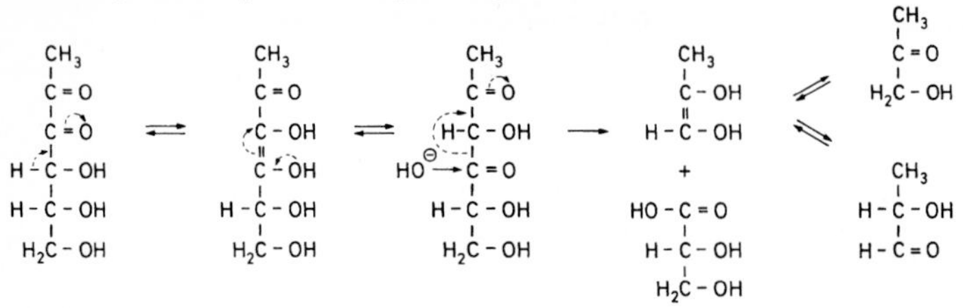

$$
\begin{array}{c}
CH_3 \\
| \\
C=O \\
| \\
C=O \\
| \\
H-C-OH \\
| \\
H-C-OH \\
| \\
H_2C-OH
\end{array}
\rightleftarrows
\begin{array}{c}
CH_3 \\
| \\
C=O \\
| \\
C-OH \\
\| \\
C-OH \\
| \\
H-C-OH \\
| \\
H_2C-OH
\end{array}
\rightleftarrows
\begin{array}{c}
CH_3 \\
| \\
C=O \\
| \\
H-C-OH \\
| \\
HO-C=O \\
| \\
H-C-OH \\
| \\
H_2C-OH
\end{array}
\rightarrow
\begin{array}{c}
CH_3 \\
| \\
C-OH \\
| \\
H-C-OH \\
+ \\
HO-C=O \\
| \\
H-C-OH \\
| \\
H_2C-OH
\end{array}
\quad
\begin{array}{c}
CH_3 \\
| \\
C=O \\
| \\
H_2C-OH \\
\\
CH_3 \\
| \\
H-C-OH \\
| \\
H-C=O
\end{array}
\tag{4.45}
$$

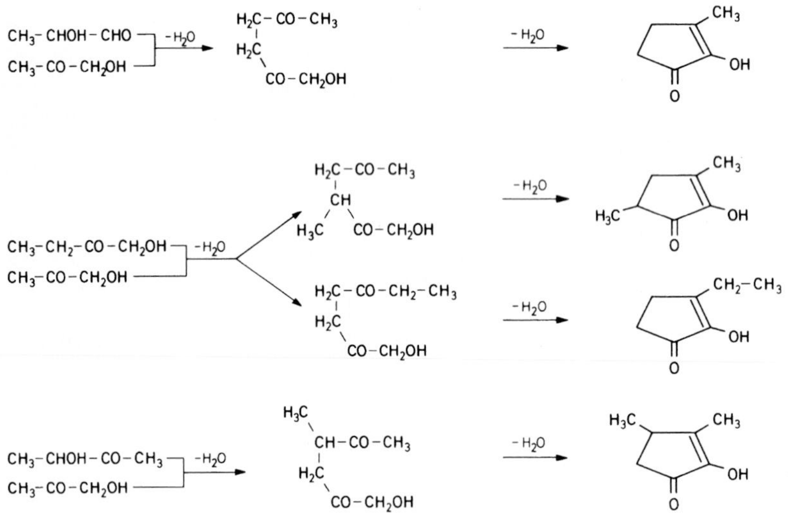

$$(4.46)$$

Since enolization is not restricted to any part of the molecule and since water elimination is not restricted in amount, even the spectrum of primary cleavage products is great. These primary products are highly reactive and provide a great number of secondary products by aldol condensations and intramolecular *Cannizzaro* reactions (cf. Reaction 4.46).

The compounds formed in fructose syrup of pH 8–10 heated for 3 h are liste in Table 4.12. Some, e. g., the cyclopentenolones, are typical caramel-like aroma substances. Their formation is assumed to follow the reaction sequence depicted under Reaction 4.46.

Saccharic acids are specific reaction products of monosaccharides in strong alkalies, particularly of alkaline-earth metals. They are obtained in each case as diastereomeric pairs by benzilic acid rearrangement of 1,2- and 2,3-deoxydicarbonyl compounds, which were mentioned in 4.2.4.3.1 as secondary products of enolization and water elimination: (cf. Reaction 4.47).

Ammonia catalyzes the same reactions. However, N-containing compounds are also formed, e. g., 1-amino-1-deoxyglycoses (glycosylamines), diglycosylamines and 1-amino-1-deoxyglyculoses. The mechanisms involved in their formation are covered in the section on N-glycosides (4.2.4.4). Reactive intermediary products can polymerize further into brown pigments, or form a number of imidazole, pyrazine and pyridine derivatives.

4.2.4.3.3 Caramelization

Brown-colored products with a typical caramel aroma are obtained by melting sugar or by heating sugar syrup in the presence of acidic and/or alkaline catalysts. The reactions involv-

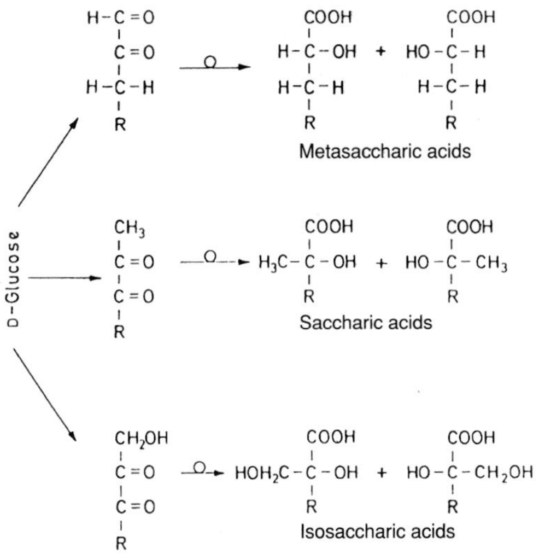

$$
\text{D-Glucose}
\begin{cases}
\\
\\
\\
\\
\\
\\
\\
\\
\end{cases}
$$

Metasaccharic acids

Saccharic acids

Isosaccharic acids

(4.47)

ed were covered in the previous two sections. The process can be directed more towards aroma formation or more towards brown pigment accumulation. Heating of saccharose syrup in a buffered solution enhances molecular fragmentation and, thereby, formation of aroma substances. Primarily dihydrofuranones, cyclopentenolones, cyclohexenolones and pyrones are formed (cf. 4.2.4.3.2). On the other hand, heating glucose syrup with sulfuric acid in the presence of ammonia provides intensively colored polymers ("sucre cou-

Table 4.12. Volatile reaction products obtained from fructose by an alkali degradation (pH 8–10)

Acetic acid
Hydroxyacetone
1-Hydroxy-2-butanone
3-Hydroxy-2-butanone
4-Hydroxy-2-butanone
Furfuryl alcohol
5-Methyl-2-furfuryl alcohol
2,5-Dimethyl-4-hydroxy-3-(2H)-furanone
2-Hydroxy-3-methyl-2-cyclopenten-1-one
3,4-Dimethyl-2-hydroxy-2-cyclopenten-1-one
3,5-Dimethyl-2-hydroxy-2-cyclopenten-1-one
3-Ethyl-2-hydroxy-2-cyclopenten-1-one
γ-Butyrolactone

leur"). The stability and solubility of these polymers are enhanced by bisulfite anion addition to double bonds:

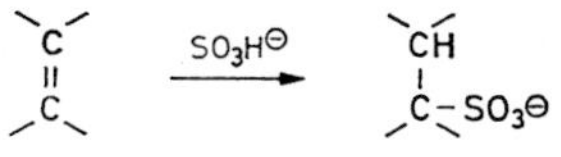

(4.48)

4.2.4.4 Reactions with Amino Compounds

In this section, the formation of N-glycosides as well as the numerous consecutive reactions classed under the term *Maillard* reaction or nonenzymatic browning will be discussed. N-Glycosides are widely distributed in nature (nucleic acids, NAD, coenzyme A). They are formed in food whenever reducing sugars occur together with proteins, peptides, amino acids or amines. They are obtained more readily at a higher temperature, low water activity and on longer storage.

On the sugar side, the reactants are mainly glucose, fructose, maltose, lactose and, to a smaller extent, reducing pentoses. On the side of the amino component, primary amines are more important than secondary amines because their concentration in foods is usually higher. Exceptions are, e.g., malt and corn products which have a high proline content. In the case of proteins, the ε-amino groups of lysine react predominantly. However, secondary products from reactions with the guanidino group of arginine are also known. In fact, imidazolones and pyrimidines, which are formed from arginine and reactive α- and β-dicarbonyl compounds obtained from sugar degradation, have been detected:

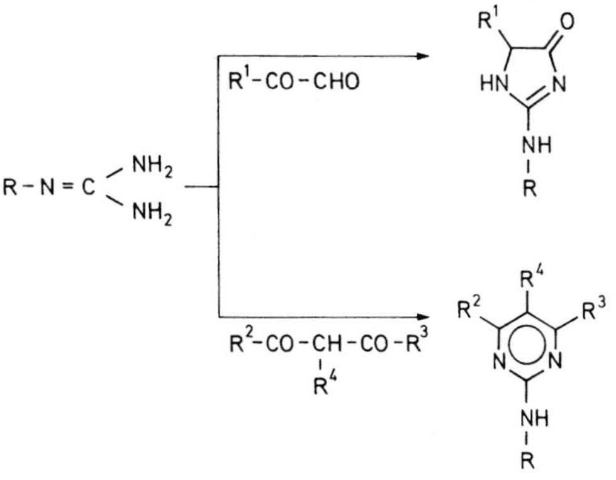

(4.49)

The consecutive reactions of N-glycosides mentioned above partially correspond to those already outlined for acid/base catalyzed conversions of monosaccharides. However, starting with N-containing intermediates, which with the nitrogen function possess a catalyst within the molecule, these reactions proceed at a high rate under substantially milder conditions, which are present in many foods.

These reactions result in:

- Brown pigments, known as melanoidins, which contain variable amounts of nitrogen and have varying molecular weights and solubilities in water. Little is known about the structure of these compounds. Studies have been conducted on fragments obtained after Curie point pyrolysis or after oxidation with ozone or sodium periodate. Browning is desired in baking and roasting, but not in foods which have a typical weak or other color of their own (condensed milk, white dried soups, tomato soup).
- Volatile compounds which are often potent aroma substances. The *Maillard* reaction is important for the desired aroma formation accompanying cooking, baking, roasting or frying. It is equally significant for the generation of off-flavors in food during storage, especially in the dehydrated state, or on heat treatment for the purpose of pasteurization, sterilization and roasting.
- Flavoring matter, especially bitter substances, which are partially desired (coffee) but can also cause an off-taste, e.g., in grilled meat or fish (roasting bitter substances).
- Compounds with highly reductive properties (reductones) which can contribute to the stabilization of food against oxidative deterioration.
- Losses of essential amino acids (cysteine, methionine).
- Compounds with mutagenic properties.
- Compounds that can cause cross-linkage of proteins. Reactions of this type apparently also play a role *in vivo* (diabetes).

4.2.4.4.1 N-Glycosides, Amadori and Heyns Compounds

The interaction of amino compounds with monosaccharides initially probably involves addition of a carbonyl group to a primary ami-

no group of an amino acid, peptide or protein, followed by water elimination, leading to an intermediary imine, which cyclizes to a glycosylamine (N-glycoside):

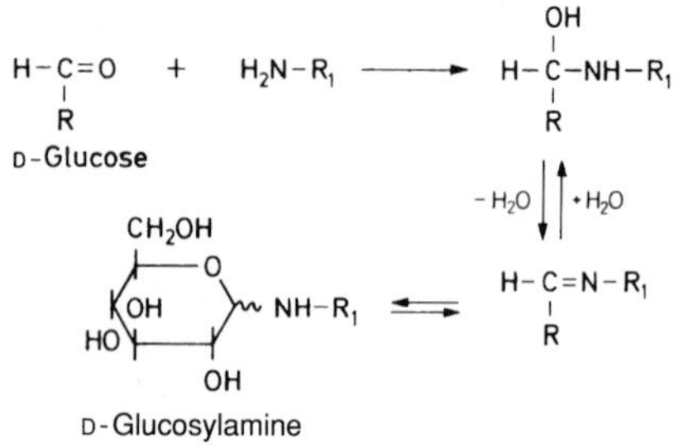

$$(4.50)$$

Unlike O-glycosides, N-glycosides exhibit mutarotation. This results from acid-catalyzed isomerization of the compound through an intermediary open-chain immonium ion:

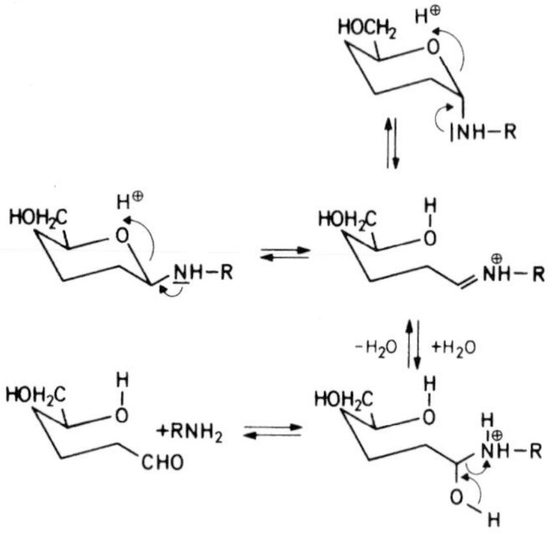

$$(4.51)$$

Aldosylamine, which is the initial product from aldose, will rearrange to a 1-amino-1-deoxyketose (an *Amadori* rearrangement). Ketosylamine can yield 2-amino-2-deoxyaldose via a *Heyns* rearrangement. Both reactions correspond to alkali-catalyzed isomerization of aldoses and ketoses, and both start with an immonium ion and proceed through an endiol-like enaminol stage:

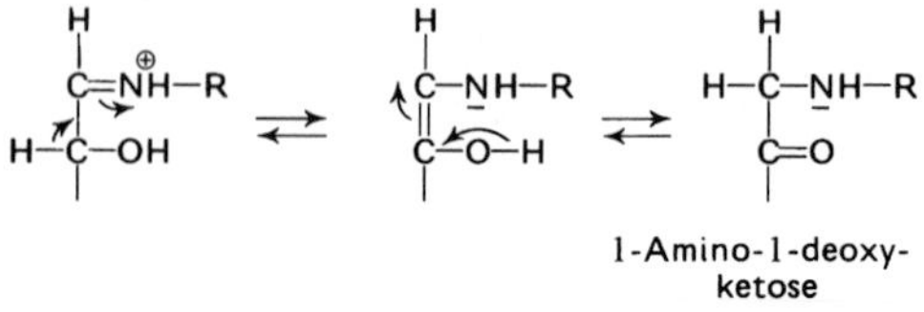

1-Amino-1-deoxy-
ketose

$$(4.52)$$

$$H-\overset{H}{\underset{C=\overset{\oplus}{N}H-R}{C}}-OH \;\rightleftharpoons\; \overset{H}{\underset{C-NH-R}{C}}\overset{\frown}{O-H} \;\rightleftharpoons\; H-\overset{H}{\underset{C-NH-R}{C}}=O$$

2-Amino-2-deoxy
aldose

$$(4.53)$$

1-Amino-1-deoxyketoses are found in various foods, such as dried fruits (peaches, apricots), dehydrated vegetables, milk powder or liver extracts.

In milk powder, the reaction occurs between the ε-amino group of protein and the disaccharide lactose (cf. Reaction 4.54). The initially formed lactosylamine is changed by an *Amadori* rearrangement into a protein-bound N-alkyl-1-amino-1-deoxylactulose which, after acidic hydrolysis, yields fructoselysine. This is then transformed into the unusual amino acids furosine and pyridosine, which occur in milk powder hydrolysate and other food hydrolysates. Both acids serve as reliable indicator compounds for revealing sugar-amine interaction.

(chemical reaction scheme)

Lactosyl – NH– (CH₂)₄ – CH $\overset{CO}{\underset{NH}{\diagdown}}$ Amadori rearrangement →

N-Alkyl-1-amino-1-deoxy-lactulose

H⊕ →

Fructoselysine

Furosine

Pyridosine

$$(4.54)$$

The *Amadori* compounds can react further with a second sugar molecule, resulting in glycosylamine formation and subsequent conversion to di-D-ketosylamino acids ("diketose amino acids") by an *Amadori* rearrangement:

(chemical structure)

Di-D-fructosylamino acid

$$(4.55)$$

4.2.4.4.2 Initial Stages of the Maillard Reaction

In the course of the *Maillard* reaction, glycosylamines and *Amadori* products are only intermediates, which are more or less stable and detectable in heated, dehydrated, and stored foods. In the pH range $4-7$, they are degraded to give the 1-, 3-, and 4-deoxydicarbonyl compounds (deoxyosones), which as reactive α-dicarbonyl compounds yield many secondary products. Formula 4.57 summarizes the formation of the *Amadori* compound from glucose and its degradation to the 1- and 3-deoxyosones ($R^1 = H$: ketosylamino acid; $R^1 = CH_2\text{-}CO\text{-}CHOH\text{-}$: diketosylamino acid; $R^2 = CHR^3\text{-}COOH$; 1, 2 : glycosyl amination, *Amadori* rearrangement, 3 : 1,2-enolization, 4 : 2,3-enolization). The spectrum of secondary products shows that the 4-deoxyosones react in the nitrogen-free and nitrogen-containing forms:

$$
\begin{array}{ll}
H_2C-OH & H_2C-NH-R \\
C=O & C=O \\
C=O & C=O \\
H-C-H & H-C-H \\
H-C-OH & H-C-OH \\
H_2C-OH & H_2C-OH
\end{array}
\qquad (4.56)
$$

Apart from the reactions of open chain *Amadori* compounds (enolization, elimination of water), there is experimental evidence for direct dehydrations of the cyclic forms. Thus, the following pathway is postulated for the conversion of a fructosylamino acid to the corresponding pyryliumbetaine (cf. Formula 4.58):

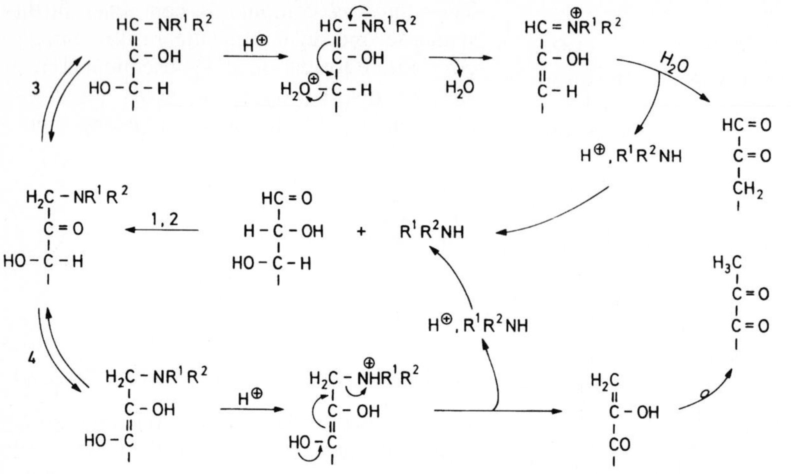

$$(4.57)$$

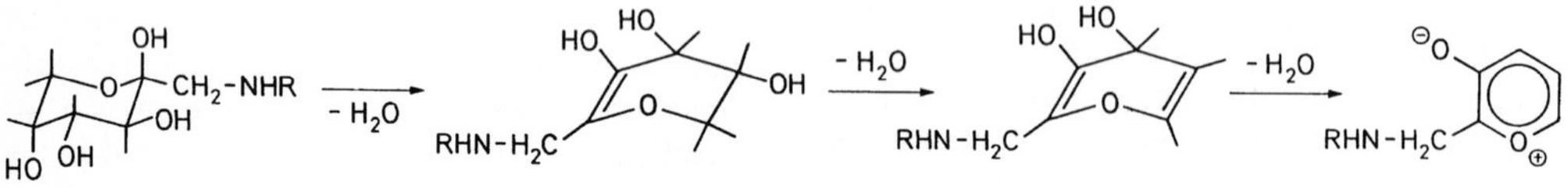

$$(4.58)$$

As in the case of the deoxyosones, the concentration of *Amadori* and *Heyns* compounds varies depending on the reaction conditions (pH value, temperature, time, type and concentration of the educts). As a result, there is a shift in the product spectrum and, consequently, in the color, taste, odor, and other properties of the foods.

In the early stages of the *Maillard* reaction, a pyrazinium radical was detected by ESR spectroscopy. It is postulated that this radical is formed from the aldimine via a retroaldol cleavage:

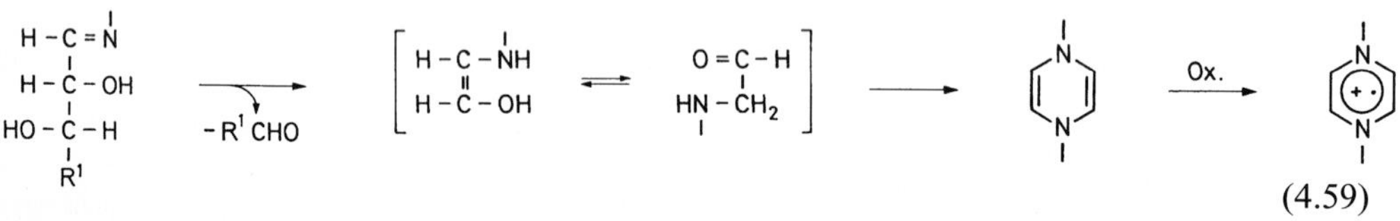

$$(4.59)$$

Further oxidation yields reactive secondary products, which are very rapidly converted to colored compounds of unknown structure. The results show that the degradation of reducing sugars can apparently proceed not only via *Amadori* compounds, but also via different pathways.

The rate of browning of a reaction system appears to depend primarily on the retroaldol reaction, which becomes more important at pH > 8, and on the C_3-fragments formed, e. g., 2-oxopropanal. Aldimines can also undergo elimination reactions at C2 to give the 2-deoxyamino compound, which finally yields 3-hydroxy-2-methylpyridine:

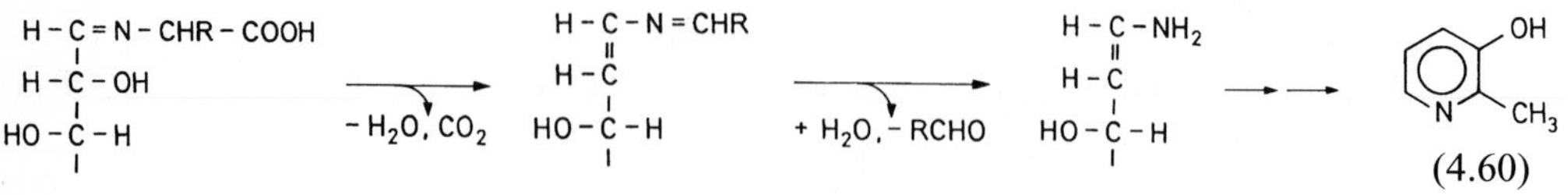

$$(4.60)$$

Oxidative cleavage of *Amadori* compounds has also been observed. It produces N-carboxymethylamino acids, e. g., ε-N-carboxymethyllysine, and can be used in vivo as an indicator for revealing the degree of *Maillard* reaction in a system:

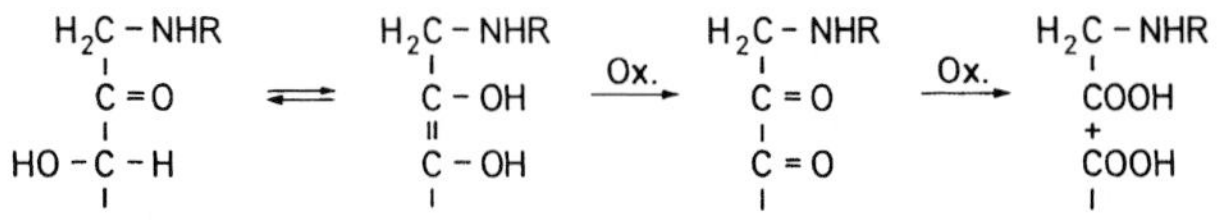

$$
\begin{array}{ccccccc}
\text{H}_2\text{C}-\text{NHR} & & \text{H}_2\text{C}-\text{NHR} & & \text{H}_2\text{C}-\text{NHR} & & \text{H}_2\text{C}-\text{NHR} \\
| & & | & & | & & | \\
\text{C}=\text{O} & \rightleftharpoons & \text{C}-\text{OH} & \xrightarrow{\text{Ox.}} & \text{C}=\text{O} & \xrightarrow{\text{Ox.}} & \text{COOH} \\
| & & \| & & | & & + \\
\text{HO}-\text{C}-\text{H} & & \text{C}-\text{OH} & & \text{C}=\text{O} & & \text{COOH} \\
| & & | & & | & & |
\end{array}
\tag{4.61}
$$

The stable secondary products of the *Maillard* reaction, that are isolated from many different reaction mixtures and have known structures, can be generally assigned to a definite deoxyosone by a series of plausible reaction steps (enolization, elimination of water, retroaldol cleavage, substitution of an amino function for a hydroxy function etc.). Moreover, deoxyosones and other reactive intermediates could be detected by trapping reactions. Thus, reactive secondary products of 3-deoxyosones, which are, e. g., involved in the formation of dyes, were isolated with o-aminoacetophenone as furoquinolines or pyranoquinolines:

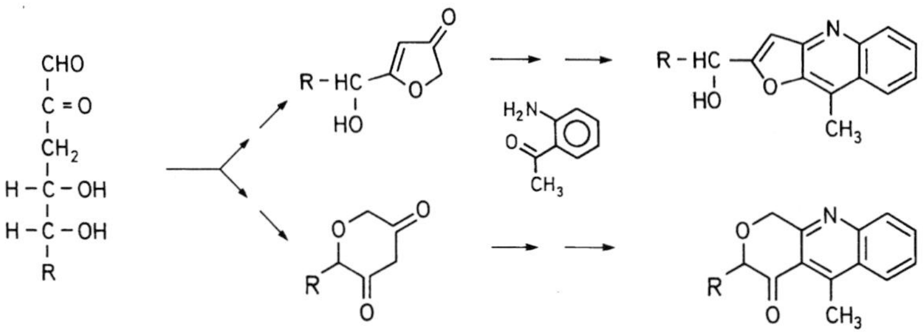

$$\tag{4.62}$$

Both 1-deoxyosone as well as the 1,4-dideoxy compounds formed via a *Strecker* reaction with α-amino acids and the secondary products obtained on elimination of water were trapped as quinoxalines with o-phenylenediamine:

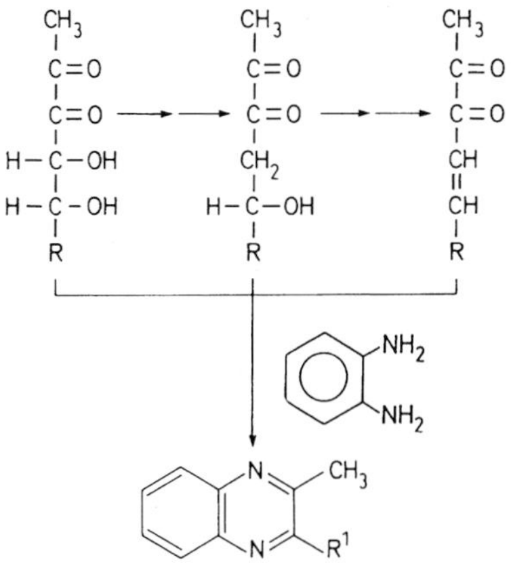

Of the large number of secondary products known today, only a few typical examples can be dealt with for each deoxyosone.

4.2.4.4.3 Secondary Products of 3-Deoxyosones

On cyclization and water elimination, the 3-deoxyosones of hexoses yield hydroxymethyl-furfural and 2-hydroxy-6-hydroxymethyl-3-pyranone, which were also trapped in a yellow colored condensation product (cf. Formula 4.64):

$$\tag{4.63}$$

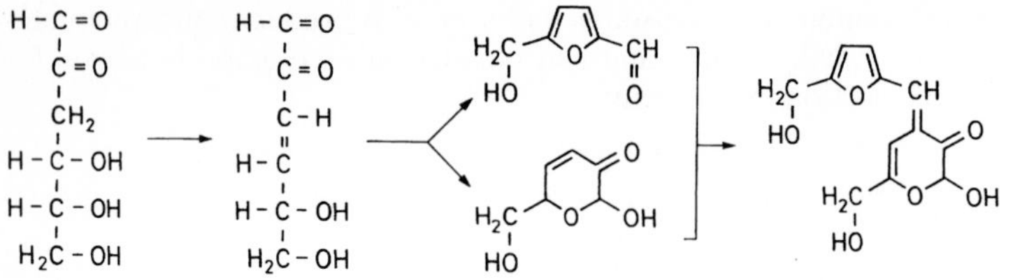

$$(4.64)$$

In the presence of higher concentrations of primary amines, the N-analogs pyrroles and pyridines, are mainly produced. Their formation can be formulated via the dideoxytriketo compound:

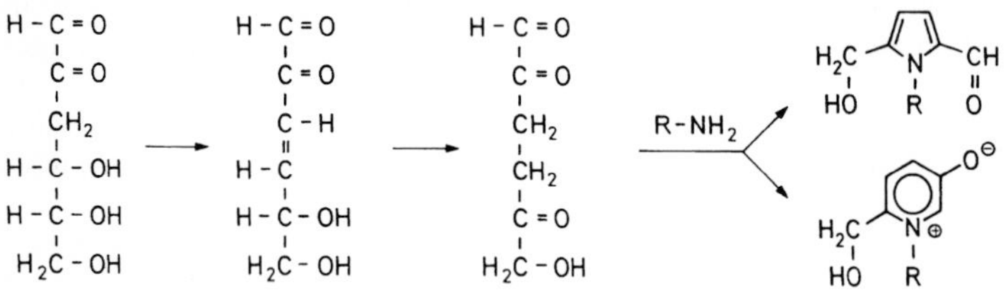

$$(4.65)$$

The corresponding pyrrole derivative of lysine was isolated from a protein, that was subjected to a *Maillard* reaction, by alkaline hydrolysis (R=HOOC–CHNH$_2$–(CH$_2$)$_4$– in Formula 4.66). Hydroxymethylformylpyrrole dimerizes very easily and could be involved in cross-linking reactions of proteins, which are possibly of importance in *in vivo* systems (diabetes).

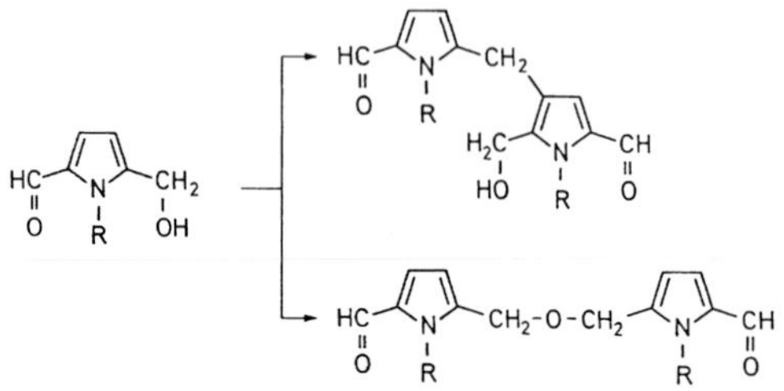

$$(4.66)$$

Starting with 3-deoxyosones, additional formylpyrroles have been detected. In fact, the products shown in Formula 4.67 must have been formed by the reaction of the deoxyosone with *Amadori* compounds on elimination of water and retroaldol cleavage. Just like those of hydroxymethylpyrroles, the α-hydroxyl groups are very easily substituted by other nucleophiles. After cleavage of the side chains with periodate, 2,3-di- and 2,3,4-triformylpyrroles are obtained from the primary reaction products.

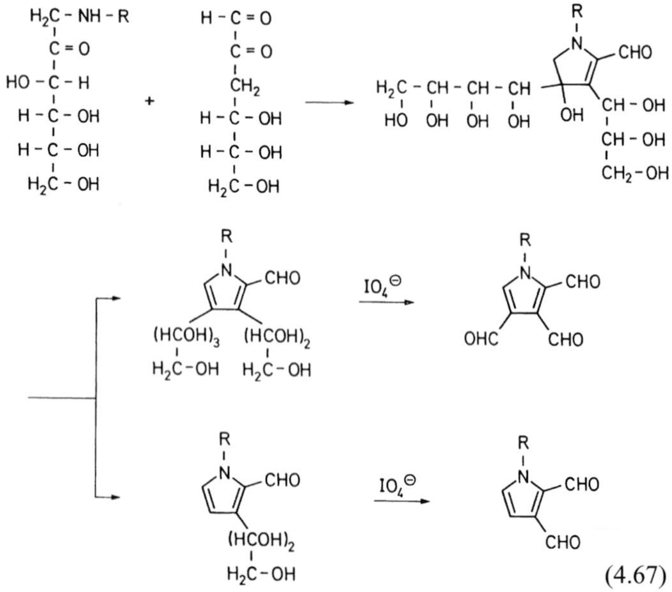

$$(4.67)$$

The isomeric 2,4-di- and 2,3,5-triformylpyrroles were also isolated from reaction mixtures of propylamine and glucose after periodate oxidation. Consequently, the reaction is very variable.

This reaction proceeds differently with secondary amines. Via enaminol as the *Strecker* product, maltoxazin is formed as the main product from proline and 3-deoxyhexosones, while 4-hydroxy-proline with a pentose yields a pyrrole derivative:

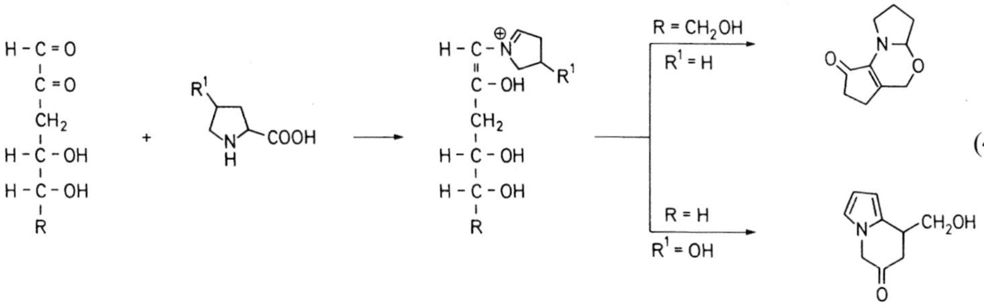

(4.68)

Even in the case of the following compounds, formation via 3-deoxyosones is assumed:

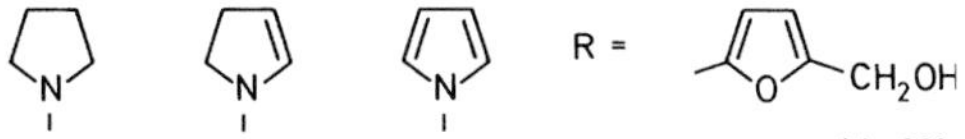

(4.69)

(4.71)

3-Deoxyosones predominantly form pyrazines and imidazoles with ammonia. The following compounds were isolated from sugar coloring:

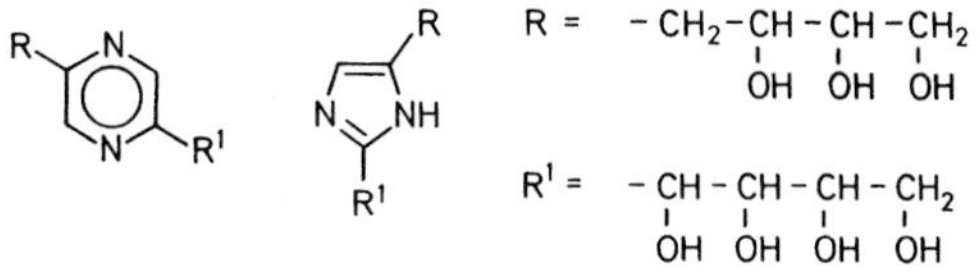

Orange dyes appear as secondary products. They are obtained from pentoses and primary or secondary amines:

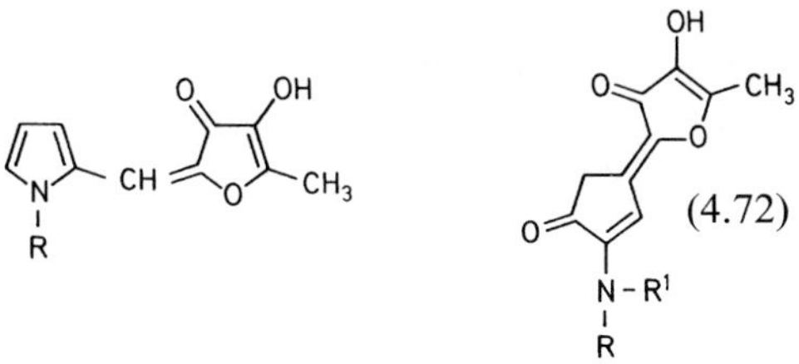

(4.70)

(4.72)

4.2.4.4.4 Secondary Products of 1-Deoxyosones

The 1-deoxyosones of pentoses, methylpentoses and hexoses give rise to furanones (R = H, CH_3, CH_2OH in Formula 4.71), which are important aroma substances (cf. 5.3.1.3).

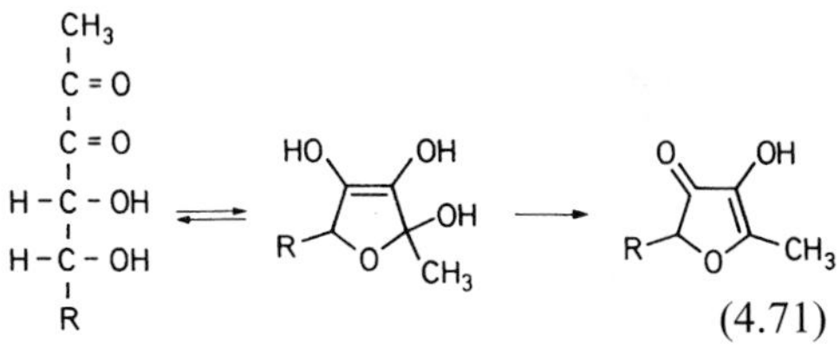

Apart from the furanones mentioned above, γ-pyranones are also formed from 1-deoxyhexosones. A second pathway leads via β-pyranone to lactyl-β-hydroxypropionic acid, which is finally hydrolyzed to give two acids (cf. Formula 4.73):

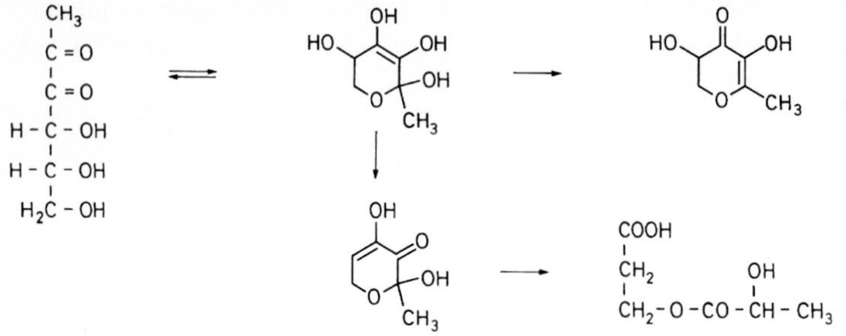

(4.73)

The formation of acetylformoin must proceed via the β-diketo compound with elimination of the hydroxyl group at C6. This very reactive compound yields pyrrolinones with primary amines and aminohexose reductones with secondary amines:

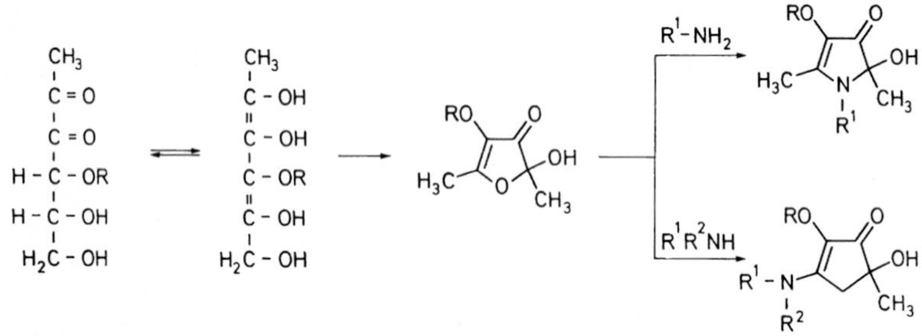

(4.74)

These reactions take place with mono- (R = H) and disaccharides (R = glycosyl). Pyrrolinone (R = H), a yellow fluorescent compound, tends to dimerize and, therefore, could be involved in cross-linking reactions of proteins.

The following reactions of 2-deoxyosones are limited to disaccharides. Cyclization and elimination of water lead to β-pyranone, which has been detected in heated milk (R = β-galactosyl-) and in red ginseng (R = α-glucosyl-). β-Pyranone, which goes back to maltose, reacts further via the pyrylium derivative to give γ-pyranone, maltol and, only to a small extent, the isomaltol derivative, which, on the other hand, is obtained from lactose as the main product:

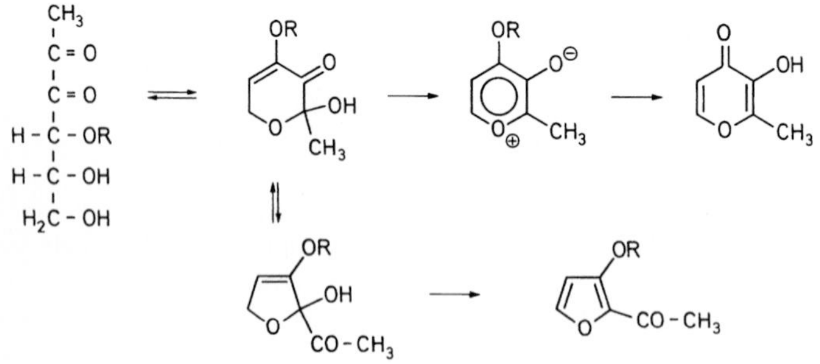

(4.75)

It is probable that the α-glucosyl residue R sterically hinders the formation of furan to a much greater extent than the β-galactosyl residue.

In the presence of primary amines, the pyrylium derivative reacts to give pyridiniumbetaine, which mainly yields the corresponding pyridinone and, to a smaller extent, pyrrole. Secondary amines give rise to various N-substituted furan derivatives (cf. Formula 4.76):

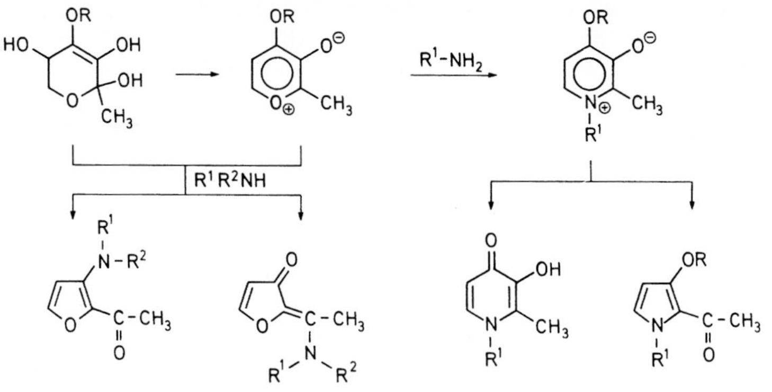

(4.76)

The following compounds were detected in reaction mixtures containing proline and hydroxyproline. Their formation must proceed via the 1-deoxyosones as well:

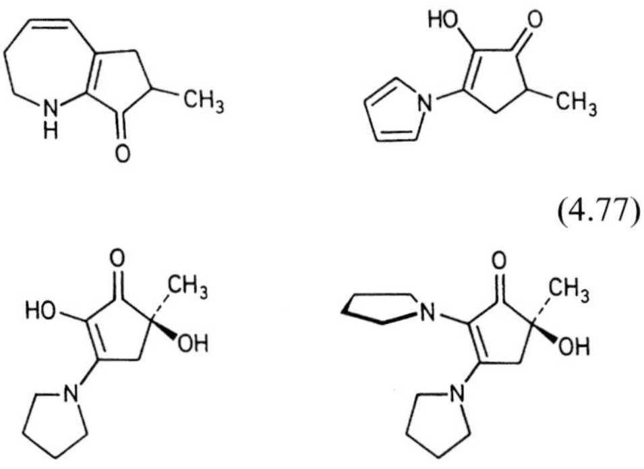

(4.77)

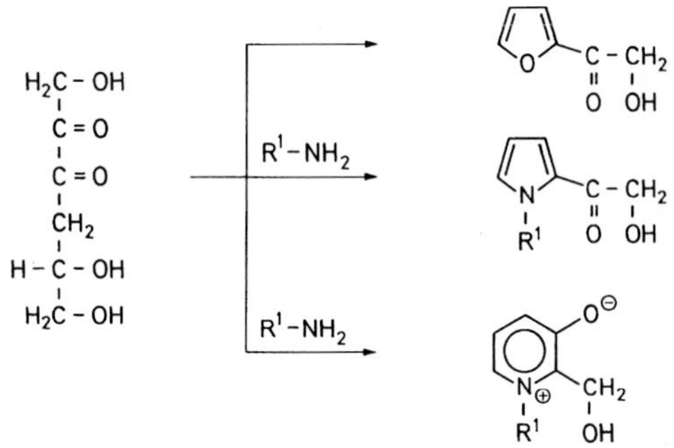

The pyrrolidino- and dipyrrolidinohexose reductones are bitter substances obtained from heated proline/saccharose mixtures (190 °C, 30 min, molar ratio 3 : 1; c_{sbi} : 0.8 and 0.03 mmol/l).

4.2.4.4.5 Secondary Products of 4-Deoxyosones

Hydroxyacetylfuran can be derived from 4-deoxyosone. In the presence of primary amines, the formation of this compound is totally suppressed in favor of the corresponding pyrrole and pyridiniumbetaine (cf. Formula 4.78):

(4.78)

As a degradation product of *Amadori* compounds, 1-amino-1,4-dideoxyosone was detected with an aminoreductone and an aminoacetylfuran as its secondary products:

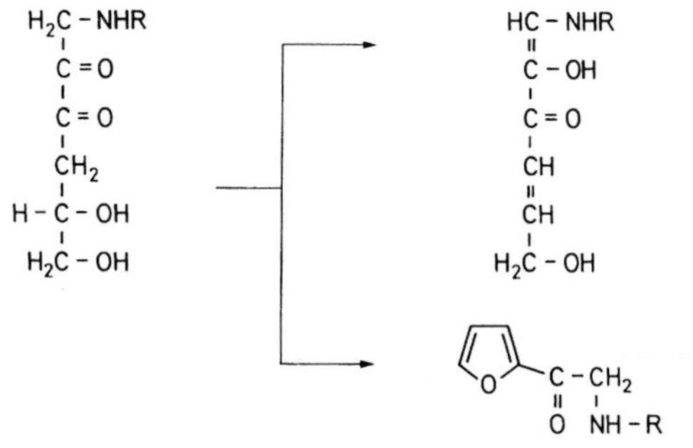

(4.79)

Aminoacetylfurans of this type were obtained until now only after acid hydrolysis of *Amadori* compounds, e. g., furosine (cf. 4.2.4.4.1).
With ammonia, aminoacetylfuran is very easily converted to 2-(2-furoyl)-5-(2-furanyl)-1H-imidazole, known as FFI, which was previously isolated from acid hydrolysates from protein/glucose reaction mixtures:

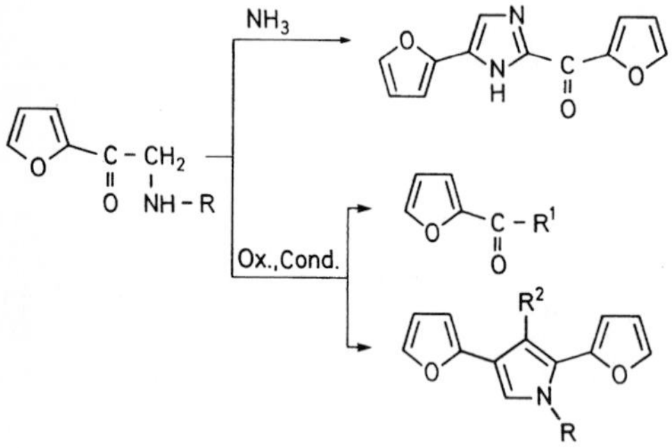

Various oxidation and condensation products (R^1 = OH, CONHR; R^2 = OH, NHR in Formula 4.80) were isolated from a heated, neutral solution. The condensation products show that aminoacetylfuran could be involved in cross-linking reactions.

4.2.4.4.6 Reactions Between Deoxyosones and Reductones

In the course of the *Maillard* reaction, deoxyosones and reductones, e. g., acetylformoin (cf. Formula 4.74), are formed. They can react to give enol and triketo compounds via an addition and disproportionation. These compounds can, in turn, give rise to a whole series of secondary products (cf. Formula 4.81).

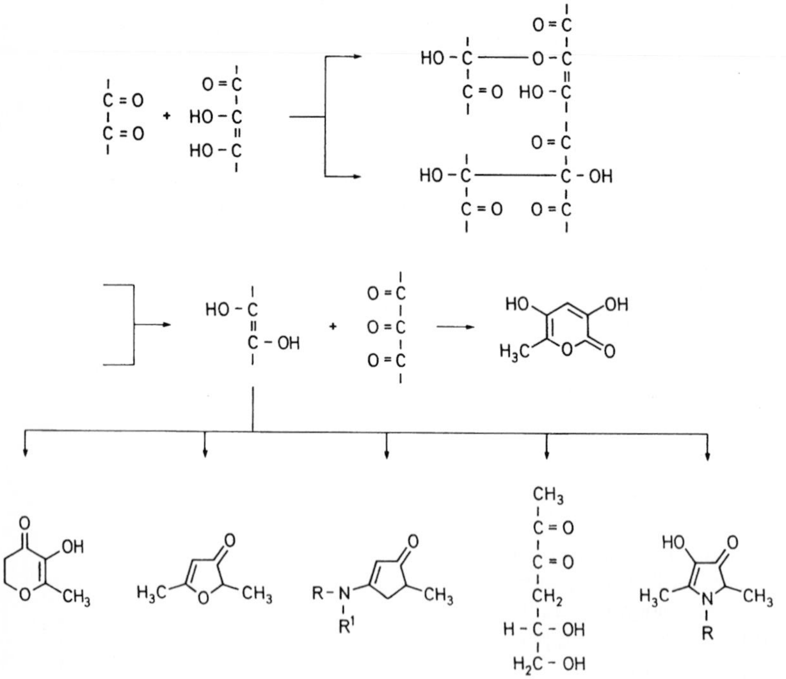

(4.80)

(4.81)

4.2.4.4.7 Strecker Reaction

The reactions between α-dicarbonyl compounds, like the deoxyosones obtained in the *Maillard* reaction, and amino acids are classed under the term *Strecker* reaction. This reaction involves transamination and gives aminoketones, aldehydes and CO_2 (cf. 1.2.4.2.4). It occurs in foods at higher concentrations of free amino acids and under more drastic reaction conditions, e.g., at higher temperatures or under pressure:

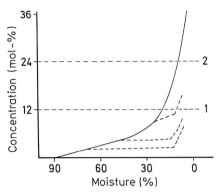

Fig. 4.10. Increase of *Amadori* compounds in two stage air drying of carrots as influenced by carrot moisture content. (— 10, 20, 30 min at 110 °C; ——— 60 °C; sensory assessment: 1) detection threshold 2) quality limit (according to *Eichner* and *Wolf*, in *Waller* and *Feather*, 1983)

(4.82)

The aldehydes formed, often called *Strecker* aldehydes, can act as food odorants (cf. 5.3.1.1). The aminoketone formed, on the other hand, can yield pyrazine derivatives (also powerful aroma constituents):

(4.83)

Other reactions have been discussed with the different deoxyosones (cf. 4.2.4.4.3–4.2.4.4.5), including those involving proline and hydroxyproline (cf. Formulas 4.68, 4.69 and 4.77).

When amino acids with functional groups in the side chain are involved, even more complex reactions are possible (cf. 5.3.1.4–5.3.1.8).

4.2.4.4.8 Inhibition of the Maillard Reaction

Measures to inhibit the *Maillard* reaction in cases where it is undesirable involve lowering of the pH value, maintenance of lowest possible temperatures and avoidance of critical water contents (cf. 0.3.2) during processing and storage, use of nonreducing sugars, and addition of sulfite. Figure 4.10 demonstrates by the example of carrot dehydration the advantages of running a two-stage process to curtail the *Maillard* reaction.

4.2.4.5 Reactions with Hydroxy Compounds (O-Glycosides)

The lactol group of monosaccharides heated in alcohol in the presence of an acid catalyst is substituted by an alkoxy or aryloxy group, denoted as an aglycone (*Fischer* synthesis), to produce alkyl- and arylglycosides. It is assumed that the initial reaction involves the open form. With the majority of sugars, the furanosides are formed in the first stage of reaction. They then equilibrate with the pyranosides. The transition from furanoside to pyranoside occurs most probably through an open carboxonium ion, whereas pyranoside isomerization is through a cyclic one (cf. Reaction 4.84). Furanosides are obtainable by stopping the reaction at a suitable time. The equilibrium state in alcohol is, as in water, dependent on conformational factors. The alcohol as solvent and its R-moiety both increase the anomeric effect and thus α-pyranoside becomes a more favorable form than was α-pyranose in

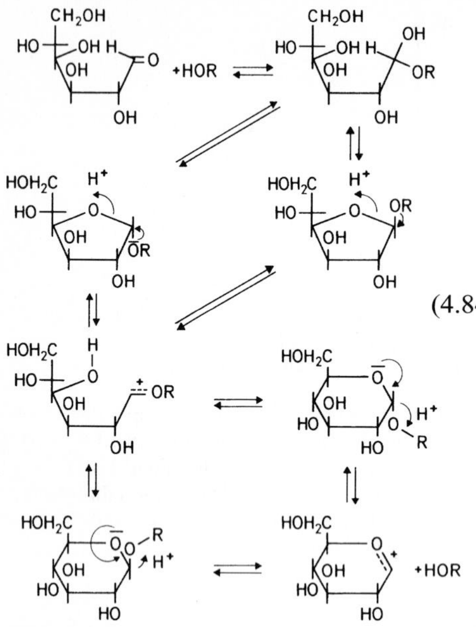

(4.84)

tion form. Due to the steric influence of the acetylated group on C-2, the 1,2-trans-glycoside is preferentially obtained, e.g., in the case of D-glucose, β-glucoside results.

Acetylglycosyl halogenides are also used for a highly stereoselective synthesis of α-glycosides. The compound is first dehalogenated into a glycal. Then, addition of nitrosylchloride follows, giving rise to 2-deoxy-2-nitrosoglycosylchloride. The latter, in the presence of alcohol, eliminates HCl and provides a 2-deoxy-2-oximino-α-glycoside. Reaction with ethanal yields the 2-oxo compound, which is then reduced to α-glycoside:

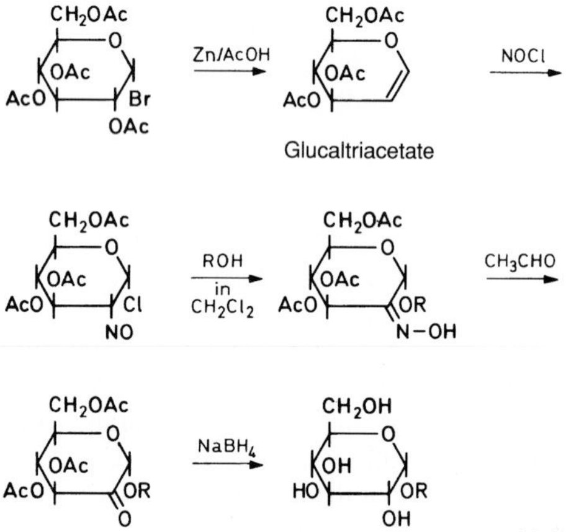

α-Glucopyranoside   (4.86)

aqueous free sugar solutions (Table 4.7). In the system D-glucose/methanol in the presence of 1% HCl, 66% of the methylglucoside is present as α-pyranoside, 32.5% as β-pyranoside, and only 0.6% and 0.9% are in α- and β-furanoside forms. Under the same conditions, D-mannose and D-galactose are 94% and 58% respectively in α-pyranoside forms.

A highly stereospecific access to glycosides is possible by C-1 bromination of acetylated sugars.

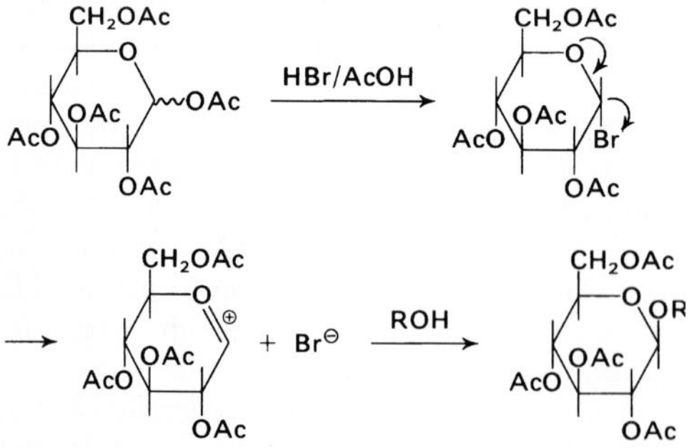

(4.85)

In the reaction of peracetylated sugar with HBr, due to the strong anomeric effect, α-halogenide is formed almost exclusively. This then reacts, probably through its glycosyl ca-

O-Glycosides are widely distributed in nature and are the constituents, such as glycolipids, glycoproteins, flavanoid glycosides or saponins, of many foods.

O-Glycosides are readily hydrolyzed by acids. Hydrolysis by alkalies is achieved only under drastic conditions which simultaneously decompose monosaccharides.

The acid hydrolysis is initiated by glycoside protonation. Alcohol elimination is followed by addition of water:

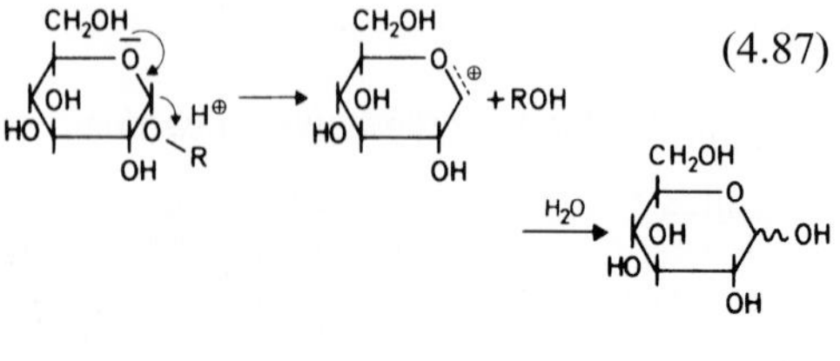

(4.87)

Table 4.13. Relative rate of hydrolysis of glycosides (a: 2 mol/l HCl, 60 °C; b: 0.5 mol/l HCl, 75 °C)

Compound	Hydrolysis condition	k_{rel}
Methyl-α-D-glucopyranoside	a	1.0
Metyhl-β-D-glucopyranoside	a	1.8
Phenyl-α-D-glucopyranoside	a	53.7
Phenyl-β-D-glucopyranoside	a	13.2
Methyl-α-D-glucopyranoside	b	1.0
Methyl-β-D-glucopyranoside	b	1.9
Methyl-α-D-mannopyranoside	b	2.4
Methyl-β-D-mannopyranoside	b	5.7
Methyl-α-D-galactopyranoside	b	5.2
Methyl-β-D-galactopyranoside	b	9.2

The hydrolysis rate is dependent on the aglycone and the monosaccharide itself. The most favored form of alkylglycoside, α-pyranoside, usually is the isomer most resistant to hydrolysis. This is also true for arylglycosides, however, due to steric effects, the β-pyranoside isomer is synthesized preferentially and so the β-isomer better resists hydrolysis.

The influence of the sugar moiety on the rate of hydrolysis is related to the conformational stability. Glucosides with high conformational stability are hydrolyzed more slowly (cf. data compiled in Table 4.13).

4.2.4.6 Esters

Esterification of monosaccharides is achieved by reaction of the sugar with an acyl halide or an acid anhydride. Acetylation, for instance with acetic anhydride, is carried out in pyridine solution:

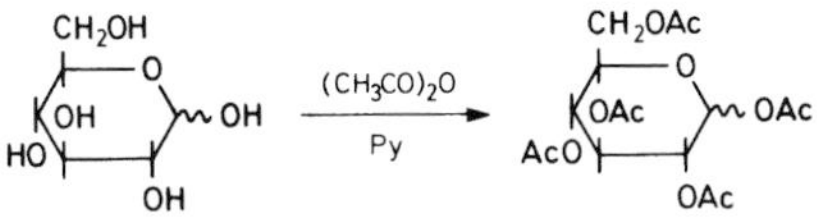

(4.88)

Acyl groups have a protective role in some synthetic reactions. Aldonitrile acetates are analytically suitable sugar derivatives for gas chromatographic separation and identification of sugars. An advantage of these compounds is that they simplify a chromatogram since there are no anomeric peaks:

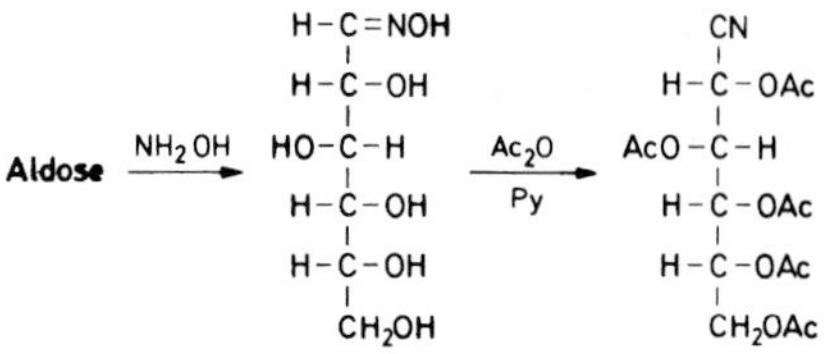

(4.89)

Selective esterification of a given HO-group is also possible. For example, glucose can be selectively acetylated in position 3 by reacting 1,2,5,6-di-O-isopropylidene-α-D-glucofuranose with acetic acid anhydride, followed by hydrolysis of the diketal:

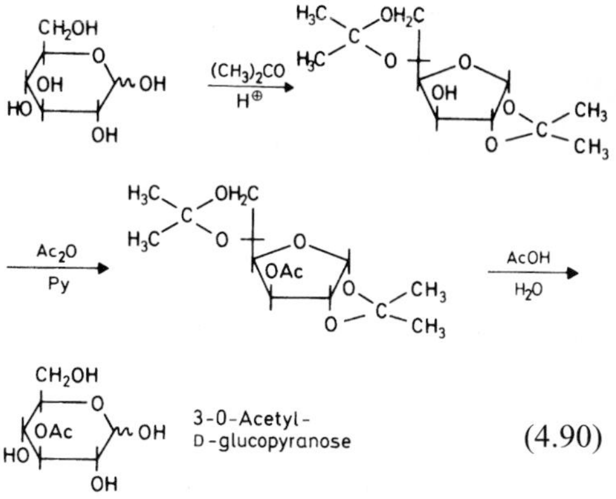

(4.90)

Hydrolysis of acyl groups can be achieved by interesterification or by an ammonolysis reaction:

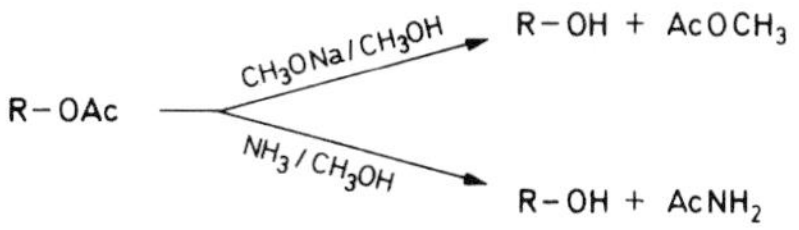

(4.91)

Sugar esters are also found widely in nature. Phosphoric acid esters are important intermediary products of metabolism, while sulfuric acid esters are constituents of some polysaccharides. Examples of organic acid esters are vacciniin in blueberry (6-benzoyl-D-glucose) and the tannintype compound, corilagin (1,3,6-trigalloyl-D-glucose):

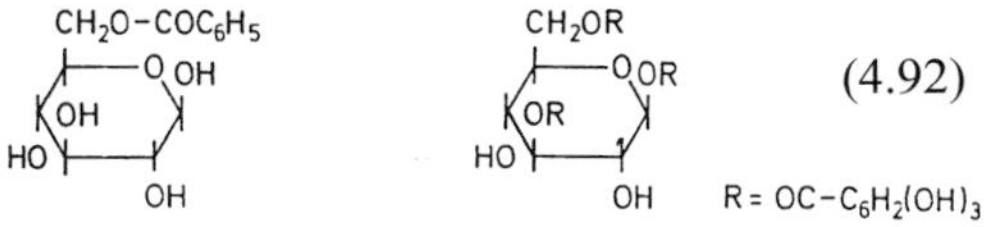

(4.92)

$R = OC-C_6H_2(OH)_3$

Sugar esters or sugar alcohol esters with long chain fatty acids (lauric, palmitic, stearic and oleic) are produced industrially and are very important as surface-active agents. These include sorbitan fatty acid esters (cf. 8.15.3.3) and those of saccharose (cf. 8.15.3.2), which have diversified uses in food processing.

4.2.4.7 Ethers

Methylation of sugar HO-groups is possible using dimethylsulfate or methyliodide as the methylating agent. Methyl ethers are of importance in analysis of sugar structure since they provide data about ring size and linkage positions.

Permethylated saccharose, for example, after acid hydrolysis provides 2,3,4,6-tetra-O-methyl-D-glucose and 1,3,4,6-tetra-O-methyl-D-fructose. This suggests the presence of a 1,2'-linkage between the two sugars and the pyranose and furanose structures for glucose and fructose, respectively:

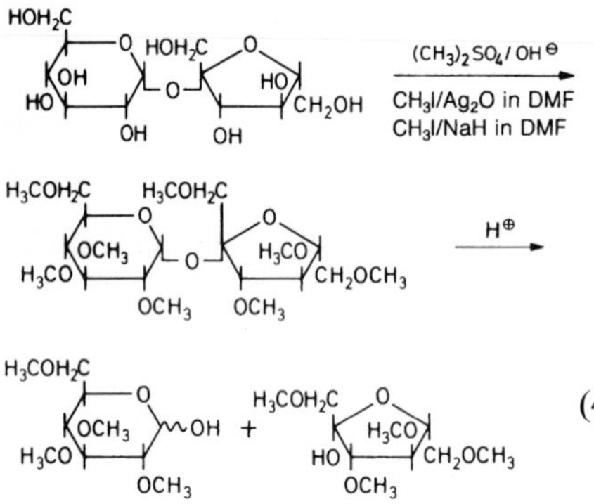

(4.93)

Trimethylsilyl ethers (TMS-ethers) are unstable against hydrolysis and alcoholysis, but have remarkable thermal stability and so are suitable for gas chromatographic sugar analysis. Treatment of a sugar with hexamethyldisilazane and trimethylchlorosilane, in pyridine as solvent, provides a sugar derivative with all HO-groups silylated:

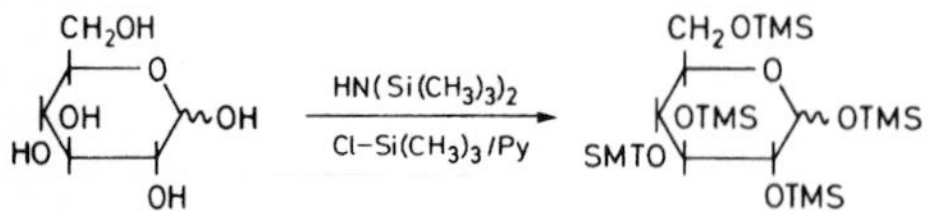

(4.94)

4.2.4.8 Halodeoxy Derivatives

Halodeoxy derivatives of sugars are of synthetic interest because they are readily convertible to amino- or deoxy- and anhydro compounds by nucleophilic substitution or reduction. On the other hand, they can be much sweeter than the sugars themselves depending on the type, number, and position of the halogen substituents. For this reason, their use as potential sweeteners is under discussion (cf. 8.8.19). Saccharose has been subjected to intensive synthetic processing. The following examples show that many different saccharose derivatives are accessible by selecting suitable halogenation agents in combination with selective blocking and deblocking of hydroxyl groups.

In 1975, the first chlorinated disaccharide with $f_{sac,g}$ (10) = 200 to be described was 4,6-dichloro-4,6-dideoxy-α-D-galactopyranosyl-1,6-dichloro-1,6-dideoxy-β-D-fructo-furanoside (4,6,1',6'-tetrachlorogalactosaccharose). With inversion at C4, saccharose reacts with sulfuryl chloride in pyridine at −30°C to yield the 4,6,6'-trichloro derivative, which is converted to the 1-sulfonate by reaction with mesitylenesulfonyl chloride. Nucleophilic substitution with LiCl in DMF gives the tetrachloro compound:

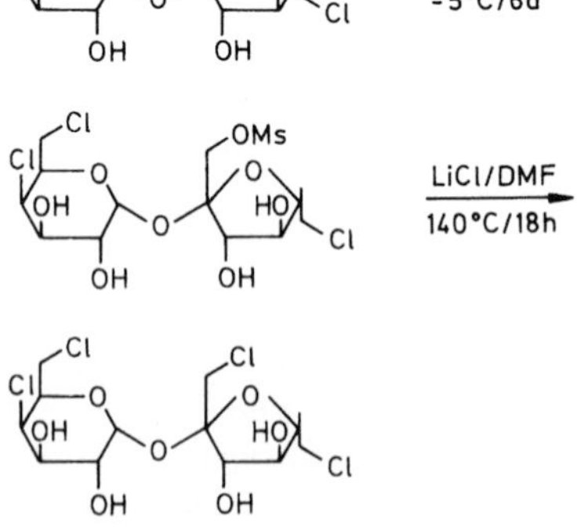

(4.95)

In the preparation of the 1'-chloro derivative ($f_{sac,g}$ (10) = 20), the hydroxyl groups in the 1'-

and 2-position are first blocked by ketalization with dimethoxydiphenylsilane. After acetylation of the remaining six hydroxyl groups with acetic anhydride, the silicon ketal is cleaved again. The primary hydroxyl group is then tritylated and the secondary acetylated. Selective elimination of the trityl group from the 1'-position is followed by chlorination with sulfuryl chloride and LiCl and subsequent deacetylation:

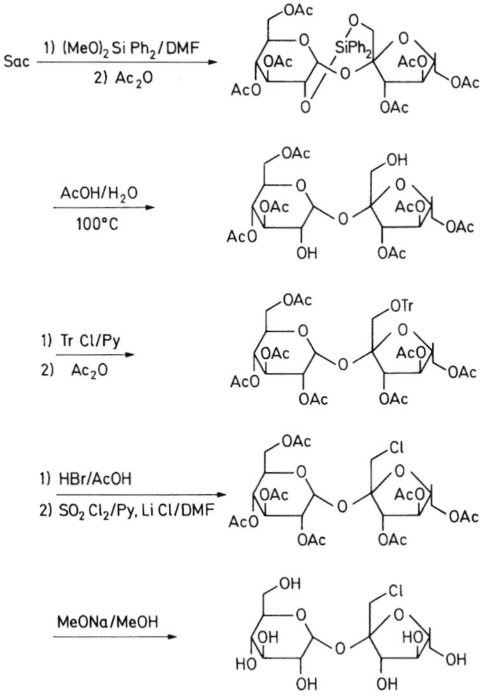

Sac $\xrightarrow[\text{2) } Ac_2O]{\text{1) } (MeO)_2 Si Ph_2/DMF}$

$\xrightarrow[100°C]{AcOH/H_2O}$

$\xrightarrow[\text{2) } Ac_2O]{\text{1) } Tr Cl/Py}$

$\xrightarrow[\text{2) } SO_2 Cl_2/Py, Li Cl/DMF]{\text{1) } HBr/AcOH}$

$\xrightarrow{MeONa/MeOH}$

(4.96)

The 6'-chloro derivative ($f_{sac,g}$ (10) = 20) is accessible through selective tritylation in the 6'-position, followed by exhaustive acetylation, detritylation, chlorination, and deacetylation:

Sac $\xrightarrow[\text{2) } Ac_2O]{\text{1) } Tr Cl/Py\ (20°C)}$

$\xrightarrow[\text{2) } SO_2Cl_2/Py]{\text{1) } HBr/AcOH}$

$\xrightarrow{MeONa/MeOH}$

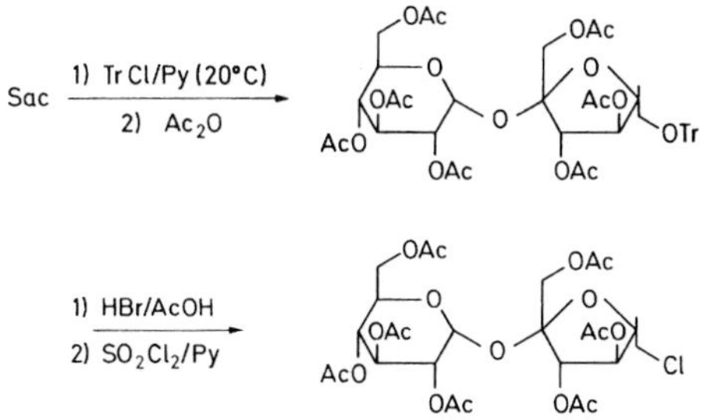

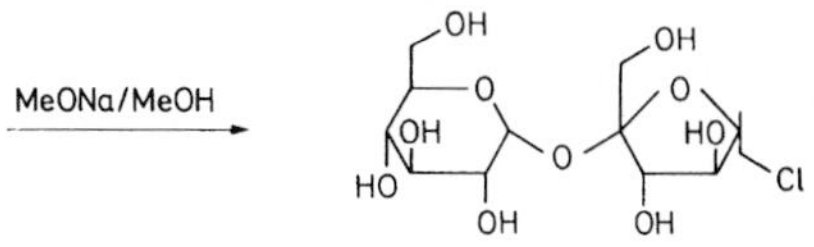

(4.97)

The pathway leading to the 1',6'-dichloro derivative ($f_{sac,g}$ (10) = 80) involves selective sulfonylation of saccharose with mesitylenesulfonyl chloride. The desired 1',6'-disulfonate is chromatographically separated from the 6,1',6'-trisulfonate formed as the main product. It is then acetylated, and subsequently chlorinated and deacetylated:

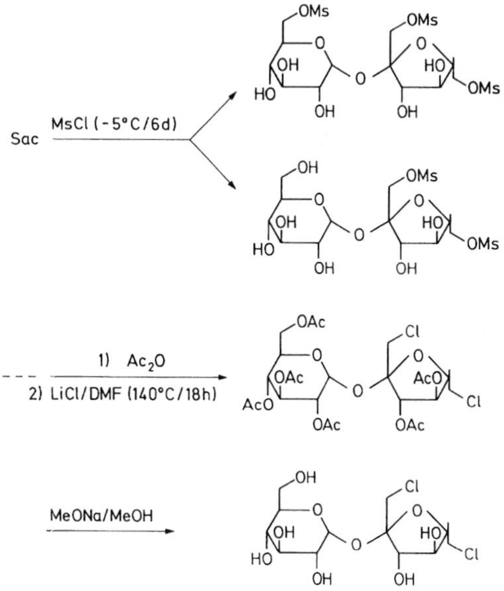

Sac $\xrightarrow{MsCl\ (-5°C/6d)}$

$\xrightarrow[\text{2) } LiCl/DMF\ (140°C/18h)]{\text{1) } Ac_2O}$

$\xrightarrow{MeONa/MeOH}$

(4.98)

Partial tritylation of saccharose with trityl chloride in pyridine at room temperature yields a mixture of trityl derivatives. The 6-trityl compound is chromatographically separated from this mixture in low yields, and acetylated. Detritylation is accompanied by migration of the acetyl group from the 4- to the 6-position. Subsequent chlorination in the 4-position and deacetylation gives 4-chloro-galactosaccharose ($f_{sac,g}$ (10) = 5) (Formula 4.99):

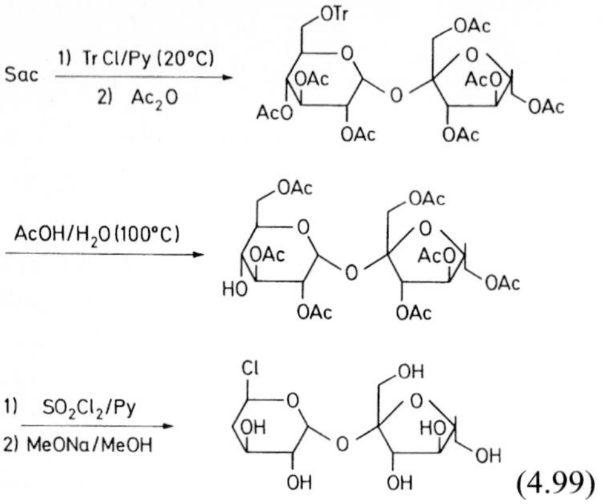

Sac 1) Tr Cl/Py (20°C)
     2) Ac₂O
→

AcOH/H₂O (100°C)
→

1) SO₂Cl₂/Py
2) MeONa/MeOH
→
(4.99)

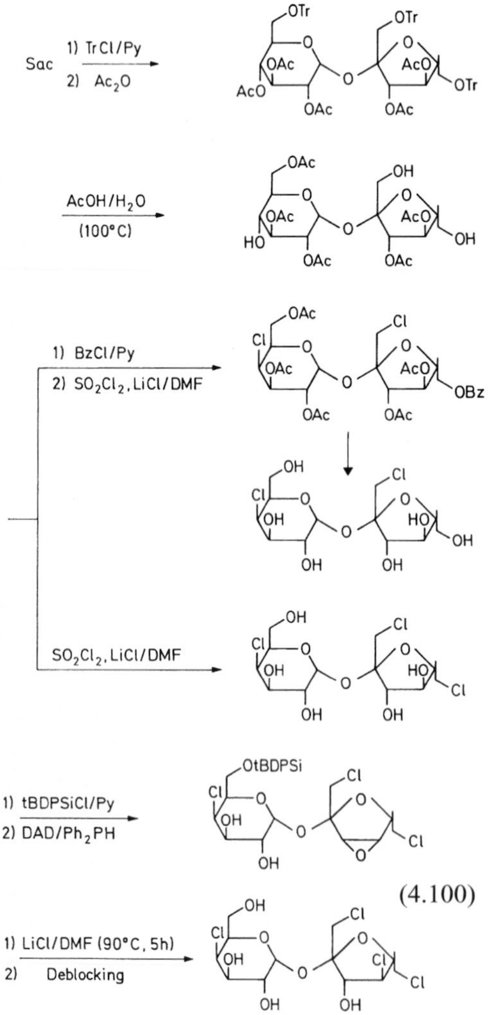

Sac 1) TrCl/Py
     2) Ac₂O
→

AcOH/H₂O
(100°C)
→

1) BzCl/Py
2) SO₂Cl₂, LiCl/DMF
→

↓

SO₂Cl₂, LiCl/DMF
→

1) tBDPSiCl/Py
2) DAD/Ph₂PH
→
(4.100)

1) LiCl/DMF (90°C, 5h)
2)   Deblocking
→

Complete tritylation of the three primary hydroxyl groups, acetylation of the secondary hydroxyl groups, and subsequent detritylation gives saccharose-2,3,6,3',4'-pentaacetate on migration of the acetyl residue from the 4- to the 6-position. Selective benzoylation in the 6'-position, followed by chlorination in the 4- and 1'-position yields 4,1'-dichlorogalactosaccharose ($f_{sac, g}$ (10) = 120) after deacetylation. If the 6'-position is not protected, chlorination of the pentaacetate gives the 4,1',6'-trichloro compound (sucralose: $f_{sac, g}$ (10) = 650), which is convertible to 4,1',4',6'-tetrachlorogalactosaccharose ($f_{sac,g}$ (10) = 2200) via several steps. First, the 6-hydroxyl group of the trichloro derivative is selectively protected by reaction with tert-butyldiphenylsilyl chloride. Subsequently, treatment with diethylazodicarboxylate and diphenylphosphine yields the 3',4'-epoxide, which is cleaved with LiCl to give the 4'-chloro derivative. Elimination of the protective group produces the desired tetrachloro derivative (cf. Formula 4.100).

The compound 2,6,1',6'-tetrachlorosaccharose is bitter. It is obtained by selective chlorination with methanesulfonyl chloride to give the 6,6'-dichloro derivative, and blocking of the 2- and 1'-hydroxyl groups with 2,2-dimethoxypropane. This is followed by acetylation of the remaining hydroxyl groups, nucleophilic substitution of chlorine for the hydroxyl groups released by hydrolysis of the ketal, which proceeds with inversion at C2, and elimination of the acetyl residues:

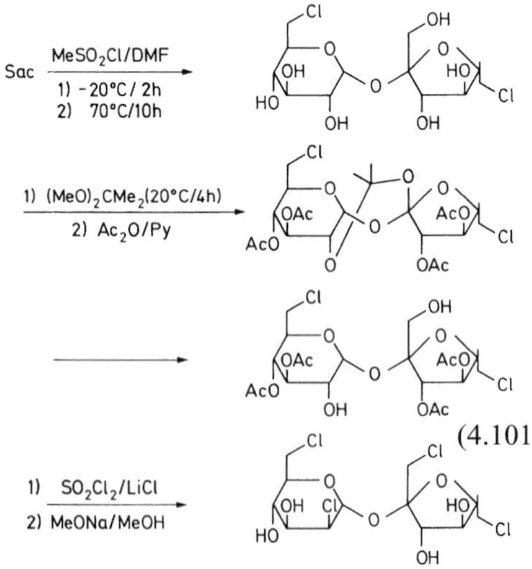

Sac MeSO₂Cl/DMF
     1) -20°C/ 2h
     2) 70°C/10h
→

1) (MeO)₂CMe₂(20°C/4h)
2) Ac₂O/Py
→

→

1) SO₂Cl₂/LiCl
2) MeONa/MeOH
→
(4.101)

Fluoro-, bromo-, and iododeoxy derivatives of saccharose and compounds with mixed halogen substitution are prepared in a similar manner. The taste properties of these compounds are discussed in Section 8.8.19.

4.2.4.9 Cleavage of Glycols

Oxidative cleavage of vicinal dihydroxy groups or hydroxy-amino groups of a sugar with lead tetraacetate or periodate is of importance for structural elucidation. Fructose, in a 5-membered furanose form, consumes 3 moles of periodate (splitting of each α-glycol group requires 1 mole of oxidant) while, in a pyranose ring form, it consumes 4 moles of periodate.

Saccharose consumes 3 moles and maltose 4 moles of periodate (cf. Reaction 4.102).

The final conclusion as to sugar linkage positions and ring structure is drawn from the periodate consumption, the amount of formic acid produced (in the case of saccharose, 1 mole; maltose, 2 moles) and the other carbonyl fragments which are oxidized additionally by bromine to stable carboxylic acids and then released by hydrolysis. The glycol splitting reaction should be considered an optional or complementary method to the permethylation reaction applied in structural elucidation of carbohydrates.

4.3 Oligosaccharides

4.3.1 Structure and Nomenclature

Monosaccharides are able to form glycosides (cf. 4.2.4.5). When this occurs between the lactol group of one monosaccharide and any HO-group of a second monosaccharide, a disaccharide results.

Compounds with up to about 10 monosaccharide residues are designated as oligosaccharides. When a glycosidic linkage is established only between the lactol groups of two monosaccharides, then a *nonreducing disaccharide* is formed, and when one lactol group and one alcoholic HO-group are involved, a *reducing disaccharide* results. The former is denoted as a glycosylglycoside, the latter as a glycosylglycose, with additional data for linkage direction

and positions. Examples are saccharose and maltose:

$$(4.102)$$

An abbreviated method of nomenclature is to use a three letter designation or symbol for a monosaccharide and suffix f or p for furanose or pyranose. For example, saccharose and maltose can be written as O-β-D-Fru$f$ $(2 \rightarrow 1)$$\alpha$-D-Glc$p$ and O-α-D-Glc$p$$(1 \rightarrow 4)$D-Glc$p$, respectively.

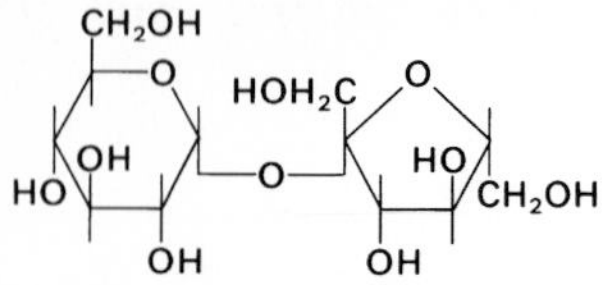

β-D-Fructofuranosyl-α-D-
glucopyranoside
(saccharose)

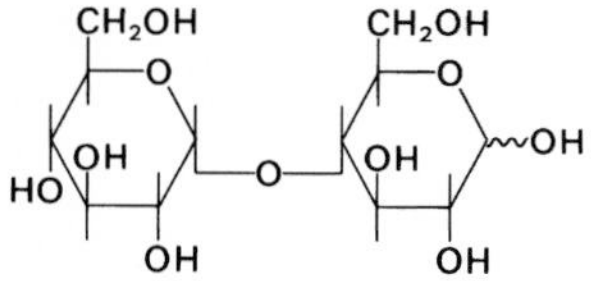

(4.103)

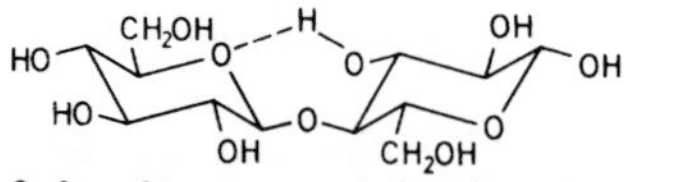

O-α-D-Glucopyranosyl-(1 → 4)-
D-glucopyranose
(maltose)

Branching also occurs in oligosaccharides. It results when one monosaccharide is bound to two glycosyl residues. The name of the second glycosyl residue is inserted into square brackets. A trisaccharide which represents a building block of the branched chain polysaccharides amylopectin and glycogen is given as an example:

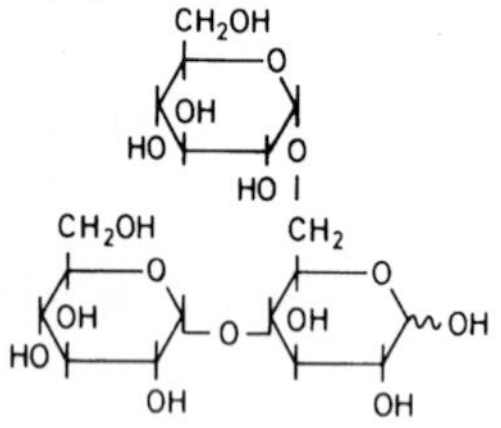

(4.104)

0-α-D-Glucopyranosyl-(1→4)-0-[α-D-
glucopyranosyl-(1→6)]-D-glucopyranose

The abbreviated formula for this trisaccharide is as follows:

α-D-Glcp (1 → 4) Glcp
                 (6
                  ↑   (4.105)
                 1)-α-D-Glcp

The conformations of oligo- and polysaccharides, like peptides, can be described by providing the angles Φ and ψ:

(4.106)

A calculation of conformational energy for all conformers with allowed Φ, ψ pairs provides a Φ, ψ diagram with lines corresponding to iso-conformational energies. The low-energy conformations calculated in this way agree with data obtained experimentally (X-ray diffraction, NMR, ORD) for oligo- and polysaccharides.

H-bonds fulfill a significant role in conformer stabilization. Cellobiose and lactose conformations are well stabilized by an H-bond formed between the HO-group of C-3 in the glucose residue and the ring oxygen of the glycosyl residue. Conformations in aqueous solutions appear to be similar to those in the crystalline state:

(4.107)

0-β-D-Glucopyranosyl-(1→4)-D-glucopyranose (cellobiose)

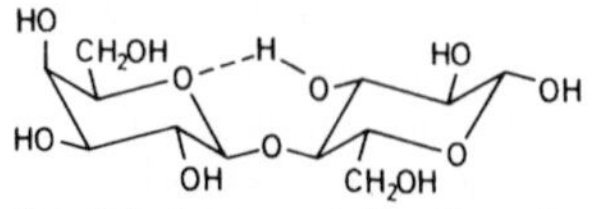

0-β-D-Galactopyranosyl-(1→4)-D-glucopyranose (lactose)

In crystalline maltose and in nonaqueous solutions of this sugar, a hydrogen bond is established between the HO-groups on C-2 of the glucosyl and on C-3 of the glucose residues (4.108). However, in aqueous solution, a conformer partially present is stabilized by H-bonds established between the CH₂OH-group of the glucosyl residue and the HO-group of C-3 on the glucose residue (4.109). Both conformers correspond to the two energy minima in the Φ, ψ diagram.

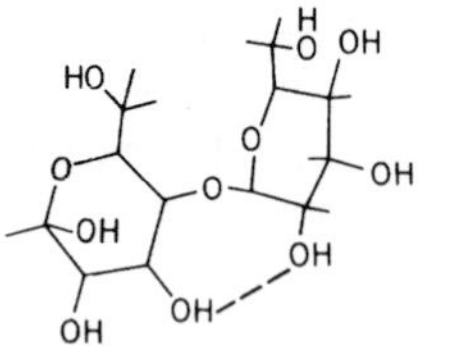

(4.108)

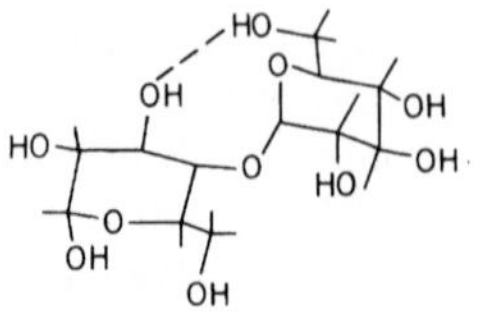

(4.109)

Two H-bonds are possible in saccharose, the first between the HO-groups on the C-1 of the fructose and the C-2 of the glucose residues, and the second between the HO-group on the C-6 of the fructose residue and the ring oxygen of the glucose residue:

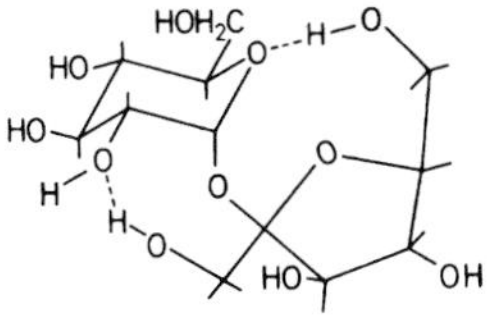

$$(4.110)$$

β -D-Fructofuranosyl -α-D-glucopyranoside (saccharose)

4.3.2 Properties and Reactions

The oligosaccharides of importance to food, together with data on their occurrence, are compiled in Table 4.14. The physical and sensory properties were covered with monosaccharides, as were reaction properties, though the difference between reducing and nonreducing oligosaccharides should be mentioned. The latter do not have a free lactol group and so lack reducing properties, mutarotation and the ability to react with alcohols and amines.

As glycosides, oligosaccharides are readily hydrolyzed by acids, while they are relatively stable against alkalies. Saccharose hydrolysis is denoted as an inversion and the resultant equimolar mixture of glucose and fructose is called invert sugar. The term is based on a change of specific rotation during hydrolysis. In saccharose the rotation is positive, while it is negative in the hydrolysate, since D-glucose rotation to the right (hence its name dextrose) is surpassed by the value of the left-rotating fructose (levulose):

$$\text{Saccharose} \xrightarrow{H^{\oplus}} \text{D-Glucose + D-Fructose}$$

$$[\alpha]_D = +66.5° \qquad [\alpha]_D = +52.7° [\alpha]_D = -92.4°$$

$$[\alpha]_D = -19.8°$$

$$(4.111)$$

Conclusions can be drawn from mutarotation, which follows hydrolysis of reducing disaccharides, about the configuration on the anomeric C-atom. Since the α-anomer has a higher specific rotation in the D-series than the

β-anomer, cleavage of β-glycosides increases the specific rotation while cleavage of α-glycosides decreases it:

$$\text{Cellobiose }(\beta) \longrightarrow 2 \text{ D-Glucose}$$

$$[\alpha]_D = +34.6° \qquad [\alpha]_D = +52.7°$$

$$\text{Maltose }(\alpha) \longrightarrow 2 \text{ D-Glucose}$$

$$[\alpha]_D = +130° \qquad [\alpha]_D = +52.7°$$

$$\text{Lactose }(\beta) \longrightarrow \text{D-Galactose + D-Glucose}$$

$$[\alpha]_D = +52.3° \qquad [\alpha]_D = +80.2° \quad [\alpha]_D = +52.7°$$

$$[\alpha]_D = +66.3°$$

$$(4.112)$$

Enzymatic cleavage of the glycosidic linkage is specified by the configuration on anomeric C-1 and also by the whole glycosyl moiety, while the aglycone residue may vary within limits.

The methods used to elucidate the linkage positions in an oligosaccharide (methylation, oxidative cleavage of glycols) were outlined under monosaccharides.

The cyclodextrins listed in Table 4.14 are prepared by the action of cyclomaltodextrin glucanotransferase (E. C. 2.4.1.19), obtained from *Bacillus macerans*, on maltodextrins. Maltodextrins are, in turn, made by the degradation of starch with α-amylase. This glucanotransferase splits the α-1,4-bond, transferring glucosyl groups to the nonreducing end of maltodextrins and forming cyclic glucosides with 6- 12 glucopyranose units. The main product, β-cyclodextrin, consists of seven glucose units and is a non-hygroscopic, slightly sweet compound:

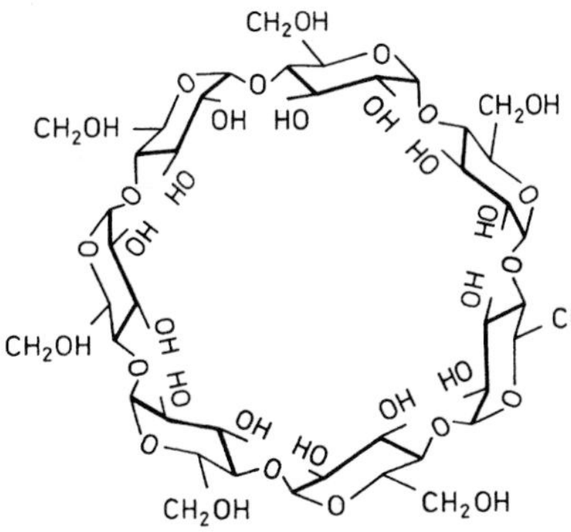

$$(4.113)$$

Table 4.14. Structure and occurrence of oligosaccharides

Name	Structure	Occurrence
Disaccharides		
Cellobiose	O-β-D-Glc*p*-(1 → 4)-D-Glc*p*	Building block of cellulose
Gentiobiose	O-β-D-Glc*p*-(1 → 6)-D-Glc*p*	Glycosides (amygdalin)
Isomaltose	O-α-D-Glc*p*-(1 → 6)-D-Glc*p*	Found in mother liquor during glucose production from starch
Lactose	O-β-D-Gal*p*-(1 → 4)-D-Glc*p*	Milk
Lactulose	O-β-D-Gal*p*-(1 → 4)-D-Fru*p*	Conversion product of lactose
Maltose	O-α-D-Glc*p*-(1 → 4)-D-Glc*p*	Building block of starch, sugar beet, honey
Maltulose	O-α-D-Glc*p*-(1 → 4)-D-Fru*f*	Conversion product of maltose, honey, beer
Melibiose	O-α-D-Gal*p*-(1 → 6)-D-Glc*p*	Cacao beans
Neohesperidose	O-α-L-Rha*p*-(1 → 2)-D-Glc*p*	Glycosides (naringin, neohesperidin)
Neotrehalose	O-α-D-Glc*p*-(1 → 1)-β-D-Glc*p*	Koji extract
Nigerose	O-α-D-Glc*p*-(1 → 3)-D-Glc*p*	Honey, beer
Palatinose	O-α-D-Glc*p*-(1 → 6)-D-Fru*f*	Microbial product of saccharose
Rutinose	O-α-L-Rha*p*-(1 → 6)-D-Glc*p*	Glycosides (hesperidin)
Saccharose	O-β-D-Fru*f*-(2 → 1)-α-D-Glc*p*	Sugar beet, sugar cane, spread widely in plants
Sophorose	O-β-D-Glc*p*-(1 → 2)-D-Glc*p*	Legumes
Trehalose	O-α-D-Glc*p*-(1 → 1)-α-D-Glc*p*	Ergot (*Claviceps purpurea*), young mushrooms
Trisaccharides		
Fucosidolactose	O-α-D-Fuc*p*-(1 → 2)-O-β-α-Gal*p*-(1 → 4)-D-Gal*p*	Human milk
Gentianose	O-β-D-Glc*p*-(1 → 6)-O-α-D-Glc*p*-(1 → 2)-β-D-Fru*f*	Gentian rhizome
Isokestose (1-Kestose)	O-α-D-Glc*p*-(1 → 2)-O-β-D-Fru*f*-(1 → 2)-β-D-Fru*f*	Product of saccharase action on saccharose as a substrate
Kestose (6-Kestose)	O-α-D-Glc*p*-(1 → 2)-O-β-D-Fru*f*-(6 → 2)-β-D-Fru*f*	Saccharose subjected to yeast saccharase activity, honey
Maltotriose	O-α-D-Glc*p*-(1 → 4)-O-α-D-Glc*p*-(1 → 4)-D-Glc*p*	Degradation product of starch, starch syrup
Manninotriose	O-α-D-Gal*p*-(1 → 6)-O-α-D-Gal*p*-(1 → 6)-D-Glc*p*	Manna
Melezitose	O-α-D-Glc*p*-(1 → 3)-O-β-D-Fru*f*-(2 → 1)-α-D-Glc*p*	Manna, nectar
Neokestose	O-β-D-Fru*f*-(2 → 6)-O-α-D-Glc*p*-(1 → 2)-β-D-Fru*f*	Product of saccharase action on saccharose as a substrate
Panose	O-α-D-Glc*p*-(1 → 6)-O-α-D-Glc*p*-(1 → 4)-D-Glc*p*	Degradation product of amylopectin, honey
Raffinose	O-α-D-Gal*p*-(1 → 6)-O-α-D-Glc*p*-(1 → 2)-β-D-Fru*f*	Sugar beet, sugar cane, widely distributed in plants
Umbelliferose	O-α-D-Gal*p*-(1 → 2)-O-α-D-Glc*p*-(1 → 2)-β-D-Fru*f*	Umbelliferae roots
Tetrasaccharides		
Maltotetraose	O-α-D-Glc*p*-(1 → 4)-O-α-D-Glc*p*-(1 → 4)-O-α-D-Glc*p*-(1 → 4)-D-Glc*p*	Starch syrup
Stachyose	O-α-D-Gal*p*-(1 → 6)-O-α-D-Gal*p*-(1 → 6)-O-α-D-Glc*p*-(1 → 2)-β-D-Fru*f*	Widespread in plants (artichoke, soybean)
Higher oligosaccharides		
Maltopentaose	[O-α-D-Glc*p*-(1 → 4)]₄-D-Glc*p*	Starch syrup
α-*Schardinger*-Dextrin, Cyclohexaglucan (α, 1 → 4)		Growth of
β-*Schardinger*-Dextrin, Cycloheptaglucan (α, 1 → 4)		*Bacillus macerans*
γ-*Schardinger*-Dextrin, Cyclooctaglucan (α, 1 → 4)		on starch syrup

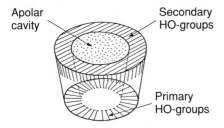

Fig. 4.11. Schematic representation of the hollow cylinder formed by β-cyclodextrin

The β-cyclodextrin molecule is a cylinder (Fig. 4.11) which has a primary hydroxyl (C6) rim on one side and a secondary hydroxyl (C2, C3) rim on the other. The surfaces made of pyranose rings are hydrophobic. Indeed, the water of hydration is very easily displaced from this hydrophobic cavity by sterically suitable apolar compounds, which are masked in this way.

In food processing, β-cyclodextrin is therefore a suitable agent for stabilizing vitamins and aroma substances and for neutralizing the taste of bitter substances

4.4 Polysaccharides

4.4.1 Classification, Structure

Polysaccharides, like oligosaccharides, consist of monosaccharides bound to each other by glycosidic linkages. Their acidic hydrolysis yields monosaccharides. Partial chemical and enzymatic hydrolysis, in addition to total hydrolysis, are of importance for structural elucidation. Enzymatic hydrolysis provides oligosaccharides, the analysis of which elucidates monosaccharide sequences and the positions and types of linkages.

Polysaccharides (glycans) can consist of one type of sugar structural unit (homoglycans) or of several types of sugar units (heteroglycans). The monosaccharides may be joined in a linear pattern (as in cellulose and amylose) or in a branched fashion (amylopectin, glycogen, guaran). The frequency of branching sites and the length of side chains can vary greatly (glycogen, guaran). The monosaccharide residue sequence may be periodic, one period containing one or several alternating structural units (cellulose, amylose or hyaluronic acid), the sequence may contain shorter or longer segments with periodically arranged residues separated by nonperiodic segments (alginate, carrageenans, pectin), or the sequence may be nonperiodic all along the chain (as in the case of carbohydrate components in glycoproteins).

4.4.2 Conformation

The monosaccharide structural unit conformation and the positions and types of linkages in the chain determine the chain conformation of a polysaccharide. In addition to irregular conformations, regular conformations are known which reflect the presence of at least a partial periodic sequence in the chain. Some typical conformations will be explained in the following discussion, with examples of glucans and some other polysaccharides.

4.4.2.1 Extended or Stretched, Ribbon-Type Conformation

This conformation is typical for 1,4-linked β-D-glucopyranosyl residues (Fig. 4.12, a), as occur, for instance, in cellulose fibers:

$$(4.114)$$

This formula shows that the stretched chain conformation is due to a *zigzag* geometry of monomer linkages involving oxygen bridging. The chain may be somewhat shortened or compressed to enable formation of H-bonds between neighboring residues and thus contribute to conformational stabilization. In the ribbon-type, stretched conformation, with the number of monomers in turn denoted as n and the pitch (advancement) in the axial direction per monomer unit as h, the range of n is from 2 to ± 4, while h is the length of a monomer unit. Thus, the chain given in Fig. 4.12a has $n = -2.55$ and $h = 5.13$ Å.

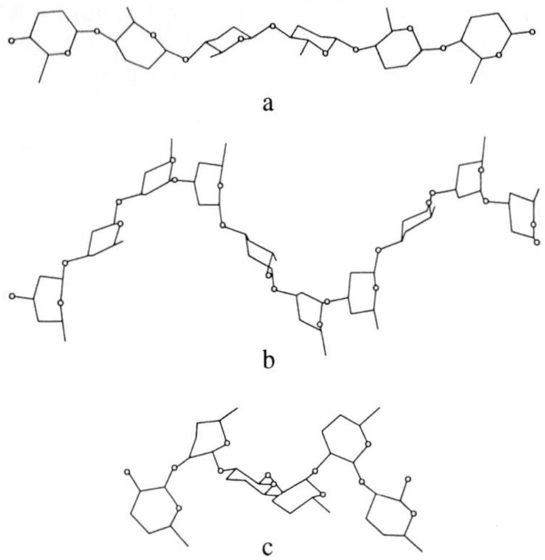

(4.116 cont.)

Ca^{2+} ions can be involved to stabilize the conformation. In this case, two alginate chains are assembled in a conformation which resembles an egg box (*egg box type of conformation*):

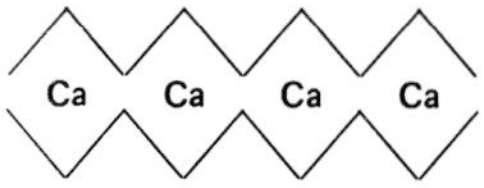

Ca   Ca   Ca   Ca

(4.117)

It should be emphasized that in all examples the linear, ribbon-type conformation has a zigzag geometry as a common feature.

a

b

c

Fig. 4.12. Conformations of some β-D-glucans. Linkages: **a** $1 \to 4$, **b** $1 \to 3$, **c** $1 \to 2$ (according to *Rees*, 1977)

A strongly pleated, ribbon-type conformation might also occur, as shown by a segment of a pectin chain (1,4-linked α-D-galactopyranosyl-uronate units):

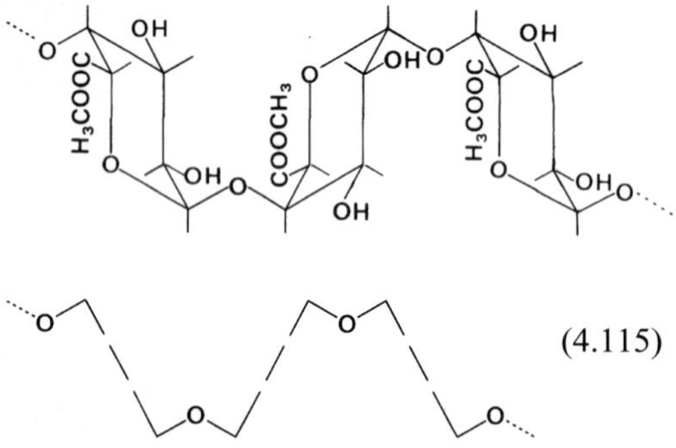

(4.115)

and the same pleated conformation is shown by an alginate chain (1,4-linked α-L-gulopyranosyluronate units):

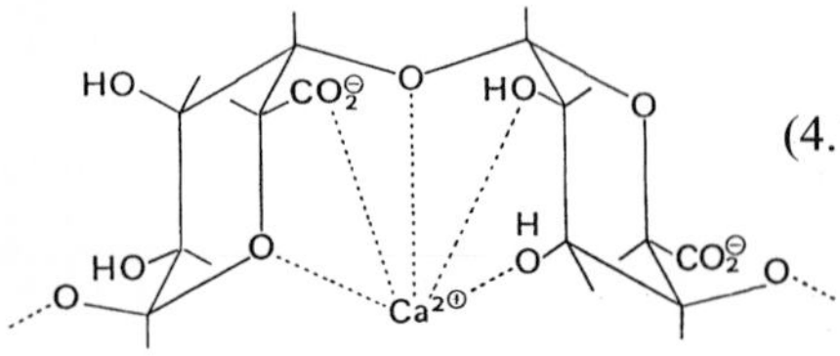

(4.116)

4.4.2.2 Hollow Helix-Type Conformation

This conformation is typical for 1,3-linked β-D-glucopyranose units (Fig. 4.12, b), as occur in the polysaccharide lichenin, found in moss-like plants (lichens):

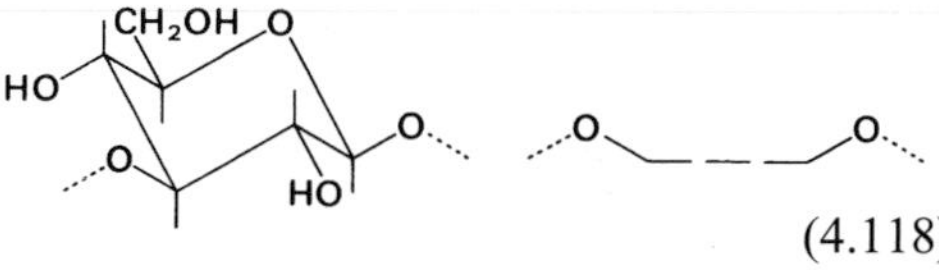

(4.118)

The formula shows that the helical conformation of the chain is imposed by a U-form geometry of the monomer linkages. Amylose (1,4-linked α-D-glucopyranosyl residues) also has such a geometry, and hence a helical conformation:

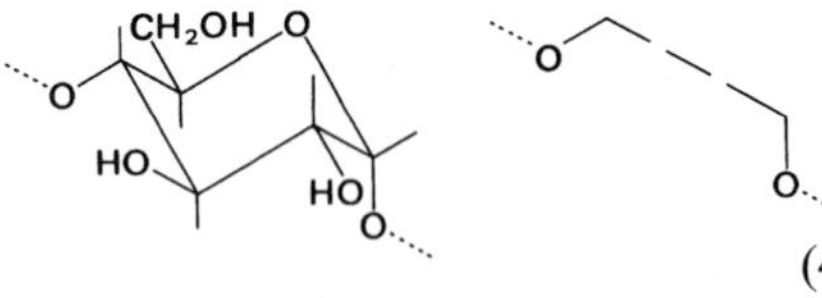

(4.119)

The number of monomers per turn (n) and the pitch in the axial direction per residue (h) is highly variable in a hollow helical conformation.

The value of n is between 2 and ± 10, whereas h can be near its limit value of 0. The conformation of a β$(1 \to 3)$-glucan, with n = 5.64 and h = 3.16 Å, is shown in Fig. 4.12, b.

a b c

Fig. 4.13. Stabilizatioin of helical conformations. **a** Clathrate compounds, **b** coiled double or triple helices, **c** "nesting" (according to *Rees*, 1977)

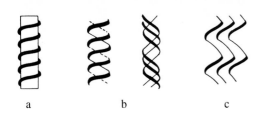

The helial conformation can be stabilized in various ways. When the helix diameter is large, inclusion (clathrate) compounds can be formed (Fig. 4.13, a; cf. 4.4.4.14.3). More extended or stretched chains, with smaller helix diameter, can form double or triple stranded helices (Fig. 4.13, b; cf. 4.4.4.3.2 and 4.4.4.14.3), while strongly-stretched chains, in order to stabilize the conformation, have a zigzag, pleated association and are not stranded (Fig. 4.13, c).

4.4.2.3 Crumpled-Type Conformation

This conformation occurs with, for example, 1,2-linked β-D-glucopyranosyl residues (Fig. 4.12, c). This is due to the wrinkled geometry of the monomer O-bridge linkages:

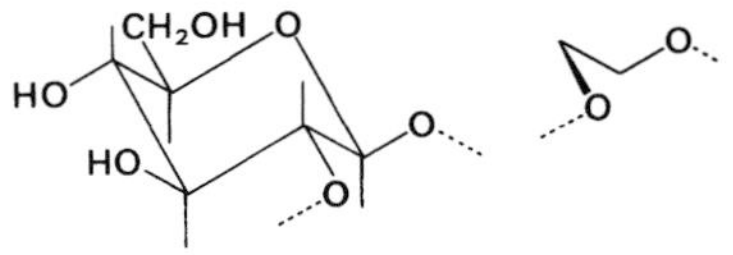

(4.120)

Here, the n value varies from 4 up to − 2 and h is 2–3 Å. The conformation reproduced in Fig. 4.12, c has n = 2.62 and h = 2.79 Å. The likelihood of such a disorderly form associating into more orderly conformations is low. Polysaccharides of this conformational type play only a negligible role in nature.

4.4.2.4 Loosely-Jointed Conformation

This is typical for glycans with 1,6-linked β-D-glucopyranosyl units, because they exhibit a particularly great variability in conformation. The great flexibility of this glycan-type conformation is based on the nature of the connecting bridge between the monomers. The bridge has three free rotational bonds and, furthermore, the sugar residues are further apart:

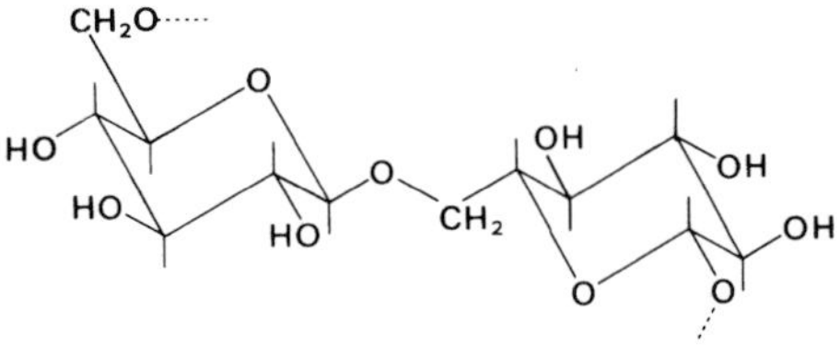

(4.121)

4.4.2.5 Conformations of Heteroglycans

The examples considered so far have demonstrated that a prediction is possible for a homoglycan conformation based on the geometry of the bonds of the monomer units which maintain the oxygen bridges. It is more difficult to predict the conformation of a heteroglycan with a periodic sequence of several monomers, which implies different types of conformations. Such a case is shown by ι-carrageenan, in which the β-D-galactopyranosyl-4-sulfate units have a U-form geometry, while the 3,6-anhydro-α-D-galactopyranosyl-2-sulfate residues have a zigzag geometry:

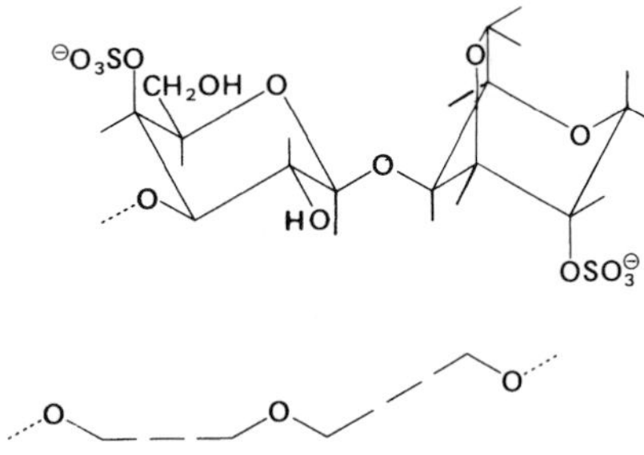

(4.122)

Calculations have shown that conformational possibilities vary from a shortened, compressed ribbon band type to a stretched helix type. X-ray diffraction analyses have proved that a stretched helix exists, but as a double stranded helix in order to stabilize the conformation (cf. 4.4.4.3.2 and Fig. 4.19).

4.4.2.6 Interchain Interactions

It was outlined in the introductory section (cf. 4.4.1) that the periodically arranged monosaccharide sequence in a polysaccharide can be interrupted by nonperiodic segments. Such sequence interferences result in conformational disorders. This will be explained in more detail with ι-carrageenan, mentioned above, since it will shed light on the gel-setting mechanism of macromolecules in general.

Initially, a periodic sequence of altering units of β-D-galactopyranose-4-sulfate (I, conformation $^4C_1$) and α-D-galactopyranose-2,6-disulfate (II, conformation $^4C_1$) is built up in carrageenan biosynthesis:

$$\text{(4.123)}$$

I

II

III

When the biosynthesis of the chain is complete, an enzyme-catalyzed reaction eliminates sulfate from most of α-D-galactopyranose-2,6-disulfate (II), transforming the unit to 3,6-anhydro-α-D-galactopyranose-2-sulfate (III, conformation $^1C_4$). This transformation is associated with a change in linkage geometry. Some II-residues remain in the sequence, acting as interference sites. While the undisturbed, ordered segment of one chain can associate with the same segment of another chain, forming a double helix, the nonperiodic or disordered segments can not participate in such associations (Fig. 4.14).

In this way, a gel is formed with a three-dimensional network in which the solvent is immobilized. The gel properties, e.g., its strength, are influenced by the number and distribution of α-D-galactopyranosyl-2,6-disulfate residues, i.e. by a structural property regulated during polysaccharide biosynthesis.

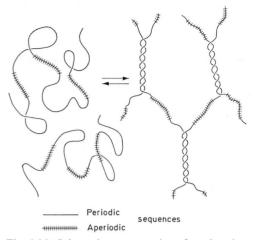

——— Periodic

++++++ Aperiodic   sequences

Fig. 4.14. Schematic representation of a gel setting process (according to *Rees*, 1977)

The example of the ι-carrageenan gel-building mechanism, involving a chain-chain interaction of sequence segments of orderly conformation, interrupted by randomly-coiled segments corresponding to a disorderly chain sequence, can be applied generally to gels of other macromolecules. Besides a sufficient chain length, the structural prerequisite for gel-setting ability is interruption of a periodic sequence and its orderly conformation. The interruption is achieved by insertion into the chain of a sugar residue of a different linkage geometry (carrageenans, alginates, pectin), by a suitable distribution of free and esterified carboxyl groups (glycuronans) or by insertion of side chains. The interchain associations during gelling (network formation), which involve segments of orderly conformation, can then occur in the form of a double helix (Fig. 4.15, a); a multiple bundle of double helices (Fig. 4.15, b); an association between stretched ribbon-type conformations, such as an egg box model (Fig. 4.15, c); some other similar associations (Fig. 4.15, d); or, lastly, forms consisting of double helix and ribbon-type combinations (Fig. 4.15, e).

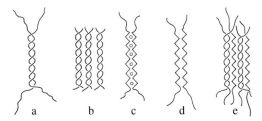

Fig. 4.15. Interchain aggregation between regular conformations. **a** Double helix, **b** double helix bundle, **c** egg-box, **d** ribbon-ribbon, and **e** double helix, ribbon interaction

4.4.3 Properties

4.4.3.1 General Remarks

Polysaccharides are widely and abundantly distributed in nature, fulfilling roles as:

- Structure-forming skeletal substances (cellulose, hemicellulose and pectin in plants; chitin, mucopolysaccharides in animals).
- Assimilative reserve substances (starch, dextrins, inulin in plants; glycogen in animals).
- Water-binding substances (agar, pectin and alginate in plants; mucopolysaccharides in animals).

As a consequence, polysaccharides occur in many food products and even then they often retain their natural role as skeletal substances (fruits and vegetables) or assimilative nutritive substances (cereals, potatoes, legumes). Isolated polysaccharides are utilized to a great extent in food processing, either in native or modified form, as: thickening or gel-setting agents (starch, alginate, pectin, guaran gum); stabilizers for emulsions and dispersions; film-forming, coating substances to protect sensitive food from undesired change; and inert fillers to increase the proportion of indigestible ballast substances in a diet (cf. 15.2.4.2). Table 4.15 gives an overview of uses in food technology.

The outlined functions of polysaccharides are based on their highly variable properties. They vary from insoluble forms (cellulose) to those with good swelling power and solubility in hot and cold water (starch, guaran gum). The solutions may exhibit low viscosities even at very high concentrations (gum arabic), or may have exceptionally high viscosities even at low concentrations (guaran gum). Some polysaccharides, even at a low concentration, set into a thermoreversible gel (alginates, pectin). While most of the gels melt at elevated temperatures, some cellulose derivatives set into a gel.

These properties and their utilization in food products are described in more detail in section 4.4.4, where individual polysaccharides are covered. Here, only a brief account will be given to relate their properties to their structures in a general way.

4.4.3.2 Perfectly Linear Polysaccharides

Compounds with a *single* neutral monosaccharide structural unit and with *one* type of linkage (as occurs in cellulose or amylose) are denoted as perfectly linear polysaccharides. They are usually insoluble in water and can be solubilized only under drastic conditions, e. g. at high temperature, or by cleaving H-bonds with alkalies or other suitable reagents. They readily precipitate from solution (example: starch retrogradation). The reason for these properties is the existence of an optimum structural prerequisite for the formation of an orderly conformation within the chain and also for chain-chain interaction. Often, the conformation is so orderly that a partial crystallinity state develops. Large differences in properties are found within these groups of polysaccharides when there is a change in structural unit, linkage type or molecular weight. This is shown by properties of cellulose, amylose or β-1,3-glucan macromolecules.

4.4.3.3 Branched Polysaccharides

Branched polysaccharides (amylopectin, glycogen) are more soluble in water than their perfectly linear counterparts since the chain-chain interaction is less pronounced and there is a greater extent of solvation of the molecules. Solutions of branched polysaccharides, once dried, are readily rehydrated. Compared to their linear counterparts of equal molecular weights and equal concentrations, solutions of branched polysaccharides have a lower viscosity. It is assumed that the viscosity reflects the "effective volume" of the macromolecule. The

Table 4.15. Examples of uses of polysaccharides in foods

Area of application/food	Suitable polysaccharides
Stabilization of emulsions/suspensions in condensed milk and chocolate milk	Carrageenan, algin, pectin, carboxymethylcellulose
Stabilization of emulsions in coffee whiteners, low-fat margarines	Carrageenan
Stabilization of ice cream against ice crystal formation, melting, phase separation; improvement of consistency (smoothness)	Algin, carrageenan, agar, gum arabic, gum tragacanth, xanthan gum, guaran gum, locust bean flour, modified starches, carboxymethylcellulose, methylcellulose
Water binding, improvement of consistency, yield increase of soft cheese, cream cheese, cheese preparations	Carrageenan, agar, gum tragacanth, karaya gum, guaran gum, locust bean flour, algin, carboxymethylcellulose
Thickening and gelation of milk in puddings made with and without heating, creams; improvement of consistency	Pectin, algin, carrageenan, guaran gum, locust bean flour, carboxymethylcellulose, modified starches
Water binding, stabilization of emulsions in meat products (corned beef, sausage)	Agar, karaya gum, guaran gum, locust bean flour
Jellies for meat, fish, and vegetable products	Algin, carrageenan, agar
Stabilization and thickening, prevention of synaeresis, freeze-thaw stability of soups, sauces, salad dressing, mayonnaise, ketchup; obtaining "body" in low-fat and low-starch products	Gum tragacanth, algin, karaya gum, xanthan gum, guaran gum, locust bean flour, carboxymethylcellulose, propylene glycol alginate, modified starches
Stabilization of protein foam in beer, whipped cream, meringues, chocolate marshmallows	Algin, carrageenan, agar, gum arabic, karaya gum, xanthan gum
Prevention of starch retrogradation in bread and cakes, water binding in dough	Agar, guaran gum, locust bean flour, carrageenan, xanthan gum
Thickening and gelation of fruit pulp (confiture, jams, jellies, fruit pulp for ice cream and yoghurt)	Pectin, algin
Gelation of jelly candies, jelly beans, glaze, icing, water-dessert jellies	Pectin, algin, carrageenan, agar, gum arabic, modified starches
Sediment stabilization in fruit juices, obtaining "body" in beverage powders	Algin, pectin, propylene glycol alginate, gum arabic, xanthan gum, guaran gum, methylcellulose
Stabilization of powdery aroma emulsions, encapsulation of aroma substances	Gum arabic, gum ghatti, xanthan gum

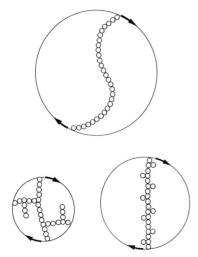

Fig. 4.16. Schematic representation of the "effective volumes" of linear, branched and linearly branched types of polysaccharides

"effective volume" is the volume of a sphere with diameter determined by the longest linear extension of the molecule. These volumes are generally larger for linear than for branched molecules (Fig. 4.16). Exceptions are found with highly pleated linear chains. The tendency of branched polysaccharides to precipitate is low. They form a sticky paste at higher concentrations, probably due to side chain-side chain interactions (interpenetration, entanglement). Thus, branched polysaccharides are suitable as binders or adhesives.

4.4.3.4 Linearly Branched Polysaccharides

Linearly branched polysaccharides, i. e. polymers with a long "backbone" chain and with many short side chains, such as guaran or alkyl cellulose, have properties which are a combination of those of perfectly linear and of branched molecules. The long "backbone" chain is responsible for high solution viscosity. The presence of numerous short side chains greatly weakens interactions between the molecules, as shown by the good solubility and rehydration rates of the molecules and by the stability even of highly concentrated solutions.

4.4.3.5 Polysaccharides with Carboxyl Groups

Polysaccharides with carboxyl groups (pectin, alginate, carboxymethyl cellulose) are very soluble as alkali salts in the neutral or alkaline pH range. The molecules are negatively charged due to carboxylate anions and, due to their repulsive charge forces, the molecules are relatively stretched and resist intermolecular associations. The solution viscosity is high and is pH-dependent. Gel setting or precipitation occurs at pH $\leq$ 3 since electrostatic repulsion ceases to exist. In addition, undissociated carboxyl groups dimerize through H-bridges. However, a divalent cation is needed to achieve gel setting in a neutral solution.

4.4.3.6 Polysaccharides with Strongly Acidic Groups

Polysaccharides with strongly acidic residues, present as esters along the polymer chains (sulfuric, phosphoric acids, as in furcellaran, carrageenan or modified starch), are also very soluble in water and form highly viscous solutions. Unlike polysaccharides with carboxyl groups, in strongly acidic media these solutions are distinctly stable.

4.4.3.7 Modified Polysaccharides

Modification of polysaccharides, even to a low substitution degree, brings about substantial changes in their properties.

4.4.3.7.1 Derivatization with Neutral Substituents

The solubility in water, viscosity and stability of solutions are all increased by binding neutral substituents to linear polysaccharide chains. Thus the properties shown by methyl, ethyl and hydroxypropyl cellulose correspond to those of guaran and locust bean gum. The effect is explained by interference of the alkyl substituents in chain interactions, which then facilitates hydration of the molecule. An increased degree of substitution increases the hydrophobicity of the molecules and, thereby, increases their solubility in organic solvents.

4.4.3.7.2 Derivatization with Acidic Substituents

Binding acid groups to a polysaccharide (carboxymethyl, sulfate or phosphate groups) also results in increased solubility and viscosity for reasons already outlined. Some derivatized polysaccharides, when moistened, have a pasty consistence.

4.4.4 Individual Polysaccharides

4.4.4.1 Agar

4.4.4.1.1 Occurrence, Isolation

Agar is a gelatinous product isolated from seaweed (red algae class, *Rhodophyceae*), e.g., *Gelidium spp., Pterocladia spp.* and *Gracilaria spp.*, by a hot water extraction process. Purification is possible by congealing the gel.

4.4.4.1.2 Structure, Properties

Agar is a heterogenous complex mixture of related polysaccharides having the same backbone chain structure. The main components of the chain are β-D-galactopyranose and 3,6-anhydro-α-L-galactopyranose, which alternate through $1 \rightarrow 4$ and $1 \rightarrow 3$ linkages:

$$(4.124)$$

The chains are esterified to a low extent with sulfuric acid. The sulfate content differentiates between the agarose fraction (the main gelling component of agar), in which close to every tenth galactose unit of the chain is esterified, and the agaropectin fraction, which has a higher sulfate esterification degree and, in addition, has pyruvic acid bound in ketal form [4,6-(1-carboxyethylidene)-D-galactose]. The ratio of the two polymers can vary greatly. Uronic acid, when present, does not exceed 1%. Agar is insoluble in cold water, slightly soluble in ethanolamine and soluble in formamide. Agar precipitated by ethanol from a warm aqueous dispersion is, in its moist state, soluble in water at 25°C, while in the dried state it is soluble only in hot water. Gel setting occurs upon cooling. Agar is a most potent gelling agent as gelation is perceptible even at 0.04%. Gel setting and stability are affected by agar concentration and its average molecular weight. A 1.5% solution sets to a gel at 32–39°C, but does not melt below 60–97°C. The great difference between gelling and melting temperatures, due to hysteresis, is a distinct and unique feature of agar.

4.4.4.1.3 Utilization

Agar is widely used, for instance in preparing nutritive media in microbiology. Its application in the food industry is based on its main properties: it is essentially indigestible, forms heat resistant gels, and has emulsifying and stabilizing activity. Agar is added to sherbets (frozen desserts of fruit juice, sugar, water or milk) and ice creams (at about 0.1%), often in combination with gum tragacanth or locust (carob) bean gum or gelatin. An amount of 0.1–1% stabilizes yoghurt, some cheeses and candy and bakery products (pastry fillings). Furthermore, agar retards bread staling and provides the desired gel texture in poultry and meat canning. Lastly, agar has a role in vegetarian diets (meat substitute products) and in desserts and pretreated instant cereal products.

4.4.4.2 Alginates

4.4.4.2.1 Occurrence, Isolation

Alginates occur in all brown algae (*Phaeophyceae*) as a skeletal component of their cell walls. The major source of industrial production is the giant kelp, *Macrocystis pyrifera*. Some species of *Laminaria, Ascophyllum* and *Sargassum* are also used. Algae are extracted with alkalies. The polysaccharide is usually precipitated from the extract by acids or calcium salts.

4.4.4.2.2 Structure, Properties

Alginate building blocks are β-D-mannuronic and α-L-guluronic acids, joined by $1 \rightarrow 4$ linkages:

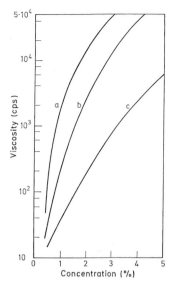

(4.125)

The ratio of the two sugars (mannuronic/guluronic acids) is generally 1.5, with some deviation depending on the source. Alginates extracted from *Laminaria hyperborea* have ratios of 0.4−1.0. Partial hydrolysis of alginate yields chain fragments which consist predominantly of mannuronic or guluronic acid, and also fragments where the two uronic acid residues alternate in a 1:1 ratio. Alginates are linear copolymers consisting of the following structural units:

[→ 4)-β-D-Man*p*A(1 → 4)-β-D-Man*p*A(1 →]$_n$

[→ 4)-α-L-Gul*p*A(1 → 4)-α-L-Gul*p*A(1 →]$_m$

[→ 4)-β-D-Man*p*A(1 → 4)-α-L-Gul*p*A(1 →]$_p$

(4.126)

The molecular weights of alginates are 32−200 kdal. This corresponds to a degree of polymerization of 180−930. The carboxyl group pK-values are 3.4−4.4. Alginates are water soluble in the form of alkali, magnesium, ammonia or amine salts. The viscosity of alginate solutions is influenced by molecular weight and the counter ion of the salt. In the absence of di- and trivalent cations or in the presence of a chelating agent, the viscosity is low ("long flow" property). However, with a rise in multivalent cation levels (e. g., calcium) there is a parallel rise in viscosity ("short flow"). Thus, the viscosity can be adjusted as desired. Freezing and thawing of a Na-alginate solution containing Ca^{2+} ions can result in a further rise in viscosity. The curves in Fig. 4.17 show the effect on viscosity of the concentrations of three alginate preparations: low, moderate and high viscosity types. These data reveal that a 1% solution, depending on the type of alginate, can have a viscosity range of 20−2,000 cps. The viscosity is unaffected in a pH range of 4.5−10. It rises at a pH below 4.5, reaching a maximum at pH 3−3.5.

Gels, fibers or films are formed by adding Ca^{2+} or acids to Na-alginate solutions. A slow reaction is needed for uniform gel formation. It is achieved by a mixture of Na-alginate, calcium phosphate and glucono-δ-lactone, or by a mixture of Na-alginate and calcium sulfate.

Depending on the concentration of calcium ions, the gels are either thermoreversible (low concentration) or not (high concentration). Figure 4.18 shows a schematic section of a calcium alginate gel.

4.4.4.2.3 Derivatives

Propylene glycol alginate is a derivative of economic importance. This ester is obtained by the reaction of propylene oxide with partially neutralized alginic acid. It is soluble down to pH 2 and, in the presence of Ca^{2+} ions, forms soft, elastic, less brittle and syneresis-free gels.

4.4.4.2.4 Utilization

Alginate is a powerful thickening, stabilizing and gel-forming agent. At a level of 0.25−0.5% it improves and stabilizes the consistency of fillings for baked products (cakes, pies), salad dressings and milk chocolates, and

Fig. 4.17. Viscosity of aqueous alginate solutions. Alginate with (a) high, (b) medium, and (c) low viscosity

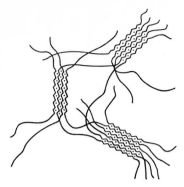

Fig. 4.18. Schematic representation of a calcium alginate gel (cross-linkage by egg box formation, cf. Formula 4.117; according to *Franz*, 1991)

Table 4.16. Building blocks of carrageenans

Carrageenan	Monosaccharide building block
ι-Carrageenan	D-Galactose-4-sulfate, 3,6-anhydro-D-galactose-2-sulfate
κ-Carrageenan	D-Galactose-4-sulfate, 3,6-anhydro-D-galactose
λ-Carrageenan	D-Galactose-2-sulfate, D-galactose-2,6-disulfate
μ-Carrageenan	D-Galactose-4-sulfate, D-galactose-6-sulfate, 3,6-anhydro-D-galactose
ν-Carrageenan	D-Galactose-4-sulfate, D-galactose-2,6-disulfate, 3,6-anhydro-D-galactose
Furcellaran	D-Galactose-D-galactose-2-sulfate, D-galactose-4-sulfate, D-galactose-6-sulfate, 3,6-anhydro-D-galactose

prevents formation of larger ice crystals in ice creams during storage. Furthermore, alginates are used in a variety of gel products (cold instant puddings, fruit gels, dessert gels, onion rings, imitation caviar) and are applied to stabilize fresh fruit juice and beer foam.

4.4.4.3 Carrageenans

4.4.4.3.1 Occurrence, Isolation

Red sea weeds (*Rhodophyceae*) produce two types of galactans: agar and agar-like polysaccharides, composed of D-galactose and 3,6-anhydro-L-galactose residues, and carrageenans and related polysaccharides, composed of D-galactose and 3,6-anhydro-D-galactose which are partially sulfated as 2-, 4- and 6-sulfates and 2,6-disulfates. Galactose residues are alternatively linked by $1 \rightarrow 3$ and $1 \rightarrow 4$ linkages. Carrageenans are isolated from *Chondrus* (*Chondrus crispus*, the Irish moss), *Eucheuma, Gigartina, Gloiopeltis* and *Iridaea* species by hot water extraction under mild alkaline conditions, followed by drying or isolate precipitation.

4.4.4.3.2 Structure, Properties

Carrageenans are a complex mixture of various polysaccharides. They can be separated by fractional precipitation with potassium ions. Table 4.16 compiles data on these fractions and their monosaccharide constituents.

Two major fractions are κ (gelling and K⁺-insoluble fraction) and λ (nongelling, K⁺-soluble).

κ-Carrageenan is composed of D-galactose, 3,6-anhydro-D-galactose and ester-bound sulfate in a molar ratio of 6 : 5 : 7. The galactose residues are essentially fully sulfated in position 4, whereas the anhydrogalactose residues can be sulfated in position 2 or substituted by α-D-galactose-6-sulfate or -2,6-disulfate. A typical sequence of κ-(or ι-)carrgeenan is:

$$^{\ominus}O_3SO \quad CH_2OH \qquad \qquad \qquad \qquad (4.127)$$

The sequence favors the formation of a double-stranded helix (Fig. 4.19). λ-Carrageenan contains as the basic building block β-D-Gal*p*-$(1 \rightarrow 4)$-α-D-Gal*p* (cf. Formula 4.128), which is joined through a 1,3-glycosidic linkage to the polymer. Position 6 of the second galactose residue is esterified with sulfuric acid as is ca. 70% of position 2 of both residues. The high sulfate content favors the formation of a zigzag ribbon-shaped conformation.

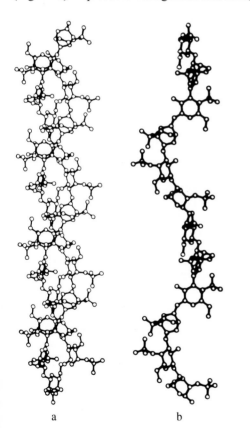

(4.128)

The molecular weights of κ- and λ-carrage-enans are 200–800 kdal. The water solubility increases as the carrageenan sulfate content increases and as the content of anhydrosugar residue decreases. The viscosity of the solution depends on the carrageenan type, molecular weight, temperature, ions present and carrageenan concentration.

As observed in all linear macromolecules with charges along the chain, the viscosity increases exponentially with the concentration (Fig. 4.20). Aqueous κ-carrageenan solutions,

in the presence of ammonium, potassium, rubidium or caesium ions, form thermally reversibly gels. This does not occur with lithium and sodium ions.

This strongly suggests that gel-setting ability is highly dependent on the radius of the hydrated counter ion. The latter is about 0.23 nm for the former group of cations, while hydrated lithium (0.34 nm) and sodium ions (0.28 nm) exceed the limit. The action of cations is visualized as a zipper arrangement between aligned segments of linear polymer sulfates, with low ionic radius cations locked between alternating sulfate residues. Gel-setting ability is probably also due to a mechanism based on formation of partial double helix structures between various chains. The extent of intermolecular double helix formation, and thus the gel strength, is greater, the more uniform the chain sequences are. Each substitution of a 3,6-anhydrogalactose residue by another residue, e.g., galactose-6-sulfate, results in a kink within the helix and, thereby, a decrease in gelling strength. The helical conformation is also affected by the position of sulfate groups. The effect is more pronounced with sulfate in the 6-position, than in 2- or 4-positions. Hence, the gel strength of κ-carrageenan is dependent primarily on the content of esterified sulfate groups in the 6-position.

The addition of carubin, which is itself non-gelling, to κ-carrageenan produces more rigid, more elastic gels that have a lower tendency

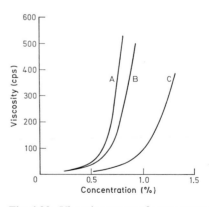

Fig. 4.19. ι-Carrageenan conformation. **a** Double helix, **b** single coil is presented to clarify the conformation (according to *Rees*, 1977)

Fig. 4.20. Viscosity curves of carrageenan aqueous solutions. A: *Eucheuma spinosum*, C: *Chondrus crispus*, B: A and C in a ratio of 2 : 1, 40 °C, 20 rpm (according to *Whistler*, 1973)

towards synaeresis. Carubin apparently prevents the aggregation of $\varkappa$-carrageenan helices.

The 6-sulfate group can be removed by heating carrageenans with alkali, yielding 3,6-anhydrogalactose residues. This elimination results in a significantly increased gel strength.

Carrageenans and other acidic polysaccharides coagulate proteins when the pH of the solution is lower than the proteins' isoelectric points. This can be utilized for separating protein mixtures.

4.4.4.3.3 Utilization

Carrageenan utilization in food processing is based on the ability of the polymer to gel, to increase solution viscosity and to stabilize emulsions and various dispersions. A level as low as 0.03% in chocolate milk prevents fat droplet separation and stabilizes the suspension of cocoa particles. Carrageenans prevent syneresis in fresh cheese and improve dough properties and enable a higher amount of milk powder incorporation in baking. The gelling property in the presence of K^+ salt is utilized in desserts and canned meat. Protein fiber texture is also improved. Protein sedimentation in condensed milk is prevented by carrageenans which, like $\varkappa$-casein, prevent milk protein coagulation by calcium ions. Carrageenans are also used to stabilize ice cream and clarify beverages.

4.4.4.4 Furcellaran

4.4.4.4.1 Occurrence, Isolation

Furcellaran (Danish agar) is produced from red sea weed (algae *Furcellaria fastigiata*). Production began in 1943 when Europe was cut off from its agar suppliers. After alkali pretreatment of algae, the polysaccharide is isolated using hot water. The extract is then concentrated under vacuum and seeded with 1–1.5% KCl solution. The separated gel threads are concentrated further by freezing, the excess water is removed by centrifugation or pressing and, lastly, the polysaccharide is dried. The product is a K-salt and contains, in addition, 8–15% occluded KCl.

4.4.4.4.2 Structure, Properties

Furcellaran is composed of D-galactose (46–53%), 3,6-anhydro-D-galactose (30–33%) and sulfated portions of both sugars (16–20%).

The structure of furcellaran is similar to $\varkappa$-carrageenan. The essential difference is that $\varkappa$-carrageenan has one sulfate ester per two sugar residues, while furcellaran has one sulfate ester residue per three to four sugar residues. Sugar sulfates identified are: D-galactose-2-sulfate, -4-sulfate and -6-sulfate, and 3,6-anhydro-D-galactose-2-sulfate. Branching of the polysaccharide chain can not be excluded. Furcellaran forms thermally reversible aqueous gels by a mechanism involving double helix formation, similar to $\varkappa$-carrageenan.

The gelling ability is affected by the polysaccharide polymerization degree, amount of 3,6-anhydro-D-galactose, and by the radius of the cations present. K^+, NH_4^+, Rb^+ and Cs^+ form very stable, strong gels. Ca^{2+} has a lower effect, while Na^+ prevents gel setting. Addition of sugar affects the gel texture, which goes from a brittle to a more elastic texture.

4.4.4.4.3 Utilization

Furcellaran, with milk, provides good gels and therefore it is used as an additive in puddings. It is also suitable for cake fillings and icings. In the presence of sucrose, it gels rapidly and retains good stability, even against food grade acids. Furcellaran has the advantage over pectin in marmalades since it allows stable gel setting at a sugar concentration even below 50–60%. The required amount of polysaccharide is 0.2–0.5%, depending on the marmalade's sugar content and the desired gel strength. To keep the hydrolysis extent low, a cold aqueous 2–3% solution of furcellaran is mixed into a hot, cooked slurry of fruits and sugar.

Furcellaran is also utilized in processed meat products, such as spreadable meat pastes and pastry fillings. It facilitates protein precipitation during brewing of beer and thus improves the final clarification of the beer.

4.4.4.5 Gum Arabic

4.4.4.5.1 Occurrence, Isolation

Gum arabic is a tree exudate of various *Acacia* species, primarily *Acacia senegal*, and is obtained as a result of tree bark injury. It is collected as air-dried droplets with diameters from 2–7 cm. The annual yield per tree averages 0.9–2.0 kg. The major producer is Sudan, with 50–60,000 t/annum, followed by several other African countries. Gum arabic has been known since ancient Egypt as "kami", an adhesive for pigmented paints.

4.4.4.5.2 Structure, Properties

Gum arabic is a mixture of closely related polysaccharides, with an average molecular weight range of 260–1,160 kdal. The main structural

units, with molar proportions for the gum exudate *A. senegal* given in brackets, are L-arabinose (3.5), L-rhamnose (1.1), D-galactose (2.9) and D-glucuronic acid (1.6). The proportion varies significantly depending on the Acacia species. Gum arabic has a major core chain built of β-D-galactopyranosyl residues linked by 1→3 bonds, in part carrying side chains attached at position 6 (cf. Formula 4.129).

Gum arabic occurs neutral or as a weakly acidic salt. Counter ions are Ca^{2+}, Mg^{2+} and K^+. Solubilization in 0.1 mol/l HCl and subsequent precipitation with ethanol yields the free acid. Gum arabic is very soluble in water and solutions of up to 50% gum can be prepared. The solution viscosity starts to rise steeply only at high concentrations (Fig. 4.21). This property is unlike that of many other polysaccharides, which provide highly viscous solutions even at low concentrations (Table 4.17).

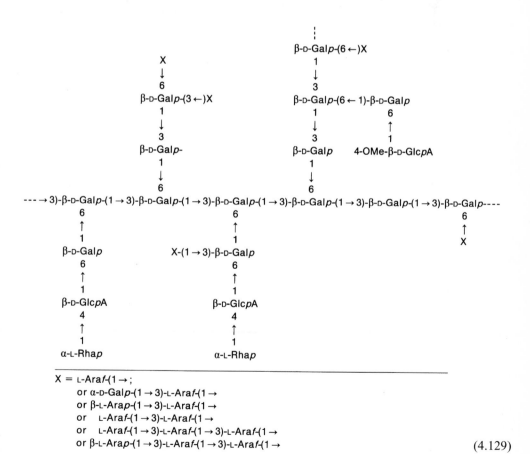

X = L-Araf-(1→ ;
  or α-D-Galp-(1→3)-L-Araf-(1→
  or β-L-Arap-(1→3)-L-Araf-(1→
  or   L-Araf-(1→3)-L-Araf-(1→
  or   L-Araf-(1→3)-L-Araf-(1→3)-L-Araf-(1→
  or β-L-Arap-(1→3)-L-Araf-(1→3)-L-Araf-(1→

(4.129)

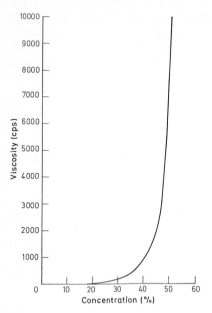

Fig. 4.21. Viscosity curve of an aqueous gum arabic solution (according to *Whistler*, 1973) (25.5 °C, Brookfield viscometer)

4.4.4.5.3 Utilization

Gum arabic is used as an emulsifier and stabilizer, e.g., in baked products. It retards sugar crystallization and fat separation in confectionery products and large ice crystal formation in ice creams, and can be used as a foam stabilizer in beverages. Gum arabic is also applied as a flavor fixative in the production of encapsulated, powdered aroma concentrates. For example, essential oils are emulsified with gum arabic solution and then spray-dried. In this process, the polysaccharide forms a film surrounding the oil droplet, which then protects the oil against oxidation and other changes.

4.4.4.6 Gum Ghatti

4.4.4.6.1 Occurrence

Gum ghatti is an exudate from the tree *Anogeissus latifolia* found in India and Ceylon.

4.4.4.6.2. Structure, Properties

The building blocks are L-arabinose, D-galactose, D-mannose, D-xylose, and D-glucuronic acid. L-Rhamnose has also been detected. The sugars are partially acetylated (5.5 % acetyl groups based on dry weight). Three characteristic structural elements have been detected (cf. Formula 4.130). This acidic polysaccharide occurs as a Ca/Mg salt. Gum ghatti is soluble in water to the extent of ca. 90 % and dispersible. Although it produces solutions that are more viscous than gum arabic, it is less soluble.

4.4.4.6.3 Utilization

Like gum arabic, gum ghatti can be used for the stabilization of suspensions and emulsions.

Table 4.17. Viscosity (cps) of polysaccharides in aqueous solution as affected by concentration (25 °C)

Concentration (%)	Gum arabic	Tragacanth	Carrageenan	Sodium alginate	Methyl cellulose 1,500 cps	Locust bean gum	Guaran gum
1		54	57	214	38.9	58.5	3,025
2		906	397	3,760	512	1,114.3	25,060
3		10,605	4,411	29,400	3,850	8,260	111,150
4		44,275	25,356		12,750	39,660	302,500
5	7.3	111,000	51,425		67,575	121,000	510,000
6		183,500					
10	16.5						
20	40.5						
30	200.0						
40	936.3						
50	4,162.5						

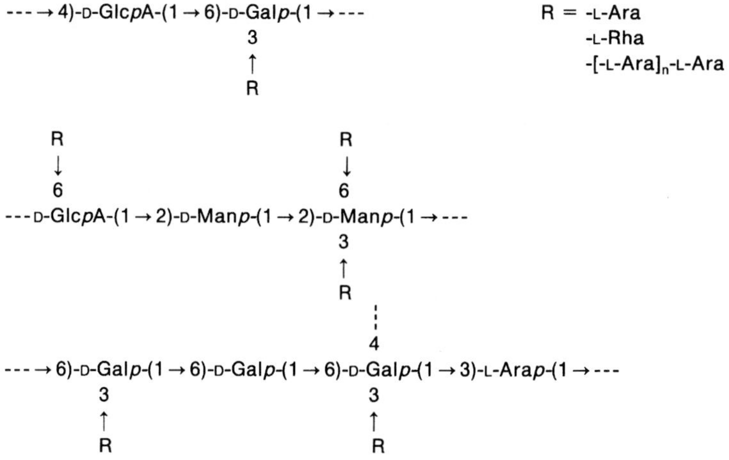

$$--- \to 4)\text{-D-Glc}p\text{A-}(1 \to 6)\text{-D-Gal}p\text{-}(1 \to ---$$
$$3$$
$$\uparrow$$
$$R$$

$$R = \text{-L-Ara}$$
$$\text{-L-Rha}$$
$$\text{-[-L-Ara]}_n\text{-L-Ara}$$

$$
\begin{array}{cc}
R & R \\
\downarrow & \downarrow \\
6 & 6
\end{array}
$$
$$---\text{D-Glc}p\text{A-}(1 \to 2)\text{-D-Man}p\text{-}(1 \to 2)\text{-D-Man}p\text{-}(1 \to ---$$
$$3$$
$$\uparrow$$
$$R$$
$$\vdots$$
$$4$$
$$--- \to 6)\text{-D-Gal}p\text{-}(1 \to 6)\text{-D-Gal}p\text{-}(1 \to 6)\text{-D-Gal}p\text{-}(1 \to 3)\text{-L-Ara}p\text{-}(1 \to ---$$
$$
\begin{array}{cc}
3 & 3 \\
\uparrow & \uparrow \\
R & R
\end{array}
$$

(4.130)

4.4.4.7 Gum Tragacanth

4.4.4.7.1 Occurrence

Gum tragacanth is a plant exudate collected from *Astragalus* species shrubs grown in the Middle East (Iran, Syria, Turkey).

4.4.4.7.2 Structure, Properties

Gum tragacanth consists of a water-soluble fraction, the so-called tragacanthic acid, and the insoluble swelling component, bassorin. Tragacanthic acid contains 43% of D-galacturonic acid, 40% of D-xylose, 10% of L-fucose, and 4% of D-galactose. Like pectin, it is composed of a main polygalacturonic acid chain which bears side chains made of the remaining sugar residues (Formula 4.131). Bassorin consists of 75% of L-arabinose, 12% of D-galactose, 3% of D-galacturonic acid methyl ester, and L-rhamnose.

Its molecular weight is about 840 kdal. The molecules are highyl elongated (450×1.9 nm) in aqueous solution and are responsible for the high viscosity of the solution (Table 4.17). As shown in Fig. 4.22, the viscosity is highly dependent on shear rate.

4.4.4.7.3 Utilization

Gum tragacanth is used as a thickening agent and a stabilizer in salad dressings (0.4–1.2%) and in fillings and icings in baked goods. As an additive in ice creams (0.5%), it provides a soft texture.

4.4.4.8 Karaya Gum

4.4.4.8.1 Occurrence

Karaya gum, also called Indian tragacanth, is an exudate from an Indian tree of the *Sterculia ureus* and other *Sterculia* species.

4.4.4.8.2 Structure, Properties

The building blocks are D-galactose, L-rhamnose, D-galacturonic acid, and L-glucuronic

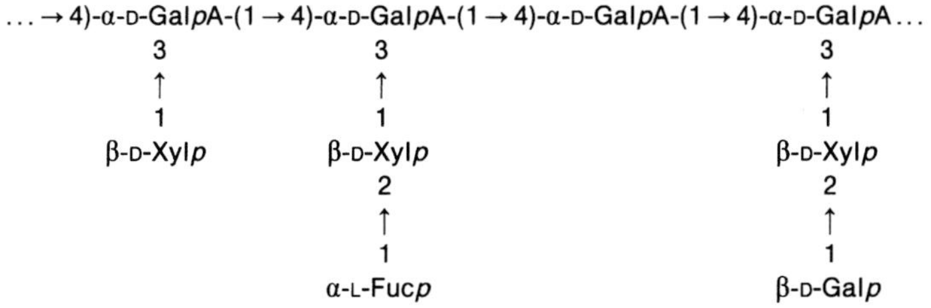

$$\ldots \to 4)\text{-}\alpha\text{-D-Gal}p\text{A-}(1 \to 4)\text{-}\alpha\text{-D-Gal}p\text{A-}(1 \to 4)\text{-}\alpha\text{-D-Gal}p\text{A-}(1 \to 4)\text{-}\alpha\text{-D-Gal}p\text{A} \ldots$$
$$
\begin{array}{ccc}
3 & 3 & 3 \\
\uparrow & \uparrow & \uparrow \\
1 & 1 & 1 \\
\beta\text{-D-Xyl}p & \beta\text{-D-Xyl}p & \beta\text{-D-Xyl}p \\
2 & & 2 \\
\uparrow & & \uparrow \\
1 & & 1 \\
\alpha\text{-L-Fuc}p & & \beta\text{-D-Gal}p
\end{array}
$$

(4.131)

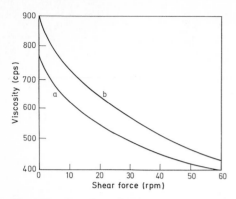

Fig. 4.22. The effect of shear rate on viscosity of aqueous tragacanth solutions. **a** Flake form tragacanth, 1%; **b** tragacanth, ribbon form, 0.5% (according to *Whistler*, 1973)

acid. The sugars are partially acetylated (13% acetyl groups based on dry weight). The molecule consists of three main chains which are polymers of different disaccharide units (cf. Formula 4.132). The main chains carry side chains and are also covalently linked via the side chains.

As a result of the strong cross-linkage, this polysaccharide is insoluble in water and resistant to enzymes and microorganisms. However, it swells greatly even in cold water. Suspensions have a pasty consistency at concentrations of more than 3%.

4.4.4.8.3 Utilization

Karaya gum is used as a water binder (soft cheese), a binding agent (meat products like corned beef, sausages), a stabilizer of protein foams (beer, whipped cream, meringues), and as a thickener (soups, sauces, salad dressings, mayonnaise, ketchup). It increases the freeze-thaw stability of products, prevents synaeresis of gels, and provides "body".

4.4.4.9 Guaran Gum

4.4.4.9.1 Occurrence, Isolation

Guar flour is obtained from the seed endosperm of the leguminous plant *Cyamopsis tetragonoloba*. The seed is decoated and the germ removed. In addition to the polysaccharide guaran, guar flour contains 10−15%

$$(4.132)$$

moisture, 5–6% protein, 2.5% crude fiber and 0.5–0.8 ash. The plant is cultivated for forage in India, Pakistan and the United States (Texas).

4.4.4.9.2 Structure, Properties

Guaran gum consists of a chain of β-D-mannopyranosyl units joined with $1 \rightarrow 4$ linkages. Every second residue has a side chain, a D-galactopyranosyl residue that is bound to the main chain by an $\alpha\ (1 \rightarrow 6)$ linkage (cf. Formula 4.133).

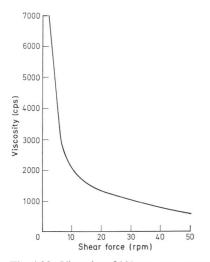

(4.133)

Guaran gum forms highly viscous solutions (Table 4.17), the viscosity of which is shear rate dependent (Fig. 4.23).

Fig. 4.23. Viscosity of 1% aqueous guar solution at 25 °C versus shear rate (rpm.). Viscometer: Haake rotovisco (according to *Whistler*, 1973)

4.4.4.9.3 Utilization

Guaran gum is used as a thickening agent and a stabilizer in salad dressings and ice creams (application level 0.3%). In addition to the food industry, it is widely used in paper, cosmetic and pharmaceutical industries.

4.4.4.10 Locust Bean Gum

4.4.4.10.1 Occurrence, Isolation

The locust bean (carob bean; St. John's bread) is from an evergreen cultivated in the Mediterranean area since ancient times. Its long, edible, fleshy seed pod is also used as fodder. The dried seeds were called "carat" by Arabs and served as a unit of weight (approx. 200 mg). Even today, the carat is used as a unit of weight for precious stones, diamonds and pearls, and as a measure of gold purity (1 carat = 1/24 part of pure gold). The locust bean seeds consist of 30–33% hull material, 23–25% germ and 42–46% endosperm. The seeds are milled and the endosperm is separated and utilized like the guar flour described above. The commercial flour contains 88% galactomannoglycan, 5% other polysaccharides, 6% protein and 1% ash.

4.4.4.10.2 Structure, Properties

The main locust bean polysaccharide is similar to that of guaran gum: a linear chain of $1 \rightarrow 4$ linked β-D-mannopyranosyl units, with α-D-galactopyranosyl residues $1 \rightarrow 6$ joined as side chains. The ratio mannose/galactose is 3 to 6; this indicates that, instead of every second mannose residue, as in guaran gum, only every 4th to 5th is substituted at the C-6 position with a galactose molecule.

The molecular weight of the galactomannan is close to 310 kdal. Physical properties correspond to those of guar gum, except the viscosity of the solution is not as high (cf. Table 4.17).

4.4.4.10.3 Utilization

Locust bean flour is used as a thickener, binder and stabilizer in meat canning, salad dressings, sausages, soft cheeses and ice creams. It also improves the water binding capacity of dough, especially when flour of low gluten content is used.

4.4.4.11 Tamarind Flour

4.4.4.11.1 Occurrence, Isolation

Tamarind is one of the most important and widely grown trees of India (*Tamarindus indica*; date of India). Its brown pods contain seeds which are rich in a polysaccharide that is readily extracted with hot water and, after drying, recovered in a powdered form.

4.4.4.11.2 Structure, Properties

The polysaccharide consists of D-galactose (1), D-xylose (2) and D-glucose (3), with respective molar ratios given in brackets. L-Arabinose is also present. The suggested structure is presented in Formula 4.134.

The polysaccharide forms a stable gel over a wide pH range. Less sugar is needed to achieve a desired gel strength than in corresponding pectin gels (Fig. 4.24). The gels exhibit only a low syneresis phenomenon.

4.4.4.11.3 Utilization

The tamarind seed polysaccharide is a suitable substitute for pectin in the production of mar-

malades and jellies. It can be used as a thickening agent and stabilizer in ice cream and mayonnaise production.

4.4.4.12 Arabinogalactan from Larch

4.4.4.12.1 Occurrence, Isolation

Coniferous larch-related woods (*Larix* species; similar to pine, but shed their needle-like leaves) contain a water-soluble arabinogalactan of 5–35% of the dry weight of the wood. It can be isolated from chipped wood by a counter-current extraction process, using water or dilute acids. The extract is then usually drum dried.

4.4.4.12.2 Structure, Properties

The polysaccharide consists of straight chain β-D-galactopyranosyl units joined by $1 \rightarrow 3$ linkages and, in part, has side chains of galactose and arabinose residues bound to positions 4 and 6. The suggested structure is given in Formula 4.135.

In general, the polysaccharide is highly branched. The molecular weight is 50–70 kdal. The molecule is nearly spherical in shape, so its aqueous solution behaves like a *Newton*ian fluid. The viscosity is exceptionally low. At a temperature of 20 °C, the viscosity of a 10% solution is 1.74 cps, a 30% solution

$$
\left[
\begin{array}{c}
\longrightarrow 4)\text{-}\beta\text{-}\text{D-Glc}p(1\longrightarrow 4)\text{-}\beta\text{-}\text{D-Glc}p(1\longrightarrow 4)\text{-}\beta\text{-}\text{D-Glc}p(1\longrightarrow 4)\text{-}\beta\text{-}\text{D-Glc}p(1\longrightarrow \\
\begin{array}{ccc}
6 & 6 & 6 \\
\uparrow & \uparrow & \uparrow \\
1 & 1 & 1 \\
\alpha\text{-}\text{D-Xyl}p & \alpha\text{-}\text{D-Xyl}p & \text{L-Ara}f \\
& 2 & \\
& \uparrow & \\
& 1 & \\
& \beta\text{-}\text{D-Gal}p &
\end{array}
\end{array}
\right]_n
\qquad (4.134)
$$

$$
\begin{array}{cccc}
\beta\text{-L-Ara}p(1\longrightarrow 3)\text{L-Ara}f & \text{L-Ara}f(1\longrightarrow 3)\text{D-Gal}p & \text{D-Gal}p(1\longrightarrow 6)\text{D-Gal}p \\
1 & 1 & 1 \\
\downarrow & \downarrow & \downarrow \\
6 & 6 & 6
\end{array}
$$

$$
\longrightarrow 3)\text{-}\beta\text{-}\text{D-Gal}p(1\longrightarrow 3)\text{-}\beta\text{-}\text{D-Gal}p(1\longrightarrow 3)\text{-}\beta\text{-}\text{D-Gal}p(1\longrightarrow 3)\text{-}\beta\text{-}\text{D-Gal}p(1\longrightarrow 3)\text{-}\beta\text{-}\text{D-Gal}p(1\longrightarrow 3)\text{-}\beta\text{-}\text{D-Gal}p(1\longrightarrow
$$

$$
\begin{array}{c}
4 \\
\uparrow \\
1 \\
\beta\text{-}\text{D-Gal}p(1\longrightarrow 6)\text{D-Gal}p
\end{array}
\qquad (4.135)
$$

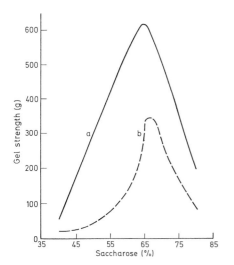

Fig. 4.24. Gel strength of (a) tamarind flour and (b) pectin from lemons versus saccharose concentration (according to *Whistler*, 1973)

7.8 cps at pH 4 or 8.15 cps at pH 11, and a 40% solution 23.5 cps. These data show that the viscosity is practically unaffected by pH. The solution acquires a thick paste consistency only at concentrations exceeding 60%.

4.4.4.12.3 Utilization

Arabinogalactan, due to its good solubility and low viscosity, is used as an emulsifier and stabilizer, and as a carrier substance in essential oils, aroma formulations, and sweeteners.

4.4.4.13 Pectin

4.4.4.13.1 Occurrence, Isolation

Pectin is widely distributed in plants. It is produced commercially from peels of citrus fruits and from apple pomace (crushed and pressed residue). It is 20−40% of the dry matter content in citrus fruit peel and 10−20% in apple pomace. Extraction is achieved at pH 1.5−3 at 60−100 °C. The process is carefully controlled to avoid hydrolysis of glycosidic and ester linkages. The extract is concentrated to a liquid pectin product or is dried by spray- or drum-drying into a powered product. Purified preparations are obtained by precipitation of pectin with ions which form insoluble pectin

salts (e.g. Al^{3+}), followed by washing with acidified alcohol to remove the added ions, or by alcoholic precipitation using isopropanol and ethanol.

4.4.4.13.2 Structure, Properties

Pectin is a chain-like polymer consisting of α-D-galacturonic structural units joined by $1 \rightarrow 4$ linkages:

R: $COO^{\ominus}$, $COOCH_3$ (4.136)

In addition the main chain contains rhamnose residues. In segments in which rhamnose is enriched, rhamnose units may be in adjacent or alternate positions. Pectin also contains small amounts of D-galactan and arabinan in its extended side chains and, to a lesser extent, fucose and xylose sugars in its short side chains (1 to 3 unit chains). These short side chains are not regarded as typical pectin constituents. The galacturonic acid carboxyl groups along the main chain are esterified to a variable extent with methanol, while the HO-groups in 2- and 3-positions may be acetylated to a small extent. Pectin stability is highest at pH 3−4. The glycosidic linkage hydrolyzes in a stronger acidic medium. In an alkaline medium, both linkages, ester and glycosidic, are split to the same extent, the latter by an elimination reaction (cf. Formula 4.137).
The elimination reaction occurs more readily with galacturonic acid units having an esterified carboxyl group, since the H-atom on C-5 is more acidic than with residues having free carboxyl groups.
At a pH of about 3, and in the presence of Ca^{2+} ions also at higher pH's, pectin forms a thermally reversible gel. The gel-forming ability, under comparable conditions, is directly proportional to the molecular weight and inversely proportional to the esterification degree. For gel formation, low-ester pectins require very low pH values and/or calcium ions, but they gelatinize in the presence of a relatively low

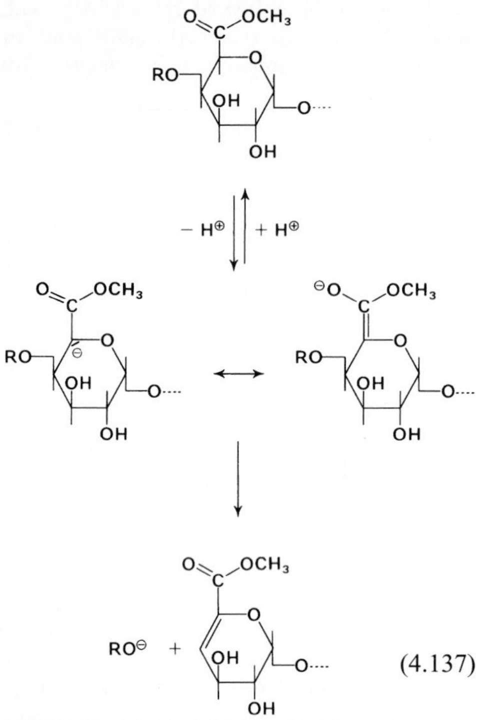

$$(4.137)$$

Table 4.18. Gelling time of pectins with differing degrees of esterification

Pectin type	Esterification degree (%)	Gelling time[a] (s)
Fast gelling	72–75	20–70
Normal	68–71	100–135
Slow gelling	62–66	180–250

[a] Difference between the time when all the prerequisites for gelling are fulfilled and the time of actual gel setting.

Table 4.19. Raw materials for starch

Raw material	Starch production 1980[a]
Raw materials of industrial importance	
Corn	77
Waxy corn	
Potato	10
Cassava	8
Wheat	4
Rice	
Waxy rice	
Other raw materials	
Sago palm	Kouzu
Sweet potato	Water chestnut
Arrowroot	Edible canna
Negro corn	Mungo bean
Lotus root	
Taro	Lentil

[a] % of the world production.

sugar content. High-ester pectins require an increasing amount of sugar with rising esterification degree. The gelsetting time for high ester pectins is longer than that for pectin products of low esterification degree (Table 4.18).

4.4.4.13.3 Utilization

Since pectin can set into a gel, it is widely used in marmalade and jelly production. Standard conditions to form a stable gel are, for instance: pectin content <1%, sucrose 58–75% and pH 2.8–3.5. In low-sugar products, low-ester pectin is used in the presence of Ca^{2+} ions. Pectin use as a stabilizer for beverages (to emulsify the essential oil components) or ice creams is also of importance.

4.4.4.14 Starch

4.4.4.14.1 Occurrence, Isolation

Starch is widely distributed in various plant organs as a storage carbohydrate. As an ingredient of many foods, it is also the most important carbohydrate source in human nutrition. In addition, starch and its derivatives are important industrially, for example, in the paper and textile industries.

Starch is isolated mainly from the sources listed in Table 4.19. Starch obtained from corn, potatoes, cassava, and wheat in the native and modified form accounted for 99% of the world production in 1980. Some other starches are also available commercially. Recently, starches obtained from legumes (peas, lentils) have become more interesting because they have properties which appear to make them a suitable substitute for chemically modified starches in a series of products.

Starches of various origin have individual, characteristic properties which go back to the shape, size, size distribution, composition, and crystallinity of the granules. The existing connections are not yet completely understood on a molecular basis.

In some cases, e.g., potato tubers, starch granules occur free, deposited in cell vacuoles; hence, their isolation is relatively simple. The plant material is disintegrated, the starch granules are washed out with water, and then sedimented from the "starch milk" suspension and dried. In other cases, such as cereals, the starch is embedded in the endosperm protein matrix, hence granule isolation is a more demanding process. Thus, a counter-current process with water at 50°C for 36−48 h is required to soften corn (steeping step of processing). The steeping water contains 0.2% SO_2 in order to loosen the protein matrix and, thereby, to accelerate the granule release and increase the starch yield. The corn grain is then disintegrated. The germ, due to its high oil content, has a low density and is readily separated by flotation. It is the source for corn oil isolation (cf. 14.3.2.2.4). The protein and starch are then separated in hydrocyclones. The separation is based on density difference (protein < starch).

The protein by-product is marketed as animal feed or used for production of a protein hydrolysate (seasoning). The recovered starch is washed and dried.

In the case of wheat flour, a dough is made first, from which a raw starch suspension is washed out. After separation of fiber particles from this suspension, the starch is fractionated by centrifugation. In addition to the relatively pure primary starch, a finer grained secondary starch is obtained which contains pentosans. The starch is then dried and further classified. The residue, gluten (cf. 15.1.5), serves, e.g., as a raw material in the production of food seasoning (cf. 12.7.3.5) and in the isolation of glutamic acid. If dried gently, it retains its baking quality and is used as a flour improver. In the case of rye, the isolation of starch is impeded by the relatively high content of swelling agents. Starch isolated from the tubers of various plants in tropical countries is available on the market under a variety of names (e.g., canna, maranta, and tacca starch). The real sago is the product obtained from the pith of the sago palm.

Starch is a mixture of two glucans, amylose and amylopectin (cf. 4.4.4.14.3 and 4.4.4.14.4).

Most starches contain 20−30% amylose (Table 4.20). New corn cultivars (amylomaize) have been developed which contain 50−80% amylose. The amylose can be isolated from starch, e.g., by crystallization of a starch dispersion, usually in the presence of salts ($MgSO_4$) or by precipitation with a polar organic compound (alcohols, such as n-butanol, or lower fatty acids, such as caprylic or capric), which forms a complex with amylose and thus enhance its precipitation.

Normal starch granules contain 70−80% amylopectin, while some corn cultivars and millet, denoted as waxy maize or waxy millet, contain almost 100% amylopectin.

4.4.4.14.2 Structure and Properties of Starch Granules

Starch granules are formed in the amyloplasts. These granules are simple or compound and consist of concentric or eccentric layers of varying density. They are of varying size (2−150 µm), size distribution, and shape (Table 4.20). In addition to amylose and amylopectin, the usually contain small amounts of proteins and lipids. They are examined by using various physical methods, including light microscopy, small-angle light scattering, electron microscopy, X-ray diffraction, small-angle neutron scattering, and small-angle X-ray scattering. On the basis of X-ray diffraction experiments, starch granules are said to have a semicrystalline character, which indicates a high degree of orientation of the glucan molecules. About 70% of the mass of a starch granule is regarded as amorphous and ca. 30% as crystalline (Table 4.20). The amorphous regions contain the main amount of amylose, but also a considerable part of the amylopectin. The crystalline regions consist primarily of amylopectin. Although this finding was surprising at first because of the branched structure of amylopectin (cf. 4.4.4.14.4), it was deduced from the fact that amylose can be dissolved out of the granule without disturbing the crystalline character and that even amylose-free starches, like waxy corn starch, are semicrystalline. The degree of crystallinity

Table 4.20. Shape, composition, and properties of different starch granules

Source	Shape[a]	Diameter (µm)	Crystallinity (%)	Gelatinization temperature (°C)	Swelling at 95°C[b]	Amylose Percentage (%)[c]	Amylose Polymerisation degree	Amylopectin Iodine binding constant[d]	Amylopectin Polymerisation degree[e]
Cereal									
Wheat	l, p	2–38	36	53–65	21	26–31	2100	0.21	19–20
Rye	l	12–40		57–70		28		0.74	26
Barley	l	2–5		56–62		22–29	1850		26
Corn	p	5–25		62–70	24	28	940	0.91	25–26
Amylomaize			20–25	67–87		52–80	1300	0.11	23
Waxy corn	p		39	63–72	64	0–1			20–22
Oats		5–15		56–62		27	1300		20
Rice	p	3–8	38	61–78	19	14–32		0.59	
Waxy rice				55–65	56	1			
Millet	p, s	4–12		69–75[g]	22[g]				
Sorghum	p, s	4–24				21–34			
Waxy sorghum				68–74	49				
Legumes									
Horsebean	s, o	17–31		64–67		32–34	1800	1.03	23
Smooth pea	n (si)	5–10		57–70[h]		33–35	1300	1.66	26
Wrinkled pea	n (c)	30–40				63–75	1100	0.91	27
Roots and tubers									
Potato	e	15–100	25	58–66		23	3200	0.58	24
Cassava	semi-s, s	5–35	38[f]	52–64		17		1.06	

[a] l = lenticular, p = polyhedral, s = spherical, o = oval, n = kidney-shaped, el = elliptical, si = simple, c = compound.
[b] Weight of swollen starch, based on its dry weight; loss of soluble polysaccharides is considered.
[c] Based on the cum of amylose and amylopectin.
[d] mg iodine/100 mg starch.
[e] Cleavage degree of polymerization, determined by degradation of branches with pullulanase or isoamylase.
[f] Tapioca.
[g] Millet.
[h] Pea.

depends on the water content. It is 24% for air-dried potato starch (19.8% of water), 29–35% for the wetted product (45–55% of water), and only 17% for starch dried via P_2O_5 and subsequently rehydrated.

On the basis of results obtained from different physical methods, the model shown in Fig. 4.25 is under discussion for the crystalline regions of the starch granule. It contains double helices of amylopectin (cf. 4.4.4.14.4), mixed amylose/amylopectin double helices, V helices of amylose with enclosed lipid (cf. 4.4.4.14.3), free amylose, and free lipid.

With the aid of X-ray diffraction diagrams, native starches can be divided into types A, B, and C. An additional form, called the V-type, occurs in swollen granules (Fig. 4.26). While types A and B are real crystalline modifications, the C-type is a mixed form. The A-type is largely present in cereal starches, and the B-type in potatoes, amylomaize, and in retro-

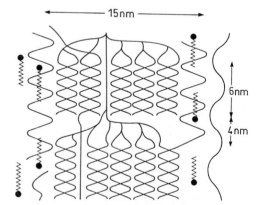

Fig. 4.25. Model of a crystalline region in a starch granule (according to *Galliard*, 1987). Amylopectin double helix ⧓⧓⧓; mixed double helix of amylose and amylopectin ⟩⟩⟩ ; V-helix of amylose and enclosed lipid ; free lipid ; free amylose ∿∿

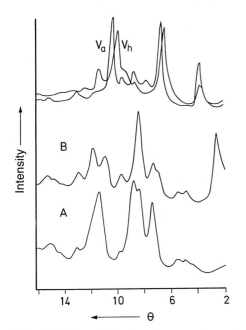

Fig. 4.26. X-ray diffraction diagrams of starches: A-type (cereals), B-type (legumes) and V-type (swollen starch, V_a: water free, V_h: hydrated) (according to *Galliard*, 1987)

tween glucose chains in the crystallites are primarily disrupted, and perhaps some of those in the amorphous regions as well. It is probable that the swelling of the amorphous regions facilitates the dissolving out of amylose from the crystallites, which are thereby destabilized. As with heating in water, the same effect occurs when starch is suspended in other solvents, e.g., liquid ammonia or dimethyl sulfoxide, or mechanically damaged, e.g., by dry grinding.

The course of gelatinization depends not only on the botanical origin of the starch and the temperature used, but also on the water content of the suspension (Fig. 4.27). Thus, dried starch with 1−3% of water undergoes only slight changes up to a temperature of 180°C, whereas starch with 60% of water completely gelatinizes at temperatures as low as 70°C.

If an aqueous starch suspension is maintained for some time at temperatures below the gelatinization temperature, a process known as tempering, the gelatinization temperature is increased, apparently due to the reorganization

graded starches. The C-type is not only observed in mixtures of corn and potato starches, but it is also found in various legume starches. When suspended in cold water, air-dried starch granules swell with an increase in diameter of 30−40%. If this suspension is heated, irreversible changes occur starting at a certain temperature, which is characteristic of each type of starch (50−70°C, cf. Table 4.20), called the gelatinization temperature. The starch granules absorb 20−40 g of water/g of starch, the viscosity of the suspension rising steeply. At the same time, a part of the amylose diffuses out of the granule and goes into solution. Finally, the granule bursts. In the first step of gelatinization, the starch crystallites melt and form a polymer network. This network breaks up at higher temperatures (ca. 100°C), resulting in a solution of amylose and amylopectin. In gelatinization, water first diffuses into the granule, crystalline regions then melt with the help of hydration, and, finally, swelling gives rise to a solution through further diffusion of water. In this process, hydrogen bridges be-

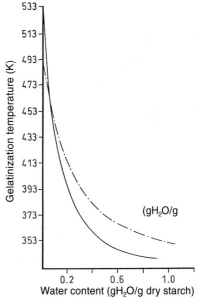

Fig. 4.27. Gelatinization temperature of differently hydrated starches (— potato starch, −·− wheat starch, determined by differential thermal analysis, differential calorimetry, and double refraction) (according to *Galliard*, 1987)

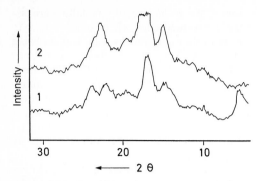

Fig. 4.28. X-ray diffraction diagrams of potato starch before (1) and after (2) thermal treatment (102 °C/16 h) at a water content of 40 %. The pattern of native starch (18.3 % of water) corresponds to the B-type and that of treated starch (24.2 % of water) to the A-type (according to *Galliard*, 1987)

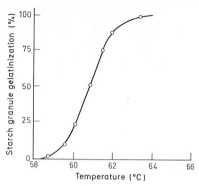

Fig. 4.29. Potato starch gelatinization curve (according to *Banks* and *Muir*, 1980)

of the structure of the granule. Treatment of starch at low water contents and higher temperatures results in the stabilization of the crystallites and, consequently, a decrease in the swelling capacity. Figure 4.28 shows the resulting change in the X-ray diffraction spectrum from type B to type A, using potato starch as an example.

The changes in the physical properties caused by treating processes of this type can, however, vary considerably, depending on the botanical origin of the starches. This is shown in Table 4.21 for potato and wheat starch. On wet heating, the swelling capacity of both starches decreases, although to different extents. On the other hand, there is a decrease in solubility only of potato starch, while that of wheat

starch clearly increases. The explanation put forward to explain these findings is that the amorphous amylose of potato starch is converted to an ordered, less soluble state, while the amylose of cereal starch, present partially in the form of helices with enclosed lipids, changes into a more easily leachable state.

A gelatinization curve for potato starch is presented in Fig. 4.29. The number of gelatinized starch granules was determined by microscopy. Another way to monitor gelatinization as a function of temperature is to measure the viscosity of a starch suspension. The viscosity curves in Fig. 4.30 show that, as mentioned above, the viscosity initially increases due to starch granule swelling. The subsequent disintegration of the swollen granule is accompanied by a drop in viscosity. The shape of the curve varies greatly for different starches. Potato starch shows a very high maximum

Table 4.21. Physicochemical properties of starches before (1) and after (2) heat treatment in the wet state (T = 100 °C, t = 16 h, H_2O = 27 %)

Property	Wheat starch		Potato starch	
	1	2	1	2
Start of gelatinization (°C)	56.5	61	60	60.5
End of gelatinization (°C)	62	74	68	79
Swelling capacity at 80 °C (ratio)	7.15	5.94	62.30	19.05
Solubility at 80 °C (%)	2.59	5.93	31.00	10.10
Water binding capacity (%)	89.1	182.6	102.00	108.7
Enzymatic vulnerability (% dissolved)	0.44	48.55	0.57	40.35

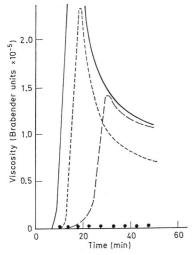

Fig. **4.30.** Gelatinization properties of various starches (according to *Banks* and *Muir*, 1980). Brabender visco-amylograph. 40 g starch/460 ml water, temperature programming: start at 50 °C, heated to 95 °C at a rate of 1.5 °C/min. Held at 95 °C for 30 min — potato, --- waxy corn, − − −normal corn, and • • • amylomaize starch

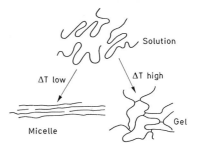

Fig. **4.31.** Behavior of amylose molecules during cooling of a concentrated aqueous solution

(∼ 4000 *Brabender* units), followed by a steep drop. Waxy corn starch exhibits similar behavior, except that the maximum is not as high. In normal corn starch, the maximum is still lower, but the following drop is slight, i.e., the granules are more stable. Under these conditions, amylomaize starch does not swell, even though ca. 35% of the amylose goes into solution. The viscosity of a starch paste generally increases on rapid cooling with mixing, while a starch gel is formed on rapid cooling without mixing.

Amylose gels tend to retrograde. This term denotes the largely irreversible transition from the solubilized or highly swollen state to an insoluble, shrunken, microcrystalline state (Fig. 4.31). This state can also be directly achieved by slowly cooling a starch paste. The tendency towards retrogradation is enhanced at low temperatures, especially near 0 °C, neutral pH values, high concentration, and by the absence of surface active agents. It also depends on the molecular weight and on the type of starch, e.g., it increases in the series potato < corn < wheat. The transitions des-

cribed from very water-deficient starting states via very highly swollen states or solutions to more or less shrunken states are linked to changes in the interactions between the glucans and to conformational changes. At present, these changes cannot be fully described because they greatly depend on the conditions in each case, e.g., even on the presence of low molecular compounds.

It is known that the gelatinization temperature is increased by polyhydroxy compounds (glycerol, sugar) and decreased by salts (NaCl, CaCl$_2$), as presented in Fig. 4.32 (top) as a function of water activity, which is lowered by the dissolved substances (a_w, cf. 0.3.1). Apart from the activity of the solvent water, if its volume fraction (v_1), which changes in reverse order to the volume fraction of the solute, is considered and if the gelatinization temperature is plotted against ln a_w/v_1 instead of a_w, the effect of the different dissolved substances is unified (Fig. 4.32, bottom). The reason is that polyhydroxy compounds cause a large change in v_e and a small change in a_w, while a small change in v_e is linked to a large change in a_w in the case of the salts.

Lipids also influence the properties of starch. Like free amino acids, monoglycerides or fatty acid esters of hydroxy acids, lipids form inclusion compounds with amylose (cf. 4.4.4.14.3). Like di- and triglycerides, they also reduce the swelling capacity and solubility by inhibiting water diffusion. Therefore, both degreasing as well as lipid addition are of importance as physical modification methods of starches.

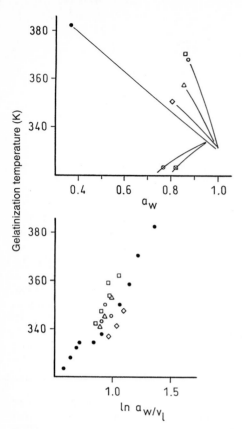

Fig. 4.32. Gelatinization temperature of potato starch as a function of water activity a_w (top) and of the natural logarithm of the quotient of activity a_w to volume fraction v_1 of water (bottom); • glycerol, ○ maltose, □ saccharose, △ glucose, ◇ ribose, ⊗ NaCl, ⊠ CaCl₂ (according to *Galliard*, 1987)

4.4.4.14.3 Structure and Properties of Amylose

Amylose is a chain polymer of α-D-glucopyranosyl residues linked 1 → 4:

$$(4.138)$$

Enzymatic hydrolysis of the chain is achieved by α-amylase, β-amylase and glucoamylase. Often, β-amylase does not degrade the molecule completely into maltose, since a very low branching is found along the chain with α (1 → 6) linkages. The molecular size of amylose is variable. The polymerization degree in wheat starch is 1,000–2,000, while in potatoes it can rise up to 4,500. This corresponds to a molecular weight of 150–750 kdal. X-ray diffraction experiments conducted on oriented amylose fibers make possible the assignment of the types of starch mentioned above to definite molecular structural elements. Oriented fibers of the A-type were obtained by cutting and stretching thin films of acetylamylose at 150 °C, deacetylation in alcoholic alkali, and conditioning at 80% relative air humidity and 85 °C. Type B fibers were obtained in a corresponding manner by conditioning the deacetylated material at room temperature for three days at 80% and for another three days at 100% relative air humidity, followed by aftertreatment in water at 90 °C for 1 h. The diffraction patterns obtained with these oriented fibers corresponded to those of types A and B given by native starch powders, allowing the development of structural models.

The structural elements of type B are double helices (Fig. 4.34a), which are packed in an antiparallel, hexagonal mode (Fig. 4.33). The central channel surrounded by six double helices is filled with water (36 H₂O/unit cell). The A-type is very similar to the B-type, except that the central channel is occupied by another double helix, making the packing more close. In this type, only eight molecules of water per unit cell are inserted between the double helices. The transition from type B to type A achieved by wet heating has been described already (4.4.4.14.2, Fig. 4.28). It is difficult to bring the postulated antiparallel arrangement of the double helices into line with the requirements of biosynthesis, where a parallel arrangement can be expected. It is possible that the present experimental data do not exclude such an arrangement.

The double helix mentioned above and shown in Fig. 4.34 can, depending on conditions, change into other helical conformations. In the presence of KOH, for instance, a more extended helix results with 6 glucose residues

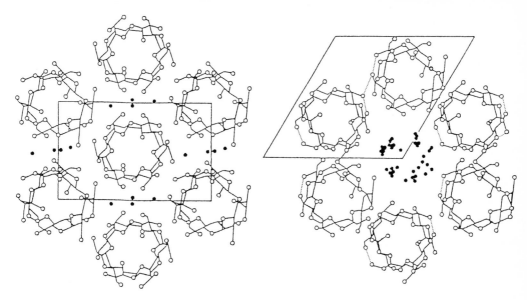

Fig. 4.33. Unit cells and arrangement of double helices (cross section) in A-amylose (left) and B-amylose (right) (according to *Galliard*, 1987)

per helical turn (Fig. 4.34, b) while, in the presence of KBr, the helix is even more stretched to 4 residues per turn (Fig. 4.34, c). Inclusion (clathrate) compounds are formed in the presence of small molecules and stabilize the V starch conformation (Fig. 4.34, d); it also has 6 glucose residues per helical turn. Stabilization may be achieved by H-bridges between O-2 and O-3 of neighboring residues within the same chain and between O-2 and O-6 of the residues i and i + 6 neighbored on the helix surface. Many molecules, such as iodine, fatty acids, fatty acid esters of hydroxycarboxylic acids (e.g., stearyllactate), monoglycerides, phenols, arylhalogenides, n-butanol, t-butanol, and cyclohexane are capable of forming clathrate compounds with amylose molecules. The helix diameter, to a certain extent, conforms to the size of the enclosed guest molecule; it varies from 13.7 Å to 16.2 Å. While the iodine complex and that of n-butanol have 6 glucose residues per turn in a V conformation, in a complex with t-butanol the helix turn is enlarged to 7 glucose residues/turn (Fig. 4.34, e). It is shown by an α-naphthol clathrate that up to 8 residues are allowed (Fig. 4.34, f). Since the helix is internally hydrophobic,

the enclosed "guest" has also to be lipophilic in nature. The enclosed molecule contributes significantly to the stability of a given conformation. For example, it is observed that the V conformation, after "guest" compound removal, slowly changes in a humid atmosphere to a more extended B conformation. Such a conformational transition also occurs during staling of bread or other bakery products. Freshly baked bread shows a V spectrum of gelatinized starch, but aged bread typically has the retrograded starch B spectrum. Figure 4.35 illustrates both conformations in the form of cylinder projections. While in V amylose, as already outlined, O-2 of residues i and O-6 of residues i + 6 come into close contact through H-bridges, in the B pattern the inserted water molecules increase the double-strand distance along the axis of progression (h) from 0.8 nm for the V helix to 1 nm for the B helix.

Cereal starches are stabilized by the enclosed lipid molecules, so their swelling power is low. The swelling is improved in the presence of alcohols (ethanol, amyl alcohol, tert-amyl alcohol). Obviously, these alcohols are dislodging and removing the "guest" lipids from the helices.

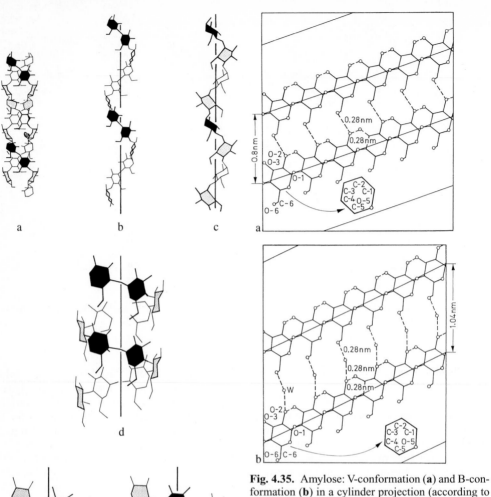

Fig. 4.35. Amylose: V-conformation (**a**) and B-conformation (**b**) in a cylinder projection (according to *Ebert*, 1980)

4.4.4.14.4 Structure and Properties of Amylopectin

Amylopectin is a branched glucan with side chains attached in the 6-position of the glucose residues of the principal chain:

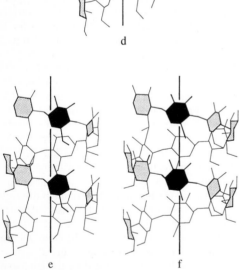

Fig. 4.34. Amylose conformation (for explanation see text) (according to *Rees*, 1977)

$$(4.139)$$

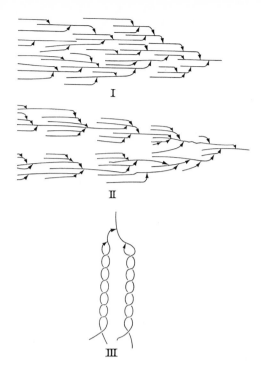

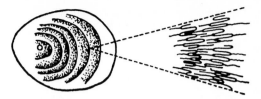

Fig. 4.37. Arrangement of amylopectin molecules in a starch granule

Fig. 4.36. Structural models (I, II) for amylopectin with parallel double helices. III is an enlarged segment of I or II (according to *Banks* and *Muir*, 1980)

An average of 15–30 glucose residues are present in short chain branches and each of these branch chains is joined by linkage of C-1 to C-6 of the next chain. The proposed structural models (Fig. 4.36) suggest that amylopectin also has double helices organized in parallel. As mentioned above, the main portion of a starch granule's crystalline structure is apparently derived from amylopectin. The structural modell II in Fig. 4.36 clearly shows from left to right the sequence of more compact (crystalline) and less compact (amorphous) sections. In this model, a distinction is made between shorter A-chains that are free of side chains and longer B chains that bear side chains. In the B chains, sections with compact successive side chains (cluster) alternate with branch-free sections.

The molecular weight of amylopectin is very high, in the range 10^7 to 7×10^8. One phosphoric acid residue is found for an average of 400 glucose residues.

The organization of amylopectin molecules in starch granules is shown in Fig. 4.37: it is radial, the reducing end being directed outwards.

Enzymatic degradation of amylopectin is similar to that of amylose. The enzyme β-amylase degrades the molecule up to the branching points. The remaining resistant core is designated as "limit-dextrin".

Amylopectin, when heated in water, forms a transparent, highly viscous solution, which is ropy, sticky and coherent. Unlike with amylose, there is no tendency toward retrogradation. There are no staling or aging phenomena and no gelling, except at very high concentrations. However, there is a rapid viscosity drop in acidic media and on autoclaving or applying stronger mechanical shear force.

4.4.4.14.5 Utilization

Starch is an important thickening and binding agent and is used extensively in the production of puddings, soups, sauces, salad dressings, diet food preparations for infants, pastry filling, mayonnaise, etc. Corn starch is the main food starch and an important raw material for the isolation of starch syrup and glucose (cf. 19.1.4.3).

A layer of amylose can be used as a protecting cover for fruits (dates or figs) and dehydrated and candied fruits, preventing their sticking together. Amylose treatment of French fries decreases their susceptibility to oxidation. The good gelling property of a dispersable amylose makes it a suitable ingredient in instant puddings or sauces. Amylose films can be used for food packaging, as edible wrapping or tubing, as exemplified by a variety of instant coffee or tea products. Amylopectin utilization is also diversified. It is used to a large extent as a thickener or stabilizer and as an adhesive or binding agent. Table 4.22 lists the range of its applications.

4.4.4.15 Modified Starches

Starch properties and those of amylose and amylopectin can be improved or "tailored" by physical and chemical methods to fit or adjust the properties to a particular application or food product.

4.4.4.15.1 Mechanically Damaged Starches

When starch granules are damaged by grinding or by application of pressure at various water contents, the amorphous portion is increased, resulting in improved dispersibility and swellability in cold water, a decrease in the gelatinization temperature by 5–10 °C, and an increase in enzymatic vulnerability. In bread dough made from flour containing damaged starch, for instance, the uptake of water is faster and higher and amylose degradation greater.

4.4.4.15.2 Extruded Starches

The X-ray diffraction diagram changes on extrusion of starch. The V-type appears first, followed by its conversion to an E-type at higher temperatures (>185 °C), and reformation of the V-type on cooling. The E-type apparently differs from the V-type only in the spacing of the V helices of amylose.
Extruded starches are easily dispersible, better soluble, and have a lower viscosity. The partial degradation of appropriately heated amylose shows that chemical changes also occur at temperatures of 185–200 °C. Apart from maltose, isomaltose, gentiobiose, sophorose, and 1,6-anhydroglucopyranose appeared.

4.4.4.15.3 Dextrins

Heating of starch (<15% of water) to 100–200 °C with small amounts of acidic or basic catalysts causes more or less extensive degradation. White and yellow powders are obtained which deliver clear or turbid, highly sticky solutions of varying viscosity. These products are used as adhesives in sweets and as fat substitutes.

Table 4.22. Utilization of amylopectin and its derivatives

Starch	Utilization
Unmodified waxy starch (also in blend with normal starch and flours)	Salad dressing, sterilized canned and frozen food, soups, broth, puffed cereals, and snack food
Pregelatinized waxy starch or isolated amylopectin	Baked products, paste (pâté) fillings, sterilized bread, salad dressing, pudding mixtures
Thin boiling waxy starch	Protective food coatings
Cross-linked waxy starch	Paste fillings, white and brown sauces, broth, sterilized or frozen canned fruit, puddings, salad dressing, soups, spreadable cream products for sandwiches, infant food
Waxy starch, hydroxypropyl ether	Sterilized and frozen canned food
Waxy starch, carboxymethyl ether	Emulsion stabilizer
Waxy starch acetic acid ester	Sterilized and frozen canned food, infant food
Waxy starch succinic- and adipic acid esters	Sterilized and frozen canned food, aroma encapsulation
Waxy starch sulfuric acid ester	Thickenig agent, emulsion stabilizer, ulcer treatment (pepsin inhibitor)

4.4.4.15.4 Pregelatinized Starch

Heating of starch suspensions, followed by drying, provides products that are swellable in cold water and form pastes or gels on heating. These products are used in instant foods, e. g., pudding, and as baking aids (cf. Table 4.22).

4.4.4.15.5 Thin-Boiling Starch

Partial acidic hydrolysis yields a starch product which is not very soluble in cold water but is readily soluble in boiling water. The solution has a lower viscosity than the untreated starch, and remains fluid after cooling. Retrogradation is slow. These starches are utilized as thickeners and as protective films (cf. Table 4.22).

4.4.4.15.6 Starch Ethers

When a 30–40% starch suspension is reacted with ethylene oxide or propylene oxide in the presence of hydroxides of alkali and/or alkali earth metals (pH 11–13), hydroxyethyl- or hydroxypropyl-derivates are obtained (R′ = H, CH_3):

$$R-OH + \underset{O}{\overset{R^1}{\triangle}} \xrightarrow{OH^{\ominus}} R-O-CH_2-\underset{\underset{OH}{|}}{CH}R^1$$

$$(4.140)$$

The derivatives are also obtained in reaction with the corresponding epichlorohydrins. The substitution degree can be controlled over a wide range by adjusting process parameters. Low substitution products contain up to 0.1 mole alkyl group/mole glucose, while those with high substitution degree have 0.8–1 mole/mole glucose. Introduction of hydroxyalkyl groups, often in combination with a small extent of cross-linking (see below) greatly improves starch swelling power and solubility, lowers the gelatinization temperature and substantially increases the freeze-thaw stability and the paste clarity of highly-viscous solutions. Therefore, these products are utilized as thickeners for refrigerated foods (apple and cherry pie fillings, etc), and heat-sterilized canned food (cf. Table 4.22).

Reaction of starch with monochloroacetic acid in an alkaline solution yields carboxymethyl starch:

$$R-OH + ClCH_2COO^{\ominus} \xrightarrow{OH^{\ominus}} R-O-CH_2-COO^{\ominus}$$

$$(4.141)$$

These products swell instantly, even in cold water and in ethanol. Dispersions of 1–3% carboxymethyl starch have an ointment-like (pomade) consistency, whereas 3–4% dispersions provide a gel-like consistency. These products are of interest as thickeners and gel-forming agents.

4.4.4.15.7 Starch Esters

Starch monophosphate ester is produced by dry heating of starch with alkaline orthophosphate or alkaline tripolyphosphate at 120–175°C:

$$R-OH \xrightarrow[POCl_3/Alkali\ phosphate]{OH^{\ominus}} R-OPO_3H^{\ominus}$$

$$(4.142)$$

Starch organic acid esters, such as those of acetic acid, longer chain fatty acids (C_6-C_{26}), succinic, adipic or citric acids, are obtained in reactions with the reactive derivatives (e.g., vinyl acetate) or by heating the starch with free acids or their salts. The thickening and paste clarity properties of the esterified starch are better than in the corresponding native starch. In addition, esterified starch has an improved freeze-thaw stability. These starches are utilized as thickeners and stabilizers in bakery products, soup powders, sauces, puddings, refrigerated food, heat-sterilized canned food and in margarines. The starch esters are also suitable as protective coatings, e.g., for dehydrated fruits or for aroma trapping or encapsulation (cf. Table 4.22).

4.4.4.15.8 Cross-Linked Starches

Cross-linked starches are obtained by the reaction of starch (R–OH) with bi- or polyfunctional reagents, such as sodium trimetaphosphate, phosphorus oxychloride, epichlorohydrin or mixed anhydrides of acetic and dicarboxylic acids (e.g., adipic acid):

$$2\,R-OH + R'CO-O-CO-(CH_2)_n-CO-O-COR'$$
$$\rightarrow R-O-CO-(CH_2)_n-CO-O-R \qquad (4.143)$$

$$R-OH + \underset{O}{\overset{CH_2Cl}{\triangle}}$$

$$\xrightarrow{OH^{\ominus}} R-O-CH_2-\underset{\underset{OH}{|}}{CH}-CH_2Cl$$

$$\xrightarrow{OH^{\ominus}} R-O-CH_2-\underset{O}{\triangle}$$

$$\xrightarrow[OH^{\ominus}]{ROH} R-O-CH_2-\underset{\underset{OH}{|}}{CH}-CH_2-O-R \qquad (4.144)$$

$$2\,R-OH \xrightarrow{(NaPO_3)_3} R-O-\overset{\overset{O}{\|}}{\underset{\underset{O^{\ominus}}{|}}{P}}-O-R \qquad (4.145)$$

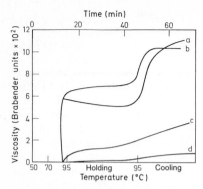

Fig. 4.38. Corn starch viscosity curves as affected by crosslinking degree. Instruments: Brabender amylograph; a control, b crosslinked with 0.05%, c 0.10%, d 0.15% epichlorohydrin (according to *Pigman*, 1970)

The starch granule gelatinization temperature increases in proportion to the extent of cross-linking, while the swelling power decreases (Fig. 4.38). Starch stability remains high at extreme pH values (as in the presence of food acids) and under conditions of shear force. Cross-linked starch derivatives are generally used when high starch stability is demanded.

4.4.4.15.9 Oxidized Starches

Starch hydrolysis and oxidation occur when aqueous starch suspensions are treated with sodium hypochlorite at a temperature below the starch gelatinization temperature range. The products obtained have an average of one carboxyl group per 25–50 glucose residues:

$$(4.146)$$

Oxidized starch is used as a lower-viscosity filler for salad dressings and mayonnaise. Unlike thinboiling starch, oxidized starch does not retrograde nor does it set to an opaque gel.

4.4.4.16 Cellulose

4.4.4.16.1 Occurrence, Isolation

Cellulose is the main constituent of plant cell walls, where it usually occurs together with hemicelluloses, pectin and lignin. Since cellulase enzymes are absent in the human digestive tract, cellulose, together with some other inert polysaccharides, constitute the indigestible carbohydrate of plant food (vegetables, fruits or cereals), referred to as dietary fiber. Cellulases are also absent in the digestive tract of animals, but herbivorous animals can utilize cellulose because of the rumen microflora (which hydrolyze the cellulose). The importance of dietary fiber in human nutrition appears mostly to be the maintenance of intestinal motility (peristalsis).

4.4.4.16.2 Structure, Properties

Cellulose consists of β-glucopyranosyl residues joined by $1 \rightarrow 4$ linkages (cf. Formula 4.147).
Cellulose crystallizes as monoclinic, rod-like crystals. The chains are oriented parallel to the fiber direction and form the long b-axis of the unit cell (Fig. 4.39). The chains are probably somewhat pleated to allow intrachain hydrogen bridge formation between O-4 and O-6, and between O-3 and O-5 (cf. Formula 4.148).
Intermolecular hydrogen bridges (stabilizing the parallel chains) are present in the direction

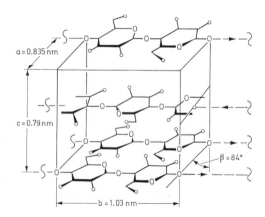

Fig. 4.39. Unit cell of cellulose (according to *Meyer* and *Misch*)

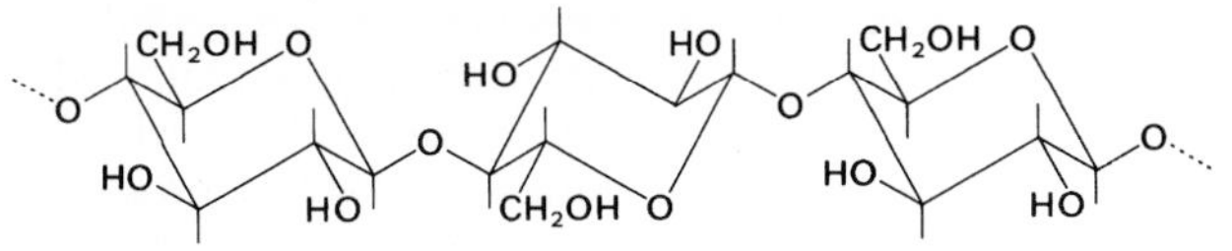

$$(4.147)$$

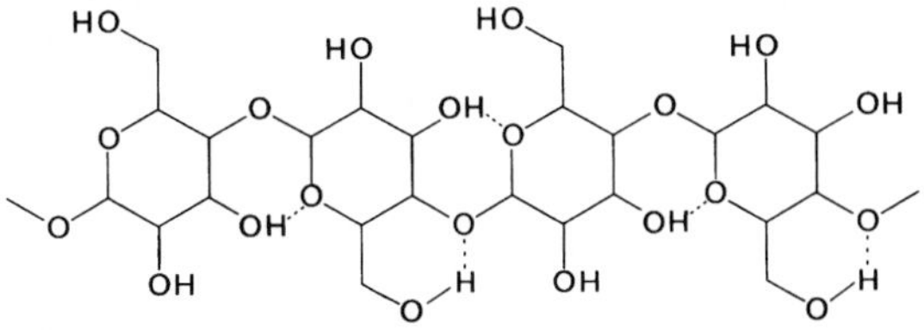

$$(4.148)$$

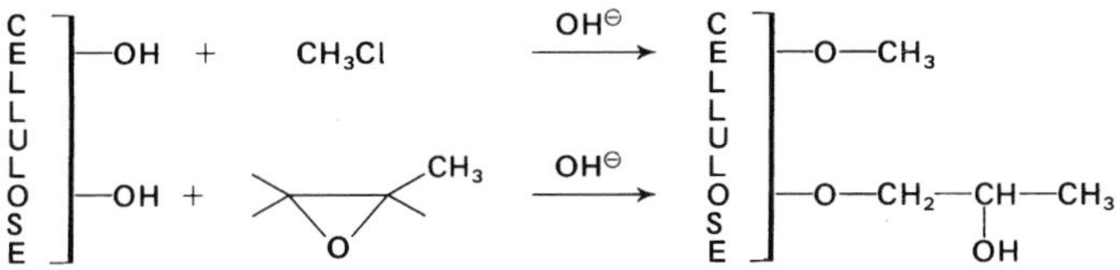

$$(4.149)$$

of the a-axis while hydrophobic interactions exist in the c-axis direction. The crystalline sections comprise an average of 60% of native cellulose. These sections are interrupted by amorphous gel regions, which can become crystalline when moisture is removed. The acid- or alkali-labile bonds also apparently occur in these regions. Microcrystalline cellulose is formed when these bonds are hydrolyzed. This partially depolymerized cellulose product with a molecular weight of 30–50 kdal, is still water insoluble, but does not have a fibrose structure.

Cellulose has a variable degree of polymerization (denoted as DP; number of glucose residues per chain) depending on its origin. The DP can range from 1,000 to 14,000 (with corresponding molecular weights of 162 to 2,268 kdal). Because of its high molecular weight and crystalline structure, cellulose is insoluble in water. Also, its swelling power or ability to absorb water, which depends partly on the cellulose source, is poor or negligible.

4.4.4.16.3 Utilization

Microcrystalline cellulose is used in low-calorie food products and in salad dressings, desserts and ice creams. Its hydration capacity and dispersibility are substantially enhanced by adding it in combination with small amounts of carboxymethyl cellulose.

4.4.4.17 Cellulose Derivatives

Cellulose can be alkylated into a number of derivatives with good swelling properties and improved solubility. Such derivatives have a wide field of application.

4.4.4.17.1 Alkyl Cellulose, Hydroxyalkyl Cellulose

The reaction of cellulose with methylchloride or propylene oxide in the presence of a strong alkali introduces methyl or hydroxypropyl groups into cellulose (cf. Reaction 4.149). The degree of substitution (DS) is dependent on reaction conditions.

Mixed substituted products are also produced, e.g., methylhydroxypropyl cellulose or methylethyl cellulose. The substituents interfere with the normal crystalline packing of the cellulose chains, thus facilitating chain solvation. Depending on the nature of the substituent (methyl, ethyl, hydroxymethyl, hydroxyethyl or hydroxypropyl) and the substitution degree, products are obtained with variable swelling powers and water solubilities. A characteristic property for methyl cellulose and double-derivatized methylhydroxypropyl cellulose is their initial viscosity drop with rising temperature, setting to a gel at a specific temperature. Gel setting is reversible. Gelling temperature is dependent on substitution type and degree. Figure 4.40 shows the dependence of gelling

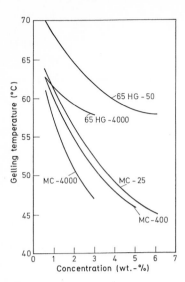

Fig. 4.40. Gelling behavior of alkyl celluloses (according to *Balser*, 1975). MC: methyl cellulose, HG: hydroxypropylmethyl cellulose with a hydroxypropyl content of about 6.5%. The numerical suffix is the viscosity (cps) of a 2% solution

temperature on the type of substitution and the concentration of the derivatives in water. Hydroxyalkyl substituents stabilize the hydration layer around the macromolecule and, thereby, increase the gelling temperature. Changing the proportion of methyl to hydroxypropyl substituents can vary the jelling temperature within a wide range. The above properties of cellulose derivatives permit their diversified application (Table 4.23). In baked products obtained from gluten-poor or gluten-free flours, such as those of rice, corn or rye, the presence of methyl and methylhydroxypropyl celluloses decreases the crumbliness and friability of the product, enables a larger volume of water to be worked into the dough and, thus, improves the extent of starch swelling during oven baking. Since differently substituted celluloses offer a large choice of gelling temperatures, each application can be met by using the most suitable derivative. Their addition to batter or a coating mix for meats (panure) decreases oil uptake in frying. Their addition to dehydrated fruits and vegetables improves rehydration characteristics and texture upon reconstitution. Sensitive

foods can be preserved by applying alkyl cellulose as a protective coating or film. Cellulose derivatives can also be used as thickening agents in low calorie diet foods. Hydroxypropyl cellulose is a powerful emulsion stabilizer, while methylethyl cellulose has the property of a whipping cream: it can be whipped into a stable foam consistency.

4.4.4.17.2 Carboxymethyl Cellulose

Carboxymethyl cellulose is obtained by treating alkaline cellulose with chloroacetic acid. The properties of the product depend on the degree of substitution (DS; 0.3–0.9) and of polymerization (DP; 500–2,000). Low substitution types (DS ≤ 0.3) are insoluble in water but soluble in alkali, whereas higher DS types (> 0.4) are water soluble. Solubility and viscosity are dependent on pH.
Carboxymethyl cellulose is an inert binding and thickening agent used to adjust or improve the texture of many food products, such as jellies, paste fillings, spreadable process cheeses, salad dressings and cake fillings and icings (Table 4.23). It retards ice crystal formation in ice cream, stabilizing the smooth and soft texture. It retards undesired saccharose crystallization in candy manufacturing and inhibits starch retrogradation or the undesired staling in baked goods. Lastly, carboxymethyl cellulose improves the stability and rehydration characteristics of many dehydrated food products.

4.4.4.18 Hemicelluloses

The term hemicelluloses refers to substances which occupy the spaces between the cellulose fibrils within the cell walls of plants. Various studies, e.g., on apples, potatoes, and beans, show that xyloglucans dominate in the class *Dicotyledoneae*. A section of the structure of a xyloglucan from runner beans is presented in Formula 4.150.
In the class *Monocotyledoneae*, the composition of the hemicelluloses in the endosperm tissue varies greatly, e.g., wheat and rye contain mainly arabinoxylans (pentosans, cf. 15.2.4.2.1), while β-glucans (cf. 15.2.4.2.2) predominate in barley and oats.

Table 4.23. Utilization of cellulose derivatives (in amounts of 0.01 to 0.8%)

Food product	Cellulose derivative [a]			Effect								
	1	2	3	A[b]	B	C	D	E	F	G	H	I
Baked products	+		+		+		+		+			
Potato products	+	+			+		+					+
Meat and fish	+		+	+		+						+
Mayonnaise, dressings	+		+	+	+			+				
Fruit jellies	+			+	+	+						
Fruit juices	+			+								
Brewery	+	+								+	+	
Wine	+	+								+	+	
Ice cream, cookies	+			+	+			+				
Diet food	+	+	+	+								

[a] 1: Carboxymethyl cellulose, Na-salt; 2: methyl cellulose; 3: hydroxypropyl methyl cellulose.
[b] A: Thickening effect; B: water binding/holding; C: cold gel setting; D: gel setting at higher temperatures; E: emulsifier; F: suspending effect; G: surface activity; H: adsorption; and I: film-forming property.

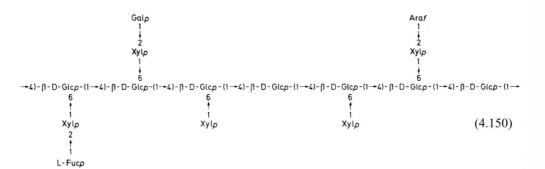

$$(4.150)$$

4.4.4.19 Xanthan Gum

4.4.4.19.1 Occurrence, Isolation

Xanthan gum, the extracellular polysaccharide from *Xanthomonas campestris* and some related microorganisms, is produced on a nutritive medium containing glucose, NH_4Cl, a mixture of amino acids, and minerals. The polysaccharide is recovered from the medium by isopropanol precipitation in the presence of KCl.

4.4.4.19.2 Structure, Properties

Xanthan gum can be regarded as a cellulose derivative. The main chain consists of 1,4 linked β-glucopyranose residues. On an average, every second glucose residue bears in the 3-position a trisaccharide of the structure β-D-

Man$p$-(1 → 4)-β-D-GlcpA(1 → 2)-α-D-Man$p$ as the side chain. The mannose bound to the main chain is acetylated in position 6 and ca. 50% of the terminal mannose residues occur ketalized with pyruvate as 4,6-O-(1-carboxyethylidene)-D-mannopyranose (cf. Formula 4.151; GlcpA: glucuronic acid).

$$[4)\text{-}\beta\text{-}\text{D-Glc}p\text{-}(1 \rightarrow 4)\text{-}\beta\text{-}\text{D-Glc}p\text{-}(1 \rightarrow]_n$$

$$\begin{array}{c} 3 \\ \uparrow \\ 1 \end{array}$$

β-D-Man$p$-(1 → 4)-β-D-GlcpA-(1 → 2)-α-D-Man$p$-6-OAc

$$(4.151)$$

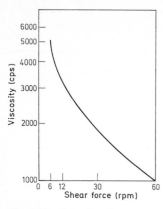

Fig. 4.41. Viscosity of aqueous xanthan gum solution as affected by shear rate (according to *Whistler*, 1973). Viscometer: Brookfield model LVF

The molecular weight of xanthan gum is $>10^6$ dal. In spite of this weight, it is quite soluble in water. The highly viscous solution exhibits a pseudoplastic behavior (Fig. 4.41). The viscosity is to a great extent, independent of temperature. Solutions, emulsions and gels, in the presence of xanthan gums, acquire a high freeze-thaw stability.

4.4.4.19.3 Utilization

The practical importance of xanthan gum is based on its emulsion-stabilizing and particle-suspending abilities (turbidity problems, essential oil emulsions in beverages). Due to its high thermal stability, it is useful as a thickening agent in food canning. Xanthan gum addition to starch gels substantially improves their freeze-thaw stability.

Xanthan gum properties might also be utilized in instant puddings: a mixture of locust bean flour, Na-pyrophosphate and milk powder with xanthan gum as an additive provides instant jelly after reconstitution in water. The pseudoplastic thixotropic properties, due to intermolecular association of single-stranded xanthan gum molecules, are of interest in the production of salad dressings, i.e. a high viscosity in the absence of a shear force and a drop in viscosity to a fluid state under a shear force.

4.4.4.20 Scleroglucan

4.4.4.20.1 Occurrence, Isolation

Sclerotium species, e.g., *S. glucanicum*, produce scleroglucan on a nutritive medium of glucose, nitrate as N-source and minerals. The polysaccharide is recovered from the nutritive medium by alcoholic precipitation.

4.4.4.20.2 Structure, Properties

The "backbone" of scleroglucan is a β-1,3-glucan chain that, on an average, has an attached glucose as a side chain on every third sugar residue (cf. Formula 4.152).

The polysaccharide has a molecular weight of about 130 kdal and is very soluble in water. Solutions have high viscosities and exhibit pseudoplastic thixotropic properties.

4.4.4.20.3 Utilization

Scleroglucan is used as a food thickener and, on the basis of its good film-forming property, is applied as a protective coating to dried foods.

4.4.4.21 Dextran

4.4.4.21.1 Occurrence

Leuconostoc mesenteroides, Streptobacterium dextranicum, Streptococcus mutans and some other bacteria produce extracellular dextran

(4.152)

from saccharose with the help of α-1,6-glucan: D-fructose-2-glucosyl transferase (dextran sucrase, EC 2.4.1.5).

4.4.4.21.2 Structure, Properties

Dextran is an α-1,6-glucan (Formula 4.153; molecular weight $M_r = 4-5 \times 10^7$ dal) with several glucose side chains, which are bound to the main chain of the macromolecule primarily through 1,3-linkages but, in part, also by 1,4- and 1,2-linkages.

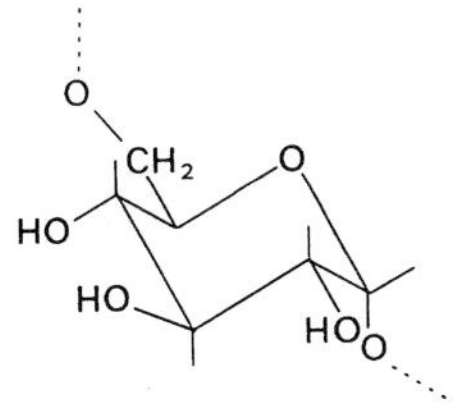

(4.153)

On an average, 95% of the glucose residues are present in the main chain. Dextran is very soluble in water.

4.4.4.21.3 Utilization

Dextran is used mostly in medicine as a blood substitute. In the food industry it is used as a thickening and stabilizing agent, as exemplified by its use in baking products, confections, beverages and in the production of ice creams.

4.4.4.22 Polyvinyl Pyrrolidone (PVP)

4.4.4.22.1 Structure, Properties

This compound is used as if it were a polysaccharide-type additive. Therefore, it is described here. The molecular weight of PVP can range from 10−360 kdal.

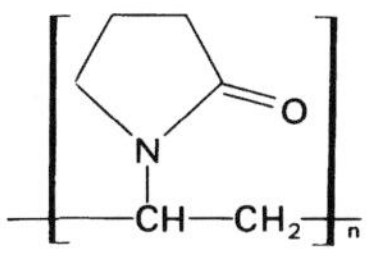

(4.154)

It is quite soluble in water and organic solvents. The viscosity of the solution is related to the molecular weight.

4.4.4.22.2 Utilization

PVP forms insoluble complexes with phenolic compounds and, therefore, is applied as a clarifying agent in the beverage industry (beer, wine, fruit juice). Furthermore, it serves as a binding and thickening agent, and as a stabilizer, e.g., of vitamin preparations. Its tendency to form films is utilized in protective food films (particle solubility enhancement and aroma fixation in instant tea and coffee production).

4.4.5 Enzymatic Degradation of Polysaccharides

Enzymes that cleave polysaccharides are of interest for plant foods. Examples are processes that occur in the ripening of fruit (cf. 18.1.3.3.2), in the processing of flour to cakes and pastries (cf. 15.2.2.1), and in the degradation of cereals in preparation for alcoholic fermentation (cf. 20.1.4). In addition, enzymes of this type are used in food technology (cf. 2.7.2.2) and in carbohydrate analysis (cf. Table 2.16 and 4.4.6). The following hydrolases are of special importance.

4.4.5.1 Amylases

Amylases hydrolyze the polysaccharides of starch.

4.4.5.1.1 α-Amylase

α-Amylase hydrolyzes starch, glycogen, and other 1,4-α-glucans. The attack occurs inside the molecule, i.e., this enzyme is comparable to endopeptidases. Oligosaccharides of 6−7 glucose units are released from amylose. The enzyme apparently attacks the molecule at the amylose helix (cf. 4.4.4.14.3) and hydrolyzes "neighboring" glycoside bonds that are one turn removed. Amylopectin is cleaved at random; the branch points (cf. 4.4.4.14.4) are overjumped. α-Amylase is activated by Ca^{2+} ions (cf. 2.3.3.1 and 2.7.2.2.2).

The viscosity of a starch solution rapidly decreases on hydrolysis by α-amylase (starch liquefaction) and the iodine color disappears. The dextrins formed at first are further degraded on longer incubation, reducing sugars appear and, finally, α-maltose is formed. The activity of the enzyme decreases rapidly with decreasing degree of polymerization of the substrate.

Catalysis is accelerated by the gelatinization of starch (cf. 4.4.4.14.2). For example, the swollen substrate is degraded 300 times faster by a bacterial amylase and 10^5 times faster by a fungal amylase than is native starch.

4.4.5.1.2 β-Amylase

This enzyme catalyzes the hydrolysis of 1,4-α-D-glucosidic bonds in polysaccharides (mechanism in Fig. 2.18), effecting successive removals of maltose units from the nonreducing end. Hydrolysis is linked to a Walden inversion at C-1, giving rise to β-maltose. This inversion, which can be detected polarimetrically, represents a definite characteristic of an exoglycanase.

In contrast to amylose, amylopectin is not completely hydrolyzed. All reaction stops even before branch points are reached.

4.4.5.1.3 Glucan-1,4-α-D-glucosidase (glucoamylase)

This glucoamylase starts at the nonreducing end of 1,4-α-D-glucans and successively liberates β-D-glucose units. In amylopectin, α-1,6-branches are cleaved ca. 30 times slower than α-1,4-bonds.

4.4.5.1.4 α-Dextrin Endo-1,6-α-glucosidase (pullulanase)

This enzyme hydrolyzes 1,6-α-D-glucosidic bonds in polysaccharides, e.g., in amylopectin, glycogen, and pullulan. Linear amylose fragments are formed from amylopectin.

4.4.5.2 Pectinolytic Enzymes

Pectins (cf. 4.4.4.13) in plant foods are attacked by a series of enzymes. A distinction is made between:

- Pectin esterases which occur widely in plants and microorganisms and demethylate pectin to pectic acid (Formula 4.155).
- Enzymes which attack the glycosidic bond in polygalacturonides (Table 4.24). These include hydrolases and lyases which cataly-

ze a transelimination reaction (see Formula 4.156). The double bond formed in the product of the last mentioned reaction results in an increase in the absorption at 235 nm.

The second group can be further subdivided according to the substrate (pectin or pectic acid) and to the site of attack (endo-/exo-enzyme), as shown in Table 4.24. The endo-enzymes strongly depolymerize and rapidly reduce the viscosity of a pectin solution.

Polygalacturonases occur in plants and microorganisms. They are activated by NaCl and some by Ca^{2+} ions as well.

Pectin and pectate lyases are only produced by microorganisms. They are activated by Ca^{2+} ions and differ in the pH optimum (pH 8.5–9.5) from the polygalacturonases (pH 5–6.5).

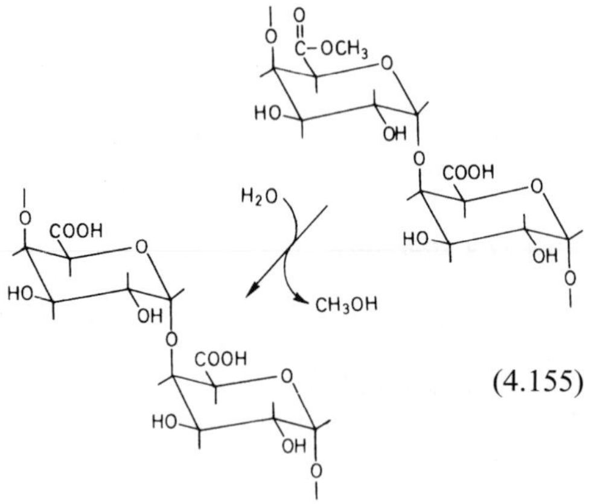

(4.155)

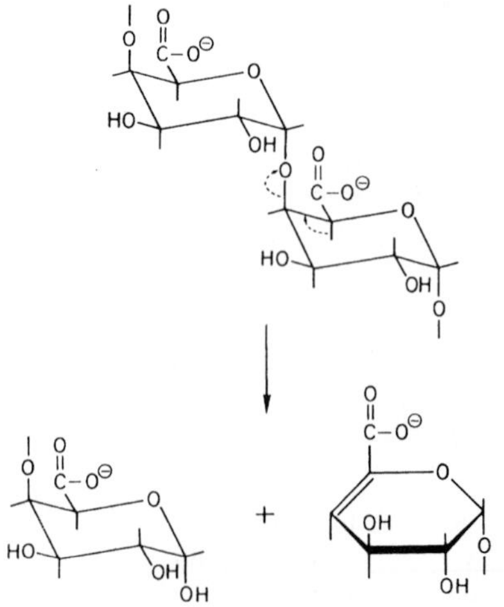

(4.156)

Table 4.24. Enzymes that cleave pectin and pectic acid

Enzyme	EC No.	Substrate
Polygalacturonase	3.2.1.15	
Endo-polymethyl galacturonase		Pectin
Endo-polygalacturonase		Pectic acid
Exo-polygalacturonase	3.2.1.67	
Exo-polymethyl galacturonase		Pectin
Exo-polygalacturonase		Pectic acid
Pectin lyase	4.2.2.10	
Endo-polymethyl galacturonlyase		Pectin
Pectate lyase	4.2.2.2	
Endo-polygalacturonate lyase		Pectic acid
Exo-polygalacturonate lyase	4.2.2.9	Pectic acid

4.4.5.3 Cellulases

Hydrolysis of completely insoluble, microcrystalline cellulose is a complicated process. For this purpose, certain microorganisms produce particles called cellusomes, (particle weight ca. 10^6). During isolation, these particles readily disintegrate into enzymes, which synergistically perform cellulose degradation, and components, which, among other things, support substrate binding. At least three enzymes are involved in the degradation of cellulose to cellobiose and glucose:

$$\text{Cellulose} \xrightarrow[C_1]{C_x} \text{Cellobiose} \xrightarrow{\text{Cellobiase}} \text{Glucose} \quad (4.157)$$

As shown in Table 4.25, the C_1 and C_x factors, which were found to be endo- and exo-1,4-β-glucanases respectively, hydrolyze cellulose to cellobiose. Since the C_1 factor is increasingly inhibited by its product, a cellobiase is needed so that cellulose breakdown is not rapidly brought to a standstill. However, cellobiase is also subject to product inhibition. Therefore, complete cellulose degradation is possible only if cellobiase is present in large excess or the glucose formed is quickly eliminated.

4.4.5.4 Endo-1,3(4)-β-glucanase

This hydrolase is also called laminarinase and hydrolyzes 1,3(4)-β-glucans. This enzyme occurs together with cellulases, e.g., in barley malt, and is involved in the degradation of β-glucans (cf. 15.2.4.2.2) in the production of beer.

4.4.5.5 Hemicellulases

The degradation of hemicelluloses also proceeds via endo- and exohydrolases. The sub-

Table 4.25. Cellulases

EC No.	Name	Synonym	Reaction
3.2.1.4	Cellulase	C_x factor CMCase[a], endo-1,4-β-glucanase	Endohydrolysis of 1,4-β-D-glucosidic bonds
3.2.1.91	Cellulose 1,4-β-cellobiosidase	C_1 factor avicellase	Exohydrolysis of 1,4-β-D-glucosidic bonds with formation of cellobiose from cellulose or 1,4-β-glucooligosaccharides. The attack proceeds from the nonreducing end.
3.2.1.21	β-Glucosidase	Cellobiase amygdalase	Hydrolysis of terminal β-D-glucose residues in β-glucans

[a] CMC: carboxymethylcellulose; the enzyme activity can be measured via the decrease in viscosity of a CMC solution.

strate specificity depends on the monosaccharide building blocks and on the type of binding, e. g., endo-1,4-β-D-xylanase, endo-1,5-α-L-arabinase. These enzymes occur in plants and microorganisms, frequently together with cellulases.

4.4.6 Analysis of Polysaccharides

The identification and quantitative determination of polysaccharides plays a role in the examination of thickening agents, balast material etc.

4.4.6.1 Thickening Agents

First, thickening agents must be concentrated. The process used for this purpose is to be modified depending on the composition of the food. In general, thickening agents are extracted from the defatted sample with hot water. Extracted starch is digested by enzymatic hydrolysis (α-amylase, glucoamylase), and proteins are separated by precipitation (e. g., with sulfosalicylic acid). The polysaccharides remaining in the solution are separated with ethanol. An electropherogram of the polysaccharides dissolved in a borate buffer provides an initial survey of the thickening agents present. It is sometimes difficult to identify and, consequently, differentiate between the added polysaccharides and those that are endogenously present in many foods. In simple cases, it is sufficient if the electropherogram is supported by structural analysis. For this purpose, the polysaccharides are first subjected to hydrolysis or methanolysis, the monomers are then qualitatively and quantitatively detected, e. g., as gluconic acid nitrile acetates (cf. 4.2.4.6) or as trimethylsilyl ethers (cf. 4.2.4.7), by gas chromatography on capillary columns. In more difficult cases, a preliminary separation of acidic and neutral polysaccharides on an ion exchanger is recommended. The linkage of the monomers is determined, if necessary. Here, free OH groups are first methylated, the polysaccharides are then acid hydrolyzed, and the methyl sugars obtained are analyzed by gas chromatography, as described above. Methanolysis or hydrolysis of polysaccharides and the detection of labile uronic acids and an-

hydro sugars are accompanied by losses and, therefore, have a critical course of analysis. The appearance of glucose with the monomers indicates, among other substances, modified glucans, e. g., modified starches or celluloses. The identification of thickening agents of this type is achieved by the specific detection of the hetero-components, e. g., acetate or phosphate.

4.4.6.2 Dietary Fibers

Gravimetric methods are the methods of choice for the determination of dietary fibers (cf. 15.2.4.2). In the defatted sample, the digestible components (1,4-α-glucans, proteins) are enzymatically hydrolyzed (heat-stable α-amylase, glucoamylase, proteinase). After centrifugation, the insoluble fibers remain in the residue. The water soluble fibers in the supernatant are isolated by precipitation with ethanol, ultrafiltration or dialysis. The protein and mineral matter still remaining with the soluble and insoluble dietary fibers is deducted with the help of correction factors.

4.5 Literature *

Angyal, S.J.: Zusammensetzung und Konformation von Zuckern in Lösung. Angew. Chem. *81*, 172 (1969)

Balser, K.: Celluloseäther. In: Ullmanns Encyklopädie der technischen Chemie, 4. edn., Vol. 9, S. 192. Verlag Chemie: Weinheim. 1975

Banks, W., Muir, D.D.: Structure and chemistry of the starch granule. In: The biochemistry of plants (Eds.: Stumpf, P.K., Conn, E.E.), Vol. 3, p. 321, Academic Press: New York. 1980

Birch, G.G.: Structural relationships of sugars to taste. Crit. Rev. Food Sci. Nutr. *8*, 57 (1967)

Birch, G.G. (Ed.): Developments in food carbohydrate-1 ff, Applied Science Publ.: London. 1977 ff.

Birch, G.G., Parker, K.J. (Eds.): Nutritive sweeteners. Applied Science Publ.: London. 1982

Birch, G.G., Parker, K.J. (Eds.): Dietary fibre. Applied Science Publ.: London. 1983

Birch, G.G. (Ed.): Analysis of food carbohydrate. Elsevier Applied Science Publ.: London. 1985

---

* Cf. 19.3.

Blanshard, J.M.V., Mitchell, J.R. (Eds.): Polysaccharides in food. Butterworths: London. 1979

Brimacombe, J.C. (Ed.): Carbohydrate chemistry, Vol. 1 ff, The Chemical Society, Burlington House: London. 1969 ff.

Davidson, R.L. (Ed.): Handbook of water-soluble gums and resins. McGraw-Hill Book Co.: New York. 1980

Ebert, G.: Biopolymere. Dr. Dietrich Steinkopff Verlag: Darmstadt. 1980

Eriksson, C. (Ed.): *Maillard* reactions in food. Chemical, physiological and technological aspects. Pergamon Press: Oxford. 1981

Galliard, T. (Ed.): Starch: Properties and Potential. John Wiley and Sons: Chichester. 1987

Guadagni, D.G., Maier, V.P., Turnbaugh, J.G.: Effect of some citrus juice constituents on taste thresholds for limonine and naringin bitterness. J. Sci. Food Agric. *24*, 1277 (1973)

Hill, R.D., Munck, L. (Eds.): New approaches to research on cereal carbohydrates. Elsevier Science Publ.: Amsterdam. 1985

Hough, L., Phadnis, S.P.: Enhancement in the sweetness of sucrose. Nature *263*, 800 (1976)

Jenner, M.R.: Sucralose. How to make sugar sweeter. ACS Symposium Series *450*, p. 68 (1991)

Ledl, F., Severin, T.: Untersuchungen zur *Maillard*-Reaktion. XIII. Bräunungsreaktion von Pentosen mit Aminen. Z. Lebensm. Unters. Forsch. *167*, 410 (1978)

Ledl, F., Severin, T.: Untersuchungen zur *Maillard*-Reaktion. XVI. Bildung farbiger Produkte aus Hexosen. Z. Lebensm. Unters. Forsch. *175*, 262 (1982)

Ledl, F., Krönig, U., Severin, T., Lotter, H.: Untersuchungen zur *Maillard*-Reaktion. XVIII. Isolierung N-haltiger farbiger Verbindungen. Z. Lebensm. Unters. Forsch. *177*, 267 (1983)

Ledl, F.: Low molecular products, intermediates and reaction mechanisms. In: Amino-carbonyl reactions in food and biological systems (Eds.: Fujimaki, M., Namiki, M., Kato, H.), p. 569, Elsevier: Amsterdam. 1986

Ledl, F., Fritul G, Hiebl, H., Pachmayr, O., Severin, T.: Degradation of *Maillard* products. In: Amino-carbonyl reactions in food and biological systems (Eds.: Fujimaki, M., Namiki, M., Kato, H.). p. 173, Elsevier: Amsterdam. 1986

Ledl, F.: Chemical Pathways of the *Maillard* Reaction. In: The *Maillard* Reaction in Food Processing, Human Nutrition and Physiology (Eds.: Finot, P.A. et al.) p. 19, Birkhäuser Verlag: Basel. 1990

Ledl, F., Schleicher, E.: Die *Maillard*-Reaktion in Lebensmitteln und im menschlichen Körper – neue Ergebnisse zu Chemie, Biochemie und Medizin. Angewandte Chemie *102*, 597 (1990)

Ledl, F., Glomb, M., Lederer, M.: Nachweis reaktiver *Maillard*-Produkte. Lebensmittelchemie *45*, 119 (1991)

Lehmann, J.: Chemie der Kohlenhydrate. Georg Thieme Verlag: Stuttgart. 1976

Loewus, F.A., Tanner, W. (Eds.): Plant carbohydrates I, II. Springer-Verlag: Berlin. 1981/82

Nedvidek, W., Noll, P., Ledl, F.: Der Einfluß des *Strecker*abbaus auf die *Maillard*-Reaktion. Lebensmittelchemie *45*, 119 (1991)

Pagington, J.S.: β-Cyclodextrin and its uses in the flavour industry. In: Developments in food flavours (Eds.: Birch, G.G., Kindley, M.G.), p. 131, Elsevier Applied Science: London. 1986

Pangborn, R.M., Leonard, S., Simone, M., Luh, B.S.: Freestone peaches. I. Effect of sucrose, citric acid and corn syrup on consumer acceptance. Food Technol. *13*, 444 (1959)

Pigman, W., Horton, D. (Eds.): The Carbohydrates. 2nd edn., Academic Press: New York. 1970–1980

Pilnik, W., Voragen, F., Neukom, H., Nittner, E.: Polysaccharide. In: Ullmanns Encyklopädie der technischen Chemie, 4. edn., Vol. 19, S. 233, Verlag Chemie: Weinheim. 1980

Preuß, A., Thier, H.-P.: Isolierung natürlicher Dickungsmittel aus Lebensmitteln zur capillargaschromatographischen Bestimmung. Z. Lebensm. Unters. Forsch. *176*, 5 (1983)

Radley, J.A. (Ed.): Starch production technology. Applied Science Publ.: London. 1976

Rees, D.A.: Polysaccharide shapes. Chapman and Hall: London. 1977

Reilly, P.J.: Xylanases: structure and function. Basic Life Sci. *18*, 111 (1981)

Salunkhe, D.K., McLaughlin, R.L., Day, S.L., Merkley, M.B.: Preparation and quality evaluation of processed fruits and fruit products with sucrose and synthetic sweeteners. Food Technol. *17*, 203 (1963)

Scherz, H., Mergenthaler, E.: Analytik der als Lebensmittelzusatzstoffe verwendeten Polysaccharide. Z. Lebensm. Unters. Forsch. *170*, 280 (1980)

Scherz, H.: Verwendung der Polysaccharide in der Lebensmittelverarbeitung. In: Polysaccharide. Eigenschaften und Nutzung (Hrsg.: Burchard, W.), S. 142, Springer-Verlag: Berlin. 1985

Schweizer, T.F.: Fortschritte in der Bestimmung von Nahrungsfasern. Mitt. Geb. Lebensmittelunters. Hyg. *75*, 469 (1984)

Shallenberger, R.S., Birch, G.G.: Sugar chemistry. AVI Publ. Co.: Westport, Conn. 1975

Shallenberger, R.S.: Advanced sugar chemistry, principles of sugar stereochemistry. Ellis Horwood Publ.: Chichester. 1982

Sutherland, I.W.: Extracellular polysaccharides. In: Biotechnology (Eds.: Rehm, H.-J., Reed, G.), Vol. 3, p. 531, Verlag Chemie: Weinheim. 1983

Von Sydow, E., Moskowitz, H., Jacobs, H., Meiselman, H.: Odor – Taste interaction in fruit juices. Lebensm. Wiss. Technol. 7, 18 (1974)

Waller, G.R., Feather, M.S. (Eds.): The *Maillard* reaction in foods and nutrition. ACS Symposium Series 215, American Chemical Society: Washington, D.C. 1983

Whistler, R.L. (Ed.): Industrial gums. 2nd edn., Academic Press: New York. 1973

Yaylayan, V.: In search of alternative mechanisms for the *Maillard* reaction, Trends Food Sci Technol 1, 20 (1990)

5 Aroma Substances

5.1 Foreword

5.1.1 Concept Delineation

When food is consumed, the interaction of taste, odor and textural feeling provides an overall sensation which is best defined by the English word "flavor". German and some other languages do not have an adequate expression for such a broad and comprehensive term. Flavor results from compounds that are divided into two broad classes: Those *responsible for taste* and those *responsible for odors*, the latter often designated as aroma substances. However, there are compounds which provide both sensations.

Compounds *responsible for taste* are generally nonvolatile at room temperature. Therefore, they interact only with taste receptors located in taste buds of the tongue. The four important basic taste perceptions are provided by: sour, sweet, bitter and salty compounds. They are covered in separate sections (cf., for example, 8.10, 22.3, 1.2.6, 1.3.3, 4.2.3 and 8.8).

Aroma substances are volatile compounds which are perceived by the odor receptor sites of the smell organ, i. e. the olfactory tissue of the nasal cavity. They reach the receptors when drawn in through the nose (nasal detection) and via the throat after being released by chewing (retronasal detection). The concept of aroma substances, like the concept of taste substances, should be used loosely, since a compound might contribute to the typical odor or taste of one food, while in another food it might cause a faulty odor or taste, or both, resulting in an off-flavor.

5.1.2 Impact Compounds of Natural Aromas

The amount of volatile substances present in food is extremely low (ca. $10-15$ mg/kg). In general, however, they comprise a large number of components. Especially foods made by

thermal processes, alone (e. g., coffee) or in combination with a fermentation process (e. g., bread, beer, cocoa, or tea), contain more than 500 volatile compounds. A great variety of compounds is often present in fruits and vegetables as well.

All the known volatile compounds are classified according to the food and the class of compounds and published in a constantly updated tabular compilation (*Maarse, H., Visscher, C. A.*, 1990). A total of 6200 compounds in more than 300 foods are listed in the 1990 edition.

Of all the volatile compounds, only a limited number are important for aroma. Compounds that are considered as aroma substances are primarily those which are present in food in concentrations higher than the odor and/or taste thresholds (cf. "Aroma Value", 5.1.4). Compounds with concentrations lower than the odor and/or taste thresholds also contribute to aroma when mixtures of them exceed these thresholds (for examples of additive effects, see 3.2.1.1 and 20.1.7.8).

Among the aroma substances, special attention is paid to those compounds that provide the characteristic aroma of the food, the character impact compounds. Examples are given in Table 5.1. Furthermore, compounds of special interest are those which produce important aroma features in complex aroma profiles, e. g., "nut-like" in hazelnuts or "roasted" in white bread (Table 5.1).

In the case of individual foods, the differentiation between aroma substances and the remaining volatile compounds has progressed to very different extents. Important findings are presented in the section on "Aroma Substances" in the corresponding chapters.

5.1.3 Threshold Value

The lowest concentration of a compound that is just enough for the recognition of its odor is

Table 5.1. Examples of character impact compounds

Compound	Aroma	Occurrence
2-trans,4-cis-Decadienoic acid ethyl ester	Pear-like	Pears
Benzaldehyde	Bitter almond-like	Almonds, cherries, plums
Neral/geranial	Lemon-like	Lemons
1-(p-Hydroxy-phenyl)-3-butanone (raspberry ketone)	Raspberry-like	Raspberries
(R)-(−)-1-Octen-3-ol	Mushroom-like	Champignons, Camembert cheese
2-trans,6-cis-Nonadienal	Cucumber-like	Cucumbers
Geosmin	Earthy	Beetroot
trans-5-Methyl-2-hepten-4-one (filberton)	Nut-like	Hazelnuts
4-Hydroxy-2,5-dimethyl-3(2H)-furanone	Caramel-like	Biscuits, beer, coffee
2-Acetyl-1-pyrroline	Roasted	White-bread crust

Table 5.2. Odor threshold values in water of some aroma compounds (20 °C)

Compound	Threshold value (mg/l)
Ethanol	100
Maltol	35
Hexanol	2.5
Furfural	3.0
Vanillin	0.02
Raspberry ketone	0.01
Limonene	0.01
Linalool	0.006
Hexanal	0.0045
2-Phenylethanal	0.004
Methylpropanal	0.001
Ethylbutyrate	0.001
(+)-Nootkatone	0.001
(−)-Nootkatone	1.0
Filberton	0.00005
Methylthiol	0.00002
2-Isobutyl-3-methoxypyrazine	0.000002
1-p-Menthen-8-thiol	0.00000002

called the odor threshold (recognition threshold). The detection threshold is lower, i.e., the concentration at which the compound is detectable but the aroma quality still cannot be unambiguously established. Threshold concentration data allow comparison of the intensity or potency of odorous substances. The examples in Table 5.2 illustrate that great differences exist between individual aroma compounds, with an odor potency range of several orders of magnitude.

In an example provided by nootkatone, an essential aroma compound of grapefruit oil (cf. 18.1.2.6.3), it is obvious that the two enantiomers (optical isomers) differ significantly in their aroma intensity (cf. 5.2.4) and, occasionally, in aroma quality or character.

The threshold concentrations (values) for aroma compounds are dependent on their vapor pressure, which is affected by both temperature and medium (cf. Table 5.3). The values are also influenced by the assay procedure and/or performance of the sensory panel. The frequent discrepancies in threshold values in the literature are basically due to such differences.

5.1.4 Aroma Value

As already indicated, compounds with high "aroma values" may contribute to the aroma of foods. The "aroma value" A_x of a compound is calculated according to the definition:

$$A_x = \frac{c_x}{a_x} \tag{5.1}$$

(c_x: concentration of compound X in the food, a_x: odor threshold (cf. 5.1.3) of compound X in the food). Methods for the identification of the corresponding compounds are desribed under Section 5.2.5.

The evaluation of volatile compounds on the basis of the aroma value provides only a rough pattern at first. The dependence of the odor intensity on the concentration must also be taken into account. In accordance with the universally valid law of *Stevens* for physiological stimuli, it is formulated as follows:

$$E = k \cdot (S - S_o)^n \tag{5.2}$$

E: perception intensity, k: constant, S: concentration of stimulant, S_o: threshold concentration of stimulant.

The examples presented in Fig. 5.1 show that the exponent n and, therefore, the dependency

Table 5.3. Comparison of threshold values[a] in water and beer

Compound	Threshold (mg/kg) in	
	Water	Beer
n-Butanol	0.5	200
3-Methylbutanol	0.25	70
Dimethylsulfide	0.00033	0.05
2-trans-Nonenal	0.00008	0.00011

[a] Odor and taste.

of the odor intensity on the concentration can vary substantially.

In addition, additive effects that are difficult to assess must also be considered. Examinations of mixtures have provided preliminary information. They show that although the intensities of compounds with a similar aroma note add up, the intensity of the mixture is usually lower than the sum of the individual intensities (cf. 3.2.1.1 and 20.1.7.8). For substances which clearly differ in their aroma note, however, the odor profile of a mixture is composed of the odor profiles of the components added together, only when the odor intensities are approximately equal. If the concentration ratio is such that the odor intensity of one component predominates, this component then largely or completely determines the odor profile.

Examples are 2-trans-hexenal and 2-trans-decenal which have clearly different odor profiles (cf. Fig. 5.2, a and 5.2, f). If the ratio of the odor intensities is approximately one, the odor notes of both aldehydes can be recognized in the odor profile of the mixture (Fig. 5.2, d). But if the dominating odor intensity is that of the decenal (Fig. 5.2, b), or of the hexenal (Fig. 5.2, e), that particular note determines the odor profile of the mixture.

The mixture in Fig. 5.2, c gives a new odor profile because definite features of the decenal (stale, paint-like, rancid) and the hexenal (like apples, almonds, sweet) can no longer be recognized in it.

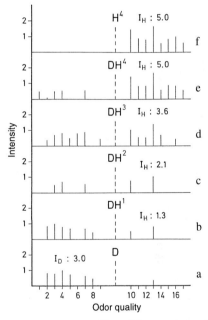

Fig. 5.2. Odor profiles of 2-trans-decenal (D), 2-trans-hexenal (H) and mixtures of both aldehydes (according to *Laing* and *Willcox*, 1983). The following concentrations (mg/kg) dissolved in di-2-ethylhexylphthalate were investigated: 50 (D); 2 (H[1]); 3.7 (H[2]); 11 (H[3]) and 33 (H[4]).

I_D and I_H: Odor intensity of each concentration of 2-trans-decenal and 2-trans-hexenal. Odor quality: **1**, warm; **2**, like clean washing; **3**, cardboard; **4**, oily, fatty; **5**, stale; **6**, paint; **7**, candle; **8**, rancid; **9**, stink-bug; **10**, fruity; **11**, apple; **12**, almond; **13**, herbal, green; **14**, sharp, pungent; **15**, sweet; **16**, banana; **17**, floral. The broken line separates the aroma qualities of 2-trans-decenal (left side) and 2-trans-hexenal

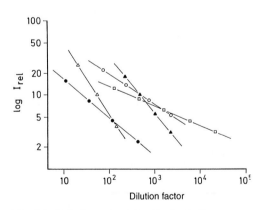

Fig. 5.1. Relative odor intensity I_{rel} (reference: n-butanol) as a function of the stimulant concentration (according to *Dravnieks*, 1977).

Air saturated with aroma substance was diluted. ●—●—● α-pinene, ○—○—○ 3-methylbutyric acid methyl ester, △—△—△ hexanoic acid, ◆—◆—◆ 2,4-hexadienal, □—□—□ hexylamine

The examples show clearly that the aroma profiles of foods containing the same aroma substances can be completely dissimilar owing to quantitative differences. For example, changes in the recipe or in the production process which cause alterations in the concentrations of the aroma substances can interfere with the balance in such a way that an aroma profile with unusual characteristics is obtained.

5.1.5 Off-Flavors, Food Taints

An off-flavor can arise through foreign aroma substances, that are normally not present in a food, loss of impact compounds, or changes in the concentration ratio of individual aroma substances. Figure 5.3 describes the causes for aroma defects in food. In the case of an odorous contaminant, which enters the food via the air or water and then gets enriched, it can be quite difficult to determine its origin if the limiting concentration for odor perception is exceeded only on enrichment. Examples of some off-flavors that can arise during food processing and storage are listed in Table 5.4. Examples of microbial metabolites wich may be involved in pigsty-like and earthy-muddy off-flavors are skatole (I; faecal-like, 10 µg/kg*), 2-methylisoborneol (II; earthy-muddy,

0.03 µg/kg*) and geosmin (III; earthy, 0.01 µg/kg*):

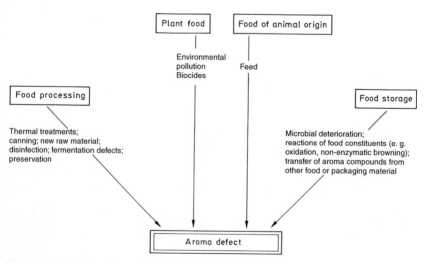

$$(5.3)$$

2,4,6-Trichloroanisole (IV) with an extremely low odor threshold (mouldy-like: 3.10^{-5} µg/kg, water) is an example of an off-flavor substance (cf. 20.2.7) which is produced by fungal degradation and methylation of pentachlorophenol fungicides.

To a certain extent, unwanted aroma substances are concealed by typical ones. Therefore, the threshold above which an off-flavor becomes noticeable can increase considerably in food compared to water as carrier, e. g., up to 0.2 µg/kg 2,4,6-trichloroanisole in dried fruits.

Fig. 5.3. The cause of aroma defects in food

---

* Odor threshold in water.

Table 5.4. "Off-flavors" in food products

Food product	Off-flavor	Cause
Milk	Sunlight flavor	Photooxidation of methionine to methional (with riboflavin as a sensitizer)
Milk powder	Bean-like	The level of O_3 in air too high: ozonolyis of 8,15- and 9,15-isolinoleic acid to 6-trans-nonenal
Milk powder	Gluey	Degradation of tryptophan to o-amino-acetophenone
Milk fat	Metallic	Autoxidation of pentaene- and hexaene fatty acids to octa-1,cis-5-dien-3-one
Milk products	Malty	Faulty fermentation by *Streptococcus lactis, var. maltigenes*; formation of phenylacetaldehyde and 2-phenylethanol from phenylalanine
Mutton-meat	Sweet, acidic	4-Methyloctanoic acid, 4-methylnonanoic acid
Peas, deep frozen	Hay-like	Saturated and unsaturated aldehydes, octa-3,5-dien-2-one, 3-alkyl-2-methoxypyrazines, hexanal
Orange juice	Grapefruit note	Metal-catalyzed oxidation or photooxidation of valencene to nootkatone 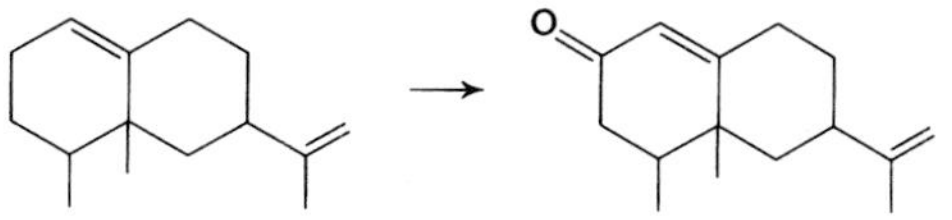
Orange juice	Terpene note	d-Limonene oxidation to carvone 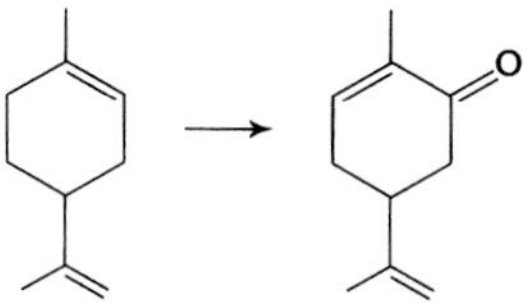
Passion fruit juice	Aroma flattening during pasteurization	Oxidation of (6-trans-2′-trans)-6-(but-2′-enyliden)-1,5,5-trimethylcyclohex-1-ene to 1,1,6-trimethyl-1,2-dihydronaphthalene: 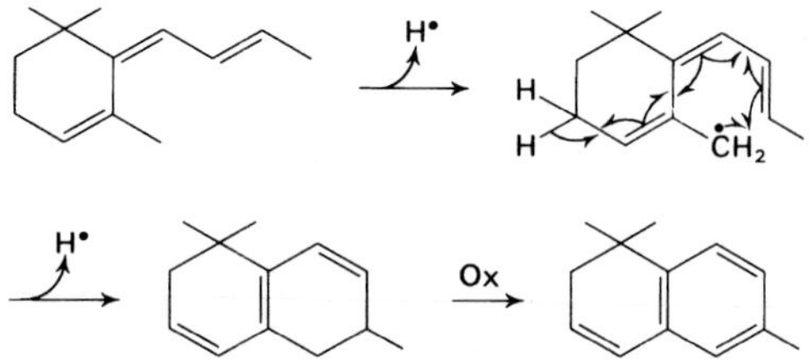
Beer	Sunlight flavor	Photolysis of humulone: reaction of one degradation product with hydrogen sulfide yielding 3-methyl-2-buten-1-thiol
Beer	Phenolic note	Faulty fermentation: hydrocinnamic acid decarboxylation by *Hafnia protea*

5.2 Aroma Analysis

The aroma substances consist of highly diversified classes of compounds, some of them being highly reactive and are present in food in extremely low concentrations. The difficulties usually encountered in qualitative and quantitative analysis of aroma compounds are based on these features. Other difficulties are associated with identification of aroma compounds, elucidation of their chemical structure and characterization of sensory properties. The results of an aroma analysis can serve as an objective guide in food processing for assessing the suitability of individual processing steps, and for assessing the quality of raw material, intermediate- and endproducts. In addition, investigation of food aroma broadens the possibility of food flavoring with substances that are prepared synthetically, but are chemically identical to those found in nature, i.e. the so-called "nature identical flavors" (cf. 5.5).

5.2.1 Aroma Isolation

The amount of starting material must be selected to detect even those aroma substances which are present in very low concentrations (ppb range), but contribute considerably to the aroma because of still lower odor thresholds. The volatile compounds should be isolated from food using different methods because each method has its own drawbacks which can result in quantitative changes in the detected spectrum of aroma substances (Table 5.5).

Additional difficulties are encountered in foods which retain fully-active enzymes, which can alter the aroma. For example, during the homogenization of fruits and vegetables, hydrolases split the aroma ester constituents, while lipoxygenase, together with hydroperoxide lyase, enrich the aroma with newly-formed volatile compounds. To avoid such interferences, tissue disintegration is done in the presence of enzyme inhibitors or, when possible, by rapid sample preparation. It is useful in some cases to inhibit enzyme-catalyzed reactions by the addition of methanol or ethanol. However, this can result in a change in

Table 5.5. Yield of volatile compounds (%) obtained by their isolation from highly diluted aqueous solutions (0.6 ppm) using the distillation and extraction (pentane) apparatus of *Likens* and *Nickerson*

C-Number	1-Alkanol	2-Alkanone	Alkanal	Alkane
3	Trace			
4	Trace	Trace	Trace	
5	93	79	101	
6	97	104	91	64
7	101	101	101	94
8	102	94	94	103
9	99	97	83	94
10		102		90
11		101		94
12				104

aroma due to the formation of esters and acetals from acids and aldehydes respectively.

At the low pH values prevalent in fruit, non-enzymatic reactions, especially reactions 4–7 shown in Table 5.6, can interfere with the isolation of aroma substances by the formation of artifacts. In the concentration of isolates from heated foods, particularly meat, it cannot be excluded that reactive substances, e.g., thiols, amines and aldehydes, get concentrated to such an extent that they condense to form heterocyclic aroma substances, among other compounds (Reaction 8, Table 5.6).

An additional aspect of aroma isolation not to be neglected is the ability of the aroma sub-

Table 5.6. Possible changes in aromas during the isolation of volatile compounds

Reaction

Enzymatic

1. Hydrolysis of esters (cf. 3.7.1)
2. Oxidative cleavage of unsaturated fatty acids (cf. 3.7.2.3)
3. Hydrogenation of aldehydes (cf. 5.3.2.1)

Non-enzymatic

4. Hydrolysis of glycosides (cf. 5.3.2.4 and 3.8.4.4)
5. Lactones from hydroxy acids
6. Cyclization of di-, tri-, and polyols (cf. 5.3.2.4)
7. Dehydration and rearrangement of tert-allyl alcohols
8. Reactions of thiols, amines, and aldehydes (cf. 5.3.1.4) in the aroma concentrate

stances to bind to the solid food matrix. Such binding ability differs for many aroma constituents (cf. 5.4).

The aroma substances present in the vapor space above the food can be very gently detected by *headspace analysis* (cf. 5.2.1.3). However, the amounts of substance isolated in this process are so small that important aroma substances, which are present in food in very low concentrations, give no detector signal after gas chromatographic separation of the sample. These substances can be determined only by sniffing the carrier gas stream. The difference in the detector sensitivity is clearly shown in Fig. 5.4, taking the aroma substances of the crust of white bread as an example. The gas chromatogram does not show, e. g., 2-acetyl-1-pyrroline and 2-ethyl-3,5-dimethylpyrazine, which are of great importance for aroma due to high FD factors in the FD chromatogram (definition in 5.2.5.2). These aroma substances can be identified only after concentration from a relatively large amount of the food.

5.2.1.1 Distillation, Extraction

The volatile aroma compounds, together with some water, are removed by vacuum distillation from an aqueous food suspension. The highly volatile compounds are condensed in an efficiently cooled trap. The aqueous distillate

is then extracted with an organic solvent to separate the organic compounds present.

An extraction combined with distillation can be achieved using an apparatus designed by *Likens-Nickerson* (Fig. 5.5). Experiments with the classes of compounds listed in Table 5.5 provide high recoveries for the C_5–C_{11} homologues. However, separation of polar compounds which are readily soluble in water is incomplete, e. g., the low-molecular homologues of the compound classes given in Table 5.5 or the 3(2H)-furanones listed in Table 5.16. On the other hand, the volatility of the compounds is reduced for molecular weights above 150 dal and, thus, their recovery is greatly decreased.

In simultaneous distillation/extraction, low-boiling solvents are usually used to make subsequent concentration of the aroma substances easier. Therefore, this process is carried out at normal pressure or slightly reduced pressure. The resulting thermal treatment of the food can lead to reactions (examples in Table 5.6) that change the aroma composition. The example in Table 5.7 shows the extent to which some aroma substances are released from glycosides during simultaneous distillation/extraction.

Recovery of the volatiles from fats and oils in a cold trap (Fig. 5.6) provides concentrates free of water.

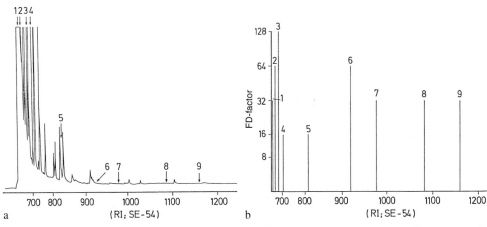

Fig. 5.4. Headspace analysis of aroma substances of white-bread crust. **a** Capillary gas chromatogram (the arrows mark the positions of the odorants), **b** FD chromatogram. Odorants: *1* methylpropanal, *2* diacetyl, *3* 3-methylbutanal, *4* 2,3-pentanedione, *5* butyric acid, *6* 2-acetyl-1-pyrroline, *7* 1-octen-3-one, *8* 2-ethyl-3,5-dimethylpyrazine, *9* (E)-2-nonenal (according to *Schieberle* and *Grosch*, 1992)

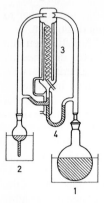

Fig. 5.5. Apparatus according to *Likens* and *Nickerson* used for simultaneous extraction and distillation of volatile compounds.
1 Flask with heating bath containing the aqueous sample, 2 flask with heating bath containing the solvent (e.g. pentane), 3 cooler, 4 condensate separator: extract is the upper and water the lower phase

Table 5.7. Isolation of odorants from cherry juice – Comparison of distillation in vacuo (I) with simultaneous distillation and extraction (II)

Odorant	I	(µg/l)	II
Benzaldehyde	202		5260
Linalool	1.1		188

The aroma substances are taken up in a highly volatile solvent and the solution is concentrated under gentle conditions because otherwise artifacts can be formed (Table 5.6). A rediluted aliquot of the sample is subjected to a sensory check to make sure that the aroma is identical to that of the starting material.

5.2.1.2 Gas Extraction

Volatile compounds can be isolated from a solid or liquid food sample by purging the sample with an inert gas (e.g., N_2, CO_2, He) and adsorbing the volatiles on a porous, granulated polymer (Tenax GC, Porapak Q, Chromosorb 105), followed by recovery of the compounds. Water is retarded to only a negligible extent by these polymers (Table 5.8). The desorption of volatiles is usually achieved stepwise in a temperature gradient. At low temperatures, the traces of water are removed by elution, while at elevated temperatures, the volatiles are released and flushed out by a carrier gas into a cold trap, usually connected to a gas chromatograph.

5.2.1.3 Headspace Analysis

The headspace analysis procedure is simple: the food is sealed in a container, then brought to the desired temperature and left for a while to establish an equilibrium between volatiles bound to the food matrix and those present in the vapor phase. A given volume of the headspace is withdrawn with a gas syringe and then injected into a gas chromatograph equipped with a suitable separation column (static headspace analysis). Since the presence of water and a large volume of headspace provides inferior separations, the sample volume has to be moderate and therefore only the major volatiles are found.

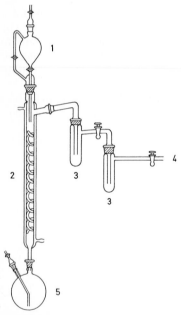

Fig. 5.6. Apparatus for separation of volatile compounds from fats and oils or other high boiling solvents (according to *Weurman*, 1969). 1 Sample, 2 glass column with heating jacket (40–60 °C) and a rotating spiral to disperse the sample over a large surface, 3 condensing traps cooled in liquid nitrogen or acetone/dry ice, 4 connection to vacuum pump, 5 receiving flask for sample depleted of volatile compounds

Table 5.8. Relative retention time (t_{rel}) of some compounds separated by gas chromatography using Porapak Q as stationary phase (Porapak: styrene divinylbenzene polymer; T: 55 °C)

Compound	t_{rel}	Compound	t_{rel}
Water	1.0	Methylthiol	2.6
Methanol	2.3	Ethylthiol	20.2
Ethanol	8.1	Dimethylsulfide	19.8
Acetaldehyde	2.5	Formic acid	
Propanal	15.8	ethyl ester	6.0

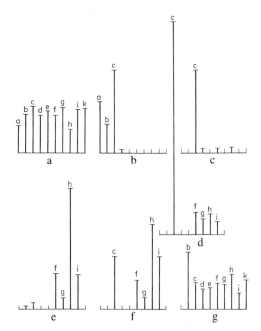

Fig. 5.7. A comparison of some methods used for aroma compound isolation (according to *Jennings* and *Filsoof*, 1977). **a** a Ethanol, b 2-pentanone, c heptane, d pentanol, e hexanol, f hexyl formate, g 2-octanone, h d-limonene, i heptyl acetate and k γ-heptalactone. **b** Headspace analysis of aroma mixture **a**. **c** From aroma mixture 10 µl is dissolved in 100 ml water and the headspace is analyzed. **d** As in **c** but the water is saturated with 80% NaCl. **e** As in **c** but purged with nitrogen and trapped by Porapak Q. **f** As in **c** but purged with nitrogen and trapped by Tenax GC. **g** As in **e** but distilled and extracted according to *Nickerson* and *Likens* (cf. Fig. 5.5)

An increase in analysis sensitivity is possible if the headspace volatiles are flushed out, adsorbed and thus concentrated in a polymer, as outlined in the previous section (dynamic headspace analysis). However, it is difficult to obtain a representative sample by this flushing procedure, a sample that would match the original headspace composition. A model system assay (Fig. 5.7) might clarify the problems. Samples (e) and (f) were obtained by adsorption on different polymers. They are different from each other and differ from sample (b), which was obtained directly for headspace analysis. The results might agree to a greater extent by varying the gas flushing parameters (gas flow, time), but substantial differences would still remain.

A comparison of samples (a) and (g) in Fig. 5.7 shows that the results obtained by the distillation-extraction procedure, as applied to a model mixture, are fairly reproducible, with ethanol as the only exception.

5.2.2 Separation

When an aroma concentrate contains phenols, organic acids and or bases, preliminary separation of these compounds from neutral volatiles by extraction with alkali or acids is advantageous.

The acidic, basic and neutral fractions are individually analyzed. The neutral fraction by itself consists of so many compounds that in most cases not even a gas chromatographic column with the highest resolving power is able to separate them into individual peaks.

Thus, separation of the neutral fraction is advisable and is usually achieved by fractional distillation, or preparative gas or high performance liquid chromatography. An example given in Fig. 5.8, which deals with an analysis of cognac aroma, may meet all the requirements for such an analysis.

5.2.3 Chemical Structure

In the structure elucidation of aroma substances, mass spectrometry has become an indispensable tool because the substance amounts eluted by gas chromatography are

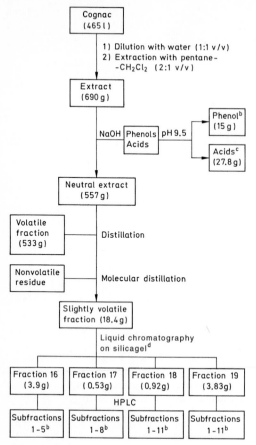

Fig. 5.8. Volatile compounds of cognac[a]. (analysis scheme according to *Ter Heide* et al., 1978)

---

[a] The analysis is limited to fractions that significantly contribute to cognac aroma
[b] GC/MS analysis identified the presence of 18 acetals, 59 alcohols, 28 aldehydes, 70 esters, 35 ketones, 3 lactones, 8 phenols, and 44 other compounds
[c] GC/MS analysis of acids as methyl esters provided 27 compounds
[d] Of 22 fractions collected, four fractions were further separated by high pressure liquid chromatography

generally sufficient for an evaluable spectrum. If the corresponding reference substance is available, identification of the aroma substance is based on correspondence of the mass spectrum, retention indices on at least two capillary columns of different polarity, and of

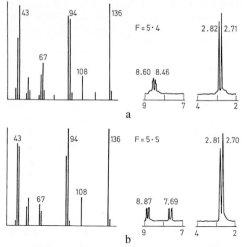

Fig. 5.9. Mass spectra and ^{1}H-NMR spectra. Excerpts from 2-acetyl-3-methylpyrazine (**a**) and 4-acetyl-2-methylpyrimidine (**b**) recordings (according to *Tressl*, 1980)

odor thresholds, which are compared by gas chromatography/olfactometry (cf. 5.2.5.2). If one of these criteria does not agree with the reference substance, the structure of the odorant must be established, e.g., by ^{1}H-NMR measurements. Indeed, ^{1}H-NMR spectra also permit the determination of substances which have ambiguous mass spectra.

As an example to be considered are the two compounds shown in Fig. 5.9. Their mass spectra are very similar. Differentiation is possible only from the ^{1}H-NMR spectra (Fig. 5.9). Wider use of ^{1}H-NMR in the elucidation of aroma compound structure is now possible since new recording techniques have been developed which are suitable for compounds also present in relatively small amounts.

It is obvious that identification of an odorant is considered final only when the proposed structure coincides in all aspects with a synthetic reference compound.

5.2.4 Enantioselective Analysis

In the case of chiral aroma substances, elucidation of the absolute configuration and determination of the enantiomeric ratio, which is usually given as the enantiomeric excess (ee), are of especial interest because the enantiomers of a compound can differ considerably in their odor quality and threshold. In addition, determination of the ee value can be used to detect aromatization with a synthetic chiral aroma substance because in many cases one enantiomer is preferentially formed in the biosynthesis of chiral aroma substances (examples in Table 5.9). In contrast to biosynthesis, chemical synthesis gives the racemate which is usually not separated for economic reasons. The addition of an aroma substance of this type can be determined by enantioselective analysis if safe data on the enantiomeric excess of the compound in the particular food are available. It should also be taken into account that the ee value can change during food processing, e.g., that of filberton decreases during the roasting of hazelnuts (cf. Table 5.9).

The method frequently applied to determine ee values is the enantioselective gas chromatographic analysis of the aroma substance on a chiral phase, e. g., peralkylated cyclodextrins. This method was used, e. g., to test raspberry fruit juice concentrates for unauthorized aromatization with trans-α-ionone. The gas chromatograms of trans-α-ionone from two different samples are shown in Fig. 5.10. The low excesses of the R-enantiomer of ee = 8% (concentrate A) and ee = 24% (B) can probably be

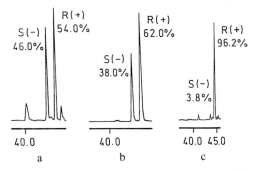

Fig. 5.10. Enantioselective gas chromatographic analysis of trans-α-ionone in aroma extracts of different raspberry fruit juice concentrates (according to *Werkhoff* et al., 1990): **a** and **b** samples with nature identical aroma, **c** natural aroma

put down to the addition of synthetic trans-α-ionone racemate to the fruit juice concentrate because in the natural aroma (C), the ee value is 92.4%.

5.2.5 Aroma Relevance

In many earlier studies on the composition of aromas, volatile compounds were equated with aroma substances. Although lists with hundreds of compounds were made for many foods, it was unclear which of them were really significant as aroma substances and to what extent important aroma substances, which are present in very low concentrations, were detected.

Investigations have for some time concentrated more and more on those compounds which appreciably contribute to aroma. Both of the following methods based on the aroma value concept (cf. 5.1.4) find application.

5.2.5.1 Calculation of Aroma Values

The composition of the volatile fraction is analyzed qualitatively and quantitatively. The aroma values are then calculated on the basis of the odor thresholds, which were determined for the compounds found in a suitable medium (e. g., water in the case of water-rich foods) (cf. 5.1.4).

As an example, the seven most important aroma substances of tomato paste determined by this method are listed in Table 5.10.

Table 5.9. Enantiomeric excess (ee) of chiral aroma substances in some foods

Aroma substance	Food	ee (%)
R(+)-γ-Decalactone	Peach, apricot, mango, strawberry pineapple, maracuya	> 80
R(+)-δ-Decalactone	Milk fat	60
R(+)-trans-α-Ionone	Raspberrry	92.4
	Carrot	90.0
	Vanilla bean	94.2
R(−)-1-Octen-3-ol	Mushroom, chanterelle	> 90
S(+)-E-5-Methyl-2-hepten-4-one (filbertone)	Hazelnut, raw	60–68
	Hazelnut, roasted	40–45
R-3-Hydroxy-4,5-dimethyl-2(5H)-furanone (sotolon)	Sherry	ca. 30

Table 5.10. Aroma substances of tomato paste [a]

Compound	Concentration (µg/kg)	Odor threshold (µg/kg)[b]	Aroma value
Dimethylsulfide	2000	0.3	6.7×10^3
β-Damascenone	14	0.002	7×10^3
3-Methylbutanal	24	0.2	1.2×10^2
1-Nitro-2-phenylethane	66	2	33
Eugenol	100	6	17
Methional	3	0.2	15
3-Methylbutyric acid	2000	250	8

[a] Dry matter: 28–30 w/w per cent.
[b] In water.

To check the result, these aroma substances were dissolved in water in the given concentrations. It was found that the odor of the model was similar to that of tomato paste. The slight deviation is probably due to the fact that dimethyltrisulfide, sotolon and 1-octen-3-one, which were subsequently identified in tomato paste as aroma substances with high aroma values, were missing in the experiments on aroma simulation. This investigation shows that of the more than 400 volatile compounds that were identified in tomato paste, only a very limited number are involved in aroma.

5.2.5.2 Aroma Extract Dilution Analysis

The example cited in 5.2.5.1 clearly shows that the collection of data required for the calculation of aroma values is difficult and not very reliable because, in principle, all the components of the volatile fraction should be identified. In addition, the threshold values of all the identified compounds must be known in order to assess their aroma relevance. Then, all the odorants must be quantitatively analyzed. In this process, serious mistakes can be made because these are volatile substances which not only occur in trace amounts, but are partly very reactive. Aroma extract dilution analysis is a considerably easier method because it involves sensory evaluation of the volatile compounds at the start of the investigation.

The concentrate of the odorants is separated by gas chromatography on a capillary column. To determine the retention times of the aroma substances, the carrier gas stream is subjected to sniffing detection after leaving the capillary column (GC/olfactometry). The sensory assessment of a single GC run, which is often reported in the literature, is not very meaningful because the perception of aroma substances in the carrier gas stream depends on limiting quantities which have nothing to do with the aroma value, e.g., the amount of food analysed, the degree of concentration of the volatile fraction, and the amount of sample separated by gas chromatography.

These limitations are eliminated by the stepwise dilution of the volatile fraction with solvent, followed by the gas chromatographic/olfactometric analysis of each dilution. The process is continued until no more aroma substance can be detected by GC olfactometry. In this way, a dilution factor is determined for each aroma substance that appears in the gas chromatogram. It is designated as the flavor dilution factor (FD factor) and indicates the number of parts of solvent required to dilute the aroma extract until the aroma value is reduced to one.

Another more elaborate variant of the dilution analysis requires, in addition, that the duration of each odor impression is entered into an EDP system and CHARM values are calculated (CHARM: acronym for combined hedonic response measurement), which are proportional to aroma values.

The result of an aroma extract dilution analysis (AEDA) can be represented as a diagram. The FD factor is plotted against the retention time in the form of the retention index (RI) and the diagram is called a FD chromatogram.

The FD chromatograms of the volatile compounds of white bread and dill are presented in Fig. 5.4 and 5.11.

The identification experiments concentrate on those aroma substances which appear in the FD chromatogram with higher FD factors. To detect all the important aroma substances, the range of FD factors taken into account must not be too narrowly set at the lower end because differences in yield shift the concentration ratios. Labile compounds can undergo substantial losses and when distillation processes are used, the yield decreases with increasing molecular weight of the aroma substances.

In the case of dill (Fig. 5.11), 16 aroma substances appearing in the FD-factor range 16–1024 were analyzed and 15 of them identified (cf. legend of Fig. 5.11).

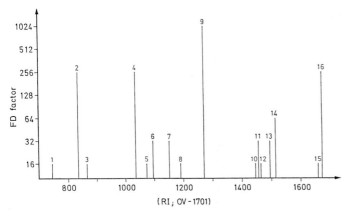

Fig. 5.11. FD Chromatogram of the volatile fraction of dill herb. The following odorants were identified: *1* methylpropionic acid methyl ester, *2* 2-methylbutyric acid methyl ester, *3* 1-hexen-3-one, *4* S(+)-α-phellandrene, *5* 1-octen-3-one, *6* Z-1,5-octadien-3-one, *7* 2-isopropyl-3-methoxypyrazine, *8* phenylacetaldehyde, *9* 3R,4S,8S-3,9-epoxy-1-p-menthene (dill ether), *10* p-anisaldehyde, *11* unknown, *12* E,E-2,4-decadienal, *13* vinylguaiacol, *14* eugenol, *15* vanillin, *16* myristicin (according to *Blank* and *Grosch*, 1991)

In GC/olfactometry, odor thresholds are considerably lower than in solution because the aroma substances are subjectied to sensory assessment in a completely vaporized state. The examples given in Table 5.11 show how great the differences can be when compared to solutions of the aroma substances in water. It should also be taken into account that additive effects (cf. 5.1.4) and masking (example in Fig. 5.2) do not come into play in GC/olfactometry because the aroma substances are sniffed individually.

The mistakes based on simplifications that are outlined here can be eliminated by simulating the aroma in question, starting with the results of a dilution analysis.

Taking dill as an example, the four compounds with the highest FD factors were quantified and dissolved in water in accordance with the concentrations found (Table 5.12). The sensory test of the model gave a very good approximation of the aroma of dill. The concentrations of the aroma substances in the solution were then varied. It was found that S(+)-α-phellandrene produces the characteristic odor note of dill and dill ether rounds it off. In comparison, myristicin and 2-methylbutyric acid methyl ester do not contribute to the total aroma. Although their concentrations in the model were 4 and 18 times the odor threshold respectively, their aromas are completely suppressed by those of the two monoterpenes (Table 5.12).

Table 5.11. Odor thresholds of aroma substances in air and water

Compound	Odor thresholds in		
	Air (a) (ng/l)	Water (b) (µg/l)	b/a
β-Damascenone	0.003	0.002	6.7×10^2
Methional	0.12	0.2	1.6×10^3
2-Methylisoborneol	0.009	0.03	3.3×10^3
2-Acetyl-1-pyrroline	0.02	0.1	5×10^3
4-Vinylguaiacol	0.6	5	8.3×10^3
Linalool	0.6	6	1.0×10^4
Vanillin	0.9	20	2.2×10^4
4-Hydroxy-2,5-dimethyl-3(2H)-furanone (furaneol)	1.0	30	3×10^4

Table 5.12. Model for investigating the typical aroma substances of dill

Compound	Concentration[a] (mg/kg, water)	Aroma value A_x
S(+)-α-Phellandrene	11.3	56
Dill ether	2.3	77
Myristicin	0.12	4
2-Methylbutyric acid methyl ester	0.007	18

[a] The amount of odorant occurring in 10 g of dill was dissolved in 1 kg of water.

5.3 Individual Aroma Compounds

A comprehensive survey of information related to the occurrence of aroma compounds in food recognizes, on the one hand, the vast diversity of the classes of compounds present and, on the other hand, reveals the frequent similarity of various food aromas.

The diversity of chemical structures suggests the involvement of numerous reactions in aroma formation. These reactions occur in many foods and are responsible for overlapping aroma spectra. Fruits and vegetables, herbs and spices contain aroma substances derived primarily as secondary products of plant metabolism. Common qualitative features of aroma are revealed in many unrelated plant families, suggesting that, in spite of their diversity, a common pattern of metabolic pathways operates in these plant cells. Thus, the wide spectrum of terpenes is found not only in citrus fruits but also in herbs and spices, red currants, cranberries, carrots and celery. The differences in aroma between these products are due to deviations which occur in terpene biosynthesis or to a dissimilarity in the secondary metabolism of these compounds.

Meat, fish, milk and cereals are, by their nature, a poor source of aroma. The specific aroma of the product is generated only through processing, such as heating and/or fermentation. When common types of carbohydrates, amino acids, peptides and/or lipids are present in food, similar or equivalent aromas are formed during food processing if the processing parameters are comparable (temperature, duration, presence of water, oxygen). Examples are the widespread formation of the intensive aroma substances methional and 1-octen-3-one in the foods mentioned above because the *Strecker* degradation of methionine as well as the autoxidation of linoleic acid can occur on heating. In the presence of the same starting compounds, the aroma typical of different foods can be formed, e. g., by the fact that in one case the formation of a character impact compound is inhibited and in another case it is not. Thus, meat and bread dough contain comparable amounts of thiamine, which on heating gives rise to the aroma substance 2-methyl-3-furanthiol, an important compound for cooked

meat (cf. 12.9). This aroma substance is not formed during the baking of bread because it or the intermediate thiols (cf. 5.3.1.4) are trapped by the carbohydrates and their degradation products that are present in large amounts.

A selection of aroma compounds, grouped according to their formation by nonenzymatic or enzymatic reactions and listed according to classes of compounds, is presented in the following sections. Some aroma compounds formed by both enzymatic and nonenzymatic reactions are covered in sections 5.3.1 and 5.3.2. It should be noted that the reaction pathways for each aroma compound are differentially established. In many cases they are dealt with by using hypothetical reaction pathways which lead from the precursor to the flavor compound. The reaction steps and the intermediates of the pathway are postulated by using the general knowledge of organic chemistry or biochemistry. So far, only in a few cases have the intermediary products been isolated and identified, the enzymes involved proven, and the proposed reaction pathway verified in a model system. Obviously, these subjects of aroma research are especially difficult since they involve, in most cases, elucidation of the side pathways occurring in chemical or biochemical reactions, which quantitatively are often not much more than negligible.

5.3.1 Nonenzymatic Reactions

Aroma changes at room temperature caused by nonenzymatic reactions are observed only after prolonged storage of food. Lipid peroxidation (cf. 3.7.2.1.5), *Strecker* degradation (cf. 4.2.4.4.7) of amino acids, carbohydrate heterolysis and the further reactions between their intermediary products, involving mostly aldehydes (cf. Fig. 3.34), all play a part. These processes are greatly accelerated during heat treatment of food.

The diversity of aroma is enriched at the higher temperatures used during roasting or frying. The food surface, once dehydrated, is subjected to pyrolysis of its main constituents: carbohydrates, proteins, lipids and other constituents and, as a consequence, a diversified aroma spectrum results.

The large number of volatile compounds formed by the degradation of only one or two constituents is characteristic of nonenzymatic reactions. For example, 41 sulfur-containing compounds, including 20 thiazoles, 11 thiophenes, 2 dithiolanes and 1 dimethyltrithiolane, are obtained by heating cysteine and xylose in tributyrin at 200 °C.

5.3.1.1 Carbonyl Compounds

The most important reactions which provide volatile carbonyl compounds were presented in sections 3.7.2.1.5 (lipid peroxidation), 4.2.4.3.3 (caramelization) and 4.2.4.4.7 (amino acid decomposition by the *Strecker* degradation mechanism).
Some *Strecker* aldehydes found in many foods are listed in Table 5.13 together with the corresponding aroma quality data. Data for carbonyls derived from fatty acid degradation are found in Table 3.31. Carbonyls are also obtained by degradation of carotenoids (cf. 3.8.4.4).

5.3.1.2 Pyranones

Maltol (3-hydroxy-2-methyl-4H-pyran-4-one) is obtained from carbohydrates as outlined in 4.2.4.4.4 and has a caramel-like odor. It has been found in a series of foods (Table 5.14),

but in concentrations that were mostly lower than the relatively high odor threshold of 35 mg/kg (water).
Maltol enhances the sweet taste of food, especially sweetness produced by sugars (cf. 8.6.3), and is able to mask the bitter flavor of hops and cola.
Ethyl maltol [3-hydroxy-2-ethyl-4H-pyran-4-one] enhances the same aroma but is 4- to 6-times more powerful than maltol. It has not been detected as a natural constituent in food. Nevertheless, it is used for food aromatization.

Table 5.14. Occurrence of maltol in food

Food product	mg/kg	Food product	mg/kg
Coffee, roasted	20–45	Chocolate	3.3
Butter, heated	5–15	Beer	0–3.4
Biscuit	19.7		

5.3.1.3 Furanones

Among the great number of products obtained from carbohydrate degradation, 3(2H)- and 2(5H)-furanones belong to the most striking aroma compounds (Table 5.15).
Compounds I–III, V and VI in Table 5.15, as well as maltol and the cyclopentenolones

Table 5.13. Some *Strecker* degradation aldehydes

Amino acid precursor	*Strecker*-aldehyde				Odor threshold value (µg/l; water)
	Name	Structure		Aroma description	
Gly	Formaldehyde	CH_2O		Mouse-urine, ester-like	50×10^3
Ala	Ethanal			Sharp, penetrating, fruity, sweet	15
Val	Methylpropanal			Malty	0.7
Leu	3-Methylbutanal			Malty	0.4
Ile	2-Methylbutanal			Malty	1.3
Phe	2-Phenylethanal			Flowery, honey-like	4

(cf. 4.2.4.3.2), have a planar enol-oxo-con-figuration

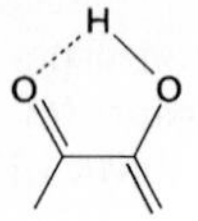

$$(5.4)$$

and a caramel-like odor. When the hydroxy group is methylated (IV), this aroma note disappears.

A list of foods in which furanone II has been identified as an important aroma substance is given in Table 5.16.

Table 5.15. Furanones in food

Structure	Substituent[a]	Aroma description	Occurrence
(I)	**A. 3(2H)-Furanones** 4-Hydroxy-5-methyl *Norfuraneol*	Roasted chicory-like, caramel	Meat broth
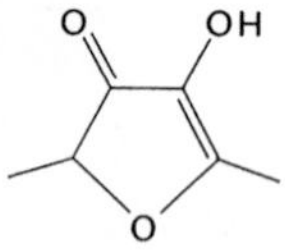 (II)	4-Hydroxy-2,5-dimethyl *Furaneol®* (nasal: 150; retronasal: 25)	Heat-treated strawberry, pineapple-like, caramel	cf. Table 5.16
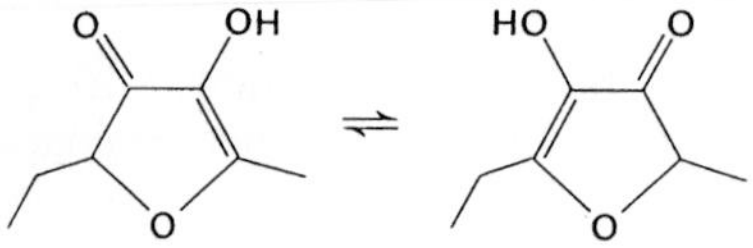 (III)	4-Hydroxy-2-ethyl-5-methyl	Sweet, pastry, caramel	Soya souce[b]
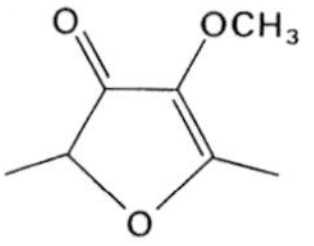 (IV)	4-Methoxy-2,5-dimethyl *Mesifuran*	Sherry-like	Strawberry, raspberry[c]
(V)	**B. 2(5H)-Furanones** 3-Hydroxy-4,5-dimethyl *Sotolon* (nasal, R-form 90, S-form 7; racemate, retronasal: 3)	Caramel, protein hydrolysate	Coffee, sherry, seasonings
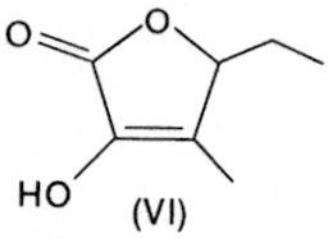 (VI)	5-Ethyl-3-hydroxy-4-methyl *Abhexon* (nasal: 30, retronasal: 3)	Caramel, protein hydrolysate	Coffee, seasonings

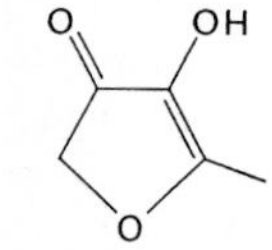

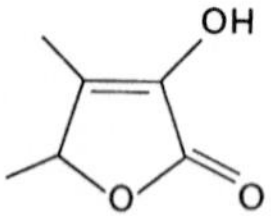

[a] Odor threshold (µg/kg) in water.
[b] Both tautomeric forms presented are identified.
[c] Arctic bramble (*Rubus arcticus*).

Table 5.16. Occurrence of 4-hydroxy-2,5-dime-thyl-3(2H)-furanone

Food	mg/kg
Beer, light	0.35
Beer, dark	1.3
White bread, crust	1.96
Coffee drink[a]	2.5–4.5
Strawberry	1–30
Pineapple	1.6–35

[a] Coffee, medium roasted, 54 g/l water.

Studies on models show that the *Maillard* reaction of rhamnose yields relatively large amounts of furaneol (Fig. 5.12). Since rhamnose occurs in the free state in foods only in very low concentrations, fructose and, to a less extent, glucose are of greater importance as precursors. Fructose can give rise to 1,6-dideoxy-2,4,5-hexotriulose (cf. 4.2.4.3.1), which yields furanone II after disproportionation (see Formula 5.6). Elimination of the OH group at C-6 of fructose, which is required for formation of the 1,6-dideoxy sugar, occurs only under relatively drastic conditions (e. g. at 150–170 °C). In comparison, this reaction is much easier with fructose-6-phosphate or fructose-1,6-biphosphate. Whether the furanone II detected in fruit, which is partly present as the β-glycoside, is formed exclusively by nonenzymatic reactions favored by the low pH is still not clear.

Furanone V (sotolon) is a significant contributor to the aroma of, e.g., sherry, French white wine, and coffee. It is a chiral compound having enantiomers that differ in their odor threshold (Table 5.15) but not in their odor quality. The precursor under discussion is 4-hydroxyisoleucine which could give rise to this furanone by oxidative deamination and cyclization. Furanone VI (abhexon) has an aroma quality similar to that of sotolon and is formed by aldol condensation of α-oxobutyric acid, a degradation product of threonine (Fig. 5.13).

Quantitative analysis of furanones is not very easy because due to their good solubility in water, they are extracted from aqueous foods with poor yields and easily decompose, e. g., sotolon (cf. Formula 5.5).

5.3.1.4 Thiols, Thioethers, Di- and Trisulfides

An abundance of sulfurous compounds is obtained from cysteine, cystine, thiamine and methionine by heating food. Some are very powerful aroma compounds (Table 5.17) and are involved in the generation of some delightful but also some irritating, unpleasant odor notes.

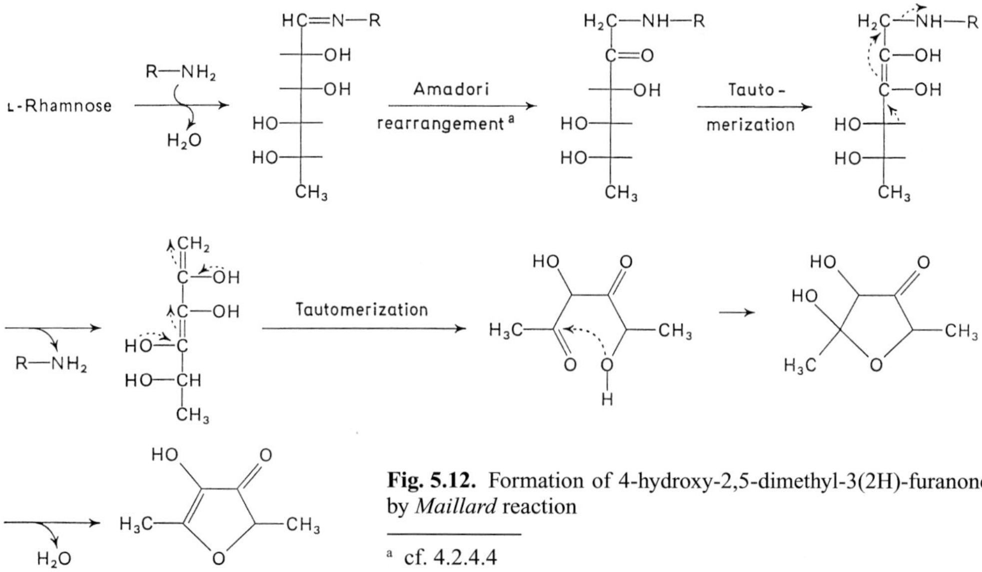

Fig. 5.12. Formation of 4-hydroxy-2,5-dimethyl-3(2H)-furanone by *Maillard* reaction

[a] cf. 4.2.4.4

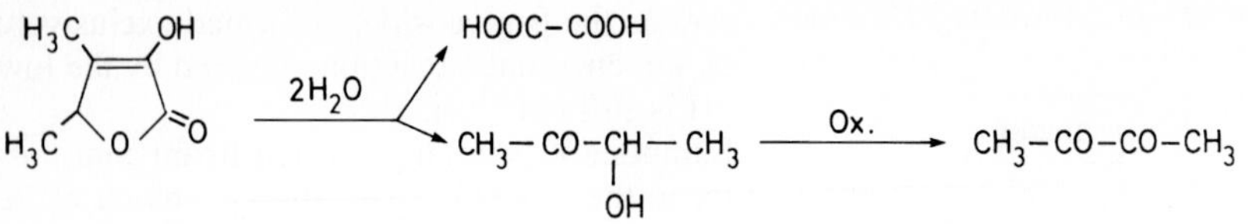

$$H_2O \quad HOOC-COOH$$

$$\xrightarrow{2H_2O} \quad CH_3-CO-CH-CH_3 \xrightarrow{Ox.} CH_3-CO-CO-CH_3 \qquad (5.5)$$
$$\underset{OH}{|}$$

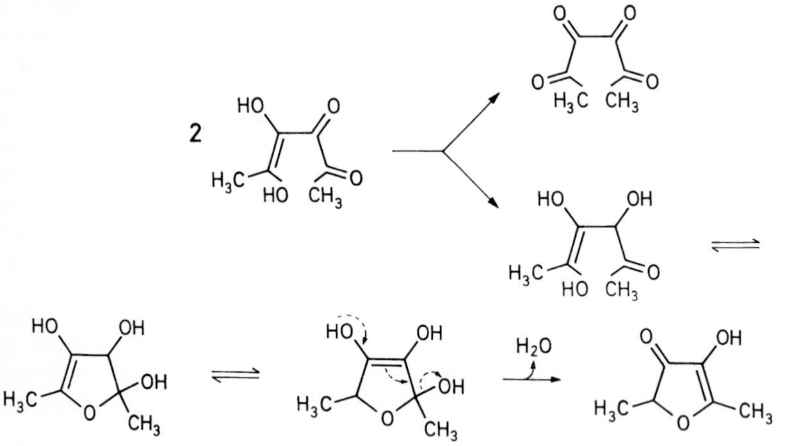

(5.6)

2 Threonine $\longrightarrow$

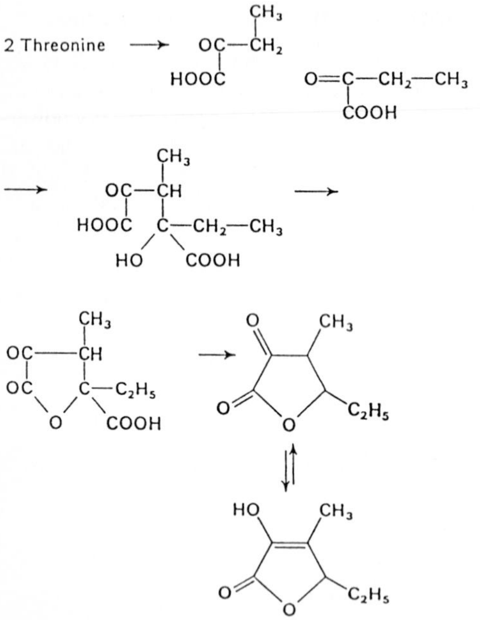

Fig. 5.13. Formation of 5-ethyl-3-hydroxy-4-methyl-2(5H)-furanone from threonine by heating

Thiols are important constituents of food aroma because of their intensive odor and their occurrence as intermediary products which can react with other volatiles by addition to carbonyl groups or to double bonds.

Hydrogen sulfide and 2-mercaptoacetaldehyde are obtained during the course of the *Strecker* degradation of cysteine (Fig. 5.14). In a similar way, methionine gives rise to methional, which releases methylthiol by β-elimination (Fig. 5.15). Dimethylsulfide is obtained by methylation during heating of methionine in the presence of pectin:

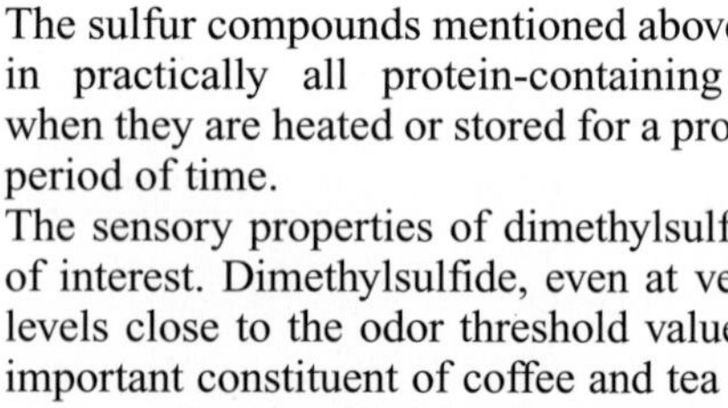

(5.7)

The sulfur compounds mentioned above occur in practically all protein-containing foods when they are heated or stored for a prolonged period of time.

The sensory properties of dimethylsulfide are of interest. Dimethylsulfide, even at very low levels close to the odor threshold value, is an important constituent of coffee and tea aroma. However, in other food, it is responsible for off-flavors designated as "crude oil" (frozen crustaceans), "onion" flavor (beer) or "feed smell" (such as occurs in milk). Bacteria are involved in the formation of dimethylsulfide in beer.

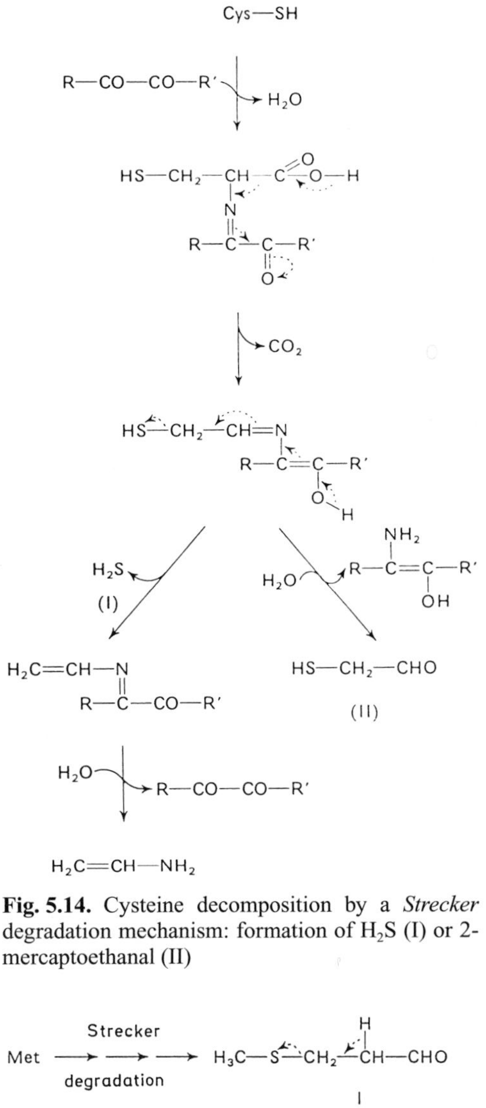

Cys—SH

R—CO—CO—R′

$\rightarrow H_2O$

HS—CH₂—CH   C   O—H
                    N

R—C—C—R′
        O

$\rightarrow CO_2$

HS—CH₂—CH=N

R—C=C—R′
        O   H

H₂S   (I) H_2O R—C=C—R′   NH₂   OH

H₂C=CH—N
R—C—CO—R′

HS—CH₂—CHO
(II)

H_2O R—CO—CO—R′

H₂C=CH—NH₂

Fig. 5.14. Cysteine decomposition by a *Strecker* degradation mechanism: formation of H₂S (I) or 2-mercaptoethanal (II)

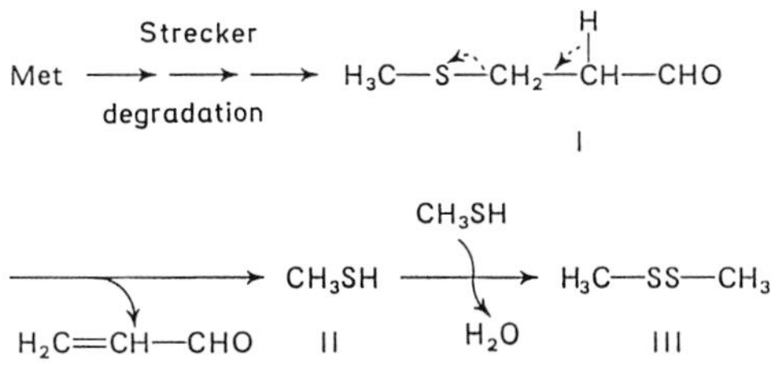

Met $\xrightarrow{\text{Strecker}}$ $\longrightarrow$ $\longrightarrow$ H₃C—S—CH₂—CH—CHO
          degradation   H

                                                                                              I

                        CH₃SH

$\longrightarrow$ CH₃SH $\longrightarrow$ H₃C—SS—CH₃
H₂C=CH—CHO   II   H₂O   III

Fig. 5.15. Methionine degradation to methional, methylthiol and dimethylsulfide

Methional is responsible for the "sunlight" flavor of milk (cf. Table 5.4) and for the typical flavor of processed potatoes and meat broth.

Table 5.17. Odor threshold values (water) of some volatile sulfur compounds

Compound	Threshold value (μg/l)
Hydrogen sulfide	10
Methylthiol	0.02
Ethylthiol	0.008
Dimethylsulfide	1.0
Dimethyldisulfide	7.6
Dimethyltrisulfide	0.01
Methional	0.2
3,4-Dimethylthiophene	1.3
5-Methyl-2-formylthiophene	1.0
3,5-Dimethyl-1,2,4-trithiolane	10
2-Methyl-3-furanthiol	0.007
Bis(2-methyl-3-furyl)disulfide	$2 \cdot 10^{-5}$
2-Furfurylthiol	0.01

The typical aroma substances of cooked meat include 2-methyl-3-furanthiol and its oxidation product bis(2-methyl-3-furyl)disulfide. Both these compounds exhibit extremely low odor thresholds (Table 5.17) and are formed by the hydrolysis of thiamine (Fig. 5.16). The postulated intermediate is the very reactive 5-hydroxy-3-mercaptopentane-2-one.

This example shows that minor constituents of food like thiamine (vitamin B₁) can play an important role as aroma precursors if they produce aroma substances with very low threshold values.

Some reaction systems, which have been described in the patent literature for the production of meat aromas, regard thiamine as precursor (cf. Table 12.23). During the cooking of meat, hydrogen sulfide released from cysteine (Fig. 5.14) or glutathione promotes the formation of 2-methyl-3-furanthiol and its disulfide

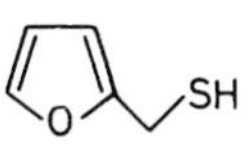

SH   (5.8)

The compound 2-furfurylthiol (see Formula 5.8), which has a roasted odor and a very low odor threshold (cf. Table 5.17), contributes to the aroma of roasted coffee, cooked meat, and popcorn.

Thiols are responsible for marked aroma defects. Besides the "sunlight" flavor defect of beer (Table 5.4), the "cat urine" taint of can-

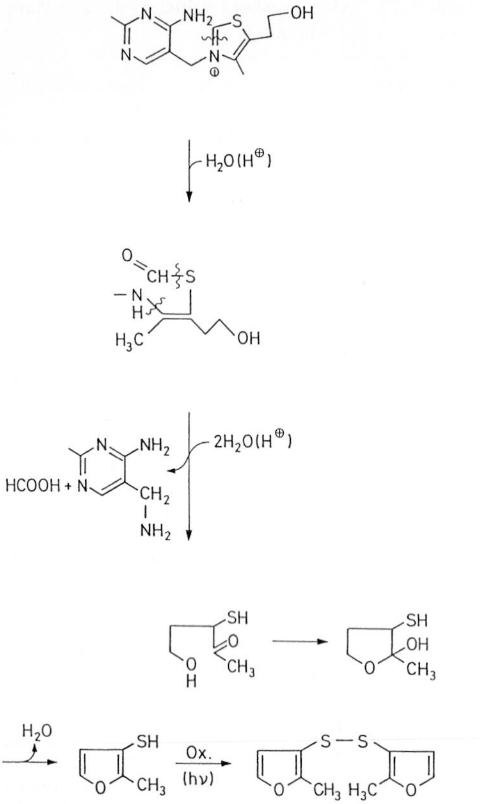

$$2\,CH_3CHO \;+\; 3\,H_2S$$

$$H_3C-CH-S-CH-CH_3$$
$$\quad\;\; SH \qquad\quad SH$$

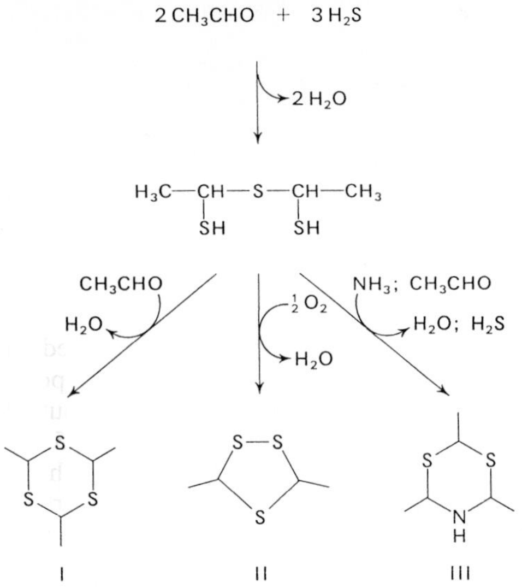

Fig. 5.17. Formation of 2,4,6-trimethyl-s-trithiane (I), 3,5-dimethyl-1,2,4-trithiolane (II) and 2,4,6-trimethyl-5,6-dihydro-1,3,5-dithiazine (III)

Fig. 5.16. Formation of 2-methyl-3-furanthiol and bis(2-methyl-3-furyl)disulfide from thiamine

ned beef should be mentioned. Here, the reaction components are mesityl oxide, probably derived from solvent contamination, and hydrogen sulfide:

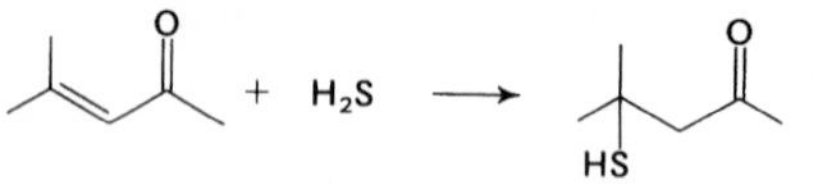

(5.9)

Thiols are readily oxidized to disulfides (Fig. 5.15), which can disproportionate to trisulfides:

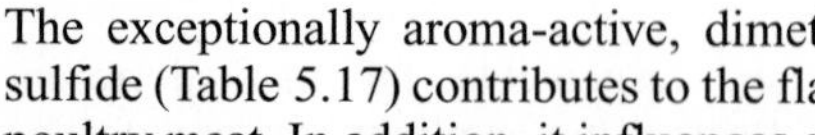

(5.10)

The exceptionally aroma-active, dimethyltrisulfide (Table 5.17) contributes to the flavor of poultry meat. In addition, it influences cooked white cabbage and cauliflower aroma.

Interaction of acetaldehyde and hydrogen sulfide provides heterocyclic compounds (Fig. 5.17).

If ammonia is present, the profile of the heterocyclic compound is broadened (Fig. 5.17).

It is unclear whether sulfides I–III in Fig. 5.17 and trithioacetone, analogous to trithioacetaldehyde (I), are really formed during the cooking of meat or whether these compounds are artifacts that are produced on concentration of the volatile fraction in the course of analysis (cf. 5.2.1).

5.3.1.5 Thiophenes

Model experiments have shown that thiophenes can also be formed from 5-hydroxy-3-mercaptopentane-2-one produced by the hydrolysis of thiamine (Fig. 5.16). Thiophenes, e.g., 2-methyl-3-thiophenethiol (see Formula 5.11), have a meat-like (cooked) odor. Whether 4-mercapto-5-methyl-3(2H)-thiophenone, which has a sweet, meat-like odor, or other thiophenes play a significant role in the aroma of cooked meat has yet to be clarified.

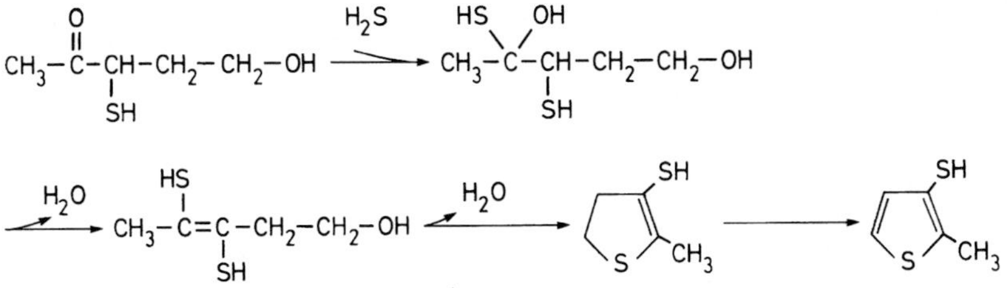

$$CH_3-\overset{\overset{\text{O}}{\|}}{C}-\underset{\underset{\text{SH}}{|}}{CH}-CH_2-CH_2-OH \longrightarrow \overset{H_2S}{}\overset{HS\quad OH}{\underset{}{\diagup\diagdown}} CH_3-\overset{\overset{}{}}{C}-\underset{\underset{\text{SH}}{|}}{CH}-CH_2-CH_2-OH$$

$$\overset{H_2O}{\longleftarrow} CH_3-\overset{\overset{HS}{|}}{C}=\underset{\underset{\text{SH}}{|}}{C}-CH_2-CH_2-OH \overset{H_2O}{\longleftarrow} \text{[ring, SH, CH}_3\text{]} \longrightarrow \text{[ring, SH, CH}_3\text{]} \qquad (5.11)$$

5.3.1.6 Thiazoles

Thiazole and its derivatives are detected in foods such as coffee, fried meat, fried potatoes, heated milk and beer. Several of about 30 compounds identified are listed in Table 5.18. Thiazole I (Table 5.18) occurs widely in heated food and contributes to the aromas of roasting. The formation of this compound is outlined in Fig. 5.18. Its precursor, 2-acetyl-2-thiazoline, has a similar odor and is more important because of its lower odor threshold (Table 5.18). This compound is one of the characteristic aroma substances in roasted meat. Thiazole III (Table 5.18) can occur in milk

when it is heated, and is responsible for a "stale" off-flavor. Thiazole IV (Table 5.18) is a significant constituent of tomato aroma. The aroma of tomato products is usually enhanced by the addition of 20–50 ppb of thiazole IV (for the biosynthesis of the compound, see Section 5.3.2.5).

5.3.1.7 Pyrroles, Pyridines

The volatile compounds formed by heating food include numerous pyrrole and pyridine derivatives. Of special interest are the N-heterocyclic compounds with the following structural feature:

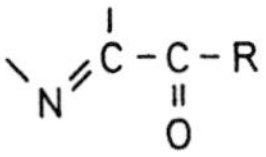

$$\underset{\text{N}}{\diagdown}\overset{}{\underset{}{C}}-\overset{\overset{\text{}}{}}{\underset{\underset{\text{O}}{\|}}{C}}-R \qquad (5.12)$$

This characteristic feature appears to be required for a roasted odor. In fact, all the pyrrolines and pyridines listed in Table 5.19 as

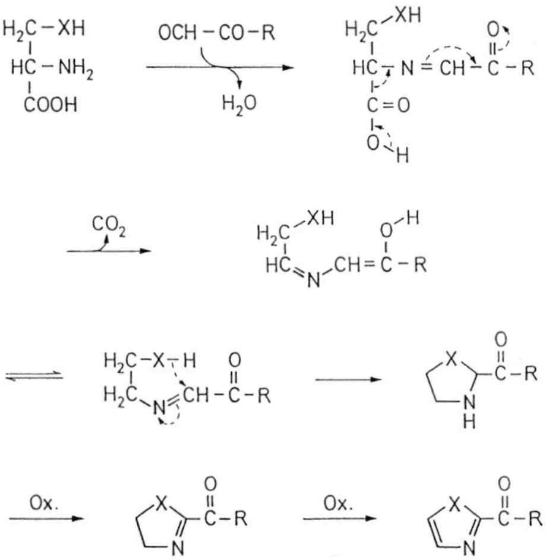

Table 5.18. Thiazoles and thiazolines in food

Name	Structure	Aroma quality	Odor threshold (µg/kg, H_2O)
2-Acetyl-thiazole	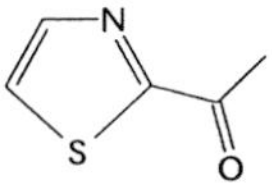 I	Cereal, popcorn	10
2-Acetyl-2-thiazoline	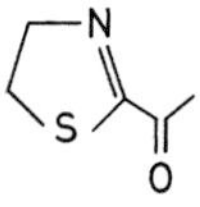 II	Popcorn	1.3
Benzo-thiazole	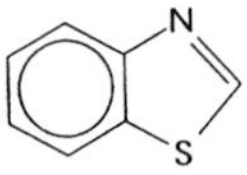 III	Quino-line, rubber	
2-Isobutyl-thiazole	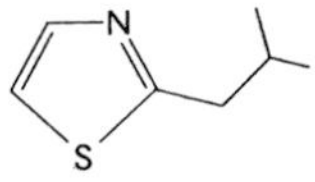 IV	Green, tomato, wine	3

Fig. 5.18. *Strecker* degradation of cysteine (x = S) with formation of 2-acetylthiazoles

Table 5.19. Pyrrole and pyridine derivatives with a roasted aroma

Name	Structure	Odor threshold (µg/kg, water)	Occurrence
2-Acetyl-1-pyrroline	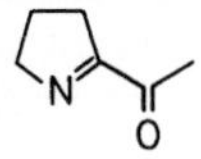	0.1	White-bread crust, rice, cooked meat, popcorn
2-Propionyl-1-pyrroline	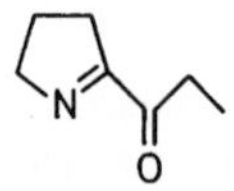	0.1	Popcorn
2-Acetyltetrahydropyridine	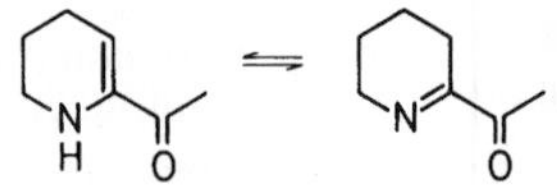	1.6	White-bread crust, popcorn
2-Acetylpyridine	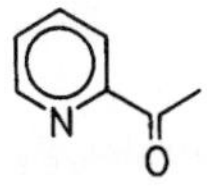	19	White-bread crust

well as 2-acetylthiazole, 2-acetylthiazoline (cf. Table 5.18) and acetylpyrazine (cf. Table 5.20) contain this structural element and have a roasted or cracker-like odor. However, the thresholds of these compounds vary greatly. The lowest values were found for 2-acetyl- and 2-propionyl-1-pyrroline.

The length of the alkanoyl group also influences the aroma quality because in the transition from 2-propionyl- to 2-butanoyl-1-pyrroline, the roasted note suddenly disappears and the odor threshold increases by several powers of ten.

2-Acetyl-1-pyrroline is responsible for the typical aroma of the crust of white bread and it produces the pleasant popcorn aroma of certain types of rice consumed mainly in Asia. In gas chromatography, 2-acetyl-1-pyrroline ap-

pears predominantly in the imine form shown in Table 5.19, whereas 2-acetyltetrahydropyridine appears as the enamine and imine tautomers.

2-Acetyl-1-pyrroline is formed from both ornithine (cf. Fig. 5.19) and proline (cf. Fig. 5.20). In the baking of white bread, ornithine comes from yeast where it is found in a concentration about four times that of free proline. In addition, triose phosphates occurring in yeast have been identified as precursors. They yield on heating, among other compounds, 2-oxopropanal (cf. Formula: 5.13) which is involved in the *Strecker* degradation. Another source of 2-oxopropanal is the retroaldol condensation of 3-deoxy-1,2-dicarbonyl compounds in the course of the *Maillard* reaction (cf. 4.2.4.4.2).

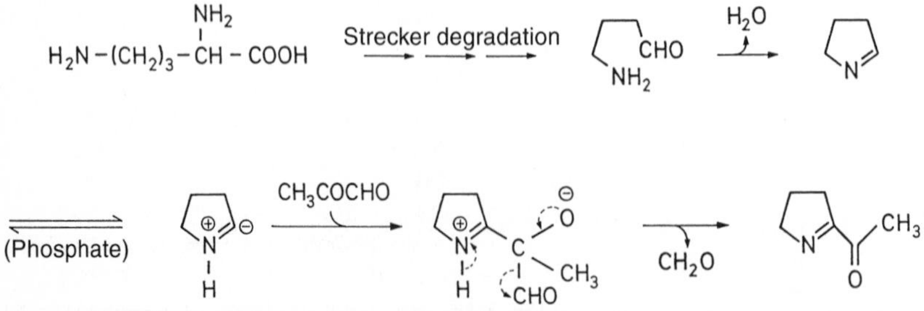

Fig. 5.19. Formation of 2-acetyl-1-pyrroline in the *Strecker* degradation of ornithine

Model studies have shown that 2-acetyl-1-pyrroline is formed from ornithine via the intermediates 4-aminobutyraldehyde and 1-pyrroline (Fig. 5.19) and that relatively large amounts are formed if 1-pyrroline and 2-oxopropanal are heated in the presence of phosphate ions.

The *Strecker* reaction of proline and 2-oxopropanal gives rise to 2-acetyltetrahydropyridine and to 2-acetyl-1-pyrroline (Fig. 5.20). However, the amount formed is less than that of the first mentioned aroma substance.

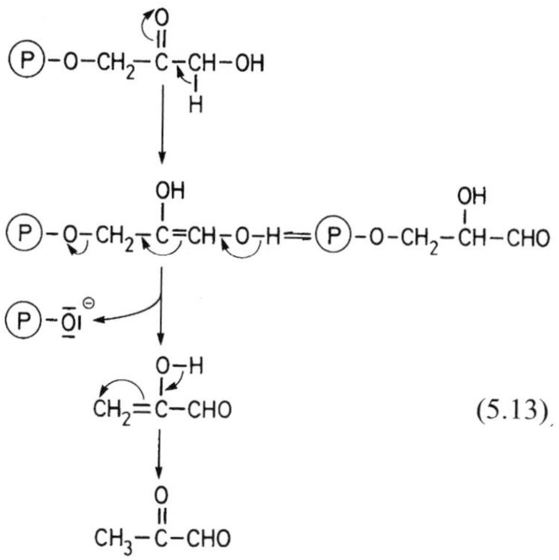

$$+ CH_3-CO-CHO \longrightarrow$$

(5.13)

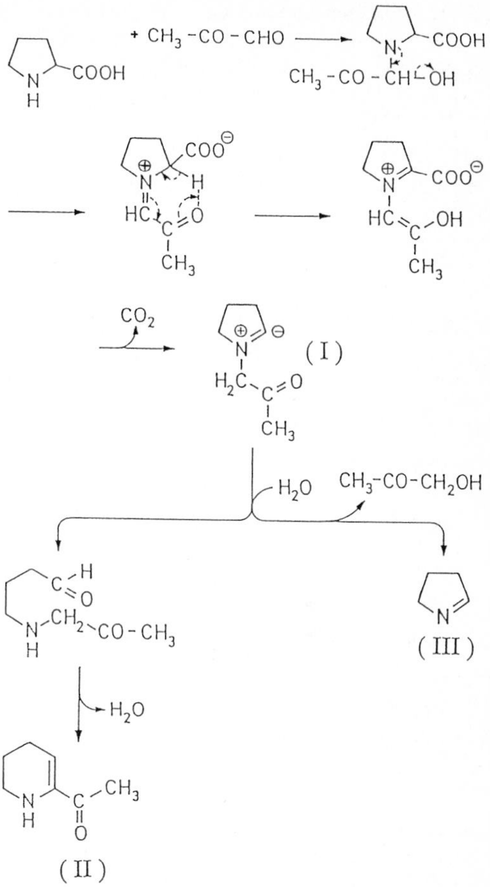

Fig. 5.20. Formation of 2-acetyl-1,4,5,6-tetrahydropyridine and 1-pyrroline in the *Strecker* reaction of proline and 2-oxopropanal

The ylide (I in Fig. 5.20) produced by the decarboxylation of N-acetonylproline is assumed to be the branch point of the reaction pathway to the two compounds. Hydrolytic cleavage of the ylide bond and enlargement of the proline ring give 2-acetyl-1,4,5,6-tetrahydropyridine (II). Hydrolysis of the ylide with release of hydroxyacetone yields 1-pyrroline (III), which can react further to give 2-acetyl-1-pyrroline, as presented in Fig. 5.19.

Although the odor threshold increases by about a factor of 10, the popcorn-like aroma note remains on oxidation of tetrahydropyridine to 2-acetylpyridine. Substantially greater effects on the aroma are obtained by the oxidation of 2-acetyl-1-pyrroline to 2-acetylpyrrole, which has an odor threshold that is more than 5 powers of ten higher and no longer smells roasted.

2-Pentylpyridine contributes to the smell of roasting lamb fat (greasy, suety odor; threshold: 0.6 µg/kg water). The precursors postulated are ammonia from the pyrolysis of proteins adhering to the fat and 2,4-decadienal:

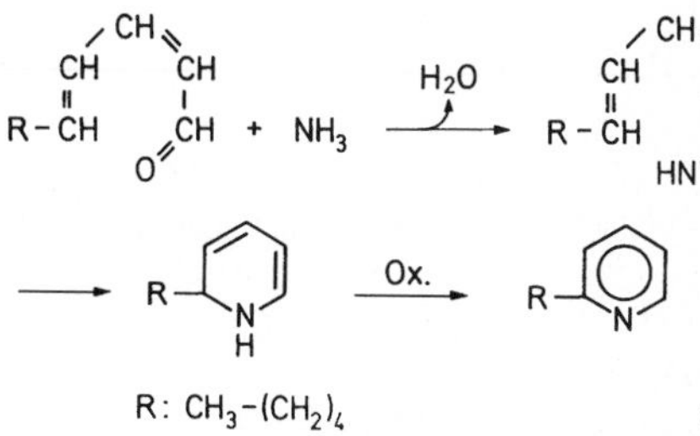

$$+ NH_3 \xrightarrow{H_2O} \xrightarrow{Ox.}$$

R: $CH_3-(CH_2)_4$

(5.14)

Table 5.20. Pyrazines in food

Structure	Substituent	Aroma quality	Odor threshold value (µg/l; water)
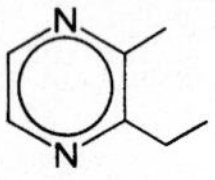 (I)	2-Methyl-3-ethyl-	Burning	130
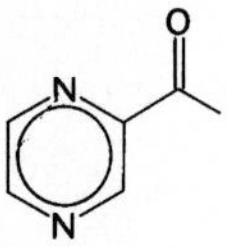 (II)	Acetyl-	Roasted corn	62
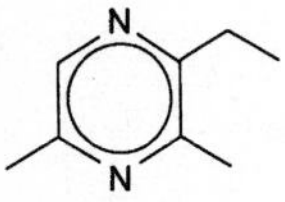 (III)	2-Ethyl-3,5-dimethyl-	Earthy, burnt	0.16
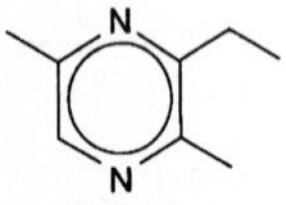 (IV)	2-Ethyl-3,6-dimethyl-	Earthy, burnt	40
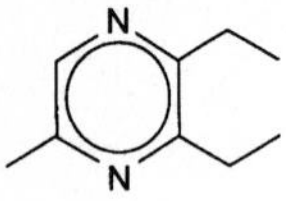 (V)	2,3-Diethyl-5-methyl-	Earthy, burnt	0.1
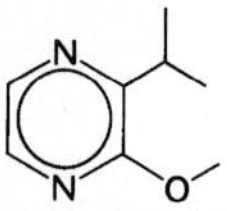 (VI)	2-Isopropyl-3-methoxy-	Potatoes	0.002
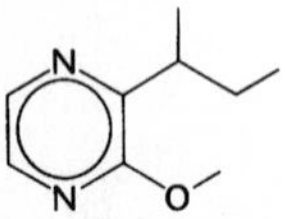 (VII)	2-sec-Butyl-3-methoxy-	Earthy	0.001
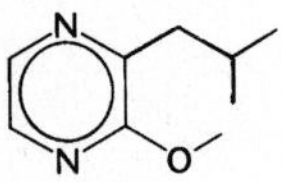 (VIII)	2-Isobutyl-3-methoxy-	Hot paprika (red pepper)	0.002

5.3.1.8 Pyrazines

Pyrazines are powerful aroma compounds. More than 80 pyrazines have been found in food. Some are presented in Table 5.20. Pyrazines are formed by the *Maillard* reaction and by pyrolysis of some amino compounds. Accordingly, they are widely distributed in heat-treated food, for example, bread, meat, roasted coffee, cocoa and roasted nuts. The examples in Table 5.20 show the aroma notes of several pyrazines (a wide aroma spectrum indeed: paprika, chocolate, coffee, potato, etc.). The comprehensive patent literature (several examples are provided in Table 5.21) serves as a convincing illustration of the efforts put into the production of pyrazines and their utilization as flavoring compounds. The odor potency can vary within this single class of compounds in a range of eight orders of magnitude. The potency is greatly affected by the nature of the molecule's side chains and positions in the ring.

Different pyrazines are formed by the *Maillard* reaction and by pyrolysis. The diversity of compounds has been studied with reference to the available N-source. Products obtained by heating glucose with N-supplying reaction components, such as various amino acids and ammonium chloride, are shown in Fig. 5.21. Pyrazines are formed as secondary products of the

Table 5.21. Food flavoring with pyrazines

Compound (mg/kg)	Food product	Aroma description
2-Ethyl-3-vinyl-pyrazine (6)	Instant coffee	Earthy
2-Ethyl-3,5-dimethyl-pyrazine (50)	Glucose syrup	Burnt almond
2-Ethyl-3,6-dimethyl-pyrazine (20)	Glucose syrup	Hazelnut
Formylpyrazine (12.5)	Instant coffee	Roasted note
2-Ethoxy-3-methyl-pyrazine	Ice cream	Roasted nuts
2-Ethyl-3-methoxy-pyrazine	Potato products	Potatoes

Strecker reaction through the condensation of two aminoketones (cf. 4.2.4.4.7). Since reductive amination of an α-dicarbonyl compound to give the aminoketone (*Strecker* reaction) requires a high activation energy, pyrazines are formed relatively quickly via these intermediates only at temperatures >100 °C. On the other hand, pyrazines are formed at room temperature from acyloins and ammonia (weakly acidic medium), e.g., 6,7-dihydro-5H-cyclopenta(b)pyrazine from cyclopentenolone/

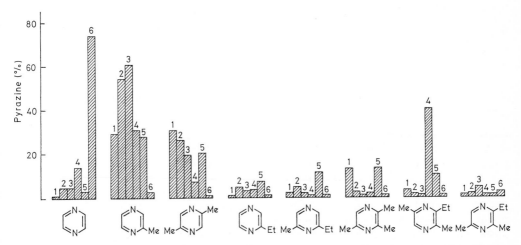

Fig. 5.21. Pyrazine formation in roasted peanuts and in *Maillard* reaction model systems (according to *Koehler* et al., 1969). Roasted peanuts (1); glucose reacted at 120 °C with asparagine (2), glutamine (3), glutamic acid (4), aspartic acid (5) and ammonium chloride (6)

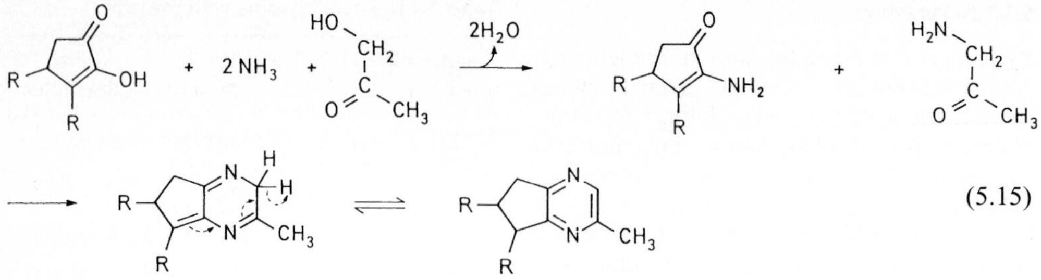

$$(5.15)$$

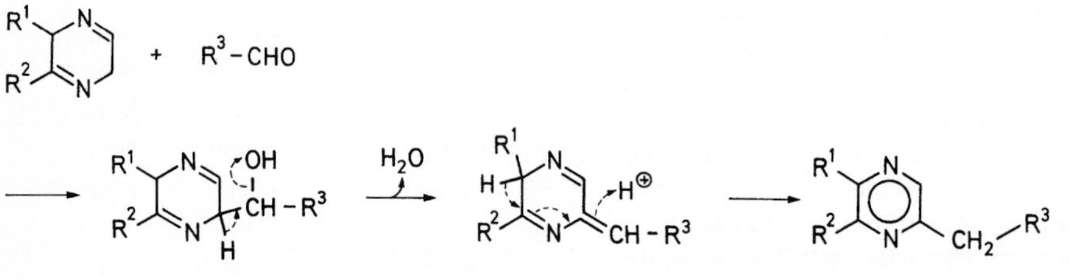

$$(5.16)$$

NH$_3$/acetol (see Formula 5.15). The formation of cyclopentenolones is described under 4.2.4.3.2.

The great variety of pyrazines that occur in heated food result from the fact that their precursors, the dihydropyrazines, can be alkylated by aldehydes (see Formula: 5.16). This reaction pathway explains the formation of the trialkylated pyrazines III–V (Table 5.20), which produce important aroma notes in roasted meat and roasted coffee and have odor thresholds considerably lower than those of the dialkylpyrazines also formed in this process, e.g., III and V in Table 5.20.

The powerfully odorous pyrazines VI–VIII (Table 5.20) appear as metabolic by-products in some plant foods and microorganisms (cf. 5.3.2.6).

5.3.1.9 Phenols

Phenolic acids and lignin are degraded thermally or decomposed by microorganisms into phenols, which are then detected in food. Some of these compounds are listed in Table 5.23.

Smoke generated by burning wood (lignin pyrolysis) is used for cold or hot smoking of meat and fish products. This is a phenol enrichment process since phenol vapors penetrate the meat or fish muscle tissue. Also, some alcoholic beverages, such as Scotch whiskey,

and also butter have low amounts of some phenols, the presence of which is needed to round-off their typical aromas. Model system studies involving pyrolysis of single phenolic acids (Table 5.24) have verified the formation of large numbers of phenols. To explain such a reaction which, for example, accompanies the process of roasting coffee or the kiln drying of malt, it has to be assumed that thermally for-

Table 5.22. Pyrazine formation by pyrolysis. Amounts obtained are recorded as very high (4), high (3), medium (2), low (1), and not detectable (0)

Pyrazine	Precursor				
	Ser	Thr	Etha-nol-amine	Glu-cose-amine	Ala
Pyrazine	3	0	4	1	0
Methylpyrazine	2	1	3	4	0
2,3-Dimethylpyrazine	1	0	0	1	0
2,5-Dimethylpyrazine	0	4	1	3	0
Ethylpyrazine	4	0	2	0	0
2-Ethyl-5-methylpyrazine	0	0	0	1	0
2-Ethyl-6-methylpyrazine	1	0	0	0	0
2,6-Dimethylpyrazine	2	0	0	0	0
3-Ethyl-2,5-dimethylpyrazine	1	3	2	0	0
Trimethylpyrazine	0	3	0	2	0

Table 5.23. Phenols in food

Name	Structure	Aroma quality	Odor threshold (μg/kg, water)	Occcurrence
p-Cresol	 (I)	Smoky	55	Coffee, sherry, milk, roasted peanuts, asparagus
4-Ethylphenol	 (II)	Woody		Milk, soya souce, roasted peanuts, tomatoes, coffee
Guaiacol	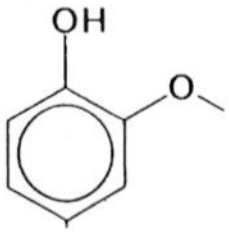 (III)	Smoky, burning, sweet	3	Coffee, milk, crispbread, meat (fried)
4-Vinylphenol	 (IV)	Harsh, smoky	10	Beer, milk, roasted peanuts
2-Methoxy-4-vinylphenol	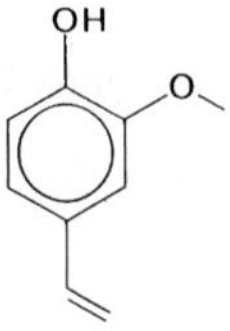 (V)	Clove-like	5	Coffee, beer, apple (cooked), asparagus
Eugenol	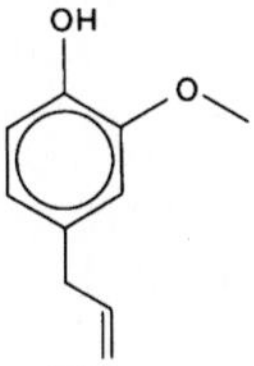 (VI)	Spicy	6	Tomato paste, brandy, plums, cherries
Vanillin	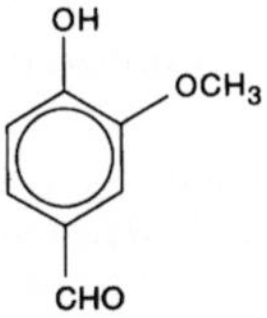 (VII)	Vanilla	20	Vanilla, rum, coffee, asparagus (cooked), butter

Table 5.24. Pyrolysis products of some phenolic acids (T: 200 °C; air)

Phenolic acid	Product	Distribution (%)
Ferulic acid	4-Vinylguaiacol	79.9
	Vanillin	6.4
	4-Ethylguaiacol	5.5
	Guaiacol	3.1
	3-Methoxy-4-hydroxy-acetophenone (Acetovanillone)	2.6
	Isoeugenol	2.5
Sinapic acid	2,6-Dimethoxy-4-vinylphenol	78.5
	Syringaldehyde	13.4
	2,6-Dimethoxyphenol	4.5
	2,6-Dimethoxy-4-ethylphenol	1.8
	3,5-Dimethoxy-4-hydroxy-acetophenone (Acetosyringone)	1.1

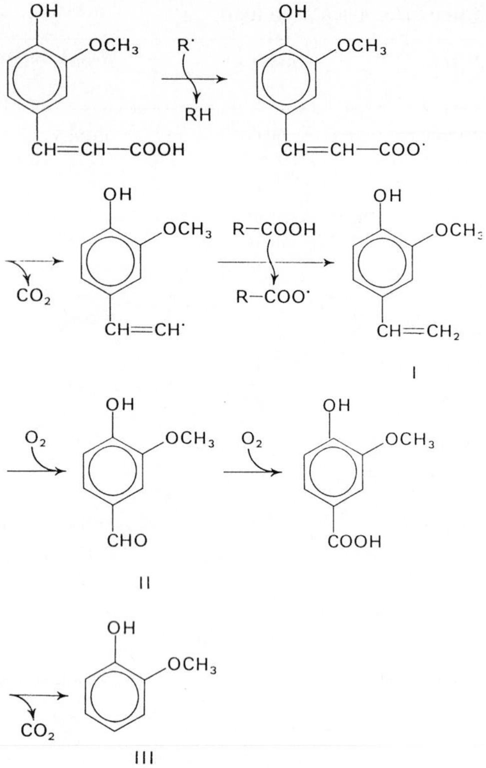

Fig. 5.22. Thermal degradation of ferulic acid. 4-Vinylguaiacol (I), vanillin (II), and guaiacol (III)

med free radicals regulate the decomposition pattern of phenolic acids (cf., for example, heat decomposition of ferulic acid, Fig. 5.22). In the pasteurization of orange juice, p-vinylguaiacol can also be formed from ferulic acid, producing a stale taste at concentrations of 1 mg/kg.

5.3.2 Enzymatic Reactions

Aroma compounds are formed by numerous reactions which occur as part of the normal metabolism of animals, plants and microorganisms. The enzymatic reactions triggered by tissue disruption, as experienced during disintegration or slicing of fruits and vegetables, are of particular importance. Enzymes can also be involved indirectly in aroma formation by providing the preliminary stage of the process, e.g. by releasing amino acids from available proteins, sugars from polysaccharides, or ortho-quinones from phenolic compounds. These are then converted into aroma compounds by further nonenzymatic reactions. In this way, the enzymes enhance the aroma of bread, meat, beer, tea and cacao.

5.3.2.1 Carbonyl Compounds, Alcohols

Fatty acids and amino acids are precursors of a great number of volatile aldehydes, while carbohydrate degradation is the source of ethanal only.

Linoleic and linolenic acids in fruits and vegetables are subjected to oxidative degradation by lipoxygenase alone or in combination with a hydroperoxide lyase, as outlined in sections 3.7.2.2 and 3.7.2.3. The oxidative cleavage yields oxo acids, aldehydes and allyl alcohols. Among the aldehydes formed, hexanal, 2-trans-hexenal, 3-cis-hexenal and/or 2-trans-nonenal, 3-cis-nonenal, 2-trans,6-cis-nonadienal and 3-cis,6-cis-nonadienal are important for aroma.

Frequently, these aldehydes appear along with some of the alcohols derived from them soon after the disintegration of the tissue in the presence of oxygen.

In comparison, lipoxygenases and hydroperoxide lyases from mushrooms exhibit a different reaction specificity. Linoleic acid, which predominates in the lipids of champignon mushrooms, is oxidatively cleaved to R(−)-1-octen-3-ol and 10-oxo-(E)-8-decenoic acid (cf. 3.7.2.3). The allyl alcohol is oxidized to a small extent by atmospheric oxygen to the corresponding ketone. Owing to an odor threshold that is about hundred times lower (cf. Table 3.31), this ketone together with the alcohol accounts for the mushroom odor of fresh champignons.

Aldehydes formed by the *Strecker* degradation (cf. 5.3.1.1; Table 5.13) can also be obtained as metabolic by-products of the enzymatic transamination or oxidative deamination of amino acids. First, the amino acids are converted enzymatically to α-keto acids and then to aldehydes by decarboxylation in a side reaction:

$$R-\underset{\underset{CO_2}{\overset{|}{NH_2}}}{\overset{|}{CH}}-COOH \longrightarrow R-\underset{\overset{\|}{O}}{\overset{}{C}}-COOH$$

$$\xrightarrow{\quad} R-CHO \qquad (5.17)$$

Unlike other amino acids, threonine can eliminate a water molecule and, by subsequent decarboxylation, yield propanal:

$$H_3C-\underset{\underset{OH}{|}}{CH}-\underset{\underset{NH_2}{|}}{CH}-COOH$$

$$\xrightarrow{H_2O} H_3C-CH=\underset{\underset{NH_2}{|}}{C}-COOH$$

$$\rightleftharpoons H_3C-CH_2-\underset{\overset{\|}{NH}}{C}-COOH$$

$$\xrightarrow[NH_3]{H_2O} H_3C-CH_2-CO-COOH$$

$$\xrightarrow[\quad]{CO_2} H_3C-CH_2-CHO \qquad (5.18)$$

Many aldehydes derived from amino acids occur in plants and fermented food.

A study involving the yeast *Saccharomyces cerevisiae* clarified the origin of 2-methylpropanal and 2- and 3-methylbutanal. They are formed to a negligible extent by decomposition but mostly as by-products during the biosynthesis of valine, leucine and isoleucine.

Figure 5.23 shows that α-ketobutyric acid, derived from threonine, can be converted into isoleucine. Butanal and 2-methylbutanal are formed by side-reaction pathways.

2-Acetolactic acid, obtained from the condensation of two pyruvate molecules, is the intermediary product in the biosynthetic pathways of valine and leucine (Fig. 5.24). However, 2-acetolactic acid can be decarboxylated in a side reaction into acetoin, the precursor of diacetyl. At α-keto-3-methylbutyric acid, the metabolic pathway branches to form 2-methylpropanal and branches again at α-keto-4-methyl valeric acid to form 3-methylbutanal (Fig. 5.24).

The enzyme that decarboxylates the α-ketocarboxylic acids to aldehydes has been detected in oranges. Substrate specificity for this decarboxylase is shown in Table 5.25.

Alcohol dehydrogenases (cf. 2.3.1.1) can reduce the aldehydes derived from fatty acid and amino acid metabolism into the corresponding alcohols:

$$R-CH_2-OH + NAD^{\oplus}$$

$$\rightleftharpoons R-CHO + NADH + H^{\oplus} \qquad (5.19)$$

Alcohol formation in plants and microorganisms is strongly favoured by the reaction equilibrium and, primarily, by the predominance of NADH over NAD^+. Nevertheless, the enzyme specificity is highly variable. In most cases aldehydes $>C_5$ are only slowly reduced; thus, with aldehydes rapidly formed by, for example, oxidative cleavage of unsaturated fatty acids, a mixture of alcohols and aldehydes results, in which the aldehydes predominate.

Table 5.25. Substrate specificity of a 2-oxocarboxylic acid decarboxylase from orange juice

Substrate	v_{rel} (%)
Pyruvate	100
2-Oxobutyric acid	34
2-Oxovaleric acid	18
2-Oxo-3-methylbutyric acid	18
2-Oxo-3-methylvaleric acid	18
2-Oxo-4-methylvaleric acid	15

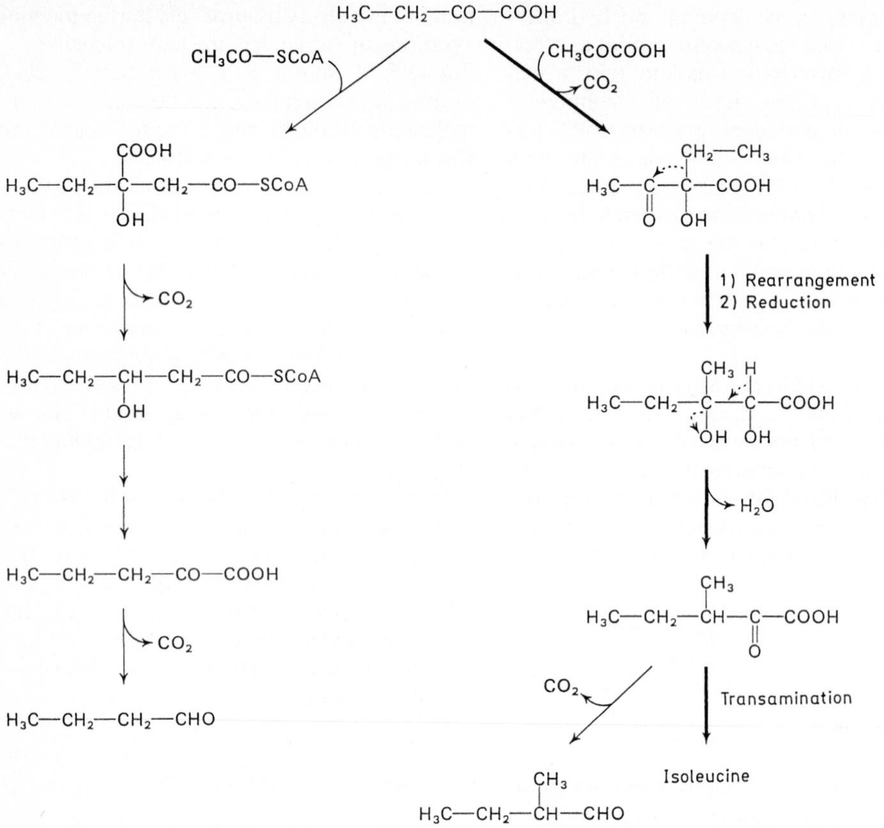

$$H_3C-CH_2-CO-COOH$$

$CH_3CO-SCoA$

$CH_3COCOOH$
CO_2

$$H_3C-CH_2-\overset{\overset{\textstyle COOH}{|}}{\underset{\underset{\textstyle OH}{|}}{C}}-CH_2-CO-SCoA$$

$$H_3C-\overset{\overset{\textstyle CH_2-CH_3}{\cdots\cdots|}}{\underset{\underset{\textstyle O \quad OH}{|\quad|}}{C}}-\overset{}{C}-COOH$$

1) Rearrangement
2) Reduction

CO_2

$$H_3C-CH_2-\underset{\underset{\textstyle OH}{|}}{CH}-CH_2-CO-SCoA$$

$$H_3C-CH_2-\overset{\overset{\textstyle CH_3 \; H}{|\quad|}}{\underset{\underset{\textstyle OH \;\; OH}{|\quad\;\;|}}{C}}\overset{}{-}\overset{}{C}-COOH$$

H_2O

$$H_3C-CH_2-CH_2-CO-COOH$$

$$H_3C-CH_2-\underset{}{CH}-\overset{\overset{\textstyle CH_3}{|}}{\underset{\underset{\textstyle O}{||}}{C}}-COOH$$

CO_2

CO_2

Transamination

$$H_3C-CH_2-CH_2-CHO$$

$$H_3C-CH_2-\overset{\overset{\textstyle CH_3}{|}}{CH}-CHO$$

Isoleucine

Fig. 5.23. Formation of aldehydes during isoleucine biosynthesis (according to *Piendl*, 1969). → main pathway → side pathway of the metabolism

5.3.2.2 Hydrocarbons, Esters

Fruits and vegetables (e.g., pineapple, apple, pear, peach, passion fruit, kiwi, celery, parsley) contain unsaturated C_{11} hydrocarbons which play a role as aroma substances. Of special interest are 1,3-trans,5-cis-undecatriene and 1,3-trans,5-cis,8-cis-undecatetraene, which with very low threshold concentrations have a balsamic, spicy, pine-like odor. It is assumed that the hydrocarbons are formed from unsaturated fatty acids by β-oxidation, lipoxygenase catalysis, oxidation of the radical to the carbonium ion and decarboxylation. The hypothetical reaction pathway from linoleic acid to 1,3-trans,5-cis-undecatriene is shown in Formula 5.21.

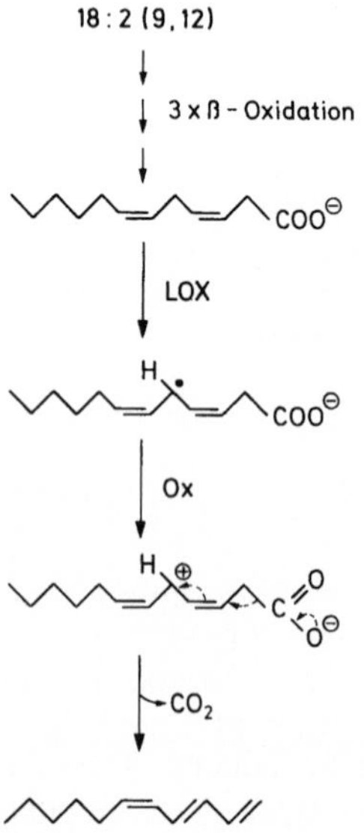

18 : 2 (9,12)

3 x ß - Oxidation

$COO^{\ominus}$

LOX

$COO^{\ominus}$

Ox

CO_2

(5.21)

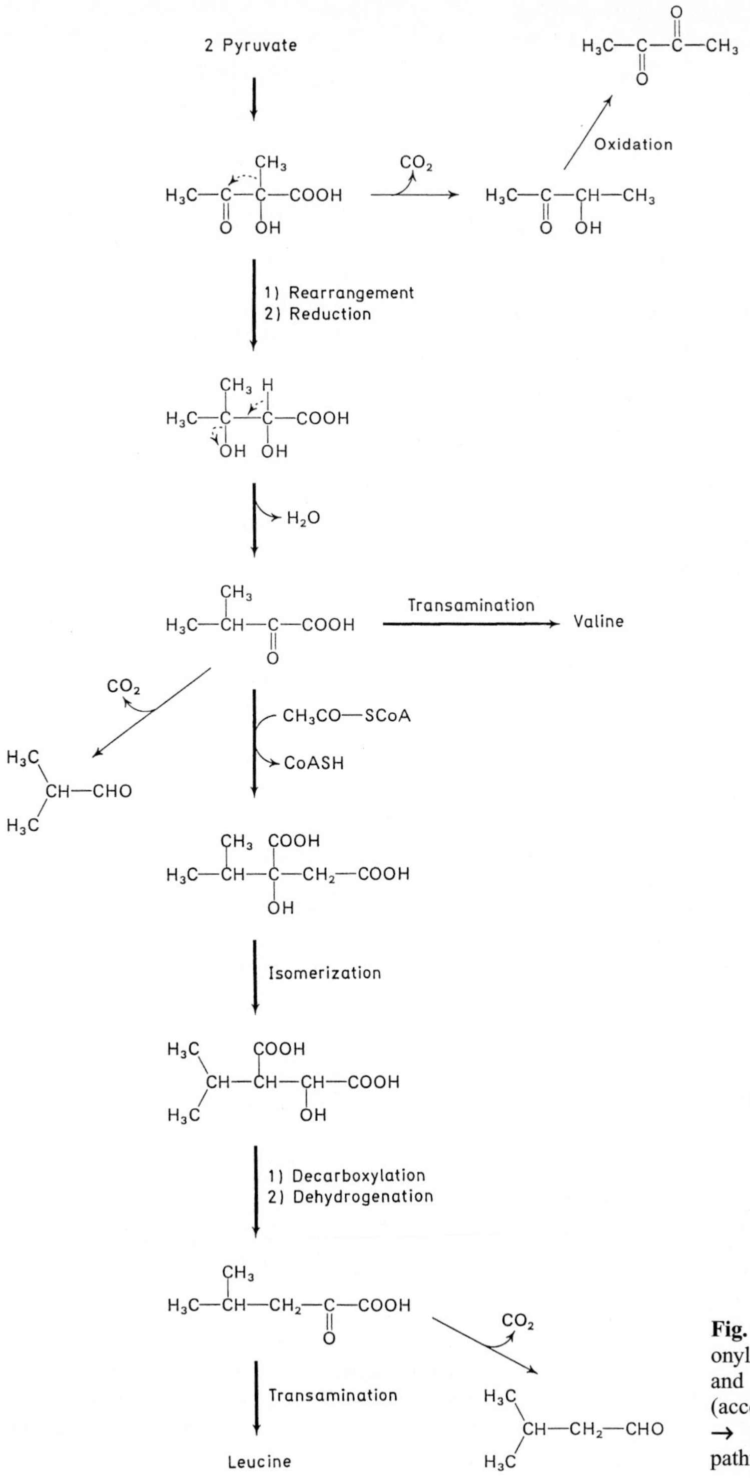

Fig. 5.24. Formation of carbonyl compounds during valine and leucine biosynthesis (according to *Piendl*, 1969). → main pathway → side pathway of the metabolism

Esters are significant aroma constituents of many fruits. They are synthetized only by intact cells:

$$R—CO—SCoA \ + \ R'—OH$$
$$\longrightarrow \ R—CO—O—R' \ + \ CoASH \qquad (5.20)$$

Acyl-CoA originates from the β-oxidation of fatty acids and also occasionally from amino acid metabolism. Figure 5.25 shows an example of how ethyl 2-trans, 4-cis-decadienoate, an important aroma constituent of pears, is synthesized from linoleic acid.

Table 5.26 gives information on the odor thresholds of some esters with a fruity aroma quality. Methyl branched esters, from the metabolism of isoleucine, were found to have the lowest values.

When fruits are homogenized, such as in the processing of juice, the esters are rapidly hydrolyzed by the hydrolase enzymes present, and the fruit aroma flattens.

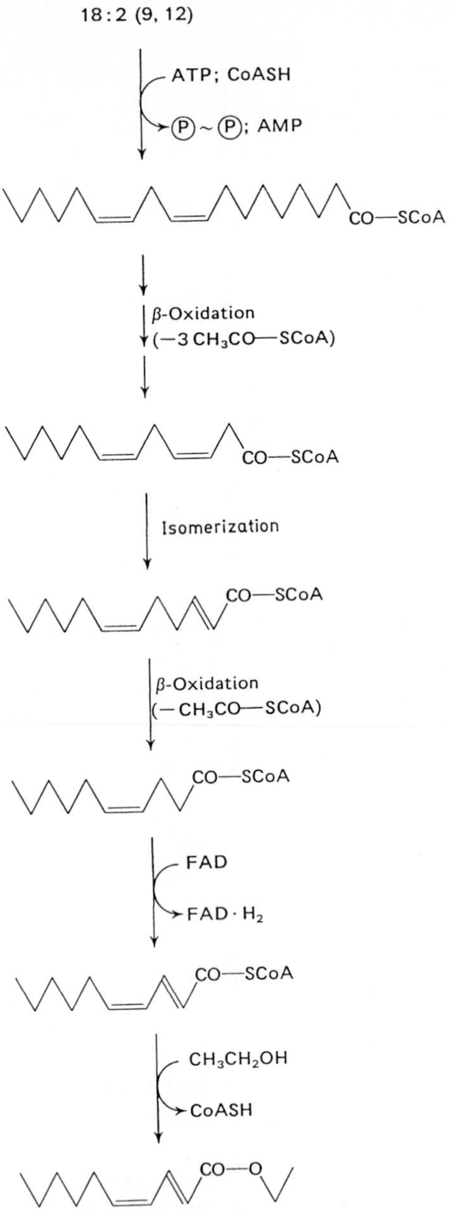

5.3.2.3 Lactones

Numerous lactones are found in food. Some of the representatives which belong to the typical aroma substances of butter, coconut oil, and various fruits are presented in Table 5.27. Since the aroma of lactones is partly very pleasant, these substances are also of interest for commercial aromatization of food. In the homologous series of γ- and δ-lactones, the odor threshold decreases with increasing molecular weight (Table 5.28).

Table 5.26. Odor thresholds of esters

Compound	Odor threshold (µg/kg, water)
2-Methylpropionic acid methyl ester	7
2-Methylbutyric acid methyl ester	0.25
Methylpropionic acid ethyl ester	0.1
2-Methylbutyric acid ethyl ester	0.1
Butyric acid ethyl ester	1
Acetic acid hexyl ester	2
Caproic acid ethyl ester	1
Benzoic acid ethyl ester	60
Salicylic acid methyl ester	40

Fig. 5.25. Biosynthesis of 2-trans, 4-cis-decadienoic acid ethyl ester in pears (according to *Jennings* and *Tressl*, 1974)

Lactones are formed from the corresponding hydroxy acids. They are chiral compounds, e.g., (R)-γ-decalactone predominates in fruits and (S)-δ-decalactone in milk fat (cf. Table 5.9).

Table 5.27. Lactones in food

Name	Structure	Aroma quality	Occurrence
4-Nonanolide (γ-nonalactone)	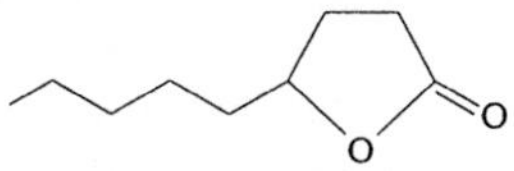	Reminiscent of coconut oil, fatty	Fat-containing food, crispbread, peaches
4-Decanolide (γ-decalactone)	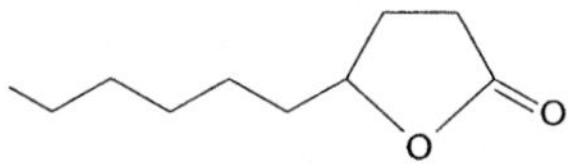	Fruity, peaches	Fat-containing food, cf. Table 5.9
5-Decanolide (δ-decalactone)	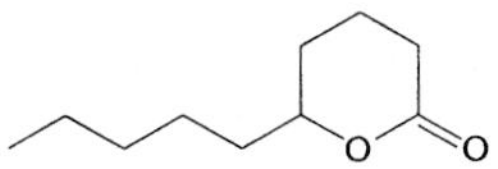	Oily, peaches	Fat-containing food, cf. Table 5.9
(Z)-6-Dodecen-γ-lactone	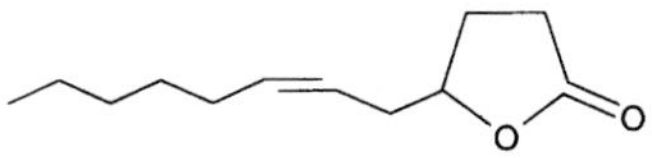	Sweet	Milk fat, peaches
3-Methyl-4-octanolide (whisky – or oak lactone)	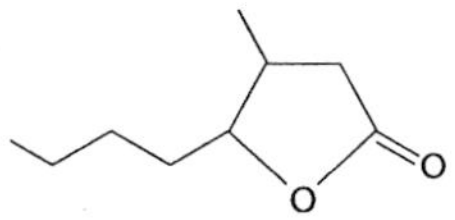	Coconut-like	Alcoholic beverages

Linoleic acid is metabolized by cows with the formation of (Z)-6-dodecen-γ-lactone as a secondary product (Fig. 5.26). Its sweetish odor enhances the aroma of butter. On the other hand, it is undesirable in meat.

The whisky or oak lactone is formed when alcoholic beverages are stored in oak barrels. 3-Methyl-4-(3,4-dihydroxy-5-methoxybenzo)octanoic acid is extracted from the wood. After elimination of the benzoic acid residue, this compound cyclizes to give the lactone. The odor thresholds of the two cis-oak lactones (3R, 4R and 3S, 4S) are about ten times lower than those of the trans diastereomers (3S, 4R and 3R, 4S).

Table 5.28. Odor thresholds of lactones

Compound	Odor threshold (μg/kg, water)
γ-Lactones	
γ-Hexalactone	1600
γ-Heptalactone	400
γ-Octalactone	7
γ-Nonalactone	30–65
γ-Decalactone	11
γ-Dodecalactone	7
δ-Lactones	
δ-Octalactone	400
δ-Decalactone	100
6-Pentyl-α-pyrone	150

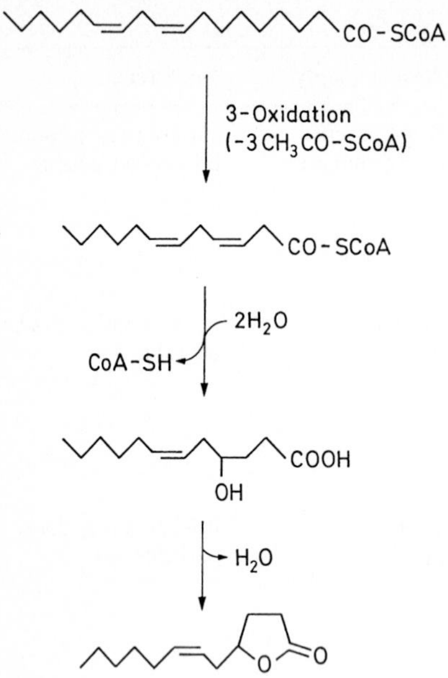

Fig. 5.26. Formation of Z-6-dodecen-γ-lactone

5.3.2.4 Terpenes

The mono- and sesquiterpenes in fruits (cf. 18.1.2.6) and vegetables (cf. 17.1.2.6), herbs and spices (cf. 22.1.1.1) and wine (cf. 20.2.6.9) are presented in Table 5.29. These compounds stimulate a wide spectrum of aromas, mostly perceived as very pleasant (examples in Table 5.30). The odor thresholds of terpenes vary greatly (Table 5.30). Certain terpenes occur in flavoring plants in such large amounts that in spite of relatively high odor thresholds, they can act as character impact compounds, e. g., S(+)-α-phellandrene in dill (cf. Table 5.12).

Monoterpenes with hydroxy groups, such as linalool, geraniol and nerol, are present in fruit juice at least in part as glycosides. Linalool-β-rutinoside (I) and linalool-6-0-α-L-arabino-furanosyl-β-D-glucopyranoside (II) have been found in wine grapes and in wine (cf. 20.2.6.9):

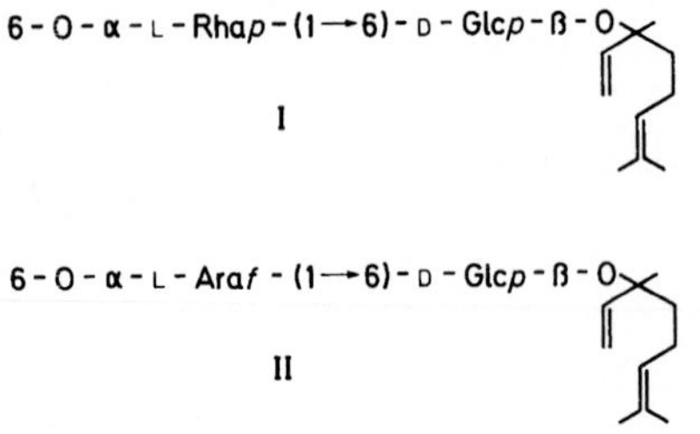

$$ \tag{5.22} $$

Table 5.29. Terpenes in food

Monoterpenes

Acyclic (including cyclic derivatives)

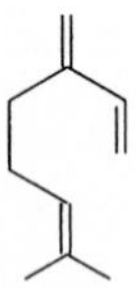

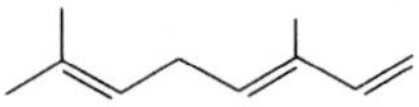

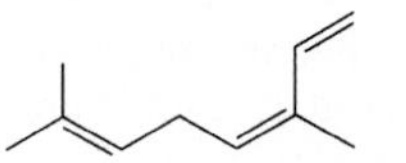

Myrcene (I)

trans-Ocimene (II)

cis-Ocimene (III)

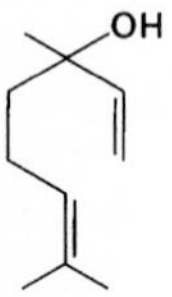

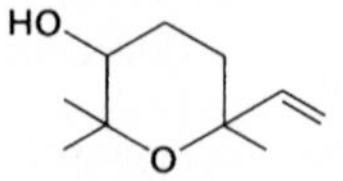

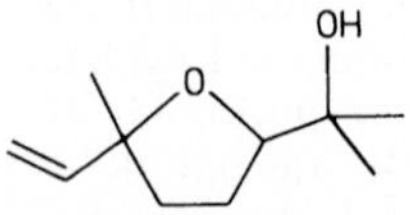

Linalool (IV)

2,6,6-Trimethyl-2-vinyl-5-hydroxytetrahydro-pyran[a] (IVa)

2-Methyl-2-vinyl-5-hydroxyisopropyltetra-hydrofuran[a] (IVb)

Table 5.29 (continued)

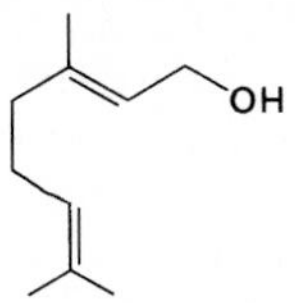

Geraniol[b] (V)

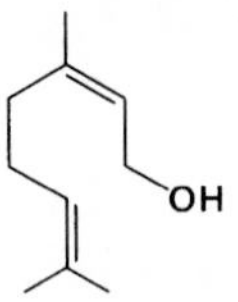

Nerol[b] (VI)

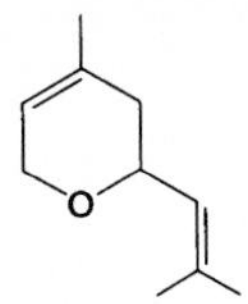

Neroloxide (VIa)

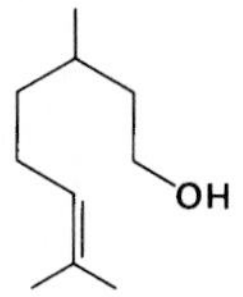

Citronellol[b] (VII)

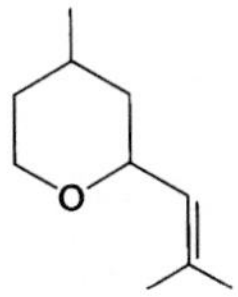

Rose oxide (VIIa)

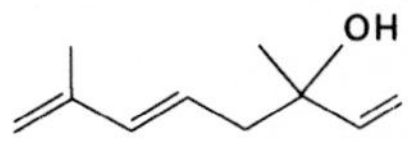

Hotrienol[c] (VIII)

Monocyclic

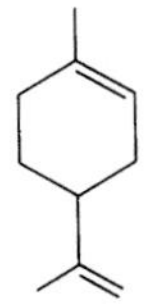

Limonene (IX)

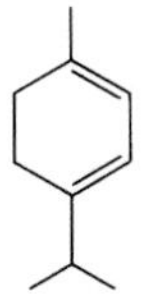

α-Terpinene (X)

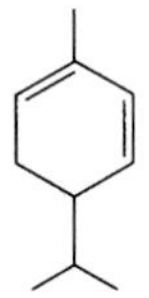

α-Phellandrene (XI)

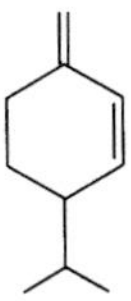

β-Phellandrene (XII)

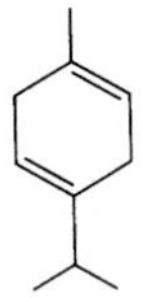

γ-Terpinene (XIII)

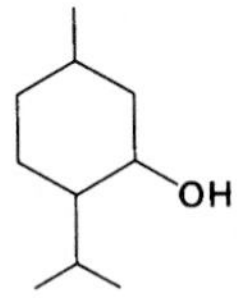

Menthol (XIV)

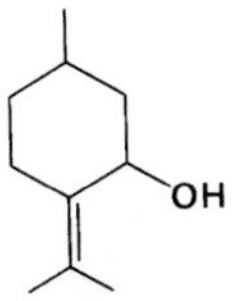

Pulegol (XV)

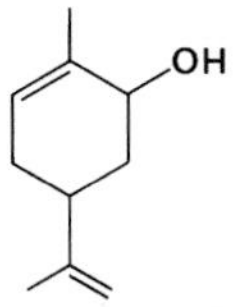

Carveol (XVI)

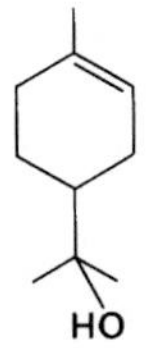

α-Terpineol (XVII)

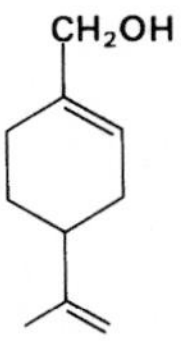

Perilla alcohol (XVIII)

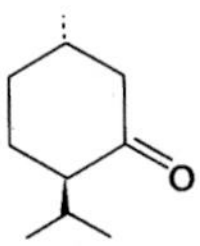

Menthone (XIX)

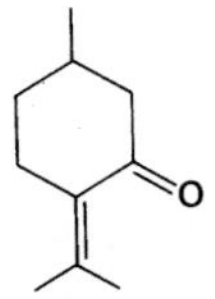

Pulegone (XX)

Table 5.29 (continued)

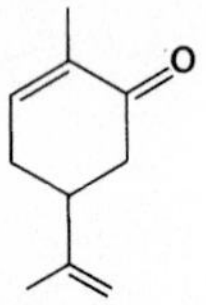

Carvone (XXI)

1,3-p-Menthadien-7-al
(XXII)

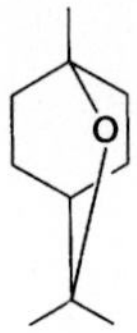

1,8-Cineole (XXIII)

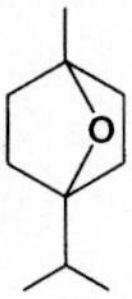

1,4-Cineole (XXIV)

Bicyclic

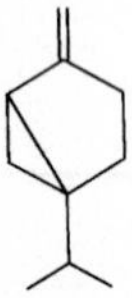

Sabinene (XXV)

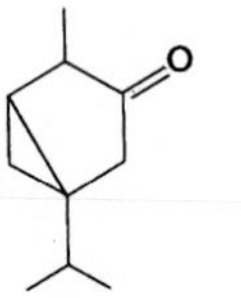

Thujone (XXVI)

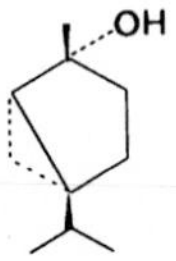

(+)-cis-Sabinene hydrate
(XXVII)

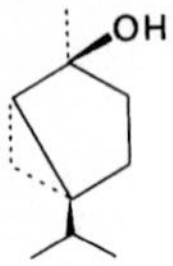

(+)-trans-Sabinene hydrate
(XXVIII)

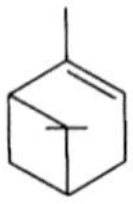

α-Pinene (XXIX)

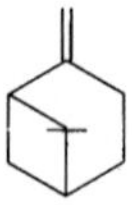

β-Pinene (XXX)

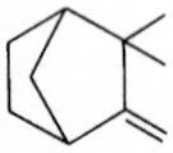

Camphene (XXXI)

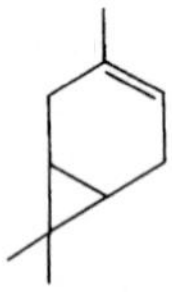

Δ³-Carene (XXXII)

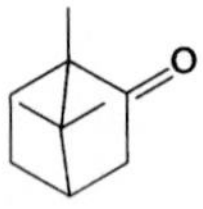

Camphor (XXXIII)

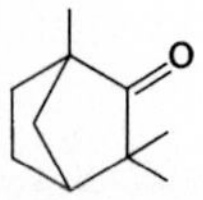

Fenchone (XXXIV)

Table 5.29 (continued)

Sesquiterpenes
Acyclic

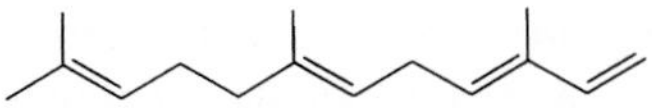

trans-α-Farnesene (XXXV) 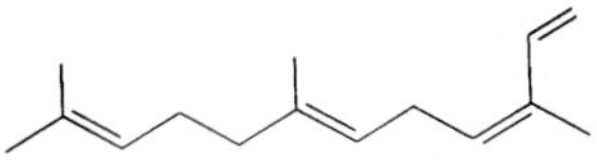   cis-α-Farnesene (XXXVI)

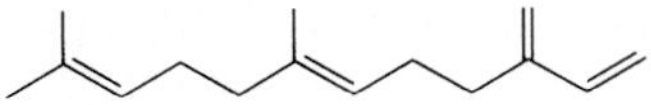

β-Farnesene (XXXVII) 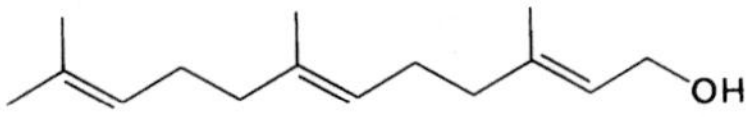   Farnesol (XXXVIII)

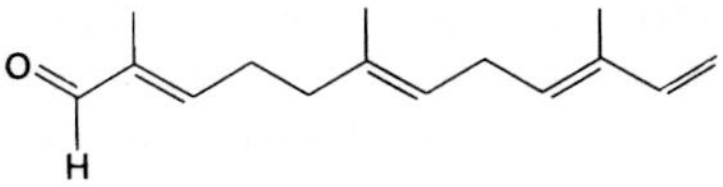

(all-trans)-α-Sinensal (XXXIX) 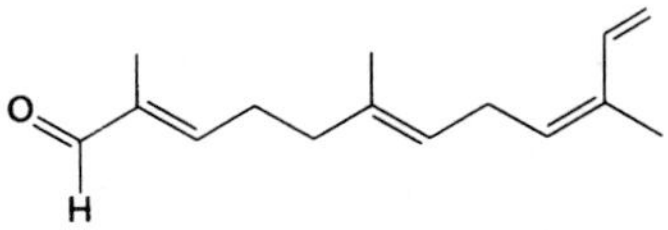   (trans,trans,cis)-α-Sinensal (XL)

Monocyclic

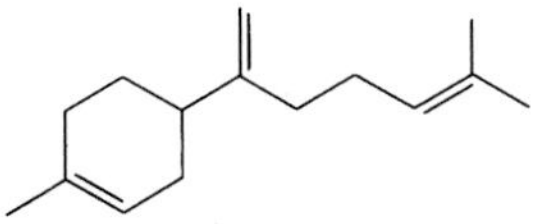

β-Bisabolene (XLI) 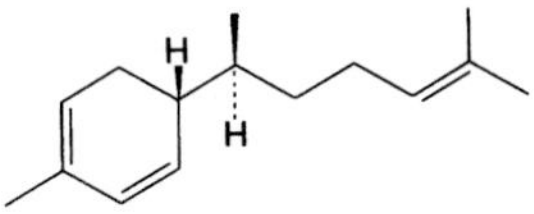   (−)-Zingiberene (XLII)

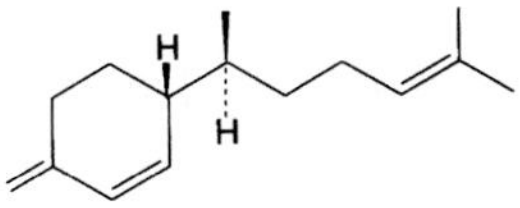

(−)-Sesquiphellandrene (XLIII) 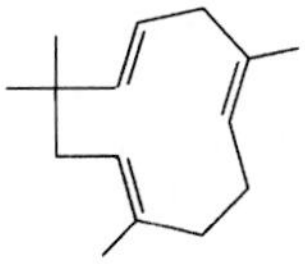   Humulene (XLIV)

Bicyclic

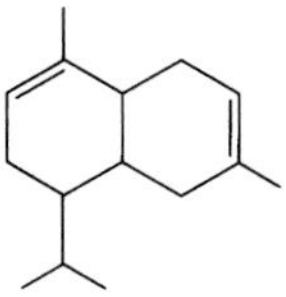

β-Cadinene (XLV) 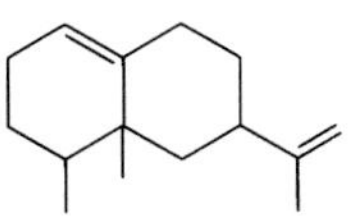   Valencene (XLVI) 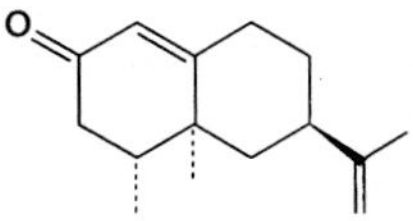   (+)-Nootkatone (XLVII)

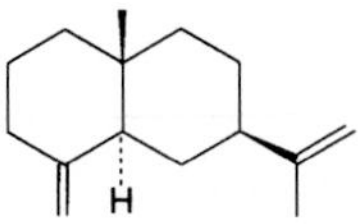

β-Selinene (XLVIII) 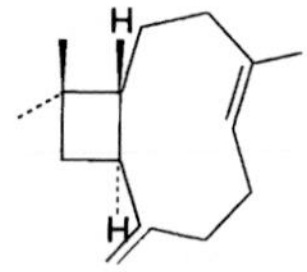   β-Caryophyllene (XLIX)

ᵃ Compounds IVa and IVb are also denoted as pyranlinalool and furanlinalool oxide, respectively.
ᵇ Corresponding aldehydes geranial (V a), neral (VI b) and citronellal (VIIa) also occur in food. Citral is a mixture of neral and geranial.
ᶜ (−)-3,7-Dimethyl-1,5,7-octatrien-3-ol (hotrienol) is found in grape, wine and tea aromas.

Table 5.30. Sensory properties of some terpenes

Compound[a]	Aroma quality	Odor threshold (μg/kg, water)
Myrcene (I)	Herbaceous, metallic	14
Linalool (IV)	Flowery	6
cis-Furanlinalool oxide (IVb)	Sweet-woody	6000
Geraniol (V)	Rose-like	40
Geranial (Va)	Citrus-like	32
Citronellol (VII)	Rose-like	40
(2S,4R)-Rose oxide (VIIa)	Geranium-like	0.5
R(+)-Limonene (IX)	Citrus-like	200
R(–)-α-Phellandrene (XI)	Terpene-like, medicinal	500
S(+)-α-Phellandrene (XI)	Dill-like, herbaceous	200
α-Terpineol (XVII)	Lilac-like, peach-like	330
1,8-Cineol (XXIII)	Spicy, camphor-like	12
(all-E)-α-Sinensal (XXXIX)	Orange-like	0.05
(–)-β-Caryophyllene (XLIX)	Spicy, dry	64

[a] The numbering of the compounds refers to Table 5.29.

Terpene glycosides hydrolyze, e.g., in the production of jams (cf. 18.1.2.6.11), either enzymatically (β-glucosidase) or due to the low pH of juices. The latter process is strongly accelerated by a heat treatment. Under these conditions, terpenes with two or three hydroxyl groups which are released undergo further reactions, forming hotrienol (IV) and nerol-oxide (V) from 3,7-dimethylocta-1,3-dien-3,7-diol (cf. Formula 5.23) in grape juice, or cis- and trans-furanlinalool oxides (VIa and VIb) from 3,7-dimethylocta-1-en-3,6,7-triol in grape juice and peach sap (cf. Formula 5.24).

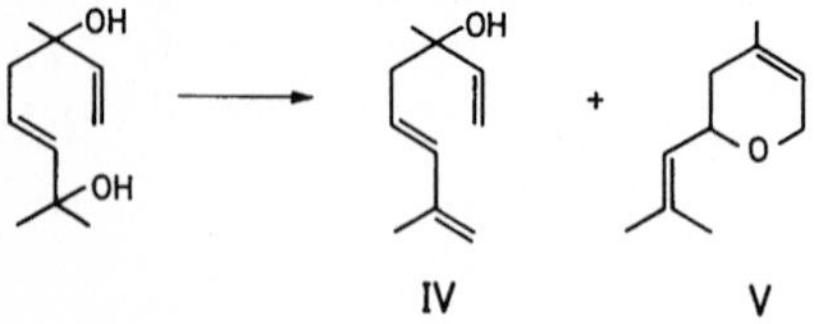

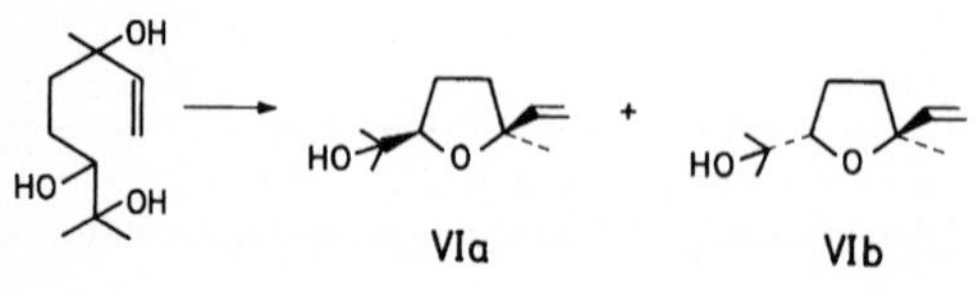

Most terpenes contain one or more chiral centers. Of several terpenes, the optically inactive form and the 1- and d-form occur in different plants. The enantiomers and diastereoisomers differ regularly in their odor characteristics. For example, menthol (XIV in Table 5.29) in the 1-form (1R, 3R, 4S) which occurs in peppermint oil, has a clean sweet, cooling and refreshing peppermint aroma, while in the d-form (1S, 3S, 4R) it has remarkable, disagreeable notes such as phenolic, medicated, camphor and musty. Carvone (XXI in Table 5.29) in the R(–)-form has a peppermint odor. In the S(+)-form it has an aroma similar to caraway. Some terpenes are readily oxidized during food storage. Examples of aroma defects resulting from oxidation are provided in Table 5.4 and Section 22.1.1.1.

Terpene biosynthesis is carried out only by plants and some microorganisms and is initiated in both cases by acetyl-CoA:

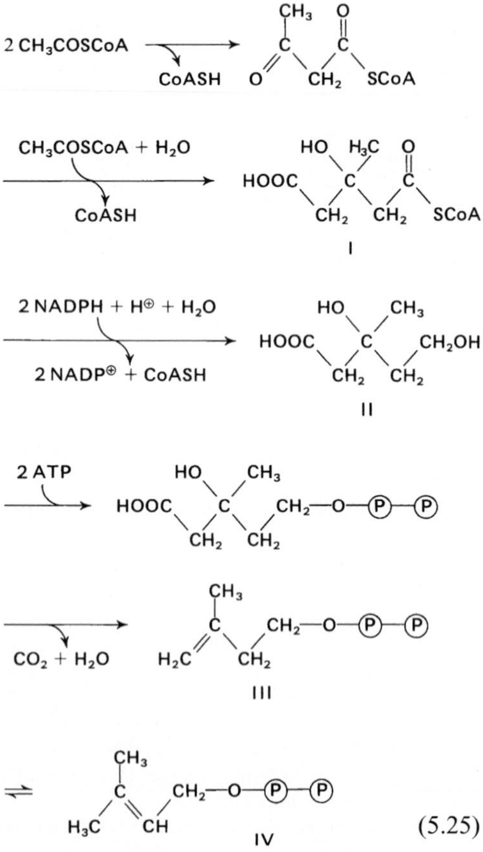

(5.23)

(5.24)

(5.25)

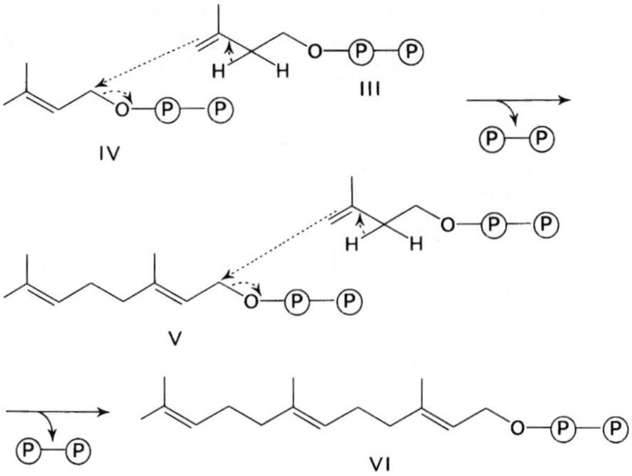

$$(5.26)$$

Three molecules of acetyl-CoA condense to form 3-hydroxy-3-methylglutaryl-CoA (I) which, after hydrolysis and reduction, is converted to mevalonic acid (II). Phosphorylation, and elimination of CO_2 and water yields isopentenyl diphosphate (III), which can partly isomerize into dimethyl-allyl pyrophosphate (IV). A cation derived from IV then reacts with the double bond of compound III (cf. Reaction 5.26).

Geranyl diphosphate (V), the parent compound for monoterpene (C_{10}) biosynthesis, is generated by elimination of a proton. Condensation of compound V with isopentenyl diphosphate (III) leads to farnesyl diphosphate (VI), which is the parent compound for the biosynthesis of all the sesquiterpenes (C_{15}).

Although acyclic monoterpenes can be generated from V (e.g. geraniol by hydrolysis of compound V), cyclization is possible. It can only occur via isomerization of compound V from the trans- to the cis-form, i.e. formation of neryl pyrophosphate (VII, Reaction 5.27). The chain elongation reaction leading to farnesyl diphosphate is interrupted by this isomerization.

It is assumed that the cyclization mechanism of VII occurs via the cation VIII, forming terpenes with a p-menthane skeleton, such as α-terpineol (IX) and limonene (X). The cation is also involved in formation of bicyclic terpenes, such as pinane (XI), bornane (XII) and thujane (XIII). The cations present as intermediary products can attract a nucleophilic HO-ion and thus provide a variety of terpene alcohols. However, the oxygen-containing moiety can also be acquired by oxidation of the complete menthane skeleton. For example, in peppermint leaves, 1(−)menthol (XIV) is synthesized from cation VIII by the following pathway:

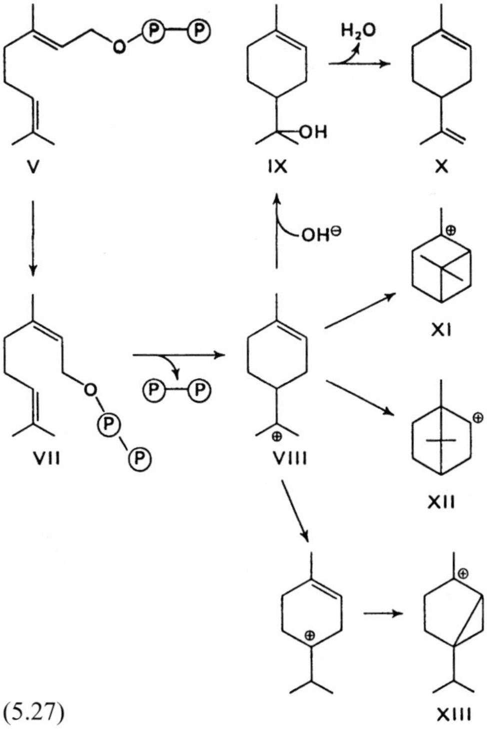

(5.27)

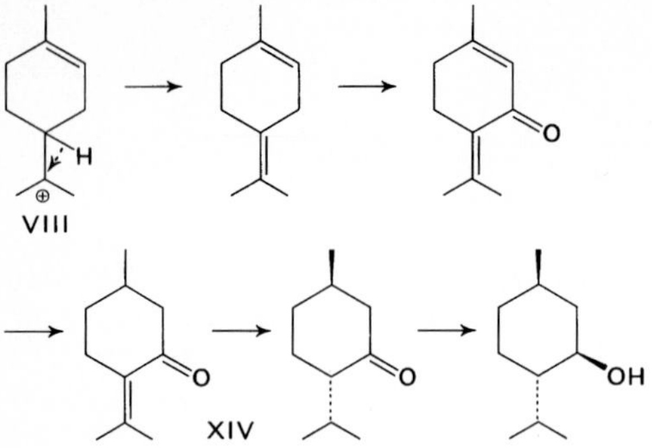

(5.28)

Similarly, sesquiterpene biosynthesis begins with a trans-to-cis isomerization. The intermediary cation (XVI) obtained from cis-farnesyl pyrophosphate (XV) provides a number of reaction possibilities due to the length of the carbon skeleton and the three double bonds within the skeleton. Only two metabolic pathways are presented here; one postulated for the biosynthesis of β-bisabolene (XVII) and the second for β-cadinene (XIX; cf. Reaction 5.29).

5.3.2.5 Volatile Sulfur Compounds

The aroma of many vegetables is due to volatile sulfur compounds obtained by a variety of enzymatic reactions. Examples are the vegetables of the plant families *Brassicacea* and *Liliaceae*; their aroma is formed by decomposition of glucosinolates or S-alkyl-cysteine-sulfoxides (cf. 17.1.2.6.7).

2-Isobutylthiazole (compound IV, Table 5.18) contributes to tomato aroma. It is probably obtained as a product of the secondary metabolism of leucine and cysteine (cf. postulated Reaction 5.30).

Isobutyric acid is the precursor of asparagus acid (1,2-dithiolane-4-carboxylic acid) found in asparagus. It is dehydrogenated to give methylacrylic acid which then adds on an unknown S-containing nucleophile (see Formula 5.31). During cooking, asparagus acid is oxidatively decarboxylated to a 1,2-dithiocyclopentene (see Formula 5.32), which contributes to the aroma of asparagus.

Volatile sulfur compounds formed in wine and beer production originate from methionine and are by-products of the microorganism's metabolism. The compounds formed are methional (I), methionol (II) and acetic acid-3-(methylthio)-propyl ester (III, cf. Reaction 5.33).

Tertiary thiols (Table 5.31) are some of the most intensive aroma substances. They have a fruity odor at the very low concentrations in

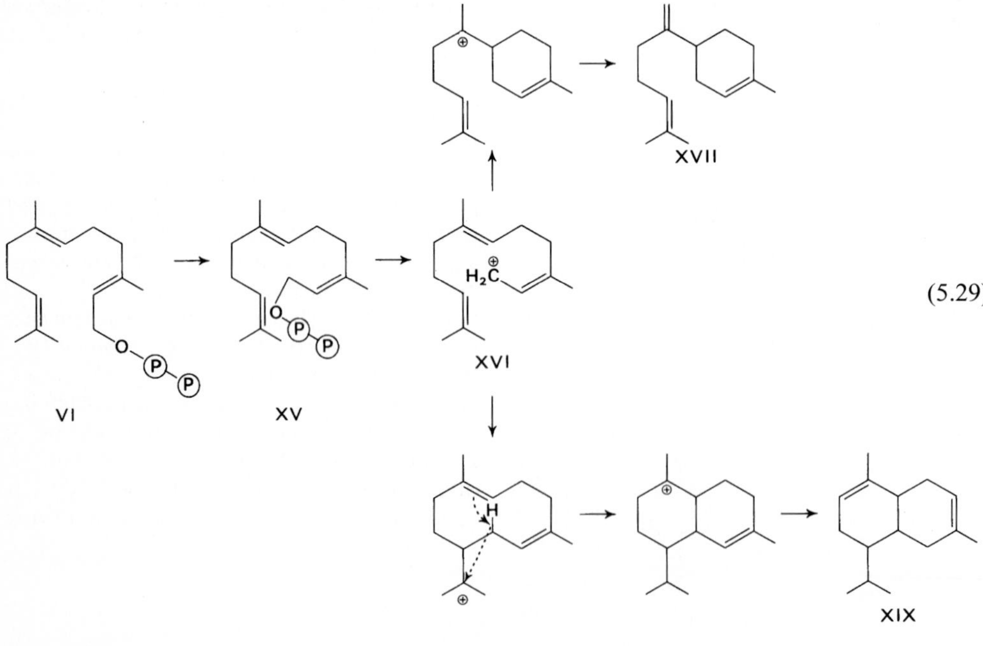

(5.29)

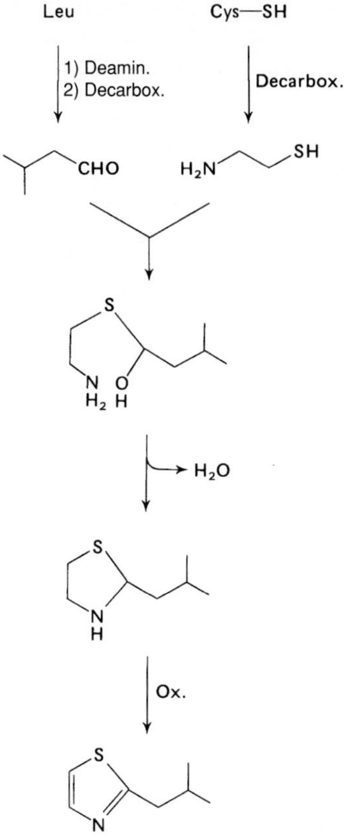
Leu
Cys—SH
1) Deamin.
2) Decarbox.
Decarbox.
CHO
H2N
SH
S
N O
H2 H
H2O
S
N
H
Ox.
S
N

(5.30)

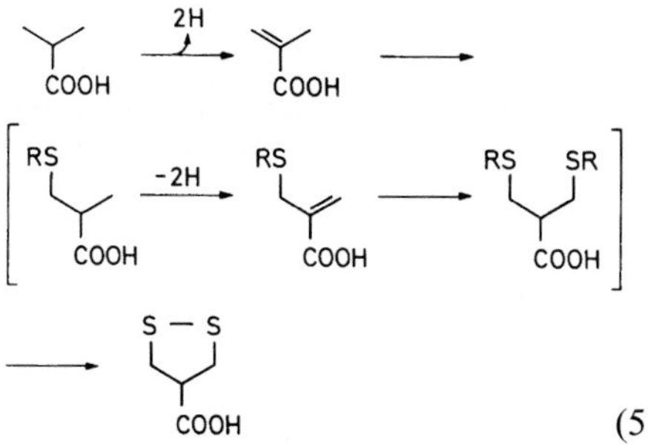
2H
COOH
COOH
RS
-2H
RS
RS SR
COOH
COOH
COOH
S — S
COOH

(5.31)

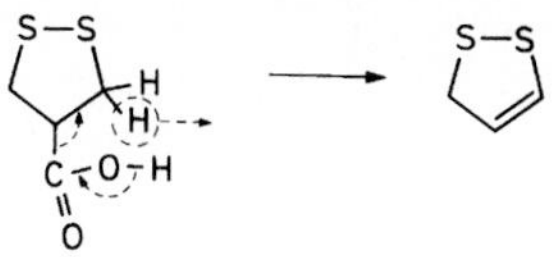
S—S
H
H
C
O
H
O
S—S

(5.32)

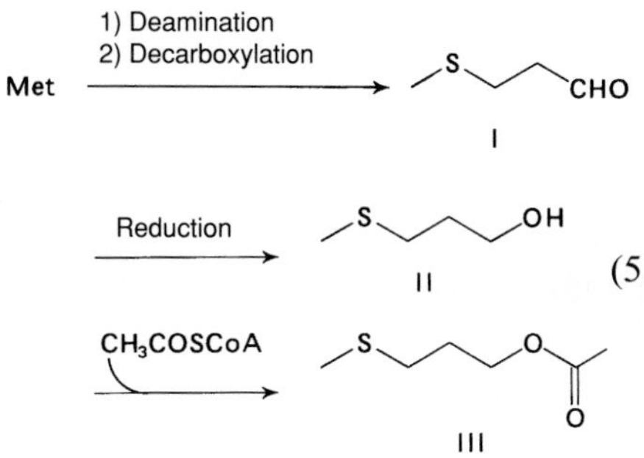
1) Deamination
2) Decarboxylation
Met
S
CHO
I
Reduction
S
OH
II
CH3COSCoA
S
O
O
III

(5.33)

which they occur in foods. With increasing concentration, they smell of cat urine and are called *catty odorants*. Tertiary thiols have been detected in fruits, olive oil and roasted coffee (Table 5.31). They make important contributions to the aroma and are possibly formed by the addition of hydrogen sulfide to metabolites of isoprene metabolism. In beer, 3-mercapto-3-methylbutylformate is undesirable because it causes off-flavor at concentrations as low as 5 ng/l. Another representative of this class of compounds, 4-mercapto-4-methyl-2-pentanone, has been identified as a reaction product of mesityl oxide (cf. 5.3.1.4).

5.3.2.6 Pyrazines

Paprika pepper (*Capsicum anuum*) and chillies (*Capsicum frutescens*) contain high concentrations of 2-isobutyl-3-methoxypyrazine (VIII in Table 5.20 for structure). Its biosynthesis from leucine is assumed to be through the following pathway:

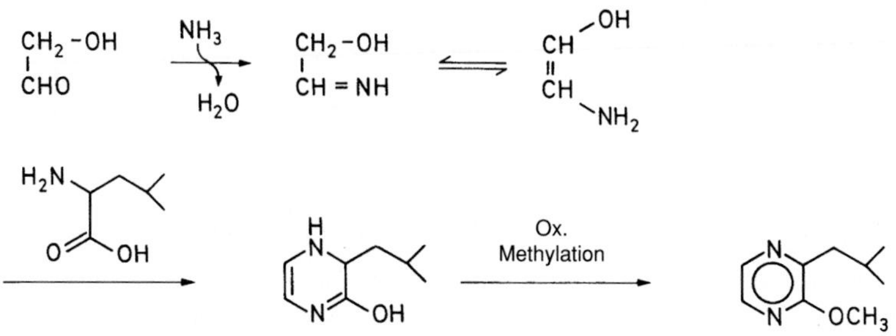
CH2-OH
CHO
NH3
H2O
CH2-OH
CH = NH
CH
OH
CH
NH2
H2N
O
OH
H
N
N
OH
Ox.
Methylation
N
N
OCH3

(5.34)

Table 5.31. Tertiary thiols in food

Name	Structure	Odor threshold (μg/kg, water)	Occurrence
4-Methoxy-2-methyl-2-butanethiol	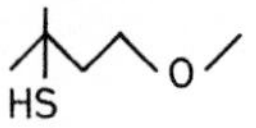	0.00002	Olive oil (cf. 14.3.2.11), black currants
3-Mercapto-3-methylbutylformate	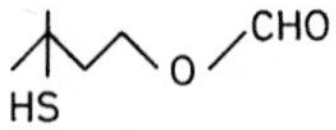	0.003	Roasted coffee
1-p-Menthen-8-thiol	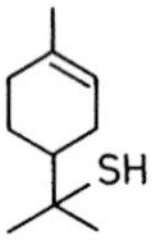	0.00002	Grapefruit

The compound 2-sec-butyl-3-methoxy-pyrazine is one of the typical aroma substances of carrots.

Pyrazines are also produced by microorganisms. For example, 2-isopropyl-3-methoxypyrazine has been identified as a metabolic by-product of *Pseudomonas perolans* and *Pseudomonas taetrolens*. This pyrazine is responsible for a musty/earthy off-flavor in eggs, dairy products and fish.

5.4 Interactions with Other Food Constituents

Aroma interactions with lipids, proteins and carbohydrates affect the retention of volatiles within the food and, thereby, the levels in the gaseous phase. Consequently, the interactions affect the intensity and quality of food aroma. Since such interactions cannot be clearly followed in a real food system, their study has been transferred to model systems which can, in essence, reliably imitate the real systems. A knowledge of the binding of aroma to solid food matrices, from the standpoint of food aromatization, aroma behavior and food processing and storage, is of great importance.

5.4.1 Lipids

In an o/w emulsion (cf. 8.15.1), the distribution coefficient, K, for aroma compounds is related to aroma activity:

$$K = \frac{C_o}{C_w} \qquad (5.35)$$

where C_o is the concentration of the aroma compound in the oil phase, and C_w the concentration of the aroma compound in the aqueous phase.

In a homologous series, e. g., n-alkane alcohols (cf. Fig. 5.27), the value of K increases with increasing chain length. The solubility in the fat or oil phase rises proportionally as the hydrophobicity imposed by chain length increases. The vapor pressure behavior is exactly the reverse; it drops as the hydrophobicity of the aroma compounds increases. The vapor pressure also drops as the volume of the oil phase increases, and the odor threshold value increases at the same time. This is well clarified in Fig. 5.28. The solubility of 2-heptanone is higher in whole milk than in skim milk which, in this case, behaves as an aqueous phase. When this phase is replaced by oil (Fig. 5.28), 2-heptanone concentration in the gas phase is the lowest.

Experiments with n-alcohols demonstrate that, with increasing chain length of volatile compounds, the migration rate of the molecules from oil to water phase increases. An increase in oil viscosity retards such migration.

5.4.2 Proteins, Polysaccharides

The sorption characteristics of various proteins for several volatile compounds are pre-

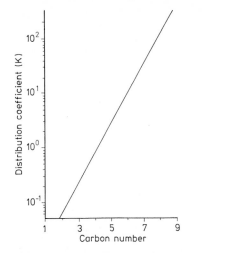

Fig. 5.27. Distribution of n-alkanols in the system oil/water (according to *McNulty* and *Karel*, 1973)

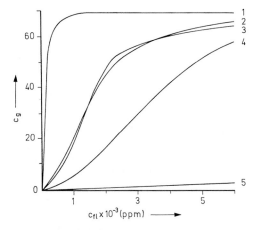

Fig. 5.28. Influence of the medium on 2-heptanone concentration in the gas phase (according to *Nawar*, 1966). 2-Heptanone alone (1), in water (2), in skim milk (3), in whole milk (4), in oil (5). c_{fl}: concentration in liquid; c_g: concentration in gas phase (detection signal height from headspace analysis)

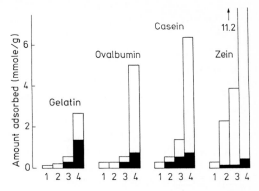

Fig. 5.29. Sorption of volatile compounds on proteins at 23 °C (according to *Maier*, 1974). Hexane (1), ethyl acetate (2), acetone (3), ethanol (4). □ plus ■: maximal sorption, ■: after desorption

When a biopolymer, B, has a group which attracts and binds the aroma molecule, A, then the following equation is valid:

$$K = \frac{[BA]}{C_f[B]} \tag{5.36}$$

where K = a single binding constant; and C_f = concentration of free aroma compound molecules.

$$[BA] = KC_f[B] \tag{5.37}$$

To calculate the average number of aroma molecules bound to a biopolymer, the specific binding capacity, r, has to be introduced:

$$r = \frac{[BA]}{([B] + [BA])} \tag{5.38}$$

The concentration of the complex BA from Equation 5.37 is substituted in Equation 5.38:

$$r = \frac{KC_f[B]}{([B] + K[B]C_f)} = \frac{KC_f}{1 + KC_f} \tag{5.39}$$

When a biopolymer binds not only one molecular species (as A in the above case) but has a number (n) of binding groups (or sites) equal in binding ability and independent of each other, then r has to be multiplied by n, and Equation 5.39 acquires the form:

$$r = \frac{nKC_f}{1 + KC_f} \tag{5.40}$$

sented in Fig. 5.29. Ethanol is bound to the greatest extent, probably with the aid of hydrogen bonds. The binding of the nonpolar aroma compounds probably occurs on the hydrophobic protein surface regions. A proposal for the evaluation of data on the sorption of aroma volatiles on a biopolymer (protein, polysaccharide) is based on the law of mass action.

$$\frac{r}{C_f} = Kn - Kr = K' - Kr \qquad (5.41)$$

where K' = overall binding constant.
The evaluation of data then follows Equation 5.41 presented in graphic form, i. e. a diagram of $r/C_f = f(r)$. Three extreme or limiting cases should be observed:

a) A straight line (Fig. 5.30, a) indicates that only one binding region on a polymer, with n binding sites (all equivalent and independent from each other) is involved. The values n and K' are obtained from the intersection of the straight line with the abscissa and the ordinate, respectively.

b) A straight line parallel to the abscissa (Fig. 5.30, b) is obtained when the single binding constant, K, is low and the value of n is very high. In this special case, Equation 5.41 has the form:

$$r = K'C_f \qquad (5.42)$$

c) A curve (Fig. 5.30, c) which in approximation is the merging of two straight lines, as shown separately (Fig. 5.30, d). This indicates two binding constants, K'_1, and K'_2, and their respective binding groups, n_1 and n_2, which are equivalent and independent of each other.

By plotting r versus C_f, values of K' are obtained from the slope of the curve. An example for a model system with two binding regions (case c in Fig. 5.30) is given by aroma binding to starch. It should be remembered that starch binds the volatiles only after gelatinization by trapping the volatiles in its helical structure, and that starch is made up of two constituents, amylose and amylopectin. The binding parameters are listed for some aroma compounds in Table 5.32. Numerous observations indicate that K'_1 and binding region n_1 are related to the inner space of the helix, while K'_2 and the n_2 region are related to the outer surface of the helix. K'_1 is larger than K'_2, which shows that, within the helix, the aroma compounds are bound more efficiently to glucose residues of the helix. The fraction $1/n$ is a measure of the size of the binding region. It decreases, as expected, with increasing molecular weight of alkyl alcohols, but it is still

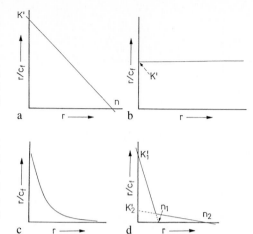

Fig. 5.30. Binding of aroma compounds by biopolymers. Graphical determination of binding parameters (according to Solms, 1975)

Table 5.32. Binding of aroma compounds by potato starch

Compounds	Binding constant			
	K'_1	n_1	K'_2	n_2
1-Hexanol	$5.45 \cdot 10^1$	0.10	–	–
1-Octanol	$2.19 \cdot 10^2$	0.05	$2.15 \cdot 10^1$	0.11
1-Decanol	$1.25 \cdot 10^2$	0.04	$1.29 \cdot 10^1$	0.11
Capric acid	$3.30 \cdot 10^2$	0.07	$4.35 \cdot 10^1$	0.19
Menthone	$1.84 \cdot 10^2$	0.012	8.97	0.045
Menthol	$1.43 \cdot 10^2$	0.007	–	–
β-Pinene	$1.30 \cdot 10^1$	0.027	1.81	0.089

larger within the helix than on the outer surface. Altogether, it should be concluded that, within a helix, the trapped compound cannot fulfill an active role as an aroma constituent.
An unlimited number of binding sites exist in proteins dissolved or dispersed in water (case b, Fig. 5.30). K' values for several aroma compounds are given in Table 5.33. The value of the constant decreases in the order of aldehydes, ketones, alcohols, while compounds such as dimethylpyrazine and butyric acid are practically unable to bind. In the case of aldehydes, it must be assumed that they can react with free amino- and SH-groups. The high values of K' can reflect other than secondary forces.

Table 5.33. Binding of aroma compounds by proteins (0.4% solutions at pH 4.5)

Aroma compound	Total binding constant $K' \cdot 10^{-3}$ ($l\,mol^{-1}$)			
	Bovine serum albumin		Soya protein	
	20°C	60°C	20°C	60°C
Butanal	9.765	11.362	10.916	9.432
Benzaldehyde	6.458	6.134	5.807	6.840
2-Butanone	4.619	5.529	4.975	5.800
1-Butanol	2.435	2.786	2.100	2.950
Phenol	3.279	3.364	3.159	3.074
Vanillin	2.070	2.490	2.040	2.335
2,5-Dimethyl pyrazine	0.494			
Butyric acid	0			

Bovine serum albumin and soya proteins are practically identical with regard to the binding of aroma compounds (Table 5.33). Since both proteins have a similar hydrophobicity, it is apparent that hydrophobic rather than hydrophilic interactions are responsible for aroma binding in proteins.

5.5 Natural and Synthetic Flavorings

Aromatized food has been produced and consumed for centuries, as exemplified by confectionery and baked products, and tea or alcoholic beverages. In recent decades the number of aromatized foods has increased greatly. In Germany, these foods account for about 15–20% of the total food consumption. A significant reason for this development is the increase in industrially produced food, which partly requires aromatization because certain raw materials are available only to a limited extent and, therefore, expensive or because aroma losses occur during production and storage. In addition, introduction of new raw materials, e. g., protein isolates, to diversify or expand traditional food sources, or the production of food substitutes is promising only if appropriate aromatization processes are

available. Aroma concentrates, essences, extracts and individual compounds are used for aromatization. They are usually blended in a given proportion by a flavorist; thus, an aroma mixture is "composed". The empirically developed "aroma formulation" is based primarily on the flavorist's experience and personal sensory assessment and is supported by the results of a physico-chemical aroma analysis. Legislative measures that regulate food aromatization differ in various countries.

At present, non-alcoholic beverages occupy the first place among aromatized foods (Table 5.34). Of the different types of aroma, citrus, mint and red fruit aromas predominate (Table 5.35).

Table 5.34. Use of aromas in the production of foods

Product group	Percentage (%)[a]
Non-alcoholic beverages	38
Confectionery	14
Savoury products[b], snacks	14
Bread and cakes	7
Milk products	6
Desserts	5
Ice cream	4
Alcoholic beverages	4
Others	8

[a] Approximate values.
[b] Salty product line like vegetables, spices, meat.

Table 5.35. Types of aroma used

Aroma type	Percentage (%)[a]
Citrus	20
Mint	15
Red fruits	11
Vanilla	10.5
Meat	10.5
Spices	8.5
Chocolate	8.5
Cheese	5.5
Nut	2.5
Others	8

[a] Approximate values.

5.5.1 Raw Materials for Essences

In Germany, up to about 60% of the aromas used for food aromatization are of plant origin and, thus, designated as "natural aroma substances". The rest of the aroma compounds are synthetic, but 99% of this portion is chemically identical to their natural counterparts. Only 1% are synthetic aroma compounds not found in nature.

5.5.1.1 Essential Oils

Essential (volatile) oils are obtained preferentially by steam distillation of plants (whole or parts) such as clove buds, nutmeg (mace), lemon, caraway, fennel, and cardamon fruits (cf. 22.1.1.1). After steam distillation, the essential oil is separated from the water layer, clarified and stored. The pressure and temperature used in the process are selected to incur the least possible loss of aroma substances by thermal decomposition, oxidation or hydrolysis.

Many essential oils, such as those of citrus fruits, contain terpene hydrocarbons which contribute little to aroma but are readily auto-oxidized and polymerized ("resin formation"). These undesirable oil constituents (for instance, limonene from orange oil) can be removed by fractional distillation. Fractional distillation is also used to enrich or isolate a single aroma compound. Usually, this compound is the dominant constituent of the essential oil. Examples of single aroma compounds isolated as the main constituent of an essential oil are: 1,8-cineole from eucalyptus, 1(−)-menthol from peppermint, anethole from anise seed, eugenol from clove, or citral (mixture of geranial and neral, the pleasant odorous compounds of lemon or lime oils) from litsea-cuba.

5.5.1.2 Extracts, Absolues

When the content of essential oil is low in the raw material or the aroma constituents are destroyed by steam distillation or the aroma is lost by its solubility in water, then the oil in the raw material is recovered by an extraction process. Examples are certain herbs or spices (cf. 22.1.1.1) and some fruit powders. Hexane,

triacetin, acetone, ethanol, water and/or edible oil or fat are used as solvents. Good yields are also obtained by using liquid CO_2. The volatile solvent is then fully removed by distillation. The oil extract (resin, absolue) often contains volatile aroma compounds in excess of 10% in addition to lipids, waxes, plant pigments and other substances extractable by the chosen solvent. Extraction may be followed by chromatographic or counter-current separation to isolate some desired aroma fractions. If the solvent used is not removed by distillation, the product is called an extract. The odor intensity of the extract, compared to the pure essential oil, is weaker for aromatization purposes by a factor of 10^2 to 10^3.

5.5.1.3 Distillates

The aroma constituents in fruit juice are more volatile during the distillation concentration process than is the bulk of the water. Hence, the aroma volatiles are condensed and collected (cf. 18.2.11). Such distillates yield highly concentrated aroma fractions through further purification steps.

5.5.1.4 Microbial Aromas

Cheese aroma concentrates offered on the market have an aroma intensity at least 20-fold higher than that of normal cheese. They are produced by the combined action of lipases and *Penicillium roqueforti* using whey and fats/oils of plant origin as substrates. In addition to C_4–C_{10} fatty acids, the aroma is determined by the presence of 2-heptanone and 2-nonanone.

5.5.1.5 Synthetic Natural Aroma Compounds

In spite of the fact that a great number of food aroma compounds have been identified, economic factors have resulted in only a limited number of them being synthesized on a commercial scale. Synthesis starts with a natural compound available in large amounts at the right cost, or with a basic chemical. Several examples will be considered below.

A most important aroma compound worldwide, vanillin, is obtained primarily by alkaline hydrolysis of lignin (sulfite waste of the wood pulp industry), which yields coniferyl alcohol.

It is converted to vanillin by oxidative cleavage:

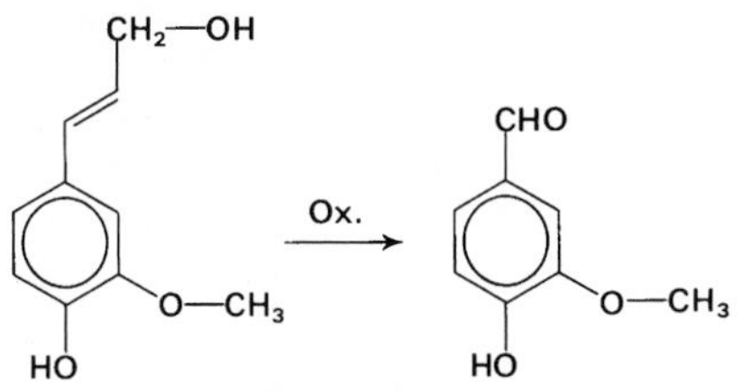

(5.43)

A distinction can be made between natural and synthetic vanillin by using quantitative ^{13}C analysis (cf. 18.4.3). The values in Table 5.36 show that the ^{13}C distribution in the molecule is more meaningful than the ^{13}C content of the entire molecule.

The most important source of citral, used in large amounts in food processing, is the steam-distilled oil of lemongrass (*Cymbopogon flexuosus*). Citral actually consists of two geometrical isomers: geranial (I) and neral (II). They are isolated from the oil in the form of bisulfite adducts:

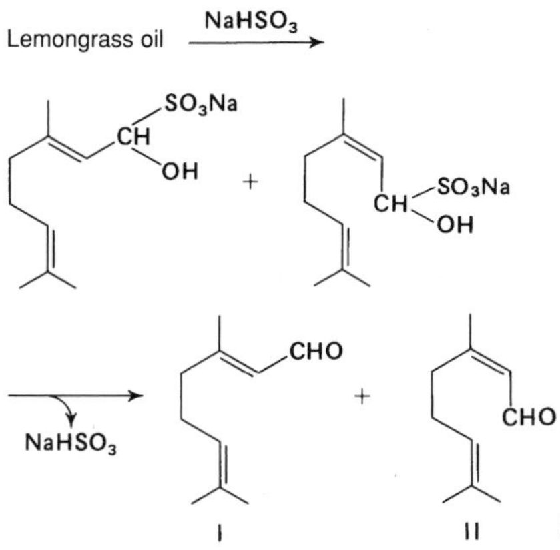

(5.44)

Table 5.36. Site-specific ^{13}C isotopic analysis of vanillin from different sources

Source	R (%)[a] in			R (%)[a] total
	CHO	Benzene ring	OCH$_3$	
Vanilla	1.074	1.113	1.061	1.101
Lignin	1.062	1.102	1.066	1.093
Guaiacol	1.067	1.102	1.026	1.089

[a] R ($^{13}C/^{12}C$) was determined by site-specific natural isotope fractionation NMR (SNIF-NMR). Standard deviation: 0.003–0.007.

The aroma compound menthol is primarily synthesized from petrochemically obtained m-cresol. Thymol is obtained by alkylation and is then further hydrogenated into racemic menthol:

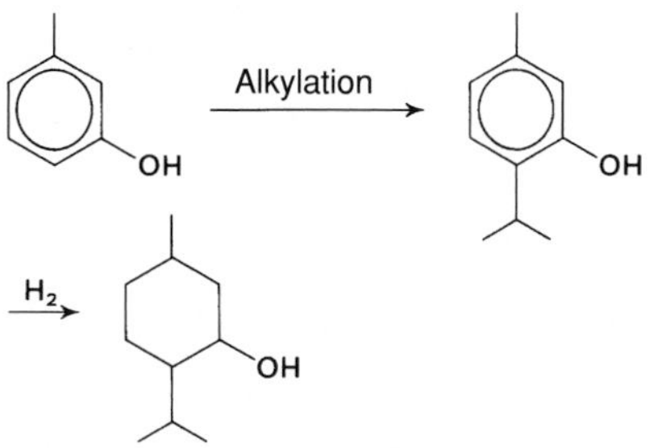

(5.45)

A more expensive processing step then follows, in which the racemic form is separated and 1(−)-menthol is recovered. The d-optical isomer substantially decreases the quality of the aroma (cf. 5.3.2.4).

The purity requirement imposed on synthetic aroma substances is very high. The purification steps usually used are not only needed to meet the stringent legal requirements (i.e. beyond any doubt safe and harmless to health), but also to remove undesirable contaminating aroma compounds. For example, menthol has a phenolic off-flavor note even in the presence of only 0.01% thymol as an impurity. This is not surprising since the odor threshold value of thymol is lower than that of 1(−)-menthol by a factor of 450.

5.5.1.6 Synthetic Aroma Compounds

Some synthetic flavorings which do not occur in food materials are compiled in Table 5.37. Of these compounds ethyl vanillin is of greater importance.

5.5.2 Essences

The flavorist composes essences from raw materials. In addition to striving for an optimal aroma, the composition of the essence has to meet food processing demands, e. g., compensation for possible losses during heating. The "aroma formulation" is an empirical one, developed as a result of long experience deal-

Table 5.37. Synthetic Flavoring Materials (not naturally occurring in food)

Name	Structure	Aroma description
Ethyl vanillin	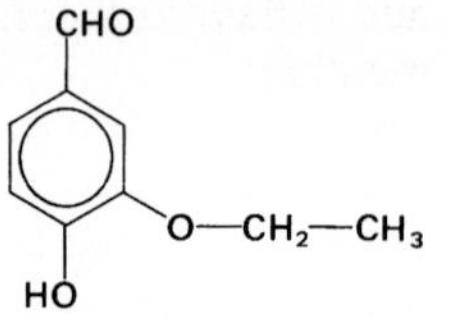	Sweet like vanilla (2 to 4-times stronger than vanillin)
Ethyl maltol	cf. 5.3.1.2	Caramel-like
Musk ambrette	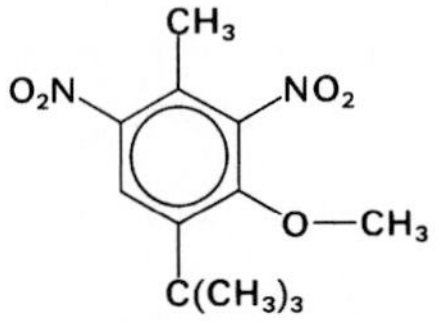	Musk-like
Allyl phenoxyacetate	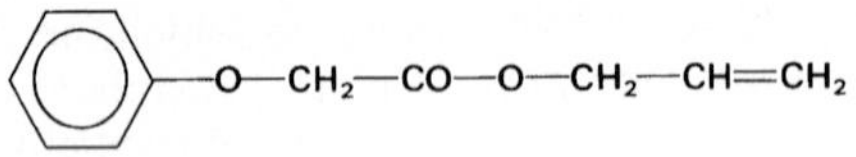	Fruity, pineapple-like
α-Amyl cinnamic-aldehyde	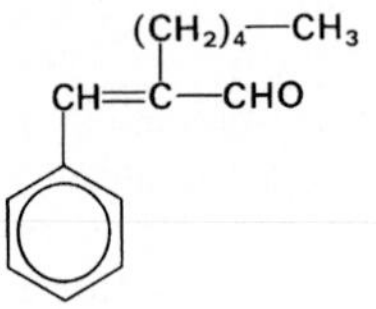	Floral, jasmin and lilies
Hydroxycitronellal	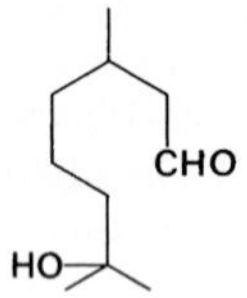	Sweet, flowery, liliaceous
6-Methyl coumarin	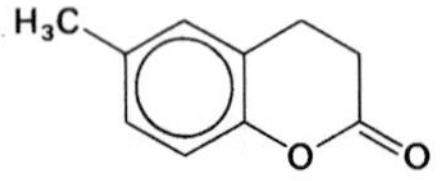	Dry, herbaceous
Propenylguaethol (vanatrope)	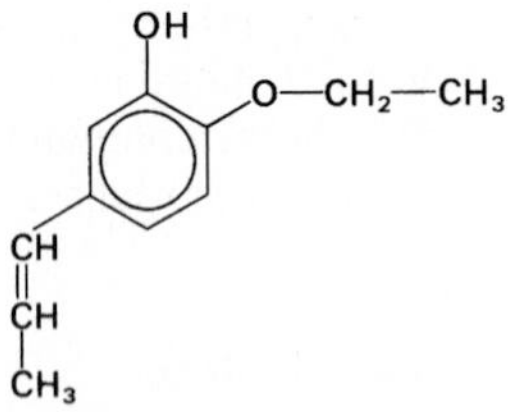	Phenolic, anise-like
Piperonyl isobutyrate	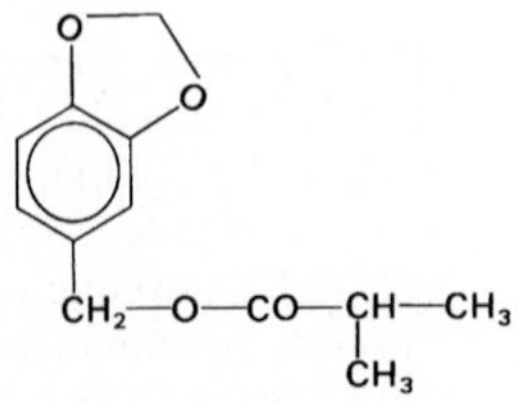	Sweet, fruity, like berry fruits

Table 5.38. Formulations for pineapple essence

From naturally occurring raw materials:	From synthetic aroma compounds (Continued):
686 g pineapple fruit juice concentrate	20.0 g diethyl sebacate
300 g pineapple shell distillate	16.4 g allyl cyclohexyl propionate
10 g orange oil[a]	16.0 g ethyl propionate
2 g oil of wine yeast[a]	13.0 g ethyl heptanoate
2 g camomile oil (*Matricaria chamomilla*)	8.0 g butyric acid
	5.6 g vanillin
	4.0 g citronellyl butyrate
1,000 g	2.5 g methyl allyl caproate
	2.0 g methyl-β-methyl thiopropionate
From synthetic aroma compounds:	2.0 g allyl phenoxy-acetate
376 g ethyl acetate	1.0 g methyl caprylate
112 g amyl butyrate	0.6 g citral
105 g butyl acetate	0.3 g cinnamyl acetate
45 g ethyl butyrate	0.1 g bornyl acetate
36 g ethyl isovalerate	162 g solvent
28.6 g amyl acetate	
22.5 g orange oil	1,000 g
21.4 g allyl caproate	

[a] Aroma compounds content approx. 1%.

ing with many problems, disappointments and failures, and is rigorously guarded after the "know-how" is acquired. Based on these facts, the example given in Table 5.38 provides a formulation for pineapple essence only in principle.

5.5.3 Aromas from Precursors

The aroma of food that has to be heated, in which the impact aroma compounds are generated by the *Maillard* reaction, can be improved by increasing the levels of precursors involved in the reaction. This is a trend in food aromatization. Some of the precursors are added directly (cf. processed flavors for meat, Table 12.23), while some precursors are generated within the food by the preliminary release of the reaction components required for the *Maillard* reaction (cf. 4.2.4.4). This is achieved by adding protein and polysaccharide hydrolases to food.

5.5.4 Stability of Aromas

Aroma substances can undergo changes during the storage of food. Aldehydes and thiols are especially sensitive because they are easily oxidized to acids and disulfides respectively. Moreover, unsaturated aldehydes are degraded by reactions which will be discussed using 2(E)-hexenal and citral as examples. These two aldehydes are important aromatization agents for leaf green and citrus notes.

In an apolar solvent, e.g., a triacetin, 2(E)-hexenal oxidizes mainly to 2(E)-hexenoic acid, with butyric acid, valeric acid and 2-penten-1-ol being formed as well. The following reaction pathway leads to the acids:

$$CH_3-(CH_2)_2-CH=CH-CHO \ (RH)$$

$$CH_3-(CH_2)_2-CH=CH-C\begin{smallmatrix}O\\O-OH\end{smallmatrix}$$

$$2\,CH_3-(CH_2)_2-CH=CH-COOH$$

$$CH_3-(CH_2)_2-CH=CH-OH$$

$$CH_3-(CH_2)_2-CH_2-CHO$$

$$CH_3-(CH_2)_2-CH_2-COOH \qquad (5.46)$$

In[•]: Start radical

At the acidic pH values found in fruit, autoxidation decreases, 2(E)-hexenal preferentially adds water with the formation of 3-hydroxy-hexanal. In addition, the double bond is isomerized with the formation of low concentrations of 3(Z)-hexenal. As a result of its low threshold value, 3(Z)-hexenal influences the aroma to a much greater extent than 3-hydroxyhexanal which has a very high threshold.

Citral is also instable in an acidic medium, e.g., lemon juice. At citral equilibrium, which consists of the stereoisomers geranial and neral in the ratio of 65 : 35, neral reacts as shown in Formula: 5.47. It cyclizes to give the

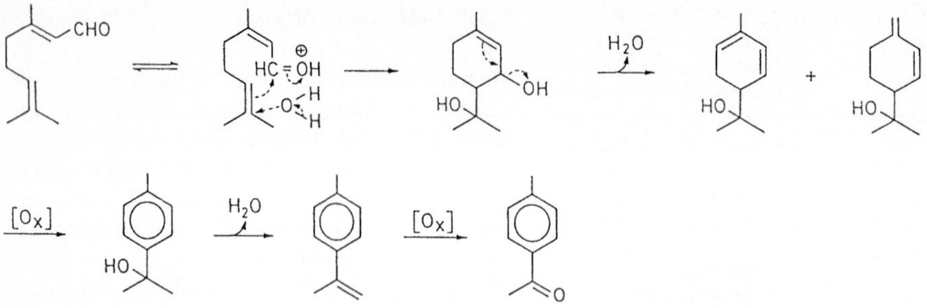

$$(5.47)$$

labile p-menth-1-en-3,8-diol which easily eliminates water, forming various p-menthadien-8-ols. This is followed by aromatization with the formation of p-cymene, p-cymen-8-ol, and α,p-dimethylstyrene. p-Methylacetophenone is formed from the last mentioned compound by oxidative cleavage of the Δ^8-double bond. Together with p-cresol, p-methylacetophenone contributes appreciably to the off-flavor formed on storage of lemon juice. Citral is also the precursor of p-cresol.

In citrus oils, limonene and γ-terpinene are also attacked in the presence of light and oxygen. Carvone and a series of limonene hydroperoxides are formed as the main aroma substances.

5.5.5 Encapsulation of Aromas

Aromas can be protected against the chemical changes described in 5.5.4 by encapsulation. Materials suitable for inclusion are polysaccharides, e.g., gum arabic, maltodextrins, modified starches, and cyclodextrins. The encapsulation proceeds via spray drying, extrusion or formation of inclusion complexes.

For spray drying, the aroma substances are emulsified in a solution or suspension of the polysaccharide, which contains solutizer in addition to the emulsifying agent.

In preparation for extrusion, a melt of wall material, aroma substances, and emulsifiers is produced. The extrusion is conducted in a cooled bath, e.g., isopropanol.

β-Cyclodextrins, among other compounds, can be used for the formation of inclusion complexes. Together with the aroma substances, they are dissolved in a water/ethanol mixture by heating. The complexes precipitate out of the cooled solution and are removed by filtration and dried.

Criteria for the evaluation of encapsulated aromas are: stability of the aroma, concentration of aroma substance, average diameter of the capsules and, amount of aroma substance adhering to the surface of the capsule.

5.6 Relationships Between Structure and Odor

5.6.1 General Aspects

The effect of stimulants on the peripheral receptors of an organism results in responses that are characterized by their *quality* and their *intensity*. The intensity is quantifiable, e.g., by determining odor threshold values (cf. 5.1.3). The quality can be described only by comparison. Odor stimulants can be grouped into those of the same or similar qualities. For example, primary quality or modality is a quality which is recognized as being homogeneous, i.e. it is not reproducible by two or more nonidentical stimulants being superimposed.

The stimulus is triggered in the olfactory zone (*Regio olfactoria*) of the nasal cavity. At the membranes of the olfactory hairs (*Cilia olfactoria*), contact occurs between odor substances and integral glycoproteins of type gp95. The resulting change in conformation of the membrane protein triggers the reaction cascade described for sweet substances (cf. 8.1.2).

5.6.2 Important Structural Elements

The question that is the subject of many studies is which structural elements control the perception of odor and taste. Several references have been made to structure-activity relationships of this type (SAR: structure-activity relationship, QSAR: quantitative structure-activity relationship) in this chapter and in other chapters on taste substances (cf. 1.2.6, 4.2.3, 8.8.1.1). Some aspects will be summarized here again.

The action of *odor substances* is influenced by:

- the geometry of the molecule and
- the functional groups.

5.6.2.1 Molecular Geometry

The significance and meaning of a compound's geometry is demonstrated, for example, by several camphor-like compounds:

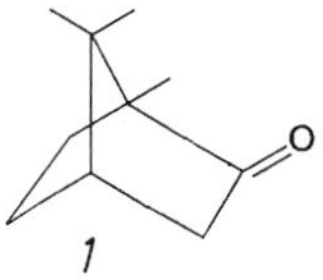

1 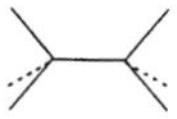   2   3   4   (5.48)

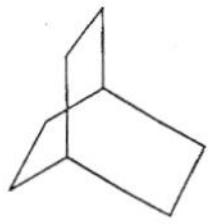

2,2,3,3-tetramethylbutane (*2*) and bicyclooctane (*3*) have an odor very close to that of camphor (*1*). Also, cyclooctane (*4*) is camphorlike, though not as clearly as *2* and *3*. These examples strongly suggest that functional groups have no decisive importance for the camphor-like odor, but the molecular shape or geometry of the molecule, which is close to that of a sphere, does.

A further indication of the great importance of the molecular geometry is the possible substitution of groups within the molecule by other groups having similar *van der Waals* radii (isosteric substitution) (e. g., H ↔ F; CH$_3$ ↔ Br). This substitution has basically no effect on the odor quality (cf. Formula 5.49).

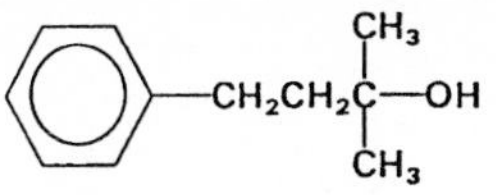

5

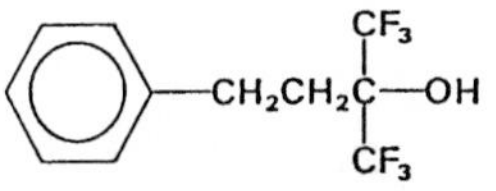

6

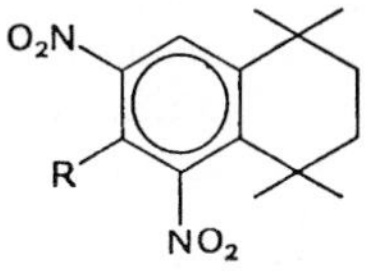

7 : R = Me
8 : R = Br

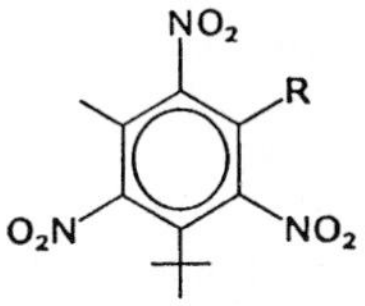

9 : R = Me
10 : R = Br

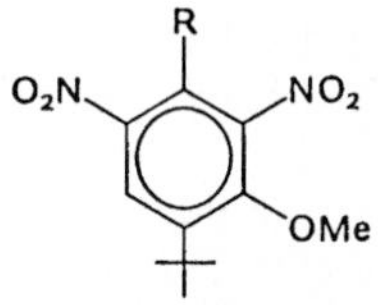

11 : R = Me
12 : R = Br

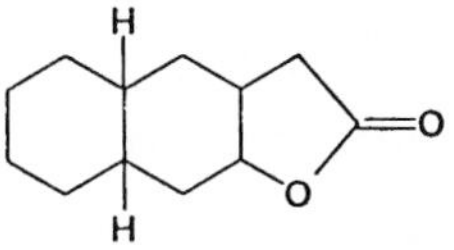

13

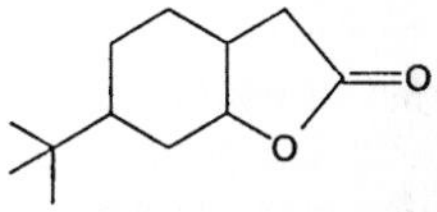

14

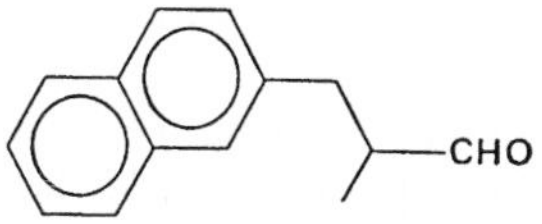

15

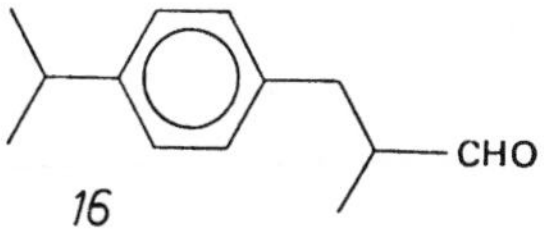

16

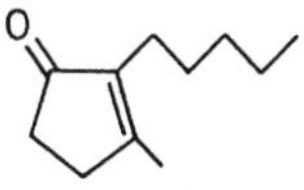

17 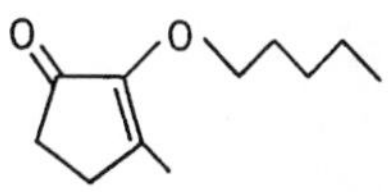   18   (5.49)

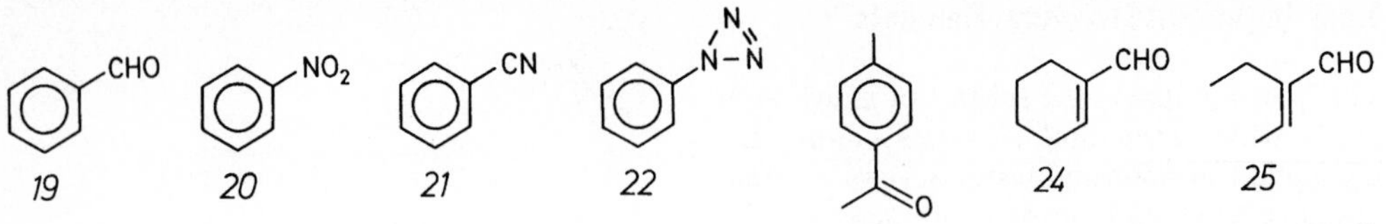

(5.50)

Both compounds, *5* and *6* are similar in odor though the fluoro-derivative *6* is slightly more herbaceous. The musky odor character is retained in compounds *7/8* and *9/10* while it is absent in compounds *11/12*. The sandalwood odor is retained by compounds *13/14* when the cis-decalin group is replaced by a t-butyl-cyclohexyl group. Also the pairs *15/16* and *17/18* (cf. Formula 5.49) have the same odors, respectively.

The bitter almond odor typical of benzaldehyde (*19* in Formula: 5.50) is also maintained when the aldehyde group is replaced by isosteric groups, e.g., nitrobenzene (*20*), benzonitrile (*21*) and benzo-1,2,3-triazole (*22*). As shown by the bitter almond-like compounds p-methylacetophenone (*23*), tetrahydrobenzaldehyde (*24*) and 3-formylpentene-2 (*25*), the H-atom of the aldehye group can be substituted by a methyl group and changes in the hydrophobic part of the molecule are also possible without losing the typical taste provided that a double bond remains in conjugation with the carbonyl group.

The need to maintain the molecule's dipole moment in such isosteric substitutions is important since the dipole moment is responsible for the required orientation of the molecule on the odor's chemoreceptor site. cis-t-Butylcyclohexyl acetate (*26*) has an intensive specific odor, while its trans-isomer (*27*) has only a weak and flat odor (cf. Formula 5.51).

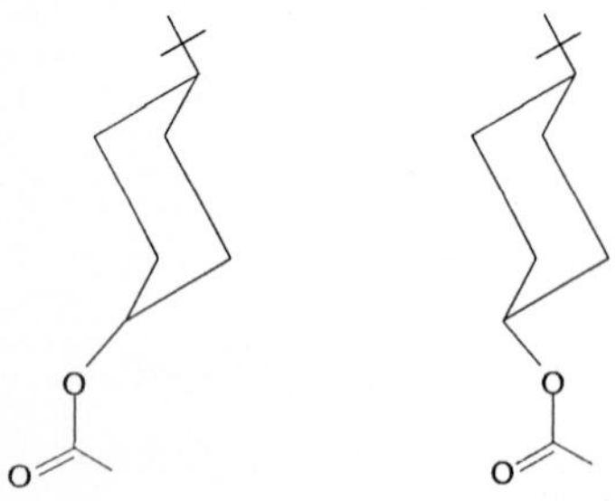

(5.51)

(Z)-3-Hexenal (*28* in Formula: 5.52) has a green odor and, on the other hand, (E)-3-hexenal (*29*) smells greasy.

(5.52)

The fact that a similar type of molecular geometry supports a similar type of odor is demonstrated in the case of limonene (*30*). Both isomers of p-methene-8 differ clearly: *31* has a characteristic orange odor (like limonene), while *32* has the flat odor of a hydrocarbon.

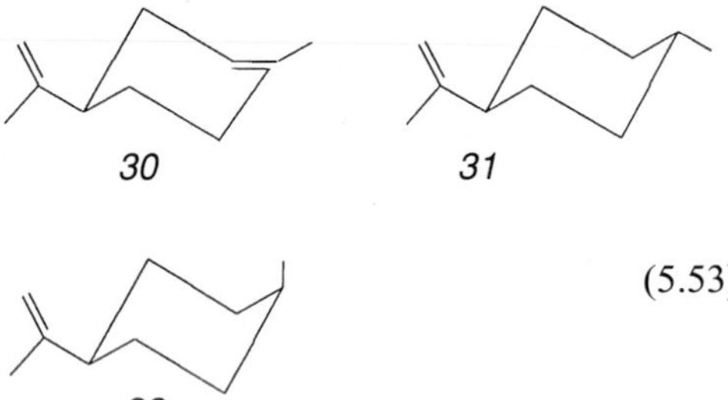

(5.53)

5.6.2.2 Functional Groups

Functional groups are not at all essential for an aroma compound, as illustrated by hydrocarbons. These compounds have a specific odor though they have no functional group, again illustrating the importance of molecular geometry. On the other hand, there are compounds, such as NH_3, H_2S, and CH_3SH, which consist of only one functional group and their smell is extremely intensive. In this case consideration of steric factors makes no sense; the functional group is obviously and solely the determinant.

An increase in molecular size can result in a decrease in the effect of the functional group and a concomitant increase in the influence of molecular orientation, due to its dipole moment, on the perception of odor quality. Actually, both odor quality and odor intensity are influenced indirectly. This is exemplified by the following compounds: R_2S, R_2NH, R_2O. For R = Me, the odor quality is distinctly different. For R^1 = Me, R^2 = PhEt, differences exist but they are not distinct, as in the case of ether and thioether, which have the same odor qualities.

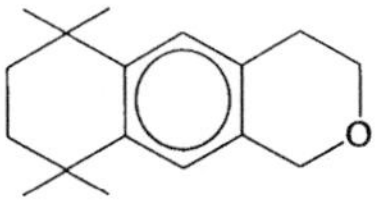

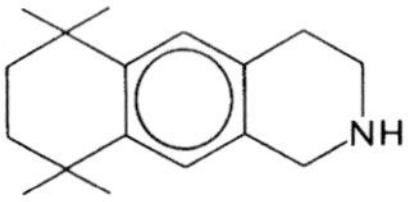

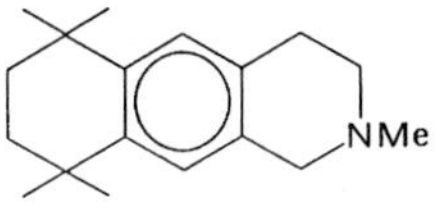

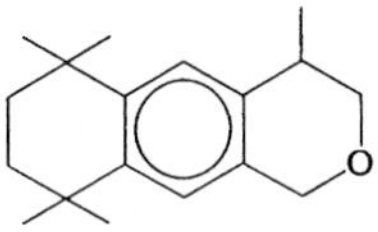

39 : R = OH   41 : R = OH
40 : R = OH   42 : R = SH
43

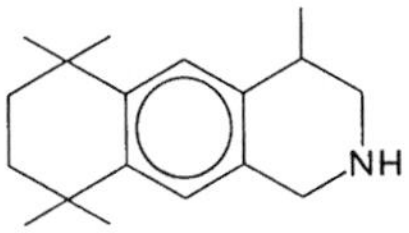

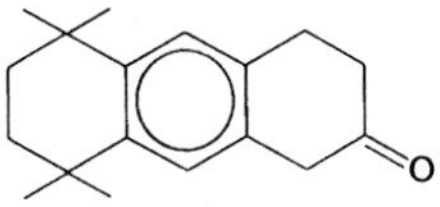

(5.55)

(5.54)

Further examples are found among the odorous substances of musk. It is possible to retain the original odor quality by substituting the oxygen in (33) with NH (34) or MeN (35); however, the odor intensity is decreased. The same is true for 36 and 37. Substitution of oxygen in 36 with CO (38) results in an odor loss. The latter is the result of a lower volatility of the new compound and change in its dipole moment.

However, replacement of oxygen by sulfur can produce a complete change in quality, even in larger molecules. Examples are the lilac-like α-terpineol (39 in Formula: 5.55), 1-p-menthen-8-thiol (40) which is fruity on dilution, minty 8-hydroxy-p-menthen-3-one (41), and the 8-thio compound that smells of black currants (42). With an altered odor profile, sila-α-terpineol (43) has the same flowery note as the carbon analog.

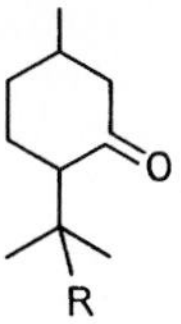

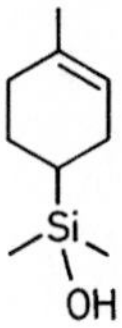

The surroundings of functional groups within a molecule are important since they influence the accessibility of these groups. Camphor odor is quite well imitated by the compound methyl isopropyl ketone. In this case, the surroundings of the functional group are very similar. Several examples of the musk-like compounds can also verify the involvement of the surroundings of the functional group as a factor in aroma perception (cf. Formula 5.56).

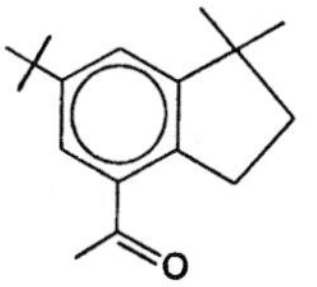

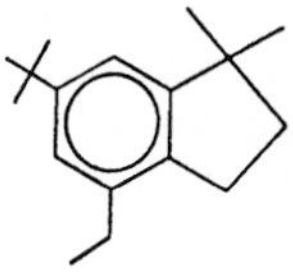

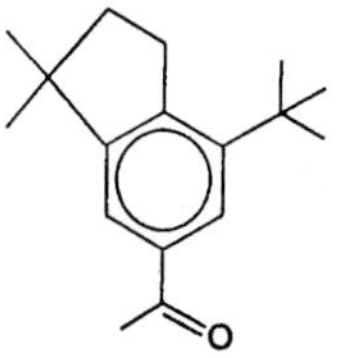

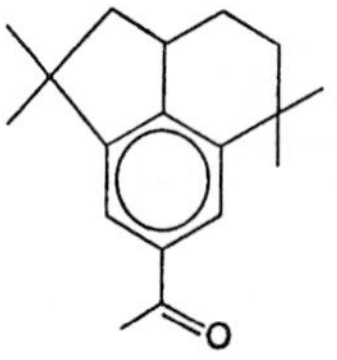

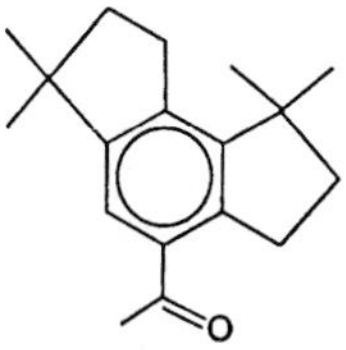

(5.56)

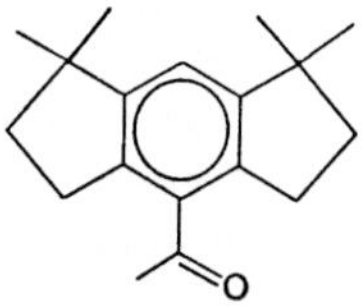

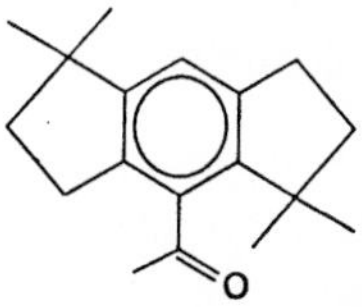

The CO group in *44* is required for a musk-like odor, since *45* does not have such an odor. The quaternary C-atoms at positions 3 and 5 are important for odor intensity. Positions 2 and 6 interfere with accessibility to the CO group and also with the coplanarity of the molecule. Compounds *46–48* have a strong musk-like odor, while *49* and *50* have a very weak musk-like odor.

Examples showing the influence of different positions of a functional group on the odor quality are β-ionone with an odor of violets (*51* in Formula 5.57) and β-damascone (*52*) with a fruity camphor-green odor.

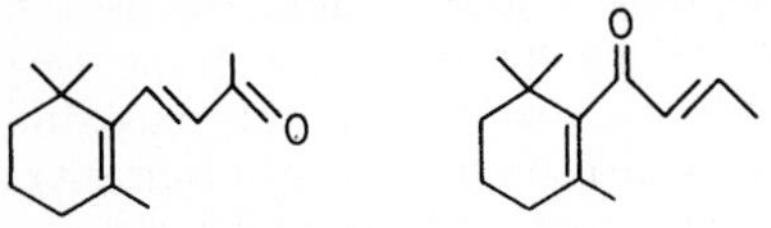

(5.57)

51   *52*

In compounds with several functional groups, their relative positions can also be of importance for odor quality. An interchange of hydroxy and methoxy groups in vanillin (*53* in Formula: 5.58) results in practically odorless isovanillin (*54*).

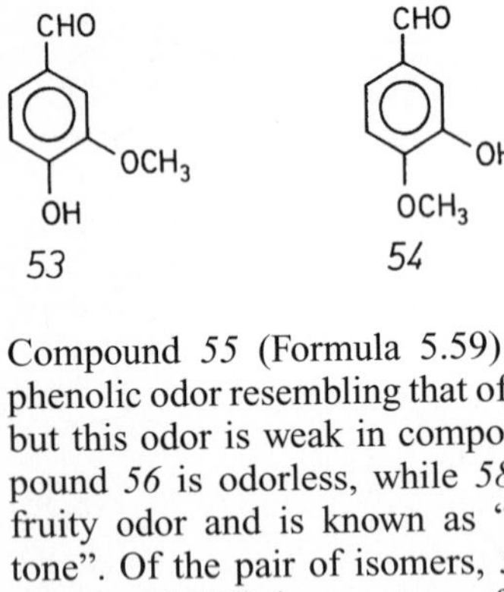

(5.58)

CHO   CHO

OCH₃

OH   OH
              OCH₃

53   *54*

Compound *55* (Formula 5.59) has a strong phenolic odor resembling that of salicylic acid, but this odor is weak in compound *57*. Compound *56* is odorless, while *58* has a strong fruity odor and is known as "raspberry ketone". Of the pair of isomers, *59/60*, the cis-compound (*59*) has a strong flowery-woody odor note, while the trans-isomer (*60*) is odorless. Very probably, both functional groups interact as a bifunctional unit with the corresponding chemoreceptor site. The OH-groups, in this case, play the role of proton donors (HA-group), while the CO-group functions as a proton acceptor (B-group). In many additional compounds with differing odor qualities, it has been found that the distance between proton donor and proton acceptor must be less than 3 Å. The odor quality in these compounds appears to be determined by the hydrophobic part of the molecule which, as the third X group, establishes the hydrophobic bond with the receptor. It is of interest that this AH/B/X-system postulated for a number of aroma compounds has stood the test for more than twenty years in the discussion of structure-activity relationships of sweet compounds (cf. 4.2.3 and 8.8.1.1).

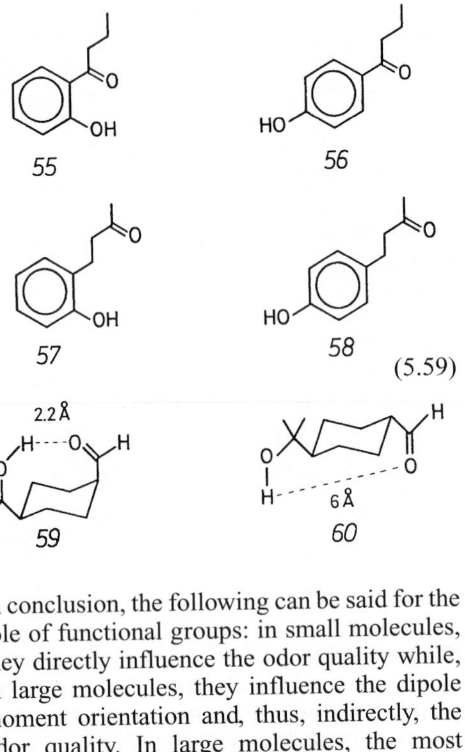

55   *56*

57   *58*

(5.59)

2.2Å

59   *60*

In conclusion, the following can be said for the role of functional groups: in small molecules, they directly influence the odor quality while, in large molecules, they influence the dipole moment orientation and, thus, indirectly, the odor quality. In large molecules, the most important influence is derived from the molecular geometry. It appears that both parameters, geometry and functional groups, cannot be separated.

5.6.2.3 Chirality

The acceptance of the above mentioned stimulant-receptor interaction over at least three points shows its usefulness in the case of chiral compounds as well. As a rule, these compounds have greatly varying aroma notes and/or threshold values. Of the numerous isomers of 2-isopropyl-5-methyl cyclohexanol, only (1R,3R,4S)-menthol (*61* in Formula: 5.60) has a fresh, minty odor which is associated with a strong cooling effect and only (S)-limonene (*62*) exhibits the typical citrus odor. Of the enantiomeric carvones, the (S)-compound (*63*) has the odor of caraway while the (R)-compound has a spicy odor. The typical aroma of yellow passion fruit is due to (2S,4R)-2-methyl-4-propyl-1,3-oxathiane (*64*) whereas the other isomers have a nonspecific sulfurous, herb-like odor. (R)-1-Octen-3-ol (*65*) is responsible for the mushroom aroma, its enrichment in champignon mushrooms ee = 97.2% and in chanterelle ee = 89%.

Of the linalool enantiomers, the (R)-compound (*66*) has a clearly lower threshold value than the (S)-compound.

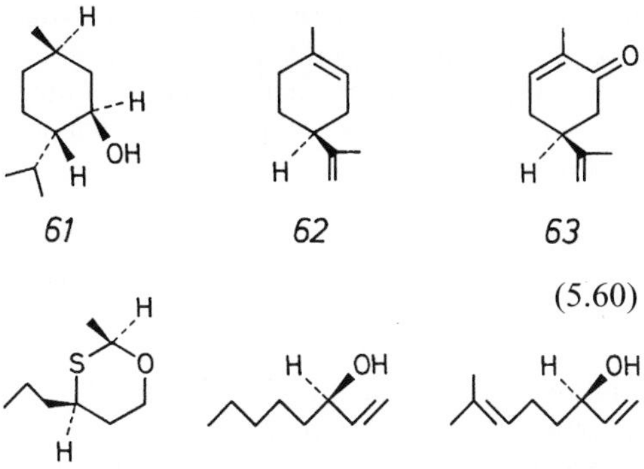

(5.60)

5.6.3 Individual Modalities or Qualities

5.6.3.1 Amber-like Compounds

The number of primary odor qualities is disputed. Some examples are reproduced in Table 5.39. Each primary quality or modality is illustrated by a compound(s) which well represents the specified odor quality.

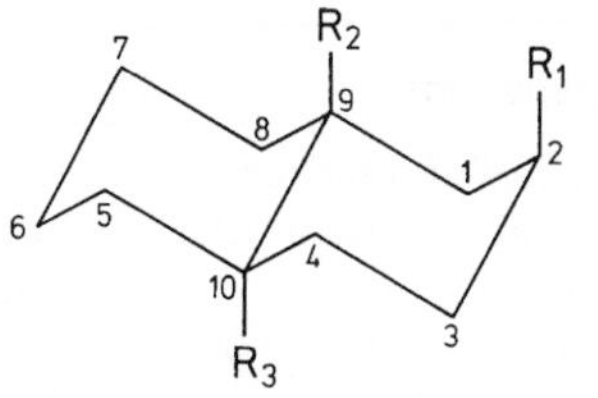

(5.61)

67

Table 5.39. Examples of primary qualities of odors

Quality	Characteristic compound
I. According to *Amoore*	
Flowery	Phenylethyl-methyl-ethyl carbinol
Ethereal	Propanol, 1,2-dichloroethane
Camphor-like	Camphor
Musk-like	Musk xylol
Mint-like	Menthone
Putrid	Dimethylsulfide
Pungent	Acetic acid, formic acid
II. According to *Harper*	
Flowery	β-Ionone
Ethereal	Propanol
Sweet	Vanillin
Rancid	Butyric acid
Sulfurous, garlic-like, onion-like	Diethylsulfide
Almond-like	Benzaldehyde
Mothball-like	Naphthalene
Ammonia-like	Ammonia
Fishy	Trimethylamine
Fruity	Benzyl acetate

In some cases, the chemical *structure-odor activity relationship* can be expressed by simple rules. An example is decaline (*67*), which has an amber tree-like odor.

The structural requirement for the amber-tree-like odor is: trans-decaline must be axially substituted (H is counted as a substituent) in positions 2, 9 and 10 and one of the substituents must be oxygen. This is the "triaxial rule" established by *Ohloff* and is supported by many examples.

5.6.3.2 Musk-like Compounds

Musk is by far very well investigated with regard to structure-activity relationships. Certain information is also available about some other modalities. On the other hand, there are still few systematic studies published on many other modalities.

Macro rings 68–70 (Formula 5.62), with n = 15 or 16, have good natural musk-like odor characteristics. For n > 16, the animal-like perfume note increases, while for ketones with n = 9–12 rings, the odor is like camphor and, for n = 13, like cedar. The nature of the atoms

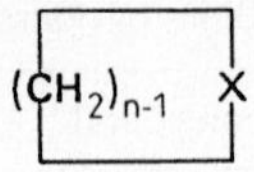

$X = CO, O, S, NH, NMe$

68

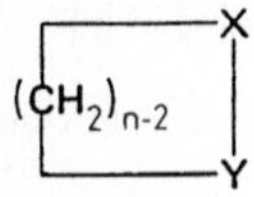

$X-Y = O-CO$ (5.62)

69

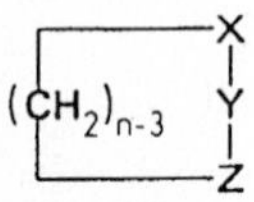

$X-Y-Z = O-CO-O$

$CO-O-CO$

$O-CH_2-O$

70

of the ring can be varied without producing any essential alteration in the odor. In the lower-membered rings, however, the odor quality depends strongly on the nature of the polar group present.

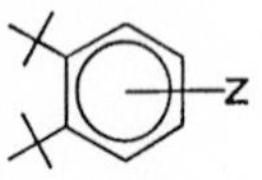

71 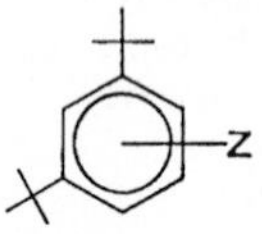   *72*

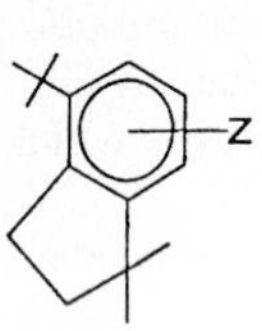

73 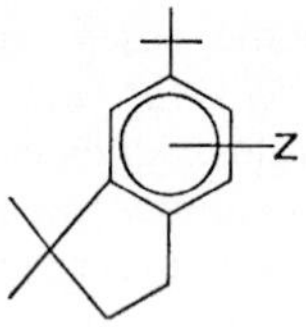   *74*   (5.63)

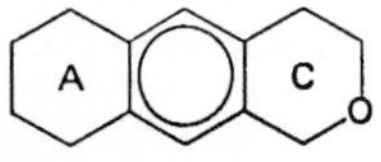

75 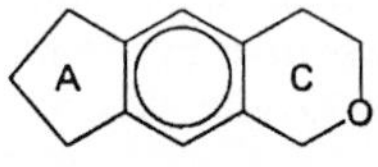   *76*   (5.64)

Aromatic ketones 71–74 (z = R – CO), as well as isochromans 75 and 76 (with A and C rings alkylated) and aromatic nitro compounds 77 and 78, likewise can have a musk-like odor.

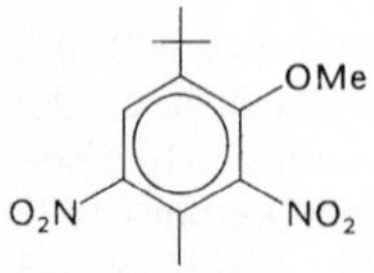

77 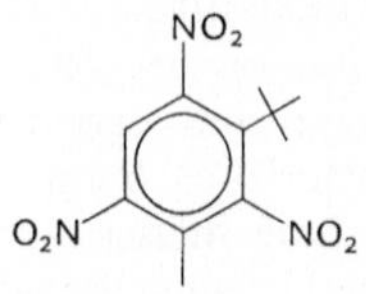   *78*   (5.65)

5.6.3.3 Camphor-like Compounds

Camphor (*1* in Formula 5.48) acts as though it has a quasi-spherical molecular shape and so do hydrocarbons and their respective polar derivatives (*2–4* in Formula 5.48; *79–82* in Formula 5.66):

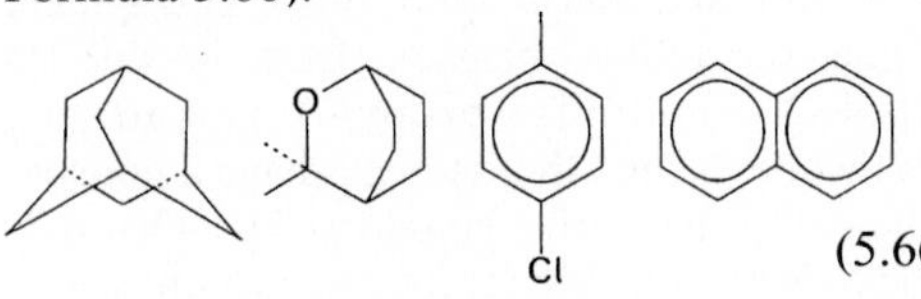

79   *80*   *81*   *82*   (5.66)

5.6.3.4 Caramel-like Compounds

The above-mentioned HA/B-system appears to be essential in the structural element (Formula: 5.67) of compounds with a caramel-like, sweet, nutty odor obtained by nonenzymatic browning of sugars present in food. Variations in odor quality are caused by the hydrophobic part of the molecule.

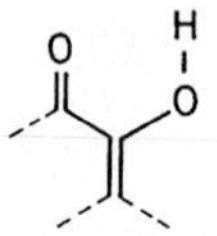

(5.67)

Maltol (*83*), isomaltol (*84*), furaneol (*85*), sotolon (*86*) and cyclotene (*87*) are typical representatives of this class of aroma compounds (Formula 5.68). Low concentrations of some thio derivatives (*88*, *89*) also have a similar odor. Any change of the bifunctional proton donor/proton acceptor unit results in an extensive change in odor quality: α-dicarbonyl compounds, e.g., 2,2,5,5-tetramethyltetrahydrofuran-3,4-dione (*90*), which have no AH/B system but two acceptor groups, are neither caramel-like nor roasted. O-methylisomaltol (*91*) is completely devoid of the typical maltol character, retaining only a weak, slightly fruity odor. The methyl ester of furaneol (*92*) has a sherry odor and not a caramel odor.

5.6.3.5 Roasted Compounds

Compounds with roasted, cracker-like aroma notes are also formed via the *Maillard* reaction. They are, however, nitrogen-containing unlike the caramel-like compounds. A common structural element is the grouping (shown in Formula 5.69):

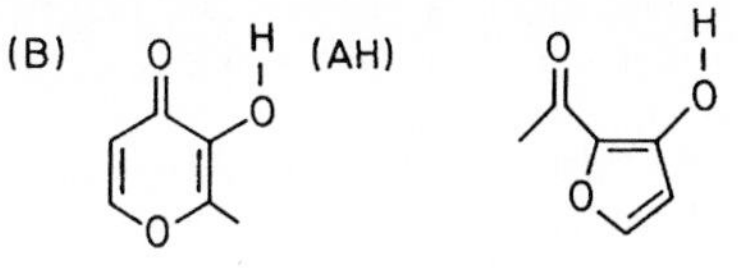

(B) (AH)

83   *84*

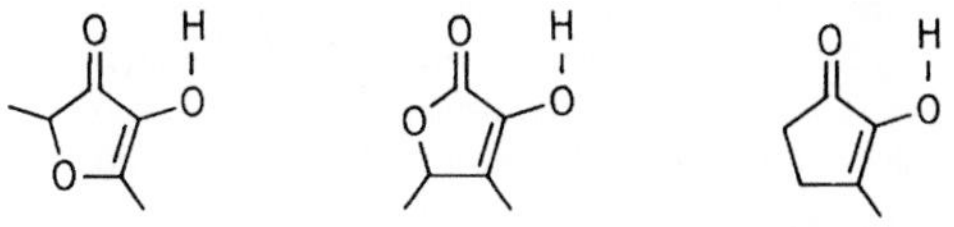

85   *86*   *87*

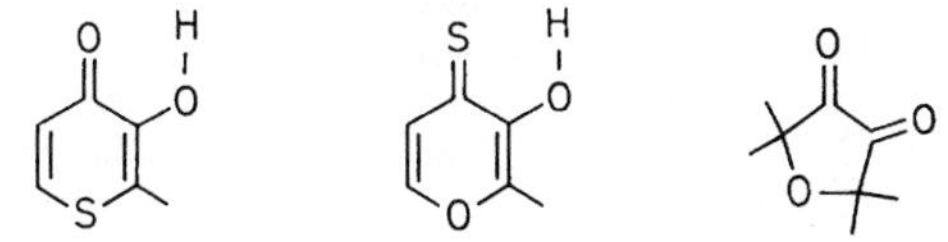

88   *89*   *90*

91   *92*   (5.68)

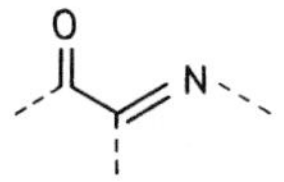

(5.69)

Typical representatives are 2-acetyl-1-pyrroline (*93* in Formula: 5.70), 2-acetyltetrahydropyridine (*94*), 2-acetylpyridine (*95*) and 2-acetylthiazoline (*96*).

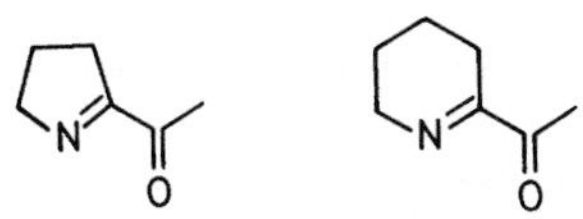

93   *94*

                              (5.70)

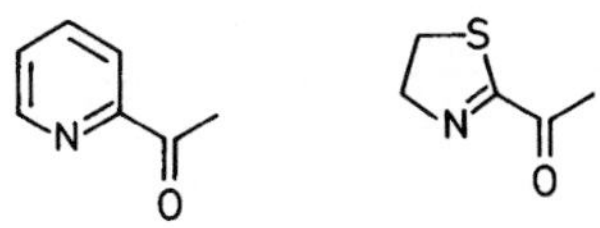

95   *96*

5.6.4 Prospects

In summary, it should be noted that the odor of chemical compounds depends on the size and geometry of the molecule, type and number of polar functional groups and their position relative to each other and to hydrophobic structural elements. A large amount of quantitative and qualitative data on odor substances are available. However, only a fraction has been systematically processed with regard to structure-activity relationships. The large number of odor qualities represent a special problem. However, computer programs available today for the production and processing of molecular models and for the determination of energetically preferred conformers increasingly allow rational processing so that in the foreseeable future a better understanding of the relationships between structure and odor can be expected.

5.7 Literature

Acree, T.E., Barnard, J., Cunningham, D.G.: A procedure for the sensory analysis of gas chromatographic effluents. Food Chem. *14*, 273 (1984)

Adda, J., Richard, H. (Eds.): International symposium on food flavors. Technique et Documentation (Lavoisier): Paris. 1983

Beets, M.G.J.: Structure – activity relationships in human chemoreception. Applied Science Publ.: London. 1978

Beyeler, M., Solms, J.: Interaction of flavour model compounds with soy protein and bovine serum albumin. Lebens. Wiss. Technol. *7*, 217 (1974)

Blank, I., Grosch, W.: Evaluation of potent odorants in dill seed and dill herb (*Anethum graveolens L.*) by aroma extract dilution analysis. J. Food Sci. *56*, 63 (1991)

Blank, I., Sen, A., Grosch, W.: Sensory study on the character-impact flavor compounds of dill herb (*Anethum graveolens L.*) Food Chem. *43*, 337 (1992)

Buttery, R.G., Haddon, W.F., Seifert, R.M., Turnbaugh, J.G.: Thiamin odor and bis(2-methyl-3-furyl)disulfide. J. Agric. Food Chem. *32*, 674 (1984)

Buttery, R.G., Teranishi, R., Ling, L.C., Turnbaugh, J.G.: Quantitative and sensory studies on tomato paste volatiles. J. Agric. Food Chem. *38*, 336 (1990)

Charalambous, G., Inglett, G.E. (Eds.): Flavor of foods and beverages. Academic Press: New York. 1978

Charalambous, G., Katz, I. (Eds.): Phenolic, sulfur, and nitrogen compounds in food flavors. ACS Symposium Series 26, American Chemical Society: Washington, D.C. 1976

Drawert, F.: Geruch- und Geschmackstoffe. Verlag Hans Carl: Nürnberg. 1975

Emberger, R.: Möglichkeiten des Aufbaus natürlicher und synthetischer Aromen. Lebensmittelchem. Gerichtl. Chem. *34*, 149 (1980)

Engel, K.-H., Flath, R.A., Buttery, R.G., Mon, T.R., Ramming, D.W., Teranishi, R.: Investigation of volatile constituents in nectarines. 1. Analytical and sensory characterization of aroma components in some nectarine cultivars. J. Agric. Food Chem. *36*, 549 (1988)

Fenaroli, G.: Handbook of flavor ingredients. 2nd edn., Vol. I and II, CRC Press: Cleveland, Ohio. 1975

Frijters, J.E.R.: A critical analysis of the odour unit number and its use. Chem. Senses Flavour *3*, 227 (1978)

Grosch, W.: Analyse von Aromastoffen. Chem. unserer Zeit *24*, 82 (1990)

Guichard, E., Fournier, N.: Enantiomeric ratios of sotolon in different media and sensory differentiation of the pure enantiomers. Poster at the EURO FOOD, CHEM VI, Hamburg, 1991

Heath, H.B., Reineccius, G.: Flavor chemistry and technology. AVI Publ. Comp., Inc., Westport, Connecticut, 1986

Jennings, W.G., Filsoof, M.: Comparison of sample preparation techniques for gas chromatographic analyses. J. Agric. Food Chem. *25*, 440 (1977)

Laing, D.G., Wilcox, M.E.: Perception of components in binary odour mixtures. Chem. Senses *7*, 249 (1983)

Land, D.G., Nursten, H.E. (Eds.): Progress in flavour research. Applied Science Publ.: London. 1979

Larsen, M., Poll, L.: Odour thresholds of some important aroma compounds in raspberries. Z. Lebensm. Unters. Forsch. *191*, 129 (1990)

Leffingwell, J.C., Leffingwell, D.: GRAS flavor chemicals-detection thresholds. Perfumer & Flavorist *16* (1), 1 (1991)

Maarse, H. (Ed.): Volatile compounds in foods and beverages, Marcel Dekker, Inc., New York. 1991

Maarse, H., Belz, R. (Eds.): Isolation, separation and identification of volatile compounds in aroma research. Akademie-Verlag: Berlin. 1981

Maarse, H., Visscher, C.A. (Eds.): Volatile compounds in food. Qualitative and quantitative data. 6th edn. with supplements 1, TNO-CIVO Food Analysis Institute: Zeist, The Netherlands. 1990

Maier, H.G.: Zur Bindung flüchtiger Aromastoffe an Proteine. Dtsch. Lebensm. Rundsch. *70*, 349 (1974)

McNulty, P.B., Karel, M.: Factors affecting flavour release and uptake in O/W-emulsions. J. Food Technol, *8*, 319 (1973)

Morton, I.D., MacLeod, A.J. (Eds.): Food Flavours. Part A. Introduction. Part B. The flavour of beverages. Part C. The flavour of fruits. Elsevier Scientific Publ. Co.: Amsterdam. 1982, 1986, 1990

Mosandl, A., Heusinger, G., Gessner, M.: Analytical and sensory differentiation of 1-octen-3-ol enantiomers. J. Agric. Food Chem. *34*, 119 (1986)

Mosandl, A.: Chirality in flavor chemistry-recent developments in synthesis and analysis. Food Rev. Internat. *4*, 1 (1988)

Mosandl, A.: Neue Ergebnisse der Aromastoff-Forschung. Lebensmittelchemie *45*, 2 (1991)

Naim, M., Striem, B.J., Kanner, J., Peleg, H.: Potential of ferulic acid as a precursor to off-flavors in stored orange juice. J. Food Sci. *53*, 500 (1988)

Nursten, H.E.: Developments in research on the flavour of foods. Res. Food Sci. Nutr. *4*, 183 (1984)

Ohloff, G.: Riechstoffe und Geruchssinn. Springer-Verlag, Berlin. 1990

Ohloff, G.: Recent developments in the field of naturally occurring aroma components. Fortschr. Chem. Org. Naturst. *35*, 431 (1978)

Ohloff, G.: Bifunctional unit concept in flavour chemistry. In: Flavour '81 (Ed.: Schreier, P.), p. 757, Walter de Gruyter: Berlin. 1981

Ohloff, G., Flament, I.: The role of heteroatomic substances in the aroma compounds of foodstuffs. Fortschr. Chem. Org. Naturst. *36*, 231 (1978)

Ohloff, G., Flament, I., Pickenhagen, W.: Flavor chemistry. Food Rev. Int. *1*, 99 (1985)

Parliment, T.H. Scarpellino, R.: Organoleptic techniques in chromatographic food flavor analysis. J. Agric. Food Chem. *25*, 97 (1977)

Piendl, A.: Brauereitechnologie und Molekularbiologie. Brauwissenschaft *22*, 175 (1969)

Reineccius, G.A.: Off-flavors in foods. Crit. Rev. Food Sci. Nutr. *29*, 381 (1991)

Risch, S.J., Reineccius, G.A. (Eds.): Flavor encapsulation. ACS Symp. Ser. *370* (1988)

Rizzi, P.: The biogenesis of food-related pyrazines. Food Rev. Internat. *4*, 375 (1988)

Schieberle, P.: Primary odorants in popcorn. J. Agric. Food Chem. *39*, 1141 (1991)

Schieberle, P.: The role of free amino acids present in yeast as precursors of the odorants 2-acetyl-1-pyrroline and 2-acetyltetrahydropyridine in wheat bread crust. Z. Lebensm. Unters. Forsch. *191*, 206 (1990)

Schreier, P.: Chromatographic studies of biogenesis of plant volatiles. Dr. Alfred Hüthig Verlag: Heidelberg. 1984

Silberzahn, W.: Aromen, in: Taschenbuch für Lebensmittelchemiker und -technologen (Frede, W., Ed.), Band 1, Springer-Verlag, Berlin. 1991

Solms, J.: Aromastoffe als Liganden. In: Geruch-
und Geschmackstoffe (Ed.: Drawert, F.), S. 201,
Verlag Hans Carl: Nürnberg. 1975

Teranishi, R., Buttery, R.G., Shahidi, F. (Eds.): Flavor chemistry – Trends and Developments. ACS
Symp. Ser. *388* (1989)

Ter Heide, R., de Valois, P.J., Visser, J., Jaegers, P.P.,
Timmer, R.: Concentration and identification of
trace constituents in alcoholic beverages. In: Analysis of foods and beverages – Headspace techniques (Ed.: Charalambous, G.), p. 249, Academic
Press: New York. 1978

Tressl, R., Helak, B., Martin, N., Kersten, E.: Formation of amino acid specific *Maillard* products
and their contribution to thermally generated
aromas. ACS Symp. Ser. *409*, 156 (1989)

Tressl, R.: Probleme bei der Beurteilung von natürlichen Aromen. Lebensmittelchem. Gerichtl.
Chem. *34*, 47 (1980)

Vernin, G. (Ed.): Chemistry of heterocyclic compounds in flavours and aromas. Ellis Horwood
Publ.: Chichester. 1982

Werkhoff, P., Bretschneider, W., Herrmann, H.-J.,
Schreiber, K.: Fortschritte in der Aromastoff-Ana-
lytik (1–9). Labor Praxis (1989) S. 306, 426, 514,
616, 766, 874, 1002, 1121 and (1990) S. 51

Werkhoff, P., Brüning, J., Emberger, R., Güntert, M.,
Köpsel, M., Kuhn, W., Surburg, H.: Isolation and
characterization of volatile sulfur-containing
meat flavor components in model systems. J.
Agric. Food Chem. *38*, 777 (1990)

Williams, P.J., Strauss, C.R., Wilson, B., Massy-
Westropp, R.A.: Novel monoterpene disaccharide
glycosides of *Vitis vinifera* grapes and wines.
Phytochemistry *21*, 2013 (1982)

Ziegler, E.: Die natürlichen und künstlichen Aromen. Dr. Alfred Hüthig Verlag: Heidelberg.
1982

6 Vitamins

6.1 Foreword

Vitamins are minor but essential constituents of food. They are required for the normal growth, maintenance and functioning of the human body. Hence, their preservation during storage and processing of food is of far-reaching importance. Data are provided in Tables 6.1 and 6.2 to illustrate vitamin losses in some preservation methods for fruits and vegetables. Vitamin losses can occur through chemical reactions which lead to inactive products, or by extraction or leaching, as in the case of water-soluble vitamins during blanching and cooking.

The vitamin requirement of the body is usually adequately supplied by a balanced diet. A deficiency can result in hypovitaminosis and, if more severe, in avitaminosis. Both can occur not only as a consequence of insufficient supply of vitamins by food intake, but can be caused by disturbances in resorption, by stress and by disease.

An assessment of the extent of vitamin supply can be made by determining the vitamin con- tent in blood plasma, or by measuring a biological activity which is dependent on the presence of a vitamin, as are many enzyme activities.

Vitamins are usually divided into two general classes: the fat-soluble vitamins, such as A, D, E and K_1, and the water-soluble vitamins, B_1, B_2, B_6, nicotinamide, pantothenic acid, biotin, folic acid, B_{12} and C.

Data on the average human requirement of some vitamins are presented by age group in Table 6.3.

6.2 Fat-Soluble Vitamins

6.2.1 Retinol (Vitamin A)

6.2.1.1 Biological Role

Retinol (I, in Formula 6.1) is of importance in protein metabolism of cells which develop from the ectoderm (such as skin or mucous-coated linings of the respiratory or digestive systems). Lack of retinol in some way nega-

Table 6.1. Vitamin losses (%) through processing/canning of vegetables

Processed/ canned product	Samples of vegetable analyzed	Vitamin losses as % of freshly cooked and drained product				
		A	B_1	B_2	Niacin	C
Frozen products (cooked and drained)	10[a]	12[c] 0–50[d]	20 0–61	24 0–45	24 0–56	26 0–78
Sterilized products (drained)	7[b]	10 0–32	67 56–83	42 14–50	49 31–65	51 28–67

[a] Asparagus, lima beans, green beans, broccoli, cauliflower, green peas, potatoes, spinach, brussels sprouts, and baby corn-cobs.
[b] As under a) with the exception of broccoli, cauliflower and brussels sprouts; the values for potato include the cooking water.
[c] Average values.
[d] Variation range.

Table 6.2. Vitamin loss (%) through processing/canning of fruits

Processed/canned product	Fruit samples analyzed	Vitamin losses as % of fresh product				
		A	B_1	B_2	Niacin	C
Frozen products (not thawed)	8[a]	37[c] 0–78[d]	29 0–66	17 0–67	16 0–33	18 0–50
Sterilized products (including the cooking water)	8[b]	39 0–68	47 22–67	57 33–83	42 25–60	56 11–86

[a] Apples, apricots, bilberries, sour cherries, orange juice concentrate (calculated for diluted juice samples), peaches, raspberries and strawberries.
[b] As under a) except orange juice and not its concentrate was analyzed.
[c] Average values.
[d] Variation range.

Table 6.3. Vitamins: daily dietary requirements

Age group (years):		< 1	1–4	4–10	10–18	> 18
Vitamin	Unit					
A	IU[a]	1,500	2,000–2,500	2,500–3,000	4,500–5,000	5,000–6,000
D	IU[b]	400	400	400	400	400
E	IU[c]	5	10	10	15	15
C	mg	35	40	40	45	45–80
B_1	mg	0.3–0.5	0.6–0.8	0.9–1.2	1.5	1.5
B_2	mg	0.4–0.6	0.8	0.9–1.2	1.3–1.8	1.2–1.8
B_6	mg	0.3–0.5	0.7–0.9	0.9–1.2	1.6–2.0	2.0–2.5
Nicotinamide	mg	5–8	9–12	12–16	14–20	12–20
Folic acid	mg	0.05	0.1–0.2	0.2–0.3	0.4	0.4–0.8
B_{12}	µg	0.3–0.4	1–1.5	1.5–2.0	2.0–3.0	3.0–4.0

[a] Requirement covers a supply in the form of vitamin A (75%) and carotenes (25%), 1 international unit $\triangleq$ 0.3 µg vitamin A, 1.8 µg β-carotene or 3.6 µg of other carotenoid with a vitamin A activity.
[b] 1 IU $\triangleq$ 0.025 µg Vitamin D_3.
[c] 1 IU $\triangleq$ 1 mg D,L-α-Tocopherol acetate.

tively affects epithelial tissue (thickening of skin, hyperkeratosis) and also causes night blindness.

Furthermore, retinol, in the form of 11-cis-retinal (II), is the chromophore component of the visual cycle chromoproteins in three types of cone cells, blue, green and red (λ_{max} 435, 540 and 565 nm, respectively) and of rods of the retina.

The chromoproteins (rhodopsins) are formed in the dark from the corresponding proteins (opsins) and 11-cis-retinal, while in the light the chromoproteins dissociate into the more

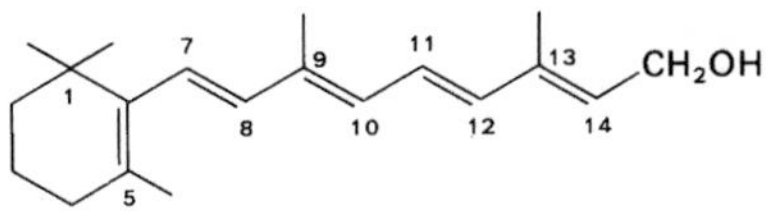

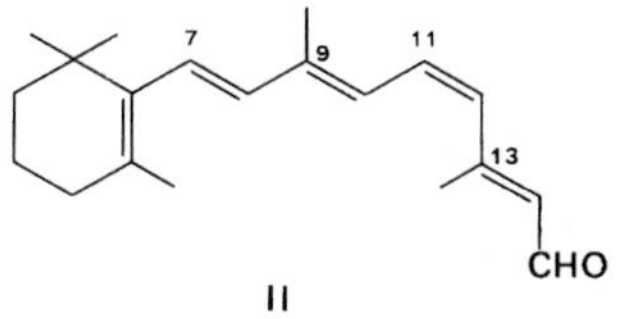

$$\text{(6.1)}$$

stable all-trans-retinal and protein. This conformational change triggers a nerve impulse in the adjacent nerve cell. The all-trans-retinal is then converted to all-trans-retinol and through an intermediary, 11-cis-retinol, is transformed back into 11-cis-retinal (see Fig. 6.1 for the visual cycle reactions).

6.2.1.2 Requirement, Occurrence

The daily adult requirement of vitamin A is 1.5–1.8 mg. Approx. 75% is provided by retinol intake (as fatty acid esters, primarily retinyl palmitate), while the remaining 25% is through β-carotene and other provitamin-active carotenoids. Due to the limited extent of carotenoid cleavage, at least 6 g of β-carotene are required to yield 1 g retinol.

Vitamin A resorption and its storage in the liver occur essentially in the form of fatty acid esters. Its content in liver is 250 µg/g fresh tissue, i.e. a total of about 240–540 mg is stored. The liver supplies the blood with free retinol, which then binds to proteins in blood. Vitamin A concentration is 45–84 µg/100 ml plasma in adults; values below 15–24 µg/100 ml indicate a deficiency.

A hypervitaminosis is known, but the symptoms disappear if the intake of retinal is decreased.

Vitamin A occurs only in animal tissue; above all in fish liver oil, in livers of mammals, in milk fat and in egg yolk. Plants are devoid of vitamin A but do contain carotenoids which yield vitamin A by cleavage of the centrally located double bond (provitamins A). Carotenoids are present in almost all vegetables but primarily in green, yellow and leafy vegetables (carrots, spinach, cress, kale, bell peppers, paprika peppers, tomatoes) and in fruit, outstanding sources being rose hips, pumpkin, apricots, oranges and palm oil, which is often used for yellow coloring. Animal carotenoids are always of plant origin, derived from feed. Table 6.7 gives the vitamin A content of some common foods. These values can vary greatly with cultivar, stage of ripeness, etc. An accurate estimate of the vitamin A content of a food must include a detailed analysis of its carotenoids.

6.2.1.3 Stability, Degradation

Food processing and storage can lead to 5–40% destruction of vitamin A and carotenoids. In the absence of oxygen and at higher temperatures, as experienced in cooking or food sterilization, the main reactions are isomerization and fragmentation. In the presence of oxygen, oxidative degradation leads to a series of products, some of which are volatile (cf. 3.8.4.4). This oxidation often parallels lipid oxidation (cooxidation process). The rate of oxidation is influenced by oxygen partial pressure, water activity, temperature, etc. Dehydrated foods are particularly sensitive to oxidative degradation.

6.2.2 Calciferol (Vitamin D)

6.2.2.1 Biological Role

Cholecalciferol (vitamin D_3, I) is formed from cholesterol in the skin through photolysis of 7-dehydrocholesterol (provitamin D_3) by ultraviolet light ("sunshine vitamin"; cf. 3.8.2.2.2). Similarly, vitamin D_2 (ergocalciferol, II; cf. Formula 6.2) is formed from ergosterol.

The active metabolites, 25-hydroxycholecalciferol and 1α,25-dihydroxycholecalciferol, are responsible for intestinal resorption of calcium and for calcium salt deposition into the organic matrix of bones by triggering the biosynthesis of the calcium-binding protein.

Vitamin D deficiency can result in an increased excretion of calcium and phosphate and, consequently, impairs bone formation through inadequate calcification of cartilage and bones (childhood rickets). Vitamin D defi-

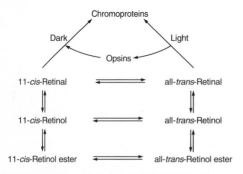

Fig. 6.1. Schematic representation of the visual cyle

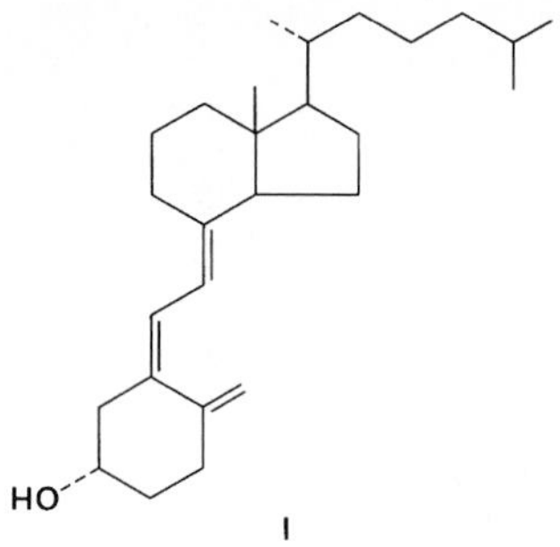

(6.2)

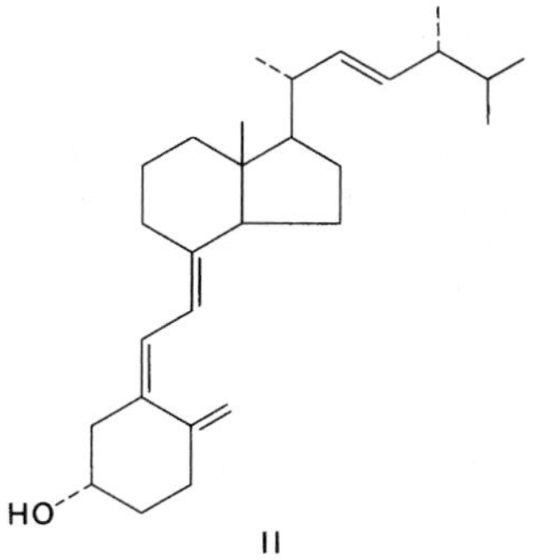

Table 6.4. Biological activity of tocopherols[a]

Tocopherol	Activity (IU/mg)
d,l-α-Tocopherol acetate[b]	1.0
d,l-α-Tocopherol	1.1
d-α-Tocopherol acetate	1.4
d-α-Tocopherol	1.5
l-α-Tocopherol	0.5
d,l-β-Tocopherol	0.3
d,l-γ-Tocopherol	0.15
d,l-δ-Tocopherol	0.01

[a] The formulas are given in Fig. 3.45.
[b] Synthetic, racemic d,l-α-tocopherol acetate is the vitamin E preparation that is used most often. It has been selected as the standard substance: 1 mg = 1 international unit (IU). Recently, d-α-tocopherol has been recommended as the standard.

ciency in adults leads to osteomalacia, a softening and weakening of bones. Hypercalcemia is a result of excessive intake of vitamin D, causing calcium carbonate and calcium phosphate deposition disorders involving various organs.

6.2.2.2 Requirement, Occurrence

The daily requirement is 10 µg. Indicators of deficiency are the concentration of the metabolite 25-hydroxycholecalciferol in plasma and the activity of alkaline serum phosphatase, which increases during vitamin deficiency.
Most natural foods have a low content of vitamin D_3. Fish liver oil is an exceptional source of vitamin D_2. The D-provitamins, ergosterol and 7-dehydrocholesterol, are widely distributed in the animal and plant kingdoms. Yeast, some mushrooms, cabbage, spinach and wheat germ oil are particularly abundant in provitamin D_2. Vitamin D_3 and its provitamin are present in egg yolk, butter, cow's milk, beef and pork liver, mollusks, animal fat and pork skin. However, the most important vitamin D source is fish oil, primarily liver oil. The vitamin D requirement of humans is best supplied

by 7-dehydrocholesterol. Table 6.7 gives data on vitamin D occurrence in some foods. However, these values can vary widely, as shown by variations in dairy cattle milk (summer or winter), caused by feed or frequency of pasture grazing and exposure to the ultraviolet rays of sunlight.

6.2.2.3 Stability, Degradation

Vitamin D is sensitive to oxygen and light. Its stability in food is not a problem, because adults usually obtain a sufficient supply of this vitamin.

6.2.3 α-Tocopherol (Vitamin E)

6.2.3.1 Biological Role

The various tocopherols differ in the number and position of the methyl groups on the ring. α-Tocopherol (Formula 6.3; the configuration at the three asymmetric centers, 2, 4′ and 8′, is R) has the highest biological activity (Table 6.4). Its activity is based mainly on its antioxidative properties, which retard or prevent lipid oxidation (cf. 3.7.3.1). Thus, it not only contributes to the stabilization of membrane structures, but also stabilizes other active agents (e.g., vitamin A, ubiquinone, hormones, and enzymes) against oxidation. Vitamin E is involved in the conversion of arachidonic acid

Table 6.5. Requirement of α-tocopherol on supply of unsaturated fatty acids

Fatty acid	α-Tocopherol (mg/g fatty acid)
Monoene acids	0.09
Diene acids	0.6
Triene acids	0.9
Tetraene acids	1.2
Pentaene acids	1.5
Hexaene acids	1.8

Table 6.6. Tocopherol stability during deep frying

	Tocopherol total (mg/100 g)	Loss (%)
Oil before deep frying	82	
after deep frying	73	11
Oil extracted from potato chips		
immediately after production	75	
after 2 weeks storage		
at room temperature	39	48
after 1 month storage		
at room temperature	22	71
after 2 months storage		
at room temperature	17	77
after 1 month kept at $-12\,^{\circ}C$	28	63
after 2 months kept at $-12\,^{\circ}C$	24	68
Oil extracted from French fries		
immediately after production	78	
after 1 month kept at $-12\,^{\circ}C$	25	68
after 2 months kept at $-12\,^{\circ}C$	20	74

to prostaglandins and slows down the aggregation of blood platelets. Vitamin E deficiency is associated with chronic disordes (sterility in domestic and experimental animals, anemia in monkeys, and muscular dystrophy in chickens). Its mechanism of action is not fully elucidated.

6.2.3.2 Requirement, Occurrence

The daily requirement is 15 mg α-tocopherol. It increases when the diet contains a high content of unsaturated fatty acids (cf. Table 6.5). A normal supply results in a concentration of 0.7–1.6 mg/100 ml in blood plasma. A level less than 0.4 mg/100 ml is considered a deficiency.
Tables 3.51 and 6.7 provide data on the tocopherol content in some foods. The main sources are vegetable oils, particularly germ oils of cereals.

6.2.3.3 Stability, Degradation

Losses occur in vegetable oil processing into margarine and shortening. Losses are also encountered in intensive lipid autoxidation, particularly in dehydrated or deep fried foods (Table 6.6).

6.2.4 Phytomenadione (Vitamin K₁)

6.2.4.1 Biological Role

The K-group vitamins are naphthoquinone derivatives which differ in their side chains. The structure of vitamin K₁ is shown in Formula 6.4. The configuration at carbon atoms 7′ and 11′ is R and corresponds to that of natural phytol. Racemic vitamin K₁ synthesized from

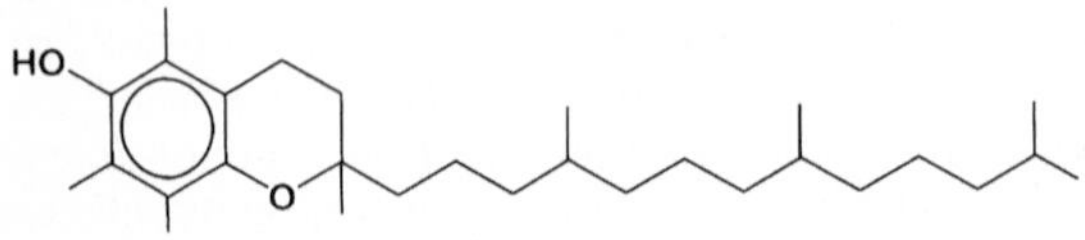

$$(6.3)$$

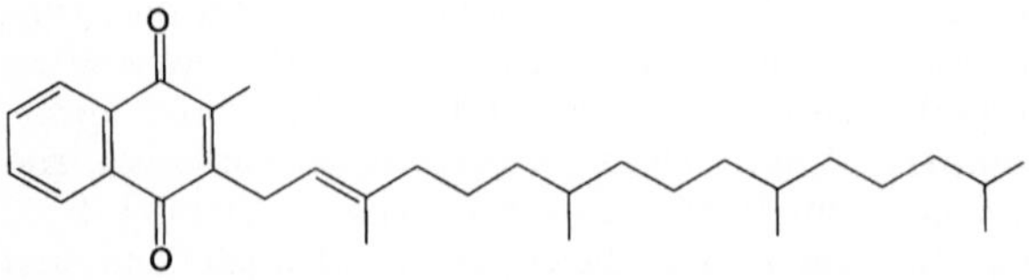

$$(6.4)$$

optically inactive isophytol has the same biological activity as the natural product. Vitamin K_1 is involved in biosynthesis of some blood-clotting factors (prothrombin, proconvertin, Christmas or Stuart factor). Deficiency of this vitamin causes a reduced activity of prothrombin and results in hemorrhage.

6.2.4.2 Requirement, Occurrence

The daily adult requirement of vitamin K_1 is estimated to be 1–4 mg. It is supplied by food, most abundantly by green leafy vegetables, and by synthesis within the body by intestinal bacteria. Their synthesizing efficiency, 1–1.5 mg/day, is impressive as it probably can meet the daily requirement.

Vitamin K_1 occurs primarily in green leafy vegetables (spinach, cabbage, cauliflower), but liver (veal or pork) is also an excellent source (Table 6.7).

6.2.4.3 Stability, Degradation

Little is known about the reactions of vitamin K_1 in foods. The vitamin K compounds are destroyed by light and alkali. They are relatively stable to atmospheric oxygen and exposure to heat.

6.3 Water-Soluble Vitamins

6.3.1 Thiamine (Vitamin B₁)

6.3.1.1 Biological Role

Thiamine, in the form of its pyrophosphate,

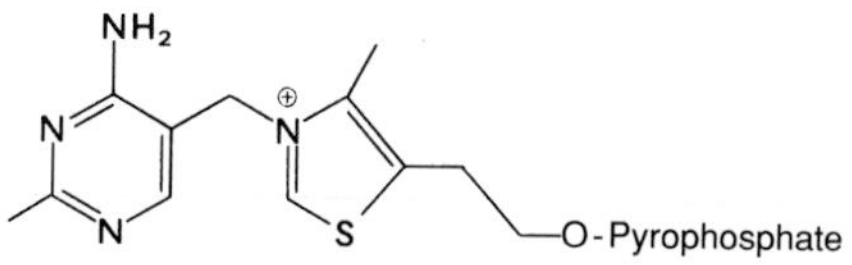

$$(6.5)$$

is the coenzyme of several important enzymes, such as pyruvate dehydrogenase, transketolase, phosphoketolase and α-ketoglutarate dehydrogenase, in reactions involving the transfer of an activated aldehyde unit (D: donor; A: acceptor):

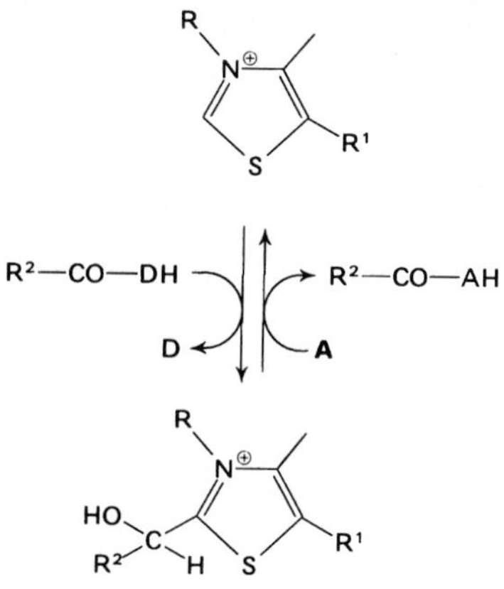

$$(6.6)$$

Vitamin B_1 deficiency is shown by a decrease in activity of the enzymes mentioned above. The disease known as beri-beri, which has neurological and cardiac symptoms, results from a severe dietary deficiency of thiamine.

6.3.1.2 Requirement, Occurrence

The daily adult requirement is 1–2 mg. Since thiamine is a key substance in carbohydrate metabolism, the requirement increases in a carbohydrate-enriched diet. The assay of transketolase activity in red blood cells or the extent of transketolase reactivation on addition of thiamine pyrophosphate can be used as indicators for sufficient vitamin intake in the diet.

Vitamin B_1 is found in many plants. It is present in the pericarp and germ of cereals, in yeast, vegetables (potatoes) and shelled fruit. It is abundant in pork, beef, fish, eggs and in animal organs such as liver, kidney, brain and heart. Human milk and cow's milk contain vitamin B_1. Whole grain bread and potatoes are important dietary sources. Since vitamin B_1 is localized in the outer part of cereal grain hulls, flour milling with a low extraction grade or rice polishing remove most of the vitamin in the bran (cf. 15.3.1.3 and 15.3.2.2.1). Table 6.7 lists data on the occurrence of thiamine.

6.3.1.3 Stability, Degradation

Thiamine stability in aqueous solution is relatively low. It is influenced by pH (Fig. 6.2), temperature (Table 6.8), ionic strength and

Table 6.7. Vitamin content of some food products[a]

Food product	Carotene[b] mg	A mg	D µg	E mg	K mg	B1 mg	B2 mg	NAM[c] mg	PAN[d] mg	B6 mg	BIO[e] µg	FOL[f] µg	B12 µg	C mg
Milk and milk products														
Bovine milk, raw	0.018	0.030	0.06	0.09		0.04	0.18	0.09	0.35	0.05	3.5	6.0	0.4	1.7
Human milk	0.024	0.054	0.05	0.52	0.003	0.02	0.04	0.17	0.21	0.01	0.6	5.0	0.05	4.4
Butter	0.38	0.59	1.3	2.2	0.06	0.005	0.02	0.03	0.05	0.005				0.2
Cheese														
Emmental	0.14	0.32	1.1	0.35		0.05	0.34	0.18	0.40	0.07	3.0	4.0	2.2	0.5
Camembert (60% fat)		0.63				0.04	0.37	1.18	0.7	0.2	2.8			
Camembert (30% fat)	0.1	0.2	0.17	0.30		0.05	0.67	1.2	0.9	0.3	5.0	66	3.1	
Eggs														
Chicken egg yolk		1.12		3.0		0.29	0.40	0.07	3.7	0.3	50	150	2.0	0.3
Chicken egg white						0.02	0.32	0.09	0.14	0.012	7	16	0.1	
Meat and meat products														
Beef, whole carcass, lean						0.08	0.18	4.9		0.5			1.3	
Pork, whole carcass, lean						0.66	0.17	3.7		0.4			0.8	
Calf liver		3.92	0.33	1.2	0.15	0.28	2.61	15.0	7.9	0.9	80	240	60	35
Pork liver		5.8		0.45	0.02	0.31	3.2	15.7	6.8	0.6	30	220	40	23
Chicken liver		11.6	1.3	0.4		0.32	2.49	11.6	7.2	0.8		380	20	28
Pork kidneys						0.34	1.8	8.4	3.1	0.6			20	16
Blood sausage	0.02					0.07	0.13	1.2					50	
Fish and fish products														
Herring		0.04	30	1.5		0.04	0.22	3.8	0.9	0.5	4.5	5	8.5	0.5
Eel		0.98	13	8		0.18	0.32	2.6	0.3			13	1	1.8
Cod-liver oil		30	330	3.26										
Cereals and cereal products														
Wheat, whole kernel	0.02			3.2		0.48	0.14	5.1	1.2	0.4	6	49		
Wheat flour, type 405				2.3		0.06	0.03	0.7	0.2	0.2	1.5	10		
Wheat flour, type 550				2.0		0.11	0.08	0.5	0.4	0.1	1.1	20		
Wheat germ				27.6	0.35	2.01	0.72	4.5	1.0	3.3	17	520		
Wheat gluten				9.1	0.08	0.65	0.51	17.7	2.5	2.5	44	400		
Rye whole kernel				3.8		0.35	0.17	1.8	1.5	0.3	4.6	42		
Rye flour, type 997				3.4		0.19	0.11	0.8						
Corn whole kernel	0.37			5.8	0.04	0.36	0.20	1.5	0.7	0.4	6	26		

Food	Vitamin A[b]	Vitamin E	B₁	B₂	Nicotinamide[c]	Pantothenic acid[d]	Vitamin B₆	Biotin[e]	Folic acid[f]	Vitamin C
Corn (breakfast cereal, corn flakes)	0.43				1.4	0.2	0.07		6	
Oat flakes	3.7		0.59	0.15	1.0	1.1	0.16	20	24	
Rice, unpolished	4.5		0.41	0.09	5.2	1.7	0.68	12	16	
Rice, polished	0.4		0.06	0.03	1.3	0.6	0.15		29	
Vegetables										
Watercress	2.68	1.94	0.09	0.17	0.7	0.07		20		51
Mushrooms (cultivated)	0.01		0.10	0.44	5.2	2.1	0.05		30	4.9
Chicory	1.29		0.05	0.03	0.24				50	10.2
Endive	1.14	0.02	0.05	0.12	0.4				50	9.4
Lamb's lettuce	3.9		0.07	0.08	0.4					35
Kale	4.1	0.09	0.1	0.25	2.1	0.4		0.5	60	105
Potatoes	0.01		0.11	0.05	1.2	0.4	0.2	0.4	7	17
Kohlrabi	0.2		0.05	0.05	1.8	0.1	0.1			63.3
Head lettuce	0.8	0.2	0.06	0.08	0.3	0.1	0.1	1.9	40	13
Lentils, dried	0.1	0.08	0.43	0.26	2.2	1.4	0.6		40	
Carrots	12	0.7	0.07	0.05	0.6	0.1		5	8	7.1
Brussels sprouts	0.4	0.6	0.11	0.14	0.7	0.4		0.4	80	114
Spinach	4.2	2.5	0.11	0.23	0.6	0.3	0.22	6.9	80	52
Edible mushroom (*Boletus edulis*)		3.1	0.03	0.37	4.9	2.7				2.5
Tomatoes	0.82	0.63	0.06	0.04	0.5	0.3	0.1	4	40	24.2
White cabbage	0.04	0.02	0.05	0.04	0.3	0.3	0.1		80	45.8
Fruits										
Orange	0.09	0.24	0.08	0.04	0.3	0.2	0.05	2.3	20	50
Apricot	1.8	0.5	0.04	0.05	0.8	0.3		4	4	9.4
Strawberry	0.05	0.22	0.03	0.05	0.5	0.3	0.06	4	20	64
Grapefruit	0.02	0.27	0.05	0.02	0.24	0.25	0.03	0.4	10	44
Rose hips										1,250
Red currants	0.04	0.21	0.04	0.03	0.23	0.06	0.05	2.6		36
Black currants	0.14	1.0	0.05	0.04	0.28	0.4	0.08	2.4		177
Sour cherries	0.3		0.05	0.06	0.4	0.4				12
Plums	0.2	0.8	0.07	0.04	0.4	0.2	0.05	0.1	2	5.4
Sea buckthorn	1.5		0.03	0.21	0.3	0.2	0.11	3.3	10	450
Yeast										
Baker's yeast, pressed			1.43	2.31	17.4	3.5	0.8	30	1,020	
Brewer's yeast, dried			12.0	3.8	44.8	7.2	4.4	20	3,200	20

[a] Values are given in mg or µg per 100 g of edible portion. [b] Total carotenoids with vitamin A activity. [c] Nicotinamide. [d] Pantothenic acid. [e] Biotin. [f] Folic acid.

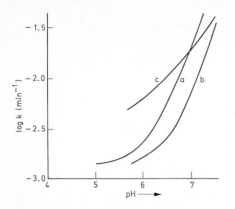

Fig. 6.2. Inactivation rate of thiamine as affected by pH **a** Thiamine in phosphate buffer, **b** thiamine in wheat or oat flour, **c** thiamine pyrophosphate in flour

metal ions. The enzyme-bound form is less stable than free thiamine (Fig. 6.2). Strong nucleophilic reagents, such as HSO_3^- or OH^-, cause rapid decomposition by forming 5-(2-hydroxyethyl)-4-methylthiazole and 2-methyl-4-amino-5(methylsulfonic acid)-pyrimidine, or 2-methyl-4-amino-5-hydroxymethylpyrimidine (see Reactions 6.7).

Thermal degradation of thiamine, which also initially yields the thiazole and pyrimidine derivatives mentioned above, is involved in the formation of meat-like aroma in cooked food. The degradation products of the thiazole, namely 2-methyl-3-furanthiol and free H_2S, are responsible for this aroma.

Table 6.8. Thiamine losses in food during storage (12 months)

Food	Thiamine loss, %	
	1.5°C	38°C
Apricots	28	65
Orange juice	0	22
Peas	0	32
Green beans	24	92
Tomato juice	0	40

Thiamine is inactivated by nitrites, probably through reaction with the amino group attached to the pyrimidine ring.

Strong oxidants, such as H_2O_2 or potassium ferricyanide, yield the fluorescent thiochrome. This reaction is often used in chemical determination of the thiamine content in food (see Reaction 6.8).

The following losses of thiamine can be expected: 15–25% in canned fruit or vegeta-

(6.8)

(6.7)

bles stored for more than a year; 0–60% in meat cooked under household conditions, depending on temperature and preparation method; 20% in salt brine pickling of meat and in baking of white bread; 15% in blanching of cabbage without sulfite and 40% with sulfite. Losses caused by sulfite are pH dependent. Practically no thiamine degradation occurs in a stronger acidic medium (e.g. lemon juice).

6.3.2 Riboflavin (Vitamin B₂)

6.3.2.1 Biological Role

Riboflavin (Formula 6.9) is the prosthetic group of flavine enzymes, which are of great importance in general metabolism and particularly in metabolism of protein.

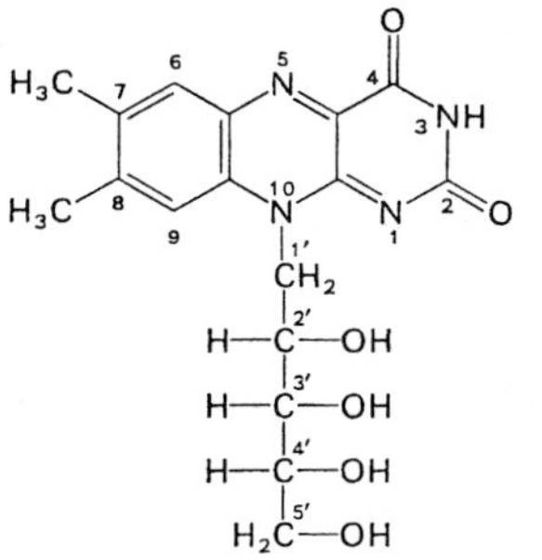

(6.9)

Riboflavin deficiency will lead to accumulation of amino acids. A specific deficiency symptom is the decrease of glutathione reductase activity in red blood cells.

6.3.2.2 Requirement, Occurrence

The daily adult requirement is 1.6–2.6 mg. Deficiency symptoms are rarely observed with a normal diet and, since the riboflavin pool in the body is very stable, even in a deficient diet it is not depleted by more than 30–50%. The riboflavin content of urine is an indicator of riboflavin supply levels. Values above 80 μg riboflavin/g creatinine are normal; 27–79 μg/g are low; and less than 27 μg/g strongly suggests a vitamin-deficient diet. Glutathione reductase activity assay can provide similar information.

The most important sources of riboflavin are milk and milk products, eggs, various vegetables, yeast, meat products, particularly variety meats such as heart, liver and kidney, and fish liver and roe. Table 6.7 provides data about the occurrence of riboflavin in some common foods.

6.3.2.3 Stability, Degradation

Riboflavin is relatively stable in normal food handling processes. Losses range from 10–15%. Exposure to light, especially in the visible spectrum from 420–560 nm, photolytically cleaves ribitol from the vitamin, converting it to lumiflavin:

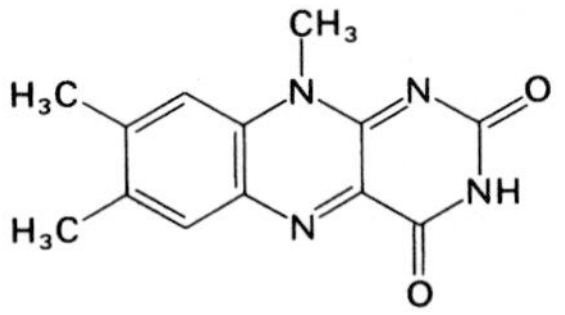

(6.10)

6.3.3 Pyridoxine (Pyridoxal, Vitamin B₆)

6.3.3.1 Biological Role

Vitamin B₆ activity is exhibited by pyridoxine (Formula 6.11) or pyridoxol (R = CH₂OH), pyridoxal (R=CHO) and pyridoxamine (R=CH₂NH₂). The metabolically active form, pyridoxal phosphate, functions as a coenzyme (cf. 2.3.2.3) of amino acid decarboxylases, amino acid racemases, amino acid dehydrases, amino transferases, serine palmitoyl transferase, lysyl oxidase, δ-aminolevulinic acid synthase, and of enzymes of tryptophan metabolism. Furthermore, it stabilizes the conformation of phosphorylases.

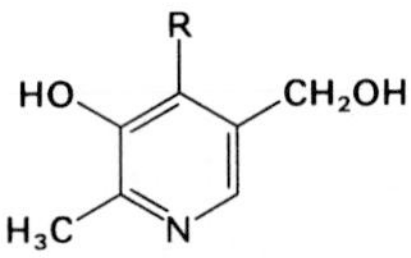

(6.11)

The intake of the vitamin occurs usually in the forms of pyridoxal or pyridoxamine.
Pyridoxine deficiency in the diet causes disorders in protein metabolism, e.g., in hemo-

globin synthesis. Hydroxykynurenine and xanthurenic acid accumulate, since the conversion of tryptophan to nicotinic acid, a step regulated by the kynureninase enzyme, is interrupted.

6.3.3.2 Requirement, Occurrence

The daily adult requirement is 2 mg. An indicator of sufficient supply is the activity of glutamate oxalacetate transaminase, an enzyme present in red blood cells. This activity is decreased in vitamin deficiency. The occurrence of pyridoxine in food is outlined in Table 6.7.

6.3.3.3 Stability, Degradation

The most stable form of the vitamin is pyridoxal, and this form is used for vitamin fortification of food. Vitamin B_6 loss is 45% in cooking of meat and 20–30% in cooking of vegetables. During milk sterilization, a reaction with cysteine transforms the vitamin into an inactive thiazolidine derivative (Formula 6.12). This reaction may account for vitamin losses also in other heat-treated foods.

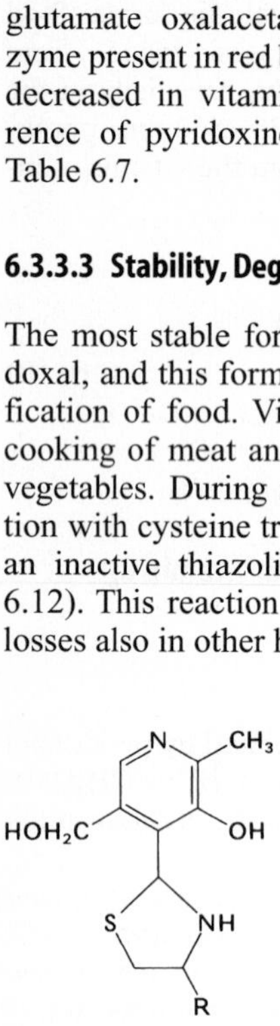

$$(6.12)$$

6.3.4 Nicotinamide (Niacin)

6.3.4.1 Biological Role

Nicotinic acid amide (I), in the form of nicotinamide adenine dinucleotide (NAD$^+$, cf. 2.3.1.1), or its phosphorylated form (NADP$^+$), is a coenzyme of dehydrogenases. Its excretion in urine is essentially in the form of N^1-methylnicotinamide (trigonelline amide, II), N^1-methyl-6-pyridone-3-carboxamide (III) and N^1-methyl-4-pyridone-3-carboxamide (IV):

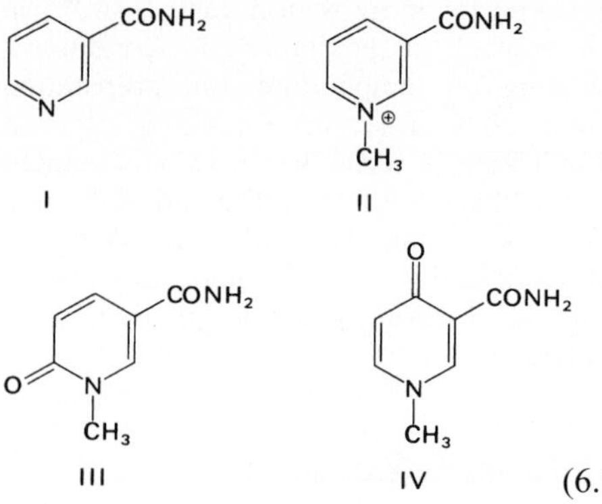

$$(6.13)$$

Vitamin deficiency is observed initially by a drop in concentration of NAD$^+$ and NADP$^+$ in liver and muscle, while levels remain normal in blood, heart and kidney. The classical deficiency disease is pellagra, which affects the skin, digestion and the nervous system (dermatitis, diarrhea and dementia). However, the initial deficiency symptoms are largely nonspecific.

6.3.4.2 Requirement, Occurrence

The daily adult requirement is 12–20 mg, with 60–70% of that covered by tryptophan intake. Hence, milk and eggs, though they contain little niacin, are good foods for prevention of pellagra because they contain tryptophan. It substitutes for niacin in the body, with 60 mg L-tryptophan equalling 1 mg nicotinamide. Indicators of sufficient supply of niacin in the diet are the levels of metabolites II (cf. Formula 6.13) in urine or III and IV in blood plasma.

The vitamin occurs in food as nicotinic acid, either as its amide or as a coenzyme. Animal organs, such as liver, and lean meat, cereals, yeast and mushrooms are abundant sources of niacin. Table 6.7 provides data on its occurrence in food.

6.3.4.3 Stability, Degradation

Nicotinic acid is quite stable. Moderate losses of up to 15% are observed (cf. Tables 6.1 and 6.2) in blanching of vegetables. The loss is 25–30% in the first days of ripening of meat.

6.3.5 Pantothenic Acid

6.3.5.1 Biological Role

Pantothenic acid (Formula 6.14) is the building unit of coenzyme A (CoA), the main carrier of acetyl and other acyl groups in cell metabolism. Acyl groups are linked to CoA by a thioester bond. Pantothenic acid occurs in free form in blood plasma, while in organs it is present as CoA. The highest concentrations are in liver, adrenal glands, heart and kidney.

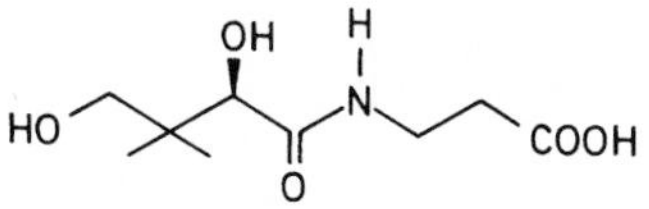

(6.14)

Only the R enantiomer occurs in nature and is biologically active. A normal diet provides an adequate supply of the vitamin.

6.3.5.2 Requirement, Occurrence

The daily adult requirement is 6–8 mg. The concentration in blood is 10–40 µg/100 ml and 2–7 mg/day are excreted in urine.
Table 6.7 lists data on pantothenic acid occurrence in food.

6.3.5.3 Stability, Degradation

Pantothenic acid is quite stable. Losses of 10% are experienced in processing of milk. Losses of 10–30%, mostly due to leaching, occur during cooking of vegetables.

6.3.6 Biotin

6.3.6.1 Biological Role

Biotin is the prosthetic group of carboxylating enzymes, such as acetyl-CoA-carboxylase, pyruvate carboxylase and propionyl-CoA-carboxylase, and therefore plays an important role in fatty acid biosynthesis and in glu-

coneogenesis. The carboxyl group of biotin forms an amide bond with the ε-amino group of a lysine residue of the particular enzyme protein. Only the (3aS, 4S, 6aR) compound, D-(+)-biotin, is biologically active:

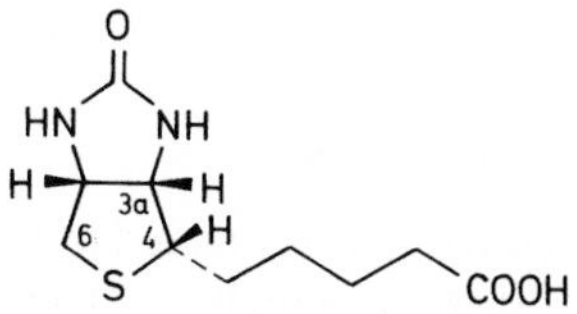

(6.15)

Biotin deficiency rarely occurs. Consumption of large amounts of raw egg white might inactivate biotin by its specific binding to avidin (cf. 11.2.3.1.9).

6.3.6.2 Requirement, Occurrence

The daily adult requirement is 150–300 µg. An indicator of sufficient biotin supply is the excretion level in the urine, which is normally 30–50 µg/day. A deficiency is indicated by a drop to 5 µg/day.
Biotin is not free in food, but is bound to proteins. Table 6.7 provides data on its occurrence in food.

6.3.6.3 Stability, Degradation

Biotin is quite stable. Losses during processing and storage of food are 10–15%.

6.3.7 Folic Acid

6.3.7.1 Biological Role

The tetrahydrofolate derivative (II) of folic acid (I in Formula 6.16) is the cofactor of enzymes which transfer single carbon units in various oxidation states, e.g., formyl or hydroxymethyl residues. In transfer reactions the single carbon unit is attached to the N^5- or N^{10}-atom of tetrahydrofolic acid.

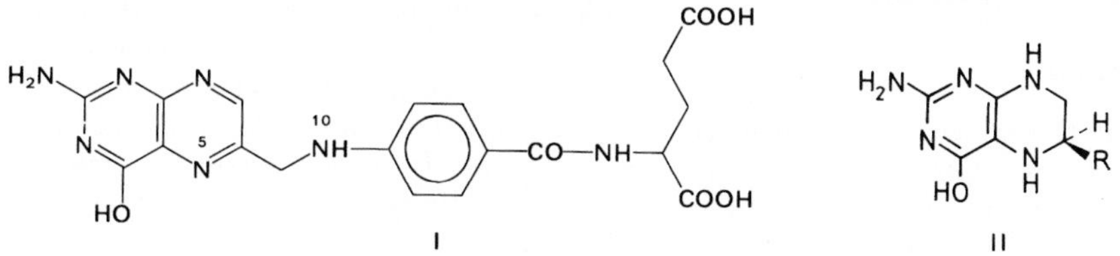

(6.16)

Folic acid deficiency can occur by insufficient supply in the diet, malfunction of resorptive processes, or therapy which uses a folic acid antagonist. Deficiency is detected by a decrease in folic acid concentration in red blood cells and plasma, and by a change in blood cell patterns.

6.3.7.2 Requirement, Occurrence

The daily adult requirement is estimated to be 0.4–0.8 mg. A sufficient supply can be monitored by the level of free folic acid in blood serum or red blood cells. Serum values of 5–20 ng/ml are normal, while less than 5 ng/ml is a deficiency level. When folic acid is lacking, there is increased excretion of formiminoglutamic acid, which is formed from histidine, since its conversion to glutamic acid as the final step in histidine degradation is dependent on folic acid.

In food folic acid is mainly bound to oligo-γ-L-glutamates made up of 1–8 glutamic acid residues. Unlike free folic acid, the resorption of this conjugated form is limited and occurs only after the glutamic acid moiety is cleaved by the enzyme folic acid conjugase, which occurs in intestinal mucosa. The folic acid content of foods varies. It is mainly present in conjugated from in vegetables, while it is in free form in liver. Data on folic acid occurrence in food are compiled in Table 6.7.

6.3.7.3 Stability, Degradation

Folic acid is quite stable. There is no destruction during blanching of vegetables, while cooking of meat gives only small losses. Losses in milk are apparently due to an oxidative process and parallel those of ascorbic acid. Ascorbate added to food preserves folic acid.

6.3.8 Cyanocobalamin (Vitamin B$_{12}$)

6.3.8.1 Biological Role

Cyanocobalamin (Formula 6.17) was isolated in 1948 from *Lactobacillus lactis*. Due to its stability and availability, it is the form in which the vitamin is used most often. In fact,

(6.17)

cyanocobalamin is formed as an artifact in the processing of biological materials. Cobalamins occur naturally as adenosylcobalamin and methylcobalamin, which instead of the cyano group contain a 5'-deoxyadenosyl residue and a methyl group respectively.

Adenosylcobalamin (coenzyme B$_{12}$) participates in rearrangement reactions in which a hydrogen atom and an alkyl residue, an acyl group or an electronegative group formally exchange places on two neighboring carbon atoms. Reactions of this type play a role in the metabolism of a series of bacteria. In mammals and bacteria a rearrangement reaction that depends on vitamin B$_{12}$ is the conversion of methylmalonyl CoA to succinyl CoA (cf. 10.2.8.3). Vitamin B$_{12}$ deficiency results in the excretion of methylmalonic acid in the urine.

Another reaction that depends on adenosylcobalamin is the reduction of ribonucleoside triphosphates to the corresponding 2'-deoxy compounds, the building blocks of deoxyribonucleic acids.

Methylcobalamin is formed, e.g., in the methylation of homocysteine to methionine with N^5-methyltetrahydrofolic acid as the intermediate stage. The enzyme involved is a cobalamin-dependent methyl transferase.

The resorption of cyanocobalamin is achieved with the aid of a glycoprotein, the "intrinsic factor" formed by the stomach mucosa. Deficiency of vitamin B$_{12}$ is usually caused by impaired resorption due to inadequate formation of "intrinsic factor" and results in pernicious anemia.

6.3.8.2 Requirement, Occurrence

The daily adult requirement of vitamin B_{12} is 3–4 µg. The plasma concentration is normally 450 pg/ml.

The ability of vitamin B_{12} to promote growth alone or together with antibiotics, for example in young chickens, suckling pigs and young hogs, is of particular importance. This effect appears to be due to the influence of the vitamin on protein metabolism, and it is used in animal feeding. The increase in feed utilization is exceptional with underdeveloped young animals. Vitamin B_{12} is of importance also in poultry operations (enhanced egg laying and chick hatching). The use of vitamin B_{12} in animal feed vitamin fortification is obviously well justified.

Liver, kidney, spleen, thymus glands and muscle tissue are abundant sources of vitamin B_{12} (Table 6.7). Consumption of internal organs (variety meats) of animals is one method of alleviating vitamin B_{12} deficiency symptoms in humans.

6.3.8.3 Stability, Degradation

The stability of vitamin B_{12} is very dependent on a number of conditions. It is fairly stable at pH 4–6, even at high temperatures. In alkaline media or in the presence of reducing agents, such as ascorbic acid or SO_2, the vitamin is destroyed to a greater extent.

6.3.9 L-Ascorbic Acid (Vitamin C)

6.3.9.1 Biological Role

Ascorbic acid (L-3-keto-threo-hexuronic acid-γ-lactone, Formula 6.18, I) is involved in hydroxylation reactions, e.g., biosynthesis of catecholamines, hydroxyproline and corticosteroids (11-β-hydroxylation of deoxycorticosterone and 17-β-hydroxylation of corticosterone). Vitamin C is fully resorbed and distributed throughout the body, with the highest concentration in adrenal and pituitary glands.

About 3% of the body's vitamin C pool, which is 20–50 mg/kg body weight, is excreted in the urine as ascorbic acid, dehydroascorbic acid (a combined total of 25%) and their metabolites, 2,3-diketo-L-gulonic acid (20%) and oxalic acid (55%). An increase in excreted oxalic acid occurs only with a very high intake of ascorbic acid.

Scurvy is caused by a dietary deficiency of ascorbic acid.

6.3.9.2 Requirement, Occurrence

The daily adult requirement is about 45–80 mg. An indicator of insufficient vitamin supply in the diet is a low level in blood plasma (0.4 mg/100 ml).

Vitamin C is present in all animal and plant cells, mostly in free form, and it is probably bound to protein as well. Vitamin C is particularly abundant in rose hips, black and red currants, strawberries, parsley, oranges, lemons (in peels more than in pulp), grapefruit, a variety of cabbages and potatoes. Vitamin C loss during storage of vegetables from winter through late spring can be as high as 70%.

Table 6.7 provides data on vitamin C occurrence in a variety of foods.

6.3.9.3 Stability, Degradation

Ascorbic acid (I) has an acidic hydroxyl group ($pK_1 = 4.04$, $pK_2 = 11.4$ at 25°C). Its UV absorption depends on the pH value (Table 6.9). Ascorbic acid is readily and reversibly oxidized to dehydroascorbic acid (II), which is present in aqueous media as a hydrated hemiketal (IV). The biological activity is lost when the dehydroascorbic acid lactone ring is irreversibly opened, giving rise to 2,3-diketogulonic acid (III), cf. Formula 6.18:

Table 6.9. Effect of pH on ultraviolet absorption maxima of ascorbic acid

pH	λ max (nm)
2	244
6–10	266
> 10	294

catalyzed oxidation, is maximal at pH 4 and minimal at pH 2. It probably proceeds through the ketoform of ascorbate, then via a keto-anion to diketogulonic acid:

$$(6.20)$$

Diketogulonic acid degradation products, xylosone and 4-deoxypentosone (Formula 6.21), are then converted into ethylglyoxal, various reductones (cf. 4.2.4.3.1), furfural and furancarboxylic acid.

$$(6.18)$$

The oxidation of ascorbic acid to dehydroascorbic acid and its further degradation products depends on a number of parameters. Oxygen partial pressure, pH, temperature and the presence of heavy metal ions are of great importance. Metal-catalyzed destruction proceeds at a higher rate than noncatalyzed spontaneous autoxidation. Traces of heavy metal ions, particularly Cu^{2+} and Fe^{3+}, result in high losses.

The principle of metal catalysis is schematically presented in Reaction 6.19 (Me = metal ion).

The rate of anaerobic vitamin C degradation, which is substantially lower than that of non-

$$(6.21)$$

$$(6.19)$$

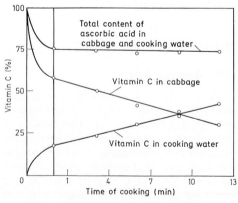

Fig. 6.3. Ascorbic acid losses as a result of cooking of cabbage (according to *Plank*, 1966)

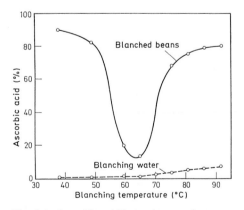

$$(6.22)$$

In the presence of amino acids, ascorbic acid, dehydroascorbic acid and their degradation products might be changed further by entering into *Maillard*-type browning reactions (cf. 4.2.4.4).

An example is the reaction of dehydroascorbic acid with amino compounds to give pigments, which can cause unwanted browning in citrus juices and dried fruits. The intermediates that have been identified are scorbamic acid (I in Formula 6.22), which is produced by *Strecker* degradation with an amino acid, and a red pigment (II).

A wealth of data is available on ascorbic acid losses during preservation, storage and processing of food. Tables 6.1 and 6.2 and Figs. 6.3 and 6.4 present several examples. Ascorbic acid degradation is often used as a general indicator of changes occurring in food.

6.4 Literature

Bässler, K. H.: On the problematic nature of vitamin E requirements: net vitamin E. Z. Ernährungswiss. *30*, 174 (1991)

Bässler, K. H., Lang, K.: Vitamine. Dr. Dietrich Steinkopff Verlag: Darmstadt. 1975

Counsell, J. N., Hornig, D. H. (Eds.): Vitamin C (Ascorbic Acid). Applied Science Publ.: London 1981

Farrer, K. T. H.: The thermal destruction of vitamin B_1 in foods. Adv. Food Res. *6*, 257 (1955)

Isler, O., Brubacher, G.: Vitamine I. Fettlösliche Vitamine. Georg Thieme Verlag: Stuttgart. 1982

Isler, O., Brubacher, G., Ghisla, S., Kräutler, B.: Vitamine II. Wasserlösliche Vitamine. Georg Thieme Verlag: Stuttgart, 1988

Körner, W. F., Völlm, J.: Vitamine. In: Klinische Pharmakologie und Pharmakotherapie. 3. Aufl. (Eds.: Kuemmerle, H. P., Garrett, E. R., Spitzy, K. H.), p. 361, Urban u. Schwarzenberg: München. 1976

Labuza, T. P., Riboh, D.: Theory and application of *Arrhenius* kinetics to the prediction of nutrient losses in foods. Food Technol. *36* (10), 66 (1982)

Fig. 6.4. Ascorbic acid losses in green beans versus blanching temperature (according to *Plank*, 1966)

Liao, M.-L., Seib, P.A.: Chemistry of L-ascorbic acid related to foods. Food Chem. *30*, 289 (1988)

Lund, D.B.: Effect of commercial processing on nutrients. Food Technol. *33* (2), 28 (1979)

Machlin, L.J. (Ed.): Handbook of Vitamins. Marcel Dekker: New York. 1984

Plank, R.: Handbuch der Kältetechnik. Bd. XIV, p. 475, Springer-Verlag: Berlin. 1966

Reiff, F., Paust, J., Meier, W., Fürst, A., Ernst, H.G., Kuhn, W., Härtner, H., Florent, J., Suter, C., Pollak, P.: Vitamine. Ullmanns Encyklopädie der technischen Chemie, 4th edn. Vol. 23, p. 621, Verlag Chemie: Weinheim. 1983

Seib, P.A., Tolbert, B.M. (Eds.): Ascorbic acid: Chemistry, metabolism, and uses. Advances in Chemistry Series 200, American Chemical Society: Washington, D.C. 1982

7 Minerals

7.1 Foreword

Minerals are the constituents which remain as ash after the incineration of plant and animal tissues. They may be divided into two categories; *main elements* (Ca, P, K, Cl, Na, Mg) and *trace elements* (Fe, Zn, Cu, Mn, I, Mo, etc.). According to their biological roles, they may also be divided into *essential elements*, for which the biological roles are known; *nonessential elements*, with unknown functions, if any; and *toxic elements*, which may be ingested through food or water or absorbed from the air.

Essential elements, including the main elements and a number of trace elements, fulfill various functions: as electrolytes, as enzyme constituents (cf. 2.3.3) and as building materials, e.g., in bones and teeth. Table 7.1 summarizes the mineral content of the human body.

Table 7.2 shows the content of sodium, potassium, calcium, iron, and phosphorus in some foods. In the same food raw material, the mineral content can fluctuate greatly depending on genetic and climatic factors, agricultural procedures, composition of the soil, and ripeness of the harvested crops, among other factors. This applies to both main and trace elements. Changes in the mineral content usually occur also in the processing of raw materials, e.g., in thermal processes and material separations. Table 7.3 shows data on mineral losses in food processing. Mineral supply depends not only on the intake in food but primarily on the bioavailability, which is essentially related to the composition of the food. Thus, the redox potential and pH value determine the valency state, solubility, and, consequently, resorption. A series of food constituents, e.g., proteins, peptides, amino acids, polysaccharides, sugars, lignin, phytin, and organic acids, bind minerals and enhance or inhibit their resorption.

The importance of minerals as food ingredients depends not only on their nutritional and physiological roles. They contribute to food flavor and activate or inhibit enzyme-catalyzed and other reactions, and they affect the texture of food.

7.2 Main Elements

7.2.1 Sodium

The sodium content of the body is 1.4 g/kg. Sodium is present mostly as an extracellular constituent and maintains the osmotic pressure of the extracellular fluid. In addition, it activates some enzymes, such as amylase. Sodium absorption is rapid; it starts 3–6 min after intake and is completed within 3 h. Daily intake of sodium ranges from 1.7–6.9 g; the adult's minimum requirement averages 460 mg/day. The intake of too little or too much sodium can result in serious disorders. From a nutritional standpoint, only excessive intake is of concern since it results in hypertension, i.e. abnormally high blood pressure. A low intake of sodium can be achieved by a nonsalty diet or by using diet salt (common salt substitutes, cf. 22.2.5). Table 7.2 provides values for the sodium content of some foods.

Table 7.1. Mineral content of the human body

Element	Content g/kg	Element	Content mg/kg
Calcium	10–20	Iron	70–100
Phosphorus	6–12	Zinc	20–30
Potassium	2–2.5	Copper	1.5–2.5
Sodium	1–1.5	Manganese	0.15–0.3
Chlorine	1–1.2	Iodine	0.1–0.2
Magnesium	0.4–0.5	Molybdenum	0.1

Table 7.2. Mineral content (Na, K, Ca, Fe, and P) of some foods

Food product	Na	K	Ca	Fe	P
Milk and dairy products					
Bovine milk, raw, high quality	48	157	120	0.046	92
Human milk	16	53	31	0.08	15
Butter	5	16	13	–	21
Cheese					
Emmental (45% fat)	450	107	1,020	0.31	636
Camembert (60% fat)	944	105	400	0.58	310
Camembert (30% fat)	900	120	600	0.17	540
Eggs					
Chicken egg yolk	51	138	140	7.2	590
Chicken egg white	170	154	11	0.2	21
Meat and meat products					
Beef, whole carcass, lean	58	342	11	2.6	170
Pork, whole carcass, lean	58	260	9	2.3	176
Calf liver	87	316	8.7	7.9	306
Pork liver	77	350	20	22.1	362
Chicken liver	68	218	18	7.4	240
Pork kidney	173	242	7	10	260
Blood sausage	680	38	6.5	6.4	22
Fish and fish products					
Herring	117	360	34	1.1	250
Eel	65	217	17	0.6	223
Cereals and cereal products					
Wheat, whole kernel	7.8	502	43.7	3.3	406
Wheat flour, type 405	2.0	108	15	1.95	–
Wheat flour, type 550	3.0	126	16	1.1	95
Wheat germ	5	837	69	8.1	1,100
Wheat gluten	2	1,390	43	3.6	1,240
Rye, whole kernel	40	530	64	–	373
Rye flour, type 997	1	240	31	2.2	–
Corn, whole kernel	6	330	15	–	256
Breakfast cereals (corn flakes)	915	139	13	2.0	59
Oat flakes	5	335	54	4.6	391
Rice, unpolished	10	150	23	2.6	325
Rice, polished	6	103	6	0.6	120
Vegetables					
Watercress	12	276	180	3.1	64
Mushrooms (cultivated)	8	422	8	1.26	123
Chicory	4.4	192	26	0.74	26
Endive	53	346	54	1.4	54
Peas, green	2	304	24	1.84	108
Lamb's lettuce	4	421	35	2.0	49
Kale	42	490	212	1.9	87
Potatoes	3.2	443	9.5	0.8	50
Kohlrabi	32	380	68	0.9	49.7
Head lettuce	10	224	37	1.1	33
Vegetables					
Lentils, dried	4	810	74	6.9	412
Carrots	60	290	41	0.66	35
Brussels sprout	7	411	31	1.1	84

Table 7.2 (continued)

Food product	Na	K	Ca	Fe	P
Vegetables					
Spinach	65	633	126	4.1	55
Edible mushroom (*Boletus edulis*)	6	486	9	1.0	115
Tomato	6	297	14	0.5	26
White cabbage	13	227	46	0.5	28
Fruits					
Apple	3	144	7	0.48	12
Orange	1.4	177	42	0.4	23
Apricots	2	278	16	0.65	21
Strawberry	2.5	147	26	0.96	29
Grapefruit	1.6	180	18	0.34	17
Rose hips	146	291	257	–	258
Currants-red	1.4	238	29	0.91	27
Currants-black	1.5	310	46	1.29	40
Cherries-sour	2	114	–	0.6	7
Plums	1.7	221	14	0.44	18
Sea buckthorn	3.5	133	42	0.44	9
Yeast					
Baker's yeast, pressed	34	649	28	4.9	605
Brewer's yeast, dried	–	1,410	–	17.6	1,900

[a] Data are in mg/100 g edible portion (average values).

Table 7.3. Mineral losses in food processing

Raw material	Product	Loss (%)						
		Cr	Mn	Fe	Co	Cu	Zn	Se
Spinach	Canned		87		71		40	
Beans	Canned						60	
Tomatoes	Canned						83	
Carrots	Canned				70			
Beetroot	Canned				67			
Green beans	Canned				89			
Wheat	Flour		89	76	68	68	78	16
Rice	Polished rice	75	26			45	75	

7.2.2 Potassium

The concentration of potassium in the body is 2 g/kg. Potassium is localized mostly within the cells. It regulates the osmotic pressure within the cell, is involved in cell membrane transport and also in the activation of a number of glycolytic and respiratory enzymes. The potassium intake in a normal diet is 2–5.9 g/day. The minimum daily requirement is estimated to be 782 mg. Potassium deficiency is associated with a number of symptoms and may be a result of undernourishment or predominant consumption of potassium-deficient foods, e. g., white bread, fat or oil. The potassium content in food is summarized in Table 7.2. Potatoes and molasses are particularly abundant sources.

7.2.3 Magnesium

The concentration of magnesium in the body is 250 mg/kg. The daily requirement is 300–350 mg. In a normal diet, the daily intake is 300–500 mg. As a constituent and activator of many enzymes, particularly those associated with the conversion of energy-rich phosphate compounds, and as a stabilizer of plasma membranes, intracellular membranes, and nucleic acids, magnesium is a life-supporting element. Because of its indispensable role in body metabolism, magnesium deficiency causes serious disorders.

7.2.4 Calcium

The total amount of calcium in the body is about 1,500 g. Because of the large amounts of calcium all over the body, it is one of the most important minerals. It is abundant in the skeleton and in some body tissues. Calcium is considered an essential nutrient because of its importance in building and maintaining bones and its role in blood clotting and muscle contraction. Therefore, calcium deficiency also causes serious disorders. The daily requirement is 0.8–1.0 g, while the average intake in a normal diet is 0.8–0.9 g. The main dietary source of calcium is milk and other dairy products. Fruits and vegetables, cereals, meat, fish and eggs have much lower amounts. Table 7.2 provides data about the occurrence of calcium in foods.

7.2.5 Chloride

The chloride content of human tissue is 1.1 g/kg body weight. Chloride intake, mostly in the form of common salt (NaCl), is 3–12 g/day. Chloride serves as a counter ion for sodium in extracellular fluid and for hydrogen ions in gastric juice. Chloride absorption is as rapid as its excretion in the urine. An accurate estimate of the daily requirement of chloride is not available.

7.2.6 Phosphate

The total phosphorus content in the body is about 700 g. The daily requirement is about 0.8–1.2 g. The Ca/P ratio in food should be about 1. Phosphorus, in the form of phosphate, free or bound as an ester or present as an anhydride, plays an important role in metabolism and, as such, is an essential nutrient. The organic forms of phosphorus in food are cleaved by intestinal phosphatases and, thereby, absorption occurs mostly in the form of inorganic phosphate. Polyphosphates, used as food additives, are absorbed only after prior hydrolysis into orthophosphate. The extent of hydrolysis is influenced by the degree of condensation of the polyphosphates. Table 7.2 includes a compilation of the phosphorus content of some foods.

7.3 Trace Elements

7.3.1 General Remarks

There are 15 essential trace elements present in hormones, vitamins, enzymes and other proteins which have distinct biological roles. In addition, numerous other elements occur in the human body and their physiological roles have not yet been determined. They are usually associated with related elements, e.g., Li with Na, or Rb with K. Table 7.4 summarizes the content of essential and nonessential trace elements in the body and their average daily intake in food. Toxic elements (e. g. Sb, Cd, Hg, Tl and Pb) may appear in food by various routes. These toxicants are dealt with separately in Chapter 9.2. A deficiency in the essential trace elements results in metabolic disorders that are primarily associated with the absence or decreased activity of metabolic enzymes.

Table 7.4. Trace elements in the human body and their daily intake[a]

Element	Content (mg/kg body weight)	Intake (mg/day)
Essential		
Fe	60	15
F	37	2.5
Zn	33	6–22
Si	ca. 14	33[b]
Cu	1.5	3.2
B	0.7	1.3–4.3
V	0.3	0.02
As	0.3	0.02–0.03
Se	0.2	0.07
Mn	0.2	2–48
I	0.2	0.2
Ni	0.1	0.4
Mo	0.1	0.3
Cr	0.1	0.005–0.2
Co	0.02	0.002–0.1
Nonessential		
Rb	4.6	1–2
Br	2.9	7.5
Al	0.9	5–35
Ba	0.3	1.3
Sn	0.2	4.0
Ti	0.1	0.9
Au	0.1	
Sb	0.1	
Te	0.1	0.2
Li	0.03	2.0
Cs	0.02	
U	0.001	
Bi	0.0004	

[a] Average values.
[b] Silicon intake is strongly influenced by overall food composition (vegetarian nutrition ≫ meat based nutrition).

7.3.2 Essential Trace Elements

7.3.2.1 Iron

The iron content of the body is 4–5 g. Most of it is present in the hemoglobin (blood) and myoglobin (muscle tissue) pigments. The metal is also present in a number of enzymes (peroxidase, catalase, hydroxylases and flavine enzymes), hence it is an essential ingredient of the daily diet. The iron requirement depends on the age and sex of the individual, it is about 1–2.8 mg/day. Iron supplied in the diet must be in the range of 5–28 mg/day in order to meet this daily requirement. The large variation in intake can be explained by different extents of absorption of the various forms of iron present in food (organic iron compounds *vs* simple salts). The most utilizable source is iron in meat, for which the extent of absorption is 20–30%. The absorption is much less from liver (6.3%) and fish (5.9%), or from cereals, vegetables and milk, from which iron absorption is the lowest (1.0–1.5%). Eggs decrease and ascorbic acid increases the extent of absorption. Bran interferes with iron absorption due to the high content of phytate. Apparently, the absorption of iron present in food is, in a healthy organism, regulated by the requirement of the organism. Nevertheless, in order to provide a sufficient supply of iron to persons who require higher amounts (children, women before menopause and pregnant or nursing women), cereals (flour, bread, rice, pasta products) fortified with iron to the extent of 55–130 mg/kg are recommended. Extensive feeding tests with chickens and rats have shown that $FeSO_4$ is the most suitable form of iron, but ferrous gluconate and ferrous glycerol phosphate are also efficiently absorbed. Two food processing problems arising from mineral fortification are the increased probability that oxidation will occur and, in the case of wheat flour, decreased baking quality. Generally, iron is an undesirable element in food processing; for example, iron catalyzes the oxidation of fat or oil, increases turbidity of wine and, as a constituent of drinking water, it supports the growth of iron-requiring bacteria. The iron content of various foods is shown in Table 7.2.

7.3.2.2 Copper

The amount of copper in the body is 100–150 mg. Copper is a component of a number of oxidoreductase enzymes (cytochrome oxidase, superoxide dismutase, tyrosinase, uricase, amine oxidase). In blood plasma, it is bound to ceruloplasmin, which catalyzes the oxidation of Fe^{2+} to Fe^{3+}. This reaction is of great significance since it is only the Fe^{3+} form in blood which is transported by the transferrin protein to the iron pool in the liver. The daily copper requirement is $1-2$ mg and it is supplied in a normal diet. Copper is even less desirable than iron during food processing and storage since it catalyzes many unwanted reactions (e. g., oxidative destruction of ascorbic acid).

7.3.2.3 Zinc

The total zinc content in adult human tissue is $2-4$ g. The daily requirement of $6-22$ mg is provided by a normal diet. Zinc is a component of a number of enzymes (e.g., alcohol dehydrogenase, lactate dehydrogenase, malate dehydrogenase, glutamate dehydrogenase, carboxypeptidases A and B, and carbonic anhydrase). Other enzymes, e.g., dipeptidases, alkaline phosphatase, lecithinase and enolase, are activated by zinc and by some other divalent metal ions. Zinc deficiency in animals causes serious disorders, while high zinc intake by humans is toxic. Zinc poisoning has been reported as a result of consumption of soured food kept in zinc-plated metal containers (e. g., potato salad from institutional catering services).

7.3.2.4 Manganese

The body contains a total of $10-40$ mg of manganese. The daily requirement, $2-48$ mg, is met by the normal daily food intake. Manganese is the metal activator for pyruvate carboxylase and, like some other divalent metal ions, it activates various enzymes, such as arginase, amino peptidase, alkaline phosphatase, lecithinase or enolase. Manganese, even in higher amounts, is relatively nontoxic.

7.3.2.5 Cobalt

The total cobalt content of the body is $1-2$ mg. Since it was discovered that vitamin B_{12} contains cobalt as its central atom, the nutritional importance of cobalt has been emphasized and it has been assigned the status of an essential element. Its requirement is met by normal nutrition.

7.3.2.6 Vanadium

The total vanadium content of the body is $17-43$ mg. Feeding tests using chickens and rats have indicated a growth promoting effect.

Obviously, this element has a biological role. The intake in food amounts to $12-30$ µg/day.

7.3.2.7 Chromium

The chromium content of the body varies considerably depending on the region; the range is $6-12$ mg. The daily intake also varies greatly from 5 to 200 µg. The supply is considered suboptimal. Chromium is important in the utilization of glucose. For instance, it activates the enzyme phosphoglucomutase and increases the activity of insulin; therefore, chromium deficiency causes a decrease in glucose tolerance. And the risk of cardiovascular disease increases. Chromium, as the chromate ion, proved to be nontoxic when used at 25 ppm in a long-term feeding experiment with rats.

7.3.2.8 Selenium

The selenium content in humans is $10-15$ mg, while the daily intake is $0.05-0.1$ mg. Depending on the region, it can vary greatly because of the varying content of selenium in the soil. Selenium is an antioxidant and can enhance tocopherol activity. The enzyme glutathione peroxidase contains selenium. It catalyzes the following reaction, protecting membranes from oxidative destruction:

$$ROOH + 2GSH \rightarrow ROH + H_2O + GSSG \tag{7.1}$$

Selenium toxicity, for example, its strong carcinogenic activity, is well known from numerous animal feeding studies and from diseases of cattle grazing in pastures on selenium-enriched soil. Serious disorders are caused by as little as $2-8$ ppm selenium in animal feed.

7.3.2.9 Molybdenum

The body contains $8-10$ mg of molybdenum. Daily intake in food is approx. 0.3 mg. It is a component of aldehyde oxidase and xanthine oxidase. The bacterial nitrate reductase involved in meat curing and pickling processes contains molybdenum. High levels of the metal are toxic, as has been shown by cattle grazing on molybdenum-enriched soil. The grass on such soil contains $20-100$ µg molybdenum/g dry matter.

7.3.2.10 Nickel

Nickel is an activator of a number of enzymes, e.g., alkaline phosphatase and oxalacetate decarboxylase, which can also be activated by other divalent metal ions. Nickel also enhances insulin activity. The essential role of nickel has been established by inducing deficiency symptoms in feeding experiments with chickens and rats. These symptoms include changes in the liver mitochondria. The daily intake in food amounts to 150–700 µg. The nickel requirement is estimated to be 35–500 µg/day.

7.3.2.11 Boron

Boron is found in humans and animals. As an ultratrace element, it apparently influences the mineral metabolism of higher animals via an effect on the activity of parathyrin. It is also important for some plants. In fact, heartrot and dry rot disease in sugar beet and browning disease in kohlrabi are due to boron deficiency. Boron is present in many foods. Based on the fresh weight, fruits contain 5–30, vegetables 0.5–2, cereals 0.5–3, eggs 0.1, and milk 0.1–0.2 ppm of boron. The intake of boron amounts to 1.3–4.3 mg/day and depends, for instance, on the amount of wine consumed, which contains 10 mg/l of boron. The requirement is estimated to be >0.4 µg/g of food. At higher concentrations, boric acid accumulates in adipose tissue and especially in the central nervous system. Since the implications of this storage pool are still unknown, it is no longer used in the preservation of food.

7.3.2.12 Silicon

Silicon, as soluble silicic acid, is rapidly absorbed. The silicon content of the body is approx. 1 g. The main source is cereal products. Silicon promotes growth and thus has a biological role. The toxicity of silicic acid is apparent only at concentrations ≥ 100 mg/kg. The intake in food amounts to 21–46 mg/day.

7.3.2.13 Fluorine

The body contains 2.6 g fluorine. It plays an essential role, as indicated by feeding experiments with rats and mice – deficient diets containing less than 2.5 ppm and 0.1–0.3 ppm, respectively, resulted in disorders in growth and reproduction. The positive effect of fluorine on teeth caries is well established. The addition to drinking water of 0.5–1.5 ppm fluorine in the form of NaF or $(NH_4)_2SiF_6$ inhibits tooth decay. Its beneficial effect appears to be in retarding solubilization of tooth enamel and inhibiting the enzymes involved in development of caries. Toxic effects of fluorine appear at a level of 2 ppm. Therefore, the beneficial effects of fluoridating drinking water are disputed by some and it is a controversial topic of mineral nutrition.

7.3.2.14 Iodine

The content of iodine in the body is about 10 mg, of which the largest portion (70–80%) is covalently bound in the thyroid gland. Iodine absorption from food occurs exclusively and rapidly as iodide and is utilized in the thyroid gland in the biosynthesis of the hormone thyroxine (tetraiodothyronine) and its less iodized form, triiodothyronine. In this process, the iodide ion is first oxidized, then iodization of the tyrosine residues of thyroglobulin occurs. Diiodotyrosine condenses with itself or with monoiodotyrosine to form thyroglobulin-bound thyroxine or triiodothyronine. Both active hormones are released from thyroglobulin by the action of a proteinase. Also released are several peptides which, however, lack activity. The iodine requirement of humans is 100–200 µg/day. Iodine deficiency results in enlargement of the thyroid gland (iodine-deficiency induced goiter). There is little iodine in most food. Good sources are milk, eggs and, above all, seafood. Drinking water contributes little to the body's iodine supply. In areas where goiter is found, the water has 0.1–2.0 µgI/l, while in goiter-free districts, 2–15 µgI/l are present in drinking water. To avoid diseases caused by low iodine supply, some countries with iodine-deficient districts employ prophylactic measures to combat the deficiency symptoms. This involves iodization of common salt with KI, with 100 µg iodine added to 1–10 g NaCl. Higher amounts of iodine are toxic and, as shown with rats, disturb the animal's normal reproduction and lactation. In humans, diseases of the thyroid can develop.

7.3.2.15 Arsenic

Arsenic was shown to be an essential trace element for the growth of chickens, rats, and goats. Its metabolic role is not yet understood. It appears to be involved in the metabolism of methionine. Choline can be replaced by arsenocholine in some of its functions. The possible human requirement is estimated to be 12–25 µg/day. The intake in food amounts to 20–30 µg/day. The main source is fish.

7.3.3 Some Nonessential Trace Elements

7.3.3.1 Tin

Tin occurs in all humans organs. Although a growth-promoting effect was detected in rats, it is disputed. The natural level of tin in food is very low, but it can be increased in the case of foods canned in tinplate cans. Very acidic foods can often dissolve substantial amounts of tin. Thus, the concentration of tin in pineapple and grapefruit juice transported in poorly tin plated cans was 2 g/l. The tin content of foods in tinplate cans is generally below 50 mg/kg and should not exceed 250 mg/kg. In the form of inorganic compounds, tin is resorbed only to a low extent and, therefore, it is only slightly toxic. In comparison, organic tin compounds can be very toxic.

7.3.3.2 Aluminum

The body contains 50–150 mg of aluminum. Higher levels are found in aging organisms. The daily average intake of aluminum is 2–10 mg. It is resorbed in only negligible amounts by the gastrointestinal tract. The largest portion is eliminated in feces. Excretion of aluminum in urine is less than 0.1 mg/day. It is not secreted in milk. Animal feeding tests with high levels of aluminum in the diet through several generations showed that aluminum is nontoxic. This seems to be true also for humans. Hence, the reluctance to use aluminum cookware in food processing is unfounded. Some recent studies, however, have revealed that a pathologically caused accumulation of aluminum in humans can cause significant damage to the cells of the central nervous system.

7.4 Minerals in Food Processing

The contribution of minerals to the nutritive/physiological value and the physical state of food has been covered in the Foreword of this Chapter and under the individual elements.

However, there are metal ions, derived from food itself or acquired during food processing and storage, which interfere with the quality and visual appearance of food. They can cause discoloration of fruit and vegetable products (cf. 18.1.2.5.7) and many metal-catalyzed reactions are responsible for losses of some essential nutrients, for example, ascorbic acid oxidation (cf. 6.3.9.3). Also, they are responsible for taste defects or off-flavors, for example, as a consequence of fat oxidation (cf. 3.7.2.1.4). Therefore, the removal of many interfering metal ions by chelating agents (cf. 8.14) or by other means is of importance in food processing.

7.5 Literature

Lang, K.: Biochemie der Ernährung, 4. Aufl., Dr. Dietrich Steinkopff Verlag: Darmstadt. 1979

Pfannhauser, W.: Essentielle Spurenelemente in der Nahrung. Springer Verlag: Berlin, 1988

Smith, K.T.: Trace Minerals in Foods. Marcel Dekker: New York, 1988

Wolfram, G., Kirchgeßner, M. (Eds.): Spurenelemente und Ernährung. Wissenschaftl. Verlagsgesellschaft: Stuttgart, 1990

8 Food Additives

8.1 Foreword

A food additive is a substance (or a mixture of substances) which is added to food and is involved in its production, processing, packaging and/or storage without being a major ingredient. Additives or their degradation products generally remain in food, but in some cases they may be removed during processing. The following examples illustrate and support the use of additives to enhance the:

- *Nutritive Value of Food*
  Additives such as vitamins, minerals, amino acids and amino acid derivatives are utilized to increase the nutritive value of food. A particular diet may also require the use of thickening agents, emulsifiers, sweeteners, etc.
- *Sensory Value of Food*
  Color, odor, taste and consistency or texture, which are important for the sensory value of food, may decrease during processing and storage. Such decreases can be corrected or readjusted by additives such as pigments, aroma compounds or flavor enhancers. Development of "off-flavor", for instance, derived from fat or oil oxidation, can be suppressed by antioxidants. Food texture can be stabilized by adding minerals or polysaccharides, and by many other means.
- *Shelf Life of Food*
  The current forms of food production and distribution, as well as the trend towards convenient foods, have increased the demand for longer shelf life. Furthermore, the world food supply situation requires preservation by avoiding deterioration as much as possible. The extension of shelf life involves protection against microbial spoilage, for example, by using antimicrobial additives and by using active agents which suppress and retard undesired

chemical and physical changes in food. The latter is achieved by stabilization of pH using buffering additives or stabilization of texture with thickening or gelling agents, which are polysaccharides.

It is implicitly understood that food additives and their degradation products should be nontoxic at their recomended levels of use. This applies equally to acute and to chronic toxicity, particularly the potential carcinogenic, teratogenic (causing a malformed fetus) and mutagenic effects. It is generally recognized that additives are applied only when required for the nutritive or sensory value of food, or for its processing or handling. The use of additives is regulated by Food and Drug or Health and Welfare administrations in most countries. The regulations differ in part from country to country but there are endeavors under way to harmonize them on the basis of both current toxicological knowledge and the requirements of modern food technology. The most important groups of additives are outlined in the following sections. No reference is made to legislated regulations or definitions provided therein. A compilation of the relative importance of various groups of additives is presented in Table 8.1.

8.2 Vitamins

Many food products are enriched or fortified with vitamins to adjust for processing losses or to increase the nutritive value. Such enrichment is important, particularly for fruit juices, canned vegetables, flour and bread, milk, margarine and infant food formulations. Table 8.2 provides an overview of vitamin enrichment of food.

Several vitamins have some desirable additional effects. Ascorbic acid is a dough improver, but can play a role similar to toco-

Table 8.1. Utilization of food additives in United States (1965 as % of total additives used)[a]

Additives, class	% of total	Additives, class	% of total
Aroma		Chelating agents	2.6
compounds	42.5	Colors	2.1
Natural aroma		Chemical	
substances	21	preservatives	1.8
Nutritional		Stabilizers	1.8
fortifiers	6.9	Antioxidants	1.7
Surface active		Maturing and	
agents		bleaching	
(tensides)	5	agents	1.4
Buffering		Sweeteners	0.5
substances,		Other	
acids, bases	3.5	additives	9.4

[a] In 1965 a total of 1696 substances (= 100 %) were utilized.

pherol as an antioxidant. Carotenoids and riboflavin are used as coloring pigments, while niacin improves the color stability of fresh and cured and pickled meat.

8.3 Amino Acids

The increase in the nutritive value of food by addition of essential amino acids and their derivatives is dealt with in sections 1.2.5 and 1.4.6.3.

8.4 Minerals

Food is usually an abundant source of minerals. Fortification is considered for iron, which is often not fully available, and for calcium, magnesium, copper and zinc. Iodization of salt is of importance in iodine deficient areas (cf. 22.2.4).

8.5 Aroma Substances

The use of aroma substances of natural or synthetic origin is of great importance (cf. Table 8.1). The aroma compounds are dealt with in detail in Chapter 5 and in individual sections covering some food commodities.

Table 8.2. Examples of vitamin fortification of food

Vitamin	Food product
B_1	Cocoa powder and its products, beverages and concentrates, confectionary and other baked products
B_2	Baked products, beverages
B_6	Baked and pasta products
B_{12}	Beverages, etc.
Pantothenic acid	Baked products
Folic acid/biotin	Not commonly used as additives
C	Fruit drinks, desserts, dairy products, flour
A	Skim milk powder, breakfast cereals (flakes), beverage concentrates, margarine, baked products, etc.
D	Milk, milk powder, etc.
E	Various food products, e.g. margarine

8.6 Flavor Enhancers

These are compounds that enhance the aroma of a food commodity, though they themselves have no distinct odor or taste in the concentrations used. An enhancer's effect is apparent to the senses as "feeling", "volume", "body" or "freshness" (particularly in thermally processed food) of the aroma, and also by the speed of aroma perception ("time factor potentiator").

8.6.1 Monosodium Glutamate (MSG)

Glutamic acid was isolated by *Ritthausen* (cf. 1.2.2.2). In 1908 *Ikeda* found that MSG is the beneficial active component of the algae *Laminaria japonica*, used for a long time in Japan as a flavor improver of soup and similarly prepared food. The consumption of MSG in 1978 was 200,000 tonnes worldwide.

In a pH range of 5–8 and at its usually applied level of 0.2–0.5%, MSG has a pleasant, slightly salty-sweet taste and a property often described as "mouth satisfaction". It affects and promotes sensory perception, particularly of meat-like aroma notes, and is frequently used as an additive in frozen, dehydrated or canned fish and meat products.

The intake of larger amounts of MSG by some hypersensitive persons can trigger a "Chinese restaurant syndrome", which is characterized by temporary disorders such as drowsiness, headache, stomach ache and stiffening of joints. These disappear after a short time.

8.6.2 5′-Nucleotides

5′-Inosine monophosphate (IMP, disodium salt) and 5′-guanosine monophosphate (GMP, disodium salt) have properties similar to MSG but heightened by a factor of 10–20. Their flavor enhancing ability at 75–500 ppm is good in all food (e. g. soups, sauces, canned meat or tomato juice). However, some other specific effects, besides the "MSG effect", have been described for nucleotides. For example, they imprint a sensation of higher viscosity in liquid food. The sensation is often expressed as "freshness" or "naturalness", the expressions "body" and "mouthfeel" being more appropriate for soups. Synergistic flavor-enhancing effects are experienced with simultaneous use of MSG and IMP or GMP (Fig. 8.1).

8.6.3 Maltol

Maltol (3-hydroxy-2-methyl-4-pyrone, cf. 5.3.1.2) has a caramel-like odor (melting point 162–164 °C). It enhances the perception of sweetness in carbohydrate-rich food (e. g. fruit juices, marmalades, fruit jelly). Addition of 5–75 ppm maltol allows a decrease of sugar content by about 15%, while retaining the sweetness intensity.

8.6.4 Other Compounds

Numerous patents have been issued or are pending for other compounds that influence, improve or enhance the flavor of food or suppress or modify unpleasant flavor notes. The significance of these compounds, with some of them active at exceptionally low levels, will obviously rise. Simultaneously, their analytical determination will become a challenge.

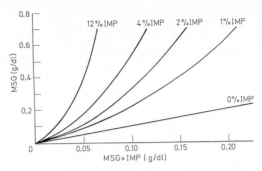

Fig. 8.1. Synergistic activities of Na-glutamate (MSG) and disodium-inosine monophosphate (IMP). The curves give the concentrations of MSG and MSG + IMP in water that are rated as being sensory equivalent by a taste panel

An example is dioctyl sodium sulfosuccinate

$$(8.1)$$

which in low concentrations provides a perception of "freshness" to sterilized milk. N,N′-di-o-tolylethylenediamine, in a concentration range of 5×10^{-7} ppb to 10 ppb,

$$(8.2)$$

enhances the butter-like aroma note of margarine and the milk flavor of reconstituted milk powder.

8.7 Sugar Substitutes

Sugar substitutes are those compounds that are used like sugars (sucrose, glucose) for sweetening, but are metabolized without the influence of insulin. Important sugar substitutes are the sugar alcohols, sorbitol, xylitol and mannitol and, to a certain extent, fructose (cf. 19.1.4.5–19.1.4.9).

8.8 Sweeteners

Sweeteners are natural or synthetic compounds which imprint a sweet sensation and
possess no or negligible nutritional value
("nonnutritive sweeteners") in relation to the
extent of sweetness.

There is considerable interest in new sweeteners. The rise in obesity in industrialized
countries has established a trend for calorie-
reduced nutrition. Also, there is an increased
discussion about the safety of saccharin and
cyclamate, the two sweeteners which were predominant for a long time. The search for new
sweeteners is complicated by the fact that the
relationship between chemical structure and
sweetness perception is not yet satisfactorily
resolved. In addition, the safety of suitable
compounds has to be certain. Some other criteria must also be met, for example, the compound must be adequately soluble and stable
over a wide pH and temperature range, have a
clean sweet taste without side or post-flavor
effects, and provide a sweetening effect as
cost-effectively as does sucrose. At present,
some new sweeteners are on the market (e.g.,
acesulfame and aspartame). The application of
a number of other compounds will be discussed here.

The following sections describe several sweeteners, irrespective of whether they are approved, banned or are just being considered for
future commercial use.

8.8.1 Sweet Taste: Structural Requirements and Molecular Biological Aspects

8.8.1.1 Structure-Activity Relationships in Sweet Compounds

A sweet taste can be derived from compounds
with very different chemical structures. *Shallenberger* and *Acree* consider that, for sweetness, a compound must contain a proton
donor/acceptor system (AH_s/B_s-system),
which has to meet some steric requirements
and which can interact with a complementary
receptor system (AH_r/B_r-system) by involvement of two hydrogen bridges (Fig. 8.2). The
expanded model of *Kier* has an additional
hydrophobic interaction with a group, X, pre-

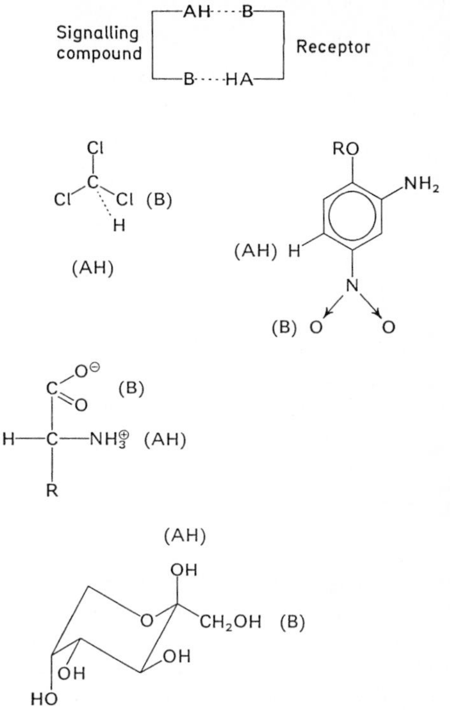

Fig. 8.2. AH/B-systems of various sweet compounds

sent at a distinct position of the molecule (Fig.
8.3). The examples in Figs. 8.2 and 8.3 show
that these models are applicable to many sweet
compounds from highly different classes. An
enlarged model substitutes a nucleophilic/
electrophilic system (n_s/e_s system) for the
AH_s/B_s system and an extended hydrophobic
contact for the localized contact with group X.
Thus, a receptor for sweet compounds is to be
depicted schematically as a hydrophobic
pocket, containing a complementary n_r/e_r
system.

It has been shown with numerous compounds
that as the hydrophobicity and the space-filling
properties of hydrophobic groups increase, the
sweetening strength increases, passes through
a maximum, and finally reaches a limit beyond
which the sweet taste is either quenched or
changes into a bitter taste.

When the e_s/n_s systems of computer-generated
molecular models of sweet and structurally
related non-sweet compounds are superim-

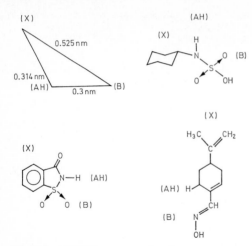

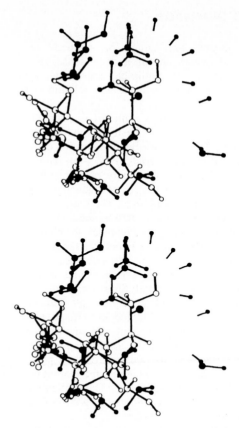

Fig. 8.3. AH/B/X-systems of various sweet compounds

posed, the hydrophobic groups of these compounds occupy different spatial regions relative to the e/n system. In this way, it is possible to deduce spatial positions that are or are not permitted to produce sweetness and, consequently, to get an idea of the size and form of the hydrophobic pockets of sweetness receptors. Figure 8.4 shows a model obtained by using this method and developed on the basis of benzisothiazolone dioxides (saccharin and related compounds), oxathiazinone dioxides (acesulfame and related compounds), amino acids, aminobenzoic acids, and halobenzoic acids. It shows in stereoscopic representation the superimposed e/n systems and the positions that are not permitted for sweetness. Sucrose is an example of a sweet compound placed in this "receptor space".

Fig. 8.4. Model of a sweetness receptor obtained by superimposing the e/n systems of sweet and structurally related non-sweet compounds. Sucrose (saccharose) is placed in the conformation of lowest energy with 3'-OH/2-OH as the e/n system. (○/● permitted/forbidden positions for sweetness; stereoscopic representation) (according to *Rohse* and *Belitz*, 1991)

It follows from present studies that the position of hydrophobic groups relative to the e/n system is of great importance for the appearance and intensity of sweetness. It is evident that at least two hydrophobic groups, which occupy different spatial regions relative to the e/n system, are characteristic of compounds with high sweetening strength. Examples are certain guanidines (cf. 8.8.14.3) or superaspartame (cf. 8.8.17) which are substantially sweeter than aspartame, saccharin or acesulfame and, unlike these compounds, contain two hydrophobic groups each. It is possible that apart from the e/n system, other polar

groups also participate in contacts with the receptor.

While *two* polar (n_s/e_s) groups should be present in sweet compounds and, when necessary, be supplemented with a hydrophobic group, a bitter tasting compound requires the presence of only *one* polar (n_s or e_s) and *one* hydrophobic group. The sweetening strength of a compound can be measured numerically and expressed as:

● Threshold detection value, c_{tsv} (the lowest concentration of an aqueous solution that can still be perceived as being sweet).

- Relative sweetening strength of a substance X, related to a standard substance S, which is the quotient of the concentrations c (w/w per cent or mol/l) of isosweet solutions of S and X:

$$f(c_s) = \frac{c_s}{c_x} \quad \text{for} \ \rightarrow \ c_s \ \text{isosweet} \ c_x \qquad (8.3)$$

Saccharose in a 2.5 or 10% solution usually serves as the standard substance ($f_{sac,g}$). Since the sweetening strength is concentration dependent (cf. Fig. 8.5), the concentration of the reference solution must always be given ($f(c_s)$). When the sweetening strength of a substance is expressed as $f_{sac,g}(10) = 100$, this means, e.g., that the substance is 100 times sweeter than a 10% saccharose solution or a 0.1% solution of this substance is isosweet with a 10% saccharose solution.

8.8.1.2 Molecular Biological Aspects of Sweetness

Experimental evidence suggests that the previously unknown receptors for sweet compounds and other receptors which are known to be coupled to G proteins (guanine nucleotide binding proteins) are homologous. These include, e.g., the photoreceptor rhodopsin, α- and β-adrenergic receptors, dopamine receptors, and other receptors for neurotransmitters and neuropeptides. As shown in Fig. 8.6, the following cascade for the transfer of taste perception is probable. The sweet compound (S) stereospecifically binds to the receptor protein (R) localized in the membrane of the taste cell sensitive to sweetness.

The receptor protein undergoes a change in conformation and interacts with a G protein of the type G_s (G), which then activates adenyl cyclase (C). This enzyme synthesizes 3′, 5′-cyclic AMP (cAMP) from ATP. As a second messenger, cAMP stimulates a cAMP-dependent protein kinase (Pk-A), resulting in the phosphorylation of proteins of a $K^{\oplus}$-channel which subsequently closes. The decreasing $K^{\oplus}$ transport into the cell causes depolarization of the membrane, $Ca^{2\oplus}$ uptake, and release of neurotransmitter at the synapse, thereby eliciting an action potential in the nerve cell.

At present, the genes that code the sweetness receptor proteins are being isolated.

8.8.2 Saccharin

Saccharin is an important sweetener ($f_{sac,g}(4) = 320$) and is mostly used in the form of the Na

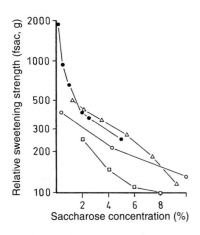

Fig. 8.5. Relative sweetening strength of some sweeteners as a function of the saccharose concentration (● neohesperidin dihydrochalcone, △ saccharin, ○ aspartame, □ acesulfame K) (according to *Bär* et al. 1990)

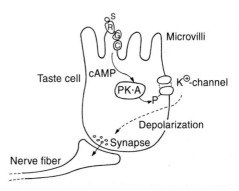

Fig. 8.6. Schematic representation of the transfer of sweetness perception (cf. text, S: sweet compound, R: receptor protein, G: G_s protein, C: adenyl cyclase, Pk-A: protein kinase. R, G, and C are localized in the apical membrane of the microvilli of the taste cell, and the $K^{\oplus}$-channel and the synapse in the basolateral membrane) (according to *Lancet* and *Ben-Arie*, 1991)

or Ca salt. At higher concentrations, this compound has a slightly bitter after-taste. The present stipulated ADI value is $0-2.5$ mg/kg of body weight. The synthesis of saccharin usually starts with toluene (*Remsen/Fahlberg* process, Formula 8.4) or sometimes with the methyl ester of anthranilic acid (*Maumee* process, Formula 8.5).

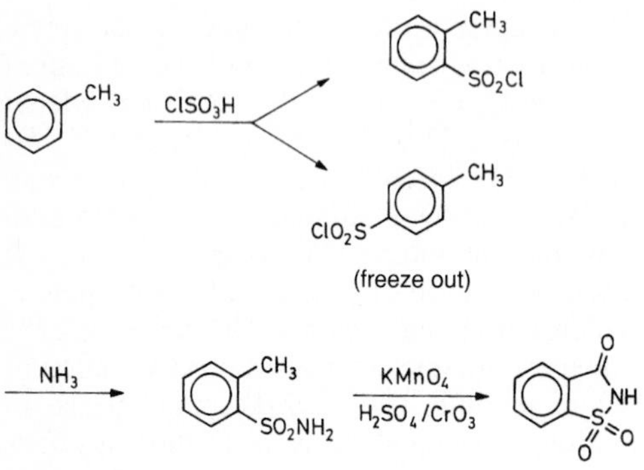

(8.4)

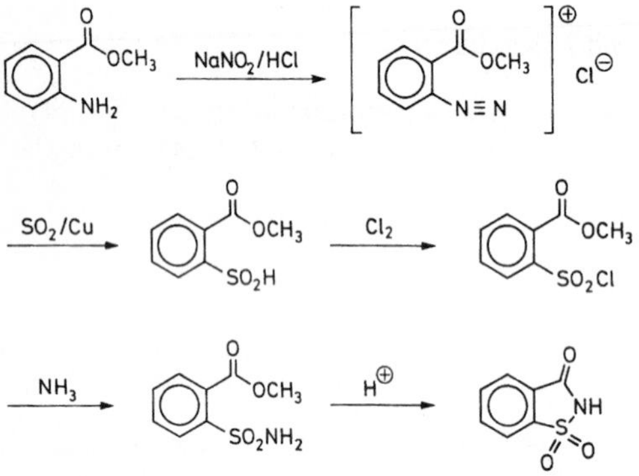

(8.5)

8.8.3 Cyclamate

Cyclamate is a widespread sweetener and is marketed as the Na- or Ca-salt of cyclohexane sulfamic acid. The sweetening strength is substantially lower than that of saccharin and is $f_{sac, g}(10) = 35$. It has no bitter after-taste. Overall, the sweet taste of cyclamate is not as pleasant as that of saccharin. The present stipulated ADI value of the acid is $0-11$ mg/kg

of body weight. The synthesis of the compound is based on sulfonation of cyclohexylamine:

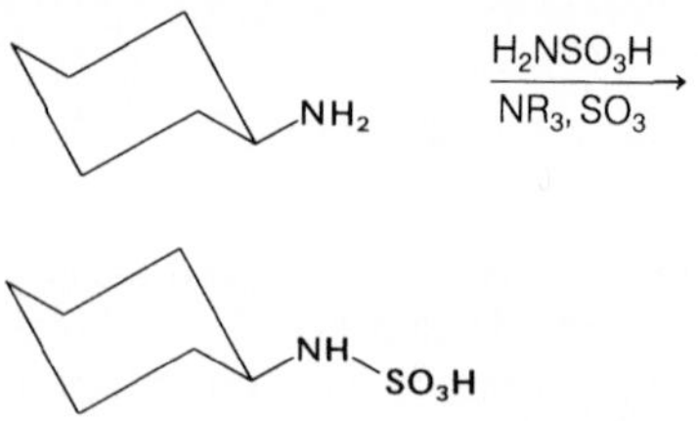

(8.6)

Table 8.3 shows several homologous compounds, illustrating the dependence of sweetness intensity on cycloalkyl ring size. The larger the ring size, the higher the sweetness, i.e. the lower the sweetness threshold value.

8.8.4 Monellin

The pulp of *Dioscoreophyllum cumminsii* fruit contains monellin, a sweet protein with a molecular weight of 11.5 kdal. It consists of two peptide chains, A and B, which are not covalently bound. Their amino acid sequences are shown in Table 8.4.
The conformation is known (Fig. 8.7 and 8.8). As a result of cross reactions with an antiserum against thaumatin (cf. 8.8.5), sequence Y (13) ASD in a β-turn is regarded as the site of contact with the sweetness receptor. It corresponds to sequence Y(57)FD of thaumatin. The separated individual chains are not sweet. When the chains are recombined, a sweet taste is restored slowly, but the sweetness intensity of the original native protein is not reached. This strongly suggests that peptide chain separation results in irreversible conformational changes. However, combination of synthesiz-

Table 8.3. Taste threshold values of some alicyclic sulfamic acids (Na-salts)

R	c_{tsw} (mmol/l)	R	c_{tsw} (mmol/l)
Cyclobutyl	100	Cycloheptyl	$0.5-0.7$
Cyclopentyl	$2-4$	Cyclooctyl	$0.5-0.8$
Cyclohexyl	$1-3$		

Table 8.4. Amino acid sequences of the A and B chains of monellin. (The sequence YASD shown in bold-type, which is localized in a β-turn, is regarded as a part of the structure responsible for the cross reaction of monellin with antibodies against thaumatin as well as for making contact with the sweetness receptor (cf. Table 8.5 and Fig. 8.7 and 8.8)

					5					10					15					20
A-chain:	Fᵃ R	E	I	K	G	Y	E	Y	Q	L	Y	V	**Y**	**A**	**S**	**D**	K	L	F	R
	A	D	I	S	E	D	Y	K	T	R	G	R	K	L	L	R	F	N	G	P
	V	P	P	P																

					5					10					15					20
B-chain	Tᵇ G	E	W	E	I	I	D	I	G	P	F	T	Q	N	L	G	K	F	A	V
	D	E	E	N	K	I	G	Q	Y	G	R	L	T	F	N	K	V	I	R	P
	C	M	K	K	T	I	Y	E	E	N										

ᵃ Ca. 10% of the A chains also contain phenylalanine at the N-terminal (Phe-A-chain).
ᵇ Ca. 19% of the B chains also contain threonine at the N-terminal (Thr-B-chain) and N-terminal glycine is absent in ca. 24% (de-gly¹-B-chain).

ed A and B chains gave a product with the same sweetening strength as natural monellin. The thermal stability of the protein was increased by covalently bonding the two peptide chains via the amino acid residues A2 and B50 (cf. Fig. 8.9). For this purpose, a synthetic gene was cloned and expressed in *E. coli* and yeast. The protein (I) thus obtained was as sweet as natural monellin (II). While the sweet taste of II was completely quenched at pH 2 by heating to 50 °C, I exhibited its full sweetness at room temperature even after being heated to 100 °C.

The threshold value is $f_{sac,g} = 3,000$. Based on its low stability, slow triggering and slow fading away of taste perception, monellin probably will not succeed as a commercial sweetener.

8.8.5 Thaumatins

The fruit of *Thaumatococcus daniellii* contains two sweet proteins: thaumatin I and II, with $f_{sac,g} \sim 2,000$. There are also low amounts of three other sweet proteins (thaumatin a, b and c). The complete amino acid sequence and the conformation (Fig. 8.7 and 8.8) of thaumatin I, a peptide chain with 207 amino acid residues, has been established (Table 8.5). Based on cross reactions with an antiserum against monellin (cf. 8.8.4), the sequence Y(57)FD located in a β-turn is regarded as the site of contact with the sweetness receptor. It corresponds to the sequence Y(A13)ASD of monellin. Successive acetylation of the 11 ε-amino groups weakens the sweet taste and, when just four groups are acetylated, it is extinguished. However, modification of up to seven ε-amino groups by reductive methylation with

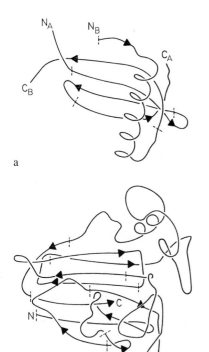

a

b

Fig. 8.7. A two dimensional representation of the conformation of the peptide chains of monellin (a) and thaumatin (b). (β-structure: ⟶ ; α-helix: ∿ ; β-turn: ⊃ ; N, C, or N_A, N_B, C_A, C_B: N and C termini of the chains) (according to *Kim* et al., 1991)

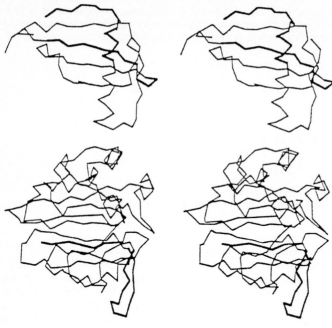

Fig. 8.8. A stereoscopic representation of the conformation of the peptide chains of monellin (top) and thaumatin (bottom) (the location of tryptic peptides that cross react with heterologous antibodies is indicated by thicker lines) (according to *Kim* et al., 1991)

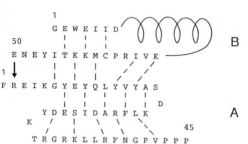

Fig. 8.9. Monellin: schematic representation of the A and B chains, showing the intra- and intermolecular hydrogen bonds (−−). Using genetic engineering techniques, the two chains were linked via a peptide bond (→) between the amino acids residues E(B50) and R (A2) (according to *Kim* et al., 1991)

HCHO/NaBH$_4$ does not decrease or modify the sweet taste intensity. Apparently, the isoelectric point of the protein plays a role in sweetness activity. Thaumatin which is regarded as toxicologically safe is used, e. g., in chewing gum and milk products. Synergistic effects have been observed when thaumatin is used in combination with saccharin, acesulfame or stevioside.

8.8.6 Curculin and Miraculin

Curculin is a sweet protein ($f_{sac, g}$ (6.8) = 550) of known sequence (Table 8.6). It occurs in the fruit of *Curculigo latifolia*. The sweet taste

Table 8.5. Amino acid sequence of thaumatin I. (Disulfide bonds: 9−204, 55−66, 71−77, 121−193, 126−177, 134−145, 149−158, 159−164; the sequence YFD shown in boldtype, which is localized in a β-turn, is regarded as a part of the structure responsible for the cross reaction of thaumatin with antibodies against monellin as well as for making contact with the sweetness receptor (cf. Table 8.4 and Fig. 8.7 and 8.8)

	5	10	15	20	25	30	35	40
	A T F E I V N R C S Y T V W A A A S K G D A A L D A G G R Q L N S G E S W T I N
41	V E P G T N G G K I W A R T D C **Y F D** D S G S G I C K T G D C G G L L R C K R F
81	G R P P T T L A I F S L N Q Y G K D Y I D I S N I K G F N V P M N F S P T T R G
121	C R G V R C A A D I V G Q C P A K L K A P G G G C N D A C T V F Q T S E Y C C T
161	T G K C G P T E Y S R F F K R L C P D A F S Y V L D K P T T V T C P G S S N Y R
201	V T F C P T A

Table 8.6. Amino acid sequence of curculin

	5	10	15	20	25	30	35	40
	D N V L L S G Q T L H A D H S L Q A G A Y T L T I Q N N C N L V K Y Q N G R Q I							
41	W A S N T D R R G S G C R L T L L S D G N L V I Y D H N N N D V N G S A C C G D							
81	A G K Y A L V L Q K D G R F V I Y G P V L W S L G P N G C R R V N G							

induced by this protein disappears after a few minutes, only to reappear with the same intensity on rinsing with water. It is assumed that $Ca^{2\oplus}$ and/or $Mg^{2\oplus}$ ions in the saliva suppress the sweet taste. Rinsing with citric acid (0.02 mol/l) considerably enhances the impression of sweetness ($f_{sac,g}(12) = 970$). Thus, like miraculin, curculin acts as a taste modifier.

Miraculin is a glycoprotein present in the fruit of *Synsepalum dulcificum* (a tropical fruit known as miracle berry). Although it is tasteless, it has the property of giving sour solutions a sweet taste and therefore it is called a taste modifier. Thus, lemon juice seems sweet when the mouth is first rinsed with a solution of miraculin. The molecular weight of this taste modifier is 42–44 kdal.

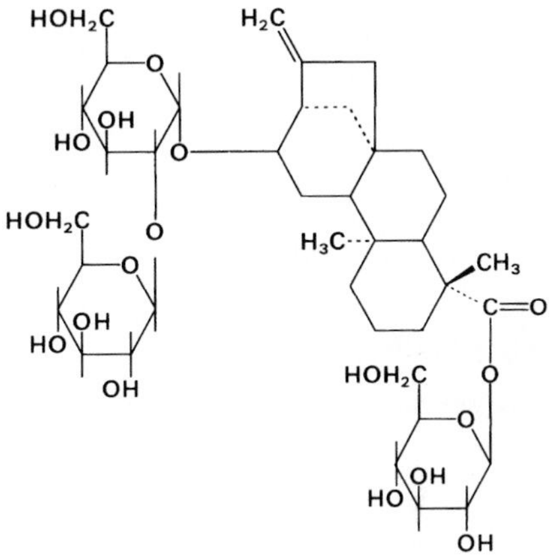

(8.7)

8.8.7 Gymnema silvestre Extract

The extract from *Gymnema silvestre* is related to the taste modifier miraculin. It has the property of eliminating the ability to perceive sweetness for a few hours, without interfering with the perception of other taste qualities. The active substance has not yet been identified.

8.8.8 Stevioside

Leaves of *Stevia rebaudiana* contain approx. 6% stevioside ($f_{sac,g}(4) \sim 300$). This compound is used as a sweetener in various countries. Its structure is shown in Formula 8.7.

8.8.9 Osladin

Osladin (Formula 8.8) occurs in rhizomes of *Polypodium vulgare*. It is very sweet ($f_{sac,g} \sim 3,000$); however, due to its toxic effect, it is not likely to become an approved sweetener.

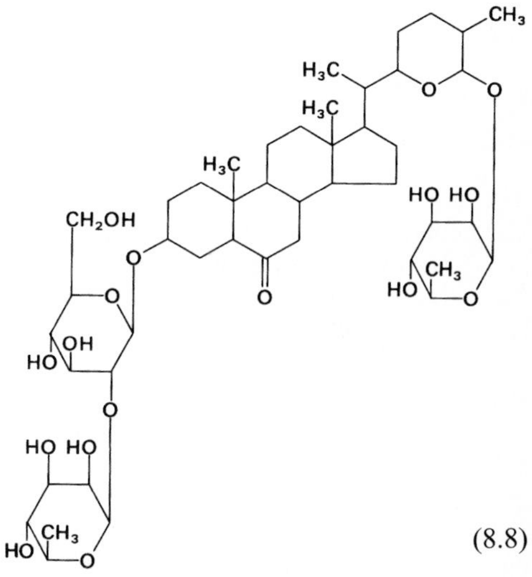

(8.8)

8.8.10 Phyllodulcin

The leaves of *Hydrangea macrophylla* contain a 3,4-dihydroisocoumarin derivative, phyllodulcin (Formula 8.9). Its sweetness matches that of dihydrochalcones and of licorice root. The taste perception builds relatively slowly and also fades away slowly. The sweetening strength is $f_{sac}(5) = 250$. Its application is being considered in the production of several brands of candy and chewing gum. A study of a number of related isocoumarin derivatives shows that taste quality and strength are very much dependent on the substitution pattern of the molecule (cf. Table 8.7).

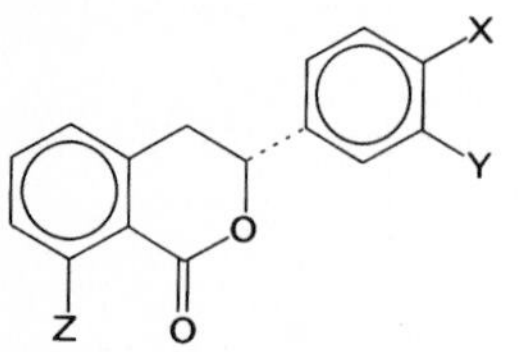

$$(8.9)$$

Z = OH, X = OMe, Y = OH

8.8.11 Glycyrrhizin

The active substance from licorice root (*Glycyrrhiza glabra*) is a β,β'-glucuronido-glucuronide of glycyrrhetic acid:

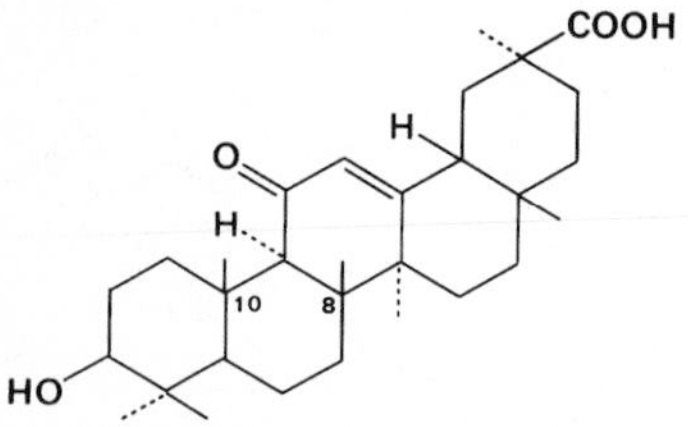

$$(8.10)$$

The sweetening strength is $f_{sac,g}(4) = 50$. The compound is utilized for production of licorice (also spelled as liquorice). Its cortisone-like side effect limits its wide application.

Table 8.7. Sensory properties of some 2,3-dihydroisocoumarins

Compound[a]			Taste
X	Y	Z	
OMe	OH	OH	very sweet
OMe	OMe	OH	bitter
OMe	OMe	OMe	no taste
OMe	OAc	OAc	slightly sweet
OH	OH	OH	no taste
OH	H	OH	no taste
OH	OH	H	no taste
OMe	OH	H	very sweet
OH	OMe	H	no taste

[a] Formula 8.9.

8.8.12 Nitroanilines

Several m-nitroanilines are potent sweeteners:

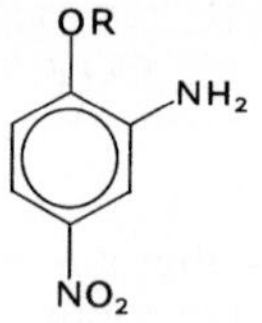

$$(8.11)$$

As shown in Table 8.8, the sweetening strength is strongly dependent on the substituent R. The propoxy derivative has an $f_{sac} = 4,100$ and was for some time used under the brand name "Ultrasweet P-4000" but, due to its toxic side effects, is no longer important.

8.8.13 Dihydrochalcones

Some dihydrochalcones are derived from flavanones (cf. 18.1.2.5.4) and have a relatively clean sweet taste that is slowly perceived but persists for some time. The sweetening strength of β-neohesperidin dihydrochalcone is $f_{sac,g} = 1100$ (threshold value) or $f_{sac,g}(10) = 667$ (R = β-neohesperidosyl in Formula 8.12).

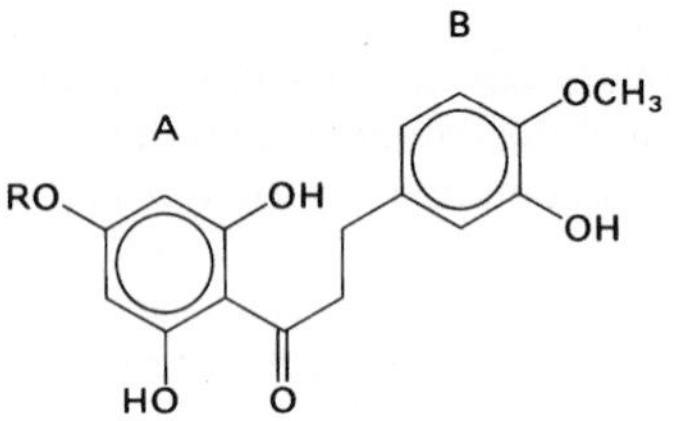

$$(8.12)$$

In different countries, this compound is used in chewing gum, mouthwashes, beverages, and various types of candy. The quality and strength of the sweet taste of dihydrochalcone are related particularly to the substitution pat-

Table 8.8. Sweetness of some 1-alkoxy-2-amino-4-nitrobenzenes (Formula 8.11)

R	f_{sac}	R	f_{sac}
H	120	C_4H_9	1,000
CH_3	220	$(CH_3)_2CH$	600
C_2H_5	950	$CH_3CH = CH$	2,000
C_3H_7	4,100		

tern in ring B. The prerequsite for a sweet taste is the presence in ring B of at least one $-OH$ group, but not three adjacent hydroxy and alkoxy substituents.

8.8.14 Ureas and Guanidines

8.8.14.1 Dulcin

Dulcin, 4-ethoxyphenylurea, is closer in taste to sucrose than saccharin. The sweetening strength is $f_{sac, g} = 109$ (5% saccharose solution), i.e. lower than that of saccharin. It was used earlier in a mix with saccharin, but now has no importance due to its toxic side effects.

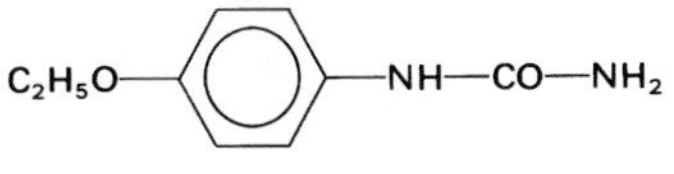

$$C_2H_5O-\langle\bigcirc\rangle-NH-CO-NH_2 \qquad (8.13)$$

8.8.14.2 Suosan

Suosan, N-[(p-nitrophenyl)carbamoyl]-β-alanine (Formula 8.14), $f_{sac, g}(2) = 700$, is clearly sweeter than dulcin and saccharin. The e/n system could be the NH/COO⁻ system of β-alanine, which corresponds to the e/n system of aspartame (cf. 1.3.3 and 8.8.17). The p-cyanophenyl compound ($f_{sac, g}(2) = 450$), the N-glycine homolog and the thiocarbamoyl compound are also sweet.

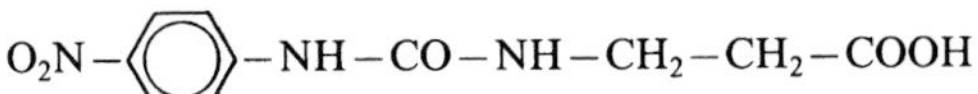

$$O_2N-\langle\bigcirc\rangle-NH-CO-NH-CH_2-CH_2-COOH$$
$$(8.14)$$

8.8.14.3 Guanidines

Derivatives of guanidinoacetic acid (Formula 8.15) are among the sweetest compounds known until now (Table 8.9). The e/n system probably consists of a nitrogen of the guanidino group and the carboxylate group. The guanidines have a sweetening strength that is two to three powers of ten higher than that of related compounds like aspartic acid derivatives (cf. 1.3.3 and 8.8.17) and suosan (cf. 8.8.14.2). This is probably due to the two hydrophobic groups which occupy a different spatial position relative to the e/n system and, if necessary, could make contact with two different hydrophobic binding sites of the receptor. It is possible that apart from the e/n system, other polar contact groups play a role.

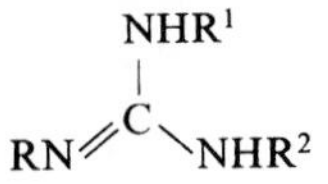

$$\begin{array}{c} NHR^1 \\ | \\ RN^{\diagup C}\diagdown NHR^2 \end{array} \qquad (8.15)$$

Replacement of the carboxyl group by a tetrazole residue results in loss of sweetening strength (Table 8.9).

Table 8.9. Taste of some guanidines (Formula 8.15)

R	R¹	R²	$f_{sac, g}(2)$
p-Cyanophenyl	H	Carboxymethyl	2700
	Benzyl		30000
	Phenylsulfonyl		45000
	1-Naphthyl		60000
	Cyclohexyl		12000
	Cyclooctyl		170000
	Cyclononyl		200000
3,5-Dichlorophenyl	Benzyl		80000
	Cyclooctyl		60000
p-Cyanophenyl	Cyclohexyl	Tetrazolylmethyl	400ᵃ
	Cyclooctyl		5000ᵇ

[a] $f_{sac, g}(4)$.
[b] $f_{sac, g}(5)$.

The guanidines can be synthesized, e.g., via the isothiocyanates:

$$R-N=C=S \xrightarrow{R^1NH_2} R-NH-\underset{\underset{S}{\|}}{C}-NHR^1$$

$$\xrightarrow{MeI} \underset{\underset{SMe}{|}}{R-N=C-NHR^1}$$

$$\xrightarrow{R^2NH_2} \underset{\underset{NHR^2}{|}}{R-N=C-NHR^1} \qquad (8.16)$$

8.8.15 Oximes

It has long been known that the anti-aldoxime of perillaldehyde (discovered in the essential oil of *Perilla nankinensis*) has an intensive sweet taste ($f_{sac, g} \sim 2,000$). For its structure see Formula 8.17 (I).

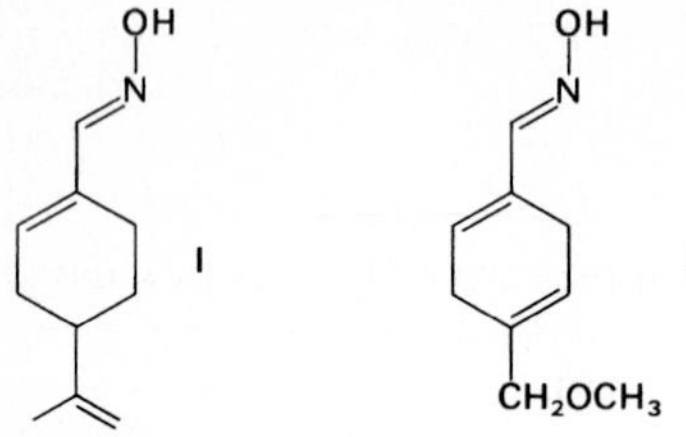

(8.17)

Its poor water solubility hinders its utilization. In the meantime, a related compound, (II), with improved solubility, has been reported, but its sweetness is just moderately high ($f_{sac,g}$ ~ 450).

8.8.16 Oxathiazinone Dioxides

These compounds (Formula 8.18) belong to a new class of sweeteners with an AH/B-system corresponding to that of saccharin. Based on their properties and present toxicological data, they are suitable for use. The sweetening strength depends on substituents R1 and R2 and is $f_{ac,g}(4) = 130$ for acesulfame K (Table 8.10).

The ADI value stipulated for the potassium salt of acesulfame is $0-9$ mg/kg of body weight.

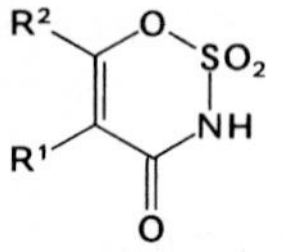

(8.18)

Oxathiazinone dioxides are obtained from fluorosulfonyl isocyanate and alkynes, or from compounds with active methylene groups, as

Table 8.10. Sweetness of some oxathiazinone dioxides (Na-salts) (Formula 8.18)

R^1	R^2	$f_{sac,g}$	R^1	R^2	$f_{sac,g}$
H	H	10	Et	H	20
H	Me	130[a]	Et	Me	250
Me	H	20	Pr	Me	30
Me	Me	130	i-Pr	Me	50
H	Et	150			

[a] Acesulfame.

exemplified by 1,3-diketones, 3-oxocarboxylic acids and 3-oxocarboxylic acid esters:

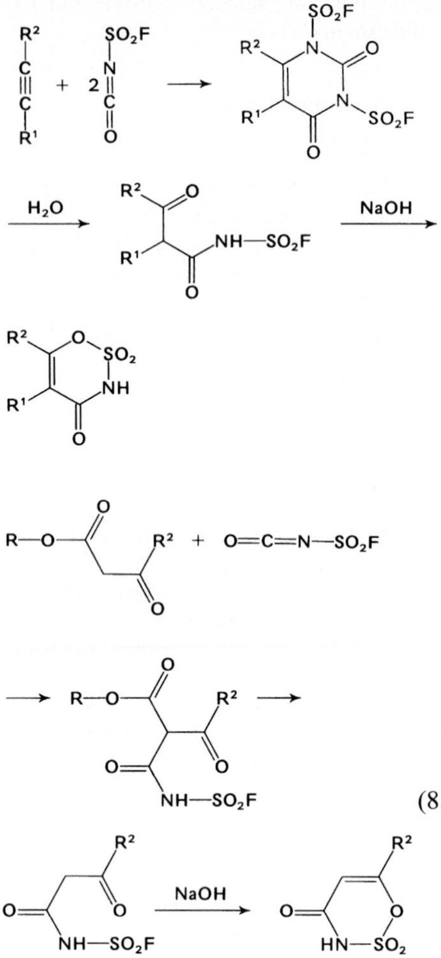

(8.19)

8.8.17 Dipeptide Esters and Amides

8.8.17.1 Aspartame

A dipeptide, L-aspartyl-L-phenylalanine methyl ester (L-Asp-L-Phe-OMe), has recently been approved for use as a sweetener in North America (aspartame, "NutraSweet"). It is as sweet as a number of other dipeptide esters of L-aspartic acid and D,L-aminomalonic acid. The relationship between chemical structure and taste among these compounds was outlined in more detail in 1.3.3. The sweetening strength relative to saccharose is concentration dependent (Table 8.11). Aspartame is used

Table 8.11. Comparison of sweetening strengths of aspartame and saccharose (concentrations of iso-sweet aqueous solutions in %)

Saccharose	Aspartame	$f_{sac, g}$
0.34[a]	0.001[a]	340
4.3	0.02	215
10.0	0.075	133
15.0	0.15	100

[a] Threshold value.

worldwide, though its stability is not always satisfactory. Unlike sweetening of drinks (coffee or tea) which are drunk immediately, problems arise in the use of aspartame in food which has to be heated or in sweetened drinks which have to be stored for longer periods of time. Possible degradation reactions involve α/β-rearrangement, hydrolysis into amino acid constituents and cyclization to the 2,5-dioxopiperazine derivative:

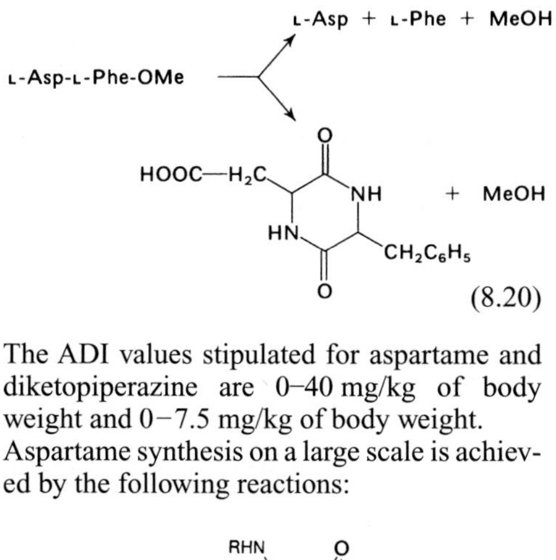

(8.20)

The ADI values stipulated for aspartame and diketopiperazine are 0–40 mg/kg of body weight and 0–7.5 mg/kg of body weight. Aspartame synthesis on a large scale is achieved by the following reactions:

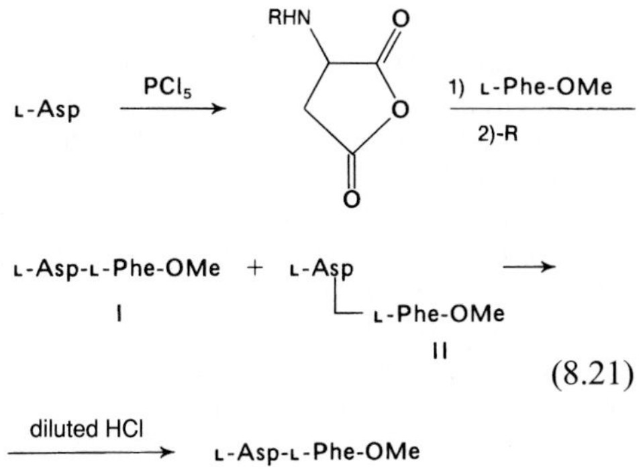

(8.21)

Separation of the two dipeptide isomers (I, II) is possible since there are solubility differences between the two isomers as a consequence of their differing isoelectric points ($IP_I > IP_{II}$).
Other possible syntheses are based on a plastein reaction (cf. 1.4.6.3.2) with an N-derivatized aspartic acid and phenylalanine methylester or on bacterial synthesis of an Asp-Phe polymer, achieved by genetic engineering techniques, enzymatic cleavage of the polymer to Asp-Phe, followed by acid or enzyme catalyzed esterification of the dipeptide with methanol.

8.8.17.2 Superaspartame

Substitution of a (p-cyanophenyl)carbamoyl residue for the free amino group of aspartame produces a compound called superaspartame (Formula 8.22), $f_{sac, g}(2) = 14000$, which is sweeter than aspartame by about two powers of ten. This molecule contains structural elements of aspartame and of cyanosuosan. The high sweetening strength is probably due to the presence of two hydrophobic groups which occupy different spatial positions relative to the e/n system.

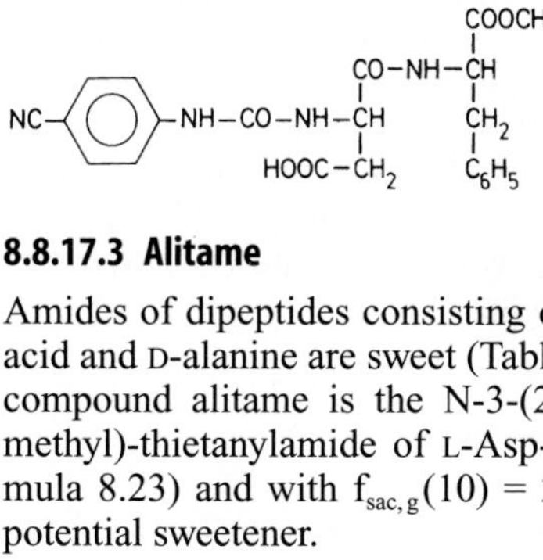

(8.22)

8.8.17.3 Alitame

Amides of dipeptides consisting of L-aspartic acid and D-alanine are sweet (Table 8.12). The compound alitame is the N-3-(2,2,4,4-tetramethyl)-thietanylamide of L-Asp-D-Ala (Formula 8.23) and with $f_{sac, g}(10) = 2000$, it is a potential sweetener.

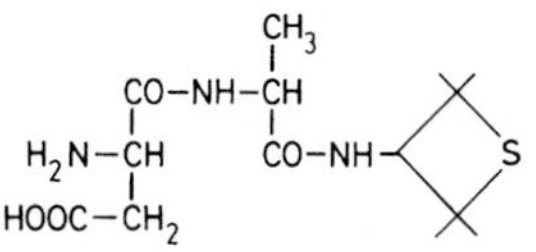

(8.23)

Since the second amino acid has a D configuration, its side chain must be small, corresponding to the structure activity relationships discussed for dipeptide esters of the aspartame type (cf. 1.3.3). On the other hand, the carbonyl group should carry the largest possible hydrophobic residue.

Table 8.12. Taste of some dipeptide amides of the type L-Asp-D-Ala-NHR

R	$f_{sac,g}(10)$
Cyclopentyl	50
Cyclohexyl	90
(2,2,5,5-tetramethyl)-cyclopentyl	800
(2,2,6,6-tetramethyl)-cyclohexyl	300
(Diethyl)-methyl	100
(Diisopropyl)-methyl	250
(Di-tert-butyl)-methyl	450
(Di-cyclopropyl)-methyl	1200
(Cyclopropyl)-(tert-butyl)-methyl	1200
(Cyclopropyl)-(methyl)-methyl	100
(2,2,4,4-Tetramethyl)-cyclobutyl	300
(2,2,4,4-Tetramethyl)-cyclobutan-3-onyl	240
(3-Hydroxy-2,2,4,4-tetramethyl)-cyclobutyl	125
3-(2,2,4,4-Tetramethyl)-thietanyl	2000[a]
3-(1-cis-Oxo-2,2,4,4-tetramethyl)-thietanyl	300
3-(1-trans-Oxo-2,2,4,4-tetramethyl)-thietanyl	350
3-(1,1-Dioxo-2,2,4,4-tetramethyl)-thietanyl	805

[a] Alitame.

The stability of dipeptide amides of the alitame type is substantially higher than that of dipeptide esters of the aspartame type. Therefore, alitame can also be used in bread and confectionery.

Like aspartame, alitame also undergoes α/β-rearrangement. Both isomers hydrolyze slowly to give L-aspartic acid and D-alanine amide, which is excreted either directly or as the glucuronide. A small part is oxidized to sulfoxides and sulfone. Cyclization to diketopiperazine which is typical of dipeptide methylesters does not occur.

8.8.18 Hernandulcin

(+)-Hernandulcin is a sweet sesquiterpene from *Lippia dulcis Trev.* (*Verbenaceae*), with the structure 6-(1,5-dimethyl-1-hydroxy-hex-4-enyl)-3-methyl-cyclohex-2-enone:

$$(8.24)$$

In comparison with sucrose, the sweetening strength of this compound is $f_{sac,mol}(0.25) = 1250$. Hernandulcin is somewhat less pleasant in taste than sucrose and exhibits some bitterness.

The racemic compound was synthesized via a directed aldol-condensation reaction by adding 6-methyl-5-hepten-2-one to a mixture of 3-methyl-2-cyclohexen-1-one and lithium diisopropylamide in tetrahydrofuran, followed by chromatographic separation of (±)-hernandulcin (I, 95%) from the diastereomeric counterpart (±)-epihernandulcin (II, 5%). Whereas I is sweet, II exhibits no sweet taste.

The carbonyl and hydroxyl groups, which are located about 0.26 nm apart in the preferred conformation, are considered as an AH/B-system. The sweet taste is lost when these groups are modified (reduction of the carbonyl group to an alcohol, or acetylation of the hydroxy group).

8.8.19 Halodeoxy Sugars

Halodeoxy sugars are sweeter than the sugars themselves, probably due to their higher hydrophobicity. Their sweetening strength depends on the type, number, and position of the substituents (Table 8.13). In fact, 4-chloro-4-deoxy-α-D-galactopyranosyl-1,6-dichloro-1,6-dideoxy-β-D-fructofuranoside (4,1′,6′-trichlorogalactosucrose), also called sucralose, has a sweetening strength of $f_{sac,g}(10) = 650$ and is under discussion as a potential sweetener because of its pleasant sweet taste and high stability.

$$(8.25)$$

The compounds listed in Table 8.13 show that chlorine in positions 4, 1′, 4′, and 6′ enhances sweetness (No. 1–6, 11), while chlorine in positions 6 (No. 5 in comparison to No. 10, No. 6/15) and 2 (No. 10/19) has a negative effect or produces a bitter taste. In the case of 4,1′,6′-tri-(No. 6–9) and 4,1′,4′,6′-tetrahalo derivatives (No. 11, 12), the sweetening strength increases greatly with increasing hydrophobicity up to the bromine derivatives (F ≪ Cl < Br ≫ I), but then it decreases in the

Table 8.13. Sweetness of halodeoxysucroses

No.	Deoxysucrose derivative	$f_{sac,g}(10)$
1	4-Chlorogalacto-	5
2	1'-Chloro	20
3	6'-Chloro-	20
4	4,1'-Dichlorogalacto-	120
5	1',6'-Dichloro-	80
6	4,1',6'-Trifluorogalacto-	40
7	4,1',6'-Trichlorogalacto-[a]	650
8	4,1',6'-Tribromogalacto-	800
9	4,1',6'-Triiodogalacto-	120
10	6,1',6'-Trichloro-	25
11	4,1',6',6'-Tetrachlorogalacto-	2200
12	4,1',4',6'-Tetrabromogalacto-	7500
13	4-Fluoro-1',4',6'-trichlorogalacto-	1000
14	4'-Iodo-4,1',6'-trichlorogalacto-	3500
15	4,6,1',6'-Tetrachlorogalacto-	200
16	6-Deoxy-4,1',6'-trichlorogalacto-	400
17	6-O-Methyl-4,1',6'-trichlorogalacto-	500
18	6-O-Isopropyl-4,1',6'-trichlorogalacto-	not sweet
19	2,6,1',6'-Tetrachloromanno-	bitter

[a] Sucralose.

case of the iodine compounds, probably for steric reasons. The synthesis of halodeoxy sugars is discussed in Section 4.3.4.8.

8.9 Food Colors

A number of natural colors are available and used to adjust or correct food discoloration or color change during processing or storage. Carotenoids (cf. 3.8.4.5) are used the most, followed by red beet pigment and brown colored caramels. The number of approved synthetic dyes is low. Table 8.14 lists the pigments of importance in food coloring. Yellow and red colors are used the most. Food products which are often colored are confections, beverages, dessert powders, cereals, ice creams and dairy products.

8.10 Acids

The acids in food fulfill a number of other functions besides flavor and antimicrobial activities. The most important acids used in food processing and storage are outlined in this section.

8.10.1 Acetic Acid and Other Fatty Acids

Acetic, propionic and sorbic acids are dealt with under antimicrobial agents (8.12). Other short chain fatty acids, such as butyric and higher homologues, are used in aroma formulations.

8.10.2 Succinic Acid

The acid ($pK_1 = 4.19$; $pK_2 = 5.63$) is applied as a plasticizer in dough making. Succinic acid monoesters with glycerol are used as emulsifiers in the baking industry. The acid is synthesized by catalytic hydrogenation of fumaric or maleic acids.

8.10.3 Succinic Acid Anhydride

This is the only acid anhydride used as a food additive. The hydrolysis proceeds slowly, hence the compound is suitable as an agent in baking powders and for binding of water in some dehydrated food products.

8.10.4 Adipic Acid

Adipic acid ($pK_1 = 4.43$; $pK_2 = 5.62$) is used in baking powders, in powdered fruit juice drinks, for improving the gelling properties of marmalades and fruit jellies, and for improving cheese texture. It is synthesized from phenol or cyclohexane (cf. Reactions 8.26).

Table 8.14. Examples of food colorants (natural and synthetic)

Number	Name	FD & C (USA)	EU No.	Color	λ max (nm) (solvent[b])	Formula[a]	Examples for utilization in food processing
1	Tartrazine	Yellow No 5	E 102	lemon-yellow (W)	426 (W)	I	Pudding powders, confectionary and candies, ice creams, pop (effervescent) drinks
2	Riboflavin		E 101	yellow (W)	445 (W)		Mayonnaise, soups, puddings, desserts, confectionary and candy products
3	Curcumin		E 100	yellow-red (E)	426 (E)	II	Mustard
4	Zeaxanthin			yellow (oil)	455–460 (CH)		Fat, hot and cold drinks, puddings, water
5	Sunset Yellow FCF	Yellow No 6	E 110	orange (W)	485 (W)	III	Beverages, fruit preserves, confectionary and candy products, honey-like products, sea salmon, crabs
6	β-Carotene		E 160a	orange (oil)	453–456 (CH)		Fat, beverages, soups, pudding, water, confectionary and candy products, yoghurt
7	Bixin		E 160b	orange (oil)	471/503 (CHCl$_3$)		Fat, mayonnaise
8	Lycopene		E 160d	orange (oil)	478 (H)		Mayonnaise, ketchup, sauces
9	Canthaxanthin	Food Orange 8	E 161g	orange (oil)	485 (CHCl$_3$)		Sea salmon, beverages, tomato products
10	Astaxanthin			orange (oil)	488 (CHCl$_3$)		Beverages, tomato products, confectionary and candy products
11	β-Apo-8'-carotenal		E 160e	orange (oil)	460–462 (CH)		Sauces, beverages, confectionary and candy products
12	Carmoisine		E 122	red with bluish tint (W)	516 (W)	IV	Beverages, confectionary and candy products, ice cream, pudding, fruit preserves
13	Amaranth	Red No 2	E 123	red with bluish tint (W)	520 (W)	V	Beverages, fruit preserves, confectionary and candy products, jams
14	Ponceau 4R		E 124	scarlet-red (W)	505 (W)	VI	Beverages, candy products (bonbons), sea salmon, cheese coatings

No.	Name		E-number	Color	Absorption max (solvent)[b]		Applications
15	Carmine		E 120	bright-red	518 (W ammonia solution)	VII	Alcoholic beverages
16	Anthocyanidin (from red grape pomace)		E 163 a–f	red-violet[c] (W)	520–546 (M + 0.01 % HCl)		Jams, pop (effervescent) drinks
17	Erythrosine	Red No 3	E 127	cherry-red (W)	527 (W)	VIII	Fruits, jams, confectionary and candy products
18	Red 2G				red with bluish tint	532 (W)	IX	Confectionary and candy products
19	Indigo Carmine (Indigotine)	Blue No 2	E 132	purple blue (W)	610 (W)	X	Also in combination with yellow colorant for confectionary and candy products and liqueurs
20	Patent Blue V		E 131	blue with a greenish tint (W)	638 (W)	XI	Mostly in combination with yellow colorants for confectionary and candy products, beverages
21	Brilliant blue FCF	Blue No 1		blue with a greenish tint (W)	630 (W)	XII	Mostly in combination with yellow colorants for confectionary and candy products, beverages
22	Chlorophyll		E 140	green	412 (CHCl₃)		Edible oils
23	Chlorophyllin copper complex		E 141	green (W)	405 (W)		Confectionary and candy products, liqueurs, jellies, cream food products
24	Green S (Brillant Green BS)		E 142	green (W)	632 (W)	XIII	Confectionary and candy products
25	Black BN		E 151	violet with bluish tint	570 (W)	XIV	Fish roe coloring, confectionary and candy products

[a] Formulas in Table 8.15; [b] Solvent W: water, CH: cyclohexane, M: methanol, H: hexane, and E: ethanol; [c] Color is pH dependent.

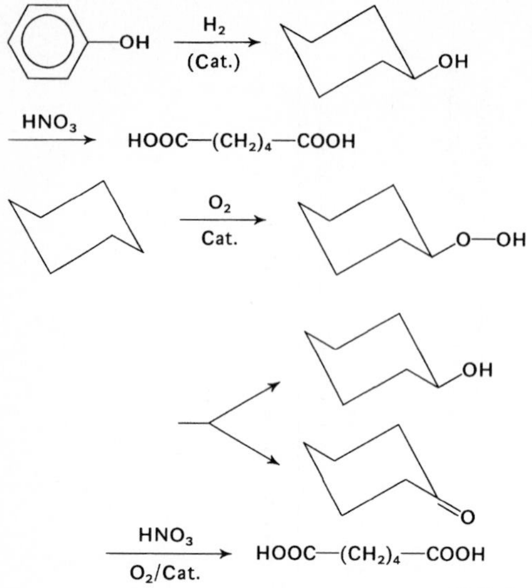

$$\xrightarrow{\text{HNO}_3} \quad \text{HOOC—(CH}_2)_4\text{—COOH}$$

$$\xrightarrow[\text{O}_2/\text{Cat.}]{\text{HNO}_3} \quad \text{HOOC—(CH}_2)_4\text{—COOH}$$

(8.26)

8.10.5 Fumaric Acid

Fumaric acid ($pK_1 = 3.00$; $pK_2 = 4.52$) increases the shelf life of some dehydrated food products (e.g. pudding and jelly powders). It is also used to lower the pH, usually together with food preservatives (e.g. benzoic acid), and as an additive promoting gel setting. Fumaric acid is synthesized via maleic acid anhydride which is obtained by catalytic oxidation of butene or benzene (cf. Reaction 8.27) or is produced from molasses by fermentation using *Rhizopus spp.*

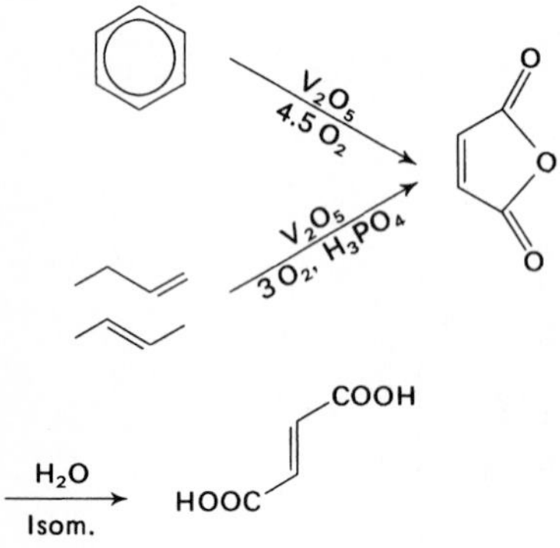

(8.27)

8.10.6 Lactic Acid

D,L- or L-lactic acid (pK = 3.86) is utilized as an 80% solution. A specific property of the acid is its formation of intermolecular esters, providing oligomers or a dimer lactide:

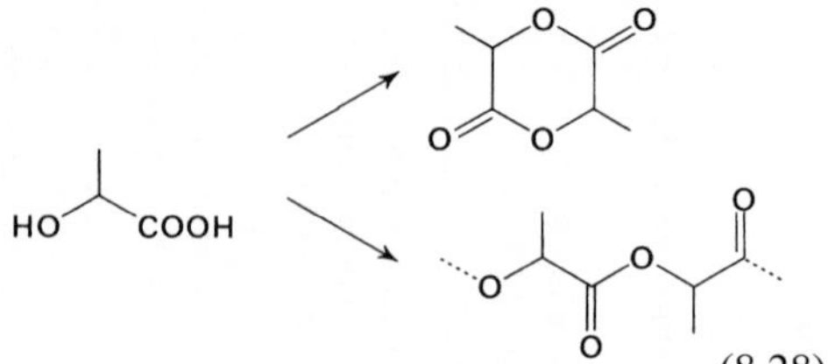

(8.28)

Such intermolecular esters are present in all lactic acid solutions with an acid concentration higher than 18%. More dilute solutions favor complete hydrolysis to lactic acid. The lactide can be utilized as an acid generator. Lactic acid is used for improving egg white whippability (pH adjustment to 4.8–5.1), flavor improvement of beverages and vinegar-pickled vegetables, prevention of discoloration of fruits and vegetables and, in the form of calcium lactate, as an additive in milk powders.

Lactic acid production is based on synthesis starting from ethanal, leading to racemic D,L-lactic acid (Formula 8.29) or on homofermentation (*Lactobacillus delbruckii, L. bulgaricus, L. leichmannii*) of carbohydrate-containing raw material, which generally provides L- but also D,L-lactic acids under the conditions of fermentation.

$$\text{CH}_3\text{CHO} \xrightarrow{\text{HCN}} \text{CH}_3\text{CHOHCN} \longrightarrow \text{Lactate}$$

(8.29)

8.10.7 Malic Acid

Malic acid ($pK_1 = 3.40$; $pK_2 = 5.05$) is widely utilized in the manufacturing of marmalades, jellies and beverages and canning of fruits and vegetables (e.g. tomato). The monoesters with fatty alcohols are effective antispattering agents in cooking and frying fats and oils.

Table 8.15. Structures of the synthetic food colorants listed in Table 8.14

I

II

III

IV

V

VI

VII

VIII

IX

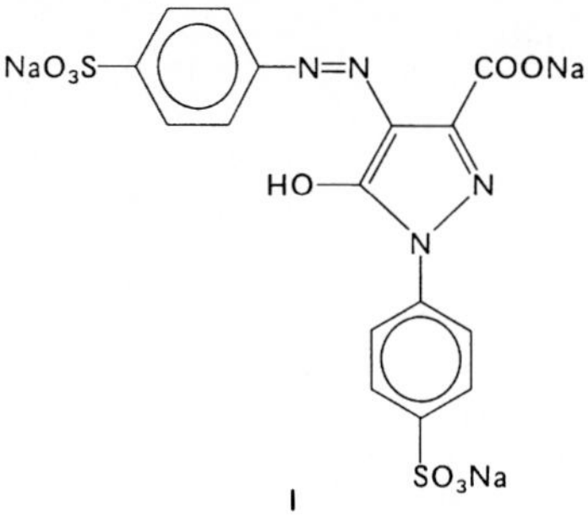

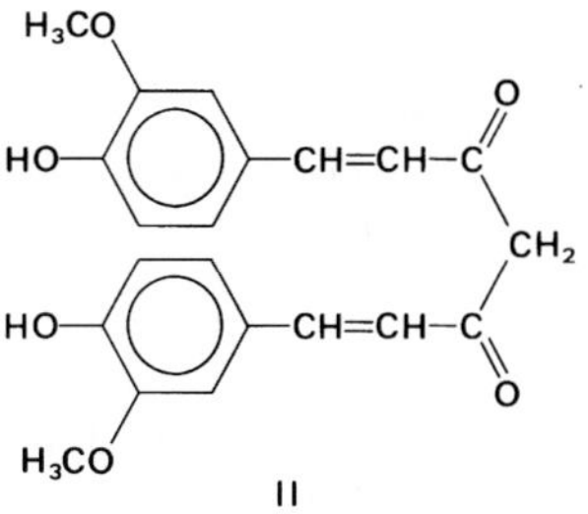

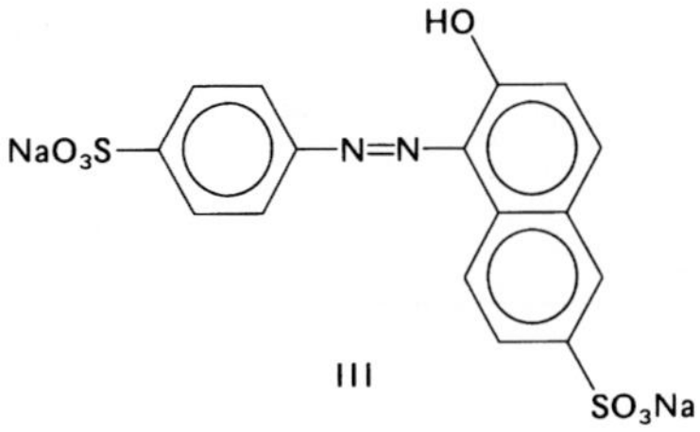

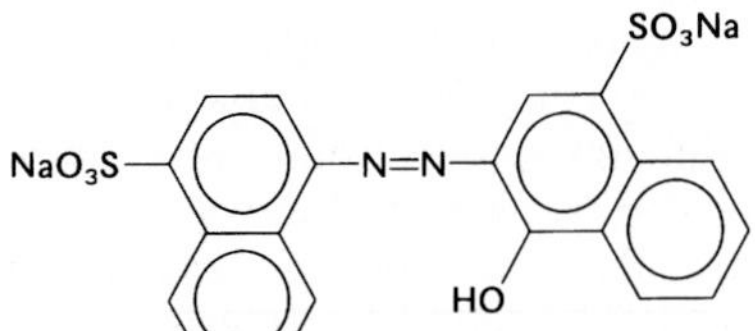

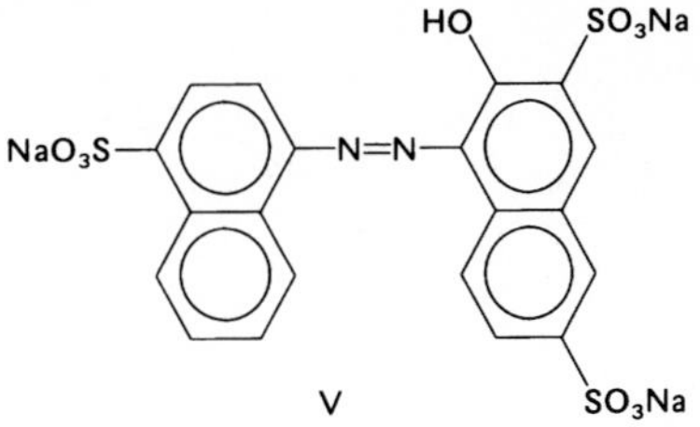

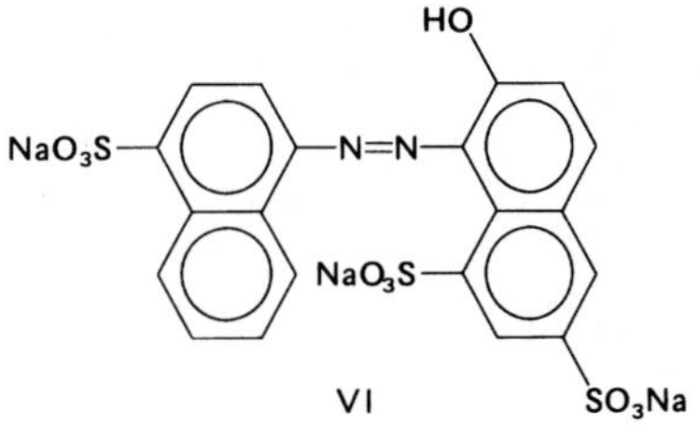

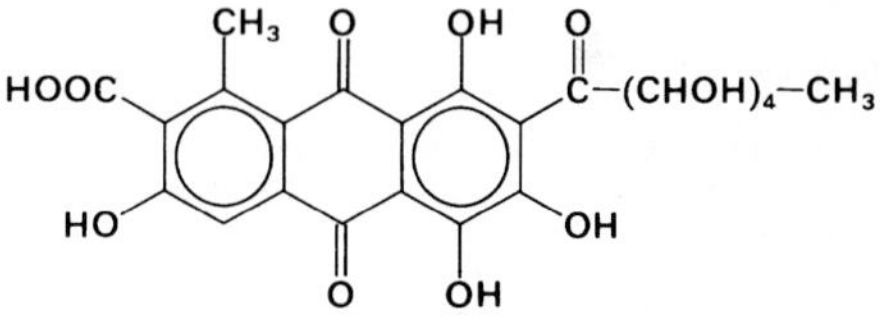

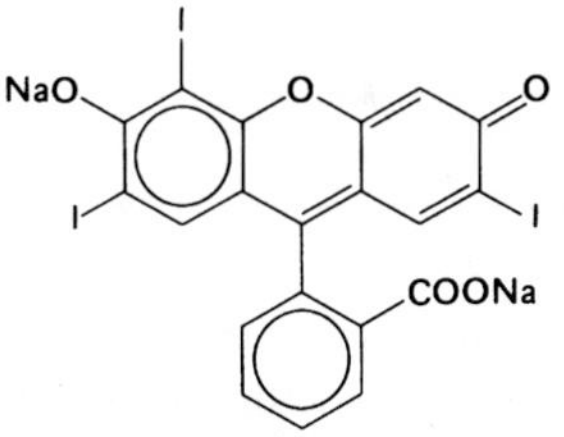

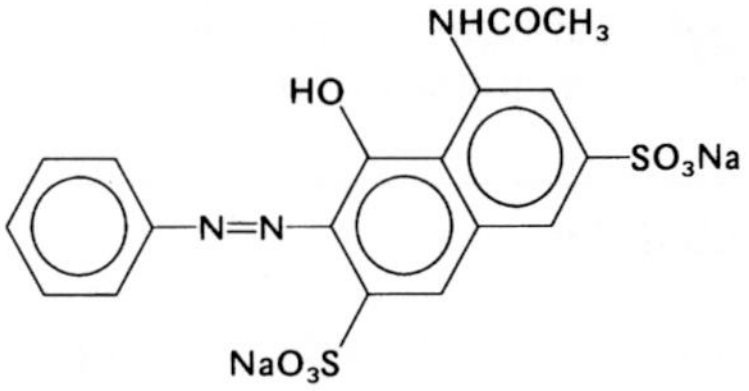

Table 8.15 (continued)

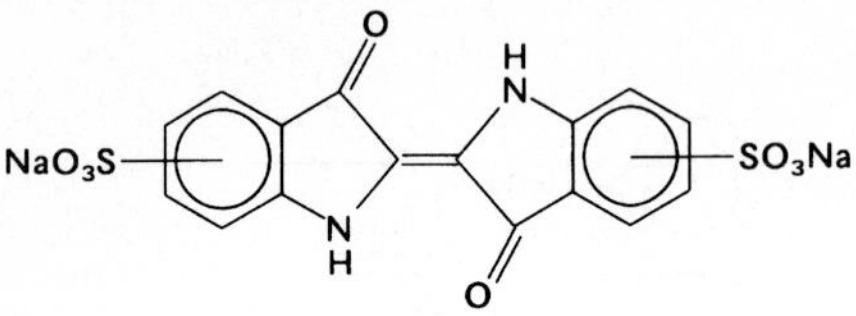

Malic acid synthesis, which provides the racemic D,L-form, is achieved by addition of water to maleic/fumaric acid. L-Malic acid can be synthesized enzymatically from fumaric acid with fumarase (*Lactobacillis brevis, Paracolobactrum spp.*) and from other C sources (paraffins) by fermentation with *Candida spp.*

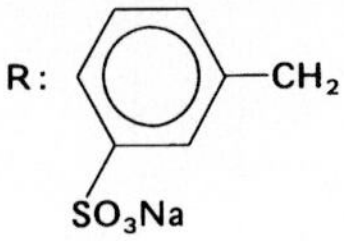

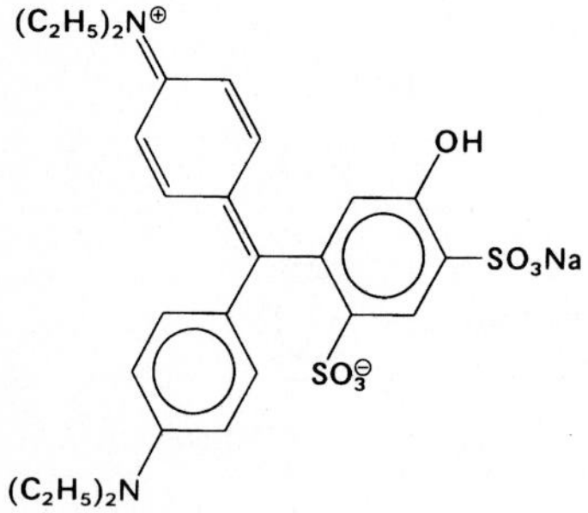

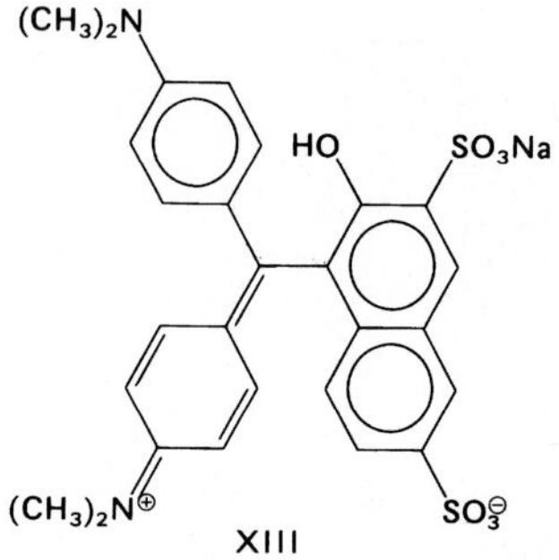

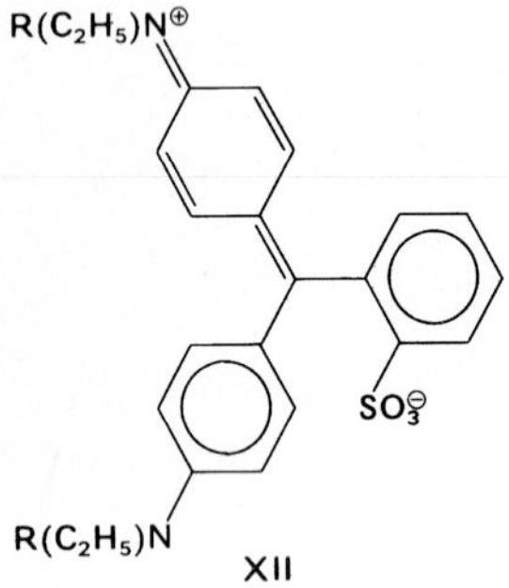

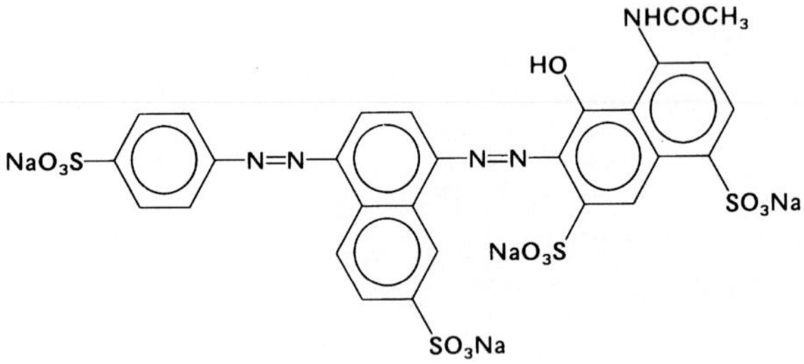

8.10.8 Tartaric Acid

Tartaric acid ($pK_1 = 2.98$; $pK_2 = 4.34$) has a "rough", "hard" sour taste. It is used for the acidification of wine, in fruit juice drinks, sour candies, icecream, baking powder, and, because of its formation of metal complexes, as a synergist for antioxidants. The production of (2R,3R)-tartaric acid is achieved from wine yeast, pomace, and cask tartar, which contain a mixture of potassium hydrogentartrate and calcium tartrate. This mixture is first converted entirely to calcium tartrate, from which tartaric acid is liberated by using sulfuric acid. Racemic tartaric acid is obtained by cis-epoxidation of maleic acid, followed by hydrolysis:

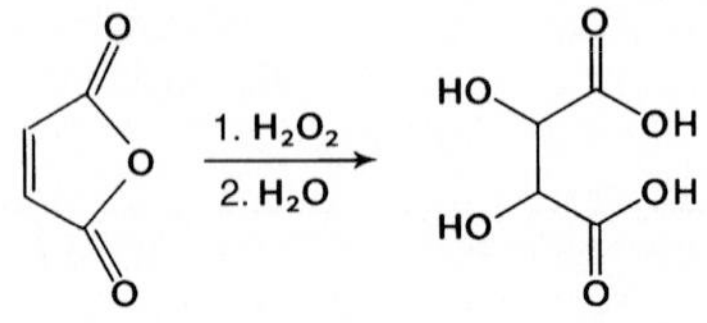

$$(8.30)$$

8.10.9 Citric Acid

Citric acid ($pK_1 = 3.09$; $pK_2 = 4.74$; $pK_3 = 5.41$) is utilized in candy production, fruit juice, ice cream, marmalade and jelly manufacturing, in vegetable canning and in dairy products such as processed cheese and buttermilk (aroma improver). It is also used to suppress browning in fruits and vegetables and as a synergetic compound for antioxidants. Its production is based on microbial fermentation of molasses by *Aspergillus niger*. The yield of citric acid is 50–70% of the fermentable sugar content.

8.10.10 Phosphoric Acid

Phosphoric acid ($pK_1 = 2.15$; $pK_2 = 7.1$; $pK_3 \sim 12.4$) and its salts account for 25% of all the acids used in food industries. The bulk of the acids (salts) used in the industry is citric acid (about 60%), while the use of other acids accounts for only 15%. The main field of use of phosphoric acid is the soft drink industry (cola drinks). It is also used in fruit jellies, processed cheese and baking powder and as an active buffering agent or pH-adjusting ingredient in fermentation processes. Acid salts, e.g., $Ca(H_2PO_4)_2 \cdot H_2O$ (fast activity), NaH_{14}-$Al_3(PO_4)_8 \cdot 4H_2O$ (slow activity) and Na_2H_2-P_2O_7 (slow activity) are used in baking powders as components of the reaction to slowly or rapidly release the CO_2 from $NaHCO_3$.

8.10.11 Hydrochloric and Sulfuric Acids

Both acids are used in starch and sucrose hydrolyses. Hydrochloric acid is also used in protein hydrolysis in industrial production of seasonings.

8.10.12 Gluconic Acid and Glucono-δ-lactone

Gluconic acid is obtained by the oxidation of glucose, which proceeds either by metal catalysis or enzymatically (*Aspergillus niger, Guconobacter suboxidans*).
Gluconic acid is used, e.g., in the production of invert sugar, beverages, and candies.

The δ-lactone is produced by evaporating a gluconic acid solution at 35–60°C. Glucono-δ-lactone slowly hydrolyzes, releasing protons. Hence, it is applied as an additive when slow acidification is needed, as with baking powder, raw sausage ripening and several sour milk products.

8.11 Bases

NaOH and a number of alkaline salts, such as $NaHCO_3$, Na_2CO_3, $MgCO_3$, MgO, $Ca(OH)_2$, Na_2HPO_4 and Na-citrate, are used in food processing for various purposes, for example:
Ripe olives are treated with 0.25–2% NaOH to eliminate the bitter flavor and to develop the desired dark fruit color.
In alkali-baked goods, molded dough pieces are dipped into 1.25% NaOH at 85–88°C in order to form the typical deep brown surface during baking.
In chocolate manufacturing, $NaHCO_3$ enhances the *Maillard* reaction, providing dark bitter chocolates.
In production of molten processed cheese, the pH rise needed to improve the swelling of casein gels is achieved by addition of alkali salts.

8.12 Antimicrobial Agents

Elimination of microflora by physical methods is not always possible, therefore, antimicrobial agents are needed. The spectrum of compounds used for this purpose has hardly changed for a long time. It is not easy to find new compounds with wider biological activity, negligible toxicity for mammals and acceptable cost.

8.12.1 Benzoic Acid

Benzoic acid activity is directed both to cell walls and to inhibition of citrate cycle enzymes (α-ketoglutaric acid dehydrogenase, succinic acid dehydrogenase) and of enzymes involved in oxidative phosphorylation.
The acid is used in its alkali salt form as an additive, since solubility of the free acid is

unsatisfactorily low. However, the undissoc-iated acid ($pK_a = 4.19$) is predominantly active – it is probably in this form that the acid acquires its ability to cross cell membranes. As in the case of sorbic acid and propionic acid, a certain activity is also attributed to the anion.

Benzoic acid usually occurs in nature as a gly-coside (in cranberry, bilberry, plum and cinna-mon trees and cloves). Its activity is primarily against yeasts and molds, less so against bac-teria. Figures 8.10 and 8.12 show the pH-dependent activity of the acid against *Escheri-chia coli, Staphylococcus aureus* and *Asper-gillus niger*.

The LD_{50} (rats; orally) is 1.7–3.7 g/kg body weight; the LD_{100} (guinea pig, cat, dog, rabbit; orally) is 1.4–2 g/kg. A daily intake of < 0.5 g Na-benzoate is tolerable for humans. No dan-gerous accumulation of the acid occurs in the body even at a dosage of as much as 4 g/day. It is eliminated by excretion in the urine as hip-puric acid while, at higher levels of intake, the glucuronic acid derivative is also excreted.

Benzoic acid (0.05–0.1%) is often used in combination with other preservatives and, on the basis of its higher activity at acidic pH's, it is used for preservation of sour food (pH 4 – 4.5 or lower), beverages with carbon dioxide, fruit salads, marmalades, jellies, fish preser-ves, margarine, paste (pâté) fillings and pick-led sour vegetables. A change in aroma, occur-ring mostly in fruit products, may result as a consequence of benzoic acid esterification.

8.12.2 PHB-Esters

The alkyl esters of p-hydroxybenzoic acid (PHB; parabens) are quite stable. Their solubi-lity in water decreases with increasing alkyl chain length (methyl → butyl). The esters are mostly soluble in 5% NaOH.

The esters are primarily antifungal agents and are also active against yeasts but less so against bacteria, especially those which are gram-negative. The activity rises with increas-ing alkyl chain length (Fig. 8.11). Neverthe-less, lower members of the homologous series are preferred because of better solubility.

The LD_{50} (mice; orally) is > 8 g/kg body weight. In a feeding experiment over 96 weeks

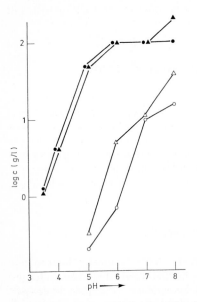

Fig. 8.10. The effect of benzoic acid on *Escherichia coli* (○ bacteriostatic. ● bactericidal activity) and *Staphylococcus aureus* (△ bacteriostatic and ▲ bac-tericidal activity)

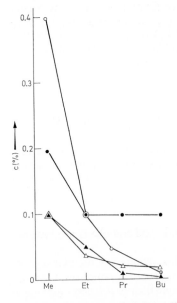

Fig. 8.11. Inhibition of *Salmonella typhosa* (●), *Aspergillus niger* (△), *Staphylococcus aureus* (○), and *Saccharomyces cerevisiae* (▲) by PHB-esters

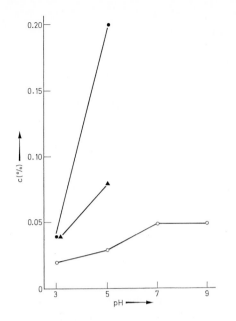

Fig. 8.12. Growth inhibitiion of *Aspergillus niger* by benzoic acid (●), p-hydroxybenzoic acid propyl ester (○) and sorbic acid (▲)

using 2% PHB-ester, no weight decrease was observed, while a slight decrease was found at the 8% level. In humans, the compounds are excreted in urine as p-hydroxybenzoic acid or its glycine or glucuronic acid conjugates.

Unlike benzoic acid, the esters can be used over a wide pH range since their activity is almost independent of pH (cf. Fig. 8.12). As additives, they are applied at 0.3–0.06% as aqueous alkali solutions or as ethanol or propylene glycol solutions in fillings for baked goods, fruit juices, marmalades, syrups, preserves, olives and pickled sour vegetables.

8.12.3 Sorbic Acid

The antimycotic effect of straight chain carboxylic acids has long been known. In particular the unsaturated acids, for example crotonic acid and its homologues, are very active. Sorbic acid (2-trans, 4-trans-hexadienoic acid; pK = 4.76) has the advantage that it is odorless and tasteless at the levels of use (0.3% or less). The acid is obtained by several syntheses:

- From parasorbic acid [(S)-2-hexen-5-olide); cf. Reaction 8.31]. The acid is present in berries of the mountain ash tree (*Sorbus aucuparia*).

$$\text{(8.31)}$$

- From ethanal:

$$3\,CH_3CHO \xrightarrow[\text{acetate}]{\text{Piperidine}} \quad \diagdown\diagup\diagdown\diagup CHO + 2\,H_2O$$

$$\xrightarrow[\text{Ag/O}_2/30°C]{\text{pH} > 12.5} \quad \diagdown\diagup\diagdown\diagup COOH \qquad \text{(8.32)}$$

- From crotonaldehyde obtained from ethanal (cf. Reaction 8.33).

The third synthesis is the most important.

The microbial activity of sorbic acid is primarily against fungi and yeasts, less so against bacteria. The activity is pH dependent (Fig. 8.12). Its utilization is recommended up to pH 6.5.

The LD_{50} (rats) of sorbic acid is ca. 10 g/kg body weight. Feeding experiments with rats for more than 90 days, with 1–8% sorbic acid in the diet, had no effect, while only 60% of the animals survive an 8% level of benzoic acid.

Sorbic acid is degraded biochemically like a fatty acid, i.e. by a β-oxidation mechanism. A small portion of the acid is degraded by ω-oxidation, yielding trans, trans-muconic acid (cf. Reaction 8.34).

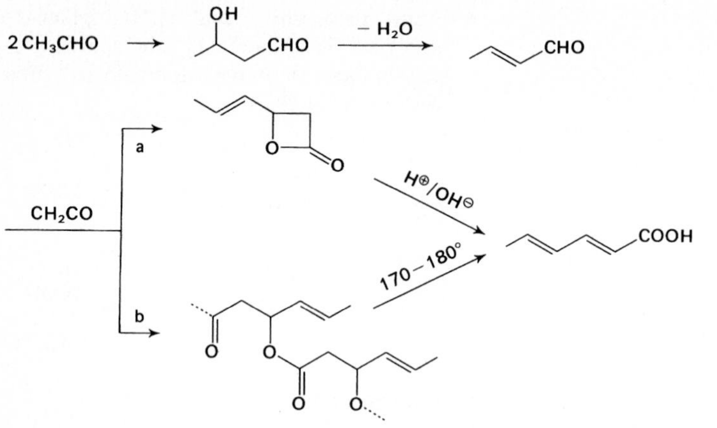

$$(8.33)$$

a: BF$_3$, T<0°C   b: Diethylene glycol, Zn-Isobutyrate, T>25°C

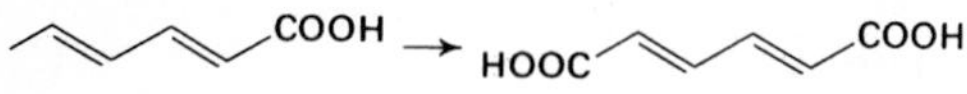

$$(8.34)$$

Some microorganisms, such as *Penicillium roqueforti*, have the ability to decarboxylate sorbic acid and thus convert it into 1,3-pentadiene, which has no antimicrobial activity and in addition may contribute to an off-flavor in cheeses:

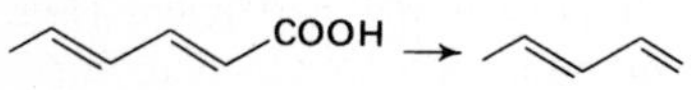

$$(8.35)$$

Sorbic acid or its salts are effective antifungal agents in baked products, cheeses, beverages (fruit juices, wines), marmalades, jellies, dried fruits and in margarine.

8.12.4 Propionic Acid

Propionic acid is found in nature where propionic acid fermentation occurs, e.g., in Emmental cheese, in which it is present up to 1%.

Its antimicrobial activity is mostly against molds, less so against bacteria. Propionic acid has practically no effect against yeast. Its activity is pH dependent. It is recommended and used up to pH 5 and only occasionally up to pH 6.

Propionic acid is practically nontoxic. It is used as an additive in baked products for inhibition of molds, and to prevent ropiness

caused by the action of *Bacillus mesentericus*. It is added to flour at 0.1–0.2% as its Ca-salt and is used in cheese manufacturing by dipping the cheese into an 8% solution of the acid.

8.12.5 Acetic Acid

The preserving activity of vinegar (cf. 22.3) has been known from ancient times. The acid has a two-fold importance: as a preservative and as a seasoning agent. It is more active against yeasts and bacteria than against molds. It is used as the free acid, Na- and Ca-salts, or as Na-diacetate (CH$_3$COOH · CH$_3$COONa · 1/2 H$_2$O), in ketchup, mayonnaise, acid-pickled vegetables, bread and other baked products.

8.12.6 SO$_2$ and Sulfite

The activity of these preserving agents covers yeasts, molds and bacteria. The activity increases with decreasing pH and is mostly derived from undissociated sulfurous acid, which predominates at a pH < 3.

Toxicity is negligible at the levels usually applied. Possible mutagenic activity is under investigation. Excretion in the urine occurs as sulfate.

Sulfite reacts with a series of food constituents, e.g., proteins with cleavage of disulfide bonds (cf. 1.4.4.4), with various cofactors like $NAD^{\oplus}$, folic acid, pyridoxal, and thiamine (cf. 6.3.1.3) and with ubiquinone:

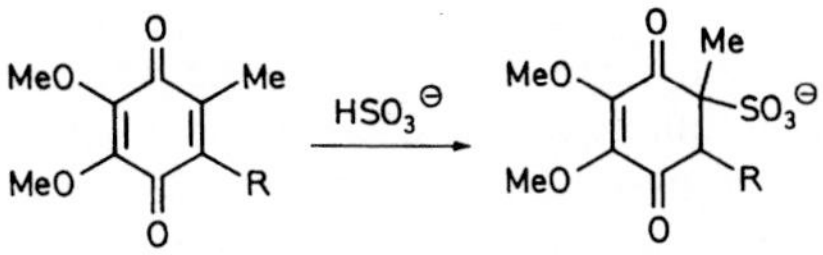

$$(8.36)$$

The pyrimidines in nucleic acids can also react, e.g., cytosine and uracil (cf. Formula 8.37). Anthocyanins are bleached (cf. 18.1.2.5.3).

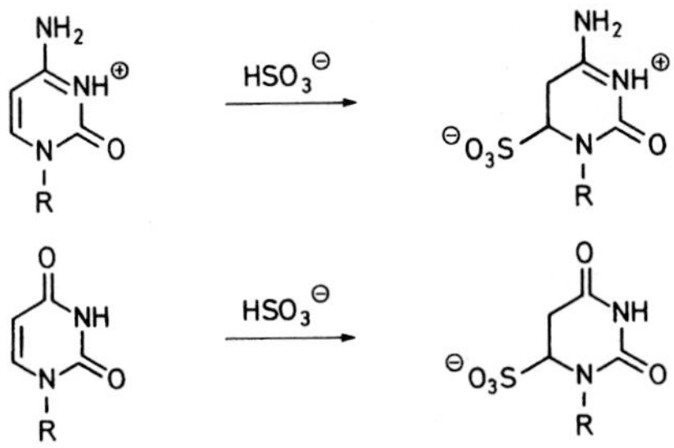

$$(8.37)$$

SO_2 is used in the production of dehydrated fruits and vegetables, fruit juices, syrups, concentrates or purée. The form of application is SO_2, Na_2SO_3, K_2SO_3, $NaHSO_3$, $Na_2S_2O_5$ and $K_2S_2O_5$ at levels of 200 ppm or less.
SO_2 is added in the course of wine making prior to must fermentation to eliminate interfering microorganisms. During wine fermentation with selected pure yeast cultures, SO_2 is used at a level of 50–100 ppm, while 50–75 ppm are used for wine storage.
SO_2 is not only antimicrobially active, but inhibits discoloration by blocking compounds with a reactive carbonyl group (*Maillard* reaction; nonenzymatic browning) or by inhibiting oxidation of phenols by phenol oxidase enzymes (enzymatic browning).

8.12.7 Diethyl (Dimethyl) Pyrocarbonate

Diethyl pyrocarbonate (DEPC or diethyl dicarbonate) is a colorless liquid of fruit-like or ester odor. Its antimicrobial activity covers yeasts (10–100 ppm), bacteria (*Lactobacilli*: 100–170 ppm) and molds (300–800 ppm).

The levels of the compound required for a clear inhibition are given in brackets. Diethyl pyrocarbonate readily hydrolyzes to yield carbon dioxide and ethanol:

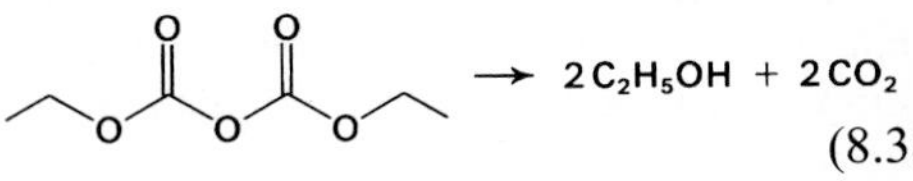

$$\longrightarrow 2\,C_2H_5OH + 2\,CO_2 \qquad (8.38)$$

or it reacts with food ingredients. In alcoholic beverages it yields a small amount of diethyl carbonate:

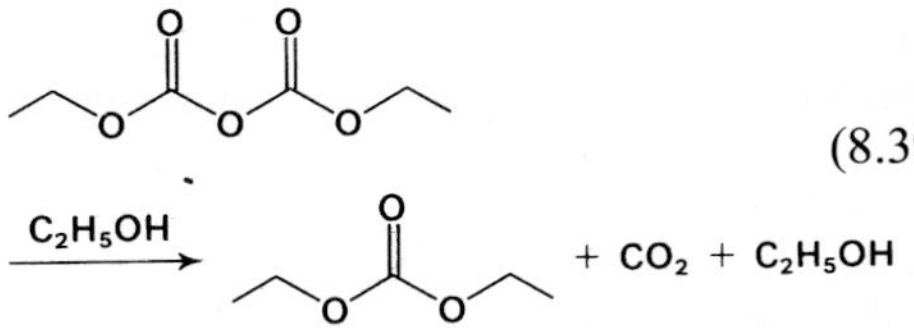

$$(8.39)$$
$$\xrightarrow{C_2H_5OH} \quad + CO_2 + C_2H_5OH$$

In the presence of ammonium salts, DEPC can form ethyl urethane in a pH-dependent reaction:

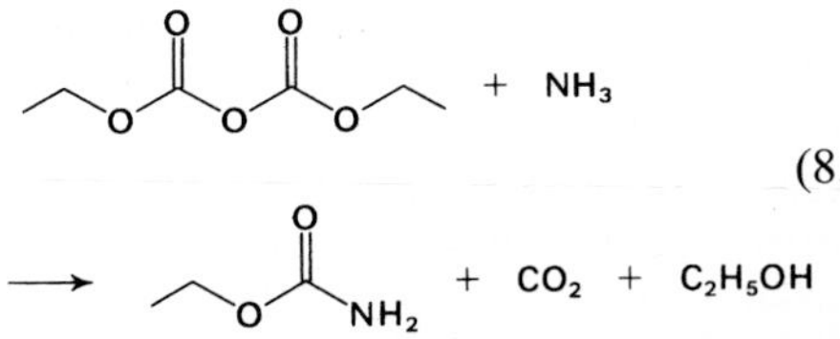

$$+ NH_3 \qquad (8.40)$$
$$\longrightarrow \quad + CO_2 + C_2H_5OH$$

Since diethyl carbonate may be a teratogenic agent and ethyl urethane is a carcinogen, the use of diethyl pyrocarbonate is discussed under toxicological aspects. The compound should be replaced by dimethyl pyrocarbonate, since methyl urethane, unlike ethyl urethane, is not carcinogenic.
Both compounds are used in cold pasteurization of fruit juices, wine and beer at a concentration of 120–300 ppm.

8.12.8 Ethylene Oxide, Propylene Oxide

These compounds are active against all microorganisms, particularly vegetative cells and spores, and also against viruses. Propylene oxide is somewhat less reactive than ethylene oxide.

Since they are efficient alkylating agents, the pure compounds are very toxic. After application, all the residual amounts must be completely removed. The glycols resulting from their hydrolysis are not as toxic (ethylene glycol: LD_{50} for rats is 8.3 g/kg body weight). Toxic reaction products can be formed, as exemplified by chlorohydrin obtained in the presence of chloride:

$$\text{(8.41)}$$

In addition, some essential food constituents react with formation of biologically inactive derivatives. Examples are riboflavin, pyridoxine, niacin, folic acid, histidine or methionine. However, these reactions are not of importance under the conditions of the normal application of ethylene oxide or propylene oxide.

Both compounds are used as gaseous sterilants (ethylene oxide, boiling point 10.7°C; propylene oxide, 35°C) against insects and for gaseous sterilization of some dehydrated foods for which other methods, e. g. heat sterilization, are not suitable. Examples are the gaseous sterilization of walnuts, starches, dehydrated foods (fruits and vegetables) and, above all, spices, in which a high spore count (and plate count in general) is often a sanitary problem. The sterilization is carried out in pressure chambers in a mixture with an inert gas (e. g. 80–90% CO_2). The need to remove the residual unreacted gas (vacuum, "gaseous rinsing") has already been emphasized. An alternative method of sterilization for the above-mentioned food products is high energy irradiation (UV-light, X-ray, or gamma irradiation).

8.12.9 Nitrite, Nitrate

These additives are used primarily to preserve the red color of meat (cf. 12.3.2.2.2). However, they also have antimicrobial activity, particularly in a mixture with common salt. Of importance is their inhibitive action, in non-sterilized meat products, against infections by *Clostridium botulinum* and, consequently,

against accumulation of its toxin. The activity is dependent on the pH and is proportional to the level of free HNO_2. Indeed, 5–20 mg of nitrite per kg are considered sufficient to redden meat, 50 mg/kg for the production of the characteristic taste, and 100 mg/kg for the desired antimicrobial effects. Acute toxicity has been found only at high levels of use (formation of methemoglobin). A problem is the possibility of the formation of nitrosamines, compounds with powerful carcinogenic activity. Numerous animal feeding tests have demonstrated tumor occurrence when the diet contained amines (sensitive to nitroso substitution) and nitrite. Consequently, the trend is to exclude or further reduce the levels of nitrate and nitrite in food. No suitable replacement has been found for nitrite in meat processing.

8.12.10 Antibiotics

The use of antibiotics in food preservation raises a problem since it might trigger development of more resistant microorganisms and thus create medical/therapeutic difficulties.

Of some importance is nisin, a polypeptide antibiotic, produced by some *Lactococcus lactis* strains. It is active against Gram-positive microorganisms and all spores, but is not used in human medicine. This heat-resistant peptide is applied as an additive for sterilization of dairy products, such as cheeses or condensed or evaporated milk (cf. 1.3.4.3). Natamycin (pimaricin, Formula 8.42), which is produced by *Streptomyces natalensis* and *S. chattanogensis*, is active at 5–100 ppm against yeasts and molds and is used as an additive in surface treatment of cheeses. It also finds application for suppressing the growth of molds on ripening raw sausages.

$$R = COOH \qquad \text{(8.42)}$$

The possibility of incorporating the wide spectrum antibiotics chlortetracycline and oxytetracycline into fresh meat, fish and poultry, in order to delay spoilage, is still under investigation.

8.12.11 Diphenyl

Diphenyl, due to its ability to inhibit growth of molds, is used to prevent their growth on peels of citrus fruits (lemon, orange, lime, grapefruit). It is applied by impregnating the wrapping paper and/or cardboard packaging material ($1-5$ g diphenyl/m^2).

8.12.12 o-Phenylphenol

This compound, at a level of $10-50$ ppm and a pH range of $5-8$, inhibits the growth of molds. The inhibition effect, which increases with increasing pH, is utilized in the preservation of citrus fruits. It is applied by dipping the fruits into a $0.5-2\%$ solution at pH 11.7.

8.12.13 Thiabendazole, 2-(4-Thiazolyl)benzimidazole

This compound (Formula 8.43) is particularly powerful against molds which cause the so-called blue mold rots, e.g., *Penicillium italicum* (blue-green-spored "contact mold") and *Penicillium digitatum* (green-spored mold). It is used for preserving the peels of citrus fruits and bananas. The application mode is by dipping or spraying the fruit with a wax emulsion containing $0.1-0.45\%$ thiabendazole.

(8.43)

8.13 Antioxidants

Since lipids are widely distributed in food and since their oxidation yields degradation products of great aroma impact, their degradation is an important cause of food deterioration by generation of undesirable aroma. Lipid oxidation can be retarded by oxygen removal or by using antioxidants as additives. The latter are mostly phenolic compounds, which provide the best results often as a mixture and in combination with a chelating agent. The most important antioxidants, natural or synthetic, are tocopherols, ascorbic acid esters, gallic acid esters, tert-butylhydroxyanisole (BHA) and di-tert-butylhydroxytoluene (BHT). They are covered in 3.7.3.2.2.

8.14 Chelating Agents (Sequestrants)

Chelating agents have acquired greater importance in food processing. Their ability to bind metal ions has contributed significantly to stabilization of food color, aroma and texture. Many natural constituents of food can act as chelating agents, e.g., carboxylic acids (oxalic, succinic), hydroxy acids (lactic, malic, tartaric, citric), polyphosphoric acids (ATP, pyrophosphates), amino acids, peptides, proteins and porphyrins.

Table 8.16 lists the chelating agents utilized by the food industry, while Table 8.17 gives the stability constants for some of their metal complexes.

Table 8.16. Chelating agents used as additives in food processing. (Compounds given in brackets are utilized only as salts or derivatives)

(Acetic acid)	Na-, K-, Ca-salts
Citric acid	Na-, K-, Ca-salts, monoisopropyl ester, monoglyceride ester, triethyl ester, monostearyl ester,
EDTA	Na-, Ca-salts
(Gluconic acid)	Na-, Ca-salts
Oxystearin	
Orthophosphoric acid	Na-, K-, Ca-salts
(Pyrophosphoric acid)	Na-salt
(Triphosphoric acid)	Na-salt
(Hexametaphosphoric acid, 10–15 residues)	Na-, Ca-salts
(Phytic acid)	Ca-salt
Sorbitol	
Tartaric acid	Na-, K-salts
(Thiosulfuric acid)	Na-salt

Table 8.17. Stability constants (pK-values) of some metal complexes

Chelating agent	Ca^{2+}	Co^{2+}	Cu^{2+}	Fe^{2+}	Fe^{3+}	Mg^{2+}	Zn^{2+}
Acetate	0.5	2.2				0.5	1.0
Glycine	1.4	5.2	8.2	4.3	10.0	3.5	5.2
Citrate	3.5	4.4	6.1	3.2	11.9	2.8	4.5
Tartrate	1.8		3.2		7.5	1.4	2.7
Gluconate	1.2		18.3			0.7	1.7
Pyrophosphate	5.0		6.7		22.2	5.7	8.7
ATP	3.6	4.6	6.1			4.0	4.3
EDTA	10.7	16.2	18.8	14.3	25.7	8.7	16.5

Traces of heavy metal ions can act as catalysts for fat or oil oxidation. Their binding by chelating agents increases antioxidant efficiency and inhibits oxidation of ascorbic acid and fat-soluble vitamins. The stability of the aroma and color of canned vegetables is substantially improved.

In the production of herb and spice extracts, the combination of an antioxidant and a chelating agent provides an improved extract quality. Chelating agents are also used in dairy products, wherein their deaggregating activity for the casein complexes is often utilized; in blood recovery processes to prevent clotting; and in the sugar industry to facilitate sucrose crystallization, a process which is otherwise retarded by sucrose-metal complexes.

8.15 Surface-Active Agents

Naturally occurring and synthetic surface-active agents (tensides), some of which are listed in Table 8.18, are used in food processing when a decrease in surface tension is required e.g., in production and stabilization of all kinds of dispersions (Table 8.19).

Dispersions include emulsions, foams, aerosols and suspensions (Table 8.20). In all cases an *outer, continuous* phase is distinct from an *inner*, discontinuous, *dispersed* phase. Emulsions are of particular importance and they will be outlined in more detail.

8.15.1 Emulsions

Emulsions are dispersed systems, usually of two immiscible liquids. When the outer phase consists of water and the inner of oil, it is con-

Table 8.18. Surfactants (surface active agents) in food

I. Naturally occurring:
A. Ions: proteins (cf. 1.4.3.6), phospholipids (lecithin, cf. 3.4.1.1), bile acids
B. Neutral substances: glycolipids (cf. 3.4.1.2), saponins

II. Synthetic:
A. Ions: stearyl-2-lactylate, Datem, Citrem (cf. Table 8.26)
B. Neutral substances: mono-, diacylglycerols and their acetic- and lactic acid esters, saccharose fatty acid esters, sorbitan fatty acid esters, polyoxyethylene sorbitan fatty acid esters

Table 8.19. Examples of surfactant utilization in the food industry

Utilization in production of	Effect
Margarine	Stabilization of a w/o emulsion
Mayonnaise	Stabilization of an o/w emulsion
Ice cream	Stabilization of an o/w emulsion, achievement of a "dry" consistency
Sausages	Prevention of fat separation
Bread and other baked products	Improvement of crumb structure, baked product volume, inhibition of starch retrogradation (bread staling)
Chocolate	Improvement of rheological properties, inhibition of "fat blooming"
Instant powders	Solubilization
Spice extracts	Solubilization

sidered as an "oil in water" (o/w) type of emulsion. When this is reversed, i.e., water is dispersed in oil, a w/o emulsion exists. Examples of food emulsions are: milk (o/w), butter (w/o) and mayonnaise (o/w).

The visual appearance of an emulsion depends on the droplet diameter. If the diameter is in the range of 0.15–100 μm, the emulsion appears milky-turbid. In comparison, microemulsions (diameter: 0.0015–0.15 μm) are transparent and considerably more stable because the sedimentation rate depends on the droplet diameter (Table 8.21).

Table 8.20. Dispersion systems

Type	Inner phase	Outer phase
Emulsion	liquid	liquid
Foam	gaseous	liquid
Aerosol	liquid or solid	gaseous
Suspension	solid	liquid

Table 8.21. Sedimentation rate (v) as a function of droplet diameter (d)

d (μm)	v (cm/24 h)
0.02	3.75×10^{-4}
0.2	3.76×10^{-2}
2	3.76
20	3.76×10^2
200	3.76×10^4

Each emulsifier can disperse a limited amount of an inner phase, i.e. it has a fixed *capacity*. When the limit is reached, further addition of outer phase breaks down the emulsion. The capacity and other related parameters differ among emulsifiers and can be measured accurately under standardized conditions.

8.15.2 Emulsifier Action

8.15.2.1 Structure and Activity

Emulsions are made and stabilized with the aid of a suitable tenside, usually called an emulsifier. Its activity is based on its molecular structure. There is a lipophilic or hydrophobic part with good solubility in a nonaqueous phase, such as an oil or fat, and a polar or hydrophilic part, soluble in water. The hydrophobic part of the molecule is generally a long-chain alkyl residue, while the hydrophilic part consists of a dissociable group or of a number of hydroxyl or polyglycolether groups.

In an immiscible system such as oil/water, the emulsifier is located on the interface, where it decreases interfacial tension. Thus, even in a very low concentration, it facilitates a fine distribution of one phase within the other. The emulsifier also prevents droplets, once

formed, from aggregating and coalescing, i.e. merging into a single, large drop (Fig. 8.13). Ionic tensides stabilize o/w emulsions in the following way (Fig. 8.14a): at the interface, their alkyl residues are solubilized in oil droplets, while the charged end groups project into the aqueous phase. The involvement of counter ions forms an electrostatic double layer, which prevents oil droplet aggregation. Nonionic, neutral tensides are oriented on the oil droplet's surface with the polar end of the

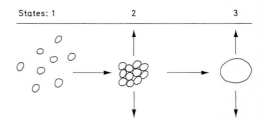

Fig. 8.13. Changes in an emulsion. 1 The droplets are dispersed in a continuous phase. 2 The droplets form aggregates. An increase in particle diameter results in acceleration of their flotation or sedimentation. 3 Coalescence: the aggregated droplets merge into larger droplets. Finally, two continuous phases are formed; the emulsion is destroyed

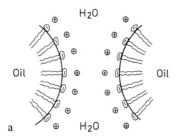

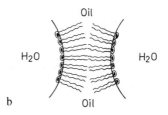

Fig. 8.14. Stabilization of an emulsion. **a** Activity of an ionic emulsifier in an o/w emulsion. **b** Activity of a nonpolar emulsifier in w/o emulsion. ○ Polar groups, ∼ apolar tails of the emulsifier

tenside projecting into the aqueous phase. The coalescence of the droplets of an o/w emulsion is prevented by an anchored "hydrate shell" built around the polar groups.

The coalescence of water droplets in a w/o emulsion first requires that water molecules break through the double-layered hydrophobic region of emulsifier molecules (Fig. 8.14b). This escape is only possible when sufficient energy is supplied to rupture the emulsifier's hydrophobic interaction.

The stability of an emulsion is increased when additives are added which curtail droplet mobility. This is the basis of the stabilization effect of hydrocolloids (cf. 4.4.3) on o/w emulsions since they increase the viscosity of the outer, aqueous phase.

A rise in temperature negatively affects emulsion stability, and can be applied whenever an emulsion has to be destroyed. Elevated temperatures are used along with shaking, agitation or pressure (mechanical destruction of interfacial films as, for example, in butter manufacturing, cf. 10.2.3.3). Other ways of decreasing the stability of an emulsion are addition of ions which collapse the electrostatic double layer, or hydrolysis to destroy the emulsifier.

8.15.2.2 Critical Micelle Concentration (CMC), Lyotropic Mesomorphism

The surface tension of an aqueous solution of an o/w emulsifier decreases down to the critical micelle concentration (CMC) as a function of the emulsifier concentration. Above this limiting value, the emulsifier aggregates reversibly to give spherical micelles, the surface tension changing only slightly. The CMC is a characteristic value of the emulsifier, which decreases as the hydrophobic part of the molecule increases. It is also influenced by the temperature, pH value, and electrolyte concentration.

The temperature at which the solubility of an emulsifier reaches the CMC is called the critical micelle temperature (T_c, *Krafft* point). Crystals, micelles, and the dissolved emulsifier are in equilibrium at the T_c (Fig. 8.15). An emulsifier cannot form micelles below the T_c which, e.g., depends on the structure of the fatty acid residues in lecithin (Table 8.22).

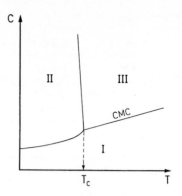

Fig. 8.15. Solubility of an emulsifier in water. Ordinate: concentration, abscissa: temperature. I: Solution, II: Crystals, III: Micelles, T_c: Critical micelle temperature

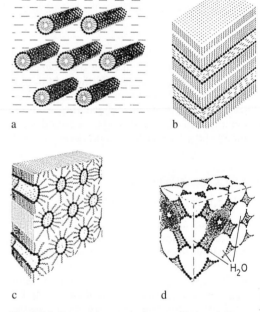

Fig. 8.16. Lyotropic mesophases of emulsifiers (according to *Schuster*, 1985). (a) Hexagonal I, (b) Lamellar, (c) Hexagonal II, (d) Cubic

●⌒⌒ Emulsifier   ⎓⎓⎓ Water

Emulsifiers are lyotropic mesomorphous, i.e., they form one of the following liquid crystalline mesophases depending on the water content and the temperature (shown schematically in Fig. 8.16):

Table 8.22. Effect of fatty acid residues on the critical micelle temperature T_c of lecithins

Fatty acid	T_c (°C)
12:0	0
14:0	23
16:0	41
18:0	58
18:1	− 20

Hexagonal I. Cylindrical aggregates of emulsifier molecules; the polar groups are oriented towards the outer water phase.

Lamellar. Emulsifier bilayers which are separated by thin water zones.

Hexagonal II (Inverse Hexagonal). Cylindrical aggregates of emulsifier molecules; the polar groups are oriented towards the inner water phase.

Cubic. Cubic space- and face-centered water aggregates in a matrix of emulsifier molecules; the polar groups are oriented towards the water.

Phase diagrams show the mesophase present as a function of water content and temperature.

In the phase diagram of the o/w emulsifier lysolecithin (Fig. 8.17, a), micelles, a lamellar, and a hexagonal phase appear. The w/o emulsifier 1-monoelaidin (Fig. 8.17, b) crystallizes at temperatures below 30 °C. The α-modification formed first is converted to the more stable β-form, which unlike the α-form, has no emulsifying properties. The melted 1-monoelaidin forms a microemulsion with little water and lamellar and cubic mesophases with much water. 1-Monoolein (Fig. 8.17, c) melts at lower temperatures and an inverse hexagonal mesophase appears.

The phases of simply constituted food emulsions, which are present at a certain temperature depending on the composition, show a ternary phase diagram, e. g., in Fig. 8.18.

8.15.2.3 HLB-Value

A tenside with a relatively strong lipophilic group and weak hydrophilic group is mainly soluble in oil and preferentially stabilizes a w/o emulsion, and vice versa. This fact led to

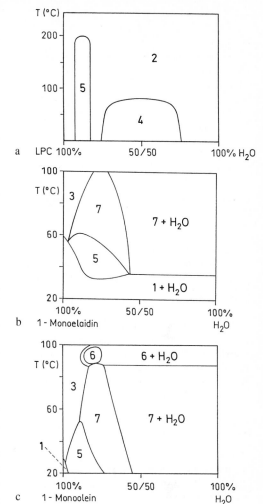

Fig. 8.17. Binary phase diagrams for the system emulsifier/water (according to *Krog*, 1990) (a) Lysolecithin, (b) 1-Monoelaidin, (c) 1-Monoolein
1 Crystals, *2* Micelles, *3* Microemulsion, *4* Hexagonal I, *5* Lamellar, *6* Hexagonal II, *7* Cubic

the development of a standard with which the relative strength or "activity" of the hydrophilic and lipophilic groups of emulsifiers can be evaluated. It is called the *HLB value* (hydrophilic-lipophilic balance). It can be determined, e. g., from dielectric constants or from the chromatographic behavior of the surface-active substance. The HLB value of the fatty

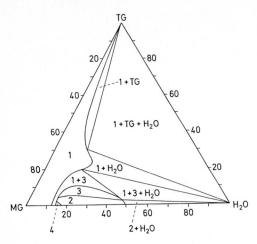

Fig. 8.18. Ternary phase diagram for the system monoglycerides (from sunflower oil) /water/soybean oil at 40°C (according to *Larsson* and *Dejmek*, 1990)
1 Microemulsion, *2* Cubic, *3* Hexagonal II, *4* Lamellar

acid esters of polyhydroxy alcohols can also be calculated as follows (SV = saponification number of the emulsifier, AV = acid value of the separated acid):

$$HLB = 20 \left(1 - \frac{SV}{AV}\right) \qquad (8.44)$$

On the basis of experimental group numbers (Table 8.23), the HLB value can be calculated using the formula:

$$HLB = \Sigma \text{ (hydrophilic group number)} -$$
$$\Sigma \text{ (hydrophobic group number)} + 7$$
$$(8.45)$$

Some examples listed in Table 8.24 show very good correspondence between calculated and experimentally found HLB values.
The HLB values indicated the first industrial applications (Table 8.25). For a detailed characterization, however, comprehensive knowledge of possible interactions of the emulsifier with the many components of a food emulsion is still lacking. Hence, emulsifiers are mainly used in accordance with empirical considerations.

Table 8.23. Group number N_H and N_L for HLB calculation

Hydrophilic group	N_H	Lipophilic group	N_L
		$\vert$	
$-OSO_3^-$, Na^+	38.7	$-CH-$	0.475
$-SO_3^-$, Na^+	37.4	$-CH_2-$	0.475
$-COO^-$, Na^+	21.1	$-CH_3$	0.475
$-COO^-$, K^+	19.1	$=CH-$	0.475
Sorbitan ring	6.8	$-CH-CH_2-O-$	0.15
Ester	2.4	$\vert$	
$-COOH$	2.1	CH_3	
$-OH$ (free)	1.9	Benzene ring	1.662
$-O-$	1.3		
$-(CH_2-CH_2-O)-$	0.33		

Table 8.24. Hydrophilic lipophilic balance (HLB) values of some surfactants

Compound	HLB-value	
	Found	Calculated
Oleic acid	1.0	
Sorbitol tristearate	2.1	2.1
Stearyl monoglyceride	3.4	3.8
Sorbitol monostearate	4.7	4.7
Sorbitol monolaurate	8.6	
Gelatin	9.8	
Polyoxyethylene sorbitol tristearate	10.5	10
Methylcellulose	10.5	
Polyoxyethylene sorbitol monostearate	14.9	
Polyoxyethylene sorbitol monooleate	15.0	15
Sodium oleate	18.0	
Potassium oleate	20.0	

Table 8.25. HLB-values related to their industrial application

HLB-range	Application
3–6	w/o-Emulsifiers
7–9	Humectants
8–18	o/w-Emulsifiers
15–18	Turbidity stabilization

It has been observed with neutral emulsifiers that the degree of hydration of the polar groups decreases with a rise in temperature and the influence of the lipophilic groups increases.

Phase inversion occurs o/w → w/o. The temperature at which inversion occurs is called the phase conversion temperature.

8.15.3 Synthetic Emulsifiers

Today, 150000–200000 t of emulsifiers are produced worldwide. Of this amount, mono- and diacylglycerides and their derivatives account for the largest part, i.e. about 75%. Synthetic emulsifiers include a series of nonionic compounds. Unlike the ionic compounds, the nonionic emulsifiers are not in danger of decreasing in interfacial activity through salt formation with food constituents.

The utilization of emulsifiers is legislated and often differently regulated in some countries. The synthetic emulsifiers described below are used worldwide.

8.15.3.1 Mono-, Diacylglycerides and Derivatives

Mono- and diacylglycerides, which are mostly used as mixtures, are produced as described in 3.3.2. Other emulsifiers with special activities are obtained by derivatization (cf. Table 8.26). As a result of the diverse reaction possibilities of the starting compounds, complex products are obtained in this process. The production of diacetyltartaric acid monoglyceride (Formula 8.46) represents an example.

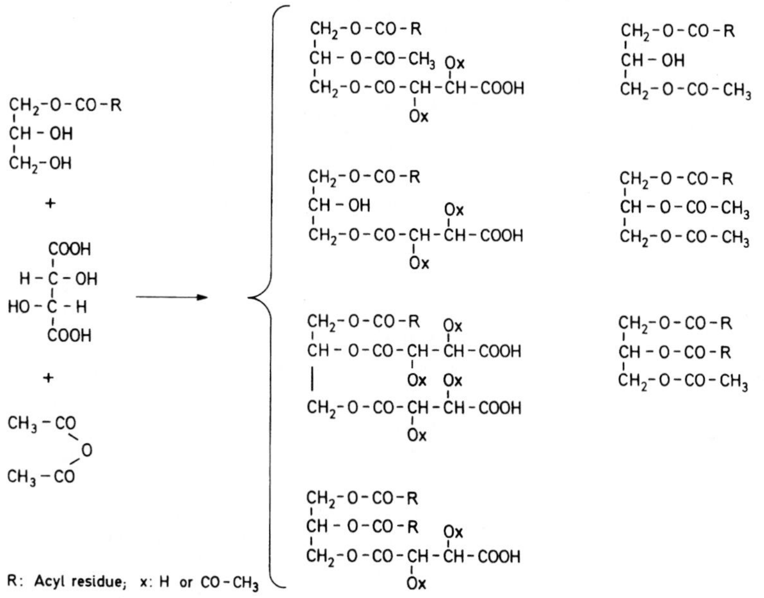

$$\text{(8.46)}$$

R: Acyl residue; x: H or $CO-CH_3$

Table 8.26. Emulsifiers from mono- and diacylglyceride mixtures

Name			Production by conversion of mixtures of mono- and
Mono- and diglycerides esterified with		EU-number	diacylglycerides with
Acetic acid (acetylated mono- and diglycerides)	*Acetem*	E 472a	Acetic anhydride
Lactic acid	*Lactem*	E472b	Lactic acid
Citric acid	*Citrem*	E472c	Citric acid
Monoacetyl- and diacetyltartaric acid	*Datem*	E472e	Tartaric acid and acetic anhydride

Unlike *acetem* and *lactem*, *datem* and *citrem* contain acidic components, e. g. *citrem*:

$$
\begin{array}{l}
CH_2\text{-}O\text{-}CO\text{-}R \\
| \\
HO\text{-}CH \\
| \\
CH_2\text{-}O\text{-}CO \\
\qquad | \\
\qquad CH_2 \\
\qquad | \\
HO\text{-}C\text{-}COOH \\
\qquad | \\
\qquad CH_2\text{-}COOH
\end{array}
\qquad (8.47)
$$

8.15.3.2 Sugar Esters

They are obtained, among other methods, by transesterification of fatty acid methyl esters (14 : 0, 16 : 0, 18 : 0 and/or 18 : 1, double bond position 9) with sucrose and lactose. The resultant mono – and diesters are odorless and tasteless. Depending on their structure, they cover an HLB range of 7–13, and are used in stabilization of o/w emulsions, or in stabilization of some instant dehydrated and powdered foods.

8.15.3.3 Sorbitan Fatty Acid Esters

Esters of sorbitan (cf. 19.1.4.6) with fatty acids (*Span*'s) serve the stabilization of w/o emulsions.

$$
\begin{array}{l}
HO \qquad OH \\
\diagdown \diagup \\
\diagup \diagdown \text{-}CHOH \\
O \quad | \\
\qquad CH_2O\text{-}COR
\end{array}
\qquad (8.48)
$$

8.15.3.4 Polyoxyethylene Sorbitan Esters

Polyoxyethylene groups are introduced into the molecules to increase the hydrophilic property of sorbitan esters:

$$
\begin{array}{l}
HO(H_4C_2O)_w \qquad (OC_2H_4)_x\,OH \\
\diagdown \diagup \\
\diagup \diagdown CH(OC_2H_4)_y\,OH \\
O \quad | \\
\qquad CH_2(OC_2H_4)_z\,O\text{-}COR
\end{array}
\qquad (8.49)
$$

Polyoxyethylene sorbitan monoesters (examples in Table 8.24) are used to stabilize o/w emulsions.

8.15.3.5 Stearyl-2-Lactylate

In the presence of sodium or calcium hydroxide, esterification of stearic acid with lactic acid gives a mixture of stearyl lactylates (Na or Ca salt), the main component being stearyl-2-lactylate:

$$
\begin{array}{c}
\qquad\qquad CH_3 \qquad\quad CH_3 \\
\qquad\qquad | \qquad\qquad\;\; | \\
CH_3\text{-}(CH_2)_{16}\text{-}CO\text{-}O\text{-}CH\text{-}CO\text{-}O\text{-}CH\text{-}COO^{\ominus}Na^{\oplus}
\end{array}
\qquad (8.50)
$$

The free acid acts as a w/o emulsifier and the salts as o/w emulsifiers. The HLB-value of the sodium salt is 8–9, and that of the calcium salt, 6–7. The sodium salt is used to stabilize an o/w emulsion which is subjected to repeated cycles of freezing and thawing.

8.16 Thickening Agents, Gel Builders, Stabilizers

A number of polysaccharides and their modified forms, even at low concentrations, are able to increase a system's viscosity, to form gels and to stabilize emulsions, suspensions or foams. These compounds are also active as crystallization inhibitors (e. g. in confections, ice creams) and are suitable for aroma encapsulation, as is often needed for dehydrated food. These properties make polysaccharides important additives in food processing and storage. The compounds of importance, together with their properties and utilization, were described in detail in the chapter on carbohydrates. Among proteins, gelatin is an important gel-forming agent used widely in food products (cf. 12.3.2.3.1).

8.17 Humectants

Some polyols (1,2-propanediol, glycerol, mannitol, sorbitol) have distinct hygroscopic properties and act as humectants, i.e. additives for retaining food moisture and softness and inhibiting crystallization. They are often required in a confectionery product. When glycerol or sorbitol is added to mashed vegetables or fruits or in the production of other powdered foods before the final drying stage, the dehydrated products have improved rehydration characteristics.

8.18 Substitutes for Fat

Low-calorie and calorie-free substitutes on a protein or carbohydrate basis or on a synthetic basis have been developed to lower the fat content.

Substitutes known by the name *Simplesse* are heat-coagulated milk and/or egg proteins. The relatively large particles which are normally formed during denaturation and feel rough in the mouth are ground by using a special technique (microparticulation). They are ground until the tongue no longer detects individual particles, but they feel like a liquid. This substitute is suitable for milk products (ice cream, desserts etc.) which are not strongly heated. In fact, 3 g of fat can be replaced by 3 g of swollen substitute (1 g of protein + 2 g of water) and 27 kcal by 4 kcal.

Non-sweet oligosaccharides (maltodextrins, DE 5), which dissolve completely in hot water, are made from corn starch. When this solution is cooled, a gel is formed, which has the texture of edible oil. It can partially replace fat, e. g., in margarine, allowing a 35% reduction of the energy content.

Synthetic substitutes are esters which are not hydrolyzed by the lipases in the gastrointestinal tract. Examples are hexa- to octa-esters of sucrose with fatty acids, called *olestra*, and polyetherpolyols formed by the reaction of glycerol with propylene oxide, followed by esterification with fatty acids. If olestra contains linoleic acid residues, dimers and trimers can be formed on heating. It is of advantage that during baking and roasting, these compounds can be heated to the same extent as cooking fats.

8.19 Anticaking Agents

Some food products, such as common salt, seasoning salt (e. g. a mixture of onion or garlic powder with common salt), dehydrated vegetable and fruit powders, soup and sauce powders and baking powder, tend to cake into a hard lump. Lumping can be avoided by using any of a number of compounds that either absorb water or provide protective hydrophobic films. Anticaking compounds include sodium, potassium and calcium hexacyanoferrate (II), calcium and magnesium silicate, tricalcium phosphate and magnesium carbonate.

8.20 Bleaching Agents

Bleaching is used primarily in flour production. The removal of yellow carotenoids by oxidation can be achieved by a number of compounds that, in addition to bleaching, improve the baking quality of flour. Examples of some approved common bleaching agents are Cl_2, ClO_2, $NOCl$, NO_2 and N_2O_4. Lipoxygenase enzyme also has an efficient bleaching activity.

8.21 Clarifying Agents

In some beverages, such as fruit juices, beer or wine, turbidity and sediment formation can occur with the involvement of phenolic compounds, pectins and proteins. These defects can be corrected by: (a) partial enzymatic degradation of pectins and proteins; (b) removal of phenolic compounds with the aid of gelatin, polyamide or polyvinyl pyrrolidone powders; and (c) by protein removal with bentonite (an aluminum silicate) or tannin.

8.22 Propellants, Protective Gases

Food sensitive to oxidation and/or microbial spoilage can be stored in an atmosphere of protective gas or a gas mixture (N_2, CO_2, CO, etc.; modified or controlled atmosphere storage). This is often a suitable method for lengthening the shelf life of food.

Liquid food can be filled into pressurized containers and, when needed, using a propellant, discharged in the form of a cream or paste (e. g. cream cheese, ketchup), a foam (whipping cream) or a mist (herb or spice extracts in oil; liquid barbecue smoke). Propellants used are N_2, N_2O, and CO_2. Octafluorocyclobutane (Freon 318) and chloropentafluoroethane (Freon 115) are being used less and less for reasons of environmental protection.

Due to its low solubility in water, fat and oil, N_2 is used preferentially as a propellant when foam formation is not desired (ketchup). On the other hand, gases such as N_2O and CO_2 are preferred at times due to their good solubility in water.

8.23 Literature

Anonymus: Fat substitute update. Food Technol. *44* (3), 92 (1990)

Ariyoshi, Y., Kohmura, M., Hasegawa, Y., Ota, M., Nio, N.: Sweet peptides and proteins: Synthetic studies. ACS Symposium Series *450*, p. 41 (1991)

Bär, A., Borrego, F., Castillo, J., del Rio, J.A.: Neohesperidin dihydrochalcone: Properties and applications. Lebensm. Wiss. Technol. *23*, 371 (1990)

Beets, M.G.J.: Structure-activity relationships in human chemoreception. Applied Science Publ.: London. 1978

Belitz, H.-D., Chen, W., Jugel H., Treleano, R., Wieser, H., Gasteiger, J., Marsili, M.: Sweet and bitter compounds: Structure and taste relationship. In: Food taste chemistry (Ed.: Boudreau, J.C.), ACS Symposium Series 115, p. 93, American Chemical Society: Washington, D.C. 1979

Belitz, H.-D., Chen, W, Jugel, H., Stempfl, W., Treleano, R., Wieser, H.: Structural requirements for sweet and bitter taste. In: Flavour '81 (Ed.: Schreier, P.), p. 741, Walter de Gruyter: Berlin. 1981

Belitz, H.-D., Chen, W., Jugel, H., Stempfl, W., Treleano, R., Wieser, H.: QSAR of bitter tasting compounds. Chem. Ind. *1983*, 23

Birch, G.G., Brennan, J.G., Parker, K.J. (Eds.): Sensory properties of foods. Applied Science Publ.: London. 1977

Branen, A.L., Davidson, P.M., Salminen, S. (Eds.): Food Additives. Marcel Dekker: New York. 1990

Buchta, K.: Lactic acid. In: Biotechnology (Eds.: Rehm, H.-J., Reed, G.), Vol. 3, p. 409, Verlag Chemic: Weinheim. 1983

Compadre, C.M., Pezzuto, J.M., Kinghorn, A.D.: Hernandulcin: an intensely sweet compound discovered by review of ancient literature. Science *227*, 417 (1985)

Deutsche Forschungsgemeinschaft: Bewertung von Lebensmittelzusatz- und -inhaltsstoffen. VCH Verlagsgesellschaft: Weinheim. 1985

Furia, T.E. (Ed.): Handbook of food additives. 2nd edn., CRC Press: Cleveland, Ohio. 1972

Gardner, D.R., Sanders, R.A.: Isolation and characterization of polymers in heated olestra and olestra/triglyceride blend. J. Am. Oil Chem. Soc. *67*, 788 (1990)

Glowaky, R.C., Hendrick, M.E., Smiles, R.E., Torres, A.: Development and uses of Alitame: A novel dipeptide amide sweetener. ACS Symposium Series *450*, p. 57 (1991)

Gould, G.W. (Ed.): Mechanisms of action of food preservation procedures. Elsevier Applied Science: London (1989)

Griffin, W.C., Lynch, M.J.: Surface active agents. In: Handbook of food additives, 2nd edn. (Ed.: Furia, T.E.), p. 397, CRC Press: Cleveland, Ohio. 1972

Haberstroh, H.-J., Hustede, H.: Weinsäure. Ullmanns Encyklopädie der technischen Chemie, 4. edn., vol. 24, p. 431. Verlag Chemie: Weinheim. 1983

Herbert, R.A.: Microbial growth at low temperatures. In: Gould, G.W. (Ed.) Mechanism of action of food preservation procedures, p. 71. Elsevier Applied Science: London (1989)

Heusch, R.: Emulsionen. In: Ullmanns Encyklopädie der technischen Chemie, 4. edn., vol. 10, p. 449, Verlag Chemie: Weinheim. 1975

Iyengar, R.B., Smits, P., Van der Ouderas, F., Van der Wel, H., Van Brouwershaven, J., Ravestein, P., Richters, G., von Wassenaar, P.D.: The complete amino-acid sequence of the sweet protein Thaumatin I. Eur. J. Biochem. *96*, 193 (1979)

Jenner, M.R.: Sucralose: How to make sugar sweeter. ACS Symposium Series *450*, p.68 (1991)

Kier, L.B.: A molecular theory of sweet taste. J. Pharm. Sci. *61*, 1394 (1972)

Kim, S.-H., Kang, C.-H., Cho, J.-M.: Sweet proteins: Biochemical studies and genetic engineering. ACS Symposium Series *450*, p. 28 (1991)

Kommission der Europäischen Gemeinschaften: Berichte des Wissenschaftlichen Lebensmittelausschusses (Sechzehnte Folge) – Süßstoffe: Brüssel. 1985

Lancet, D., Ben-Arie, N.: Sweet taste transduction. A molecular-biological analysis. ACS Symposium Series *450*, p. 226 (1991)

Lange, H., Kurzendorfer, C.-P.: Zum Mechanismus der Stabilisierung von Emulsionen. Fette Seifen Anstrichm. *76*, 120 (1974)

Larsson, K., Friberg, S.E.: Food emulsions. 2. edn., Marcel Dekker, Inc.: New York, 1990

Lück, E.: Chemische Lebensmittelkonservierung. Springer Verlag: Berlin. 1977

Maga, J. A.: Flavor potentiators. Crit. Rev. Food Sci. Nutr. *18*, 231 (1982/83)

Owens, W. H., Kellogg, M. S., Klade, C. A., Madigan, D. L., Mazur, R. H., Muller, G. W.: Tetrazoles as carboxylic acid surrogates: High-potency sweeteners. ACS Symposium Series *450*, p. 100 (1991)

Powrie, W. D., Tung, M. A.: Food dispersions. In: Principles of food science, part I (Ed.: Fennema, O. R.), p. 539, Marcel Dekker, Inc.: New York. 1976

Röhr, M., Kubicek, C. P., Kominek, J.: Citric acid. In: Biotechnology (Eds.: Rehm, H.-J., Reed, G.), Vol. 3, p. 419, Verlag Chemie: Weinheim. 1983

Röhr, M., Kubicek, C. P., Kominek, J.: Gluconic acid. In: Biotechnology (Eds.: Rehm, H.-J., Reed, G.), Vol. 3, p. 455, Verlag Chemie: Weinheim. 1983

Rohse, H., Belitz, H.-D.: Shape of sweet receptors studied by computer modeling. ACS Symposium Series *450*, p. 176 (1991)

Rosival, L., Engst, R., Szokolay, A.: Fremd- und Zusatzstoffe in Lebensmitteln. VEB Fachbuchverlag: Leipzig. 1978

Schuster, G. (Ed.): Emulgatoren für Lebensmittel. Springer-Verlag: Berlin. 1985

Schwall, H.: Milchsäure. In: Ullmanns Encyklopädie der technischen Chemie, 4. ed., vol. 17, p. 1, Verlag Chemie: Weinheim. 1979

Shallenberger, R. S., Acree, T. E.: Chemical structure of compounds and their sweet and bitter taste. In: Handbook of sensory physiology, Vol. IV, Part 2 (Ed.: Beidler, L. M.), p. 221, Springer-Verlag: Berlin. 1971

Souci, S. W., Mergenthaler, E.: Fremdstoffe in Lebensmitteln. J. F. Bergmann Verlag: München. 1958

Tinti, J.-M., Nofre, C.: Design of sweeteners: A rational approach. ACS Symposium Series *450*, p. 88 (1991)

Walters, D. E., Orthoefer, F. T., DuBois, G. E. (Eds.): Sweeteners – Discovery, Molecular Design and Chemoreception. ACS Symposium Series *450*, American Chemical Society: Washington, D.C. 1991

Van der Wel, H., Bel, W. J.: Effect of acetylation and methylation on the sweetness intensity of Thaumatin I. Chem. Senses Flavor *2*, 211 (1976)

Yamashita, H., Theerasilp, S., Aiuchi T., Nakaya, K., Nakamura, Y., Kurihara, Y.: Purification and complete amino acid sequence of a new type of sweet protein with taste-modifying activity; curulin. J. Biol. Chem. *265*, 15770 (1990)

9 Food Contamination

9.1 General Remarks

Special attention must be paid to the possibility of contamination of food with toxic compounds. They may be present incidentally and may be derived in various ways. Examples of such contaminants are:

- Pollutants derived from burning of fossil fuels, radionuclides from fallout, or emissions from industrial processing (toxic trace elements, radionuclides, polycyclic aromatic hydrocarbons, dioxins).
- Components of packaging material, and of other frequently used products (monomers, polymer stabilizers, plasticizers, polychlorinated biphenyls, cleansing/washing agents and disinfectants).
- Toxic metabolites of microorganisms (enterotoxins, mycotoxins).
- Residues of agricultural chemicals used for crop protection (pesticides, such as insecticides, fungicides and herbicides, anti-sprouting agents, plant hormones).
- Residues from livestock and poultry husbandry (veterinary medicinals and feed additives).

Toxic food contaminants might also be formed within the food itself or within the human digestive tract by reactions of some food ingredients and additives (e. g. nitrosamines).

Measures required to prevent contamination include:

- Extensive analytical control of food.
- Determination of the sources of contamination.
- Legislation (legal standards to permit, ban, curtail or control the use of potent food contaminants, and the processes associated with them) to establish permissible levels of contaminants.

Such efforts have paid off, as shown in a report of the German Society for Nutrition released in 1980 on chemical/toxicological aspects of food.

This report, which is still valid today, stated:

"When the overall situation in food processing and distribution is compared to that described in the nutrition report of 1976, we have to say that no new health hazard source has been discovered. Where contamination problems existed, appropriate measures were taken to reduce or avoid them; consequently, decreases in residues hazardous to health have already been observed. Where residue problems can not be avoided, our duty is to inform the consumer not to include the food as a regular part of the daily diet but to less frequently consume some of the more contaminated foods, such as some wild mushrooms, kidney, liver, specialty meats and freshwater fish.

Where man is obviously responsible for the creation of residue problems by the use of pesticides at improper times, banned pesticides, or illegal animal drugs, or by not observing waiting times, only greater vigilance and control can be of help. Even individual producers of milk and eggs, growers of certain fruits and vegetables, and especially importers of certain food commodities should be controlled more carefully".

These remarks are supplemented with the final observations of the nutrition report, 1988:

"The situation presented in previous nutrition reports regarding the presence of residues and contaminants in food has been essentially confirmed in the time up to 1987. The burden on the consumer due to agricultural pesticides, especially persistent organochlorine compounds, has continued to decrease slightly. However, the concentrations of polychlorinated biphenyls (PCB) have remained unchanged. This is due to the worldwide industrial application and the high chemical and biological persistence of these substances. Residues of veterinary drugs and feed additives play no special role. Their presence indicates a disregard of application regulations."

Toxicological assessment of a contaminant may, for various reasons, be a difficult task. Firstly, sufficient data are not available for all compounds. Also, the possibility of synergistic effects of various substances, often including their degradation products, should not be excluded. Further, the hazard might be influenced by age, sex, state of health and by habitual consumption. Based on these considerations, any nutritional statement about the "tolerable concentration" must take sufficient safety factors into account.

Toxicity assay involves the determination of:

- Acute toxicity, designated as LD_{50} (the dose that will kill 50% of the animals in a test series).
- Subacute toxicity, determined by animal feeding tests lasting four weeks.
- Chronic toxicity, assessed by animal feeding tests lasting for 6 months to 2 years.

In chronic toxicity tests attention is especially given to the occurrence of carcinogenic, mutagenic and teratogenic symptoms. The tests are conducted with at least two animal species, one of which is not a rodent.

The upper dosage level for a substance, fed to test animals over their life span and observed for several generations, which does not produce any effect, is designated as the "No Observed Effect Level" (NOEL, mg/kg body weight of the animal tested or mg/kg feed). This level can be used as a basis for estimating the hazard for humans in all cases in which a correlation between dose and effect has been observed. The NOEL is multiplied by an empirical uncertainty factor of 10^{-1} to 5×10^{-4}, but usually 10^{-2}, to obtain the toxicologically acceptable dose. It is expressed in mg/kg body weight/day and is called the Reference Dose (RfD) previously also known as the Acceptable Daily Intake (ADI). These dose calculations also take into account the extreme deviations of data from the mean, as found with particularly sensitive organisms.

Taking into consideration the consumption habits, a toxicologically Permissible Concentration (PC) can be calculated from the Reference Dose for a contaminant present in a given food:

$$PC = \frac{NOEL \times FV}{SF} \times \frac{BW}{CA \times ASF}$$

In this formula, PC is the toxicological permissible concentration for a particular food (expressed in mg/kg food); NOEL, no observed effect level (mg/kg feed); FV, daily intake of feed by test animals (kg feed/kg body weight); SF, uncertainty factor (10–2,000, but usually 100); BW, body weight of an adult (50–80 kg); CA, amount in kg consumed per day of food for which the PC is being calculated; and ASF, additional safety factor for particularly sensitive persons, such as children or the sick (the accepted ASF is 2.5). The maximum concentrations of contaminants allowed by legislation are often still well below toxicological tolerance concentrations, because other parameters such as "good agricultural practice", are taken into account.

9.2 Toxic Trace Elements

9.2.1 Mercury

Mercury poisoning caused by food intake is derived from organomercury compounds, e.g., dimethyl mercury (CH_3–Hg–CH_3), methyl mercury salts (CH_3–Hg–X; X = chloride or phosphate), and phenyl mercury salts (C_6H_5–Hg–X; X = chloride or acetate). These highly toxic compounds are lipid soluble, readily absorbed and accumulate in erythrocytes and the central nervous system. Some are used as fungicides and for treating seeds (seed dressing). Methyl mercury compounds are also synthesized by microflora from inorganic mercury salt sediments found on lake and river bottoms. Hence, the content of these compounds might rise in fish and other organisms living in water.

The natural mercury level in the environment appears to have stabilized in the last 50 years. Poisonings recorded in Japan appear to have been caused by consumption of fish caught in waters heavily contaminated by mercury-containing industrial waste water, and in Iraq by milling and consuming seed cereals dressed with mercury, which were intended for sowing. The permitted level or tolerance dose for an adult of 70 kg is 0.35 mg Hg per week, of which a maximum of 0.2 mg may be derived from the highly toxic methyl mercury. The mercury content in some foods is compiled in Table 9.1. The average mercury intake with food, most of which consumed fish, is estimated at 0.12 mg/capita/week in FR Germany.

9.2.2 Lead

The contamination of the environment with lead is increased by industrialization and by emissions from cars running on leaded gasoline. Tetraethyllead [$(C_2H_5)_4Pb$], an antiknocking additive used to increase the octane value of gasoline, is converted by combustion into PbO, $PbCl_2$ and other inorganic lead compounds. The major part of these compounds is found in an approx. 30 m wide band along roads or highways; the lead level sharply decreases beyond this distance. At a distance of 100 m from a road with heavy traffic, the lead level in the atmosphere decreases by a factor of 10 and that in soil and plants by a factor of 20 from the level found at or close to the road. A decrease in the level of lead in gasoline and increased use of unleaded gasoline has resulted in a drop in the extent of contamination. Environmental lead contamination has not, however, significantly increased the level of lead in food. The lead in soil is rather immobilized; thus the increase in the lead level of plants is not proportional to the extent of soil contamination. Vegetables with larger surface areas (spinach, cabbage) may contain higher levels of lead when cultivated near the lead emission source. When contaminated plants are fed to animals, the body does not absorb much lead since most is excreted in feces.

Further sources of contamination are lead-containing tin cookware and soldered metal cans and lead-containing enamels. This is particularly so in contact with sour food. These sources of contamination are of lesser importance.

3.5 mg of lead are considered as the temporary acceptable weekly intake for adults of 70 kg.

Table 9.1. Mercury, lead, and cadmium levels in food (FR Germany report for 1979)[a]

Food	Mercury			Lead			Cadmium		
	1	2	3	1	2	3	1	2	3
Milk				0.001–0.0835	0.019	0.05	0.001–0.007	0.001	0.002
Chicken eggs	0.000 8–0.24	0.011	0.03	0.0002–0.8689	0.074	0.2	0.0005–0.0871	0.024	0.05
Beef/veal meat	0.000 5–0.105	0.003	0.02	0.001–0.967	0.070	0.3	0.001–0.32	0.016	0.1
Pork meat	0.001–0.18	0.006	0.05	0.01–0.600	0.061	0.3	0.001–0.099	0.009	0.1
Beef/veal liver	0.002 5–0.879	0.015	0.1	0.01–3.31	0.278	0.8	0.001–4.1	0.127	0.5
Pork liver	0.001–1.434	0.058	0.1	0.007–1.488	0.149	0.8	0.002 5–1.61	0.165	0.8
Fish (fresh water)	0.000 5–2.74	0.257	1.0	0.000 5–1.08	0.124	0.5	0.000 5–0.8035	0.020	0.05
Fish (sea)	0.003 5–1.78	0.128	1.0						
Fish products	0.002–1.6	0.189	1.0						
Leafy vegetables	0.000 25–0.033	0.004		0.002 5–9.136	0.620	1.2	0.011–0.3875	0.044	0.1
Sprout vegetables	0.000 25–0.025	0.003		0.000 5–0.55	0.101	1.2	0.000 5–0.09	0.019	0.1
Fruit vegetables	0.000 25–0.012	0.003		0.001 5–1.91	0.120	0.2	0.000 5–0.166	0.020	0.1
Root vegetables				0.001 5–1.24	0.205	0.5	0.001–0.104	0.023	0.05
Pomme fruits	0.000 25–0.0125	0.002		0.000 5–1.54	0.171	0.5	0.000 5–0.116	0.010	0.05
Stone fruits	0.000 25–0.0099	0.001		0.007 5–1.349	0.142	0.5	0.000 5–0.076	0.014	0.05
Berries	0.000 25–0.00167	0.002		0.000 2–2.08	0.245	0.5	0.000 1–0.101	0.018	0.05
Fruit juice				0.01–0.20	0.057	0.2	0.004–0.015	0.007	0.02
Cereals	0.000 5–0.0642	0.004	0.03	0.01–0.61	0.041	0.5	0.004–0.80	0.035	0.1
Potatoes	0.000 5–0.0154	0.006	0.02	0.001 5–0.319	0.075	0.2	0.001–0.202	0.050	0.1
Wine				0.005–3.08	0.173	0.3	0.000 5–0.03	0.003	0.1
Drinking water	0.000 02–0.002	0.0003	0.004	0.002 1–0.022 5	0.009	0.04	0.000 4–0.0044	0.001	0.006

[a] All values given are in mg/kg or mg/l, column 1: variation range, column 2: an arithmetic average, column 3: a guiding value which should not be exceeded, or in some cases legally set as the highest permissible level.

The lead content of food is compiled in Table 9.1. The average intake of lead in FR Germany is 0.98 mg/week, of which in excess of 90% is derived from food. A decrease in lead intake is desirable.

Hair and bone analyses have revealed that lead contamination of humans in preindustrialized times was apparently higher than today. This might be due to the use in those days of lead pipes for drinking water, lead-containing tinware, and excessive use of lead salts for heavily glazed pottery used as kitchenware.

9.2.3 Cadmium

Cadmium ions, unlike Pb^{2+} and Hg^{2+}, are readily absorbed by plants and distributed uniformly in all their tissues, thus decontamination by dehulling or by removal of outer leaves, as with lead, is not possible. Some wild mushrooms accumulate higher levels of cadmium. In food of animal origin, cadmium is found primarily in internal organs, such as liver and kidney, and in milk. The contamination sources are industrial waste water and the sludge from plant clarifiers, which is often used as fertilizer. The cadmium content of food is compiled in Table 9.1.

A prolonged intake of cadmium results in its accumulation in the human organism, primarily in liver and kidney. A level of 0.2–0.3 mg Cd/g kidney cortex causes damage of the tubuli. The tolerance dose of cadmium is estimated at 0.5 mg/week. Its intake varies, but is estimated at 0.19 mg/week in FR Germany and some other countries; 95% of which is derived from the diet. This intake suggests that no acute health hazard exists with cadmium. Nevertheless, cadmium contamination, like lead and mercury, still presents a problem and, hence, a decrease in cadmium intake is a justified aim.

9.2.4 Radionuclides

It is estimated that the average radiation exposure in FR Germany in 1975 was 172 mrad, of which 21 mrad were ascribed to internal radiation by natural radionuclides incorporated in the body (about 90% from ^{40}K, the rest from ^{14}C) and less than 1 mrad by nuclides acquired as a result of atmospheric fallout from nuclear explosion tests (50% from ^{137}Cs, a radionuclide with a half life of 30 years, but quickly excreted by the body; approx. 50% from ^{90}Sr, a most dangerous radioisotope, capable of inducing leukemia and bone cancer; and traces of ^{14}C and tritium). ^{137}Cs and ^{90}Sr are the escort elements of potassium and calcium, respectively. Food contamination with radionuclides in FR Germany had its peak in 1964/65, when the intake in food per day per person was 240 pCi of ^{137}Cs and 30 pCi of ^{90}Sr. Up to the Chernobyl reactor accident in April 1986, the intake was less than 10% of previous values as a result of the moratorium on atmospheric testing of atomic weapons. Radionuclide residues in food were not a health hazard.

For 1986, the accident in Chernobyl caused an additional intake of radionuclides with food that is estimated at (children up to the age of 1 year/adults) 1779/4598 Bq/year of ^{131}J, 986/1758 Bq/year of ^{134}Cs, and 1849/3399 Bq/year of ^{137}Cs. The resulting additional effective equivalent dose for people in the FR Germany is estimated at 0.06–0.22 mSv. In comparison, natural radiation exposure is about 2 mSv per year, of which 0.38 mSv/year is caused by radionuclides in food. As a precaution, maximum activity values of 500 Bq/l and 250 Bq/kg have been stipulated for milk and vegetables respectively.

The level of tritium infiltrating the biosphere is expected to rise from 28 MCi to 250 MCi due to nuclear plant operation on a long-term basis.

9.3 Toxic Compounds of Microbial Origin

9.3.1 Food Poisoning by Bacterial Toxins

Most (60–90%) cases of food poisoning are bacterial in nature. They are distinguished by food intake causing:

- Intoxication (poisoning, e.g., by *Clostridium botulinum, Staphylococcus aureus*).
- Diseases caused by massive pollution with facultative pathogenic spores, e.g., *Clostridium perfringens, Bacillus cereus*.
- Infections by *Salmonella* spp. or *Shigella* spp.

- Diseases of unclear etiology, such as those from *Proteus* spp., *Escherichia coli*, *Pseudomonas* spp.

The harmful activity of these bacteria in the digestive tract is ascribed to enterotoxins, which are classified into two groups: exotoxins (toxins excreted by microorganisms into the surrounding medium) and endotoxins (retained by the microorganism cells but released when the cell disintegrates).
Exotoxins are released primarily by gram-positive bacteria during their growth. They consist mostly of proteins which are antigenic and very poisonous. They become active after a latent period. This group includes the toxins released by *Clostridium botulinum* (botulin toxin, a globular protein neurotoxin), *Cl. perfringens* and *Staphylococcus aureus*. Table 9.2 gives some important data for these microorganisms, including harmful effects. Intoxications with *St. aureus* are the most frequent cause of food poisoning. Symptoms are vomiting, diarrhea and stomach ache and are caused

primarily by food of animal origin (meat and meat products, poultry, cheese, potato salad, pastry).
Endotoxins are produced primarily by gram-negative bacteria. They act as antigens, are firmly bound to the bacterial cell wall and are complex in nature. They have protein, polysaccharide and lipid components. Endotoxins are relatively heat stable and are in general active without a latent period. The toxins causing typhoid and paratyphoid fevers, salmonellosis and bacterial dysentery are in this group. Salmonellosis is very serious. It is an infection by toxins of about 300 different but closely related organisms. The infection is characterized by enteric fever, gastroenteritis and salmonella septicemia. Sources of infections are egg products, frozen poultry, ground or minced beef, confectionery products and cocoa.

9.3.2 Mycotoxins

There are more than 200 mycotoxins produced under certain conditions by about 120 fungi or

Table 9.2. Food poisoning by bacterial toxins

Microorganism:	*Staphylococcus aureus*	*Clostridium botulinum*	*Bacillus cereus*	*Clostridium perfringens*
Growth conditions				
temperature range	10–45°C	4–35°C	10–45°C	12–52°C
pH range	4.5	5		5–8.5
Toxin				
type	Protein	Protein	Lipid (?)	Protein
effective amount	0.5–1 µg	0.1–1 µg	10^8 spores/g	10^6 spores/g
stability	Relatively thermostable	Thermolabile, inactivation at 80°C/30 min or 100°C/5 min		
Incubation time	2–6 h	1–3 days	1–12 h	8–24 h
Duration of illness	1–3 days	Death after 1–8 days, with survivors ill 6–8 months	0.5–1 day	0.5–1 day
Symptoms	Vomiting, diarrhea, abdominal pain	Paralysis of the nerve centres of the *medulla oblongata*	Abdominal pain, nausea, diarrhea, vomiting	Diarrhoe, abdominal convulsions, nausea, loss of appetite
Foods usually accounting for poisoning	Cold meat and cheese slices, mildly acidic salads (meat, poultry, sausage, cheese, potatoes), mayonnaise, cream fillings in baked products	Homemade canned meat, hind bony ham, sliced sausages, trout fillets, canned green beans	Institutional/ community catering. Heated and warmed dishes, cereal containing dishes (corn, rice)	Institutional/ communal catering. Heated and warmed meat dishes, warm desserts, puddings, cream fillings in baked products

molds. Table 9.3 presents data on mycotoxins of particular interest to food preservation and storage. The chemical structures of these toxins are presented in Fig. 9.1.

Infections of rye and, to a lesser extent, of other cereal grains with *Claviceps purpurea* (ergot, or rooster's spur) are responsible for the disease called ergotism (symptoms: gangrene and convulsions). The disease was important in the past when bread from infected rye grain was eaten. It has practically ceased to exist due to seed treatment with fungistatic agents and grain cleaning prior to milling.

Most mycotoxin data are on the genera *Aspergillus* and the aflatoxins they produce during growth. These are the most common and highly toxic fungal toxins, e.g., Aflatoxin B_1, the most powerful carcinogen known. In animal feeding tests with rats, its carcinogenic effect is revealed at a daily dose of only 10 µg/kg body weight. In a comparative study, the carcinogenic property of the highly toxic dimethyl-nitrosamine was revealed only at a daily dose of 750 µg/kg body weight. It is primarily plant material (particularly peanuts, peanut butter, rice, maize) that is contaminated with aflatoxins. Aflatoxin passes from moldy feed to animal products, primarily milk. The dairy cow's metabolism converts the B-group aflatoxins to those of the M-group ("M" stands for metabolite), which are also carcinogenic. Nephrotoxic ochratoxin A passes from fodder cereals mainly to the blood and kidney tissue of pigs, but it is also found in the muscles, liver and adipose tissue.

Thorough analytical research and strict legislative measures have significantly reduced aflatoxin contamination and kept it under control. Only *one* peanut out of 10^4 is contaminated with aflatoxin. Aflatoxins M_1 and M_2, obtained by hydroxylation of aflatoxins B_1 and B_2 respectively, average $3-8$ ng/l of milk.

Analytical data are also increasingly available on the occurrence of the other mycotoxins

Table 9.3. Mycotoxins

Fungus/mold	Toxin[a]	Toxicity[b]	Effect	Occurrence
Claviceps purpurea	Ergot alkaloids (I)		Ergotism (gangrenous convulsions)	Mainly rye, to a lesser extent wheat
Aspergillus flavus *A. parasiticus*	Aflatoxins (II)	7.2 mg/kg (rat, orally)	Liver cirrhosis, liver cancer	Groundnuts and other nuts (almond, Brasil nut) corn and other cereals, animal feed, milk
Aspergillus versicolor *A. nidulans*	Sterigmatocystin (III)	120 mg/kg (rat, orally)	Liver cancer	Corn, wheat, animal feed
Penicillium expansum *P. urticae* *Byssochlamis nivea,* *B. fulva*	Patulin (IV)	35 mg/kg (mouse, orally)	Cellular poison	Putrifying fruits, fruit juices
Aspergillus ochraceus *A. melleus*	Ochratoxin A (V)	20 mg/kg (rat, orally)	Fatty liver and kidney damage	Barley, corn
Fusarium graminearum	Zearalenone (VI) (Fusariotoxin F_2)	0.1 mg/kg over 5 days (swine, orally[c])	Estrogen, infertility	Corn and other cereals, animal feed
Fusarium oxysporum *F. tricinctum*	Fusariotoxin T_2 (VII)	3.8 mg/kg (rat, orally)	Toxic aleukia, hemorrhagic syndrome	Cereals, animal feed
Fusarium roseum *F. graminearum*	Vomitoxin (VIII)	70 mg/kg (mouse)	Vomiting	Cereals, animal feed

[a] Roman numerals refer to the structural formulas in Fig. 9.1.
[b] Acute toxicity (LD_{50}).
[c] Estrogenic activity.

listed in Table 9.2. A quantitatively important household source of mycotoxins is mold infections of fruit, bread and other baked products, meat and processed meat products.

9.4 Pesticides

9.4.1 General Remarks

The term pesticides includes the compounds used in agricultural food production to protect cultivated plants from plant- and insect-caused diseases, parasites or weeds, or from detrimental microorganisms. The most important groups of pesticides are: (1) *herbicides* to protect the plant from weeds; (2) *fungicides* to suppress the growth of undesired fungi or molds; and (3) *insecticides* to protect the plants from damage caused by insects. In addition to these main groups, there are *acaricides* to control mites, *nematocides* to control worms or nematodes, *molluscicides* to protect the plant from snails and slugs, *rodenticides* to control rodents (mice or rats), and plant *growth regulators* (plant growth hormones [cf.

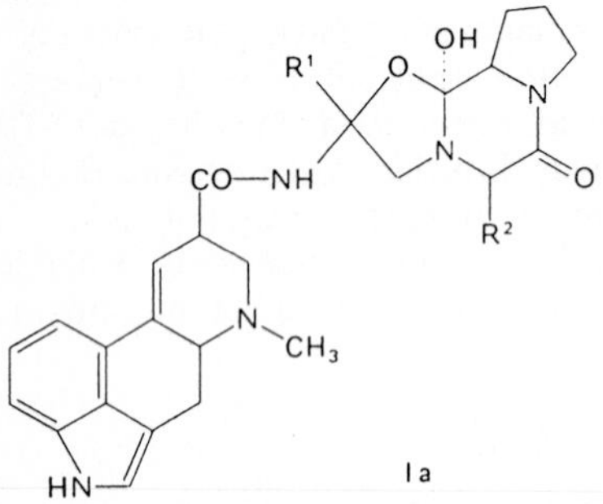

I: Alkaloids of Ergot
Ia: Ergocristine (R^1 = CH(CH$_3$)$_2$, R^2 = H$_2$C–C$_6$H$_5$), Ergostine (R^1 = C$_2$H$_5$, R^2 = H$_2$C–C$_6$H$_5$), Ergotamine (R^1 = CH$_3$, R^2 = H$_2$C–C$_6$H$_5$), Ergocryptine (R^1 = CH(CH$_3$)$_2$, R^2 = H$_2$C–CH(CH$_3$)$_2$), Ergocornine (R^1 = CH(CH$_3$)$_2$, R^2 = CH(CH$_3$)$_2$), Ergosine (R^1 = CH$_3$, R^2 = CH$_2$–CH(CH$_3$)$_2$) Ib: Ergometrine

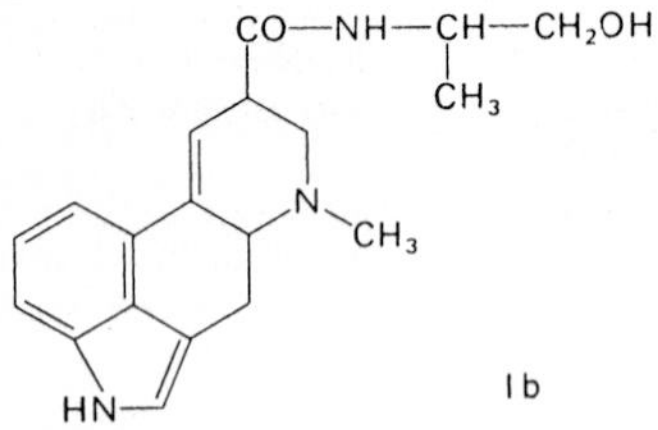

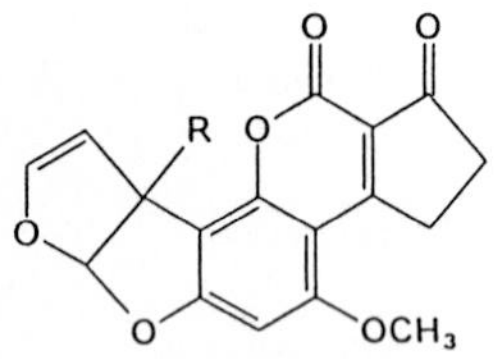

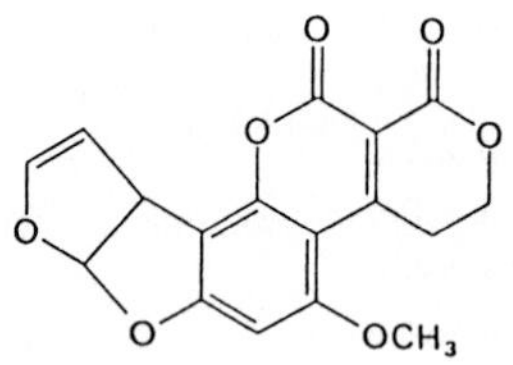

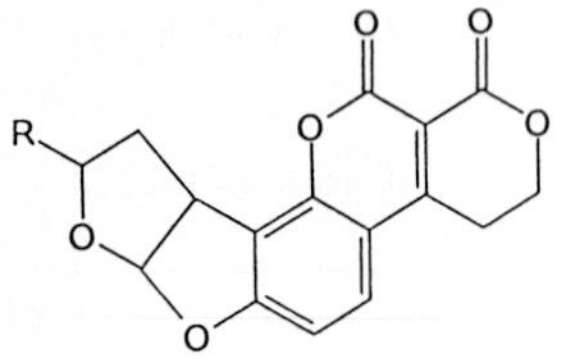

II: Aflatoxins
IIa: Aflatoxin B$_1$ (R=H), Aflatoxin M$_1$ (R=OH), IIb: Aflatoxin B$_2$ (R, R^1=H), Aflatoxin M$_2$ (R=OH, R^1=H), Aflatoxin B$_{2a}$ (R=H, R^1=OH), IIc: Aflatoxin G$_1$, IId: Aflatoxin G$_2$ (R=H), Aflatoxin G$_{2a}$ (R=OH)

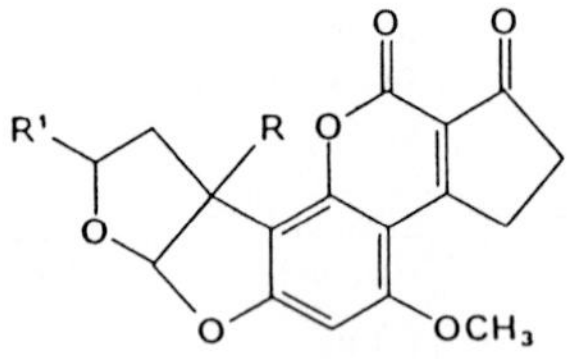

Fig. 9.1. Structures of some mycotoxins (cf. Table 9.3)

Fig. 9.1 (continued)

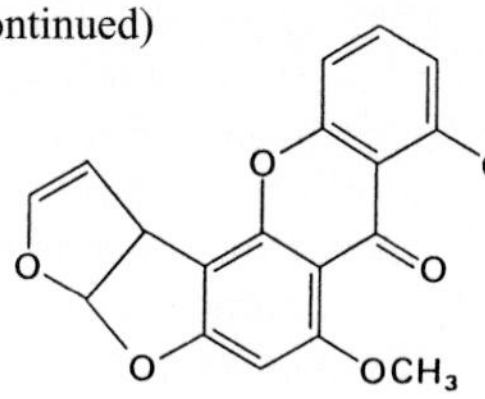

III: Sterigmatocystine
IV: Patuline

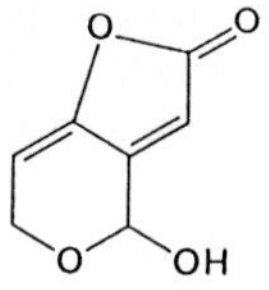

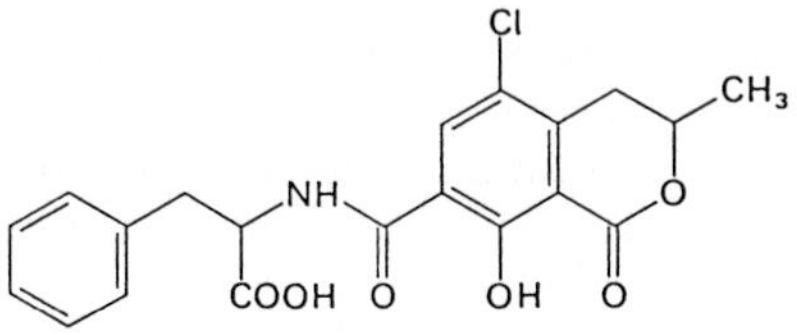

III

IV

V

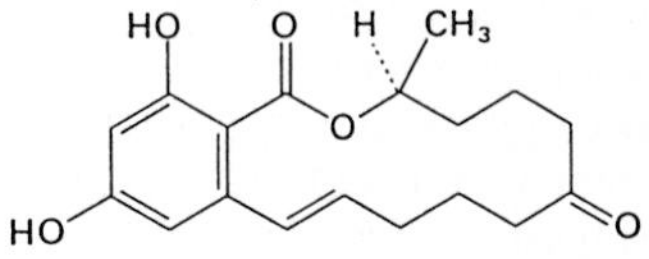

VI

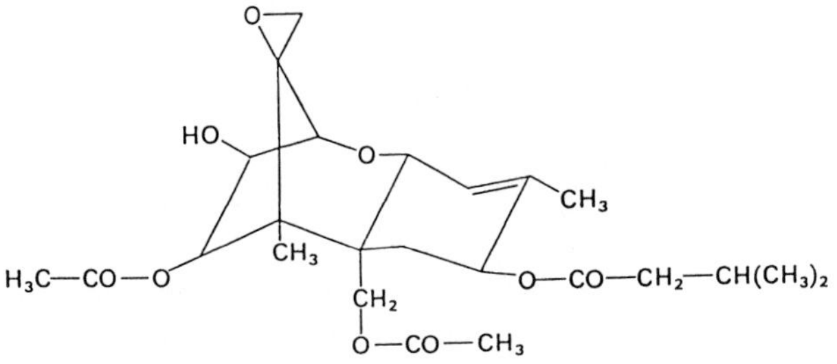

VII

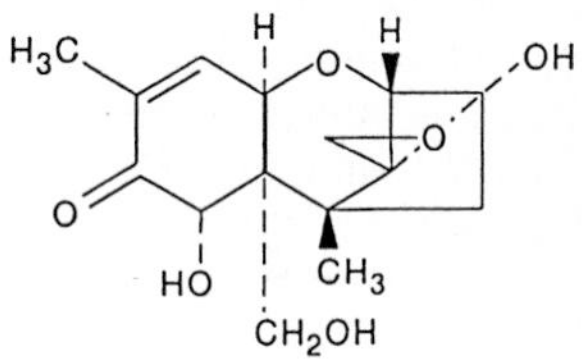

VIII

V: Ochratoxin A
VI: trans-Zearalenone
VII: Fusariotoxin T_2
VIII: Vomitoxin

18.1.4] and sprout inhibitors). Table 9.4 provides data on the global share of various groups of pesticides. In the FR Germany, about 1700 preparations based on ca. 300 active agents were approved for use in 1988. Indeed, 44% of these products were herbicides, 21% insecticides and acaricides, 16% fungicides, and the rest were other groups. Of the amount of active agents sold in the country in 1987, 56.8% were herbicides, 30.8% fungicides, and only 4% were insecticides.

Pesticide use is rewarding since it reduces losses in crop yield and stocks. It has also contributed to the control or eradication of insect-spread diseases such as malaria. Nevertheless, crop harvest losses average 35% woldwide (Europe, 25%; Asia, 43%).

Pesticides are applied in various forms and by various means: dusting as powder, fumigation, spraying as a liquid, or pad or furrow irrigation. The strict observance of directions for

use, waiting the recommended time between final application and harvest, and restricting the application to the necessary dose, is re-

Table 9.4. World utilization of pesticides

	Pesticide group[a]				Total[b]
	Insecti-cides	Fungi-cides	Herbi-cides	Others	
Western Europe	21	27	47	5	20.6
FR Germany	11	19	64	6	2.8
France	19	25	50	6	7.2
Italy	28	45	22	5	3.2
Eastern Europe	22	28	45	5	18.3
USA/Canada	28	6	65	2	23.1
Latin America	58	13	28	1	12.0
Africa	60	17	16	7	4.9
Asia/Middle East	49	22	26	2	19.4
Australia/ New Zealand	30	13	56		1.7

[a] Percent of the total value (wholesale price).
[b] Percent of the world market (wholesale price).

quired to maintain the residual pesticide levels in food at a minimum. These requirements and recommendations, as set by regulations, are listed in an application booklet.

Contamination of food of plant origin can occur directly by treating the crop before storage and distribution (fruit and vegetable treatment with fungicides, cereal treatment with insecticides). It can occur indirectly by uptake from the soil of residual pesticides by the subsequent crop, from the atmosphere or drifting from neighboring fields, or from a storage space pretreated with pesticide.

Contamination of food of animal origin occurs by ingestion of feed containing stall- and barn-cleansing agents (fungicides, insecticides), by coming in contact with wooden studs and boards preserved with fungistatic agents, and veterinary medicines and, occasionally, by use of disinfected corn as fodder.

Table 9.5 and Fig. 9.2 review the important pesticides and their application methods. Table 9.6 gives data on residues in FR Germany.

9.4.2 Insecticides

The most important insecticide classes are chlorinated hydrocarbons, organophosphoric acid esters (organophosphates) and carbamates (Table 9.5; Fig. 9.2). The greatest attention is given to chlorinated hydrocarbons. They are very stable, persistent in the environment, and fat soluble and thus deposit in human adipose tissue and in human milk fat. Direct analysis of human fat for these residues is a good indicator of the overall contamination by this class of insecticide.

The contamination of food with chlorinated organic compounds is decreasing because the use of these compounds is steadily declining in favor of thiophosphoric acid esters, carbamates, pyrethrum compounds, and pyrethroids. Since these compounds are substantially degraded over the span of a recommended waiting period, residue problems do not arise.

In the residue analyses shown in Table 9.6, positive results involve the compounds listed in Tables 9.7 and 9.8. Foods of animal origin contained almost without exception low amounts of chlorinated organic compounds (HCH, HCB, DDT, DDE, dieldrin and PCB). In comparison with previous years, however, a clear trend away from agrochemicals and towards polychlorinated biphenyls can be observed. As a result of the concentration of chlorinated organic compounds in human milk (Table 9.9), ADI values are clearly exceeded in the case of breast-fed children. Nevertheless, the expert opinion is that the advantages of infant feeding with mother's milk during a 3–6 month period outweigh the health hazard of the residue levels.

In Fig. 9.3 fresh dairy milk is used as an example to demonstrate the difference in results for chlorinated hydrocarbon residues in blended milk samples and in milk delivered by individual dairy farmers. The data emphasize the importance of taking random samples from individual dairy farmers.

9.4.3 Herbicides

Herbicides, used to protect the growth of cultivated plants from weeds, are divided into compounds with broad or selective activity. The former group includes chlorates, copper sulfate, calcium cyanamide and chlorinated fatty acid derivatives. The group with selective activity includes growth regulators, such as aryloxy fatty acids, carbamic acid and urea derivatives, triazines and pyridines (cf. Table 9.5 and Fig. 9.2).

The main field of herbicide application is corn (maize), other cereals and beet cultivation. Residue problems are almost nonexistent. The toxicity of the compounds used is generally low in warmblooded organisms.

A possible side effect of herbicides which is not to be ignored or underestimated is their influence on the arthropods and the natural microflora of soil.

9.4.4 Fungicides

These compounds are used to protect plants against diseases caused by fungi or molds, e.g., potato and tomato rots, flour dew and fruit scabs.

Important fungicides, in addition to inorganic compounds (copper oxychloride, sulfur, sulfurlime broth), are dithiocarbamates and or-

Table 9.5. Some selected pesticide trade names and applications

Number	Name	Application	Number	Name	Application
Synthetic Insecticides			Natural insecticides		
I	Lindane (γ-HCH), BHC, Gammexane	Seed dressing (cereals, beets, legumes), vegetables and fruits	XXVIa	Nicotine	Vegetables, fruits
			XXVIb	Pyrethrins (from *Chrysanthemum cinerariiaefolium*)	Fruits, vegetables, potatoes, rapeseed, beets, mushroom, vines
Ia	β-HCH	In Germany not permitted (the same is the case with α-HCH)	XXVIc	Rotenone, Derris	Fruits, vegetables, potatoes, rapeseed, beets, vines
II	p,p′-DDT	Fight against malaria, in Germany not permitted	Acaricides		
III	Methoxychlor	Fruits, vegetables, potatoes, cabbage, wheat, rye	XXVII	Ethion	Fruits
			XXVIII	Kelthane, Dicofol	Fruits, vegetables, vines
IV	Endosulfan, Thiodan	Vegetables, fruits, potatoes, rapeseed, beets, field beans, corn, animal feed	XXIX	Tetradifon, Tedion	Fruits, vines
			XXX	Tetrasul, Animert	Cucumbers, tomatoes
V	Aldrin, HHDN	Grape (vine) cultivation	Fungicides		
VI	Dieldrin, HEOD	In Germany not permitted	XXXI	Copperoxichloride	Fruits, vegetables, rapeseed, beets, potatoes, vines
VII	Heptachlor	Seed dressing (beets)	XXXII	Sulfur	Fruits, vegetables, vines, potato seeds
VIII	Heptachlor-epoxide (A, B)	Metabolites from VII	XXXIII	Thiram, TMTD, Pomarsol	Pomme fruits, strawberries, chicory, head lettuce, vegetables, vines
IX	Azinphosethyl Gusathion H	Potatoes, cereals, rapeseed, beets, vegetables, animal feed	XXXIV	Ferbam	Pomme fruits
X	Azinphosmethyl Gusathion	Fruits, vegetables (asparagus), vines	XXXV	Ziram	Pomme fruits
			XXXVI	Maneb, Dithane M-22	Potatoes, beets, vegetables, asparagus, berry fruits, vines, pomme fruits
XI	Dimethoate, Rogor, Perfekthion	Vegetables, fruits, vines, potatoes, beets, cereals, animal feed	XXXVII	Zineb, Dithane, Z-78	Pomme fruits, stone fruits, asparagus, vines
XIa	Disulfoton	Seed potatoes, hops, tropical cultures	XXXVIII	Mancozeb (complex consisting of XXXVI and XXXVII), Dithane ultra	Pomme fruits, stone fruits, cucumbers, tomatoes, vegetables, vines
XIb	Phosalon	Rapeseed, fruit			
XIc	Ethion				
XII	Fenitrothion	Vegetables, fruits, potatoes, cereals, clover, beets, vines	XXXIX	Benomyl, Benlate	Pomme fruits, cucumbers, head lettuce, strawberries, cereals (seed dressing)
XIII	Methidathion	Fruits, potatoes, beets, vines	XXXIXa	Iprodion	Vines, strawberries, vegetables, rapeseed, barley
XIV	Mevinphos, Phosdrin, PD 5	Fruits, vegetables, cereals potatoes, rapeseed, beets, animal feed			
XIVa	Chlorpyrifos		XXXIXb	Imazalil	
XV	Malathion	Cereals, potatoes, vegetables, fruits	XXXIXc	Vinclozolin	Vines, strawberries, vegetables, rapeseed
XVI	Parathion (-ethyl), E 605, Eftol	Vegetables, fruits cereals, beets, rapeseed, potatoes, vines, animal feed	XXXIXd	Procymidon	Vines, fruits, vegetables, hops
			XL	Captafol, Difolatan	Pomme fruits, peaches, vegetables, legumes, potatoes, vines
XVII	Parathionmethyl, ME 605	As compound XVI			
XVIII	Phosphamidon	Vegetables, fruits, cereals, potatoes, rapeseed, beets, clover, vines	XLI	Captan, Orthocide-406	Pomme fruits, cherries, peaches, vegetables, corn, legumes, head lettuce, vines
XIX	Bromophos, Nexion	Vegetables, fruits, cereals, potatoes, beets, rapeseed, animal feed	XLII	Dichlofluanid, Euparen	Head lettuce, tomatoes, berry fruits, pomme fruits, peaches, vines
XX	Dibrom, Naled	As compound XIX	XLIII	Folpet, Phaltan	Pomme fruits, cherries, strawberries, beans, peas, cucumbers, vines
XXI	Chlorfenvinphos, Birlane	Vegetables, potatoes, beets, corn			
XXII	Dichlorvos, DDVP, Vapona	Vegetables, fruits, cereals, rapeseed, beets, potatoes, mushroom, vines, animal feed	XLIV	Hexachlorobenzene (HCB)	In Germany not permitted
			XLV	Quintozene, PCNB, Brassicol (contains up to 3% XLIV)	Seed dressing (cereals, potato seed)
XXIII	Carbaryl, Sevin	Fruits, vegetables, potatoes, vines			
XXIV	Methomyl, Lannate	Vegetables, vines	XLVa	Chlorthalonil	
XXV	Promecarb, Carbamult	Pomme fruits, vegetables, potatoes, rapeseed, beets	XLVI	Thiabendazole, Tecto	Pomme fruits, surface treatment of fruits (bananas, citrus fruits)
XXVI	Propoxur, Unden	Vegetables, fruits, cereals, potatoes, beets	XLVII	Fentin acetate, Brestan	Carrots, potatoes

Table 9.5 (continued)

Number	Name	Application
XLVIII	Methoxyethyl-mercuric chloride, Ceresan	Wheat, rye, barley oats
IL	Phenylmercuric acetate, PMA	As compound XLVIII
L	Phenylmercuric chloride PMC	As compound XLVIII
Herbicides		
LI	Alachlor	Corn, cabbages
LII	Amitrole	For all crops and pomme fruits
LIII	Atrazine	Corn, asparagus
LIV	Bromacil	Pomme fruits
LV	Buturone	Berry fruits, vines, winter wheat cultivars
LVI	Chlorbufam	Vegetables (carrots, red beet)
LVII	Chloropropham, CIPC	Potatoes (antisprouting agent)
LVIII	Chloroxuron, Tenoran	Vegetables, carrots
LIX	Chlorotoluron	Winter wheat cultivars
LX	2,4-D	Cereals
LXI	Disquat	Potatoes (vine killing) vegetables, strawberries
LXII	Desmetryn	Cabbages
LXIII	Diallate, DDTC	Peas, red beets, beets
LXIV	Diuron	Asparagus, berry fruits
LXV	Lenacil	Strawberries, spinach, beets
LXVI	Linuron	Peas, beans, carrots, celery, asparagus, potatoes, vines
LXVII	MCPA	Pomme fruits, barley, vines
LXVIII	Metobromuron	Lamb's lettuce, potatoes
LXIX	Metoxuron	Carrots, cereals
LXX	Metribuzin	Tomatoes, asparagus, potatoes
LXXI	Monalid	Carrots, parsley, celery
LXXII	Monolinuron	Bush snap beans, asparagus, potatoes, cereals, vines
LXXIII	Paraquat, Gramoxone	Vegetables, fruits, straw-berries, vines and all forage crops

Number	Name	Application
LXXIV	Pentanochlor	Carrots, parsley, tomatoes, celery, peppermint
LXXV	Prometryn	Carrots, leek, celery
LXXVI	Propham, IPC	Caraway, potatoes (antisprouting agent)
LXXVII	Propyzamide	Chicory, head lettuce, pomme fruits, rapeseed
LXXVIII	Simazire, Gesatop	Pomme and stone fruits, strawberries, vines, peas, corn, field beans, sugar beet
LXXIX	2,4,5-T	Cereals, vines
LXXX	TCA	Sugar beet
LXXXI	Terbutryne, Igran	Peas, corn
LXXXII	Triallate, Avadex	Cereals, sugar beet
LXXXIII	Trifluoralin	Cauliflower, turnips, winter cereal cultivars
Nematicides		
LXXXIV	Dazomet	Fruits, vegetables, forage crops, potatoes
LXXXV	1,3-Dichloro-propane/ 1,2-dichlorpropene, DD mixture	Potatoes, fruits, vegetables, vines, forage crops
LXXXVI	Methyl bromide, Terabol	Vegetables, fruits, forage crops, potatoes, beets
LX XXVII	Methyl isocyanate	As compound LXXXVI
LXXXVIII	Zinophos	Vegetables
Molluscicides		
LXXXIX	Metaldehyde	Vegetables, strawberries, cereals
XC	Mercapto-dimethur	Cereals
Rodenticides		
XCI	Endrin	Pomme- and stone fruits, currants
XCII	Toxaphene (Chlorinated camphene)	Cereals

ganometallic compounds (cf. Table 9.5 and Fig. 9.2). Residues of HCB and dithiocarbamates are observed in green vegetables, particularly in lettuce. Quintozene, which contains HCB as an impurity, also occurs as a residue.

9.5 Veterinary Medicines and Feed Additives

9.5.1 Foreword

The current practice in animal husbandry is the wide use of veterinary medicines, which serve not only for therapy, but to a large extent for prophylaxis and economic aims (e.g. to shorten animal growth or feeding time; to abate the risk of losses). Veterinary preparation residues in food are ingested by humans in low amounts but continuously and, hence, could be a health hazard. This possibility was, for a long time, not carefully examined. Therefore, as in the field of pesticides, supporting and maintaining appropriate measures (printing suitable directions for use, analytical control or supervision, elucidation of toxicological problems) has the ultimate aim of protecting human health.

A brief outline of some important groups of veterinary medicines follows. Table 9.10 and Fig. 9.4 provide a review of their use and chemical structures.

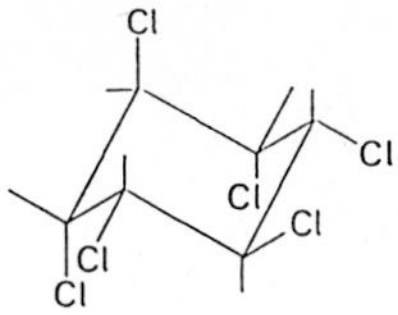

I

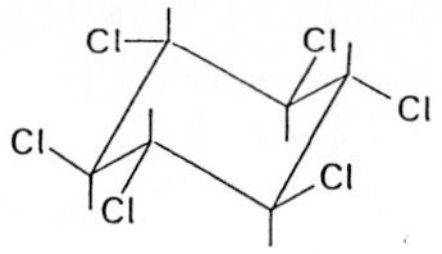

Ia

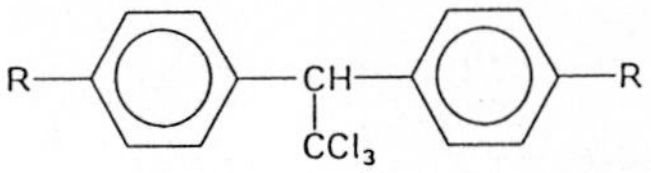

II: R = Cl,   III: R = OCH₃

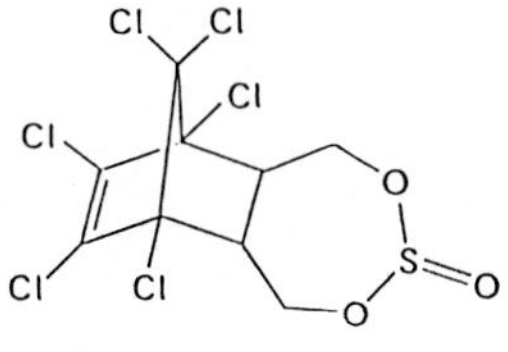

IV

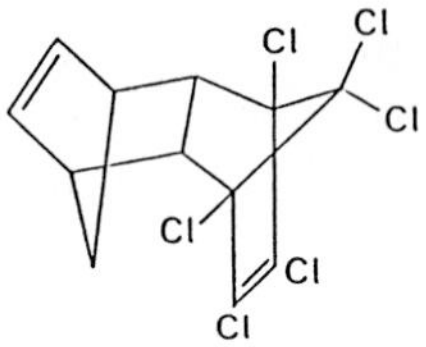

V

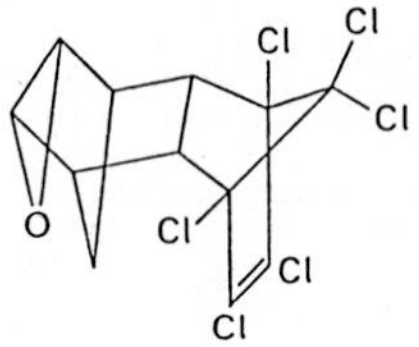

VI

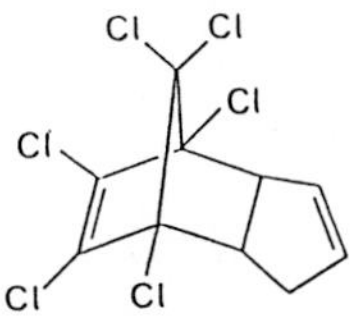

VII

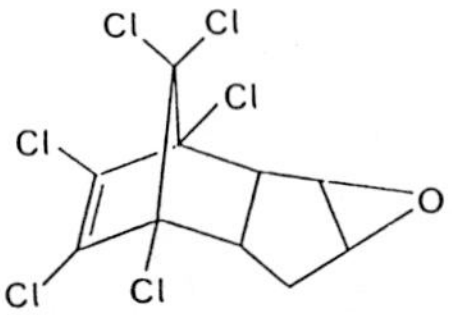

VIII

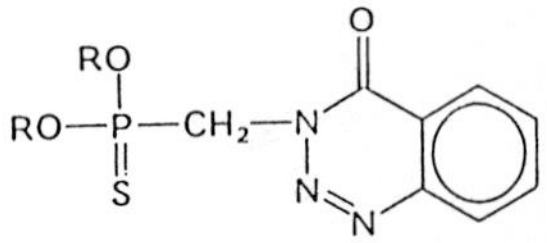

IX: R = C₂H₅,   X: R = CH₃

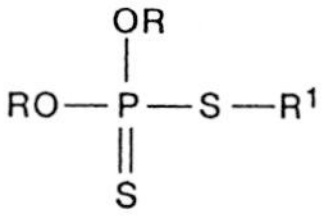

$$RO-\underset{\underset{S}{\|}}{\overset{\overset{OR}{|}}{P}}-S-R^1$$

R = CH₃,   R¹ = CH₂—CO—NH—CH₃

XI

R = C₂H₅,   R¹ = CH₂—CH₂—S—C₂H₅,

XI a

R = C₂H₅,   R¹ = CH₂—N

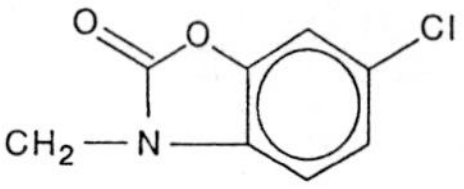

XI b

R = C₂H₅,   R¹ = CH₂—N—S—P(OC₂H₅)(OC₂H₅)

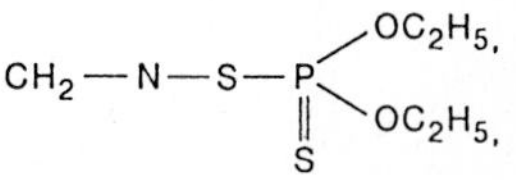

XI c

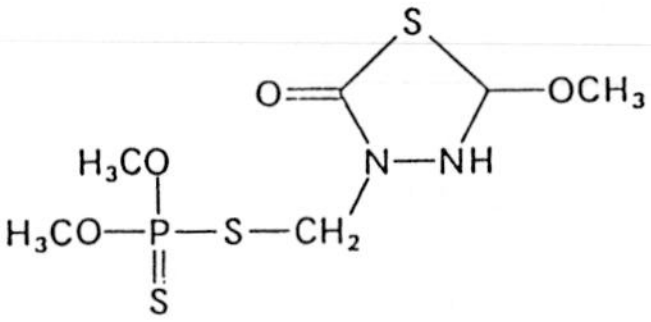

XIII

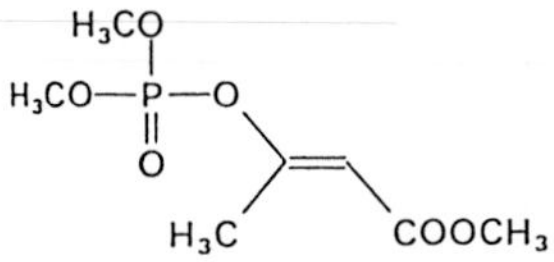

XIV

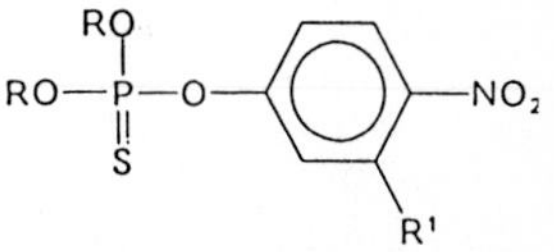

XII : R = CH₃, R¹ = CH₃
XVI : R = C₂H₅, R¹ = H
XVII: R = CH₃, R¹ = H

Fig. 9.2. Structures of some selected pesticides. Part 1. The Roman numerals refer to Table 9.5

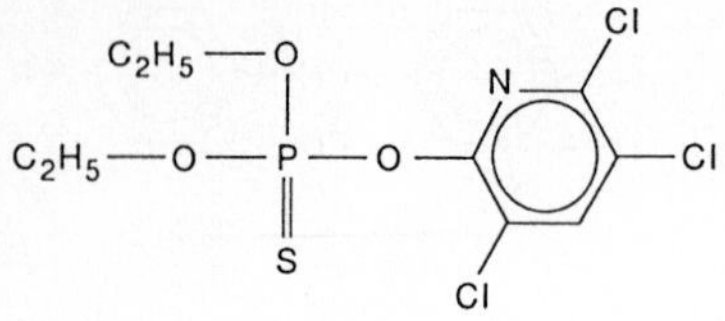

XIV a

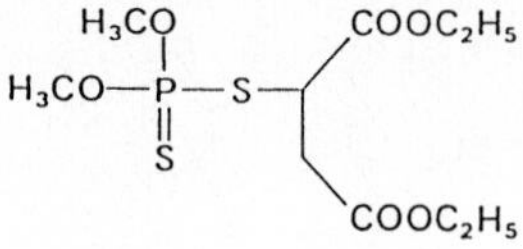

XV

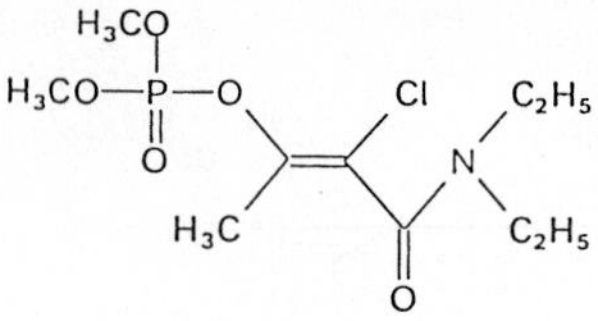

XVIII

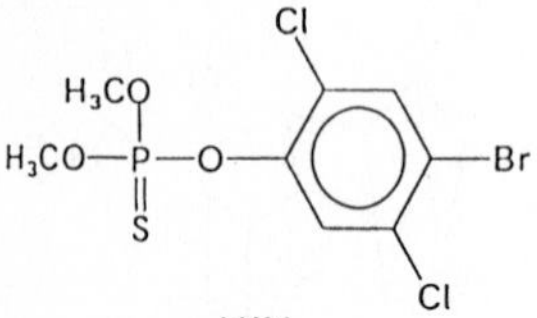

XIX

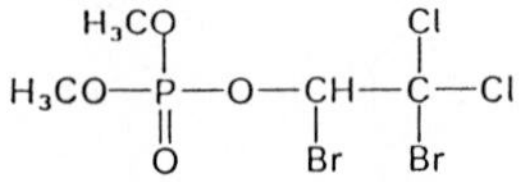

XX

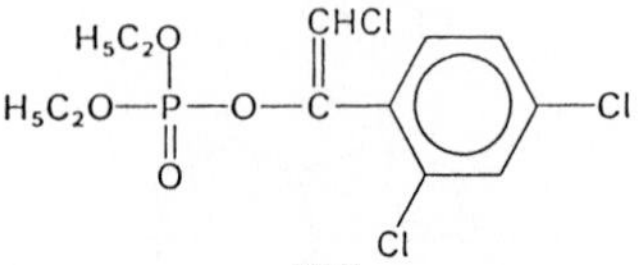

XXI

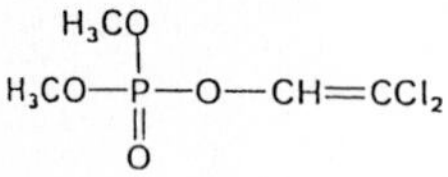

XXII

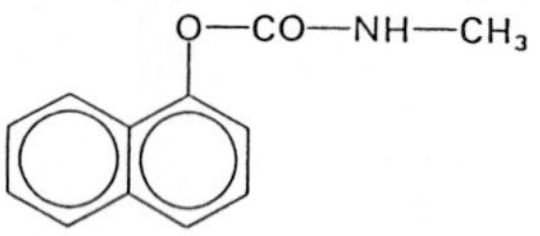

XXIII

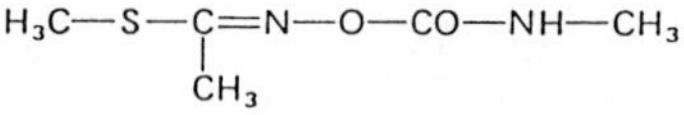

XXIV

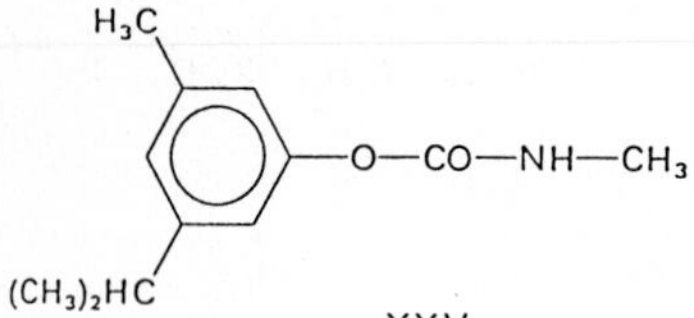

XXV

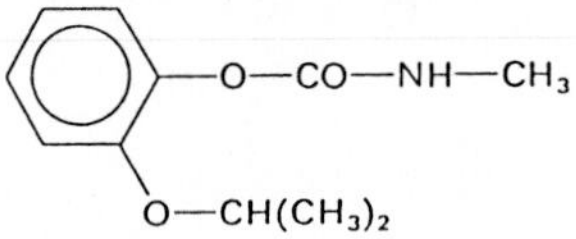

XXVI

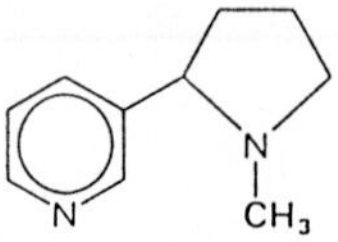

XXVIa

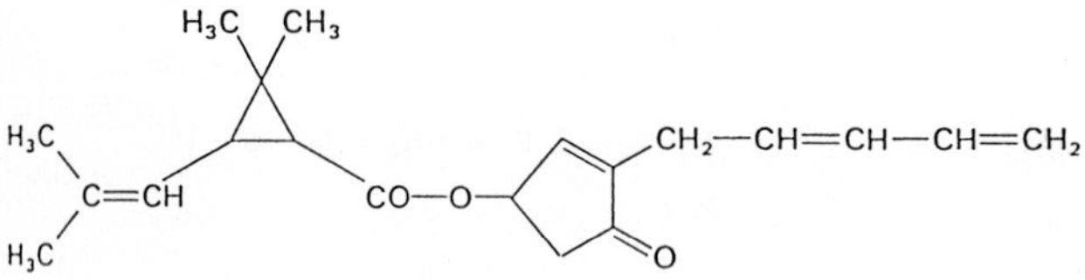

XXVIb (Pyrethrin I)

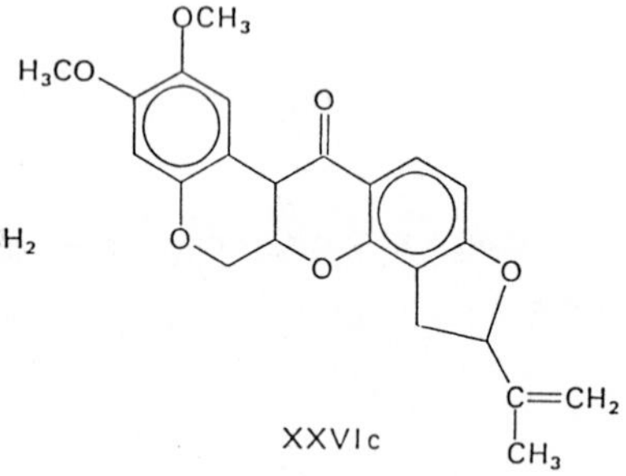

XXVIc

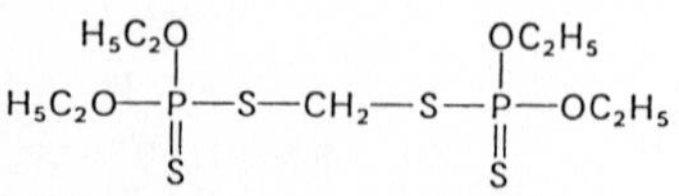

XXVII

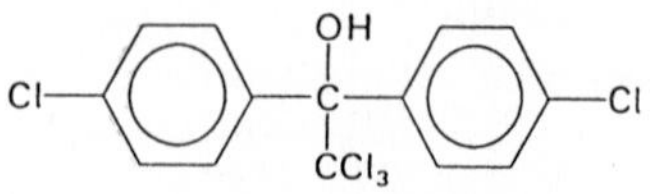

XXVIII

Fig. 9.2. Structures of some selected pesticides. Part 2. The Roman numerals refer to Table 9.5

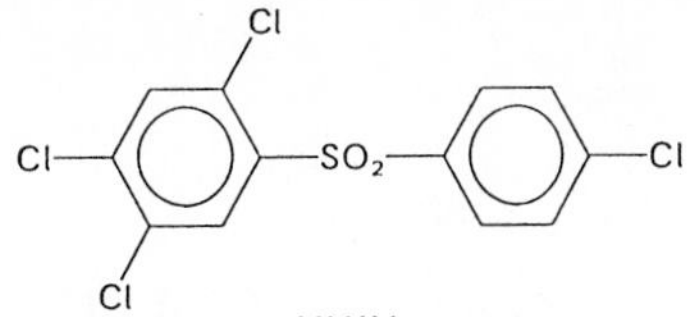

XXIX

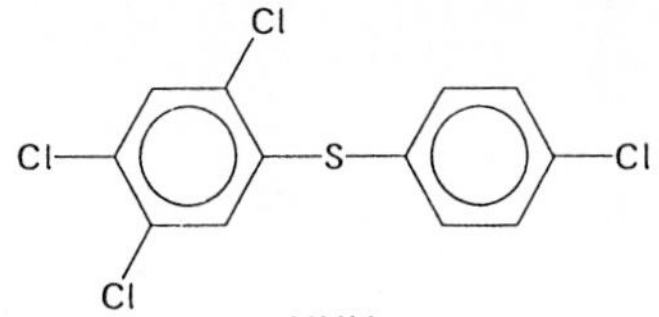

XXX

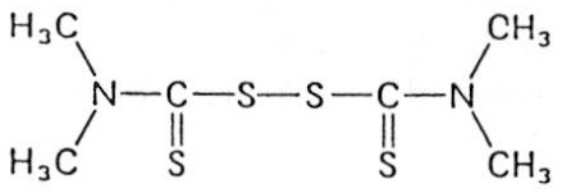

XXXIII

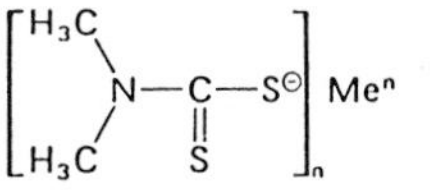

XXXIV: n = 3, Me = $Fe^{3\oplus}$
XXXV: n = 2, Me = $Zn^{2\oplus}$

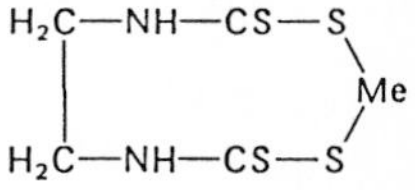

XXXVI: Me = $Mn^{2\oplus}$
XXXVII: Me = $Zn^{2\oplus}$

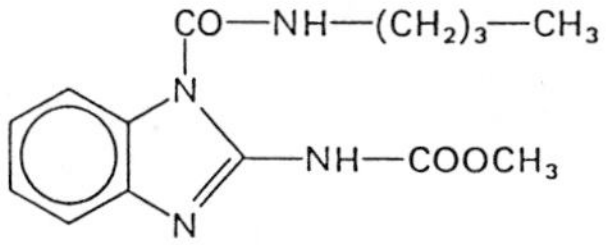

XXXIX

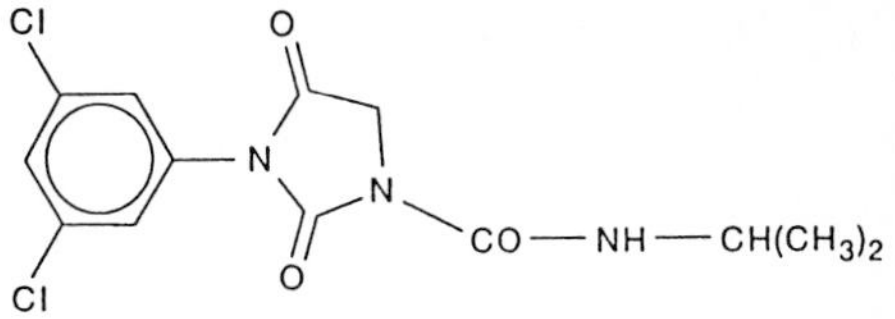

XXXIX a

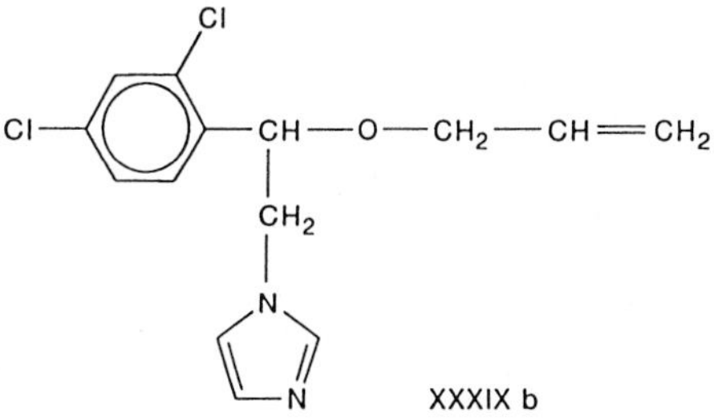

XXXIX b

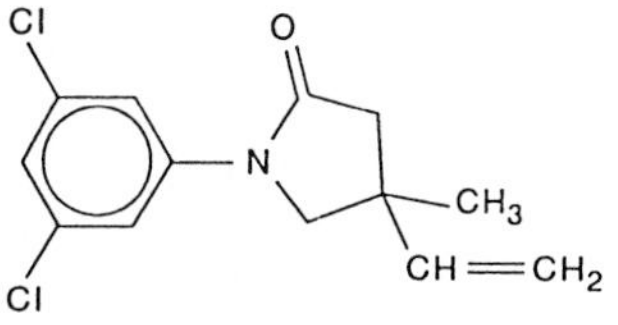

XXXIX c

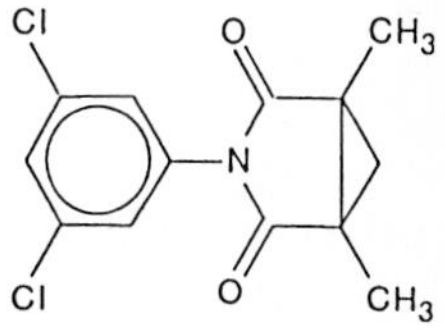

XXXIX d

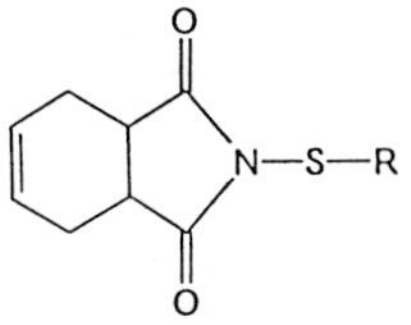

XL: R = CCl_2—$CHCl_2$
XLI: R = CCl_3

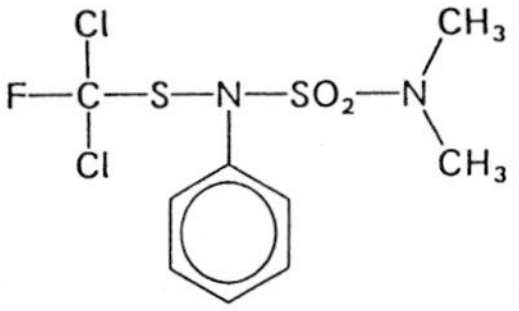

XLII

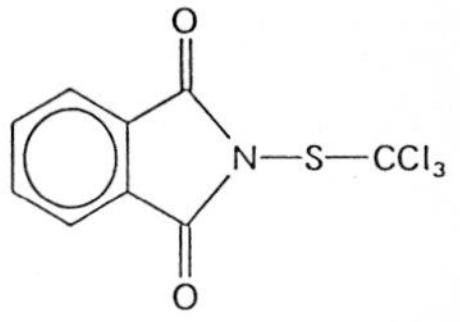

XLIII

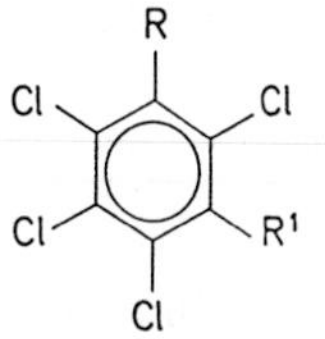

XLIV: R, R^1 = Cl
XLV: R = Cl, R^1 = NO_2
XLV a: R, R^1 = CN

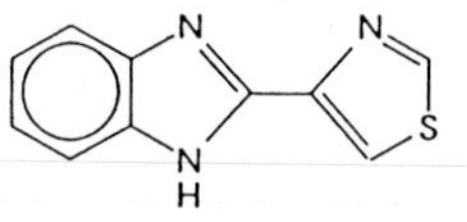

XLVI

R_nMeX

XLVII:	n = 3, Me = $Sn^{4\oplus}$, X = $OCOCH_3$, R = C_6H_5
XLVIII:	n = 1, Me = $Hg^{2\oplus}$, X = Cl, R = $CH_3O(CH_2)_2$
IL:	n = 1, Me = $Hg^{2\oplus}$, X = $OCOCH_3$, R = C_6H_5
L:	n = 1, Me = $Hg^{2\oplus}$, X = Cl, R = C_6H_5

Fig. 9.2. Structures of some selected pesticides. Part 3. The Roman numerals refer to Table 9.5

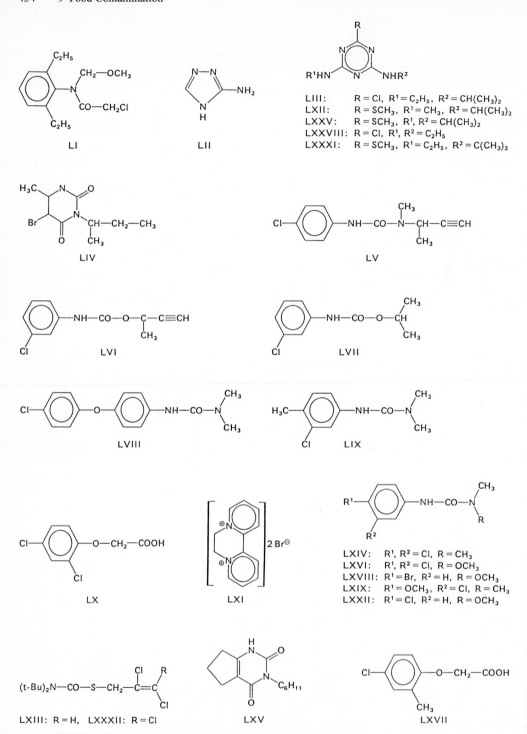

Fig. 9.2. Structures of some selected pesticides. Part 4. The Roman numerals refer to Table 9.5

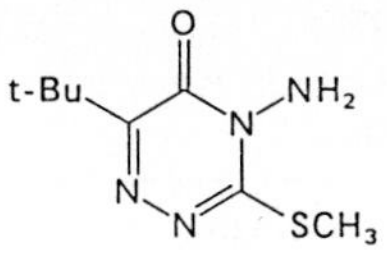

LXX

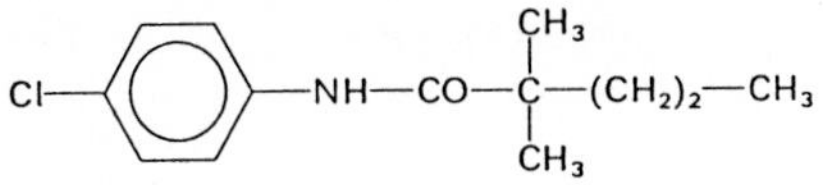

LXXI

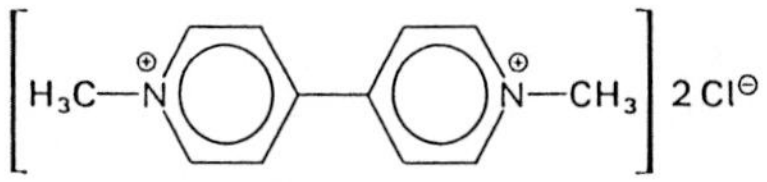

LXXIII

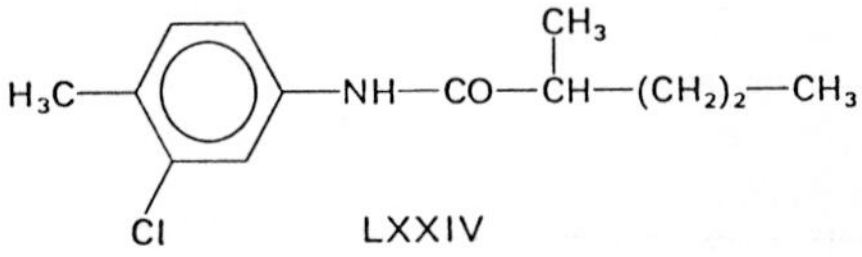

LXXIV

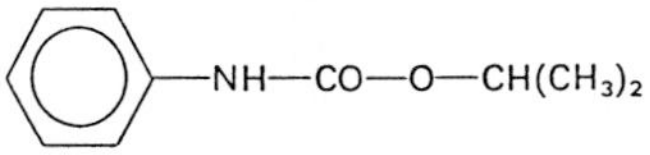

LXXVI

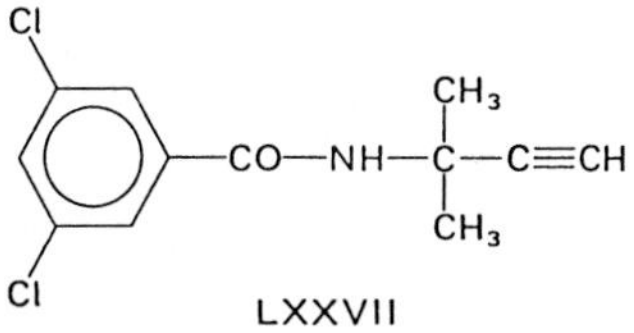

LXXVII

CCl₃COOH

LXXX

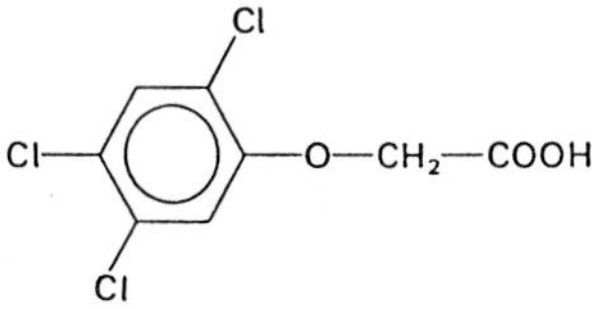

LXXIX

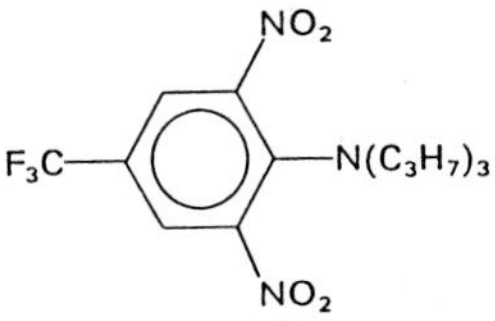

LXXXIII

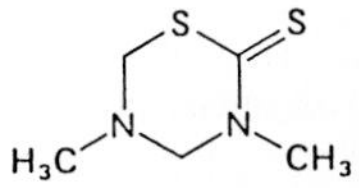

LXXXIV

HC≡CCl—CH₂Cl

ClCH=CH—CH₂Cl

LXXXV

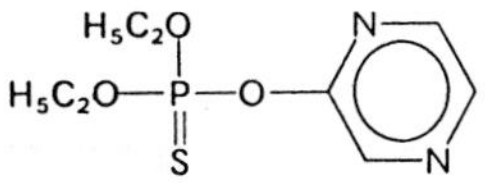

LXXXVIII

CH₃CHO

LXXXIX

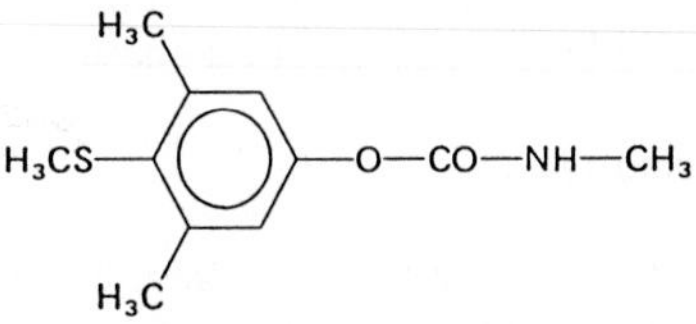

XC

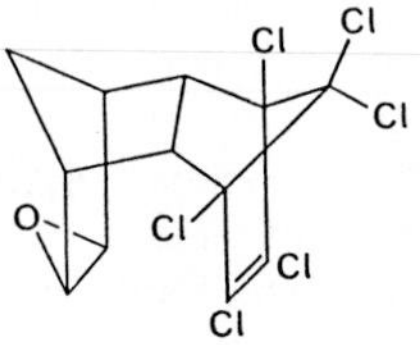

XCI

Fig. 9.2. Structures of some selected pesticides. Part 5. The Roman numerals refer to Table 9.5

Table 9.6. Pesticide residues in food (FR Germany, 1984–1986[a]; Nutrition Report, 1988)

Product group	No. of samples	Samples with residues	
		< Maximum permissible amount (%)[b]	> Maximum permissible amount (%)[b]
Foods of plant origin	18762	36.0	2.6
Fruits (excluding citrus fruits), total	4497	42.9	1.7
Domestic origin	1652	52.4	0.3
Foreign origin	1909	39.1	3.5
Citrus fruits	1657	33.2	1.1
Vegetables (excluding potatoes), total	5876	24.7	3.2
Domestic origin	2201	12.3	2.1
Foreign origin	2305	33.6	5.3
Potatoes	428	11.5	0.2
Cereals, cereal products	1479	54.0	1.0
Bread, cookies, confectionery	72	75.0	0
Pasta	111	60.4	0.9
Tea, tea-like products	568	82.9	5.8
Infant and baby food	568	35.3	0
Spices, condiments	155	68.4	3.9
Legumes, oil seeds, nuts	768	61.4	11.6
Jams, preserves	16	25.0	0
Honey	380	31.6	0.3
Coffee, coffee substitute	22	4.5	0
Cocoa, chocolate	14	50.0	7.10
Fruit juices, nonalcoholic drinks	134	3.7	0
Wine	345	76.2	0
Beer	36	2.8	0

[a] Summarized results of the official food control.
[b] Based on the number of samples (= 100%).

9.5.2 Antibiotics

Antibiotics are used for therapy and as growth-promoting agents, since they improve feed utilization and, thus, animal growth (calves, hogs and poultry). Residues may be found in eggs and milk (e.g. after treatment for mastitis). A constant intake of antibiotics, even at low doses, is a risk to human health since some microorganisms may become resistant and allergic reactions may develop. Therefore, the trend in current practice is to use only those antibiotics as feed additives, which are not used for human therapy or in treatment of animal diseases.

9.5.3 Glucocorticoides

Preparations of the hormones of the adrenal gland cortex, e.g., cortisone, are used in situations when animals are stressed. These preparations have a broad activity and should not have uncontrolled access to the food chain.

9.5.4 Sex Hormones

In addition to their use in therapy, these compounds are used as growth or feed utilization promoters due to their anabolic activity. These include testicular hormones, such as testosterone and estradiol, as well as synthetic com-

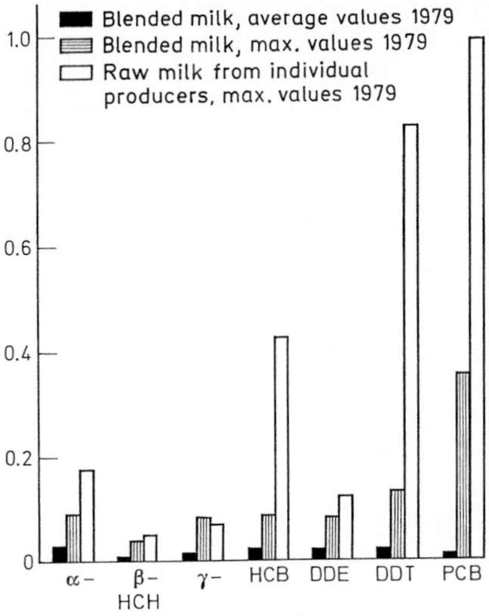

Fig. 9.3. Chlorinated hydrocarbons in cow's milk, mg/kg fat (according to *Acker*, 1981). For abbrevations see Table 9.9

pounds with corresponding activity: esters of estradiol and testosterone, trenbolone, diethylstilbestrol and zeranol.
While residues of the endogenic hormones are assumed not to present a health hazard, the other compounds mentioned are generally stable and have a long residence time in the body. Diethylstilbestrol has teratogenic and carcinogenic side effects.

9.5.5 Psychosomatic Agents

Compounds of this group (e.g. librium, valium) are used as sedatives to lessen the irritation or excitement of the animal and are given prior to vaccination or slaughter. The anabolic conversion of feed by a tranquil animal may be indirectly improved as well. However, the use of sedatives might impede the recognition of sickness in an animal.

9.5.6 Thyreostatica

These compounds lower the basal metabolic rate, thus effecting an increase in body weight (muscle and adipose tissue deposition) and improving the quality of the muscle. Thiouracil and mercaptobenzimidazole are the most commonly used. Thiourea derivatives, found among degradation products, have carcinogenic activity.

9.5.7 Coccidiostatica

The compounds of this class are added to animal feed to combat coccidiosis diseases (such as enteritis or cachexie) caused by protozoans living as parasites in intestines. Poultry and rabbits are the animals most often affected. Residues may be found in eggs.

9.5.8 Other Compounds

Antiparasitica are used, for example, against the cattle liver leech, which enters and infects the body after intake of snails or slugs during grazing. Antiparasitica residues may occur in milk.
Antioxidants, e.g., ethoxyquin (6-ethoxy-1,2-dihydro-2,2,4-trimethylquinoline), are used as an additive to protect animal feed from spontaneous combustion during transport.

9.6 Polychlorinated Biphenyls (PCB's)

Chlorination of biphenyl yields a mixture of compounds with differing chlorine contents. A widely used commercial product, Clophen A60, with 60% bound chlorine (corresponding in North America to Aroclor 1260, a compound with an average chlorine content of 63% and an average molecular weight of 372), consists of about 55 components of differing toxicities. The toxicity is influenced by the number of chlorine atoms and their substitution pattern in the biphenyl molecule, and by the overall composition of the product.
PCB's are widely used in industry as plasticizers in synthetic polymers, in paints, in heat exchange media, in hydraulic presses and as a dielectric fluid in transformers. Although external use has been abandoned, PCB's are still widely found as environmental contaminants (soil, atmosphere and water) and as a consequence of this as residues in food (eggs,

Table 9.7. Relative frequency[a] of pesticides in Germany according to product groups

Product group	Active agent	Total (%)	> Maximum permissible amount (%)
Fruits (excluding citrus fruits)	Vinclozolin	100	6.5
	Dichlofluanid	90	
	Dithiocabamates	33.2	30.4
	Procymidon	30.8	
	Captan	24.4	6.5
	Lindane	19.0	
	Chlorpyrifos	16.4	100
	Endosulfan	11.3	
	Dicofol	11.3	
	Parathion	9.1	4.3
	Ethion		17.4
	Pyrazophos		4.3
	Iprodion		4.3
Vegetables (excluding potatoes)	Vinclozolin	100	46.6
	Lindane	50.2	16.6
	Iprodion	46.4	
	Bromine-containing fumigants	39.0	66.6
	Dithiocarbamates	28.4	86.6
	Procymidon	28.0	
	Dieldrin	26.4	100
	Hexachlorbenzene	24.3	
	Endosulfan	18.8	
	Chlorpyrifos	9.2	43.3
	Chlorthalonil	8.8	36.7
	Dicofol	8.8	
	Pentachlorophenol		13.3
	Quintozene		13.3
	Profenofos		10.0
	Fonofos		10.0
Citrus fruits	Methidathion	100	
	Parathion	46.5	
	Chlorpyrifos	40.3	
	Dicofol	39.5	
	Ethion	38.7	
	Fenithrothion	27.9	20.0
	Anzinphos methyl	19.4	
	Pentachlorophenol	17.1	100
	Parathion methyl	16.3	
	Chlorfenvinphos	16.3	
	Disulfoton	16.3	
	Malathion	14.0	
	Carbophenothion	4.6	
	Imazalil	1.1	
	Bromopropylat	3.1	

[a] Based on the most frequently occurring compound (= 100%)

Table 9.8. Frequency[a] of pesticide residues in fruits and vegetables according to compound classes (Germany)

Fungicide	Insecticide	Others
Vinclozolin	Lindane	Dicofol (kelthan)
Captan	Dieldrin	Endosulfan
Procymidin	Chlorpyrifos	Tetradifon
Dithiocarbamates	Ethion	HCH isomers (excluding lindane)
Dichlofluanid	Fenitrothion	Bromine-containing fumigants
Folpet	Anzinphos methyl	
Chlorthalonil	Malathion	
Iprodion (glycophen)	Disulfoton	
Quintozene	Phosalon	
Imazalil	Methidathion	
Hexachlorobenzene (HCB)	Parathion	
	Parathion methyl	
	Chlorfenvinphos	

[a] Arranged in order of decreasing frequency.

Table 9.9. Chlorinated hydrocarbons in human and bovine milk ($\mu g/kg$ fat)

		α-HCH	β-HCH	γ-HCH	HCB
Bovine milk	1979	28	9	31	30
	1985/86	11	2	24	17
Human milk	1974/75	32	560	87	2650
	1981–86		230		650
Acceptable daily dose[a]		5	1	10	0.6

		DDT	HCE	DIE	PCB
Bovine milk	1979	33			110
	1985/86	18	0.3	4	80
Human milk	1974/75	3510			6500
	1981–86	1250			2640
Acceptable daily dose[a]		5	0.5	0.1	1

[a] $\mu g/kg$ body weight.
HCH: hexachlorocyclohexane, HCB: hexachlorobenzene, DDT; 4,4-dichlorodiphenyltrichloroethane + degradation products DDE [1,1-dichloro-2,2-bis(4-chlorophenyl)-ethylene] and DDD [1,1-dichloro-2,3-bis(4-chlorophenyl)-ethane], HCE: heptachloroepoxide, DIE: dieldrin, PCB: polychlorinated biphenyls (as Chlophen A 60)

wild game, fish, etc.). The concentrations, averaging 12 $\mu g/kg$, are indeed low but, as with all persistent lipophilic compounds, the level is constantly being increased through the nutritional chain (plant-animal-human).

9.7 Polycyclic Aromatic Hydrocarbons

Burning of organic materials, such as wood (wood smoke and its semi-dry distillation product, the wood smoke vapor phase), coal or fuel oil, results in pyrolytic reactions which yield a great number of polycyclic aromatic hydrocarbons with more than three linearly or

Table 9.10. Animal medicines and feed additives (selected structural formulas are presented in Fig. 9.4)

Number	Compound	Application
Antibiotics, Sulfonamides		
I	Penicillins (*Penicillium notatum*)	Therapeutics, feed additives
II	Streptomycin (*Streptomyces griseus*)	Therapeutics
III	Tetracyclines (*Streptomyces spp.*)	Therapeutics, feed additives
IV	Chloramphenicol (*Streptomyces venezuelae*)	As III
V	Oleandomycin (*Streptomyces antibioticus*)	Feed additive for poultry and swine
VI	Spiramycin (*S. ambofaciens*)	As V and calves, sheep, goats
VII	Tylosin (*S. fradiae*)	As V, for swine
VIII	Flavophospholipol	As V, for poultry, calves, swines
IX	Virginiamycin (*S. virginae*)	As VIII
X	Zinc-Bacitracin (*Bacillus subtilis*)	As VI
XI	Sulfonamides	Therapeutics
XIa	Nitrofurans	Therapeutics
XII	Carbadox	Therapeutic for animal breeding
Steroid hormones and other compounds with estrogenic activity		
XIII	Corticosteroids	Therapeutics
XIV	17-β-Estradiol	
XV	Oestrone	
XVI	17-α-Estradiol	
XVII	Estriol	
XVIII	Progesterone	
XIX	Testosterone	
XIXa	19-Nortestosterone	
XX	17-β-Estradiol-3-benzoate	Therapeutics, anabolic agents
XXI	Estradiol-17-monopalmitate	
XXII	Testosterone propionate	
XXIII	Trenbolone	
XXIV	Trenbolone acetate	
XXV	Diethylstilbestrol	
XXVI	Genisterin	
XXVII	Hexestrol	
XXVIII	Coumestrol	
XXIX	Dienestrol	
XXX	Zeranol	Anabolic agents
Psychopharmaceuticals		Therapeutics (Sedatives)
XXXI	Meprobamate (Aneural, Miltaun)	
XXXII	Hydroxyzine (Atarax, Marmoran)	
XXXIII	Chlordiazepoxide (Librium)	
XXXIV	Diazepam (Valium)	

Table 9.10 (Continued)

Number	Compound	Application
XXXV	Oxazepam (Adumbran)	
XXXVI	Chlorpromazine (Megaphen)	
XXXVII	Promazine (Verophen)	
XXXVIII	Azepromazine (Plegicil)	
XXXIX	Xylazine (Rompun)	
XL	Azaperone (Stresnil)	
XLI	Reserpine	
Thyreostatica		Therapeutics, anabolic agents
XLII	Methylthiouracil (Thyreostat)	
XLIII	Thiamazole (Favistan)	
XLIV	Carbimazole (Neo-Thyreostat)	
Coccidiostatica		Poultry feed additive for prophylaxis against Cocciodiose
XLV	Amprolium	
XLVI	Decoquinate	
XLVII	DOT, Dinitolmide	
XLVIIa	Clopidol	
XLVIIb	Nicarbazine	
Antiparasitica		
XLVIII	Niclofolan (Menichlopholan)	A remedy against liver leech
IL	Oxyclozanide	As XLVIII
L	Trichlorphon (Metrifonate)	
Other medicines		
LI	Niclosamide (Masonil)	A remedy against worms
LII	Nitarsone	Therapy (poultry, swine diarrhea), growth promoter
Antioxidants		Feed additives
LIII	Ethoxyquin	
LIV	Butylated hydroxytoluene (BHT)	

angularly fused benzene rings, that are carcinogenic to varying extents (1,2-benzanthracene, benzo[α]pyrene, chrysene, fluoranthene, pyrene, etc.). The quantity and diversity of compounds generated is affected by the conditions of the burning process. Benzo[α]pyrene (Formula 9.1) usually serves as an indicator compound.

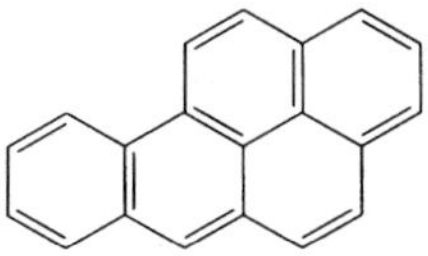

$$(9.1)$$

Contamination of food with polycyclic compounds can occur by deposition from the atmosphere (as often occurs with fruit and leafy vegetables in industrial districts), by direct drying of cereals with combustion gases, by smoking or roasting of food (barbecuing or charcoal broiling; smoking of sausage, ham or fish; roasting of coffee). The content in meat and processed meat products should not exceed 1 µg/kg end-product measured as benzo[α]pyrene. Most smoked products (about 98%) are within this limit. The highest concentrations of polycyclic aromatic hydrocarbons in smoked fish have been found in eels and salmon.

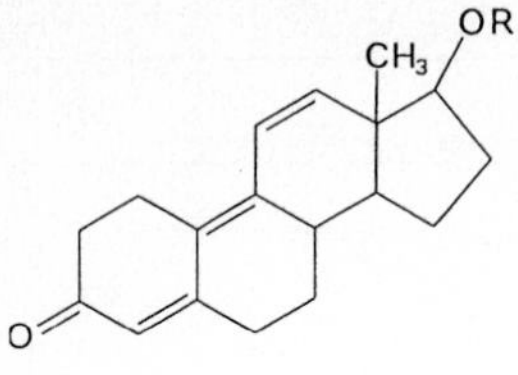

XXIII: R = H
XXIV: R = COCH₃

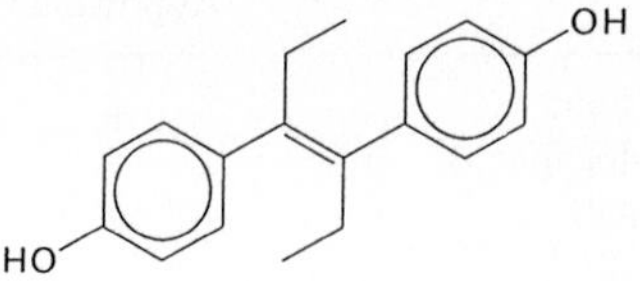

XXV

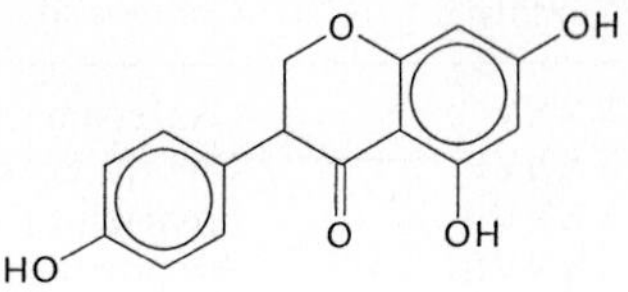

XXVI

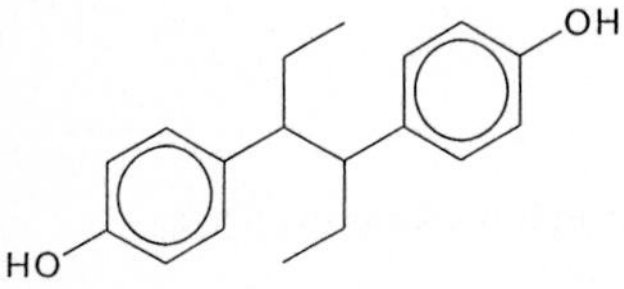

XXVII

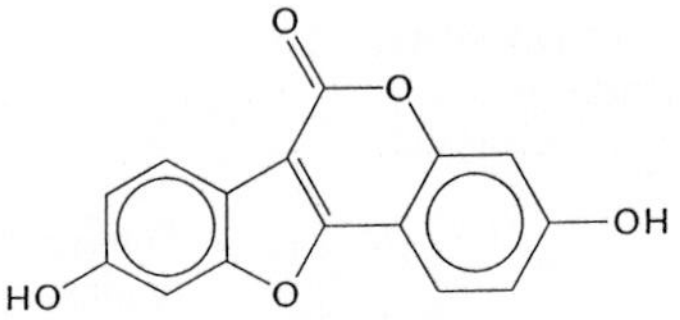

XXVIII

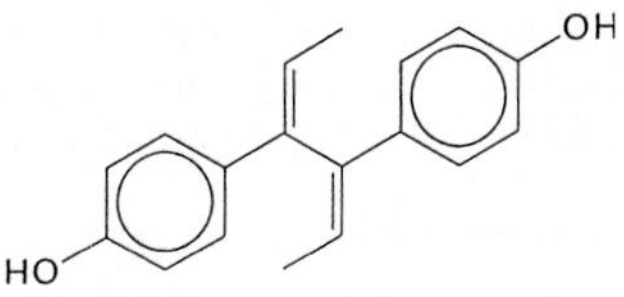

XXIX

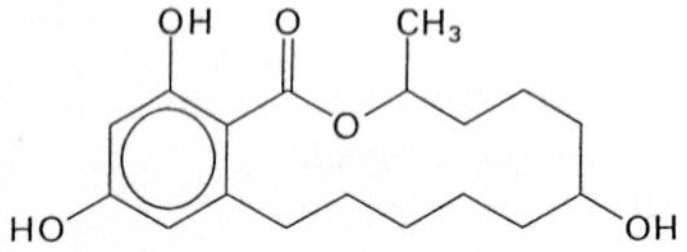

XXX

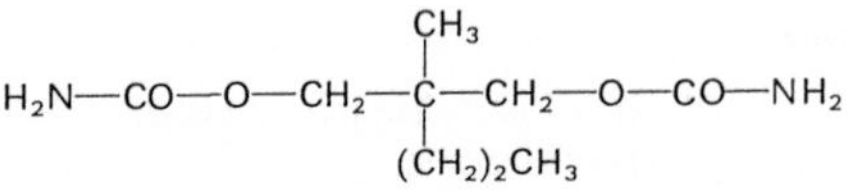

XXXI

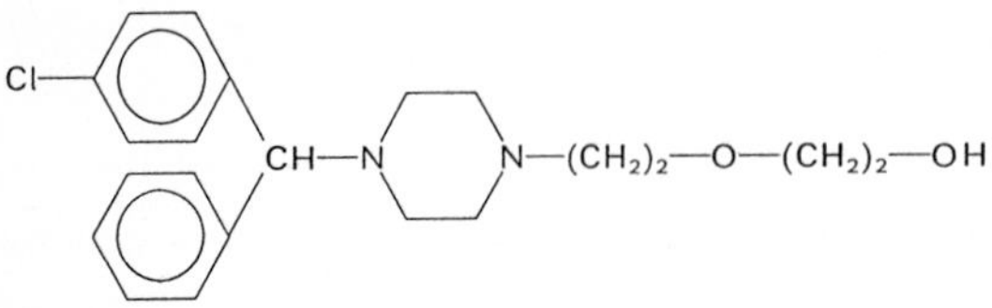

XXXII

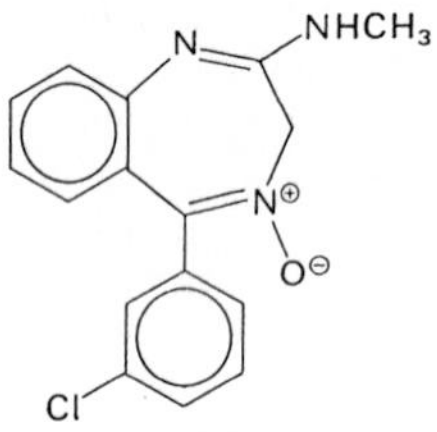

XXXIII

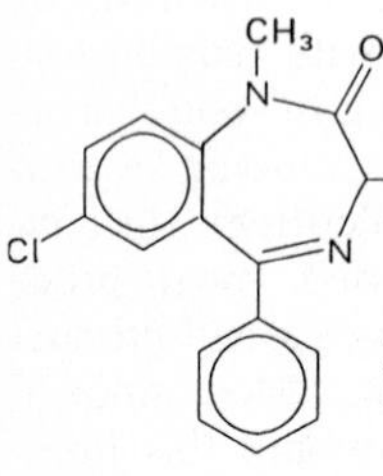

XXXIV: R = H
XXXV: R = OH

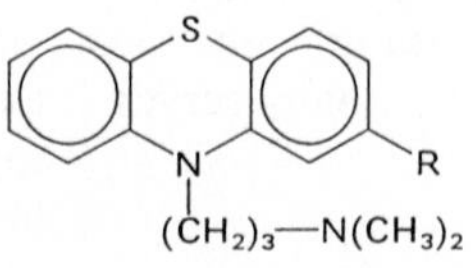

XXXVI: R = Cl
XXXVII: R = H
XXXVIII: R – COCH₃

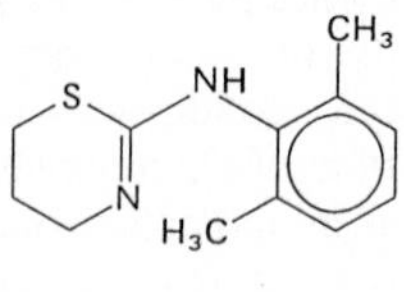

XXXIX

Fig. 9.4. Structures of some selected veterinary medicines and feed additives. The Roman numerals refer to Table 9.10

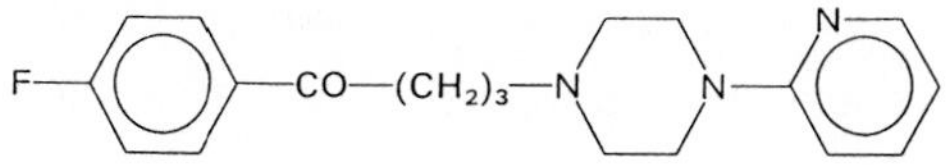

XL

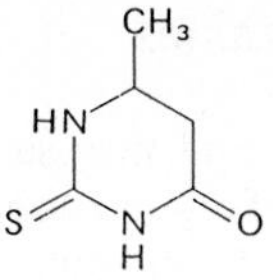

XLII

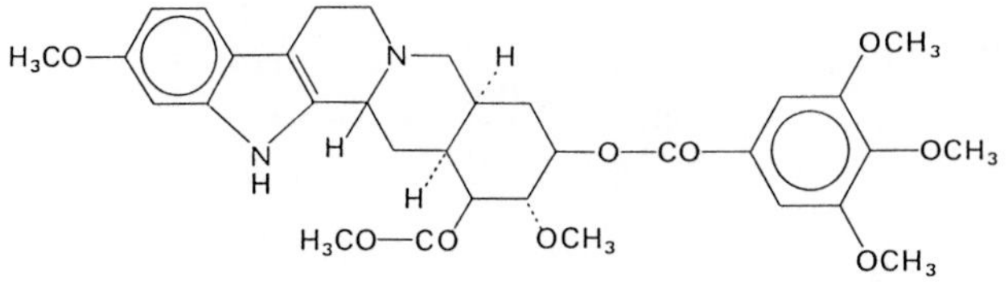

XLI

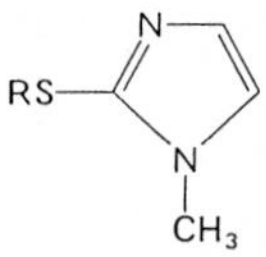

XLIII: R = H
XLIV: R = OCOCH3

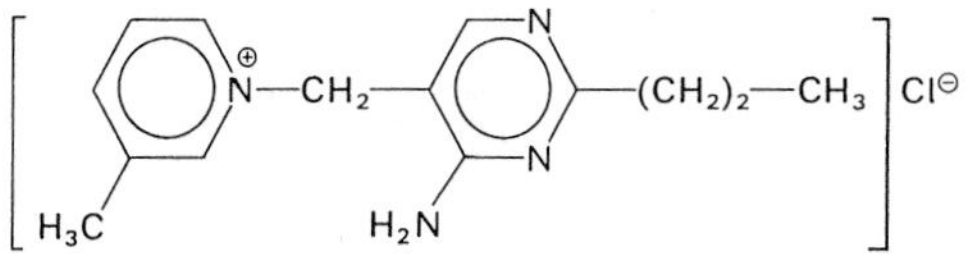

XLV

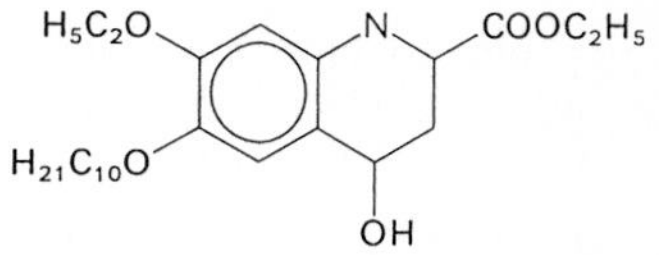

XLVI

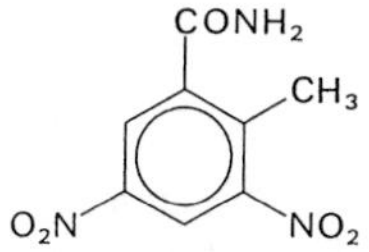

XLVII

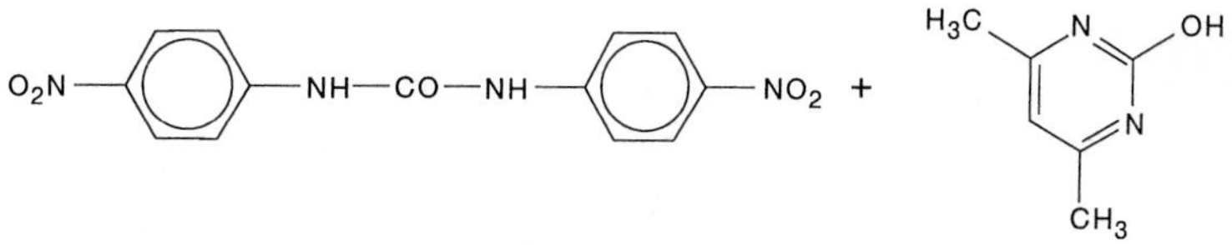

XLVII b

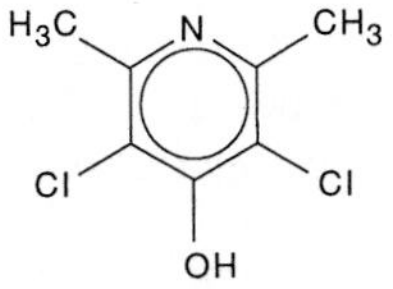

XLVII a

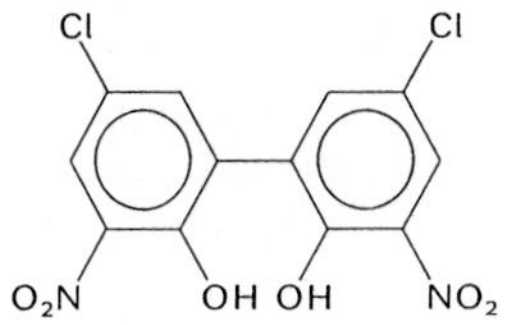

XLVIII

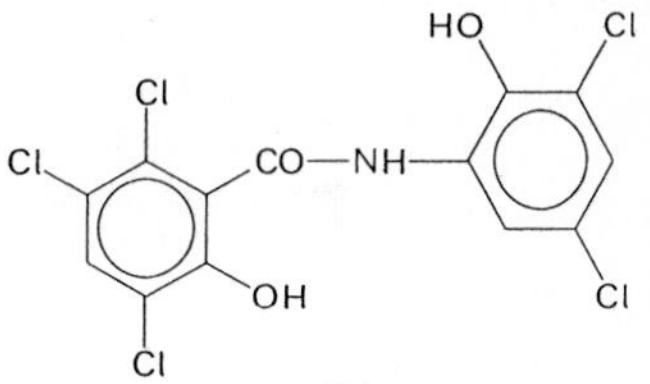

IL

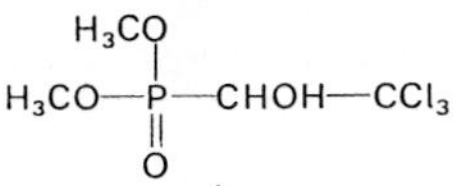

L

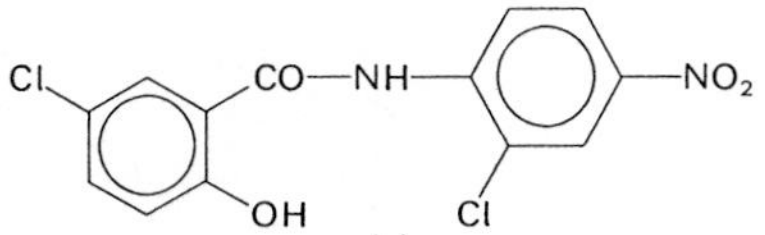

LI

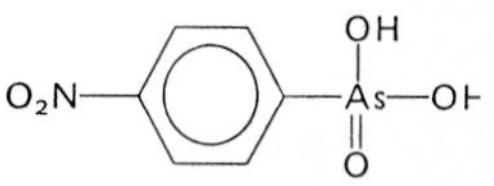

LII

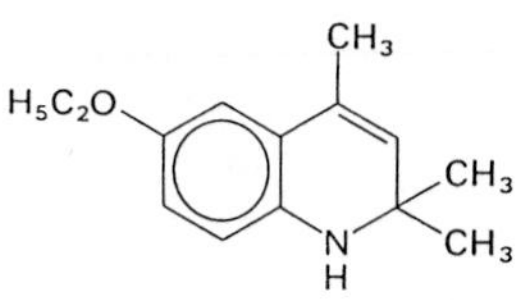

LIII

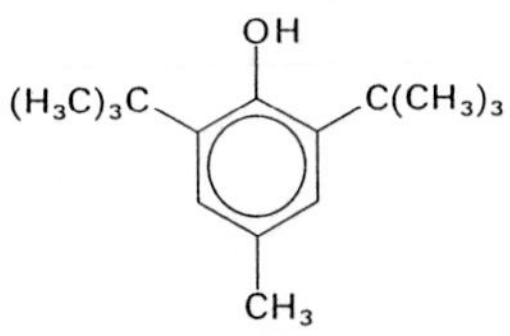

LIV

Fig. 9.4 (continued)

9.8 Nitrosamines

Nitrosamines and nitrosamides are powerful carcinogens. They are obtained from secondary amines, N-substituted amides and nitrous acid:

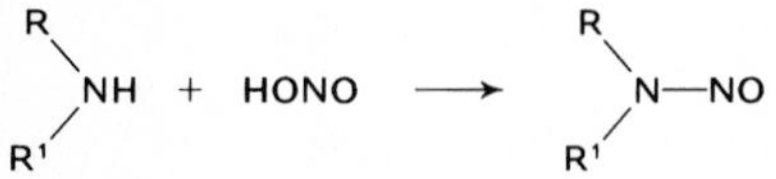

$$R\!\!\diagdown\!\!NH + HONO \longrightarrow R\!\!\diagdown\!\!N\!\!-\!\!NO \qquad (9.2)$$

with R^1 substituents.

$$R\!-\!CO\!-\!NHR^1 + HONO \longrightarrow R\!-\!CO\!-\!\underset{NO}{N}\!-\!R^1 \qquad (9.3)$$

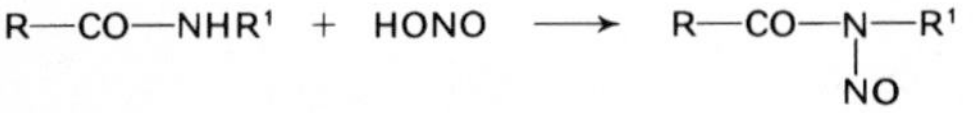

The nitrosonium ion, NO^+, or a nitrosyl halogenide, XNO, is the reactive intermediate:

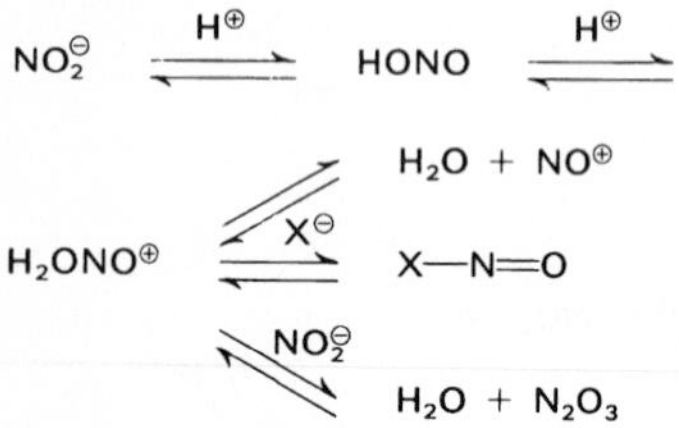

$$NO_2^{\ominus} \underset{H^{\oplus}}{\overset{H^{\oplus}}{\rightleftharpoons}} HONO \overset{H^{\oplus}}{\rightleftharpoons}$$

$$H_2ONO^{\oplus} \begin{array}{c} \nearrow \; H_2O + NO^{\oplus} \\ \overset{X^{\ominus}}{\rightleftharpoons} X\!-\!N\!=\!O \\ \underset{NO_2^{\ominus}}{\searrow} \; H_2O + N_2O_3 \end{array} \qquad (9.4)$$

Nitrosamine formation is also possible from primary amines:

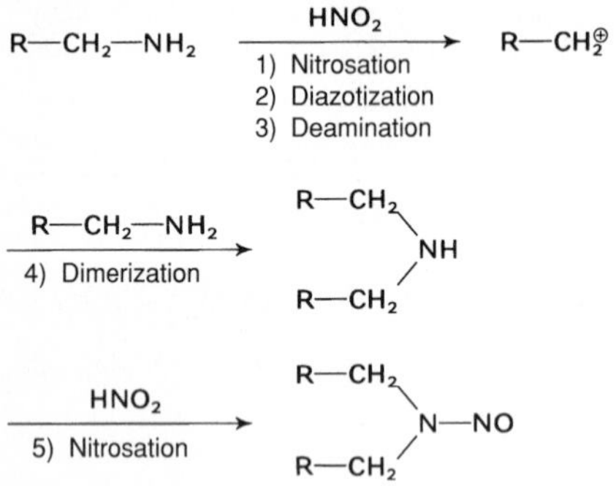

$$R\!-\!CH_2\!-\!NH_2 \xrightarrow[\substack{1)\ Nitrosation \\ 2)\ Diazotization \\ 3)\ Deamination}]{HNO_2} R\!-\!CH_2^{\oplus}$$

$$R\!-\!CH_2\!-\!NH_2 \xrightarrow[4)\ Dimerization]{} \overset{R\!-\!CH_2}{\underset{R\!-\!CH_2}{\diagup}}\!\!NH$$

$$\xrightarrow[5)\ Nitrosation]{HNO_2} \overset{R\!-\!CH_2}{\underset{R\!-\!CH_2}{\diagup}}\!\!N\!-\!NO \qquad (9.5)$$

from diamines:

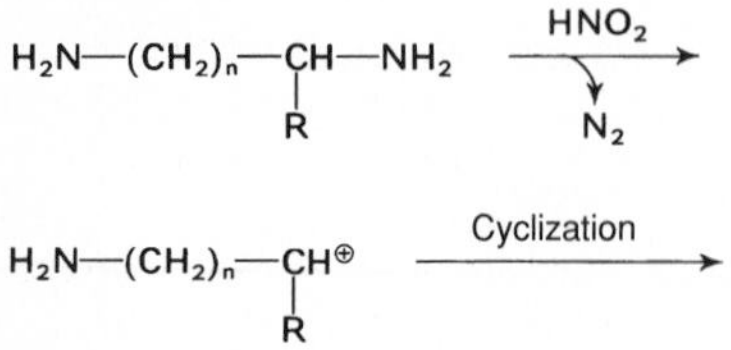

$$H_2N\!-\!(CH_2)_n\!-\!\underset{R}{CH}\!-\!NH_2 \xrightarrow[N_2]{HNO_2}$$

$$H_2N\!-\!(CH_2)_n\!-\!\underset{R}{CH^{\oplus}} \xrightarrow{Cyclization} \qquad (9.6)$$

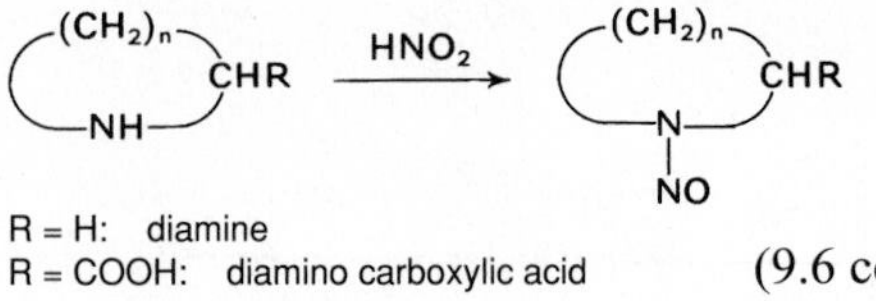

$$\overset{-(CH_2)_n-}{\underset{-NH}{\diagdown}}\!\!CHR \xrightarrow{HNO_2} \overset{-(CH_2)_n-}{\underset{\underset{NO}{N}}{\diagdown}}\!\!CHR$$

R = H: diamine
R = COOH: diamino carboxylic acid   (9.6 cont.)

and from tertiary amines:

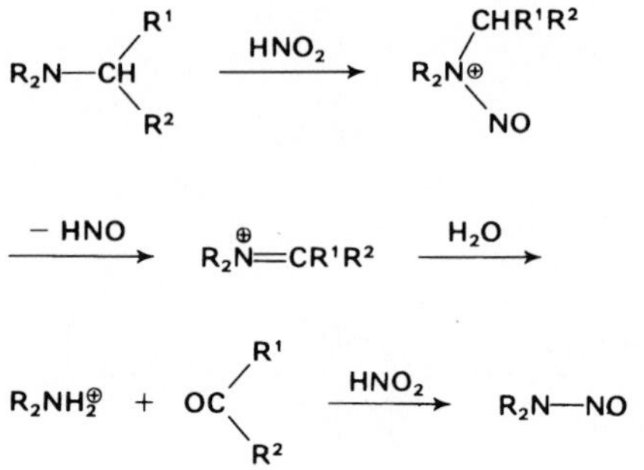

$$R_2N\!-\!\underset{R^2}{\overset{R^1}{CH}} \xrightarrow{HNO_2} R_2\overset{\oplus}{\underset{NO}{N}}\!\!\overset{CHR^1R^2}{}$$

$$\xrightarrow{-\,HNO} R_2\overset{\oplus}{N}\!=\!CR^1R^2 \xrightarrow{H_2O}$$

$$R_2NH_2^{\oplus} + OC\!\!\overset{R^1}{\underset{R^2}{\diagdown}} \xrightarrow{HNO_2} R_2N\!-\!NO \qquad (9.7)$$

$$2\,HNO \longrightarrow H_2N_2O_2 \longrightarrow H_2O + N_2O \qquad (9.8)$$

Nitrosamines are detected in variable amounts in many foods (Table 9.11). The most common compound is dimethylnitrosamine, which is also a most powerful carcinogen. Some activity has been ascribed to nitrosopiperidine and nitrosopyrrolidine. In meat products cured and treated with pickling salt, 30% of the samples contained nitrosodimethylamine (NDMA; 0.5–15 µg/kg) and 13% nitrosopyrrolidine (NPYR; >0.5 µg/kg). About 25% of the cheese samples analyzed were contaminated (0.5–4.9 µg/kg).

Nitrosopyrrolidine is formed from the amino acid proline by nitrosation followed by decarboxylation at elevated temperatures, such as in roasting or frying:

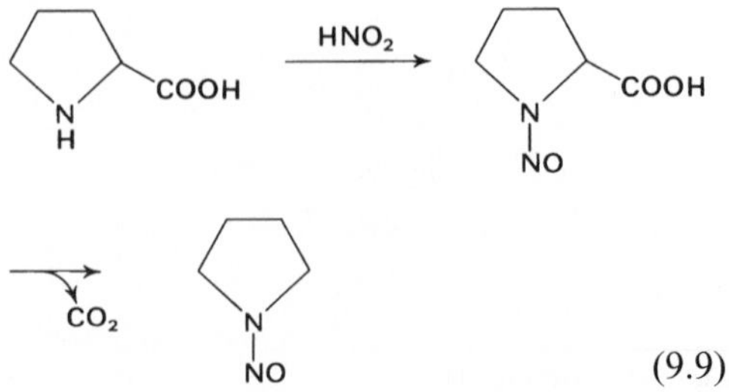

$$\underset{H}{\overset{}{\text{pyrrolidine-COOH}}} \xrightarrow{HNO_2} \underset{NO}{\overset{}{\text{pyrrolidine-COOH}}}$$

$$\xrightarrow{-CO_2} \underset{NO}{\overset{}{\text{pyrrolidine}}} \qquad (9.9)$$

Table 9.11. Nitrosamines in food

Food product	Compound[a]	Content µg/kg	Year of analysis
Frankfurter (hot dog)	NDMA	0–84	1972
Fish (raw)	NDMA	0–4	1971, 1972
Fish, smoked and pickled with nitrites or nitrates	NDMA	4–26	1971
Fish, fried	NDMA	1–9	1972
Cheese (Danish, Blue, Gouda, Tilsiter, goatmilk cheese)	NDMA	1–4	1972
Salami	NDMA	10–80	1972
Bacon (hog's hind leg) smoked meat	NDMA	1–60	1975
Pepper-coated ham, raw and roasted	NPIP	4–67	1975
	NPYR	1–78	1975

[a] NDMA: N-Nitrosodimethylamine, NPIP: N-nitrosopiperidine, NPYR: N-nitrosopyrrolidine.

The nitrosopyrrolidine (1.5 µg/kg) in meat products increases almost ten fold (to 15.4 µg/kg) during roasting and frying. An estimate of the average daily intake of nitrosamines ranges from 0.1 µg nitrosodimethylamine and 0.1 µg nitrosopyrrolidine to a total of 1 µg.

An endogenic dose should be included in addition to the above exogenic dose. It may result from ingestion of amines, and of nitrate ions, both of which are abundant in food. Table 9.12 presents data on the forms of amines present in food. The occurrence of nitrate is very high in some vegetables, and nitrate is occasionally found in drinking water (Table 9.13).

The minimum nitrate intake is estimated at 75 mg/person/day. Nitrate is reduced to nitrite by microflora of mouth saliva. This is a prerequisite for a nitrosation reaction in the gastric acid medium. Nitrosation may also occur with a number of medicinal products. Dimethylamino or diethylamino compounds, which might be degraded with release of dimethyl or diethyl nitrosamines, are most commonly involved.

Inhibition of a nitrosation reaction is possible, e.g., with ascorbic acid, which is oxidized by nitrite to its dehydro form, while nitrite is

reduced to NO. Similarly, tocopherols and some other food constituents inhibit substitution reactions. Representative suitable measures to decrease exo- and endogenic nitrosamine hazards are:

- Decreasing nitrite and nitrate incorporation into processed meat. However, to completely relinquish the use of nitrite is a great health hazard due to the danger from bacterial intoxication (especially by botulism).
- Addition of inhibitors (ascorbic acid, tocopherols).
- Decreasing the nitrate content of vegetables.

9.9 Cleansing Agents and Disinfectants

Residues introduced by large-scale animal husbandry and use of milking machines are gaining importance. The residues in meat and processed meat products originate from the surfaces of processing equipment, while residues in milk arise from measures involved in disinfection of the udder. Iodine-containing disinfectants, including udder dipping or soaking agents, may be an additional source of contamination of milk with iodine.

9.10 Polychlorinated Dibenzodioxins (PCDD) and Dibenzofurans (PCDF)

Polyhalogen-dibenzo-p-dioxins (Hal_xDD) and polyhalogen-dibenzofurans (Hal_xDF), informally called "dioxins", occur as companion compounds or impurities in a large number of bromine- and chlorine-containing chemicals.

$$(9.10)$$

Futhermore, they are formed in many thermal processes ($600°C > T \geq 200°C$) in the presence of chlorine or other halogens in inorganic or organic form. Consequently, they are widely distributed in the environment. The number of isomers (congeners) is large, For rodents,

Table 9.12. Amines in food (mg/kg)

Compound	Spinach	Cabbage (kale) red	Cabbage (kale) green	Carrots	Red beet	Celery	Lettuce	Rhubarb	Herring salted	Herring smoked	Herring in oil	Cheese Tilsiter	Cheese Camembert	Cheese Limburger
Ammonia	18.280	11.060	15.260	3.970	8.800	19.600	10.260	6.340	2.928	270	–	164.400	–	–
Methylamine	12	22.7	16.6	3.8	30	64	37.5	–	3.4	–	7	–	12	3
Ethylamine	8.4	1.3	–	1	–	–	3.3	–	0.1	0.4	–	–	4	1
Dimethylamine	–	2.8	5.5	–	–	51	7.2	–	7.8	6.3	45	–	–	–
Methylethylamine	–	0.9	–	7	–	–	7.5	–	–	–	1	–	–	2
n-Propylamine	–	–	–	–	–	–	–	–	–	–	–	8.7	2	2
Diethylamine	15	–	–	–	–	–	–	–	1.9	5.2	–	–	–	–
n-Butylamine	–	–	7	–	–	–	–	–	–	–	–	3.7	–	–
i-Butylamine	–	–	–	–	–	–	–	–	–	0.3	–	–	0.2	0.2
Pyrroline	–	–	–	–	–	–	–	–	–	–	–	–	–	–
n-Pentylamine	0.3	0.6	0.4	–	–	0.8	3	–	–	–	17	1.2	0.2	tr[a]
i-Pentylamine	3.8	–	0.5	–	–	–	–	3.9	–	–	–	–	–	0.1
Pyrrolidine	2.5	–	–	–	–	0.4	–	–	–	–	–	19.9	1	–
Di-n-propylamine	–	–	–	–	–	–	–	–	–	–	–	8.4	–	–
Piperidine	–	–	–	–	–	–	–	–	0.7	0.2	–	–	–	tr
Aniline	–	1.0	0.7	30.9	0.6	0.7	0.6	5	–	–	–	–	–	–
N-Methylaniline	3.4	0.3	–	0.8	–	0.5	–	–	–	–	–	37.9	–	–
N-Methylbenzylamine	–	–	–	16.5	–	–	10	–	–	–	2	–	–	–
Toluidine	–	–	1.1	7.2	–	1.1	–	–	–	–	–	–	–	–
Benzylamine	6.1	3.3	3.8	2.8	0.1	3.4	11.5	2.9	–	–	–	–	–	–
Phenylethylamine	1.1	8.6	3	2	0.5	–	–	3.2	–	–	–	–	–	–
N-Methylphenyl-ethylamine	2.4	3.7	2	2	0.4	0.5	0.4	2.6	–	–	0.1	2.6	–	–

[a] Traces.

Table 9.13. Nitrate and nitrite in food (mg/kg fresh substance) analyzed in Germany

Food	Nitrate			Nitrite		
	n[a]	Mean	Variation	n[a]	Mean	Variation
Milk	16	1.4	1.0–4.1			
Cheese				39	0.3	0.2–1.3
Meat	110	7.6	1.0–49.5			
Uncooked smoked pork ribs	73	68.6	5.0–425.5	47	27.9	0.2–94.1
Uncooked smoked black forest ham	23	351.0	21.6–1384.3	20	12.3	1.2–80.2
Uncooked smoked ham				23	10.7	0.9–44.2
Uncooked sausages, firm	20	208.4	7.0–1042.0			
Cooked smoked shoulder ham				44	15.7	0.8–91.0
Salami				76	5.1	0.3–48.7
Fresh soft sausage				35	6.9	0.2–45.6
Fried sausage				108	3.5	0.2–41.5
Finely minced pork sausage				32	7.8	0.2–18.6
Calf-liver sausage, finely grained				19	5.4	1.9–12.3
Salted herring filet	154	27.4	1.0–405.0			
Herring titbit	103	74.7	19.0–276.0			
Cereals	75	7.2	0.3–19.0	10	0.5	0.3–1.0
Potatoes	270	93.3	10.0–463.0	160	0.2	< 0.1–15.6
Fresh vegetables	3776	720.6	< 0.1–6798.0	2044	0.1	< 0.1–19.6
Lettuce	526	1489.2	10.0–5570.0			
Corn salad	163	1434.8	10.0–4125.0			
White cabbage	102	451.2	10.0–1790.0			
Spinach	117	964.8	10.0–3894.0			
Cress	24	2326.3	10.0–5364.0			
Fennel	19	1541.4	129.0–5893.0			
Tomatoes	169	27.2	0.4–747.0			
Carrots	65	232.6	14.8–841.6			
Radish	203	2030.0	10.0–6684.0			
Beetroot	108	1630.2	10.0–6798.0			
Fresh fruits	532	70.1	1.0–3291.0	155	0.3	0.2–1.0
Strawberries	67	138.6	2.5–425.0			
Grapes, white	23	8.2	1.0–30.0			
Apples	99	18.8	1.0–688.0			
Pears	24	13.5	2.0–49.0			
Rhubarb	19	986.3	90.0–3291.0			
Wine	735	13.7	0.8–62.9			
Beer	39	23.5	0.4–53.4			
Infant food	588	81.0	2.0–453.0	273	0.4	0.3–1.1

[a] Number of samples.

Table 9.14. Polychlorodibenzo-p-dioxins (Cl$_x$DD) and polychlorodibenzofurans (Cl$_x$DF) in bovine milk and human milk (ng/kg fat)[a]

Dioxin/furan	Structure	TEF[b]	Bovine milk (n[c] = 8)	Human milk (n[c] = 30)
D 48	2,3,7,8-Cl$_4$DD	1	< 0.2	3.4
D 54	1,2,3,7,8-Cl$_5$DD	0.1	0.8	15
D 66	1,2,3,4,7,8-Cl$_6$DD	0.1	0.3	12
D 67	1,2,3,6,7,8-Cl$_6$DD	0.1	1.2	59
D 70	1,2,3,7,8,9-Cl$_6$DD	0.1	0.4	11
D 73	1,2,3,4,6,7,8-Cl$_7$DD	0.01	< 20	61
D 75	OCDD/Cl$_8$DD	0.001	< 10	530
F 83	2,3,7,8-Cl$_4$DF	0.1	0.7	2.5
F 94	1,2,3,7,8-Cl$_5$DF	0.05	< 0.2	< 1
F 114	2,3,4,7,8-Cl$_5$DF	0.5	1.4	20
F 118	1,2,3,4,7,8-Cl$_6$DF	0.1	0.8	8.5
F 121	1,2,3,6,7,8-Cl$_6$DF	0.1	0.8	7.8
F 124	1,2,3,7,8,9-Cl$_6$DF	0.1	n.d.	n.d.
F 130	2,3,4,6,7,8-Cl$_6$DF	0.1	0.7	3.0
F 131	1,2,3,4,6,7,8-Cl$_7$DF	0.01	< 0.5	8.5
F 134	1,2,3,4,7,8,9-Cl$_7$DF	0.01	n.d.	n.d.
F 135	OCDF/Cl$_8$DD	0.001	< 1	3

[a] FR Germany, 1986.
[b] Toxicity equivalent factors (2,3,7,8-Cl$_4$DD $\triangleq$ 1).
[c] Number of samples.
n.d., not determined.

2,3,7,8-tetrachlorodibenzodioxin (2,3,7,8-TCDD, "Seveso dioxin") has proved to be expecially toxic (LD$_{50}$ = 0.6 μg/kg, guinea-pig) and carcinogenic. A NOEL (cf. 9.1) of 1 ng/kg of body weight and day was determined in experiments over a period of two years. The toxicity of the other compounds is lower and is generally expressed as toxicity equivalent factors (TEFs), based on 2,3,7,8-TCDD (TEF = 1). With the help of these values, 2,3,7,8-TCDD equivalents (TEQ) can be calculated, which are a measure of the total exposure to corresponding compounds (cf. Tables 9.14 and 9.15).

The daily intake of dioxin in industrial countries is estimated at 1–2 pg TEQ/kg of body weight. With a half life of 8 years, this leads to the measured concentrations of ~30 ppt TEQ in body fat. Some data on the contamination of bovine and human milk are presented in Table 9.14. The different compound patterns indicate different contamination pathways. An estimation of the dioxin intake with food is presented in Table 9.15.

Table 9.15. Average daily intake of 2,3,7,8-tetra-chlorodibenzo-p-dioxin (2,3,7,8-Cl$_4$DD) and related compounds with the food (pg/day)[a]

	2,3,7,8-Cl$_4$DD	Σ TEQ[b]
Meat products (including poultry)	7	23.5
Milk	6.2	28.5
Eggs	0.8	4.2
Fish	8.6	33.3
Vegetable oil	< 0.2[c]	< 0.6
Vegetables	< 2.4[c]	< 2.4[c]
Fruits	< 1.4[c]	< 2.6[c]
Sum:	24.6	93.5[d]

[a] Based on an "average food basked".
[b] Sum of the compounds taken in, expressed as toxicity equivalents TEQ (cf. text).
[c] These numbers are included in the sum with 50%.
[d] At present, the ADI value is 1 pg/kg body weight and day. In outdoor air that is not directly contaminated, it can be assumed that the intake through breathing is 0.03 pg TEQ/kg body weight and day. The intake with food is 1.3 pg TEQ/kg body weight and day.

9.11 Literature

Acker, L.: Die Rückstandssituation in der Bundesrepublik Deutschland – Versuch einer Bestandsaufnahme. Lebensmittelchem. Gerichtl. Chem. *35*, 1, (1981)

Ballschmiter, K.: Chemie und Vorkommen der halogenierten Dioxine und Furane. Nachr. Chem. Tech. Lab. *39*, 988 (1991)

Berg, H. W., Diehl, J. F., Frank, H.: Rückstände und Verunreinigungen in Lebensmitteln. Dr. Dietrich Steinkopff Verlag: Darmstadt. 1978

Bund für Lebensmittelrecht und Lebensmittelkunde e.V.: Wie sicher sind unsere Lebensmittel? Schriftenreihe Heft 102, B. Behr's Verlag: Hamburg. 1983

Bundesforschungsanstalt für Ernährung: Essentielle und toxinale Inhaltsstoffe in der täglichen Gesamtnahrung. Bericht BFE-R–83-02, Karlsruhe, Dezember 1983

Bundesforschungsanstalt für Ernährung: Radioaktivität in Lebensmitteln – Tschernobyl und die Folgen. Bericht BFE-R–86-04, Karlsruhe, Dezember 1986

Cikryt, P.: Die Gefährdung der Menschen durch Dioxine und verwandte Verbindungen. Nachr. Chem. Tech. Lab. *39*, 648 (1991)

Deutsche Forschungsgemeinschaft: Rückstände in Gebäck und Getreideprodukten. Harald Boldt Verlag KG: Boppard. 1981

Deutsche Forschungsgemeinschaft: Rückstände in Lebensmitteln tierischer Herkunft. Verlag Chemie: Weinheim. 1983

Deutsche Forschungsgemeinschaft: Das Nitrosamin-Problem. Verlag Chemie: Weinheim. 1983

Deutsche Forschungsgemeinschaft: Rückstände und Verunreinigungen in alkoholfreien Getränken. Verlag Chemie: Weinheim. 1983

Deutsche Forschungsgemeinschaft: Polychlorierte Biphenyle. Verlag Chemie: Weinheim. 1988

Eisenbrand, G.: N-Nitrosoverbindungen in Nahrung und Umwelt. Wissenschaftliche Verlagsgesellschaft mbH: Stuttgart. 1981

Ernährungsbericht 1980. Deutsche Gesellschaft für Ernährung. e.V.: Frankfurt/Main. 1980

Ernährungsbericht 1984. Deutsche Gesellschaft für Ernährung. e.V.: Frankfurt/Main. 1984

Ernährungsbericht 1988. Deutsche Gesellschaft für Ernährung e.V.: Frankfurt/Main. 1988

Großklaus, D. (Ed.): Rückstände in von Tieren stammenden Lebensmitteln. Verlag Paul Parey, Berlin. 1989

Hassall, K. A.: The chemistry of pesticides. Verlag Chemie: Weinheim. 1982

Hathway, D. E.: Molecular aspects of toxicology. The Royal Society of Chemistry, Burlington House, London. 1984.

Heitefuß, R.: Pflanzenschutz, 2. ed. Georg Thieme Verlag: Stuttgart, 1987

Heyns K.: Über die endogene Nitrosamin-Entstehung beim Menschen. Landwirtschaftliche Forschung, Sonderheft 36, S. 145, Kongreßband 1979 Gießen, J.D. Sauerländer Verlag: Frankfurt/Main. 1980

Lindner, E.: Toxikologie der Nahrungsmittel. 2. ed., Georg Thieme Verlag: Stuttgart. 1979

Macholz, R., Lewerenz, H.: Lebensmitteltoxikologie. Springer-Verlag: Berlin. 1989

National Research Council (U.S.A.): Regulating Pesticides in Food. National Academy Press: Washington, D.C. 1987

Sinell, H.-J.: Einführung in die Lebensmittelhygiene. 2. ed. Verlag Paul Parey: Berlin. 1985

Teufel, M., Niessen, K.-H.: Rückstände in Muttermilch. Ernährungs-Umschau *38*, 142 (1991)

Thier, H.-P., Frehse, H.: Rückstandsanalytik von Pflanzenschutzmitteln. Georg Thieme Verlag: Stuttgart. 1986

Wegler, R. (Hrsg.): Chemie der Pflanzenschutz- und Schädlingsbekämpfungsmittel. Bde. 1–8, Springer-Verlag: Berlin. 1970–1982

10 Milk and Dairy Products

10.1 Milk

Milk is the secreted fluid of the mammary glands of female mammals. It contains nearly all the nutrients necessary to sustain life. Since the earliest times, mankind has used the milk of goats, sheep and cows as food. Today the term "milk" is synonymous with cow's milk. The milk of other animals is spelled out, e.g., sheep milk or goat milk, when supplied commercially.

The yield of milk per cow in Germany more than doubled within this century as a result of dairy cattle breeding and improvement in feed quality. The yield was 1,260 kg per cow in 1812, 2,163 kg in 1926, 3,800 kg in 1970 and 4,181 kg in 1977.

Milk production in various countries, its processing into dairy products and its consumption are summarized in Tables. 10.1–10.3.

10.1.1 Physical and Physico-Chemical Properties

Milk is a white or yellow-white, opaque liquid. The color is influenced by scattering and absorption of light by milk fat globules and

Table 10.1. World production of milk, 1996 (1,000 t)

Continent	Cow milk	Buffalo milk	Sheep milk	Goat milk
World	466317	51884	7756	10144
Africa	16141	1600	1554	2078
America, North-, Central-	89910	–	–	148
America, South-	41880	–	33	184
Asia	83693	50193	3250	5577
Europe	215696	91	2919	2156
Oceania	18997	–	–	–

Table 10.1 (continued)

Country	Cow milk	Country	Buffalo milk
USA	69975	India	31990
Russian Fed.	35445	Pakistan	14800
India	33000	China	2200
Germany	28621	Egypt	1600
France	25668	Nepal	665
Brazil	19845	Iran	160
Ukraine	15704	Turkey	108
UK	14600	Myanmar	100
Poland	11430	Italy	79
Netherlands	11188	Sri Lanka	67
Italy	10674		
New Zealand	9934	Σ (%)[a]	99
Argentina	9176		
Turkey	9133		
Australia	8986		
Japan	8290		
Mexico	8059		
Canada	8000		
China	5838		
Colombia	5000		
Σ (%)[a]	75		

Country	Sheep milk	Country	Goat milk
Turkey	922	India	2010
Italy	800	Bangladesh	1152
China	680	Pakistan	720
Greece	670	Sudan	645
Sudan	510	Greece	460
Syria	450	France	430
Iran	438	Somalia	415
Somalia	430	Iran	412
Romania	394	Spain	321
Spain	313	Turkey	280
Algeria	220	Russian Fed.	268
		Ukraine	197
Σ (%)[a]	75	Chile	195
		Indonesia	192
		Σ (%)[a]	76

[a] World production = 100%.

Table 10.2. World production of dairy products, 1996, (1,000 t)

Continent	Cheese	Butter[a]	Condensed milk	Whole milk powder	Skim milk powder[b]	Whey powder
World	14994	6565	4553	2386	3477	1709
Africa	517	179	41	28	15	1
America, North-, Central-	4019	687	1263	143	604	578
America, South-	641	181	229	482	500	504
Asia	1177	2326	941	110	299	–
Europe	7327	2770	1972	1187	2058	1064
Oceania	497	422	107	437	449	66

Country	Cheese	Country	Butter[a]	Country	Condensed milk
USA	3607	India	1350	USA	949
France	1679	USA	539	Germany	550
Germany	1420	Germany	478	India	440
Italy	899	France	470	Netherlands	330
Netherlands	688	Pakistan	394	Russian Fed.	225
Russian Fed.	477	Russian Fed.	380	UK	204
Argentina	370	New Zealand	278	Malaysia	187
UK	355	Ukraine	155	Peru	157
Poland	350	Poland	150	Mexico	137
Egypt	349	Australia	142	Estonia	130
Canada	300	Ireland	139	Czech Rep.	108
Denmark	299	Netherlands	125		
Greece	221	UK	125	Σ (%)[c]	75
China	202	Iran	119		
		Turkey	117		
Σ (%)[c]	75				
		Σ (%)	76		

Country	Whole milk powder	Country	Skim milk powder[b]	Country	Whey powder
New Zealand	326	USA	500	USA	504
France	260	Germany	415	France	490
Germany	221	France	402	Netherlands	278
Brazil	200	Russian Fed.	306	Germany	154
Argentina	120	Australia	227	Canada	74
Netherlands	120	New Zealand	222	Australia	47
Australia	111	Japan	190	Ireland	39
Russian Fed.	110	Ukraine	128	UK	37
Denmark	105	Ireland	122	Denmark	32
UK	86	Poland	120	New Zealand	19
Mexico	75				
Belgium/ Luxembourg	70	Σ (%)[c]	76	Σ (%)[c]	98
Σ (%)[c]	76				

[a] Includes fat from buffalo milk (ghee).
[b] Includes powder from buttermilk recovered from butter churning.
[c] World production = 100%.

Table 10.3. Consumption of milk and dairy products in FR Germany (in kg/capita and year)

	64/65	66/67	68/69	70/71	72/73	74/75	76/77
Fluid milk	95.7	93.4	92.0	92.5	86.5	82.9	83.8
Cream	2.5	2.8	3.2	3.5	3.7	3.9	4.1
Cheese	8.4	9.1	9.4	10.2	11.1	11.7	12.4
Butter	8.5	8.6	8.5	8.3	7.3	7.0	6.4

protein micelles. Therefore, skim milk also retains its white color. A yellowish, i.e. yellow-green, color is derived from carotene (ingested primarily during pasture grazing) present in the fat phase and from riboflavin present in the aqueous phase. Milk tastes mildly sweet, while its odor and flavor are normally quite faint.

Milk fat occurs in the form of droplets or globules, surrounded by a membrane and emulsified in milk serum (also called whey). The fat globules (called cream) separate after prolonged storage or after centrifugation. The fat globules float on the skim milk. Homogenization of milk so finely divides and emulsifies the fat globules that cream separation does not occur even after prolonged standing.

Proteins of various sizes are dispersed in milk serum. They are called micelles and consist mostly of calcium salts of casein molecules. Furthermore, milk contains lipoprotein particles, also called milk microsomes, which consist of the residues of cell membranes, microvilli, etc., as well as somatic cells, which are mainly leucocytes (10^8/l of milk). Some of the properties of the main structural elements of milk are listed in Table 10.4.

Various proteins, carbohydrates, minerals and other ingredients are solubilized in milk serum. The specific density of milk decreases with increasing fat content, and increases with increasing amounts of protein, milk sugar and salts. The specific density of cow's milk ranges from 1.029 to 1.039 (15°C). Defatted (skim) milk has a higher specific density than whole milk. From the relationships given by *Fleischmann*:

$$m = 1.2\,f + \frac{266.5\,(s-1)}{s} \qquad (10.1)$$

and by *Richmond*:

$$m = 0.25\,s + 1.21\,f + 0.66 \qquad (10.2)$$

the dry matter content of milk, m, in percent, can be calculated from the percent fat content (f), knowing the specific density (s).

The freezing point of milk is -0.53 to -0.55°C. This rather constant value is a suitable test for detection of watering of milk.

The pH of fresh milk is 6.5–6.75, while the acid degree according to *Soxhlet-Henkel* (°SH) is 6.5–7.5.

The refractive index (n_D^{20}) is 1.3410–1.3480, and the specific conductivity at 25°C is 4–5.5 $\times 10^{-3}$ ohm^{-1} cm^{-1}.

The measurement of redox potentials of milk and its products can also be of value. The redox potential is $+0.30$ V for raw and $+0.10$ V for pasteurized milk, $+0.05$ V for processed cheese, -0.15 V for yoghurt and -0.30 V for Emmental cheese.

Table 10.4. Main structural elements of milk

Name	Type of dispersion	Percentage	Number (l^{-1})	Diameter (mm)	Surface (m^2/l milk)	Specific density[a] (g/ml)
Fat globules	Emulsion	3.8	10^{13}	100–10000	70	0.92
Casein micelles	Suspension	2.8	10^{17}	10–300	4000	1.11
Globular proteins (whey proteins)	Colloidal solution	0.6	10^{20}	3–6	5000	1.34
Lipoprotein particles	Colloidal suspension	0.01	10^{17}	10	10	1.10

[a] 20°C.

10.1.2 Composition

The composition of dairy cattle milk varies to a fairly significant extent. Table 10.5 provides some data. In all cases water is the main ingredient of milk at 63–87%. In the following sections, only cow's milk will be dealt with in detail since it is the main source of our dairy foods.

Table 10.5. Composition of human milk and milk of various mammals (%)

Milk	Protein	Casein	Whey protein	Sugar	Fat	Ash
Human	0.9[a]	0.4	0.5	7.1	4.5	0.2
Cow (bovine)	3.2	2.6	0.6	4.6	3.9	0.7
Donkey	2.0	1.0	1.0	7.4	1.4	0.5
Horse	2.5	1.3	1.2	6.2	1.9	0.5
Camel	3.6	2.7	0.9	5.0	4.0	0.8
Zebu	3.2	2.6	0.6	4.7	4.7	0.7
Yak	5.8			4.6	6.5	0.9
Buffalo	3.8	3.2	0.6	4.8	7.4	0.8
Goat	3.2	2.6	0.6	4.3	4.5	0.8
Sheep	4.6	3.9	0.7	4.8	7.2	0.9
Reindeer	10.1	8.6	1.5	2.8	18.0	1.5
Cat	7.0	3.8	3.2	4.8	4.8	0.6
Dog	7.4	4.8	2.6			
Rabbit	10.4					

[a] After the 15-th day of the breast feeding period the protein content is increased to 1.6%.

10.1.2.1 Proteins

In 1877 *O. Hammarsten* distinguished three proteins in milk: casein, lactalbumin and lactoglobulin. He also outlined a procedure for their separation: skim milk is diluted then acidified with acetic acid. Casein flocculates, while the whey proteins stay in solution. This established a specific property of casein: it is insoluble in weakly acidic media. It was later revealed that the milk protein system is much more complex. In 1936 *Pedersen* used ultracentrifugation to demonstrate the nonhomogeneity of casein, while in 1939 *Mellander* used electrophoresis to prove that casein consists of three fractions, i. e. α-, β- and γ-casein. The most important proteins of milk are listed in Table 10.7. The casein fraction forms the main portion. Major constituents of whey proteins, β-lactoglobulin A and B and α-lactalbumin, can be differentiated genetically. Other

Table 10.6. Amino acid composition (g AA/100 g protein) of the total protein, casein, and whey protein of bovine milk

Amino acid	Total protein	Casein	Whey protein
Alanine	3.7	3.1	5.5
Arginine	3.6	4.1	3.3
Aspartic acid	8.2	7.0	11.0
Cystine	0.8	0.3	3.0
Glutamic acid	22.8	23.4	15.5
Glycine	2.2	2.1	3.5
Histidine	2.8	3.0	2.4
Isoleucine	6.2	5.7	7.0
Leucine	10.4	10.5	11.8
Lysine	8.3	8.2	9.6
Methionine	2.9	3.0	2.4
Phenylalanine	5.3	5.1	4.2
Proline	10.2	12.0	4.4
Serine	5.8	5.5	5.5
Threonine	4.8	4.4	8.5
Tryptophan	1.5	1.5	2.1
Tyrosine	5.4	6.1	4.2
Valine	6.8	7.0	7.5

protein constituents, e.g., enzymes, are present in much lower quantities; they are not listed in Table 10.7.

The amino acid composition of the total protein, casein, and whey protein of bovine milk is presented in Table 10.6.

10.1.2.1.1 Casein Fractions

The main constituents of this milk protein fraction have been fairly well investigated. Their amino acid sequences are summarized in Table 10.8. Data showing the genetic variations are provided in Table 10.9.

α_s-*Caseins*. The B variant of α_{s1}-casein consists of a peptide chain with 199 amino acid residues and has a molecular weight of 23 kdal. The sequence contains 8 phosphoserine residues, 7 of which are localized in positions 43–80, and these positions have an additional 12 carboxyl groups. Thus these positions are extremely polar acidic segments along the peptide chain. Proline is uniformly distributed along the chain and apparently to a great extent hinders the formation of a regular structure. A portion of the chain, up to 30%, is assumed to have regular conformations.

Table 10.7. Bovine milk proteins

Fraction	Genetic variants	Portion[a]	Isoionic point	Molecular weight[b] (kdal)	Phosphorus content (%)
Caseins		80	–	–	0.9
α_{s1}-Casein	A, B, C, D, E	34	4.92–5.35	23.6[f]	1.1
α_{s2}-Casein	A, B, C, D	8		25.2[g]	1.4
$\varkappa$-Casein	A, B	9	5.77–6.07	19[h]	0.2
β-Casein	A^1, A^2, A^3, B, C, D, E	25	5.20–5.85	24	0.6
γ-Casein		4	5.8–6.0	12–21	0.1
γ_1-Casein	A^1, A^2, A^3, B			20.5	
γ_2-Casein	$A^1/A^2, A^3$, B			11.8	
γ_3-Casein	$A^1/A^2/A^3$, B			11.6	
Whey proteins		20	–	–	
β-Lactoglobulin	A, B, C, D, E, F, G	9	5.35–5.41	18.3	
α-Lactalbumin	A, B, C	4	4.2–4.5[e]	14.2	
Serum albumin	A	1	5.13	66.3	
Immunoglobulin		2			
IgG1			5.5–6.8	162	
IgG2			7.5–8.3	152	
IgA			–	400[c]	
IgM			–	950[d]	
FSC(s)[i]				80	
Proteose-Peptone		4	3.3–3.7	4–41	

[a] As % of skim milk total protein, [b] monomers, [c] dimer, [d] pentamer, [e] isoelectric point, [f] Variant B, [g] Variant A, [h] Variant A^2, [i] Free secretory component.

Amino acid residues 100–199 are distinctly apolar and are responsible for strong association tendencies, which are limited by the repulsing forces of phosphate groups. In the presence of Ca^{2+} ions, in the levels found in milk, α_{s1}-casein forms an insoluble Ca-salt. In the A variant of the molecule, amino acid residues 14–26 are missing; in the C variant the glutamic acid in position 192 (Glu-192) is replaced by Gly-192; and in the D variant Pth-53 (phosphothreonine) replaces Ala-53.

α_{s2}-*Casein* has a pronounced dipolar structure with a concentration of anionic groups in the region of the N-terminus and cationic groups in the region of the C-terminus. With Ca^{2+}, it is more easily precipitable than α_{s1}-casein.

β-*Caseins*. The A^2 variant is a peptide chain consisting of 209 residues and has a molecular weight of 24.5 kdal. Five phosphoserine residues are localized in positions 1–40; these positions contain practically all of the ionizing sites of the molecule. Positions 136–209 con-

tain mainly residues with apolar side chains. The molecule has a structure with a "polar head" and an "apolar tail", thus resembling a "soaplike" molecule. Indeed, CD measurements have shown that β-casein contains about 9% of α-helix structure and about 25% of β-structure. An increase in temperature results in an increase in the β-structure at the cost of the aperiodic part. The self-association of β-casein is an endothermic process. The protein precipitates in the presence of Ca^{2+} ions at the levels found in milk. However, at temperatures at or below 1°C the calcium salt is quite soluble.

$\varkappa$-*Caseins*. The B variant consists of a peptide chain with 169 residues and has a molecular weight of 18 kdal. This monomer is accessible only under reducing conditions. Normally, $\varkappa$-casein occurs as a trimer or as a higher oligomer in which the formation of disulfide bonds is probably involved. The protein contains varying amounts of carbohydrates (average values: 1% galactose, 1.2% galactosamine,

Table 10.8. Amino acid sequences of bovine milk proteins

α_{s1}-*Casein B-8P*

R	P	K	H	P	I	K	H	Q	G	L	P	Q	E	V	L	N	E	N	L
L	R	F	F	V	A	P	F	P	Q	V	F	G	K	E	K	V	N	Q	L
S	K	D	I	G	S^a	E	S^a	T	E	D	Q	A	M	E	D	I	K	E	M
E	A	E	S^a	I	S^a	S^a	S^a	E	E	I	V	P	N	S^a	V	E	Q	K	H
I	Q	K	E	D	V	P	S	E	R	Y	L	G	Y	L	E	Q	L	L	R
L	K	K	Y	K	V	P	Q	L	E	I	V	P	N	S^a	A	E	E	R	L
H	S	M	K	E	G	I	H	A	Q	Q	K	E	P	M	I	G	V	N	Q
E	L	A	Y	F	Y	P	E	L	F	R	Q	F	Y	Q	L	D	A	Y	P
S	G	A	W	Y	Y	V	P	L	G	T	Q	Y	T	D	A	P	S	F	S
D	I	P	N	P	I	G	S	E	N	S	E	K	T	T	M	P	L	W	

α_{s2}-*Casein A-11P*

K	N	T	M	E	H	V	S^a	S^a	S^a	E	E	S	I	I	S^a	Q	E	T	Y
K	Q	E	K	N	M	A	I	N	P	S	K	E	N	L	C	S	T	F	C
K	E	V	V	R	N	A	N	E	E	E	Y	S	I	G	S^a	S^a	S^a	E	E
S^a	A	E	V	A	T	E	E	V	K	I	T	V	D	D	K	H	Y	Q	K
A	L	N	E	I	N	E	F	Y	Q	K	F	P	Q	Y	L	Q	Y	L	Y
Q	G	P	I	V	L	N	P	W	D	Q	V	K	R	N	A	V	P	I	T
P	T	L	N	R	E	Q	L	S^a	T	S^a	E	E	N	S	K	K	T	V	D
M	E	S^a	T	E	V	F	T	K	K	T	K	L	T	E	E	E	K	N	R
L	N	F	L	K	K	I	S	Q	R	Y	Q	K	F	A	L	P	Q	Y	L
K	T	V	Y	Q	H	Q	K	A	M	K	P	W	I	Q	P	K	T	K	V
I	P	Y	V	R	Y	L													

β-*Casein A^2-5P*

R	E	L	E	E	L	N	V	P	G	E	I	V	E	S^a	L	S^a	S^a	S^a	E
E	S	I	T	R	I	N	K	K	I	E	K	F	Q	S^a	E	E	Q	Q	Q
T	E	D	E	L	Q	D	K	I	H	P	F	A	Q	T	Q	S	L	V	Y
P	F	P	G	P	I	P	N	S	L	P	Q	N	I	P	P	L	T	Q	T
P	V	V	V	P	P	F	L	Q	P	E	V	M	G	V	S	K	V	K	T
A	M	A	P	K	H	K	E	M	P	F	P	K	Y	P	V	Q	P	F	T
E	S	Q	S	L	T	L	T	D	V	E	N	L	H	L	P	P	L	L	L
Q	S	W	M	H	Q	P	H	Q	P	L	P	P	T	V	M	F	P	P	Q
S	V	L	S	L	S	Q	S	K	V	L	P	V	P	E	K	A	V	P	Y
P	Q	R	D	M	P	I	Q	A	F	L	L	Y	Q	Q	P	V	L	G	P
V	R	G	P	F	P	I	I	V											

$\varkappa$-*Casein B-1P*

Z^d	E	Q	N	Q	E	Q	P	I	R	C	E	K	D	E	R	F	F	S	D
K	I	A	K	Y	I	P	I	Q	Y	V	L	S	R	Y	P	S	Y	G	L
N	Y	Y	Q	Q	K	P	V	A	L	I	N	N	Q	F	L	P	Y	P	Y
Y	A	K	P	A	A	V	R	S	P	A	Q	I	L	Q	W	Q	V	L	S
D	T	V	P	A	K	S	C	Q	A	Q	P	T	T	M	A	R	H	P	H
P	H	L	S	F	M	A	I	P	P	K	K	N	Q	D	K	T	E	I	P
T	I	N	T	I	A	S	G	E	P	T	S	T^b	P	T	I	E	A	V	E
S	T	V	A	T	L	E	S^a	P	E	V	I	E	S	P	P	E	I	N	
T	V	Q	V	T	S	T	A	V											

Table 10.8 (continued)

α-Lactalbumin B [c]

E	Q	L	T	K	C	E	V	F	R	E	L	K	D	L	K	G	Y	G	G
V	S	L	P	E	W	V	C	T	T	F	H	T	S	G	Y	D	T	E	A
I	V	E	N	N	Q	S	T	D	Y	G	L	F	Q	I	N	N	K	I	W
C	K	N	D	Q	D	P	H	S	S	N	I	C	N	I	S	C	D	K	F
L	N	N	D	L	T	N	N	I	M	C	V	K	K	I	L	D	K	V	G
I	N	Y	W	L	A	H	K	A	L	C	S	E	K	L	D	Q	W	L	C
E	K	L																	

β-Lactoglobulin B [e]

L	I	V	T	Q	T	M	K	G	L	D	I	Q	K	V	A	G	T	W	Y
S	L	A	M	A	A	S	D	I	S	L	L	D	A	Q	S	A	P	L	R
V	Y	V	E	E	L	K	P	T	P	E	G	D	L	E	I	L	L	Q	K
W	E	N	G	E	C	A	Q	K	K	I	I	A	E	K	T	K	I	P	A
V	F	K	I	D	A	L	N	E	N	K	V	L	V	L	D	T	D	Y	K
K	Y	L	L	F	C	M	E	N	S	A	E	P	E	Q	S	L	A	C	Q
C	L	V	R	T	P	E	V	D	D	E	A	L	E	K	F	D	K	A	L
K	A	L	P	M	H	I	R	L	S	F	N	P	T	Q	L	E	E	Q	C
H	I																		

[a] The serine residue is phosphorylated.
[b] The threonine residue is glycosylated.
[c] Disulfide bonds: 6–120, 28–111, 61–77, 73–91.
[d] Pyrrolidone carboxylic acid.
[e] Disulfide bonds: 66–160 and apparently either 106–119 or 106–121.
  Accordingly, the free thiol group is either Cys-119 or Cys-121.

2.4% N-acetyl neuramic acid) that are bound to the peptide chain through Thr-131, 133, 135 or (in variant A) 136. *κ*-Casein is separated electrophoretically into various components that have the same composition of amino acids, but differ in their carbohydrate moiety, e.g., per protein molecule they contain 0–3 moles N-acetyl neuramic acid, 0–4 moles galatose and 0–3 moles galactosamine. Three different glycosyl residues could be isolated, one of which has the structure shown in Formula 10.3.

In the other two oligosaccharide units, one of the two N-acetylneuraminic acid residues is lacking in each case.

κ-Casein is the only main constituent of casein which remains soluble in the presence of Ca^{2+} ions in the concentrations found in milk (Fig. 10.1). Aggregation of $α_{s1}$- and β-caseins with *κ*-casein prevents their coagulation in the presence of Ca^{2+} ions (Fig. 10.2). This property of *κ*-casein is of utmost importance for formation and maintenance of stable casein complexes and casein micelles, as occur in milk.

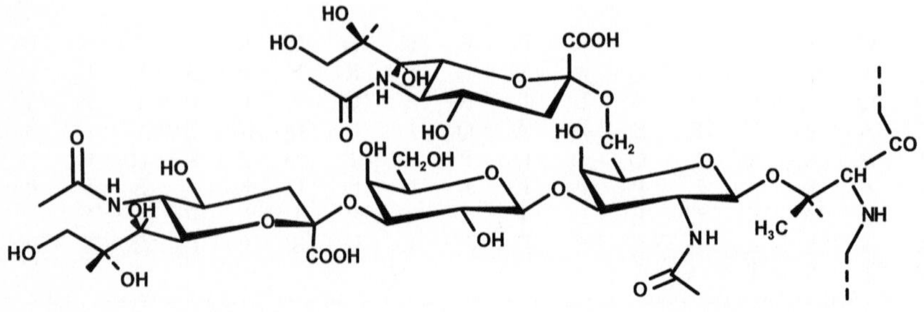

(10.3)

Table 10.9. Amino acid sequences[a] of genetic variants of bovine milk proteins

Protein	Variant	Frequency[b]	Positions of the substitutions						
			14–26	53	59	192			
α_{s1}-Casein	A	s	are lacking						
	B	w		Ala	Glu	Glu			
	C	i				Gly			
	D	s		ThrP					
	E	s			Lys	Gly			
			18	35	36	37	67	106	122
β-Casein	A¹						His		
	A²	w, i	SerP	SerP	Glu	Glu	Pro	His	Ser
	A³							Gln	
	B	s							Arg
	C	s		Ser		Lys			
	D	s	Lys						
	E	s			Lys				
			136	148					
$\varkappa$-Casein	A	w, i	Thr	Asp					
	B	x	Ile	Ala					
			10						
α-Lactalbumin	A	i	Gln						
	B	w	Arg						
			45	59	64	118	158		
β-Lactoglobulin	A	x			Asp	Val			
	B	w, i	Glu	Gln	Gly	Ala	Glu		
	C	s		His					
	D	s	Gln						
	E(D$_\text{Yak}$)	s					Gly		

[a] cf. Table 10.8.
[b] w: predominant in the western world (*Bos taurus*), i: predominant in India (*Bos indicus, Bos grunniens*), s: rare, x: not predominant, but not rare.

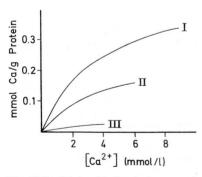

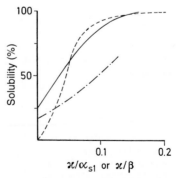

Fig. 10.1. Calcium binding by I: α_{s1}-casein (0.38), II: β-casein (0.21) and III: $\varkappa$-casein (0.05). The bound phosphate residues in mmol/g of casein are given in brackets (according to *Walstra* and *Jenness*, 1984)

Fig. 10.2. Influence of $\varkappa$-casein on the solubility of α_{s1}-casein (– – – 2.5 mg/ml) and β-casein (—— 1.5 mg/ml; —·—·— 6 mg/ml) at pH 7.0, 30 °C, 100 mmol/l CaCl$_2$ (according to *Walstra* and *Jenness*, 1984)

Chymosin (rennet, rennin cf. 1.4.5.2) selectively cleaves the peptide chain of $\varkappa$-casein at – Phe^{105} – Met^{106} – into two fragments: para-$\varkappa$-casein and a glycopeptide (Pyg = pyroglutamic acid, i.e. pyrrolidone carboxylic acid):

$$
\begin{array}{cccc}
1 & 105 & 106 & 169 \\
\text{Pyg} \cdots & \text{Phe} - \text{Met} & \cdots\cdots & \text{Val} \\
 & \multicolumn{2}{c}{\varkappa\text{-Casein}} & \\
\end{array}
$$

$$
\xrightarrow{\text{Lab}}
\begin{array}{cccc}
1 & 105 & 106 & 169 \\
\text{Pyg} \cdots & \text{Phe} + \text{Met} & \cdots\cdots & \text{Val} \\
\multicolumn{2}{c}{\text{para-}\varkappa\text{-Casein}} & \text{Glycopeptide} & \\
\end{array}
$$

$$(10.4)$$

The released glycopeptide is soluble, while para-$\varkappa$-casein precipitates in the presence of Ca^{2+} ions. In this way $\varkappa$-casein loses its protective effect; the casein complexes and casein micelles coagulate (curdle formation) from the milk. The specificity of rennin is high, as is shown by data given in Table 10.10. The sugar moiety of $\varkappa$-casein is not essential for rennin action, nor for the stabilizing property of its protein portion. However, the sugar moiety delays protein cleavage by rennin. Also, it appears that the stability of α_s- and $\varkappa$-casein mixtures in the presence of Ca^{2+} ions is influenced by the carbohydrate content of $\varkappa$-casein.

γ-*Caseins*. These proteins are degradation products of the β-caseins, formed by milk proteases, e.g., γ_1-casein is obtained by cleavage of the residues 1–28. The peptide released is identical to the proteose-peptone PP8F which has been found in milk. Correspondingly, γ_2- and γ_3-caseins are formed by hydrolysis of the amino acid residues 1–105 and 1–107

Table 10.10. Chymosin specificity: relative rate of hydrolysis of $\varkappa$-casein and peptides

Substrate	k_{rel}
$\varkappa$-Casein	
105 106 169	
(Pyg—Leu-Ser-Phe-Met-Ala—Val)	1
Phe-Met	0
Phe-Met-OMe	0
Z-Phe-Met-OMe	0
Leu-Ser-Phe-Met-Ala-OMe	0.001 to 0.01

respectively. According to more recent nomenclature recommendations, β-casein fragments should be described by the position numbers. Thus, γ_1-casein from any β-casein variant X is called, e.g., β-casein X (f 29–209) and the corresponding proteose peptone PF8F β-casein X (f 1–28).

λ-*Caseins*. The λ-casein fraction consists mainly of fragments of the α_{s1}-caseins. In vitro the λ-caseins are formed by incubation of the α_{s1}-caseins with bovine plasmin.
The molar ratio of the main components $\alpha_{s1}/\beta + \gamma/\varkappa/\alpha_{s2}$ is on an average 8/8/3/2. All casein forms contain phosphoric acid, which always occurs in a tripeptide sequence pattern (Pse = phosphoserine):

$$\text{Pse-X-Glu} \quad \text{or} \quad \text{Pse-X-Pse} \qquad (10.5)$$

in which X is any amino acid, including phosphoserine and glutamic acid. Examples are:

$$
\begin{array}{ll}
\alpha_{s1}\text{-Casein:} & \text{Pse-Glu-Pse} \\
 & \text{Pse-Ile-Pse-Pse-Pse-Glu} \\
 & \text{Pse-Val-Glu} \\
 & \text{Pse-Ala-Glu} \qquad (10.6) \\
\beta\text{-Casein:} & \text{Pse-Leu-Pse-Pse-Pse-Glu} \\
 & \text{Pse-Glu-Glu} \\
\varkappa\text{-Casein:} & \text{Pse-Pro-Glu} \\
\end{array}
$$

Most probably this regular pattern originates from the action of a specific protein kinase.
The various distribution of polar and apolar groups of the individual proteins outlined above are summarized in Table 10.11. The hydrophobicity values listed are average hydrophobicity values $\bar{H}$ of the amino acid side chains present in the sequence of the given segments, and are calculated as follows:
A measure of the hydrophobicity of a compound is the free energy, F_t, needed to transfer the compound from water into an organic solvent, and is given as the ratio of the compound's solubility in water (N_w, as mole fraction) and in the organic solvent (N_{org}, as mole fraction), involving the activity coefficients (γ_w, γ_{org}):

$$\Delta F_t = RT \ln \frac{N_w \cdot \gamma_w}{N_{org} \cdot \gamma_{org}} \qquad (10.7)$$

Table 10.11. Distribution of amino acid residues with ionizing side chains (net charge) and with non-polar side chains (hydrophobicity) in α_{s1}-casein and β-casein

Residue	α_{s1}-Casein		Residue	β-Casein	
	1	2		1	2
1–40	+3	1,340	1–43	−16	783
41–80	−22.5	641	44–92	−3.5	1,429
81–120	0	1,310	93–135	+2	1,173
121–160	−1	1,264	136–177	+3	1,467
161–199	−2.5	1,164	178–209	+2	1,738

1 Net charge,
2 Hydrophobicity $\bar{H}$ (Cal/mole; cf. text).

The corresponding free energy of transfer of the side chain of an amino acid $H\Phi_i$ is obtained from the following relationship:

$$H\Phi_i = \Delta F_t \text{ (amino acid i)} - \Delta F_t \text{ (glycine)}$$

The average hydrophobicity of a sequence segment of a polypeptide chain with n amino acid residues is then:

$$\bar{H} = \frac{\Sigma H\Phi_i}{n} \qquad (10.8)$$

The higher the $H\Phi_i$, i.e. $\bar{H}$, the higher is the hydrophobicity of individual side chains, i.e. the sequence segment. Data provided in Table 10.11 are related to the ethanol/water system.

10.1.2.1.2 Micelle Formation

Only up to 10% of the total casein fraction is present as monomers. They are usually designated as serum caseins and the concentration ratio $c_\beta > c_\varkappa > c_{\alpha s1}$ is quite valid. However, the main portion is aggregated to casein complexes and casein micelles. This aggregation is regulated by a set of parameters, as presented in Fig. 10.3. Dialysis of casein complexes against a chelating agent might shift the equilibrium completely to monomers, while against high Ca^{2+} ion concentrations the shift would be to large micelles.

From Fig. 10.4 it follows that the diameter of the micelles in skim milk varies from 50–300 nm, with a particle distribution peak at

Monomers (soluble caseins)

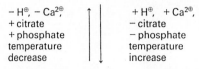

Casein complex

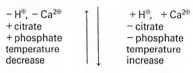

Micella
(calcium caseinate + calcium phosphate)

Fig. 10.3. Casein complex and casein micelle formation

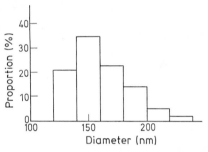

Fig. 10.4. Particle size distribution of casein micelles in skim milk (fixation with glutaraldehyde)

150 nm. Using an average diameter of 140 nm, the micelle volume is 1.4×10^6 nm^3 and the particle weight is 10^7–10^9 dal. This corresponds to 25,000 monomers per micelle. Casein micelles are substantially smaller than fat globules, which have diameters between 0.1–10 µm. Scanning electron micrographs of micelles are shown in Fig. 10.5 and compositional data are provided in Table 10.12.
The ratio of monomers in micelles varies to a great extent (Table 10.13), depending on dairy

Table 10.12. Composition of casein micelles (%)

Casein	93.2	Phosphate	
Ca	2.9	(organic)	2.3
Mg	0.1	Phosphate	
Na	0.1	(inorganic)	2.9
K	0.3	Citrate	0.4

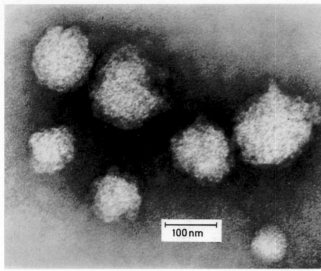

Fig. 10.5. Electron micrograph of the casein micelles in skim milk (according to *Webb*, 1974). The micelles are fixed with glutaraldehyde and then stained with phosphomolybdic acid

Table 10.13. Typical distribution of components in casein micelles

Component	Ratio numbers			
α_{s1}	3	6	9	12
β	1	1	4	4
γ		1	1	1
$\varkappa$	1	3	3	3

Table 10.14. Composition and size of casein micelles isolated by centrifugation

Centrifugation time (min)[a]	Composition of the sediment (%)			
	α_{s1}	β	$\varkappa$	Others
0[b]	50	32	15	3
0–7.5	47	34	16	3
7.5–15	46	32	18	4
15–30	45	31	20	4
39–60	42	29	26	3
Serum casein	39	23	33	5

[a] Centrifugation speed $10^5 \times g$.
[b] Isoelectric casein.

cattle breed, season and fodder, and is influenced also by micellular size (Table 10.14). The micelles are not tightly packed and so are of variable density. They are strongly solvated (1.9 g water/g protein) and hence are porous. The monomers are kept together with:

- Hydrophobic interactions that are minimal at a temperature less than 5 °C.
- Electrostatic interactions, mostly as calcium or calcium phosphate bridges between phosphoserine and glutamic acid residues (Fig. 10.6).
- Hydrogen bonds.

On a molecular level different micelle models have been proposed which to a certain extent explain the experimental findings. The most probable model is shown in Fig. 10.7. This model comprises subunits (submicelles, M_r ~ 760,000) which consist of ca. 30 different casein monomers and aggregate to large micelles via calcium phosphate bridges. Two types of subunits apparently exist: one type contains $\varkappa$-casein and the other does not. The $\varkappa$-casein molecules are arranged on the surface of the corresponding submicelles. At various positions, their hydrophilic C-termini protrude like hairs from the surface, preventing aggregation. Indeed, aggregation of the submicelles proceeds until the entire surface of the forming micelle is covered with $\varkappa$-casein, i.e., covered with "hair", and, therefore, exhibits steric repulsion. The effective density of the hair layer is at least 5 nm. A small part of the $\varkappa$-casein is also found inside the micelle.

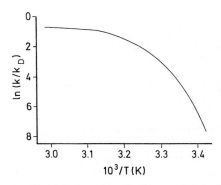

(Structure: P E P T I D E ... chain with calcium ion bridging)

$$-CH_2-O-\overset{\overset{\displaystyle OH}{|}}{\underset{\underset{\displaystyle O}{\|}}{P}}-O^{\ominus}Ca^{\oplus\oplus}\left[^{\ominus}O-\overset{\overset{\displaystyle OH}{|}}{\underset{\underset{\displaystyle O}{\|}}{P}}-O^{\ominus}Ca^{\oplus\oplus}\right]_{o-n}{}^{\ominus}O-\overset{\overset{\displaystyle OH}{|}}{\underset{\underset{\displaystyle O}{\|}}{P}}-O-CH_2-$$

Fig. 10.6. Peptide chain bridging with calcium ions

10.1.2.1.3 Gel Formation

The micelle system, can be destabilized by the action of rennin or souring. Rennin attacks ϰ-casein, eliminating not only the C-terminus in the form of the soluble glycopeotide 106–169, but also the cause of repulsion. The remaining paracasein micelles first form small aggregates with an irregular and often long form, which then assemble with gel formation to give a three dimensional network with a pore diameter of a few μm. The fat globules present are included in this network with pore enlargement. It is assumed that dynamic equilibria exist between casein monomers and submicelles, dissolved and bound calcium phosphate, and submicelles and micelles.

The rate of gel formation increases with increasing temperature (Fig. 10.8). It is slow at $T < 25\,^\circ C$ and proceeds almost under diffusion control at $T \sim 60\,^\circ C$. It follows that hydrophobic interactions, especially due to the very hydrophobic para-ϰ-casein remaining on the surface after the action of rennin, are the driving force for gel formation. In addition, other temperature-dependent reactions play a role, like the binding of calcium ions and of β-casein to the micelles, and the change in solubility of colloidal calcium phosphate.

Hydrophilic shell
Hydrophobic core

COO^-
COO^-
$PO_4^=$
$PO_4^=$
COO^-
$PO_4^=$
COO^-
COO^-
COO^-
$PO_4^=$

a

b

○ Submicele
)~ Protruding chain
— Calcium phosphate

Fig. 10.7. Schematic model of a casein micelle; (**a**) a subunit consisting of α_{s1}-, β-, γ-, ϰ-caseins, (**b**) Micelle made of subunits bound by calcium phosphate bridges. (according to *Webb*, 1974)

Fig. 10.8. Temperature dependency of the aggregation rate of para-casein micelles (rate constant k in fractions of the diffusion-controlled rate k_D; according to *Dalgleish*, 1983)

(Graph: y-axis $\ln(k/k_D)$ from 0 to 8; x-axis $10^3/T$ (K) from 3.0 to 3.4)

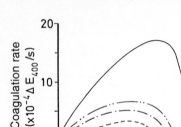

Fig. 10.9. Rate of coagulation of casein micelles as a function of temperature and pH value (— 25°C, —··— 15°C, —·— 10°C, − − − 5°C, according to *Bringe, Kinsella*, 1986)

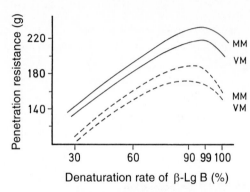

Fig. 10.10. Firmness of yoghurt as a function of the rate of denaturation of β-lactoglobulin B (the final value of the penetration resistance of a conical test piece in stiff yoghurt is given; heating temperature 85°C: —, 130°C: − − −, WM: whole milk with 3.5% fat, SM: skim milk; according to *Kessler* 1988)

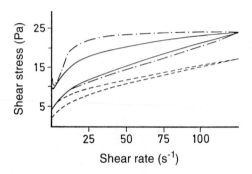

Fig. 10.11. Flow curves of stiff skim-milk yoghurt subjected to defined prestirring as a function of the rate of denaturation of β-lactoglobulin B (temperature/time/denaturation rate 90°C/2.2 s/10%: − − −, 90°C/21 s/60%: —, 90°C/360 s/99%: —··—; according to *Kessler*, 1988)

Acid coagulation of casein is also definitely caused by hydrophobic interactions, as shown by the dependency of the coagulation rate on the temperature and pH value (Fig. 10.9). On acidification, the micelle structure changes due to the migration of calcium phosphate and monomeric casein. Since the size of the micelle remains practically constant, this migration of components must be associated with swelling. During coagulation, dissolved casein reassociates with the micelles, forming a gel network.

The gel structure can be controlled via changes in the hydrophobicity of the micelle surface. A decrease in hydrophobicity is possible, e.g., by heating milk (90°C/10 min). Covalent bonding of denatured β-lactoglobulin to ϰ-casein (cf. 10.1.3.5) occurs, burying hydrophobic groups. Due to weaker interactions, stable, rigid gels with a chain-like structure are formed on acidification. These gels exhibit no syneresis and are desirable, e.g., in yoghurt (10.2.1.2). Figure 10.10 shows that the firmness of stiff yoghurt is highest when the denaturation of β-lactoglobulin is 90–99%. If this rate of denaturation is achieved at lower temperatures (e.g., 85°C), gels are formed that are more rigid and coarser than those formed by heating to higher temperatures (e.g., 130°C), which results in a soft, smooth gelatinous mass. The gel stability of whole-milk yoghurt is lower than that of skim-milk yoghurt because the protein network is interrupted by included fat globules.

Flow curves of skim-milk yoghurt as a function of the rate of denaturation of β-lactoglobulin are presented in Fig. 10.11. The yield point is a measure of the elastic properties of the gel and the area enclosed by the hysteresis loop is a measure of the total energy required to destroy the gel. Both parameters increase with increasing rate of denaturation, which is a sign of increasing gel stability.

In contrast to yoghurt production, syneresis of the gel is desirable in the production of cottage cheese, so that the typical texture is attained. For this reason, the milk is only slightly heat treated and the surface hydrophobicity is increased by the addition of chymosin before acidification.

10.1.2.1.4 Whey Proteins

β-*Lactoglobulin* occurs in genetic variants A, B and C of the Jersey dairy cattle breed, and variant D of the Montbeliarde dairy cow. Two other A_{Dr} and B_{Dr} variants of Australian drought master cows are identical to variants A and B apart from the carbohydrate content. Table 10.9 shows the corresponding changes in the amino acid composition and Table 10.8 shows the amino acid sequence of β-lactoglobulin.

The monomeric β-lactoglobulin, with a molecular weight of 18 kdal, shows a reversible, pH-dependent oligomerization, as represented by the equation:

$$A \rightleftharpoons A_2 \rightleftharpoons (A_2)_4 \rightleftharpoons A_2 \rightleftharpoons A$$

$$\text{pH} < 3.5 \qquad 3.7 < \text{pH} < 5.1 \qquad \text{pH} > 7.5$$

$$(10.9)$$

Hence, the monomer is stable only at a pH less than 3.5 or above 7.5. The octamer occurs with variant A, but not with variants B and C. Irreversible denaturation occurs at a pH above 8.6 as well as by heating or at higher levels of Ca^{2+} ions. β-Lactoglobulin contains one thiol group that in the native protein is buried within the molecule. This group is exposed on partial denaturation and can participate either in protein dimerization via disulfide bridge formation or in reactions with other milk proteins, especially with ϰ-casein and α-lactalbumin, which proceed during the heating of milk.

α-*Lactalbumin*. This protein exists in two genetic forms, A and B (Gln → Arg). Its amino acid sequence, which is similar to that of lysozyme, has been elucidated. α-Lactalbumin has a biological function since it is the B subunit of the enzyme lactose synthetase. The enzyme subunit A is a nonspecific galactosyltransferase; the subunit B promotes selective galactose transfer to glucose in lactose biosynthesis.

10.1.2.2 Carbohydrates

The main sugar in milk is lactose, an O-β-D-galactopyranosyl-(1 → 4)-D-glucopyranose, which is 4−6% of milk.

The most stable form is α-lactose monohydrate, $C_{12}H_{22}O_{11} \cdot H_2O$. Lactose crystallizes in this form from a supersaturated aqueous solution at T < 93.5 °C. The crystals may have a prism- or pyramid-like form, depending on conditions. Vacuum drying at T > 100 °C yields a hygroscopic α-anhydride. Crystallization from aqueous solutions above 93.5 °C provides water-free β-lactose (β-anhydride, cf. Formula 10.10). Rapid drying of a lactose solution, as in milk powder production, gives a hygroscopic and amorphous equilibrium mixture of α- and β-lactose.

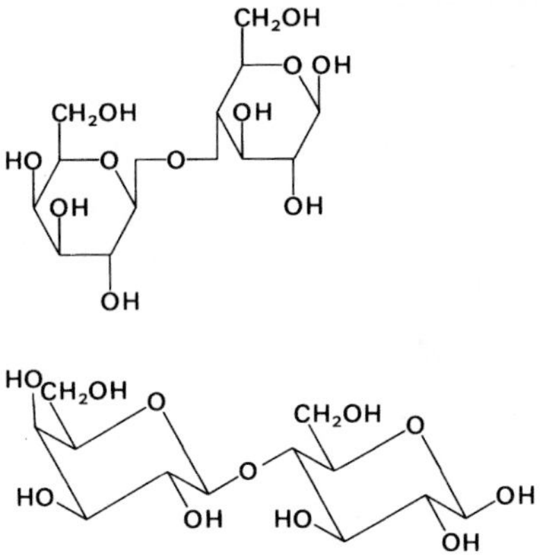

$$(10.10)$$

β-Lactose

Some physical data of lactose are summarized in Table 10.15. The ratio of anomers is temperature dependent. As temperature increases, the β-form decreases. The mutarotation rate is temperature ($Q_{10} = 2.8$) and pH dependent (Fig. 10.12). The rise in mutarotation rate at pH < 2 and pH > 7 originates from the rate-determining step of ring opening, which is catalyzed by both H^+ and OH^- ions:

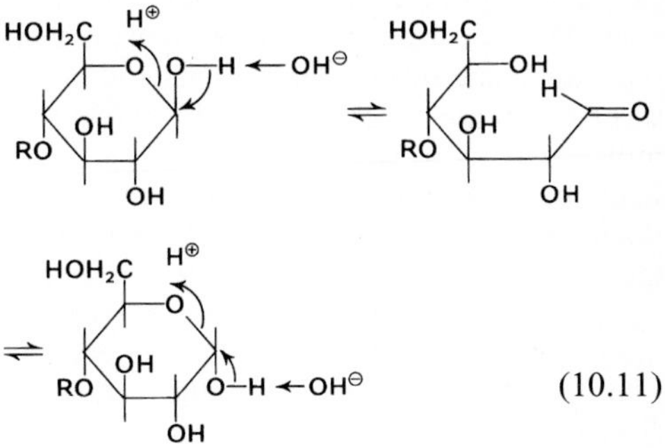

$$(10.11)$$

Table 10.15. Some physical characteristics of lactose

	α-Lactose	β-Lactose	Equilibrium mixture
Melting point (°C)	201.6[a]	252.2	
Spec. rotation $[\alpha]_D^{20}$	89.4	35.0	

Equilibrium in aqueous solution[b]		
0 °C	1.00	1.80
20 °C	1.00	1.68
50 °C	1.00	1.63

Solubility in water[c]			
0 °C	5.0	45.1	11.9
25 °C	8.6		21.6
39 °C	12.6		31.5
100 °C	70	94.7	157.6

[a] Hydrate. [b] Relative concentration.
[c] g Lactose/100 g water.

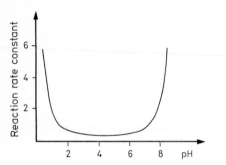

Fig. 10.12. Mutarotation rate of lactose as affected by pH

The great solubility difference between the two anomers is noteworthy. The sweetness of lactose is significantly lower than that of fructose, glucose or sucrose (Table 10.16). Glucose and some other amino sugars and oligosaccharides are present in small amounts in milk.

Lactulose is found in heated milk products. It is a little sweeter and clearly more soluble than lactose. For example, condensed milk contains up to 1% of lactulose, corresponding to an isomerization of ca. 10% of the lactose present. The formation proceeds via the *Lobry de Bruyn-van Ekenstein* rearrangement (cf.

Table 10.16. Relative sweetness of saccharose, glucose, fructose and lactose[a]

Saccharose	Glucose	Fructose	Lactose
0.5	0.9	0.4	1.9
5.0	8.3	4.2	15.7
10.0	12.7	8.7	20.7
20.0	21.8	16.7	33.3

[a] Results are expressed as concentration % for isosweet aqueous sugar solutions.

4.2.4.3.2) or via *Schiff* base. Traces of epilactose (4-O-β-D-glactopyranosyl-D-mannose) are also formed on heating milk.

10.1.2.3 Lipids

The composition of milk fat is presented in Table 10.17. Milk fat contains 95–96% triglycerols. Its fatty acid composition is given in Table 10.18. The relatively high content of low molecular weight fatty acids, primarily of butyric acid, is characteristic of milk. The content of unsaturated acids varies with season and fodder. The level can be increased by incorporating highly unsaturated fats, in protected encapsulated forms, into feed. The disadvantage of such a nutritionally/physiologically interesting approach is the changed physico-chemical properties of the dairy product, e.g., an increased susceptibility to oxidation and the formation of unsaturated lactones (γ-dodec-cis-6-enolactone from linoleic acid) which influences the flavor of milk and meat. In addition to the main straight-chain fatty

Table 10.17. Milk lipids

Lipid fraction	Percent of the total lipid
Triacylglycerols	95–96
Diacylglycerols	1.3–1.6
Monoacylglycerols	0.02–0.04
Keto acid glycerides	0.9–1.3
Hydroxy acid glycerides	0.6–0.8
Free fatty acids	0.1–0.4
Phospholipids	0.8–1.0
Sphingolipids	0.06
Sterols	0.2–0.4

Table 10.18. Fatty acid composition of milk fat[a]

Fatty acid	Weight-%
Saturated, straight chain	
Butyric acid	2.79
Caproic acid	2.34
Caprylic acid	1.06
Capric acid	3.04
Lauric acid	2.87
Myristic acid	8.94
Pentadecanoic acid	0.79
Palmitic acid	23.8
Heptadecanoic acid	0.70
Stearic acid	13.2
Nonadecanoic acid	0.27
Arachidic acid	0.28
Behenic acid	0.11
Saturated, branched chain	
12-Methyltetradecanoic acid	0.23
13-Methyltetradecanoic acid	0.14
14-Methylpentadecanoic acid	0.20
14-Methylhexadecanoic acid	0.23
15-Methylhexadecanoic acid	0.36
3,7,11,15-Tetramethylhexadecanoic acid	0.12–0.18
Unsaturated	
9-Decenoic acid	0.27
9-cis-Tetradecenoic acid	0.72
9-cis-Hexadecenoic acid	1.46
9-cis-Heptadecenoic acid	0.19
8-cis-Octadecenoic acid	0.45
Oleic acid	25.5
11-cis-Octadecenoic acid	0.67
9-trans-Octadecenoic acid	0.31
10-trans-Octadecenoic acid	0.32
11-trans-Oxtadecenoic acid	1.08
12-trans-Octadecenoic acid	0.12
13-trans-Octadecenoic acid	0.32
14-trans-Octadecenoic acid	0.27
15-trans-Octadecenoic acid	0.21
16-trans-Octadecenoic acid	0.23
Linoleic acid	2.11
Linolenic acid	0.38

[a] Only acids with a content higher than 0.1% are listed.

Table 10.19. Melting characteristics of butterfat

Temperature (°C)	Solid content (%)	Temperature (°C)	Solid content (%)
5	43–47	30	6–8
10	40–43	35	1–2
20	21–22	40	0

Table 10.20. Membrane composition of milkfat globules

Constituent	Proportion (%)
Protein	41
Phospho- and glycolipids	30
Cholesterol	2
Neutral glycerides	14
Water	13

Butterfat melting properties, as affected by season and fodder, are listed in Table 10.19.

Milk fat is very finely distributed in plasma. The diameter of fat globules is 0.1–10 μm. During homogenization, milk at 50–75 °C is forced through small passages under pressure of 25–350 bar, the diameter of the globules lowers to <1 μm, depending on the pressure. The fat droplets are surrounded by a membrane that consists of phospholipids and a double layer of proteins and accounts for about 2% of the total mass of the droplet. The layer thickness is on average 8–9 nm, but is not uniform. Membrane compositional data are given in Table 10.20.

Membrane proteins are specific in nature. Casein proteins enter and participate in membrane formation when the fat globule surface area is expanded 4- to 6-fold during homogenization of milk. In addition to lipoproteins, the membrane also contains enzymes, including a very active lipoprotein lipase (a glycoprotein, 8.3% carbohydrates, molecular weight 48.3 kdal). However, if the milking and storage procedures are appropriate, the raw milk can be kept for several days without the development of a rancid off-flavor. It is likely that the membranes of the fat globules prevent lipolysis. Disruption of the organized structure of the membrane, for instance by homogenization, allows the lipase to bind to the fat globu-

acids, small amounts of odd-C-number, branched-chain and oxo-fatty acids (cf. 3.2.1.3) are present.

Phospholipids are 0.8–1.0% of milk fat and sterols, mostly cholesterol, are 0.2–0.4%.

les and to hydrolyze the triacylglycerols at a high rate (1 µmole fatty acid per min per ml milk, pH 7, 37 °C). The milk becomes unpalatable within a few minutes. Therefore the lipoprotein lipase has to be inactivated by pasteurization prior to milk homogenization.

The small amounts of gangliosides that occur in milk (5.6 µmol/l, calculated as ganglioside-bound sialic acid) are of interest for the analytical differentiation of skim-milk and butter-milk powder. As structural elements of the membrane of the fat globules, the cream gets enriched with gangliosides during skimming and only about 8% remain in the skimmed milk. During butter-making, the membrane of the fat globules is mechanically destroyed and the highly polar gangliosides pass almost completely into the buttermilk. Therefore, unlike skim-milk powder, butter-milk powder is rich in gangliosides (ca. 480 µmol/kg, calculated as sialic acid).

10.1.2.4 Organic Acids

Citric acid (1.8 g/l) is the predominant organic acid in milk. During storage it disappears rapidly as a result of the action of bacteria. Other acids (lactic, acetic) are degradation products of lactose. The occurrence of orotic acid (73 mg/l), an intermediary product in biosynthesis of pyrimidine nucleotides, is specific for milk:

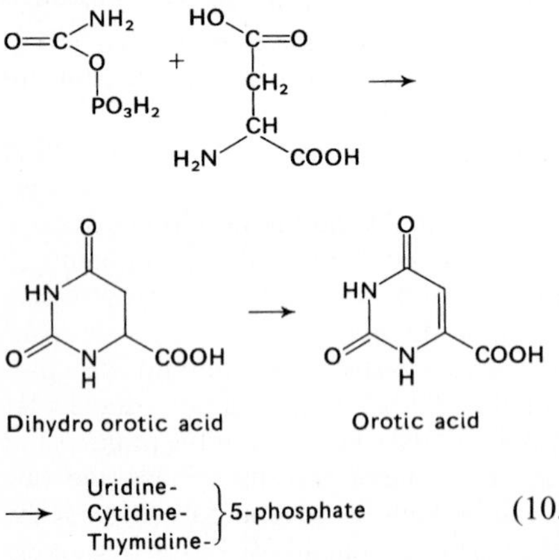

Dihydro orotic acid   Orotic acid

$$\longrightarrow \left. \begin{array}{l} \text{Uridine-} \\ \text{Cytidine-} \\ \text{Thymidine-} \end{array} \right\} \text{5-phosphate} \qquad (10.12)$$

Orotic acid as well as total creatinine and uric acid are suitable indicators for the determination of the proportion of milk in foods. The average values for whole-milk and skim-milk powder given in Table 10.21 can serve as reference values.

10.1.2.5 Minerals

Minerals, including trace elements, in milk are compiled in Table 10.22.

10.1.2.6 Vitamins

Milk contains all the vitamins in variable amounts (Table 10.23). During processing, the fat-soluble vitamins are retained by the cream, while the water-soluble vitamins remain in skim milk or whey.

10.1.2.7 Enzymes

Milk contains a great number of enzymes which are not only of analytical importance for the detecton of heat-treated milk, but can also

Table 10.21. Indicators for the proportion of milk in foods

Compound	Whole milk powder	Skim milk powder
Orotic acid		
photometric	50.6	66.4
polarographic	46.6	58.1
Total creatinine	66.3	84.4
Uric acid	12.4	15.3

Expressed as mg/100 g solids.

Table 10.22. Mineral composition of milk

Constituent	mg/l	Constituent	µg/l
Potassium	1,500	Zinc	4,000
Calcium	1,200	Aluminum	500
Sodium	500	Iron	400
Magnesium	120	Copper	120
		Molybdenum	60
Phosphate	3,000	Manganese	30
Chloride	1,000	Nickel	25
Sulfate	100		
		Silicon	1,500
		Bromine	1,000
		Boron	200
		Fluorine	150
		Iodine	60

Table 10.23. Vitamin content of milk

Vitamin	mg/l	Vitamin	mg/l
A (Retinol)	0.4	Nicotinamide	1
D (Calciferol)	0.001	Pantothenic acid	3.5
E (Tocopherol)	1.0	C (Ascorbic acid)	20
B_1 (Thiamine)	0.4	Biotin	0.03
B_2 (Riboflavin)	1.7	Folic acid	0.05
B_6 (Pyridoxine)	0.6		
B_{12} (Cyanocobal-amine)	0.005		

Table 10.24. Enzymes in bovine milk

EC Number	Name	Localization[a]
1.1.1.27	L-Lactate dehydrogenase	P
1.1.1.37	Malate dehydrogenase	
1.1.3.22	Xanthine oxidase	F
1.4.3.6	Amine oxidase (copper-containing)	
1.6.99.3	NADH dehydrogenase	F
1.8	Sulfhydryl oxidase[b]	S
1.8.1.4	Dihydrolipoamide dehydrogenase	F
1.11.1.6	Catalase	L
1.11.1.7	Peroxidase	S
1.15.1.1	Superoxide dismutase	
2.3.2.2	γ-Glutamyltransferase	F
2.4.1.22	Lactose synthase	S
2.4.99.1	β-Galactoside-α-2,6-sialyltransferase	
2.6.1.1	Aspartate aminotransferase	P
2.6.1.2	Alanine aminotransferase	
2.7.1.26	Riboflavin kinase	
2.7.1.30	Glycerol kinase	
2.7.7.2	FMN adenyl transferase	
2.8.1.1	Thiosulfate sulfurtransferase	
3.1.1.1	Carboxylesterase	S
3.1.1.2	Arylesterase	
3.1.1.7	Acetylcholine esterase	F
3.1.1.8	Choline esterase	S
3.1.1.34	Lipoproteinlipase	C
3.1.3.1	Alkaline phosphatase	F
3.1.3.2	Acid phosphatase	F
3.1.3.5	5'-Nucleotidase	F
3.1.3.9	Glucose-6-phosphatase	F
3.1.3.16	Phosphoprotein phosphatase	P
3.1.4.1	Phosphodiesterase I	F
3.1.27.5	Pancreatic ribonuclease	S
3.2.1.1	α-Amylase	S
3.2.1.2	β-Amylase	
3.2.1.17	Lysozyme	S
3.2.1.24	α-Mannosidase	
3.2.1.31	β-Glucuronidase	
3.2.1.52	β-N-Acetylhexosaminidase	
3.4	Peptidases	
3.4.21.7	Plasmin	C
3.6.1.3	Adenosinetriphosphatase	F
3.6.1.9	Nucleotide pyrophosphatase	
4.1.2.13	Fructosebiphosphate aldolase	
5.3.1.9	Glucose-6-phosphate isomerase	

[a] C: casein micelle, F: fat-globule membrane, L: leucocytes. P: plasma, S: serum.
[b] Not thiol oxidase EC 1.8.3.2.

influence the processing properties. The rate of heat inactivation of the enzymes indicates the type and extent of heating (cf. 2.5.4 and Fig. 2.37).
Hydrolases identified include: amylases, lipases, esterases, proteinases and phosphatases. Proteinase inhibitors have also been found. Important oxidoreductases in milk are aldehyde dehydrogenase (xanthine oxidase), lactoperoxidase and catalase. A general idea of the occurrence and localization of enzymes in bovine milk is given in Table 10.24.

10.1.3 Processing of Milk

Only a small amount of milk is sold to the consumer without processing (certified raw milk). The main part is subjected to a processing procedure shown schematically in Fig. 10.13.

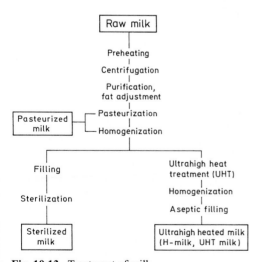

Fig. 10.13. Treatment of milk

10.1.3.1 Purification

The milk is usually delivered in the tank of a milk truck. For purification, it is fed into a clarifier (self-cleaning disk separator) via a deaerating vessel. These separators can process either cold or warm milk (40 °C) at speeds of 4500–8400 rpm with throughput capacities of up to 50,000 l/h.

10.1.3.2 Creaming

After heating to about 40°C (increase in creaming efficiency by lowering the viscosity), the milk is separated into cream and skimmed milk in a cream separator. Cream separators have a nominal capacity of up to 25,000 l/h at speeds of 4700–6500 rpm. The fat content of the milk can be standardized by careful back-mixing.

10.1.3.3 Heat Treatment

The fluid milk is heated to improve its durability and to kill disease-causing microorganisms. Heat treatments used are (cf. Fig. 10.14):

- Pasteurization.
  The milk is treated: at high temperature (85°C for 2–3 s) in a short-time, flash process (72–75°C/15–30 s) in plate heaters; or by the low temperature or holder process, in which it is heated at 62–65°C for at least 30–32 min, with stirring, and is then cooled.
- Ultrahigh Temperature (UHT) Treatment.
  The process involves indirect heating by coils or plates at 136–138°C for 5–8 s, or direct heating by live steam injection at 140–145°C for 2–4 s, followed by aseptic packaging.
  To prevent dilution or concentration of the milk, the amount of injected steam must be controlled in such a way that it corresponds to the amount of water withdrawn during expansion under vacuum.
- Bactotherm Process.
  This is a combination of centrifugal sterilization in bactofuges (65 to 70°C) and UHT heating of the separated sediment (2–3% of the milk), followed by recombination. Since the total amount of milk is not heated in this process, the taste is improved. The storability is ca. 8–10 days.
- Sterilization.
  Milk in retail packages is heated in autoclaves at 107–115°C/20–40 min, 120–130°C/8–12 min.

10.1.3.4 Homogenization

Homogenization is conducted to stabilize the emulsion milk by reducing the size of the fat

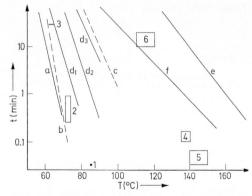

Fig. 10.14. Heating of milk. *1–3* Pasteurization: 1 high temperature treatment, 2 short time and 3 long time heat treatment; *4* and *5* UHT treatment: 4 indirect and 5 direct; *6* sterilization. a: Killing pathogenic microorganisms (*Tubercle bacilli* as labelling organism), b/c: inactivation of alkaline/acid phosphatase. d_1, d_2, d_3 denaturation (5, 40, 100%) of whey proteins. e: casein heat coagulation, f: start of milk browning

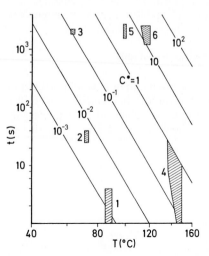

Fig. 10.15. Chemical reactions in heat-treated milk. ("Chemical effect" C* = 1: losses of approx. 3% thiamine and approx. 0.7% lysine and formation of approx. 0.8 mg/l HMF); commonly used heat treatments: 1 high heat, 2 short time heating, 3 prolonged heating, 4 UHT treatment, 5 boiling, 6 sterilization (according to *Kessler*, 1983). HMF: Hydroxymethylfurfural

globules. This is achieved by high-pressure homogenization (up to 350 bar, 50–75°C). In principle, the high-pressure homogenizer is a high-pressure pump which presses the product through a homogenizing valve. The fat globules are reduced in size by turbulence, cavitation, and shear forces.

10.1.3.5 Reactions During Heating

Heat treatment affects several milk constituents. Casein, strictly speaking, is not a heat-coagulable protein; it coagulates only at very high temperatures (cf. Fig. 10.14). Heating at 120 °C for 5 h dephosphorylates sodium or calcium caseinate solutions (100% and 85%, respectively) and releases 15% of the nitrogen in the form of low molecular weight fragments.

However, temperature and pH strongly affect casein association and cause changes in micellular structure (cf. 10.1.2.1.2 and 10.1.2.1.3). An example of such a change is the pH-dependent heat coagulation of skim milk. The coagulation temperature drops with decreasing pH (Fig. 10.16 and 10.9). Salt concentration also has an influence, e. g., the heat stability of milk decreases with a rise in the content of free calcium.

All pasteurization processes supposedly kill the pathogenic microorganisms in milk. The inactivation of the alkaline phosphatase is used in determining the effectiveness of pasteurization. At higher temperatures or with longer heating times, the whey proteins start to denature – this coincides with the complete inactivation of acid phosphatase. Denatured whey proteins, within the pH-range of their isoelectric points, cease to the soluble and

coagulate together with casein due to souring or chymosin action of the milk. Such coprecipitation of the milk proteins is of importance in some milk processing (as in cottage cheese production). The thermal stability of whey proteins is illustrated in Fig. 10.17.

Heat treatment of milk activates thiol groups; e.g., a thiol-disulfide exchange reaction occurs between $\varkappa$-casein and β-lactoglobulin. This reaction reduces the vulnerability of $\varkappa$-casein to chymosin, resulting in a more or less strong retardation of the rennet coagulation of heated milk.

Further changes induced by heating of milk are:

- Calcium phosphate precipitation on casein micelles.
- *Maillard* reactions between lactose and amino groups (e. g. lysine) which, in a classical sterilization process, causes browning of milk and formation of hydroxymethyl furfural (HMF).
- δ-Lactone and methyl ketone formation from glycerides esterified with hydroxy- or ketofatty acids.
- Thiamine degradation.

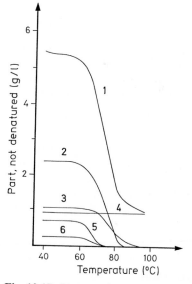

Fig. 10.17. Denaturation of whey proteins by heating at various temperatures for 30 min. 1 Total whey protein, 2 β-lactoglobulin, 3 α-lactalbumin, 4 proteose peptone, 5 immunoglobulin, 6 serum albumin

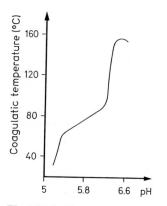

Fig. 10.16. Thermal coagulation of skim milk

- Changes in membranes of milk fat globules, which affect the cream separation property of the globules.

Detailed studies have shown that the rate of several reactions which occur during heating of milk, e.g., thiamine and lysine degradations, formation of HMF and nonenzymic browning, can be calculated over a great temperature-time range (including extended storage) by application of a second-order rate law. Assuming an average activation energy of $E_a = 102$ kJ/mole, a "chemical effect" $C^* = 1$ has been defined which gives a straight line in a $\log t$ vs. T^{-1} diagram, from which the thiamine loss is seen to be approx. 0.8 mg/l (Fig. 10.15). Other lines in Fig. 10.15 represent a power of ten of heat treatment and chemical reactions ($C^* = 10^{-1}, 10^{-2}, ...,$ or $10^1, 10^2, ...$). The pigments formed in a browning reaction become visible only in the range of $C^* = 10$.

Quality deterioration in the form of nutritional degradation, changes in color or development of off-flavor have also been predicted for other foods by application of a suitable mathematical model. In most cases the loss of quality fits a zero- or first-order rate law. Knowledge of the rate constant allows one to predict the extent of reaction for any time.

The influence of temperature on the reaction rate follows the *Arrhenius* equation (cf. 2.5.4). Thus by studying a reaction and measuring the rate constants at two or three high temperatures, one could then extrapolate with a straight line to a lower temperature and predict the rate of the reaction at the desired lower temperature. However, these data allow only a prediction of the shelf life when the physical and chemical properties of the components of a food do not alter with temperature. For example, as temperature rises a solid fat goes into a liquid state. The reactants may be mobile in the liquid fat and not in the solid phase. Thus, shelf life will be underestimated for the lower temperature.

10.1.4 Types of Milk

Milk is consumed in the following forms:

- *Raw fluid milk* (high quality milk), which has to comply with strict hygienic demands.

- *Whole milk* is heat-treated and contains at least 3% fat. It can be a standardized whole milk adjusted to a predetermined fat content, in which case the fat content has to be at least 3.5%.
- *Low-fat milk* is heat-treated and the cream is separated. The fat content is 1.5–2%.
- *Skim milk* is heat-treated and the fat content is less than 0.3%.
- *Reconstituted milk* is most common in regions where milk production is not feasible (e.g. many Japanese cities). It is made from whole or skim milk powders to which cream or butter fat is added.
- *Filled* or *imitation milk* is a fat-substituted milk produced from nonfat milk solids by addition of vegetable fats (coconut, safflower or corn oil).
- *Toned milk* is a blend of a fat-rich fresh milk and reconstitued skim milk in which the nonfat solids are "toned up". Addition of water "tones down" the fat and nonfat solids.

10.2 Dairy Products

Milk processing is illustrated schematically in Fig. 10.18.

10.2.1 Fermented Milk Products

All sour milk products have undergone fermentation, which can involve not only lactic-acid bacteria, but also other microorganisms, e.g., yeasts. To the lactic acid bacteria count the genera *Lactobacillus*, *Lactococcus*, *Leuconostoc*, and *Pediococcus*. The most important species are presented in Table 10.25.

Depending on the microorganisms involved, fermentation proceeds via the glycolysis pathway with the almost exclusive formation of lactic acid (homofermentation), via the pentose phosphate pathway with formation of lactic acid, acetic acid (ethanol), and possibly CO_2 (heterofermentation) or via both pathways. These metabolic pathways are shown in Fig. 10.19. Organisms that are obligatorily homofermentative have fructose-bisphosphate

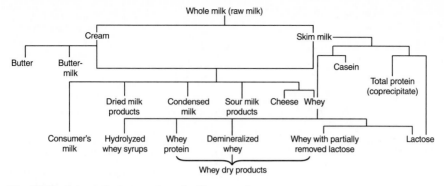

Fig. 10.18. Schematic presentation of milk processing

Table 10.25. Lactic acid bacteria

Microorganisms	L-Lactic acid[a] (%)	Remarks
Lactobacillus bulgaricus	0.6–4	thermophilic,
L. lactis	0	homofermentative,
L. leichmanii		D-, L- or
L. delbrueckii		D,L-Lactic acid
L. helveticus	70	
L. jugurti		
L. acidophilus	60	
L. casei subsp. casei		mesophilic,
L. casei subsp. alactosus		homofermentative,
L. casei subsp. pseudo		D-, L- or
plantarum		D,L-Lactic acid
L. casei subsp. rhamnosus		
L. casei subsp. fusiformis		
L. casei subsp. tolerans		
L. plantarum		
L. curvatus		
L. fermentum		heterofermentative,
L. cellobiosus		D,L-Lactic acid
L. brevii		
L. hilgardii		
L. vermiformis		
L. reuteri		
Streptococcus thermophilus	99	thermophilic,
		homofermentative
S. faecium		homofermentative
Lactococcus lactis		mesophilic,
subsp. lactis	92–99	homofermentativ
Lactococcus lactis		
subsp. cremoris	99	
Leuconostoc cremoris		heterofermentative,
L. mesenteroides		D-Lactic acid
L. dextranicum		
L. lactis		
Pediococcus acidilactici		thermophilic,
		homofermentative,
		D,L-Lactic acid

[a] Orientation values; the proportion of L-lactic acid depends on the bacterial strain and on the culture conditions.

aldolase, but not glucose 6-phosphate dehydrogenase and 6-phosphogluconate dehydrogenase. However, organisms that are obligatorily heterofermentative have both dehydrogenases, but no aldolase. Facultatively homofermentative organisms have all three enzymes and can use both metabolic pathways.

Apart from the type of fermentation, the configuration of the lactic acid formed also depends on the microorganisms involved. As shown in Table 10.25, both enantiomers are formed in varying amounts. Table 10.26 lists the content of total lactic acid and L-lactic acid in various dairy products.

In human metabolism, L-lactic acid is formed exclusively and only one L-lactate dehydrogenase is available. Therefore, the intake of larger amounts of D-lactic acid can result in

Table 10.26. Total lactic acid and L-lactic acid in some dairy products

Product	Total lactic acid (%)	L-Lactic acid (%)[b]
Sour milk	0.97	88
Buttermilk	0.86	87
Sour cream	0.86	96
Joghurt	1.08	54
Curd	0.59	94
Cottage cheese	0.34	92
Emmental	0.27	76
Sbrinz	1.53	58
Tilset cheese	1.27	52

[a] Average values.
[b] Based on total lactic acid.

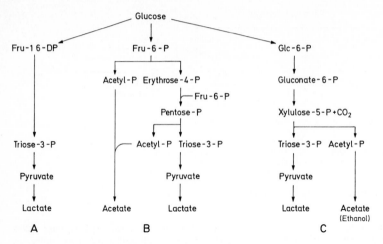

Fig. 10.19. Glucose metabolism in lactic acid bacteria. A: homofermentation, B: Bifidus pathway, and C: heterofermentation (6-phosphogluconate pathway)

enrichment in the blood and hyperacidity of the urine. For this reason, the WHO recommends a limitation of the intake of D-lactic acid to 100 mg per day and kg of body weight. Apart from the main products mentioned, various aroma substances are formed in the course of fermentation (cf. 10.3.3). In addition, proteolytic and lipolytic processes occur to a certain extent.

According to the consistency, a distinction is made between stiff, gel-like products, stirred, creamy products, and drinkable, flowable products. The thermal pretreatment of milk influences the rheological properties of the products as described in section 10.1.2.1.3. The keeping time of sour milk products can be increased if they are produced and filled under aseptic conditions or produced under normal conditions but subsequently heat treated.

10.2.1.1 Sour Milk

Sour milk is the product obtained by the fermentation of milk, which occurs either by spontaneous souring caused by various lactic-acid-producing bacteria or on addition of mesophilic microorganisms (*Lactococcus lactis, L. cremoris, L. diacetylactis, Leuconostoc cremoris*) to heated milk at 20 °C. As fermentation proceeds, lactose is transformed into lactic acid, which coagulates casein at pH 4–5. The thick, sour-tasting curdled milk is manufactured from whole milk (at least 3.5% milk fat), low-fat milk (1.5–1.8% fat) or from

skim milk (at most 0.3% fat), often by blending with skim milk powder to increase the total solids content and to improve the resultant protein gel structure. Sour milk contains 0.5–0.9% of lactic acid. In some countries sheep, water buffalo, reindeer or mare's milk are also processed. Sour cream is produced by a process very similar to that used in sour milk manufacture except that coffee grade cream is used as the raw material.

10.2.1.2 Yoghurt

The production of yoghurt is presented schematically in Fig. 10.20. Yoghurt cultures consist of thermophilic lactic acid bacteria that live together symbiotically (*Streptococcus thermophilus* and *Lactobacillus bulgaricus*). Incubation is conducted on addition of 1.5–3% of the operating culture at 42–45 °C for about 3 h. The final product has a pH value of about 4–4.2 and contains 0.7–1.1% of lactic acid. A special yoghurt, "acidophilus milk", is produced by *Lactobacillus acidophilus* (37–38 °C). It contains 0.5–0.7% of lactic acid.

The addition of fruit or fruit pastes and sugar yields special products (fruit yoghurts).

An essential part of the specific yoghurt aroma comes from carbonyl compounds, predominantly acetaldehyde.

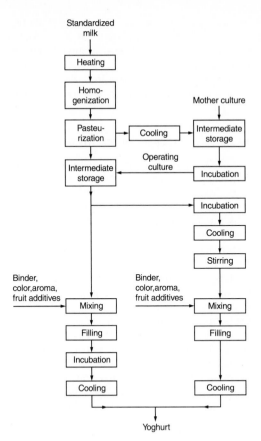

Fig. 10.20. Production of different types of yoghurt

10.2.1.3 Kefir and Kumiss

Kefir and kumiss are sparkling, carbonated alcoholic beverages. The microflora of kefir include *Torula* yeast (responsible for alcoholic fermentation) and *Lactobacillus brevis, L. casei, Leuconostoc mesenteroides, Streptococcus durans, Saccharomyces delbrueckii, S. cerevisiae* and *Acetobacter aceti*. The kefir bacillus causes a buildup of "kefir grains", which resemble cauliflower heads when wet and brownish seeds when dry, and are particles of clotted milk plus the kefir organisms. Their addition to fluid milk produces kefir. Kumiss is made of mare's or goat's milk fermented by the obligatory pure kumiss culture.

Both dairy beverages are indigenous to the Caucasus and steppes of Turkestan. Kefir

contains lactic acid (0.5–1.0%), noticeable amounts of alcohol (0.5–2.0%) and carbon dioxide, and some products of casein degradation resulting from proteolytic action of yeast. Normal kumiss contains 1.0–3.0% of alcohol. The production is similar to that of yoghurt.

10.2.1.4 Taette Milk

Taette (Lapp's milk) is a specially fermented, sour cow's milk product consumed in Sweden, Norway and Finland. Its thread-like, viscous structure is due to the formation of slimy substances at the low fermentation temperatures used. Mesophilic microorganisms (*Lactococcus* and *Leuconostoc spp.*) are involved in this process.

10.2.2 Cream

Milk is practically completely defatted (remaining fat content 0.03–0.06%) in hermetic, self-cleaning or hermetic/self-cleaning creaming separators. The cream products are subsequently standardized by back-mixing. Whipping cream contains at least 30% milk fat, coffee cream at least 10% and butter cream 25–82%. Cream is utilized in many ways, either by direct consumption or for production of butter and ice creams.

Whippability and stability of the whipped foam products are necessary whipping cream properties. For the best quality cream, a volume increase of at least 80% is expected and a standard cone with 100 g load must penetrate 3 cm deep in >10 s. No serum separation should occur at 18 °C after 1 h.

Fat droplets accumulate during whipping on the surface of large air bubbles which form the froth. An increased build-up of smaller bubbles tears apart the membrane of the droplet and enlarges the fat interphase area, thus resulting in gel setting of the lamella separating the individual air bubbles. Sour cream is the product of progressive lactic acid fermentation of cream.

10.2.3 Butter

Butter is a water-in-oil emulsion (w/o emulsion) made from cream by phase inversion occurring in the butter-making process. According to its manufacturing process, three types exist:

- Butter from sour cream (cultured-cream butter).
- Butter from nonsoured, sweet cream (sweet cream butter).
- Butter from sweet cream, which is soured in a subsequent step (soured butter).

Butter contains 81–85% fat, 14–16% water and 0.5–2.0% fat-free solids. The composition generally must meet legal standards. Butter is an emulsion with a continuous phase of liquid milk fat in which are trapped crystallized fat grains, water droplets and air bubbles. A freeze-fracture micrograph of butter showing the continuous fat phase with included fat globules and water droplets is shown in Fig. 10.21. Butter consistency is determined by the ratio of free fluid fat to that of solidified fat. Due to seasonal variations in the unsaturated fatty acid content of milk fat, the solid/fluid fat ratio fluctuates at 24 °C between 1.0 in summer and 1.5 in winter. Equalization of these ratios is achieved by a preliminary cream-tempering step in a cream-ripening process, then churning and kneading the cream, which influences the extent of fluid fat inclusion into

the solidified "fat grains". Figure 10.22 shows the crystalline shell of a cut fat grain, from which the liquid fat was removed during preparation.

A general idea of the most important processing steps involved in butter making is given in Fig. 10.23.

10.2.3.1 Cream Separation and Treatment

Cream is separated from whole milk by high-efficiency separators (cf. 10.1.3.2 and 10.2.2). The cream, depending on the subsequent churning process, might contain 25–82% milk fat. The cream is then pasteurized at 90–110 °C (smaller creameries use batch methods; larger plants use flash pasteurization at 73–85 °C for 16–25 s or vacuum pasteurization, heating the cream to 94 °C with steam).

Cream ripening and souring are the most important steps in the production of sour cream butter. The process is performed in a cream ripener or vat, with suitable mixing and temperature control. Soon after the cream has filled the ripener, a "starter culture" is added to the cream to achieve souring. The "starter culture" consists of various strains of lactic acid bacteria (primarily *Lactococcus lactis subsp. lactis, Lactococus lactis subsp. cremoris, Lactococus lactis subsp. diacetylactis* and *Lacto-*

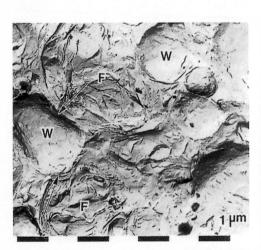

Fig. 10.21. Freeze-fracture micrograph of butter (F: fat globule, W: water droplet; according to *Juriaanse* and *Heertje*, 1988)

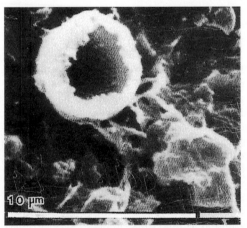

Fig. 10.22. Crystalline shell of a fat grain, as found in butter, which was obtained by eliminating the included oil; (according to *Juriaanse* and *Heertje*, 1988)

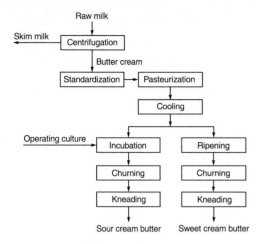

Fig. 10.23. Production of butter

coccus lactis subsp. cremoris bv. citrovorum). The subsequent ripening at 8–19 °C proceeds for up to 24 h.

In addition to build-up of acids and some aroma substances, suitable temperature control gives a partial crystallization of fat. This has a significant influence on butter consistency. The souring step is omitted in sweet cream butter production. Accelerated fat crystallization is achieved by additional storage of cream at 6–10 °C for 3–29 h.

10.2.3.2 Churning

Churning is essentially strong mechanical cream shearing which tears the membranes of the fat globules and facilitates coalescence of the globules. The cream "breaks" and tiny granules of butter appear. Prolonged churning results in a continuous fat phase. Foam build-up is also desirable since the tiny air bubbles, with their large surface area, attract some membrane materials. Some membrane phospholipids are transferred into the aqueous phase. Buttermilk, a milky, turbid liquid, separates out initially (it is later drained off), followed by the butter granules of approx. 2 mm diameter. These granules still contain 30% of the aqueous phase. This is reduced to 15–19% by churning. The finely distributed water droplets (diameter 10 µm or less) are retained by the fat phase.

Churning can be conducted in a discontinuous process, using cream with a fat content of 25–35%. The stainless steel vessels employed have various forms and rotate nonsymmetrically. Continuously operated churns are also used with cream having a fat content of 32–38% (sour cream butter) or 40–50% (sweet cream butter). The machines are divided into churning, separation, and kneading compartments. In the churning compartment, a rotating impact wave causes butter granule formation. The separation compartment is divided into two parts. The butter is first churned further, resulting in the formation of butter granules of a larger diameter. Subsequently, the buttermilk is separated and the butter is washed, if necessary. The cooled kneading compartment consists of transport screws and kneading elements for further processing the butter. Both kneading compartments are operated under vacuum conditions to reduce the air content of the butter to less than 1%. The final salt and water content of the butter is adjusted by apportioning.

In the continuous *Alfa*-process the phase conversion is achieved in a screw-type cooler, using an 82% cream and repeated chilling, without the aqueous phase being separated.

The *Booser* process and the *NIZO* process allow a subsequent souring of butter from sweet cream. Both processes are of economic interest, because they yield a more aromatic sour butter and sweet buttermilk, which is a more useful by-product than sour buttermilk.

During the *Booser* process 3–4% of starter cultures are incorporated into the butter granules (water content: 13.5–14.5%) obtained from sweet cream.

Lactic acid and a flavor concentrate are obtained by separate fermentations during the first step of the *NIZO* process. In a second step they are mixed and incorporated into the butter granules from sweet cream.

Lactobacillus helveticus cultivated on whey produces the lactic acid, which is then separated by ultrafiltration and concentrated in vacuum up to about 18%. The flavor concentrate is obtained by growing starter cultures and *Lactococcus lactis subsp. diacetylactis* on skim milk of about 16% dry matter.

10.2.3.3 Packaging

After the butter is formed, it is cut by machine into rectangular blocks and is wrapped in waxed or grease-proof paper or metallic (aluminium) foil laminates (coated within with polyethylene).

10.2.3.4 Products Derived from Butter

- Melted butter consists of at least 99.3% milk fat. The aqueous phase is removed by decantation of the melted butter or by evaporation.
- Fractionated butterfat. The butter is separated by fractional crystallization into high- and low-melting fractions, and is utilized for various purposes (e.g. consistency improvement of whipping creams and butters).
- Spreadable blends with vegetable oils ("butterine").

10.2.4 Condensed Milk

Condensed milk is made from milk by the partial removal of water and addition of saccharose, if necessary (sweetened condensed milk). It is used, diluted or undiluted, like milk. Nonsweetened condensed milk is mainly

available with a fat content of 4%, 7.5%, or 10%.

The production process (Fig. 10.24) starts with milk of the desired fat content. The milk is first heated to 85 to 100 °C for 10–25 minutes to separate albumin, kill germs, and reduce the danger of delayed thickening. Subsequently, it is evaporated in a continuously operated vacuum evaporator at 40 to 80 °C. In comparison with previously used circulation, riser, and flat-plate evaporators, film evaporators are mainly employed today. Several units (up to seven stages) are usually connected in series, each unit being heated by the vapor from the previous stage. The temperature and pressure decrease from stage to stage. Optimal energy utilization is achieved by mechanical or thermal vapor compression. The resulting evaporated milk, with a solid content of 25–33% or more, is homogenized, poured into lacquer (enamel)-coated cans made of white metal sheets, and is sterilized in an autoclave at 115–120 °C for 20 min. Continuous flow sterilization followed by aseptic packaging is also used. To prevent coagulation during processing and storage, Na-hydrogen carbonate, disodium phosphate and trisodium citrate are incorporated into the condensed milk. These additives have a dual effect: pH correction and adjustment of free Ca^{2+} ion concentration, both aimed at preventing casein aggregation (cf. Fig. 10.3).

In the production of *sweetened condensed milk*, after a preheating step (short-time heating at 110–130 °C), sucrose is added to a concentration of 45–50% of the weight of the end-product. Homogenization and sterilization steps are omitted.

To avoid graininess caused by lactose crystallization – the solubility limit of lactose is exceeded after sucrose addition – the condensed milk is cooled rapidly, then seeded with finely pulverized α-lactose hydrate. Seeding ensures that the lactose crystal size is 10 μm or less.

The critical quality characteristics of condensed milk are the degree of heat damage (lysine degradation), prevention of separation during the storage life, absence of coarse crystallized lactose, as well as color and taste. These criteria are influenced not only by the process management (heat treatment during evapora-

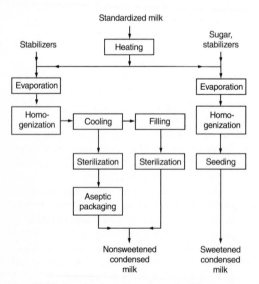

Fig. 10.24. Production of condensed milk

tion and sterilization and suitable selection of the homogenization temperature and pressure), but also by the source of the milk (feed) and the producer's ability to maintain hygienic conditions.

10.2.5 Dehydrated Milk Products

Skim milk powder and whole milk powder are used either for the reconstitution of milk in countries that for climatic reasons have no dairy farming or as intermediate products for further processing into infant milk products, milk chocolate etc. The quality of these instant products depends on the durability, redissolution capacity (cold and warm), taste, microbiological characteristics, and preservation of essential constituents (proteins, vitamins) during production.

The main drying process used is spray drying. However, drum drying (with and without vacuum) and fluid-bed drying (foaming with inert gas N_2 or CO_2) are used for special purposes. Freeze drying offers no particular advantages over the less expensive spray drying process and is only of interest for special products.

Using film evaporating systems, the milk is first preconcentrated to 40–50% solids.

In drum drying, the liquid is applied in a thin layer to a heated drying cylinder and, after a defined residence time (rotation), removed with a doctor knife. The liquid film can be applied in various ways. In drum drying, relatively large particles are obtained. The thermal exposure (temperature, time) is considerably higher than in spray drying. The solubility is poor due to the denaturation of whey proteins. The product is clearly brown owing to the *Maillard* reaction.

In spray drying, the milk is finely dispersed in the spray tower by centrifugal atomization or by nozzle atomization and dried with hot air (150–200 °C) cocurrently or countercurrently. Multistage processes have been favored in the past years. The first step involves the actual spray drying; fluid-bed drying in a vibration fluid bed forms the second step. This subsequent use of vibration fluid beds provides better energy utilization in the spraying tower (highest possible temperature difference between entry and exit) and gives agglomerized dry products with improved wetting properties. New spray drying systems operate with a fluid bed integrated into the spray-drying chamber (compact system).

Particles with a diameter in the range of 5 to 100 μm consist of a continuous mass of amorphous lactose and other low-molecular components, which includes fat globules, casein micelles, whey proteins and usually vacuoles. When the powder absorbs water, lactose crystallizes at $a_w > 0.4$, causing agglomeration. During drying, the temperature of the particles normally does not rise above 70 °C. Therefore, the whey proteins do not denature and remain soluble. Many enzymes are still active. Storage problems are caused by the *Maillard* reaction and by fat oxidation in the case of fat-containing powders. Foam dried products can have excellent properties (aroma, solubility).

Other dehydrated dairy products, in addition to whole milk or skim milk powders, are manufactured by similar processes. Products include dehydrated malted milk powder, spray- or roller-dried creams with at least 42% fat content of their solids and a maximum 4% moisture, and butter or cream powders with 70–80% milk fat. Dehydrated buttermilk and lactic acid-soured milk are utilized as children's food.

Adaptation of infant milk product formulation to approximate mothers' milk can be achieved, for example, by addition of whey proteins, sucrose, whey or lactose, vegetable oil, vitamins and trace elements and by reduction of minerals, i.e., by a shift of the Na/K ratio.

The compositions of some dehydrated dairy products are illustrated in Table 10.27.

Table 10.27. Composition of dried milk products (%)

	1	2	3	4	5
Water	3.5	4.3	4.0	3.1	7.1
Protein	25.2	35.0	21.5	33.4	12.0
Fat	26.2	1.0	40.0	2.3	1.2
Lactose	38.1	51.9	29.5	54.7	71.5
Minerals	7.0	7.8	5.0	6.5	8.2

1: Whole milk powder, 2: skim milk powder, 3: cream powder, 4: buttermilk powder, and 5: whey powder.

10.2.6 Coffee Whitener

Coffee whiteners are products that are available in liquid, but more often in dried instant form. They are used like coffee cream or condensed milk. A formulation typical of these products is shown in Table 10.28. In contrast to milk products, plant fats are used in the production of coffee whiteners. Caseinates are usually the protein component. The most important process steps in the production are: preemulsification of the constituents at temperatures of up to 90 °C, high-pressure homogenization (cf. 10.1.3.4), spray drying, and instantization (cf. 10.2.5).

Table 10.28. Typical formulation of coffee whiteners

Constituent	Amount (%)
Glucose syrup	52.6
Fat	30.0
Sodium caseinate	12.0
Water	3.15
Emulsifiers	1.6
K_2HPO_4	0.6
Carrageenan	0.05
Color and aroma substances	

10.2.7 Ice Cream

Ice cream is a frozen mass which can contain whole milk, skim milk products, cream or butter, sugar, vegetable oil, egg products, fruit and fruit ingredients, coffee, cocoa, aroma substances and approved food colors. A typical formulation is 10% milk fat, 11% fat-free milk solids, 14% saccharose, 2% glucose syrup-solids, 0.3% emulsifiers, 0.3% thickener, and 62% water. The thickeners, mostly polysaccharides (cf. Table 4.15), increase the viscosity and the emulsifiers destabilize the fat globules, favoring their aggregation during the freezing process.

For the production of ice cream, the mixture of components is subjected to high-temperature short-time pasteurization (80–85 °C, 20–30 s), high-pressure homogenization (150–200 bar) and cooling to ca. 5 °C. Air is then mixed into the mixture (60–100 vol%) while

it is frozen at temperatures of up to – 10 °C and then hardened at – 15 to – 25 °C. The freezers used are mainly cylindrical, ammonia-cooled, scraped-surface heat exchangers, similar to those used in the production of margarine. To foam up the ice cream, more air is mixed into the mixture inside the cylinder.

The structural elements of ice cream are ice crystals (~ 50 µm), air bubbles (60–150 µm), fat globules (< 2 µm), and aggregated fat globules (5–10 µm). The fat is mostly attached to the air bubbles. The air bubbles have a three fold function: they reduce the nutritional value, soften the product, and prevent a strong cold sensation during consumption.

10.2.8 Cheese

Cheese is obtained from curdled milk by removal of whey and by curd ripening in the presence of special microflora (Table 10.29). The great abundance of cheese varieties, about 2000 worldwide, can be classified from many viewpoints, e.g., according to:

- Milk utilized (cow, goat or sheep milk).
- Curd formation (using acids, rennet extract or a combination of both).
- Texture or consistency, or water content (%) in fat-free cheese. Following the latter criterion, the more important cheese groups are (water content, %, in brackets):
    Very hard cheese for grating (< 47);
    Hard cheese (< 56);
    Cutting (slicing) cheese (54–63);
    Semi-solid slicing cheese (61–69);
    Soft cheese (67–76);
    Fresh cheese (73–87).
- Fat content (% dry matter). By this criterion, the more important groups are:
    Double cream cheese (60–85% fat);
    Cream cheese (≥ 50);
    Whole fat cheese (≥ 45);
    Fat cheese (≥ 40);
    Semi fat cheese (≥ 20);
    Skim cheese (max. 10).

Within each group, individual cheeses are characterized by aroma. A small selection of the more important cheese varieties is listed in Table 10.30. Cheese manufacturing essentially consists of curd formation and ripening (Fig. 10.25).

Table 10.29. Characteristic microflora of some types of cheese

Type of cheese	Starter cultures	Other species
Parmigiano-Reggiano	*Streptococcus thermophilus* *Lactobacillus helveticus* *L. bulgaricus*	
Emmental	*Lactococcus lactis* subsp. *lactis* *Lactococcus lactis* subsp. *cremoris* *S. thermophilus* *Lactobacillus helveticus* *L. bulgaricus*	*Propionibacterium freudenreichii* *P. freudenreichii* subsp. *shermanii*
Cheddar	*Lactococcus lactis* subsp. *cremoris* *(Lactococcus lactis* subsp. *lactis)*	*None*
Roquefort	*Lactococcus lactis* subsp. *lactis* *Lactococcus lactis* subsp. *cremoris* *Lactococcus lactis* subsp. *diacetylactis* *Leuconostoc cremoris*	*Penicillium roqueforti*
Limburger	*Lactococcus lactis* subsp. *lactis* *Lactococcus lactis* subsp. *cremoris*	*Brevibacterium linens* *Micrococcus spp.* *Yeasts*
Edamer, Gouda	*Lactococcus lactis* subsp. *lactis* *Lactococcus lactis* subsp. *cremoris* *Lactococcus lactis* subsp. *diacetylactis* *Leuconostoc cremoris*	*Brevibacterium linens* *Micrococcus spp.* *Yeasts*
Camembert, Brie	*Lactococcus lactis* *Lactococcus lactis* subsp. *cremoris* *Lactococcus lactis* subsp. *diacetylactis*	*Penicillium camemberti* *P. caseicolum* *Brevibacterium linens* *Micrococcus spp.* *Yeasts*

Table 10.30. Cheese Varieties

Unripened Cheeses (F: < 10–70, T: 39–44, R: unripened)

Quark. Neuchâtel, Petit Suisse, Demi Sel, Cottage Cheese
Schichtkäse (layers of different fat content)
Rahm-, Doppelrahmfrischkäse, Demi Suisse, Gervais, Carré-frais, Cream Cheese
Mozzarella (plastic curd by heating to > 60 °C within the whey). Scamorze

Table 10.30 (continued)

Ripened Cheese

Hard Cheeses (F: 30–50, T: 58–63, R: 2–8 M)

Chester, Cheddar, Cheshire, Cantal
Emmental, Alpkäse, Bergkäse, Gruyère, Herrgårdskäse, Samsoe
Gruyère (Greyerzer), l'Emmental française, Beaufort, Gruyère de Comte,
Parmigiano-Reggiano (granular structure, very hard, grating type), Grana, Bagozzo, Sbrinz
Provolone (plastic curd by heating to > 60 °C in the whey: Pasta filata), Cacciocavallo

Slicing Cheeses (F: 30–60, T: 44–57, R: 3–5 W)

Edam, Geheimratskäse, Brotkäse, Molbo, Thybo Gouda, Fynbo, Naribo
Pecorino (from ewe's milk), Aunis Brinsenkäse
Port Salut, St. Paulin, Esrom, Jerome, Deutscher Trappistenkäse
Tilsiter, Appenzeller, Danbo, Steppenkäse, Svecia-Ost

Semi-solid slicing Cheeses (F: 30–40, T: 44–55, R: 3–5 W)

Butterkäse, Italico, Bel Paese, Klosterkäse
Roquefort (from ewe's milk), Bleu d'Auvergne, Bresse Bleu, Bleu du Haut-Jura, Gorgonzola, Stracchino, Stilton, Blue Dorset, Blue Cheese, Danablue Steinbuscher
Weißlacker, Bierkäse

Soft Cheeses (F: 20–60, T 35–52, R: 2 W)

Chevre (from goat's milk), Chevret, Chevretin, Nicolin, Cacciotta, Rebbiola, Pinsgauer Käse
Brie, Le Coulommiers
Camembert, veritable Camembert, Petit Camembert
Limburger, Backsteinkäse, Allgäuer Stangenkäse
Münsterkäse, Mainauer, Mondseer, Le Munster, Gérômè
Pont l'Eveque, Angelot, Maroilles
Romadour, Kümmelkäse, Weinkäse, Limburger

Low-fat Cheeses (F: < 10, T: 35, R: 1–2 W)

Harzer Käse, Mainzer Käse (ripened with *Bact. linens,* different cocci and yeasts)
Handkäse, Korbkäse, Stangenkäse, Spitzkäse (ripened with *Bact. linens,* different cocci and yeasts, or with *Penicillium camemberti*), Gamelost

Cooking cheese (from Cottage Cheese by heating with emulsifying agents, F: < 10–60)

[a] Related types are grouped together. For the classes average values are given for
– fat content in the dry matter: F (%)
– dry matter: T(%)
– ripening time: R in months (M) or weeks (W).

10.2.8.1 Curd Formation

The milk fat content is adjusted to a desired level and, when necessary, the protein content is also adjusted. Additives include calcium salts to improve protein coagulation and cheese texture, nitrates to inhibit anaerobic spore-forming microflora, and color pigments. The prepared raw or pasteurized milk is mixed at 18–50°C in a vat with a starter culture (cf. Table 10.29) (lactic acid or propionic acid bacteria; molds, such as *Penicillium camemberti, P. candidum, P. roqueforti;* red- or yellow-smearing cultures, such as *Bacterium linens* with *cocci* and yeast). The milk coagulates into a soft, semi-solid mass, the curd, after lactic acid fermentation (sour milk cheese, pH 4.9–4.6), or by addition of rennet (sweet milk cheese, pH 6.6–6.3), or some other combination, the most common being combined acid and rennet treatment. This protein gel is cut into cubes while being heated and is then gently stirred. The whey is drained off while the retained fat-containing curd is subjected to a firming process (syneresis). The firming gets more intense as the mechanical input and the applied temperature increase. The process and the starter culture (pH) determine the curd properties. When the desired curd consistency has been achieved, curd and whey separation is accomplished either by draining off the whey or by pressing off the curd while simultaneously molding it.

New methods of cheese making aim at including the whey proteins in the curd, instead of removing them with the whey. Apart from giving higher yields (12–18%), these processes help to economize on waste water costs or elaborate whey treatments (cf. 10.2.10). The use of ultrafiltration steps as compared with conventional cheese making is shown in Fig. 10.25. Alternatively, conventionally produced whey can be concentrated by ultrafiltration and then added to the curd or milk can be soured with starter culture and/or rennet addition and then concentrated by ultrafiltration. To reduce the cost of enzymes in the casein precipitation step with chymosin (rennet or usually microbial rennet substitutes), processes using carrier-bound enzymes are being tested. Here, the enzyme reaction proceeds in the cold and precipitation occurs subsequently

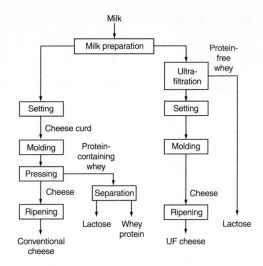

Fig. 10.25. Cheese making (conventional or with ultrafiltration)

on heating the milk. In this way, clogging of the enzyme bed is avoided.

The individual process steps in cheese making are being increasingly mechanized and automated. The equipment used includes discontinuously operated cheesemakers (vats or tanks with stirring and cutting devices) and coagulators for continuous curd formation with subsequent fully automatic whey separation and molding.

10.2.8.2 Unripened Cheese

Unripened cheeses have a soft (quark), gelatinous (layer cheese), or grainy (cottage cheese) consistency. In the production of quark, the whey is usually separated after souring. Cottage cheese is generally produced in continuously operated coagulators with special temperature regulation. After whey separation via a filter band, the curd grain can be washed in a screw vat, cooled, and dried via another drying band.

10.2.8.3 Ripening

The molded cheese mass is placed in a salt bath for some time, dried, and then left to ripen in air-conditioned rooms. Ripening, or curing, is dependent on cheese mass composition, particularly the water content, the microflora and the external conditions, such as temperature and humidity in the curing rooms.

The ripening of soft cheeses proceeds inwards, so in the early stages there is a ripened rind and an unripe inner core. This nonuniform ripening is due to the high whey content which causes increased formation of lactic acid and a pH drop at the start of ripening. In the rind, special molds that grow more favorably at higher pH values contribute to a pH increase by decarboxylating amino acids.

Ripening in hard cheeses occurs uniformly throughout the whole cheese mass. Rind formation is the result of surface drying, so it can be avoided by packaging the cheese mass in suitable plastic foils before curing commences. The duration of curing varies and lasts several days for soft cheeses and up to several months or even a couple of years for hard cheeses. The yield per 100 kg fluid milk is 8 kg for hard cheeses and up to 12 kg for soft cheeses.

All cheese ingredients are degraded biochemically to varying extents during curing.

Lactose is degraded to lactic acid by homofermentation. In cheddar cheese, for example, the pH drops from 6.55 to 5.15 from the addition of the starter culture to the end of mold pressing. In the presence of propionic acid bacteria (as in the case of Emmental cheese), lactic acid is converted further to propionic and acetic acids and CO_2, according to the reaction:

$$3\ CH_3CHOHCOOH \rightarrow 2\ CH_3CH_2COOH + CH_3COOH + CO_2 + H_2O \quad (10.13)$$

The ratio of propionic to acetic acid is influenced by the redox potential of the cheese, and in the presence of nitrates, for example, the ratio is lower. Propionic acid fermentation is shown in Fig. 10.26. The crucial step is the reversible rearrangement of succinyl-CoA into methylmalonyl-CoA:

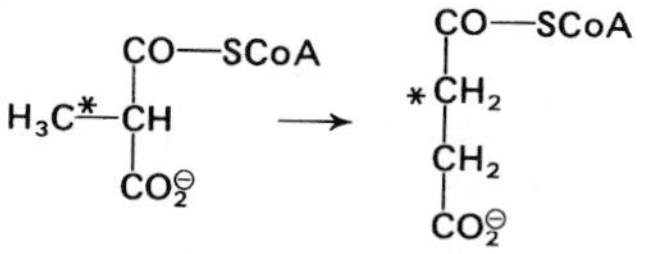

$$(10.14)$$

The catalysis is mediated by adenosyl-B_{12}, which is a coenzyme for transformations of the general type:

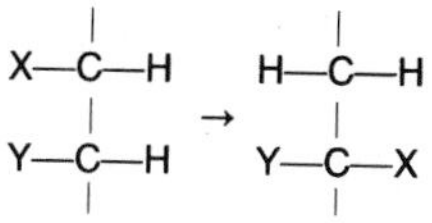

$$(10.15)$$

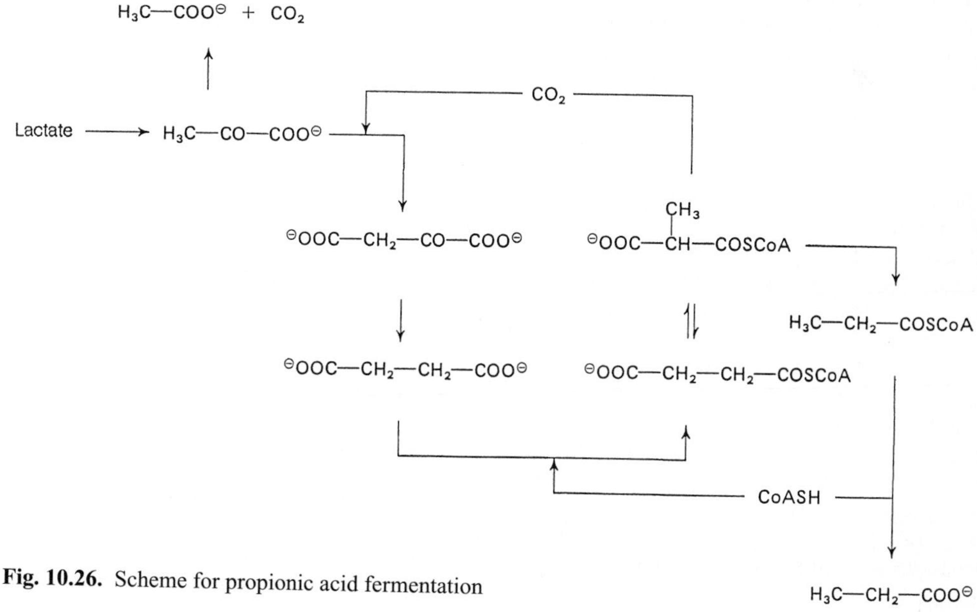

Fig. 10.26. Scheme for propionic acid fermentation

Based on a study of a coenzyme B_{12}-analogue, it is obvious that a nonclassical carbanion mechanism is involved:

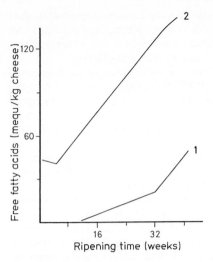

Fig. 10.27. Lipolysis during ripening of blue cheese: 1 untreated milk, 2 homogenized milk

(10.16)

The mode and extent of milk *fat* degradation depend on the microflora involved in cheese ripening. In most types of cheese, as little lipolysis as possible is a prerequisite for good aroma. Exceptions are varieties like Roquefort, Gorgonzola, and Stilton, that are characterized by a marked degradation of fat.

Lipolysis is strongly enhanced by homogenization of the milk (Fig. 10.27). The release of fatty acids, especially those that affect cheese aroma, depends on the specificity of the lipases (Table 10.31). In addition to free fatty acids, 2-alkanones and 2-alkanols are formed as by-products of the β-oxidation of the fatty acids (cf. 3.7.5).

Molds, particularly *Penicillium roqueforti*, utilize β-ketoacyl-CoA deacylase (thiohydrolase) and β-ketoacid decarboxylase to provide the compounds typical for the aroma of semi-soft cheeses e.g., the blue-veined cheese (Roquefort, Stilton, Gorgonzola, cf. Table 10.32).

Table 10.31. Substrate specificity of a lipase from *Penicillium roqueforti*

Substrate	Hydrolysis (V_{rel})
Tributyrin	100
Tripropionin	25
Tricaprylin	75
Tricaprin	50
Triolein	15

Table 10.32. 2-Alkanones in blue cheese

2-Alkanone n^a	mg/100 g cheese (dry matter)
3	0.5–0.8
5	1.4–4.1
7	3.8–8.0
9	4.4–17.6
11	1.2–5.9

[a] Number of C-atoms.

Protein degradation to amino acids occurs through peptides as intermediary products. Depending on the cheese variety, 20–40% of casein is transformed into soluble protein derivatives, of which 5–15% are amino acids. A pH range of 3–6 is optimum for the activity of peptidases from *Penicillium roqueforti*. Proteolysis is strongly influenced by the water and salt content of the cheese. The amino acid content is 2.6–9% of the cheese solids and is of importance to cheese aroma. Ripening defects can produce bitter-tasting peptides.

The amino acids are transformed further. In early stages of cheese ripening, at a lower pH, they are decarboxylated to amines. In later stages, at a higher pH, oxidation reactions prevail:

$$R-\underset{\underset{COOH}{|}}{CH}-NH_2 \quad\diagup\quad \begin{array}{l} R-CH_2-NH_2 \;+\; CO_2 \\[2em] R-CO-COOH \end{array}$$

$$\qquad\qquad\qquad (10.17)$$

$$\downarrow$$

$$R-CHO \;+\; CO_2$$

Proteolysis contributes not only to aroma formation, but it affects cheese texture. In overripening of soft cheese, proteolysis can proceed almost to liquefaction of the entire cheese mass.

The progress of proteolysis can be followed by electrophoretic and chromatographic methods, e.g., via the peptide pattern obtained with the help of RP-HPLC (Fig. 10.28) and via changes in concentration of individual peptides which correspond to certain casein sequences (Table 10.33) and can serve as an indicator of the degree of cheese ripening.

Table 10.33. Amino acid sequences of some small peptides from Cheddar cheese

Peptide[a]	Sequence	Corresponding casein sequence	
30	A P F P E	α_{s1}B	26–30[b]
37	D K I (H) P F	βA²	47–52
39	L P Q E (V L)	α_{s1}B	11–16
46	L Q D K I (H) P (F)	βA²	45–52
58	Y P F P G P I P N	βA²	60–68
60	A P F P E (V F)	α_{s1}B	26–32[b]

[a] Numbering cf. Fig. 10.28.
[b] In the literature, Q represents position 30 of α_{s1}-casein B, and E the corresponding position of the precursor protein. () Added on the basis of the amino acid composition.

The content of biogenic amines in some types of cheese is presented in Table 10.34. These values can fluctuate greatly depending on the degree of ripening.

10.2.8.4 Processed Cheese

Processed (or melted) cheese is made from natural, very hard grating or hard cheeses by shredding and then heating the shreds to 75–95 °C in the presence of 2–3% melting salts (lactate, citrate, phosphate) and, when required, utilizing other ingredients, such as milk powder, cream, aromas, seasonings and vegetable and/or meat products. The cheese can be spreadable or made firm and cut as

Table 10.34. Biogenic amines in cheese (mg/100 g)

Cheese	Phenylethyl-amine	Tyramine	Tryptamine	Histamine	Putrescine	Cadaverine
Cheddar	0–30	13–39	0–0.2	2.4–140	0–100	0–88
Emmentaler	0–23.4	3.3–40	0–1.3	0.4–250	0–15	0–8
Gruyere		6.4–9.9		0–20		
Parmesan		0.4–2.9		0–58		
Provolone					1–20	2–20
Edamer	0–1.3		0–0.4	1.4–6.5		0.5–9.4
Gouda		0–110		5.6–7.4	2–20	2.5
Tilsiter	0–14.8	0–78	0–7.1	0–95.3	0–31.3	0–31.8
Gorgonzola					0–75	0–430
Roquefort		2.7–110	0–160	1–16.8	1.5–3.3	7.1–9.3
Camembert		2–200	2	0–48	0.7–3.3	1.2–3.7

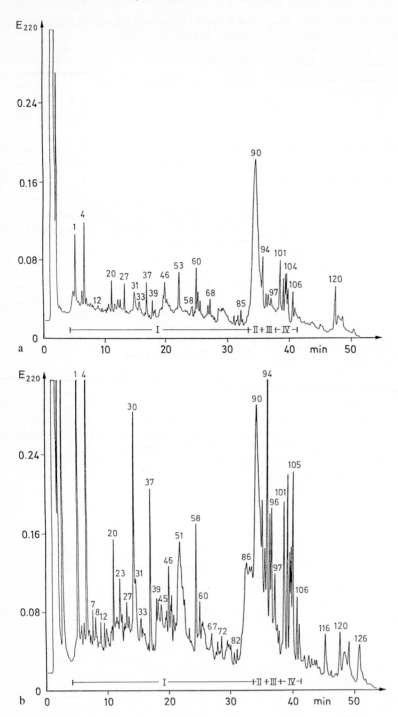

Fig. 10.28. RP-HPLC of the pH 4.6-soluble fraction of a citrate extract of Cheddar cheese after 3 (a) and 24 (b) weeks; ripening at 10°C (according to *Kaiser* et al., 1992)

desired. The shelf life of processed cheese is long due to thermal killing of microflora.

The heating process is carried out batchwise by steam injection in a double-walled pressure vessel equipped with a mixer, usually under a slight vacuum. Continuous processes are conducted in double-walled cylinders with agitator shafts.

10.2.8.5 Imitation Cheese

These products are mainly found in North America (USA 1981: 954000 t). They are made of protein (mostly milk protein), fat (mostly hardened vegetable fat), water, and stabilizers by using processed cheese technology. A typical formulation is shown in Table 10.35.

10.2.9 Casein, Caseinates, Coprecipitate

The production of casein, caseinates, and coprecipitate is shown schematically in Fig. 10.29.

Coagulation and separation of casein from milk is possible by souring the milk by lactic acid fermentation, or by adding acids such as HCl, H_2SO_4, lactic acid or H_3PO_4. Another

Table 10.35. Typical formulation of imitation cheese (Mozarella type)

Component	Amount (%)
Water	51.1
Ca/Na caseinate	26.0
Vegetable oil (partially hydrogenated)	18.0
Glucono-δ-lactone	2.8
Salt	2.0
Color and aroma substances	

way to achieve coagulation is to add proteinase enzymes, such as chymosin and pepsin. The acid coagulation is achieved at 35–50°C and pH 4.2–4.6 (isoelectric point of casein is pH 4.6–4.7). Casein precipitates out as coarse grains and is usually separated in sedimentation centrifuges, washed, and dried (whirlwind drier). The enzymatic process involves heating to 65°C after precipitation in whey.

Increasing the level of Ca^{2+} ions (addition to milk of 0.24% $CaCl_2$) causes casein and whey proteins to coagulate when the temperature is at 90°C. Joint coagulation of proteins can also be achieved by first heat-denaturing the whey proteins, then acidifying the milk. Washing

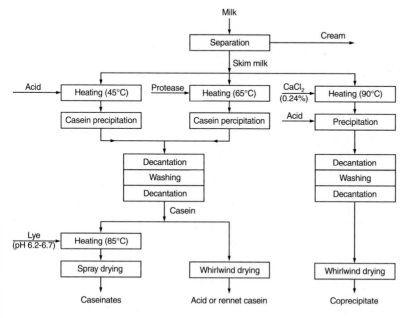

Fig. 10.29. Production of casein, caseinates, and coprecipitate

followed by drying of the curd gives a copreci-pitate which contains up to 96% of the total proteins of the milk. When casein dispersions, 20–25%, are treated wtih alkali [NaOH or Ca(OH)$_2$, alkali or alkaline-earth carbonates or citrates] at 80–90 °C and pH 6.2–6.7, and then the solubilized product is spray-dried, a soluble or readily dispersable casein product is obtained (caseinate, disintegrated milk pro-tein).

Casein and caseinate are utilized as food and also have nonfood uses. In food manufactur-ing they are used for protein enrichment and/or to achieve stabilization of some phy-sical properties of processed meats, baked products, candies, cereal products, ice creams, whipping creams, coffee whiteners, and some dietetic food products and drugs.

The nonfood uses involve wide application of casein/caseinate as a sizing (coating) for better quality papers (for books and journals, with a surface suitable for fine printing), in glue manufacturing, as a type of waterproof glue (alkali caseinate with calcium components as binder); in the textile industry (dye fixing, water-repellent impregnations); and for casein paints and production of some plastics (knobs, piano keyboards, etc.).

10.2.10 Whey Products

Whey accumulates in considerable amounts in the production of cheese and casein (Table 10.37). Increasing waste water costs and al-tered processing structures made the develop-ment of recovery processes necessary.

The composition of whey and whey products is presented in Table 10.36. Whey and whey products are used in animal feed, dietetic foods (infant food), bread, confectionery, candies, and beverages.

10.2.10.1 Whey Powder

In dairy farming, two process variants are applied for the drying of whey:

- Preliminary concentration of the whey to 50–55% dry matter in falling-film eva-porating systems (thermal or mechanical vapor compression), followed by spray dry-ing (one step or two step with subsequent vibraton fluid bed).

- Preliminary concentration of the whey to 21–25% dry matter by reverse osmosis (hyperfiltration), followed by concentration to 50–55% dry matter via falling-film evaporators and spray drying.

The second variant is more advantageous as it is energetically favored.

10.2.10.2 Demineralized Whey Powder

The production of demineralized whey pow-der proceeds via ion exchange or, preferential-ly, electrodialysis (1.5–4.5 V/cell; current density 5–20 mA/cm^2 membrane area, Fig. 10.30).

Table 10.36. Protein, lactose and mineral contents of whey products[a]

Product	DM[b] (%)	Protein (%)	Lactose (%)	Minerals (%)
Skim milk	9.0	36	53	7
Whey (from coagulating with rennet)	6.0–6.4	13	75	8
Whey (from coagulating with acid)	5.8–6.2	12	67	14
Demineralized whey powder	12–13	85		1–2
Whey protein powder[c]				
I		47	44	9
II		74	20	6

[a] Average values are expressed as % of dry matter.
[b] Dry matter.
[c] After one (I) and two (II) ultrafiltrations.

Table 10.37. Accumulation of whey (in 1000 t of dry matter)

Region	1976	1980
Western Europe	1617	1987
FR Germany	258	323
USA	838	968
World	2807	3362

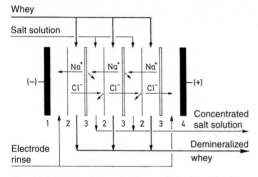

Fig. 10.30. Principle of electrodialysis of whey. *1* cathode, *2* cation membrane, *3* anion membrane, *4* anode

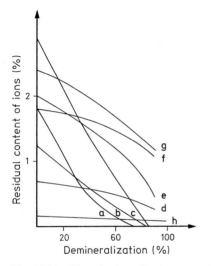

Fig. 10.31. Whey demineralization. Ions of *a* chloride, *b* sodium, *c* potassium, *d* calcium, *e* phosphate, *f* lactate, *g* citrate, and *h* magnesium

10.2.10.3 Partially Desugared Whey Protein Concentrates

In the ultrafiltration of whey, protein concentrates depleted of lactose to various extents are obtained, depending on the number of stages and amount of wash water. Another, less gentle method involves the heating of whey (95 °C, 3–4 min) by direct steam injection, followed by precipitation of the denatured proteins at pH 4.5, separation in a sedimentation centrifuge (2000–4000 min^{-1}), and drying.

10.2.10.4 Hydrolyzed Whey Syrups

The production of sweet whey syrups is becoming increasingly important due to the use of carrier-bound lactase (β-galactosidase, EC. 3.2.1.23). In these syrups, lactose is hydrolyzed to glucose and galactose. Concentration to 60–75% solids is achieved by evaporation.

10.2.11 Lactose

For lactose production the whey is evaporated to 55–65% solid content, and the concentrate is then seeded and cooled slowly to induce sugar crystallization. The raw lactose (food quality) is recrystallized to yield a raffinade (pharmaceutical-grade lactose). Lactose is used in manufacturing of drugs (tablet filler), dietetic food products, baked products, dehydrated foods, cocoa products, beverages and ice creams.

10.2.12 Cholesterol-Reduced Milk and Milk Products

In the production of milk products with a reduced cholesterol content, more than 90% of the cholesterol is removed from water-free milk fat by extraction with supercritical carbon dioxide or by steam distillation. The fat is then recombined with skim milk to give low-cholesterol milk, which is used to make the usual milk products. The extent of cholesterol reduction in a series of products is listed in Table 10.38.

Recombined milk does not have the same properties as the original milk because, e.g., the membrane composition of the fat globule changes in the process. Cheese made from milk of this type can exhibit texture defects.

Since skim milk with a fat content of 0.2% still contains about 18 mg/l of cholesterol, skim milk must also be freed of cholesterol for the production of cholesterol-free products.

Table 10.38. Effects of a 90% reduction of cholesterol in butter oil on the cholesterol content of recombined milk and its products

Food	Fat (%)	Cholesterol (mg/kg) I[a]	II[a]
Whole milk	3.3	135	26
Butter	81	2400	300
Yoghurt	3.5	124	26
Ice cream	10.8	450	41
Cottage cheese	4.6	150	12
Mozzarella	21.6	786	68
Brie	20.8	1000	75
Camembert	24.6	714	57
Roquefort	30.6	929	107
Cheddar	33.1	1071	114

[a] Product before (I) and after (II) cholesterol reduction.

10.3 Aroma of Milk and Dairy Products

10.3.1 Milk

Raw or gently pasteurized milk has a mild, but characteristic taste. In concentrates of aroma substances (1–100 mg/kg of milk), more than 400 volatile compounds, which contribute to aroma to different extents, have been identified. Some compounds which determine the aroma of gently pasteurized milk (73 °C, 12 s) and UHT milk (indirect heating, 142 °C, 4.6 s) are listed in Table 10.39.

Milk pasteurized at low temperatures (e.g., 73 °C, 12 s) does not have a cooked flavor. Typical aroma substances are dimethylsulfide, diacetyl, 2-methylbutanol, 4-cis-heptenal, 3-butenylisothiocyanate, and 2-trans-nonenal. A cooked flavor appears at slightly higher pasteurization temperatures (83 °C, 10 s), mainly due to hydrogen sulfide and other sulfur compounds, which are primarily formed from proteins of the fat globule membrane. Furthermore, methyl ketones formed by the thermal decarboxylation of β-keto acids as well as lactones formed from γ- and δ-hydroxy fatty acids also contribute to the cooked flavor.

The aroma of UHT milk is also determined by 2-alkanones, lactones, and sulfur compounds, including 2-heptanone, 2-nonanone, dimethylsulfide, diacetyl, 2-hexanone, 4-cis-heptenal,

Table 10.39. Volatile compounds which contribute to the aroma of milk[a]

Compound	Contribution to aroma[b] GPM	UHT
Hydrogen sulfide	0	2
Ethanol	1	0
Methylthiol	0	1
Dimethylsulfide	3	3
Diacetyl	2	3
3-Methylbutanal	1	2
2-Methylbutanal	1	1
2-Pentanone	0	1
Pentanal	0	1
i-Butylthiol	0	1
Methylisothiocyanate	0	1
Dimethyldisulfide	0	2
4-Pentennitrile	1	0
2-Methylbutanol	2	1
2-Hexanone	1	3
Ethyl isothiocyanate	0	1
Hexanal	1	2
Ethyl butyrate	1	0
Furfural	0	1
2,4-Dithiapentane	1	1
2-Heptanone	0	4
4-cis-Heptenal	2	3
Heptanal	1	1
Benzaldehyde	0	1
2,3,4-Trithiapentane	0	2
3-Butenylisothiocyanate	2	0
Benzonitrile	1	1
1-Octen-3-one	1	1
2-Octanone	0	1
1-Octen-3-ol	1	1
Octanal	0	1
Acetophenone	0	1
2-Nonanone	0	4
Nonanal	1	1
p-Cresol	1	1
2-trans-Nonenal	2	2
Naphthalene	0	1
2-trans,4-trans-Nonadienal	1	0
Benzothiazole	0	1
γ-Octalactone	0	1
δ-Octalactone	0	1
Decanol	0	1
2-Undecanone	0	2
γ-Decalactone	0	1
δ-Decalactone	1	2
2-Tridecanone	0	1
γ-Dodecalactone	0	2
δ-Dodecalactone	1	3

[a] Gently pasteurized milk (GPM): 12 s, 73 °C; UHT milk (UHT): indirect heating, 4.6 s, 142 °C.
[b] Score: 0–4: no, weak, moderate, strong, very strong contribution to aroma.

δ-dodecalactone, hydrogen sulfide, 3-methyl butanal, dimethyldisulfide, hexanal, 2,3,4-trithiapentane, 2-trans-nonenal, 2-undecanone, δ-decalactone, and γ-dodecalactone.

A still higher thermal exposure of milk, e.g., by sterilization, allows the accumulation of *Maillard* products, such as maltol, isomaltol, 5-hydroxymethylfurfural, 4-hydroxy-2,5-dimethyl-3(2H)-furanone, and 2,5-dimethylpyrazine.

10.3.2 Condensed Milk, Dried Milk Products

During the concentration and drying of milk, reactions that are similar to those described for heat-treated milk (cf. 10.1.3.5 and 10.3.1) occur, but to a greater extent. Therefore, like the aroma of UHT milk (cf. 10.3.1 and Table 10.39), the aroma of condensed milk is also caused by ketones, lactones, and *Maillard* reaction products. Typical compounds include 2-alkanones, benzaldehyde, acetophenone, maltol, 5-hydroxymethylfurfural, 4-hydroxy-2,5-dimethyl-3(2H)-furanone, 2,5-dimethylpyrazine, furfuryl alcohol, δ-decalactone, benzothiazole, and o-aminoacetophenone. The stale flavor that appears when condensed milk is stored for longer periods is due to higher concentrations of these compounds, especially concentrations of ≥ 1 µg/kg of o-aminoacetophenone formed on degradation of tryptophan. A rubbery aroma defect results from higher concentrations of benzothiazole.

Lactones, 2-alkanones and *Maillard* reaction products are also characteristic of the taste of milk powder. The development of aroma defects during the storage of whole milk powder is due to products of lipid peroxidation.

10.3.3 Sour Milk Products, Yoghurt

Metabolic products of lactic acid bacteria, such as diacetyl, ethanal, dimethylsulfide, acetic acid, lactic acid, and various other aldehydes, ketones, and esters, are characteristic of this aroma. Although CO_2 is not an aroma substance, it appears to be important. In good sour milk products, the concentration ratio of diacetyl/ethanal should be ca. 4. At values of ≤ 3, a green taste appears, which is to be regarded as an aroma defect.

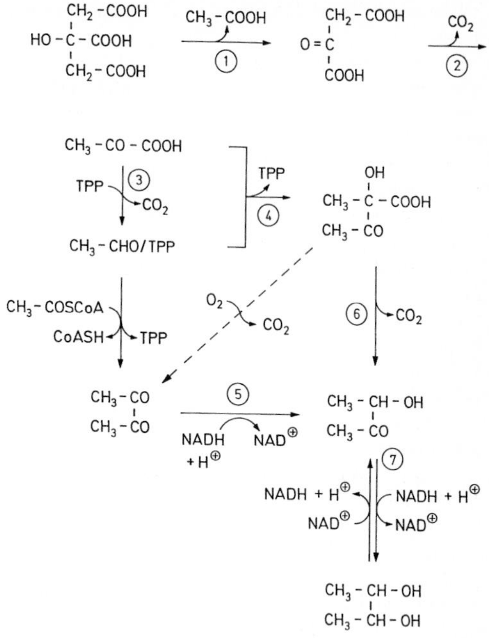

Fig. 10.32. Formation of diacetyl and butanediol from citrate by Streptococci. *1* citratase, *2* oxaloacetate decarboxylase, *3* pyruvate decarboxylase, *4* α-acetolactate synthase, *5* diacetyl reductase, *6* α-acetolactate decarboxylase, *7* 2,3-butanediol dehydrogenase

Diacetyl is formed from citrate (Fig. 10.32). The conversion of acetolactate to diacetyl is disputed. It should occur spontaneously or be catalyzed by an α-acetolactate oxidase.

Ethanal greatly contributes to the aroma of yoghurt. Concentrations of $13-16$ µg/kg are characteristic of good products.

10.3.4 Cream, Butter

The compounds that contribute to the aroma of sweet cream butter include free fatty acids (C_{10}, C_{12}), δ-lactones (C_8, C_{10}, C_{12}), dimethylsulfide, 4-cis-heptenal, indole, and skatole. For sour cream butter, metabolic products of the starter cultures are also of importance, especially diacetyl, lactic acid, and acetic acid. A formulation suitable for the aromatization of sweet cream butter is presented in Table 10.40.

Table 10.40. Formulation of sweet cream butter aroma

Compound	Concentration[a]
Diacetyl	4
3-Methylbutanal	0.01
4-cis-Heptenal	0.006
2-Phenylethanal	0.002
Acetic acid	57
Valeric acid	0.15
Phenol	0.01
p-Cresol	0.005
Guaiacol	0.002
Ethyl butyrate	0.002
γ-Decalactone	3
δ-Decalactone	3.6
δ-Dodecalactone	6.9
Hydrogen sulfide	0.01
Methylthiol	0.01
Dimethylsulfide	0.06
Indole	0.006
Skatole	0.67
Sodium glutamate	1.0
pH 4.6 adjusted by lactic acid	

[a] mg/kg of sweet cream butter.

If butter contains lipases, fatty acids are released on storage. Above certain limiting concentrations (cf. 3.2.1.1), these fatty acids cause a rancid off-flavor. Table 3.23 shows the composition of free fatty acids in butter samples which taste perfect or rancid.

Rancid, soapy aroma defects, which occur in butter samples with very low concentrations of free fatty acids, can be due to contamination with anionic detergents (sodium dodecyl sulfate, sodium dodecyl benzosulfonate). Detergents of this type are used to disinfect the udder and the milking machine.

10.3.5 Cheese

A very large number of compounds contribute to the aroma of cheese. Apart from qualitative differences, the quantitative balance seems to be especially important for the typical aroma note of each variety of cheese.

Table 10.41 shows some formulations with which the aroma of blue cheese (Roquefort) can be simulated. Apart from many other com-

Table 10.41. Formulation of blue cheese aroma (mg/kg cheese) 1: amounts found in blue cheese (+: detected), 2–4: artificial formulations

Compound	1	2	3	4
Acetic acid	825	8	550	
Butyric acid	1,500	6	950	
Propionic acid				126
Isobutyric acid		2		
Valeric acid		3		
Isovaleric acid		3		
Caproic acid	900	265	600	
Heptanoic acid		5.4		
Caprylic acid	770	89	515	
Pelargonic acid		4.5		
Capric acid	2,000	133		
Lauric acid	3,000			
Glyoxylic acid		71		
Cinnamic acid		0.03		
Pyruvic acid		20		
2-Keto isovaleric acid		17		
2-Keto isocaproic acid		43		
Oxalacetic acid		51		
2-Ketoglutaric acid		42		
Butyric acid ethyl ester	+	0.5	1.5	
Caproic acid methyl ester			6	
Caproic acid ethyl ester		0.5		
Caprylic acid methyl ester			6	
Caprylic acid ethyl ester		0.5		
Cinnamic acid methyl ester				4
2-Pentanol	0.5	6	1	
2-Heptanol	6.0	1	12	28
2-Nonanol	3.5	10	7	
1-Octen-3-ol		22		4
Phenylethanol	+		2	
Ethanal		5		
Propanal		2		
Butanal		2		
Pentanal		2		
Phenylethanal		2		
Methional		0.1		
Acetone	3		6	
2-Pentanone	15	8	30	
2-Heptanone	35	65	70	1.6
2-Nonanone	33	116	65	36
2-Undecanone	8	10	17	
2-Tridecanone		2		
δ-Decalactone		0.7		
δ-Dodecalactone		0.7		
Indole		0.1		

pounds, high concentrations of lower fatty acids as well as 2-alkanones and 2-alkanols from lipid metabolism are especially typical of this cheese.

A character impact compound of Camembert is 1-octen-3-ol, which is present in higher concentrations and is responsible for the mushroom aroma note. The compounds 2-phenylethanol and 2-phenylethylacetate are involved in the flowery note and 1,3-dimethoxybenzene and cinnamic acid methyl ester in the hazelnut note. The garlic note of ripe Camembert is due to sulfur compounds, e. g., 2,4-dithiapentane, 2,4,5-trithiahexane, and 3-methylthio-2,4-dithiapentane.

In cheeses with bacterial surface ripening, e. g., Pont l'Eveque, phenol, cresol and acetophenone as well as methylthio esters of lower fatty acids (C_2, C_3, C_4, methyl C_4) are of importance for the aroma.

Methyl thioacetate has also been found in cheeses which undergo propionic acid fermentation, e. g., Emmentaler and Gruyere. Apart from propionic acid, other lower fatty acids and some keto acids, various alcohols and esters, as well as monoamines and alkylpyrazines appear to play a role in the generation of aroma.

It is probable that not only peptides, but also other amides are responsible for the bitter taste of cheese. For example, the presence of bitter N-isobutyl acetamide has been detected in Camembert cheese.

10.3.6 Aroma Defects

As already indicated, aroma defects can arise in milk and milk products either by absorption of aroma substances from the surroundings or by formation of aroma substances via thermal and enzymatic reactions.

Exogenous aroma substances from the feed or cowshed air enter the milk primarily via the respiratory or digestive tract of the cow. Direct absorption apparently plays only a minor role. Metabolic disorders of the cow can cause aroma defects, e. g., the acetone content of milk is increased in ketosis.

The oxidation of lipids is involved in the endogenous formation of aroma defects. While very low concentrations of certain carbonyl compounds, e. g., 4-cis-heptenal (1 µg/kg), 1-octen-3-one, and hexanal, appear to contribute to the full creamy taste, increased concentrations of these and other compounds produce cardboard-like, metallic, and green aroma notes. In butter, for instance, the phospholipids of the fat globule membrane are especially susceptible to oxidation. The subsequent products get distributed in the entire fat fraction and cause taste defects which range from metallic to fatty and from fishy to tallowy.

Heat-induced changes in aroma have already been discussed. It is characteristic that with increasing heat exposure, hydrogen sulfide and other volatile sulfur compounds, 2-alkanones, lactones, and *Maillard* compounds become increasingly important. Light can cause the degradation of methionine to 3-methylthiopropanal via riboflavin as sensitizer. Together with other sulfides and methylthiol, this sulfur compound produces the aroma defect of milk and milk products called "light taste".

A series of aroma defects are caused by enzymatic reactions. These include:

- An unclean taste due to an increased concentration of dimethylsulfide produced by psychotropic microorganisms.
- A fruity taste due to the formation of ethyl esters produced by psychotropic microorganisms, e. g., *Pseudomonas fragii*.
- A malty taste due to increased formation of 3-methylbutanal, 2-methylbutanal, and methylpropanal by *Strept. lactis var. maltigenes*.
- A phenolic taste due to spores of *Bacillus circulans*.
- A rancid taste due to the release of lower fatty acids (C_4–C_{12}) by milk lipases or bacterial lipases.
- A bitter taste can occur due to proteolytic activity, e. g., on storage of UHT milk. The milk proteinase plasmin is inactivated on intensive heating ($142\,°C$, > 16 s). However, some bacterial proteinases can still be active even after much longer exposure to heat ($142\,°C$, 6 min).

10.4 Literature

Adda, J., Roger, S., Dumont, J.-P.: Some recent advances in the knowledge of cheese flavor. In: Flavor of foods and beverages (Eds.: Charalambous, G., Inglett, G.E.), p. 65, Academic Press: New York. 1978

Aimutis, W.R., Eigel, W.N.: Identification of λ-casein as plasmin-derived fragments of bovine α-casein. J. Dairy Sci. *65*, 175 (1982)

Andrews, A.T.: The formation and structure of some proteose-peptone components. J. Dairy Res. *46*, 215 (1979)

Badings, H.T., De Jong, C., Dooper, R.P.M., De Nijs, R.C.M.: Rapid analysis of volatile compounds in food products by purge- and cold-trapping/capillary gas chromatography. In: Progress in flavor research 1984 (Ed.: Adda, J.), p. 523, Elsevier Science Publ., Amsterdam. 1985

Bringe, N.A., Kinsella, J.E.: The effects of pH, calcium chloride and temperature on the rate of acid coagulation of casein micelles. Am. Dairy Sci., Annual Meeting, 23–26 June, Davis, CA. (1986)

Bottazzi, V.: Other fermented dairy products. In: Biotechnology (Eds.: Rehm, H.-J., Reed, G.), Vol. 5, p. 315, Verlag Chemie: Weinheim. 1983

Creamer, L.K., Richardson, T., Parry, D.A.D.: Secondary structure of bovine α_{s1}-and β-casein in solution. Arch. Biochem. Biophys. *211*, 689 (1981)

Dalgleish, D.G.: Coagulation of renneted bovine casein micelles: dependence on temperature, calcium ion concentration and ionic strength. J. Dairy Res. *50*, 331 (1983)

Davies, F.L., Law, B.A.: Advances in the microbiology and biochemistry of cheese and fermented milk. Elsevier Appl. Science Publ., London. 1984

Dirks, U., Reimerdes, E.H.: Analytik von Gangliosiden unter besonderer Berücksichtigung der Milch. Z. Lebensm. Unters. Forsch. *186*, 99 (1988)

Eigel, W.N., Butler, J.E., Ernstrom, C.A., Farrell Jr., H.M., Harwalkar, V.R., Jenness, R., Whitney, R.McL.: Nomenclature of Proteins of Cow's Milk: Fifth Revision, J. Dairy Sci. *67*, 1599 (1984)

Graham, E.R.B., Malcolm, G.N., McKenzie, H.A.: On the isolation and conformation of bovine β-casein A[1]. Int. J. Biol. Macromol. *6*, 155 (1984)

Henle, T., Walter, H., Krause, I., Klostermeyer, H.: Efficient determination of individual *Maillard* compounds in heat-treated milk products by amino acid analysis. Int. Dairy Journal *1*, 125 (1991)

Jolles, P., Fiat, A.-M.: The carbohydrate portion of milk glycoproteins. J. Dairy Res. *46*, 187 (1979)

Juriaanse, A.C., Herrtje, J.: Microstructure of shortenings, margarine and butter – a review. Food Microstructure 7, 181 (1988)

Kaiser, K.-P., Belitz, H.-D., Fritsch, R.J.: Monitoring Cheddar cheese ripening by chemical indices of proteolysis. 2. Peptide mapping of casein fragments by reverse-phase high-performance liquid chromatography. Z. Lebensm. Unters. Forsch. *195*, 8 (1992)

Kessler, H.G.: Lebensmittel- und Bioverfahrenstechnik. Molkereitechnologie. 3. Auflage. Verlag A. Kessler: Freising. 1988

Kessler, H.G.: Leidet der ernährungsphysiologische Wert industriell bearbeiteter Lebensmittel – am Beispiel Milch? Deutsche Molkerei-Zeitung *104*, 482 (1983)

Kiermeier, F., Lechner, E.: Milch und Milcherzeugnisse. Verlag Paul Parey: Berlin. 1973

Kinsella, J.E., Hwang, D.H.: Enzymes of Penicillium roqueforti involved in the biosynthesis of cheese flavor. Crit. Rev. Food Sci. Nutr. *8*, 191 (1977)

Kosikowski, F.V.: "Cholesterol-free" milks and milk products: Limitations in production and labeling. Food Technol. *44* (11), 130 (1990)

Labuza, T.P., Riboh, D.: Theory and application of *Arrhenius* kinetics to the prediction of nutrient losses in foods. Food Technol. *36* (10), 66 (1982)

Nemitz, G.: Technologie der Milch-Nebenprodukte. Landwirtschaftliche Forschung, Sonderheft 33/II, p. 300, Kongreßband 1976 Oldenburg, J.D. Sauerländer's Verlag: Frankfurt/Main. 1977

Olivecrona, T., Bengtsson, G.: Lipase in milk. In: Lipases (Eds.: Borgström, B., Brockman, H.L.), p. 205, Elsevier Science Publ., Amsterdam. 1984

Payens, T.A.J.: Casein micelles: the colloid-chemical approach. J. Dairy Res. *46*, 291 (1979)

Plasser, A.J., Klostermeyer, H.: NPN-Komponenten der Milch als analytische Indikatoren in Lebensmitteln. Z. Lebensm. Unters. Forsch. *187*, 552 (1988)

Stein, J.L., Imhof, K.: Milch und Milchprodukte. In: Ullmanns Encyklopädie der technischen Chemie, 4. Ed., vol. 16, p. 689, Verlag Chemie: Weinheim. 1978

Swaisgood, H.E.: The Caseins. Crit. Rev. Food Technol. *3*, 375 (1972/73)

Timm, F.: Speiseeis. Verlag Paul Parey: Berlin. 1985

Vedamuthu, E.R., Washam, C.: Cheese. In: Biotechnology (Eds.: Rehm, H.-J., Reed, G.), Vol. 5, p. 231. Verlag Chemie: Weinheim. 1983

Walstra, P., Jenness, R.: Dairy chemistry and physics. John Wiley & Sons. New York. 1984

Webb, B.H., Johnson, A.H., Alford, J.A. (Eds.): Fundamentals of dairy chemistry. 2nd edn., AVI Publ. Co.: Westport, Conn. 1974

Woo, A.H., Lindsay, R.C.: Anionic detergent contamination detected in soapy-flavored butters. J. Food Prot. *45*, 1232 (1982)

11 Eggs

11.1 Foreword

Eggs have been a human food since ancient times. They are one of nature's nearly perfect protein foods and have other high quality nutrients. Eggs are readily digested and can provide a significant portion of the nutrients required daily for growth and maintenance of body tissues. They are utilized in many ways both in the food industry and the home. Chicken eggs are the most important. Those of other birds (geese, ducks, plovers, seagulls, quail) are of lesser significance. Thus, the term "eggs", without a prefix, generally relates to chicken eggs and is so considered in this chapter. Table 11.1 gives some data on the production of eggs.

11.2 Structure, Physical Properties and Composition

11.2.1 General Outline

The egg (Fig. 11.1) is surrounded by a 0.2–0.4 mm thick calcareous and porous shell. Shells of chicken eggs are white-yellow to brown, duck's are greenish to white, and those

Table 11.1. World production of eggs, 1996 (1,000 t)[a]

Continent	Chicken eggs	Other eggs
World	43,159	4,179
Africa	1,697	6
America, North-, Central-	6,502	1
America, South-	2,575	25
Asia	22,911	4,075
Europe	9,269	70
Oceania	203	2

Country	Chicken eggs	Country	Other eggs
China	13,995	China	3,453
USA	4,501	Thailand	287
Japan	2,562	Indonesia	128
Russian Fed.	1,747	Viet Nam	80
India	1,540	Philippines	60
Brazil	1,400	Bangladesh	28
Mexico	1,266	Brazil	24
France	1,018	Romania	23
Germany	836	Russian Fed.	17
Spain	696	Malaysia	14
Italy	680		
UK	614	Σ (%)[b]	98
Netherlands	593		
Turkey	560		
Iran	520		
Σ (%)[b]	75		

[a] Including eggs for hatching.
[b] World production = 100 %.

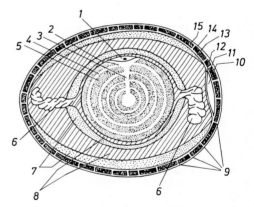

Fig. 11.1. Cross-section of a chicken egg – a schematic representation. Egg yolk: 1 germinal disc (blastoderm), 2 yolk membrane, 3 latebra, 4 a layer of light colored yolk, 5 a layer of dark colored yolk, 6 chalaza, 7 egg white (albumen) thin gel, 8 albumen thick gel, 9 pores, 10 air cell, 11 shell membrane, 12 inner egg membrane, 13 shell surface cemented to the mammillary layer, 14 cuticle, and 15 the spongy calcareous layer

of most wild birds are characteristically spotted. The inside of the shell is lined with two closely adhering membranes (inner and outer). The two membranes separate at the large end of the egg to form an air space, the so-called air cell. The air cell is approx. 5 mm in diameter in fresh eggs and increases in size during storage, hence it can be used to determine the age of eggs. The egg white (albumen) is an aqueous, faintly straw-tinted, gel-like liquid, consisting of three fractions that differ in viscosity. The inner portion of the egg, the yolk, is surrounded by albumen. A thin but very firm layer of albumen (chalaziferous layer) closely surrounds the yolk and it branches on opposite sides of the yolk into two chalazae that extend into the thick albumen.

The chalazae resemble two twisted rope-like cords, twisted clockwise at the large end of the egg and counterclockwise at the small end. They serve as anchors to keep the yolk in the center. In an opened egg the chalazae remain with the yolk. The germinal disc (blastoderm) is located at the top of a clubshaped latebra on one side of the yolk. The yolk consists of alternate layers of dark- and light-colored material arranged concentrically.

The average weight of a chicken egg is 58 g. Its main components are water ($\sim 74\%$), protein ($\sim 12\%$), and lipids ($\sim 11\%$). The proportions of the three main egg parts, yolk, white and shell, and the major ingredients are listed in Table 11.2. Table 11.3 gives the amino acid composition of whole egg, white and yolk.

Table 11.2. Average composition of chicken eggs

Fraction	Percent of the total weight (%)	Dry matter (%)	Pro- tein (%)	Fat (%)	Carbo- hydrates (%)	Minerals (%)
Shell	10.3	98.4	3.3[a]			95.1
Egg white	56.9	12.1	10.6	0.03	0.9	0.6
Egg yolk	32.8	51.3	16.6	32.6	1.0	1.1

[a] A protein mucopolysaccharide complex.

Table 11.3. Amino acid composition of whole egg, egg white and yolk (g/100 g edible portion)

Amino acid	Whole egg	Egg white	Egg yolk
Ala	0.71	0.65	0.82
Arg	0.84	0.63	1.13
Asx	1.20	0.85	1.37
Cys	0.30	0.26	0.27
Glx	1.58	1.52	1.95
Gly	0.45	0.40	0.57
His	0.31	0.23	0.37
Ile	0.85	0.70	1.00
Leu	1.13	0.95	1.37
Lys	0.68	0.65	1.07
Met	0.40	0.42	0.42
Phe	0.74	0.69	0.72
Pro	0.54	0.41	0.72
Ser	0.92	0.75	1.31
Thr	0.51	0.48	0.83
Trp	0.21	0.16	0.24
Tyr	0.55	0.45	0.76
Val	0.95	0.84	1.12

11.2.2 Shell

The shell consists of calcite crystals embedded in an organic matrix or framework of interwoven protein fibers and spherical masses (protein-mucopolysaccharide complex) in a proportion of 50:1. There are also small amounts of magnesium carbonate and phosphates.

The shell structure is divided into four parts: the cuticle or bloom, the spongy layer, the mammillary layer and the pores. The outermost shell coating is an extremely thin (10 μm), transparent, mucilaginous protein layer called the cuticle, or bloom. The spongy, calcareous layer, i.e. a matrix comprising two-thirds of the shell thickness, is below the thin cuticle. The mammillary layer consists of a small layer of compressed, knob-like particles, with one side firmly cemented to the spongy layer and the other side adhering closely to the outer surface of the shell membrane. The shell membrane is made of two layers (48 and 22 μm, respectively), each an interwoven network of protein polysaccharide fibers. The outer layer adheres closely to the mammillary layer. Tiny pore canals which extend through the shell are seen as minute pores or round openings (7,000–17,000 per egg). The cuticle protein partially seals the pores, but they remain permeable to gases while restricting penetration by microorganisms.

11.2.3 Albumen (Egg White)

Albumen is a 10% aqueous solution of various proteins. Other components are present in very low amounts. The thick, gel-like albumen differs from thin albumen (cf. Fig. 11.1) only in its approx. four-fold content of ovomucin. Albumen is a pseudoplastic fluid. Its viscosity depends on shearing force (Fig. 11.2). The surface tension (12.5% solution, pH 7.8, 24 °C) is 49.9 dynes cm^{-1}. The pH of albumen of freshly laid egg is 7.6–7.9 and rises to 9.7 during storage due to diffusion of solubilized CO_2 through the shell. The rise is time and temperature dependent. For example, a pH of 9.4 was recorded after 21 days of storage at 3–35 °C.

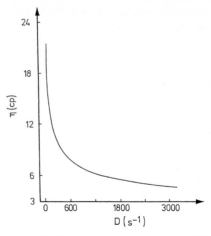

Fig. 11.2. Egg white viscosity, η, as affected by shear rate D, at 10°C. (according to *Stadelman*, 1977)

11.2.3.1 Proteins

Table 11.4 lists the most important albumen proteins in order of their abundance in egg white.
The carbohydrate moieties of the glycoprotein constituents are presented in Table 11.5. Several albumen proteins have biological activity (Table 11.4), i.e., as enzymes (e.g., lysozyme), enzyme inhibitors (e.g., ovomucoid, ovoinhibitor) and complex-forming agents for some coenzymes (e.g., flavoprotein, avidin). The biological activities may be related to protection of the egg from microbial spoilage. Egg white protein separation is relatively easy:

Table 11.4. Proteins of egg white

Protein	Percent of the total protein[a]	Denaturation temperature (°C)	Molecular weight (kdal)	Isoelectric point (pH)	Comments
Ovalbumin	54	84.5	44.5	4.5	
Conalbumin (Ovotransferrin)	12	61.5	76	6.1	binds metal ions
Ovomucoid	11	70.0	28	4.1	proteinase inhibitor
Ovomucin	3.5		$5.5-8.3\times10^6$	4.5–5.0	inhibits viral hemagglutination
Lysozyme (Ovoglobulin G$_1$)	3.4	75.0	14.3	10.7	N-acetylmuramidase
Ovoglobulin G$_2$	4	92.5	30–45	5.5 ⎫	good foam builders
Ovoglobulin G$_3$	4			5.8 ⎭	
Flavoprotein	0.8		32	4.0	binds riboflavin
Ovoglycoprotein	1.0		24	3.9	
Ovomacroglobulin	0.5		760–900	4.5	
Ovoinhibitor	0.1		49	5.1	proteinase inhibitor
Avidin	0.05		68.3[b]	9.5	binds biotin
Ficin inhibitor	0.05		12.7	5.1	inhibits cysteine peptidases

[a] Average values are presented.
[b] Four times 15.6 kdal + approx. 10% carbohydrate.

the albumen is treated with an equal volume of saturated ammonium sulfate; the globulin fraction precipitates together with lysozyme, ovomucin and other globulins; while the major portion of the egg white remains in solution. This albumen fraction consists of ovalbumin, conalbumin and ovomucoid. Further separation of these fractions is achieved by ion-exchange chromatography.

11.2.3.1.1 Ovalbumin

This is the main albumen protein, crystallized by *Hofmeister* in 1890. It is a glycophosphoprotein with 3.2% carbohydrates (Table 11.5) and 0–2 moles of serine-bound phosphoric acid per mole of protein (ovalbumin components A_3, A_2 and A_1, approx. 3, 12 and 85%, respectively). Ovalbumin contains 4 thiol and 1 disulfide group. During storage of eggs, heat-stable S-ovalbumin is formed from native protein, probably by a thiol-disulfide exchange reaction.

The carbohydrate moiety is bound to Asn-292 in the sequence:
–Glu–Lys–Thr–Asn–Leu–Thr–Ser– with a probable structure as follows:

$$(\beta GlcNAc)_{0-1} \rightarrow \alpha Man \rightarrow (\alpha Man)_3(1 \rightarrow$$

$$\uparrow$$
$$(\alpha Man)_{0-1}$$

$$\rightarrow 3)\beta GlcNAc(1-4)\beta GlcNAc \rightarrow Asn$$
$$4$$
$$\uparrow$$
$$1$$
$$\beta Man$$
$$\uparrow$$
$$(\beta GlcNAc)_{0-2} \qquad (11.1)$$

Ovalbumin is relatively readily denatured, for example, by shaking or whipping its aqueous solution. This is an interphase denaturation which occurs through unfolding and aggregation of protein molecules.

11.2.3.1.2 Conalbumin (Ovotransferrin)

Conalbumin and serum transferrin are identical in the chicken. This protein, unlike ovalbumin, is not denatured at the interphase but coagulates at lower temperatures. Conalbumin consists of one peptide chain and contains one

Table 11.5. Carbohydrate composition of some chicken egg white glycoproteins

Protein	Carbo-hydrate (%)	Components (moles/mole protein)				
		Gal	Man	GlcN	GalN	Sialic acid
Ovalbumin	3.2		5	3		
Ovomucoid	23	2	7	23		1
α-Ovomucin[a]	13	21	46	63	6	7
Ovoglyco-protein	31	6	12	19		2
Ovoinhibitor (A)	9.2		10[b]	14		0.2
Avidin[c]	10		4(5)	3		

[a] In addition to carbohydrate, it contains 15 moles of esterified sulfuric acid per mole protein.
[b] Sum of Gal + Man.
[c] Data per subunit (16 kdal).

Table 11.6. Metal complexes of conalbumin

Metal ion	λ max (nm)	ε (1 mol^{-1} cm^{-1})	Complex color
Fe^{3+}	470	3,280	pinkish
Cu^{2+}	440	2,500	yellow
	670	350	
Mn^{3+}	429	4,000	yellow

oligosaccharide unit made of four mannose and eight N-acetylglucosamine residues.

Binding of metal ions (2 moles Mn^{3+}, Fe^{3+}, Cu^{2+} or Zn^{2+} per mole of protein) at pH 6 or above is a characteristic property of conalbumin. Table 11.6 lists the absorption maxima of several complexes. The occasional red discoloration of egg products during processing originates from a conalbumin-iron complex. The complex is fully dissociated at a pH less than 4. Tyrosine and histidine residues are involved in metal binding. Alkylation of 10 to 14 histidine residues with bromoacetate or nitration of tyrosine residues with tetranitromethane removes its iron-binding ability. Conalbumin has the ability to retard growth of microorganisms.

11.2.3.1.3 Ovomucoid

Ion-exhcange chromatography or electrophoresis reveals 2 or 3 forms of this protein, which apparently differ in their sialic acid contents. The carbohydrate moiety (Table 11.5) consists of three oligosaccharide units bound to protein through asparagine residues. The protein has 9 disulfide bonds and, therefore, stability against heat coagulation. Hence, it can be isolated from the supernatants of heat-coagulated albumen solutions, and then precipitated by ethanol or acetone. Ovomucoid inhibits bovine but not human trypsin activities. The proportion of regular structural elements is high (26% of α-helix, 46% of β-structure, and 10% of β-turn).

11.2.3.1.4 Lysozyme (Ovoglobulin G1)

Lysozyme is widely distributed and is found not only in egg white but in many animal tissues and secretions, in latex exudates of some plants and in some fungi. This protein, with three known components, is an N-acetylmuramidase enzyme that hydrolyzes the cell walls of Gram-positive bacteria (murein; AG = N-acetyl-glucosamine; AMA = N-acetylmuramic acid; → = lysozyme attack):

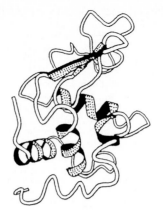

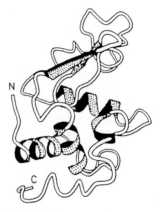

$$-\beta AG(1-4)\beta AMS(1-4)\beta AG(1\rightarrow$$
$$\mid$$
$$\text{Peptide}$$
$$\mid$$
$$\rightarrow 4)\beta AMS(1-4)\beta AG-$$
$$\mid$$
$$\text{Peptide}$$
$$\mid \qquad (11.2)$$

Fig. 11.3. Stereoscopic representation of the tertiary structure of lysozyme from chicken egg white (according to *McKenzie* and *White*, 1991)

Lysozyme consists of a peptide chain with 129 amino acid residues and four disulfide bonds, Its primary (Table 11.7) and tertiary structures have been elucidated (Fig. 11.3).

11.2.3.1.5 Ovoglobulins G2 and G3

These proteins are good foam builders.

11.2.3.1.6 Ovomucin

This protein, of which three components are known, can apparently form fibrillar structures and so contribute to a rise in viscosity of albumen, particularly of the thick, gel-like egg white (see egg structure, Fig. 11.1), where it occurs in a four-fold higher concentration than in fractions of thin albumen.

Ovomucin has been separated into a low-carbohydrate (carbohydrate content ca. 15%) α-fraction and a high-carbohydrate (carbohydrate content ca. 50%) β-fraction. It appears to be associated with polysaccharides. The compositions of its carbohydrate moieties are given in Table 11.5.

Ovomucin is heat stable. It forms a water-insoluble complex with lysozyme. The dissociation of the complex is pH dependent. Presumably it is of importance in connection with the thinning of egg white during storage of eggs.

11.2.3.1.7 Flavoprotein

This protein binds firmly with riboflavin and probably functions to facilitate transfer of this coenzyme from blood serum to egg.

Table 11.7. Amino acid sequences of avidin (1) and lysozyme (2)

1)		Ala	Arg	Lys	*Cys*	Ser	*Leu*	Thr	Gly	Lys	Trp
2)	Lys	Val	Phe	Gly	Arg	*Cys*	Glu	*Leu*	Ala	Ala	Ala Met
1)		Thr	Asn	Asp	Leu	Gly	Ser	*Asn*[a]	Met	Thr	Ile
2)		Lys	Arg	His	Gly	Leu	Asp	*Asn*[a]	Tyr	Arg	Gly
1)		Gly	Ala	Val	Asn	Ser	Arg	Gly	Glu	Phe	Thr
2)		Tyr	Ser	Leu	Gly	Asn	Trp	Val	Cys	Ala	Ala
1)		Gly	Thr	Tyr	Ile	Thr	Ala	Val	*Thr*	Ala	Thr
2)		Lys	Phe	Glu	Ser	Asn	Phe	Asn	*Thr*	Glu	Ala
1)		Ser	*Asn*	Glu	Ile	Lys	Glu	Ser	Pro	Leu	His
2)		Thr	*Asn*	Arg	Asn	Thr	Asp	Gly	Ser	Thr	Asp
1)		Gly	Thr	Glu	Asn	Thr	*Ile*	Asn	Lys	*Arg*	Thr
2)		Tyr	Gly	Ile	Leu	Glu	*Ile*	Asn	Ser	*Arg*	Trp
1)		Gln	Pro	Thr	Phe	*Gly*	Phe	*Thr*	Val	Asn	Trp
2)		Trp	Cys	Asn	Asp	*Gly*	Arg	*Thr*	Pro	Gly	Ser
1)		Lys	Phe	Ser	Glu	Ser	Thr	Thr	Val	Phe	Thr
2)		Arg	Asn	Leu	Cys	Asp	Ile	Pro	Cys	Ser	Ala
1)		Gly	Gln	Cys	Phe	Ile	Asp	Arg	Asn	Gly	Lys
2)		Leu	Leu	Ser	Ser	Asp	Ile	Thr	Ala	Ser	Val
1)		Glu	Val	Leu	*Lys*	Thr	Met	Trp	Leu	Leu	Arg
2)		Asn	Cys	Ala	*Lys*	Lys	Ile	Val	Ser	Asp	Gly
1)		Ser	Ser	Val	*Asn*	Asp	Ile	Gly	Asp	Asp	*Trp*
2)		Asp	Glu	Met	*Asn*	Ala	Trp		Val	Ala	*Trp*
1)		Lys	Ala	Thr	*Arg*	Val	*Gly*	Ile	Asn	Ile	Phe
2)		Arg	Asn	Arg	*Cys*	Lys	*Gly*	Thr	Asp	Val	Gln
1)		Thr	Arg	Leu	*Arg*	Thr	Gln	Lys	Glu		
2)		Ala	Trp	Ile	*Arg*	Gly	Cys	Arg	Leu		

[a] Binding site for carbohydrate.
Italics: Identical amino acid in 1 and 2.

11.2.3.1.8 Ovoinhibitor

This protein is, like ovomucoid, a proteinase inhibitor. It inhibits the activities of trypsin, chymotrypsin and some proteinases of microbial origin. Its carbohydrate composition is given in Table 11.5.

11.2.3.1.9 Avidin

Avidin is a basic glycoprotein (Table 11.5). Its amino acid sequence has been determined. Noteworthy is the finding that 15 positions (12% of the total sequence, Table 11.7) are identical with those of lysozyme. Avidin is a tetramer consisting of four identical subunits, each of which binds one mole of biotin. The dissociation constant of the avidin-biotin complex at pH 5.0 is $k_{-1}/k_1 = 1.3 \times 10^{-15}$ mol/l, i.e., it is extremely low. The free energy and free enthalpy of complex formation are $\Delta G = -85$ kJ/mole and $\Delta H = -90$ kJ/mole, respectively. Avidin, in its form in egg white, is practically free of biotin, and presumably fulfills an antibacterial role. Of interest is the occurrence of

a related biotin-binding protein (streptavidin) in *Streptomyces* spp., which has antibiotic properties.

11.2.3.1.10 Ficin Inhibitor (Cystatin C)

Chicken egg cystatin C consists of one peptide chain with a ca. 120 amino acid residues ($M_r \sim$ 13,000). The two isomers known differ in their isoelectric point (pI 5.6 and pI 6.5) and their immunological properties. This inhibitor inhibits ficin and papain, but not bromelain. In fact, cathepsins B, H, and L and dipeptidyl peptidase I are also inhibited. Cystatin does not act on serine proteinases (trypsin, chymotrypsin, and microbial enzymes).

11.2.3.2 Other Constituents

11.2.3.2.1 Lipids

The lipid content of albumen is negligible (0.03%).

11.2.3.2.2 Carbohydrates

Carbohydrates (approx. 1%) are partly bound to protein (approx. 0.5%) and partly free (0.4–0.5%). Free carbohydrates include glucose (98%) and mannose, galactose, arabinose, xylose, ribose and deoxyribose, totaling 0.2–2.0 mg/100 g egg white. There are no free oligosaccharides or polysaccharides. Bound carbohydrates were covered previously with proteins (cf. 11.2.3.1 Table 11.5). Mannose, galactose and glucosamine are predominant, and sialic acid and galactosamine are also present.

11.2.3.2.3 Minerals

The mineral content of egg white is 0.6%. Its composition is listed in Table 11.8.

Table 11.8. Mineral composition of eggs

	Egg white (%)	Egg yolk (%)
Sulfur	0.195	0.016
Phosphorus	0.018	0.543–0.980
Sodium	0.161–0.169	0.070–0.093
Potassium	0.145–0.167	0.112–0.360
Magnesium	0.009	0.032–0.128
Calcium	0.008–0.02	0.121–0.262
Iron	0.0009	0.0053–0.011

11.2.3.2.4 Vitamins

Data on vitamins found in egg white are summarized in Table 11.12.

11.2.4 Egg Yolk

Yolk is a fat-in-water emulsion with about 50% dry matter content, and consisting of proteins (one-third) and lipids (two-thirds). Water transfer from egg white drops the solid content of the yolk by 2–4% during storage for 1–2 weeks. Yolk contains particles of differing size that can be classified into two groups:

- *Yolk droplets* of highly variable size, with a diameter range of 20–40 μm. They resemble fat droplets, consist mostly of lipids, and some have protein membranes. They are a mixture of lipoproteins with a low density (LDL, cf. 3.5.1.2).
- *Granules* that have a diameter of 1.0–1.3 μm, i.e., they are substantially smaller than yolk droplets, and are more uniform in size but less uniform in shape. They have a substructure and consist of proteins but also contain lipids and minerals.

Older methods of yolk separation, which included at least partial defatting with various solvents (ether, ethanol, butanol), led to lipoprotein destruction and through it to artifacts of varying composition. Yolk studies are now based on ultracentrifugation, when necessary in the presence of electrolytes, which provides native yolk fractions.

Figure 11.4 schematically presents such a fractionation. The granules are separated from the plasma by ultracentrifugation of diluted yolk solution. After NaCl addition, the granules are separated further into a low density lipoprotein fraction (LDL granules' fraction) and a lipovitellin-phosvitin complex. The latter can be separated into its constituents by chromatographic techniques. In the presence of NaCl the plasma can be further separated by centrifugation into a floating, low density lipoprotein fraction (LDL-fraction, lipovitellenin, cf. 3.5.1) and a water-soluble livetin fraction. Table 11.9 provides compositional data on granules, plasma and some of their constituents. The values given are calculated from different literature data and so should be considered only as guiding values. The data available deviate significantly due to methodological difficulties with yolk separation.

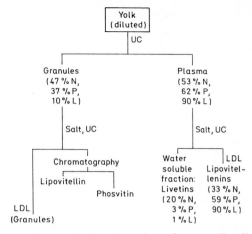

Fig. 11.4. A fractionation scheme for egg yolk. All percentages refer to the total content of the yolk. N: nitrogen, P: phosphorus, L: lipids, UC: ultracentrifugation, LDL: low density lipoprotein

Egg yolk is a pseudoplastic non-*Newton*ian fluid with a viscosity which depends on the shear forces applied. Its surface tension is 0.044 Nm^{-1} (25 °C), while its pH is 6.0 and, unlike egg white, increases only slightly (to 6.4–6.9) even after prolonged storage.

Table 11.9. Composition of egg yolk granules and plasma fractions

Fraction	Lipid	Protein	Minerals
Egg yolk	63.5	32.4	2.1
Granules	*6.9*	*16.1*	*1.4*
Lipovitellins (HDL)[b]	3.5	12.3	
Phosvitin		4.6	
LDL[c]	2.5	0.3	
Plasma	*59.3*	*13.9*	*1.5*
Livetins		10.6	
LDL[c]	59.4	6.6	

[a] Data are expressed as percentage of egg yolk dry matter.
[b] High density lipoprotein.
[c] Low density lipoprotein.

11.2.4.1 Proteins of Granules

11.2.4.1.1 Lipovitellins

The lipovitellin fraction represents high density lipoproteins (HDL). Its lipid moiety is 22% of dry matter and consists of 35% triglycerides, approx. 60% phospholipids and close to 5% cholesterol and cholesterol esters (cf. 3.5.1). The lipovitellins can be separated by electrophoretic and chromatographic methods into their α- and β-components, which differ in their protein-bound phosphorus content (0.50 and 0.27% P, respectively). The protein components probably also contain carbohydrates. At a pH < 7 lipovitellins occur as dimers with a molecular weight of 400 kdal. The amino acid composition is shown in Table 11.10. The protein moiety of this lipoprotein fraction has not been well elucidated. It can be assumed that protein subunits of 30 kdal are present. Lipovitellins occur in the yolk as a complex with phosvitin.

11.2.4.1.2 Phosvitin

Phosvitin is a glycophosphoprotein with an exceptionally high amount of phosphoric acid bound to serine residues. It consists of 2 components, α- and β-phosvitin, which are protein aggregates with molecular weights of 160,000 and 190,000. α-Phosvitin consists of three different subunits (M_r = 37,500, 42,500 and 45,000) and β-phosvitin mainly of one subunit (M_r = 45,000). Its amino acid composition is given in Table 11.10. The partial specific volume of 0.545 ml/g is very low, probably due to the large repulsive charges of the molecule. The frictional ratio suggests the presence of a long, mostly stretched molecular form.

A partial review of its amino acid sequence shows that sequences of 6–8 phosphoserine residues, interrupted by basic and other amino acid residues, are typical of this protein:

$$\cdots Asp—(Pse)_6—Arg—Asp\cdots$$
$$\cdots His—His—Arg—(Pse)_6—Arg—His—Lys\cdots$$

$$(11.3)$$

The carbohydrate moiety is a branched oligosaccharide, consisting of mannose (3 residues), galactose (also 3 residues), N-acetylglucosamine (5) and N-acetylneuraminic acid (2). The oligosaccharide is bound by an N-glycosidic linkage to asparagine. The amino acid sequence in the vicinity of the linkage position is:

$$\cdots Ser—Asn—Ser—Gly—(Pse)_8—Arg—Ser—$$
$$|$$
$$Carbohydrate$$

$$—Val—Ser—His—His\cdots \qquad (11.4)$$

There are indications that phosvitin contains a phosphothreonine residue and that 5–7 serine residues per mole are in free rather than esterified form.

Phosvitin efficiently binds metal ions. Intermolecular complexes are formed through cross linkages in the presence of Ca^{2+} and Mg^{2+}. The Fe^{3+} ion forms a monomeric complex and phosvitin is saturated with iron at a molar ratio of Fe/P = 0.5; this strongly suggests formation of a chelate complex involving two phosphate groups from the same peptide chain per iron. It can be assumed that metal complexing is one of the biological roles of phosvitins.

Table 11.10. Amino acid composition of phosvitin and α- and β-lipovitellins (mole %)

Amino acid	Phosvitin[a]	α-Lipovitellin	β-Lipovitellin
Gly	2.7	5.0	4.6
Ala	3.6	8.0	7.5
Val	1.3	6.2	6.6
Leu	1.3	9.2	9.0
Ile	0.9	5.6	6.2
Pro	1.3	5.5	5.5
Phe	0.9	3.2	3.3
Tyr	0.5	3.3	3.0
Trp	0.5	0.8	0.8
Ser	54.5	9.0	9.0
Thr	2.2	5.2	5.6
Cys	0.0	2.1	1.9
Met	0.5	2.6	2.6
Asx	6.2	9.6	9.3
Glx	5.8	11.4	11.6
His	4.9	2.2	2.0
Lys	7.6	5.7	5.9
Arg	5.3	5.4	5.6

[a] The phosphoric acid content amounts to 50–55 mole-%.

11.2.4.2 Plasma Proteins

11.2.4.1.2 Lipovitellenin

Lipovitellenin is obtained as a floating, low density lipoprotein (LDL) by ultracentrifugation of diluted yolk. Several components with varying densities can be separated by fractional centrifugation. The lipid moiety represents 84–90% of the dry matter and consists of 74% triglycerides and 26% phospholipids. The latter contain predominantly phosphatidyl choline (approx. 75%), phosphatidyl ethanolamine (approx. 18%) as well as sphingomyelin and lysophospholipids (approx. 8%). The molecular weight of lipovitellenin is several million dal. The individual components of this plasma protein are not well characterized.

11.2.4.2.2 Livetin

The water-soluble globular protein fraction can be separated electrophoretically into α-, β- and γ-livetins. These have been proven to correspond to chicken blood serum proteins, i.e. serum albumin, α_2-glycoprotein and γ-globulin.

11.2.4.3 Lipids

Egg yolk contains 32.6% of lipid whose composition is given in Table 11.11. These lipids occur as the lipoproteins described above and, as such, are closely associated with the proteins occurring in yolk.

Table 11.11. Egg yolk lipids

Lipid fraction	a	b
Triacylglycerols	66	
Phospholipids	28	
Phosphatidyl choline		73
Phosphatidyl ethanolamine		15.5
Lysophosphatidyl choline		5.8
Sphingomyelin		2.5
Lysophosphatidyl ethanolamine		2.1
Plasmalogen		0.9
Phosphatidyl inositol		0.6
Cholesterol, cholesterol esters and other compounds	6	

[a] As percent of total lipids.
[b] As percent of phospholipid fraction.

11.2.4.4 Other Constituents

11.2.4.4.1 Carbohydrates

Egg yolk carbohydrates are about 1% of the dry matter, with 0.2% bound to proteins. The free carbohydrates present in addition to glucose are the same monosaccharides identified in egg white (cf. 11.2.3.2.2).

11.2.4.4.2 Minerals

The minerals in egg yolk are listed in Table 11.8.

11.2.4.4.3 Vitamins

The vitamins in egg yolk are presented in Table 11.12.

Table 11.12. Vitamin content of whole egg, egg white and yolk (mg/100 g edible portion)

Vitamin	Whole egg	Egg white	Egg yolk
Retinol (A)	0.22	0	1.12
Thiamine	0.11	Trace	0.29
Riboflavin	0.30	0.27	0.44
Niacin	0.1	0.1	0.1
Pyridoxine (B_6)	0.12	Trace	0.3
Pantothenic acid	1.59	0.14	3.72
Biotin	0.025	0.007	
Folic acid	0.051	0.016	0.15
Tocopherols	1.0	0	3.0
α-Tocopherol	0.46		

11.2.4.4.4 Aroma Substances

The typical aroma substances of egg white and egg yolk are still unknown. The "fishy" aroma defect that can occur in eggs is caused by trimethylamine, which has an odor threshold of 25 µg/kg (pH 7.8). Trimethylamine is formed by the microbial degradation of choline, e.g., on feeding fish meal or soy meal.

11.3 Storage of Eggs

A series of changes occurs in eggs during storage. The diffusion of CO_2 through the pores of the shell, which starts soon after the egg is laid, causes a sharp rise in pH, especially in egg white. The gradual evaporation of water through the shell causes a decrease in density (initially approx. 1.086 g/cm³; the daily reduction coefficient is about 0.0017 g/cm³) and the air cell enlarges. The viscosity of the egg white drops. The yolk is compact and upright in a fresh egg, but it flattens during storage. After the egg is cracked and the contents are released onto a level surface, this flattening is expressed as yolk index, the ratio of yolk height to diameter. Furthermore, the vitellin membrane of the yolk becomes rigid and tears readily once the egg is opened. Of importance for egg processing is the fact that several properties change, such as egg white whipping behavior and foam stability. In addition, a "stale" flavor develops.

These changes are used for determination of the age of an egg, e.g., floating test (change in egg density), flash candling (egg yolk form and position), egg white viscosity test, measurement of air cell size, refractive index, and sensory assay of the "stale" flavor (performed mostly with softboiled eggs). The lower the storage temperature and the lower the losses of CO_2 and water, the lower the quality loss during storage of eggs. Therefore, cold storage is an important part of egg preservation. A temperature of 0 to $-1.5\,°C$ (common chilled storage or subcooling at $-1.5\,°C$) and a relative humidity of 85–90% are generally used. A coating (oiling) of the shell surface with light paraffin-base mineral oil quite efficiently retards CO_2 and vapor escape, but a tangible benefit is derived only if oil is applied soon (1 h) after laying, since at this time the CO_2 loss is the highest. Controlled atmosphere storage of eggs (air or nitrogen with up to 45% CO_2) has been shown to be a beneficial form of egg preservation. Cold storage preserves eggs for 6–9 months, with a particularly increased shelf life with subcooled storage at $-1.5\,°C$. Egg weight loss is 3.0–6.5% during storage.

11.4 Egg Products

11.4.1 General Outline

Egg products, in liquid, frozen or dried forms, are made from whole eggs, white or yolk. They are utilized further as semi-end products in the manufacturing of baked goods, noodles, confectionery, pastry products, mayonnaise and other salad dressings, soup powders, margarine, meat products, ice creams and egg liqueurs. Figure 11.5 gives an overview of the main processing steps involved in manufacturing of egg products.

11.4.2 Technically-Important Properties

The many uses of egg products are basically a result of three properties of eggs: coagulation when heated; foaming ability (whippability); and emulsifying properties. The coloring ability and aroma of egg should also be mentioned.

11.4.2.1 Thermal Coagulation

Egg white begins to coagulate at $62\,°C$ and egg yolk at $65\,°C$. The coagulation temperature is influenced by pH. At a pH at or above 11.9 egg white gels or sets even at room temperature, though after a while the gel liquefies. All egg proteins coagulate, except ovomucoid and phosvitin. Conalbumin is particularly sensitive, but can be stabilized by complexing it with metal ions. Due to their ability to coagulate, egg products are important food-binding agents.

11.4.2.2 Foaming Ability

Whipping of egg white builds a foam which entraps air and hence is used as a leavening agent in many food products (baked goods, angel cakes, biscuits, soufflés, etc.).

Due to a large surface area increase in the liquid/air interphase, proteins denature and aggregate during whipping. In particular, ovomucin forms a film of insoluble material between the liquid lamella and air bubble, thereby stabilizing the foam. Egg globulin also contributes to this effect by increasing the fluid viscosity and by decreasing the surface tension, both effects of importance in the initial stage of the whipping process. In angel cake, egg white without ovomucin and globulins leads to long whipping times and cakes with reduced volumes. An excessive ovomucin content decreases the elasticity of the ovomucin film and thus decreases the thermal stability (expansion of air bubbles) of the foam.

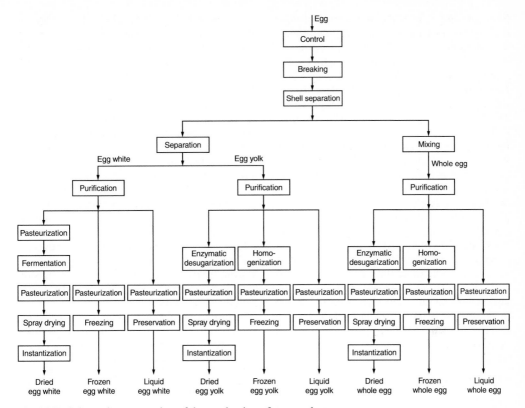

Fig. 11.5. Schematic presentation of the production of egg products

The whippability of egg white can be assayed by measurement of foam volume and foam stability (amount of liquid released from the foam in a given time).

11.4.2.3 Emulsifying Effect

The emulsifying effect of whole egg or egg yolk alone is utilized, for example, in the production of mayonnaise and of creamy salad dressings (made by beating a mixture of egg yolk, olive oil, lemon juice or vinegar, and seasonings: cf. 14.4.6) LD-Lipoproteins and proteins are responsible for the emulsifying action of eggs.

11.4.3 Dried Products

The liquid content of eggs is mixed or churned either immediately or only after egg white and yolk separation. This homogenization is followed by a purification step using centrifuges (separators), and then by a pasteurization step (Fig. 11.5).

The sugars are removed prior to egg drying to prevent reaction between amino components (proteins, phosphatidyl ethanolamines) and reducing sugars (glucose), thereby avoiding undesired brown discoloration and faulty aroma.

Sugars are removed from egg white after pasteurization (cf. 11.4.5), usually by microbiological sugar fermentation. The pasteurized egg liquid is adjusted to pH 7.0–7.5 using citric or lactic acid, and then is incubated at 30–33 °C with suitable microorganisms (*Streptococcus* spp., *Aerobacter* spp.). The sugar in whole egg homogenate or yolk is removed in part by yeasts (e.g. *Saccharomyces cerevisiae*) or mainly by glucoseoxidase/catalase enzymes (cf. 2.7.2.1.1 and 2.7.2.1.2), which oxidize glucose to gluconic acid. Addition of hydrogen peroxide releases oxygen and accelerates the process.

Spray drying with a jet or centrifugal spray drier is the most important egg drying process. The dispersed egg pulp meets a current of warm air, which enters at 120–230 °C. This rapidly reduces the pulp moisture to 5% or less. Whole egg or egg yolk powders are then rapidly cooled. Other egg drying processes, e.g., freeze drying, are rarely applied commercially.

Dried instant powder can be made in the usual way: rewetting and additionally drying the agglomerated particles. Egg white agglomerization is facilitated by addition of sugar (sucrose or lactose).

The shelf life of dried egg white is essentially unlimited. Whole egg powder devoid of sugar has a shelf life of approx. 1 year at room temperature, while sugarless yolk lasts 8 months at 20–24 °C and more than a year in cold storage. The shelf-life of powders containing egg yolk is limited by aroma defects which develop gradually from oxidation of yolk fat. The compositions of dried egg products are given in Table 11.13.

11.4.4 Frozen Egg Products

The eggs are pretreated as described above (cf. 11.4.3 and Fig. 11.5). The homogenate is pasteurized at 63 °C for 1 min (cf. 11.4.5) to lower the microflora count and is then frozen between –23 and –25 °C. The shelf-life of the frozen eggs is 8–10 months at a storage temperature of –15 to –18 °C.

Frozen egg white thickens negligibly after thawing, while the viscosity of egg yolk rises irreversibly when freezing and storage temperatures are below –6 °C (Fig. 11.6). The egg yolk has a gel-like consistency after thawing, which hampers further utilization by dosage metering or mixing. Thawed whole egg gels can cause similar problems, but to a lesser extent than yolk.

Pretreatment of yolk with proteolytic enzymes, such as papain, and with phospholipase A prevents gel formation. Mechanical treatments after thawing of yolk can result in a drop in viscosity. Gel formation can also be prevented by adding 2–10% common salt or 8–10% sucrose to egg yolk (Fig. 11.7). Although salted and sugar-sweetened yolk is of limited acceptability to some manufacturers, this process is of great importance.

The consistency of the frozen egg products is influenced by the temperature gradients during freezing and thawing, and also by storage duration and temperature. Rapid freezing and thawing are best.

The molecular events leading to gel formation by freezing are poorly understood. Apparently, the formation of ice crystals causes a partial dehydration of protein, coupled with a rearrangement of lipoprotein. This probably induces formation of entangled protein strands.

Table 11.13. Composition of dried egg products (values in %)

Constituent	Whole egg	Egg white	Egg yolk
Moisture[a]	5.0	8.0	5.0
Fat[b]	40.0	traces	57.0
Protein[b]	45.0	80.0	30.0
Ash	3.7	5.7	3.4
Reducing sugars[a]	0.1	0.1	0.1

[a] Maximum values.
[b] Minimum values.

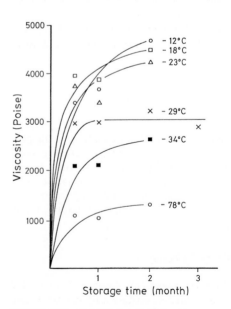

Fig. 11.6. Egg yolk viscosity after frozen storage. (According to *Palmer* et al., 1970)

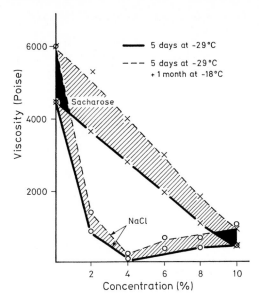

Fig. 11.7. Egg yolk viscosity on addition of NaCl or saccharose and after frozen storage. (according to *Palmer* et al., 1970)

The whippability of egg white can be enhanced by various additives, such as glycerol, starch syrup and triethyl citrate. Typical compositional data for frozen egg products are provided in Table 11.14.

11.4.5 Liquid Egg Products

Eggs are pretreated as described earlier (cf. 11.4.3 and Fig. 11.5). Despite sanitary conditions at plants, eggs cannot be entirely protected from microorganisms. Pasteurization is difficult due to the heat sensitivity of egg protein and the need to kill the pathogens under specific conditions. It is especially important to eliminate *Salmonella* spp., which have varying resistances to heat. The most resistant are *S. senftenberg*, *S. oranienburg* and *S. paratyphi B*. Inactivation of α-amylase occurs as the temperature lethal to *S. senftenberg* is approached; hence, this enzyme can be used as an indicator to monitor the adequacy of the heat treatment. The heating conditions differ for different liquid egg products.

Table 11.14. Composition of frozen and liquid egg products (values in %)

Constituent	Whole egg	Egg white	Egg yolk
Moisture	75.3	88.0	57.0
Fat	11	< 0.03[a]	27.2
Protein	12	10.5	13.5
Reducing sugars	0.7	0.8	0.7

[a] Proportion of egg yolk (weight-%).

A generally adequate heating regimen for whole egg and egg yolk is 64–65 °C for 2.5–3 min, though milder or more drastic conditions are suggested. The conditions for egg white can be milder, since the heat resistance of *Salmonella* spp. is lower at higher pH's. However, at a higher pH protein coagulation occurs at a lower temperature. Addition of salt or sucrose to liquid eggs increases the resistance of microflora to heat.

Most of the egg white proteins are relatively stable at pH 7, so normal pasteurization conditions do not negatively affect processing properties such as whippability. An exception is conalbumin, but addition of metal ions (e.g. Al-lactate) can stabilize even this protein. Addition of Na-hexametaphosphate can also improve the stability of conalbumin.

Pasteurized liquid egg products are generally also preserved by chemical means, e.g., addition of sorbic or benzoic acid.

The compositions of liquid egg products are presented in Table 11.14.

11.5 Literature

Chang, P.K., Powrie, W.D., Fennema, O.: Effect of heat treatment on viscosity of yolk. J. Food Sci. *35*, 864 (1970)

Green, N.M.: Avidin. Adv. Protein Chem. *29*, 85 (1975)

Janssen, H.J.L.: Neuere Entwicklungen bei der Fabrikation von Eiprodukten. Alimenta *10*, 121 (1971)

Maga, J.A.: Egg and egg product flavour. J. Agric. Food Chem. *30*, 9 (1982)

McKenzie, H.A., White jr., F.H.: Lysozyme and α-lactalbumin: Structure, function, and interrelationships. Adv. Protein Chem. *41*, 174 (1991)

Palmer, H.H., Ijichi, K., Roff, H.: Partial thermal reversal of gelation in thawed egg yolk products. J. Food Sci. *35*, 403 (1970)

Spiro, R.G.: Glycoproteins. Adv. Protein Chem. *27*, 349 (1973)

Stadelmann, W.J., Cotterill, O.J. (Eds.): Egg science and technology. 2nd edn., AVI Publ. Co.: Westport, Conn. 1977

Taborsky, G.: Phosphoproteins. Adv. Protein Chem. *28*, 1 (1974)

Tunmann, P., Silberzahn, H.: Über die Kohlenhydrate im Hühnerei. I. Freie Kohlenhydrate. Z. Lebensm. Unters. Forsch. *115*, 121 (1961)

12 Meat

12.1 Foreword

Much evidence from many civilizations has verified that the meat of wild and domesticated animals has played a significant role in human nutrition since ancient times. In addition to the skeletal muscle of warm-blooded animals, which in a strict sense is "meat", other parts are also used: fat tissue, some internal organs and blood. Definitions of the term "meat" can vary greatly, corresponding to the intended purpose. From the aspect of food legislation for instance, the term meat includes all the parts of warm-blooded animals, in fresh or processed form, which are suitable for human consumption. In the colloquial language the term meat means skeletal muscle tissue containing more-or-less adhering fat. Some data concerning meat production and consumption are compiled in Tables 12.1–12.3.

12.2 Structure of Muscle Tissue

12.2.1 Skeletal Muscle

Skeletal muscle tissue consists of long, thin, parallel cells arranged into fiber bundles. Each of these muscle fibers exists as a separate entity surrounded by connective tissue, the endomysium. Numbers of these primary muscle fibers are held together in a bundle which is surrounded by a larger sheet of thin connective tissue, the perimysium. Many such primary bundles are then held together and wrapped by an outer, large, thick layer of connective tissue called the epimysium. Figure 12.1 shows a cross-section of rabbit *Psoas major* muscle in which the endomysium and perimysium are readily recognized.

The membrane surrounding each individual muscle fiber is called the sarcolemma. It consists of three layers: the endomysium, a middle amorphous layer and an inner plasma membrane. The individual myofibrils, the contractile units of the muscle fiber, are within the muscle fiber and are surrounded and imbed-

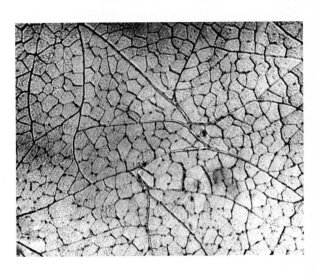

Fig. 12.1. A cross-section of *M. psoas* rabbit muscle. (From *Schultz, Anglemier*, 1964)

Table 12.1. World meat production in 1996 (1,000 t)[a]

Continent	Beef/Veal	Buffalo	Mutton/Lamb	Goat
World	53956	2704	7289	3562
Africa	3418	150	939	670
America, North-, Central-	14654	–	169	51
America, South-	9760	–	301	70
Asia	10374	2552	3247	2642
Europe	13330	1	1542	117
Oceania	2419	–	1089	13

Continent	Pork	Horse	Poultry	Meat, grand total
World	85761	508	58122	215169
Africa	753	13	2293	9328
America, North-, Central-	10197	154	17566	43002
America, South-	2579	92	7163	20079
Asia	47272	113	20046	87077
Europe	24534	113	10450	51073
Oceania	425	24	603	4611

Country	Beef/Veal	Country	Buffalo	Country	Mutton/Lamb
USA	11757	India	1204	China	1250
Brazil	4960	Pakistan	531	Australia	555
China	4604	China	392	New Zealand	534
Russian Fed.	2500	Egypt	150	UK	346
Argentina	2471	Nepal	105	Iran	280
Australia	1702	Viet Nam	97	Pakistan	268
France	1686	Thailand	63	Turkey	260
Germany	1407	Indonesia	45	Russian Fed.	240
Mexico	1349	Philippines	44	Spain	217
India	1292	Myanmar	23	Kazakstan	180
Italy	1185			India	179
Ukraine	1037	Σ (%)[b]	98	Algeria	175
Canada	1015			France	141
UK	712			USA	122
Colombia	698			Afghanistan	117
New Zeland	692			South Africa	111
Japan	600			Syria	98
Netherlands	580			Ireland	91
				Mongolia	91
Σ (%)[b]	75			Sudan	85
				Brazil	84
				Greece	83
				Σ (%)[b]	76

Table 12.1 (continued)

Country	Goat	Country	Pork (swine)	Country	Horse
China	1104	China	40570	Mexico	79
India	490	Germany	3700	China	70
Pakistan	463	France	2160	Italy	56
Nigeria	130	Spain	2100	Argentina	50
Bangladesh	105	Poland	1881	USA	49
Iran	100	Russian Fed.	1700	Mongolia	28
Indonesia	66	Netherlands	1619	Australia	22
Ethiopia	62	Brazil	1520	Brazil	18
Turkey	61	Denmark	1495	Canada	14
Philippines	59	Italy	1430	Chile	11
Greece	48	Japan	1260		
		Philippines	1095	Σ (%)[b]	78
Σ (%)[b]	75	Viet Nam	1052		
		UK	995		
		Mexico	920		
		Korea Rep.	870		
		Σ (%)[b]	75		

Country	Poultry	Country	Meat, grand total
USA	14672	China	60095
China	11647	USA	34564
Brazil	4350	Brazil	10965
France	2027	France	6326
UK	1319	Germany	5840
Mexico	1263	Russian Fed.	5272
Japan	1255	India	4295
Italy	1091	Italy	4072
Indonesia	955	Spain	3800
Canada	886	Mexico	3682
Spain	857	UK	3385
Russian Fed.	795	Canada	3169
Malaysia	701	Japan	3126
Thailand	687	Australia	3109
Iran	672	Netherlands	2870
Netherlands	651	Poland	2691
		Ukraine	2086
Σ (%)[b]	75	Indonesia	2039
		Σ (%)[b]	75

[a] Data refer to slaughtered animals irrespective of the possibility of being imported as live animals.
[b] World production = 100%.

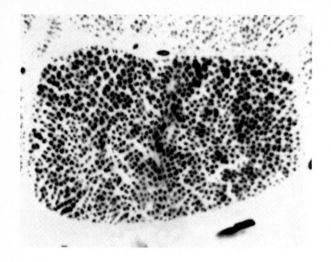

Fig. 12.2. A cross-section of a muscle fiber. (from *Schultz, Anglemier*, 1964)

Table 12.2. Annual meat consumption in FR Germany (kg/person)

Year	Beef/Veal	Pork	Poultry	Total
1960	18.7	29.3	4.2	59.0
1964/65	19.2	33.9	6.0	66.5
1970	22.9	36.1	7.4	72.0
1972/73	20.5	42.0	9.0	79.0
1974/75	21.0	44.6	8.8	82.5
1976/77	21.6	45.5	9.1	84.9

Table 12.3. Meat consumption in selected countries (kg/person/year)

Country/region	Year	Beef/Veal	Pork	Poultry	Total
USA	1960	41.5	29.5	17.2	95.3
	1970	48.6	30.6	18.9	99.9
EEC (European Economic Community)	1960	19.9	19.2	5.2	52.2
	1970	25.2	23.7	8.9	65.7
France	1960	29.2	19.8	8.6	74.9
	1970	35.9	24.8	11.3	89.3
Italy	1960	12.9	7.2	3.6	30.0
	1970	18.9	9.1	9.2	43.5

White muscle (birds, poultry), which has a high ratio of myofibrils to sarcoplasm, contracts rapidly but tires quickly. It can be distinguished from red muscle, which is poor in myofibrils but rich in sarcoplasm. Red muscles are used for slow, long-lasting contractions and do not tire quickly. Figure 12.2 shows a cross-section of a muscle fiber with numerous myofibrils. A greatly magnified, oblique view of a fiber of this type is presented in Fig. 12.3 and Fig. 12.4 shows separated myofibrils.

The organization of the muscle contractile apparatus is revealed in a longitudinal section of the muscle fiber. The characteristic cross-bondings ("striations"; Fig. 12.5) of skeletal

Fig. 12.3. An oblique view of a fractured muscle fiber; scanning electron microscopy at − 180 °C (*Sargent*, 1988)

ded in a homogeneous matrix, the sarcoplasm, as are other subcellular particles, such as nuclei, mitochondria and the sarcoplasmic reticulum.

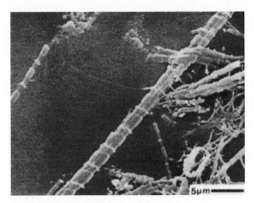

Fig. 12.4. Separated myofibrils; scanning electron microscopy at – 180 °C (*Sargent*, 1988)

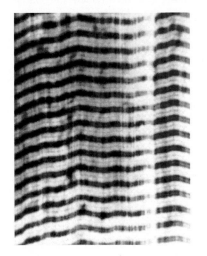

Fig. 12.5. A longitudinal section of two adjacent muscle fibers. (from *Schultz* and *Anglemier*, 1964)

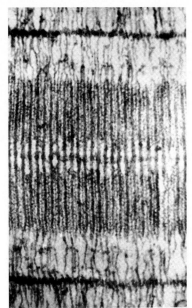

Fig. 12.6. A longitudinal section of a sarcomere. (from *Schultz, Anglemier*, 1964)

muscle are due to the regularly overlapping anisotropic A bands, which double refract polarized light, and the isotropic I bands. The dark bands, the Z line, are in the middle of the light I bands and perpendicular to the axis of the fiber. The dark A bands are crossed in the middle with light H bands, while the dark M line is situated in the middle of the H bands (Fig. 12.6). A single contractile unit of a myofibril, called the sarcomere, stretches from one Z line to the next and consists of thick and thin filaments. The thick filaments are formed from the protein myosin. They stretch through the entire A band and are fixed in a hexagonal

arrangement by the bulge at the center (M line) (Fig. 12.10a). Thin filaments consist mainly of actin. They originate from the Z line and pass across the I band and between the thick filaments to the edge of the H zone, where they penetrate the A bands (Figs. 12.6 and 12.7). During muscle contraction, the mechanism of which is explained in section 12.3.2.1.5, the thick filaments penetrate into the H zones and the Z lines move closer to each other. Thus, the width of the I band gradually decreases and finally disappears. Figure 12.7 schematically presents these changes which take place during muscle contraction.

12.2.2 Heart Muscle

The structure of heart muscle is similar to striated skeletal muscle but has significantly more mitochondria and sarcoplasm.

12.2.3 Smooth Muscle

The smooth muscle cells are distinguished by their centrally located cell nuclei and optically

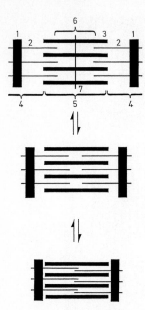

Fig. 12.7. A schematic representation of a sarcomere in a relaxed and contracted state. 1 Z-line, 2 thin filament, 3 thick filament, 4 I-band, 5 A-band, 6 H-zone, and 7 M-line

uniform myofibrils which do not have cross-striations. Smooth muscles occur in mucous linings, the spleen, lymphatic glands, epidermis and intestinal tract. Smooth muscle fibers are useful in the examination of meat products; preferentially for the detection of pharynx (esophagus), stomach or calf pluck (heart, liver and lungs).

12.3 Muscle Tissue: Composition and Function

12.3.1 Overview

Muscles freed from adhering fat contain on the average 76% moisture, 21.5% N-substances, 1.5% lipids and 1% minerals. In addition, variable amounts of carbohydrates (0.05–0.2%) are present. Table 12.4 provides data on the average composition of some cuts of beef, pork and chicken.

12.3.2 Proteins

Muscle proteins can be divided into three large groups (cf. Table 12.5):

- Proteins of the contractile apparatus, extractable with concentrated salt solutions (actomyosin, together with tropomyosin and troponin).
- Proteins soluble in water or dilute salt solutions (myoglobin and enzymes).
- Insoluble proteins (connective tissue and membrane proteins).

12.3.2.1 Proteins of the Contractile Apparatus and Their Functions

About 20 different myofibrillar proteins are known. Myosin and actin quantitatively predominate, acounting for 65–70% of the total protein. The remaining proteins are the tro-

Table 12.4. Average composition of meat (%)

Meat	Cut	Moisture	Protein	Fat	Ash
Pork	Boston butt (*M. subscapularis*)	74.9	19.5	4.7	1.1
	Loin (*M. psoas maior*)	75.3	21.1	2.4	1.2
	Cutlets, chops[a]	54.5	15.2	29.4	0.8
	Ham	75	20.2	3.6	1.1
	Side cuts	40	11.2	48.2	0.6
Beef	Shank	76.4	21.8	0.7	1.2
	Sirloin steak[a]	74.6	22.0	2.2	1.2
Chicken	Hind leg (thigh + drum stick)	73.3	20.0	5.5	1.2
	Breast	74.4	23.3	1.2	1.1

[a] With adhering adipose tissue.

Table 12.5. Muscle proteins

Proteins	Percentage[a]	
Myofibrillar proteins	60.5	
Myosin (H-, L-meromyosin, various associated components)		29
Actin		13
Connectin		3.7
Tropomyosins		3.2
Troponins C, I, T		3.2
α, β-, γ-Actinins		2.6
Myomesin, N-line proteins etc.		3.7
Desmin etc.		2.1
Sarcoplasma proteins	29	
Glyceraldehydephosphate dehydrogenase		6.5
Aldolase		3.3
Creatine kinase		2.7
Other glycolytic enzymes		12.0
Myoglobin		1.1
Hemoglobin, other extracellular proteins		3.3
Connective tissue proteins		
Proteins from organelles	10.5	
Collagen		5.2
Elastin		0.3
Mitochondrial proteins (including cytochrome c and insoluble enzymes)		5.0

[a] Average percentage of the total protein of a typical mammalian muscle after rigor mortis and before other post mortem changes.

pomyosins and troponins, which are important for contraction, and various cytoskeletal proteins, which are involved in the stabilization of the sarcomere.

12.3.2.1.1 Myosin

Myosin molecules form the thick filaments and make up 50–60% of the total proteins present in the contractile apparatus. Myosin can be isolated from muscle tissue with a high ionic strength buffer, for example, 0.3 mol/l KCl/0.15 mol/l phosphate buffer, pH 6.5. The molecular weight of myosin is approx. 500 kdal. Myosin consists of two very long, identical peptide chains (2 × 140 nm) (Fig. 12.8 a). The two peptide chains form a long, double-stranded α-helical rod with a double head of

globular protein, both heads being joined at the same end of the coil (head dimensions, 5 × 20 nm). The myosin ATPase activity is localized in the heads and is required for the interaction of the heads with actin, the protein constituent of the thin filaments. Myosin is cleaved by trypsin into two fragments: light (LMM, M_r 150000) and heavy meromyosin (HMM, M_r 340000). The HMM fraction contains the globular-headed region and has the ATPase activity and the ability to react with actin. Further proteolysis of HMM yields two subfragments S1 and S2, which correspond to

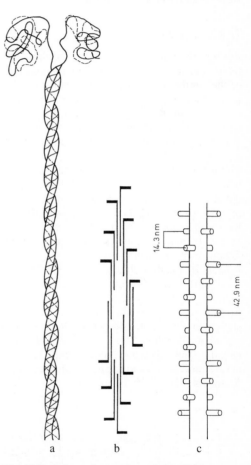

Fig. 12.8. Schematic representations of (**a**) a myosin molecule (according to *Lehninger*, 1975), (**b**) arrangement of myosin molecules in a thick filament (according to *Huxley*, 1963), and (**c**) a thick filament (according to *Lehninger*, 1975)

the actual head and neck. In addition, four other noncovalently bound peptide chains (M_r = 20000) are found in the region of the head.

The individual myosin molecules in the thick filaments are arranged as presented schematically in Fig. 12.8 (b, c). By bringing the tails together, a major cord is formed and on its surface the heads are spirally located. The distance between two adjacent heads on such a spiral is 14.3 nm, and that between the two repeating heads in the same row or line is 42.9 nm. Each filament contains approx. 400 myosin molecules. Their association is reversible under certain conditions.

12.3.2.1.2 Actin

Actin is the main constituent of the thin filament. It makes up 15–30% of the total protein of the contractile apparatus. It is substantially less soluble than myosin, probably because it is fixed to substances in the Z line. Actin can be isolated, for example, by extraction of pulverized, acetone-dried muscle tissue with an aqueous ATP solution.

The actin monomer has a globular shape and hence is designated as G-actin. It has a molecular weight of approx. 46 kdal and binds to myosin. In the presence of salts or ATP and Mg^{2+} (ATP hydrolyzes to ADP which remains bound to actin), actin polymerizes into its fibrous form, F-actin.

F-actin in the thin filaments (1 ~ 1000 nm, d ~ 8 nm) is in the form of a double-stranded helix in which the G-actin beads are stabilized by two tropomyosin fibrils (cf. 12.3.2.1.3), as shown in Fig. 12.9. Altogether, it is a four-stranded filament. Six F-actin strands surround a thick filament; consequently, each F-actin strand adheres to the heads of three thick filaments (Fig. 12.10, b).

12.3.2.1.3 Tropomyosin and Troponin

Tropomyosin is a highly elongated molecule (2 × 45 nm) with a molecular weight of about 70 kdal, and is assumed to be a double-stranded α-helix. The monomer readily forms polymeric fibrils which are bound to F-actin on the thin filament.

Troponin sits on the actin filaments (cf. Fig. 12.9) and controls the contact between the filaments of myosin and actin during muscle contraction by means of a $Ca^{2\oplus}$ concentration-dependent change in conformation (cf. 12.3.2.1.5). It is a complex of three components, T, I, and C. Troponin T consists of a peptide chain with 259 amino acid residues and binds to tropomyosin. Troponin I (179 amino acid residues) binds to actin and inhibits various enzyme activities (ATPase). Troponin C (158 amino acid residues) binds $Ca^{2\oplus}$ ions reversibly through a change in conformation.

12.3.2.1.4 Other Myofibrillar Proteins

Apart from the main components of sarcomeres, myosin and actin, a series of cytoskeletal proteins exist that are responsible for the stabilization of the structure of the sarcomeres. The most important component is *connectin*, an insoluble protein (M_r = 700 000 – 1 000 000) capable of forming fine filaments (g-filaments, d = 2 nm) which start at the Z line and proceed between the thick filaments of neighboring sarcomeres. These g-filaments greatly contribute to the firmness of meat.

Another protein of the cell skeleton is *myomesin* (subunit: M_r = 165 000). As the main component of the M line, myomesin is involved in fixing the thick filaments of the A band and in connecting neighboring myofibrils. Since myomesin strongly binds to myosin, it is possibly involved in the packing and cohesion of

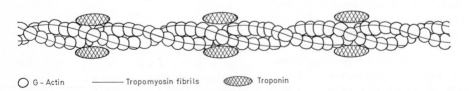

○ G - Actin   ——— Tropomyosin fibrils   ⬡ Troponin

Fig. 12.9. A schematic representation of a thin filament. (according to *Karlsson*, 1977)

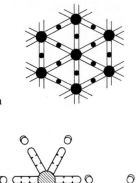

a

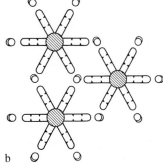

b

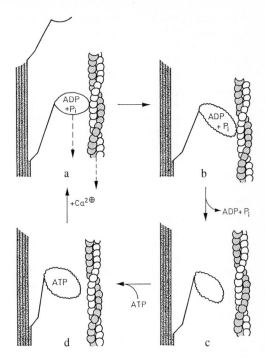

Fig. 12.10. Schematic cross-sections of myofibrils. **a** Hexagonal arrangement of thick filaments (myosin filaments, ●) in the region of the M line; cross linking through M filaments (•) and M bridges (=), which consist of myomesin, among other substances (according to *Lawrie*, 1985). **b** Hexagonal arrangement of thick and thin filaments (myosin and actin filaments) (according to *Schultz, Anglemier*, 1964)

Fig. 12.11. Molecular processes involved in muscle contraction (see text; according to *Karlsson*, 1977)

the myosin molecules in the thick filaments as well.

Among other proteins, α-*actinin* (M_r = 200 000), *desmin* (M_r = 55 000), *vimentin* (M_r = 58 000) and *synemin* (M_r = 23 000) are localized in the Z lines. Desmin appears to connect neighboring myofibrils.

A *N line protein* (M_r = 60 000) has been isolated from the N lines which run parallel to the Z lines on both sides and through the I bands.

12.3.2.1.5 *Contraction and Relaxation*

Muscle stimulation by a nerve impulse triggers depolarization of the outer membrane of the muscle cell and thus release of Ca^{2+} ions from the sarcoplasmic reticulum. The Ca^{2+} concentratioin in the sarcoplasm of the resting muscle increases quickly from 10^{-7} to 10^{-5} mole/l. The binding of this Ca^{2+} to the troponin complex causes a conformational

change in this protein. As a consequence, displacement of the tropomyosin fibrils occurs along the F-actin filament. Thus, the sterically hindered sites on the actin units are exposed for interaction with the myosin heads. The energy required for the shifting of the unbound myosin heads is obtained from the hydrolysis of ATP. The hydrolysis products of ATP, ADP and inorganic phosphate (P_i), remain on the myosin heads, which then bind to the actin monomers (Fig. 12.11a). Consequently, the myosin heads, now bound to actin, are forced to undergo a conformational change, which forces the thin filament to move relative to the thick filament (Fig. 12.11b).

The thin filaments and the heads of the thick filaments reverses half way between the Z lines. Therefore, the two thin filaments which interact with one thick filament are drawn toward each other, resulting in a shortening of the distance between the Z lines.

When the myosin heads release ADP and P_1 and become detached from the thin filaments

(Fig. 12.11c), the heads are ready to take up a fresh charge of ATP (Fig. 12.11d). If the Ca^{2+} concentration in the sarcoplasm remains high, the ATP will again hydrolyze and the interaction of the myosin heads with the thin filament is repeated (Fig. 12.11a). However, if the Ca^{2+} concentration drops in the meantime, no ATP hydrolysis occurs, tropomyosin again blocks the access of myosin heads to the actin binding sites and the muscle returns to its resting state. The decrease in Ca^{2+} concentration when muscle excitation has ceased, as well as the increase in Ca^{2+} during stimulation, i.e. the flow of calcium ions, is controlled by the sarcoplasmic reticulum. The Ca^{2+} concentration is low in the sarcoplasm of the resting muscle, while it is high within the sarcoplasmic reticulum. When the ATP level is low, detachment of the myosin and actin filaments does not occur. The muscle remains in a stiff, contracted state called rigor mortis (cf. 12.4). Hence, relaxation of muscle depends on the presence of regenerated ATP.

12.3.2.1.6 Actomyosin

Solutions of F-actin and myosin at high ionic strength ($\mu = 0.6$) *in vitro* form a complex called actomyosin. The formation of the complex is reflected by an increase in viscosity and occurs in a definite molar ratio: 1 molecule of myosin per 2 molecules of G-actin, the basic unit of the double-helical F-actin strand. It appears that a spike-like structure is formed, which consists of myosin molecules embedded in a "backbone" made of the F-actin double helix. Addition of ATP to actomyosin causes a sudden drop in viscosity due to dissociation of the complex. When this addition of ATP is followed by addition of Ca^{2+}, the myosin ATPase is activated, ATP is hydrolyzed and the actomyosin complex again restored after the ATP concentration decreases.

Upon spinning of an actomyosin solution into water, fibers are obtained which, analogous to muscle fibers, contract in the presence of ATP. Glycerol extraction of muscle fibers removes all the soluble components and abolishes the semipermeability of the membrane. Such a model muscle system shows all the reactions of *in vivo* muscle contraction after the readdition of ATP and Ca^{2+}.

This and similar model studies demonstrate that the muscle contraction mechanism is understood in principle, although some molecular details are still not clarified.

12.3.2.2 Soluble Proteins

Soluble proteins make up 25–30% of the total protein in muscle tissue. They consist of ca. 50 components, mostly enzymes and myoglobin (cf. Table 12.5). The high viscosity of the sarcoplasm is derived from a high concentration of solubilized proteins, which can amount to 20–30%. The glycolytic enzymes are bound to the myofibrillar proteins *in vivo*.

12.3.2.2.1 Enzymes

Sarcoplasm contains most of the enzymes needed to support the glycolytic pathway and the pentosephosphate cycle. Glyceraldehyde-3-phosphate dehydrogenase can make up more than 20% of the total soluble protein. A series of enzymes involved in ATP metabolism, e.g., creatine phosphokinase and ADP-deaminase (cf. 12.3.6 and 12.3.8) are also present.

12.3.2.2.2 Myoglobin

Muscle tissue dry matter contains an average of 1% of the purple-red pigment myoglobin. However, the amounts in white and red meat vary considerably.

Myoglobin consists of a peptide chain (globin) of molecular weight of 16.8 kdal. It has known primary and tertiary structures (Fig. 12.12). The pigment component is present in a hydrophobic pocket of globin and is bound to a histidyl (His^{93}) residue of the protein. The pigment, heme, is the same as that in hemoglobin (blood pigment), i.e. Fe^{2+}-protoporphyrin (Fig. 12.13).

Myoglobin supplies oxygen because of its ability to bind oxygen reversibly. Comparison of the oxygen binding curves for hemoglobin and myoglobin (Fig. 12.14) shows that at low p_{O_2}, such as exists in muscle, hemoglobin releases oxygen to myoglobin. The sigmoidal shape of the O_2-binding curve for hemoglobin is due to its quaternary structure. It consists of four polypeptide chains, with one pigment molecule bound to each. The binding of O_2 to the four pigment molecules occurs cooperatively

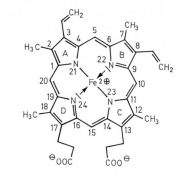

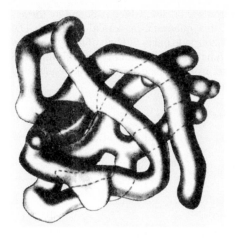

a

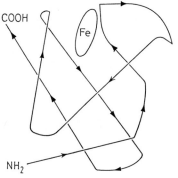

b

Fig. 12.12. Molecular model of myoglobin (**a**) and a schematic representation of peptide chain course (**b**). (from *Schormueller*, 1965)

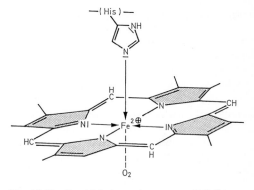

Fig. 12.13. Octahedral environment of Fe^{2+}-protoporphyrin with the imidazole ring of a globin histidine residue and oxygen (according to *Karlsson*, 1977)

because of allosteric effects. Therefore, the degree of saturation, S, is expressed by the following equation (p_{O_2} = oxygen partial pressure; k = dissociation constant for the O_2-protein complex):

$$S = \frac{k \cdot p_{O_2}^n}{1 + k \cdot p_{O_2}^n} \qquad (12.1)$$

For hemoglobin, n ~ 2.8 (sigmoidal saturation curve), and for myoglobin, n = 1 (hyperbolic saturation curve). The efficiency of O_2 transfer from hemoglobin to myoglobin is further enhanced by a decrease in pH since oxygen binding is pH-dependent (the *Bohr* effect).

While in the living animal approx. 10% of the total iron is bound to myoglobin, 95% of all the iron in well-bled beef muscle is bound to

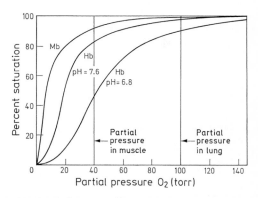

Fig. 12.14. Oxygen binding curves of myoglobin and hemoglobin

myoglobin. Unlike myoglobin, hemoglobin contributes little to the color of meat. The contribution of other pigments, such as the cytochromes, is negligible.

However, attention must be paid to the fact that the visual appearance of a cut of meat is influenced not only by the light absorption of pigments, i.e. primarily myoglobin, but also by light scattering by the surface of muscle fiber. A bright red color is obtained when the coefficient of absorption is high and that of light scattering is low.

Myoglobin (Mb) is purple (λ_{max} = 555 nm); oxymyoglobin (MbO$_2$), a covalent complex of ferrous Mb and O$_2$, is bright red (λ_{max} = 542 and 580 nm); and the oxidation product of Mb in the ferric state, metmyoglobin (MMb$^+$), is brown (λ_{max} = 505 and 635 nm). Some other ligands, such as electron pair donors (e.g. CO, NO, N$_3^-$, CN$^-$), like O$_2$, bind covalently, giving rise to low-spin complexes with similar absorption spectra and hence to a color similar to MbO$_2$. Figure 12.15 shows several absorption spectra of myoglobins.

Heme devoid of globin (free heme, Fe^{2+}-protoporphyrin) does not form the O$_2$-adduct, but oxidizes rapidly to hemin (Fe^{3+}-protoporphyrin). A prerequisite for reversible O$_2$ binding is the presence of an effective donor ligand on the iron's axial site, which is bound by formation of a quadratic-pyramidal complex. The imidazole side chain of His^{93} of myoglobin has this function. Upon interaction with this fifth ligand, iron is raised above the heme plane by about 0.05 nm:

$$(12.2)$$

Binding of the sixth ligand moves the iron to its original position in the heme plane. Since the Fe–N bond distance (His^{93}) remains constant, dislocation of the fifth ligand occurs (His^{93}, proximal His), i.e. a conformational change of the globin takes place.

The basicity of the fifth ligand affects the binding of the sixth ligand. The imidazole ring of His^{93} is a good π-donor and, hence, stabilizes the O$_2$-adduct. A weaker base would enhance oxidation of the iron rather than adduct formation, while a stronger base would increase the stability of the adduct and diminish the probability of iron oxidation. From a biochemical viewpoint, the latter effect is rated as (O$_2$ supplier) negative; while from a food science point of view, it is desirable and positive (stable, bright red meat color).

As mentioned above, His^{93} is located in a hydrophobic pocket of the myoglobin molecule. The electron density and, therefore, the

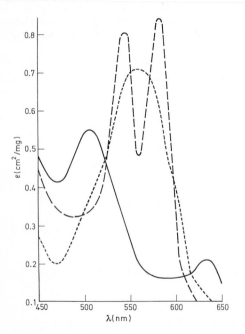

Fig. 12.15. Absorption spectra of myoglobin (----), oxymyoglobin (– – –) and metmyoglobin (——). (according to *Fennema*, 1976)

oxidation state of the iron are regulated by protonation and deprotonation of the imidazole ring. With an increase in pH, there is an increase in basicity and, hence, an increase in binding of O_2 (the *Bohr* effect; cf. Fig. 12.14). A second histidine residue of myoglobin, His^{64} (distal His), contributes to heme-O_2-complex stabilization by formation of a hydrogen bridge or ionic bond between N and O (cf. Formula 12.2).

The color of fresh meat is determined by the ratios of myoglobin (Mb), oxymyoglobin (MbO$_2$) and metmyoglobin (MMb$^+$):

$$\begin{array}{ccc} & +O_2 & \\ Mb & \rightleftharpoons & MbO_2 \\ & -O_2 & \\ & \searrow \quad \nearrow & \\ & MMb^\oplus & \end{array} \qquad (12.3)$$

Stable MbO$_2$ is formed at a high partial pressure of oxygen. Fresh cuts of meat, to a depth of about 1 cm, acquire a bright cherry-red color which is considered a mark of quality. A slow and continuous oxidation to MMb$^+$ occurs at a low partial pressure of O_2. The change of $Fe^{2+} \rightarrow Fe^{3+}$ is reflected in the change in color from red to brown. MMb$^+$ does not form an O_2-adduct, since Fe^{3+} appears to be a less efficient π donor than Fe^{2+}. With better donor ligands than O_2 (CN$^-$, NO, N$_3^-$), low-spin complexes are formed, the spectra of which are similar to those of MbO$_2$.

The change of $Fe^{2+} \rightarrow Fe^{3+}$ is designated as autoxidation:

"Dioxygen-Fe"   "Fe$^{3\oplus}$- superoxide"
(ionic)

$$(12.4)$$

The oxygen molecule dissociates from the heme, taking along an electron from the iron, after protonation of its outer, more negative oxygen atom to form a hydroperoxy radical, the conjugate acid of the superoxide anion (cf.

3.7.2.1.4). The proton may originate from the distal histidine residue or other globin residues or from the surrounding medium. Autoxidation is accelerated by a drop in pH. The reason is the increased dissociation of the protein-pigment complex:

$$Globin + Heme \underset{k_{-1}}{\overset{k_1}{\rightleftharpoons}} Myoglobin \qquad (12.5)$$

Soon after slaughter, the meat has a pH at or near 7, at which the equilibrium constant of the above reaction is $K = k_1/k_{-1} = 10^{12}-10^{15}$ mole^{-1}. Since, during post-rigor, glycolysis decreases the pH of the meat to 5–6, myoglobin becomes increasingly susceptible to autoxidation.

The stability of MbO$_2$ is also highly dependent on temperature. Its half-life, τ, at pH 5 is 2.8 h at 25°C and 5 days at 0°C. Fresh meat has a system which can reduce MMb$^+$ back to Mb. This system appears to be related to that which reduces methemoglobin in erythrocytes. The slow formation of MMb$^+$ can be reversed at the low partial pressure of O_2 which is found inside the cut of meat or in packaged, sealed meat. Therefore, for color stabilty, packaging of meat in O_2-permeable materials is not suitable since, after a time, its reduction capacity is fully exhausted. A non-O_2-permeable material is suitable for packaging meat. All of the pigment is present as Mb and is transformed to the bright red MbO$_2$ only when the package is opened.

Stabilization of the color of meat is also possible under controlled atmosphere packaging. A gaseous mixture of CO and air appears to be advantageous.

Copper ions promote autoxidation of heme to a great extent, while other metal ions, such as Fe^{3+}, Zn^{2+} or Al^{3+}, are less active.

Color stabilization by the addition of nitrate or nitrite (meat curing) plays an important role in meat processing. Nitrite initially oxidizes myoglobin to metmyoglobin:

$$Mb + NO_2^\ominus \longrightarrow MMb^\oplus + NO \qquad (12.6)$$

The resulting NO forms bright-red, highly stable complexes with Mb and MMb$^+$, MbNO and MMb$^+$NO:

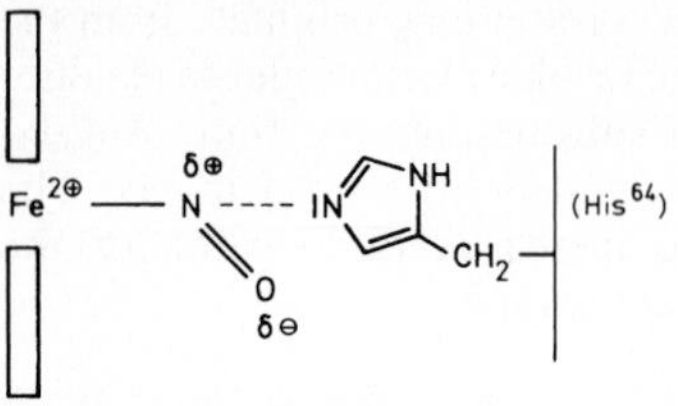

$$Fe^{2\oplus} \underset{\substack{\| \\ O \\ \delta\ominus}}{\overset{\substack{\delta\oplus}}{—N}} ---- N\begin{array}{c} \text{NH} \\ \| \\ \text{CH}_2 \end{array} \Big| (His^{64}) \qquad (12.7)$$

Reducing agents, such as ascorbate, thiols or NADH, accelerate the formation of a red color by reducing nitrite to NO and MMb^+ to Mb. It appears that MMb^+NO, after reduction of iron, is partially converted to MbNO. It is possible that this nitrosylmyoglobin is the sole end product of NO interaction with Mb and MMb^+.

MbNO is highly stable when O_2 is absent. However, in the presence of O_2, the NO released by dissociation of MbNO is oxidized to NO_2.

The color of cured meat is heat stable. Denatured nitrosomyoglobin is present in heated meat, or, due to dissociation of the protein-pigment complex, heme occurs with NO ligands present in both axial binding sites:

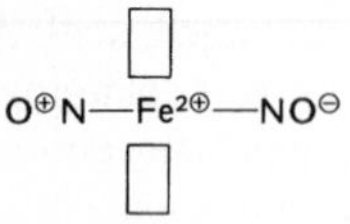

$$O^{\oplus}N—Fe^{2\oplus}—NO^{\ominus} \qquad (12.8)$$

A color change to brown is observed when noncured meat is heated. A Fe^{3+} complex is present which has its fifth and sixth coordina-tion sites occupied by histidine residues of denatured meat proteins.

The myoglobin reactions relevant to meat color are presented schematically in Fig. 12.16.

12.3.2.3 Insoluble Proteins

The main fraction of proteins insoluble in water or salt solutions are the proteins of connective tissue. Membranes and the insoluble portion of the contractile apparatus are included in this group (cf. 12.3.2.1.4).

Connective tissue contains various types of cells. These cells synthesize many intercellular amorphous substances (carbohydrates, lipids, proteins) in which the collagen fibers are embedded.

Lipoproteins are present mostly in membranes. The lipids make up 3–4% of muscle tissue and are located in membranes. They consist of phospholipids, triacylglycerols and cholesterol. The phospholipid portion varies greatly: it makes up 50% of the plasma membrane and 90% of the mitochondrial membrane.

12.3.2.3.1 Collagen

Collagen constitutes 20–25% of the total protein in mammals. Table 12.6 shows data on its amino acid composition. The high contents of glycine and proline and the occurrence of 4-hydroxyproline and 5-hydroxylysine are char-

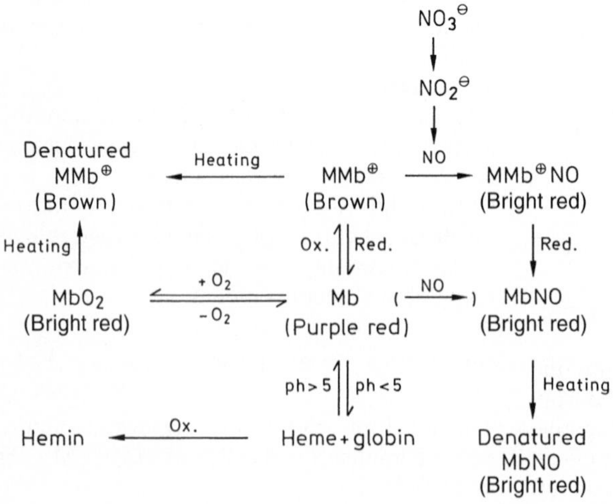

Fig. 12.16. Myoglobin reactions (Mb: myoglobin, MMb⁺: metmyoglobin, MbO₂: oxymyoglobin, MbNO: nitrosomyoglobin, MMb⁺NO: nitrosometmyoglobin)

Table 12.6. Amino acid composition of muscle proteins (values are in g/16 g N)

Amino acid	Beef muscle total	Poultry muscle total [a]	Myosin	Actin	Collagen (calf skin)	Elastin
Aspartic acid	9.7–9.9	9.7–11.0	10.9	10.4	5.4	1.0
Threonine	4.8	3.5–4.5	4.7	6.7	2.1	1.1
Serine	4.1–4.5	–	4.1	5.6	2.9	0.9
Glutamic acid	15.8–16.2	16–18	21.9	14.2	9.7	2.4
Proline	3.0–4.1	–	2.4	4.9	13.0	11.6
Hydroxyproline					10.5	1.5
Glycine	4.6–6.1	4.6–6.7	2.8	4.8	22.5	25.5
Alanine	6.1–6.3	–	6.7	6.1	8.2	21.1
Cystine	1.3–1.5	–	1.0	1.3	0	0.3
Valine	4.8–5.5	4.7–4.9	4.7	4.7	2.9	16.5
Methionine	4.1–4.5	–	3.1	4.3	0.7	Trace
Isoleucine	5.2	4.6–5.2	5.3	7.2 ⎫	4.8 [b]	3.7
Leucine	8.1–8.7	7.3–7.8	9.9	7.9 ⎭		8.6
Tyrosine	3.8–4.0	–	3.1	5.6	1.2	1.3
Phenylalanine	3.8–4.5	3.7–3.9	4.5	4.6	2.2	5.9
Lysine	9.2–9.4	8.3–8.8	11.9	7.3	3.9	0.5
Hydroxylysine					1.1	–
Histidine	3.7–3.9	2.2–2.3	2.2	2.8	0.7	0.1
Arginine	5.3–5.5	5.7–6.1	6.8	6.3	7.6	1.2
Tryptophan	–	–	0.8	2.0	0	–

[a] Chicken, duck, turkey: average values.
[b] Sum of isoleucine and leucine.

acteristic. Since the occurrence of hydroxyproline is confined to connective tissue, its determination may provide quantitative data on the extent of connective tissue incorporation into a meat product.

Collagen also contains carbohydrates (glucose and galactose). These are attached to hydroxylysine residues of the peptide chain by O-glycosidic bonds. The presence of 2-O-α-D-glucosyl-O-β-D-galactosyl-hydroxylysine and of O-β-D-galactosyl-hydroxylysine has been confirmed.

Various types of collagen are known. They are characteristic of different organs and also of different connective tissue layers of muscular tissue (cf. 12.2.1). An overview is presented in Table 12.7. The amino acid sequence of an α^1-chain of collagen type I of mammalian skin is shown in Table 12.8. It is typical that every third residue in this sequence is glycine. Deviations from this regularity have been observed only at the ends of a chain. A frequently recurring sequence is:

–Gly–Pro–Hyp–.

As a result of the specificity of the hydroxylating enzyme in vertebrates, hydroxyproline is always located, as shown in the sequence (Table 12.8) before glycine.

Collagen consists of three peptide chains which can be different or identical, depending on the type (cf. Table 12.7). The three peptide chains, each of which has a helical structure, form together a triple-stranded helix which has a structure corresponding to that of polyglycine II. A triple helix of this type is shown in Fig. 12.17.

The basic structural unit of collagen fibers is called tropocollagen. It has a molecular weight of approx. 30 kdal. With a length of approx. 280 nm and a diameter of 1.4–1.5 nm, collagen is one of the longest proteins. Tropocolla-

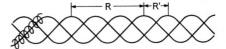

Fig. 12.17. Schematic representation of the conformation of tropocollagen (period R = 8.7 nm, pseudoperiod R′ = 2.9 nm)

Table 12.7. Types of collagen

Type	Peptide chains[a]	Molecular composition	Occurrence
I	α^1, α^2	$[\alpha^1(I)]_2\,\alpha^2(I)$	Skin, tendons, bones, muscle (epimysium)
II	α^1	$[\alpha^1(II)]_3$	Cartilage
III	α^1	$[\alpha^1(III)]_3$	Fetal skin, cardiovascular system, synovial membranes, inner organs, muscle (perimysium)
IV	α^1, α^2	$[\alpha^1(IV)]_3\,(?)^b$	Basal membranes, capsule of lens, glomeruli
		(?)	Placental membrane, lung, muscle (endomysium)
V	$\alpha A, \alpha B,$ $\alpha C\ (?)$	$[\alpha B]_2\,\alpha A$ or $(\alpha B)_3+$ $(\alpha A)_3$ or $(\alpha C)_3\,(?)$	Placental membrane, cardiovascular system, lung, muscle (endomysium), secondary component of many tissues

[a] Since the α chains of various types of collagen differ, they are called α^1 (I), α^1 (II), αA etc.
[b] (?) Not completely elucidated.

Fig. 12.18. Build up of a collagen fiber (**b**) from tropocollagen (**a**) molecules

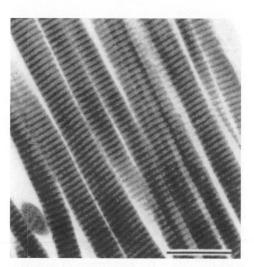

Fig. 12.19. Collagen fibers of bovine muscle (*Extensor carpi radialis*); transmission electron microscopy; sample fixed with 2% glutaraldehyde/paraformaldehyde, bar: 0.5 μm. (according to *Elkhalifa* et al., 1988)

gen fibers associate in a specific way to form collagen fibers, as presented schematically in Fig. 12.18. The association of adjacent rows is not in register, but is displaced by about one-fourth of the tropocollagen length (a "quarter staggered" array). This is responsible for cross-striations in the collagen fibers.
Figure 12.19 shows an electron micrograph of collagen fibers of bovine muscle.
During maturation or aging, collagen fibers strengthen and are stabilized, primarily by covalent cross-linkages. Thus, cross-links confer mechanical strength to collagen fibers.
Cross-link formation involves the following reactions:

- Enzymatic oxidation of lysine and hydroxy-lysine to the corresponding ω-aldehydes.
- Conversion of these aldehydes to aldols and aldimines.
- Stabilization of these primary products by additional reduction or oxidation reactions.

It appears that only certain residues undergo reaction, mainly in the terminal, non-helical regions of the peptide chains.

Lysine and hydroxylysine residues within the peptide chain are oxidized by an enzyme that requires Cu^{2+} and pyridoxal phosphate for its activity and which is related to amine oxidase. This reaction yields an α-aminoadipic acid semialdehyde residue bound to the existing peptide chain (R = H or OH):

$$
\begin{array}{l}
\diagup \\
NH \\
\mid \\
CH\!-\!CH_2\!-\!CH_2\!-\!CHR\!-\!CH_2NH_2 \\
\mid \\
CO \\
\diagdown
\end{array}
$$

$$
\longrightarrow
\begin{array}{l}
\diagup \\
NH \\
\mid \\
CH\!-\!CH_2\!-\!CH_2\!-\!CHR\!-\!C\!\!\diagup^{\textstyle O}_{\textstyle H} \\
\mid \\
CO \\
\diagdown
\end{array}
\qquad (12.9)
$$

Table 12.8. Sequence of mammalian skin collagen, α^1-chain[a]

```
                                          Z* M  S  Y  G  Y  D  E  K  S  A  G  V  S  V  P
G  P  M  G  P  S  G  P  R  G  L  P* G  P  P* G  A  P* G  P  Q  G  F  Q  G  P  P* G  E  P*
G  E  P* G  A  S  G  P  M  G  P  R  G  P  P* G  P  P* G  K  N  G  D  D  G  E  A  G  K  P
G  R  P* G  Q  R  G  P  P* G  P  Q  G  A  R  G  L  P* G  T  A  G  L  P* G  M  K* G  H  R
G  F  S  G  L  D  G  A  K  G  N  T  G  P  A  G  P  K  G  E  P* G  S  P* G  Z  B  G  A  P*
G  Q  M  G  P  R  G  L  P* G  E  R  G  R  P* G  P  P* G  S  A  G  A  R  G  D  D  G  A  V
G  A  A  G  P  P* G  P  T  G  P  T  G  P  P* G  F  P* G  A  A  G  A  K  G  E  A  G  P  Q
G  A  R  G  S  E  G  P  Q  G  V  R  G  E  P* G  P  P* G  P  A  G  A  A  G  P  A  G  N  P*
G  A  D  G  Q  P* G  A  K  G  A  N  G  A  P* G  I  A  G  A  P* G  F  P* G  A  R  G  P  S
G  P  D  G  P  S  G  A  P* G  P  K  G  N  S  G  E  P* G  A  P* G  N  K  G  D  T  G  A  K
G  E  P* G  P  A  G  V  Q  G  P  P* G  P  A  G  E  E  G  K  R  G  A  R  G  E  P* G  P  S
G  L  P* G  P  P* G  E  R  G  G  P* G  S  R  G  F  P* G  A  D  G  V  A  G  P  K  G  P  A
G  E  R  G  S  P* G  P  A  G  P  K  G  S  P* G  E  A  G  R  P* G  E  A  G  L  P* G  A  K
G  L  T  G  S  P* G  S  P* G  P  D  G  K  T  G  P  P* G  P  A  G  Q  D  G  R  P* G  P  A
G  P  P* G  A  R  G  Q  A  G  V  M  G  F  P* G  P  K  G  A  A  G  E  P* G  K  A  G  E  R
G  V  P* G  P  P* G  A  V  G  P  A  G  K  D  G  E  A  G  A  Q  G  P  P* G  P  A  G  P  A
G  E  R  G  E  Q  G  P  A  G  S  P* G  F  Q  G  L  P* G  P  A  G  P  P* G  E  A  G  K  P*
G  E  Q  G  V  P* G  D  L  G  A  P* G  P  S  G  A  R  G  E  R  G  F  P* G  E  R  G  V  E
G  P  P* G  P  A  G  P  R  G  A  N  G  A  P* G  N  D  G  A  K  G  D  A  G  A  P* G  A  P*
G  S  Q  G  A  P* G  L  Q  G  M  P* G  E  R  G  A  A  G  L  P* G  P  K  G  D  R  G  D  A
G  P  K  G  A  D  G  A  P  G  K  D  G  V  R  G  L  T  G  P  I  G  P  P* G  P  A  G  A  P*
G  D  K  G  E  A  G  P  S  G  P  A  G  T  R  G  A  P* G  D  R  G  E  P* G  P  P* G  P  A
G  F  A  G  P  P* G  A  D  G  Q  P* G  A  K  G  E  P* G  D  A  G  A  K  G  D  A  G  P  P*
G  P  A  G  P  A  G  P  P* G  P  I  G  N  V  G  A  P* G  P  K* G  A  R  G  S  A  G  P  P*
G  A  T  G  F  P* G  A  A  G  R  V  G  P  P* G  P  S  G  N  A  G  P  P* G  P  P* G  P  A
G  K  E  G  S  K  G  P  R  G  E  T  G  P  A  G  R  P* G  E  V  G  P  P* G  P  P* G  P  A
G  E  K  G  A  P* G  A  D  G  P  A  G  A  P* G  T  P  G  P  Q  G  I  A  G  Q  R  G  V  V
G  L  P* G  Q  R  G  E  R  G  F  P* G  L  P* G  P  S  G  E  P* G  K  Q  G  P  S  G  A  S
G  E  R  G  P  P* G  P  M  G  P  P* G  L  A  G  P  P* G  E  S  G  R  E  G  A  P* G  A  E
G  S  P* G  R  D  G  S  P* G  A  K  G  D  R  G  E  T  G  P  A  G  A  P* G  P  P* G  A  P*
G  A  P* G  P  V  G  P  A  G  K  S  G  D  R  G  E  T  G  P  A  G  P  I  G  P  V  G  P  A
G  A  R  G  P  A  G  P  Q  G  P  R  G  B  K* G  Z  T  G  Z  Z  G  B  R  G  I  K* G  H  R
G  F  S  G  L  Q  G  P  P* G  P  P* G  S  P* G  E  Q  G  P  S  G  A  S  G  P  A  G  P  R
G  P  P* G  S  A  G  S  P* G  K  D  G  L  N  G  L  P* G  P  I  G  P*P* G  P  R  G  R  T
G  D  A  G  P  A  G  P  P* G  P  P* G  P  P* G  P  P* G  P  P  S  G  G  Y  D  L  S  F  L
P  Q  P  P  Q  Q  Z  K  A  H  D  G  G  R  Y  Y
```

Z*: Pyrrolidone carboxylic acid, P*: 4-hydroxyproline, K*: 5-hydroxylysine.
[a] The sequence is derived from very similar sequences of skin collagen of various mammals.

The two aldehyde-containing chains may interact through an aldol condensation followed by elimination of water, forming a cross-link:

```
Protein         Protein
 |               |
H2C—CHO    +    H2C—CHO    →
 |               |
 I               I

Protein  Protein          Protein Protein
 |        |                |       |
H2C—CHOH—CH—CHO     →     H2C—CH=C—CHO
                 |                II
```
(12.10)

A polypeptide chain with an aldehyde residue (I) can interact with a lysine residue of the adjacent chain to form an aldimine, which can be further reduced to peptide-bound lysinonorleucine (III):

```
Protein           Protein          Protein    Protein
 |                 |                 |          |
H2C—CHO     +     H2N—CH2    →     H2C—CH=N—CH2
 |
 I

        Protein      Protein
         |            |
   →    H2C—CH2—NH—CH2
```
III
(12.11)

If hydroxylysine is involved, an aldimine formed initially can be converted to a more stable β-aminoketone by *Amadori* rearrangement (cf. 4.2.4.4.1):

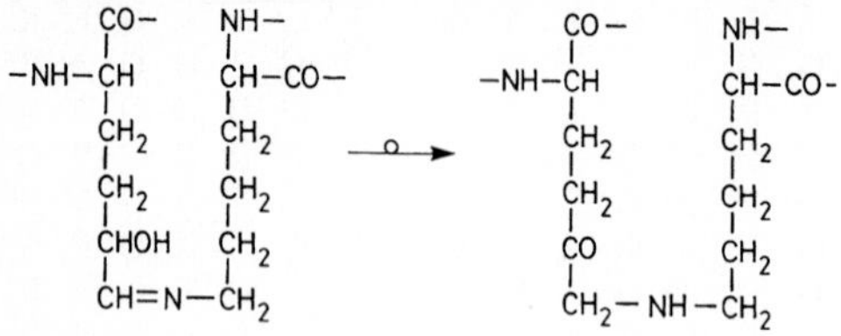

$$\text{(12.12)}$$

Likewise, aldehyde II can interact with one lysine residue through the intermediary dehydromerodesmosine (IV) to merodesmosine (V) and, thus, provide cross-links between the three adjacent polypeptide chains:

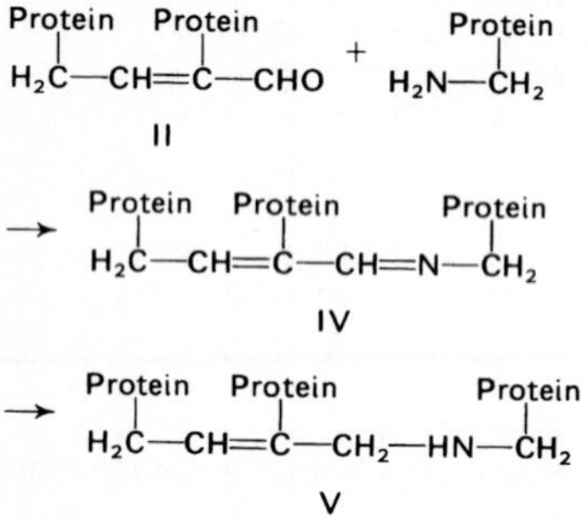

$$\text{(12.13)}$$

During the reaction of three aldehyde molecules of type I with a lysine residue (actually a total of four lysine side chains are involved), a pyridine derivative is formed which, depending on the extent of reduction, yields desmosine (VI), dihydro- (VII) and tetrahydrodesmosine (VIII):

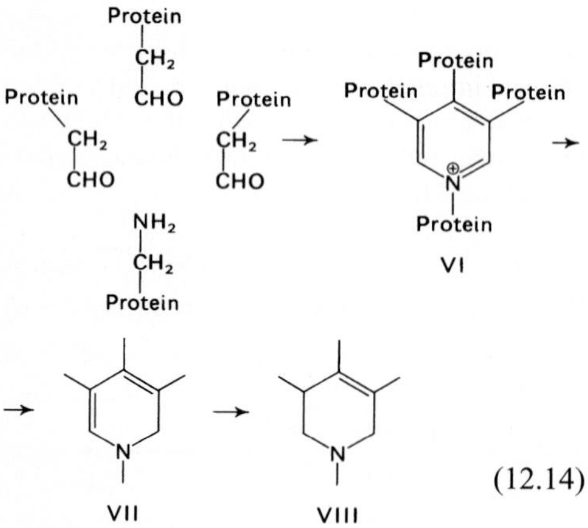

$$\text{(12.14)}$$

Depending on the kind of condensation, in addition to desmosine VI, designated as an A-type condensation product, rings with other substitution patterns are observed, i.e. B- and C-type condensation products:

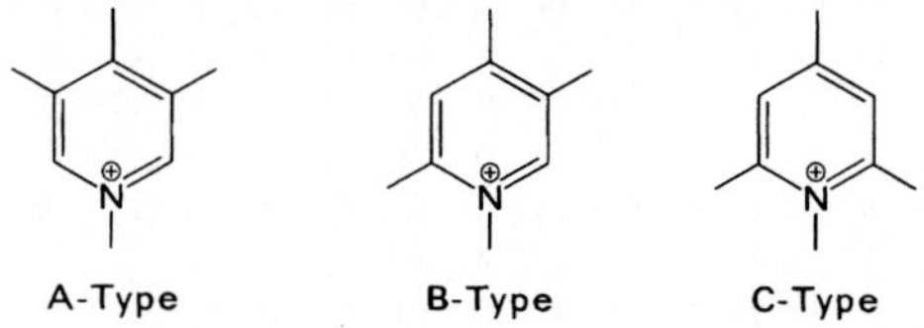

A-Type   B-Type   C-Type

$$\text{(12.15)}$$

Pyridinolines have also been detected. They are probably formed from β-aminoketones and the ω-aldehyde of a hydroxylysine residue:

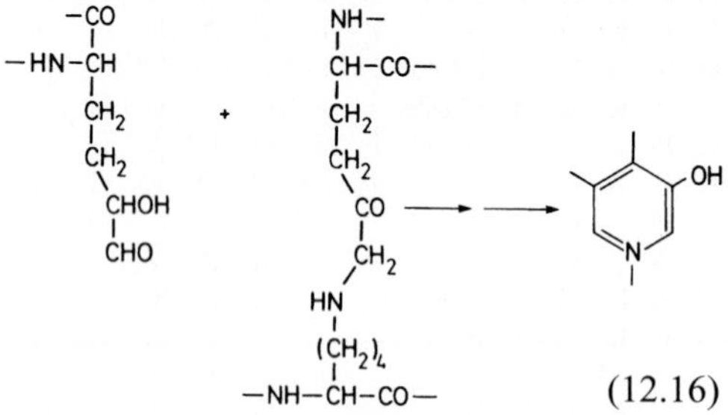

$$\text{(12.16)}$$

Studies of bovine muscle collagens have shown that the pyridinoline content increases with increasing age of the animal and, like the collagen content, negatively correlates with the tenderness. In intensively fattened cows, the pyridinoline content was higher than in extensively fattened animals.

Histidine can also be involved in cross-linking reactions, as shown by the detection of histidino-hydroxylysino-norleucine:

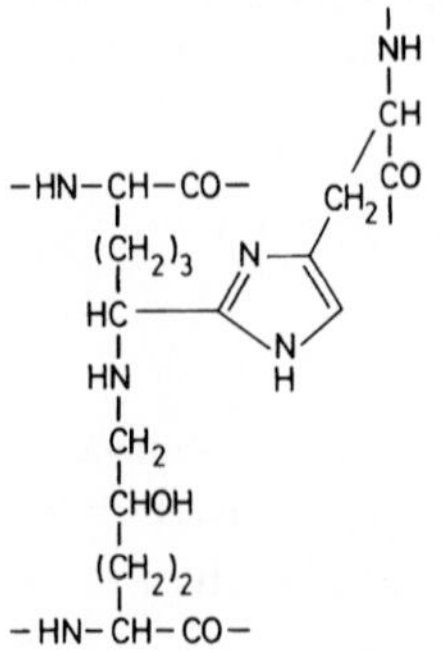

$$\text{(12.17)}$$

The amino acid pentosidine was also obtained from collagen, which indicates the bonding of lysine and arginine with the participation of a pentose:

$$(12.18)$$

The outlined reactions can also occur with hydroxylysine residues present on collagen fibers. Of all the compounds mentioned, hydroxylysino-norleucine and dihydroxyly-sino-norleucine have been isolated from collagen in significant amounts.

In the case of type I, collagen biosynthesis (Fig. 12.20a–h) involves first the synthesis of pro-α^1- and pro-α^2-precursor chains. The N-terminus of these precursors contains up to 25% of extended α^1- and α^2-chains (a). Imme-

diately after the chains are released from polysomes, hydroxylation of the proline and lysine residues occurs (cf. reactions under 12.19).

Realignment of the chains follows: two strands of pro-α^1 and one chain of pro-α^2 are joined to form a triple-stranded helix (b–d). The extended peptides at the N-terminus appear to play a distinct role in these reactions. Di-

Proline hydroxylase (Fe²⊕, ascorbate)

+ 2-oxoglutarate ⟶

$$(12.19)$$

Lysine hydroxylase (Fe²⊕, ascorbate)

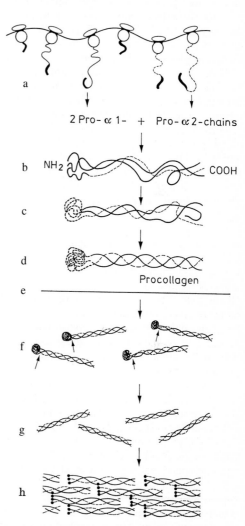

2 Pro-α1- + Pro-α2-chains

Procollagen

Fig. 12.20. Collagen biosynthesis (according to *Bornstein*, 1974). **a** Polysome, **b** hydroxylation, **c** chain straightening, **d** disulfide bond formation, **e** cell membrane, **f** membrane crossing, **g** a limited hydrolysis to tropocollagen, **h** collagen fiber formation, cross-linking

sulfide bridging occurs between the strands at this stage in order to stabilize the structure. The procollagen thus formed will cross the membrane of the cell in which it was synthesized (e). The N-terminal peptides are removed by limited proteolysis (f) and the procollagen is converted to tropocollagen (g). Finally, the tropocollagen is realigned to form collagen fibers (h). At this stage, collagen maturation, which coincides with strengthening of collagen fibers by covalent cross-linking along the peptide strands, begins. The maturation is initiated by oxidation of lysine and is followed by the reactions described above.

Collagen swells but does not solubilize. Enzymatically, it can be hydrolyzed to various extents with a series of collagenases from different sources and with different specificities. A vertebrate animal collagenase, which is a metal proteinase, splits a special bond in native collagen while the collagenase from *Clostridium histolyticum*, also a metal proteinase, cleaves collagen preferentially at glycine residues, forming tripeptides:

$$-Pro—X—Gly—Pro—X—Gly—Pro- \qquad (12.20)$$

Collagenase enzymes which are serine proteinases are also known.

Denatured collagen, as formed post-mortem by the action of lactic acid, can also be cleaved by lysosomal enzymes, e. g., lysosomal collagenase and cysteine proteinase cathepsin B_1. Thermally denatured collagen is attacked by pepsin and trypsin.

One characteristic of the intact collagen fiber is that it shrinks when heated (cooking or roasting). The shrinkage temperature (T_s) is different for different species. For fish collagen, the T_s is 45 °C and for mammals, 60–65 °C. When native or intact collagen is heated to $T > T_s$, the triple-stranded helix is destroyed to a great extent, depending on the cross-links. The disrupted structure now exists as random coils which are soluble in water and are called gelatin. Depending on the concentration of the gelatin solution and of the temperature gradient, a transition into organized structures occurs during cooling. Figure 12.21 schematically summarizes these transitions. At low concentrations, intramolecular back-pleating

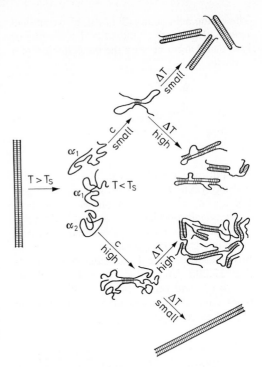

Fig. 12.21. Collagen conversion into gelatin. (according to *Traub* and *Piez*, 1971). T_s: shrinkage temperature, T: temperature, c: concentration; see text)

occurs preferentially with single-strands. At higher concentrations and slow rates of cooling, a structure is rebuilt which resembles the original native structure. At even higher concentrations and rapid cooling, structures are obtained in which the helical segments alternate with randomly coiled portions of the strand. All these structures can immobilize a large amount of water and form gelatin gels.

The transition of collagen to gelatin outlined above occurs during the cooking and roasting of meat. The extent of gelatinization is affected by the collagen cross-linking as determined by the age of the animal and the amount of heat applied (temperature, time, pressure).

Gelatin plays a role as a gelling agent. It is produced on a large scale from animal bones or skin by treatment with alkali or acid, followed by a water extraction step. Depending on the process, products are obtained which differ in molecular weight and, consequently, in their gelling properties. Some brands are used as food gelatins, others play an important role in industry (film emulsions, glue manufacturing).

12.3.2.3.2 Elastin

Elastin is found in lower amounts in connective tissue along with collagen. It is a nonswelling, highly stable protein which forms elastic fibers. The protein has rubber-like properties. It can stretch and then return to its original length or shape. Large amounts of elastin are present in ligaments and the walls of blood vessels. The ligament located in the neck of grazing animals is an exceptionally rich source of this protein. Table 12.6 shows that the amino acid composition is different from that of collagen. The amount of basic amino acids is low, while the content of acids rich in nonpolar aliphatic residues (alanine, valine, leucine, isoleucine) is markedly higher. This may account for the lack of swelling of elastin when it is heated in water.

The elastic properties are put down to the presence of a pentapeptide unit, VPGVG. As a polymer, this unit forms a special conformation, a β-spiral. This is a sequence of β-turns connected by glycine residues, which are wound to give an α-helix:

$$
\begin{array}{ccccccc}
-G-V & V-G-V & V-G-V & V- \\
\quad | \quad | & \quad | \quad | & \quad | \quad | \\
P-G & P-G & P-G
\end{array} \quad (12.21)
$$

For steric reasons, the glycine residue connecting the β-turns cannot be replaced by any other amino acid residue.

Elastin is hydrolyzed by the serine proteinase elastase, which is excreted by the pancreas. This enzyme preferentially cleaves peptide bonds at sites where the carbonyl residue has a nonaromatic, nonpolar side chain.

12.3.3 Free Amino Acids

Fresh beef muscle contains 0.1–0.3% free amino acids (fresh weight basis). All amino acids are detectable in low amounts (< 0.005), with alanine (0.01–0.05%) and glutamic acid (0.01–0.05%) being most predominant.

The free amino acid fraction also contains 0.02–0.1% taurine (I). As such, taurine should be regarded as a major constituent of this fraction. It is obtained biosynthetically from cysteine through cysteic acid and/or from a side pathway involving cysteamine and hypotaurine (II):

$$
\begin{array}{ccc}
\begin{array}{l} CH_2-SH \\ | \\ CH-NH_2 \\ | \\ COOH \end{array}
&\longrightarrow&
\begin{array}{l} CH_2-SO_2H \\ | \\ CH-NH_2 \\ | \\ COOH \end{array}
&\longrightarrow&
\begin{array}{l} CH_2-SO_3H \\ | \\ CH-NH_2 \\ | \\ COOH \end{array}
\end{array}
$$

$$
\downarrow CO_2 \qquad \downarrow CO_2 \qquad \downarrow CO_2
$$

$$
\begin{array}{ccc}
\begin{array}{l} CH_2-SH \\ | \\ CH_2-NH_2 \end{array}
&\longrightarrow&
\begin{array}{l} CH_2-SO_2H \\ | \\ CH_2-NH_2 \end{array}
&\longrightarrow&
\begin{array}{l} CH_2-SO_3H \\ | \\ CH_2-NH_2 \end{array} \\
& II & & I
\end{array}
$$

$$(12.22)$$

The biochemical role of taurine includes derivatization of bile acids (taurocholic and taurodeoxycholic acids). A neurotransmitting function has also been ascribed to this compound.

12.3.4 Peptides

The characteristic β-alanyl histidine peptides, carnosine, anserine and balenine, of muscle are described in section 1.3.4.2.

12.3.5 Amines

Methylamine in fresh beef muscle is present at 2 mg/kg, while the other volatile aliphatic amines (dimethyl-, trimethyl-, ethyl-, diethyl- and isopropylamine) are detected only in trace amounts. Biogenic amines, such as histamine (from histidine), tyramine (tyrosine), tryptamine (tryptophan) and colamine (serine), are obtained by amino acid decarboxylation. These amines are present at 10 mg/kg fresh muscle. Putrescine (from ornithine), cadaverine (from lysine), spermine [1,4-bis-(3′-aminopropylamino)-butane] and spermidine (3′-aminopropyl-1,4-diaminobutane) are present in very low amounts. There is a significant increase in the content of biogenic amines (1–20 g/kg) in muscle due to autolysis, to be described later, and, starting bacterial degradation.

12.3.6 Guanidine Compounds

Creatine and creatinine (I and II, respectively; cf. Formula 12.23) are characteristic constituents of muscle tissue and their assay is used to detect the presence of meat extract in a food product. Creatine is present in fresh beef at 0.3–0.6% and creatinine at 0.02–0.04%.

In living muscle, 50–80% of creatine is in the phosphorylated form, creatine phosphate (III, cf. Formula 12.23), which is in equilibrium with ATP. The reaction rate is highly influenced by the enzyme creatine phosphokinase. Creatine phosphate serves as an energy reservoir (free energy of hydrolysis, $\Delta G^0 = -42.7$ kJ/mole; of ATP: $\Delta G^0 = -29.7$ kJ/mole). Creatine phosphate has a higher phosphoryl group transfer potential than ATP. Hence, when muscle is stimulated for a prolonged period in the absence of glycolysis or respiration, the supply of creatine phosphate will become depleted within a couple of hours by maintaining the ATP concentration. This is especially the case in post-mortem muscle, when the ATP supply has declined significantly through oxidative respiration.

12.3.7 Quaternary Ammonium Compounds

Choline and carnitine are present in muscle tissue at 0.02–0.06% and 0.05–0.2%, respectively (on a fresh weight basis). Choline is synthesized from serine with colamine as an intermediary product (cf. Reactions 12.24) and carnitine is obtained from lysine through ε-N-trimethyllysine and butyrobetaine (cf. Reactions 12.25).

$$
\begin{array}{ccc}
CH_2OH & CH_2OH & CH_2OH \\
CHNH_2 & \longrightarrow & CH_2NH_2 & \longrightarrow & CH_2N^{\oplus}(CH_3)_3 \\
COOH & CO_2 & &
\end{array}
$$
(12.24)

$$
\begin{array}{cccc}
CH_2{-}NH_2 & CH_2N^{\oplus}(CH_3)_3 & CH_2N^{\oplus}(CH_3)_3 \\
CH_2 & CH_2 & CH_2 \\
CH_2 & \longrightarrow & CH_2 & \longrightarrow & CH_2 & \longrightarrow \\
CH_2 & CH_2 & COOH \\
CHNH_2 & CHNH_2 \\
COOH & COOH
\end{array}
$$

$$
\xrightarrow[\text{Fe}^{2\oplus}, \text{ Ascorbate}]{O_2, \text{ 2-Ketoglutarate}}
\begin{array}{c}
CH_2N^{\oplus}(CH_3)_3 \\
CHOH \\
CH_2 \\
COOH
\end{array}
$$
(12.25)

The carnitine fatty acid esters, which are in equilibrium with long chain acyl-CoA molecules in living muscle tissue, are of biochemical importance. The carnitine fatty acid ester, but not the acyl-CoA ester, can traverse the inner mitochondrial membrane. After the fatty acid is oxidized within the mitochondria, carnitine is instrumental in transporting the generated acetic acid out of the mitochondria.

12.3.8 Purines and Pyrimidines

The total content of purines in fresh beef muscle tissue is 0.1–0.25% (on a fresh weight basis). ATP, present predominantly in living tissue, breaks down to inosine-5′-monophosphate (IMP) in the post-mortem stages. The breakdown rate is influenced by the condition

$$
\begin{array}{ccc}
& NH_2 & & & NH{-}PO_3H_2 \\
HN{=} & & + \text{ ATP} \rightleftharpoons & HN{=} & & + \text{ ADP} \\
& N{-}CH_2{-}COOH & & & N{-}CH_2{-}COOH \\
& CH_3 \quad \text{I} & & & CH_3 \quad \text{III}
\end{array}
$$

$$
\begin{array}{c}
HN{=} \\
\begin{array}{c}
H \\
N \quad O \\
\\
N \\
CH_3 \quad \text{II}
\end{array}
\end{array}
$$
(12.23)

Table 12.9. Purines and pyrimidines in fresh-beef muscle

Compound	Content (%)
Inosine-5'-phosphate	0.02–0.2[a]
Inosine	Trace
Hypoxanthine	0.01–0.03
Adenosine-5'-phosphate	0.001–0.01
Adenosine-5'-diphosphate ⎱	
Adenosine-5'-triphosphate ⎰	<0.3[b]
Nicotinamide-adenine-dinucleotide	0.1
Guanosine-5'-phosphate	0.002
Cytidine-5'-phosphate	0.001
Uridine-5'-phosphate	0.002

[a] Until approx. 1 h post-mortem no IMP is found in muscle.
[b] There is a fairly rapid decrease in post-mortem concentration influenced by cooling and other muscle handling conditions.

Table 12.10. Post-mortem changes in the concentration of some constituents of rabbit muscle (*M. psoas*)

Compound	μmol/g Fresh tissue	
	living muscle	post-rigor muscle
Total acid-soluble phosphorus	68	68
Inorganic phosphorus	<12	>48
Adenosine triphosphate (ATP)	9	<1
Adenosine diphosphate (ADP)	1	<1
Inosine monophosphate	<1	9
Creatine phosphate	20	<1
Creatine	23	42
NAD/NADP	2	1
Glycogen	50	<10
Glucose-1-phosphate	<1	<1
Glucose-6-phosphate	5	6
Fructose-1,6-bisphosphate	<1	<1
Lactic acid	10	100

of the animal and by temperature. IMP is then slowly decomposed through successive steps to hypoxanthine, with inosine as an intermediary product:

ATP $\rightleftarrows$ ADP + P$_{in}$ (Myosin-ATPase)

2 ADP $\rightleftarrows$ ATP + AMP (Myokinase)

AMP $\longrightarrow$ IMP + NH$_3$ (Adenylate-deaminase)

IMP $\longrightarrow$ Inosine + P$_{in}$ (5'-Nucleotidase)

Inosine $\longrightarrow$ Hypoxanthine + Ribose (Nucleosidase)

$$(12.26)$$

Post-mortem data on the *Psoas major* rabbit muscle are given in Table 12.10. They relate to nucleotide breakdown and to other important muscle tissue constituents.

The changes in water holding capacity of meat resulting from ATP transition to IMP are dealt with in Section 12.5. Unlike purines, pyrimidine nucleotide content in muscle is very low (Table 12.9).

12.3.9 Organic Acids

The predominant acid in muscle tissue is the lactic acid formed by glycolysis (0.2–0.8% on a fresh meat weight basis), followed by glycolic (0.1%) and succinic acids (0.05%). Other acids of the *Krebs* cycle are present in negligible amounts.

12.3.10 Carbohydrates

The glycogen content of muscle varies greatly (0.02–1.0% on a fresh tissue weight basis) and is influenced by the age and condition of the animal prior to slaugher. The rate of the post-mortem decrease in glycogen als varies. Sugars are only 0.1–0.15% of the weight of fresh muscle, of which 0.1% is shared by glucose-6-phosphate and other phosphorylated sugars. The free sugars present are glucose (0.009–0.09%), fructose and ribose.

12.3.11 Vitamins

Table 12.11 provides data on water-soluble vitamins in beef muscle.

Table 12.11. Vitamins in beef muscle

Compound	mg/kg Fresh tissue
Thiamine	0.6–1.6
Riboflavin	1–3
Nicotinamide	40–120
Pyridoxine, pyridoxal, pyridoxamine	1–4
Pantothenic acid	4–10
Folic acid	0.1–0.3
Biotin	0.05
Cyanocobalamine (B$_{12}$)	0.01–0.02

12.3.12 Minerals

Table 12.12 provides data on minerals in meat. Table 12.13 provides data on the occurrence of soluble and insoluble iron in meat of different animals. The other trace elements, which are 1 mg/kg fresh meat tissue, are not listed individually.

12.4 Post-Mortem Changes in Muscle

Immediately after death, the muscle is soft, limp, and dry and can be reversibly extended by using a low load (5–15 kPa). Cadaveric rigidity (rigor mortis) occurs after a few hours. The muscle can then be extended only by using a heavy load (> 200 kPa) and becomes moist or wet. Rigor can occur in various stages of contraction or stretching. It subsides after some time and the muscle can be easily extended, but irreversibly. Proteolytically caused morphological changes result in a tender consistency (maturation, aging).

12.4.1 Rigor Mortis

Cessation of blood circulation ends the O_2 supply to muscle. Anaerobic conditions start to develop. The energy-rich phosphates, such as creatine phsophate, ATP and ADP, are degraded. The glycolysis process, which is pH and temperature dependent and which is influenced by the presence of glycogen, is the sole remaining energy source. The lactic acid formed remains in the muscle, thereby decreasing the muscle pH from 6.5 to less than 5.8.

Table 12.10 gives an example of post-mortem changes in rabbit *Psoas major* as related to concentrations of some of the more important muscle tissue constituents. The data shown in Fig. 12.22 illustrate the post-mortem decreases in pH, creatine phosphate and ATP in beef *Longissimus dorsi* and *Psoas major* muscles and emphasizes that the changes are dependent on the type of muscle.

Although muscle tissue is soft and flexible and dry on the surface immediately following death, its flexibility or extensibility is lost very rapidly. ATP breaks down (Fig. 12.22). The muscle tissue becomes stiff and rigid (death's stiffening, rigor mortis; cf. 12.3.2.1.5 and 12.3.2.1.6) and, as the rigor proceeds, the muscle tissue surface becomes wetter (the drip or muscle exudate increases).

The onset of rigor mortis occurs in beef muscle within 10–24 h; in pork, 4–18 h; and in chicken, 2–4 h.

The rate of decrease in pH and the final pH value of meat are of significance for water holding capacity and, therefore, for meat quality. Figure 12.23 shows that a more rapid and intensive cooling of the post-mortem muscle results in meat with a noticeably higher water holding capacity than that of muscle cooled slowly.

Table 12.12. Minerals in beef muscle

Element	% in fresh tissue	Element	% in fresh tissue
K	0.25–0.4	Zn	0.001–0.008
Na	0.07–0.2	P (as	
Mg	0.015–0.035	P_2O_5)	0.30–0.55
Ca	0.005–0.025	Cl	0.04–0.1
Fe	0.001–0.005		

Table 12.13. Occurrence of iron in meat of different animal species

Animal species	Concentration (µg/g)[a]		Distribution of soluble iron (%)[a]			
	Insoluble Fe	Soluble Fe	Ferritin	Hemoglobin	Myoglobin	Free Fe
Beef (rump steak)	5.9	20.0	1.6	6.0	89.0	3.4
Pork (loin)	3.0	3.6	8.4	22.2	64.0	5.4
Lamb (loin)	5.9	12.3	7.3	13.0	74.0	5.7
Chicken (leg)	4.7	3.4	26.4	55.7	12.1	5.8

[a] Average value of three meat samples.

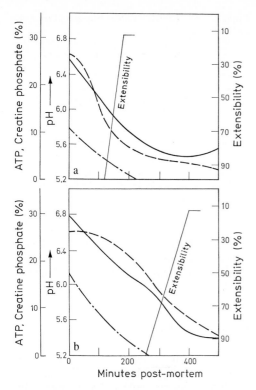

Fig. 12.22. Post-mortem changes in beef muscle.
a *M. longissimus dorsi*; **b** *M. psoas*; —— : pH value;
– – – : ATP as % of the total acid-soluble phosphate;
–·–·– : creatine phosphate as % of the total acid
soluble phosphate. (according to *Hamm*, 1972)

12.4.2 Defects (PSE and DFD Meat)

Rapid drops in ATP and pH (Fig. 12.24) cause pork muscle to become pale and soft and to undergo extensive drip loss because of lowered water holding capacity (PSE meat: pale, soft, exudative). PSE meat has a low tensile strength and loses a substantial amount of weight when hung and, when thawed, drip losses occur. Such defects are typical of hogs with an inherited sensitivity towards stress, such as fear prior to slaughter, anxiety during transport, exposure to temperature changes, etc. Immediately prior to or during slaughtering, an abnormally rapid ATP breakdown occurs and, consequently, the rate of glycolysis is accelerated. A lower pH is thus achieved at a relatively high temperature. As a result, pre-

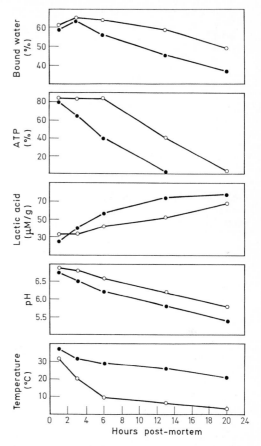

Fig. 12.23. The effect of temperature on post-mortem changes in beef muscle.
M. semimembranosus •–•: normal cooling, animal carcass kept for the first hour post-mortem at $2-4\,°C$ then posterior hind quarters cut and kept at $14\,°C$ for 10 h followed by $2\,°C$; ◦—◦: cooling in ice, hind quarters 11 h in crushed ice, followed by $2\,°C$. Temperature measurement of the meat at 4 cm depth; bound water as percent of total water; lactic acid results are on fresh weight basis and ATP expressed as percent of total nucleotides. (according to *Disney* et al., 1967)

cipitation of sarcoplasmic proteins occurs on the surface of myofibrils and, hence, their swelling properties are changed (Fig. 12.25). The phenomena described above occur preferentially with light-colored muscle tissue, for example, *Longissimus dorsi* or *Glutaeus medius*, while dark muscles of the same animal, e. g., *Rectus abdomini*, are in a normal state.

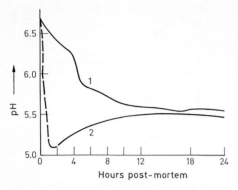

Fig. 12.24. Post-mortem pH-fall in normal pork (1) and in PSE-meat (2); *M. longissimus dorsi* (according to *Briskey* and *Wismer-Pedersen*, 1961)

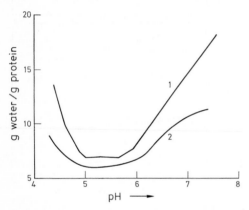

Fig. 12.25. Water holding capacity of washed myofibrils from normal (1) and exudative (2) pork (PSE-meat) (according to *Bendall* and *Wismer-Pedersen*, 1962)

The occurrence of dark and firm pork meat (DFD meat: dark, firm and dry) is likewise characteristic of a stress-impaired hog. In contrast to the PSE effect, it seems that lactate and hydrogen ions move from DFD muscle to blood prior to or soon after slaughtering. Thus, DFD meat, in contrast to PSE meat, has a high pH and a low level of lactic acid 45 min postmortem.

Data relating to normal and faulty cuts of meat are summarized in Table 12.14. Both defects mentioned may occur in different muscles of the same animal. The PSE effect is not significant in beef muscle tissue since energy is available from fat oxidation so glycogen break-

Table 12.14. Some differences between normal and faulty meat[a]

Quality	pH (1 h)	pH (24 h)	ATP	Glycogen	Lactate
Normal meat		6.5	5.8	2.2	6.2	4.7
PSE-Meat	Pale, exudative, loose soft texture	5.6	5.6	0.3	1.9	9.0
DFD-Meat	Dark, sticky, firm texture	6.5	6.3	1.1	1.5	4.0

[a] Pork meat: *M. longissimus dorsi*. Values are averages expressed as mg/g muscle 1 h post-mortem; pH 1 h (initial) and pH 24 h (final) values post-mortem.

down can occur slowly. These meat defects may be avoided in hog muscles by careful handling of stress-sensitive animals and by rapid cooling of carcasses.

12.4.3 Aging of Meat

Rigor mortis in beef muscle tissue is usually resolved 2–3 days post-mortem. By this time, the meat again becomes soft and tender. Further aging of the meat to improve tenderness and to form aroma requires various amounts of time, depending on the temperature. Beef requires 14 days at 0 °C, 6 days at 8–10 °C and 4 days at 16–18 °C. A slight rise in pH is observed with aging, the water holding capacity is increased somewhat and, also, fluid loss from heat-treated meat is slightly decreased. Maturation or aging is accompanied by morphological changes which primarily affect the cytoskeleton. The Z lines get disorganized, the g-filaments soften and disintegrate, and the lateral adhesion between neighboring myofibrils is lost.

The processes involved are proteolytic in nature. For instance, the binding of α-actinin to the Z lines is weakened and, consequently, the fixation of actin filaments as well. The degradation of desmin weakens the assembly of adjacent myofibrils in the region of the Z lines. Soluble proteins, e.g., fibronectins, are also involved in the adhesion of muscle cells. Their concentration decreases with aging. The g-filaments which consist of connectin are also proteolytically degraded.

Various endogenous enzymes are involved in these processes. A protein known as CASF (calcium activated sarcoplasmic factor) is primarily responsible for the attack on the Z line material. The post mortem release of calcium from the endoplasmic reticulum activates this enzyme. The pH optimum of CASF is around 7 and distinct activity is observed starting at pH 6. In vivo, CASF degrades desmin, connectin, M line proteins, troponin T and tropomyosin, but has no effect on myosin and actin.

Lysosomal enzymes also contribute to proteolysis, e. g., cathepsins B, D, and L, which attack troponin T, myosin, and actin, among other proteins.

12.5 Water Holding Capacity of Meat

Muscle tissue contains 20–25% protein and approx. 74–76% water, i.e. 350–360 g water per 100 g protein. Of this total water not more than 5% is bound directly to hydrophilic groups on the proteins. The rest of the water in the muscle tissue, i. e. 95%, is held by capillary forces between the thick and thin filaments. When a larger amount of water is bound to the network, the muscle is more swollen and the meat is softer and juicier. Hence, water holding capacity, protein swelling and meat consistency are intimately interrelated. The extent of water holding by the protein gel network depends on the abundance of cross-links among the peptide chains. These links may be hydrophobic bonds, hydrogen and ionic bonds and may involve divalent metal ions. A decrease in the number of these cross-linkages results in swelling, whereas an increase in the number of cross-linkages results in shrinkage (syneresis) of the protein gel.

The transversal swelling of the myofibrils caused by NaCl has been visualized by phase-contrast microscopy. On washing with 0.6–1.0 mol/L of NaCl, first the centers of the myosin polymer A bands (thick filaments) (cf. 12.3.1 and 12.3.2.1.1) are extracted, and, with increasing concentration, the entire bands are extracted. There is a 2.5 fold increase in the diameter of the myofibrils, corresponding to a 6 fold increase in volume. The cause of these

changes is attributed to the depolymerization of the thick filaments to give soluble myosin molecules and the weakening of the bonding of myosin heads to actin. Furthermore, weakening of transversal structural elements (M line, Z line, cf. 12.3.2.1.4) probably occurs, which facilitates the extension of myofibrils.

The water holding capacity of meat is of great practical importance for meat processing and is affected by pH and the ion environment of the proteins (cf. 1.4.3.1 and 1.4.3.3).

The total charge on the proteins and, hence, their electrostatic interactions are the highest at their isoelectric points. Therefore, meat swelling is minimal in the pH range of 5.0–5.5 (Fig. 12.26).

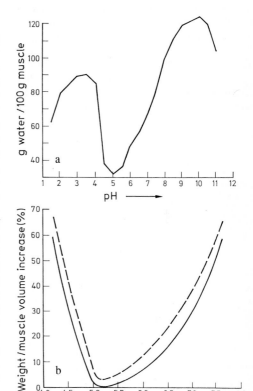

Fig. 12.26. Swelling of meat as affected by pH. **a** Beef muscle homogenate, 5 days post-mortem, **b** beef muscle cut in cubes 3 mm edge length, – – – weight increase, — volume increase (according to *Hamm*, 1972)

Addition of salt to meat shifts the isoelectric point and, hence, the corresponding swelling minimum to lower pH values, due to the preferred binding of the anion. This means that, in the presence of salts, water holding is increased at all pH's higher than the isoelectric point of the unsalted meat (Fig. 12.27).

The water holding capacity of muscle tissue soon after slaughter is high because the muscle is still warm and due to the presence of high concentrations of ATP. After the onset of rigor mortis ATP breaks down, the rigidity of the tissue increases and the water holding capacity starts to decrease (Fig. 12.28). Addition of ATP to muscle tissue homogenates prior to the onset of rigor mortis brings about a rise in tissue swelling (Fig. 12.29). Addition of low levels of ATP (to about 1×10^{-3} molar) during post-rigor brings about tissue contraction or shrinkage, while higher levels of ATP cause tissue swelling (Fig. 12.29). This influence on swelling, however, is of short duration since, as ATP breaks down, contraction and shrinkage take place. Nevertheless, these studies amply illustrate the softening effect of ATP and, as already mentioned, the ability of ATP to dissociate actinmyosin complexes (cf. 12.3.2.1.5 and 12.3.2.1.6). Thus, because of high ATP levels and high pH, the slaughtered muscle which is still warm has a high water holding capacity, whereas post-rigor meat, with low ATP and low pH, has a low water holding capacity.

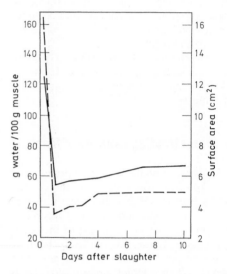

Fig. 12.28. Water holding capacity and rigidity of beef muscle. — Water holding capacity, – – – rigidity (stiffness) expressed as the surface area acquired by homogenate after being pressed between filter papers, under standardized conditions (according to *Hamm*, 1972)

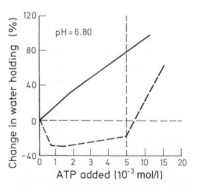

Fig. 12.29. Swelling of meat as affected by ATP addition. Beef muscle homogenate; pH 6.8; — 2 h post-mortem, – – – 4 days post-mortem (according to *Hamm*, 1972)

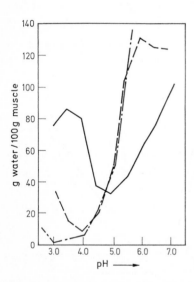

Fig. 12.27. Swelling of meat as affected by salts. Beef muscle homogenate; the ionic strength of the salt added to homogenate is $\mu = 0.20$; — control, – – – NaCl, – · – · – NaSCN (according to *Hamm*, 1972)

12.6 Kinds of Meat, Storage, Processing

Modern slaughterhouses are higly automated. After the delivered animals are stunned either electrically, or by using a bolt apparatus, or with CO_2, they are bled. The blood can be processed into plasma and blood concentrate (hemoglobin). The animal bodies are then passed to skinning machines via scalding vats and unhairing machines. Subsequently, the animals are disemboweled, the red organs and the stomach/intestine package are separated for further processing. The animals sides are passed through a shock tunnel (-4 to $-10\,°C$, $1-2$ h). They are stored in the cold until they are cut up on conveyor belts. During processing, accumulating fat is fed to the grease boiler. All discarded materials and bones are processed into meat and bone meal in carcass processing plants. The waste water is treated using specific processes.

12.6.1 Kinds of Meat, By-Products

12.6.1.1 Beef

Depending on the age of the animal and breed of beef cattle, meat is bright cherry red or dark red and lean or marbled with fat deposits between the connective tissue membranes of the muscle fibers. The average amount of waste from slaughterhouse oxen is $40-55\%$; that from cows, $42-66\%$. Beef carcasses are hung for $4-8$ days before being cut up for soup meat, and $10-14$ days for roasts or steaks.

12.6.1.2 Veal

Veal is white to pale red, with fine, tender fibers. The muscle tissue is generally limp and sticky. The characteristic odor of veal is sour and aromatic and is influenced by the presence of lactic acid. The best age at which to slaughter the calf for veal is $4-14$ weeks. Young calves yield meat with a water content exceeding 80%. The meat is hung for 8 days before use.

12.6.1.3 Mutton and Lamb

Mutton (sheep older than 1 year) and lamb (young sheep) meats are interspersed with fat tissue. The best quality meat is obtained from $2-4$ year old sheep slaughtered in autumn. The flavor and taste are characteristic and are milder for lamb than mutton.

12.6.1.4 Goat Meat

Goat meat is lighter in color than mutton. The male goat (ram) readily gives off a sex aroma if slaughtered improperly.

12.6.1.5 Pork

Pork is the meat from pigs. It has fine fibers, a soft, tender consistency and is pale pink or pink, or greyish-white in color, thus making it different from all other meats. The meat is interspersed with and surrounded by fat, which may be as much as $20-40\%$. The carcasses are hung for $3-4$ days before butchering.

12.6.1.6 Horse Meat

The meat of a young horse is bright red, whereas that of older horses is dark or reddish-brown or, when exposed to air, darkens to a reddish-black color. The consistency of the meat is firm and compact and the muscle tissue is not marbled with fat. During cooking, the white fat (melting point $30\,°C$) appears as yellow droplets on the surface of the broth. The characteristic sweet flavor and taste of the meat are derived from the high glycogen content. In addition to the determination of glycogen, an immunoassay (cf. 2.6.3) and fatty acid analysis can be used to detect horse meat. Horse fat is characterized by a higher content of linolenic acid than beef or pork lard.

12.6.1.7 Poultry

The color of poultry meat differs according to age, breed and body part (breast meat is light, thighs and drumsticks are dark). Species of poultry which have dark meat (geese, ducks, pigeons) can be distinguished from those with light meat (chickens, turkeys, peacocks). The age, breed and feeding of the bird influence meat quality. Poultry fat tends to become rancid because of its high content of unsaturated fatty acids.

12.6.1.8 Game

Wild game can be divided into fur-bearing animals: deer (antelope, caribou, elk, white-tailed deer), wild boars (wild pigs) and other wild game (hare, rabbit, badger, beaver, bear); and birds or fowl (heathcock, partridge, pheasant, snipe, etc.). The meat of wild game consists of fragile fibers with a firm consistency. The meat remains red to red-brown in color. It has low amounts of connective and adipose tissues. The taste and flavor of each type of wild meat is characteristic. Aging of the meat requires a longer time than meat from domestic animals because of the thick and compact muscle tissue structure. The flavor of the meat then becomes pleasantly piquant or pungent and the meat is colored dark-brown to black-red.

12.6.1.9 Variety Meats

Meats of various animal organs are called variety meats. They include tongue, heart, liver, kidney, spleen, brains, retina, intestines, tripe (the first and second stomachs of ruminants), bladder, pork crackling (skin), cow udders, etc. Many of these variety meats, such as liver, kidney or heart, are highly-valued foods because they contain vitamins and trace elements as well as high quality protein. Liver contributes the specific aroma of liver sausage and pastes (goose liver). Liver is also consumed as such. Heart, kidney, lungs, pork or beef stomach, calf giblets and cow's udders are incorporated into less expensive sausages: spleen is also made into sausage. Tongues are cooked, pickled and smoked, used for the production of better-quality sausages, and canned or sold as fresh meat. Calf brain and sweetbreads (thymus glands) are especially valued as food for patients. The compositions of some variety meats are shown in Table 12.15.

Intestines, with their high content of elastin, make excellent natural sausage casings. These and beef stomach are specialty dishes.

Table 12.15. Average composition of some internal organs and blood (g/100 g edible portion)

Organ	Moisture	Protein	Fat	Carbo-hydrate	Caloric value (kJ)
Heart					
beef	75.5	16.8	6.0	0.56	556
pork	76.8	16.9	4.8	0.4	510
Kidney					
beef	76.1	16.6	5.1	–	510
pork	76.3	16.5	5.2	0.80	523
Liver					
beef	69.9	19.7	3.1	5.90	590
pork	71.8	20.1	5.7	1.14	615
Spleen					
beef	76.7	18.5	2.9		456
pork	77.4	17.2	3.6	–	464
Tongue, beef	66.8	16.0	15.9	0.4	933
Lung, pork	79.1	13.5	6.7	–	510
Brain, veal	79.4	9.8	8.6	0.8	536
Thymus, veal	77.7	17.2	3.4	–	452
Blood					
beef	80.5	17.8	0.13	0.065	335
pork	79.2	18.5	0.11	0.06	372

Pork cracklings play a role in the production of jellied meat and blood sausage, as is the case with collared pork (head). They are also consumed directly and are a good source of vitamin D. Cartilage and bones contain tendons and ligaments which are collagen- and elastin-type proteins. Cartilage and bones are similar in composition, with the exception of their mineral content; the former contains 1% minerals and the later averages 22% minerals, ranging from 20–70%. The fat content of bones can be as high as 30% and commonly varies between 10–25%. Spinal cord and ribs, when boiled in water, release gelatin-type substances and fat and, therefore, both are used in soup preparations (bouillon, clear broth or bouillon cubes or concentrated stock).

12.6.1.10 Blood

The blood which drains from a slaughtered animal is, on the average, about 5% of the live weight (oxen, cows, calves) but is particularly high for horses (9.98%) and low for hogs (3.3%). Blood has been used since ancient times for making blood and red sausages and other food products.

Blood consists of protein-rich plasma in which the cells or corpuscles are suspended. They are the red and white blood cells (erythrocytes and leucocytes, respectively) and the platelets (thrombocytes). The red blood cells do not have nuclei and are flexible round or elliptical discs with indented centers. The diameters of red blood cells vary (in µm: 4 in goat; 6 in pig; 10 in whale; and up to 50 or more in birds, amphibians, reptiles and fish). Red blood cells contain hemoglobin, the red blood pigment. White blood cells contain nuclei but no pigments, are surrounded by membranes, are $4-14$ µm in diameter and are fewer in number than red blood cells. In addition to salts (potassium phosphate, sodium chloride and lesser amounts of Ca-, Mg- and Fe-salts), various proteins, such as albumins, globulins and fibrinogen, are present in blood.

The N-containing low molecular weight substances ("residual nitrogen") of blood comprise primarily urea and lesser amounts of amino acids, uric acid, creatine and creatinine. During coagulation or clotting of blood, the soluble fibrinogen in the plasma is converted to insoluble fibrin fibers which separate as a clot. Coagulation is a complex reaction catalyzed by the enzyme thrombin, the precursor of which is prothrombin. Thrombin reacts with fibrinogen to form insoluble fibrin. The mesh of long fibrin fibers traps and holds blood cells (platelets, erythrocytes and leucocytes). Hence, the clot is colored red. The remaining fluid, which contains albumins and globulins, is the serum. Blood plasma contains 0.3–0.4% fibrinogen and 6.5–8.5% albumin plus globulin in the ratio of 2.9 : 2.0.

The composition of blood is given in Table 12.15. Blood clotting requires the presence of Ca^{2+} ions. Hence, Ca^{2+}-binding agents, such as citrate, phosphate, oxalate and fluoride, prevent blood coagulation. In the processing of blood into food, coagulation is occasionally retarded by stirring the blood with metal rods onto which the fibrin deposits. Currently, blood clotting is inhibited by using Ca^{2+}-complexing salts. After centrifugation, blood stabilized in this way yields about 70% of plasma containing 7–8% protein. The proteins can be processed further by spray-drying into powdered plasma. Recovery of liquid plasma is permitted only from the blood of cattle (excluding calves) and hogs. Addition of dried and liquid plasma to processed meats is legal.

Citrate and/or phosphate are used as calcium-binding agents.

12.6.1.11 Glandular Products

Animal glands, such as the adrenal, pancreas, pineal, mammary, ovary, pituitary and thyroid glands, provide useful by-products for pharmaceutical use. Some of these products are adrenalin, cortisone, epinephrine, insulin, progesterone, trypsin and thyroid gland extract.

12.6.2 Storage and Preservation Processes

Meat must be appropriately treated to allow storage.

12.6.2.1 Cooling

Refrigeration (cooling or freezing the meat) is an important process for prolonged preservation of fresh meat. Carcasses in the form of sides or quarters are cooled. Cooling is performed slowly (e.g., with a blast of air at 0.5 m/s at 4 °C) or quickly (e.g., stepwise for 3 h with a 3.5 m/s blast of air at -10 °C, for 19 h with a blast of air at 1.2 m/s at 2 °C, and over 18 days with air at 4 °C). The shelf-life of meat at 0 °C is 3 to 6 weeks. Weight loss due to moisture evaporation is low at high relative humidities, and decreases as the water holding capacity increases.

If meat is cooled to cold storage temperatures (<10 °C) before rigor occurs, it shrinks and becomes tough. This is due to the fact that at low temperatures, binding of $Ca^{2\oplus}$ by the sarcoplasmic reticulum and mitochondria is reduced, the $Ca^{2\oplus}$ concentration in the intracellular space is increased, inducing contraction (cf. 12.3.2.1.5). To prevent this phenomenon, meat is kept at $15-16$ °C for $16-24$ h and cooled after rigor has occurred. Electrical stimulation is also possible. This process causes rigor by accelerating glycolysis and a decline in pH. The same effect is achieved by stunning the animals with CO_2.

As long as the meat is stored in the cold in large cuts, lipid oxidation is very slow. Only chopping or mincing or warming of the muscle tissue causes a high rate of peroxidation. Muscle disintegration results in a low but significant release of highly unsaturated membrane phospholipids and Fe^{2+} ions from myoglobin. This non-heme iron is an effective catalyst of lipid peroxidation. Its concentration increases during cooking, as shown for beef in Table 12.17. Even after short cold storage of heated meat, a rancid off-flavor may develop (warmed over flavor, WOF), which is primarily due to a rapid increase in hexanal (e. g., from 0.3 to more than 10 mg/kg of cooked beef in 48 h at 4 °C) and the formation of trans-4,5-epoxy-(E)-2-decenal. In addition, important sulfur-containing aroma substances of cooked meat are lost on cold storage, so that the aroma defect becomes more evident, especially in reheated meat.

The WOF is inhibited by additives which bind Fe ions, e.g., polyphosphates, phytin, and EDTA. On the other hand, antioxidants are almost ineffective. Therefore, it is assumed that a site specific mechanism is involved in the formation of WOF. The Fe ions released during cooking are bound by phospholipids, via the negatively charged phosphate residues, and are then adjacent to the unsaturated acyl groups of these lipids. Radicals from the *Fenton* reaction of Fe ions with hydroperoxides (cf. 3.7.2.1.4) are not trapped by non-lipids, but attack only the unsaturated acyl groups and initiate their peroxidation. This hypothesis can also explain the observation that multivalent ions ($Ca^{2\oplus}$, $Al^{3\oplus}$) inhibit WOF because they probably displace the Fe ions from the phospholipids. Curing also prevents WOF. Myoglobin is stabilized by nitrite, therefore, no additional non-heme iron is formed during cooking (Table 12.17). Lipid peroxidation does not occur and new aroma substances are formed that are characteristic of cured meat.

12.6.2.2 Freezing

The shelf life of meat is substantially lengthened by freezing. Freezing can be performed in a single step (direct freezing) or in a two step process (initial cooling followed by freezing) using an air blast freezer with an air temperature of -40 °C and an air stream velocity of $3-10$ m/s. The shelf life for storage at -18 °C to -20 °C and 90% relative humidity is 9 to 15 months. The shelf life of frozen chicken, as affected by storage temperature, is presented in Fig. 12.30, while Table 12.16 shows the

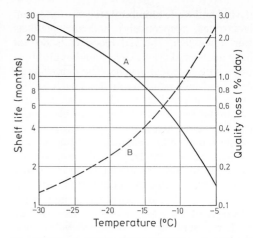

Fig. 12.30. Shelf life of frozen chicken as affected by storage temperature. — Shelf life A, – – – quality loss B = 100/A (according to *Gutschmidt*, 1974)

deterioration of frozen chicken as it is shipped from producer to consumer. The shelf life is largely determined by oxidative changes affecting the lipids, which take place more readily in poultry (ducks, geese, chickens) and pork than in beef or mutton.

The water holding capacity of frozen meat increases as the freezing temperature decreases. Water holding capacity also remains high when freezing is performed rapidly. Under these conditions the formation of large ice crystals is suppressed and damage to mem-

Table 12.16. Loss of quality of frozen chicken from producer to consumer[a]

Frozen food chain	Average storage temperature (°C)	Shelf-life (day)	Quality loss (% per day)	Average storage time (day)	Quality loss (%)
Producer	-23	540	0.186	40	7.5
Transport	-20	420	0.239	2	0.5
Wholesaler	-22	520	0.196	190	37.1
Transport	-16	370	0.370	1	0.4
Retailer	-20	420	0.239	30	7.2
	-14[b]	210	0.476	3	1.4
Transport	-7	60	1.67	0.17	0.3
Consumer	-12	150	0.666	14	9.3
Σ				280	63.7

[a] For definition cf. Fig. 12.30.
[b] A temperature estimate for food storage on the surface of open freezers.

Table 12.17. Oxidative fat deterioration in cooked beef

	Beef			
	Raw	Cooked	Cured[a]	Cured and cooked[a]
Non-heme iron (µg/g)	6.62	10.8	6.65	6.80
Storage at 4°C	Malonic aldehyde (mg/kg)[b]			
0. Day			0.58	0.56
5. Day			1.55	0.48
12. Day			2.78	0.47
21. Day			2.83	0.54

[a] Cured with 156 mg/kg nitrite.
[b] Determined with the thiobarbituric acid test.

branes and the irreversible change in myofibrillar proteins caused by temporary high salt concentrations are avoided.

Freezing meat immediately after slaughtering, without precooling (single-stage process), causes substantial shortening and high fluid losses on thawing, if the meat freezes completely before rigor occurs. However, this is possible only with smaller cuts. The reason for this phenomenon is an extremely fast ATP breakdown with a corresponding decrease in the water holding capacity. The sudden high rate of ATP breakdown is initiated by release of Ca^{2+} ions from the sarcoplasmic reticulum, which triggers the high activity of myosin-ATPase ("thaw rigor"). This "thaw rigor", which is associated with toughness, can be avoided if the warm meat is frozen and then minced in the frozen state after addition of NaCl. Thaw rigor may also be avoided by disintegrating warm meat in the presence of NaCl and then freezing it.

Freezing matured meat results in lower fluid losses than freezing meat in a prerigor or rigor state. However, this process is not widely used for economic reasons. Rigor can be induced before freezing by using electrical stimulation. Long storage of frozen meat results in a decrease in water holding capacity. Solubility changes and shifts in the isoelectric point of proteins of the sarcoplasm and contractile apparatus are observed.

Slow thawing of frozen meat is generally considered more favorable than rapid thawing, al-

though some opposing data exist. Obviously, freezing, storage and thawing should be considered as related process steps, which should be coordinated.

12.6.2.3 Drying

Drying is an ancient method of meat preservation. Drying is frequently used in combination with salting, curing, and smoking. Some processes are: drying in a stream of hot air (40–60°C), drying in vacuum under variable conditions, e.g., in hot fat, and freeze-drying, the most gentle process. The moisture content of the end product is usually 3–10%. Important quality criteria of such dried meat products are the rehydration capacity, which can be determined by water uptake under standard conditions, and the fraction of firmly-bound water. The drying process should not affect the water holding and aroma characteristics of the meat. The shelf life of dried meat products is limited by the development of off-flavors due to fat oxidation and by discoloration due to the *Maillard* reaction. Dried beef and chicken are important ingredients of many soup powders. In addition to pieces of meat, minced meat, with or without binders, and processed meats, e.g., meat balls or dumplings, are also dried for this purpose.

12.6.2.4 Salt and Pickle Curing

Salt in high concentrations inhibits the growth of microorganisms and curtails activity of meat enzymes. Hence, salt is considered as a meat preservative. Salting meat at a level up to 5% NaCl causes swelling (cf. Fig. 12.27). Higher salt concentrations (10–20%) induce shrinkage in meat and its products, causing a decrease in moisture to a level below that of untreated meat. The meat retains its natural color, usually dark red, since the myoglobin concentration increases due to the moisture loss. The color of such meat changes upon cooking to grayish-brown.

Salting by the addition of sodium nitrite and/or nitrate (curing or pickling) produces products with highly stable color (cf. 12.3.2.2.2). Since nitrite reacts faster and less is required for color stabilization, it is widely used in place of nitrate. Salt curing is done either by rubbing

salt on the meat surface (dry curing or pickling), by submerging the meat in 15–20%
brine (wet pickle curing), or by injection of
brine in special automats.

Additives, such as sugar or spices, which
favorably affect the red color and formation of
meat aroma, are often added to pickling salts.
The aroma of cured meat is specific and differs
from that of noncured meat. Aroma formation
is enhanced by the microflora (*Micrococcus*
spp. and *Achromobacter* spp.) of curing brine,
which are simultaneously involved in reduction of nitrate (NO_3^-) and nitrite (NO_2^-) ions
and thereby contribute to the stabilization of
the pinkish or red color of cured meat.

12.6.2.5 Smoking

Smoking of meat is usually associated with
salting. Depending on the smoking procedure,
the moisture drops 10–40%. Compounds present in smoke with bactericidal and antioxidative properties are deposited on and penetrate
into the meat. Important smoke ingredients
include phenols, acids, and carbonyl compounds. The concentration of polycyclic
hydrocarbons in smoke depends on the type of
smoke generation and can be largely suppressed by suitable process management, e.g., by
external smoke generation with cleaning of
the smoke via cold traps, showers, or filters. A
distinction is made between hot smoking
(≤70 °C) over a period ranging from less than
one hour to several hours (e.g., used for cooked and boiling sausages) and cold smoking
(16–28 °C) over a period ranging from two
days to several weeks (e.g., used for raw sausage and ham). Special smoking processes
include wet smoking processes, electrostatic
processes, and the use of smoke condensates.

12.6.2.6 Heating

Heat treatment is an important finishing process and also serves for the production of canned meat. Typical changes involved in heat treatment are: development of grayish-brown
color, protein coagulation, release of juices
due to decrease in water holding capacity
(Fig. 12.31), increase in pH, development of a
typical cooked or roasted meat aroma and,
finally, softening induced by the shrinking and

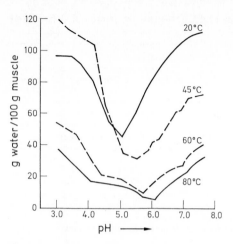

Fig. 12.31. Water holding capacity of beef muscle
versus heat treatment and pH. (according to *Hamm*,
1972)

partial conversion of collagen to gelatin (cf.
12.3.2.3.1).

Refrigerated storage of heated meat and re-
heating may lead to WOF (cf. 12.6.2.1).

12.6.2.7 Tenderizing

Plant enzyme preparations (ficin, papain,
bromelain) are used to tenderize meat. These
substances are either sprayed onto the meat
cuts or are distributed via the blood vessels of
the animal either shortly before or after
slaughtering.

12.7 Meat Products

Canned meat, ham, and sausages, and meat
extracts are produced from meat.

12.7.1 Canned Meat

Examples of canned meat are beef and pork in
their own juice, corned beef, luncheon meat,
cooked sausages, jellied meat, and cured and
pickled hams. In order to achieve sterile
canned meat, the required heating time and
temperature depend on the size and content
of the can since heat penetration is highly
variable (Table 12.18). Prolonged heating leads
to inferior aroma, texture and appearance

Table 12.18. Effect of can size and product on required heating time of canned meat (time in min to reach 121 °C at the center of the can)

Canned meat	400 g	850 g	2,500 g
Beef	47	57	80
Pork	58	98	120
Liver sausage	90	130	
Blood sausage	106	113	130

(separation of fat and jelly), hence the current processing trend for canning of meat is to use high temperatures and short times.

Apart from fully preserved sterile cans (shelf life ca. 4 years at 25 °C), which spoil abiotically by means of chemical reactions, semipreserved (e. g., cured ham, heating temperature in the center 65–75 °C, shelf life up to 6 months at 5 °C) and three-quarter preserved cans (cooked and boiling sausages, heating temperature at the center 110 °C, shelf life 9–12 months at 15 °C) with limited shelf lives are also produced.

12.7.2 Ham, Sausages, Pastes

12.7.2.1 Ham, Bacon

12.7.2.1.1 Raw Smoked Hams

After the center ham has been cut (longitudinal or circular), *ham on the bone* is dry, then wet cured (4–7 weeks), matured (reddened) for 2–3 weeks by dry storing, followed by washing, drying and exposure to cold smoke for 4–7 weeks. In *rolled ham*, the bone is taken out, and it is subsequently processed like ham on the bone, except that the curing time is shorter. *Lightly-salted lean hams* are made from cutlet or chop meats by a mild curing process, filled into casings and warm smoked.

12.7.2.1.2 Cooked Ham

Bone-free ham is cured for 2–3 weeks, stored dry to mature, washed and warm smoked. It is subsequently cooked by gently simmering. In the cook-in process, the cured meat is first packed in foil that is resistant to boiling, then cooked and smoked. *Ham sausage* is cured by rubbing the curing solution into the ham in

special machines (possibly under vacuum). *Praha ham* is a special cooked ham which is often baked in a bread dough.

12.7.2.1.3 Bacon

Back fat from the pig is salted, washed, dried, and cold smoked.

12.7.2.2 Sausages

Sausage manufacturing consists of grinding, mincing or chopping the muscle tissue and other organs and blending them with fat, salts, seasonings (herbs and spices) and, when necessary, with binders or extenders. The sausage mix or dough is then stuffed into cylindrical synthetic or cellulose casings or tubings of traditional sausage shape or, often, natural casings, such as hog or sheep intestines or the hog's bun (for liver sausage) are used. They are sold as raw, precooked or cooked, and/or smoked sausages. Some sausage formulations are presented in Table 12.19 and the composition of ham and sausage products is shown in Table 12.20.

The different types of sausages have in common that a continuous, hydrophilic salt/protein/water matrix stabilizes a disperse phase (coarse meat/fat particles, fat globules, insoluble proteins, connective tissue, and seasoning particles). The stability of systems of this type is influenced by the pH value, ionic strength, melting range of the lipids, and by the protein content. In finely ground systems with emulsion character, the grinding temperature is also important for stability. A temperature of 14 °C is regarded as optimal, unstable products resulting at T > 20 °C.

In the emulsions mentioned above, a monomolecular protein film is formed around the fat globules present (Fig. 12.32). The importance of the different protein components as film formers decreases in the following order: myosin > actomyosin > sarcoplasma proteins > actin. The hydrophobic heads of the myosin molecules evidently dip into the fat globules, while the tails interact with actomyosin in the continuous phase. The monomolecular myosin layer formed in this way should have a thickness of ~130 nm. On the outside, there is probably a multimolecular actomyosin layer

Table 12.19. Types of Sausages

Sausage Variety	Formulation/Production [a, b]
Raw Fermented Sausages	
Coarse mettwurst	Beef (24%), pork trimmings, lean (48%), pork backfat (24%), salt (37%)[b], sodium nitrate (0.03%), white pepper (0.4%). Dried and exposed to cold smoke until slicing texture is achieved.
Tea Sausage	Veal (30%), light beef (30%), pork trimmings (30%), unsalted pork backfat (10%), salt, sodium nitrate, white pepper, nutmeg. Stored until curing is finished and then smoked.
Salami (Italian style)	Beef trimmings (84%), pork backfat (16%), salt, sodium nitrate, ginger, cloves. Dried and moderately-warm smoked.
Salami (Hungarian style)	Presalted pork (75%), pork backfat (25%), salt, sugar, pepper, paprika. Dried and smoked.
Cervelat Sausage	Beef trimmings (20%), lean pork trimmings (50%), pork backfat (30%), salt, sodium nitrate and whole white pepper. Dried and then exposed to cold smoke.
Plockwurst	Beef trimmings (50%), pork trimmings (25%), pork backfat (25%), salt, sodium nitrate, sugar, ground and whole white pepper. Dried and smoked until slicing consistency is achieved.
Landjager Sausage	Lean pork trimmings (40%), pork backfat (30%), beef trimmings (30%), salt, sodium nitrate, sugar, pepper, mustard seed. Molded into flat rod-shaped matching pairs of small sausages. Dried and warm smoked.
Saxony Knackwurst	Beef (50%), pork (33%), pork backfat (17%), salt, sodium nitrate, sugar, pepper, caraway.
Thuringian Knackwurst	Marbled pork (100%), salt, caraway, pepper, marjoram. Dried and exposed to cold smoke.
Frying Sausage	
Thuringian Bratwurst	Marbled pork (70%), steer[c] meat (30%), salt, pepper, nutmeg, caraway. To be roasted on a grate over fire (barbecued).
Nuremberg Bratwurst	Lean pork trimmings (70%), pork backfat (30%), salt, pepper, marjoram.
Munich Wollwurst	Steer meat (70%), milk (30%), salt, pepper, lemon peel.
Boiling Sausages	
Wieners	Steer meat (49%), lean pork (19.5%), pork backfat (20%), nitrite pickle brine (2%), sugar (0.2%), white pepper (0.2%), nutmeg (0.03%), coriander (0.02%). Stuffed into sheep casings and warm smoked until a yellow-pink color is attained.
Halberstaedter Sausages	Steer meat (60%), lightly-marbled pork (10%), pork backfat (30%), nitrite pickle brine, sugar, white pepper, nutmeg, ginger, grainy meat broth. Stuffed into animal casings and smoked.
Frankfurters	Lean pork (70%), pork backfat (30%), salt, nitrite pickle brine, sugar, nutmeg, pepper, cardamon. Stuffed into animal casings and warm smoked.
Hunter's Sausage	Lean pork (60%), beef (30%), pork backfat (10%), nitrite pickle brine, sugar, pepper, coriander, nutmeg, grainy meat broth. Ripened to a red color, and hot smoked.

Table 12.19 (continued)

Sausage Variety	Formulation/Production[a, b]
Mortadella	Steer meat (50%), marbled pork (30%), pork backfat (20%), nitrite pickle brine, sugar, pepper, cardamon, ginger, pistachio nuts, grainy meat broth. Smoked hot and, after cooling, further exposed to cold smoke.
Goettingers	Lean beef (40%), lean pork (30%), pork backfat (30%), nitrite pickle brine, sugar, white pepper, ginger, rum, meat broth. Left to ripen until red colored, then hot smoked, cooled and exposed further to cold smoke.
Gelbwurst (yellow sausage)	Steer meat (70%), marbled pork (30%), salt, sugar, pepper, mace: stuffed and heated to about 67 °C and colored yellow with saffron.
Munich Weisswurst (white sausage)	Steer meat (50%), marbled pork (30%), veal head trimmings (20%), salt, sugar, pepper, mace, ginger, lemon peel, onions, parsley.
Bavarian Sausage Loaf	Steer meat (50%), marbled pork (40%), pork backfat (10%), some beef liver, nitrite pickle brine, sugar, pepper, mace, ginger, onions. It is then roasted in molds.

Cooked Sausages

Veal Liver Sausage, economy grade (top grade product is Braunschweiger)	Pork liver (34%), cooked veal (19%), cooked marbled pork (44%), salt (2.5%)[b], pepper (0.3%), mace (0.1%), cloves (0.04%), ginger (0.04%). Cooked in boiling water, cooled, dried and moderately smoked.
Homemade Liver Sausage	Liver (30%), meat and fat trimmings (50%), fresh onions (5%), pork backfat (15%), salt, pepper, pimentos, marjoram, thyme. Cooked in boiling water, cooled, dried and moderately smoked.
Goose Liver Sausage (Strasbourg Style)	Pork liver (22%), pork pouch (32%), liver of fattened geese (35%), tongue (6%), truffle garnish (5%), salt, onion, mace, ginger, white pepper, pistachio. Cooked, cooled, dried and moderately smoked.
Liver Paste (Pâté)	Pork liver (50%), pork backfat (50%), salt, mushrooms, nutmeg, ginger, white pepper, onion. Cooked in molds or baked in the oven.
Rotwurst (Blood Sausage)	Diced pork backfat (50%), meat trimmings, heart (15%), cooked pork skin (15%), pork blood (20%), salt, black pepper, marjoram, clove, pimentos. Cooked, cooled, dried and smoked.
Meat-Blood Sausage	Marbled pork (50%), meat trimmings, heart, kidney (25%), pork skin (15%), pork blood (10%), salt, white pepper, clove, marjoram. Cooked, then cold smoked.
Blood and Tongue sausage	Pork, veal and beef tongue (40%), pork backfat (28%), cooked pork skin (14%), pork blood (18%), salt, black pepper, clove, pimentos, thyme, onions, marjoram. Cooked, dried and cold smoked.
Hannover Cooked Mettwurst (white sausage)	Fat pork (100%), salt, white pepper, mace, onions. Stuffed and then cooked.
Bulk Headcheese	Pork knuckles, pork back, other meat, pork skin, salt, pepper, clove, cara-way. The cooked broth is chilled in molds, then wrapped in cellulose or rectangular plastic casings.

[a] The formulations listed should be considered only as examples. They may deviate greatly within a group or between related sausage groups, depending on country, region, district or market situation, and on degree of quality.
[b] Data for the amount of salt and seasonings (herbs and spices) used are given for only one sausage variety within a group.
[c] A steer is a young castrated bull.

Table 12.20. Protein and fat content of ham and sausage products

Product	Moisture %	Protein %	Fat %	Caloric value (kJ) (kJ/100 g)
Salami (German style)	27.7	17.8	49.7	2303
Cervelat sausage	34.8	16.9	43.2	2028
Knackwurst	50.1	11.9	33.7	1559
Bratwurst (pork)	52.7	12.7	32.4	1522
Hunter's sausage	52.5	15.6	29.2	1448
Gelbwurst	53.1	11.8	32.7	1520
Munich Weisswurst (white sausage Munich style)	65.2	11.1	21.7	1067
Bockwurst	59.1	12.3	25.3	1232
Liver sausage	42.9	12.4	41.2	1881
Rotwurst	45.5	13.3	38.5	1775
Ham, raw	43.4	18.0	33.3	1665
Ham, cooked	62.0	21.4	12.8	903
Bacon, marbled	20.0	9.1	65.0	2751

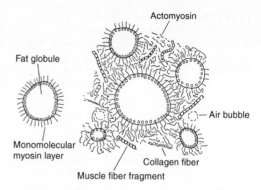

Fig. 12.32. Schematic representation of a sausage emulsion (according to *Morrissey* et al., 1987)

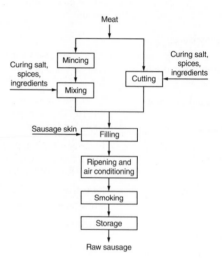

Fig. 12.33. Production of raw sausage

which binds water and contributes to the stabilization of the emulsion because of its viscous, elastic, and cohesive properties. Higher temperatures, which lead to destabilization (see above), probably cause increased protein/protein interactions in the actomyosin layer which, in turn, result in a decrease in the water binding, elasticity losses, and disturbances in the myosin film.

While the formation of myosin films on fat globules is responsible for the stabilization of raw sausages with emulsion character, protein/protein interactions and gel formation are important for the stabilization of fat and water in the system in the case of cooked and boiling sausages.

12.7.2.2.1 Raw Sausages

Raw sausages are made of raw, ground, skeletal muscle tissue, fat, and spices. Typical products are Cervelat sausage, salami, and the German Mettwurst. The production of raw sausage is schematically presented in Fig. 12.33. Depending on the desired grain size of the final product, grinding is performed in a mincer (coarse) or cutter (fine). When a mincer is used, the coarsely ground meat material is fed to a cutting set with a perforated disc via a feed screw. By using separators, the mincer can simultaneously separate meat from hard parts (tendons, sinews, cartilage, and bone).

Cutters are cutting mixers in which the following steps are conducted: cutting, distribution, mixing, very fine grinding, and emulsification. A cutter consists essentially of a dish rotating around a high-speed cutter head. Products of varying texture can be produced by varying the form and speed of the cutter, the size of the cutting chamber (possible insertion of stowing rings), plate speed, and cutting duration. Operation under a vacuum has advantages. Apart from batch cutters with a dish content of up to 750 l, continuously operating dish cutters are also available today.

In the case of firm types of raw sausages, frozen material is used for grinding (−20 °C) and the temperature is kept below 4 °C during the grinding process by cooling. After the mass has been stuffed, the sausage is ripened in air conditioned rooms (30–180 °C, 95–75% relative air humidity) and smoked.

The specific aroma is formed in the course of ripening by microorganisms present (Micrococci and Lactobacilli, often added in the form of starter cultures). Here, too, maturing plays a big role. The drop in pH due to lactic acid formation (5.2–4.8) results in shrinkage of the protein gel. The sausages become firm and suitable for slicing after vaporization of the water released (20–35% weight loss). Accelerated ripening is made possible by souring with the help of glucono-δ-lactone. The white layer on various types of salami is due to mold mycelia or, in cheaper products, a layer of lime milk dip.

12.7.2.2.2 Cooked Sausages

Cooked sausages are made from cooked starting materials. Typical products are liver sausage, blood sausage, and jellied sausage. The production of liver sausage is shown schematically in Fig. 12.34. Modern plants generally use cooking cutters in which the following steps are conducted in one machine: preliminary grinding, cooking, mixing, and cutting.

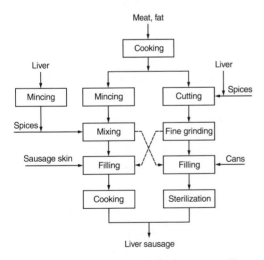

Fig. 12.34. Production of cooked sausage (liver sausage)

12.7.2.2.3 Boiling Sausages

Boiling sausages (German "Brühwurst") are made in a cutter from ground raw meat, usually with the addition of water. If no freshly slaughtered or freshly slaughtered pre-salted meat is available, the water holding capacity is increased by the addition of common salt and cutter aids (condensed phosphates, lactate, acetate, tartrate, and citrate). The swelling resulting from added salts is caused by an increase in the pH of the meat slurry and by the complexing of divalent cations which, in free form, suppress swelling. The temperature during grinding/chopping has to be kept low (addition of ice or ice-cold water) since higher temperatures decrease the water holding capacity. Water retention increases as the fat component of the meat slurry is increased as long as the fat:protein ratio does not exceed 2.8 to 1. As a consequence the salt concentration is increased.

After chopping and stuffing, the sausages may be smoked at temperatures less than 80 °C (internal product temperature of 68 °C). At this temperature, coagulation of protein gel, which holds the water, forms the broken texture so typical of these sausages.

Typical products are bockwurst, wieners, white and hunter's sausage and mortadella. Figure 12.35 schematically shows the production of boiling sausages.

12.7.2.3 Meat Paste (Pâté)

12.7.2.3.1 Pastes

Meat pastes are delicately cooked meat products made primarily from meat and fat of calves and hogs and, often, from poultry (e.g. goose liver paste) or wild animal meat (hare, deer or boar). Unlike sausages, pastes contain quality meat and are free of slaughter scrapings or other inferior by-products. A portion of meat or the whole meat used is present as finely comminuted spreadable paste.

12.7.2.3.2 Pains

Pains usually consist of larger pieces of meat (especially game and poultry), which are processed into a cooked sausage-like mass with fat, truffle, and various spices.

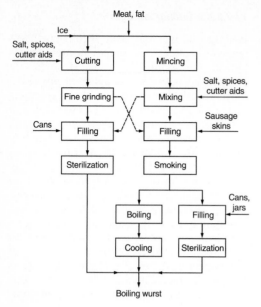

Fig. 12.35. Production of boiling sausage ("Brüh-wurst")

12.7.3 Meat Extracts and Related Products

12.7.3.1 Beef Extract

Meat extract is a concentrate of water-soluble beef ingredients devoid of fat and proteins. Its preparation dates back to *Liebig's* work in Munich in 1847. Comminuted beef is counter-currently extracted with water at 90°C. After removal of fat by separators and subsequent filtration, the extract containing 1.5–5% solids is concentrated to 45–65% solids in a multiple stage vacuum evaporator which operates in a decreasing temperature gradient (a range of 92 to 46°C). The final evaporation to 80–83% solids is then carried out under atmospheric pressure at a temperature of 65°C or higher or under vacuum on a belt dryer.

In the same way, the cooking water recovered during the production of corned beef can be processed into meat extract. Only this latter source of meat extract is of economic significance. The yield is 4% of fresh meat weight. The composition of the extract is given in Table 12.21. For addition to soup powders and sauce powders, the thick pasty meat extract is blended with a carrier substance and vacuum- or spray-dried.

Table 12.21. Chemical composition of beef extract

	%
Organic matter	56–64
Amino acids, peptides	15–20
Other N-compounds	10–15
Total creatinine	5.4–8.2
Ammonia	0.2–0.4
Urea	0.1–0.3
N-free compounds	10–15
Total lipids	> 1.5
Pigments	10–20
Minerals	18–24
Sodium chloride	2.5–5
Moisture	15–23

pH-value of a 10% aqueous solution	approx. 5.5

12.7.3.2 Whale Meat Extract

This product is obtained from meat of various whales (blue, finback, sei, humpback and sperm) in a process similar to that used for beef extract.

12.7.3.3 Poultry Meat Extract

Chicken extract is obtained by evaporation of chicken broth or by extraction of chicken halves with water at 80°C, followed by a concentration step under vacuum to an end-product of 70–80% solids.

12.7.3.4 Yeast Extract

Yeast cells (*Saccharomyces* and *Torula* spp.) are forced to undergo shrinking of protoplasm by addition of salt, which causes loss of cell water and solutes (plasmolysis), or the cells are steamed or subjected to autolysis. Cells treated in this way are extracted with water and the extract is concentrated to yield a brown paste. Yeast extract is rich in the B-vitamins. The concentrations of thiamine and thiamine diphosphate are above their taste threshold values and may contribute to the product's unpleasant flavor. On the other hand, the spicy flavor of the paste is essentially due to 5'-nucleotides freed during hydrolysis and to amino acids, particularly glutamic acid.

12.7.3.5 Hydrolyzed Vegetable Proteins

The production of this protein hydrolysate is schematically presented in Fig. 12.36. According to the given formulation, the different plant protein-containing raw materials, such as wheat and rice gluten and roughly ground soybeans, palm kernels or peanuts, are automatically delivered from raw material silos, weighed, and fed to a hydrolysis boiler (double-walled, pressure-stable stirred tank). Hydrolysis proceeds at temperatures above 100 °C and the appropriate pressure with hydrochloric acid or sulfuric acid (salt-free seasoning).

The hydrolysate is subsequently neutralized to pH 5.8 with sodium or calcium carbonate or with sodium hydroxide solution. In this process, the pH range of 2.5−4 must be passed through as quickly as possible to repress the formation of pyrrolidone carboxylic acid from glutamic acid.

The hydroysate is filtered and the filtrate (seasoning) stored. The filtration residue is washed with water and refiltered, if necessary. The diluted filtrate is evaporated and added to the seasoning obtained in the first step.

The seasoning is subsequently stored; it is filtered several times before filling. Apart from liquid food seasoning, seasoning in paste and powder form and mixtures for use in dry soups and sauces are produced. These products are partly bleached with activated carbon and the taste is neutralized.

The compound 3-hydroxy-4,5-dimethyl-2(5H)-furanone (cf. 5.3.1.3) is responsible for the intensive, typical seasoning aroma. The products have a meat- or bouillon-like odor and taste. It was found in 1978 that genotoxic compounds are formed in hydrochloric acid hydrolysates of protein-containing raw materials. Thus, 3-chloropropane-1,2-diol, 2-chloropropane-1,3-diol, 1,3-dichloropropane-2-ol, 1,2-dichloropropane-3-ol, and 3-chloropropane-1-ol have been identified as secondary products of lipids in amounts of 0.1 to >100 ppm in commercial protein hydrolysates and products derived from them. In feeding experiments on rats, these dichloro compounds were found to be carcinogenic. The testing of the monochloro compounds is still in progress. The chlorinated glycerols, which are partly also present as fatty acid esters, have half life periods of several hundred days in the hydrolysates. The N-(2,3-dihydroxypropyl) derivatives of the amino acids serine and threonine as well as 3-aminopropane-1,2-diol have been detected as aminolysis products.

Chlorinated steriods, e.g., 3-chloro-5-cholestene (Formula 12.27, a), 3-chloro-24-methyl-5,22-cholestadiene (Formula 12.27b) and 3-

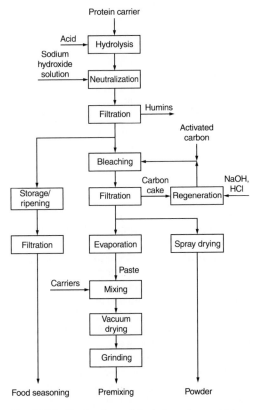

Fig. 12.36. Production of hydrolyzed vegetable protein

a R = (12.27)

b R = , R¹ = Me

c R wie b, R¹ = Et

chloro-24-ethyl-5,22-cholestadiene (Formula 12.27c), have been identified in the insoluble residue of the corresponding products.

Moreover, there have been indications of the presence of chlorinated *Maillard* compounds in hydrochloric acid hydrolysates, e.g., 5-(chloromethyl)furfural.

To avoid or minimize the unwanted compounds mentioned above, the production process has been or is being modified, e.g., in the form of an additional alkali treatment of the hydrochloric acid hydrolysate. Thus, concentrations of <1 ppm of 3-chloro-1,2-propanediol were found in the majority of samples tested in 1990, which is clearly less than it was in previous years.

12.8 Dry Soups and Dry Sauces

Meat extract, hydrolysates of vegetable proteins, and yeast autolysate are used to a large extent in the production of dry soups amd dry sauces. For this reason, these substances will be described here. The industrial production of these products for use in home and canteen kitchens has become increasingly important in the past 20 years. In particular, a special pretreatment of the raw materials made possible the development of products which, after quick rehydration, give ready-to-consume complete meals (dry stews), snacks between meals (dry soups, instant soups), or sauces.

12.8.1 Main Components

Not only meat extracts, protein hydrolysates, and yeast autolysates, but also glutamate, ribonucleotides (inosinate/guanylate), and reaction products of amino and carbonyl compounds with a meaty aroma are used as the taste-bearing substance (cf. Table 12.23). These substances are dried with and without a carrier (belt vacuum drying, spray drying). Flour (wheat, rice, corn), legume flour (peas, lentils, beans), and starches (potato, rice and corn) serve as binding agent. Apart from native flour or starch, swelling flour or instant starch that is pregelatinized by drum drying or boil extrusion is used. In fact, especially good swelling and dispersing properties are achiev-

ed by agglomeration. Legumes are precooked in pressure vessels for up to several hours before drying. The rehydration time can be reduced to 4–5 minutes by freeze drying. Standard products are normally air dried on belt dryers. Pasta is subjected to a precooking process by means of steam and/or water or used in a fat dried form, like in the Far East.

Rice is added in a pre-cooked, freeze-dried form or as reformed rice (dried rice flour extrudate). After the appropriate pretreatment (e.g., blanching), vegetables and mushrooms are dried (drum, spray, and freeze drying). Products with instant character are obtained by centrifugal fluidized bed drying. In this process, which is used on a large scale for carrots and rice, the products in a perforated and basket-shaped rotating cylinder are dried with hot air of ca. 130 °C with simultaneous puffing. The fats used are mainly beef fine tallow, hardened plant fats, chicken fat, and milk fat. These fats are often applied in powder form (cf. 14.4.7). The meat additives are primarily beef and chicken which are air dried or freeze dried. To perfect the taste, salt and spices are used as ground natural spices or in the form of spice extracts.

To improve the technological properties, dry soups and sauces contain a series of other ingredients, e.g., milk products, egg products, sugar, and maltodextrin, acids, soybean protein, sugar coloring, and antioxidants.

12.8.2 Production

The production of dry soups and sauces essentially involves mixing the preproduced raw materials. The process steps are shown in Fig. 12.37.

Weighing of individual components from the raw material silos and their pneumatic dosing into the mixer are conducted automatically. In soup mixtures that contain breakable components, such as pasta and dry vegetables, a basic mixture of the powdery components (binder, fat powder, extract powder etc.) is first produced in high-speed mixers. The breakable components are gently mixed in a second slow mixing step. The mixtures are agglomerated for special uses (instant soups and sauces); they generally have no coarse components. This is usually conducted in batchwise or con-

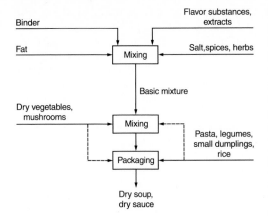

Fig. 12.37. Production of dry soups and sauces

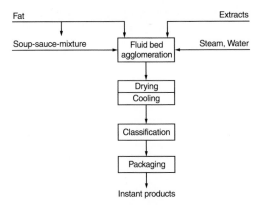

Fig. 12.38. Production of instant products by agglomeratiion

tinuously operated fluid bed spray granulators. In continuous agglomeration plants (Fig. 12.38), extract substances and fat are dosed in separated systems. Alternatively, finished soup/sauce mixtures are agglomerated by back wetting with steam or water and dried via a separate fluid bed. The packaging materials used protect the dry mixture from light, air, and moisture.

12.9 Meat Aroma

Raw meat has only a weak aroma. Numerous intensive aroma variations arise from heating,

the character of the aroma being dependent on the type of meat and the method of preparation (stewing, cooking, pressure cooking, roasting or broiling-barbecuing). The preparation effects are based on reaction temperatures and reactant concentrations. Thus, a carefully dried, cold aqueous meat extract provides a roasted meat aroma when heated, while an extract heated directly, without drying, provides a bouillon aroma. When meat is fried in hot oil, the frying oil is an additional reactant. Meat aroma consists of: (a) nonvolatile flavor substances, (b) flavor enhancers and (c) volatile aroma constituents. The latter compounds or their precursors originate essentially from the water-soluble fraction. The constituents listed in Table 12.22 have been identified as the flavor substances of beef broth. A solution of these substances in the given concentrations (Table 12.22) gives the typical taste profile, which is composed of sweet, sour, salty, and bouillon-like notes.

The compounds 2-methyl-3-furanthiol, its disulfide, 2-furfurylthiol, and methional contribute with high aroma values to the odor of bouillon. 2-Methyl-3-furanthiol, which has the pleasant odor of cooked meat near the very low odor threshold (cf. 5.3.1.4), is formed on degradation of thiamine (cf. 5.3.1.4) and by a reaction of ribose with cysteine.

Table 12.22. Flavor substances of beef broth

Compound	Concentration	
	(mmol/l)	(mg/l)
Aspartic acid	0.05	7
Glutamic acid	0.3	45
5′-AMP	0.14	51
5′-IMP	0.4	136
Carnosine	6.2	1402
Anserine	0.7	212
Carnitine	2.0	322
Lactic acid	25.6	2306
Pyroglutamic acid	2.6	336
Sodium	2.3	53
Potassium	3.1	1216
Calcium	1.0	40
Magnesium	3.0	73
Chloride	3.1	110
Phosphate	10.1	959

The roasted, caramel-like and earthy odor notes in the aroma profile of roasted beef are produced by 2-acetyl-2-thiazoline, 4-hydroxy-2,5-dimethyl-3(2H)-furanone, guajacol, 2-ethyl-3,5-dimethylpyrazine and 2,3-diethyl-5-methylpyrazine. Diacetyl and methional are also included in the important aroma substances. The S-containing furans, which are essential for boiled meat, are of minor importance for the aroma of roasted meat.

The differences in the aroma of meat from various animal species are largely caused by the fat, because lean meat (beef, pork, mutton) heated by itself always produces the beef note. The animal-specific aroma appears only on addition of fatty tissue. Chicken contains about ten times as much linoleic acid as beef. When chicken broth is boiled, this fatty acid is peroxidized, producing (E,E)-2,4-decadienal as the most intensive odor substance, which modifies to "chicken" the aroma of the sulfur-containing furans mentioned above.

The generation of a meat-like aroma is also possible by heating cystine, cysteine, methionine or thiamine with reducing sugars, by reaction of H_2S with alkenals and hydroxy-dihydrofurans, or by other reactions. Many formulations for these "reaction aromas" have been proposed in the patent literature (examples in Table 12.23). In most cases, a relatively inexpensive protein hydrolysate is used as the amino acid source and other important precursors of meat aromas like thiamine and monosaccharide phosphates are applied in the form of yeast autolysates.

In a model system for the production of vegetable-based roast gravy, which also uses amino acids and sugar as the precursors of the thermally formed aroma substances, it could be shown that with the same heating time and temperature, the type of aroma and the depth of color depend only on the solids of the reaction mixture (Fig. 12.39).

Table 12.23. Examples of patents on aromas of processed meat

Inventor	Company	Number	Year	Reactions
May and Morton	Unilever	Brit. 858,660	1961	Amino acid source (must contain cysteine) + aldehyde + pentoses in water: reflux
Hack and Königsdorf	Corn Products Co.	US 3,480,447	1969	Amino acids (cysteine-free) + hydrolyzed vegetable protein + reducing sugar + taurine: heat for 15–20 h to 110 °C
Giacino	IFF	US 3,519,437	1970	Taurine + thiamine + hydrolyzed vegetable protein + water or fat + heat or flash heating
Kitada et al.	Ajinomoto	US 3,620,772	1971	Cysteine + reducing sugar 1–10 h at 50–120 °C and 70–200 kg/cm²
van Pottelsberghe, de la Potterie	Nestle	US 3,716,379	1973	Hydrolyzed vegetable protein (cysteine-free) + thiamine + one mono – or polysaccharide + water + heat
Lee	General Foods	US 3,741,775	1973	Cysteine + 3-deoxyhexose + thiamine
Baugher and Township	Procter & Gamble	US 3,930,046	1975	Hydrolyzed whey protein + cysteine + xylose + thiamine
Eguchi	Ajinomoto	US 4,066,793	1978	Yeast autolysate + 5'-nucleotides + NaCl + potassium salt ($[K^{\oplus}] > 0.5 (Na^{\oplus})$)
Corbett	Stauffer	US 4,165,391	1979	HVP + yeast autolysate + whey components

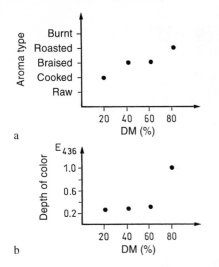

a

b

Fig. 12.39. Model system for the production of roast gravy: aroma type (**a**) and depth of color (**b**) as a function of the solids (dry matter, DM) in %; formulation (%): vegetable (fresh tissue apple, onion, carrot) 70, spices, yeast extract 15, salt 5, sugar 5, fat 5; the mixture was adjusted to different solid contents; heating: T = 95 °C, t = 4.5 min; according to *Biller*, 1989)

A typical aroma substance of seasonings is sotolon (cf. 5.3.1.3). The aroma value of 5-ethyl-3-hydroxy-4-methyl-2(5H)-furanone (abhexon), which was first regarded as a characteristic aroma substance, is considerably lower because of the lower concentration.
Homocysteic acid, cysteine-S-sulfonic acid, tricholomic and ibotenic acids are flavor enhancers with an effect similar to that of glutamic acid.

12.10 Meat Analysis

12.10.1 Meat

The determination of the kind of animal, the origin of meat, differentiation of fresh meat from that kept frozen and then thawed, and the control of veterinary medicines is of interest. The latter include antibiotics (penicillin, streptomycin, tetracyclines, etc.) used to treat dairy cattle infected with mastitis, and other chemicals, including diethyl stilbestrol, used for cattle to increase the efficiency of conversion of feed into meat.

12.10.1.1 Animal Origin

The animal origin of the meat can be determined by immunochemical and/or electrophoretic methods of analysis.

12.10.1.1.1 Serological Differentiation

This analysis is performed in the same way as described for soya and milk proteins (cf. 12.10.2.3.2), i.e. by an antibody-antigen reaction, when antibody is available from the animal being investigated.

12.10.1.1.2 Electrophoresis

To determine the animal or plant origin of the food, electrophoretic procedures have often proved to be valuable when the electropherograms of the protein extracts reveal protein zones or bands specific for the protein source. Thus, in meat analysis, such a method allows for the differentiation between more than 40 animal species, e.g., beef, pork, horse, buffalo, sheep, game, and poultry (cf. Fig. 12.40).
To carry out an analyis, the sarcoplasm proteins are extracted with water. The electrophoretic separation is predominantly conducted on polyacrylamide gels, previously also on starch and agarose gels. The application of a pH gradient (isoelectric focusing) provides excellent protein patterns. The first assignment is achieved directly after the electrophoretic separation on the basis of two red myoglobin zones (Fig. 12.40a). The ratio of the intensities of these zones, which represent met- and oxymyoglobin, changes with the storage time of the meat or extract and is not important for evaluation. The myoglobin and hemoglobin zones can be intensified by treatment with o-dianisidine/H_2O_2 (Fig. 12.40b) and subsequent staining with coomassie brilliant blue makes all the proteins visible (12.40c).
Some animal species can be recognized via the myoglobin bands (e.g., beef, buffalo, pork, horse, red and grey kangaroo) and others are assigned to groups. The identification is achieved with electropherograms stained with coomassie blue. A differentiation within the families Cervidae (deer) and Bovidae (horned

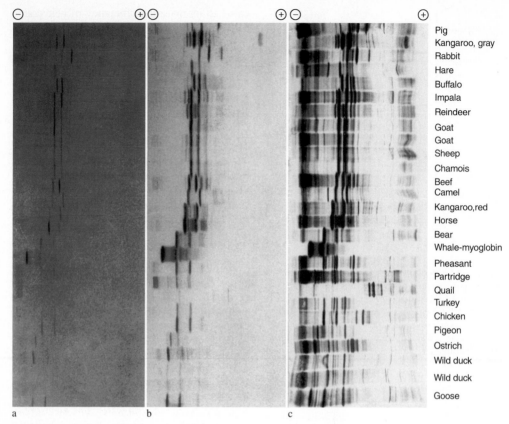

Fig. 12.40. Separation of sarcoplasm proteins of various warm blooded animals (mammals and fowl) by isoelectric focusing on polyacrylamide gels (PAGIF, PAGplate, pH range 3.5–9.5). (according to *Kaiser*, 1988). **a** Myoglobin (and hemoglobin) zones without staining; **b** Myoglobin and hemoglobin zones after treatment with o-dianisidine/H_2O_2; **c** Protein zones after staining with coomassie brilliant blue

animals) is difficult, with the exception of the subfamily Bovinae (cattle), e.g., between roe deer, fallow buck, elk, reindeer, kudu, springbok, impala, sheep, goat, and chamois. Here, the hemoglobins can be used if the meat contains sufficient blood components, as is usually the case with game, or if blood is separately available (Fig. 12.41 a).

The analyses mentioned above are largely limited to raw meat because protein denaturation occurs in heat treated meat. Denaturation increases with temperature and time and makes the immunochemical and electrophoretic identification more and more difficult.

From the intensities of the indicator zones in an electropherogram, it is possible to estimate the proportion of one kind of meat in a meat mix. This is illustrated in Fig. 12.42 using a mixture of ground beef and pork.

12.10.1.2 Differentiation of Fresh and Frozen Meat

The isoenzyme patterns of cell organelles, for instance mitochondria and microsomes, differ often from those of cytoplasm. When the organelle membranes are damaged by a physical or chemical process, isoenzyme blending will occur in the cytoplasm.

Such membrane damage has been observed by freezing and thawing of tissue, for example, of muscle tissue, in which the isoenzymes of glutamate oxalacetate transaminase (GOT) bound to mitochondrial membranes are parti-

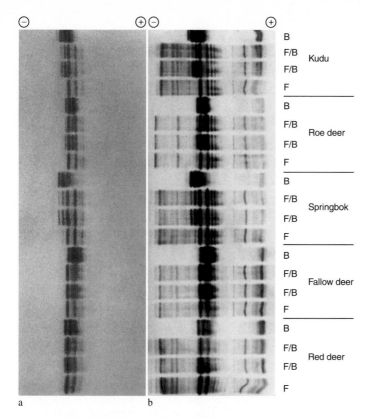

Fig. 12.41. Animal species with the same myoglobin patterns. Separation by isoelectric focusing of water soluble muscle proteins [F], blood [B], and mixtures of both [F/B] (cf. Fig. 12.40, according to *Kaiser*, 1988). **a** Myoglobin and hemoglobin zones after treatment with o-dianisidine/H$_2$O$_2$; **b** Protein zones after staining with coomassie brilliant blue

ally released and found in the sarcoplasm. The pressed sap collected from fresh unfrozen meat has only sarcoplasm enzymes, while the frozen and thawed meat has, in addition, the isoenzymes derived from mitochondria. The GOT isoenzymes can be separated by electrophoresis (Fig. 12.43). This procedure is also applicable to fish.

12.10.1.3 Pigments

Pigment analysis is carried out for the evaluation of meat freshness. The individual pigments, such as myoglobin (purple-red), oxymyoglobin (red) and metmyoglobin (brown), are determined.

12.10.1.4 Treatment with Proteinase Preparations

Proteinases injected intramuscularly or through blood vessels degrade the structural proteins and, hence, proteolytic enzymes can be used to soften or tenderize meat. The enzymes are of plant or microbial origin and are used in the meat and poultry industries, while some are also used in the household as meat tenderizers. Analytical determination of proteinases is relatively difficult.

A possible assay may be based on disc gel electrophoresis of meat extracts, prepared in the presence of urea and SDS. The band intensities of the lower molecular weight collagen fragments increase in proteinase-treated meat.

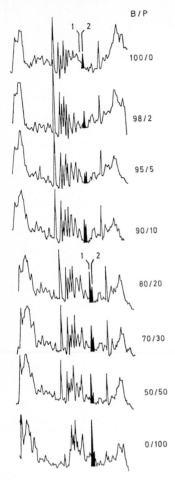

Fig. 12.42. Blended beef and pork meat: Densitograms of sarcoplasm proteins separated by PAGIF and PAG-plate; pH range 3.5–9.5. B/P (beef/pork) blend ratios in weight %. (according to *Kaiser*, 1980b)

12.10.1.5 Anabolic Steroids

Anabolic compounds present in animal feed as an additive increase muscle tissue growth. Owing to a potential health hazard, some of these compounds are banned in many countries. Their detection can be achieved by the mouse uterus test or by a radioimmunoassay. Special receptor proteins which have the property of binding strongly to estrogens are isolated from rabbit or cattle uterus. The hormone-receptor complex is in equilibrium with its components:

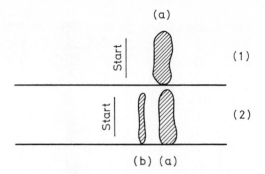

Fig. 12.43. Differentiation of fresh liver (1) from frozen and thawed liver (2) by electrophoretic separation of glutamate-oxalacetate transaminases (a) GOT sarcoplasm, (b) GOT mitochondria (according to *Hamm* and *Mašić*, 1975)

$$Receptor + estrogen \rightleftharpoons Receptor\text{-}estrogen\ complex \qquad (12.28)$$

The nonlabelled estrogens bound to receptor in the test sample will be competitively displaced by the addition of 17-β-estradiol labelled with tritium for radiochemical assay.

To reach equilibrium, a suitable amount of receptor protein and a constant amount of labelled estradiol are incubated together with the test sample. The amount of the radioactive ^{3}H-estradiol receptor complex will decrease in the presence of competitive estrogens from the meat extract. The binding affinity of the estrogen receptor depends on the type of estrogen present (Fig. 12.44). Hence, detection limits differ and range from 0.3 to 50 ppb (mg per metric ton).

Anabolic compounds can be further separated by gas-liquid chromatography after derivatization of the polar functional groups, and identified by mass spectrometry. This method allows the determination of weak or nonestrogenic components too, but in the past it suffered from high losses in sample preparation and could not compete with radioimmunoassay in sensitivity. In the meantime disadvantages of the method have been eliminated.

12.10.1.6 Antibiotics

Antibiotics are used as part of therapy to treat animal diseases and, sometimes in low concentrations, as constitutents of animal feed to

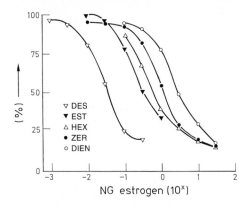

Fig. 12.44. Relative binding affinity of estrogen compounds to estrogen receptor. 50% Binding achieved by: 0.034 ng diethylstilbestrol (DES), 0.33 ng 17-β-estradiol (EST), 0.6 ng hexestrol (HEX), 1.2 ng zeranol (ZER), 2.9 ng dienestrol (DIEN). (according to *Ingerowski* and *Stan*, 1978)

increase feed utilization and to accelerate animal growth.

Detection of antibiotics is usually achieved by the inhibition of the growth of bacteria ("inhibitor test"). *Bacillus subtilis*, strain BGA, is one of the recommended test organisms.

Chemical methods must be used in order to identify and quantify the antibiotics and other veterinary medical residues. The principal method is chromatographic separation coupled with mass spectrometry. The tetracyclines, which are common antibiotics, can be determined relatively easily by fluorometric measurement of adequately prepared and purified meat extracts.

12.10.2 Processed Meats

Besides the estimation of the animal species and the control of additives, the analysis of processed meats is associated with verifying composition. Here the emphasis is on the content of extraneous added water, carbohydrate-containing thickeners and binders, nonmeat protein additives and fat. In addition, the determination of nitrites, nitrates, nitrosamines and, for enhancing the pinkish-red color of processed meat, L-ascorbic acid is of importance in pickle-cured meat products. Other analytical problems involve the detection of

condensed phosphates, citric acid and glucono-δ-lactone, as well as the detection of polycyclic aromatic compounds in smoked meats, of mycotoxins in products with desirable or undesirable growth of molds and of chlorine compounds in seasonings.

12.10.2.1 Main Ingredients

The first insight into the composition of processed meat, i.e. whether it contains an excess of fat or carbohydrate, which would lower the protein content and thus lower the value of the processed meat, is obtained by proximate analysis of the product's main ingredients: moisture, raw protein, fat and ash content. If their sum is less than $100 \pm 0.5\%$ of the sample weight, then the presence of carbohydrate binders should be verified. A positive finding should be further investigated since incorporation of liver into processed meat may provide glycogen. Hence, thorough carbohydrate analysis is required.

12.10.2.2 Added Water

Moisture content is related to protein content and is relatively constant. *Feder's* method of analysis of water added to chopped or ground meat or to emulsion-type sausages is based on these findings. The method uses the empirical equation:

$$\text{Water added (\%)} = \text{Moisture (\%)} - \text{Protein (\%)} \times F \qquad (12.29)$$

F for beef, emulsion-type sausage = 4.0;
F for pork = 4.5

This indirect method for assessing the amount of added water has been repeatedly criticized. In spite of this, no better method has yet been developed. Moreover, the calculated water content is never used alone to evaluate a meat product. Other significant data, such as muscle protein content and the proportion of fat to protein, are also included.

12.10.2.3 Lean Meat Free of Connective Tissue

A measure of meat quality is expressed as the amount of lean meat free of connective tissue, which corresponds to meat proteins devoid of connective tissue protein (MPDCP). To obtain this value, the meat sample is analyzed for connective tissue proteins (CP), added extraneous protein (EP) and nonprotein-nitrogen (NPN), e.g., glutamate, purine and pyrimidine derivatives, urea. These values are then deducted from the value for total protein (TP):

$$MPDPC = TP - (CP + EP + NPN) \qquad (12.30)$$

Another method still being tested is based on drastic treatment (heating to 130 °C at pH 9) of a meat sample. Under these conditions, extraneous proteins, collagen and blood plasma proteins solubilize, while the residual protein is calculated as MPDCP using a constant factor.

12.10.2.3.1 Connective Tissue Protein

The amino acid 4-hydroxyproline is a marker compound for connective tissue. It occurs only in connective tissue protein. The amount of 4-hydroxyproline is determined in the acid-hydrolysate of the sample or the isolated protein using an amino acid analyzer, or colorimetrically using a specific color reaction. The latter, accepted widely in practice, is a direct photometric procedure based on the oxidation of hydroxyproline in alkaline solution by H_2O_2 or N-chloro-p-toluenesulfonamide (chloramine-T). The oxidation yields a pyrrole derivative which is then condensed with p-dimethylaminobenzaldehyde to form a red pigment. The connective tissue content of meat is calculated by multiplying the hydroxyproline value by a factor of 8, which corresponds to an average of 12.4% hydroxyproline content of connective tissue.

12.10.2.3.2 Added Protein

In order to extend or improve the water holding capacity of processed meat, the product may contain milk, egg or soy proteins. These proteins can be detected immunochemically with high sensitivity using a simple and rapid diffusion test on agar gel (Fig. 12.45).

Intensive heat treatment as, for example, in emulsion-type canned meat (Table 12.24, may cause severe changes in the solubility and the antigenic properties. In such cases, no quantitative assessment is possible. An additional difficulty arises when the antibody reacts with several related proteins. This so-called "cross reaction" has been observed, for example

Fig. 12.45. Immunochemical detection of nonmeat proteins in a boiling sausage made at 85 °C. (according to *Guenther*, 1969). Agar-agar gel-coated glass plate has reservoir cavities (wells). A diffusion process occurs against antibody solution in the centrally located well with soya protein (reference) from wells 1, 3, 5 and boiling sausage extract from wells 2, 4, 6. After staining with Amido Black two precipitation zones are revealed for soya protein and one for boiling sausage extract

Table 12.24. Determination of soja protein in boiling sausages

Soya protein (%)		Temperature (°C)[a]
added	found	
0.75	0.5 ± 0.1	115
1.50	1.4 ± 0.1	115
3.0	2.0	115
0.75	0.6 ± 0.1	121
1.50	1.0 ± 0.1	121
3.0	0.9 ± 0.1	121

[a] Temperature recorded in center of can.

when egg white antibody precipitated not only egg white proteins, but also those of milk. In such cases, instead of an agar gel diffusion, the immunoelectrophoresis method is suitable. In addition to precipitation, the positions of the protein bands in an electropherogram provide the qualitative data needed to unequivocally identify the extraneous protein in the processed meat.

Another very sensitive method of detection of soybean in meat products is based on the electrophoretic separation of extracts, followed by protein transfer to membranes and immunochemical identification. For this purpose, the meat and soybean proteins are extracted in the presence of sodium dodecyl sulfate (SDS) to increase their solubility. Electrophoretic separation is conducted on polyacrylamide gels. The separated proteins are transferred by diffusion- or electro-blotting to nitrocellulose or immobilon P membranes and stained immunochemically. A specific glycoprotein stain, in which concanavalin A attaches to the con-glycinins of the soybean proteins (7S fraction), has proved useful. Furthermore, glycoprotein peroxidase binds to the remaining free binding sites of concanavalin A and can be detected by a color reaction. In this way, the addition of even 0.1% soybean isolate can be detected in unheated sausage dough and in boiling sausage ("Brühwurst") (75 °C). In meat products which are heated to 120 °C for 60 minutes, the detection limit increases to ca. 2.5% (Fig. 12.46).

Another immunochemical staining method, immunogold/silver staining (IGSS), is more sensitive. In the first step, the soybean antigen reacts with rabbit antiserum against soybean protein. In the second step, gold-labelled anti-(rabbit-IgG)-IgG from the goat is bound to the first antibody. The barely visible gold colored zones are intensified to strong gold colored to brown zones by treatment with silver solution.

In raw sausage dough and in boiling sausage (75 °C), the detection limit of soybean isolate is ca. 0.02–0.05% based on dough or 0.2% based on protein. In meat products which are heated to 120 °C for 60 minutes, even 0.1% of soybean isolate can be detected with this method (Fig. 12.47). The higher sensitivity of this method is due to the detection of glycinins

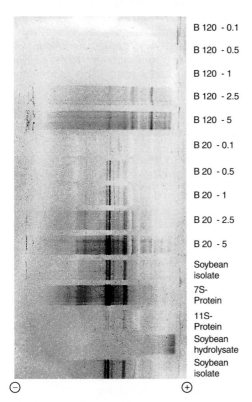

Fig. 12.46. Separation of extracts of soybean containing meat products with SDS-PAGE on ExcelGel SDS, diffusion blotting on immobilon P membrane and glycoprotein staining (according to *Kaiser* and *Krause*, 1991). B20; B75; B120: sausage dough unheated or heated to 75 °C and to 120 °C, with the addition of 0; 0.1; 0.5; 1.0; 2.5; and 5.0% of soybean isolate. 7S-, 11S-proteins: soybean fractions

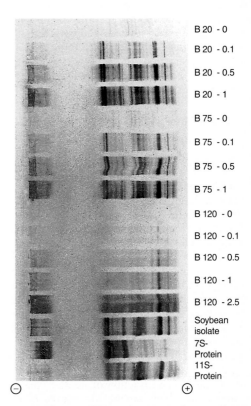

Fig. 12.47. Separation of extracts from soybean containing meat products with SDS-PAGE on Excelgel SDS, diffusion blotting on immobilon P membrane and immunogold/silver staining (according to *Kaiser* and *Krause*, 1991). B20; B75; B120: sausage dough unheated or heated to 75 °C and 120 °C, with the addition of 0; 0.1; 0.5; 1.0; 2.5; and 5.0% of soybean isolate. 7S-, 11S-proteins: soybean fractions

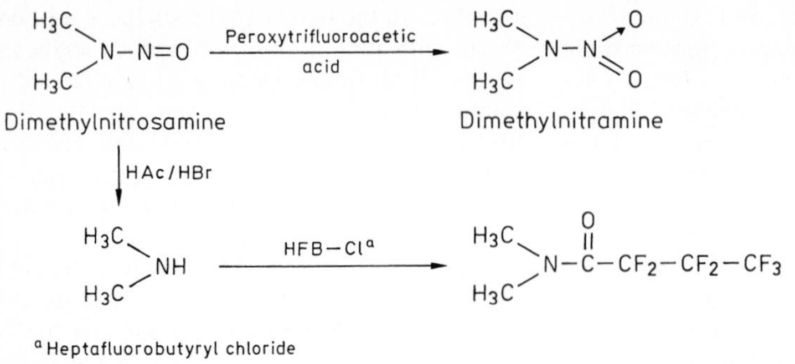

H₃C
   \
    N—N=O $\xrightarrow{\text{Peroxytrifluoroacetic acid}}$
   /
H₃C

Dimethylnitrosamine

| HAc/HBr

H₃C
   \
    NH $\xrightarrow{\text{HFB—Cl}^{a}}$
   /
H₃C

H₃C
   \
    N—N
   /   \
H₃C   O

Dimethylnitramine

H₃C   O
   \   ‖
    N—C—CF₂—CF₂—CF₃
   /
H₃C

^{a} Heptafluorobutyryl chloride

Fig. 12.48. Gas chromatographic detection of nitrosamines after derivatization (electron capture detector)

(11S fraction), which are present in larger amounts in soybean protein, in addition to conglycinins. Furthermore, this method has the advantage that soybean additives can still be detected when only certain protein fractions (7S or 11S fraction) or soybean hydrolysates (as long as they are electrophoretically detectable) are used.

12.10.2.4 Nitrosamines

Not only does the question of the content of nitrite or nitrate in pickle-cured meat arise, but also whether nitrosamines are formed and to what extent they occur in meat (cf. 9.8).

In general, nitrosamines arise only in very low concentrations. Since some of these compounds are a great health hazard, they should be detectable below 0.1 ppm in food for human consumption. The same procedures are available for identifying volatile nitrosamines which have been described earlier for the analysis of aroma constituents (cf. 5.2). However, precautions should be taken during the isolation step. Isolation of nitrosamines should not proceed at low pH since an acid medium in the presence of residual meat nitrites promotes further *de novo* synthesis of nitrosamines. Since the isolated fraction of neutral volatile compounds, which also includes nitrosamines, is highly complex in composition, reliable nitrosamines identification by gas chromatographic retention data is not possible. Additional mass spectrometric data are needed to verify the chemical structure.

The limit of detection of nitrosamines can be further improved when quantitative analysis involves prior derivatization to a product which is readily detected by the highly sensitive electron capture detector (ECD). Two methods are outlined in Fig. 12.48: nitrosamine is oxidized to nitramine, or the nitroso group is eliminated, and the corresponding amine is converted into heptafluorobutyric acid amide.

12.11 Literature

Bailey, A. J. (Ed.): Recent advances in the chemistry of meat. The Royal Society of Chemistry, Burlington House: London. 1984

Bailey, A. J.: The Chemistry of Collagen Cross-links and their role in meat texture. Proc. 42nd, Annual Reciprocal Meat Conf., 127 (1989) publ. 1990

Beck, G., Mantel, Th.: Der biologische Hemmstofftest, ein Weg zur Klärung des Rückstandsproblems in Fleisch. Fleischwirtschaft *52*, 1605 (1972)

Bendall, J. R., Wismer-Pedersen, J.: Some properties of the fibrillar proteins of normal and watery pork muscle. J. Food Sci. *27*, 144 (1962)

Biller, F.: Küchenmäßige Zubereitung und industrielle Praxis – Beispiele der Anwendung der *Maillard*-Reaktion. In: Lebensmittelqualität: Wissenschaft und Technik (Stute, R., Ed.) VCH Verlagsgesellschaft, Weinheim, 1989

Bornstein, P.: The biosynthesis of collagen. Annu. Rev. Biochem. *43*, 567 (1974)

Bornstein, P., Sage, H.: Structurally distinct collagen types. Annu. Rev. Biochem. *49*, 957 (1980)

Brander, J., Eyring, G., Richter, B.: Würzen. In: Ullmanns Encyklopädie der technischen Chemie, 4. edn., vol. 24, p. 507, Verlag Chemie: Weinheim. 1983

Briskey, E. J., Wismer-Pedersen, J.: Biochemistry of pork muscle structure; I. Rate of anaerobic glycolysis and temperature change versus the apparent structure of muscle tissue. J. Food Sci. *26*, 297 (1961)

Cerny, C., Grosch, W.: Evaluation of potent odorants in roasted beef by aroma extract dilution analysis. Z. Lebensm. Unters. Forsch. *194*, 322 (1992)

Disney, J.G., Follet, M.J., Ratcliff, P.W.: Biochemical changes in beef muscle post mortem and the effect of rapid cooling in ice. J. Sci. Food Agric. *18*, 314 (1967)

Drumm, T.D., Spanier, A.M.: Changes in the content of lipid autoxidation and sulfur-containing compounds in cooked beef during storage. J. Agric. Food Chem. *39*, 336 (1991)

Elkhalifa, E.A., Marriott, N.G., Grayson, R.L., Graham, P.P., Perkins, S.K.: Ultrastructural and textural properties of restructured beef treated with a bacterial culture and splenic pulp. Food Microstructure *7*, 137 (1988)

Fennema, O.R. (Ed.): Principles of food science. Part I: Food chemistry. Marcel Dekker, Inc.: New York 1976

Flament I., Willhalm, B., Ohloff, G.: New developments in meat aroma research. In: Flavor of foods and beverages (Eds.: Charalambous, G., Inglett, G.E.), p. 15, Academic Press: New York. 1978

Gasser, U., Grosch, W.: Primary odorants of chicken broth. A comparative study with meat broths from cow and ox. Z. Lebensm. Unters. Forsch. *190*, 3 (1990)

Giddings, G.G.: The basis of color in muscle foods. Crit. Rev. Food Sci. Nutr. *9*, 81 (1977)

Grau, R.: Fleisch und Fleischwaren. Verlag Paul Parey: Berlin. 1969

Graf, E., Panter, S.S.: Inhibition of WOF development by polyvalent cations. J. Food Sci. *56*, 1055 (1991)

Grüttner, F., Lienhop, E.: Taschenbuch der Fleischwarenherstellung. 6. edn., Günter Hempel: Braunschweig. 1962

Günther, H.: Bestimmung von Fremdeiweiß in Fleischwaren. Arch. Lebensmittelhyg. *20*, 97 (1969)

Gutschmidt, J.: The storage life of frozen chicken with regard to the temperature in the cold chain. Lebensm. Wiss. Technol. *7*, 139 (1974)

Hamm, R.: Kolloidchemie des Fleisches. Verlag Paul Parey: Berlin. 1972

Hamm, R., Masic, D.: Routinemethode zur Unterscheidung zwischen frischer Leber und aufgetauter Gefrierleber. Fleischwirtschaft *55*, 242 (1975)

Harrington, W.F., Rao, N.V.: Collagen structure in solution. I. Kinetics of helix regeneration in single-chain gelatins. Biochemistry *9*, 3714 (1970)

Hawkes, R.: Identification of Concanavalin A-binding proteins after sodium dodecyl sulfate-gel electrophoresis and protein blotting. Anal. Biochem. *123*, 143 (1982)

Hermann, Ch., Merkle, Ch., Kotter, L.: Zur Problematik serologischer Nachweisreaktionen für Fremdeiweiße in erhitzten Fleischerzeugnissen. Fleischwirtschaft *53*, 97 and 249 (1973)

Herrmann, Ch., Thoma, H., Kotter, L.: Zur direkten Bestimmung von Muskeleiweiß in Fleischerzeugnissen. Fleischwirtschaft *56*, 87 (1976)

Huxley, H.E.: Electron microscope studies on the structure of natural and synthetic protein filaments from striated muscle. J. Mol. Biol. *7*, 281 (1963)

Igene, J.O., Yamauchi, K., Pearson, A.M., Gray, J.I., Aust, S.D.: Mechanisms by which nitrite inhibits the development of warmed-over flavour (WOF) in cured meat. Food Chem. *18*, 1 (1985)

Ingerowski, G.H., Stan, H.-J.: Nachweis von Östrogen-Rückständen in Fleisch mit Hilfe des cytoplasmatischen Östrogenrezeptors aus Rinderuterus. Dtsch. Lebensm. Rundsch. *74*, 1 (1978)

Kaiser, K.-P., Belitz, H.-D.: Tierartidentifizierung mit elektrophoretischen Methoden: Untersuchungen von Sarkoplasma- und Blutproteinen in rohem Fleisch. Deutsche Forschungsanstalt für Lebensmittelchemie (DFA) Report 1988, p. 34

Kaiser, K.-P., Matheis G., Kmita-Dürrmann, Ch., Belitz, H.-D.: Identifizierung der Tierart bei Fleisch, Fisch und abgeleiteten Produkten durch Proteindifferenzierung mit elektrophoretischen Methoden. I. Rohes Fleisch und roher Fisch. Z. Lebensm. Unters. Forsch. *170*, 334 (1980a)

Kaiser, K.-P., Matheis G., Kmita-Dürrmann, Ch., Belitz, H.-D.: Proteindifferenzierung mit elektrophoretischen Methoden bei Fleisch, Fisch und abgeleiteten Produkten. II. Qualitative and quantitative Analyse roher binärer Fleischmischungen durch isoelektrische Fokussierung in Polyacrylamidgel. Z. Lebensm. Unters. Forsch. *171*, 415 (1980b)

Karlson, P.: Kurzes Lehrbuch der Biochemie. 10. edn., Georg Thieme Verlag: Stuttgart. 1977

Ladikos, D., Lougovois, V.: Lipid oxidation in muscle foods: a review. Food Chem. *35*, 295 (1990)

Lawrie, R.A.: The conversion of muscle to meat. In: Recent advances in food science (Eds.: Hawthorn, J., Leitch, J.M.), Vol. I, p. 68, Butterworths: London. 1962

Lawrie, R.A. (Ed.): Developments in meat science-1 ff, Applied Science Publ.: London. 1980 ff

Lawrie, R.A.: Meat science, 4th edn., Pergamon Press: Oxford. 1985

Ledward, D. A.: A note on the nature of the haematin pigments present in freeze dried and cooked beef. Meat Sci. *21*, 231 (1987)

Lehninger, A. L.: Biochemie. 2nd edn., Verlag Chemie: Weinheim. 1977

MacLeod, G., Seyyedain-Ardebili, M.: Natural and simulated meat flavors (with particular reference to beef). Crit. Rev. Food Sci. Nutr. *14*, 309 (1981)

Moody, W. G.: Beef flavor – a review. Food Technol. *37*, 227 (1983)

Mottram, D. S.: Some observations on the role of lipids in meat flavour. In: Sensory quality in foods and beverages (Eds.: Williams. A. A., Atkin, R. K.), p. 394, Ellis Horwood Ltd.: Chichester. 1983

Pearson, A. M., Young, R. B.: Muscle and meat biochemistry. Academic Press: San Diego, CA. 1989

Piasecki, A., Ruge, A., Marquardt, H.: Malignant transformation of mouse M2-fibroblasts by glycerol chlorohydrines contained in protein hydrolysates and commercial food. Arzneim.-Forsch./Drug. Res. *40*, 1054 (1990)

Ohloff, G., Flament, I: Heterocyclic constituents of meat aroma. Heterocycles *11*, 663 (1978)

Potthast, K., Hamm, R.: Biochemie des DFD-Fleisches. Fleischwirtschaft *56*, 978 (1976)

Price, J. F., Schweigert, B. S. (Eds.): The science of meat and meat products. 2nd edn., W.H. Freeman: San Francisco. 1971

Rich, A., Davies, D. R., Crick, F. H. C., Watson, J. D.: The molecular structure of polyadenylic acid. J. Mol. Biol. *3*, 71 (1961)

Sargent, J. A.: The application of cold stage scanning electron microscopy to food research. Food Microstructure *7*, 123 (1988)

Scanlan, R. A.: N-Nitrosamines in foods. Crit. Rev. Food Technol. *5*, 357 (1974)

Schormüller, J. (Ed.): Handbuch der Lebensmittelchemie. Vol. I, Springer-Verlag: Berlin. 1965

Schultz, H. W., Anglemier, A. F. (Eds.): Proteins and their reactions. AVI Publ. Co.: Westport, Conn. 1964

Silhankova, L., Smid, F., Cerna, M., Davidek, J., Velisek, J.: Mutagenicity of glycerol chlorohydrines and of their esters with higher fatty acids present in protein hydrolysates. Mutation Research *103*, 77 (1982)

Steinhart, H.: Chemische Kriterien zur Beurteilung der Fleischqualität. Lebensmittelchemie *46*, 61 (1992)

Sulser, H.: Die Extraktstoffe des Fleisches. Wissenschaftliche Verlagsgesellschaft mbH: Stuttgart. 1978

Thanh, V. H., Shibasaki, K.: Major proteins of soybean seeds. A straightforward fractionation and their characterization. J. Agric. Food Chem. *24*, 1117 (1976)

Tóth, L.: Chemie der Räucherung. Verlag Chemie: Weinheim. 1982

Traub, W., Piez, K. A.: The chemistry and structure of collagen. Adv. Protein Chem. *25*, 243 (1971)

Van Rillaer, W., Beernaert, H.: Determination of residual 1,3-dichloro-2-propanol in protein hydrolysates by capillary gas chromatography. Z. Lebensm. Unters. Forsch. *188*, 343 (1989)

Velisek, J., Davidek, J., Davidek, T., Hamburg, A.: 3-Chloro-1,2-propanediol derived amino alcohol in protein hydrolysates. J. Food Sci. *56*, 136 (1991)

Velisek, J., Davidek, J., Davidek, T., Kubelka, V., Viden, I.: 3-Chloro-1,2-propanediol derived amino acids in protein hydrolysates. J. Food Sci. *56*, 139 (1991)

Velisek, J., Davidek, J., Hajslova, J., Kubelka, V., Janisek, G., Mankova, B.: Chlorohydrins in protein hydrolysates. Z. Lebensm. Unters. Forsch. *167*, 241 (1978)

Velisek, J., Davidek, J., Kubelka, V., Janisek, G., Svobodava, Z., Simicova, Z.: New chlorine-containing organic compounds in protein hydrolysates. J. Agric. Food Chem. *28*, 1142 (1980)

Velisek, J., Davidek, J., Kubelka, V.: Formation of $\Delta^{3,5}$-diene and 3-chloro Δ^5-ene analogues of sterols in protein hydrolysates. J. Agric. Food Chem. *34*, 660 (1986)

Warendorf, T., Belitz, H.-D.: Zum Geschmack von Fleischbrühe. 2. Sensorische Analyse der Inhaltsstoffe und Imitation einer Brühe. Z. Lebensm. Unters. Forsch. *195*, 209 (1992)

Wittkowski, R.: Phenole im Räucherrauch. VCH Verlagsgesellschaft: Weinheim. 1985

Wittmann, R.: Chlorpropandiole in Lebensmitteln. Lebensmittelchemie *45*, 89 (1991)

Wykle, B., Gillett, T. A., Addis, P. B.: Myoglobin heterogeneity in pigs with PSE and normal muscle by an improved isoelectric focusing technique. J. Animal. Sci. *47*, 1260 (1978)

13 Fish, Whales, Crustaceans, Mollusks

13.1 Fish

13.1.1 Foreword

Fish and fish products fulfill an important role in human nutrition as a source of biologically valuable proteins, fats and fat-soluble vitamins. Fish can be categorized in many different ways, e.g., according to:

- The environment in which the fish lives: sea fish (herring, cod, saithe) and freshwater fish (pike, carp, trout), or those which can live in both environments, e.g., eels and salmon. Sea fish can be subdivided into groundfish and pelagial fish.
- The body form: round (cod, saithe) or flat (common sole, turbot or plaice).

Commercial fishing takes place in the open sea, coastal and freshwater areas. Conservation programs and hatcheries to rebuild stocks play important roles in the management of fresh and saltwater fish resources.

The fishing industry catch has risen sharply in tonnage during this century. In 1900 the catch was approx. 4 million t, while it had increased to 102 million t by 1996. Table 13.1 shows the catch in tonnage of the leading countries engaged in fishing. This includes shellfish products, i.e. lobsters and crustaceans such as crabs, crayfish and shrimps, and mollusks such as clams, oysters, scallops, squid, etc., which are not true fish but are harvested from the sea by the fishing industry. Table 13.2 lists the catch in the same year of the chief kinds of fish and shellfish. A review of the forms of fish products entering the market it given in Table 13.3.

13.1.2 Food Fish

Table 13.4 shows the important food fish. In general, a predatory fish is better tasting than a nonpredaceous fish, a fatty fish better than a nonfatty fish. The fishbone-rich species, such as carp, perch, pike and fench, are often less in demand than fish with fewer bones.

Some important food fish will be described in more detail.

Table 13.1. World catch of fish, crustaceans and mollusks (1996)

Continent	1,000 t	Country	1,000 t
World	102,178	China	14,222
Africa	3,336	Peru	9,515
America, North-,		Chile	6,692
Central-	7,610	Japan	5,964
America, South-	19,466	USA	5,000
Asia	38,896	Russian Fed.	4,675
Europe	11,691	Indonesia	3,729
Former USSR	5,152	India	3,492
Oceania	76	Thailand	3,138
		Norway	2,638
		Korea,	
		Republic	2,413
		Phillipinen	1,790
		Korea, Volks-	
		republic	1,725
		Denmark	1,681
		Mexico	1,422
		Argentina	1,237
		Malaysia	1,130
		Spain	1,055
		Canada	900
		UK	876
		Bangladesh	874
		Vietnam	811
		Myanmar	804
		Brazil	798
		Ecuador	684
		Σ (%)[a]	76

[a] World production $\triangleq$ 100%.

Table 13.2. Catch of fish, crustaceans and mollusks (1996)

	1,000 t
Freshwater fish[a]	
Carps, barbels etc.	835
Tilapias	517
Sturgeons	4
River eels	13
Salmons, trouts, smelts, etc.	1,029
Sea fish	
Flounders, halibuts, sole, etc.	920
Cods, hakes, haddocks etc.	10,711
Redfishes, basses, congers, etc.	6,605
Jacks, mullets, sauries, etc.	11,135
Herrings, sardines, anchovis	22,323
Tunas, bonitos, billfishes, etc.	4,584
Mackerels, snoeks, cutlassfishes, etc.	5,136
Sharks, rays, chimaeras, etc.	758
Others	11,993
Crustaceans	
Fresh water crustaceans	475
Sea-spiders, crabs, etc.	1,219
Lobsters	208
Shrimps	2,470
Other sea crustaceans	1,093
Mollusks	
Freshwater mollusks	560
Mussels	203
Oysters	156
Octopuses, squids, cuttlefishes	3,037
Other sea mollusks	1,045

[a] Includes fish species journeying between the sea and freshwater lakes or rivers.

Table 13.3. World market for fish and fish products (1989)

Marketed	Amount (%)[a]
Fresh fish	21.8
Frozen fish	23.9
Salted, smoked, marinated	11.0
Canned	12.8
	69.5
Fish meal, fish oil	29.0
Others	1.5

[a] As % of total catch (99.5 mill. t $\triangleq$ 100%).

13.1.2.1 Sea Fish

13.1.2.1.1 Sharks

Dogfish (*Squalus acanthias*) about 1 m long are marinated or smoked before marketing. In North America fish of the family *Squalidae* are generally referred to as dogfish sharks. Other names are spiny, spring or piked dogfish and rock salmon. Trade names used in th U.K. are flake, huss or rig. In Germany the name used is Dornhai (Dornfisch), and the smoked dorsal muscle is sold as "Seeaal", while the hot-smoked, skimmed belly walls are called "Schillerlocken". Mackerel sharks of the family *Lamnidae* are also in this group. The main species of this family are: (a) porbeagle, blue dog or Beaumaris shark (*Lamna nasus*); and (b) salmon, (c) mako and (d) white sharks. The blue shark is found in the Atlantic Ocean and the North Sea as an escort of herring schools. It possesses a meat similar to veal and is known in the trade as sea or wild sturgeon, or calf-fish. Due to a high content of urea (cf. 13.1.4.3.6), the meat of these fish is often tainted with a mild odor of ammonia. Endeavors to popularize shark and related fish meat are well justified since the meat is highly nutritious; however consumer acceptance would be hampered by the word "shark", therefore other terms are commonly preferred in the trade. Shark fins are a favorite dish in China and are imported to Europe as a specialty food.

13.1.2.1.2 Herring

The herring (*Clupea harengus*) is one of the most processed and most important food fish and is a source of raw materials for meal and marine oil. Herrings are categorized according to the season of the catch (spring or winter herring), spawning time or stage of development (e.g. matje, the young fatty herring with roe only slightly developed, is cured and packed in half barrels), or according to the ways in which the fish were caught: drag or drift nets, trawling nets, gill and trammel nets or by seining (purse seining), the most important form of snaring. Electrofishing methods have proven to be particularly economical on the high seas. The main fishing time for the German fleet is July, and for the English and

Table 13.4. Major commercial fish species – quality and utilization

Name	Family	Genus, sp.	Comments on quality and processing
Sea fish			
Pleurotremata (sharks)			
Dogfish	Squalidae	*Squalus acanthias (Acanthias vulgaris)*	
Rajiformes (skates)			
Skates, e.g., thornback, common skate	Rajidae	*Raja clavata, R. batis*	Used are the wing shaped body widenings, the pectoral part and breast fins as a delicacy; it is fried, smoked or jellied
Acipenseriformes (sturgeons)			
Sturgeon	Acipenseridae	*Acipenser sturio*	Exceptionally delicate when smoked, caviar is made from its roe
Clupeiformes (herrings)			
Herring	Clupeidae	*Clupea harengus*	Valuable fish with fine white meat, fried and grilled; industrially processed, for example into Bismarck herrings, rollmops and brat-herring
Sprat	Clupeidae	*Sprattus sprattus*	Mostly cold or warm smoked; anchovies
Sardine	Clupeidae	*Sardina pilchardus*	Mostly steam cooked and canned in oil; along sea coast grilled and fried
Anchovy (Anchovis)	Engraulidae	*Engraulis encrasicolus*	Pleasant, aromatic fragrant, cured in brine, made into rings and paste
Lophiiformes (anglers)			
Angler, allmouth	Lophiidae	*Lophius piscatorius*	White, good and firm meat, poached
Gadiformes (cods)			
Ling	Gadidae	*Molva molva*	Tasty firm white meat
Cod	Gadidae	*Gadus morhua*	Meat is prone to fracturing, used fresh, filleted, salted and frozen, dried (stock- and klipfish), cooked, poached; oil is produced from liver
Haddock	Gadidae	*Melanogrammus aeglefinus*	Very fine in taste, processed as fresh, pickled, or marinated, smoked, fried, roasted, cooked or poached, or used for fish salad
Coalfish pollack, black coor or Boston bluefish	Gadidae	*Pollachius virens, P. pollachius*	Meat is lightly tinted grayish-brown, it is filleted, smoked, sliced as cutlets or chops and processed in oil (used for salmon substitute)
Whithing	Merlangius	*Merlangius merlangius*	Good meat, easily digested, but very sensitive, fried or deep fried roasted or smoked, used for fish stuffings
Hake	Merluccidae	*Merluccius merluccius*	Fresh or frozen, all processing methods are used

Table 13.4 (Continued)

Name	Family	Genus, sp.	Comments on quality and processing
Scorpaeniformes			
Red fish, ocean perch	Scorpaenidae	*Sebastes marinus*	Tasty meat, fattier than cod, it is filleted or smoked
Gurnard, sea robin (gray gurnard, red gurnard)	Triglidae	*Trigla gurnardus, T. lucerna*	White firm meat (red sp. is of higher quality), used fresh or smoked
Lumpfish, sea hen	Cyclopteridae	*Cyclopterus lumpus*	Smoked, its roe is processed into caviar substitute
Perciformes (percoid fishes)			
Red mullet	Mullidae	*Mullus barbatus*	White fine and a piquant delicius meat, mostly grilled
Catfish, wolffish	Anarhicha-didae	*Anarhichas lupus, A. minor*	Fine white fragrant meat, poached grilled, dough-type crust coated
Mackerel	Scombridae	*Scomber scombrus*	Highly valued fish, tasty reddish meat, fried, grilled, smoked or canned; its meat is not readily digestible
Tuna	Scombridae	*Thunnus thynnus*	Reddish meat of exceptional taste, it is fried, roasted, smoked, or canned in oil or processed into paste sausages or rolls
Pleuronectiformes (flat fishes)			
Turbot (butt or britt)	Scopthalmidae	*Psetta maxima (Rhombus maximum)*	Apart from common sole, the highest valued flat fish, meat is snow-white firm and piquant, it is cooked, grilled or poached
Halibut	Pleuronectidae	*Hippoglossus h. hippoglossus*	Tasty meat, it is poached, fried or smoked
Plaice	Pleuronectidae	*Pleuronectes platessa*	Tasty meat, fried or filleted and poached
Flounder	Pleuronectidae	*Platichthys flesus (Pleuronectes flesus)*	Good white meat, it is poached, fried or smoked
Common sole	Soleidae	*Solea solea*	It is the finest flat fish, it is poached, fried, grilled or roasted
Freshwater fish			
Petromyzones (lampreys)			
Lamprey	Petromyzoni-dae	*Lampetra fluviatilis*	It is industrially processed
Anguilliformes (eel sp.)			
Eel	Angullidae	*Anguilla anguilla*	Tasty meat, good quality when not exceeding 1 kg in weight, the fresh fish is fried, roasted, or it is smoked, marinated of jellied
Salmoniformes (salmons)			
Salmon	Salmonidae	*Salmo salar*	High quality fish (5–10 kg), it is poached, grilled, cured or smoked, also pickled

Table 13.4 (Continued)

Name	Family	Genus, sp.	Comments on quality and processing
River trout	Salmonidae	*Salmo trutta*	High quality fish, no fishbone, bluish tinted when cooked, roasted a la meuniere
Rainbow trout (lake- or steelhead trout)	Salmonidae	*Salmo gairdnerii*	
Brook trout	Salmonidae	*Salvelinus fontinalis*	A worthy fish, meat is pale pinkish, processed as trout but mostly fried
Whitefish	Salmonidae	*Coregonus sp.*	Processed as trout
Coregonus, whitefish	Salmonidae	*Coregonus sp.*	White tender and tasty meat, though somewhat dry, it is fried or deep fried
Smelt	Osmeridae	*Osmerus eperlanus*	A fishbone rich meat which is mostly deep fried
Pike (jackfish)	Esocidae	*Esox lucius*	Young pikes (best quality 2–3 kg) are soft tender and tasty, the meat is well rated though it is bone rich, steam cooked, cooked or fried
Cypriniformes (carps)			
Roach	Cyprinidae	*Rutilus rutilus*	It has a tasty meat though fishbone rich
Bream	Cyprinidae	*Abramis brama*	Tasty meat but fishbone rich
Tench	Cyprinidae	*Tinca tinca*	Tender fatty meat, tasty, bluish when cooked, mostly steam cooked
Carp	Cyprinidae	*Cyprinus carpio*	Soft meat, readily disgestible, a valuable fish food, bluish when cooked
Crucian	Cyprinidae	*Carassim carassim*	A good food fish, but not as good as carp, the meat is bone rich
Perciformes (perchoid fishes)			
Perch	Percidae	*Perca fluviatilis*	Firm, white and very tasty meat, best quality is below 1 kg (25–40 cm), it is fried, filleted and/or steam cooked
Zander	Percidae	*Stizostedion lucioperca*	White, soft, juicy and tasty meat, 40–50 cm, fried or steam cooked, it is the best quality freshwater fish
Ruffle	Percidae	*Gymnocephalus cernua*	Exceptionally tasty meat

Norwegian fishing fleets it is October to December.

The herring averages 12–35 cm in length and migrates in large swarms or schools throughout the nothern temperate and cold seas.

Herring is marketed cold or hot smoked (kippers, buckling), frozen, salt-cured, dried and spiced, jellied, marinated and canned in a large variety of sauces, creams, vegetable oils, etc. Sprat (*Sprattus sprattus phalericus*), called brisling in Scandinavia and Sprotte in Germany, is processed into an "Appetitsild" (skinned fillets or spice-cured sprats packed in vinegar, salt, sugar and seasonings). Canned brisling is packed in edible oil, tomato and mustard sauces, etc. and is sold as brisling sardines. Brisling is often lightly smoked and marketed as such. Sprats are also processed into a delicatessen product called "Anchosen", which consists mostly of small sprats, sometimes mixed with cured matje, and preserved in salt and sugar, with or without spices and sodium nitrate.

Anchovies (*Engraulis encrasicolus*; German term "Sardellen"; found in Atlantic and Pacific Oceans) should also be included in the herring group. Anchovies are usually salted (cured in brine in barrels until the flesh has reddened). They are also canned in glass jars, marketed as a paste or cream, smoked or dried.

Sardines (*Sardina pilchardus*), from the Mediterranean or Atlantic (France, Spain, Portugal) or from Africa's west coast, are often marketed steamed, fried or grilled, or canned in oil or tomato sauce. The fully grown sardine is known in the trade as pilchard (Californian, Chilean, Japanese). It is also salt-cured and pressed in barrels or canned in edible oil or in sauce. "Russian sardines" or "Kronsardine" are actually marinated small herrings or sprats caught in the Baltic Sea. Also in the herring group is the allis shad (*Alosa alosa* or *Clupea alosa*), which is sold fresh, smoked or canned.

13.1.2.1.3 Cod Fish

These nonfatty fish (from the North Sea, Iceland or Greenland) are usually marketed fresh, whole and gutted, and many have the head and/or skin removed or be filleted. The Atlantic cod (*Gadus morhua*) is considered the most important food fish of Northern Europe. Classified according to size, designations are: small codling, codling, sprag and cod in the U.K. and Iceland; and scrod in the United States.

Saithe is also known as coalfish or pollack (*Pollachius virens*) or by names such as black cod or Boston bluefish. After salt-curing and slicing, it is lightly smoked and packed in edible oil. Saithe is marketed in Germany as a salmon substitute called "Seelachs". Rolled in balls and canned, it is called "side boller" in Norway.

Whiting (*Merlangius merlangius*), known as merlan in France, is a North Atlantic, North Sea fish, marketed in many forms.

Haddock (*Melanogrammus aeglefinus*) is a North Atlantic and Arctic Sea fish. Small haddock are called gibbers or pingers, and large ones are jumbos. The annual haddock catch is lower than those of the above-mentioned fish, i.e. anchovy, herring, cod, sardine and pollack. Hake (*Merluccius merluccius*) is an Atlantic and North Sea fish. Its various subspecies are the Cape, Chilean, Northeast Pacific, Mediterranean and North American east coast white hake. The annual catch is somewhat higher than that of haddock. Even higher than both is the catch of menhaden (*Brevoortia tyrannus*), which accounts for almost 38% of the fish tonnage in the United States.

13.1.2.1.4 Scorpaenidae

Red fish of the North Atlantic and arctic regions (*Sebastes marinus* and other species), which are known as red fish or ocean perch (U.K.) or rose-fish or Norway haddock (U.S.A.), have gained in importance in recent decades. Red fish meat is rich in vitamins and fat. It is marketed fresh or frozen, whole or as fillets; as cold or hot smoked steaks; and roasted or cooked.

13.1.2.1.5 Perch-like Fish

The bluefin tuna (*Thunnus thynnus*) is one of the several *Thunnus* spp. It has a red, beef-like muscle tissue and is caught primarily in the North and Mediterranean Seas and the Atlantic Ocean. It is marketed salted and dried, smoked, or canned in edible oil, brine or tomato sauce. Tuna meat is also a common delicatessen item (tuna paste, sausages, rolls, etc.).

The Atlantic mackerel (*Scomber scombrus*) is of great importance, as are the chub or Pacific mackerel (*S. japonicus*) and the blue Australian mackerel (*S. australasicus*). Mackerel are sold whole, gutted or ungutted; or filleted, frozen, smoked, salted, pickle-salted (Boston mackerel), etc.

13.1.2.1.6 Flat Fish

This group includes: plaice or hen fish (*Pleuronectes platessa*); flounder (*Platichthys flesus*, also known as fluke); Atlantic halibut or butt (*Hippoglossus hippoglossus*); common dab (*Pleuronectes limanda*); brill (*Rhombus laevis*); Atlantic and North Sea common sole (*Solea solea*, "Dover" sole); and turbot (*Psetta maxima*, also called butt or britt). These fish and haddock (cf. 13.1.2.1.3) are the sea fish most popular with consumers.

13.1.2.2 Freshwater Fish

Some important freshwater fish are; eels; carp; tench; roach; silver bream; pike, jackfish or pickerel; perch, pike-perch or blue pike; salmon; rainbow, river or brown trout; and pollan (freshwater herring or white fish). Unlike sea fish, the catch of freshwater fish is of little economic importance (cf. Table 13.2), although it does offer an important source of biologically valuable proteins.

13.1.2.2.1 Eels

Freshwater and sea eels (*Anguilla anguilla*, *A. rostrata*, *Conger conger*, etc.) are sold fresh, marinated, jellied, frozen or smoked as unripe summer (yellow or brown eel) or ripe winter eels (bright or silver eel). Due to their high fat content (approx. 25% fresh weight), eels are not readily digestible.

13.1.2.2.2 Salmon

Salmon (*Salmo salar*) and sea trout (*Salmo trutta*) are migratory. Salted or frozen fish are supplied to the European market by Norway and by imports from Alaska and the Pacific Coast of Canada. Also included in this group are: river trout (*Salmo trutta f. fario*) and lake trout (*Salmo gairdnerii*), which is commonly called steelhead trout in North America when it journeys between the sea and inland lakes.

13.1.3 Skin and Muscle Tissue Structure

As in other backboned animals, fish skin consists of two layers: the outer epidermis and the inner derma (cutis or corium). The outer epidermis is not horny but is rich in water, has numerous gland cells and is responsible for the slimy surface. Mucopolysaccharides are major components of this mucous, with galactosamine and glucosamine as the main sugars. The derma is permeated with connective tissue fibers and has various pigment cells, among them guanophores, which contain silvery-white glistening guanine crystals. Scales protrude from the derma. Their number, size and kind differ from species to species. This is of importance in fish processing since it determines whether a fish can be processed with or without skin. The nature and state of fish skin affects shelf life and flavor. The spreading of skin microflora after death is the main cause of the rapid decay of fish. The skin contains numerous spores resistant to low temperatures; they can grow even at $<-10\,°C$ (psychrophiles or psychrotolerant microorganisms). The decay is also enhanced by bacteria present in fish intestines.

The fish body is fully covered by muscle tissue. It is divided dorsoventrally by spinous processes and fin rays and in the horizontal direction by septa. Corresponding to the number of vertebra, the rump muscle tissue is divided into muscle sections (myomeres) which are separated from each other by connective tissue envelopes. The transversal envelopes are called myocommata, the horizontal ones myosepta. While myosepta are arranged in a straight line, myocommata are pleated in a zig-zag fashion. Since cooking gelatinizes the connective tissue, the muscle tissue is readily disintegrated into flake-like segments.

As in mammals, fish muscle fibers are striated. Depending on the myoglobin content, fish flesh is dark or light colored. In some fish, such as herring and mackerel, the portion of dark-colored tissue is very high (10%), while in others, such as cod, it is limited to a thin layer immediately below the derma.

13.1.4 Composition

13.1.4.1 Overview

The edible portion of a fish body is less than in warm-blooded animals. The total waste might approach 50% and 10%–15% after head removal. Fish meat and that of land animals are readily digestible, but fish is digested substantially faster and has therefore a much lower nutritive saturation value. The cooking loss is approx. 15% with fish, which is significantly less than that of beef. The biological value of fish proteins is similar to that of land animals. While the crude protein content of fish is about 17–20%, the fat and water contents vary widely. Some are distinctly nonfatty, with fat contents of only 0.1–0.4% (haddock

Table 13.5. Average chemical composition of fish

Fish	Moisture[a]	Protein[a]	Fat[a]	Minerals[a]	Edible portion[b]
Freshwater fish					
Eel	61	13	26	1.0	70
Perch	80	18	0.8	1.3	38
Zander	78	19	0.7	1.2	50
Carp	72	19	7	1.3	55
Tench	77	18	0.8	1.8	40
Pike	80	18	0.9	1.1	55
Salmon	66	20	14	1.0	64
River trout	78	19	2	1.2	50
Smelt	80	17	1.7	0.9	48
Sea fish					
Cod	82	17	0.4	1.0	56
Haddock	81	18	0.1	1.1	57
Ling	79	19	0.6	1.0	68
Hake	81	17	0.9	1.1	58
Red fish (ocean perch)	78	19	3	1.4	52
Catfish	80	16	3	1.1	52
Plaice	81	17	0.8	1.4	56
Flounder	81	17	0.7	1.3	45
Common sole	80	18	1.4	1.1	71
Halibut (butt)	75	19	5	1.0	75
Turbot (britt)	80	17	1.7	0.7	46
Herring (Northern Sea)	63	17	18	1.3	67
Herring (Baltic Sea)	71	18	9	1.3	65
Sardine	74	19	5		59
Mackerel	68	19	12	1.3	62
Tuna	62	22	16	1.1	61

[a] As % of edible portion.
[b] As % of the whole fish weight.

or cod), while some are very fatty (eels, herring or tuna), with fat contents of 16–26%. Many fish species have fat contents between these extreme values. Table 13.5 provides data on the basic composition of fish.

13.1.4.2 Proteins

The protein-N content of fish muscle tissue is between 2–3%. The amino acid composition, when compared to that of beef or milk casein (Table 13.6), reveals the high nutritional value of fish proteins. The sarcoplasma protein accounts for 16–22% of the muscle tissue total protein. The contractile apparatus accounts for 75% protein; the connective tissue of teleosts is 3%; and of elasmobranchs, such as sharks and rays (skate or rocker), is up to 10%. The individual protein groups and their functions in muscle tissue of mammals (cf. 12.3.2) also apply to fish.

13.1.4.2.1 Sarcoplasma Proteins

Fish sarcoplasma proteins consist largely of enzymes. The enzymes correspond to those of mammalian muscle tissue. When these proteins are separated electrophoretically, specific patterns are obtained for each fish species. This is a helpful chemical means of fish taxonomy. The content of pigments (myoglobin, cytochromes) varies greatly, but is never as high as in mammalian muscle. In strongly pigmented fish (e.g. tuna), pigment degradation reactions can induce meat discoloration

Table 13.6. Amino acid composition of fish muscle, beef muscle and casein (amino acid-N as % of total-N)

	Casein	Beef muscle	Cod muscle
Aspartic acid	4.7	4.0	6.8
Threonine	3.6	3.7	3.4
Serine	5.3	4.6	3.6
Glutamic acid	13.3	9.3	8.8
Proline	7.5	4.3	3.4
Glycine	3.2	6.0	5.8
Alanine	3.0	4.9	5.9
Cystine	0.2	0.8	2.5
Valine	5.4	3.7	2.5
Methionine	1.8	2.2	2.0
Isoleucine	4.1	4.2	2.7
Leucine	6.1	5.1	5.1
Tyrosine	3.0	2.1	1.7
Phenylalanine	2.7	2.7	2.1
Tryptophan	1.0	1.2	1.1
Lysine	9.8	9.8	11.7
Histidine	5.3	4.9	3.5
Arginine	8.2	14.5	13.2

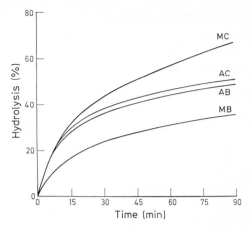

Fig. 13.1. Tryptic hydrolysis of myofibrils (M) and actin (A) from cod fish (C) and beef (B) under the same conditions. (according to *Connell*, 1964)

(e.g. observable "greening" in canned tuna meat).

13.1.4.2.2 Contractile Proteins

The proportion of contractile proteins in fish total protein is higher than in mammalian muscle tissue, however the proportions among individual components are similar. The heat stability of fish proteins is lower than that of mammals, the protein denaturation induced by urea occurs more readily, and protein hydrolysis by trypsin is faster (Fig. 13.1). These properties provide additional evidence of the good digestibility of fish proteins.

13.1.4.2.3 Connective Tissue Protein

The content of connective tissue protein in fish muscle is lower than in mammalian flesh. The shrinkage temperature, T_s, is about 45 °C in fish collagen, i.e. much lower than in mammalian collagen (60–65 °C). These two factors make fish meat more tender than mammalian meat.

13.1.4.2.4 Serum Proteins

The congealing temperature of the blood serum of polar fish of Arctic or Antarctic regions (e.g. *Trematomus borchgrevinski, Dissostichus mawsoni, Boreogadus saida*) is about − 2 °C and thus is significantly lower than that of other fish (− 0.6 to − 0.8 °C). Antifreeze glycoproteins account for such low values. The amino acid sequence of this class of proteins is characterized by high periodicity:

$$[Ala—Ala— Thr]_n—Ala—Ala^a \qquad (13.1)$$

[a] C-terminal has one or two alanine residues

The molecular weight range is 10.5–27 kdal, while the conformation is generally stretched, with several *a*-helical regions. These glycoproteins are hydrated to a great extent in solution. The antifreezing effects are attributed to the disaccharide residues as well as to the methyl side chains of the peptide moiety.

13.1.4.3 Other N-Compounds

The nonprotein-N content is 9–18% of the total nitrogen content in teleosts and 33–38% in elasmobranchs.

13.1.4.3.1 Free Amino Acids, Peptides

Histidine is the predominant free amino acid in fish with dark-colored flesh (tuna, mackerel). Its content in the flesh is 0.6–1.3% fresh weight and can even exceed 2%. During bacterial decay of the flesh, a large amount of histamine is formed from histidine. Fish with light colored flesh contain only 0.005–0.05% free histidine. Free 1-methylhistidine is also present in fish muscle tissue. Anserine and carnosine contents are 25 mg/kg fresh tissue. Taurine content is high (500 mg/kg).

13.1.4.3.2 Amines, Amine Oxides

Sea fish contain 40–120 mg/kg of trimethylamine oxide, which is involved in the regulation of the osmotic pressure. After death, this compound is reduced by bacteria to "fishy" smelling trimethylamine. On the other hand, fresh-water fish contain only very low amounts of trimethylamine (0–5 mg/kg). On storage of fish, a part of the trimethylamine is enzymatically broken down to dimethylamine and formaldehyde. The latter then undergoes cross-linking reactions with proteins, which make the fish tougher. In addition to trimethylamine, the amine fraction contains dimethyl- and monomethylamines and ammonia, and some other biogenic amines derived from amino acid decarboxylation. The concentration of volatile nitrogen bases increases after death, the increase being influenced by storage duration and conditions. The level of volatile amines can be used as an objective measure of fish freshness (Fig. 13.2).

13.1.4.3.3 Guanidine Compounds

Guanidine compounds, such as creatine, are 600–700 mg/kg fresh fish muscle tissue. In crustaceans, the role of creatine is taken over by arginine.

13.1.4.3.4 Quaternary Ammonium Compounds

Glycine betaine and γ-butyrobetaine are present in low amounts in fish flesh.

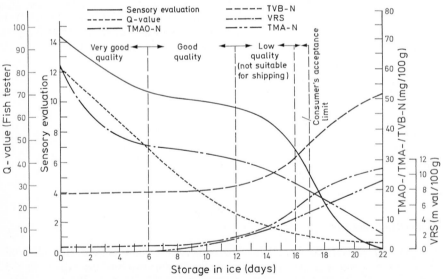

Fig. 13.2. Cod fish quality change during storage. (according to *Ludorff*, 1973). Sensory evaluation: in total 15 points are given, 5 for visual appearance and 10 for odor, taste and texture; Q-value: electric resistance of the fish tissue as recorded by a "fish tester"; Q40: quality class S, Q = 30–40: A, Q = 20–30: B, Q20: C and worse; TMAO-N: trimethylamine oxide-N; TVB-N: total volatile base-N; VRS: volatile reducing substances, TMA-N: trimethylamine-N

13.1.4.3.5 Purines

The purine content in fish muscle tissue is about 300 mg/kg.

13.1.4.3.6 Urea

The fairly high content of urea in muscle tissue (1.3–2.1 g/kg) is characteristic of elasmobranchs (rays, sharks). The compound is decomposed to ammonia by bacterial urease during fish storage.

13.1.4.4 Carbohydrates

Glycogen is the principal carbohydrate. Its content (up to 0.3%) is generally lower than in mammalian muscle tissue.

13.1.4.5 Lipids

The fat (oil) content of fish is highly variable. It is influenced not only by the kind of fish but by the maturity, season, food availability and feeding habit. Fat deposition occurs in muscle tissue (e.g. carp, herring), in liver (cod, haddock, saithe) or in intestines (blue pike, pike, perch).

Fish is an important source of ω-3-polyenic acids with 5 and 6 double bonds (cf. Table 13.7), which are considered valuable from a physiological and nutritional viewpoint. In contrast to the high content of unsaturated fatty acids, the level of antioxidatively active tocopherols is relatively low. Therefore, the

Table 13.7. The content of ω-PUFA in fish (g/100 g of fillet)

Type of fish	EPA (20:5)[a]	DHA (22:6)[a]
Mackerel	0.65	1.10
Salmon (Atlantic)	0.18	0.61
Salmon (red)	1.30	1.70
Trout	0.22	0.62
Tuna	0.63	1.70
Cod	0.08	0.15
Flounder	0.11	0.11
Perch	0.17	0.47
Haddock	0.05	0.10
Sole	0.09	0.09

[a] Structure in Table 3.7.
PUFA: polyunsaturated fatty acid.

lipids of fish represents a major problem in preservation because of its easy peroxidation (cf. 13.1.4.8).

13.1.4.6 Vitamins

Fish fat and liver (liver oil) are significant sources of fat-soluble vitamins, A and D. Also present are vitamins E (tocopherol) and K. The water-soluble vitamins, thiamine, riboflavin and niacin, occur in relatively high amounts, while others are present only in low amounts.

13.1.4.7 Minerals

The average content of major minerals in fish muscle tissue is compiled in Table 13.8.

Table 13.8. Minerals in fish muscle

Element	Content (mg/kg)	Element	Content (mg/kg)
Ca	48–420	Fe	5–248
Mg	240–310	Cu	0.4–1.7
P	1,730–2,170	I	0.1–1.0

13.1.4.8 Aroma Substances

Aroma substances are formed by the enzymatic oxidative degradation of the highly unsaturated fatty acids with the participation of lipoxygenases of varying specificity. These aroma substances contribute to the mild "green-metallic" aroma of fresh fish: hexanal, 2-trans-hexenal, 3-cis-hexenal, 1-octen-3-ol, 1-octen-3-one, 1,5-cis-octadiene-3-ol, 1,5-cis-octadien-3-one, 2-trans,6-cis-nonadienal.

The compound 2,6-dibromophenol, which has a very low aroma threshold of 0.5 ng/kg, also contributes to the aroma of fresh sea fish. In higher concentrations, it produces an iodoform-like odor defect which has been observed in shrimps. The meaty aroma note of cooked tuna is caused by the formation of 2-methyl-3-furanthiol (cf. 12.9).

A fishy odor of train oil can be formed very quickly. This aroma defect is due to the reduction of trimethylamine oxide (cf. 13.1.4.3.2) and/or autoxidation of ω-3-fatty acids with three and more double bonds. These PUFAs yield 1,5-octadien-3-one which in combination with methional (precursor methionine) cause the "fishy" aroma defect of boiled fish and also of other food containing these precursors, e.g., cooked potatoes.

13.1.4.9 Other Constituents

More than 500 tropical fish species (barracuda, sting ray, fugu, globefish), including some valuable food fish, are known to be passively poisonous. Poisoning can occur as a result of their consumption. The toxicity can vary with the season, and can extend to the whole body or be localized in individual organs (gonads, i.e. ovaries and testicles, liver, intestines, blood). Cooking can destroy some of these toxic substances. They consist of peptides, proteins and other compounds. Some of their structures have been elucidated. There are also actively poisonous fish, with prickles or tiny needle-like spines used as the poisoning apparatus. These are triggered as a weapon in defence or attack. This group of fish includes the species *Dasytidae*, *Scorpaenidae* and *Trachinidae*. The latter, known as lesser weever (*Trachinus vipera*), is a fish of the Atlantic Ocean and the Mediterranean Sea.

13.1.5 Post-mortem Changes

After death, fish muscle tissue is subjected to practically the same spontaneous reactions as mammalian muscle tissue. Due to the low glycogen content of fish muscle, its pH drop is small. Generally, pH values of 6.2 are obtained. Rigor mortis in cold-blooded animals is much shorter than in warm-blooded animals, a fact of great importance for preservation of fish quality.

Fish exhausted by lengthy struggle in a trawling net give meat of low keeping quality. The duration of rigor mortis is shortened as under these conditions the pH remains high. Therefore, to have an extended rigor the current trend in fishing is to avoid fish exhaustion. Fish muscle tissue differs basically from that of a land animal in that its maturation time is short while a land animal's is prolonged.

Because of the particular structure of fish muscle, the tendency to generate an alkaline pH reaction in muscle, and a high probability of microbial infection during fishing and fish dressing, conditions are highly favorable for rapid spoilage of fish. Therefore, bacteriological supervision and control, from the market to processing plants and during distribution, are of utmost importance. Fish muscle autolysis is low, since fish proteinases, the cathepsins, have a pH optimum at 4.3, so they are not active at the fish muscle pH of 6–7.

There are various physical and chemical criteria for assessing fish meat freshness.

The pH of fresh fish is 6.0–6.5. The suitability limit for consumption is pH 6.8, while spoiled fish meat has a pH of 7.0 or above.

The specific resistance of fish muscle changes with storage duration. Soon after catching it is 440–460 ohms, after 4-days approx. 280 ohms, and after 12 days it drops to 260 ohms. The suitability limit for consumption is reached after 16 days, when the resistance is 220 ohms.

The refractive index (n) of fish eye fluid is affected by storage duration. In cod of very good quality, n ranges from 1.3347 to 1.3366. Fish with a n of 1.3394 or higher is not suitable for marketing. The decrease in TMAO concentration and a concomitant increase in volatile N-containing substances, such as trimethylamine and several volatile reducing compounds, are chemical criteria for fish quality assessment.

Figure 13.2 provides data on the usefulness of some quality criteria, with cod stored in ice taken as an example. In addition to chemical and physical data, sensory evaluation data are included. A method is considered to be more suitable if it is highly sensitive during the first 12 days of storage, because it is during this time that the fish quality changes from very good to tolerable.

13.1.6 Storage and Processing of Fish and Fish Products

13.1.6.1 General Remarks

Today, the fishing grounds are not only further and further away and the fishing trips longer, but the fishing ships must also be economically utilized. Therefore, as a result of the easy deterioration of fish, it has become increasingly necessary to process the fish on accompanying factory ships. An overview of the

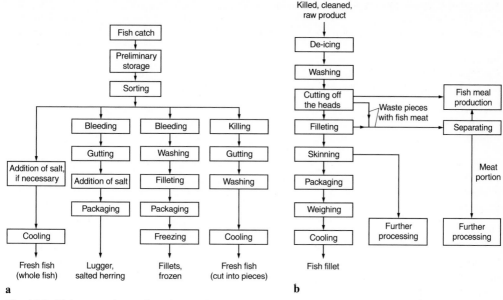

Fig. 13.3. Fish processing on board (**a**) and on land (**b**)

steps involved in fish processing is given in Fig. 13.3. The manual operation steps of the past, such as bleeding, gutting, washing, cutting off the heads, skinning, and filleting, have now been replaced to a significant extent by machines.

The fish waste that accumulates on processing, which accounts for up to 50% of the whole fish, is economically utilized by processing into fish meal on board and on land (cf. 13.1.6.13).

The ready decomposition or spoilage of fish flesh is the result of the special structure of the muscle tissue and the diverse ways in which microbial contamination occurs while handling fish, from catching, through processing and during distribution. From the earliest times, fish handling methods, like those for land animals, have been designed to increase the shelf life or storage ability.

Fish are usually initially cooled or frozen, or are dried, salted and smoked, followed by pickling in vinegar or in gelatin with vinegar added. They may also be deep fried in oil, or pickled with or without vinegar and soaked in a sauce in an airtight, sealed container. The expected shelf life of such products determines if they are considered fully preserved,

canned or semi-preserved products. Semi-preserves may contain additives against microbial spoilage. The compositions of some fish products are given in Table 13.9.

13.1.6.2 Cooling and Freezing

Preservation of freshness by refrigeration is the most modern and effective way to retain the wholesomeness and nutritional value of food. Refrigeration also enables fishing fleets to range the oceans for months in search of fish. Refrigeration permits stockpiling of fish, thus making fish processing plants more economical and better able to respond to market demand and supply.

Fish deteriorates rapidly at temperatures only slightly above 0°C. Therefore, immediately after catching fish are packed in ice on board the ship. The ice used may be sprinkled with a bactericidal substance. Freezing, which may also be used on ships, is suitable for whole fish (gutted or ungutted, with or without head or skin removal), as is the case with flat fish, tuna, mackerel or herring, or for fish fillets (cod, haddock, saithe, red fish).

Only quick freezing is used (−30 to −40°C; cf. Fig. 13.4), so the critical temperature range of

Table 13.9. Average chemical composition of processed fish

Product	Moisture[a]	Protein[a]	Fat[a]	NaCl[a]	Edible portion[b]
Salted fish					
Matje herring	54	18	18	10	68
Salt cured herring	48	21	16	15	68
Dried fish					
Stockfish	15	79	2.5	3	64
Klipfish	34	45	0.7	13	99
Smoked fish					
Buckling	58	23	16	3	62
Smoked sprats	62	17	20	2	60
Eel	53	19	26	1	73
Mackerel	61	21	16	1	70
Schillerlocken					
(smoked haddock filet)	53	21	24		100
Semi-preserved fish					
Bismarck herring	60	20	17	3	95
Bratherring (fried and					
pickled herring)	62	17	15	4	92
Herring, jellied	56	29	13		55
Anchovies	69	13	5	1	100
Herring tidbit	62	15	10	3	100

[a] As % of edible portion. [b] As % of the whole fish weight.

−0.5 to −5°C is rapidly passed over. Apart from air and contact freezing processes, cryogen frosters are being increasingly used, especially for sensitive and high quality products (crustaceans).

In air freezing, freezing takes place in a cold current of air in differently arranged, usually continuously operated systems (tunnel, spiral band etc.). In the contact freezing processes used, the fish are pressed and frosted between two contact plates that are cooled by a flow of coolant. The blocks obtained by this process can be portioned into slabs or sticks using band saws. They can be sold to the consumer as such or breaded and pre-fried (170°C, 20 s). Waste pieces (8–12%) are used for fishburgers and similar products. In comparison with conventional freezing systems, the food comes into direct contact with the refrigerant (liquid nitrogen or liquid carbon dioxide) in cryogenic frosters. The spatial arrangement of the freezing systems essentially corresponds to those in air freezing.

During freezing of fish, problems associated with drip or sap losses, discoloration and rancidity or lipid peroxidation and, consequently, fish weight loss, poor visual appearance and flat taste may arise and must be avoided by suitable processing. Cold storage should proceed at high air humidity (90%) and with stationary, noncirculating air. Data on the storage properties of some frozen fish are provided in Table 13.10.

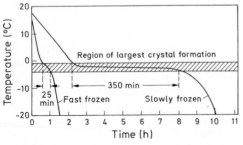

Fig. 13.4. Temperature course during fish fillet freezing process

Table 13.10. Shelf life of frozen fish, crustaceans and mollusks as influenced by storage temperature

Product	Shelf life (months) at		
	−18 °C	−25 °C	−30 °C
Fatty fish	4	8	12
Nonfatty fish	8	18	24
Lobsters, cray- and crawfish	6	12	15
Crabs	6	12	12
Oysters	4	10	12

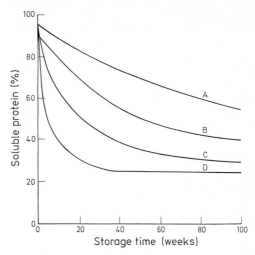

Fig. 13.5. Changes in fish muscle protein solubility as a result of cold storage (−14 °C). A: plaice, B: halibut, C: dogfish, D: cod. (according to *Connell*, 1964)

Thawing of fish in the home often is done either at room temperature (20 °C) or under a stream of running tap water (15 °C). It is important that the refrigeration chain from processor to distributor to consumer be maintained. Fish must be consumed shortly after thawing, otherwise juices are rapidly lost by dripping and the meat begins to decay. Fish muscle enzymes have noticeable activity even at −10 °C. Excessively long storage or insufficient cooling, especially of fatty fish, results in a rancid off-flavor and an unattractive, yellow colored fish surface. Antioxidants and associated synergistic compounds, such as ascorbic and citric acids, are used to inhibit fat deterioration. Changes in muscle texture are primarily due to changes in protein solubility (Fig. 13.5).

13.1.6.3 Drying

Fish can be preserved by drying naturally in the sun or in drying installations.

Stock fish, primarily nonfatty fish (cod, saithe, haddock, ling or tuck, which is often called cusk in North America) with head removed, split and gutted, is spread outdoors to dry in sea air (water content ≤18%). It is an unsalted fish product that is consumed in Southern Europe and in tropical countries.

Alternatively, machine-cut, headless and tailless fish which has been belly-clipped ("clipped" fish) is salted, either directly or in brine, and then put through a drying process (salt content 18–20%, water content <40%). This is most often done with cod or other nonfatty fish species. The main consumers of dried salted fish are Italy, Portugal, Spain and South American countries.

13.1.6.4 Salting

Salted fish (whole or parts) are obtained by salting fresh, deep frozen or frozen fish. Salt is the most important and oldest preservative for fish. Rubbing or sprinkling of fish with salt or immersion in brine, often followed by smoking, is called fish curing. Pickling in vinegar might be an additional preservation step. Salted products include: herring, anchovies, saithe, cod, salmon, tuna, and roe or caviar. It should be taken into account that if salting is used as the sole method of preservation without further processing (like matje, marinaded fish, smoked products etc.), complete bacterial protection is not provided because halophilic microorganisms can cause spoilage.

In dry salting, fish and salt are alternately stacked in open piles and the resulting brine can drain off.

In wet salting, the fish are put into more or less concentrated salt solutions. In heavily salted fish there are at least 20 g of salt in 100 g tissue fluid; in a medium-salted fish the salt content is 12–20 g.

Salting of herring is of special importance. There are the mildly salted matje (8–10% NaCl), the medium-salted "scotch cure", and the heavily salted herring, i.e. dry salting up to 25% NaCl. Herrings are also dressed, salted and packaged, both at sea and on land. The shelf life of salted herring is several months. Matje herring (immature sea herring, often wrongly called "sardines") must be consumed soon after they are removed from refrigeration. Salting might provide a finished end-product, but it is often used as a form of fast, temporary preservation, yielding semi-finished products which are later to be processed further. After salting, herring pass through a maturation process which generates a typical flavor. The proteolytic enzyme of the fish are involved in such "gibbed" herring maturation. During gibbing (i. e. a process of removing gills, long gut and stomach), the milt (male fish) or roe (female fish) and some of the pyloric cacea are left in the fish. These organs release enzymes which contribute to the maturation of the fish. If all organs are removed, no maturation occurs. Salting causes protein denaturation and cell shrinkage, wherein the glossy, sticky and nearly transparent fish muscle tissue starts to ripen. Similar in importance to herring are salted cod (Atlantic, Pacific and Greenland cod), which are salted dry or in brine as split or boneless fillets, and salted saithe, pollack (Dover hake) and some other saltwater fish of the *Gadus* species. Salted achovies (Mediterranean or Scandinavian) are also of importance. They are salted headless, gutted or whole ungutted, packed in barrels and ripened for several months, or are just salted and sold as a semi-preserved product.

13.1.6.5 Smoking

Smoked fish are obtained from fresh, deep frozen, frozen or salted fish which have been dressed in various ways. The whole fish body or fish portions are exposed to freshly generated sawdust smoke.

Cold smoking is performed for 1–3 days below 30 °C, generally at 18–26 °C, and is most often used with salt-cured fish (large size herring, salmon, cod, tuna). A product called "Buckling" is a large, fatty herring, sometimes nobbed (with head), that has been smoked. Delicatessen Buckling are made from gutted herring. Kippered herring (Newcastle Kipper) are obtained from fresh, fatty herring with the back split from head to tail. The dressed fish are then lightly brined and cold smoked. In the

United States the term "kipper" corresponds to products hot smoked on trays (e. g. kippered black cod). Salted or frozen salmon are smoked in North America. The shelf life is 2 weeks.

Hot smoking is performed for 1–4 h at temperatures between 70 and 90 °C. It is used with whole, gutted or descaled fish such as herring (Buckling), sprat ("Kieler Sprotten" in Germany), plaice, flounder, halibut (with or without skin), eels, mackerel, tuna, haddock, whiting (merlan), saithe, cod, red fish (ocean perch), dog-fish, sturgeon and shad. Unlike cold-smoked fish, hot-smoked fish have only a limited shelf life, 3–10 days, which can be extended only by cold storage. Hot-smoked caviar (cod or saithe) is available. Smoking of fish in traditional kilns or in batch-type smoke houses is being increasingly replaced by large continuous smoking installations.

13.1.6.6 Marinated, Fried and Cooked Fish Products

Marinade is vinegar or wine, or a mixture of both, usually spiced and salted, in which fish are soaked or steeped before use or before being pickled and stored for a longer time. The fish used might be fresh, deep frozen or frozen, or salted whole fish or fish portions. Marinating tenderizes muscle tissue without heat treatment. Fish preservation by pickling in this manner is based on the combined action of salt and vinegar. Vinegar-packed herring, called simply marinades, are a popular German fish food packaged in retail glass jars. Pickled fish can be packaged together with some plant extracts, sauces, gravy, creams, mayonnaise or related products, or they can be immersed in an edible oil (although oilpacked fish are not called pickled fish). Some of these products might contain chemical preservatives.

Marinated fish are packaged in cans, jars, etc., and may be handled without packaging. Fish marinades have only a limited shelf life (they are semi-preserves), and even chemical preservatives can not prevent their eventual decay. Marinated fish considered as delicatessen items are "Kronsardines", Bismarck herring, rollmops and pickled herring.

Fried fish products are prepared from variously dressed fresh, deep frozen or frozen whole fish or fish portions, with or without further dressing in eggs and bread crumbs (batter formulations, such as "shake and bake"). They are then made tender by frying, baking, roasting or barbecuing. These products may be packaged or canned in the presence of vinegar, sauces, gravy or an edible oil, often with a chemical preservative added. Examples of these products are fried marinated fish sticks, "Brathering", "Bratrollmops", balls, etc.

Cooked fish products are processed in a similar manner. Tenderization is achieved by cooking or steaming. Processing also involves the use of vinegar or wine, addition of salt and the use of a preservative. The cooked products can be solidified, with or without plant ingredients, into a jelly (herring in jelly) or packaged with other extracts, sauces or gravy. Cooked fish products include herring marinades, rollmops, bacon rollmops in jelly, sea eel (dogfish) in jelly, or broths made from disintegrated saltwater fish meat. The occasional liquefaction of cooked fish jelly ("jelly disease") indicates microbial proteolysis.

13.1.6.7 Saithe

Saithe, often called coalfish, coley, pollack or Boston bluefish (trade name "Dover hake"), are processed into fillets, saltcured, dyed or tinted, and smoked. They are then cut into slices or cutlets and covered with edible oil. The product has a good shelf life.

13.1.6.8 Anchosen

Anchosen are made from fresh, frozen or deep-frozen small sprats and herring, preserved with salt in the presence of added sugar or sugars derived from starch saccharification, spiced and biologically ripened with sodium nitrate. Flavors are also added. Proteinases are also used to accelerate ripening.

Anchosen can be packed in sauce (gravy), creams or in edible oils, garnished with plant ingredients and a chemical preservative may be added. Examples of anchosen are appetitsild, cut spiced herring and spiced herring. Appetitsild is a product consisting of skinned fillets of spice-cured sprats, cured and packed in vinegar, salt, sugar and spices.

13.1.6.9 Pasteurized Fish Products

Pasteurized fish products made of fresh, deep-frozen or frozen fish or fish portions have shelf lives, even without cold storage, of at least 6 months. These products are prepared by prolonged heat treatment of fish at temperatures below 100 °C. They are then tightly sealed in a container. Such products are salted or soaked in vinegar prior to pasteurization.

13.1.6.10 Fish Products with an Extended Shelf Life

Canned fish products of extended shelf life are made by steam retorting of fresh or frozen whole fish or fish portions, followed by packaging in vacuum-sealed, air-tight containers. Their shelf life, without special cold storage, is at least 1 year. Cans are usually wrapped in paper for labelling. Special can materials have to be chosen when the fish is canned with corrosive ingredients such as tomato or mustard sauce, vinegar or lemon juice. The can is usually made of a lacquer-coated tinplate or inert aluminum.

Products with extended shelf lives are in their own juice or in added oil, or in some sauce or cream (e.g. "sardine" pilchards, *Sardina pilchardus*, packed in olive or soya oil, tomato mustard, or lemon juice). Also available are fish paste, meat balls or "Frikadellen" (Germany), i.e. flesh of white fish made into rissoles using flour, eggs and spices, which are then roasted, deep fried and used ready-to-serve, as hors d'œuvres, and fish salad. The latter products are canned or packed in glass jars, and may be packed under controlled atmosphere.

13.1.6.11 Other Fish Products

These include ready-to-serve or instant fish foods, usually garnished with vegetables and fruit, e.g., fish cakes, dumplings, sausages (including cod liver sausage), pastes of anchovy, herring or cod liver, or salmon paste which may include prawn or shrimp and, occasionally, butter. There are also many fish salads and butters, such as anchovy and salmon.

13.1.6.12 Fish Eggs and Sperm

13.1.6.12.1 Caviar

Specially prepared sturgeon eggs (roe) are called caviar. The roe ("hard roe") are detached from the fish ovary gland. The roe are washed in cold water, salted and left to ripen until they become transparent. They are then drained from the brine slime and are marketed for the wholesale market in small metal or glass containers or in barrels. Occasionally, the caviar is pasteurized. Two basic types are marketed: grainy caviar, where eggs are readily detached from roe, and pressed caviar, where the ovarian membrane and the excess fluid are removed by gentle pressing. Caviar is made from various sturgeon species (beluga, stoer or sevruga). The roe of these sturgeon species caught in winter, when mildly salted (below 6% NaCl), give a high quality caviar called "Malossol". The beluga (the largest of the three sturgeons mentioned) provides the most valuable caviar.

Pressed caviar is obtained from all species. Salmon caviar (such as Amur and Keta caviars from Siberian salmon roe) is processed using less than 8.5% salt. American whitefish caviar is a mixture of roe from salmon, whitefish, carp and some other fish. Scandinavian caviar is from cod and lumpfish.

Sturgeon caviar is gray or brown to black in color. Salmon caviar is yellow-red or red. Most caviar is imported from the Soviet Union and Iran (Caspian Sea caviars). It readily decays and so must be kept refrigerated. A medium-size beluga sturgeon can provide 15–20 kg caviar.

13.1.6.12.2 Caviar Substitutes

Caviar substitutes are made of roe of various sea and freshwater fish. Germany produces the dyed caviar of lumpfish (lumpsuckers), and also cod and herring caviars. The roe are soured, salted, spiced, dyed black, treated with traganth gum and, occasionally, a preservative is added.

13.1.6.12.3 Fish Sperm

Fish sperm are a product of the gonads of male fish and are often called milt or soft roe. Salted sperm from sea and freshwater fish, particularly herring, are most commonly marketed.

13.1.6.13 Some Other Fish Products

These include the nutritional products and seasonings derived from fish protein hydrolysates; insulin from shark pancreas; proteins recovered from saltwater fish fillet cutting; fish meal used as feed for young animals, poultry and pond fish; and, lastly, fish fat (oil), as mentioned in 14.3.1.2. Of increasing importance is the production of fish protein concentrates and, when necessary, their modified products (cf. 1.4.6.3.2 and Table 1.44).

13.2 Whales

Although a whale is in a true sense a mammal and not a fish, it will be covered here. The blue (*Balaenoptera muculus*) and the finback whale (*B. physalus*) are the two most important whales, each growing up to 30 m in length and up to 150 tons in weight. Also caught are the humpback (*Megaptera nodosa*), the sperm (*Physeter macrocephalus*) and the sei whale (*Balaenoptera borealis*). Whale meat is similar to big game meat or beef. It has long and coarse muscle fibers arranged in bundles and colored gray-reddish. The color of the meat is affected by the age of the whale, and may be bright red or dark red, while frozen whale meat becomes dark black-brown in color. Freezing imparts a rough, firm texture to the meat.

The fresh meat has a pleasant flavor but, due to the fast rate of fat oxidation, the shelf life is very short. For this reason bulk whale meat is not readily accepted by food wholesalers and the retail market. Whale meat extracts are also produced (cf. 12.7.3.2).

13.3 Crustaceans

Crustaceans have no backbone; their body is divided into sections, each bearing a pair of joint-legs. An armor-like shell covers and protects the body. Included are shrimp, crayfish (also called crabfish), crabs (e.g. freshwater, edible green shore crab) and lobster. Compositional data are provided in Table 13.11).

Table 13.11. Average chemical composition of crustaceans and mollusks

Crustaceans/ mollusks	Mois-ture[a]	Pro-tein[a]	Fat[a]	Min-erals[a]	Edible por-tion[b]
Shrimps	78	19	2		41
Lobsters	80	16	2	2.1	36
Crayfish	83	15	0.5	1.3	23
Oysters	83	9	1.2	2.0	10
Scallop	80	16	0.1	1.4	44
Mussel	83	10	1.3	1.7	18

[a] As % of edible portion.
[b] As % of the whole fish weight.

13.3.1 Shrimps

The most important shrimps are the common or brown shrimp from the North Sea (*Crangon crangon*), the Baltic Sea shrimp (*Palaemon adspersus fabricii*), the deep sea shrimp (*Pandalus borealis*) and the larger species in tropical waters, such as blue Brasilian (*Penaeus* spp.) or the royal red shrimps (*Hymenopenaeus robustus*). Larger species are called prawn.

Shrimps are marketed soon after catch as: live, fresh with shell, with or without head, cooked in brine, or cooked without shell. They have very short shelf lives. Shrimps are also sold canned, deep frozen or as an extract or a salad ingredient. Canned shrimps are heated (pasteurized) at just 80–90 °C so as not to affect their flavor; hence, they are semi-preserves with a limited shelf life.

13.3.2 Crabs

Crabs live in shallow or deep water along the sea coast or in freshwater. Blue crab (*Callinectes sapidus*) is the most common crab of the Atlantic coast of North America. Other important species are the common shore crab (also called green shore crab); the edible crab of Europe (*Cancer pagurus*), which lives in sandy, shallow water; the king crab of Alaska (*Pralithodes camchaticus*), also called Japanese crab; and the dungeness crab (*Cancer magister*) from the shallow waters from Cali-

fornia to Alaska. These crabs differ in shape and size of their big claws, but all have no tail. The color and shape of the body varies, as does the ability to swim or to run sideways.

When crabs shed their shell and the new shell has not yet hardened, they are at their tastiest and are marketed as "extra choice soft" crabs. The forms sold are: live, fresh, frozen and canned. Crab paste, canned soup and crab cakes similar to deepfried fish cakes are delicatessen sea foods. In the trade, the term crab meat means white muscle meat, colored red only in leg muscles and chelae, and is distinguished from brown crab meat obtained from crab liver and gonads. The latter are usually processed into crab paste. All crab products are of limited shelf life.

13.3.3 Lobsters

The European lobster (*Homarus gammarus*) caught in the Atlantic is the largest in Europe. It reaches a length of 35–90 cm and a maximum weight of 10 kg. The major area of catch is Helgoland, the north and west sea coast of Europe, the Mediterranean and the Black Sea. The tastiest lobster meat is that from the breast shell. The American or northern lobster (*Homarus americanus*) is closely related to the European lobster. Lobsters are marketed live (remain alive up to 36 h after catch), whole boiled, or canned as cooked meat in its own juice or as soup (cream of lobster, lobster chowder). Lobster paste is also available. Cooking of lobster changes its color to red. The color change involves the release of astaxanthin from ovoverdin, a brown-green chromoprotein (cf. 3.8.4.1.2).

The Norway lobster (*Nephrops norvegicus*; also called Langoustine) also belongs to the lobster family. It is marketed fresh, frozen, semi-preserved, as in salad, or canned, as soup, paste or mildly-brined meat in its own juice.

13.3.4 Crayfish, Crawfish

Crayfish are freshwater crustaceans considered as a delicacy in Europe. The major crayfish of Europe belongs to *Astacus* spp. (*Astacus*

astacus or *fluviatilis*). Its meat is the most tasteful in May–August when it sheds its shell and the new shell is still soft. The eastern part of North America has the freshwater crayfish of *Cambarus* spp. The Australian crayfish belongs to *Enastacus serratus.*

The cray(craw)fish die when they are dropped into boiling water. Their tail curls up – this is a sign that they were cooked fresh or alive. For the color change during cooking see above (cf. 13.3.3).

The seawater species of crayfish are called crawfish. They include *Palinurus*, *Panulirus* and *Jasus* spp. The most important crawfish are the European spiny lobster (*Palinurus vulgaris*), the Pacific North American counterpart (*Panulinus interruptus*) and others from Africa, Australia and Japan. The European spiny lobster is 30–40 cm long, up to 6 kg in weight, has rudimentary front legs shaped into sharp claws and has a knobby shell covering the body. It is often caught in the Mediterranean Sea, the west and south coasts of England, and along the coast of Ireland. The rock lobster (*Jasus lalandei*) and the Mediterranean crawfish (*Palinurus elephas*) are also available on European markets. The meat of these crawfish is rather coarse and fiberlike and is colored yellow to yellow-red.

The cray- and crawfishes are marketed fresh live, raw or cooked, and canned in different forms: meat, butter (precooked meat mixed with butter), soup and soup powders, soup extracts (these are crawfish butter, spiced, salted and blended with flour) and crayfish bisque (French purée or thick soup of crayfish and lobsters).

13.4 Mollusks (*Mollusca*)

13.4.1 Mollusks (*Bivalvia*)

The bivalve mollusks include clams, oysters, mussels and scallops. The common oyster (also called flat native or European oyster) and the blue or common mussel are the most often processed molluscan shellfish.

Oysters (*Ostreidae*, e. g. the European oyster, *Ostrea edulis*) live in colonies along the sea coast or river banks, or are cultivated in ponds

("oyster farms") which are often connected with the sea. Oysters are usually sold unshelled. Only the adductor muscle is consumed; the pleated gills and the digestive system are discarded.

In addition to the common oyster, the Portuguese oyster (*Gryphea angulata*) and the American blue point oyster, (*Crassostrea virginica*), used most commonly for canning, are of importance. The best meat is obtained from oysters harvested when they are 3–5 years old, with the top quality harvested between September and April (an old saying is: oysters should be eaten in months which have "r" in their names).

The blue or common mussel (*Mytilus edulis*) lives in shallow, sandy freshwater, while the sea mussel lives in ocean water or is cultivated in ponds or lakes. The shell, 7–15 cm long, is bluish black and the body meat is yellowish. The meat is rich in protein (16.8%) and also in vitamin A and the vitamin B-complex. The meat is eaten cooked, fried or marinated. The major mussel growing areas in Germany are the Kiel Bay and the East Friesian Islands.

In addition to common mussel, numerous other mussels are eaten, mostly canned in vegetable oil, e. g., Pacific Bay or Cape Cod scallops (*Pectinidae*) and cockles (*Cardidae*). Due to rapid spoilage, mussels are marketed live or canned. They are eaten soon after being caught or after the can is opened, and are avoided in warm seasons. Moreover, they should originate from uncontaminated clear waters.

13.4.2 Snails

Snails are univalve mollusks, i. e. they have only a single, coiled shell. They are eaten preferentially in Italy, France and Germany, and are nearly exclusively the large Helix garden snail (*Helix pomatia*). Snails are sometimes collected wild in South or Central Germany and in France, but most are supplied by snail gardens and feeding lots where lettuce and cabbage leaves are the food source, or in damp shady cellars, where wheat bran and leafy vegetable leaves (e. g. cabbage) are used as a feed. The meat is considered a delicacy. Since the shelf life of the meat is very limited, snails are marketed live (with the shell plugged) or canned.

Marine snails of various kinds are fried, steamed, baked or cooked in soups, and are also considered a delicacy.

13.4.3 Octopus, Sepia, Squid

Octopus, sepia and calmar (*Cephalopoda*) are softbodied mollusks with eight or ten arms, and without an outside shell.
The sepia or cuttlefish (*Sepia officinalis*), the squid or calmar (*Loligo loligo*) and the octopus or devilfish (*Octopus vulgaris*) are caught in the Mediterranean region, mostly in Italy, and other parts of the world (Atlantic and Pacific Oceans, e.g. the North American poulp, Japanese *Polypus* spp., etc.). They are consumed deep fried in oil, baked, cooked in wine, pickled in vinegar after being boiled, cooked in soups, in salads, stewed or canned.

13.5 Turtles

Turtles, tortoises or terrapin (for American fresh-water turtles) are reptiles with a shell used as a "house". The logger head and green sea turtles are caught commercially for their meat. In Germany turtle is mostly eaten in soup or stew. The meat of the so-called soup turtle (*Chelonia mydas*) is faintly red to bright red, and is marketed canned. An imitation or mock turtle soup is prepared from edible parts of heads of calves and has no relation to turtles except for the name.

13.6 Frogdrums

The thigh portion (frogdrum) of a frog's hinged leg is sold as a delicacy. Frogs providing frogdrums are the common bullfrog (*Rana catesboniana*), the leopard from (*Rana pipiens*) and others (*Rana arvalis, Rana tigrena, Rana esculenta*). The meat is soft in texture, white in color and tasty; however, it has a very limited shelf life as it readily deteriorates. Frogdrums are eaten cooked, roasted or stewed.

13.7 Literature

Borgstrom, G. (Ed.): Fish as food. Vol. I–III. Academic Press: New York. 1961–1965

Connell, J.J.: Fish muscle proteins and some effects on them of processing. In: Proteins and the reactions (Eds.: Schultz, H.W., Anglemier, A.F.), p. 255, AVI Publ. Co.: Westport, Conn. 1964

Connell, J.J. (Ed): Advances in fish science and technology. Fishing New Books Ltd: Farnham, Surrey, England. 1980

Feeney, R.E., Yin Yeh: Antifreeze proteins from fish bloods. Adv. Protein Chem. *32*, 191 (1978)

Habermehl, G.: Gift-Tiere und ihre Waffen. 2. edn., Springer-Verlag: Berlin. 1977

Lindsay, R.C.: Fish flavors. Food Reviews International *6*, 437 (1990)

Ludorff, W., Meyer, V.: Fische und Fischerzeugnisse. 2. edn., Verlag Paul Parey: Berlin. 1973

Multilingual dictionary of fish and fish products (prepared by the OECD), 2nd edn., Fishing News Books Ltd., Farnham, Surrey, England. 1978

Whitfield, F.B.: Flavor of prawns and lobsters. Food Reviews International *6*, 505 (1990)

14 Edible Fats* and Oils

14.1 Foreword

Most fats and oils consist of triacylglycerides (recently also denoted as triacylglycerols; cf. 3.3.1) which differ in their fatty acid compositions to a certain extent. Other constituents which make up less than 3% of fats and oils are the unsaponifiable fraction (cf. 3.8) and a number of acyl lipids; e.g., traces of free fatty acids, mono- and diacylglycerols.

The term "fat" generally designates a solid at room temperature and "oil" a liquid. The designations are rather imprecise, since the degree of firmness is dependent on climate and, moreover, many fats are neither solid nor liquid, but are semi-solid. Nevertheless, in this chapter, unless specifically emphasized, these terms based on consistency will be retained.

14.2 Data on Production and Consumption

Data on the production of oilseeds and other crops are summarized in Table 14.0. The world production of these crops has doubled (Table 14.1) since the Second World War. There has been a significant rise in production since 1964 of soybean, palm and sunflower oils, while marine oil production has declined steadily. Soybean oil, butter and edible beef fat and lard are most commonly produced in FR Germany (Table 14.1). The per capita consumption of edible fats (Table 14.2) in FR Germany more than doubled when compared to the world average. The slight drop recorded from 1970–75 was due to a temporary reduction in butter consumption.

---

* Butter is dealt with in Chapter 10.2.3.

14.3 Origin of Individual Fats and Oils

14.3.1 Animal Fats

14.3.1.1 Land Animal Fats

The depot fats and organ fats of domestic animals, such as cattle and hogs, and milk fat, which was covered in Chapter 10, are important animal raw materials for fat production. The role of sheep fat, however, is not significant. The major fatty acids of these three sources are oleic, stearic and palmitic (Table 14.3).

It should be noted that the fatty acid composition of individual fat samples may vary greatly. The fat composition of land animals is affected by the kind and breed of animal and by the feed. The composition of plant fats depends on the cultivar and growth environment, i.e. climate and geographical location of the oilseed or fruit plant (cf. Fig. 3.10). Therefore only average values are given in the following tables dealing with fatty acid composition.

In contrast to oil from plant tissue, the recovery of animal fat is not restricted by rigid cell walls or sclerenchyma supporting tissue. Only heating is needed to release fat from adipose tissue (dry or wet rendering with hot water or steam). The fat expands when heated, tearing the adipose tissue cell membrane and flowing freely. Further fat separation is simple and does not pose a technical problem (Fig. 14.1).

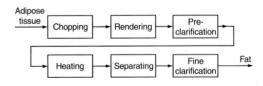

Fig. 14.1. Steps involved in wet rendering

Table 14.0. Production of major oilseeds, 1996 (1,000 t)[a]

Continent	Castorbean	Sunflower seed	Rapeseed	Sesame seed
World	1,418	25,101	30,385	2,527
Africa	42	1,124	182	526
America, North-, Central-	2	1,687	5,286	101
America, South-	70	5,577	66	36
Asia	1,301	4,124	15,500	1,863
Europe	2	12,502	8,730	–
Oceania	–	87	621	–

Continent	Linseed	Safflower seed	Cottonseed	Copra
World	2,320	839	54,132	4,774
Africa	57	35	4,585	207
America, North-, Central-	884	291	11,729	262
America, South-	102	9	3,293	38
Asia	930	471	32,153	4,018
Europe	333	6	1,357	1
Oceania	15	27	1,015	248

Continent	Palm kernel	Palm oil	Olives	Olive oil
World	5,098	17,046	12,980	2,260
Africa	909	1,756	2,425	357
America, North-, Central-	64	204	175	2
America, South-	321	812	131	10
Asia	3,735	13,998	2,477	358
Europe	–	–	7,770	1,533
Oceania	69	276	1	–

Country	Castor-bean	Country	Sunflower seed	Country	Rapeseed
India	1,000	Argentina	5,300	China	9,000
China	260	Russian Fed.	3,000	India	6,000
Brazil	46	Ukraine	2,123	Canada	5,037
Paraguay	17	France	1,997	France	2,904
Thailand	15	USA	1,627	Germany	1,949
Ethiopia	14	India	1,450	UK	1,410
Pakistan	9	China	1,360	Australia	616
Ecuador	7	Spain	1,208	Czech Rep.	531
Philippines	7	Romania	1,090	Poland	440
Sudan	7	Hungary	893	Denmark	251
Σ (%)[b]	97	Σ (%)[b]	80	Σ (%)[b]	93

Table 14.0 (continued)

Country	Sesame seed	Country	Linseed	Country	Safflower seed
India	620	Canada	843	India	450
China	580	China	500	USA	191
Myanmar	351	India	340	Mexico	100
Sudan	160	UK	86	Ethiopia	35
Uganda	72	Argentina	80	Australia	27
Nigeria	60	Germany	78	China	20
Bangladesh	49	Russian Fed.	50	Argentina	9
Pakistan	40	Bangladesh	46	Russian Fed.	6
Mexico	38	USA	41	Pakistan	1
Egypt	34	France	33		
Turkey	34			Σ (%)[b]	100
		Σ (%)[b]	90		
Σ (%)[b]	81				

Country	Cotton-seed	Country	Copra	Country	Palm seed
China	11,250	Philippines	1,800	Malaysia	2,490
USA	10,606	Indonesia	1,150	Indonesia	1,115
India	7,651	India	635	Nigeria	548
Pakistan	4,597	Viet Nam	208	Brazil	185
Uzbekistan	3,300	Mexico	204	Colombia	78
Turkey	2,089	Papua-New Guinea	120	Thailand	75
Argentina	1,326			Zaire	72
Greece	1,100	Mozambique	74	Papua-New Guinea	62
India	1,084	Malaysia	70		
Mexico	1,084	Thailand	70	Cameroon	56
		Sri Lanka	58	Guinea	62
Σ (%)[b]	81				
		Σ (%)[b]	92	Σ (%)[b]	93

Country	Palm oil	Country	Olives	Country	Olive oil
Malaysia	8,385	Italy	3,000	Spain	602
Indonesia	4,998	Spain	2,856	Italy	554
Nigeria	776	Greece	1,600	Greece	333
Colombia	446	Turkey	1,500	Tunisia	218
Thailand	400	Tunisia	1,050	Turkey	194
Côte d'Ivoire	267	Morocco	800	Syria	129
Papua-New Guinea	250	Syria	600	Morocco	86
		Algeria	313	Algeria	48
Ecuador	234	Portugal	230	Portugal	36
Zaire	181	Egypt	210	Jordan	14
Cameroon	161				
		Σ (%)[b]	94	Σ (%)[b]	98
Σ (%)[b]	94				

[a] Soybean and peanuts are presented in Table 16.1.
[b] World production = 100%.

Table 14.1. World production of fats and oils (in 1,000 t)

Fat/oil	1935/39	1965	1981	1991
Soya oil	1,229	4,860	12,495	
Sunflower oil	562	2,375	4,515	
Cottonseed oil	1,560	2,570	3,245	
Peanut oil	1,506	3,165	2,945	
Rapeseed (canola) oil	1,207	1,465	3,750	
Palmkernel- and palm oil	1,334	1,595	6,155	10,871
Coconut oil	1,932	2,225	2,925	
Olive oil	871	1,951[a]	1,325[b]	1,911
Other oils of plant origin	95	1,720	1,940	
Butter (butter fat)	3,611	4,735	5,515	7,635
Lard	2,495	4,375	4,615	
Edible tallow	1,442	4,305	5,980	
Marine oils	975	1,075	1,195	1,626
	18,819	36,416	56,000	

[a] Production data for 1964.
[b] An estimate for 1982.

Table 14.2. Consumption of edible fats and oils in FR Germany (kg per capita per year)

Year	Butter[a]	Animal fat	Oils[b]	Total
1969/70	7.3	6.1	12.8	26.2
1970/71	7.0	6.4	13.2	26.6
1971/72	6.2	6.3	13.4	25.9
1972/73	6.2	6.1	13.4	25.7
1973/74	6.0	6.1	13.3	25.4
1974/75	5.9	6.3	12.9	25.1
1975/76	5.5	6.4	13.2	25.1
1976/77	5.3	6.3	14.0	25.6

[a] Butter fat; butter weight is higher by 19.5%.
[b] It includes vegetable and marine oils.

14.3.1.1.1 Edible Beef Fat

Edible beef fat is obtained from bovine adipose tissue covering the abdominal cavity and surrounding the kidney and heart and from other compact, undamaged fat tissues. The fat is light-yellow due to carotenoids derived from animal feed. It is of a friable, brittle consistency and melts between 45 and 50 °C.
The fatty acid composition of beef fat (Table 14.3) is not influenced greatly by feed intake, but that of hog fat (lard) is. The composition of edible beef fat triacylglycerols is given in Table 3.13).
The following commercial products are prepared from beef fat: *Prime Beef Fat ("premier*

Table 14.3. Average fatty acid composition of some animal fats (weight-%)

Fatty acid	Beef tallow	Sheep tallow	Lard	Goose fat
12 : 0	0	0.5	0	0
14 : 0	3	2	2	0.5
14 : 1 (9)	0.5	0.5	0.5	0
16 : 0	26	21	24	21
16 : 1 (9)	3.5	3	4	2.5
18 : 0	19.5	28	14	6.5
18 : 1 (9)	40	37	43	58
18 : 2 (9, 12)	4.5	4	9	9.5
18 : 3 (9, 12, 15)	0	0	1	2[a]
20 : 0	0	0.5	0.5	0
20 : 1 ⎫ 20 : 2 ⎭	0	0.5	2	
Others	3	3	0	0

[a] It includes fatty acid 20 : 1.

jus") is obtained by melting fresh and selected fat trimmings in water heated to 50–55 °C. The acid value resulting from lipolytic action (cf. 14.5.3.1) is not allowed to exceed 1.3 (corresponding to approx. 0.65% free fatty acid). This beef fat, when heated to 30–34 °C, yields two fractions: oleomargarine (liquid) and oleostearine (solid). Oleomargarine is a soft fat with a consistency similar to that of melted butter. It is used by the margarine and baking industries. Oleostearine (pressed tallow) has a high melting point of 50–56 °C and is used in the production of shortenings (cf. Table 14.18). *Edible Beef Fat (secunda beef fat)* is obtained by melting fat in water at 60–65 °C, followed by a purification step. It has a typical beef fat odor and taste and a free fatty acid content not exceeding 1.5%. Lower quality tallow has only industrial or technical importance, for example, as raw material for the soap and detergent industries.

14.3.1.1.2 Sheep Tallow

The unpleasant odor adhering to sheep tallow is difficult to remove, hence it is not used as an edible fat. Sheep tallow is harder and more brittle or friable than beef tallow. The fatty acid composition of sheep tallow is presented in Table 14.3.

14.3.1.1.3 Hog Fat (Lard)

Hog (swine) fat, called lard, is obtained from fat tissue covering the belly (belly trimmings) and other parts of the body. The back fat is mainly utilized for manufacturing bacon. After tallow and butter, lard is currently the animal fat which is consumed the most (Table 14.1). Its grainy and oily consistency is influenced by the breed and feeding of hogs.
Some commercial products are:
Lard obtained exclusively from belly trimmings (abdominal wall fat). This is the highest quality *neutral lard*. It has a mild flavor, is white in color and its acid value is not more than 0.8.
Lard from other organs and from the back is rendered using steam. The maximum acid value is 1.0.
Lard obtained from all the dispersed fat tissues, including the residues left after the recovery of neutral lard, is rendered in an autoclave with steam (120–130 °C). This type of lard has a maximum acid value of 1.5.
In contrast to the composition of triacylglycerols found in beef fat (Table 3.13), lard contains fewer triacylglycerols of the type SSS and more of the types SUU, USU and UUU (S = saturated; U = unsaturated fatty acid). As a consequence, lard melts at lower temperatures and over a range of temperatures rather than sharply at a single temperature, and its shelf life is not particularly long.

14.3.1.1.4 Goose Fat

As the only kind of poultry fat produced, goose fat is a delicacy. Its production is insignificant in quantity. The fatty acid composition of goose fat is given in Table 14.3.

14.3.1.2 Marine Oils

Sea mammals, whales and seals, and fish of the herring family serve as sources of marine oils. These oils typically contain highly unsaturated fatty acids with 4–6 allyl groups, such as (double bond positions are given in brackets): 18 : 4 (6, 9, 12, 15); 20 : 5 (5, 8, 11, 14, 17); 22 : 5 (7, 10, 13, 16, 19); and 22 : 6 (4, 7, 10, 13, 16, 19) (Table 14.4). Since these acids are readily susceptible to autoxidation, marine oils are not utilized directly as edible

Table 14.4. Average fatty acid composition of some marine oils (weight-%)

Fatty acid	Blue whale	Seal	Herring (*Clupea harengus*)	Pilchard[a]	Menhaden (*Brevoortia tyrannus*)
14 : 0	5	4	7.5	7.5	8
16 : 0	8	7	18	16	29
16 : 1	9	16	8	9	8
18 : 0	2	1	2	3.5	4
18 : 1	29	28	17	11	13
18 : 2	2	1	1.5	1	1
18 : 3	0.5		0.5	1	1
18 : 4	0.4		3	2	2
20 : 1	22	12	9.5	3	1
20 : 4	0.5		0.5	1.5	1
20 : 5	2.5	5	9	17	10
22 : 1	14	7	11	4	2
22 : 5	1.5	3	1.5	2.5	1.5
22 : 6	3	6	7.5	13	13

[a] Trade name of grown sardines (*Sardinóps caerulea*)

oils, but only after hydrogenation of double bonds and refining.
Of analytical interest is the occurrence in marine oils of about 1% branched methylated fatty acids, for example, 12-methyl- and 13-methyltetradecanoic acids or 14-methylhexadecanoic acid. These acids are also readily detectable in hardened marine oils.

14.3.1.2.1 Whale Oil

There are two suborders of whales: Baleen whales which have horny plates rather than teeth, and whales which have teeth. The blue and the finback whales, both live on plankton, and belong to the Baleen suborder. The oils from these whales do not differ substantially in their fatty acid compositions.
A blue whale, weighing approx. 130 t, yields 25–28 t of oil, which is usually recovered by a wet rendering process. The ruthless exploitation of the sea has nearly wiped out the whale population, hence their raw oil has become a rare product.

14.3.1.2.2 Seal Oil

The composition of seal oil is similar to that of whale oil (Table 14.4).

14.3.1.2.3 Herring Oil

The following members of the herring fish family are considered to be satisfactory sources of oil: herring, sardines (Californian or Japanese pilchards, etc.), sprat or brisling, anchovies (German Sardellen or Swedish sardell) and the Atlantic menhaden. The fatty acid compositions of the various fish oils differ from each other (Table 14.4).

14.3.2 Oils of Plant Origin

All the edible oils (with the sole exception of oleomargarine-type products) are of plant origin. With regard to the processes used to recover plant oils, it is practical to divide them into fruit and oilseed oils. While only two fruits are of economic importance in oil production, the number of oilseed sources is enormous.
The oils are sold and consumed as pure oil from a single oilseed plant or fruit plant, for example, olive, sunflower or corn oils, or are marketed and used as blended oils, which are generally designated as edible, cooking, frying, table or salad oil.

14.3.2.1 Fruit Pulp Oils

The oils obtained from the fruits of the olive tree and several oil palm species are of great economic importance. The fatty acid compositions of the oils of these fruits are summarized in Table 14.5. Due to the high enzymatic activity in fruit pulp, particularly of lipases, the shelf life of fruit oil is severely limited.

14.3.2.1.1 Olive Oil

Olive oil is obtained from the pulp of the stone fruit of the olive tree (*Olea europaea sativa*). More than 90% of the world's olive harvest comes from the Mediterranean region, primarily in Italy and Spain (cf. Table 14.0). Olive tree plantations are found to a smaller extent in Japan, Australia, California and South America. Altogether, olive oil production is stagnant. The laborious, painstaking harvesting, which has not yet been mechanized, is especially to blame for this situation.

Table 14.5. Characteristics of olives (fruits/oil) and oil palm

	Olive (*Olea europaea sativa*)	Oil palm (*Elaeis guineensis*)
Fruits		
Length (cm)	2–3	3–5
Width (cm)	2–3	2–4
Fruit pulp (weight-%)	78–84	35–85
Fruit seed (weight-%)	14–16	15–65
Fruit pulp (mesocarp)		
Oil (weight-%)	38–58	30–55
Moisture (weight-%)	to 60	35–45
Fruit pulp oil		
Solidification point (°C)	– 5 to – 9	27–38

Average fatty acid composition (weight-%)

	Olive oil	Palm oil
14 : 0	0	1
16 : 0	11.5	43.8
16 : 1	1.5	0.5
18 : 0	2.5	5
18 : 1 (9)	75.5	39
18 : 2 (9, 12)	7.5	10
18 : 3 (9, 12, 15)	1.0	0.2
20 : 0	0.5	0.5

Oil Production. The disintegrated fruit is kneaded to release the oil droplets from the pulp, occasionally by adding common salt. The oil is then pressed out or separated by gravity decantation. The initial cold pressing provides virgin oil (provence oil). This is then followed by warm pressing at about 40 °C.
In addition to the conditions used for oil recovery, the quality of olive oil is affected by the ripeness of the fruit (overripe fruit is not preferred) and length of storage. In virgin oils there is a relationship between sensory properties and the content of free fatty acids:

- *Extra vierge*: pleasant aromatic flavor with the free fatty acid content not exceeding 1% (calculated as oleic acid).
- *Fine vierge:* slightly less aromatic in flavor, with a free fatty acid content not exceeding 1.5%.
- *Semi-fine vierge* (courante grade): cold pressed oil, less aromatic in flavor and with a free fatty acid content up to 3%.

Oil that is obtained by pressing and contains more than 1% of free acid or has an imperfect aroma is called "*lampante*" oil in Italy. After slight refining, and in mixtures with cold-pressed oil, it is sold as "olive oil" or "pure olive oil".

The oil cake is subsequently extracted with a solvent and the resulting extraction oil ("*sansa*" oil) is refined. It accounts for about one quarter of the production.

From time to time, the expensive *extra vierge* oils are adulterated with refined "lampante" oils or extraction oils. In particular, the concentrations of waxes, sterol esters, and of the triterpene alcohols erythrodiol (I) and uvaol (II, cf. Formula 14.1) are used for an analytical differentiation (Table 14.6 and 14.5.2.4).

$$ (14.1) $$

$$ I : R^1 : H, \ R^2 : CH_3 $$
$$ II : R^1 : CH_3, \ R^2 : H $$

The aroma of natural oils is of special interest. The most important aroma substances of two *extra vierge* olive oils with different aromas are shown in Table 14.7. In oil I, green, fruity, and fatty notes are predominant while in oil II, the compounds with a green odor are concealed by an aroma substance with a "black-currant" odor, possibly due to the higher degree of ripeness of the olives. The compound involved is the extremely intensive odorant 4-methoxy-2-methyl-2-butanethiol (cf. Table 5.31), which has the highest aroma value of all the volatile compounds in oil II (Table 14.7). The fatty acid composition of tea seed oil is very similar to that of olive oil. However, these two oils can be differentiated by using the *Fitelson* Test (cf. Table 14.21).

14.3.2.1.2 Palm Oil

This oil is obtained from oil palm, the utilization of which is constantly increasing (cf. Table 14.1). Palm plantations are found primarily in western Malaysia and Indonesia (cf. Table 14.0). The fruits provide two different oils, the first from the pulp and the second from the seeds.

Oil Production. The fruit cluster, which contains about 3,000–6,000 fruits, is first steam-treated to inactivate the high lipase activity and to separate the pulp from the seed. The oil is recovered by pressing the disintegrated pulp. The crude oil is then clarified by centrifugation. Washing with hot water, followed by drying, provides a crude oil product that has a high carotene content (cf. 3.8.4.5) and, hence, the color of the oil is yellow to red. During refining (cf. 14.4.1), the palm oil color is destroyed by bleaching and the free fatty acids are removed. Palm fruit characteristics and oil composition are given in Table 14.5.

Table 14.6. Olive oils: concentrations of the minor constituents

Type	Alcohols[a]	Waxes[a, b]	Sitosterol[a]		Erythrodiol[a]	Erythrodiol + Uvaol (%)[c]
			free	esterified		
Extra vierge oil	67	40	914	219	13	1
"Lampante" oil, raw	84	292	945	877	10	0.6
"Lampante" oil, refined	44	180	692	544	8	0.8
Extraction oil, raw	725	3,294	1,234	2,702	283	13.5
Extraction oil, refined	75	3,277	659	2,624	116	5.6

[a] Values in mg/kg.
[b] Sum of the wax esters C_{40}–C_{46}.
[c] Percentage of the sum of sterols and triterpene dialcohols.

Table 14.7. Important aroma substances of two *extra vierge* olive oils[a]

Aroma substance	Aroma quality	Oil I		Oil II	
		C	A_x	C	A_x
Isobutyric acid ethyl ester	Fruity	4.9	7	14	19
2-Methylbutyric acid ethyl ester	Fruity	3.9	5	14	19
Cyclohexanoic acid ethyl ester	Fruity	1.6	4.2	4.3	11
(Z)-3-Hexenal	Green	33	12	53	19
(Z)-2-Nonenal	Green, fatty	9	15	10	17
Acetic acid	Like vinegar	10,490	10	6 680	6
4-Methoxy-2-methyl-2-butanethiol	Like blackcurrants	n.d.		1.8	40

[a] Oil I is from Italy, oil II from Spain; concentration C is in µg/kg and the aroma value A_x is calculated on the basis of the odor thresholds (retronasal) in an oil; n.d.: not detectable (C < 0.05 µg/kg).

Adulteration of palm oil with palm stearin increases the ratio of the triacylglycerides PPP to MOP, which is usually between 3.5 and 4.5.

14.3.2.2 Seed Oils

Some oilseeds have acquired great significance in the large-scale industrial production of edible oils. After a general review of their production, some individual oils, grouped together according to their characteristic fatty acid compositions, will be discussed.

14.3.2.2.1 Production

Conditioning. The ground or flaked seeds are heated with live steam of about 90 °C to facilitate oil recovery. This treatment ruptures all the cells, partly denatures the proteins and inactivates most of the enzymes. The temperature is regulated to avoid formation of undesirable colors and aromas.

After conditioning and moisture adjustment to about 3%, the oil is obtained by pressing and/or solvent extraction. The choice between these two processes depends on the oil content of the seed. Solvent extraction is the only economic choice for seeds with an oil content below 25%.

Pressing. The oil is removed by pressure from an expeller or screw press. The residual oil in the resultant meal flakes is 4–7%. It is, however, more economical to apply lower pressures and to leave 15–20% of the oil in the flakes, and then to remove this oil by a solvent extraction process ("prepress solvent extraction" process).

Extraction. The ground seeds are rolled into thin flakes by passing them between smooth steel rollers. This flaking step provides the enlarged surface area needed for efficient solvent extraction. The extraction is performed using petroleum ether, i.e. technical hexane, as a solvent (boiling point 60–70 °C). In addition to n-hexane, it contains 2- and 3-methylpentane and 2,3-dimethylbutane and is free of aromatic compounds.

Solvent removal from the raw oil-solvent mixture, called miscella, is achieved by distillation. The maximum amount of solvent remaining in the oil is 0.1%. The oil-free flakes are then steamed to remove the solvent ("desolventizing") and, after dry heating ("toasting"), cooled and sold as protein-rich feed for cattle. The production of oil from soybeans is schematically shown in Fig. 14.2.

The crude oil obtained either by pressing or solvent extraction contains suspended plant debris, protein and mucous substances. These impurities are removed by filtration.

14.3.2.2.2 Oils Rich in Lauric and Myristic Acids

The most important representatives of this group of oils are coconut, palm seed and babassu oils. The acceptable shelf life stability of these oils is reflected in their fatty acid compositions (Table 14.8). Since linoleic acid is present in negligible amounts, autoxidative changes in these oils do not occur. However, when these oils are used in preparations containing water, microbiological deterioration may occur; this involves release of free

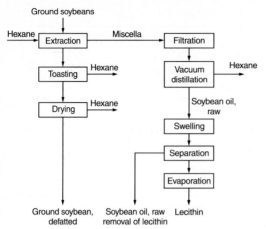

Ground soybeans

Fig. 14.2. Production of oil and lecithin from soybeans

Table 14.8. Characteristics of palm kernel oils

	Oil palm (*Elaeis guineensis*)	Coconut palm (*Cocos nucifera*)	Babassu palm (*Orbignya speciosa*)
Kernel oil content (weight-%)	40–52	63–70	67–69
Fat/oil melting range (°C)	23–30	20–28	22–26
Average fatty acid composition (weight-%)			
8 : 0	6	8	4.5
10 : 0	4	6	7
12 : 0	47	47	45
14 : 0	16	18	16
16 : 0	8	9	7
18 : 0	2.5	2.5	4
18 : 1 (9)	14	7	14
18 : 2 (9, 12)	2.5	2.5	2.5

C_8–C_{12} fatty acids and their partial degradation to methyl ketones ("perfume scent rancidity", cf. 3.7.6).

Coconut and palm seed oils are important ingredients of vegetable margarines which are solid at room temperature. However, they melt in the mouth with a significant heat uptake, producing a cooling effect.

Coconut oil is obtained from the stone fruit of the coconut palm, which grows throughout the tropics. The moisture content of the oil-containing endosperm, when dried, decreases from 50% to about 5–7%. Such crushed and dried coconut endosperm is called "copra" and is sold under this name as a raw material for oil production around the world.

Palm kernel oil is obtained from the kernels of the fruit of the oil palm. The kernels are separated from the fruit pulp, then removed from the stone shells and dried prior to recovery of the oil. Babassu oil is obtained from seeds of the babassu palm, which is native to Brazil. This oil is rarely found on the world market and is mainly consumed in Brazil.

14.3.2.2.3 Oils Rich in Palmitic and Stearic Acids

Cocoa butter and fats (solid at room temperature) belong to this group, with the latter referred to as cocoa butter substitutes ("cocoa butter interchangeable fats"). They are relatively hard and can crystallize in several polymorphic forms (cf. 3.3.1.2). Their melting points are between 30 and 40°C. The relatively narrow melting range for cocoa butter, as well as for some other types of butter, is to be expected (Table 14.9). When cocoa butter melts in the mouth, a pleasant, cooling sensation is experienced (cf. 14.3.2.2.2). This is characteristic of only a few types of triacylglycerols present in fats which contain predominantly palmitic, oleic and stearic acids. This fatty acid composition is also reflected by the resistance of these fats to autoxidation and microbiological deterioration (Table 14.9). These fats are utilized preferentially in the manufacturing of chocolates, candy and confections.

Cocoa butter is the fat from cocoa beans. The seed germ contains up to 50–58% of the fat, which is recovered as a by-product during cocoa manufacturing (cf. 21.3.2.7). It is light yellow and has the pleasant, mild odor of cocoa. Cocoa butter contains 1,3-dipalmito-2-olein, 1-palmito-3-stearo-2-olein, and 1,3-distearo-2-olein in an almost constant ratio of 22 : 46 : 31 (% peak area). Since cocoa butter substitutes clearly differ in the content of these TGs, the amount of cocoa butter can be determined by HPLC of the TGs. Bromination of the double bonds (cf. 3.3.1.4) improves the separation of the three TGs. In addition, the

Table 14.9. Fatty acid composition of cocoa butter and cocoa butter substitutes

Trade name	Cocoa butter	Illipè butter (Mowrah butter)	Borneo tallow (Tengkawang fat, Illipè butter)	Shea butter (Kerité fat)
Source	Cacao tree (*Theobroma cacao*)	*Madhuca longifolia*	*Shorea stenoptera*	*Butyrospermum parkii*
Fat, melting range (°C)[a]	28–36	24.5–28.5	28–37	23–42
Average fatty acid composition (weight-%)				
16 : 0	25	28	20	7
18 : 0	37	14	42	38
18 : 1 (9)	34	49	36	50
18 : 2 (9, 12)	3	9	<1	5

[a] The melting ranges reflect a pronounced fat polymorphism (cf. 3.3.1.2); the highest temperature given represents the melting point of the stable fat modification.

ratio of stigmasterol/campesterol is characteristic of cocoa butter.

The denotation of the "cocoa butter interchangeable fats" may be confusing since fats from diverse sources are sometimes marketed under a collective name such as Illipè butter. Confusion can be avoided by providing the Latin name of the plant, i.e. the source of the fat.

Shea Butter (Kerité Fat) is obtained from seeds of a tree which grows in western Africa and the cultivation of which appears to be uneconomical. The high content of unsaponifiable matter (up to 11%) in this kind of butter is of interest.

Borneo Tallow (Illipè Butter) is obtained from the seeds of a plant native to Java, Borneo, the Philippines and India. It serves as a valuable edible fat in the Tropics. *Mowrah butter* (often marketed as Illipè butter) is derived from a different plant (*Madhuca longifolia*) and is also indigenous to the Asian tropics.

14.3.2.2.4 Oils Rich in Palmitic Acid

Oils in this group contain more than 10% palmitic acid along with oleic and linoleic acids (cf. Table 14.10).

Cottonseed Oil is obtained from seeds of many cotton plant cultivars. The plant is widely cultivated (cf. Table 14.0). The raw oil is dark, usually dark red, and has a unique odor. It contains a poisonous phenolic, gossypol,

(14.2)

which is removed during refining. Another substance present in this oil is malvalia acid,

$$CH_3-(CH_2)_7-C=C-(CH_2)_6-COOH$$
$$\diagdown / \atop CH_2$$ (14.3)

which survives refining, but not hydrogenation of the oil. This substance is responsible for detection or identification of the oil by the *Halphen* reaction (cf. Table 14.21).

At temperatures below + 8°C, cottonseed oil becomes turbid due to crystallization of high melting point triacylglycerols. Such undesirable low temperature characteristics are avoided using a "winterization" process (cf. 14.4.4).

Cereal Germ Oils. All cereals contain significant amounts of oil in the germ. It is available after the germ is separated during grain processing. Corn (maize) oil is the most important. Germ separation is achieved during dry

Table 14.10. Oils rich in palmitic acid

	Cottonseed (*Gossypium*)	Corn germ (*Zea mays*)	Wheat germ[a] (*Triticum aestivum*)	Pumpkin seed (*Cucurbita pepo*)
Seed oil content (weight-%)	22–24	3.5–5[b]		35
Solidification point (°C)	0 to +4	−10 to −18		−15 to −16
Average fatty acid composition (weight-%)				
14 :0	1.5	0	0	0
16 :0	22	10.5	17	16
18 :0	5	2.5	1	5
20 :0	1	0.5	0	0
16 :1(9)	1.5	0.5	0	0.5
18 :1(9)	16	32.5	20	24
18 :2(9, 12)	55	52	52	54
18 :3(9,12,15)	0	1	10	0.5

[a] Oil content in germ amounts to 8–11 weight-%.
[b] Of the seed oil content 80% is located in germ and the rest in seed endosperm.

or wet processing of the kernels into corn meal and starch (cf. 4.4.4.14.1). The oil is recovered from the germ collected as a by-product by pressing and solvent extraction. After crude oil refining, the corn waxes which originate from the skin-like coating of the epidermis (the cuticle), are removed by a winterization process (cf. 14.4.4).

Corn oil is suitable for manufacture of margarine and mayonnaise (creamy salad dressing), but is used preferentially as salad and cooking oil.

The oil present in wheat and rice is also concentrated in the germ. This oil can be recovered by pressing and/or solvent extraction of the germ. Wheat germ oil has a high content of tocopherol and, therefore, additional nutritive value. Rice germ oil is consumed to a minor extent in Asia. Pumpkin oil is obtained by pressing dehulled pumpkin seeds. In southern Europe it is utilized as an edible oil. It is brown in color and has a nut-like taste.

14.3.2.2.5 Oils Low in Palmitic Acid and Rich in Oleic and Linoleic Acids

A large number of oils from diverse plant families belong to this group (cf. Table 14.11). These oils are important raw materials for manufacturing margarine.

The sunflower is the most cultivated oilseed plant in Europe. Data on the production of the sunflower in regions and countries are given in Table 14.0. Prepressing of dehulled sunflower seeds yields a light yellow oil with a mild flavor. The oil is suitable for consumption once it is clarified mechanically. Refined oils are used in large amounts as salad oil or as frying oil and as a raw material for margarine production. The refining of the oil includes a wax-removal step.

Two legume oils, soybean and peanut (or ground nut), are of great economic significance (cf. Table 14.1). Soybean oil (fatty acid composition in Table 14.11) is currently at the top of the world production of edible oils of plant origin. It is cultivated mostly in the United States, Brazil and China. The refined oil is light yellow and has a mild flavor. It contains in low concentrations (Table 3.9) branched furan fatty acids which are rapidly oxidized on exposure to light. In fact, two of these fatty acids, which differ only in the length of the carboxyl ends (see Formula 3.3), produce the intensive aroma substance 3-methyl-2,4-nonandione in a side reaction with singlet oxygen (cf. Fig. 3.25). This aroma substance and diacetyl are significantly involved in the "bean-like, buttery, hay-like" aroma defect called *reversion flavor*.

In the case of the soybean oils listed in Table 14.12, the two furan fatty acids were almost

Table 14.11. Oils low in palmitic acid and rich in oleic and linoleic acids

	Sunflower (*Helianthus annuus*)	Soya (*Glycine max.*)	Peanut[a] (*Arachis hypogaea*)	Rapeseed[b] (*Brassica napus*)	Sesame (*Sesamum indicum*)	Safflower (*Carthamus tinctorius*)	Linseed (*Linum usitatissimum*)	Poppy (*Papaver somniferum*)	Walnut (*Juglans regia*)
Seed oil content (weight-%)	25–30	18–23	42–52	ca. 40	45–55	32–43	32–43	40–51	58–71
Solidification point (°C)	−18 to −20	−8 to −18	−2 to +3	0 to −2	−3 to −6	−13 to −20	−18 to −27	−15 to −20	−15 to −20
	Average fatty acid composition (weight-%)								
16:0	6.5	10	10	4	8.5	6	6.5	9.5	8
18:0	5	5	3	1.5	4.5	2.5	3.5	2	2
20:0	0.5	0.5	1.5	0.5	0.5	0.5	0	0	1
22:0	0	0	3	0	0	0	0	0	0
18:1 (9)	23	21	41	63	42	12	18	10.5	16
18:2 (9, 12)	63	53	35.5	20	44.5	78	14	76	59
18:3 (9, 12, 15)	<0.5	8	0	9	0	0.5	58	1	12
20:1 & 20:2	1	3.5	1	1	0	0.5	0	0	0
22:1 (13)	0	0	–	0.5	0	0	0	0	0

[a] African peanut oil. [b] Canola-type oil (practically free of erucic acid).

Table 14.12. Oxidation of furan fatty acids I and II and formation of 3-methyl-2,4-nonandione in three refined soybean oils[a]

Compound[a]	Time[b]	Soybean oil		
		A	B	C
		(mg/kg)		
Furan fatty acid I	0 h	143	148	131
Furan fatty acid I	48 h	5	5	3
Furan fatty acid II	0 h	152	172	148
Furan fatty acid II	48 h	5	5	5
		(µ/kg)		
MND	0 h	< 1	< 1	< 1
MND	48 h	89	3.4	< 1
MND	30 d	721	204	43

[a] I: 10,13-Epoxy-11,12-dimethyloctadeca-10,12-dienoic acid;
II: 12,15-Epoxy-13,14-dimethyleicosa-12,14-dienoic acid. MND: 3-Methyl-2,4-nonandione.
[b] The soybean oils were stored at room temperature at a window facing north.

completely oxidized after 48 hours. However, the amounts of 3-methyl-2,4-nonandione formed were very different. This is put down to differences in the stability of the intermediate hydroperoxide (cf. Fig. 3.25).

Other experiments have shown that the hydroperoxides formed from furan fatty acids on exposure to light fragment to the dione, even if the soybean oil is subsequently stored in the dark. In the complete absence of light, soybean oil is relatively stable. The shelf life of the oil is also improved significantly by partial hydrogenation to give a melting point range of 22–28 °C or 36–43 °C. Such oils are utilized as raw materials for the manufacture of margarine and shortening (semi-solid vegetable fats used in baked products, such as pastry, to make them crisp or flaky).

The fatty acid composition of peanut oil is greatly influenced by the region in which the peanuts are grown. In contrast to the peanut oils produced in Africa (Senegal or Nigeria), the peanut oils from South America are enriched in linoleic acid (41% vs 25%, w/w; see fatty acid composition, Table 14.11) at the expense of oleic acid (37% vs 55%, w/w). The contents of arachidic (20 : 0), eicosenoic (20 : 1), behenic (22 : 0), erucic (22 : 1) and ligno-ceric (24 : 0) acids are characteristic of peanut oil. Their glycerols readily crystallize below 8 °C.

Peanut Butter is a spreadable paste made from roasted and ground peanuts by the addition of peanut oil and, occasionally, hydrogenated peanut oil.

Rapeseed Oil. This oil is produced from seeds of two *Brasica* species: *B. napus*, known as Argentinian rape or, in Europe, as summer or Swedish rape, and *B. campestris* (Polish or summer turnip rape). The latter plants yield slightly less oil, are shorter (approx. 80 cm), but mature more quickly. They are more tolerant to frost and have improved resistance to pest and diseases. Old rape and turnip rape cultivars contained high levels of erucic acid (45–50% by weight), which is hazardous in human nutrition. "Zero"erucic acid cultivars (22 : 1 < 5% by weight), called *Canola*, have been developed and, recently, "double zero" cultivars, with low levels of erucic in the oil and goitrogenic compounds in the seed meal, have been developed. The major rapeseed-cultivating regions and countries are listed in Table 14.0.

The above-mentioned plants, such as *Brassicacea*, contain mustard oil glucosides (glucosinolates, cf. 17.1.2.6.5) which, immediately after seed crushing, are hydrolyzed to esters of iso-thiocyanic acid. The hydrolysis is dependent on seed moisture and is catalyzed by a thioglucosidase enzyme called myrosinase (EC 3.2.3.1). In the presence of the enzyme, some of the isocyanates are isomerized into thiocyanates (esters of normal thiocyanic acid or rhodanides) and, in part, are decomposed into nitrile compounds which do not contain sulfur. All these compounds are volatile and, when dissolved in oil, are hazardous to health and detrimental to oil flavor. Moreover, they interfere with hydrogenation of the oil by acting as Ni-catalyst poisons (cf. 14.4.2.2). Therefore, in the production of rapeseed oil, a dry conditioning step is used (without live steam) to inactivate the myrosinase enzyme and only then is the seed ground and subjected to prepress and solvent extraction processes.

Despite these precautions, small amounts of volatile sulfur compounds are formed. However, they are removed during the refining process. Irrespective of technical achievements in rapeseed production and processing, the selection and breeding of rapeseed "double zero" cultivars is being continued.

Rapeseed (Canola) Oil is used as an edible oil. It is susceptible to autoxidation because of its relatively high content of linolenic acid. It is saturated by hyrogenation to a melting point of 32–34 °C and, with its stability and melting properties, resembles coconut oil.
Turnip rape oil has practically the same composition as the *B. napus* oil.

Sesame Oil is obtained from an ancient oilseed crop (*Sesamum indicum*, L.), which is widely cultivated in India, China, Burma and east Africa (cf. Table 14.0). In its refined form the oil is nearly crystal clear and has a good shelf life. In addition to a considerable amount of tocopherols, it contains another phenolic antioxidant, sesamol, which is derived from hydrolysis of sesamolin (Fig. 14.3).
Sesame oil can be readily identified with great reliability (cf. Table 14.21). Therefore, in some countries, blending this oil into margarine is required by law in order to identify the product as margarine.

Safflower Oil is obtained from a thistle-like plant (*Carthamus tinctorius*) grown in the arid regions of North America and India (cf. Table 14.0). New cultivars have been bred with oil compositions which deviate greatly from those listed in Table 14.11. These new cultivars contain 80% by weight oleic acid (18 : 1) and 15% by weight linoleic acid (18 : 2; 9, 12).

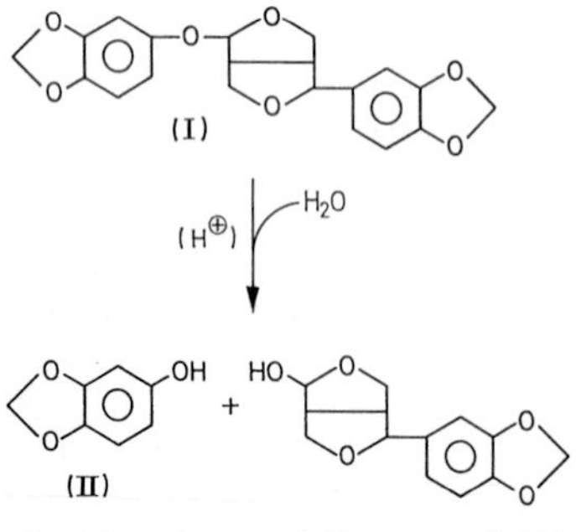

Fig. 14.3. Sesame oil: sesamol (II) formation by sesamolin (I) hydrolysis

Linseed Oil. Flax, used for fiber and seed production and the subsequent processing of the seed into linseed oil, is grown mainly in Canada, China and India (cf. Table 14.0). Due to its high content of linolenic acid (cf. Table 14.11), the oil readily autoxidizes, one of the processes by which some bitter substances are created. Since autoxidation involving polymerization reactions proceeds rapidly, the oil solidifies ("fast drying oil"). Therefore, it is used as a base for oil paints, varnishes and linoleum manufacturing, etc. A comparatively negligible amount, particularly of the cold-pressed oil, is utilized as an edible oil.

Poppy Oil is very rich in linoleic acid (Table 14.11). The cold-pressed oil from flawless seeds is colorless to light yellow and can be used directly as an edible oil.

Walnut Oil has a pleasant odor and a nut-like taste. It contains relatively high concentrations of linolenic acid (Table 14.11) and, consequently, has a very limited shelf life.

14.4 Processing of Fats and Oils

14.4.1 Refining

Apart from some oils obtained by cold pressing (examples in 14.3.2.1), most of the oils obtained using expeller, screw or hydraulic presses, solvent extraction or by melting at elevated temperatures are not suitable for immediate consumption. Depending on the raw material and the oil recovery process, the oil contains polar lipids, especially phospholipids, free fatty acids, some odor- and taste-imparting substances, waxes, pigments (chlorophyll, carotenoids and their degradation products), sulfur-containing compounds (e.g. thioglucosides in rapeseed oils), phenolic compounds, trace metal ions, contaminants (pesticides or polycyclic hydrocarbons) and autoxidation products. In a refining process comprising the following steps:

- Lecithin removal
- Degumming
- Free fatty acid removal
- Bleaching
- Deodorization,

all the undesired compounds and contaminants are removed. An overview of the refining

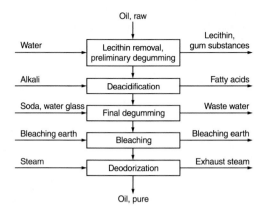

Water → Lecithin removal, preliminary degumming → Lecithin, gum substances

Alkali → Deacidification → Fatty acids

Soda, water glass → Final degumming → Waste water

Bleaching earth → Bleaching → Bleaching earth

Steam → Deodorization → Exhaust steam

Oil, raw → ... → Oil, pure

Fig. 14.4. Refining of oils

process is given in Fig. 14.4. In practice the refining steps used depend on the quality of the crude oil and its special constituents (e.g. carotene in palm oil or gossypol in cottonseed oil). The following precautionary measures are taken during refining in order to avoid undesirable autoxidation and polymerization reactions:

- Absence of oxygen (also required during transport or storage)
- Avoidance of heavy metal contamination
- Maintaining the processing temperatures as low and duration as short as possible.

14.4.1.1 Removal of Lecithin

This processing step is of special importance for rapeseed and soybean oils. Water (2–3%) is added to crude oil, thereby enriching the phospholipids in the oil/water interface. The emulsion thus formed is heated up to 80 °C and then separated or clarified by centrifugation. The "crude lecithin" (cf. 3.4.1.1) is isolated from the aqueous phase and is recovered as crude vegetable lecithin after evaporating the water in a vacuum.

14.4.1.2 Degumming

Finely dispersed protein and carbohydrates are coagulated in oil by addition of phosphoric acid (0.1% of oil weight). A filtering aid is then added and the oil is clarified by filtration. This also removes the residual phospholipids from the previous processing step.

14.4.1.3 Removal of Free Fatty Acids (Deacidification)

Several methods exist for deacidification of fat or oil. The choice depends on the amount of free fatty acids present in crude fat or oil.

The removal of fatty acids with 15% sodium hydroxide (alkali refining) is the most frequently used method. Technically, this is not very simple since fat hydrolysis has to be avoided and, moreover, the sodium soap (the "soapstock"), which tends to form stable emulsions, has to be washed out by hot water. After vacuum drying, the fat or oil may contain only about 0.05% free fatty acids and 60 to 70 ppm of sodium soaps. When the fat or oil is treated with diluted phosphoric acid, the content of sodium soaps decreases to 20 ppm and part of the trace heavy-metal ions is removed.

Fats (oils) with a high content of free fatty acids require relatively high amounts of alkali for extraction, resulting in an unavoidably high loss of neutral fats (oils) due to alkaline hydrolysis. Therefore, extraction with alkali is frequently replaced by deacidification by distillation in these cases (14.4.1.5).

In special cases, a selective fluid/fluid extraction is of interest. Ethanol extracts free fatty acids (above a level of 3%) from triacylglycerols in crude oils – this is a suitable way to treat oils with exceptionally high amounts of free acids. At a given temperature, furfural can extract only the polyunsaturated triacylglycerols. On the other hand, propane under pressure preferentially solubilizes the saturated triacylglycerols and leaves behind the unsaturated ones, together with unsaponifiable matter. Pressurized propane is utilized in marine oil fractionation, e.g., in the production of vitamin A concentrates.

14.4.1.4 Bleaching

In order to remove the plant pigments (chlorophyll, carotenoids) and autoxidation products, the fat or oil is stirred for 30 min in the presence of Al-silicates (bleaching or *Fuller's* earth) in a vacuum at 90 °C. The silicate has to be activated prior to use – a suspension in water is treated with hydrochloric acid, followed by thorough washing with water, then drying. The amount of silicate used is 0.5–2%

of the fat (oil) weight. It is often used together with 0.1–0.4% activated charcoal.

The bleached oil is removed from the adsorbent by filtration. The oil retained by the absorbent can be recovered by hexane extraction and recycled into the refining process. The residual alkali soaps, gums, part of the unsaponifiable matter and the heavy metal ions are also removed during the bleaching process.

After bleaching, some oils or fats which contain polyunsaturated fatty acids show an increase in absorbance at 270 nm. This is due to decomposition of hydroperoxides, formed by autoxidation, into oxo-dienes (I) and fatty acids with three double bonds (II):

$$R_1-CH=CH-CH=CH-\overset{\overset{\displaystyle O}{\|}}{C}-R_2$$

(I) ↓ ⟶ H₂O

$$R_1-CH=CH-CH=CH-\underset{\underset{\displaystyle OOH}{|}}{CH}-R_2$$

(Hydroperoxide)

↓ ⟶ •OH

$$R_1-CH=CH-CH=CH-\underset{\underset{\displaystyle O\cdot}{|}}{CH}-R_2$$ (14.4)

↓ ⟵ RH
↓ ⟶ R•

$$R_1-CH=CH-CH=CH-\underset{\underset{\displaystyle OH}{|}}{CH}-R_2$$

↓ ⟶ H₂O

$$R_1-CH=CH-CH=CH-CH=CH-R_3$$

(II)

As shown in Table 14.13 with rapeseed oil, most of the chlorophylls and their degradation products are removed during bleaching. However, the rest, which is in the range of 70–1200 µg/kg in refined plant oils, could still accelerate photooxidation (Fig. 14.5). As a result of their high stability on exposure to light, the pheophytins are stronger photooxidants than the chlorophylls. As shown in Table 14.13, the pheophytins are the predominating pigments.

Table 14.13. Removal of chloro pigments in the refining of rapeseed oil (values in mg/kg)

	Raw oil	Amounts after		
		Deacid-ification	Bleaching	Deodor-ization
Chlorophyll A	2.62	0.89	0.028	0.007
Chlorophyll B	2.92	0.08	0.059	0.023
Pheophytin A	35.6	31.5	0.235	0.108
Pheophytin B	4.99	6.85	0.071	0.036
Sum	46.1	39.3	0.393	0.174

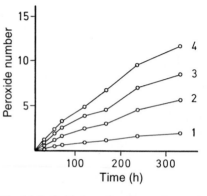

Fig. 14.5. Oxidation of soybean oil on exposure to room light (according to *R. Usuki* et al., 1984). (1), (2), (3), and (4) contain 39, 233, 425, and 623 µg/kg respectively of the mixture chlorophyll A/chlorophyll B/pheophytin A/pheophytin B (1 : 3 : 10 : 3)

14.4.1.5 Deodorization

Deodorization is essentially vacuum steam distillation (190–230 °C, 0.5–10 mbar). The volatile compounds, together with undesirable odorants present in the fat or oil, are separated in this refining step. Deodorization takes from 20 min to 6 h, depending on the type of fat or oil and the content of volatile compounds.

The processing loss in this refining step is 0.2%. This is negligible since the fat or oil droplets carried by the steam are caught by baffles or are intercepted by an external trap system.

Deodorization can been combined with deacidification by distillation when the oil has a low content of accompanying substances or if they have been largely removed by degumming and bleaching, e.g., after reduction of the phospholipids to less than 5 mg/kg. Since fatty acids are less steam-volatile than the odorous substances, higher temperatures (up to 270 °C) are used than in deodorization. Carotenoids are decomposed, so that, e.g., palm oil is thermally bleached in this process.

The combination of deodorization with distillative deacidification is called physical refining; it is the process of choice in the case of higher acid concentrations (> 0.7–1%). The relatively large amounts of waste water which have to be disposed off after alkaline extraction are avoided here. In addition, the fatty acids accumulated as by-products are of higher quality then the "refining fatty acids" obtained by alkali extraction (cf. 14.4.1.3).

In steaming or physical refining, the double bonds of linoleic and linolenic acid isomerize to a small extent. For this reason, an HPLC determination of isomeric linoleic acids is used to distinguish between refined and natural plant oils (cf. 14.5.3.4).

14.4.1.6 Product Quality Control

In addition to sensory evaluation, free fatty acid analysis (the content is usually below 0.05%) and analysis of possible contaminants are carried out. The data given in Table 14.14 illustrate the amounts of pesticides and polycyclic aromatic compounds removed by deodorization. However, this refining step also removes the highly desirable aroma substances which are characteristic of some cold-pressed oils such as olive oil.

Table 14.14. Removal of endrin and polycyclic hydrocarbons during edible oil refining (µg/kg)

Compound	Content in raw oil	Content in oil after		
		deacid-ification	bleaching	steaming
Endrin	620[a]	590	510	< 30
Anthracene	10.1[b]	5.8	4.0	0.4
Phenanthrene	100[b]	68	42	15
1,2-Benz-anthracene	14[b]	7.8	5.0	3.1
3,4-Benzpyrene	2.5[b]	1.6	1.0	0.9

[a] Soybean oil.
[b] Rapeseed oil.

The composition of phytosterols and tocopherols does not change appreciably during refining. Therefore, an analysis of these compounds is suitable for the identification of the type of fat. On the other hand, cholesterol can increase during steaming, e.g., due to the cleavage of glycosides. In palm oil, for instance, the percentage of cholesterol in the sterol fraction increased from 2.8% to 8.8%.

14.4.2 Hydrogenation

14.4.2.1 General Remarks

Liquid oils are supplied mostly from natural sources. However, a great demand exists for fats which are solid or semi-solid at room temperature. To satisfy this demand, *W. Normann* developed a process in 1902 to convert liquid oil into solid fat, based on the hydrogenation of unsaturated triacylglycerols using Ni as catalyst; a process designated as "fat hardening". The process rapidly gained great economic importance; even marine oils became suitable for human consumption after the hardening process. Today more than 4 million tons of fat per year are produced worldwide by hydrogenation of oils; most is consumed as food.

The unsaturated triacylglycerols can be fully hydrogenated, providing high-melting cooking, frying and baking fats or partially hydrogenated, providing products such as:

- Oils rich in fatty acids with one double bond. They are stable and resistant to autoxidation and have a shelf life similar to olive oil. They are used as salad oil or as shortening.
- Products in which the linolenic acid is hydrogenated, but most of the essential fatty acid, linoleic acid, is left intact. An example is soybean oil, hydrogenated selectively to increase its stability against oxidation.
- Fats that melt close to 30 °C and have a plastic or spreadable consistency at room temperature.

Fully or partially hydrogenated oils are important raw materials for margarine manufacturing.

14.4.2.2 Catalysts

The principle of the heterogeneous catalytic hydrogenation of unsaturated acylglycerols was outlined under 3.2.3.2.4. The most widely used catalyst is carrier-bound nickel. Raney nickel, copper, and noble metals serve special purposes. The choice of catalyst is made according to:

- Reaction specificity.
- Extent of trans-isomer formation
- Duration of activity and cost.

To determine the specificity of a catalyst, the reaction rates for each individual hydrogenation step must be determined. Simplified, there are three reaction rate constants (k) involved (AG = acylglycerol):

$$\text{Triene-AR} \xrightarrow{k_3} \text{Diene-AR} \xrightarrow{k_2} \text{Monoene-AR}$$

$$\downarrow k_1 \quad (14.5)$$

AR-acylresidue   Saturated-AR

The catalytic reactions considered here require that $k_3 > k_2 > k_1$. The following equations determine the specificity "S":

$$s_{32} = \frac{k_3}{k_2}; \quad s_{21} = \frac{k_2}{k_1}; \quad s_{31} = \frac{k_3}{k_1}; \quad (14.6)$$

That means, the greater the value of "S", the faster the hydrogenation at this step. Therefore, specificity (or selectivity) is proportional to the value of "S". For the three catalysts mentioned, Table 14.15 shows that the hydrogenation of diene → monoene by Ni_3S_2 and the hydrogenation of triene → monoene by copper become accelerated with marked specificity. Copper is particularly suitable for decreasing the linolenic acid content in soybean and rapeseed oils. However, copper catalysts are not sufficiently economical, since they can not be used more than twice and their complete removal is relatively tedious.

Although noble metals are up to 100 times more effective than nickel catalysts, they are not popular because of their high costs. It is of great advantage that the nickel catalyst can be used repeatedly for up to 50 times under the following conditions: the plant oil must be

Table 14.15. Properties of hydrogenation catalysts

Catalyst	Selectivity		trans-Fatty acids (weight-%)[a]
	S_{32}	S_{21}	
Nickel-contact	2–3	40	40
Ni_3S_2-contact	1–2	75	90
Copper-contact	10–12	50	10

[a] trans-Fatty acids as monoenoic acids total content is calculated as elaidic acid.

deacidified, freed of gum ingredients and contain no sulfur compounds (cf. rapeseed oil, 14.3.2.2.5). The favorable ratio of duration of activity to cost places the nickel catalysts ahead with advantages not readily surpassed by any other catalyst. For the production of nickel-carrier catalysts, kieselguhr is impregnated with nickel hydroxide, which is precipitated out of a solution of nickel nitrate with sodium hydroxide or carbonate. After drying, nickel hydroxide is reduced to nickel with hydrogen at 350–500 °C.

For the production of carrier-free nickel catalysts, nickel formate is suspended in a fat and then decomposed:

$$\text{Ni(HCOO)}_2 \xrightarrow[(200-250\,°C)]{} \text{Ni} + 2\,CO_2 + H_2$$
$$(14.7)$$

The Ni produced in this way is a finely dispersed pyrophoric metal. For this reason it is embedded in fat and handled and marketed in this form.

To evaluate the catalysts, calculation programs were developed for the determination of the actual selectivity of a catalyst based on the fatty acid composition of the starting material and of the hydrogenated product.

During hydrogenation, linolenic acid yields, among others, isolinoleic and isooleic acids (cf. Reaction 14.8).

The diversity of the reaction products present in partially hydrogenated fat is increased further by the positional- and stereoisomers of the double bonds. Hydrogenation of soybean oil in the presence of a copper catalyst gives, for example, a number of trans-monoene fatty acids (Table 14.16). The extent of isomeriza-

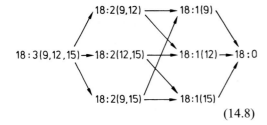

$$(14.8)$$

tion is affected, among other factors, by the type of catalyst used in hydrogenation.

Although extensive nutritional/physiological studies have provided no evidence of possible deleterious effects of elaidic or iso-elaidic acids in fat metabolism, the current processing trend is still to lower the accumulation of trans-fatty acids in hydrogenated fats. Hydrogenated fat is readily distinguished from nonhydrogenated fat by the presence of trans-fatty acids. The trans-acids are routinely revealed and quantitatively determined by infra-red (IR) spectroscopy or by gas chromatography (cf. 14.5.2.3).

A further drawback of the partial hydrogenation of an oil is the pattern of linoleic acid isomers formed. The two isomers formed during hydrogenation, linolelaidic acid 18 : 2 (9 trans, 12 cis) and 18 : 2 (9 cis, 12 trans) are, unlike linoleic acid, not essential fatty acids (cf. 3.2.1.2).

Table 14.16. Fatty acid composition of a soya oil before and after hydrogenation with a copper catalyst

Fatty acid	Hydrogenation	
	before (weight-%)	after (weight-%)
16 : 0	10.0	10.0
18 : 0	4.2	4.2
18 : 1 (9)	26.0	30.4
18 : 1 [a]	0	5.5
18 : 2 (9, 12)	52.5	42.5
18 : 2 (conjugated) [b]	0	0.7
18 : 2 [c]	0	5.2
18 : 3 (9, 12, 15)	7.3	0.7

[a] This fraction contains eight trans fatty acids: 18 : 1 (7 tr) – 18 : 1 (14 tr); major components are 18 : 1 (10 tr) and 18 : 1 (11 tr).
[b] It consists of various conjugated fatty acids.
[c] Isolinoleic and isolinolelaidic acids.

14.4.2.3 The Process

The hydrogen required can be obtained by electrolysis of dilute aqueous KOH, through water-to-gas conversion:

$$H_2O + C \longrightarrow H_2 + CO$$
$$CO + H_2O \longrightarrow H_2 + CO_2 \qquad (14.9)$$

or by the decomposition of natural gas with steam:

$$CH_3(CH_2)_x CH_3 + H_2O \longrightarrow H_2 + CO$$
$$CO + H_2O \longrightarrow H_2 + CO_2$$
$$(14.10)$$

In the latter two processes, the poisonous by-products, H_2S and CO, have to be completely removed.

Oil hydrogenation is performed in an autoclave equipped with a stirrer under hydrogen gas pressure of 1–5 bar and a temperature of 150–250 °C. A newer hydrogenation process uses a recycling hydrogenation unit equipped with a spraying nozzle, external heat exchanger and recycling pump.

The process conditions have a significant effect on the composition and therefore on the consistency of the end-product. Selective hydrogenation of double bonds is favored by a high concentration of catalyst (which, depending on the Ni activity, is 200–800 g Ni/t fat), a high temperature and low pressure of hydrogen gas. After hydrogenation, the fat is filtered, then deacidified, bleached and deodorized during further refining (cf. 14.4.1.3– 14.4.1.5).

Some constituents of the unsaponifiable matter are also affected by the hydrogenation process. Carotenoids, including vitamin A, are hydrogenated extensively. Some of the chlorine-containing pesticide contaminants are hydrogenated. Sterols, under the usual operating conditions, are not affected. The ratios and levels of tocopherols are essentially unchanged.

14.4.3 Interesterification

Natural fats and oils are subjected to extensive interesterification during processing. This involves a rearrangement or randomization of acyl residues in triacylglycerols and thus provides fats or oils with new properties. By choosing the raw material and processing parameters, the interesterification can be controlled to obtain a fat with melting characteristics and consistency which match the intended use ("tailored fats").

The basics of the interesterification process are outlined under 3.3.1.3. Sodium methylate is used almost exclusively as the catalyst. The dried and deacidified fat (or oil) is stirred at 80–100°C in the presence of alcoholate (0.1–0.3% of fat weight). When the reaction is completed, the catalyst is destroyed by adding water, then the degraded catalyst together with the resultant soaps are removed from the fat (oil) by repeated washing with water. The interesterified product is then bleached (cf. 14.4.1.4) and deodorized (cf. 14.4.1.5).

Table 14.17 illustrates the changes in triacylglycerols brought about by the process and its influence on fat melting points.

The baking properties of lard (improvement of volume and softness of the baked goods) are improved by interesterification. The uniform distribution of palmitic acid in the triacylglycerols accounts for such an improvement.

Furthermore, interesterification is of importance in the manufacturing of different varieties of margarine with a given composition, for example:

- Vegetable margarine with 30% w/w of 18 : 2 (9, 12) fatty acid may be produced by interesterification of partially hardened sunflower oil blended with its natural liquid oil.
- Interesterification of palm oil with palm seed or coconut oil (2 : 1) and the use of 6 parts of this product with 4 parts of sunflower oil provides a margarine which contains 20–25% w/w of linoleic acid and does not contain hydrogenated fat.

14.4.4 Fractionation

The undesirable fat ingredients are removed or the desirable triacylglycerols (TG) are enriched by fractional crystallization. The rising demand of food processors for special fats with standardized properties has led to large-scale isolation of special fractions, particularly from palm oil and the fats and oils listed under 14.3.2.2.2. The following procedures are used for the fractional crystallization of fats: The melted fat is slowly cooled until the high melting TG selectively crystallize, i.e. without forming mixed crystals of low and high melting TG. A sharp separation into two or more fractions is assumed to be satisfactory when the melting points of the fractions differ by at least 10°C. The separated crystals are either removed by filtration or are washed out with a tenside solution. In the latter case, the fat crystals adsorb a water soluble surface-active agent, such as sodium dodecyl sulfate, and thus acquire hydrophilic properties. The crystals are then transferred to the aqueous phase. The isolated aqueous suspension is then heated and the TG recovered as liquid fat.

An even sharper fractional crystallization procedure may be achieved by solubilization of fat in hexane or some other suitable solvent. However, solvent distillation and recovery are rather time consuming, so the use of this procedure is justified only in very special cases.

In the processing step of "winterization" of rapeseed (Canola), cottonseed or sunflower oil, small amounts of higher melting TG or waxes are removed which would otherwise cause turbidity during refrigeration. The basis of winterization is the fractional crystallization by slowly cooling the oil, as outlined

Table 14.17. Changes in the pattern of triacylglycerols in a partially hydrogenated palm oil by interesterification

Melting point	Prior to interester-ification	Single phase interester-ification	Directed interester-ification
Melting point (°C)	41	47	52
	Triacylglycerols[a] in mole-%		
S$_3$	7	13	32
S$_2$U	49	38	13
SU$_2$	38	37	31
U$_3$	6	12	24

[a] S: Saturated, U: unsaturated fatty acids.

above. Other procedures for the production of cold-stable oils are based on the use of crystallization inhibitors. These are mono- and diacylglycerols, esters of succinic acid, etc.

The application of the fractional extraction of fat or oil, instead of crystallization, has been outlined under 14.4.1.3.

14.4.5 Margarine – Manufacturing and Properties

The inventor of margarine, *H. Mège Mouries*, described in his patent issued in 1869 a process for the production of spreadable fat from beef fat which would substitute for and imitate the scarce and costly dairy butter. Based on the assumption that margaric acid (17 : 0) is the predominant fatty acid of beef fat, the name "margarine" was suggested for the new product. The assumption was, however, proven to be incorrect (cf. Table 14.3). Nevertheless, the name remained.

Margarine, which is produced worldwide in amounts exceeding 7 million t/a, is a water in oil emulsion. Its stability is achieved by an increase in viscosity of the continuous fat phase due to partial crystallization and through emulsifiers. The fat crystals form a three dimensional network. They should be present in the β'-modification; a conversion β' → β is undesirable because the β-modification causes a "sandy" texture defect. Hydrogenated fats, which are frequently used as raw materials, crystallize in the β'-modification when the lengths of the acyl residues differ.

The erucic acid-rich and partially hydrogenated rapeseed fat used in the past crystallizes in the β'-form. The cultivation of rapeseed with a low content of erucic acid at first produced a fat that, after partial hydrogenation, consisted to almost 90% of 18 : 0 and 18 : 1 and, as a result of this homogeneity, crystallized in the β-form. By means of cultivation, 16 : 0 was increased from 5 to 12% at the cost of 18 : 1, which is sufficient for the stabilization of the β'-form.

14.4.5.1 Composition

The properties of margarine, such as nutritional value, spreadability, plasticity, shelf life

and melting properties, resemble those of butter and are influenced essentially by the varieties and properties of the main fat ingredients. Since choice of ingredients is large, numerous varieties of margarine are produced (cf. Table 14.18).

The fat in margarine, which by regulation is 80% by weight (diet margarine is 39–41% fat), contains about 18% of emulsified water. The W/O emulsion is stabilized by a mixture of mono- and diacylglycerols (approx. 0.5%) and crude lecithin (approx. 0.25%). Diet margarines have higher levels of emulsifiers. Skim milk or skim milk powder suspended in water (milk proteins, 1%; 2% in semi-fat margarine) is added in the production of high quality retail brands of margarine. The casein assists the action of the emulsifiers and, together with lactose, provides the desired browning when heated.

The aqueous phase of the margarine acquires a pH of 4.2–4.5 by addition of citric and lactic acids. This not only affects the flavor, but protects against microbial spoilage. In addition, traces of heavy metal ions are complexed. Margarine also contains the aroma substances typical of butter, which can be produced by microbiological souring (cf. 10.2.3.2). Readily available synthetic compounds, such as diacetyl, butyric acid, lactones of C_8–C_{14} hydroxy-fatty acids (cf. 5.3.1.4) and 4-cis-heptenal, may also be used for aromatization. Common salt (0.1–0.2%) is used to round-off the flavor. Margarine is colored with β-carotene or with gently refined, unbleached palm oil. Attention is also given to maintaining the presence of 1 mg of α-tocopherol per g of linoleic acid. High quality products are vitaminized by the addition of about 25 IU/g vitamin A and 1 IU/g vitamin D_2. The authenticity of margarine is verified in some countries by an indicator substance added to it. This is required by legislation. Gently refined sesame oil (for its detection, see Table 14.21) is one of these substances.

14.4.5.2 Manufacturing

Margarine is manufactured continuously by a process consisting essentially of three steps:

- Emulsification of water within the continuous oil phase.

- Chilling and mechanical handling of the emulsion.
- Crystallization, preserving the type of w/o emulsion by efficient removal of the released heat of crystallization.

The triacylglycerols should preferentially crystallize in their β′-forms (cf. 3.3.1.2). The higher melting β-forms are not desired since they cause a "sandy" texture. The transitioin β′ → β-form is inhibited by addition of 1% saturated diacylglycerols.

14.4.5.3 Varieties of Margarine

The characteristic features of some varieties of margarine are summarized in Table 14.18.

14.4.6 Mayonnaise

Mayonnaise is an "oil in water" or o/w emulsion (cf. 8.15.1) consisting of 50–85% edible oil, 5–10% egg yolk, vinegar, salt and seasonings (cf. 11.4.2.3). The emulsion is stabilized by egg yolk phospholipids. Products with a lower oil content (<50%) may contain thick-ening agents such as starch, pectin, traganth, agar-agar, alginate, carboxymethylcellulose, milk proteins or gelatin. Sorbic acid, benzoic acid or the ethyl ester of p-hydroxybenzoic acid are added as preservatives. The stable emulsion is produced in a combinator with a homogenizer and then packed.

14.4.7 Fat Powder

In contrast to fats and oils, fat powders have better stability against autoxidation and, in some food products such as dehydrated soup powders or broths, are easier to handle. They are manufactured from natural or hardened plant fats, sometimes with the addition of emulsifiers and protein carriers. Butter and cream powders are also produced.
Two basic flow diagrams of the production of fat powders are shown in Fig. 14.6.
In a cold-spray process, the melted fat is sprayed under high pressure into a cooled ($-35\,°C$) airblast crystallization chamber, where the fat particles solidify. After being recrystallized, the particles are coated to avoid clumping.

Table 14.18. Examples of margarine types

Type	Comments
A. Household margarine	
Standard product	At least 50% of the fat is vegetable oil, the rest being animal fat.
Vegetable margarine	At least 98% of the fat is vegetable oil; contains at least 15% linoleic acid.
Linoleic acid enriched margarine	At least 30% linoleic acid, otherwise as vegetable margarine.
B. Semi-fat margarine	The fat content is halved. This type is not suitable for baking and frying.
C. Molten or fused margarine	Practically free of water and protein. It is aromatized with diacetyl and butyric acid; soft consistency; with large TG crystals it has a grainy structure; applied in cooking, frying and baking.
D. Special types for industrial processing	
Baking margarine	Strongly aromatized with heat stable compounds that contribute to baked products' aroma; mainly moderately melting TG's.
Margarine for pastry production	This margarine is strongly aromatized; its high melting TG's are embedded in oil phase; suitable for dough extension into thin sheets ("strudel dough") used in flaky pastry production.
Creamy margarine	It is not or only slightly aromatized; has a soft consistency; contains high content of coconut oil and approx. 10 vol-% of air.

TG: Triacylglycerol.

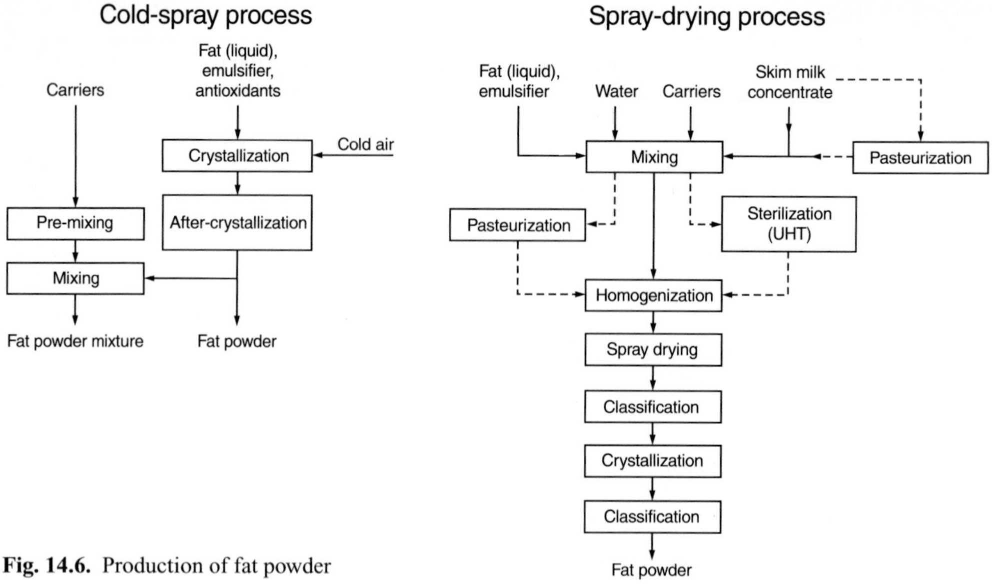

Cold-spray process

Spray-drying process

Fig. 14.6. Production of fat powder

In a spray-drying process, the fat is homogenized with emulsifiers, water and skim milk, spray dried and subsequently crystallized.

14.5 Analysis

14.5.0 Scope

The problems and scope of fat or oil analysis include identification of the type, determination of the composition of the blend, detection of additives, antioxidants, color pigments, and extraneous contaminants (solvent residues, pesticides, trace metals, mineral oils, plasticizers). In addition, the scope of analysis encompasses determination of other quality parameters, such as the extent of lipolysis, autoxidation or thermal treatments. Also of interest is the extent of refining which the fats and oils have been subjected to as well as detection of hardened fat and products which were interesterified.

14.5.1 Determination of Fat in Food

The methods used for determination of fat or oil in food are often based on extraction with either ethyl ether or petroleum ether and gravimetric determination of the extraction residue. These methods may provide unreliable or incorrect results, particularly with food of animal origin. As shown in Table 14.19, where a corned beef sample was analyzed, the amount and composition of fatty acids in the fat residue were influenced greatly by the analytical methods used. In addition to the accessible free lipids, the emulsifiers present and the changes induced by autoxidation affect the amount of extractable lipids and the lipid-to-nonlipid ratio in the residue. The use of a standard method still does not eliminate the disadvantages shown by analytical methods of fat analysis. Therefore, in questionable cases, quantitative determination of fatty acids and/or glycerol is recommended.

A rapid and accurate determination of fats or oils in food is achieved by IR- (cf. 15.3.1) and ^{1}H-NMR spectrometry. The method is based on the fact that hydrogen nuclei in fluids respond to substantially higher magnetic resonance effects than do immobilized hydrogen atoms of solid substances. Thus, the ^{1}H-NMR signal of a fluid, such as an oil, differs from that of a nonoil matrix, such as carbohydrate, protein or firmly-bound water. The intensity of the signal is directly proportional to the oil

Table 14.19. Determination of the fat content of canned corned beef

Analytical method	Fat content (%)[a]	Fatty acid composition (g/100 g)			
		Saturated acids	18 : 1 (9)	18 : 2 (9, 12)	18 : 3 (9, 12, 15)
1. Dried sample is extracted with ethyl ether	7.9	3.98	2.06	0.05	0.08
2. Sample is homogenized in 95% ethanol and then extracted with ether	15.8	4.0	2.60	0.77	0.32
3. Sample is hydrolyzed with 4 mol/l HCl (at 60°C for 30 min), then extracted with ether	12.3	5.66	3.94	0.95	0.71
4. Sample is hydrolyzed with conc. HCl (at 100°C for 1 h), methanol added and then extracted with carbon tetrachloride	13.9	2.45	1.68	0.34	0.21
5. Sample is homogenized in chloroform methanol mixture (2 : 1 v/v), washed with water and then the chloroform phase recovered	11.2	4.89	3.31	0.85	0.39

[a] The fat is determined gravimetrically after the solvent is evaporated.

content. This method is also of great value in oilseed selection or breeding research, since it permits determination of the oil content in a single kernel without damaging it by grinding or drying, i.e. retaining its ability to germinate.

The proportion of solid to fluid triacylglycerols in fat can also be determined using ^{1}H-NMR spectrometry.

14.5.2 Identification of Fat

14.5.2.1 Chemical Constants

For both, the identification and the determination of the quality of a fat or oil, the older lipid chemistry defines a series of so-called chemical constants in which the reagent uptake is used to quantitatively estimate the selected functional groups or calculate the constituents of a fat or oil. The introduction of new analytical methods, such as gas chromatography of fatty acids and the HPLC of triacylglycerols (cf. 3.3.1.4), has made many of these constants obsolete. The constants which are still used to differentiate fats or oils are:

Saponification Number (SN). This is the weight of KOH (in mg) needed to hydrolyze 1 g of fat or oil under standardized conditions. The higher the SN, the lower the average molecular weight of the fatty acids in the triacylglycerols (for examples, see Table 14.20).

Acid Value (AV). This value is important for a first quick characterization of the quality of a fat. It is the number of milligrams of KOH needed to neutralize the organic acids present in 1 g of fat.

Iodine Number (IN). This number is the number of grams of halogen, calculated as iodine, which bind to 100 g fat (cf. 3.2.3.2.1). The halogen uptake by fat or oil is affected by the contents of oleic (IN: 89.9), linoleic (IN: 181) and linolenic (IN: 273) acids. Examples of iodine numbers are provided in Table 14.20.

Hydroxyl Number (OHN). This number reflects the content of hydroxy fatty acids, fatty alcohols, mono- and diacylglycerols and free glycerol.

Table 14.20. Iodine (IN) and saponification numbers (SN) of various edible fats and oils

Oil/fat	IN	SN	Oil/fat	IN	SN
Coconut	256	9	Rapeseed		
Palm kernel	250	17	(turnip)	225	30
Cocoa	194	37	Sunflower	190	132
Palm	199	55	Soya	192	134
Olive	190	84	Butter	225	30
Peanut	192	156			

Table 14.21. Color reactions for fat and oil identification

Reaction according to [a]	Identification of
Baudouin (furfural and hydrochloric acid)	Sesame oil
Halphen (sulfur and carbondisulfide)	Cottonseed oil
Fitelson[b] (sulfuric acid and acetic acid anhydride)	Teaseed oil

[a] Reagents are listed in brackets.
[b] It is a modification of *Liebermann-Burchard* reaction for sterols (cf. 3.8.2.4).

Table 14.22. Fatty acid indicators suitable for determination of fat and oil origin

Fatty acid	Content (%)[a]	Indicator of
4 : 0	3.7	Milk fat
12 : 0	45	Coconut-, palm kernel-, and babassu fat
18 : 1 (9)	65–85[b]	Teaseed-, olive- and hazelnut oil
18 : 3 (9, 12, 15)	9	Soya-, rapeseed (also erucic acid free) oil
18 : 2 (9, 12)	50–70	Sunflower-, corn germ-, cottonseed-, wheat germ-, and soya oil
22 : 0	3	Peanut oil
20 : 4 (5, 8, 11, 14)	0.1–0.6	Animal fat
18 : 1 (9, 12-OH)	80	Castor bean oil
Trans-fatty acids		Partially or fully hydrogenated oil/fat[c]
Methyl-branched fatty acids	0.2–1.6	Animal fat[d]

[a] When value range is omitted fatty acid percentage composition is given as an average value.
[b] A high percentage of this acid is a characteristic indicator.
[c] Here precautions are needed: animal fat, e. g. from beef, might contain up to 10% trans fatty acids.
[d] It is relatively high in marine oils (approx. 1%).

14.5.2.2 Color Reactions

Some oils give specific color reactions caused by particular ingredients. Examples are summarized in Table 14.21. Since many specific nonfat components are removed from oils by refining, these tests are negative when applied to refined oils.

14.5.2.3 Composition of Fatty Acids and Triacylglycerides

The acyl residues of an acylglycerol are released as methyl esters (cf. 3.3.1.3) and are analyzed as such by gas chromatography. However, free fatty acid analysis is also possible by using specially selected stationary solid phases. Capillary-column gas chromatography should be used to differentiate between cis and trans fatty acids, which is required for the detection of partially hydrogenated fats. The fatty acids indicative of the identity or type of fat or oil are summarized in Table 14.22. An enrichment step must precede gas chromatographic separation when fatty acids of analytical significance are present as minor constituents.

Prior to the enrichment step, specific techniques such as "argentation" chromatography (cf. 3.2.3.2.3) or fractionation by urea-adduct formation (cf. 3.2.2.3) are carried out in addition to the usual preparative chromatographic procedure. The methyl branched fatty acids in marine oils are an example. These acids are enriched by the urea-adduct inclusion method since, unlike straight-chain acids, they are unable to form inclusion compounds. These branched-chain fatty acids do not change during hydrogenation, hence they can be used as marine oil indicators, i. e. to reveal the presence of marine oil in hydrogenated vegetable oils such as margarine. Another example is the use of gas chromatography to determine furan fatty acids in soybean oils (cf. Table 3.9), which is possible only after enrichment in an urea filtrate.

In the interpretation of the results of fatty acid analyses, it should be taken into account that the fatty acid composition is subject to considerable variations. It depends on the breed and feed in the case of animal fats, and on the plant variety, geographic location of the area of cultivation, and the climate in the case of plant fats. Therefore, guide values have been set for individual oils and fats (cf. Table 14.23), which can differ from country to country.

The ratio of the content of a fatty acid in position 2 of the triacylglycerides to its total content (E factor, E = enrichment) is independent of the origin of the plant oil. After hydrolysis of the fat with pancreatic lipase, separation of the 2-monoglycerides, and their methanolysis, the concentration in position 2 is determined by gas chromatography and the E-factor calculated (examples for linoleic acid are shown in Table 14.24). Adulteration of olive oils with ester oils is shown by an increased E-factor for palmitic acid (cf. 3.3.1.4).

In many cases, the triacylglyceride pattern is more expressive than the fatty acid composition. This pattern can be quickly and easily determined with the help of HPLC and GC (cf. 3.3.1.4). An example is the detection of for-

eign fat in milk fat. From extensive data on the triacylglyceride composition (GC differentiation according to the C-number), formulas have been developed which permit the detection of all important plant and animal fats up to a limiting value of 2–5 percent by weight. The older method, which is based on a decrease in the butyric acid concentration due to the foreign fat, does not safely detect an addition of 20 percent by weight because of the biological variation (3.5–4.5 w/w percent 4:0).

14.5.2.4 Minor Constituents

Some fats which can not be unequivocally distinguished by their fatty acid or triacylglyceride composition may be identified by analysis of the unsaponifiable minor constituents. Examples are given in Table 14.25.

The detection of adulteration of oils and fats has been improved further by coupled HPLC and GC of the minor constituents. The saponification of the sample is not required, free and esterified compounds being detected separately.

An example is the differentiation between the olive oil qualities "*extra vierge*" and "*lampante*". After esterification of the free OH-groups with pivalic acid, the free fat alcohols, wax esters, free acids, triterpene alcohols and

Table 14.23. Fatty acid composition of sunflower oil

Fatty acid	Per cent by weight	
	Average	Variation range [a]
16:0	6.2	3.0–10.0
16:1	0.08	< 0.1
18:0	4.75	1.0–10.0
18:1 (9)	19.8	14–65
18:2 (9, 12)	67.0	20–75
18:3 (9, 12, 15)	0.08	< 0.7
20:0	0.34	< 1.5
20:1	0.15	< 0.5
22:0	0.89	< 1.0

[a] German guide values.

Table 14.24. E-factor of various oils for linoleic acid

Oils/fats	E-factor
Sunflower	1.2
Corn	1.3
Soybean	1.3
Rapeseed	1.7
Peanut	1.7

Table 14.25. Fat or oil identification by analysis of unsaponifiable constituents

Analysis	Identification
Squalene	Olive or rice oil and fish liver oil
Campesterol/stigmasterol[a] (cf. 3.8.2.3.1)	Cocoa butter substitutes
Carotene (cf. 3.8.4.5)	Raw palm oil
γ-/β-Tocopherol[b] (cf. Table 3.51)	Corn oil
γ-Tocopherol (cf. Table 3.51)	Wheat germ oil
α-/γ-Tocopherol[b] (cf. Table 3.51)	Sunflower oil
γ-/δ-Tocopherol[b] (cf. Table 3.51)	Soybean oil
Cholesterol[c] (cf. 3.8.2.2.1)	Animal fat

[a] Concentration ratios are characteristic.
[b] Concentratiion of individual compounds and their concentration ratios are characteristic.
[c] Cholesterol concentration must exceed by 5% the total sterol fraction.

esters are eluted in a relatively narrow fraction in HPLC and separated from the triacylglycerides. The eluate is transferred to a gas chromatograph and analyzed on an apolar capillary column. As shown in Fig. 14.7, a clear distinction is made between *"lampante"* oils and *"extra vierge"* oils because the former have high contents of wax and sterol esters (sitosterol, 24-methylene cycloartenol) (cf. 14.3.2.1.1).

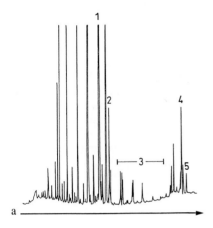

a

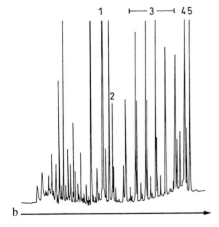

b

Fig. 14.7. On-line HPLC-GC of sterol and wax fractions of olive oils. **a** *"Extra vierge"* oil, **b** *"lampante"*oil. Peak 1: sitosterol, peak 2: 24-methylene cycloartenol, peak group 3: wax esters, peak 4: sitosterol ester, peak 5: 24-methylene cycloartenolester (according to *Grob* et al., 1991)

14.5.2.5 Melting Points

In addition to specific density, index of refraction, color and viscosity, the melting properties can be used to identify fats and oils.

The composition and the crystalline forms (cf. 3.3.1.2) of triacylglycerols present in fat determine the melting points and the temperature range over which melting occurs. The onset, flow point and end point of melting are of interest. They are determined by standardized procedures.

The melting properties of fat are more accurately determined by differential thermal analysis. The temperature difference is measured between the fat sample and a blank, i.e. a thermally inert substance, as a function of the heating temperature (Fig. 14.8). In this way the temperatures at which polymorphic transitions of fat occur are detectable. In addition, the content of solid triacylglycerols can be assessed from the heat absorbed during melting at various temperatures. Thus, the solid triacylglycerol (TG) portion of coconut oil at $-3\,^{\circ}$C can be calculated using data from the recorded curve (Fig. 14.8) and the following formula:

$$\% \text{ Coconut (solid TG)} = \frac{\text{Area (BCDE)}}{\text{Area (AEDA)}} \cdot 100$$
$$(14.11)$$

The solid: liquid ratio of acylglycerols is of importance in fat hydrogenation and interesterification processes (cf. 14.4.3). This ratio can also be assessed using the Solid Fat Index by measuring the expansion of the fat, i.e. the volume increase of a fat during its transition from solid to liquid.

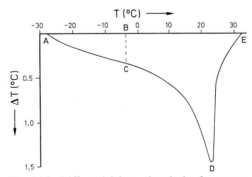

Fig. 14.8. Differential thermal analysis of a coconut fat

14.5.3 Detection of Changes During Processing and Storage

Processes used in recovery and refining and subsequent storage conditions are the main factors affecting the quality of edible fat or oil. A number of analytical methods are available for assessing the quality and deterioration of fat or oil.

14.5.3.1 Lipolysis

The extent of lipolysis (cf. 3.7.1) is determined by the free fatty acid content (FFA or Acid Value). Oils with FFA content exceeding 1% are commonly designated as crude oils, while lard with this level of free acids is considered spoiled. An exception is olive oil, which is still considered suitable for direct consumption even with a 3% FFA content. The FFA content is lowered to less than 0.1% by refining of oil or fat.

There is no relationship between the sensory perception of quality deterioration and the levels of FFA in fats which contain low-molecular acyl residues (e.g., milk, coconut, and palm kernel fats) because among the free fatty acids, the sensory-relevant compounds (C number < 14) usually take second place. A better correlation is provided by the analysis of the low-molecular free fatty acids (cf. 3.7.1.1). They are first separated from the fat, e.g., with an ion exchanger, released from the exchanger by esterification with ethyl iodide, and then determined by gas chromatography.

14.5.3.2 Oxidative Deterioration

Fats and oils deteriorate rather rapidly by autoxidation of their unsaturted acyl residues (cf. 3.7.2.1). A number of analytical methods have been developed to determine the extent of such deterioration and to predict the expected shelf life of a fat or oil.

14.5.3.2.1 Oxidation State

Peroxide Value. The method for determination of peroxide concentration is based on the reduction of the hydroperoxide group with HI or Fe^{2+}. The result of the iodometric titration is expressed as the peroxide value. The Fe^{2+} method is more suitable for determining a low hydroperoxide concentration since the amount of the resultant Fe^{3+}, in the form of the ferrithiocyanate (rhodanide) complex, is deter-

mined photometrically with high sensitivity (Fe-test in Table 14.26). The peroxide concentration reveals the extent of oxidative deterioration of the fat, nevertheless, no relationship exists between the peroxide value and aroma defects, e.g. rancidity (already existing or anticipated). This is because hydroperoxide degradation into odorants is influenced by so many factors (cf. 3.7.2.1.5) which make its retention by fat or oil or its further conversion into volatiles unpredictable.

Carbonyl Compounds. The analysis of the compounds responsible for the rancid aroma defect of a fat or oil is of great value. Volatile carbonyls (cf. 3.7.2.1.5) are among such compounds.

In a simple test, such as benzidine, anisidine or heptanal values, the volatile aldehydes are not separated from fat or oil, rather the reaction with the group-specific reagents is carried out in the fat or oil. In addition to the odorous aldehydes, the flavorless oxo-acylglycerols and oxo-acids can be determined. Since the proportion of aroma-active and sensory neutral carbonyls is not known, any correlation found between the carbonyl value and aroma defects is clearly coincidental.

The *thiobarbituric acid test* (TBA) is a preferred method for detecting lipid peroxidation in biological systems. However, the reaction is nonspecific since a number of primary and secondary products of lipid peroxidation form

Table 14.26. Analytical aspects related to the determination of the extent of oxidation of unsaturated fatty acids: relative sensitivities of spectrophotometric procedures[a]

Method		Autoxidized fatty acid methyl esters	
		18 : 2 (9, 12)[b]	18 : 3 (9, 12, 15)[c]
UV-Absorption	234 nm	1.0	1.0
	270 nm	0.1	0.3
Fe^{2+}/Thiocyanate (rhodanide)		9.4	6.3
Thiobarbituric acid test	452 nm	0.1	0.5
	530 nm	0.1	1.0
Kreis-test		0.1	0.1
Anisidine value		0.3	0.75
Heptanal value		0.1	0.1

[a] Related to UV absorption at 234 nm.
[b] Peroxide value: 475.
[c] Peroxide value: 450.

malonaldehyde which in turn reacts in the TBA test. In food containing oleic and linoleic acids, the TBA-test is not as sensitive as the Fe^{2+}-test outlined above.

The *gas chromatographic determination* of individual carbonyl compounds appears to be a method suitable for comparison with findings of sensory panel tests. Analytical methods for the odorants causing aroma defects is still in the early stages of development because only a few fats or fat-containing foods have been examined in such detail that the aroma substances involved are clearly identified.

The well studied warmed-over flavor of cooked meat (cf. 12.6.2.1) is an example. It can be controlled relatively easily because the easy-to-determine hexanal has been identified as the most important off-flavor substance. On the other hand, the easily induced rancid aroma defect of rapeseed oil is primarily caused by the volatile hydroperoxides (1-octen-3-hydroperoxide, (Z)-1,5-octadiene-3-hydroperoxide) and (Z)-2-nonenal which are all relatively difficult to quantify. This limitation also applies to 3-methyl-2,4-nonandione, which appears as the most important off-flavor substance in soybean oil on exposure to light. To simplify the analytical procedure, individual aldehydes (e. g., hexanal, 2,4-decadienal), which are formed in larger amounts during lipid peroxidation, have been proposed as indicators. In most of the cases, however, it was not tested whether the indicator increases proportionally to the off-flavor substances which cause the aroma defect.

14.5.3.2.2 Shelf Life Prediction Test

To estimate susceptibility to oxidation, the fat or oil is subjected to an accelerated oxidation test under standardized conditions so that the signs of deterioration are revealed within several hours or days. Examples of such tests are the *Schaal test* (fat maintained at 60 °C) and the *Swift stability test* (fat kept at 97.8 °C and aerated continuously). The extent of oxidation is then measured by sensory and chemical tests such as peroxide value (cf. 3.7.2), ultraviolet absorption (suitable for fats and oils containing linoleic or linolenic acids) or oxygen uptake. There are also methods based on the fact that in the process of triglycerol

oxidation, when the initiation period is terminated, large amounts of low molecular weight acids are released. They are then determined electrochemically. During oxidation of a given fat or oil sample, a good correlation exists between the length of the induction period and the shelf-life.

14.5.3.3 Heat Stability

The behavior of a frying oil, when heated, is assessed from the content of oxidized fatty acids which are insoluble in petroleum ether and from the smoke point (cf. 3.7.4) of the fat or oil. The smoke point of a fat or oil is the temperature at which its triacylglycerols start to decompose in the presence of air. Smoke is the sign of decomposition. The smoke point of a fat or oil is normally in the range of 200–230 °C during prolonged frying, and it decreases in the presence of decomposition products. When it falls below 170 °C, the fat is considered to be spoiled. At this point, the amount of fatty acids which are insoluble in petroleum ether exceeds 0.7%. However, this petroleum ether method is not reproducible. Fat separation by column chromatography is more reliable. The heated fat or oil is separated into a polar and a nonpolar fraction using silicic acid as an adsorbent. The value of 0.7% oxidized petroleum ether-insoluble fat corresponds, in this chromatographic separation, to 73% nonpolar and 27% polar fractions.

14.5.3.4 Refining

The addition of a refined oil to natural plant oil is detected by the determination of substances which can be formed during bleaching and deodorization (cf. Table 14.27).

Table 14.27. Indicators of refined oils and fats

Refining step	Indicator	Remarks
Bleaching (cf. 14.4.1.4)	a. Fatty acids with conjugated triene systems	Determination of "b" is more reliable than the UV measurement of "a"
	b. Disterylether (> 0.5 mg/kg)	
Deodorization (cf. 14.4.1.5)	a. Dimeric and oligomeric triacylglycerides	Unlike "a", the indicators "b" appear even on relatively gentle deodorization
	b. Position and substitution isomers of linoleic acids	

14.6 Literature

Baltes, J.: Gewinnung und Verarbeitung von Nahrungsfetten. Verlag Paul Parey: Berlin. 1975

DGF-Einheitsmethoden. Ed. Deutsche Gesellschaft für Fettwissenschaft e.V., Münster/Westf., Wissenschaftliche Verlagsgesellschaft mbH: Stuttgart. 1950–1979

Gertz, C.: Native und nicht raffinierte Speisefette und -öle. Fat Sci. Technol. *93*, 545 (1991)

Grob, K., Artho, A., Mariani, C.: Gekoppelte LC-GC für die Analyse von Olivenölen. Fat Sci. Technol. *93*, 494 (1991)

Grosch, W., Tsoukalas, B.: Analysis of fat deterioration – Comparison of some photometric tests. J. Am. Oil Chem. Soc. *54*, 490 (1977)

Guhr, G., Waibel, J.: Untersuchungen an Fritierfetten; Zusammenhänge zwischen dem Gehalt an petrolätherunlöslichen Fettsäuren und dem Gehalt an polaren Substanzen bzw. dem Gehalt an polymeren Triglyceriden. Fette Seifen Anstrichm. *80*, 106 (1978)

Hamilton, R.J., Rossell, J.B. (Eds.): Analysis of oils and fats. Elsevier Applied Science Publ.: London. 1986

Hoffmann, G.: The chemistry and technology of edible oils and fats and their high fat products. Academic Press: London. 1989

Kiritsakis, A., Markakis, P.: Olive oil – a review. Adv. Food Res. *31*, 453 (1987)

Kroll, S.: Margarine und Backfette. In: Ullmanns Encyklopädie der technischen Chemie, 4. edn., vol. 16, p. 481, Verlag Chemie: Weinheim. 1978

Official methods and recommended practices of the American Oil Chemists' Society, Firestone, D. (Ed.), 4th edn. American Oil Chemists' Society, Champaign, 1989

Pardun, H.: Analyse der Nahrungsfette. Verlag Paul Parey: Berlin. 1976

Precht, D.: Detection of adulterated milk fat by fatty acid and triglyceride analysis. Fat Sci. Technol. *93*, 538 (1991)

Schulte, E., Weber, N.: Disterylether in gebleichten Fetten und Ölen – Vergleich mit den konjugierten Trienen. Fat Sci. Technol. *93*, 517 (1991)

Sheppard, A.J., Hubbard, W.D., Prosser, A.R.: Evaluation of eight extraction methods and their effects upon total fat and gas liquid chromatographic fatty acid composition analyses of food products. J. Am. Oil Chem. Soc. *51*, 417 (1974)

Stansby, M.E.: Fish oils. AVI Publ. Co.: Westport, Conn. 1967

Swern, D. (Ed.): Bailey's industrial oil and fat products. 4th edn., Vol. 1, John Wiley and Sons: New York. 1979

Thomas, A.: Fette and Öle. In: Ullmanns Encyklopädie der technischen Chemie, 4. edn., vol. 11, p. 455, Verlag Chemie: Weinheim. 1976

Usuki, R., Suzuki, T., Endo, Y., Kaneda, T.: Residual amounts of chlorophylls and pheophytins in refined edible oils. J. Am. Oil Chem. Soc. *61*, 785 (1984)

15 Cereals and Cereal Products

15.1 Foreword

15.1.1 Introduction

Cereal products are amongst the most important staple foods of mankind. Nutrients provided by bread consumption in industrial countries meet close to 50% of the daily requirement of carbohydrates, one third of the proteins and 50–60% of vitamin B. Moreover, cereal products are also a source of minerals and trace elements.

The major cereals are wheat, rye, rice, barley, millet and oats. Wheat and rye have a special role since only they are suitable for bread-making.

15.1.2 Origin

The genealogy of the cereals begins with wild grasses (*Poaceae*), as shown in Fig. 15.1. Barley (*Hordeum vulgare*), probably one of the first cereals grown systematically, was known as early as 5000 B.C. in Egypt and Babylon. Also, the bearded wheat cultivars from the groups Einkorn (*Triticum monococcum*) and Emmer (*T. dicoccum*), with diploid (genome formula: AA, 2n = 14) and tetraploid (AABB, 2n=28) sets of chromosomes, (the chromosome number of the wheat genome is n=7), were found among cultivated plants that were widely spread in temperate zones of Euroasia during the neolithic period. These cultivars are becoming extinct. Only the durum form of Emmer (*T. turgidum durum*, hard wheat, AABB), at 10% of the total wheat grown, has a significant role. The hexaploid (AABBDD, 2n = 42) bare wheats (*T. aestivum* L., soft, bread wheats) derived from Dinkel (*T. spelta* L.) are grown worldwide. Archeological studies have shown that *T. aestivum* originated about 10,000 years ago. Its A genome is closely related to that of Einkorn (*T. monococcum*). The origin of the B genome is unknown. It probably comes from species of the genus *Aegilops* and the D genome from *Aegilops squarrosa*.

Rice (*Oryza sativa*) and corn (*Zea mays*) have been cultivated for 5,000 years, first in tropical Southeast Asia and then in Central and South America. Cereals designated as millet have had a role from antiquity in subtropical and tropical regions of Asia and Africa. True millet from the subfamilies *Eragrostoideae* and *Panicoideae*, to which many regionally important cultivars belong (for instance, *Eragrostis tef, Eleusine coracan, Echinochloa frumentacea, Pennisetum glaucum, Setaria italica*), is distinguished from sorghum (*Sorghum bicolor*), which belongs to the subfamily *Andropogonoideae* and is cultivated worldwide.

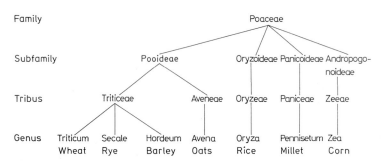

Family			Poaceae			
Subfamily		Pooideae		Oryzoideae	Panicoideae	Andropogonoideae
Tribus		Triticeae	Aveneae	Oryzeae	Paniceae	Zeeae
Genus	Triticum Wheat	Secale Rye	Hordeum Barley	Avena Oats	Oryza Rice	Pennisetum Millet	Zea Corn

Fig. 15.1. Evolutionary development (phylogeny) of cereals

Rye (*Secale cereale*) and oats (*Avena sativa*) are so-called secondary culture plants. Initially hardy and unwanted escorts of cultivated plants, they prospered and established themselves in northern regions with unfavorable climates. Their high tolerance for unfavorable climates surpassed that of both wheat and barley. Rye and oats have been cultivated for millenia.

Breeders have for many years attempted to combine the baking quality of wheat with the hardiness of rye. *Triticale*, the man-made hybrid of wheat and rye, does not yet fulfill this aim, hence its economic significance is low.

15.1.3 Production

Cereals are of great importance as raw materials for production of food and feed. Accordingly, they are grown on close to 60% of cultivated land in the world. Wheat production takes up the greatest part of land cultivated with cereals (Table 15.1) and wheat is produced in the largest quantity (Table 15.2). Wheat surplus producers are the USA, Canada, Argentina, Australia, and France. In the Federal Republic of Germany (FRG), winter wheat (92%) and spring wheat (8%) are both cultivated. The rise of cereal production in the world is shown in Table 15.3. The yields per hectare vary greatly from one country to another (Table 15.4). Due to an intensive effort in breeding and crop production programs, the yields per hectare in the FRG are very high and are surpassed by only a few countries, e.g., Holland. The FRG utilized 25.7×10^6 tons of cereals in 1976/77, of which 38% was bread and 62% feed cereals.

Table 15.1. Land cultivated with a cereal crop as % of the world total area under cereal cultivation (1979: 7.6×10^8 ha)

Cereal	1966	1976	1984	1988	1990	1996
Wheat	30.6	31.5	34.5	31.2	32.7	32.4
Rice	18.8	19.2	21.9	20.9	20.6	21.2
Corn	15.5	15.9	19.3	18.2	18.3	19.7
Millet	15.4[a]	15.6[a]	12.3[a]	5.7	5.3	5.1
Sorghum				6.6	6.3	6.6
Barley	12.2	11.9	11.7	10.8	10.1	9.4
Oats	4.5	3.8	3.8	3.2	3.1	2.4
Rye	2.4	2.1	2.6	2.3	2.3	1.6

[a] Sum of millet and sorghum.

Table 15.2. Cereal production in 1996 (1,000 t)

Continent	Wheat	Rice	Barley	Corn	Rye
World	584,874	562,259	155,261	576,821	23,156
Africa	22,801	16,029	8,486	43,982	28
America, North-, Central-	96,181	9,879	25,169	263,885	551
America, South-	21,234	18,282	1,358	48,746	50
Asia	242,106	513,733	24,205	155,962	987
Europe	178,793	3,366	89,583	63,764	21,519
Oceania	23,759	970	6,460	482	21

Continent	Oats	Millet	Sorghum	Cereals, grand total
World	30,967	29,563	69,000	2,049,578
Africa	243	12,881	20,852	127,584
America, North-, Central-	6,721	180	25,774	493,378
America, South-	705	45	3,508	94,049
Asia	1,375	15,795	16,679	973,122
Europe	20,263	635	632	389,960
Oceania	1,661	27	1,556	35,485

Table 15.2 (continued)

Country	Wheat	Country	Rice	Country	Barley
China	109,005	China	190,100	Canada	15,912
India	62,620	India	120,100	Russian Fed.	15,900
USA	62,099	Indonesia	51,165	Germany	12,074
France	35,946	Bangladesh	28,008	Spain	10,660
Russian Fed.	34,900	Vietnam	26,300	France	9,463
Canada	30,495	Thailand	21,800	USA	8,640
Australia	23,497	Myanmar	20,865	Turkey	8,000
Germany	18,922	Japan	13,000	UK	7,765
Turkey	18,515	Philippines	11,284	Australia	6,075
Pakistan	16,907	Brazil	10,035	Ukraine	5,714
UK	16,031			Denmark	4,196
Argentina	15,200	Σ (%)[a]	88	China	4,100
				Morocco	3,831
Σ (%)[a]	76			Poland	3,583
				Σ (%)[a]	75

Country	Corn	Country	Rye	Country	Oats
USA	236,064	Poland	5,927	Russian Fed.	8,570
China	119,350	Russian Fed.	5,900	Canada	4,374
Brazil	31,975	Germany	4,214	USA	2,253
Mexico	17,300	Belarus	1,794	Germany	1,650
France	14,449	Ukraine	1,093	Australia	1,616
Argentina	10,466	China	700	Poland	1,588
South Africa	10,351	Denmark	369	Finland	1,261
Romania	9,600	Lithuania	324	Sweden	987
Indonesia	8,925	Canada	322	Ukraine	739
Italy	8,712	Spain	272	Belarus	707
Σ (%)[a]	81	Σ (%)[a]	90	Σ (%)[a]	77

Country	Millet	Country	Sorghum	Country	Cereals, grand total
India	10,500	USA	20,379	China	435,654
Nigeria	5,681	India	10,500	USA	337,667
China	4,501	Nigeria	7,084	India	214,082
Niger	1,832	China	5,098	Russian Fed.	68,030
Burkina Faso	785	Mexico	4,817	France	62,488
Mali	707	Sudan	4,104	Indonesia	60,090
Senegal	650	Argentina	2,132	Canada	59,407
Uganda	640	Ethiopia	1,980	Brazil	46,101
Russian Fed.	500	Australia	1,555	Germany	42,102
Sudan	491	Burkina Faso	1,314	Australia	34,602
				Argentina	29,554
Σ (%)[a]	89	Σ (%)[a]	85	Bangladesh	29,445
				Turkey	29,342
				Vietnam	27,296
				Mexico	26,846
				Thailand	26,426
				Σ (%)[a]	75

[a] World production = 100 %.

Table 15.3. World production of cereals 1948–1996 (10^6 t)

Year	Amount	Year	Amount
1948	683	1984	1,802
1956	789	1988	1,742
1964	1,019	1989	1,881
1968	1,180	1990	1,955
1976	1,456	1996	2,050

15.1.4 Anatomy – Chemical Composition, a Review

Cereals, in contrast to forage grasses, form a relatively large fruit, termed a caryopsis, in which the fruit shell is strongly bound to the seed shell. The kernel size, which is expressed as grams per 1000 kernels (Table 15.5), is not only dependent on the kind of cereal but on the cultivar and crop production techniques, hence it varies widely.

In oats, barley, and rice the front and back husks are fused together with the fruit. In contrast, threshing separates wheat and rye kernels from the husks as bare seed.

The major constituents of seven kinds of cereal are fairly uniform (Table 15.6). Noteworthy variations are the higher lipid content of oats and a lower starch content in oats, barley and rye. Instead of starch, in these cereals the "other carbohydrates" are increased and consist mainly of nonstarchy polysaccharides (cf. 15.2.4.2). These cereals also differ in their vitamin B content (Table 15.6).

Table 15.5. Average thousand kernel weight of cereals (g)

Wheat	37	Oats	32
Rye	30	Barley	37
Corn	285	Millet	23
Rice	27		

Table 15.6. Chemical composition of cereals (average values)

	Wheat	Rye	Corn	Barley	Oats	Rice	Millet
	weight %						
Moisture	13.2	13.7	12.5	11.7	13.0	13.1	12.1
Protein (N×6.25)	11.7	11.6	9.2	10.6	12.6	7.4	10.6
Lipids	2.2	1.7	3.8	2.1	5.7	2.4[a]	4.05
Starch	59.2	52.4	62.6	52.2	40.1	70.4	64.4
Other carbohydrates	10.1	16.6	8.4	19.6	22.8	5.0	6.3
Crude fiber	2.0	2.1	2.15	1.55	1.56	0.67	1.1
Minerals	1.5	1.9	1.30	2.25	2.85	1.2	1.6
	mg/kg						
Thiamine	5.5	4.4	4.6	5.7	7.0	3.4	4.6
Niacin	63.6	15.0	26.6	64.5	17.8	54.1	48.4
Riboflavin	1.3	1.8	1.3	2.2	1.8	0.55	1.5
Pantothenic acid	13.6	7.7	5.9	7.3	14.5	7.0	12.5

[a] Polished rice: 0.8%.

Fruit and seed coats enclose the nutrient tissue (endosperm) and germ in the kernel (Fig. 15.2). Botanically the endosperm consists of the starchy endosperm (70–80% of the kernel; Table 15.7) and the aleurone layer, which, with exception of barley, is a single cell layer. The

Table 15.4. Yield per hectare in 1989–91/1994/1995/1996 (100 kg/ha)

	Wheat	Rye	Corn	Rice
Germany	62.4/67.6/68.9/72.9	41.1/47.8/52.3/52.1	70.5/70.8/73.7/78.6	–.–/ –.–/ –.–/ –.–
Argentina	19.9/21.7/19.2/23.0	8.3/ 9.6/ 7.8/ 7.8	34.7/42.4/45.2/40.1	41.2/43.0/50.3/50.5
Australia	15.6/11.4/17.7/21.4	6.9/ 6.7/ 6.4/ 6.4	41.2/46.6/48.3/52.0	86.5/83.6/85.4/67.9
China	31.1/34.3/35.4/37.6	17.3/20.0/14.0/14.0	43.3/47.0/49.2/51.7	56.1/58.3/60.2/60.6
France	65.0/66.7/65.1/71.3	36.4/39.1/40.8/45.6	67.2/79.1/77.2/83.6	57.8/45.6/49.7/52.3
India	22.2/23.8/25.6/24.9	–.–/ –.–/ –.–/ –.–	15.1/14.9/14.8/14.1	26.2/28.8/27.8/28.1
Russian Fed.	–.–/14.5/12.6/14.0	–.–/15.4/12.7/13.3	–.–/17.0/27.0/13.6	–.–/27.1/27.0/27.6
Turkey	20.0/17.9/19.2/19.8	14.1/13.4/16.4/16.6	40.9/38.1/36.9/36.4	49.7/49.4/50.0/51.5
USA	23.4/25.3/24.1/24.4	16.7/17.5/16.4/16.3	71.8/87.1/71.2/79.8	63.6/66.9/63.0/68.5
World	24.6/24.5/24.7/25.4	21.4/20.6/21.7/20.5	36.7/41.4/37.9/41.2	35.1/36.7/36.9/37.3

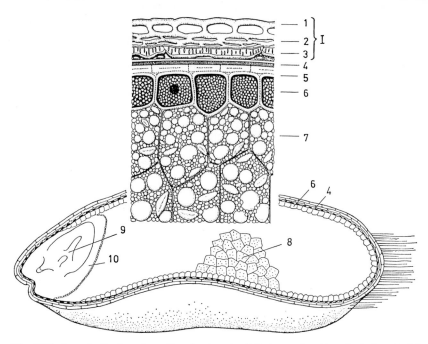

Fig. 15.2. Longitudinal section of a wheat grain. I Pericarp, 1 epidermis (epicarp), 2 hypodermis, 3 tube cells, 4 seed coat (testa), 5 nucellar tissue, 6 aleurone layer, 7 outer starchy endosperm cells, 8 inner starchy endosperm cells, 9 germ and 10 scutellum

Table 15.7. Fractions of various cereals separated by milling (average weight-%)

Cereal variety	Husk	Bran	Germ	Endosperm
Wheat	0	15.0	2.0	83.0
Corn	0	7.2	11.0	81.8
Oats	20	8.0	2.0	70.0
Rice	20	8.0	2.0	70.0
Millet	0	7.9	9.8	82.3

aleurone layer is rich in protein and also contains high amounts of fat, enzymes and vitamins (Table 15.8 and 15.9). The proteins, of which half are water-soluble, appear as granules in the aleurone cells. They have no influence on the baking properties of wheat. Millers regard the aleurone layer as part of the bran.

The starchy endosperm is the source of flour. Its thin-walled cells are packed with starch

Table 15.8. Chemical composition of anatomical parts of a wheat kernel (average weight-% on dry weight basis)

	Ash	Crude protein (N × 6.25)	Lipids	Crude fiber[a]	Cellulose	Pentosans	Starch
Longitudinal cells	1.3	3.9	1.0	27.7	32.1	50.1	–
Cross- and tube cells	10.6	10.7	0.5	20.7	22.9	38.9	–
Fruit and seed coatings	3.4	6.9	0.8	23.9	27.0	46.6	–
Aleurone cells[b]	10.9	31.7	9.1	6.6	5.3	28.3	–
Germ[b]	5.8	34.0	27.6	2.4	–	–	–
Starchy endosperm	0.6	12.6	1.6	0.3	0.3	3.3	80.4

[a] Crude fiber includes parts of cellulose and pentosans.   [b] Data for carbohydrates are incomplete.

Table 15.9. Mineral and vitamin distribution as % in kernel fractions of wheat

Fraction	Min-erals	Thia-mine	Ribo-flavin	Niacin	Pyri-doxal phos-phate	Panto-thenic acid
Fruit coat	7	1	5	4	12	9
Germ	12	64	26	2	21	7
Aleurone layer	61	32	37	82	61	41
Starchy endosperm	20	3	32	12	6	43

granules which lie imbedded in a matrix which is largely protein. A portion of these proteins, the gluten proteins, is responsible for the baking properties of wheat. The concentrations of the proteins and some other constituents (vitamins and minerals) decrease from outer to inner cells of the endosperm. The germ is separated from the endosperm by the scutellum. The germ is rich in enzymes and lipids (Table 15.8). Table 15.9 shows that wheat milling, when starchy endosperm cells are separated from germ and bran, results in a substantial loss of B-vitamins and minerals.

15.1.5 Special Role of Wheat – Gluten Formation

After addition of water a viscoelastic cohesive dough can be kneaded only from wheat flour. The resulting gluten, which can be isolated as a residue after washing out the dough with water, removing starch and other ingredients, is responsible for plasticity and dough stability.

Gluten consists of 90% protein (cf. 15.2.1.3), 8% lipids and 2% carbohydrates. The latter are primarily the water-insoluble pentosans (cf. 15.2.4.2.1), which are able to bind and hold a significant amount of water, while the lipids (cf. 15.2.5) form a lipoprotein complex with certain gluten proteins. In addition, enzymes such as proteinases and lipoxygenase are detectable in freshly isolated gluten.

The gluten proteins, in association with lipids, are responsible for the cohesive and viscoelastic flow properties of dough. Such rheological properties give the dough gas-holding

capacity during leavening and provide a porous, spongy product with an elastic crumb after baking.

Rye and other cereals can not form gluten. The baking quality of rye is due to pentosans and to some proteins which swell after acidification (cf. 15.4.2.2) and contribute to gas-holding properties.

15.1.6 Celiac Disease

Wheat, rye and barley can cause celiac disease (celiac sprue, or gluten-induced enteropathy); the role of oats in this disease is uncertain. Celiac disease affects both infants and adolescents, but rarely adults. It is associated with a loss of villous structure of the intestinal mucosa; epithelial cells exhibit degenerative changes and nutrient absorption functions are severly impaired. Incidence of the disease varies, e.g., 0.05% of the children are affected in central Europe and 0.33% in Ireland. The prolamin fractions of wheat, barley or rye are the cause of the disease, which is therefore eliminated by a change of diet to rice, millet or corn.

15.2 Individual Constituents

The role of constituents is of particular interest in the processing of wheat and rye into bakery products.

15.2.1 Proteins

15.2.1.1 Differences in Amino Acid Composition

The proteins of different cereal flours vary in their amino acid composition (Table 15.10). Lysine content is low in all cereals. Methionine is also low, particularly in wheat, rye, barley, oats and corn. Both amino acids are significantly lower in flour than in muscle, egg or milk proteins. By breeding, attempts are being made to improve the content of all essential amino acids. This approach has been successful in the case of high-lysine barley and several corn cultivars.

Table 15.10. Amino acid composition of the total proteins (mole-%) of flours from various cereals

Amino acid	Wheat	Rye	Barley	Oats	Rice	Millet	Corn
Asx	4.2	6.9	4.9	8.1	8.8	7.7	5.9
Thr	3.2	4.0	3.8	3.9	4.1	4.5	3.7
Ser	6.6	6.4	6.0	6.6	6.8	6.6	6.4
Glx	31.1	23.6	24.8	19.5	15.4	17.1	17.7
Pro	12.6	12.2	14.3	6.2	5.2	7.5	10.8
Gly	6.1	7.0	6.0	8.2	7.8	5.7	4.9
Ala	4.3	6.0	5.1	6.7	8.1	11.2	11.2
Cys	1.8	1.6	1.5	2.6	1.6	1.2	1.6
Val	4.9	5.5	6.1	6.2	6.7	6.7	5.0
Met	1.4	1.3	1.6	1.7	2.6	2.9	1.8
Ile	3.8	3.6	3.7	4.0	4.2	3.9	3.6
Leu	6.8	6.6	6.8	7.6	8.1	9.6	14.1
Tyr	2.3	2.2	2.7	2.8	3.8	2.7	3.1
Phe	3.8	3.9	4.3	4.4	4.1	4.0	4.0
His	1.8	1.9	1.8	2.0	2.2	2.1	2.2
Lys	1.8	3.1	2.6	3.3	3.3	2.5	1.4
Arg	2.8	3.7	3.3	5.4	6.4	3.1	2.4
Trp	0.7	0.5	0.7	0.8	0.8	1.0	0.2
Amide group	31.0	24.4	26.1	19.2	15.7	22.8	19.8

15.2.1.2 A Review of the Osborne Fractions of Cereals

In 1907 *T. B. Osborne* separated wheat proteins, on the basis of their solubility, into four fractions. Sequential extraction of a flour sample yielded: water-soluble albumins, salt-soluble (e.g., 0.4 mol/l NaCl) globulins, and 70% aqueous ethanol-soluble prolamins. The glutelins remained in the flour residue. They can be separated into two sub-fractions. For this purpose, all the proteins remaining in the residue are first dissolved in 50% aqueous 1-propanol at 60 °C with reduction of the disulfide bonds, e.g., with dithioerythritol. The high-molecular (HMW) subunits (cf. 15.2.1.3.1) precipitate out on increasing the propanol concentration to 60%, while the low-molecular (LMW) subunits (cf. 15.2.1.3.3) remain in solution.

Further separation of the *Osborne* fractions and subfractions into the components is possible analytically with electrophoretic methods (cf. Fig. 15.3, 15.4) and analytically and preparatively with RP-HPLC (cf. Fig. 15.5–15.8).

In the literature, *Osborne* fractions derived from different cereals are often designated by special names (cf. review Table 15.11). The various designations may result in confusion and incorrect conclusions with regard to pro-

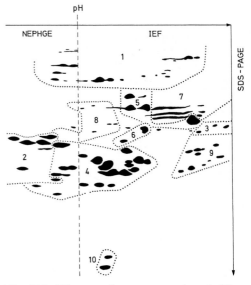

Fig. 15.3. Wheat endosperm proteins (cultivar *"Chinese Spring"*). Simplified schematic representation of a two-dimensional electrophoretic separation.

1st. Dimension: isoelectric focussing (IEF) and non-equilibrium pH gradient electrophoresis (NEPH-GE). The electropherograms obtained by both methods are put together at the broken line in such a way that a continuous pH gradient is formed.

2nd. Dimension: polyacrylamide gel electrophoresis in the presence of sodium dodecyl sulfate and mercaptoethanol (SDS-PAGE).

The following protein fractions can be recognized: high-molecular glutenin subunits (1); basic (2) and acidic (3) low-molecular glutenin subunits; α-, β-, γ- (4) and ω-gliadins (5); subunits of the triplet band (6, 10); high-molecular albumins (7); globulins (8) and nonreserve proteins (9). (according to *Payne* et al., 1985)

tein homogeneity. Therefore, it may be better to use the general designations of the *Osborne* fractions and specify the protein source, e.g., wheat glutelin instead of glutenin.

Albumins and globulins are derived mostly from cytoplasmic residues and other subcellular fractions which are part of the kernel. Thus, enzymes are present in the first two *Osborne* fractions. Prolamins and glutelins, on the other hand, are storage proteins.

Cereals contain variable levels of *Osborne* fractions (Table 15.12). Wheat has the highest content of prolamin, corn has the second high-

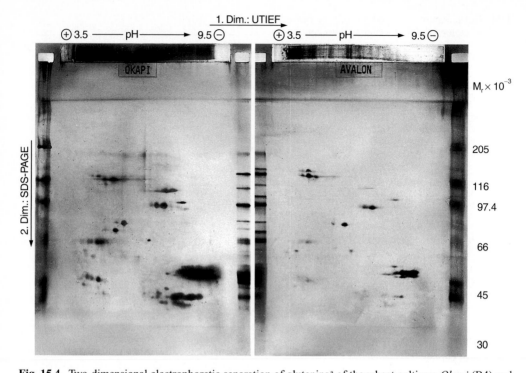

Fig. 15.4. Two-dimensional electrophoretic separation of glutenins[a] of the wheat cultivars *Okapi* (B4) and *Avalon* (A6)[b]. (according to *Krause* et al., 1988)
1st. Dimension: isoelectric focussing in ultrathin (0.25 mm) layer (UTIEF), pH 3.5–9.5; 8 mol/l urea.
2nd. Dimension: polyacrylamide gel electrophoresis in the presence of sodium dodecyl sulfate and mercaptoethanol (SDS-PAGE)

[a] Residue after extraction of defatted flour with water, salt solution, and aqueous ethanol.
[b] The baking quality class is given in brackets after the variety (bread volume yield for A6: average to high, for B4: low to average).

Table 15.11. Designations of *Osborne*-fractions

Fraction	Wheat	Rye	Oats	Barley	Corn	Rice	Millet
Albumins	Leukosin						
Globulins	Edestin		Avenalin				
Prolamins	Gliadin	Secalin	Avenin	Hordein	Zein	Oryzin	Cafirin
Glutelins	Glutenin	Secalinin		Hordenin	Zeanin	Oryzenin	

est. The albumin fraction is the highest in rye and the lowest in corn. The content of albumin in oats is comparable to that in rye. Oats and rice have a higher content of glutelin than wheat, while rye, millet and corn have a much lower glutelin content. The amino acid composition of only the prolamins (Table 15.13) can

be correlated to the botanical genealogy of cereals as shown in Fig. 15.1. In general, the amino acid composition is similar for wheat, rye and barley. The prolamin composition of oats is intermediate between *Triticeae* and the other cereals. The amount of glutamic acid in oat prolamins is similar to that of the *Triticeae*,

Table 15.12. Protein distribution (%)[a] in *Osborne*-fractions[b]

Fraction	Wheat	Rye	Barley	Oats	Rice	Millet	Corn
Albumins	14.7	44.4	12.1	20.2	10.8	18.2	4.0
Globulins	7.0	10.2	8.4	11.9	9.7	6.1	2.8
Prolamins	32.6	20.9	25.0	14.0	2.2	33.9	47.9
Glutelins[c]	45.7	24.5	54.5	53.9	77.3	41.8	45.3

[a] Calculated from amino acid analyses.
[b] Ash content of the flours (% based on dry weight), wheat (0.55), rye (0.97), barley (0.96), oats (1.87), rice (1.0), millet (1.10), and corn (0.33).
[c] Protein residue after extraction of prolamins.

whereas amounts of proline and leucine in oat prolamins are lower and higher, respectively, than those found in the *Triticeae*; this is also the case in comparison with rice, millet and corn. The amino acid compositions of rice, millet and corn are not related to the *Pooideae*.

The *Triticeae*, in which the prolamin amino acid compositions are closely related, can also cause *Celiac* disease (cf. 15.1.6). In comparison to other cereals, *Triticeae* prolamins contain significantly higher levels of glutamic acid and proline. This suggest that the difference in prolamin composition, induced by these amino acids, may be responsible for *Celiac* disease.

15.2.1.3 Protein Components of Wheat Gluten

Wheat protein fractionation by the *Osborne* method provides prolamins and glutelins in a ratio of 2 : 3. Both fractions, in hydrated form, have different effects on the rheological characteristics of dough: prolamins are responsible, preferentially, for viscosity, and glutelins for dough elasticity.

The genes for the gluten proteins occur at nine different complex loci in the wheat genome. The high molecular weight glutenin subunits are coded by the loci Glu-A1, Glu-B1 and Glu-D1, which are carried on the long arms of the chromosomes 1A, 1B and 1D. The low molecular weight glutenin subunits, the ω-gliadins and the γ-gliadins are coded by the loci Gli-A1, Gli-B1 and Gli-D1, which occur on the short arms of the chromosomes 1A, 1B and 1D. The α- and β-gliadins are coded by the loci Gli-A2, Gli-B2 and Gli-D2 on the short

Table 15.13. Amino acid composition (mole %) of the *Osborne* fractions of various cereals

Amino acid	Albumins						
	Wheat	Rye	Barley	Oats	Rice	Millet	Corn
Asx	9.7	8.8	10.2	10.2	9.9	11.0	16.7
Thr	3.8	4.0	4.7	4.4	4.6	5.0	4.4
Ser	6.2	6.2	6.4	8.9	6.5	6.3	6.2
Glx	20.9	22.1	13.8	12.4	14.2	12.1	12.4
Pro	9.3	12.0	7.4	6.1	4.6	5.1	8.6
Gly	6.9	6.6	9.7	12.6	9.8	10.0	9.7
Ala	6.9	6.5	8.2	7.6	9.4	10.5	10.0
Cys	3.2	2.3	3.8	6.8	1.9	1.4	1.8
Val	6.0	5.2	6.3	4.7	6.3	6.4	4.8
Met	1.6	1.3	2.0	1.2	1.7	2.0	1.1
Ile	3.3	3.4	3.3	2.7	3.7	3.3	3.0
Leu	6.4	6.3	5.9	5.5	6.9	6.5	5.1
Tyr	2.8	2.4	3.4	3.2	3.1	3.0	3.8
Phe	3.1	3.9	2.6	2.7	3.2	3.1	2.0
His	1.8	1.7	1.7	1.6	2.3	2.3	2.1
Lys	3.0	2.9	4.3	4.5	5.0	5.6	3.9
Arg	4.0	3.6	4.5	3.7	6.1	5.7	3.9
Trp	1.1	0.8	1.8	1.2	0.8	0.7	0.5
Amide groups	21.3	23.4	14.0	14.4	11.9	13.4	20.4

Table 15.13. (continued)

Amino acid	Globulins						
	Wheat	Rye	Barley	Oats	Rice	Millet	Corn
Asx	7.7	6.8	8.6	7.9	6.5	7.8	9.1
Thr	4.6	4.6	4.8	4.3	2.9	4.5	5.2
Ser	6.6	6.9	6.5	6.9	7.0	8.1	7.5
Glx	15.2	17.0	12.9	16.0	14.6	12.1	10.7
Pro	6.9	7.8	6.8	5.3	5.6	5.2	5.6
Gly	8.3	8.7	9.5	9.4	10.2	9.3	10.3
Ala	7.5	7.6	8.3	7.4	8.0	9.7	10.7
Cys	3.6	2.1	3.0	2.4	4.1	3.5	3.2
Val	6.8	6.3	6.8	6.5	6.1	6.3	6.2
Met	2.0	1.5	1.4	1.3	4.4	0.9	1.5
Ile	3.8	3.9	3.1	4.1	2.6	3.6	4.1
Leu	7.3	6.9	7.5	7.0	6.2	6.8	6.5
Tyr	2.9	2.3	2.7	2.7	3.7	3.0	2.6
Phe	3.1	3.6	3.3	4.0	2.8	3.3	3.2
His	2.4	2.5	2.2	2.4	2.2	2.9	2.3
Lys	4.0	4.3	4.7	4.4	2.4	4.0	4.6
Arg	6.4	6.5	7.0	7.3	9.8	8.2	6.0
Trp	0.9	0.7	0.9	0.7	0.9	0.8	0.7
Amide groups	13.9	14.6	9.6	14.5	10.4	11.3	11.2

Amino acid	Prolamins						
	Wheat	Rye	Barley	Oats	Rice	Millet	Corn
Asx	2.7	2.4	1.7	2.3	7.3	6.8	4.9
Thr	2.3	2.6	2.1	2.3	2.9	3.8	3.1
Ser	5.9	6.6	4.6	3.8	7.5	6.4	6.9
Glx	37.1	35.4	35.3	34.1	19.6	21.8	19.4
Pro	16.6	18.4	23.0	10.2	5.1	7.8	10.2
Gly	2.9	4.5	2.2	2.7	5.8	1.5	2.6
Ala	2.8	3.0	2.3	5.5	9.1	13.5	13.6
Cys	2.2	2.2	1.9	3.3	0.8	1.1	1.0
Val	4.2	4.4	3.9	7.7	6.9	6.4	4.0
Met	1.1	1.0	0.9	2.1	0.5	1.7	1.1
Ile	4.1	3.0	3.6	3.3	4.6	5.2	3.9
Leu	6.9	5.8	6.1	10.6	11.8	13.4	18.5
Tyr	2.0	1.7	2.3	1.7	6.1	2.1	3.6
Phe	4.6	4.5	5.8	5.3	4.8	4.9	4.9
His	1.7	1.2	1.2	1.1	1.5	1.3	1.1
Lys	0.8	1.0	0.5	1.0	0.5	0.0	0.0
Arg	1.7	1.9	2.0	2.7	4.7	0.8	1.2
Trp	0.4	0.4	0.6	0.3	0.5	1.5	0.0
Amide groups	37.5	34.7	34.9	31.6	23.3	28.6	23.0

Table 15.13. (continued)

Amino acid	Glutelins[a]						
	Wheat	Rye	Barley	Oats	Rice	Millet	Corn
Asx	3.7	7.1	4.9	9.3	9.5	7.6	5.5
Thr	3.6	4.7	4.2	4.2	4.2	5.1	4.2
Ser	7.3	6.9	6.7	6.6	6.7	5.9	6.1
Glx	30.1	19.7	24.2	19.0	15.5	16.8	16.0
Pro	11.9	9.4	14.2	5.5	5.1	8.4	11.1
Gly	7.9	9.2	6.4	7.9	7.4	6.9	6.9
Ala	4.4	7.3	5.6	6.5	7.9	10.1	9.4
Cys	1.4	0.8	0.5	1.2	1.2	1.7	1.8
Val	4.8	5.9	7.2	6.2	7.0	6.6	6.1
Met	1.3	1.6	1.3	1.3	2.4	1.6	2.8
Ile	3.5	3.7	4.0	4.6	4.5	4.1	3.4
Leu	6.9	7.4	7.5	7.8	8.4	9.1	10.9
Tyr	2.4	2.3	1.7	2.8	3.6	2.9	2.9
Phe	3.6	3.8	4.0	4.8	4.3	3.7	3.3
His	1.8	2.0	2.0	2.4	2.1	2.3	3.3
Lys	2.1	4.0	2.8	3.2	3.3	3.1	2.4
Arg	2.7	3.8	2.5	6.0	6.1	3.5	3.2
Trp	0.6	0.4	0.3	0.7	0.8	0.6	0.3
Amide groups	31.0	21.3	23.6	20.2	16.6	17.0	16.4

[a] Protein residue after extraction of prolamins.

arms of the group G chromosomes. It is presumed that the variation seen in different varieties is due to the presence of allelic genes at each of the nine storage protein loci. The relative importance of different alleles for gluten quality seems to be Glu-1 > Gli-1 > Gli-2.

A fractionation of gluten proteins is possible by two-dimensional electrophoresis. Fig. 15.3 provides a schematic overview of the position of the most important protein groups in a two-dimensional electropherogram. The pattern of glutenins of two wheat cultivars are shown in Fig. 15.4.

Gluten proteins can be separated into their components on an analytical and micropreparative scale by using RP-HPLC. In general, this separation starts with the *Osborne* fractions or subfractions.

In this way, the *prolamines* of wheat can be separated into ω-, α-, γ-gliadins (Fig. 15.5), different varieties of wheat giving different patterns, e. g., the cultivars Clement and Maris Huntsman known to produce sticky dough have a characteristically high ω-gliadin content.

The prolamin patterns of other cereals (Fig. 15.6) differ greatly from that of wheat. In *rye*, the hydrophilic ω-secalins are followed by the hydrophobic γ-secalins. And unlike wheat (α-gliadins), the area of moderate hydrophobicity is not occupied. In *barley*, a hydrophilic fraction is missing: the C-hordeins eluted in the middle area are followed by the hydrophobic B-hordeins. The chromatogram of *oats* is characterized by two hydrophobic fractions that are close to each other.

The *low-molecular subunits of wheat glutelins* also give a chromatogram rich in components (Fig. 15.7). This chromatogram also contains the ω5-, ω1,2-, and γ-gliadins, which are not separated during pre-fractionation because of varying solubilities (cf. 15.2.1.2).

The *high-molecular subunits of wheat glutelins* show a protein pattern typical of the cultivar (Fig. 15.8).

Based on the data available on the structure of gluten proteins, three main groups can be formed which consist of several subunits. A *high-molecular group* with the HMW subunits of glutenins, a *group of intermediate molecular weights* with the ω5- and ω1,2-gliadins, and a *low-molecular group* with the α- and γ-gliadins as well as the LMW subunits of the glutenins. The amino acid composition of these protein groups is shown in Table 15.14.

15.2.1.3.1 High-Molecular Group (HMW Subunits of Glutenin)

As shown in Table 15.14, the HMW subunits of glutenin are the only gluten proteins in which Gly (ca. 19%), and not Pro (ca. 12%), takes second place in the order of amino acids after Glx (ca. 36%). Furthermore, the proteins are characterized by the highest content of Tyr

(ca. 6%) and Thr (ca. 3.5%) and the lowest content of Phe (ca. 0.3%) and Ile (ca. 0.8%).

In the N-terminal amino acid sequence presented in Table 15.15, the sequence EGEAS-RQLQC is valid for all known HMW subunits and varies only in position 6 (E, K, G). From the total sequences known until now, it can be deduced that the HMW subunits consist of three domains (Table 15.16). The N- and C-terminal domains contain no recurring sequences and are characterized by the occurrence of Cys and amino acids with charged side chains. The middle domain consists of recurring sequences with the peptide unit QQPGQG as the backbone and insertions with

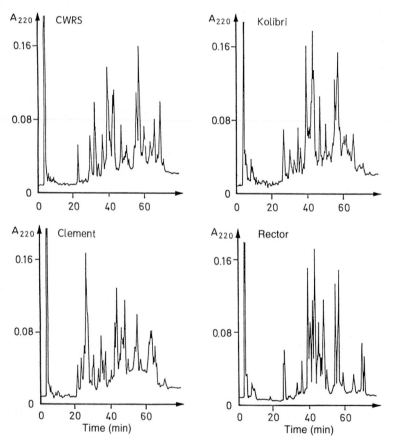

Fig. 15.5. RP-HPLC of the gliadin fractions of various wheat cultivars[a] on Synchro Pac C$_{18}$ (50 °C, aqueous 2-propanol/trifluoroacetic acid/acetonitrile; 22–34 min: ω-gliadins, 33–51 min: α-gliadins, 52–72 min: γ-gliadins; according to *Wieser* et al., 1987)

[a] CWRS (Canadian Western Red Spring) is a mark of origin.

the sequences YYPTSP, QQG, and QPG. It largely determines the unusual amino acid composition (high Gly and Tyr content). The individual HMW subunits differ mainly in the substitution of individual amino acid residues and in the number and arrangement of recurring peptide units.

The relative molecular masses (M_r) derived from SDS-PAGE are 90,000–124,000. The molecular masses calculated from the known total sequences are about 10–15% lower. Based on typical differences in the N-terminals and middle sequence segments, the HMW

subunits can be assigned to two subgroups (x-type, M_r = 104,000–120,000; y-type, M_r = 90,000–102,000) (Table 15.16).

15.2.1.3.2 Intermediate Molecular Weight Group (ω5-Gliadins, ω1,2-Gliadins)

This group of ω-gliadins is characterized by high values of Glx, Pro, and Phe (Table 15.14). The proportion of most of the other amino acids is less than in the other groups and the sulfur-containing amino acids Cys and Met are either absent or present only in traces. Total

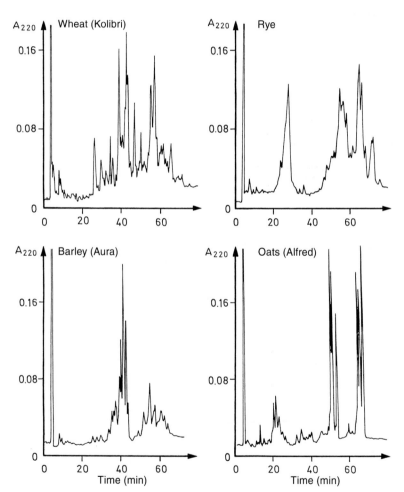

Fig. 15.6. RP-HPLC of the prolamin fractions of different varieties of cereal (conditions as in Fig. 15.5. *Wheat* 26–30 min: ω-gliadins, 32–50 min: α-gliadins, 54–71 min: γ-gliadins. *Rye* 21–37 min: ω-secalins, 45–77 min: γ-secalins. *Barley* 32–44 min: C-hordeins, 46–66 min: B-hordeins. *Oats* 49–55 min/62–69 min: avenins; according to *Wieser, Belitz* et al., 1989)

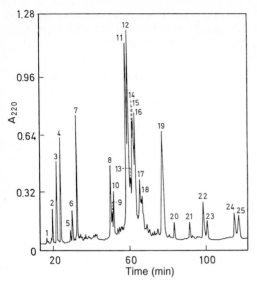

Fig. 15.7. RP-HPLC of low-molecular (LMW) subunits of wheat glutelin of the cultivar Rektor on Nucleosil C_{18} (60 °C, aqueous 2-propanol/trifluoroacetic acid/acetonitrile. After the prolamins, the protein fraction was extracted from the residue with 70% aqueous ethanol/0.5% dithioerythritol at pH 7.6 and 4 °C; peaks 2–4, 5–7: ω5-, ω1,2-gliadins. Peaks 8–17: LMW subunits, peaks 20–25: γ-gliadins or related proteins, according to *Wieser* et al., 1990)

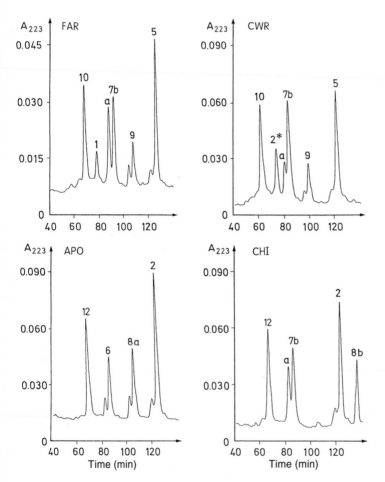

Fig. 15.8. RP-HPLC of the high-molecular (HMW) subunits of the glutelins of different varieties of wheat on Nucleosil C_8 (60 °C, urea/trifluoroacetic acid/acetonitrile/dithioerythritol; numbering of the peaks as in Fig. 15.9; (FAR: Farmer, CWR: Canadian Western Red Spring, APO: Apollo, CHI: Chinese spring; according to *Seilmeier* et al., 1991)

Table 15.14. Amino acid composition[a] of protein groups of wheat gluten (cultivar Rektor)

	HMW subunits of glutenin	ω5-Gliadins	ω1,2-Gliadins	LMW subunits of glutenin	α-Gliadins	γ-Gliadins
Asx	0.7–0.9	0.3–0.5	0.5–1.3	0.7–1.5	2.7–3.3	1.9–4.0
Thr	3.2–3.8	0.4–0.6	0.8–2.3	1.8–2.9	1.5–2.3	1.6–2.4
Ser	6.4–8.4	2.6–3.3	5.8–6.3	7.7–9.5	5.3–6.6	4.9–6.8
Glx	35.9–37.0	55.4–56.0	42.5–44.9	38.0–41.9	35.8–40.4	34.2–39.1
Pro	11.2–12.8	19.7–19.8	24.8–27.4	14.0–16.2	15.0–16.6	15.8–18.4
Gly	18.2–19.8	0.6–0.8	0.9–2.1	2.3–3.2	1.9–3.2	2.0–3.0
Ala	2.9–3.5	0.2–0.3	0.3–1.3	1.7–2.3	2.6–4.1	2.8–3.5
Cys	0.6–1.3	0.0	0.0	1.9–2.6	1.9–2.2	2.2–2.8
Val	1.6–2.7	0.3	0.6–1.4	3.8–5.3	4.2–4.9	4.4–5.4
Met	0.1–0.3	0.0	0.0–0.3	1.2–1.6	0.4–0.9	1.2–1.6
Ile	0.7–1.1	4.3–4.7	1.9–3.5	3.6–4.4	3.6–4.6	4.0–4.6
Leu	3.1–4.3	2.7–3.3	3.9–5.3	5.3–7.5	6.5–8.7	6.4–7.3
Tyr	5.1–6.4	0.6–0.7	0.8–1.5	0.9–1.9	2.3–3.2	0.6–1.4
Phe	0.2–0.5	9.0–9.5	7.6–8.1	3.8–5.5	2.9–3.9	4.7–5.6
His	0.8–1.9	1.3–1.4	0.6–1.1	1.3–1.8	1.4–2.8	1.1–1.5
Lys	0.7–1.1	0.4–0.5	0.3–0.6	0.2–0.6	0.2–0.6	0.4–0.9
Arg	1.6–2.1	0.5–0.6	0.5–1.4	1.5–2.1	1.7–2.9	1.2–2.9

[a] mol % (without Trp).

sequences have not yet been published, but some information is available on partial sequences. The ω-gliadins can be divided into two subgroups, the ω5-type and the ω1,2-type. This nomenclature is based on the varying mobility on acidic PAGE.

The *ω5-gliadins* are characterized by an extremely high content of Glx (ca. 56%) and a relatively high content of Phe (ca. 9%). Although the content of Pro (ca. 20%) is lower than in the ω1,2-type, it is clearly higher than in the other groups. These three amino acids account for about 85% of the total protein. ω5-Gliadins are free of sulfur-containing amino acids and the content of the remaining amino acids is comparatively low. It is noticeable that only this protein type has more Ile (ca. 4.5%) than Leu (ca. 3%).

The N-terminal sequences unanimously consist of the sequence SRLLSPRGKELHT and are typical of this type of protein. Recurring sequences with the peptide unit PQQQF evidently start from position 14. This represents a clear difference to the ω1,2-type, which explains the typical differences in the amino acid composition of the two subgroups.

The ω5-gliadins have a higher mobility than the ω1,2-gliadins in acidic PAGE and a lower

Table 15.15. N-terminal sequences of the protein groups of wheat gluten

	Position 5	10

HMW subunits of glutenin	E G E A S R Q L Q C
	      E
	      G
	      K	

ω5-Gliadins	S R L L S P R G K E	
ω1,2-Gliadins	– – – – – – – – K E L Q S – R Q L N P S D Q A N	
LMW subunits of glutenin	S C I S G L E R P W R P	

α-Gliadins	V R V P V P Q L Q P
	    F
	                P
	                    L	

γ-Gliadins	N M Q V D P S G Q V
	  I
	                S
	        A	

mobility in SDS-PAGE, M_r being between 64,000 and 79,000.

In comparison with the ω5-gliadins, *ω1,2-gliadins* have lower values for Glx (ca. 43%), Phe (ca. 7.5%) and Ile (ca. 3%) (Table 15.14). Most of the other values are higher, especially the content of Pro (ca. 26%), which is the highest within the gluten groups of proteins.

In the case of the N-terminal sequences, three basic variants apparently exist a: ARQLNPS-NKELQS; b: RQLNPSDQELQS, c: KELQS, which with varying length are homologous and lead into a recurring sequence. The variant a was found more frequently in ω2-gliadins and the variants b and c more frequently in ω1-gliadins. The N-terminal sequence of the

ω5-gliadins corresponds to that of variant a in 6 positions. The immediately following recurring sequence consists of the peptide unit PQQPY, while the dominating recurring peptide unit in this protein type seems to be the sequence PQQPFPQQ.

In acidic PAGE, ω1,2-gliadins have the lowest mobility. Indeed, M_r ranges of 54,000–64,000 have been obtained in SDS-PAGE.

15.2.1.3.3 Low-Molecular Group (α-Gliadins, γ-Gliadins, LMW Subunits of Glutenin)

The quantitatively predominant low-molecular protein group in gluten has the best balanced amino acid composition. Most of the values lie

Table 15.16. Sequence comparison of the HMW subunits of glutenin

a) x-Typ (HMW-subunit 1Dx5)*

N-terminal sequence			
1	E G E A S E Q L Q C E R E L Q E L Q E R E L K A C Q Q V MD Q Q L R D I S P E C		
41	H P V V V S P V A G Q Y E Q Q I V V P P K G G S F Y P G E T T P P Q Q L Q Q R I		
81	F WG I P A L L K R		

Recurring sequences				
91	Y Y P S V T C P Q – Q V S	448		Q Q P G Q G Q Q G
103	Y Y P G Q A S P Q – R P G Q G	457		Q Q P G Q G Q Q G
117	Q – Q P G Q G Q Q G	466		Q Q P G Q G Q P G
126	Y Y P – – T S P Q – Q P G Q W	475	Y Y P T S P	Q Q S G Q G
138	Q – Q P E Q G Q P R	487		Q Q P G Q W
147	Y Y P – – T S P Q – Q S G Q L	493		Q Q P G Q G Q P G
159	Q – Q P A Q G	502	Y Y P T S P L	Q P G Q G Q P G
165	Q – Q P G Q G Q Q G	517	Y D P T S P	Q Q P G Q G
174	Q – Q P G Q G Q P G	529		Q Q P G Q L
183	Y Y P – – T S S Q L Q P G Q L	535		Q Q P A Q G Q Q G
196	Q – Q P A Q G Q Q G	544		Q Q L A Q G Q Q G
205	Q – Q P G Q A Q Q G	553		Q Q P A Q V Q Q G
214	Q – Q P G Q G	562		Q R P A Q G Q Q G
220	Q – Q P G Q G Q Q G	571		Q Q P G Q G Q Q G
229	Q – Q P G Q G	580		Q Q L G Q G Q Q G
235	Q – Q P G Q G Q Q G	589		Q Q P G Q G Q Q G
244	Q – Q L G Q G Q Q G	598		Q Q P A Q G Q Q G
253	Y Y P – – T S L Q – Q S G Q G Q P G	607		Q Q P G Q G Q Q G
268	Y Y P – – T S L Q – Q L G Q G Q S G	616		Q Q P G Q G Q Q G
283	Y Y P – – T S P Q – Q P G Q G	625		Q Q P G Q G
295	Q – Q P G Q L	631		Q Q P G Q G Q P W
301	Q – Q P A Q G	640	Y Y P T S P	Q E S G Q G
307	Q – Q P G Q G Q Q G	652		Q Q P G Q W
316	Q – Q P G Q G Q Q G	658		Q Q P G Q G Q P G
325	Q – Q P G Q G	667	Y Y L T S P L	Q L G Q G Q Q G
331	Q – Q P G Q G Q P G	682	Y Y P T S L	Q Q P G Q G
340	Y Y P – – T S P Q – Q S G Q G Q P G	694		Q Q P G Q W
355	Y Y P – – T S S Q – Q P T Q S	700		Q Q S G Q G Q H W
367	Q – Q P G Q G Q Q G	709	Y Y P T S P	Q L S G Q G
376	Q – Q V G Q G Q Q A	721		Q R P G Q W
385	Q – Q P G Q G	727		L Q P G Q G Q Q G
391	Q – Q P G Q G Q P G	736	Y Y P T S P	Q Q P G Q G
400	Y Y P – – T S P Q – Q S G Q G Q P G	748		Q Q L G Q W
415	Y Y L – – T S P Q – Q S G Q G	754		L Q P G Q G Q Q G
427	Q – Q P G Q L	763	Y Y P T S L	Q Q T G Q G
433	Q – Q S A Q G Q K G	775		Q Q S G Q G Q Q G
442	Q – Q P G Q G	784	Y Y	

C-terminal sequence		
786	S S Y H V S V E H Q A A S L K V A K A Q Q L A A Q L P A MC R L E G G D A L	
824	S A S Q	

Table 15.16 (continued)

b) γ-Type (HMW subunit 1By9)*

N-terminal sequence		
1	E G E A S R Q L Q C E R E L Q E S S L E A C R Q V V D Q Q L A G R L P WS T G L	
41	Q M R C C Q Q L R D V S A K C R P V A V S Q V V R Q Y E Q T V V P P K G G S F Y	
81	P G E T T P L Q Q L Q Q V I F W G T S S Q T V Q G	

Recurring sequence				
106	Y Y P S V S S P Q Q G P	408		Q Q T R Q G
118	Y Y P G Q A S P Q Q P G Q G	414		Q Q L E Q G
132	Q Q P G K W	420		Q Q P G Q G
138	Q E L G Q G Q Q G	426		Q Q T R Q G
147	Y Y P - - T S L H Q S G Q G Q Q G	432		Q Q L E Q G
162	Y Y P - - S S L Q Q P G Q G	438		Q Q P G Q G Q Q G
174	Q Q I G Q G Q Q G	447	Y Y P T S P	Q Q S G Q G
183	Y Y P - - T S L Q Q P G Q G	459		Q Q P G Q S
195	Q Q I G Q G Q Q G	465		Q Q P G Q G Q Q G
204	Y Y P - - T S P Q H P G Q R	474	Y Y S S S L Q Q P G Q G L Q G	
216	Q Q P G Q G	489	H Y P A S L Q Q P G Q G	
222	Q Q I G Q G	501		H P G Q R
228	Q Q L G Q G	506		Q Q P G Q G
234	R Q I G Q G	512		Q Q P E Q G
240	Q Q S G Q G Q Q G	518		Q Q P G Q G Q Q G
249	Y Y P - - T S P Q Q L G Q G	527	Y Y P T S P	Q Q P G Q G
261	Q Q P G Q W	539		K Q L G Q G Q Q G
267	Q Q S G Q G Q Q G	548	Y P T S P Q Q P G Q G	
276	Y Y P - - T S Q Q Q P G Q G Q Q G	560		Q Q P G Q G Q Q G
291	Q Y P - - A S Q Q Q P G Q G Q Q G	569	H C P T S P	Q Q T G Q A
306	Q Y P - - A S Q Q Q P G Q G Q Q G	581		Q Q P G Q G
321	Q Y P - - A S Q Q Q P G Q G Q Q G	587		Q Q I G Q V
336	H Y L - - A S Q Q Q P G Q G Q Q R	593		Q Q P G Q G Q Q G
351	H Y P - - A S L Q Q P G Q G Q Q G	602	Y Y P I S L	Q Q S G Q G
366	H Y T - - A S L Q Q P G Q G Q Q G	614		Q Q S G Q G
381	H Y P - - A S L Q Q V G Q G	620		Q Q S G Q G
393	Q Q I G Q L G Q R	626		H Q L G Q G
402	Q Q P G Q G	632		Q Q S G Q E Q Q G
		641	Y D	

C-terminal sequence		
643	N P Y H V N T E Q Q T A S P K V A K V Q Q P A T Q L P I MC R ME G G D A L	
681	S A S Q	

The numbers give the positions that the amino acids at the beginning of the line occupy in the total sequence. The segments of recurring sequences are arranged according to the best possible homology. (–: space to maximize homology)
* The numbering of the HMW subunits (5 and 9) corresponds to Fig. 15.8 and 15.9.

between those of the high and intermediate molecular groups. Only the content of Cys, Val, Met, and Leu is higher (Table 15.14). A large number of partial and total sequences are found in the literature. With one exception (A-gliadin), the total sequences have all been derived from the corresponding nucleic acids. Based on present data, the low-molecular gluten proteins can be assigned to three subgroups (α-gliadins, γ-gliadins, LMW subunits of glutenin). As shown in Table 15.17 with three examples, the total sequences consist of at least six differently structured segments. The individual subgroups differ in the N- and C-terminal sequences, in the recurring sequences, and in the Gln-rich sequences. On the other hand, they exhibit long homologous sequence segments that are low in Pro. These segments are characterized by the frequent occurrence of amino acids with charged side chains. In addition, with a few exceptions, they contain all the sulfurous amino acids. The M_r of this group of proteins lies in the range of 28,000–39,000.

As shown in Table 15.14, the amino acid composition of the α-gliadins differs on the whole only slightly from that of the γ-gliadins and LMW subunits of glutenin. In the case of individual amino acids, however, significant differences are exhibited. The content of Tyr (ca. 3%) is considerably higher and the content of Met (ca. 0.7%) and Phe (ca. 3.4%) is lower.

The N-terminal sequences determined directly by *Edman* degradation correspond to those derived from the nucleic acids. Apart from a

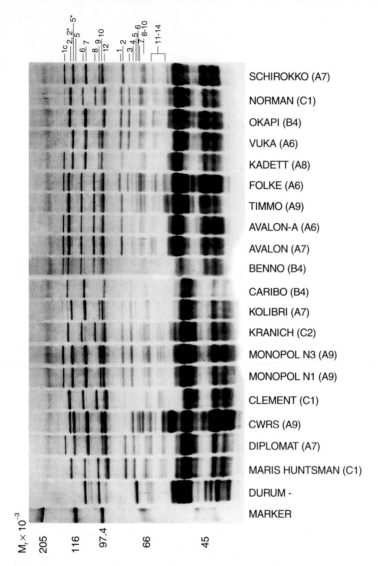

Fig. 15.9. Electrophoretic separation of the glutenins[a,b] of various wheat cultivars[c] on polyacrylamide gel in the presence of sodium dodecyl sulfate and mercaptoethanol (SDS-PAGE) (according to *Krause* et al., 1988)

[a] Residue after the extraction of defatted flour with water, salt solution, and aqueous ethanol.

[b] The numbering of the HMW subunits of glutenin differs from that in the original publication, *Payne* et al., 1981b.

[c] After each variety, the baking quality class is given in brackets (bread volume yield very high: A9, high to very high: A8, high: A7, average to high: A6, average: B5, low to average: B4, low: B3, very low to low: C2, very low: C1).

few variations, the individual amino acid residues VRVPVPQLQPQN have been found for these N-terminal sequences (Table 15.15). The recurring sequences consist of the peptide unit QPQPFPPQQPYP, which usually occurs five times and varies in individual amino acids (Table 15.17). The balanced Tyr/Phe ratio of the α-gliadins is based on this domain. Deviat-

Table 15.17. Sequence comparison[a] of an α-gliadin (clone pGli-A42), γ-gliadin (clone pW1621), and a LMW subunit of glutenin (clone LMWG-1D1)

N-terminal sequences	α-	1	V R V P V P Q L Q P Q N P S Q Q Q P Q E Q V P L V Q Q Q Q F P G					
	γ-	1	N I Q V D P S G Q V Q W L Q					
	LMW	1	R C I P G L E R P W					

			α-		γ		LMW	
Recurring sequences		33	Q Q Q Q Q F P Q P Q Q P Y P	15	Q Q L V P Q	11	Q Q Q P L P P	
		45	Q P Q P – F P – S Q Q P Y L	21	L Q Q P L S Q	18	Q Q T – F P	
		57	Q L Q P – F P –	28	Q P Q Q T F P Q	23	Q Q P L F S	
		63	Q P Q P – F P – P L – P Y P	36	P Q Q T F P H	29	Q Q Q Q Q Q L – F P	
		74	Q P Q S – F P – P Q Q P Y P	43	Q P Q Q Q V P Q	38	Q Q P S F S	
		86	Q Q Q P – Y L – P Q Q P I S	51	P Q Q P Q Q P F L Q	44	Q Q Q P P F W	
				61	P Q Q P F P Q	51	Q Q Q P P F S	
				68	Q P Q Q P F P Q	58	Q Q Q P I L P	
				76	T Q Q P Q Q P F P Q	65	Q Q P P F S	
				86	Q P Q Q P F P Q	71	Q Q Q Q L V L P	
				94	T Q Q P Q Q P F P Q	79	Q Q P P F S	
				104	Q P Q Q P F P Q	85	Q Q Q Q P V L P	
				112	T Q Q P Q Q P F P Q	93	P Q Q S P F P	
				122	L Q Q P Q Q P F P Q	100	Q Q Q Q Q H	
				132	P Q Q Q L P Q	106	Q Q L V	
				139	P Q Q P Q Q S F P Q	110	Q Q Q I P	
				149	Q Q R P			

Poly-Gln sequences	α-	100	Q Q Q A Q Q Q Q Q Q Q Q Q Q Q Q Q Q Q Q					

Sequences I low in Pro	α-	119	I L Q Q I L Q Q Q L I P C R D V V L Q Q H N I A H A S S Q V L – – – – – Q Q S T Y Q L L Q Q					
	γ-	153	F I Q P S L Q Q Q L N P C K N L L L Q Q C R P V S L V S S L W– S I I W P Q S D C Q V M R Q					
	LMW	115	V V Q P S I L Q Q L N P C K V F L Q Q Q C S P V A M P Q R L A R S Q M L Q Q S S C H V M Q Q					
	α-	160	L C C Q Q L L Q I P E Q S Q C Q A I H N V A H A I I M					
	γ-	198	Q C C Q Q L A Q I P Q Q L Q C A A I H S V V H S I I M					
	LMW	161	Q C C Q Q L P Q I P Q Q S R Y E A I R A I I Y S I I L					

Sequences high in Gln	α-	187	H Q Q Q Q Q Q E Q K Q Q L Q Q Q Q Q Q Q Q Q L Q Q Q Q Q Q Q Q Q Q P S S Q V S F Q Q P Q					
	γ-	225	Q Q Q Q Q Q Q Q Q Q G I D I F L P L S Q H E Q V G Q G S L V					
	LMW	188	Q E Q Q Q V Q G S I Q S Q Q Q Q P Q Q L G Q C V S Q P Q Q Q S Q Q Q L G Q Q P Q Q Q Q					
	α-	232	Q Q Y P S S Q V S F Q P S Q L N P					

Sequences II low in Pro	α-	249	Q A Q G S – V Q P Q Q L P Q F A E I R N L A L Q T L P A M C N V Y I P P H C S T T I A P F G					
	γ-	255	Q G Q G I – I Q P Q Q P A Q L É A I R S L V L Q T L P S M C N V Y V P P E C S I M R A P F A					
	LMW	231	L A Q G T F L Q P H Q I A Q L E V M T S I A L R I L P T M C S V N V P L Y R T T T S V P F G					

C-terminal sequences	α-	294	I S G T N					
	γ-	300	S I V A G I G G Q					
	LMW	277	V G T G V G A Y					

[a] The segments of the recurring and low-proline sequences are arranged according to the best possible homology (–: space to maximize homology).

ing from the γ-gliadins and the LMW subunits of glutenin, α-gliadin contains a poly-Gln sequence between the recurring and low-Pro sequence segments.

In comparison with the α-gliadins, the γ-*gliadins* exhibit higher values for Phe (ca. 5%) and Met (ca. 1.4%) and lower values for Tyr (ca. 1%) (Table 15.14).

The most common N-terminal sequence that is directly determined or derived from the nucleic acids is NMQVDPSGQV. Individual positions are modified, e.g., position 2 with I (Table 15.15). The recurring sequences consist of the peptide units PQQPFPQ, in which Q,

TQQ, LQQ or PQQ can be inserted. There are up to 15 repetitions of such peptide units which can vary in individual residues (Table 15.17). The absence of Tyr in the recurring sequence segments shifts the Tyr/Phe ratio to ca. 1.5 (Table 15.14).

The *LMW subunits of glutenin* differ from the α- and γ-gliadins by having higher values for Ser (ca. 9%) and lower values for Ala (ca. 2%) and Asx (ca. 1%). The values for the other amino acids coincide (Table 15.14).

The N-terminal sequences of the LMW subunits were found to be SCISGLERPW and RCIPGLERPW. The three known total sequences show that the LMW subunits of glutenin have typical N-terminal, C-terminal, Gln-rich and recurring sequence segments (Table 15.17). The remaining sequence segments correspond largely to those of α- and γ-gliadins. The recurring peptide units usually consist of the sequence Q_nPPFS with $n = 2-10$. These units are repeated up to 20 times and the hydrophobic tripeptide PPF is partly varied (e.g., by PVL, PLP). In comparison with the α- and γ-gliadins, the high Ser content in the total composition is due to the recurring sequences.

15.2.1.4 Structure of Wheat Gluten

Although there is much known about the protein components of gluten, there is little information available on the type and extent of polymerization or aggregation of subunits or on other interactions in native gluten. Many different gluten models have been postulated in the literature, but they are all only inadequately supported by experiments.

It has been proved that ω-, α-, γ-gliadins are present in gluten as monomeric proteins. On the other hand, the high-molecular (HMW) and low-molecular (LMW) subunits obtained on reduction of glutenin are present only to a smaller extent as monomers and to a larger extent in the polymerized or aggregated form. Apart from hydrophobic bonds, H bridges, and ionic bonds, intermolecular disulfide bonds are involved in the formation of these polymers or aggregates. Only in a few cases have these disulfide bonds been detected by the isolation and structure elucidation of cystine peptides from enzymatic glutenin hydrolysates. Thus, a linkage of two HMW subunits of type y via two disulfide bonds between adjacent cysteine residues (positions 44 and 45 in Table 15.6) could be detected in the peptide:

$$
\begin{array}{l}
C\ C\ Q\ Q\ L \\
|\ \ | \\
C\ C\ Q\ Q\ L
\end{array}
\qquad (15.1)
$$

Further studies have to show whether HMW and LMW subunits form homogeneous and/or heterogeneous oligomers/polymers via disulfide bridges. Model experiments have shown that in aqueous solution both protein types can be oxidatively polymerized separately. It is interesting that this does not apply to the

γ-gliadins that are structurally related to the LMW subunits (cf. Table 15.14 and 15.17). Like the α-gliadins, they remain in the monomeric state after reduction and subsequent re-oxidation under the same conditions, i.e., they form only intramolecular disulfide bonds.

Glutenin, consisting of HMW and LMW subunits more or less extensively polymerized via disulfide bonds, probably forms an insoluble, but swellable skeleton in native gluten. The stability of this skeleton appears to depend on the type of subunits involved and on the degree of polymerization. Rheological measurements (Fig. 15.10) indicate that the gliadins added to the glutenins soften the gluten formed. The gliadins act as plasticizer.

The parameters mentioned above are definitely responsible also for the technological properties of wheat doughs, which vary with the variety of wheat and the cultivation conditions. However, these properties are not yet understood in detail on a molecular level.

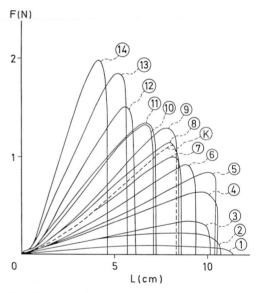

Fig. 15.10. Tensile tests of glutens with varying content of gliadin (gluten K from retail wheat flour was extracted with 70% aqueous ethanol. The extracted gliadin and the remaining glutenin were freeze dried, remixed in different proportions, and then hydrated. Gliadin content of the glutens: K) 33.9%, 1) 55.9%, 14) 22.6%; the gliadin contents of the other samples are in between, according to *Kim* et al., 1988)

15.2.2 Enzymes

Of the enzymes present in cereal kernels, those which play a role in processing or are involved in the reactions which are decisive for the quality of a cereal product have been thoroughly investigated.

15.2.2.1 Amylases

α- and β-amylases (for their reactions, see 4.4.5.1) are present in all cereals. Wheat and rye amylases are of particular interest; their optimum activities are desirable in dough making in the presence of yeast (cf. 15.4.1.4.8). In mature kernels, α-amylase activity is minimal, while it increases abruptly during sprouting or germination. Unlike the situation with wheat, dormancy in rye is not very pronounced. Unfavorable harvest conditions (high moisture and temperature) favor premature germination ("sprouting"), not visible externally. During this time, α-amylase activity rises, resulting in extensive starch degradation during the baking process. Bread faults appear, as mentioned under 15.4.1.2.

Two α-amylases, α-AI and α-AII, have been isolated from wheat by affinity chromatography and chromatofocussing. These two enzymes produce a series of multiple forms on SDS-PAGE electrophoresis. The ratio of the concentrations of the two α-amylases depends on the stage of development. After flowering, α-AI appears first in the outer layers of the kernel, then decreases with increasing ripeness. Low activities of α-AII are detectable even before dormancy, but they greatly increase during germination. The two α-amylases differ in their pH optimum, molar mass, and isoelectric point (Table 15.18). α-AII is more temperature resistant.

The pH optimum of α-amylase in germinating rye lies in a range similar to that of α-AII of wheat. Therefore, α-amylase is partially inhibited by the decrease in pH in sour dough (cf. 15.4.2.2).

The properties of wheat β-amylase are shown in Table 15.18.

15.2.2.2 Proteinases

Acid proteinases with pH optima of 4–5 occur in wheat, rye and barley. Their substrate specificity has been determined. The possibility that the wheat proteinases are involved in cleavage of gluten bonds, thereby affecting softening or mellowing of gluten during baking, is still disputed.

15.2.2.3 Lipases

These enzymes occur in various concentrations in all cereals. Carboxylester hydrolase, readily isolated from wheat germ, is not considered a lipase but an esterase (cf. 3.7.1.1). The activity in dormant seeds is low, but increases greatly on germination and can be detected with great sensitivity with a fluorochrome substrate, e.g., fluorescein dibutyrate. Therefore, this forms the basis of a method for the quick detection of "sprouting" in wheat and rye.

In addition to the esterase, a wheat lipase occurs enriched in the bran. A rise in free fatty acids observable during flour storage also involves lipases from metabolism of microorganisms present in flour.

In comparison to other cereals, oats contain a significant level of lipase. Its high activity is released once the oat kernel is disintegrated, crushed or squeezed. Linoleic acid is released from the acyl lipids that are present. It is then

Table 15.18. Amylases in wheat

Properties	α-Amylase I	α-Amylase II	β-Amylase
pH optimum	3.6–5.75	5.5–5.7	5.4–6.2
Molar mass	37,000[a]	21,000[a]	64,200[b]
Isoelectric point	4.65–5.11	6.05–6.20	4.1–4.9

[a] gel chromatography, [b] ultracentrifugation.

converted into hydroxy fatty acids by lipoxygenase and hydroperoxidase enzymes, giving
rise to off-flavors (Fig. 15.11). All these enzymes are inactivated by heat treatment and
thus quality deterioration can be avoided (cf.
15.3.2.2.2).
It should be taken into account in lipid extraction that the phospholipase D activities are
relatively high in ripe cereals and this enzyme
transfers the phosphatidyl residue of phospholipids to alcohols, which are used to extract
lipids (cf. 3.7.1.2.1). The enzyme is inactivated during extraction with boiling water-
saturated butanol. A phospholipase that hydrolyzes both acyl residues in the lecithin molecule ("phospholipase B") has been found in
germinating cereal. It influences the foam
stability in beer (cf. 20.1.7.9).
In the production and storage of egg dough
products, phopholipases B and D can lower the
phospholipid content.

15.2.2.4 Phytase

Close to 70% of the phosphorus in wheat is
bound to phytin, which is 1% of the kernels. A
major portion of this is hydrolyzed during
dough making by microbial phytases:

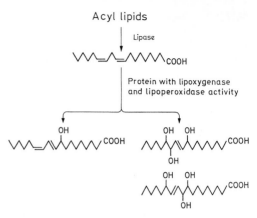

Fig. 15.11. Formation of bitter tasting compounds
in oats. (taste threshold values in Table 3.35)

Phytin

meso-Inositol   (15.2)

The reaction is nutritionally and physiologically desirable since phytin inhibits, by formation of water-insoluble salts, the intestinal
absorption of calcium and iron ions.

15.2.2.5 Lipoxygenases

Cereals contain lipoxygenases (cf. 3.7.2.2)
which, with the exception of the enzyme from

rye, oxidize linoleic acid preferentially to 9-
hydroperoxy acids. The rye lipoxygenase
forms mainly the 13-hydroperoxide isomer.
Though the enzyme from wheat belongs to
type I (cf. Table 3.32) and thus cooxidizes
carotenoids at a slow rate, it can still bring
about a loss of yellow color in pasta products.
This is the reason for inactivation of wheat
lipoxygenase during preparation of pasta
products (cf. 15.5).
The involvement of endogenous lipoxygenase
in the baking of wheat flour is not clear. However, by addition of lipoxygenase-active soy
flour, a significant improvement of the flour
quality is achieved (cf. 15.4.1.4.3).
As shown in Fig. 15.11, oats contain a lipoxygenase with lipoperoxidase activity. This
activity reduces the hydroperoxides initially
formed, in the presence of phenolic compounds as H-donors, to the corresponding
hydroxy acids:

$$R_1: \quad HOOC-(CH_2)_7; \qquad R_2: \quad H_3C-(CH_2)_4 \qquad (15.3)$$

15.2.2.6 Peroxidase, Catalase

Both enzymes are widely distributed among
cereals. The pH-activity curves of the enzymes
from wheat show that at the normal pH values
of a dough, about 6.3, catalase still has

40–50% and peroxidase less than 10% of its activity at the pH optimum (peroxidase pH 4.5; catalase pH 7.5). Therefore, it is unlikely that the oxidative cross-linkage of pentosans (Fig. 15.14), which is catalyzed by peroxidase, plays an essential role in dough.

As heme catalysts they accelerate the non-enzymatic oxidation of ascorbic acid to the dehydro form. The involvement of both enzymes in the action of ascorbic acid as an improver will be discussed (cf. 15.4.1.4.1).

15.2.2.7 Glutathione Dehydrogenase

This enzyme catalyzes the oxidation of glutathione (GSH) in the presence of dehydroascorbic acid as an H-acceptor:

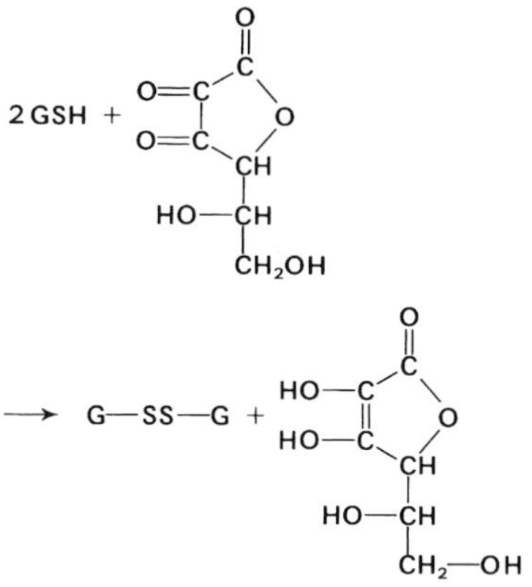

$$(15.4)$$

It has been purified from wheat flour in which its activity is relatively high. The enzyme is highly specific for the H-donor (Table 15.19) because it oxidizes only GSH and neither cysteinyl glycine nor cysteine, which also occur in wheat flour. The specificity for the H-acceptor is not so pronounced. As shown in Table 15.19, all four diastereomeric forms of dehydroascorbic acid are converted, but with different velocities. The substrate specificity corresponds to the varying activity of the diastereomeric dehydroascorbic acids in flour improvement (cf. 15.4.1.4.1).

15.2.2.8 Polyphenoloxidases

In cereals, polyphenoloxidases preferably occur in the outer layers of the kernels. Wheat enzymes that exhibited cresol activity only (cf.

Table 15.19. Substrate specificity of wheat glutathione dehydrogenase

Substrate	Relative activity (%)
H-Donor	
Glutathione (reduced form)	100
Cysteine	0
Cysteinyl glycine	0
H-Acceptor	
Dehydro ascorbic acid	
L-threo	100
L-erythro	67
D-erythro	60
D-threo	16

2.3.3.2) and were known as tyrosinases have been separated from polyphenoloxidases by chromatography and preparative gel electrophoresis.

Polyphenoloxidases can cause browning in whole-meal flours.

15.2.3 Other Nitrogen Compounds

Wheat contains glutathione and cysteine in the free state as thiol compounds (GSH, CSH), in the oxidized forms (GSSG, CSSC) and in the protein-bound forms (GSSProt and CSSProt) (Table 15.20). Reduction of GSS-Prot and CSSProt releases GSH and CSH respectively, e. g., with dithioerythritol.

Table 15.20. Glutathione and cysteine in wheat flour [a]

Wheat cultivar	Ash (w/w%)	Glutathione (nmol/g)[a]		Cysteine (nmol/g)[a]	
		GSH	Total[b]	CSH	Total[c]
CWRS[d]	0.54	16	172		
	0.71	35	348		
	1.44	60	575		
DNS[d]	0.59	41	175		
	0.78	110	345	13	159
	1.57	215	657		
Maris	0.55	20	185		
Huntsman	0.68	94	273	9	145
	1.73	210	435		

[a] Calculated as GSH or CSH based on the dry weight.
[b] Sum of GSH, GSSG, and GSSProt.
[c] Sum of CSH, CSSC, and CSSProt.
[d] Marks of origin: Canadian Western Red Spring (CWRS), Dark Northern Spring (DNS).

It has been shown that glutathione is predominantly localized in the germ and in the aleurone layer. Therefore its concentration in flour increases as the extraction grade increases (Table 15.20).

During dough making, GSH reacts very quickly undergoing disulfide interchange with flour proteins:

$$GSH + ProtSSProt = ProtSSG + ProtSH \quad (15.5)$$

If high-molecular gluten proteins are cleaved, the viscosity of the dough drops. Rheological measurements of flour/water doughs (Fig. 15.12) show that even very low concentrations of GSH can have a considerable effect. GSSG and CSSC are rheologically inactive, however, they are converted to rheologically active thiols by reduction or SS/SH interchange.

15.2.4 Carbohydrates

15.2.4.1 Starch

The major carbohydrate storage form of cereals, starch (cf. Table 15.6) occurs only in the endosperm cells. The size and form of the starch granules is specific for different cereals. The polysaccharide molecules in starch granules are radially organized. Due to the presence of alternating water-deficient and water-enriched layers, differences in indices of refraction can be observed under a microscope.

Starch granules swell when heated in water suspension. At the end of swelling, they lose their form; i.e. they gelatinize. The temperature range in which these changes occur and also the extent of swelling at a given temperature are specific (cf. Table 4.20) and may be used for starch source identification.

Cereal starches consist of about 25% amylose and 75% amylopectin (cf. Table 4.20). The chemical structures of these polysaccharides are presented in 4.4.4.14.3 and 4.4.4.14.4. Starch granules in some cultivars, for instance waxy corns, contain only amylopectin, while some cultivars are rich in amylose (cf. Table 4.20). Waxy corn starch swells considerably on heating, while granules with amylose swell only slightly (cf. Table 4.20 and Fig. 4.31).

Lipids (Table 15.21) and proteins (about 0.5%) are among the heterogeneous constituents of starch granules. Lipids are enclosed within the amylose helices. In wheat starch,

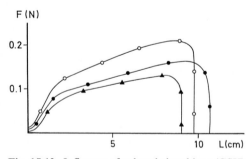

Fig. 15.12. Influence of reduced glutathione (GSH) on the rheological properties of wheat doughs (according to *Sarwin* et al., 1992). Micro-scale extensigrams with flour doughs (cultivar *Kanzler*, 0.6% ash in dry weight, 35 nmole/g, GSH) water, 2% NaCl and the following GSH additives (nmole/g flour): ○—○ without, ●—● 20, ▲—▲ 30. F: force, L: distance

Table 15.21. Lipids in various cereal starches

	Wheat	Corn[a] (maize)	Amylo-maize[a]	Waxy maize[a]
	(% or mg/100 g)[b]			
Nonpolar lipids	6%	60%	73%	88%
Sterol esters	2	3	9	7
Triacylglycerols	15	5	16	12
Diacylglycerols	7	3	16	6
Monoacylglycerols	8	12	13	5
Free fatty acids	27	380	650	105
Glycolipids	5%	1%	5%	6%
Sterol glycosides	3	7	13	3
Monogalactosyldiacyl-glycerols	4			1
Monogalactosylmono-acylglycerols	10		18	
Digalactosyldiacyl-glycerols	11			2
Digalactosylmono-glycerols	24		17	3
Phospholipids	89%	39%	22%	6%
Lyso-phosphatidyl ethanolamines	104	17	16	1
Lyso-phosphatidyl glycerols	23	6	7	trace
Lyso-phosphatidyl cholines	783	226	183	8
Lyso-phosphatidyl serines Lyso-phosphatidyl inositols	26	8	6	trace
Total lipids	1,047	667	964	153

[a] Amylose content in starch amounts to 23% (corn), 70% (amylomaize) and 5% (waxy maize cultivars).
[b] Results for lipid classes are expressed as % of total lipids present in starch, and for individual lipid compounds as mg/100 g starch dry matter.

they consist predominantly of lysolecithins (Table 15.21). They are extractable from partially gelatinized starch by using hot water-saturated butanol. During extraction, the lipid in the amylose helix is replaced by butanol.

The lipids complexed within the starch granules retard swelling and increase their gelatinization temperatures; thus they influence the baking behavior of cereals and the properties of the baked products.

15.2.4.2 Polysaccharides Other than Starch

Cereals contain polysaccharides other than starch. In endosperm cells their content is much less than that of starch (cf. Table 15.22). They include hemicelluloses, pentosans, cellulose, β-glucans and glucofructans. These designations are not uniformly applied in the literature. This is partly due to the fact that analytical criteria are not yet reliable for unequivocal classification. These polysaccharides are primarily constituents of cell walls, and are more abundant in the outer portions than the inner portions of the kernel. Therefore, their content in flour increases as the degree of fineness increases (cf. rye as an example in Table 15.29).

From a nutritional and physiological viewpoint, soluble and insoluble polysaccharides other than starch and lignin (cf. 18.1.2.5.1) are also called dietary fiber. The most important fiber sources are cereals and legumes, while their content in fruits and vegetables is relatively low.

15.2.4.2.1 Pentosans

The pentosan content of cereals varies. Rye flour is exceptionally rich (6–8%) in comparison to wheat flour (2–3%). A portion of pentosans, 1–1.5% in wheat and 15–25% in rye, is water-soluble. Some authors have designated the water-insoluble pentosans as hemicelluloses.

Unlike the water-soluble proteins of cereals, the soluble pentosans are able to absorb 15–20 times more water and thus can form highly viscous solutions. This soluble fraction consists mainly of a linear arabinoxylan, which is insoluble in 80% ethanol, and a soluble highly branched arabinogalactan peptide. A chain of D-xylopyranose units is typical of the structure of arabinoxylan. The OH groups in the 2- and 3-position of this chain are glycosidically linked to L-arabinofuranose (e.g. 3-position in Fig. 15.13). The arabinose residues can be cleaved by mild acid hydrolysis or treatment with an α-L-arabinofuranosidase, giving water-insoluble xylan. The backbone of the arabinogalactan peptide is made of β$(1 \rightarrow 3)$ and β$(1 \rightarrow 6)$ linked galactopyranose units. It is α-glycosidically bound and contains, in addition, arabinofuranose residues. The bonding to the peptide is achieved via 4-trans-hydroxyproline.

The insoluble portion of pentosans swells extensively in water. This portion is responsible for the rheological properties of dough and the baking behavior of rye, and increases the crumb juiciness and chewability of baked products. An optimum starch-pentosan ratio is 16:1 (by weight) for rye flour. Pentosans also play an important role in wheat baking quality, since they also participate in gluten formation (cf. 15.1.5).

Insoluble pentosans are solubilized by alkali treatment. Pentosan preparations from rye, when prepared and additionally purified by

Table 15.22. Distribution of carbohydrates in wheat (%)

	Endosperm	Germ	Bran
Pentosans and hemicelluloses	2.4	15.3	43.1
Cellulose	0.3	16.8	35.2
Starch	95.8	31.5	14.1
Sugars	1.5	36.4	7.6

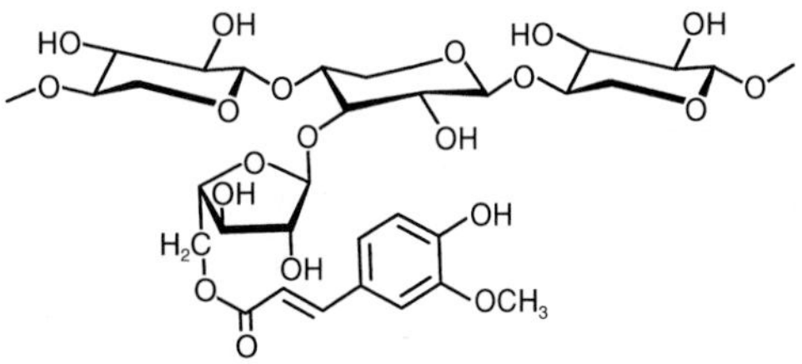

Fig. 15.13. Structures of a portion of a $(1 \rightarrow 4)$-β-xylan chain substituted at the 3 position of one residue with a 5-0-trans-feruloyl-α-L-arabinofuranosyl substituent

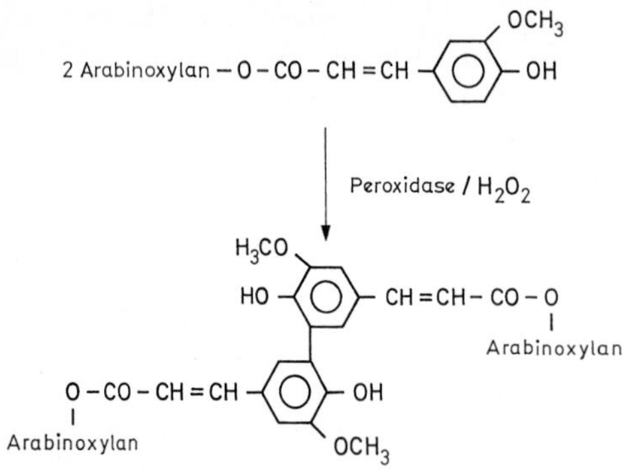

Fig. 15.14. Oxidative crosslinking of cereal pentosans

electrodialysis and added to wheat flour (2%), improve flour baking quality, in particular bread volume and crumb texture, but decrease the freshness. Additionally, pentosans allow wheat flour to be blended with 5–10% of nonbread cereals (corn, millet, soy) without affecting its baking quality. Such flour mixtures are designated as "composite flours".

Pentosan solutions gel when treated with hydrogen peroxide/peroxidase. This is due to the presence of low levels of ferulic acid (ca. 0.2%) (cf. Fig. 15.13). An enzymic phenol oxidation occurs (cf. Fig. 15.14), which causes polymerization. This results in build-up of a network which, along with the low content of branched arabinofuranose, is responsible for the lack of solubility of most pentosans.

15.2.4.2.2 β-Glucan

The β-glucan content of cereals varies: barley 3–7%, oats 2.2–4.2%, wheat and rye kernels only 0.5–2%. These are linear polysaccharides with D-glucopyranose units joined by β-1,3 and β-1,4 linkages. Polysaccharides of the β-glucan type are also called lichenins. At 38°C, 38–69% of the β-glucans of barley dissolve in 2 hours and 65–90% of the β-glucans of oats. β-Glucans are slimy mucous substances which provide a high viscosity to water solutions. In beer production from barley β-glucans can interfere in wort filtration.

15.2.4.2.3 Glucofructans

Wheat flour contains 1% water-soluble, nonreducing oligosaccharides of molecular weight up to 2 kdal. They consist of D-glucose and D-fructose. Glucofructan, which predominates in durum wheats, probably has the following structure:

$$\text{ß-D-Fru}_f\,(2{\to}6)\text{ß-D-Fru}_f\,(2{\to}6)\text{ß-D-Fru}_f\,(2{\to}1)\alpha\text{-D-Glc}_p$$
$$\overset{\displaystyle 1}{\underset{\displaystyle 2}{\uparrow}}$$
$$\text{ß-D-Fru}_f \qquad\qquad (15.6)$$

15.2.4.2.4 Cellulose

Cellulose is a minor constituent of the carbohydrate fraction obtained from starchy endosperm cells (cf. Table 15.22).

15.2.4.3 Sugars

Mono-, di- and trisaccharides, as well as other low molecular weight degradation products of starch, occur in wheat and other cereals in relatively low concentrations (Table 15.23). When starch degradation occurs during dough making, their levels increase (cf. 15.4.2.5). Mono-, di- and trisaccharides are of importance for dough leavening in the presence of yeast (cf. 15.4.1.6.1).

15.2.5 Lipids

Cereal kernels contain relatively low levels of lipids; nevertheless, differences occur among cereals (cf. Table 15.6). The endosperm cells of oats contain a higher level of lipids (6–8%) than wheat (1.6%). For this reason, the overall lipid content of oats is higher than in wheat and in other cereals.

Table 15.23. Mono- and oligosaccharides in wheat flour

Compound	(%)
Raffinose	0.05–0.17
Glucodifructose	0.20–0.30
Maltose	0.05–0.10
Saccharose	0.10–0.40
Glucose	0.01–0.09
Fructose	0.02–0.08
Oligosaccharides[a]	1.2 –1.3

[a] Fraction soluble in 80% ethanol.

Table 15.24. Average fatty acid composition of acyl lipids of cereals (weight-%)

	14:0	16:0	16:1	18:0	18:1	18:2	18:3
Wheat		20	1.5	1.5	14	55	4
Rye		18	<3	1	25	46	4
Corn		17.7		1.2	29.9	50.0	1.2
Oats	0.6	18.9		1.6	36.4	40.5	1.9
Barley	2	22	<1	<2	11	57	5
Millet		14.3	1.0	2.1	31.0	49.0	2.7
Rice	1	<28	6	2	35	39	3

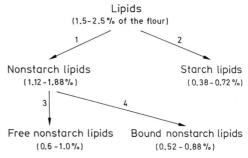

Fig. 15.15. Differentiation of wheat flour lipids by their solubility. 1 Flour extraction with water-saturated butanol (WSB) at room temperature, 2 with WSB at 90–100°C, 3 with petroleum-ether, and subsequently, 4 with WSB

The lipids are preferentially stored in the germ which, in the case of corn and wheat, serves as a source for oil production (cf. 14.3.2.2.4). Lipids are stored to a smaller extent in the aleurone layer.

Cereal lipids do not differ significantly in their fatty acid composition (Table 15.24). Linoleic acid always predominates. Close attention has been given to wheat lipids since they greatly influence baking quality and they have therefore been studied thoroughly.

A wheat kernel weighs 30–42 mg and contains 0.92–1.24 µg of lipid. The germ and the aleurone cells are rich in triglycerides, which are present as spherosomes, while phospholipids and glycolipids predominate in the endosperm.

Wheat flour contains 1.5–2.5% lipids, depending on milling extraction rate. Part of this lipid is nonstarch lipid. This portion is extracted with a polar solvent, water-saturated butanol, at room temperature. Nonstarch lipid comprises about 75% of the total lipid of flour (Fig. 15.15). The residual lipids (25%) are bound to starch (cf. 15.2.3.1).

Nonstarch- and starch-bound lipids in wheat differ in their composition (cf. Table 15.25 and Table 15.21). In nonstarch-bound lipids the major constituents are the triacylglycerides and digalactosyl diacylglycerides, while in starch-bound lipids, the major constituents are lysophosphatides in which the acyl residue is located primarily in position 1. A decrease in amylose content is accompanied by a decrease in the lipid content (Table 15.21). The ratios of nonstarch-bound lipid classes are dependent on the flour extraction grade. An increase in extraction grade increases the triacylglyceride content, since more of the germ is transferred into the flour.

The rheological dough properties are affected by nonstarch-bound lipids which are separated into free and bound lipids when extracted with solvents of different polarity. The free lipid fraction contains 90% of the total nonpolar lipids and 20% of the total polar lipids listed in Table 15.25. By kneading the flour into dough, the glycolipids become completely bound to gluten, while other lipids are only 70–80% bound. The extent of binding of triacylglycerides depends on dough handling. Intensive oxygen aeration and, particularly, addition of lipoxygenase (cf. 15.4.1.4.3) increase the fraction of free lipids.

The increased binding of lipids in the transition of flour to dough, which is expressed in their decreasing extractability, is explained by the following hypothesis.

The neutral lipids are present in flour in the form of spherosomes and their membranes are formed by a part of the phospholipids. The spherosomes can be extracted with nonpolar solvents. The other phospholipids and all the glycolipids form inverse hexagonal phases (cf. 8.15.2.2), which are only partly extractable. During dough making, the water added results in the conversion of the inverse hexagonal to a laminar phase, which in turn stabilizes a microemulsion of the neutral lipids. The microemulsion vesicles are enclosed by the network of gluten proteins and, consequently, difficult to extract. If the dough is suspended in water, the lipids appear in the aqueous phase that separates on ultracentrifugation only when the framework of gluten proteins

Table 15.25. Nonstarch lipids in wheat flour

	mg/100 g[a]
Nonpolar lipids (59%)	
Sterol lipids	43
Triacylglycerols (TG)	909
Diacylglycerols (DG)	67
Monoacylglycerols (MG)	53
Free fatty acids (FFA)	64
Glycolipids (26%)	
Sterol glycosides	18
Monogalactosyldiacylglycerols (MGDG)	115
Monogalactosylmonoacylglycerols (MGMG)	17
Digalactosyldiacylglycerols (DGDG)	322
Digalactosylmonoacylglycerols (DGMG)	52
Phospholipids (15%)	
N-Acyl-phosphatidyl ethanolamines	95
N-Acyl-lyso-phosphatidyl ethanolamines	33
Phosphatidyl ethanolamines	
Phosphatidyl glycerols	19
Phosphatidyl cholines	96
Phosphatidyl serines	
Phosphatidyl inositols	9
Lyso-phosphatidyl glycerols	5
Lyso-phosphatidyl cholines	29

[a] Based on dry matter.

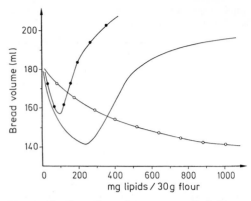

Fig. 15.16. The effect of free nonstarch lipids on the baking quality of defatted wheat flour (according to *W.R. Morrison*, 1976). — Lipids (total), −o−o− nonpolar lipids, −•−•− polar lipids

has been destroyed by reduction, e.g., with dithiothreitol.

Other hypotheses which explain the decreasing extractability of free lipids by selective binding, e.g., of glycolipids to starch and gluten, have not been confirmed.

The gas-holding capacity of doughs and, after passing through a minimum, the baking volume (Fig. 15.16) are positively influenced by polar lipids. Two effects are assumed in explanation. The polar lipids get concentrated in the boundary layer gas/liquid and stabilize the gas bubbles against coalescence. In addition, the lipid vesicles seal the pores which are formed in the protein films on kneading. On the other hand, the nonpolar lipids generally negatively influence the backing result with most varieties of wheat (Fig. 15.16).

Carotenoids and tocopherols belong to the minor components of the cereal lipid fraction.

Wheat flour has a carotenoid content averaging 5.7 mg/kg. In durum wheats, which have a more intense yellow color, the carotenoids are 7.3 mg/kg of flour.

The major carotenoid, lutein (cf. 3.8.4.1.2), is present in free or esterified form (either mono- or diester) with the fatty acids listed in Table 15.24. The following carotenoid pigments are also present: β-carotene, β-apo-carotenal, cryptoxanthin, zeaxanthin and antheraxanthin (for structures see 3.8.4.1). Carotenoid content of corn, depending on the cultivar, is 0.6– 57.9 mg/kg, with lutein and zeaxanthin being the major constituents.

The composition of the tocopherols of wheat (Table 15.26) show that the proportions of germ and aleurone lipids in nonstarch lipids can be determined by using β-T and β-T-3 as markers. Values of ca. 25% have been found, but they can fluctuate greatly depending on the milling process and extraction grade.

Table 15.26. Tocopherol content of parts of the wheat kernel

Part of kernel	Tocopherols in mg/kg			
	α-T	β-T	α-T-3	β-T-3
Germ	256	114	n.d.	n.d.
Aleurone layer	0.5	n.d.	10	69
Endosperm	0.07	0.10	0.45	13.5

n.d., not detectable.

15.3 Cereals – Milling

15.3.1 Wheat and Rye

Quality control of the raw materials and milling products usually includes the determination of water, protein, and minerals. The absorption bands of food in the near infrared region (0.8–2.6 μm) are suitable for a quick basic analysis (water, protein, fat, carbohydrates etc.).

The overtones of CH, OH, and NH valence vibrations appear in the near IR region. Therefore, foods give a large number of absorption bands that can be assigned to definite components and have intensities that correlate with the amounts of the constituents. As an example, Fig. 15.17 shows the absorption of wheat in the near IR region. The sample containing water absorbs at 1.94 μm in addition. Therefore, after subtraction of the absorption of dried wheat and after calibration, the water content can be determined. Other consituents which can be determined in food by near-infrared (NIR) spectrophotometry are listed in Table 15.27.

In the development of methods for these materials, the measurement of IR reflection was at first given the most attention because it is technically easier to perform. Since reproducible

Table 15.27. Examples of quantitative analysis of foods by near-infrared (NIR) spectrophotometry

Component	Food
Water	Meat, cereals, control of fruit and vegetable drying processes, chocolate, coffee
Protein	Meat, cereals, milk and milk products
Fat	Meat, cereals, milk and milk products, oil seeds
Minerals	Cereals, meat
Starch	Cereals
Pentosans	Wheat
β-Glucans	Barley
Lysine	Wheat, barley

results can be obtained only if the surface and granulation of the samples are constant, sources of error arise here. In the meantime, however, technical improvements allow food, e.g., cereal kernels, to be irradiated in the range of 0.8–1.1 μm. Thus, the water and protein content in unground samples can also be determined by measuring the transmission. In food technology, measurements in the near IR region are widely used for the quick quality control of raw materials (Table 15.27).

15.3.1.1 Storage

Cereals can be stored without loss of quality for 2 to 3 years, provided that the kernel moisture content, which is 20–24% after threshing, is reduced to at least 14%. The low moisture content prevents microbial spoilage, especially by mycotoxin-forming organisms, and it also lowers kernel respiration, i.e., metabolism.

The water is slowly removed from grains by ripple-type dryers in a stream of hot air or burned gas at 60–80 °C (to the extent of 4% per passage) to avoid damage to kernels by uncontrolled shrinkage. Grains with high moisture content can be stored for short periods of time in the cold without quality deterioration. Stored grains are fumigated for pest control. Aluminum and magnesium phosphides are introduced. At 20 °C and 75% relative humidity, they decompose into gaseous PH_3. HCN or ethylene oxide fumigants are also used.

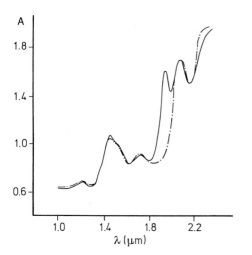

Fig. 15.17. Absorption of ground wheat in the near IR region. Sample dried (–·–·–) and with 9 w/w % water (——)

Wheat and rye are suitable for the production of bakery products, especially bread, and are called bread cereals. Other cereals serve only as additives for bakery products and are mainly used in other ways, e. g., for porridge and pancakes.

15.3.1.2 Milling

The aim of milling is to obtain preferentially a flour in which the constituents of the endosperm cells predominate. The outer part of the kernel, including the germ and aleurone layer (cf. Fig. 15.2) is removed. Such a requirement is not easy to accomplish since the kernel's groove and the unequal sizes of aleurone cells in cereals do not facilitate simple dehulling. Therefore, the grain has to be carefully broken, the particles sorted and separated by size and, only then, further disintegrated.

In a preliminary step to milling, the grain is cleaned of impurities such as weed seeds, straw, soil particles, spoiled decayed grains, dust, etc. This cleaning step is based on the cereal's kernel size and specific gravity. Washing with water is rarely done, since it promotes the growth of microorganisms.

The next step is grain wetting or steeping in water for 3–24 h, since an increased moisture content to 15–17% facilitates the separation of starchy endosperm cells from germ and hull. An alternative procedure is wheat conditioning at elevated temperatures up to 65 °C; it is faster than steeping and also favorably affects the baking quality. The kernels are disintegrated stepwise. Each passage through rollers involves particle size reduction by pressure and shear forces, followed by flour separation according to particle size using

sieves in the form of flat sifters (Fig. 15.18). Rollers are matched to the product needed. Their size, surface flutes, rotation velocity, gap between pairs of rollers rotating in opposite directions at dissimilar speeds – all can be selected or adjusted. Wheat and rye are milled differently because of structural differences in the kernels. The wheat kernel is rather brittle; the rye kernel is gluey or sticky. Therefore, rye is less suitable for coarse grist milling than wheat. The wheat milling process can be adjusted so that the first passages provide the grist and the following ones provide the flour.

The germ of the rye kernel, because of its loose attachment, falls off readily during the cleaning step, while the wheat germ is removed only on sifters. The hull and a substantial part of the aleurone layer is removed in form of bran.

A portion (ca. 5–8%) of the starch granules is mechanically damaged during milling. The extent depends both on the type and intensity of milling and on the hardness of the kernel. The harder the structure of the kernel, the greater the damage. Since the rate of water absorption during dough making and the enzymatic degradation of starch increase with increasing damage, they are important for the baking process and desirable to a limited extent. To measure starch damage, the amylose extractable with a sodium sulfate solution is determined. Alternatively, the amount of starch degradable without gelatinization, e. g., at 30 °C by α- and/or β-amylase is determined. The starch damage expected during the milling process can also be estimated by determining the hardness of the kernel, e. g., by NIR reflectance spectrophotometry (cf. Table 15.27).

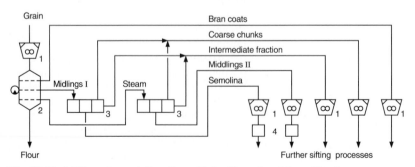

Fig. 15.18. Milling of cereal (1: roller mill, 2: sifters, 3 and 4: purifiers)

15.3.1.3 Milling Products

A miller distinguishes the end-products of milling on the basis of particle size or diameter, e.g., >500 µm for grist; 200–500 µm for semolina from durum or farina from bread wheats; 120–200 µm for "dunst"; and 14–120 µm for flour. The larger flour particles can be felt between the fingers (graspable flour), as opposed to smooth or polished flours in which the average particle size is 40–50 µm.

Differently milled flours vary considerably in baking quality. Flours obtained also differ greatly from cultivar to cultivar. This is especially the case with wheat cultivars (cf. 15.4.1.1). In addition, quality depends on whether the milled flour comes from the inner or outer parts of the endosperm. Therefore, milled flour is controlled in the plant for its baking properties and blended or mixed to yield a commercial product based on present market standards (see also below). The characteristics of a few milling products and their applications are listed in Table 15.28.

The chemical composition of the flour depends on the milling extraction rate, e.g., flour weight obtained from 100 parts by weight of grain. Examples are given in Table 15.29. Increasing the rate of flour extraction decreases the proportion of starch and increases the amount of kernel-coating consti-

tuents such as minerals, vitamins and crude fiber (cf. Tables 15.8 and 15.9). Comparing products of the same extraction rate, rye flour contains higher proportions of both minerals and vitamins than wheat flour (Fig. 15.19). It should be pointed out that in the case of some B-vitamins, such as niacin, this difference is well-balanced by the higher concentrations in wheat in comparison to rye kernels (cf. Table 15.6). Consequently the concentrations of such vitamins are similar in rye and wheat flour.

Bread flours are standardized on the basis of their ash content in Europe and, particularly, Germany.

The type of flour = ash content (weight %) × 1000 corresponds to the extraction grade. Examples are provided in Table 15.29 for wheat and rye flours and their chemical composition is detailed. Protein and starch contents are also related to flour particle size (cf. Table 15.30).

Because of the variable particle sizes and densities of protein and starch, a flour sample can be separated by air classification into a fraction enriched in protein and starch. These are the so-called special purpose flours.

Table 15.28. Wheat and rye milling products

All purpose flour	Commercially available (retail market) flour for household preparations of baked products.
Special flour	It is used for special baked products, e.g., strong gluten wheat flour for toast bread, wheat flour with weak gluten for baked goods of loose tender or crispy structure as pastry etc.
Compounded (ready to use) flour	Special flour that contains other ingredients such as milk or egg powder, sugar etc., required by formulation of a selected baked product.
Groats (grist)	Coarsely ground dehulled cereal (devoid of germ and seed hull).
Whole grain groats	Ground from whole kernel (including germ).

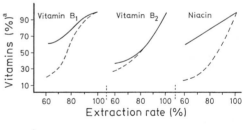

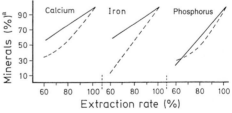

Fig. 15.19. Content of B vitamins and minerals in flour as affected by milling extraction rate (according to Lebensmittellexikon, 1979). —— rye, – – – wheat

[a] Calculated as percent of the total content present in grain.

Table 15.29. Average composition of wheat and rye flours[a]

A. Wheat flour

	Type				
	405	550	812	1050	1700[b]
	Flour extraction rate[c]				
	40–56%	64–71%	76–79%	82–85%	100%
Starch	84.2	81.8	78.4	78.2	66
Protein (N × 6.25)	11.7	12.3	13.0	13.3	14.8
Lipids	1.0	1.2	1.5	1.9	2.3
Dietary fiber[d]	3.7	3.7		4.9	10.9
Minerals (ash)	0.41	0.55	0.81	1.05	1.7

B. Rye flour

	Type				
	815	997	1150	1370	1740
	Flour extraction rate[c]				
	69–72%	75–78%	79–83%	84–87%	90–95%
Starch	77.5	74.6	72.2	69.3	62.8
Protein (N × 6.25)	9.6	10.1	10.6	11.2	12.4
Lipids	1.1	1.1	1.3	1.5	1.5
Insoluble pentosans	3.8	4.3	4.8	5.2	6.5
Soluble pentosans	1.4	1.5	1.6	1.7	1.9
Minerals (ash)	0.82	1.0	1.15	1.37	1.74

[a] Weight-% per dry matter of wheat and rye flours. Flour average moisture content is 13 weight-%.
[b] Whole wheat flour.
[c] Approximate data.
[d] Indigestible carbohydrates (water soluble and insoluble), lignin.

Table 15.30. Protein content of wheat flours as affected by flour particle size

Particle size (µm)	As portion of flour (weight %)	Protein content (weight %)
0–13	4	19
13–17	8	14
17–22	18	7
22–28	18	5
28–35	9	7
> 35	43	11.5

The commercial product semolina ("griess") is made from endosperm cells of hard durum wheats. Semolina keeps its integrity during cooking and is used mostly for pasta production. Since semolina is a milled flour of low extraction rate, it contains few minerals and vitamins.

15.3.2 Other Cereals

15.3.2.1 Corn

Corn endosperm, with the germ removed, is ground to grist for corn porridge (Polenta) and into corn flour for flat cakes (tortillas). Corn flakes are made from cooked and sweetened corn slurry, by drying, flaking and toasting. Similar products are made from millet, rice and oats.

15.3.2.2 Hull Cereals

Dehulling of rice, oats and barley requires special processes (cf. 15.1.4).

15.3.2.2.1 Rice

Rice milling involves the following processing steps: rough rice (paddy rice) → hull removal

→ brown rice → polishing to remove the bran coats (fruit and seed coats), the silvery cuticle, the germ and the aleurone layer → rubbing-off or rice polishing to obtain the end-product, white rice. Undamaged rice (45–55%), broken kernels or flour (20–35%) and a husk/hull fraction (20–24%) are obtained.

Polished white rice is made from this cleaned rice by additional treatment of the kernels with talc (a magnesium silicate) and 50% glucose solution. This imparts a glossy, transparent coating to the kernels.

White rice, in comparison to rough or brown rice, is low in vitamin content (cf. Table 15.31) and in minerals. A nutritionally improved product may be obtained by a parboiling process, originally developed to facilitate seed coat removal. About 25% of the world's rice harvest is treated by the following process: raw rice → steeping in hot water, steaming in autoclaves, followed by drying and polishing → parboiled rice.

This treatment causes the following changes: the starch gelatinizes, but partly retrogrades again during drying. Enzymes are inactivated by the heat, causing inhibition of the enzymatic hydrolysis of lipids during storage of rice. The oil droplets (cf. 3.3.1.5) are broken and lipids partly migrate from the endosperm to the outer layers of the rice kernels. Since antioxidants are simultaneously destroyed, parboiled rice is more susceptible to lipid peroxidation. In contrast, minerals and vitamins diffuse from the outer layers to the inner endosperm and remain there after the separation of the aleurone layer (Table 15.31). The changes in starch mentioned above result in reduced cooking time.

Unlike in Europe and USA, some rice varieties popular in Asia develop a popcorn-like aroma on cooking. This is due to the formation of 2-acetyl-1-pyrroline, which is present in concentrations of 550–750 µg/kg in aromatic varieties of rice (cooked) and <8 µg/kg in low-aroma varieties.

15.3.2.2.2 Oats

Oat flakes are produced by the following processing steps: the kernels are steamed and then the moisture content is decreased to 5% by heating at 75 °C for 60–90 min. The hull (fruit and seed coats) is removed, i.e., the kernel is polished. This is followed by repeated steaming, squeezing between drum rollers, and drying of the moist flakes. The yield is 55–65%. This hydrothermic process also inactivates the oat enzymes involved in off-flavor development.

15.3.2.2.3 Barley

Removal of hull (fruit and seed coatings) yields groats which, after grinding, provide marketable products of large or fine particle size.

15.4 Baked Products

Baked products (for a review, see Table 15.32) are made from milled wheat, rye and, to a lesser extent, other cereals by the addition of water, salt, a leavening agent and other, ingredients (shortening, milk, sugar, eggs, etc.). The following operations are involved:

Table 15.31. Vitamin content of raw, white and parboiled rice

	B-vitamins (mg/kg)		
	Thiamine	Riboflavin	Niacin
Raw rice	3.4	0.55	54.1
White rice	0.5	0.19	16.4
Parboiled rice	2.5	0.38	32.2

Table 15.32. Classification of baked products

Bread including small baked products (rolls, buns)	Made entirely or mostly from cereal flours; moisture content on average 15%. Addition of sugar, milk and/or shortenings amounts to less than 10%. Small baked products differ from bread only by their size, form and weight.
Fine baked goods, including long term or extended shelf life products such as biscuits, crackers, cookies etc.	Made of cereal flours with at least 10% shortening and/or sugar, as well as other added ingredients. In baked goods for long shelf life the moisture content is greatly reduced.

- Selection and preparation of the raw materials
- Dough making and handling
- Baking
- Measures for quality preservation

15.4.1 Raw Materials

Among the ingredients involved in a formulation, only flour and those additives which affect dough rheological and/or baking properties will be covered. Flour improvers and dough leavening agents will be emphasized.
Characterization of the raw materials and additives is, in practice, made by assessing the dough rheological properties and by baking tests. Basic research endeavors to understand the nature of flour constituents and the reactions which affect their behavior in dough handling and baking.

15.4.1.1 Wheat Flour

A flour of optimal baking properties is required and chosen to match the quality of the desired product (cf. Table 15.28). The baking quality of wheat is strongly influenced by the cultivar (cf. Table 15.34) and also by conditions of growth and cultivation (climate, location), and subsequently by flour storage conditions and duration. Prior quality control is of importance to assess the overall baking quality of wheat flour. Flour particle size and color are assessed by sensory analysis. Graspable flours (cf. 15.3.1.3) are made from hard gluten-rich cultivars. Water uptake is slow when compared to smooth flour, and they make dry doughs. The color difference is important, and is assessed with a wetted flour sample on a black background (*Pekar*-test).

15.4.1.1.1 Chemical Assays

Flour acidity (ml of 0.1 mol/l NaOH/10 g, titrated in the presence of phenolphthalein) depends upon the extraction rate of the flour and ranges between 2.0 ml/g (flour type 450) and 5.5 ml/g (flour type 1800). Too low acidity often reflects poorly aged flour. Acidity above 7.0 suggests microbial spoilage.
The content of gluten, which is the residue left after the dough is washed (10 g flour kneaded

into a dough with 6 ml of 2% NaCl, then washed with tap water), provides an indication of flour quality. A very low gluten content (< 20%) frequently results in dough deterioration when machine-handled and also in baking faults. A higher content of gluten will not guarantee good baking quality (see "Maris Huntsman" cultivar, Table 15.34). Gluten swelling power is assessed by a *sedimentation value* as recommended by *Zeleny*. The flour is mixed in a solution of lactic acid in aqueous isopropanol and left to stand. The higher the volume of the sedimented gluten, the better should be the baking quality of the flour.
For a given wheat cultivar, grown under similar climatic and soil conditions, the baking volume correlates with the protein content of the flour (Fig. 15.20). A similar linear relationship is not readily attainable for flours from different cultivars, as evidenced by the very different slopes of the regression lines.
The parameters involved here are definitely those described in Section 15.2.1.4 as respon-

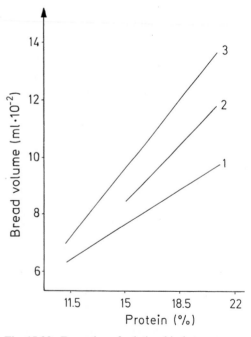

Fig. 15.20. Examples of relationship between protein content of flour and bread volume (according to *Y. Pomeranz*, 1977). United States winter wheat cultivars: 1 Chiefkan, 2 Blackhull, 3 Nebred. The regression lines are based on numerous sample analyses

sible for the properties of gluten. These include type, amount, and degree of polymerization of the HMW and LMW subunits of glutenin as well as the ratio gliadin/glutenin. Numerous attempts have been made to correlate these parameters, either individually or in a suitable combination, with the baking properties of flours with the purpose of making predictions possible. Figure 15.21 shows, for example, that the resistance to extension of doughs from different varieties of wheat correlates well with the concentration of the HMW subunits of glutenin of type-x (r = 0.89), but not with that of type-y (r = 0.29).
On the whole, the structure of wheat gluten has not been studied enough to be able to safely describe variety-specific differences in technological properties.

Wheat cultivars differ in the content of their thiol and disulfide groups (Table 15.33). This implies that the stability of a dough may be strongly influenced by a SH/SS exchange between a low molecular weight SH-peptide and gluten proteins (cf. 15.4.1.4.1). This also implies that a positive correlation between the contents of SH- and -SS-groups in flour, or their ratios, would be reflected in baking quality. However, low correlation coefficients of about 0.6 have been found, suggesting that such relationships are much more complex than assumed in a model system. Hence, SH- and -SS-data, obtainable by simple analysis, are not suitable for flour quality assessment.
Of all enzymes in flour, quality control is aimed at the determination of amylase activity. The Falling Number test (*Hagberg* and *Perten*)

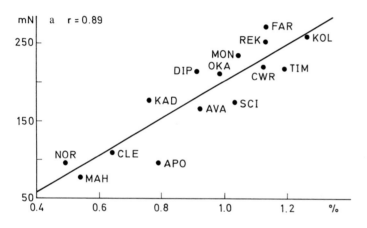

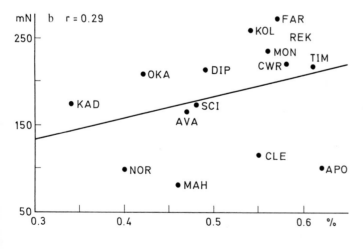

Fig. 15.21. Correlation between the maximal resistance to stretching of dough from various wheat cultivars in a micro-scale extension test and the concentration of HMW subunits (% based on flour) of x- (a) and y-type (b). (according to *Wieser* et al., 1992)

Table 15.33. Concentration of SH- and SS-groups in flour of different wheat cultivars

Cultivar	SH	SS	SS/SH
	μmole per g flour		
Kolibri	1.15	12.5	10.9
Caribo topfit	0.88	12.2	13.9
Strong Canadian wheats	0.95	13.4	14.1
Inland wheat I [a]	0.75	10.2	13.6
Inland II [a]	1.05	12.6	12.0
Canadian Western Red Spring Wheat (CWRS)	1.26	12.9	10.2

[a] Marketed flour blendings.

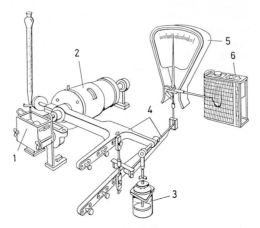

Fig. 15.22. Farinograph (according to *M. Rohrlich* and *B. Thomas*, 1967). The apparatus consists of a thermostated mixer or kneader (1), its blades are driven by an electromotor (2). The reaction torque acts through a lever system (4) of analytical balance precision on the indicator scale (5) simultaneously recorded on a strip chart recorder (6). The movement of the lever system is damped by an oil dash pot (3). The farinogram is a diagram of force versus time

serves this aim. A piston-type mixer falls through an aqueous flour paste. The falling time of the piston is measured for a given distance under standard conditions. The results are related, among other things, to starch granule stability in the presence of amylase enzymes. Dextrin values should be determined to assess amylase activity specifically. In a method developed by *Lemmerzahl*, the extent of standard dextrin hydrolysis in the presence of flour extract is measured. The fermentation power of a flour (cf. 15.4.1.6.1) involves determination of the maltose value (diastatic activity). This is a quantitative determination of reducing sugars prior to and after incubation of a flour suspension at 27 °C for 1 h. Flours with a maltose content of <1.0% are regarded as weak fermentation promoters; values above 2.5% are flours from sprouted kernels. They provide poor baking quality.

15.4.1.1.2 Physical Assays

The instruments widely used in practice for the determination of the rheological properties of dough can be divided into recording dough kneaders and tensile testers. Dough development is followed with a Brabender farinograph (Fig. 15.22), with which measurement is made of the volume of water absorbed by the flour in order to make a dough of predetermined consistency (normal consistency). A plot of dough consistency versus time is recorded, as shown in Fig. 15.23.
In addition to the water absorption, the shape of the farinogram is used to characterize a

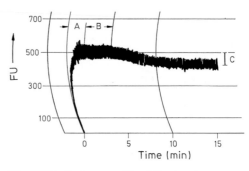

Fig. 15.23. Farinogram. The following data are pertinent for quality assessment of flour: A dough development time, B dough stability (dough consistency does not change), C decrease in dough consistency after a given time, here 12 min. FU: farinogram units

flour. Various indices have been defined (cf. Fig. 15.23); usually they refer to doughs with a maximum consistency of 500 FU.
Flours with strong gluten absorb more water and show longer dough development and stability times than do flours with weak gluten (Table 15.34). Corresponding results are obtained with the Swanson and Working mixographs.

A standardized piece of dough is stretched with the hook of a Brabender extensograph until the piece breaks (Fig. 15.24). As shown in Fig. 15.26, a graph of force (resistance to extension) versus stretching distance (extensibility) provides information about the stability of a dough, its gasholding capacity and fermentation tolerance. Of the examples given in Table 15.34, the "Monopol" cultivar obviously has strong gluten. The "Nimbus" cultivar has short gluten, as reflected by its low extensibility. The "Maris Huntsman" cultivar has a very weak gluten, as shown by the low resistance of its dough to extension and also by its low extensibility, and very small extension area.

Similar results are obtained with the Chopin extensograph or alveograph used widely in France. A piece of dough mounted on a perforated plate is blown into a ball. The pressure in the ball of dough is plotted against the time (cf. Fig. 15.25). In contrast to the Brabender extensograph, the dough is extended in two dimensions. As in the extensogram, the resistance of the dough to extension and its extensibility are obtained from the maximal height and width of the alveogram.

15.4.1.1.3 Baking Tests

Straightforward and reliable information about the baking quality of a flour is obtained from baking tests under standardized conditions. Baking volume (cf. Table 15.34), form, crumb structure and elasticity, and the taste of the baked product are evaluated. A baking test is performed with at least 100 g flour for each product.

When the effects of expensive and not readily available flour constituents and/or additives are tested or a new cultivar is assessed, of which only several hundred kernels are available, a "micro baking test" is used, with 10 g flour for each baked product (cf. Fig. 15.30).

Table 15.34. Baking quality data of some wheat flours

	Wheat cultivar[a]		
	Monopol	Nimbus	Maris Huntsman
Protein (% dry matter)[b]	13.2	11.6	11.8
Wet gluten (%)	35.1	24.7	34.3
Farinogram[c]			
Water absorption (%)	59.2	54.8	59.8
Dough development time (min)	5.0	1.0	2.0
Dough stability (min)	5.0	1.5	0.5
Mixing Tolerance Index[d] (FU)	30	80	130
Extensogram[e]			
Area (dough strength, cm^2)	143	75	17
Resistance of the dough to extension (EU)	700	680	110
Extensibility (mm)	170	92	100
Baking test			
Dough surface	somewhat wet to normal	normal	wet, gluey
Dough elasticity	normal	somewhat short	weak
Baking volume (ml)	738	630	510

[a] Wheat cultivars with breadmaking quality corresponding to very good ("Monopol"), average ("Nimbus") and poor ("Maris Huntsman").
[b] Factor N × 5.7.
[c] Explanation in Fig. 15.23; dough consistency: 500 FU.
[d] Measured after 10 min in Farinogram units (FU).
[e] Explanation in Fig. 15.26.

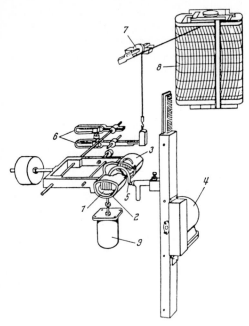

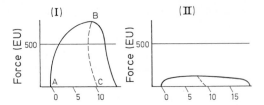

Fig. 15.26. Extensograms of a normal (I) and weak dough (II). For quality assessment the following parameters are determined: resistance to extension, height of the curve at its peak (B–C) given in extensogram units (EU); extensibility, abscissa length between A–C in mm; extension area (A–B–C–A, cm^2) is related to energy input required to reach the maximum resistance; extensogram number (overall dough quality) is the ratio of extension resistance to extensibility

Fig. 15.24. Extensograph (according to *M. Rohrlich* and *B. Thomas*, 1967). The cylindrical piece of dough (1) is fixed by dough clamps (3) and placed on the balance fork (2). The motor (4) of the stretching unit (5) is then started. The arm moves downward into the dough and extends it at constant speed. Simultaneously, the forces opposing the stretching action are transmitted through the lever system (6) to the balance system (7). This is coupled to a recording arm of the strip chart recorder (8). The fork of the balance system is coupled to an oil damper (9) to reduce the recoil

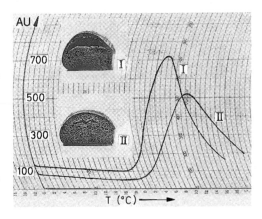

Fig. 15.27. Amylograms of two rye flours (according to *H. Stephan*, 1976)

	Gelatinization maximum (peak)	Gelatinization temperature	α-Amylase
Flour I	720 AU	67 °C	high
Flour II	520 AU	73.5 °C	low

AU: amylogram units.

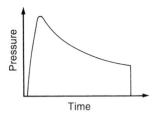

Fig. 15.25. Alveogram (cf. text)

15.4.1.2 Rye Flour

The Falling Number test (cf. 15.4.1.1.1) and an amylographic assay are the most important tests to assess the baking properties of rye flour. These tests depend to a great extent on gelatinization properties of starches and the presence of α-amylase. The higher the α-amylase activity, the lower the Falling Number.

An amylograph is a rotational torsion viscometer. It measures the viscosity change of an aqueous suspension of flour as a function of temperature. The recorded curve, called an amylogram (Fig. 15.27), shows that with increasing temperature there is an initial small fall followed by a steep rise in viscosity to a maximum value. The steep rise is due to intensive starch gelatinization. The viscosity value and temperature at maximum viscosity (i.e., the temperature reflecting the end of gelatinization) are then read.

In rye flour with balanced baking properties, an optimal relationship should exist between α-amylase activity and starch quality. The extent of enzymatic starch degradation influences the stabilty of the gas-cell membranes which are formed by gas released in the dough and which consolidate during baking into an elastic crumb structure. These membranes contain pentosans, proteins and intact starch granules in addition to gelatinized and partially hydrolyzed starch. High α-amylase activity in rye or a large difference between the temperatures needed for enzyme inactivation (close to 75°C) and those required for termination of starch gelatinization will produce poor bread since too much starch will be degraded during breadmaking. The gas-cell membranes are liquefied to a great extent; so the gas can escape. This gas will then be trapped in a hollow space below the bread crust (I in Fig. 15.27). Low α-amylase activity, especially in conjunction with low starch gelatinization, leads to a firm and brittle crumb structure.

15.4.1.3 Storage

Rye flour acquires optimal baking properties after 1–2 weeks of storage after milling. Wheat flour requires 3–4 weeks. This storage period is the flour "maturation time". In wheat the time is needed for oxidative processes to occur and thus provide a stronger (shorter) gluten. Also, the concentration of endogenous glutathione will be lowered by oxidation to its disulfide (Table 15.20), hence gluten stability during dough handling will be enhanced (cf. 15.2.3).

Flour with a moisture content of <12% may be stored at 20°C and a relative humidity of <70% for more than 6 months without significant change in baking quality.

Flour fumigation with Cl_2, ClO_2, NOCl, N_2O_4 or NO, or treatment with dibenzoyl or acetone peroxide results in carotenoid destruction. The flour becomes bleached. Other reactions, not yet elucidated, are involved with Cl_2, NOCl, ClO_2 and acetone peroxide treatment since they provide simultaneous improvement in baking quality of flours which have weak gluten.

15.4.1.4 Influence of Additives/Minor Ingredients on Baking Properties of Wheat Flour

The baking properties of wheat flours differ widely (cf. Table 15.34). In small traditional plants, a baker can use his experience to compensate for changes in the quality of raw materials: flexibility in formulations, dough handling and baking – all these parameters can be adjusted in order to obtain the desired end-product.

In a large-scale automated bakery, economic production demands uniform raw materials with uniform properties. Additives are used when necessary to adjust the flour characteristics to match the baking process (for instance, shortened dough handling time with low energy input). Additives are also used to ensure that the end-product meets existing standards. Incorporation of ascorbic acid, alkali bromates or enzyme-active soy flour improves the quality of weak gluten flour – e.g., in bread or bun baking. In these cases the dough becomes drier and there are increases in dough resistance to extension, mixing tolerance and fermentation stability. In addition, baking volume will increase and the crumb structure will improve. Ascorbic acid and lipoxygenase require oxygen for their actions; hence their beneficial role is very dependent on the intensity of dough mixing, which traps oxygen from the air.

In contrast, opposite effects may be observed by adding cysteine or proteinases, the result being gluten softening. Biscuits are made from such mellowed, softened doughs, which are made with little energy input. Additives which affect the rheological quality of the dough and/or the quality of baked products include emulsifiers, shortenings, salt, milk, soy flour, α-amylase and proteinase preparations and starch syrups.

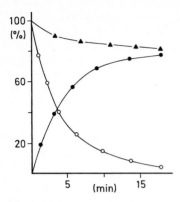

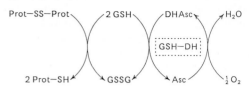

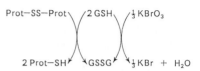

Fig. 15.28. Oxidation of ascorbic acid in dough making from wheat flour (according to *Nicolas* et al., 1980) ○—○ ascorbic acid, ●—● dehydroascorbic acid, ▲—▲ sum of ascorbic and dehydroascorbic acids

Fig. 15.29. Reactions involved in flour improvement by ascorbic acid or potassium bromate (according to *G. Mair* and *W. Grosch*, 1979). Asc: L-threo-ascorbic acid, DHAsc: L-threo-dehydroascorbic acid, GSH: reduced glutathione, GSSG: oxidized glutathione, Prot-SS-Prot: gluten proteins (compare 15.2.1.4), GSH-DH: glutathione dehydrogenase (see 15.2.2.7)

15.4.1.4.1 Ascorbic Acid

The improver effect of ascorbic acid (Asc) was recognized by *Jorgensen* as early as 1935. He found that small amounts (2–6 g Asc per 100 kg flour) caused a pronounced increase in both dough strength and bread volume. The oxidation product of Asc, dehydroascorbic acid (DHAsc) is also effective (Table 15.35), but its use would be uneconomical. The improver action in conventional doughmaking is different from that in the continuous process, in which Asc lowers the requirement for mixing and DHAsc is inactive. The four diastereomers of Asc are differently active as improvers in conventional doughmaking: L-threo-Asc (vitamin C) enhances most strongly the handling and baking characteristics (Table 15.35), both D- and L-erythro-Asc are less active while D-threo-Asc is inactive. This ranking was also shown in rheological measurements. The dehydro forms of the four ascorbic acids show the same ranking as improvers. This specificity suggests that at least one enzyme is involved in the action of ascorbic acid in the conventional process.

The order of substrate specificity of the enzyme glutathione dehydrogenase (GSH-DH) towards the four DHAsc stereoisomers (cf. 15.2.2.7) corresponds with that observed for the improver action of dough. This parallel suggests that GSH-DH is involved in the improver action of Asc via its dehydro-form.

The effect of Asc takes place in two steps (Fig. 15.29). First, Asc is oxidized to DHAsc by oxygen present in the dough. The reaction is catalyzed by an ascorbic acid oxidase and also nonenzymatically by traces of heavy metal ions and heme-containing compounds, e. g., peroxidase or catalase. In the subsequent reaction, endogenous glutathione is withdrawn from the SH/SS interchange with gluten proteins (cf. 15.2.3). The SH/SS interchange, which starts immediately on dough making and leads to depolymerization of gluten proteins and, consequently, to weakening of gluten, is inhibited. According to this hypothesis, the content of free GSH during dough making with L-threo-Asc decreases much faster than during dough making without L-threo-Asc or with D-erythro-Asc (Table 15.36).

15.4.1.4.2 Bromate, Azodicarbonamide

Addition of alkali bromates to flour also prevents excessive softening of gluten during dough making. The reaction involves oxidation of endogenous glutathione to its disulfide (Fig. 15.29). During baking, bromates are completely reduced to bromides with no bromination of flour constituents.

Azodicarbonamide is of interest as a flour improver

$$H_2N-CO-N=N-CO-NH_2 \qquad (15.7)$$

since it improves not only the dough properties of weak gluten flour, but also lowers the energy input in dough mixing (cf. Fig. 15.35). Details of the reactions involved are unknown.

15.4.1.4.3 Lipoxygenase

The addition of a small amount of enzyme-active soy flour to a wheat dough increases the mixing tolerance, improves the rheological properties and may increase the bread volume. The effect on dough rheology is shown only with high-power mixing in the presence of air. The carotenoid pigments of wheat flour are bleached by the addition of enzyme-active soy flour. This is desirable in the production of white bread. The amount of enzyme-active soy flour is restricted to approximately 1% since higher levels may generate off-flavors.

Table 15.35. Effect of additives on the rheological properties of wheat dough

Additive (0.15 μmol/g flour)	Resistance to extension (%)	Extensibility[a] (%)
Control (without additive)	100	100
Cysteine	63	106
Glutathione (reduced form)	56	105
L-threo-Ascorbic acid	147	58
D-erythro-Ascorbic acid	122	86
L-erythro-Ascorbic acid	118	93
D-threo-Ascorbic acid	94	88
L-threo-Dehydroascorbic acid	145	56

[a] Relative values.

Table 15.36. Changes in the GSH content during dough making from wheat flour

Flour/dough[a]	Additive (30 mg/kg)	GSH (nmol/g)
Flour	–	100
Dough	–	44
Dough	L-threo-Ascorbic acid	20
Dough	D-erythro-Ascorbic acid	39

[a] Flour from DNS (0.78% ash) was kneaded for 3 min and then freeze dried.

It was demonstrated that type II lipoxygenase (cf. 3.7.2.2) is responsible for the improver action (Fig. 15.30) and the bleaching effect caused by the enzyme-active soy flour. This enzyme, in contrast to endogenous wheat flour lipoxygenase, releases peroxy radicals which cooxidize carotenoids and other flour constituents.

15.4.1.4.4 Cysteine

Cysteine, in its hydrochloride form, softens gluten due to a SH/SS interchange with gluten

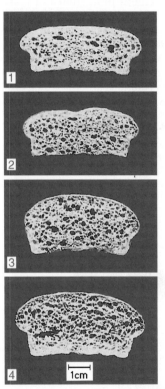

Fig. 15.30. Wheat flour quality improvement by lipoxygenase type-II enzyme of soybean[a] (according to *R. Kieffer* and *W. Grosch*, 1979). Additions: **1** control (no addition, bread volume 31 ml), **2** extract of defatted soya meal in which lipoxygenase was thermally inactivated (31 ml), **3** extract of a defatted soya meal with 290 units of lipoxygenase[b] (35 ml), **4** purified type-II enzyme with 285 activity units (37 ml)

[a] Results in small-scale baking, 10 g flour cv. Clement.
[b] One enzyme unit = 1 μmole · min^{-1} oxygen uptake with linoleic acid as substrate.

proteins, as outlined for GSH under 15.4.1.4.1. Decreases in dough development time and dough stability, as shown in farinograms (Fig. 15.31), clearly reveal the addition of cysteine. Flours with strong gluten and with optimum levels of cysteine also show a favorable increase in baking volume since, prior to baking, the gas trapped within the dough can develop a more spongy dough. The action of sodium sulfite is similar to that of cysteine.

15.4.1.4.5 Proteinases (Peptidases)

Proteinase preparations of microbial or plant origin are used for dough softening (cf. 2.7.2.2.1). Their action involves protein hydrolysis, i.e., gluten-protein endo-hydrolysis. Their effect on dough rheology, therefore, depends on the nature of the enzymes and the activity of the preparations towards gluten proteins. This is shown in Fig. 15.32. Despite equal hydrolase activities with hemoglobin as a test substrate, a fungal proteinase degrades gluten to a lesser extent and consequently causes a smaller decrease in dough resistance to extension in comparison to a bacterial enzyme preparation. Also, the latter is more effective than papain.

Fungal proteinases, because of their low enzyme activity and, therefore, high dosage tolerance, are suitable for optimization of flours containing strong gluten, used for bread and buns. However, bacterial enzymes are preferred in production of biscuits and wafers since they degrade gluten to a greater extent, providing accurate flat dough pieces with high form stability. Bacterial enzymes are also preferred for the desirable end product qualities of porosity and breaking strength.

Data are shown in Table 15.37 for white bread prepared with and without papain. There is a rise in the content of both free amino acids in the crumb and volatile carbonyl compounds in the crust when proteinase is used. As long as proteinases are active in a baking process, they release amino acids from flour proteins, which are then changed via *Strecker* degradation (cf. 4.2.4.4.7) into volatile carbonyl compounds in the crust. Bread aroma is enhanced, as is the crust color, by a build-up of melanoidin compounds from nonenzymatic browning reactions.

15.4.1.4.6 Salt

The taste of bread is rounded-off by the addition to dough of about 1.5% NaCl. As with other salts with small cations (e.g., sodium fumarate or phytate), the addition of NaCl increases dough stability. It is assumed that this is due to the ions masking the repulsion

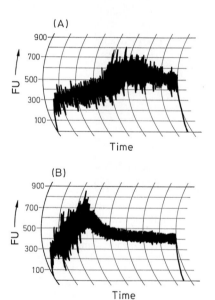

Fig. 15.31. Farinograms. Effect of L-cysteine hydrochloride on a flour with strong gluten (according to *K.F. Finney* et al., 1971). **A** control (no addition), **B** cysteine added (120 ppm)

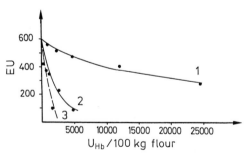

Fig. 15.32. The effect of a proteinase preparation on resistance to extension (in extensogram units) of a wheat flour dough (according to *B. Sproessler*, 1980). Proteinase preparation: 1 fungal, 2 papain, and 3 bacterial. U_{Hb} proteinase activity units determined with hemoglobin as a substrate

Table 15.37. Effects of papain addition in white bread making (values in µmole/g dry matter)

Constituent		Without papain	With papain
Free amino acids	Dough	183	186
	Crumb	182	272
	Crust	10	15
Volatile carbonyl compounds	Crust	158	217

Table 15.38. The effect of shortening on baking volume

Wheat flour	Baking volume (ml)[a]	
	Without shortening	With 3% shortening
I	64.5	81.0
II	73.3	71.8
III [b]	51.6	46.3

[a] Baking test performed on a small scale (10 g flour).
[b] Flour of poor baking quality.

between one charged gluten protein molecule and another of like charge. This allows a sufficiently close approach of one molecule to another, thus hydrophobic and hydrophilic interactions can occur.

15.4.1.4.7 Emulsifiers, Shortenings

Flour baking quality is positively correlated to the content of polar lipids, particularly glycolipids (cf. 15.2.5). Further improvements in dough properties, baking results and end-product freshness or shelf life (cf. 15.4.4) are gained by adding emulsifiers to the dough, e.g., crude lecithin (cf. 3.4.1.1), mono- and diacylglycerides or their derivatives in which the OH-group(s) is esterified with acetic, tartaric, lactic, monoacetyl or diacetyl tartaric acid (cf. 3.3.2 and 8.15.3.1). The hypothesises presented in 15.2.5 are under discussion to explain this effect in the baking process.

Addition of triacylglycerides (shortenings) generally reduces the end-product volume, but there are exceptions depending on the wheat variety. As illustrated by flour I in Table 15.38, addition of 3% shortening provides a substantial increase in baking volume.

15.4.1.4.8 α-Amylase

Flours contain very small amounts of sugars which are metabolizable by yeast (cf. Table 15.23). Addition of sucrose or starch syrup at 1–2% to dough is advisable to maintain favorable growth of yeast and therefore to provide CO_2 needed for dough leavening. Uniform leavening over an extended time improves the quality of many baked end-products; the crumb structure acquires finer and more uniform porosity, while the crust has greater elasticity.

Flours derived from wheat without sprouted grains have some β- but very little α-amylase activity (cf. 15.2.2.1). Thus, only a small amount of starch is degraded to fermentable maltose by handling dough. An insight into the extent of starch degradation is provided by the maltose value (cf. 15.4.1.1.1). Addition of α-amylase in the form of malt flour or as a microbial preparation increases the flour capacity to hydrolyze the starch.

The activity of α-amylase as well as the levels of maltose and glucose increase in the germination of cereals; hence, addition of flour from malted grains enhances the growth of yeast in dough. However, the addition of malt to flours with weak gluten may not be expedient because of the proteolytic activity of the malt. α-Amylase preparations free of proteolytic activity are available from microorganisms (cf. 2.7.2.2.2).

Examples in Table 15.39 illustrate the effects of α-amylse from various sources on baking quality. While malt and fungal amylases show similar effects, the heat-stable α-amylase from *Bacillus subtilis*, with its prolonged activity even in the oven, may be easily used to excess. Products formed by the activities of α- and β-amylases are also available as reactants for nonenzymatic browning reactions. This favorably affects the aroma and color of the crust.

15.4.1.4.9 Milk and Soy Products

Dairy products such as skim milk, buttermilk, whey and casein are added to flour in combination with the ingredients or additives mentioned so far. These dairy products are used in

Table 15.39. The effect of α-amylase preparations on baking results

α-Amylase preparation		White bread		
Origin	Activity[a] (units)	Volume (ml)	Crumb	
			pores	structure
Without addition		2,400	average	average
Wheat malt	140	2,790	good	good
	560	3,000	good	good
	1,120	2,860	average	good
Aspergillus	140	2,750	very good	very good
oryzae	560	2,900	good	good
	1,120	2,950	average	average
Bacillus	7	2,600	good	good
subtilis	35	2,600	good	average
	140	2,640	poor	very poor

[a] α-Amylase units in 700 g flour.

either powdered or liquid form as well as either whole or in the form of defatted powder. In such cases, the proteins added to the dough increase its water binding capacity and provide a juicy crumb.

15.4.1.5 Influence of Additives on Baking Properties of Rye Flour

Rye flour often requires an improved water binding capacity. For this purpose, 2–4% of pregelatinized flour is added. In addition, artificial acidification of the rye dough is practiced; hence both aspects will be covered.

15.4.1.5.1 Pregelatinized Flour

Pregelatinized flour is made from ground cereals such as wheat, rye, rice, millet, etc. by cooking and steaming in autoclaves followed by drying and repeated milling. Such pregelatinized flours are sometimes blended with guar flour or alginates.

15.4.1.5.2 Acids

Rye flour is used in bread baking with sour dough fermentation (cf. 15.4.2.2).
Artificial acidification can be achieved by the addition of lactic, acetic, tartaric or citric acid to rye dough or by adding acidic forms of sodium and calcium salts of ortho- and/or pyrophosphoric acids.

Other preparations for acidification, the so-called dry or instant acids, consist of pregelatinized flour blended with a sour dough concentrate or of cereal mash prefermented by lactic bacteria. The acid values (for definitions see 15.4.1.1.1) vary from 100–1000.

15.4.1.6 Dough Leavening Agents

Dough consisting only of flour and water gives a dense flat cake. Baked products with a porous crumb, such as bread, are obtained only after the dough is leavened. This is achieved for wheat dough by addition of yeast while, for fine baked products, baking powders are used. Rye dough leavening is achieved by a sour dough formulation which includes lactic and acetic acid bacteria.

5.4.1.6.1 Yeast

A given amount (Table 15.40) of surface-fermenting yeast, *Saccharomyces cerevisiae*, is used. While normal yeasts preferentially degrade sucrose rather than maltose, special rapidly fermenting yeasts are used which metabolize both disaccharides at the same rate, shortening the fermentation time.
Yeasts differ in their growth temperature optima (24–26 °C) and their fermentation temperature optima (28–32 °C). The optimum pH for growth is 4.0–5.0. In addition to CO_2 and ethanol, which raise the dough, the yeast forms a variety of aroma compounds (cf. 5.3.2.1). Whether other compounds released by the growth of yeast would affect the dough rheology is unclear; there appears to be no effect of yeast proteinase and GSH.

Table 15.40. Amount of yeast used in bread and other baked products

Baked product	Yeast added[a] (%)
Rye bread	0.5–1.5
Rye mix bread	1.0–2.0
Wheat mix bread	1.5–2.5
Wheat bread	2.0–4.0
Breakfast rolls	4.0–6.0
Rusk ("Zwieback")	6.0–10.0

[a] Based on flour content.

15.4.1.6.2 Chemical Leavening Agents

The interaction of water, acid, heat and chemical leavening agents (baking powders) releases CO_2. The release of gas may occur in the dough prior to or during oven baking. The agents consist of a CO_2-generating source, usually sodium bicarbonate, and an acid constituent such as tartaric, citric or adipic acid or their sodium or calcium salts, disodium hydrogen phosphate or aluminum sulfate. Glucono delta lactone is recommended since it hydrolyzes quickly in the dough to produce gluconic acid. In baking powder, the two reactive constituents are blended with a filler which consists of corn, rice, wheat or tapioca starch or just wheat flour. The filler content in the powder is up to 30%. The role of the filler is to prevent premature release of CO_2. The market also offers baking powders flavored with vanillin or ethyl vanillin.

For every 500 g of flour, baking powder should develop 2.35–2.85 g CO_2, equivalent to about 1.25 liters.

$NaHCO_3$ alone is used for some flat shelf-stable cookies. Ginger beer and honey cookies are sometimes leavened by ammonium carbonate together with potassium carbonate. The former salt is often a mixture of ammonium hydrogencarbonate and ammonium carbamate ($H_2NCOONH_4$). Both decompose above 60 °C to NH_3, CO_2 and water.

15.4.2 Dough Preparation

15.4.2.1 Addition of Yeast

15.4.2.1.1 Direct Addition

Flour, water, yeast, salt and other ingredients are directly mixed into the dough.

15.4.2.1.2 Indirect Addition

Yeast is propagated at 25–27 °C in a well-aerated liquid pre-ferment which contains flour, water and some sugar. After a given time, the liquid is blended with the bulk of flour and water and other ingredients and then made into a dough in a mixer.

For continuous indirect addition of yeast, special liquid starters (sponges) with a pH of 5.0–5.3 are also used with incubation at 38 °C

to develop aroma. Such matured fermented sponge is then metered continuously into a kneader which handles the dough.

15.4.2.2 Sour Dough Making

In sour dough making (lowering the pH to 4.0–4.3) rye flour acquires the aroma and taste properties so typical of rye bread (cf. 15.1.5).

Yeast (*Saccharomyces cerevisae, Saccharomyces minor* and others), which are mainly responsible for dough leavening, and a complex bacterial flora in which lactic acid-forming organisms dominate (*Lactobacillus plantarum* and *Lactobacillus brevis*) are present in sour dough.

Sour dough is prepared by various procedures which differ considerably in the length of time required (Fig. 15.33). A three-stage procedure takes into account the optimum temperature and humidity needs of yeast and bacteria. Yeast prefer to grow at 26 °C, while the bacteria of interest grow best at 35 °C.

In setting up a three-stage process, initially an aqueous flour suspension is inoculated. This is the first "full sour" build-up stage (Fig. 15.33). After maturation, further amounts of flour and water are added and the process is continued with a "basic sour" stage at 35 °C and then, in a similar way, continued with an additional "full sour" third stage at 26 °C.

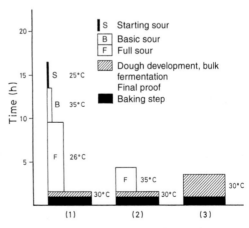

Fig. 15.33. Time requirement for various sour dough development methods (according to *M. Rothe*, 1974). 1 A three step process, 2 short sour, 3 dough souring agents used

The incubation conditions given in Fig. 15.33 are only the essential outline. Temperature deviations influence the spectrum of fermentation products. At warmer temperatures (30–35 °C) lactic acid is preferentially formed (Fig. 15.34), while at cooler temperatures (20–25 °C) more acetic acid is produced. The desirable lactic acid: acetic acid ratio, called the "fermentation ratio", is close to 80 : 20. A ratio with a higher acetic acid concentration gives too sharp an acid taste. The portion of rye flour in the end-product determines the amount of rye sour (full sour) to be added to the dough in the preparation stage. Thus, for rye bread the sour dough to be added is 35–45%, while for a rye mix bread it is 40–60% (on the basis of rye flour). In the short sour method the growth of yeast is negligible. Only a single sour stage, which lasts about 3 h, is involved, yeast is added and the dough is ready for use (Fig. 15.33). However, this short method requires a relatively high content of starter saved from a previous ripe sour. Additional time can be saved by using dough acidifiers (cf. 15.4.1.5.2 and Fig. 15.33). In short sour processes all the organic acids needed for the sour taste of the rye end-product are present. However, there is a lack of aroma compounds and precursors from which odorants can be generated during baking. In a three-stage rye sour procedure, part of the flour proteins is hydrolyzed by proteinases of the microflora into free amino acids which then participate in *Maillard* reactions during baking, providing the more intense aroma.

15.4.2.3 Kneading

The kneading process is characterized by the following stages: mixing of the ingredients and seasonings; dough development and dough plastification.

The energy input into dough kneading, the dough properties and baking volumes are interrelated. For each dough the baking volume passes through an optimum which is dependent on kneading energy input (Fig. 15.35). This optimum shifts towards lower energy input with a flour of weak gluten content and towards higher energy input with flours of strong gluten content; and, as expected, the position of the optimum can be influenced by flour improvers. Increased additions, especially of azodicarbonamide, to the dough result in a successive drop in kneading energy input (Fig. 15.35).

As the kneading energy moves away from the optimum, the dough becomes wetter, it starts to stick to trough walls and its gasholding ability drops (cf. 15.4.2.5 and Fig. 15.39, *14* and *56*). Dough development of wheat flours requires close to double the kneading time of rye flours.

The machines used for kneading are grouped according to their performance based on kneading time: fast, intensive, and high power kneaders and mixers (Table 15.41). However, the groups are not sharply divided. As the

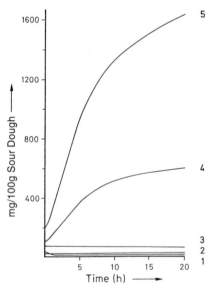

Fig. 15.34. Acid formation in sour dough versus time at 30 °C (according to E. Rabe, 1980). 1 Malate, 2 pyruvate, 3 citrate, 4 acetate, and 5 lactate

Table 15.41. Examples for kneading conditions in white bread dough making

Dough mixer/ kneader	Speed (rpm)	Kneading time (min)	Dough heat[a] ΔT (°C)
Rapid kneader	60–75	20	2
Intensive kneader	120–180	10	5
High power kneader	450	3–5	
Mixer	1,440	1	9
Mixer	2,900	0.75	14

[a] Temperature rise during kneading time.

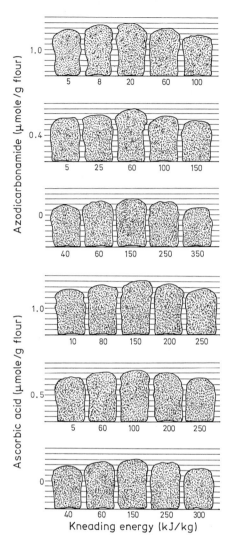

Fig. 15.35. Bread volume as affected by kneading energy input (according to *P. J. Frazier* et al., 1979)

kneading speed increases, the temperature of the dough rises (Table 15.41). Hence, cooling must be used during kneading to keep the temperature at 22–30 °C or, with high speed mixers, at 26–33 °C. The mixer, in a true sense, does not knead the dough, but rips or ruptures it. This could reduce the stability of the dough to such an extent that it could be baked only as panbread (in which case the pan walls support the dough) but not as bread made from self-supporting dough.

15.4.2.4 Fermentation

Dough passes through several stages of fermentation in the presence of growing yeast, a biological leavening agent. After initial fermentation, the dough is divided and scaled, then the dough pieces are rounded-off. A short fermentation is followed by sheeting and moulded dough fermentation. The dough acquires its enlarged final volume in the oven. The yeast produces CO_2 and ethanol which, as long as they do not dissolve in the aqueous phase of the dough, expand the air bubbles (10^2–10^5/mm^3) that arise in the dough during kneading. The volume of a square white loaf increases 4 to 5 fold and more during initial, intermediate, and moulded dough fermentation and 5 to 7 fold during oven fermentation. The length of time of the fermentation varies. It depends on flour type (cf. Fig. 15.37), seasonings incorporated, the amount of yeast and oven temperature. The flour character determines the fermentation tolerance, i.e. the minimum or maximum time after which the fermentation has to be stopped and the dough loaded into the oven. Dough fermentation of a weak gluten flour is rapid, but its fermentation tolerance is low.

The main dough fermentation step (cf. Fig. 15.36) can be substantially shortened by kneading the dough energetically and/or by incorporating fast-acting additives (for example, a mixture of bromates, ascorbic acid and cysteine) into the dough. This provides a favorable dough structure, able to accommodate large amounts of yeast. This is the basis for "no-time" dough making procedures, which provide a continuous flow of dough.

In continuously operated baking processes, the resting times required during the working of dough (intermediate and final fermentation) are realized in air conditioned fermentation rooms. The resting dough forms pass through these rooms with a defined speed.

15.4.2.5 Events Involved in Dough Making

Bread dough is prepared by mixing water and flour (70:30 w/w). Water uptake, which depends on flour type, predetermines most of the subsequent reactions. A high water uptake favors the mobility of all the constituents involved in reactions, e. g., enzymatic degra-

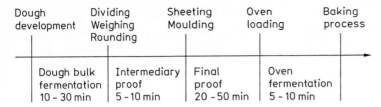

Dough development	Dividing Weighing Rounding	Sheeting Moulding	Oven loading	Baking process
Dough bulk fermentation 10 - 30 min	Intermediary proof 5 - 10 min	Final proof 20 - 50 min	Oven fermentation 5 - 10 min	

Fig. 15.36. Fermentation process for biologically leavened dough; temperature 26–32 °C (according to *H. Bueskens*, 1978)

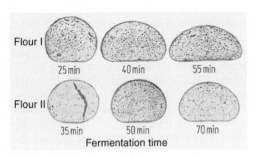

Fig. 15.37. The effect of fermentation time on baking results. (rye-mix bread with two flours which differ in baking quality; according to *H. Bueskens*, 1978)

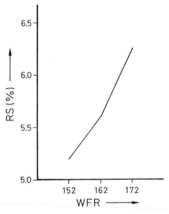

Fig. 15.38. The reducing sugar content in wheat bread crumb as affected by water content of dough (according to *L. Wassermann* and *H. H. Doerfner*, 1971).

$$\text{Water flour ratio (WFR)} = \frac{(\text{flour} + \text{water}) \times 100}{\text{flour}}.$$

RS: Reducing sugar expressed as maltose

dation of starch into reducing sugars (Fig. 15.38).

Observation of wheat dough development by light or scanning electron microscopy reveals that a sequence of forceful changes occurs in the arrangement of the flour proteins.

When a light microscope is used to look at a wheat flour particle under water, practically no protein structure is discernible (Fig. 15.39, *1a*). If the particle is stretched in one direction by moving the slide cover glass against the microscope slide, numerous protein strands with inserted starch granules become visible. These strands are oriented in the direction of stretching (*1b, 1c*) and partially adhere to the glass at one end (*2*). If circular movements are made with the cover glass, the protein strands are two-dimensionally stressed and most of the starch is released (*3*). Protein films are spread between the strands, which are then bent (*4*). As a result of the stickiness of the protein, the strands can be easily aggregated to a ball by further rotary movements. Another way of representing the protein structure is to spread flour particles on the water surface (*5*).

Under a scanning electron microscope at higher magnification, a flour particle, after the removal of starch with amylase, looks like a protein sponge (*6*) in which starch granules were inserted. One-dimensional stretching gives strands (*7*) and on two-dimensional stressing, protein films (*8*) appear, that can also be looked at under the light microscope (*4, 5*). All the structures observed with individual flour particles also occur in dough. From the start of dough making, a continuous network (*10*) with branched protein strands (*11*) is present in the dough. The kneading process presses the sticky flour particles together and promotes their agglomeration. The shear forces act over the entire structure and the network is two-dimensionally extended (*12*) with beginning film formation at the branch points

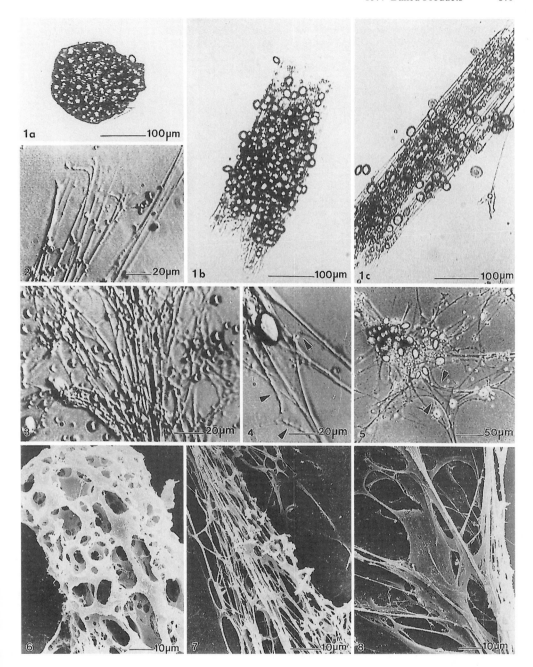

Fig. 15.39. *1−5 Light microscopy. 1a*: Individual flour particles in water. *1b*: Flour particles, slightly extended by moving the cover glass. *1c*: Flour particles, highly extended; *2*: Extended protein strands with one end adhering to the glass. *3*: Network of protein strands after two-dimensional extension of a flour particle. *4*: Protein film (arrows) between bent protein strands. *5*: Flour particles stretched on the water surface. Protein films between bent protein strands (arrows).
6−17, 56−57 Scanning electron microscopy. 6: Flour particle unstretched, holes from enzymatically

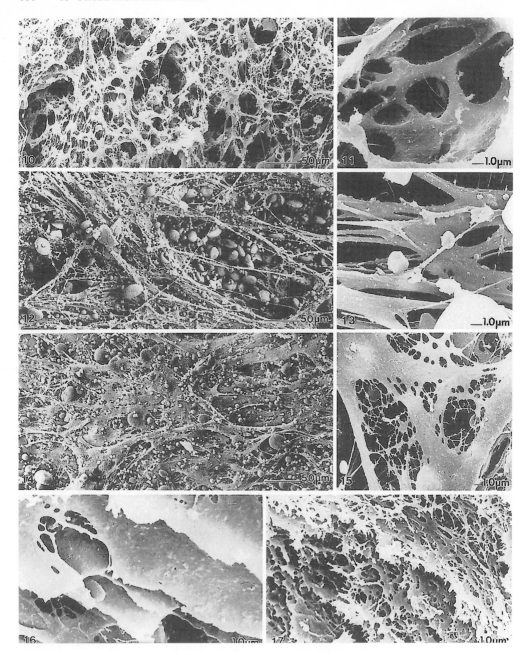

Fig. 15.39 (continued 1)

removed starch. *7*: Flour particle, extended (starch removed). *8*: Flour particle, two dimensionally extended (starch removed), protein films between strands. *10*: Dough after addition of water (starch removed), not kneaded, practically unstretched, connected protein network. *11*: Detail from *10*. *12*: Kneaded dough with extended network. *13*: Detail from *12*, beginning film formation between protein strands. *14*: Optimally kneaded dough, complete formation of protein films by two-dimensional extension. *15*: Detail from *14* with partially perforated protein film. *56*: Highly overkneaded dough with

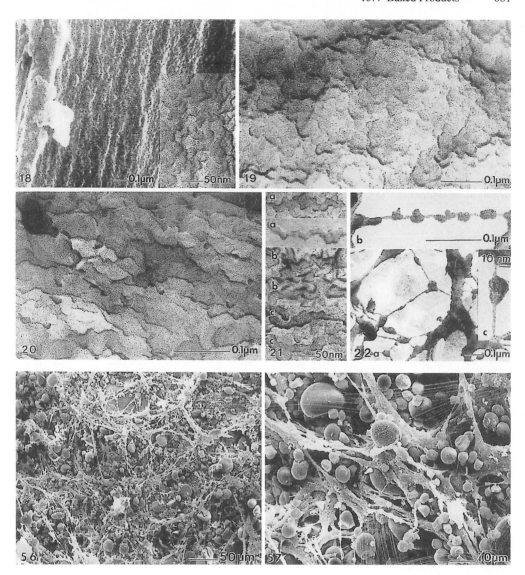

Fig. 15.39 (continued 2)

short irregular protein strands. *57*: Detail from *56*: *16*: Protein films of a gluten membrane arranged in layers. *17*: Perforated protein films of a highly extended gluten membrane.
18–22: Transmission electron microscopy. 18: Slightly extended gluten protein strand with rough surface, enlarged section. *19*: Protein strand, more extended and with smoother surface compared with *18*. *20*: Platelet-like structures on a highly extended gluten film. *21*: Protein threads on a gluten film in a) water, b) triethylamine solution, c) dithioerythritol solution. *22*: a) Highly extended protein filbrils with thickenings, b, c) enlarged

(13). Further kneading of the dough up to optimal development increases the formation of films which, apart from the strands still present, are now the main structural element *(14)* and already start to perforate *(15)*. This initiates the re-formation of the films into short, irregular protein strands, which are the characteristic structural element of a highly overkneaded dough *(56, 57)*.

Washed out gluten exhibits structures comparable to those of dough. Thus, gluten extended to form a membrane consists of protein films arranged in layers *(16)*. Overstretching of this membrane causes perforation of the layers, which are converted to two-dimensional networks.

With the help of transmission electron microscopy, it can be shown at still higher magnification that the surface of unstretched protein strands has an irregular globular structure *(18)*. As a result of the washing out of gliadin with a large excess of water, these strands should essentially consist of glutenin. On two-dimensional stretching, the globular surface is flattened *(19)* and platelet-like forms appear *(20)* which are arranged parallel to the plane of stretching and are less than 10 nm in thickness. The globular surface structures are probably highly tangled, strand-shaped proteins which are unfolded due to mechanical stress and are stabilized in the form of superimposed layers due to intermolecular interactions.

These protein threads have a diameter of 10–30 nm and all look similar, irrespective of the type of preparation, e.g., in water *(21a)*, in triethylamine *(21b)* or in dithioerythritol solution *(21c)*. One-dimensional stretching causes individual protein threads to be partly stretched into fibrils, which, including the metal layers vapor deposited for stabilization, have a diameter of only 3 nm *(22a, b, c)*.

Based on the microscope pictures, dough formation can be summarized as follows. The individual flour particles consist of a sponge-like protein matrix in which starch is embedded. After addition of water, the matrix protein becomes sticky and causes the flour particles to form a continuous structure on kneading. At the same time, the protein matrix is extended and protein films are formed at the branch points of the strands. In an optimally kneaded dough, the protein films are the predominant

structural element and should contribute to the gas-holding capacity. Further kneading causes increased perforation of the films with formation of short, irregular protein strands, which are characteristic of overkneaded dough.

15.4.3 Baking Process

15.4.3.1 Conditions

The oven temperature and time of baking for some baked products are summarized in Table 15.42. Conditions for baking of rye and rye mix bread sometimes deviate from these values. They are prebaked at higher temperatures, for instance at 400°C for 1–3 min, and then post-baked at 150°C (for the effect on quality see Table 15.45). In a continuous process, tunnel-type ovens with circulation heaters are used. Gratings frequently serve as the conveyor band.

In an oven with the temperatures given in Table 15.42, since heat transfer occurs slowly in dough, there is a steep temperature gradient, $200 \rightarrow 120°C$, inward from the crust of the dough piece. By the end of baking, a temperature of 96°C is attained within the product. Higher temperatures up to 106°C are found when the crust is able to resist the rise in inner steam pressure. The water evaporates only in the crust region during dough baking. Water diffusion towards the center of the bread can give the fresh crumb a higher moisture content than the dough. The steam concentration in the

Table 15.42. Baking times and temperatures

Baked product	Weight (g)	Baking time (min)	Oven temperature (°C)
Buns, rolls and other small baked products	45	18–20	250–240
Wheat bread (self-supported dough)[a]	500	25–30	240–230
Wheat bread (pan-baked)[b]	500	35–40	240–230
Wheat bread (self-supported dough)	1,000	40–50	240–220
Rye mix bread (self-supported dough)	1,500	55–65	250–200
Rye bread (self-supported dough)	1,500	60–70	260–200
Pumpernickel (pan-baked)	3,000	16–14 hrs.	180–100

[a] Hearth bread.
[b] Pan bread.

oven also affects the baking results. A steam header is provided in most oven designs to regulate oven moisture.

A baking weight loss is experienced as a result of water evaporation during crust formation. The extent of the loss is related to the form and size of the baked bread and is 8–14% of the fresh dough weight.

15.4.3.2 Chemical and Physical Changes

15.4.3.2.1 Texture

The foamy texture of dough is changed into the spongy texture of crumb by baking. The following processes are involved in this conversion.

Up to ca. 50 °C, yeast produces CO_2 and ethanol at a rate that initially increases. At the same time, water and ethanol evaporate and, together with the liberated CO_2, expand the exisiting gas bubbles, further increasing the volume of the baked product. Parallel to this, the viscosity of the dough falls rapidly in the lower temperature range, reaches a minimum at ca. 60 °C, and then increases rapidly (Fig. 15.40). The increase is caused, on the one hand, by the swelling of starch and the accompanying release of amylose and, on the other hand, by protein denaturation. These processes result in a tremendous increase in the tensile

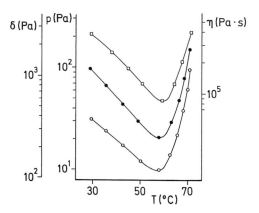

Fig. 15.40. Viscosity (η, □—□) and tensile stress (δ, ○—○) of a wheat dough as well as pressure (p, ●—●) in the gas bubbles as a function of temperature during the baking process (according to *Bloksma*, 1990)

stress of the dough and in the pressure in the gas bubbles at temperatures above ca. 60 °C (Fig. 15.40). The membranes give way and become permeable, allowing CO_2, ethanol, and water vapor to escape. The baking volume decreases slightly until the denatured proteins, with swollen and partially gelatinized starch, form a stable crumb framework, which contains pores down to 3 μm in diameter.

Thin-walled membranes which can stand a greater increase in temperature on stretching, without becoming gas permeable; this is the prerequisite for a baked product with a large volume and uniform fine pores. A relatively large amount of high-molecular glutelins in gluten has a favorable effect because it leads to gastight and highly extensible membranes. Dough made from wheat varieties with poor baking properties becomes gas permeable at a relatively low temperature and the baking volume remains correspondingly low. The extent of starch swelling depends on the available water. The water in dough is preferentially bound by prolamins, glutelins and pentosans. Part of this water becomes available to swell the starch during baking. Limited starch swelling results in a brittle crumb, whereas extensive swelling makes the crumb greasy or gluey.

In contrast to the crumb, the starch granules of the crust surface gelatinize almost completely. This is expecially the case when the oven humidity is high, e.g., when baking occurs below a steam header. Investigations involving gluten and starch mixtures to which the emulsifier stearyl-2-lactylate was added revealed that lipid transfer occurs from gluten to starch during heating of the mixture above 50 °C (Table 15.43). Apparently, the high swelling and gelatinization of the starch granules, which occurs above 50 °C (cf. Table 4.20), promotes lipid binding.

The specific volume of white bread is higher than that of rye bread (Table 15.44). The rye crumb is stronger and less elastic, suggesting that the pentosans can not fully compensate for the lack of texturizing quality of rye proteins (cf. 15.1.5). Heating of a dough accelerates enzymatic reactions, e.g., starch degradation (cf. 15.4.2.4). Above the "temperature optimum" (cf. 2.5.4.3) the reactions are inhibited by denaturation of the enzymes.

Table 15.43. The effect of temperature on stearyl-2-lactylate (SSL) binding in a blend of gluten and starch[a]

T (°C)	SSL free[b]	SSL bound[b] to	
		gluten	starch
30	22.0	64	14
40	20.0	66	14
50	22.0	62	16
60	20.0	6	74
70	16.0	6	78
80	12.0	8	80
90	12.0	2	86

[a] Blends of 17.9 g starch, 2.7 g gluten and 0.103 g SSL.
[b] Values in % of total SSL.

Table 15.44. Specific volumes[a] of bread

Bread variety	ml/g
Toast bread	3.5–4.0
White bread	3.3–3.7
White mix bread[b]	2.5–3.0
Rye mix bread[b]	2.1–2.6
Rye bread	1.9–2.4

[a] Specific volume = volume/weight.
[b] cf. Table 15.51.

The vitamins of the B group are lost to different extents during baking. In white bread, the losses amount to 20% (flour type 500) – 50% (flour type 1150) of thiamine, 6–14% of riboflavin and 0–15% of pyridoxine.

Starch degrades to dextrins, mono- and disaccharides at the relatively high temperatures to which the outer part of the dough is exposed. Caramelization and nonenzymatic browning reactions also occur, providing the sweetness and color of the crust. The thickness of the crust is dependent on temperature and baking time (Table 15.45) and type of baked product (Table 15.46). The composition of some types of bread is presented in Table 15.47.

15.4.3.2.2 Aroma

Substances that have high aroma values and are of importance for white-bread crust and crumb are presented in Table 15.48. In the

Table 15.45. The effect of baking time and temperature on the quality of rye whole meal bread

Baking time (min)	90	180	270
Baking temperature (°C)	240–210	210–185	185–160
Bread yield (ml)	142	142	140
Crust strength (mm)	4	5	6
Taste	raw, slightly aromatic	aromatic	strongly aromatic

Table 15.46. Crumb and crust ratios in different bread varieties

Bread variety	Crumb (%)	Crust (%)
Buns, rolls (50 g)	72.5	27.5
Stick (French) white bread	68.5	31.5
White bread, pan-baked (500 g)	75.0	25.0
White bread (self-supported dough, 500 g)	73.8	26.2
Rye mix bread (self-supported dough, 1000 g)	73.3	26.7
Rye mix bread (pan-baked, 1000 g)	84.5	15.5

crust, two heterocyclic compounds as well as furaneol, 2- and 3-methylbutanal are responsible for the roasty, malty and caramel notes, while autoxidation products of linoleic acid, methional, and diacetyl are involved in the aroma of the crumb. If the dough is fermented for a longer time, 3-methylbutanol and 2-phenylethanol, which are formed by yeast, increase rapidly in the crumb and are responsible for the "yeasty" flavor impression.

Important precursors of 2-acetyl-1-pyrroline are ornithine and 2-oxopropanal (cf. 5.3.1.7), which mainly originate from yeast metabolism. The concentrations of 2-acetyl-1-pyrroline and 2-acetyltetrahydropyridine in the crumb are less than those in the crust by a factor of about 30. The reason being that only in the crust area is the temperature high enough to release aroma substances or their precursors from yeast.

The differences in the aroma of the crust of white bread and rye bread are due to the fact that 2-acetyl-1-pyrroline (cf. 5.3.1.9) is formed in white bread in considerably higher concentrations (Table 15.49). This aroma sub-

Table 15.47. Chemical composition of some types of bread

Bread	Water (%)	Protein (%)	Digestible carbohydrate (%)	Fiber (%)	Lipid (%)	Minerals (%)
White bread	38.3	8.2	49.7	2.9	1.2	1.6
White mix bread	37.6	6.7	41.6	4.1	1.1	1.5
Rye mix bread	39.1	6.9	41.0	4.9	1.1	1.8
Rye bread	38.1	6.7	39.4	5.5	1.0	1.6
Rye whole grain bread	42.0	7.3	36.3	7.2	1.2	1.5
Crisp bread	7.0	10.1	63.2	14.6	1.4	2.3

stance has the lowest threshold value of all known crusty aroma substances. On the other hand, rye bread contains higher concentrations of 3-methylbutanal, diacetyl, 1-octen-3-one, 2,5-dimethyl-3-ethylpyrazine and phenylacetaldehyde (Table 15.49). In comparison with rye flour breads, rye whole meal breads have less pyrazines which produce the crusty notes (Table 15.49).

15.4.4 Changes During Storage

Bread quality changes rapidly during storage. Due to moisture adsorption, the crust loses its

crispiness and glossyness. The aroma compounds of freshly baked bread evaporate or are entrapped preferentially by amylose helices which occur in the crumb. Repeated heating of aged bread releases these compounds. Very labile aroma compounds also contribute to the aroma of bread, e.g., 2-acetyl-1-pyrroline. They decrease rapidly on storage due to oxidation or other reactions (Table 15.50).
The crumb structure also changes, although at a lower rate. The crumb becomes firm, its elasticity and juiciness are lost, and it crumbles more easily. The so-called staling defect of the crumb is basically a starch retrogradation phenomenon (cf. 4.4.4.14.2), which proceeds at different rates with amylose and amylopectin. On cooling bread, the high-molecular amylose very rapidly forms a three-dimensional net-

Table 15.48. Important aroma substances in white-bread crust and crumb

Crust	Crumb
2-Acetyl-1-pyrroline	(E)-2-Nonenal
6-Acetyltetrahydropyridine	(E,E)-2,4-Decadienal
2- and 3-Methylbutanal	trans-4,5-Epox-(E)-2-decenal
Methylpropanal	
(E)-2-Nonenal	1-Octen-3-one
Furaneol	(Z)-2-Nonenal
	Methional
	Diacetyl

Table 15.50. Decrease in 2-acetyl-1-pyrroline in the crust of white bread during storage

Time (h)	2-Acetyl-1-pyrroline (%)
0	100
3	53
24	23
168	11

Table 15.49. "Crusty" aroma substances in the crusts of three types of bread

Compound [a]	White bread	Rye bread [b]	Rye whole meal bread
2-Methyl-3-ethylpyrazine	66	94	18
Acetylpyrazine	19	80	10
5-Methyl-(5H)cyclopenta(b)pyrazine	15	37	7
2-Acetyl-1-pyrroline	19	4	< 0.5

[a] Values in µg/kg.
[b] Three-stage sour dough process.

work and the crystalline states of order of amylose/lipid complexes increase. These processes stabilize the crumb.

On the other hand, the amylopectin is in an amorphous state because the crystalline regions present in flour melt on baking. This is in contrast to the behavior of crystalline amylose/lipid complexes. Thermograms of an aqueous starch suspension (Fig. 15.41) show the differences in the melting points. In comparison with native starch (I), the endotherm peak *a* at 60 °C caused by the melting of crystalline amylopectin is absent in the thermogram of gelatinized starch (II). However, the melting point of amylose/lipid complexes (ca. 110 °C, peak *b* in curve II) is not reached in the crumb on baking.

Staling of white-bread crumb begins with the formation of crystalline structures in amylopectin. The endotherm peak at 60 °C appears again in the thermogram of stored white bread (Fig. 15.42). A state of order arises which corresponds to that of B starch (cf. 4.4.14.2) and binds up to 27% of crystal water, which is withdrawn from amorphous starch and proteins. The crumb loses its elasticity and becomes stale. On storage of white bread, the amount of water that can freeze decreases corresponding to the conversion to non-freezing crystal water (Fig. 15.43).

The formation of crystal nuclei, which proceeds very rapidly at 0 °C and does not occur at temperatures below −5 °C (Fig. 15.44), determines the rate of amylopectin retrograda-

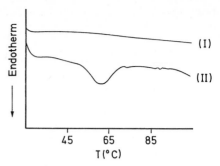

Fig. 15.42. DSC thermograms of white bread: I: fresh from the oven, II: after storage for 1 week at room temperature (according to *Slade, Levine*, 1991)

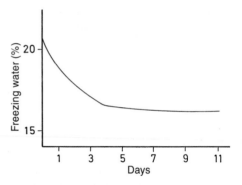

Fig. 15.43. Decrease in freezing water in the storage of white bread. The bread was stored at room temperature encapsulated to prevent drying (according to *Slade, Levine*, 1991)

tion. The nuclei grow most rapidly shortly before the melting point (60 °C) is reached (Fig. 15.44). The ageing process resulting from these events reaches a maximum at ca. 14 °C. As a result of this course, the ageing of white-bread crumb can be prevented by storage at <−5 °C. But the temperature must very quickly fall below the critical temperature for nucleation.

Temperatures above 14 °C also inhibit staling, e.g., increasing the storage temperature from 21 to 35 °C decreases the rate of amylopectin retrogradation by a factor of 4 and improves freshness of the crumb, but the aroma is dissipated. Increased protein or pentosan content slows retrogradation. A choice – actually a rule – to extend the shelf life or freshness of the baked product is the use of emulsifiers,

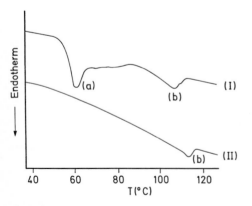

Fig. 15.41. DSC thermograms of wheat starch in water (45:55, g/g) I: native starch, II: gelatinized starch (according to *Slade, Levine*, 1991)

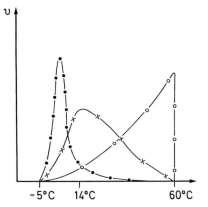

Fig. 15.44. Rate of crystallization of B starch as a function of temperature, ($-\bullet-\bullet$) formation of crystal nuclei, ($-\circ-\circ$) crystal growth, ($-\times-\times-$) total crystal formation (according to *Slade, Levine*, 1991)

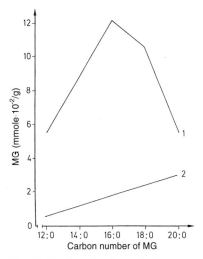

Fig. 15.45. Complex formation between monoacylglycerides (MG) and amylose (1) or amylopectin (2) (according to *W.H. Knightly*, 1977). x-axis: carbon number of the saturated acyl residue. y-axis: tendency of MG to form a complex with amylose or amylopectin (mmole · 10^{-2} MG/g polysaccharide)

such as monoacylglycerides or stearyl-2-lactylate. During baking the emulsifier will be complexed with both starch constituents, though to a different extents (Table 15.43 and Fig. 15.45). Such complexes retard starch retrogradation. Fewer carbohydrates can be extracted from starch/monoacylglyceride

adducts than from starch alone. This effect probably contributes to the increase in the cooking stability of pasta after addition of monoacylglycerides (cf. 15.5).

The staling of crumb is also delayed by bacterial α-amylase. From amylopectin, this enzyme cleaves branched oligosaccharides that consist of 19–24 glucose units. Consequently, the formation of crystalline structures in amylopectin is hindered.

15.4.5 Bread Types

Only those bread types of significant economic importance are listed in Table 15.51. Corresponding data on chemical composition are given in Table 15.47.

Crisp bread (Knaeckebrot) and Pumpernickel are special rye breads.

The flat crisp bread is produced mostly from whole rye meal with low α-amylase activity. The dough is ice-cooled and mixed using compressors until foaming occurs, then sheeted and baked for 8–10 min in a tunnel-type oven. Additional drying reduces moisture by 10–20% to a level of 5%. In addition to this mechanically leavened bread, made by mixing air or nitrogen into the dough, there are crisp breads in which biological leavening (yeast or rye sour) is used. Flat bread is also produced in fully automatic cooker-extruders. The heart of these systems is represented by single-screw or double-screw extruders with co- or counter-

Table 15.51. Bread varieties

No.	Bread variety	Formulation
1.	Wheat bread (white bread)	At least 90% wheat; middlings less than 10%; occasionally with addition of dairy products, sugar, shortenings.
2.	Wheat mix bread	50–89% wheat, the rest rye milling products and other ingredients as under 1.
3.	Rye mix bread	50–89% rye, the rest wheat milling products and others as under 1.
4.	Rye bread	At least 90% rye flour, up to 10% wheat flour; other ingredients as under 1.
5.	Rye whole grain bread	From whole rye meal including also whole kernels, other rye and wheat products less than 10%.

rotating screws. This is mainly a high-temperature, short-time heating process. The material is degraded to some extent (partial starch
gelatinization amongst others) by a combination of pressure, temperature, and shear forces
and then deformed by the nozzle head plate.
The sudden drop in pressure at the nozzle
mouth results in expansion. Water then evaporates and causes the formation of the desired
light and bubbly structure.

Pumpernickel bread originates from West-
phalia. The sour rye dough, heated in sealed
ovens, is more steam-cooked than baked (cf.
Table 15.42). Prolonged heating considerably
degrades the starch into dextrins and maltose,
which are responsible for the sweet taste. The
increased buildup of melanoidin pigments
accounts for the dark color.

15.4.6 Fine Bakery Products

Until a few years ago, the production of fine
bakery products was the domain of confec-
tioners. Today, the importance of the industrial
production of these products has grown substantially. In general, the process techniques
described for the production of bread can be
adapted for fine bakery products. Thus, the
relevant machine-building companies offer
practically automatic production lines for
various fine and stable bakery products.

15.5 Pasta Products

15.5.1 Raw Materials

Pasta products are made of wheat semolina
and grist (cf. 15.3.1.3), in which the flour
extraction grade is less than 70%, and may
incorporate egg. The preferred ingredient is
durum wheat semolina rather than the soft
wheat counterpart (farina) since the former
has better cooking and biting strengths and
also has a higher content of carotenoids (cf.
15.2.5) which provide the yellow color of
pasta products. In wheat mixtures, the soft
wheat characteristics emerge when the soft
wheat content is higher than 30%. In egg-
pasta products (chemical composition in Table
15.52), 2–4 eggs/kg semolina provide a pasta
with improved cooking strength and color.

Table 15.52. Composition of pasta products containing eggs (4 eggs per 1 kg flour)

Constituent	%	Constituent	%
Water	11.1	Carbohydrates	70.0
Protein	14.5	Crude fiber	0.5
Fat	2.9	Minerals	1.0

15.5.2 Additives

Cysteine hydrochloride (about 0.01%) lowers
the mixing/kneading time by 15–20% (cf.
15.4.1.4.4). The cysteine also inhibits melanoidin build-up due to nonenzymatic browning, and suppresses the greyish-brown pig-
mentation. Addition of monoglycerides (about
0.4%) brings about amylose and amylopectin
complexing, thereby increasing cooking
strength (cf. 15.4.4). Through competitive
inhibition, ascorbic acid prevents the action of
lipoxygenase (Fig. 15.46). Although the enzyme is a type I lipoxygenase (cf. 3.7.2.2) and
only slowly cooxidizes carotenoids, the low
enzyme activity can still destroy the pigments
because pasta production is a relatively long
process. Addition of ascorbic acid inhibits this
cooxidation (Fig. 15.47).

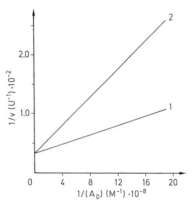

Fig. 15.46. Competitive inhibition of wheat lipoxygenase by ascorbic acid (according to *D. E. Walsh* et
al., 1970). Activity assay with linoleic acid as a substrate (1) without, and (2) in the presence of ascorbic
acid ($2 \cdot 10^{-6}$ mol/l)

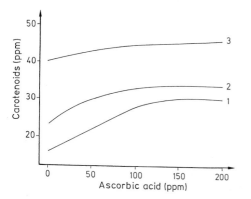

Fig. 15.47. Carotenoid stability in pasta products made of three Durum wheat cultivars as affected by added ascorbic acid (according to *D. E. Walsh* et al., 1970). 1–3 wheat cv. Durum

15.5.3 Production

Pasta products are manufactured continuously by a vacuum extruder, which consists of a mixing trough and press segments. The vacuum is used to retard oxidative degradation of carotenoids.

The semolina and added water (30%) and, when necessary, egg or egg powder are mixed in a mixing trough to form a crumb dough (diameter 1–3 mm), pressed at 150–200 bar into a uniform paste and then pressed through an extruder pressure head die to provide the familiar pasta strings.

Drying is the most demanding stage of pasta manufacturing. The surface of a pasta product must not be allowed to harden prior to the interior core, otherwise cracks, fractures or bursts develop. The freshly extruded strings are initially dried from the outside until they are no longer sticky, then drying is continued at 45–60 °C, either very slowly or stepwise. The moisture content drops to 20–24% after such a predrying process. The moisture is then allowed to equilibrate between inner and outer parts, which brings the content of the final dried product to 11–13%.

15.6 Literature

Acker, L.: Die Lipide der Stärken – ein Forschungsgebiet zwischen Kohlenhydraten und Lipiden. Fette Seifen Anstrichm. *79*, 1 (1977)

Acker, L.: Phospholipases of cereals. In: Advances in cereal science and technology (Ed.: Pomeranz, Y.), Vol. VII, p. 85, American Association of Cereal Chemists: St. Paul. Minn. 1985

Ali, M. R., D'Appolonia, B. L.: Einfluß von Roggenpentosanen auf die Teig- und Broteigenschaften. Getreide Mehl Brot *33*, 334 (1979)

Aman, P., Graham, H.: Analysis of total and insoluble mixed-linked (1 → 3), (1 → 4)-β-D-Glucans in barley and oats. J. Agric. Food Chem. *35*, 704 (1987)

Amend, T., Belitz, H.-D.: Gluten formation studied by the transmission electron microscope. Z. Lebensm. Unters. Forsch. *191*, 184 (1990)

Amend, T., Belitz, H.-D.: Electron microscopic studies on protein films from wheat and other sources at the air/water interface. Z. Lebensm. Unters. Forsch. *190*, 217 (1990)

Amend, T., Belitz, H.-D.: The formation of dough and gluten – a study by scanning electron microscopy. Z. Lebensm. Unters. Forsch. *190*, 401 (1990)

Amend, T., Belitz, H.-D.: Mikroskopische Untersuchungen von Mehl/Wasser-Systemen. Getreide Mehl Brot *43*, 296 (1989)

Amend, T., Belitz, H.-D.: Microstructural studies of gluten and a hypothesis on dough formation. Food Structure *10*, 277 (1991)

Autorenkollektiv: Lebensmittel-Lexikon. VEB Fachbuchverlag: Leipzig. 1979

Barnes, P. J. (Ed.): Lipids in cereal technology. Academic Press: New York. 1983

Bernardin, J. E.: Gluten protein interaction with small molecules and ions – the control of flour properties. Bakers Dig. *52*, Aug. 1978, p. 20

Bhattacharya, K. R., Ali, S. Z.: Changes in rice during parboiling, and properties of parboiled rice. In: Advances in cereal science and technology (Ed.: Pomeranz, Y.), Vol. VII, p. 105, American Association of Cereal Chemists: St. Paul, Minn. 1985

Bietz, J. A., Huebner, F. R., Sanderson, J. E., Wall, J. S.: Wheat gliadin homology revealed through N-terminal amino acid sequence analysis. Cereal Chem. *54*, 1070 (1977)

Bietz, J. A., Huebner, F. R., Wall, J. S.: Glutenin – the strength protein of wheat flour. Bakers Dig. *47*, Febr. 1973. p. 26

Bietz, J. A., Shepherd, K. W., Wall, J. S.: Singlekernel analysis of glutenin: use in wheat genetics and breeding. Cereal Chem. *52*, 513 (1975)

Bietz, J.A., Wall, J.S.: Identity of high molecular weight gliadin and ethanol-soluble glutenin subunits of wheat: relation to gluten structure. Cereal Chem. *57*, 415 (1980)

Bietz, J.A., Wall, J.S.: Wheat gluten subunits: molecular weights determined by sodium dodecyl sulfate polyacrylamide gel electrophoresis. Cereal Chem. *49*, 416 (1972)

Biliaderis, C.G., Tonogai, J.R.: Influence of lipids on the thermal and mechanical properties of concentrated starch gels. J. Agric. Food Chem. *39*, 833 (1991)

Bloksma, A.H.: Dough structure, dough rheology, and baking quality. Cereal Foods World *35*, 237 (1990)

Bloksma, A.H.: Rheology of the breadmaking process. Cereal Foods World *35*, 228 (1990)

Bolling, H., Meyer, D.: Zur Verarbeitungsqualität neu zugelassener Weizensorten 1976. Getreide Mehl Brot *30*, 291 (1976)

Bushuk, W. (Ed.): Rye: Production, chemistry, and technology. American Association of Cereal Chemists: St. Paul, Minn. 1976

Büskens, H.: Die Backschule – Fachkunde für Bäcker. 7th edn., W. Girardet: Essen. 1978

Buttery, R.G., Ling, L.C., Mon, T.R.: Quantitative analysis of 2-acetyl-1-pyrroline in rice. J. Agric. Food Chem. *34*, 112 (1986)

D'Appolonia, B.L., Kim, S.K.: Wheat flour pentosans. Bakers Dig. *50*, June 1976, p. 45

Finney, K.F., Tsen, C.C., Shogren, M.D.: Cysteine's effect on mixing time, water absorption, oxidation requirement, and loaf volume of Red River 68. Cereal Chem. *48*, 540 (1971)

Fox, P.F., Mulvihill, D.M.: Enzymes in wheat, flour, and bread. In: Advances in cereal science and technology (Ed.: Pomeranz, Y.), Vol. V, p. 107, American Association of Cereal Chemists: St. Paul, Minn. 1982

Frazier, P.J.: Lipoxygenase action and lipid binding in breadmaking. Bakers Dig. *53*, Dec. 1979, p. 8

Grosch, W., Schieberle, P.: Bread. In: Volatile compounds in foods and beverages (Ed.: Maarse, H.), Marcel Dekker, Inc., New York, 1991

Hoseney, R.C.: Principles of cereal science and technology. American Association of Cereal Chemists St. Paul, Minn. 1986

Houston, D.F. (Ed.): Rice, chemistry and technology. American Association of Cereal Chemists: St. Paul, Minn. 1972

Huebner, F.R., Wall, J.S.: Fractionation and quantitative differences of glutenin from wheat varieties varying the baking quality. Cereal Chem. *53*, 258 (1976)

ICC-Standards: Standard-Methoden der Internationalen Gesellschaft für Getreidechemie (ICC). Verlag Moritz Schäfer: Detmold

Inglett, G.E. (Ed.): Maize, recent progress in chemistry and technology. Academic Press: New York. 1982

Inglett, G.E., Munck, L. (Eds.): Cereals for food and beverages, recent progress in cereal chemistry and technology. Academic Press: New York. 1980

Jackson, E.A., Holt, L.M., Payne, P.I.: Characterization of high-molecular weight gliadin and low-molecular weight glutenin subunits of wheat endosperm by two-dimensional electrophoresis and the chromosomal location of their controlling genes. Theor. Appl. Genet. *66*, 29 (1983)

Juliano, B.O.: Rice. Chemistry and technology. American Association of Cereal Chemists: St. Paul. Minn. 1985

Kasarda, D.D., Bernardin, J.E., Nimmo, C.C.: Wheat proteins. In: Advances in cereal science and technology (Ed.: Pomeranz, Y.), Vol. I, p. 158, American Association of Cereal Chemists: St. Paul, Minn. 1976

Kasarda, D.D., Okita, T.W., Bernardin, J.E., Baecker, P.A., Nimmo, C.C., Lew, E.J.-L., Dietler, M.D., Greene, F.C.: Nucleic acid (cDNA) and amino acid sequences of α-type gliadins from wheat (*Triticum aestivum*). Proc. Natl. Acad. Sci. USA *81*, 4712 (1984)

Khan, K., Bushuk, W.: Glutenin: Structure and functionality in breadmaking. Bakers. Dig. *52*, April 1978, p. 14

Kieffer, R., Belitz, H.-D.: Molekulargewichte von Kleberproteinen des Weizens. Z. Lebensm. Unters. Forsch. *172*, 96 (1981)

Kieffer, R., Grosch, W.: Verbesserung der Backeigenschaften von Weizenmehlen durch die Typ II-Lipoxygenase aus Sojabohnen. Z. Lebensm. Unters. Forsch. *170*, 258 (1980)

Kieffer, R., Kim, J.-J., Walther, C., Laskawy, G., Grosch, W.: Influence of glutathione and cysteine on the improver effect of ascorbic acid stereoisomers. J. Cereal Sci. *11*, 143 (1990)

Krause, J., Müller, U., Belitz, H.-D.: Charakterisierung von Weizensorten durch SDS-Polyacrylamid-Elektrophorese (SDS-PAGE) und zweidimensionale Elektrophorese (2D-PAGE) der Glutenine. Z. Lebensm. Unters. Forsch. *186*, 398 (1988)

Kosmina, N.P.: Biochemie der Brotherstellung. VEB Fachbuchverlag: Leipzig. 1977

Kruger, J.E., Marchylo, B.A.: A comparison of the catalysis of starch components by isoenzymes from the two major groups of germinated wheat-α-amylase. Cereal Chem. *62*, 11 (1985)

Lawrence, G.J., Shepherd, K.W.: Variation in glutenin protein subunits of wheat. Aust. J. Biol. Sci. *33*, 221 (1980)

Lorenz, K.: The history, development, and utilization of Triticale. Crit. Rev. Food Technol. *5*, 175 (1974)

MacRitchie, F.: Physicochemical aspects of some problems in wheat research. In: Advances in cereal science and technology (Ed.: Pomeranz, Y.), Vol. III, p. 271, American Association of Cereal Chemists: St. Paul, Minn. 1980

Maga, J. A.: Bread flavour. Crit. Rev. Food Technol. *5*, 55 (1974)

Maga, J. A.: Bread staling. Crit. Rev. Food Technol. *5*, 443 (1974)

Marchylo, B. A., Kruger, J. E., Hatcher, D. W.: Quantitative reversed-phase high-performance liquid chromatographic analysis of wheat storage proteins as a potential quality prediction tool. J. Cereal Sci. *9*, 113 (1989)

Menger, A.: Einfluß von Rohstoffen und Prozeßfaktoren auf die Teigwarenqualität. Getreide Mehl Brot *30*, 149 (1976)

Morrison, W. R.: Lipide in Mehl, Teig und Brot. Getreide Mehl Brot *30*, 244 (1976)

Morrison, W. R.: Cereal lipids. In: Advances in cereal science and technology (Ed.: Pomeranz, Y.), Vol. II, p. 221, American Association of Cereal Chemists: St. Paul, Minn. 1978

Morrison, W. R.: Recent progress in the chemistry and functionality of flour lipids. In: Wheat end-use properties (Ed.: Salovaara, H.) Proc. Symp. ICC '89. University of Helsinki, Helsinki, pp. 131–149 (1989)

Orth, R. A., Bushuk, W.: A comparative study of the proteins of wheat of diverse baking qualities. Cereal Chem. *49*, 268 (1972)

Payne, P. I., Corfield, K. G., Blackman, J. A.: Identification of a high-molecular-weight subunit of glutenin, whose presence correlates with bread-making-quality in wheats of related pedigree. Theor. Appl. Genet. *55*, 153 (1979)

Payne, P. I., Harris, P. A., Law, C. N., Holt, L. M., Blackman, J. A.: The high-molecular-weight subunits of glutenin: structure, genetics and relationship to bread-making-quality. Ann. Technol. Agric. *29*, 309 (1980)

Payne, P. I., Corfield, K. G., Holt, L. M., Blackman, J. A.: Correlation between the inheritance of certain high-molecular-weight subunits of glutenin and bread-making quality in progenies of six crosses of bread wheat. J. Sci. Food Agric. *32*, 51 (1981a)

Payne, P. J., Holt, L. M., Lano, C. N.: Structural and genetical studies on the high-molecular weight subunits of wheat glutenin. Part 1: Allelic variation in subunits amongst varieties of wheat (*Triticum aestivum*). Theor. Appl. Genet. *60*, 229 (1981b)

Payne, P. I., Holt, L. M., Jackson, E. A., Law, C. N.: Wheat storage proteins: their genetics and their potential for manipulation by plant breeding. Philos. Trans. R. Soc. London Ser. B *304*, 359 (1984)

Payne, P. I., Holt, L. M., Jarvis, M. G., Jackson, E. A.: Two-dimensional fractionation of the endosperm proteins of bread wheat (*Triticum aestivum*): biochemical and genetic studies. Cereal Chem. *62*, 319 (1985)

Pomeranz, Y.: Dispersibility of wheat proteins in aqueous urea solution – a new parameter to evaluate bread-making potentialities of wheat flours. J. Sci. Food Agric. *16*, 586 (1965)

Pomeranz, Y.: Theorie und Praxis der Bestimmung unterschiedlichen Einflusses einzelner Weizenmehlkomponenten auf das Backverhalten. Getreide Mehl Brot *31*, 147 (1977)

Pomeranz, Y. (Ed.): Wheat: Chemistry and technology. Volumes I and II. American Association of Cereal Chemists: St. Paul, Minn. 1988

Pomeranz, Y.: Wheat is unique. American Association of Cereal Chemists: St. Paul. Minn. 1989

Rabe, E.: Organische Säuren in unterschiedlich geführten Broten. Getreide Mehl Brot *34*, 90 (1980)

Rohrlich, M.: Kleberforschung – Ein historisch-wissenschaftlicher Versuch. Verlag Moritz Schäfer: Detmold. 1969

Rohrlich, M.: Getreideenzyme. Verlag Paul Parey: Berlin. 1969

Rohrlich, M., Brückner, G.: Das Getreide. I. und II. Teil, 2. Aufl., Verlag Paul Parey: Berlin. 1966/1967

Rohrlich, M., Eßner, W.: Menge und Verteilung der SH-Gruppen und SS-Bindungen in Weizenmahlprodukten. Brot Gebäck *20*, 4 (1966)

Rohrlich, M., Thomas, B.: Getreide und Getreidemahlprodukte. In: Handbuch der Lebensmittelchemie (Ed.: Schormüller, J.), Vol. V/1st Part, Springer-Verlag: Berlin. 1967

Sarwin, R., Walther, C., Laskawy, G., Butz, B., Grosch, W.: Determination of free reduced and total glutathione in wheat flours by an isotope dilution assay. Z. Lebensm. Unters. Forsch. *195*, 27 (1992)

Sarwin, R., Walther, C., Kieffer, R., Grosch, W.: Quantitative Analyse von Glutathion in Weizenmehlen und Teigen. Getreide, Mehl, Brot, *46*, 294 (1992)

Sauer, D. B.: Storage of cereal grains and their products. American Association of Cereal Chemists, Inc., St. Paul, Minn., 1992

Schieberle, P., Grosch, W.: Quantitative analysis of aroma compounds in wheat and rye bread crusts using a stable isotope dilution assay. J. Agric. Food Chem. *35*, 252 (1987)

Schieberle, P., Grosch, W.: Potent odorants of the wheat bread crumb. Differences to the crust and effect of a longer dough fermentation. Z. Lebensm. Unters. Forsch. *192*, 130 (1991)

Seibel, W. (Ed.): Feine Backwaren. Paul Parey: Berlin, 1991

Seilmeier, W., Belitz, H.-D., Wieser, H.: Separation and quantitative determination of high-molecular subunits of glutenin from different wheat varieties and genetic variants of the variety Sicco. Z. Lebensm. Unters. Forsch. *192*, 124 (1991)

Shewry, P.R., Miflin, B.J.: Seed storage proteins of economically important cereals. In: Advances in cereal science and technology (Ed.: Pomeranz, Y.), Vol. VII, p. 1, American Association of Cereal Chemists: St. Paul, Minn. 1985

Shukla, T.P.: Cereal proteins: Chemistry and food applications. Crit. Rev. Food. Sci. Nutr. *6*, 1 (1975)

Slade, W., Levine, H.: Beyond water activity – Recent advances based on an alternative approach to the assessment of food quality and safety. Crit. Rev. Food Sci. Nutr. *30*, 115 (1991)

Sprößler, B.: Wirkung von Proteinasen beim Zusatz zum Mehl. Getreide Mehl Brot *35*, 60 (1981)

Stear, C.A.: Handbook of breadmaking technology. Elsevier Applied Science, London. 1990

Stephan, H.: Roggenmehl. Getreide Mehl Brot *30*, 77 (1976)

Sugiyama, T., Rafalski, A., Peterson, D., Söll, D.: A wheat HMW glutenin subunit gene reveals a highly repeated structure. Nucleic Acids Res. *13*, 8729 (1985)

Sumner-Smith, M., Rafalski, J.A., Sugiyama, T., Stoll, M., Soell, D.: Conservation and variability of wheat α/β-gliadin genes. Nucleic Acids Res. *13*, 3905 (1985)

Tan, S.L., Morrison, W.R.: The distribution of lipids in the germ, endosperm, pericarp and tip cap of amylomaize, LG-11 hybrid maize and waxy maize. J. Am. Oil Chem. Soc. *56*, 531 (1979)

Tatham, A.S., Shewry, P.R.: The conformation of wheat gluten proteins. The secondary structures and thermal stabilities of α-, β-, γ- and ω-gliadins. J. Cereal Sci. *3*, 103 (1985)

Walsh, D.E., Youngs, V.L., Gilles, K.A.: Inhibition of durum wheat lipoxygenase with L-ascorbic acid. Cereal Chem. *47*, 119 (1970)

Wassermann, L., Dörffner, H.H.: Der Einfluß des Wasser-Mehl-Verhältnisses in Brotteigen auf die Zusammensetzung und Eigenschaften der daraus hergestellten Brote. Brot Gebäck *25*, 148 (1971)

Watson, S.A., Ramstad, P.E.: Corn. Chemistry and technology. American Association of Cereal Chemists: St. Paul, Minn. 1987

Webster, F.H. (Ed.): Oats: Chemistry and technology. American Association of Cereal Chemists: St. Paul, Minn. 1986

Wieser, H., Seilmeier, W., Belitz, H.-D.: Vergleichende Untersuchungen über partielle Aminosäuresequenzen von Prolaminen und Glutelinen verschiedener Getreidearten. I. Proteinfraktionierung nach Osborne. Z. Lebensm. Unters. Forsch. *170*, 17 (1980)

Wieser, H., Seilmeier, W., Belitz, H.-D.: Vergleichende Untersuchungen über partielle Aminosäuresequenzen von Prolaminen und Glutelinen verschiedener Getreidearten. VII. Aminosäuresequenzen von Prolaminpeptiden. Z. Lebensm. Unters. Forsch. *184*, 366 (1987)

Wieser, H., Mödl, A., Seilmeier, W., Belitz, H.-D.: High-performance liquid chromatography of gliadins from different wheat varieties: amino acid composition and N-terminal amino acid sequences of components. Z. Lebensm. Unters. Forsch. *185*, 371 (1987)

Wieser, H., Belitz, H.-D.: Amino acid compositions of avenins separated by reversed-phase high-performance chromatography. J. Cereal Sci. *9*, 221 (1989)

Wieser, H., Seilmeier, W., Belitz, H.-D.: Characterization of ethanol-extractable reduced subunits of glutenin separated by reversed-phase high-performance liquid chromatography. J. Cereal Sci. *12*, 63 (1990)

Wieser, H., Seilmeier, W., Belitz, H.-D.: Use of RP-HPLC for a better understanding of the structure and the functionality of wheat gluten proteins. In: High-Performance Liquid Chromatography of Cereal and Legume Proteins (Eds.: Kruger, J.E., Bietz, J.A.), p. 235, American Association of Cereal Chemists: St. Paul. Minn., 1994

Wood, P.J., Siddiqui, I.R., Paton, D.: Extraction of high-viscosity gums from oats. Cereal Chem. *55*, 1038 (1978)

Youngs, V.L., Peterson, D.M., Brown, C.M.: Oats. In: Advances in cereal science and technology (Ed.: Pomeranz, Y.), Vol. V. p. 49, American Association of Cereal Chemists: St. Paul, Minn. 1982

16 Legumes

16.1 Foreword

Ripe seeds of the plant family *Fabaceae*, known commonly as "legumes" or "pulses", are an important source of proteins for much of the world's population*. The extent of the production of major legumes is illustrated in Table 16.1. Legumes contain relatively high amounts of protein (Table 16.2). Hence, they are an indispensable supply of protein for the "third world". Soybeans and peanuts are oil seeds (cf. 4.3.2.2.5) and, even in industrialized countries, are used as an important source of raw proteins.

With regard to the biological value, legume proteins are somewhat deficient in the S-containing amino acids (Table 16.3 and 1.8).

Some legumes contain toxic substances (e.g., cyanogenic glycosides and nonprotein amino acids) and antinutritive factors (e.g., proteinase inhibitors, lectins) which, when necessary, are destroyed by suitable procedures, for example, heating.

16.2 Individual Constituents

16.2.1 Proteins

About 80% of the proteins from soybean can be extracted at pH 6.8. A large number of these proteins can be precipitated by acidification at pH 4.5. (cf. Figure 1.54) This pH-dependent solubility is used in large-scale preparations of soy proteins.

Fractionation of legume proteins using solubility procedures, as applied to cereals by *T. B. Osborne* (cf. 15.2.1.2), yields three fractions: albumins, globulins, and glutelins, with globulins being predominant (Table 16.4).

---

* Semi-ripe peas and beans are considered as vegetables (cf. Chapter 17).

The high content of globulins in seeds indicates that they function mostly as storage proteins, which are mobilized during the course of germination.

The globulin fraction can be separated by ultracentrifugation or chromatography into two major components present in all the legumes: *vicilin* (~7S) and *legumin* (~11S). Legumin from soybeans is called glycinin and from peanuts is called arachin. Molecular weights and sedimentation coefficients for the 7S and 11S globulins isolated from various legumes are presented in Table 16.5.

The 11S globulins originate from a protein precursor ($M_r \sim 60,000$) which is split into an acidic α-polypeptide (pI ~5) and a basic β-polypeptide (pI ~8.2) by cleaving the peptide bond between Asn (417) and Gly (418) (cf. Table 16.6). These two polypeptides are connected by a disulfide bridge between Cys (92) and Cys (424) and are regarded as one subunit. Six such subunits join to give 11S globulin, the hydrophobic β-polypeptides evidently forming the core of the subunits and of the entire structure. Little is known about the tertiary and quaternary structure.

On the other hand, the amino acid sequences of the subunits of the 11S globulins of a number of legumes are known. They were mainly derived from the nucleotide sequences of the coding nucleic acids. As an example, Table 16.6 shows the sequences of legumin J from the pea (*Pisum sativum*) and glycinin A_2B_{1a} from the soybean (*Glycine max*). Homology exists between the 11S globulins of different legumes. Variable regions are primarily found in the acidic α-polypeptide, while the basic β-polypeptide is conservative with slight variability in the region of the C-terminal. Conserved residues are uniformly distributed throughout the sequence. The α/β cleavage site is conserved in all the proteins studied until now (cf. Table 16.7). Thus, cleavage of the protein precursor evidently occurs through

Table 16.1. World production of seed legumes, 1996 (1,000 t)

Continent	Legumes total[a]	Beans[b]	Broad beans	Peas
World	56,774	18,639	3,531	10,945
Africa	7,551	2,218	936	364
America, North-, Central-	5,541	3,314	77	1,402
America, South-	3,770	3,505	91	95
Asia	28,265	8,807	1,997	1,874
Europe	9,380	760	307	6,769
Oceania	2,267	35	123	441

Continent	Chick peas	Lentils	Soybeans	Peanuts[c]
World	8,908	2,819	130,302	29,196
Africa	292	56	621	6,208
America, North-, Central-	152	495	67,521	1,835
America, South-	25	24	39,288	686
Asia	8,041	2,123	21,011	20,399
Europe	120	69	1,795	18
Oceania	278	51	73	49

Country	Legumes, grand total	Country	Beans	Country	Broad beans
India	15,414	India	4,140	China	1,900
China	4,979	Brazil	2,837	Egypt	442
Brazil	2,862	China	1,711	Ethiopia	281
France	2,623	Mexico	1,495	Australia	123
Australia	2,186	USA	1,241	Morocco	100
Canada	1,852	Myanmar	935	Italy	97
Ukraine	1,845	Indonesia	810	Germany	77
Turkey	1,831	Ethiopia	390	Turkey	52
Nigeria	1,700	Uganda	387	Sudan	45
Mexico	1,688	Belarus	310	Tunisia	37
USA	1,431				
Russian Fed.	1,300	Σ (%)[d]	76	Σ (%)[d]	89
Myanmar	1,289				
Ethiopia	1,108				
Pakistan	883				
Σ (%)[d]	76				

Country	Peas	Country	Chick peas	Country	Lentils
France	2,570	India	6,000	India	792
Ukraine	1,700	Turkey	732	Turkey	615
Canada	1,260	Pakistan	638	Canada	430
China	1,150	Iran	360	Bangladesh	170
Russian Fed.	1,000	Australia	278	Syria	152
India	561	Mexico	152	China	115
Australia	379	Ethiopia	126	Iran	110
Denmark	270	Myanmar	121	Nepal	110
UK	240	Spain	99	USA	55
Belarus	181	Bangladesh	62	Australia	38
Σ (%)[d]	85	Σ (%)[d]	96	Σ (%)[d]	92

Table 16.1 (continued)

Country	Soy-beans	Country	Peanuts
USA	64,840	China	9,692
Brazil	23,211	India	8,000
China	13,310	Nigeria	1,723
Argentina	12,654	USA	1,653
India	4,200	Indonesia	1,240
Paraguay	2,395	Senegal	816
Canada	2,250	Zaire	580
Indonesia	1,968	Myanmar	568
Bolivia	858	Argentina	464
Thailand	412	Sudan	430
Σ (%)[d]	97	Σ (%)[d]	86

[a] Without soybeans and peanuts.
[b] Without broad beans.
[c] With hull included.
[d] World production = 100%.

Table 16.2. Chemical composition of legumes[a]

Name	Systematic name	Crude protein (%)	Lipid (%)	Digestible carbohydrate (%)	Crude fiber (%)	Minerals (%)
Soybeans	*Glycine max*	39.0	19.6	7.6	16.6	5.5
Peanuts	*Arachis hypogaea*	27.4	50.7	9.1	7.5	2.7
Peas	*Pisum sativum*	25.7	1.4	53.7	18.7	3.0
Garden beans	*Phaseolus vulgaris*	24.1	1.8	54.1	19.2	4.4
Runner beans	*Phaseolus coccineus*	23.1	2.1	n.a.	n.a.	3.9
Black gram	*Phaseolus mungo*	26.9	1.6	46.3	n.a.	3.6
Green gram (mungo beans)	*Phaseolus aureus*	26.7	1.3	51.7	21.7	3.8
Lima beans	*Phaseolus lunatus*	25.0	1.6	n.a.	n.a.	3.9
Chick peas	*Cicer arietinum*	22.7	5.0	54.6	10.7	3.0
Broad beans	*Vicia faba*	26.7	2.3	n.a.	n.a.	3.6
Lentils	*Lens culinaris*	28.6	1.6	57.6	11.9	3.6

[a] The result are average values given as weight-%/dry matter.
n.a.: not analyzed.

Table 16.3. Essential amino acids in legumes (g/16 g N)

Amino acid	Soybean	Broad bean
Cystine	1.3	0.8
Methionine	1.3	0.7
Lysine	6.4	6.5
Isoleucine	4.5	4.0
Leucine	7.8	7.1
Phenylalanine	4.9	4.3
Tyrosine	3.1	3.2
Threonine	3.9	3.4
Tryptophan	1.3	n.a.
Valine	4.8	4.4

n.a.: not analyzed.

Table 16.4. Legumes: protein distribution (%) by *Osborne* fractions

Fraction	Soy-beans	Peanuts	Peas	Mungo beans	Broad beans
Albumin	10	15	21	4	20
Globulin	90	70	66	67	60
Glutelin	0	10	12	29	15

Table 16.5. Molecular weight and sedimentation coefficient of the 7S and 11S globulins from legumes

Legume	7S globulin		11S globulin	
	Sedimentation coefficient	Mol. weight (kdal)	Sedimentation coefficient	Mol. weight (kdal)
Soybeans	7.9 ($S_{20,w}$)	193	12.3 ($S_{20,w}$)	360
Peanuts	8.7 (S_{20})	190	13.2 ($S_{20,w}$)	340
Peas	8.1 (S_{20})		13.1 (S_{20})	398
Garden beans	7.6 ($S_{20,w}$)	140	11.6 ($S_{20,w}$)	340
Broad beans	7.1 ($S_{20,w}$)	150	11.4 ($S_{20,w}$)	328

the same, very specific, but not yet characterized proteinase. With a few exceptions, the 11S globulins are not glycosylated.

The 7S globulins are made of three subunits ($M_r \sim 50{,}000$) which can be identical or different (homo- and heteropolymeric forms). There is little information available on the tertiary and quaternary structure. The subunits can consist of up to three polypeptides (α, β, γ) which are formed from the intact subunit (precursor protein) by proteolysis. Since the amino acid sequences of the α/β (239/240 in Table 16.8) and β/γ (376/377 in Table 16.8) cleavage sites are variable (Table 16.9), intact subunits and subunits with only one cleaved bond are observed, unlike the behavior of the 11S globulin subunits. Thus, the bond between N (376) and D (377) in vicilin 47k is cleaved, but corresponding ED bonds in other vicilins are evidently not split (cf. Table 16.9).

The amino acid sequences of the 7S globulin subunits of a number of legumes are known and were mainly derived from the nucleotide sequences of the coding nucleic acids. Table 16.8 shows the sequences of phaseolin from the garden bean (*Phaseolus vulgaris*), vicilin

from the pea (*Pisum sativum*), and β-conglycinin (β) from the soybean (*Glycine max*). Sequence homology, which is more pronounced than in the 11S globulins, exists between the proteins of various legumes. Variable domains are found in the N- and C-terminal regions, but not inside the structure.

The 7S globulins are glycosylated to different extents. The carbohydrate content is 0.5–1.4% in vicilin from peas, 1.2–5.5% in phaseolin from garden beans, and 2.7–5.4% in β-conglycinin from soybeans. The structures of the oligosaccharide residues are partly known. In β-conglycinin, for example, 6–8 mannose residues are bound to Asn in a branched structure via two N-acetylglucosamine residues.

Under non-denaturing conditions, the 11S and 7S globulins exhibit a tendency towards reversible dissociation/association, which greatly depends on the pH value and the ionic strength. According to their behavior, they can be attributed to different types. The 11S globulins are relatively more stable than the 7S globulins. They noticeably associate only in the region of the isoelectric point, isoelectric precipitation occurring at low ionic strength (cf. 16.3.1.2.1).

If at pH 7.6, the ionic strength is reduced from $\mu = 0.5$ to $\mu < 0.1$, soybean 11S globulin dissociates stepwise (α, β: acidic and basic proteins):

$$11\,S(6\,\alpha\beta) \rightarrow 7.5\,S(3\,\alpha\beta) \rightarrow 3\,S(\alpha\beta) \quad (16.1)$$

Complete dissociation occurs when the disulfide bonds are reduced in the presence of protein-unfolding agents, such as urea or SDS:

$$(\alpha\beta) \rightarrow \alpha + \beta \quad (16.2)$$

Soybean 7S globulin has similar properties, as illustrated in Fig. 16.1. Hence, its molecular

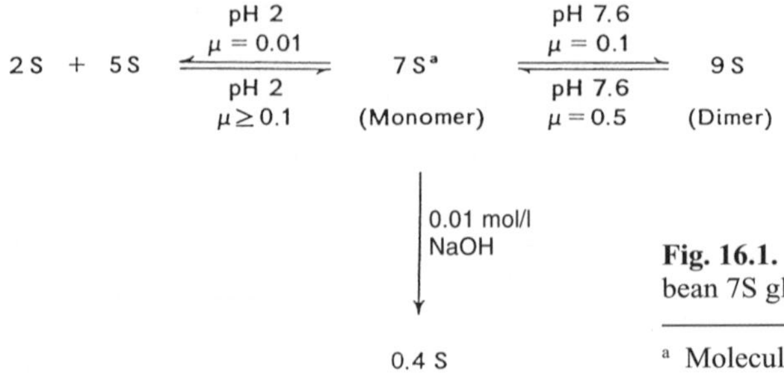

$$2\,S + 5\,S \; \underset{\substack{pH\ 2 \\ \mu \geq 0.1}}{\overset{\substack{pH\ 2 \\ \mu = 0.01}}{\rightleftharpoons}} \; 7\,S^a \; \underset{\substack{pH\ 7.6 \\ \mu = 0.5}}{\overset{\substack{pH\ 7.6 \\ \mu = 0.1}}{\rightleftharpoons}} \; 9\,S$$

(Monomer)   (Dimer)

↓ 0.01 mol/l NaOH

0.4 S

Fig. 16.1. Dissociation and aggregation of the soybean 7S globulin

[a] Molecular weight: 193 kcal.

Table 16.6. Amino acid sequences of the α/β subunits[a] of 11S globulins, 1) legumin J (*Pisum sativum*) and 2) glycinin A₂B₁ₐ (*Glycine max*)

```
       1                  10                20                30               40                50              60
1  L A T S S E F D R L - - - N Q C Q L D S I N A L E P D H R V V S E A G L T E T W N P N H P E L K C A G V S L I R R T I D
2  - - - - L R E Q A Q Q N E C Q I Q K L N A L K P D N R I E S E G G F I E T W N P N N K P F Q C A G V A L S R C T L N

      61        70        80        90       100       110      120
1  P N G L H L P S F S P S P Q L I F I I Q G K G V L G L S F P G C P E T Y E E P R S S Q S R Q - - - - E S R Q - - - Q
2  R N A L R R P S Y T N G P Q E I Y I Q Q G N G I F G M I F P G C P S T Y Q E P Q E S Q - - Q R G - R S Q - - - R - - P Q

     121       130       140       150      160       170      180
1  Q G - - - - - - - - - - - - - - - - - - - - - - - - - - D S H Q K V R R F R K G D I I A I P S G I P Y W T Y N H G
2  - - - - - - - - - - - - - - - - - - - - - - - - - - - D R H Q K V H R F R E G D L I A V P T G V A W W M Y N N E

     181       190       200       210      220       230      240
1  D E P L V A I S L L D T S N I A N Q L D S T P R V F Y L G G N P E T E F P E T Q E E Q Q G R H R - Q K H S Y P V G R R S
2  D T P V V A V S I I D T N S L E N Q L D Q M P R R F Y L A G N Q E Q E F - L K Y Q Q Q Q Q G - - - - G S Q S Q K

     241       250       260       270      280       290      300
1  G H H - Q Q E E E S E Q Q N E G N S V L S G F S S E F L A Q T F N T E E D T A K R L R S P R D E - R S - Q I V R V E G G
2  G K - - Q Q E E - - - N E G S N I L S G F A P E F L K E A F G V N M Q I V R N L Q G E N E E E D S G A I V T V K G G

     301       310       320       330      340       350      360
1  L R I I K G - - R - - T - - - - - - - - - - E E - E K - E Q - - S H - - - - S H S H R E E K E E - - - - - - -
2  L R V T A P A M R K P - - - Q Q - - - - - E - E D D D D E E Q - P - - Q - C V E T - D K G C Q - - - - - - -

     361       370       380       390      400       410      420
1  - - - - - - - - - - - - E E E E E E D E E E - - - - - - - - - E E - R - - - - - K N G L E
2  R - - - O S K R S R - - - - - - - - - - - - - K Q - R S - - - - - - E E - R - - - - - N G I D

     421       430       440       450      460       470      480
1  E T I C S A K I R E N I A D A A R A D L Y N P R A G R I S T A N S L T L P V L R Y L R L S A E Y V R L Y R N G I Y A P H
2  E T I C T M R L R Q N I G Q N S S P D I Y N P Q A G S I T T A T S L D F P A L W L L K L S A Q Y G S L R K N A M F V P H

     481       490       500       510      520       530      540
1  W N I N A N S L L Y V I R G E G R V R I R S C E L P T N T M F D N K L R K G H L V V V P Q N F V V A E Q A G E E E G L E
2  Y T L N A N S I I Y A L N G R A L V Q V V N C N - G - E R V F D G E L Q E G G V L I V P Q N F A V A A K S Q S D N - F E

     541       550       560       570      580       590      600
1  Y V V F K T N D R A A V S - H V Q Q - - - V F R A T P S E V L A N A F G L R Q R Q V T E L K L S G N R G P M - V H - P -
2  Y V S F K T N D R P - S I G N L A G A N S L L N A L P E E V I Q H T F N L K S Q Q A R Q V K - - N N N - P F S F L V P P

     601       610       620
1  R - S Q S Q S H
2  Q E S Q - - R R A V A
```

[a] The α-polypeptide ends with Asn (417) and the β-polypeptide starts with Gly (418). The two polypeptides are connected by a disulfide bond between Cys (92) and Cys (424). The two remaining Cys residues of the α-polypeptide (16, 49) probably form an intramolecular disulfide bridge. "-": space to maximize homology.

Table 16.7. Amino acid sequences in the vicinity of the α/β cleavage site (417/418 in Table 16.6) of subunits of various 11S globulins (−: space to maximize homology, ··· sequence not known)

Protein	420
Legumin J (*Pisum sativum*)	− − K N G L E E T I C S
Legumin A (*Pisum sativum*)	D − − N G L E E T V C T
Glycinin A_2B_{1a} (*Glycine max*)	− − − N G I D E T I C T
Glycinin $A_5A_4B_3$ (*Glycine max*)	E T R N G V E E N I C T
Glycinin A_3B_4 (*Glycine max*)	Q T R N G V E E N I C T
Glycinin $A_{1a}B_{1b}$ (*Glycine max*)	− − − N G I D E T I C T
Glycinin $A_{1b}B_2$ (*Glycine max*)	· · · N G I D E T I C T
Cruciferin (*Brassica napus*)	− − − N G L E E T I C S
Legumin $β_1$ (*Vicia faba*)	D − − N G L E E T V C T
Legumin B (*Vicia faba*)	− − R N G L E E T I C S
Avenin (*Avena sativa*)	− − − N G L E E N F C D
Glutelin (*Oryza sativa*)	N G L D E T F C T

weight is also strongly dependent on pH and ionic strength.

The thermal stability of the 11S and 7S globulins varies. While 7S globulin coagulates in a 10% salt solution at 99°C, 11S globulin remains in solution. The opposite is true at $μ = 0.001$. Since dissociated proteins are more easily coagulable thermally than associated proteins, it follows that 11S globulin is destabilized by dissociation at low ionic stringth, as shown above. Under these conditions, however, the 7S globulin is stabilized by association.

The amino acid compositions of both major soybean proteins, with the exception of methionine, are very similar (Table 16.10). However, large differences exist in their carbohydrate contents. The 7S globulin contains 5% carbohydrate and the 11S globulin less than 1% carbohydrate.

Legume proteins exhibit a marked gelling capacity. The gel properties depend on the protein used and on the production conditions (pH value, ionic species, ionic strength, and temperature). They are suitable for the production of foams and emulsions.

In the pH range of 4−10, the 7S globulin is a better emulsifier than the 11S globulin, when the capacity (Fig. 16.2) and the stability of an o/w emulsion are compared. Partial acid hydrolysis improves the emulsifier properties.

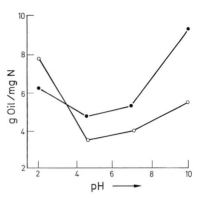

Fig. 16.2. Soybean globulin as an emulsifier. (According to *H. Aoki* et al., 1980). The capacity of an o/w-emulsion after addition of 11S globulin (−○−) and 7S globulin (−●−) is plotted versus pH

16.2.2 Enzymes

Various forms of lipoxygenase (cf. 3.7.2.2) are of interest in food chemistry since they strongly affect the legume aroma.

Urease, which hydrolyses urea,

$$NH_2 - CO - NH_2 + H_2O \rightarrow CO_2 + 2NH_3 \quad (16.3)$$

occurs in soybeans in relatively high concentration. Heat treatments of soy preparations can be detected by measuring the activity of this enzyme.

16.2.3 Proteinase and Amylase Inhibitors

16.2.3.1 Occurrence and Properties

Inhibitors of hydrolases, themselves proteins, form stoichiometric inactive complexes with the enzymes and are distributed in microorganisms, plants, and animals. Apart from the thoroughly examined group of proteinase inhibitors, some proteins that inhibit amylases are known.

Of the large number of known proteinase inhibitors, only those compounds found in foods are of interest in food chemistry. These include in particular the inhibitors in egg white, plant seeds, and plant nodules. Table 16.11 shows the most important sources of proteinase inhibitors, which have molecular weights between 6,000 and 50,000. The specificity for proteinases varies considerably. Some inhibitors inhibit only trypsin, many act on both trypsin

Table 16.8. Amino acid sequences of subunits[a] of the 7S globulins, 1) β-conglycinin α′ (*Glycine max*), 2) phaseolin (*Phaseolus vulgaris*) and 3) vicilin (*Pisum sativum*)

```
     1         10        20        30        40        50
1  - - - - DEDE E QDKE S QE S EGS E S QRE P R RHKNKNP FHF NS  KR-  FQT L FKNQ
2  AT S L REE E E- - - - S QD- - - - - - - - - - - - - - - - NP FYF NS  DNS WNT L FKNQ
3  - - - - - - - - - - R- - - - - - - S DP Q- - - - - - - - - NP FI FKS NK-  FQT L FENE
    51        60        70        80        90        100
1  YGHV RVL QR F NKR S QQL QNL R DY R I  L E F NS  KP NT L L L P HHADADY L I  VI  L
2  YGHI RVL QR F DQQS  KR L QNL E DY R L VE F R S  KP E T L L L P QQADAE L L L VVR
3  NGHI R L L QK F DQR S  KI  F E NL QNY R L L E Y KS  KP HT I  F L P QHT DADY I  L VVL
    101       110       120       130       140       150
1  NGTAI  L T L VNNDDR DS  Y- NL QS  GDA-  L - - - - - R VP AGT T  FYVVNP DNDE N
2  S GS AI  L VL VKP DDR R E Y F F L T QGDNP I  F S  DNQK I  P- - - - - - - - - - - - - -
3  S GKAI  L T V L KP DDR NS  F - NL E R GDT -  I  - - - - - K L P- - - - - - - - - - - - - -
    151       160       170       180       190       200
1  L RMI AGT T  FYVVNP DNDE NL RMI  T L AI  P VNKP E R F E S  F F L S  S T QAQQS  Y L
2  - - - - AGT I  FY L VNP DP KE DL R I  I  QL AMP VNNP Q-  I  HE F F L S  S T E AQQS  Y L
3  - - - - AGT I  AY L VNR DDNE E L RV L DL AI  P VNR P GQL QS  F L L S  GNQNQQNY L
    201       210       220       230       240       250
1  QGF S  KNI  L E AS  Y DT KF E E I  NKV L F GR E E GQQ-  Q- - - G- - -  E E R - - - - L QE
2  QE F S  KHI  L E AS  F NS  KF E E I  NRV L F -  E E E GQQ- - - - - - - - - - - -  E E GQQE
3  S GF S  KNI  L E AS  F NT DY E E I  E KV L L -  E E HE KE T QHR R S  L KD-  KR -  QQS  QE E
    251       260       270       280       290       300
1  S V I  VE I  S KKQI  R E L S  KHAKS  S  S R KT I  S  S  E DKP F NL GS  R DP I  Y S  NKL GKL F
2  GV I  VNI  DS  E QI  E E L S  KHAKS  S  S R KS  HS  - - - KQ-  D- - - -  NT I  -  GNE F GNL T
3  NV I  VKL S  R GQI  E E L S  KNAKS  T S  KKS  VS  S  E S  E P F NL R S  R GP I  Y S  NE F GKF F
    301       310       320       330       340       350
1  E I  T -  QR N-  P QL R DL DV F L S  VV DMNE GAL F L P HF NS  KAI  VV L V I  NE GE ANI
2  E R T -  D-  N- - - - S  -  L NV L I  S  S  I  E MKE GAL F VP HY Y S  KAI  V I  L VVNE GE AHV
3  E I  T P E KN-  P QL QDL DI  F VNS  VE I  KE GS  L L L P HY NS  R AI  V I  VT VNE GKGDF
    351       360       370       380       390       400
1  E L VGI  K- - - - - - - - - -  E QQQR QQQ- - -  E E Q- -  P - - - - - L E VR KY R AE L S
2  E L VGP K- - - - - - - - - -  GN- - - -  KE - - - - - - - -  T -  L E F E S  Y R AE L S
3  E L VGQR - - - - - - - - -  NE NQQE QR KE DDE E E QGE E E I  NKQVQNY KAKL S
    401       410       420       430       440       450
1  E QDI  F V I  P AGY P VMVNAT S  DL NF -  F -  AF G- - - - - I  NAE NNQR NF L AGS  KD
2  KDDV F V I  P AAY P VAI  KAT S  NVNF - -  T GF G- - - - - I  NANNNNR NL L AGKT D
3  S GDV F V I  P AGHP VAL KAS  S  NL DL -  L -  GF G- - - - - I  NAE NNQR NF L AGDE D
    451       460       470       480       490       500
1  NV I  S QI  P S  QV- -  QE - - -  L AF P R S  AKDI  E NL I  KS  Q-  S  E S  Y F VDA- - - - QP Q
2  NV I  S  S  I  GR AL DGKDV L GL T F S  GS  GE E VMKL I  NKQ-  S  GS  Y F VDGHHHQQE Q
3  NV I  S QVQR P V- -  KE - - -  L AF P GS  AQE VDR I  L E NQ-  KQS  HF ADA- - - - QP Q
    501       510       520       530
1  QKE E -  GN- - - - - -  KGR KGP L S  S  I  L R AF -  Y
2  QK- - -  GS  HQQE QQKGR KG- - - - - - -  AF VY
3  QR E -  R GS  - -  R E -  T R DR - - -  L S  S  V
```

[a] α/β cleavage site: 239/240, β/γ cleavage site: 376/377; – – space to maximize homology.

and chymotrypsin, and others inhibit microbial or plant proteinases as well, e.g., subtilisin or papain. Proteinase inhibitors are often located in, but not limited to, the seeds of plants. The seeds of legumes (soybeans ca. 20 g/kg, white beans ca. 3.6 g/kg, chick peas ca. 1.5 g/kg, mungo beans ca. 0.25 g/kg), the tubers of Solanaceae (potatoes ca. 1−2 g/kg), and cereal grains (ca. 2−3 g/kg) contain especially high concentrations. The inhibitor content greatly depends on the variety, degree of ripeness, and storage time.

Table 16.9. Amino acid sequences in the vicinity of the α/β cleavage site (239/240 in Table 16.8) and the β/γ cleavage site (376/377 in Table 16.8) of subunits of various 7S globulins (−: space to maximize the homology)

Protein	α/β 240	β/γ 377
Phaseolin *(Phaseolus vulgaris)*	− −	− −
Vicilin *(Pisum sativum)*	K D	E D
Convicilin *(Pisum sativum)*	R D	E D
Vicilin 50k *(Pisum sativum)*	R D	E D
Vicilin 47k *(Pisum sativum)*	K D	N D
β-Conglycinin α' *(Glycine max)*	− −	− −
β-Conglycinin β *(Glycine max)*	− −	− −
α-Gossypulin B *(Gossypium sp.)*	− −	− −

Table 16.10. Amino acid composition of 7S and 11S globulins from soybeans

Amino acid	g amino acid/100 g protein	
	7S globulin	11S globulin
Asx	11.18	13.10
Thr	3.14	3.37
Ser	4.79	4.16
Glx	17.54	18.03
Pro	5.21	5.40
Gly	3.37	3.97
Ala	3.66	3.55
Cys	1.52	1.44
Val	4.68	5.05
Met	0.43	1.84
Ile	4.99	4.69
Leu	8.15	7.17
Tyr	3.51	4.05
Phe	5.55	5.73
His	2.32	2.22
Lys	6.26	4.88
Arg	7.37	7.75

Often, several different inhibitors are found in plant materials. They differ in their isoelectric points and also in their specificity for proteinases, specific activities and thermal stabilities. For example, in the more than 30 legumes analyzed so far, nine inhibitors have been identified and five partially purified.

Figure 16.3 provides an example of raw protein separation from potato sap by isoelectric focussing. More than ten inhibitors for bovine trypsin and chymotrypsin have been found.

Food which contains inhibitors might cause nutritional problems. For example, feeding rats and chickens with raw soymeal leads to reversible pancreatic blistering. A consequence of excessive secretion of pancreatic juice is increased secretion of nitrogen in the feces. Furthermore, growth inhibition occurs which can be eliminated by incorporating methionine, threonine and valine into the diet. These findings indicate that the poor growth rate might be due to some amino acid deficiencies, which are a result of increased N-excretion. All the possible effects of proteinase inhibitors are not fully understood.

16.2.3.2 Structure

Many proteinase inhibitors have been isolated and their structures elucidated. The active center often contains a peptide bond specific for the inhibited enzyme, e.g., Lys-X or Arg-X in trypsin inhibitors and Leu-X, Phe-X or Tyr-X in chymotrypsin inhibitors (Table 16.12, 16.13). In addition, inhibitors are known that inhibit trypsin and chymotrypsin and contain only one trypsin-specific peptide bond at the active center, e.g., *Kunitz* inhibitors from bovine pancreas and soybeans (cf. Table 16.13). Some double-headed inhibitors contain two different active centers, which, e.g., are both directed towards trypsin or towards trypsin and chymotrypsin. An example of the latter type is represented by the *Bowman-Birk* inhibitors found in legumes (cf. Table 16.13). Their reactive centers are localized in two homologous domains of the peptide chain, each of which form a 29 membered ring via a disulfide bridge (cf. 1.4.2.3.2). In this way, the centers are exposed for contact with the enzyme. An active center can also be exposed by another suitable conformation, as is the case with the *Kunitz* inhibitor from soybeans.

X-ray analyses of the trypsin inhibitor complex show that 12 amino acid residues of the inhibitor are involved in enzyme contact, including the sequence Ser (61) − Phe (66) with the active center Arg (63) − Ile (64).

The double-headed *Bowman-Birk* inhibitor from soybeans was cleaved into two fragments by cyanogen bromide (Met (27) − Arg (28)) and pepsin (Asp (56) − Phe (57)) (cf. Fig. 1.25). Each of these fragments contained an

Table 16.11. Proteinase inhibitors of aninal and plant origin

Source/Inhibitor	Molecular weight	T	CT	P	Bs	AP	SG	PP
Animal tissues								
Bovine pancreas								
Kazal inhibitor	6,153	+	−	−				
Kunitz inhibitor	6,512	+	+	−	−	−	+	
Chicken egg								
Ovomucoid	27–31,000	+	−		−			
Ovoinhibitor	44–52,000	+	+	−	+	+		
Ficin-papain-inhibitor	12,700	−	−	+	−			
Plant tissues								
Cruciferae								
Raphanus sativus[b]	8–11,200	+	±	−	+	+		
Brassica juncea[b]	10–20,000	+	±					
Leguminosae								
Arachis hypogaea[b]	7,500–17,000	+	+					
Cicer arietinum[b]	12,000	+	+					
Dolichos lablab[b]	9,500–23,500	+	+	−				
Glycine max								
Kunitz inhibitor	21,500	+	+	−	−			
Bowman-Birk inhibitor	8,000	+	+	−	+			
Lathyrus odoratus	11,800	+	+					
L. sativus				+				
Phaseolus aureus[b]	8–18,000	+	±	+				
P. coccineus[c]	8,800–10,700	+	+					
P. lunatus[c]	8,300–16,200	+	+	−	−	±		
P. vulgaris[c]	8–10,000	+	+	−	−			
Pisum sativum[b]	8–12,800	+	+					
Vicia faba[b]	23,000	+	+					
Vigna sinensis	9,500–13,300	+	+					
Convolvulaceae								
Ipomoea batatas[b]	23–24,000	+	−		−	−	−	
Solanaceae								
Solanum tuberosum[b]	22–42,000[c]	±	±	−	±	±	±	±
Bromeliaceae								
Ananas comosus[b]	5,500	+	+					
Gramineae								
Hordeum vulgare[b]	14–25,000	±	−	−	±	±	±	
Oryza sativa[b]		±	−	+				
Secale cereale[b]	10–18,700	+	+	−				
Triticum aestivum[b]	12–18,500	±	−	−				
Zea mays[c]	7–18,500	+	+	−				

[a] T: trypsin, CT: α-chymotrypsin, P: papain, Bs: Bacillus subtilis proteases, AP: Aspergillus spp. proteases, SG: Streptomyces griseus proteases, PP: Penicillium spp. proteases, +: inhibited, −: not inhibited, ±: inhibited by some inhibitors of the particular source.
[b] The properties of different inhibitors are combined.
[c] Subunits 6–10,000.

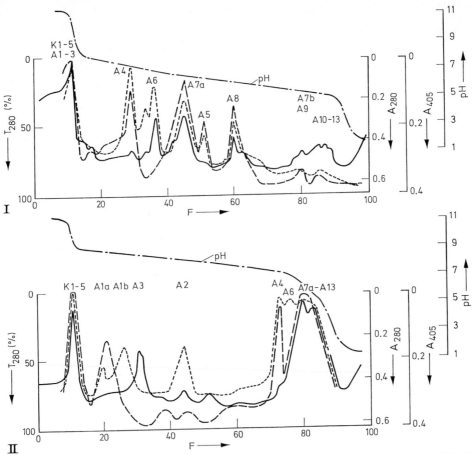

Fig. 16.3. Isoelectric focussing of proteins from potato tubers. (according to *H.-D. Belitz* et al., 1971). **I**: pH range 5–8, **II**: pH range 7–10, F: 1 ml-fractions: — protein, transmittance scanning at 280 nm; ---- trypsin inhibitor activity: inhibition of 20 µg bovine trypsin by 0.025 ml aliquots, substrate N$^\alpha$-benzoyl-L-arginine-p-nitroanilide (BAPA). Absorbance at 405 nm; ––– chymotrypsin inhibitor activity: inhibition of 12 µg bovine chymotrypsin by 0.025 ml aliquots, substrate casein. Absorbance 280 nm; –·–·– pH gradient

active center and, therefore, inhibited only one enzyme with remaining activities of 84% (trypsin) and 16% (chymotrypsin) compared with the native inhibitor.

Modifications of the active center of an inhibitor result in changes in the properties. For example, Arg (63) of the *Kunitz* inhibitor from soybeans can be replaced by Lys without changing the inhibitory behavior, while substitution by Trp abolishes the inhibition of trypsin and increases the inhibition of chymotrypsin. Indeed, Ile (64) can be replaced by Ala, Leu, or Gly without change in activity,

while the insertion of an amino acid residue, e.g., Arg (63)-Glu (63 a)-Ile (64), abolishes all inhibition and makes the inhibitor a normal substrate of trypsin.

The inhibitors from potatoes mentioned above (Fig. 16.3) exclusively inhibit serine proteinases, but differ in their specificity (Table 16.14).

16.2.3.3 Physiological Function

The biological functions of most proteinase inhibitors of plant origin are unknown. During

Table 16.12. Active centers in *Bowman-Birk* type inhibitors

Inhibited enzyme	Active center	Occurrence
Trypsin	Lys-X	Adzuki bean (API II) Chickpea Garden bean (GBI I) Lima bean (LBI IV) Soybean (BBI) Wisteria (inhibitor II)
	Arg-X	Garden bean (GBI II) Soybean (inhibitor C-II) Soybean (inhibitor D-II)
α-Chymotrypsin	Leu-X	Lima bean (LBI IV) Soybean (BBI)
	Tyr-X	Adzuki bean (API II) Chickpea
	Phe-X	Garden bean (PVI 3) Lima bean (LBI IV')
Elastase	Ala-X	Garden bean (GBI II) Soybean (inhibitor C-II)

germination of seeds or bulbs, an increase as well as a decrease in the inhibitor concentration has been observed, but only in a few cases were endogenous enzymes inhibited. It is probable that the inhibitors act against damage to plants by higher animals, insects, and microorganisms. This is indicated by the inhibition of proteinases of the genera *Tribolium* and *Tenebrio* and by the increase in inhibitor concentration in tomato and potato leaves after infection with the potato bug or its larvae. Proteinase inhibitors from the potato also inhibit the proteinases of microorganisms found in rotting potatoes, e.g., *Fusarium solani*.

16.2.3.4 Action on Human Enzymes

Inhibitor activity is normally determined with commercial animal enzymes, e.g., bovine

Table 16.13. Amino acid sequences in the region of the active centers of proteinase inhibitors

Inhibitor	Sequences at the active center[a]	Inhibited enzyme[b]
Bovine pancreas *Kazal* inh.	18 NGCP<u>RI</u>YNPVCG	T
Kunitz inh.	15 TGPC<u>KA</u>RIIRYF	T, CT
Soybean *Kunitz* inh.	63 SPSY<u>RI</u>RFIAEG	T, CT
Bowman-Birk inh.	16 CACT<u>KS</u>NPPQCR	T
	43 CICA<u>LS</u>YPAQCF	CT
Lima bean inhibitor IV	26 C$_A^L$CT<u>KS</u>IPPQCR	T
	53 CICT<u>LS</u>IPAQCV	CT
Potato, subunit A	60 PVVG<u>MD</u>FRCDRV	CT
Corn	25 GIPG<u>RL</u>PPLZKT	T

[a] Active center underlined.
[b] T: trypsin, CT: α-chymotrypsin.

Table 16.14. Inhibition of serine proteinases by proteinase inhibitors from potatoes[a]

Enzyme	Inhibitor[c]									
	K-group	Ala	Alb	A2	A4	A5	A6	A7a	A7b	A8
Trypsin[b]	+	+	+	+	+	+	+	+	+	+
α-Chymo-trypsin[b]	+	+	0	0	+	+	0	+	+	+
B. subtilis[d]	+	−	0	0	+	+	0	0	+	0
Asp. oryzae[d]	+	−	0	0	+	+	0	0	+	0
Pronase E	+	−	0	0	+	+	0	0	+	0
Proteinase K	+	−	0	0	+	+	0	0	+	0

[a] cv. Maritta. [b] Bovine. [c] +: inhibition, 0: no inhibition, −: not analyzed.
[d] Proteinase of the given microorganism.

trypsin or bovine chymotrypsin. The evaluation of a potential effect of the inhibitors on human health assumes that the inhibition of human enzymes is known. Present data (cf. Fig. 16.4) show that inhibitors from legumes generally inhibit human trypsin to the same extent or a little less than bovine trypsin. On the other hand, human chymotrypsin is inhibited to a much greater extent by most legumes. Ovomucoid and ovoinhibitor from egg white as well as the *Kazal* inhibitor from bovine pancreas do not inhibit the human enzymes. The *Kunitz* inhibitor from bovine pancreas inhibits human trypsin but not chymotrypsin. The data obtained greatly depend not only on the substrate used, but also on the enzyme preparation and the reaction conditions, e.g., on the ratio enzyme/inhibitor. The stability of an inhibitor as it passes through the stomach must also be

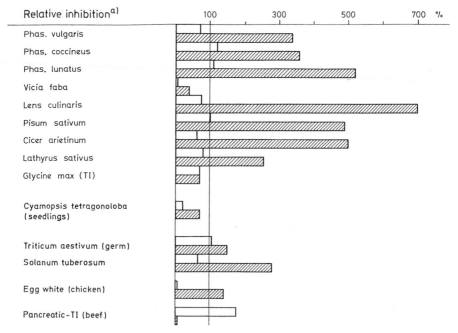

Fig. 16.4. The extent of inhibition of enzymes present in human intestinal exudates[b] as compared to that of bovine trypsin and α-chymotrypsin

[a] Related to
　▭ Inhibition of bovine trypsin = 100%,
　▨ Inhibition of bovine α-chymotrypsin = 100%.
[b] Substrate: N^α-benzoyl-L-arginine-p-nitroanilide (BAPA) for trypsin and N-glutaryl-L-phenylalanine-p-nitroanilide (Gluphepa) for chymotrypsin.

Table 16.15. Resistance of inhibitors[a] to pepsin at pH 2

Source/Inhibitor	Remaining activity[b] (%)
Soybean, *Kunitz* inhibitor	0
Bowman-Birk inhibitor (BBI)	100
extract	30–40
Lima bean, BBI-type inhibitor	70–93
Kidney bean, BBI-type inhibitor	100
Kintoki bean, BBI-type inhibitor	100
Lentil, BBI-type inhibitor	83–100
Chick pea, inhibitors	100
Broad bean, trypsin-chymotrypsin inhibitor	100
Moth bean, trypsin inhibitor	91
Broad bean, trypsin inhibitor	100

[a] Different incubation times.
[b] Against bovine and human trypsin and chymotrypsin.

taken into account in the evaluation of a potential effect (cf. Table 16.15). The *Kunitz* inhibitor of soybeans, for example, is completely inactivated by human gastric juice, but the *Bowman-Birk* inhibitor from the same source

is not. The available data show that the average amount of trypsin and chymotrypsin produced by humans per day can be completely inhibited by extracts from 100 g of raw soybeans or 200 g of lentils or other legumes.

16.2.3.5 Inactivation

The inactivation of proteinase inhibitors in the course of food processing has been the subject of many studies. In general, the inhibitors are thermolabile and can be more or less extensively inactivated by suitable heating processes. In these processes, both the starting material as well as the process parameters are of great importance (time, temperature, pressure, and water content of the sample) (Table 16.16). Steaming of soybeans for 9 minutes at 100 °C causes an 87% destruction of inhibitors (Table 16.17).

A decrease in the inhibitor activity can also be achieved by soaking. A thermal step can then

Table 16.16. Destruction of trypsin inhibitors by heating

Sample	Process	Destruction (%)
Soy flour	Live steam, 100 °C, 9 min	87
Soy bean	10% Ca(OH)$_2$, 80 °C, 1 h	100
Navy bean	Autoclaving, 121 °C, 5 min	80
	Autoclaving, 121 °C, 30 min	100
	Dry roasting, 196–204 °C, 20–25 s	75
Pinto bean	Cooking, 97 °C, 30 min	96
	Autoclaving, 121 °C, 7.5 min	100
Navy bean	Pressure cooking, 15 min	89
Winged bean	Autoclaving	92
	Soaking + autoclaving	95
Chickpeak	Autoclaving	54
Broad bean	Autoclaving, 120 °C, 20 min	90
Horse bean	Autoclaving	100
Black gram	Cooking, 100 °C, 10 min	15
	Autoclaving, 108 °C, 10 min	27
	Autoclaving, 116 °C, 10 min	38
Cow pea	Cooking, 90–95 °C, 45 min	52
	Autoclaving, 121 °C, 15 min	11
	Toasting, 210 °C, 30 min	44
	Toasting, 240 °C, 30 min	22
	Extrusion cooking	19
Peanut	Moist heat, 100 °C, 15 min	100

follow under gentler conditions. Although the processing of soybeans into protein isolates, textured protein, or meat surrogates causes a decrease in the inhibitor activity against trypsin, noticeable activity can still be present (Table 16.18).

Soybeans promote the growth of rats to the same extent as casein when about 90% of the inhibitor activity is eliminated (Table 16.19).

Table 16.17. Inactivation of soybean trypsin inhibitors by steaming (100°C)

Steaming (min)	Trypsin inhibitor	
	Concentration (mg/g soy meal)	Inactivation (%)
0	40	0
3	30	25
6	16.5	59
9	5.2	87
12	1.7	96
15	0.9	98

Table 16.18. Inhibition of bovine trypsin activity by some soya products[a]

Product	Extracted with	
	$0.125\,mol/l\,H_2SO_4$	$0.01\,mol/l\,NaOH$
Untreated soybean (cv. Caloria)	51.5	33.7
Supro G 10	6.8	15.6
Soyflour	1.1	8.7
TVP U 110 chunks	0	4.1
Flocosoya	0	1.9

[a] A 50% inhibition of mg trypsin/g product; substrate: N^α-benzoyl-L-arginine-p-nitroanilide.

Table 16.19. Effect of soybean trypsin inhibitors on the growth of rats

Trypsin inhibitor Amount (mg/100 g diet)	Inactivation (%)	Body weight (g)	Protein efficiency ratio (PER)
887	0	79	1.59
532	40	111	2.37
282	68	121	2.78
119	87	148	3.08
Control (casein)		145	3.35

16.2.3.6 Amylase Inhibitors

Relatively thermostable proteins, which have an inhibitory effect on pancreatic amylase, are found in aqueous extracts of navy beans, wheat, and rye. As a result of the high thermostabilty, inhibitor activity is also detectable in breakfast cereals and bread.

The amylase inhibitor of navy beans is instable in the stomach and becomes active only after preincubation with the enzyme in the absence of starch. As a result, it has no measurable influence on the digestion of starch by human beings. Moreover, the average amounts of inhibitor ingested with the food are small compared to the amylase activity present.

16.2.3.7 Conclusions

In summary, it can be concluded that many foods in the raw state contain inhibitors of hydrolases. The heating processes normally used in the home and in industry generally inactivate the inhibitors more or less completely, so that damage to human health is not to be expected. As a result of the greatly varying thermal stability of the inhibitors, constant and careful control of raw materials and products is required, especially when new materials and processes are applied.

16.2.4 Hemagglutinins (Lectins)

Some proteins or glycoproteins which are able to attach themselves to erythrocytes and cause their precipitation or agglutination are widely distributed in foods of plant origin. These substances are therefore designated as hemagglutinins or by the selectivity in their reactions with red cells of human blood groups as lectins (Latin *legere* = to choose, select, pick out). In erythrocytes, certain glycan sites function as receptors for lectins. Accordingly, any biopolymer will be precipitated which possesses a glycosidic residue recognizable by lectins (cf. data for specificity in Table 16.20).

Most lectins are glycoproteins. When their molecular weight exceeds 30 kdal, they consist of several subunits (Table 16.20), which readily dissociate by a change in pH or ionic strength. A characteristic feature of their amino acid composition is the high content of

Table 16.20. Occurrence of phytohemagglutinins (lectins) in food

Source	Molecular weight (kdal)	Subunits	Glycan-component		Specificity[a]
			% Carbo-hydrate	Building blocks	
Soybean	110	4	5.0	D-Man, D-GlcNAc	D-GalNAc
Garden beans	98–138	4	4.1	GlcN, Man	D-GalNAc
Jack beans[b]	112	4	0		α-D-Man
Lentils	42–69	2	2.0	GlcN, Glc	α-D-Man
Potato	20		5.2	Ara	D-GlcNAc
Wheat	26		4.5	Glc, Xyl, Hexosamine	D-GlcNAc

[a] Precipitates biopolymers that contain the given building blocks (polysaccharides, glycoproteins, lipopolysaccharides).
[b] *Canavalia ensiformis.*

acidic and hydroxy amino acids and the absence or low content of methionine.

Among the well-characterized lectins is concanavalin A, a protein isolated from jack beans (*Canavalia ensiformis*). This lectin interacts preferentially with polysaccharides containing D-mannose, but is able to bind to biopolymers consisting exclusively of α-D-glucopyranose, such as dextrins.

Lectins occur mostly in legume seeds. Animal tests have demonstrated that their toxicity often does not parallel hemagglutination activity. Thus, both lectins from soybeans and garden beans are toxic, but not that from peas. These and other observations suggest that it is not the hemagglutination activity but other activities of lectins which are responsible for their toxicity. Most observations point to the binding of lectins to epithelial cells on the intestinal wall, causing a deleterious nutritional effect by interfering with nutrient absorption, while some other lectins act as inhibitors of protein biosynthesis.

After prolonged cooking or dry heating, the activities of legume lectins and the associated toxic effects are destroyed.

16.2.5 Carbohydrates

The carbohydrates which are present in legumes are listed in Table 16.21. The major carbohydrate is starch, amounting to 75–80%. In soybeans, the presence of starch has not been proven. Instead, the presence of arabinoxylane and galactane (3.6 and 2.3% respectively) has been confirmed. In peanuts, about one-third of the total carbohydrate is starch.

Oligosaccharides in legumes are present in higher concentration than in cereals. Predominant in this fraction are sucrose, stachyose and verbascose (Table 16.21).

After legume consumption, oligosaccharides might cause flatulency, a symptom of gas accumulation in the stomach or intestines. It is a result of the growth of anaerobic microorganisms in the intestines, which hydrolyze the oligo- into monosaccharides and cause their further degradation to CO_2, CH_4 and H_2. Model feeding tests have demonstrated that phenolic ingredients, such as ferulic and syringic acids, inhibit microorganism metabolism and the related flatulency.

Table 16.21. Carbohydrates in legume flours[a]

Flour	Glucose	Saccharose	Raffinose	Stachyose	Verbascose	Starch
Garden beans	0.04	2.23	0.41	2.59	0.13	51.6
Broad beans	0.34	1.55	0.24	0.80	1.94	52.7
Lentils	0.07	1.81	0.39	1.85	1.20	52.3
Green gram (mungo beans)	0.05	1.28	0.32	1.65	2.77	52.0
Soybean[b]	0.01	4.5	1.1	3.7		

[a] Weight-% of the dry matter.
[b] Defatted flour.

Table 16.22. Cyanogenic glycosides in fruit and some field crops

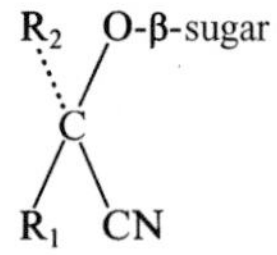

Glycoside Name	Structure		Sugar	Amino acid precursor	Occurrence (seeds)
	R_1	R_2			
Linamarin	CH_3	CH_3	Glucose	Val	Lima bean Linseed (flax) Cassava
(R)-Lotaustralin	C_2H_5	CH_3	Glucose	Ile	like Linamarin
(R)-Prunasin	Phenyl	H	Glucose	Phe	Prunes
(R)-Amygdalin	Phenyl	H	Gentiobiose	Phe	Bitter almond Apricots Peaches Apples
(S)-Dhurrin	HO–Phenyl	H	Glucose	Tyr	Sorghum sp.

Table 16.23. Amount of glycoside-bound hydrocyanic acid in food

Food	HCN (mg/100 g)
Lima bean[a]	210–310
Bitter almond	250
Sorghum sp.	250
Cassava	110
Pea	2.3
Bean	2.0
Chick pea	0.8

[a] In the United States new cultivars have been developed that contain only 10 mg HCN/100 g seed.

16.2.6 Cyanogenic Glycosides

Cyanogenic glycosides (Table 16.22) are present in lima beans and in some other plant foods. Precursors of cyanogenic glycosides are the amino acids listed in Table 16.22. As in the biosynthesis of glucosinolates (cf. 17.1.2.6.5), an aldoxime is initially formed, which is then transformed into a cyanogenic glycoside by means of the postulated reaction pathway shown in Fig. 16.5.

Seeds are ground and moistened in order to detoxify them. This initiates glycoside degradation with formation of HCN (cf. Table 16.23) which, after incubation, is expelled by heating.

The cyanogenic glycoside degradation is initiated by β-glucosidase (Fig. 16.6) which in the cells is separated from its substrate. Once the cell structure is ruptured by seed grinding, the enzyme and the substrate are brought together and the reaction starts.

The substrate specificity of β-glucosidase is governed by an aglycon moiety. Thus, the enzymes present in "emulsin", a glycosidase mixture from bitter almonds, hydrolyze not only amygdalin but also other cyanogenic glycosides which are derived from phenylalanine or tyrosine, but not linamarin.

As shown in Fig. 16.6, β-glucosidase hydrolysis produces an unstable hydroxynitrile which slowly degrades into the corresponding carbonyl compound and HCN. However, most legume seeds contain a hydroxynitrile lyase which accelerates this reaction.

16.2.7 Lipids*

With the exception of soybeans and peanuts, the lipid content of most legumes is so low (cf. Table 16.2) that they can not be considered as a source of fats or oils. Examples of their fatty acid composition are listed in Table 16.24.

---

* The composition of soy and peanut lipids is covered in Chapters 3 and 14.

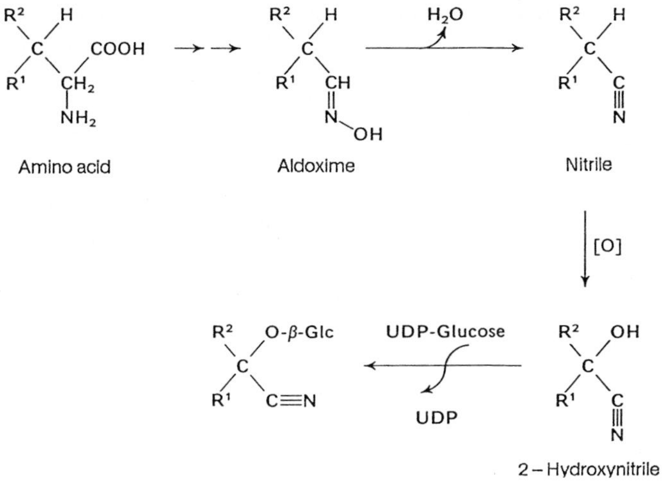

R² H   C   COOH   → →   R² H   C   CH   N–OH   H₂O   R² H   C   C≡N

R¹ CH₂ NH₂

Amino acid   Aldoxime   Nitrile

[O]

R² O-β-Glc   C   C≡N   ←   UDP-Glucose / UDP   R² OH   C   C≡N

R¹

2 – Hydroxynitrile

Fig. 16.5. Biosynthesis of cyanogenic glucosides

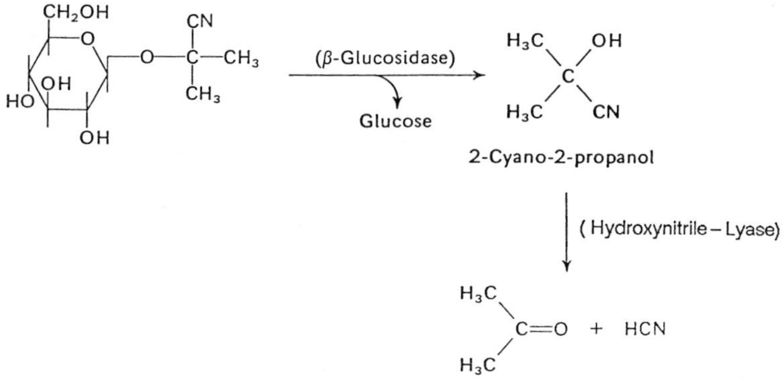

CH₂OH ... O–C(CN)(CH₃)–CH₃   (β-Glucosidase) / Glucose   H₃C OH   C   H₃C CN

2-Cyano-2-propanol

(Hydroxynitrile – Lyase)

H₃C \ C=O + HCN / H₃C

Fig. 16.6. Lima beans: linamarin degradation, resulting in a release of hydrocyanic acid

Table 16.24. Fatty acid composition of legume lipids (weight-%)[a]

Fatty acid	Garden beans	Chick peas	Broad beans	Lentils
14:0	0.22	1.3	0.6	0.85
16:0	21.8	8.9	9.3	23.2
18:0	4.7	1.6	4.9	4.6
20:0	0.53	0.03	0.7	2.3
22:0	2.9	0	0.42	2.7
24:0	1.1	0	0	0.85
16:1 (9)	0.21	0.05	0	0.15
18:1 (9)	11.6	35.4	33.8	36.0
18:2 (9, 12)	29.8	51.1	42.1	20.6
18:3 (9,12,15)	27.4	1.7	6.4	1.6
20:1	0.02	0	0.7	1.9

[a] In Table 14.11 the fatty acid compositions are provided for soya oil and peanut butter.

16.2.8 Vitamins and Minerals

Vitamin and mineral content of some legumes is presented in Table 16.25. In addition to B-vitamins, the two oilseeds are rich in vitamin E.

16.2.9 Coumestrol

Among the phenolic constituents of legumes, coumestrol is of interest because of its estrogenic effect:

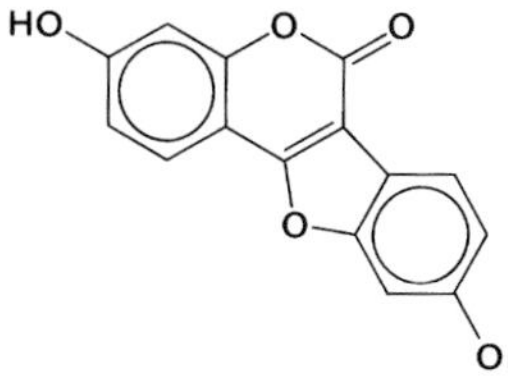

(16.4)

Table 16.25. Vitamin and mineral composition of legumes[a]

	Soybeans	Peas	Garden beans	Peanuts
Vitamins				
E	127	12		202
B_1	8.2	1.2	4.5	9.0
B_2	4.3	0.64	1.6	1.5
Nicotinamide	20.8	9.5	20.8	153
Pantothenic acid	15.9	2.9	9.7	26
B_6	9.9	0.64	2.8	3.0
Minerals				
Na	33	8.0	20	52
K	1.4×10^4	1.2×10^3	1.3×10^4	7.1×10^3
Mg	2.1×10^3	132	1.3×10^3	1.6×10^3
Ca	2.1×10^3	96	1.1×10^3	590
Fe	71	7.4	60.4	21.1
Zn	8.3	10.6		30.7
P	4.9×10^3	432	4.3×10^3	3.7×10^3
Cl	58	160	248	70

[a] Results are given in mg/kg.

It occurs mostly in soybean hulls at levels of 0.05–30 ppm.

16.2.10 Saponins

Legumes contain a number of saponins. Saponin has been isolated and identified from soy and other beans (cf. Formula 16.5). The suggestion that these compounds impart a bitter taste and might develop during storage of soy products has been proven wrong. On the other hand, the insecticidal and/or fungicidal effect of saponins is still of practical interest.

16.2.11 Other Constituents

The meadow pea, *Lathyrus sativus*, which is cultivated in India in periods of drought, contains β-N-oxalyl-α,β-diaminopropionic acid (cf. XXXVII in Table 17.5). Possibly due to its structural similarity to glutamic acid, this compound causes the disease known as neurolathyrism, which is characterized by paralysis of the lower limbs. More than 100,000 cases of this disease were described in 1975 alone. The diaminopropionic acid derivative can be largely eliminated by cooking the seeds in excess water, which is then discarded, or by soaking the seeds overnight, followed by steaming, roasting, or drying in the sun. The flour obtained from dried seeds has 24–28% of protein and a high lysine content. It can be used to make unleavened Indian bread ("chapatis"). The horse bean, *Vicia faba*, contains the glucosides vicin (Formula 16.6, I) and convicin (II). The aglycones of these compounds divicin (III) and isouramil (IV) can be released by the β-glycosidases of the digestive tract. In the oxidized form, they cause quick oxidation of glutathione in erythrocytes (cf. Formula 16.6) which have a hereditary deficiency of glucose-6-phosphate dehydrogenase. Consequently, these erythrocytes are incapable of re-producing reduced glutathione with the help of glutathione reductase for lack of NADPH. The lack of reduced glutathione causes a hemolytic anemia called favism. This genetic defect is found especially in people from the Middle East. Since *Vicia faba* plays a big role in the protein supply of people in this region, attempts are being made to cultivate variants which do not contain these toxic glucosides or to develop suitable methods for its removal (soaking, heating).

16.3 Processing

16.3.1 Soybeans and Peanuts

16.3.1.1 Aroma Defects

Preparation and storage of products from both oilseeds is often inhibited by rancidity and bitter aroma defects caused mostly by volatile

α-Rhap (1 → 2) β-Galp (1 → 2)-β-GlcAp (1—O   (16.5)

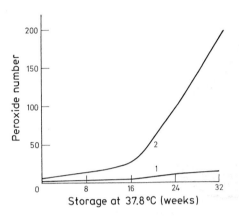

$$\text{(16.6)}$$

I : R = NH_2
II : R = OH

III (ox.) : R = NH_2
IV (ox.) : R = OH

III (red.) : R = NH_2
IV (red.) : R = OH

carbonyl compounds, particularly hexanal derived from oxidized linoleic acid. The rancidity-causing compounds are formed through lipid peroxidation, accelerated by the enzyme lipoxygenase and/or by hem(in) proteins (cf. 3.7.2.2). One way to increase quality is to thermally inactivate enzymes or hem(in) catalysts. Table 16.26 illustrates steam heating of peanuts for a prolonged time in order to inactivate peroxidase activity. Lipoxygenase denaturation, under the conditions given in Table 16.26, occurs after 2 min, but this alone does not yield a satisfactory storage stability. Peroxidase and probably other catalysts should be excluded as well (Fig. 16.7).

The complete removal of lipids is used as an additional precautionary measure in order to obtain an off-flavor-free product, particularly in the case of production of protein isolates. For example, the lipid residue which remains in soy flakes after hexane solvent extraction (cf. 14.3.2.2.1) is removed by extraction with hexane-ethanol 82 : 18 v/v.

16.3.1.2 Individual Products

Protein preparations and milk-like products are processed from soybeans and peanuts. Alone or together with cereals, soybeans are

Fig. 16.7. Storage stability of peanut flakes. (according to *J. H. Mitchel* and *R. K. Malphrus*, 1977). Peanut flakes treated with steam at 100 °C for 30 min (1) and 5 min (2)

processed into a large number of fermented products in Asia. The following products are made from soybeans.

16.3.1.2.1 Soy Proteins

Figure 16.8 gives an overview of the most important process steps in soybean processing. Soy protein concentrate is usually obtained from the flaked and defatted soy meal that is left after oil extraction (cf. 14.3.2.2.1). The process involves soaking of flakes in water, acidification of the aqueous extract to pH 4−5 (cf. 16.2.1) and separation of the precipitate from solubilized ingredients by centrifugation followed by washing and drying of the sediment collected.

Soy meal isolates enriched in protein are obtained by a preliminary extraction of soluble soy constituents with water or diluted alkali, pH 8−9, followed by protein precipitation from the aqueous extract by adjusting the pH to 4−5. Such protein isolates, texturized and flavored (cf. 1.4.7) are used as meat substitutes.

Table 16.26. Thermal inactivation of lipoxygenase and peroxidase in peanuts

Heat treatment			Enzyme activity (%)	
Type	°C	Time (min)	Per-oxidase	Lipoxy-genase
Control			100	100
Dry heat	110	60	48	7
Steam	100	2	35	0
Steam	100	6	8	0
Steam	100	30	1	0

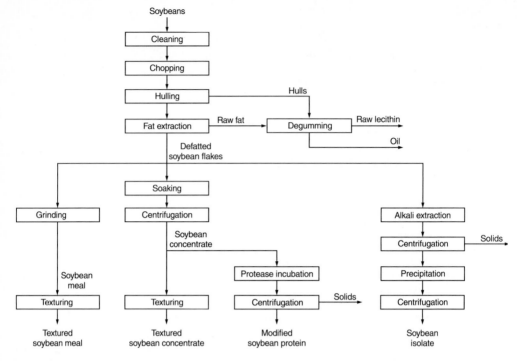

Fig. 16.8. Soybean processing

The compositions of protein concentrates and isolates are compared in Table 16.27. For both products, the essential amino acid content corresponds to that of soybeans (cf. Table 16.3). Soy protein is added as an ingredient to baked and meat products and to baby food preparations to raise their protein level and to improve their processing qualities, such as increased water binding capacity or stabilization of o/w emulsions. These properties are required for processing at higher temperatures. The addition of soy protein to beverages at a pH of 3 results in better solubility of beverage constituents. Soy protein market value may be increased by its partial hydrolysis with papain (cf. 2.7.2.2.1).

Table 16.27. Composition of soya protein concentrate and isolate (%)

Product	Protein	Crude fiber	Ash
Concentrate	72	3.5	5.5
Isolate	95.6	0.2	4.0

16.3.1.2.2 Soy Milk

Soybeans are swollen and ground in the presence of a 10-fold excess of water. Heating the suspension close to its boiling point for 15–20 min pasteurizes the suspension and inactivates lipoxygenase enzyme and proteinase inhibitors. A soy milk preparation enriched with calcium and vitamins is of importance in infant nutrition as a replacement for cow's milk, which close to 7% of infants in the USA are unable to tolerate.

16.3.1.2.3 Tofu

When calcium sulfate (3 g salt/kg milk) is added to soy milk at 65 °C, a gel (called soy "curd") slowly precipitates. The curd is separated from excess fluid by gentle squeezing in a special wooden filter box. A washing procedure then follows. The water content of the product is about 88%. Tofu contains 55% protein and 28% fat dry weight. In China and some other Asian countries, tofu is the largest source of food protein. It is consumed fresh or

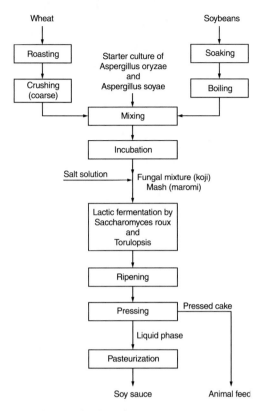

Fig. 16.9. Production of soy sauce

dried, or fried in fat and seasoned with soy sauce.

16.3.1.2.4 Soy Sauce (Shoyu)

Defatted soy meal is used as a starting material in the production of this seasoning sauce. The meal is moistened, then mixed with roasted and crushed wheat and heated in an autoclave for 45 min. The mix ratio in Japan is fixed at 1:1, while in China it varies up to 4 : 1. Increasing the amount of soy decreases the quality of the endproduct. The mix, with a water content of 26%, is then inoculated with *Aspergillus oryzae* and *Aspergillus soyae*. Initial incubation is at 30 °C for 24 h and then at 40 °C for an additional 48 h. This fermentation starter, called "koji", is then salted to 18% by addition of 22.6% NaCl solution. Inoculation with *Lactobacillus delbrueckii* and with *Hansenula* yeast species results in lactic acid

fermentation, which proceeds under gentle aeration in order to prevent the growth of undesired anaerobic microorganisms. It is a long and tedious fermentation carried out in stepwise fashion: for example, starting at 15 °C for one month, followed by 28 °C for four months, and finishing at 15°C for an additional month. Highly-valued products ripen for several years. After the fermentation is completed, the soy sauce of pH 4.6 is filtered, pasteurized at 65–80°C and preserved with benzoic acid for the export market.

During fermentation the microorganisms produce extracellular hydrolases which decompose the main components of the raw material: proteins, carbohydrates and nucleic acids. Soy sauce contains 1.5% N (of which 60% corresponds to amino-N) and 4.4% reducing sugar. The N-containing fraction consists of 40–50% amino acids (glutamic acid predominates at 1.2% of the product), 40–50% peptides, 10–15% ammonia and less than 1% protein. In addition, soy sauce contains by-products of microorganism metabolism, such as ethanol (1.2%) and lactic, succinic and acetic acids.

Soy sauce products of lower quality are blended with spices and are prepared by acid hydrolysis of the above mix of raw materials (cf. 12.7.3.5).

The compound 2(5)-ethyl-4-hydroxy-5(2)-methyl-3(2H)-furanone (EHMF) is responsible for the sweetish caramel-like aroma note. It is formed by the yeast *Zygosaccharomyces rouxii* from D-sedoheptulose-7-phosphate, which originates from the pentose phosphate cycle. Apart from EHMF, furaneol and sotolon contribute to the aroma. Abhexone is also present, but is of secondary importance because of its lower concentration compared to sotolon.

16.3.1.2.5 Miso

Miso is a fermented soybean paste. To produce this substance, rice is soaked, heated, and incubated with *Aspergillus oryzae* at 28–35°C for 40–50 hours. At the same time, whole soybeans are soaked, heated, and mixed with the incubated rice (60:30) with the addition of salt (4–13%). The mixture is allowed to ferment for several months at 25–30°C in the presence of lactic acid bacteria and yeasts. The product is then pasteurized and packed. The aroma of miso can be enhanced by the addition of EHMF (cf. 16.3.1.2.4).

16.3.1.2.6 Natto

Various types of natto, a fermented soybean product, are known. For production (Itohiki type), soybeans are soaked in water, boiled and after cooling, incubated with *Bacillus nato*, a variant of *Bacillus subtilis*, for 16−20 hours at 40−45 °C. The surface of natto has a characteristic viscous texture caused by a polyglutamic acid produced by *B. natto*.

16.3.1.2.7 Sufu

Sufu is soy cheese made from tofu. Tofu is cut into cubes (3 cm edge length), treated with an acidified salt solution (6% NaCl, 2.5% citric acid), heated (100 °C, 15 min) and inoculated with *Actinomucor elegans*. After incubation at 12−25 °C for 2−7 days, sufu is placed in a 5−10% salt solution which contains fermented soybean paste and ethanol, if necessary, and allowed to ripen for 1−12 months.

16.3.2 Peas and Beans

Peas and beans are consumed only when cooked. In order to shorten the cooking time which, even after preliminary soaking in water overnight (preliminary swelling), is several hours, the legumes are precooked or parboiled by the process described in 15.3.2.2.1.

Additionally, seed hull removal provides about a 40% reduction in cooking time which, for peas, involves seed steaming at 90 °C, followed by drying and subsequent dehulling.

The softening of legumes during cooking is due to the disintegration of the cotyledonous tissue in individual cells. This is caused by the conversion of native protopectin to pectin, which quickly depolymerizes on heating. The middle lamella of the cell walls, which consists of pectins and strengthens the tissue, disintegrates in this process.

Conversely, the hardening of legumes during cooking is due to cross linkage of the cell walls. The following reactions which can start even during storage at higher temperatures are under discussion as the cause of cross linkage. Calcium and magnesium phytates included in the middle lamellae are hydrolyzed by the phytase present (cf. 15.2.2.4). Apart from meso-inositol and phosphoric acid, Ca^{2+} and Mg^{2+} ions also released cross link the pectic acids and thus strengthen the middle lamellae. Pectin esterases, which demethylate pectin to the acid, promote the hardening of the tissue. In the case of legumes that are relatively rich in phenolic compounds and polyphenol oxidases, the formation of complexes between proteins and polyphenols should contribute to the strengthening of the tissue.

Similar to soybeans, a number of beans are processed into fermented products in Asia.

16.4 Literature

Angelo, A.J.S., Ory, R.L.: Effects of lipoperoxides on proteins in raw and processed peanuts. J. Agric. Food Chem. *23*, 141 (1975)

Aoki, H., Taneyana, O., Inami, M.: Emulsifying properties of soy protein: characteristics of 7S and 11S proteins. J. Food Sci. *45*, 534 (1980)

Badley, R.A., Atkinson, D., Hauser, H., Oldani, D., Green, J.P., Stubbs, J.M.: The structure, physical and chemical properties of the soybean protein glycinin. Biochim. Biophys. Acta *412*, 214 (1975)

Belitz, H.-D.: Vegetable proteins as human food. FEBS 11th Meeting Copenhagen 1977, Vol. 44, Symposium A3, Pergamon Press: Oxford-New York. 1978

Belitz, H.-D., Kaiser, K.-P., Santarius, K.: Trypsin and chymotrypsin inhibitors from potatoes: isolation and some properties. Biochem. Biophys. Res. Commun. *42*, 420 (1971)

Belitz, H.-D., Weder, J.K.P.: Protein inhibitors of hydrolases in plant foodstuffs. Food Rev. Int. *6*, 151 (1990)

Beuchat, L.R.: Indigenous fermented foods. In: Biotechnology (Eds.: Rehm, H.-J., Reed, G.), Vol. 6, p. 477, Verlag Chemie: Weinheim. 1983

Boulter, D., Derbyshire, E.: The general properties, classification and distribution of plant proteins. In: Plant proteins (Ed.: Norton G.), p. 3, Butterworths: London. 1978

Derbyshire, E., Wright, D.J., Boulter, D.: Legumin and vicilin, storage proteins of legume seeds. Phytochemistry *15*, 3 (1976)

Friedman, M. (Ed.): Nutritional and toxicological significance of enzyme inhibitors in foods. Adv.

Exp. Med. Biol. 199, Plenum Press: New York. 1986

Gallaher, D., Schneeman, B. O.: Nutritional and metabolic response to plant inhibitors of digestive enzymes. Adv. Exp. Med. Biol. *177*, 299 (1984)

Liener, I. E.: Phytohemagglutinins: their nutritional significance. J. Agric. Food Chem. *22*, 17 (1974)

Liener, I. E.: Protease inhibitors and other toxic factors in seeds. In: Plant proteins (Ed.: Norton, G.), p. 117, Butterworths: London. 1978

Liener, I. E.: Significance for humans of biologically active factors in soybeans and other food legumes. J. Am. Oil Chem. Soc. *56*, 121 (1979)

Liener, I. E. (Ed.): Toxic constituents of plant foodstuffs. 2nd. ed. Academic Press: New York. 1980

Lookhart, G. L., Jones, B. L., Finney, K. F.: Determination of coumestrol in soybeans by high-performance liquid and thin-layer chromatography. Cereal Chem. *55*, 967 (1978)

Mitchell, J. H., Malphrus, R. K.: Lipid oxidation in spanish peanuts: the effect of moist heat treatments. J. Food Sci. *42*, 1457 (1977)

Mossor, G., Skupin, J., Romanowska, B.: Plant inhibitors of proteolytic enzymes. Nahrung *28*, 93 (1984)

Naivikul, O., D'Appolonia, B. L.: Comparison of legume and wheat flour carbohydrates. I. Sugar analysis. Cereal Chem. *55*, 913 (1978)

Pernollet, J.-C., Mossé, J.: Structure and location of legume and cereal seed storage proteins. In: Seed proteins (Eds.: Daussant, J., Mossé, J., Vaughan, J.), p. 155, Academic Press: London. 1983

Rackis, J. J., Sessa, D. J., Steggerda, F. R., Shimizu, T., Anderson, J., Pearl, S. L.: Soybean factors relating to gas production by intestinal bacteria. J. Food Sci. *35*, 634 (1970)

Salunkhe, D. K., Kadam, S. S. (Eds.): CRC Handbook of World Food Legumes: Nutritional Chemistry, Processing Technology and Utilization, Vol. I–III. CRC Press: Boca Raton, FL, 1989

Sasaki, M., Nunomura, N., Matsudo, T.: Biosynthesis of 4-hydroxy-2(or5)-ethyl-5(or2)-methyl-3-(2H)-furanone by yeast. J. Agric. Food Chem. *39*, 934 (1991)

Sathe, S. K., Salunkhe, D. K.: Technology of removal of unwanted components of dry beans. CRC Crit. Rev. Food Sci. Nutr. *21*, 263 (1984)

Smith, A. K., Circle, S. J. (Eds.): Soybeans: Chemistry and technology. Vol. 1, AVI Publ. Co.: Westport, Conn. 1972

Stanley, D. W., Aguilera, J. M.: A review of textural defects in cooked reconstituted legumes – The influence of structure and composition. J. Food Biochem. *9*, 277 (1985)

Warchalewski, J. R.: Present-day studies on cereals protein nature α-amylase inhibitors. Nahrung *27*, 103 (1983)

Whitaker, J. R., Feeney, R. E.: Enzyme inhibitors in foods. In: Toxicants occurring naturally in foods, 2nd edn., p. 276, National Academy of Sciences: Washington, D.C. 1973

Wolf, W. J.: Chemistry and technology of soybeans. In: Advances in cereal science and technology (Ed.: Pomeranz, Y.), Vol. I, p. 325, American Association of Cereal Chemists: St. Paul, Minn. 1976

Wright, D. J.: The seed globulins. In: Development of Food Proteins-5; (Ed.: Hudson, B. J. F.), p. 81, Elsevier Applied Science: London. 1987

Wright, D. J.: The seed globulins – Part. II. In: Development in Food Proteins-6; (Ed.: Hudson, B. J. F.), p. 119, Elsevier Applied Science: London. 1987

17 Vegetables and Vegetable Products

17.1 Vegetables

17.1.1 Foreword

Vegetables are defined as the fresh parts of plants which, either raw, cooked, canned or processed in some other way, provide suitable human nutrition. Fruits of perennial trees are not considered to be vegetables. Ripe seeds are also excluded (peas, beans, cereal grains, etc.). From a botanical point of view, vegetables can be divided into algae (seaweed), mushrooms, root vegetables (carrots), tubers (potatoes, yams), bulbs and stem or stalk (kohlrabi, parsley), leafy (spinach), inflorescence (broccoli), seed (green peas) and fruit (tomato) vegetables. The most important vegetables, with data relating to their botanical classification and use, are presented in Table 17.1. Information

about vegetable production follows in Tables 17.2 and 17.3.

17.1.2 Composition

The composition of vegetables can vary significantly depending upon the cultivar and origin. Table 17.4 shows that the amount of dry matter in most vegetables is between 10 and 20%. The nitrogen content is in the range of 1–5%, carbohydrates 3–20%, lipids 0.1–0.3%, crude fiber about 1%, and minerals close to 1%. Some tuber and seed vegetables have a high starch content and therefore a high dry matter content. Vitamins, minerals, flavor substances and dietary fibers are important secondary constituents.

Table 17.1. List of some important vegetables

Number	Common name	Latin name	Class, order, family	Consumed as
Mushrooms (cultivated or wildly grown edible species)				
1	Ringed boletus	*Suillus luteus*	Basidiomycetes/Boletales	
2	Saffron milk cap	*Lactarius deliciosus*	Basidiomycetes/Agaricales	
3	Field champignon	*Agaricus campester*	Basidiomycetes/Agaricales	
4	Garden champignon	*Agaricus hortensis*	Basidiomycetes/Agaricales	
5	Cep	*Xerocomus badius*	Basidiomycetes/Boletales	
6	Truffle	*Tuber melanosporum*	Ascomycetes/Tuberales	Steamed, fried, dried,
7	Chanterelle	*Cantharellus cibarius*	Basidiomycetes/Aphyllophorales	pickled or salted
8		*Xerocomus chrysenteron*	Basidiomycetes/Boletales	
9	Morel	*Morchella esculenta*	Ascomycetes/Pezizales	
10	Edible boletus	*Boletus edulis*	Basidiomycetes/Boletales	
11	Goat's lip	*Xerocomus subtomentosus*	Basidiomycetes/Boletales	
Algae (seaweed)				
12	Sea lettuce	*Ulva lactuca*		Eaten raw as a salad, cooked in soups (Chile, Scotland, West Indies)
13	Sweet tangle	*Laminaria saccharina*		Eaten raw or cooked (Scotland)
14		*Laminaria sp.*		Eaten dried ("combu") or as a vegetable (Japan)
15		*Porphyra laciniata*		Eaten raw in salads, cooked as a vegetable (England, America)
16		*Porphyra sp.*		Dried or cooked ("nari" products, Japan and Korea)
17		*Undaria pinnatifida*		Eaten dried ("wakami") and as a vegetable (Japan)

Table 17.1 (continued)

Num-ber	Common name	Latin name	Class, order, family	Consumed as
Rooty vegetables				
18	Carrot	*Daucus carota*	Apiaceae	Eaten raw or cooked
19	Radish (white elong-ated fleshy root)	*Raphanus sativus var. niger*	Brassicaceae	The pungent fleshy root eaten raw, salted
20	Viper's grass, scorzonera	*Scorzonera hispanica*	Asteraceae	Cooked as a vegetable
21	Parsley	*Petroselinum crispum ssp. tuberosum*	Apiaceae	Long tapered roots cooked as a vegetable, or used for seasoning
Tuberous vegetables (sprouting tubers)				
22	Arrowroot	*Tacca leontopetaloides*	Taccaceae	Cooked or milled into flour for breadmaking
23	White (Irish) potato	*Solanum tuberosum*	Solanaceae	Cooked, fried or deep fried in many forms, or unpeeled baked, also for starch and alcohol production
24	Celery tuber	*Apium graveolens, var. rapaceum*	Apiaceae	Cooked as salad, and cooked and fried as a vegetable
25	Kohlrabi, turnip cabbage	*Brassica oleracea convar. acephala var. gongylodes*	Brassicaceae	Eaten raw or cooked as a vegetable
26	Rutabaga	*Brassica napus var. naprobrassica*	Brassicaceae	Cooked as a vegetable
27	Radish (reddish round root)	*Raphanus sativus var. sativus/var. niger*	Brassicaceae	The pungent fleshy root is eaten raw, usually salted
28	Red beet, beetroot	*Beta vulgaris spp. vulgaris var. conditiva*	Chenopodiaceae	Cooked as a salad
Tuberous (rhizomatic) vegetables				
29	Sweet potatoes	*Ipomoea batatas*	Convolvulaceae	Cooked, fried or baked
30	Cassava (manioc)	*Manihot esculenta*	Euphorbiaceae	Cooked or roasted
31	Yam	*Dioscorea*	Dioscoreaceae	Cooked or roasted
Bulbous rooty vegetables				
32	Vegetable fennel	*Foeniculum vulgare var. azoricum*	Apiaceae	Eaten raw as salad, cooked as a vegetable
33	Garlic	*Allium sativum*	Liliaceae	Raw, cooked as seasoning
34	Onion	*Allium cepa*	Liliaceae	Eaten raw, fried as seasoning, cooked as a vegetable
34a	Leek	*Allium porrum*	Liliaceae	The pungent succulent leaves and thick cylindrical stalk are cooked as a vegetable
Stem (shoot) vegetables				
35	Bamboo roots	*Bambusa vulgaris*	Poaceae	Cooked for salads
36	Asparagus	*Asparagus officinalis*	Liliaceae	Young shoots cooked as a vegetable or eaten as salad
Leafy (stalk) vegetables				
37	Celery	*Apium graveolens var. dulce*	Apiaceae	Leafy crispy stalks eaten raw as salad, or are cooked as vegetable
38	Rhubarb	*Rheum rhabarbarum, Rheum rhaponticum*	Polygonaceae	Large thick and succulent petioles are cooked as preserves or baked; used as a pie filling
Leafy vegetables				
39	Watercress	*Nasturtium officinale*	Brassicaceae	Moderately pungent leaves are eaten raw in salads or used as garnish
40	Endive (escarole, chicory)	*Cichorium intybus L. var. foliosum*	Cichoriaceae	Eaten raw as a salad, or is cooked as a vegetable
41	Chinese cabbage	*Brassica chinensis*	Brassicaceae	Eaten raw in salads, or is cooked as a vegetable
42	Lamb's salad (lettuce or corn salad)	*Valerianella locusta*	Valerianaceae	Eaten raw in salads
43	Garden cress	*Lepidium sativum*	Brassicaceae	Eaten raw in salads
44	Kale (borecole)	*Brassica oleracea convar. acephala var. sabellica*	Brassicaceae	Cooked as a vegetable
45	Head lettuce	*Lactuca capitata var. capitata*	Cichoriaceae	Juicy succulent leaves are eaten raw in salads
46	Mangold (mangel-wurzel, beet root)	*Beta vulgaris spp. vulgaris var. vulgaris*	Chenopidiaceae	Cooked as a vegetable

Table 17.1 (continued)

Num-ber	Common name	Latin name	Class, order, family	Consumed as
47	Chinese (Peking) cabbage	*Brassica pekinensis*	Brassicaceae	Cooked as a vegetable
48	Brussels sprouts	*Brassica oleracea convar. oleracea var. gemmifera*	Brassicaceae	Cooked as a vegetable
49	Red cabbage	*Brassica oleracea convar. capitata var. capitata f. rubra*	Brassicaceae	Eaten raw in salads or is cooked as a vegetable
50	Romaine lettuce	*Lactuca capitata var. crispa*	Cichoriaceae	Eaten raw as a salad
51	Spinach	*Spinacia oleracea*	Chenopodiaceae	Cooked as a vegetable or is eaten raw as a salad
52	White (common) cabbage	*Brassica oleracea convar. capitata var. capitata f. alba*	Brassicaceae	Juicy succulent leaves are eaten raw in salads, or are fermented (sauerkraut), steamed or cooked as a vegetable
53	Winter endive	*Cichoricum endivia*	Cichoriaceae	Eaten raw as a salad
54	Savoy cabbage	*Brassica oleracea convar. capitata, var. sabauda*	Brassicaceae	Cooked as a vegetable
Flowerhead (calix) vegetables				
55	Artichoke	*Cynara scolymus*	Asteraceae	Flowerhead is cooked as a vegetable
56	Cauliflower	*Brassica oleracea convar. botrytis var. botrytis*	Brassicaceae	Cooked as a vegetable or used in salads (raw or pickled)
57	Broccoli	*Brassica oleracea convar. botrytis var. italica*	Brassicaceae	The tight green florets are cooked as a vegetable
Seed vegetables				
58	Chestnut	*Castanea sativa*	Fagaceae	Cooked as a vegetable, roasted, or milled into a flour and used in soups and bread doughs
59	Green beans	*Phaseolus vulgaris*	Fabaceae	The immature pod is cooked as a vegetable or is steamed or pickled for salads
60	Green peas	*Pisum sativum ssp. sativum*	Fabaceae	The rounded smooth or (wrinkled) green seeds are cooked as a vegetable or are steamed/cooked for salads
Fruity vegetables				
61	Eggplant	*Solanum melongena*	Solanaceae	Steamed as a vegetable
62	Garden squash	*Cucurbita pepo*	Cucurbitaceae	Cooked as a compote or as a vegetable
63	Green bell pepper	*Capsicum annuum*	Solanaceae	Eaten raw in salads, or is cooked, steamed or baked
64	Cucumber	*Cucumis sativus*	Cucurbitaceae	Eaten raw in salads, cooked as a vegetable or pickled
65	Okra	*Abelmoschus esculentus*	Malvaceae	Its mucilaginous green pods are cooked as a vegetable in soups or stewed, or eaten as a salad
66	Tomato	*Lycopersicon lycopersicum*	Solanaceae	The reddish pulpy berry is eaten raw, in salads, cooked as a vegetable, used as a paste or seasoned puree; immature green tomatoes are pickled and then eaten as salad
67	Zucchini	*Cucurbita pepo convar. giromontiina*	Cucurbitaceae	The cylindrical dark green fruits are peeled and cooked as a vegetable

17.1.2.1 Nitrogen Compounds

Vegetables contain an average of $1-3\%$ nitrogen compounds. Of this, $35-80\%$ is protein, the rest is amino acids, peptides and other compounds.

17.1.2.1.1 Proteins

The protein fraction consists to a great extent of enzymes which may have either a beneficial or a detrimental effect on processing. They may contribute to the typical flavor or to formation of undesirable flavors, tissue softening and discoloration. Enzymes of all the main groups are present in vegetables:

- *Oxidoreductases* such as lipoxygenases, phenoloxidases, peroxidases;
- *Hydrolases* such as glycosidases, esterases, proteinases;

Table 17.2. Production of vegetables in 1996 (1,000 t)

Continent	Vegetables + melons, grand total	Cabbages	Artichokes	Tomatoes
World	565,523	46,656	1,150	84,873
Africa	35,478	780	90	9,591
America, North-, Central-	45,646	2,208	51	14,899
America, South-	16,623	526	94	5,671
Asia	376,119	28,758	25	36,250
Europe	88,727	14,265	891	17,947
Oceania	2,930	119	–	515

Continent	Cauliflower	Squash	Cucumbers	Eggplants
World	12,725	9,822	23,051	11,981
Africa	180	1,174	391	550
America, North-, Central-	438	497	1,478	83
America, South-	62	828	65	5
Asia	9,419	5,236	17,667	10,816
Europe	2,516	1,915	3,427	524
Oceania	110	171	22	2

Continent	Chilies[a]	Onions, air dried	Garlic	Green beans
World	14,068	35,644	10,401	3,620
Africa	1,839	2,252	197	193
America, North-, Central-	1,575	3,082	308	240
America, South-	248	2,434	203	99
Asia	7,905	21,341	9,075	2,059
Europe	2,476	6,288	618	979
Oceania	25	246	1	51

Continent	Green peas	Carrots	Watermelons	Cantaloups and other melons (muskmelons)
World	5,214	16,477	39,725	16,190
Africa	257	820	1,979	1,039
America, North-, Central-	1,213	2,327	2,509	2,045
America, South-	146	664	1,118	295
Asia	1,274	5,911	30,168	10,234
Europe	2,184	6,511	3,871	2,502
Oceania	140	246	267	75

- *Transferases* such as transaminases;
- *Lyases* such as glutamic acid decarboxylase, alliinase, hydroperoxide lyase.
- *Ligases* such as glutamine synthetase.

Enzyme inhibitors are also present, e. g., potatoes contain proteins which have an inhibitory effect on serine proteinases, while proteins from beans and cucumbers inhibit pectolytic

Table 17.2 (continued)

Country	Vegetables + melons, grand total	Country	Cabbages	Country	Artichokes
China	202,155	China	14,214	Italy	517
India	64,672	Russian Fed.	5,035	Spain	277
USA	34,393	India	3,300	Argentina	75
Turkey	20,769	Korea Rep.	3,200	France	72
Japan	13,589	Japan	2,702	USA	49
Italy	13,555	Poland	1,891	Morocco	34
Russian Fed.	11,099	USA	1,730	Egypt	29
Korea Rep.	10,562	Indonesia	1,500	Greece	24
Spain	10,524	Korea Democr. People's Rep.	900	Tunisia	21
Iran	9,900			Chile	17
Egypt	9,377	Germany	877		
France	7,927			Σ (%)[b]	97
Mexico	6,740	Σ (%)[b]	76		
Brazil	6,165				
Nigeria	6,039				
Σ (%)[b]	75				

Country	Tomatoes	Country	Cauliflower	Country	Squash
China	13,632	India	4,800	China	2,668
USA	11,700	China	3,858	Ukraine	925
Turkey	7,300	France	531	Argentina	400
Italy	5,156	Italy	487	Turkey	350
Egypt	5,038	UK	384	Mexico	325
India	4,800	Spain	307	Egypt	315
Spain	2,788	USA	305	Italy	286
Brazil	2,639	Poland	244	South Africa	280
Iran	2,150	Pakistan	190	Spain	250
Mexico	2,145	Germany	163	Japan	242
Greece	1,990			Pakistan	230
Chile	1,370	Σ (%)[b]	89	Iran	220
Russian Fed.	1,350			Thailand	215
Ukraine	1,077			Morocco	200
Uzbekistan	970			Indonesia	180
				Korea Rep.	165
Σ (%)[b]	76			Sri Lanka	165
				Σ (%)[b]	76

Country	Cucumbers	Country	Eggplants	Country	Chilies[a]
China	11,851	China	8,325	China	5,522
Iran	1,250	Turkey	810	Turkey	1,100
Turkey	1,150	Japan	478	Nigeria	970
USA	992	Egypt	355	Mexico	904
Japan	826	Italy	234	Spain	835
Ukraine	550	Syria	165	USA	597
Netherlands	480	Indonesia	155	Indonesia	460
Poland	379	Iraq	150	Italy	308
Iraq	340	Philippines	140	Korea Rep.	300
Korea Rep.	340	Spain	110	Bulgaria	263
Σ (%)[b]	79	Σ (%)[b]	91	Σ (%)[b]	80

Table 17.2 (continued)

Country	Onions, air dried	Country	Garlic	Country	Green beans
China	8,230	China	7,674	China	950
India	4,058	Korea Rep.	360	Turkey	440
USA	2,793	India	350	Indonesia	235
Turkey	1,900	USA	232	Spain	224
Japan	1,278	Spain	221	Italy	158
Iran	1,200	Indonesia	130	USA	141
Pakistan	1,082	Egypt	120	Egypt	109
Spain	1,007	Thailand	112	France	103
Brazil	943	Turkey	93	Thailand	87
Russian Fed.	700	Korea Democr. People's Rep.	86	Belgium/ Luxembourg	80
Netherlands	630	Ukraine	86	Netherlands	80
Indonesia	625			Greece	75
Poland	607	Σ (%)[b]	91	Japan	75
Korea Rep.	570				
Ukraine	500			Σ (%)[b]	76
Argentina	460				
Σ (%)[b]	75				

Country	Green peas	Country	Carrots	Country	Water-melons
USA	1,100	China	3,956	China	17,308
China	819	USA	1,740	Turkey	3,600
France	580	Russian Fed.	1,390	Iran	2,650
UK	535	Poland	793	USA	1,852
India	270	UK	751	Korea Rep.	880
Belgium/ Luxembourg	165	Japan	724	Spain	795
Russian Fed.	160	France	640	Georgia	750
Italy	130	Netherlands	430	Egypt	730
Hungary	120	Italy	396	Japan	617
Egypt	107	Ukraine	350	Italy	611
		Germany	340		
Σ (%)[b]	76	Canada	331	Σ (%)[b]	75
		Morocco	300		
		Spain	300		
		Σ (%)[b]	76		

Country	Cantaloups and other melons
China	5,262
Turkey	1,800
Iran	1,215
USA	965
Spain	943
Mexico	680
Romania	639
Egypt	470
Morocco	415
Japan	400
Σ (%)[b]	79

[a] Data including other Capsicum species.
[b] World production = 100%.

Table 17.3. Production of starch containing roots, rhizomes and tubers in 1996 (1,000 t)

Continent	Tubers + rhizomes grand total	Potato	Sweet potato	Casava (manioc)
World	635,337	294,834	134,244	162,942
Africa	137,733	7,796	7,411	85,041
America, North-, Central-	31,250	28,190	1,146	1,039
America, South-	44,986	12,067	1,305	30,600
Asia	261,970	89,167	123,765	46,065
Europe	156,274	156,209	57	–
Oceania	3,124	1,404	560	197

Country	Tubers + rhizomes grand total	Country	Potato	Country	Sweet potato
China	166,111	China	46,034	China	115,196
Nigeria	56,085	Russian Fed.	38,529	Indonesia	2,424
Russian Fed.	38,529	USA	22,549	Uganda	2,250
Brazil	28,156	Poland	22,500	Vietnam	1,700
India	25,070	Ukraine	18,400	Japan	1,181
USA	23,169	India	17,942	India	1,128
Poland	22,500	Germany	13,600	Rwanda	1,100
Indonesia	19,014	Belarus	10,677	Burundi	670
Zaire	18,861	Netherlands	8,081	Brazil	655
Ukraine	18,400	UK	7,219	Kenia	635
Thailand	16,208	France	6,462		
Germany	13,600	Turkey	4,950	Σ (%)[a]	95
Belarus	10,677	Spain	4,032		
Ghana	10,493				
Netherlands	8,081	Σ (%)[a]	75		
Σ (%)[a]	75				

Country	Cassava (manioc)
Nigeria	31,500
Brazil	24,587
Zaire	18,000
Thailand	16,000
Indonesia	15,438
Ghana	6,899
India	6,000
Tanzania	5,912
Mozambique	4,733
China	3,501
Σ (%)[a]	81

[a] World production = 100%.

Table 17.4. Average composition of vegetables (as % of fresh edible portion)

Vegetable	Dry matter	N-Com-pounds	Carbo-hydrates	Lipids	Crude fiber	Ash
Mushrooms						
Champignon (cultivated *Agaricus arvensis, campestris*)	10.0	4.8	3.5	0.2	0.8	0.8
Chanterelle	8.5	2.6	3.5	0.8	1.0	0.7
Edible boletus (*Boletus edulis*)	13.0	5.4	5.2	0.4	1.0	1.0
Rooty vegetables						
Carrots	11.8	1.1	8.7	0.2	1.0	0.8
Radish (*Raphanus sativus*, elongated white fleshy root)	5.5	1.0	2.9	0.2	0.7	0.8
Viper's grass, *scorzonera*	21.4	1.4	16.3	0.4	2.3	1.0
Parsley	12.0	2.9	2.3	0.6		1.6
Tuberous vegetables (sprouting tubers)						
White (Irish) potato	22.2	2.0	18.9[a]	0.2	0.8	1.1
Celery (root)	11.6	1.8	7.2	0.3	1.3	1.0
Kohlrabi	9.7	2.0	5.6	0.1	1.0	1.0
Rutabaga	13.0	1.1	9.9	0.2	1.1	0.8
Radish (*Raphanus sativus*, reddish fleshy root)	5.6	1.1	3.5	0.1		0.9
Red beet, beetroot	12.7	1.6	9.1	0.1	0.8	1.1
Tuberous root vegetables						
Sweet potato	30.8	1.6	26.6[b]	0.6	0.9	1.1
Cassava (manioc)	35.0	0.9	32.0	0.4	0.8	0.4
Yam	28.0	1.8	23.8	0.2	0.7	1.0
Bulbous root vegetables						
Onion	10.9	1.5	8.1	0.3	0.6	0.6
Leek	14.6	2.2	9.9	0.3	1.3	0.9
Vegetable fennel	14.0	2.4	9.1	0.3	0.5	1.7
Stem (shoot) vegetables						
Asparagus	8.3	2.5	4.3	0.1	0.7	0.6
Leafy (stalk) vegetables						
Rhubarb	5.2	0.6	3.0	0.2	0.7	0.8
Leafy vegetables						
Endive (escarole)	5.6	1.3	2.3	0.2	0.9	1.0
Kale (curly cabbage)	17.3	6.0	7.5	0.9		1.5
Head lettuce	5.1	1.6	1.7	0.3	0.7	0.9
Brussels sprouts	14.8	4.9	6.7	0.6	1.6	1.2
Red cabbage	9.8	2.0	5.9	0.2	1.0	0.7
Spinach	9.3	3.2	3.7	0.4	0.6	1.5
Common (white) cabbage	7.6	1.3	4.6	0.2	0.8	0.7
Flowerhead (calix) vegetables						
Artichoke	14.5	2.9	8.2	0.1	2.4	0.8
Cauliflower	9.0	2.7	4.2	0.3	1.0	0.9
Broccoli	10.9	3.6	4.4		1.5	1.1

[a] Starch content 14.1%.   [b] Starch and saccharose contents 19.6 and 2.8%, respectively.

Table 17.4 (continued)

Vegetable	Dry matter	N-Com-pounds	Carbo-hydrates	Lipids	Crude fiber	Ash
Seed vegetables						
Chestnut	49.9	2.9	42.8	1.9	1.4	1.2
Green beans	9.9	1.9	6.1		1.0	0.7
Green peas	22.0	6.3	12.4		2.0	0.9
Fruity vegetables						
Eggplant	7.6	1.2	4.7	0.2	0.9	0.6
Squash	8.7	1.1	5.5	0.1	1.2	0.8
Green bell pepper	6.6	1.2	3.4	0.3	1.4	0.4
Cucumber	5.9	0.9	2.8	0.2	0.6	0.5
Tomato	6.5	1.1	4.2	0.2	0.5	0.5

enzymes. Protein and enzyme patterns, as obtained by electrophoretic separation, are often characteristic of species or cultivars and can be used for analytical differentiation. Figure 17.1 shows typical protein and pro-teinase inhibitor patterns for several potato cultivars.

17.1.2.1.2 Free Amino Acids

In addition to protein-building amino acids, nonprotein amino acids occur in vegetables as well as in other plants. Tables 17.5 and 17.6 present data on the occurrence and structure of these amino acids. Information about their biosynthetic pathways is given below.

The higher homologues of amino acids, such as homoserine, homomethionine and amino-adipic acid, are generally derived from a re-action sequence which corresponds to that of oxalacetate to ketoglutarate in the *Krebs* cycle:

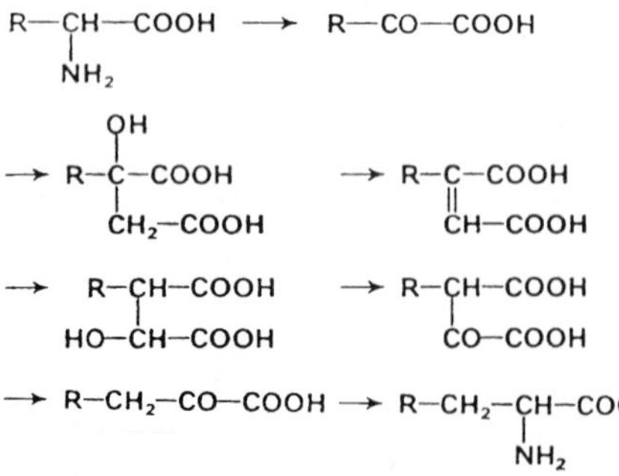

$$R—CH—COOH \longrightarrow R—CO—COOH$$
$$| \atop NH_2$$

$$\longrightarrow R—\underset{\underset{CH_2—COOH}{|}}{\overset{\overset{OH}{|}}{C}}—COOH \qquad \longrightarrow R—\underset{\underset{CH—COOH}{||}}{C}—COOH$$

$$\longrightarrow R—\underset{\underset{HO—CH—COOH}{|}}{CH}—COOH \qquad \longrightarrow R—\underset{\underset{CO—COOH}{|}}{CH}—COOH$$

$$\longrightarrow R—CH_2—CO—COOH \longrightarrow R—CH_2—\underset{\underset{NH_2}{|}}{CH}—COOH \qquad (17.1)$$

4-Methyleneglutamic acid (Table 17.5: XXXI) is formed from pyruvic acid:

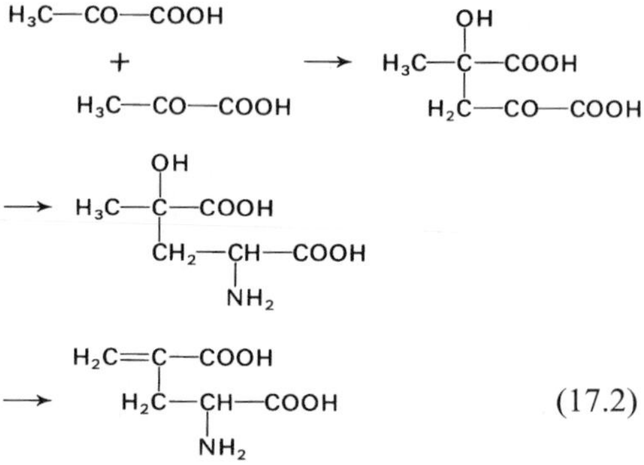

$$H_3C—CO—COOH$$
$$+ \qquad\qquad \longrightarrow H_3C—\underset{\underset{H_2C—CO—COOH}{|}}{\overset{\overset{OH}{|}}{C}}—COOH$$
$$H_3C—CO—COOH$$

$$\longrightarrow H_3C—\underset{\underset{\underset{NH_2}{|}}{CH_2—CH—COOH}}{\overset{\overset{OH}{|}}{C}}—COOH$$

$$\longrightarrow \underset{\underset{\underset{NH_2}{|}}{H_2C—CH—COOH}}{H_2C=C—COOH} \qquad (17.2)$$

The important precursors of onion flavor, the S-alkylcysteine sulfoxides, are formed as fol-lows:

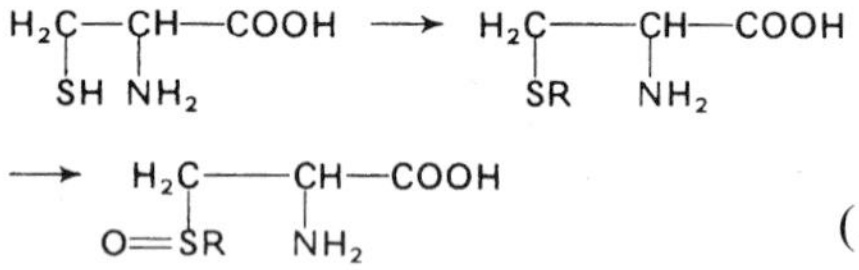

$$\underset{\underset{SH\ \ NH_2}{|\ \ \ |}}{H_2C—CH—COOH} \longrightarrow \underset{\underset{SR\ \ \ NH_2}{|\ \ \ \ |}}{H_2C—CH—COOH}$$

$$\longrightarrow \underset{\underset{O=SR\ \ \ NH_2}{|\ \ \ \ \ \ \ |}}{H_2C—CH—COOH} \qquad (17.3)$$

2,4-Diaminobutyric acid and some other com-pounds are derived from cysteine (cf. Reaction (17.4).

The aspartic acid semi-nitrile formed initially can be decarboxylated to β-amino propioni-trile which, just as its γ-glutamyl derivative, is responsible for osteolathyrism in animals.

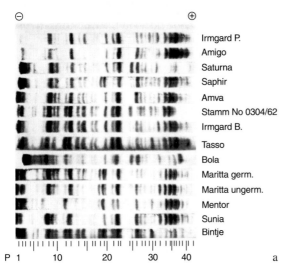

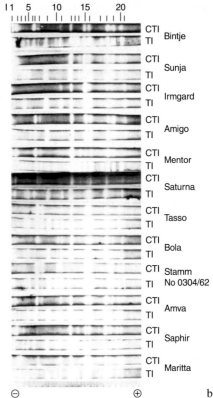

Fig. 17.1. Protein patterns of different potato cultivars obtained by isoelectric focussing on polyacrylamide gel pH 3–10. **a** Protein bands stained with Coomassie Blue; **b** Staining of trypsin and chymotrypsin inhibitors (TI, CTI): Incubation with trypsin or chymotrypsin, N-acetylphenylalanine-β-naphthyl ester and diazo blue B: inhibitor zones appear white on a red-violet background. (according to *Kaiser, Bruhn* and *Belitz*, 1974)

Hydrolysis of the semi-nitrile yields aspartic acid, hydrolysis and reduction yield 2,4-diaminobutyric acid, the oxalyl derivative of which, like oxalyldiaminopropionic acid, is a human neurotoxin. The main symptoms of neurolathyrism are paralysis of the limbs and muscu-

$$\text{Cysteine} \xrightarrow[\substack{\downarrow \\ H_2S}]{HCN} NC{-}CH_2{-}\underset{\underset{NH_2}{|}}{CH}{-}COOH$$

$$-CO_2 \qquad\qquad +H_2O \qquad\qquad Red.$$

$$NC{-}CH_2{-}CH_2{-}NH_2 \qquad H_2NOC{-}CH_2{-}\underset{\underset{NH_2}{|}}{CH}{-}COOH \qquad H_2N{-}CH_2{-}CH_2{-}\underset{\underset{NH_2}{|}}{CH}{-}COOH$$

γ-Glutamyl-β-aminopropionitrile

$$OHC{-}CH_2{-}\underset{\underset{NH_2}{|}}{CH}{-}COOH$$

XXI   (17.4)

Table 17.5. Occurrence of nonprotein amino acids in plants (the Roman numerals refer to Table 17.6)

Amino acid		Plant		Family
Neutral aliphatic amino acids				
I	2-(Methylenecyclopropyl)-glycine	litchi	*Litchi chinensis*	Sapidaceae
II	3-(Methylenecyclopropyl)-L-alanine (Hypoglycine A)	akee	*Bligia sapida*	Sapidaceae
III	3-Cyano-L-alanine	common vetch	*Vicia sativa*	Fabaceae
IV	L-2-Aminobutyric acid	garden sage	*Salvia officinalis*	Lamiaceae
V	L-Homoserine	garden pea	*Pisum sativum*	Fabaceae
VI	O-Acetyl-L-homoserine	garden pea		
VII	O-Oxalyl-L-homoserine	vetchling	*Lathyrus sativum*	Fabaceae
VIII	5-Hydroxy-L-norvaline	jackbean	*Canavalia ensiformis*	Fabaceae
IX	4-Hydroxy-L-isoleucine	fenugreek	*Trigonella foenum-graecum*	Fabaceae
X	1-Amino-cyclopropane-1-carboxylic acid	apple pear	*Malus sylvestris* *Pyrus communis*	Rosaceae Rosaceae
Sulfurcontaining amino acids				
XI	S-Methyl-L-cysteine	garden bean	*Phaseolus vulgaris*	Fabaceae
XII	S-Methyl-L-cysteinesulfoxide	radish, cabbage cauliflower, broccoli	*Brassica oleracea*	Brassicaceae
XIII	S-(Prop-1-enyl)cysteine	garlic	*Allium sativum*	Liliaceae
XIV	S-(Prop-1-enyl)cysteinesulfoxide	onion	*Allium cepa*	Liliaceae
XV	γ-Glutamyl-S-(prop-1-enyl)cysteine	chive	*Allium schoenoprasum*	Liliaceae
XVI	S-(Carboxymethyl)cysteine	radish	*Raphanus sativus*	Brassicaceae
XVII	3,3′-(Methylenedithio)dialanine (Djenkolic acid)	djenkol bean	*Pithecolobium lobatum*	Fabaceae
XVIII	3,3′-(-2-Methylethenyl-1,2-dithio)-dialanine (as γ-Glutamyl derivative)	chive	*Allium schoenoprasum*	Liliaceae
XIX	S-Methylmethionine	jackbean white cabbage asparagus	*Canavalia ensiformis* *Brassica oleracea* *Asparagus officinalis*	Fabaceae Brassicaceae Liliaceae
XX	Homomethionine	white cabbage	*Brassica oleracea*	Brassicaceae
Imino acids				
XXI	Azetidine-2-carboxylic acid	sugar beet	*Beta vulgaris ssp.*	Chenopodiaceae
XXII	tr-4-Methyl-L-proline	apple	*Malus sylvestris*	Rosaceae
XXIII	cis-4-Hydroxymethyl-L-proline	apple peel	*Malus sylvestris*	Rosaceae
XXIV	trans-4-Hydroxymethyl-L-proline	loquat	*Eriobotrya japonica*	Rosaceae
XXV	trans-4-Hydroxymethyl-D-proline	loquat	*Eriobotrya japonica*	Rosaceae
XXVI	4-Methylene-D,L-proline	loquat	*Eriobotrya japonica*	Rosaceae
XXVII	cis-3-Amino-L-proline	morel	*Morchella esculenta*	Ascomycetes
XXVIII	Pipecolic acid	many plants		
XXIX	3-Carboxy-6,7-dihydroxy-1,2,3,4-tetrahydroisoquinoline	cowage	*Mucuna sp.*	Fabaceae
XXX	1-Methyl-3-carboxy-6,7-dihydroxy-1,2,3,4-tetrahydroisoquinoline	cowage	*Mucuna sp.*	Fabaceae
Acidic amino acids and related compounds				
XXXI	4-Methyleneglutamic acid	peanut	*Arachis hypogaea*	Fabaceae
XXXII	4-Methyleneglutamine	peanut	*Arachis hypogaea*	Fabaceae
XXXIII	N^5-Ethyl-L-glutamine (L-Theanine)	tea	*Thea sinensis*	Theaceae
XXXIV	L-threo-4-Hydroxyglutamic acid			
XXXV	3,4-Dihydroxyglutamic acid	garden cress rhubarb carrot currant spinach longwort	*Lepidium sativum* *Rheum rhabarbarum* *Daucus carota* *Ribis rubrum* *Spinacia oleracea* *Angelica archangelica*	Brassicaceae Polygonaceae Apiaceae Saxifragaceae Chenopodiaceae Apiaceae
XXXVI	L-2-Aminoadipic acid	many plants		
Basic amino acids and related compounds				
XXXVII	N^2-Oxalyl-diaminopropionic acid	vetchling	*Lathyrus sativus*	Fabaceae
XXXVIII	N^3-Oxalyl-diaminopropionic acid	vetchling	*Lathyrus sativus*	Fabaceae
XXXIX	2,4-Diaminobutyric acid (as N^4-Lactyl compound)	sugar beet	*Beta vulgaris ssp.*	Chenopodiaceae
XL	2-Amino-4-(guanidinooxy)butyric acid (Canavanine)	jackbean soybean	*Canavalia ensiformis* *Glycine max*	Fabaceae Fabaceae
XLI	4-Hydroxyornithine	common vetch	*Vicia sativa*	Fabaceae
XLII	L-Citrulline	watermelon	*Citrullus lanatus*	Cucurbitaceae

Table 17.5 (continued)

Amino acid		Plant		Family
XLIII	Homocitrulline	horse bean	*Vicia faba*	Fabaceae
XLIV	4-Hydroxyhomocitrulline	horse bean	*Vicia faba*	Fabaceae
XLV	4-Hydroxyarginine	common vetch	*Vicia sativa*	Fabaceae
XLVI	4-Hydroxylysine	garden sage	*Salvia officinalis*	Lamiaceae
XLVII	5-Hydroxylysine	lucern	*Medicago sativa*	Fabaceae
XLVIII	N^6-Acetyl-L-lysine	sugar beet	*Beta vulgaris*	Chenopodiaceae
XLIX	N^6-Acetyl-allo-5-hydroxy-L-lysine	sugar beet	*Beta vulgaris*	Chenopodiaceae
Heterocyclic amino acids				
L	3-(2-Furoyl)-L-alanine	buck wheat	*Fagopyrum esculentum*	Polygonaceae
LI	3-Pyrazol-1-ylalanine	watermelon	*Citrullus lanatus*	Cucurbitaceae
LII	1-Alanyluracil (Willardin)	cucumber	*Cicumis sativus*	Cucurbitaceae
		garden pea	*Pisum sativum*	Fabaceae
LIII	3-Alanyluracil (Isowillardin)	garden pea	*Pisum sativum*	Fabaceae
LIV	3-Amino-3-carboxypyrrolidine	musk melon	*Cucurbita monlata*	Cucurbitaceae
LV	3-(2,6-Dihydroxypyrimidine-5-yl)-alanine	garden pea	*Pisum sativum*	Fabaceae
LVI	3-(Isoxazoline-5-one-2-yl)alanine	garden pea	*Pisum sativum*	Fabaceae
LVII	3-(2-β-D-Glucopyranosyl-isoxazoline-5-one-4-yl)alanine	garden pea	*Pisum sativum*	Fabaceae
Aromatic amino acids				
LVIII	N-Carbamoyl-4-hydroxy-phenylglycine	horse bean	*Vicia faba*	Fabaceae
LIX	L-3,4-Dihydroxyphenylalanine	horse bean	*Vicia fabea*	Fabaceae
		cowage	*Mucuna sp.*	Fabaceae
Other amino acids				
LX	γ-Glutamyl-L-β-phenyl-β-alanine	adzuki bean	*Phaseolus angularis*	Fabaceae
LXI	Saccharopine	yeast	*Saccharomyces cerevisiae*	Saccharomycetaceae

lar rigidity. 2,4-Diaminobutryic acid can be converted via the aspartic acid semialdehyde into 2-azetidine carboxylic acid (XXI), which occurs, for example, in sugar beets (Table 17.5).

Freshly harvested mushrooms contain aprox. 0.1% agaritin, β-N-(γ-L(+)-glutamyl)-4 hydroxymethylphenylhydrazine. Enzymes present can hydrolyze agaritin and oxidize the released 4-hydroxymethyl-phenylhydrazine to the diazonium salt.

17.1.2.1.3 Amines

The presence of amines has been confirmed in various vegetables; e.g., histamine, N-acetyl-histamine and N,N-dimethylhistamine in spinach; and tryptamine, serotonin and tyramine in tomatoes and eggplant (3−4 mg/100 g dry matter).

17.1.2.2 Carbohydrates

17.1.2.2.1 Mono- and Oligosaccharides, Sugar Alcohols

The predominant sugars in vegetables are glucose and fructose (0.3−4%) as well as sucrose (0.1−12%). Other sugars occur in small amounts; e.g. glycosidically bound apiose in *Umbelliferae* (celery and parsley); 1^F-β- and 6^G-β-fructosylsaccharose in the allium group (onions, leeks); raffinose, stachyose and verbascose in *Fabaceae*; and mannitol in *Brassicaceae* and *Cucurbitaceae*.

17.1.2.2.2 Polysaccharides

Starch occurs widely as a storage carbohydrate and is present in large amounts in some root and tuber vegetables. In *Compositae* (e.g., artichoke, viper's grass, bot. *Scorzonera*), inulin, rather than starch, is the storage carbohydrate.

Other polysaccharides are cellulose, hemicelluloses and pectins. The pectin fraction has a distinct role in the tissue firmness of vegetables. Tomatoes become firmer as the total pectin

Table 17.6. Structures of nonprotein amino acids in plants (structures and Roman numerals refer to Table 17.5)

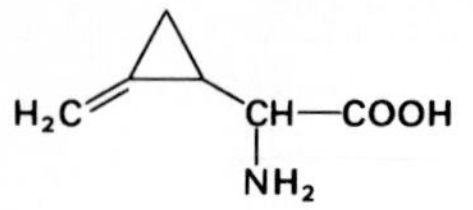

I

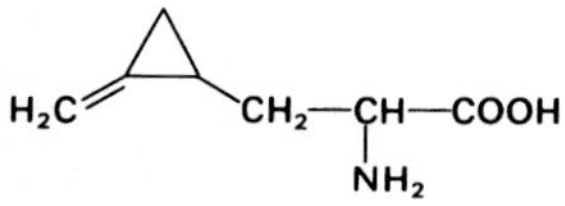

H_2C=⟨cyclopropane⟩—CH—COOH, NH₂

II

H_2C=⟨cyclopropane⟩—CH₂—CH—COOH, NH₂

III

NC—CH₂—CH—COOH, NH₂

IV

H_3C—CH₂—CH—COOH, NH₂

$ROCH_2$—CH₂—CH—COOH, NH₂

V: R = H
VI: R = H_3C—CO
VII: R = HOOC—CO

VIII

HO—CH₂—CH₂—CH₂—CH—COOH, NH₂

IX

H_3C—CH—CH—CH—COOH, with OH and CH₃, NH₂

X

⟨cyclopropane⟩—NH₂, COOH

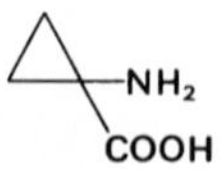

XI: R = CH₃
XIII: R = H_3C—CH=CH
XVI: R = HOOC—CH₂

R—S—CH₂—CH—COOH, NH₂

R—S(=O)—CH₂—CH—COOH, NH₂

XII: R = CH₃
XIV: R = H_3C—CH=CH

XV

H_3C—CH=CH—S—CH₂—CH—COOH; HOOC—CH—(CH₂)₂—CO—NH, NH₂

XVII

HOOC—CH—CH₂—S—CH₂—S—CH₂—CH—COOH, NH₂, NH₂

XVIII

HOOC—CH—CH₂—S—CH₂—CH—S—CH₂—CH—COOH, NH₂, CH₃, NH₂

XIX

H_3C—S⁺—CH₂—CH₂—CH—COOH, CH₃, NH₂

XX

H_3C—S—CH₂—CH₂—CH₂—CH—COOH, NH₂

XXI

⟨azetidine ring⟩—COOH, N, H

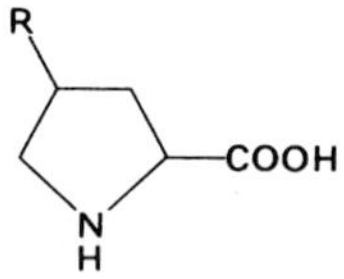

XXII: R = CH₃
XXIII, XXIV, XXV: R = HOCH₂

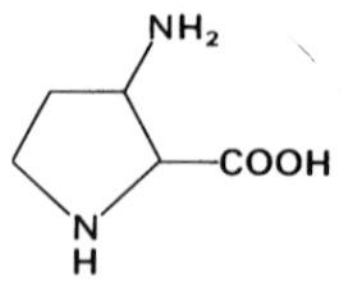

⟨pyrrolidine ring with R⟩—COOH, N, H

XXVII

⟨pyrrolidine ring with NH₂⟩—COOH, N, H

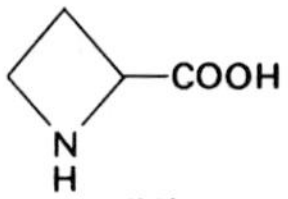

XXVI

⟨pyrrolidine ring with H₂C=⟩—COOH, N, H

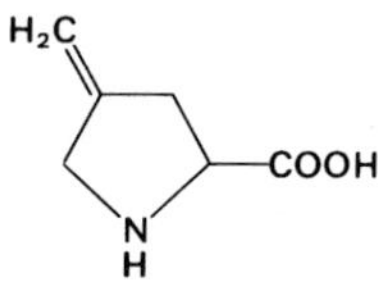

Table 17.6 (continued)

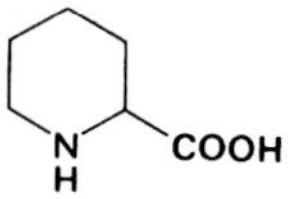

XXVIII

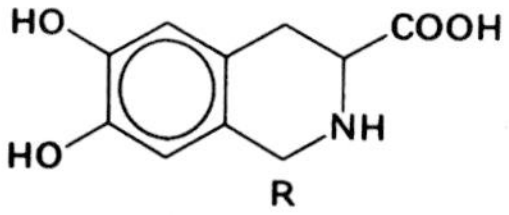

XXIX: R = H
XXX: R = CH$_3$

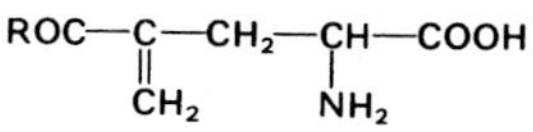

$H_2N-\overset{\underset{\displaystyle O}{\|}}{C}-NH-(CH_2)_n-\overset{\underset{\displaystyle R}{|}}{CH}-CH_2-\overset{\underset{\displaystyle NH_2}{|}}{CH}-COOH$

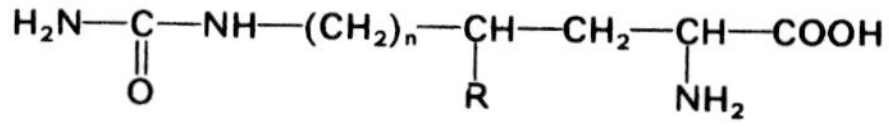

XLII: n = 1, R = H
XLIII: n = 2, R = H
XLIV: n = 2, R = OH

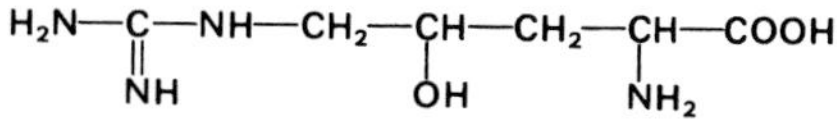

$ROC-\overset{\underset{\displaystyle CH_2}{|}}{C}-CH_2-\overset{\underset{\displaystyle NH_2}{|}}{CH}-COOH$

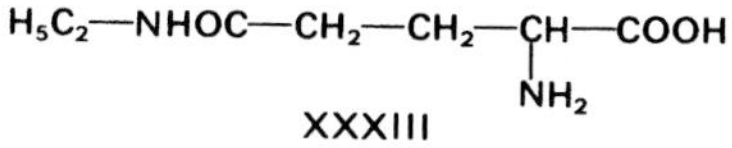

XXXI: R = OH   XXXII: R = NH$_2$

$H_2N-\overset{\underset{\displaystyle NH}{\|}}{C}-NH-CH_2-\overset{\underset{\displaystyle OH}{|}}{CH}-CH_2-\overset{\underset{\displaystyle NH_2}{|}}{CH}-COOH$

XLV

$H_5C_2-NHOC-CH_2-CH_2-\overset{\underset{\displaystyle NH_2}{|}}{CH}-COOH$

XXXIII

$R^2-HN-CH_2-\overset{\underset{\displaystyle R^1}{|}}{CH}-\overset{\underset{\displaystyle R}{|}}{CH}-CH_2-\overset{\underset{\displaystyle NH_2}{|}}{CH}-COOH$

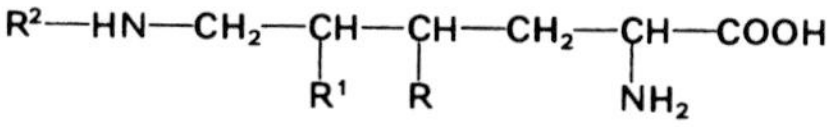

XLVI: R = OH, R^1, R^2 = H
XLVII: R^1 = OH, R, R^2 = H
XLVIII: R, R^1 = H, R^2 = CH$_3$CO
XLIX: R = H, R^1 = OH, R^2 = CH$_3$CO

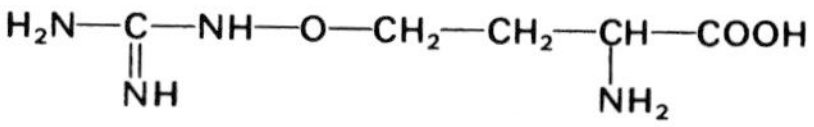

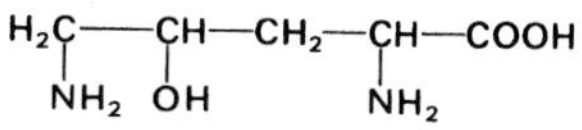

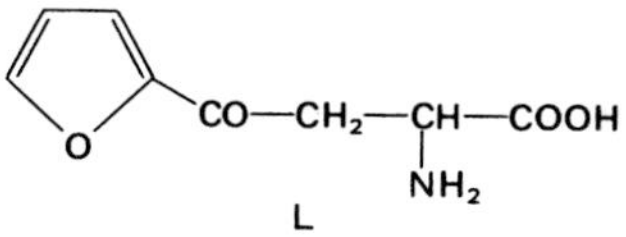

$HOOC-\overset{\underset{\displaystyle R}{|}}{CH}-\overset{\underset{\displaystyle R^1}{|}}{CH}-\overset{\underset{\displaystyle NH_2}{|}}{CH}-COOH$

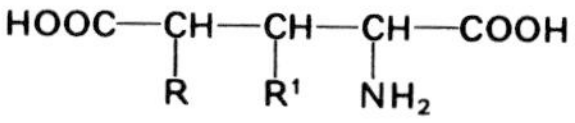

XXXIV: R = HO, R^1 = H
XXXV: R, R^1 = HO

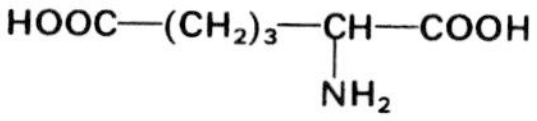

L

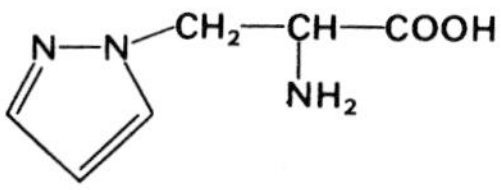

$HOOC-(CH_2)_3-\overset{\underset{\displaystyle NH_2}{|}}{CH}-COOH$

XXXVI

LI

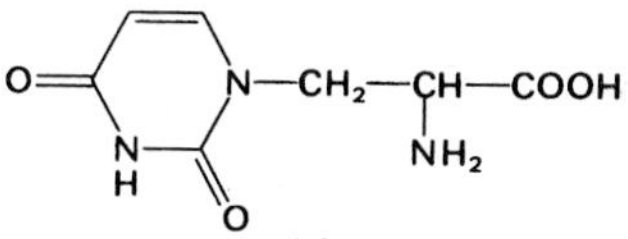

$H_2\overset{\underset{\displaystyle NHR^1}{|}}{C}-\overset{\underset{\displaystyle NHR}{|}}{CH}-COOH$

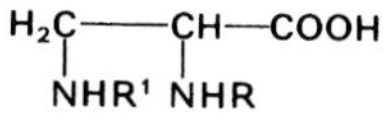

XXXVII: R = HOOC—CO, R^1 = H
XXXVIII: R^1 = HOOC—CO, R = H

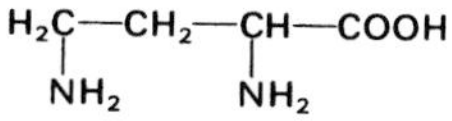

LII

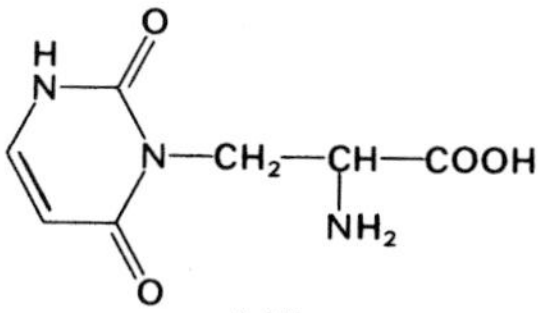

$H_2\overset{\underset{\displaystyle NH_2}{|}}{C}-CH_2-\overset{\underset{\displaystyle NH_2}{|}}{CH}-COOH$

XXXIX

LIII

$H_2N-\overset{\underset{\displaystyle NH}{\|}}{C}-NH-O-CH_2-CH_2-\overset{\underset{\displaystyle NH_2}{|}}{CH}-COOH$

XL

$H_2\overset{\underset{\displaystyle NH_2}{|}}{C}-\overset{\underset{\displaystyle OH}{|}}{CH}-CH_2-\overset{\underset{\displaystyle NH_2}{|}}{CH}-COOH$

XLI

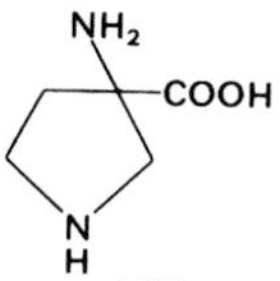

LIV

Table 17.6 (continued)

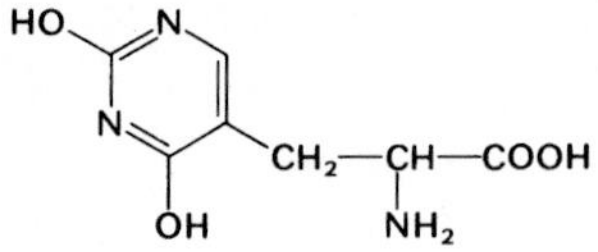

LV: HO–(pyrimidine with N, N, OH)–CH_2—CH—COOH with NH_2

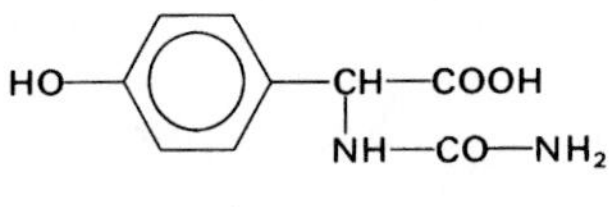

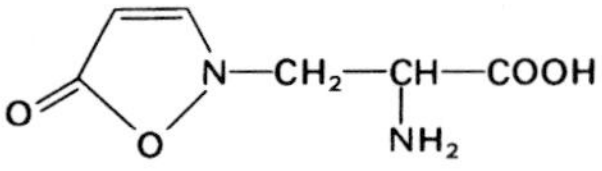

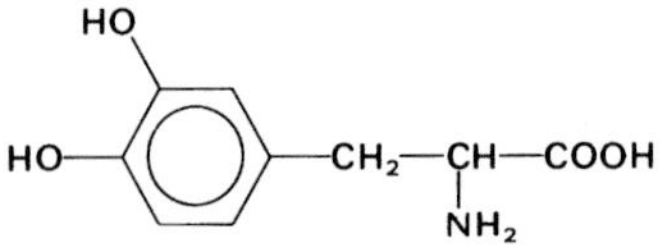

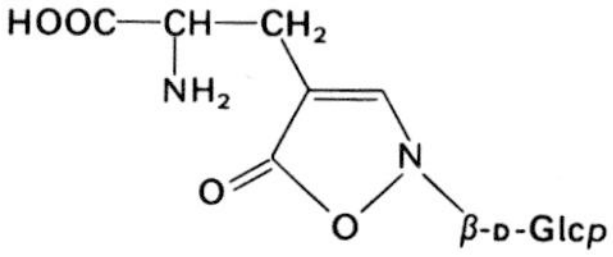

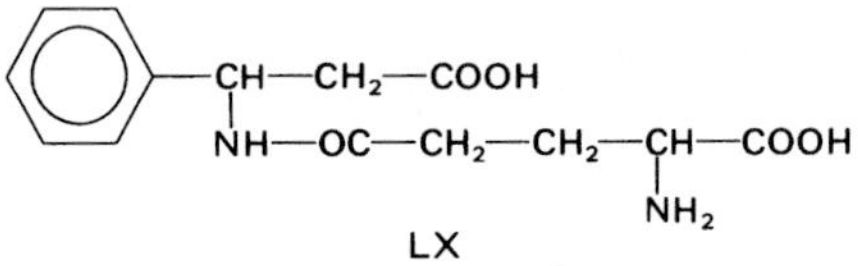

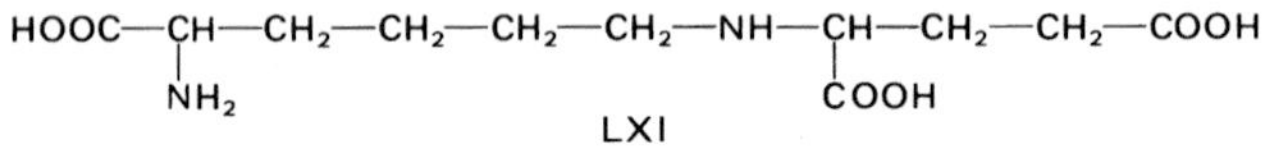

LVI: (isoxazolone ring O, N)—N—CH_2—CH—COOH with NH_2

LVII: HOOC—CH—CH_2 with NH_2, (isoxazolone ring O, N, O)—β-D-Glc$p$

LVIII: HO—(benzene ring)—CH—COOH with NH—CO—NH_2

LIX: HO / HO—(benzene ring)—CH_2—CH—COOH with NH_2

LX: (benzene ring)—CH—CH_2—COOH with NH—OC—CH_2—CH_2—CH—COOH with NH_2

LXI: HOOC—CH—CH_2—CH_2—CH_2—CH_2—NH—CH—CH_2—CH_2—COOH with NH_2 and COOH

content and the content of some minerals (Ca, Mg) increases, and as the degree of esterification of the pectin decreases. In processing cauliflower (cf. 17.2.3), 70 °C is favorable for preserving tissue firmness. The reason for this effect is the presence of pectinmethylesterase which, in vegetables, is fully inactivated only at temperatures above 88 °C, while at 70 °C it is active and provides a build-up of insoluble pectates. For the conversion of protopectin to pectin during plant tissue maturation or ripening see 18.1.3.3.1.

17.1.2.3 Lipids

The lipid content of vegetables is generally low (0.1–0.9%). In addition to triacylglycerides, glyco- and phospholipids are present. Carotenoids are occasionally found in large amounts (cf. 18.1.2.3.2). Table 17.7 provides data on carotenoid compounds in green bell and paprika peppers, tomato and watermelon. For the occurrence of bitter cucurbitacins in *Cucurbitaceae*, see 18.1.2.3.3

Table 17.7. Carotenoids in vegetables[a]

	Green bell pepper	Red pepper (paprika)	Tomato	Water-melon
Total carotenoids[b]	0.9–1.1	12.7–28.4	5.1–6.3	2.5
Phytoene (I)	1.4	1.7	5.3	2.1
Phytofluene (II)	0.2	1.1	2.8	1.4
α-Carotene (VI)	0.4	0.2	0.03	0.06
β-Carotene (VII)	13.4	11.6	3.7	4.1
γ-Carotene (V)			1.2	0.4
ζ-Carotene (III)	0.4	1.5	0.9	1.6
Lycopene (IV)			78.7	81.3
α-Cryptoxanthin	1.2	1.0		
β-Cryptoxanthin	0.5	6.7		
Lutein (IX)	40.8		2.0	
Zeaxanthin (VIII)	0.6	2.3	0.08	
Violaxanthin (XIII)	13.8	9.9	0.8	
Luteoxanthin (XIV)	6.9	0.9	0.1	
Capsanthin (X)		34.7		
Neoxanthin (XX)	15.1	0.7	0.7	

[a] Roman numerals refer to structural formula presented in Chapter 3.8.4.1.

[b] mg carotene/100 g fresh weight.

17.1.2.4 Organic Acids

The organic acids present in the highest concentration in vegetables are malic and citric acids (Table 17.8). The content of free titratable acids is 0.2−0.4 g/100 g fresh tissue, an amount which is low in comparison to fruits. Accordingly, the pH, with several exceptions such as tomato or rhubarb, is relatively high (5.5−6.5). Other acids of the citric acid cycle are present in negligible amounts.

Oxalic acid occurs in larger amounts in some vegetables (Table 17.8).

17.1.2.5 Phenolic Compounds

The phenolic compounds in plant material are dealt with in detail in 18.1.2.5. Hydroxybenzoic and hydroxycinnamic acids, flavones and flavonols also occur in vegetables. Table 17.9 provides data on the occurrence of anthocyanins in some vegetables.

17.1.2.6 Aroma Substances

Characteristic aroma compounds of several vegetables will be dealt with in more detail. The number following each vegetable corresponds to that given in Table 17.1. For aroma biosynthesis see 5.3.2.

Table 17.8. Organic acids in vegetables (mg/100 g fresh weight)

Vegetable	Malic acid	Citric acid	Oxalic acid
Artichoke	170	100	
Eggplant	170	−	
Cauliflower	390	210	
Green beans	112	34	20−45
Broccoli	120	210	
Green peas	75	142	
Kale	50	350	13−125
Carrot	240	90	0−60
Leek		59	0−89
Rhubarb	910	137	230−500
Brussels sprouts	200	240	37
Red beet	−	110	30−138
Sorrel			360
White common cabbage	100	140	
Onion	170	20	

Table 17.9. Anthocyanins in vegetables

Vegetable	Anthocyanin
Eggplant	Delphinidin-3-(p-coumaroyl-L-rhamnosyl-D-glucosyl)-5-D-glucoside
Radish	Pelargonidin-3-[glucosyl(1 → 2)-6-(p-coumaroyl)-β-D-glucosido]-5-glucoside Pelargonidin-3-[glucosyl(1 → 2)-6-(feruloyl)-β-D-glucosido]-5-glucoside
Red cabbage	Cyanidin-3-sophorosido-5-glucoside (sugar moiety esterified with sinapic acid, 1−3 moles)
Onion (red shell)	Cyanidin glycoside Peonidin-3-arabinoside

17.1.2.6.1 Mushrooms (4)

The aroma in champignons originates from (R)-1-octen-3-ol derived from enzymatic oxidative degradation of linoleic acid (cf. 3.7.2.3). A small part of the alcohol is oxidized to 1-octen-3-one in fresh champignons. This compound has a mushroom-like odor when highly diluted and a metallic odor in higher concentrations. It contributes to the mushroom odor because its threshold value is lower by two powers of ten. Heating of chamignons results in the complete oxidation of the alcohol to the ketone. The mushroom *Lentium ediodes*, which is widely consumed in China and Japan, has a very intense aroma. The presence of 1,2,3,5,6-pentathiepane (lenthionine) has been confirmed, and it is a typical impact compound:

$$(17.5)$$

Its threshold values are 0.27−0.53 ppm (in water) or 12.5−25 ppm (in edible oil) It is derived biosynthetically from an S-alkyl cysteine sulfoxide, lentinic acid. Truffles, edible potato-shaped fungi, contain approx. 50 ng/g 5α-androst-16-ene-3α-ol, which has a musky odor that contributes to the typical aroma (cf. 3.8.2.2.1).

17.1.2.6.2 Parsley Roots (21)

Parsley aroma is derived from various mono-terpene hydrocarbons, including myrcene, α-pinene, β-pinene, α-thujone, camphene, sabinene, 3-carene, α- and β-phellandrene, (-)-limonene, γ-terpinene, p-cymene and terpinolene.

17.1.2.6.3 Potatoes (23)

Of great importance for the aroma of raw potatoes is 2-isopropyl-3-methoxypyrazine. Addition of this compound or of the related compound 2-methoxy-3-ethylpyrazine to potato products (e. g., dehydrated mashed potatoes) in amounts of 0.1–0.2 ppm significantly enhances the typical potato aroma. 2,5-Dimethylpyrazine, which possesses an earthy aroma, also plays a significant role. Numerous carbonyl compounds and alcohols have also been identified.

17.1.2.6.4 Celery Tubers (24)

Celery aroma is due to the occurrence of phthalides and dihydrophthalides, namely 3-n-butylphthalide (I) and 3-n-butyl-4,5-dihydrophthalide (II), in leaves, root, tuber and seeds. Their content in essential oil from celery seeds is 3–10%.

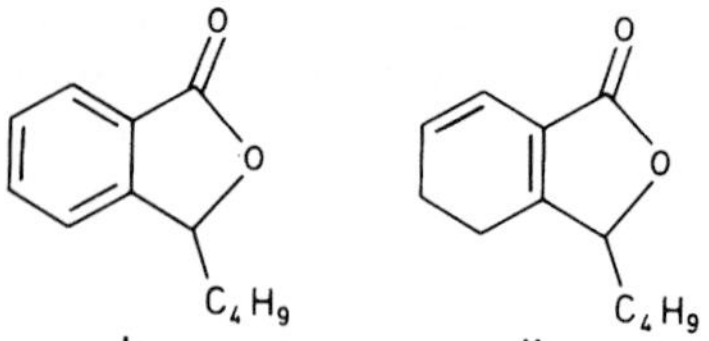

$$\text{(17.6)}$$

17.1.2.6.5 Radishes (27)

The sharp taste of the radish is due to 4-methylthio-trans-3-butenyl-isothiocyanate, which is released from the corresponding glucosinolate after the radish is sliced.

Glucosinolates are widely distributed among *Brassicaceae, Capparaceae* and some other plant families (Table 17.10). Glucosinolates are hydrolyzed by myrosinase, a thioglucosidase enzyme, into the corresponding isothiocyanates (mustard oils) on disintegration of

Table 17.10. Examples of glucosinolates and their corresponding mustard oils or oxazolidin-2-thiones

Glucosinolate	Occurrence	Biosynthesis from	Mustard oil
Sinigrin	Seeds of black mustard (*Brassica nigra*)	Homomethionine	Allyl-
Sinalbin	Seeds of white mustard (*Sinapis alba*)	Tyrosine	p-Hydroxybenzyl-
Gluconapin	Rapeseed (*Brassica napus*)		3-Butenyl-
Glucobrassicanapin	Rapeseed		4-Pentenyl-
Glucotropeolin	Garden cress (*Lepidium sativum*), Nasturtium (*Tropaeolum majus*)	Phenylalanine	Benzyl-
Gluconasturtiin	Watercress (*Nasturtium officinale*), Winter cress (*Barbaraea praecox*)	Homophenylalanine	Phenylethyl-
Glucoibervirin		Homomethionine	ω-Methylthiopropyl-
Glucobrassicin		Tryptophan	Indolylmethyl-
			Oxazolidin-2-thione
Glucoconringiin	*Conringia orientalis*	Leucine/4-hydroxyleucine	5,5-Dimethyl-
Glucobarbarin		2-Amino-4-phenyl-4-hydroxybutyric acid	5-Phenyl-

[a] Many glucosinolates are widely spread; the listed sources are just a few examples.

the tissue. The 'R' radical represents a branched or unbranched alkyl-, alkenyl- or an ω-methylthioalkyl-, monoketoalkyl-, aryl-alkyl- or heterocyclic moiety of the glucosinolate molecule:

$$R-C\begin{smallmatrix} S-Glc \\ \\ N-OSO_3^{\ominus} \end{smallmatrix} \longrightarrow$$

$$\xrightarrow[\text{Thioglucosidase}]{H_2O} R-C\begin{smallmatrix} S^{\ominus} \\ \\ N-OSO_3^{\ominus} \end{smallmatrix} + \text{Glucose}$$

$$\longrightarrow R-N=C=S + HSO_4^{\ominus} \qquad (17.7)$$

The decomposition corresponds to a *Lossen's* rearrangement of a hydroxamic acid. In addition to isothiocyanates, rhodanides and nitriles have been observed among the reaction products.

The isothiocyanates can react further, e.g., with hydroxy compounds or thiols, to form thiourethanes or dithiourethanes. In the presence of amines, thioureas result; while hydrolysis yields the corresponding amines and releases CO_2 and H_2S:

$$R-N=C=S \begin{cases} \xrightarrow{R'OH} R-NH-CS-OR' \\ \xrightarrow{R'NH_2} R-NH-CS-NHR' \\ \xrightarrow{H_2O} R-NH_2 + CO_2 + H_2S \end{cases}$$

$$(17.8)$$

Biosynthesis of glucosinolates (reaction 17.9) starts from the corresponding amino acids, and proceeds via an oxime (I) and thiohydroximic acid (III). The intermediate reactions between steps I and III are not yet clarified. Tests with ^{14}C- and ^{35}S-labelled compounds suggest that the aci-form of the corresponding nitro-compound (II) functions as a thiol acceptor. Cysteine may be involved as a thiol donor. The sulfation is achieved by 3'-phosphoadenosine-5'-phosphosulfate (PAPS). The biosynthetic pathway for cyanogenic glycosides branches at the aldoxime (I) intermediate (cf. 16.2.6).

$$R-CH-COOH \longrightarrow R-CH-COOH$$
$$\quad\;\; NH_2 \qquad\qquad\qquad NHOH$$

$$\longrightarrow R-CH \longrightarrow \left[R-CH_2 \rightleftarrows R-CH\right.$$
$$\qquad\quad NOH \qquad\quad NO_2 \qquad HO-N^{\oplus}-O^{\ominus}$$
$$\qquad\qquad I \qquad\qquad\qquad\qquad\qquad\qquad II$$

$$\xrightarrow{R'S^{\ominus}} R-C-SR^1 \longrightarrow \left. R-C-S^{\ominus}\right.$$
$$\;\; OH^{\ominus} \qquad NOH \qquad R^1 \qquad\qquad NOH$$
$$\qquad\qquad\qquad\qquad\qquad\qquad\qquad\qquad III$$

$$\xrightarrow{UDPGlc} R-C-SGlc \xrightarrow{PAPS} R-C-SGlc$$
$$\quad UDP \quad NOH \qquad\qquad PAP \quad NOSO_3^{\ominus}$$

$$(17.9)$$

17.1.2.6.6 Red Beets (28)

Geosmin is the character impact compound of the red beet:

$$(17.10)$$

17.1.2.6.7 Garlic (33) and Onions (34)

The compound which causes tears (the lachrymatory factor) is (Z)-propanethial-S-oxide (II) which, once the onion bulb is sliced, is derived from trans-(+)-S-(1-propenyl)-L-cysteine sulfoxide (I) by the action of the enzyme alliinase. Alliinase has pyridoxalphosphate as its coenzyme (cf. reaction sequence 17.11). Chopping of onions releases in addition 3,4-dimethyl-2,5-thiophendione which with an odor threshold of 7 μg/kg (water) smells like hydrogen sulfide.

Alkylthiosulfonates (III) are responsible for the aroma of raw onions, while propyl- and propenyl disulfides (IV) and trisulfides are predominant in the aroma of cooked onions. The aroma of fried onions is derived from dimethylthiophenes.

Precursors of importance for the aroma of onions, other than compound I, are S-methyl- and S-propyl-L-cysteine sulfoxide. Precursor I is biosynthesized from valine and cysteine (cf. reaction sequence 17.12).

The key precursor for garlic aroma is S-allyl-L-cysteine sulfoxide (alliin) which, as in onions, occurs in garlic bulbs together with S-methyl- and S-propyl-compounds. The allyl- and propyl-compounds are assumed to be synthesized from serine and corresponding thiols:

$$R-SH + HO-CH_2-\underset{NH_2}{CH}-COOH$$

$$\longrightarrow R-S-CH_2-\underset{NH_2}{CH}-COOH$$

$$\longrightarrow R-S-CH_2-\underset{NH_2}{\underset{O}{CH}}-COOH \qquad (17.13)$$

Diallylthiosulfinate (allicin) and diallyldisulfide are formed from the main component by means of the enzyme alliinase. Both are character impact compounds.

17.1.2.6.8 Watercress (39)

Phenylethylisothiocyanate is responsible for the aroma of this plant of the mustard family (*Brassicaceae*). Decomposition of the corresponding glucosinolate gives phenylpropionitrile, the main component, and some other nitriles, e. g., 8-methylthiooctanonitrile and 9-methylthiononanonitrile.

17.1.2.6.9 White Cabbage, Red Cabage and Brussels Sprouts (52, 49, 48)

Mustard oil is more than 6% of the total volatile fraction of cooked white and red cabbages. Major constituents are 2-propenyl-, 3-butenyl- and 2-phenylethyl isothiocyanates. Of these substances, only the last mentioned compound has an odor theshold (6 µg/kg; water) low enough to contribute to the odor. A great number of other sulfur-containing compounds have been identified, including dimethylsulfide and dimethyltrisulfide. It also appears that 3-alkyl-2-methoxypyrazine plays a role in cabbage aroma.

The total impact of the aroma in cooked frozen Brussels sprouts is less satisfactory than in cooked fresh material. In the former case, analysis has revealed comparatively little allyl mustard oil and more allylnitrile. Isothiocyanates in low concentrations are pleasant and appetite-stimulating, while nitriles are reminiscent of garlic odor. The shift in the concentration ratio of the two compounds is attributed to myrosinase enzyme inactivation during blanching prior to freezing. As a consequence of this, allylglucosinolate in frozen Brussels sprouts is thermally degraded only on subsequent cooking, preferentially forming nitriles. Goitrin is responsible for the bitter taste that can occur in Brussels sprouts (cf. 17.1.2.9.3).

17.1.2.6.10 Cauliflower (56), Broccoli (57)

In cooked cauliflower and broccoli, the aroma compounds of importance are the sulfur compounds mentioned for white cabbage as well as nonanal. 3-Methylthiopropylisothiocyanate contributes to the typical aroma of cauliflower and 4-methylthiobutylisothiocyanate to the aroma of broccoli.

During blanching of these vegetables, cystathionine-β-lyase (EC 4.4.1.8, cystine lyase) must be inactivated because this enzyme, which catalyzes the reaction shown in formula 17.14, produces an aroma defect. The undesirable aroma substances are formed by the degradation of the homocysteine released.

$$
\begin{array}{l}
\text{COOH} \\
| \\
\text{H}_2\text{N}-\text{CH} \\
| \\
\text{CH}_2 \\
| \\
\text{CH}_2 \\
| \\
\text{S} \\
| \\
\text{CH}_2 \\
| \\
\text{CH}-\text{NH}_2 \\
| \\
\text{COOH}
\end{array}
\longrightarrow
\begin{array}{l}
\text{COOH} \\
| \\
\text{H}_2\text{N}-\text{CH} \\
| \\
\text{CH}_2 \\
| \\
\text{CH}_2-\text{SH}
\end{array}
+
\begin{array}{l}
\text{COOH} \\
| \\
\text{C}=\text{O} \\
| \\
\text{CH}_3
\end{array}
+ \text{NH}_3
$$

$$(17.14)$$

17.1.2.6.11 Green Peas (60)

The aroma of green peas is derived from aldehydes and pyrazines (3-isopropyl-, 3-*sec*-butyl- and 3-isobutyl-2-methoxypyrazine).

17.1.2.6.12 Cucumbers (64)

The following aldehydes play an important role in cucumber aroma: trans-2,cis-6-nonadienal and trans-2-nonenal. Linoleic and linolenic acids, as shown in Fig. 3.31, are the precursors for these and other aldehydes (cis-3-hexenal, trans-2-hexenal, trans-2-nonenal).

17.1.2.6.13 Tomatoes (66)

Among a large number of volatile compounds, 3(Z)-hexenal, β-ionone, hexanal, β-damascenone, 1-penten-3-one, and 3-methylbutanal are of special importance for the aroma of tomatoes (cf. Table 17.11).

In tomato paste, for example (cf. Table 5.10), it was found that the changes in aroma caused by heating are primarily due to the formation of dimethylsulfide, the increase in β-damascenone and β-ionone, and a substantial decrease in 3(Z)-hexenal and hexanal.

17.1.2.7 Vitamins

Table 17.12 provides data on the vitamin content of some vegetables. The values given may vary significantly with vegetable cultivar and climate. In spinach, for example, the ascorbic acid content varies from $40-155$ mg/100 g fresh weight.

Table 17.11. Aroma substances in tomatoes

Compound	$A_x{}^a$
3(Z)-Hexenal	5×10^4
β-Ionone	6.3×10^2
Hexanal	6.3×10^2
β-Damascenone	5×10^2
1-Penten-3-one	5×10^2
3-Methylbutanal	1.3×10^2
(E)-Hexenal	16
2-Isobutylthiazole	10
1-Nitro-2-phenylethane	8
2(E)-Heptenal	5
Phenylacetaldehyde	4
6-Methyl-5-hepten-2-one	2.5
3(Z)-Hexenol	2
2-Phenylethanol	2

[a] The aroma values A_x were calculated on the basis of odor thresholds in water.

Table 17.12. Vitamin content in vegetables (mg/100 g fresh weight)

Vegetable	Ascorbic acid	Thi- amine	Ribo- flavin	Nicotinic acid	Folic acid
Artichoke	12	0.08	0.05	1.0	
Eggplant	5	0.05	0.05	0.6	
Cauliflower	78	0.11	0.10	0.7	0.02
Broccoli	113	0.10	0.23	0.9	
Kale	186	0.16	0.26	2.1	
Cucumber	11	0.03	0.04	0.2	0.02
Head lettuce	10	0.06	0.09	0.4	
Carrot	8	0.06	0.05	0.6	
Green bell pepper	128	0.08	0.08	0.5	
Leek	17	0.11	0.06	0.5	
Radish	26	0.03	0.03	0.3	
Brussels sprouts	102	0.10	0.16	0.9	
Red beet	10	0.03	0.05	0.4	
Red cabbage	61	0.09	0.06	0.4	0.04
Celery	8	0.05	0.06	0.7	
Asparagus	33	0.18	0.20	1.5	
Spinach	51	0.10	0.20	0.6	0.08
Tomato	23	0.06	0.04	0.7	0.01

17.1.2.8 Minerals

Table 17.13 reviews the mineral content of some vegetables. Potassium is by far the most abundant constituent, followed by calcium, sodium and magnesium. The major anions are phosphate, chloride and carbonate. All other elements are present in much lower amounts. For nitrate content see Table 9.13.

17.1.2.9 Other Constituents

Plant pigments other than carotenoids and anthocyanins, e. g., chlorophyll and betalains, are also of great importance in vegetables and are covered in this section together with goitrogenic compounds occurring in *Brassicaceae*.

17.1.2.9.1 Chlorophyll

The green color of leaves and unripe fruits is due to the pigments chlorophyll a (blue-green)

and chlorophyll b (yellow-green), occurring together in a ratio of about $3:1$ (see Formula 17.15). Figure 17.2 shows the absorption spectra of chlorophylls a and b. Removal of magnesium from the chlorophylls gives pheophytins a and b, both of which are olive-brown. Replacing magnesium by metal ions such as Sn^{2+} or Fe^{3+} likewise yields greyish-brown compounds, while copper or zinc ions retain the green color. Upon removal of the phytol group, for example by the action of the chlorophyllase enzyme, the chlorophylls are converted into chlorophyllides a and b, while the hydrolysis of pheophytins yields pheophorbides a and b.

Chlorophylls and pheophytins are lipophilic due to the presence of the phytol group, while chlorophyllides and pheophorbides, without phytol, are hydrophilic. Conversion of chloro-

Chlorophyll a: $R^1 = CH_3$
Chlorophyll b: $R^1 = CHO$ (17.15)

Table 17.13. Minerals in vegetables (mg/100 g fresh weight)

Vegetable	K	Na	Ca	Mg	Fe	Mn	Co	Cu	Zn	P	Cl	F	I
Cauliflower	328	16	20	17	0.6	0.2		0.1	0.2	54	29	0.01	0.1
Green beans	256	1.7	51	26	0.8	0.5		0.1		37	36	0.01	
Green peas	296	2	26	33	1.9	0.7	0.003	0.2	3	119	40	0.02	0.004
Cucumber	141	8.5	15	8	0.5	0.2		0.1	0.2	23	37	0.02	0.003
Red beet	336	86	29	1.4	0.9	1.0	0.01	0.2	0.6	45	0.4	0.02	
Tomato	297	6.3	14	20	0.5	0.1	0.01	0.1	0.2	26	60	0.02	0.002
White common cabbage	227	13	46	23	0.5	0.1	0.01	0.1	0.8	28	37	0.01	0.005

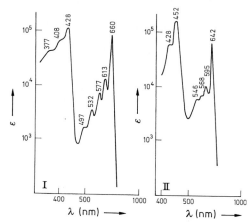

Fig. 17.2. Absorption spectra of chlorophylls a (I) and b (II). Solvent: diethyl ether (I) or diethyl ether +1% CCl_4 (II)

On stronger heating (sterilization, drying), a part of the pheophytins undergoes hydrolysis, releasing carbonic acid monomethylester which decomposes into CO_2 and methanol:

$$CO_2, CH_3OH \qquad (17.16)$$

The corresponding pyropheophytins are formed which can be determined next to the pheophytins by using HPLC (Fig. 17.3). For example, Table 17.15 shows the changes in the chloro-pigments of spinach as a function of the duration of heat sterilization.

A change in color occurs during storage of dried vegetables, its extent increases with increasing water content. The conversion of chlorophylls to pheophytins continues in blanched vegetables even during frozen storage. In beans and Brussels sprouts, immediately after blanching (2 min at 100 °C), the pheophytin content amounts to 8–9%, while after storage for 12 months at −18 °C it increases to 68–83%. Pheophytin content rises from 0% to only 4–6% in paprika peppers and peas under the same conditions.

phylls to pheophytins, which is accompanied by a color change, occurs readily upon heating plant material in weakly acidic solutions and, less readily, at pH 7. Color changes are encountered most visibly in processing of green peas, green beans, kale, Brussels sprouts and spinach. Table 17.14 shows that higher temperatures and shorter heating times provide better color retention than prolonged heating at lower temperatures.

Chlorophyllase is mostly inactivated when vegetables are blanched, hence chlorophyllides and pheophorbides are rarely detected. However, in the fermentation of cucumbers, chlorophyllase is active. The result is a color change from dark-green to olive-green, caused by large amounts of pheophorbides.

17.1.2.9.2 Betalains

Pigments known as betalains occur in centrospermae, e.g., in red beet and also in some mushrooms (the red cap of fly amanita). They

Table 17.14. Changes in the chlorophyll fraction during processing (values in % of the total pigment content of unprocessed vegetables)

Vegetable	Process	Chlorophylls		Chlorophyllides		Pheophytins		Pheophorbides	
		a	b	a	b	a	b	a	b
Green beans	Untreated	49	25	0	0	18	8	0	0
	Blanched, 4 min/100 °C	37	24	0	0	19	10	0	0
Cucumbers	Untreated	51	30	0	0	15	5	0	0
	Blanched, 4 min/100 °C	34	24	6	3	22	1	5	7
Cucumbers	Untreated	67	33	0	0	0	0	0	0
	Fermented (pickled), 6 days	4	7	3	5	10	3	47	15
	Fermented (pickled), 24 days	0	0	0	0	16	7	57	28

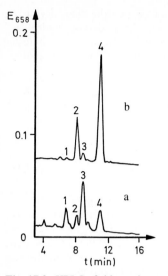

E_{658}

0.2

0.1

0

4   8   12   16

t (min)

Fig. 17.3. HPLC of chloro-pigments from sterilized cans. Green beans (a), spinach (b) (according to *Schwartz* and *von Elbe*, 1983). **1** Pheophytin b, **2** pyropheophytin b, **3** pheophytin a, **4** pyropheophytin a

Table 17.15. Effects of the heat sterilization of spinach on the composition of chloropigments (mg/g solids)

Heating to 121 °C (min)	Chlorophyll		Pheophytin		Pyropheophytin	
	a	b	a	b	a	b
Control	6.98	2.49	0	0	0	0
2	5.72	2.46	1.36	0.13	0	0
4	4.59	2.21	2.20	0.29	0.12	0
7	2.81	1.75	3.12	0.57	0.35	0
15	0.59	0.89	3.32	0.78	1.09	0.27
30	0	0.24	2.45	0.66	1.74	0.57
60			1.01	0.32	3.62	1.24

consist of red-violet betacyanins (λ_{max} ~ 540 nm) and yellow betaxanthins (λ_{max} ~ 480 nm). They have the general structure:

(17.17)

About 50 betalains have been identified. The majority have an acylated sugar moiety. The acids involved are sulfuric, malonic, caffeic, sinapic, citric and *p*-coumaric acids. All betacyanins are derived from two aglycones: betanidin (I) and isobetanidin (II), the latter being the C-15 epimer of betanidin:

(17.18)

Betanin is the main pigment of red beet. It is a betanidin 5-0-β-glucoside. The betaxanthins have only the dihydropyridine ring in common. The other structural features are more variable than in betacyanins. Examples of betaxanthins are natural vulgaxanthins I and II, also from red beet (*Beta vulgaris*):

(17.19)

I: R=OH
II: R=NH$_2$

Betalain biosynthesis starts with dopa by opening of its benzene ring, followed by cyclization to a dihydropyridine. The (S)-betalamic acid which is formed undergoes condensation with (S)-cyclodopa to betacyanins or with some other amino acids to betaxanthins (cf. reaction sequence 17.20).

Red betanin is water soluble and is used to color food. Its application is, however, limited because it hydrolytically decomposes into the colorless cyclodopa-5-O-β-glucoside and the yellow (S)-betalamic acid. This reaction is reversible. Since the activation energy of the forward reaction $(72 \text{ kJ} \times \text{mol}^{-1})$ greatly exceeds that of the back reaction $(2.7 \text{ kJ} \times \text{mol}^{-1})$, a part of the betanin is regenerated at higher temperatures. Betanin is also sensitive to oxygen.

17.1.2.9.3 Goitrogenic Substances

Brassicaceae contain glucosinolates which decompose enzymatically, e.g., into rhodanides. For example, in savoy cabbage the rhodanide content is 30 mg/100 g fresh weight, while in cauliflower it is 10 mg and in kohlrabi 2 mg. Since rhodanide interferes with iodine uptake by the thyroid gland, large amounts of cabbage together with low amounts of iodine in the diet may cause goiter.

Oxazolidine-2-thiones are also goitrogenic. They occur as secondary products in the enzymatic hydrolysate of glucosinolates when the initially formed mustard oils contain a hydroxy group in position 2:

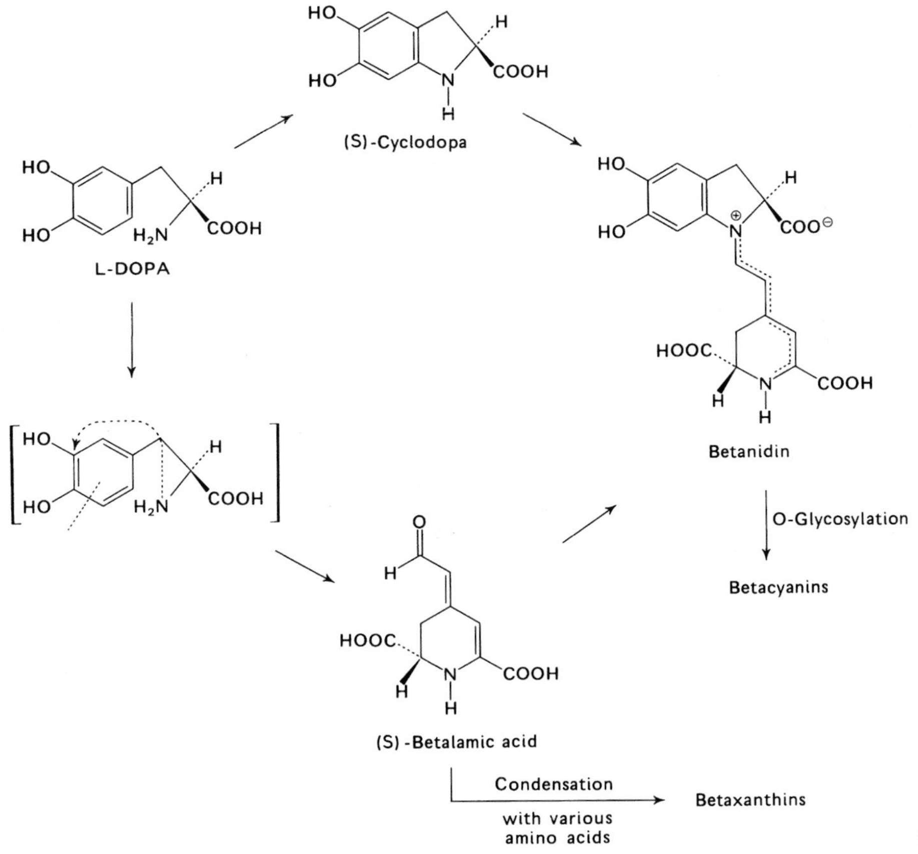

(17.20)

Glucosinolate $\xrightarrow[\text{sinase}]{\text{Myro-}}$ R—CH—CH$_2$—N=C=S

$$\text{R—CH—CH}_2\text{—N=C=S}$$
$$\text{|}$$
$$\text{OH}$$

R: CH$_2$ = CH   (17.21)

The levels of the corresponding glucosinolates are up to 0.02% in yellow and white beets and up to 0.8% in seeds of *Brassicaceae* (all members of the cabbage family; kohlrabi, turnip; rapeseed). The leaves contain only negligible amounts of these compounds.

There are 3−15 mg/kg of 5-vinyloxazolidine-2-thione in sliced turnips. Direct intake of thiooxazolidones by humans is unlikely since the vegetable is generally consumed in cooked form. Consequently, the myrosinase enzyme is inactivated and there is no release of goitrogenic compounds. However, brussels sprouts are exceptions, as higher amounts (70−110 mg /kg) of bitter tasting goitrin is formed from progoitrin during cooking. An indirect intake is possible through milk when such plants are used as animal feed, resulting in a goitrogenic compound content of 50−100 μg/l of milk. The oxazolidine-2-thiones inhibit the iodination of tyrosine, an effect unlike that of rhodanides, which may be offset not by intake of iodine but only by intake of thyroxine.

17.1.3 Storage

The storability of vegetables varies greatly and depends mostly on type, but also on vegetable quality. While some leafy vegetables, such as lettuce and spinach as well as beans, peas, cauliflower, cucumbers, asparagus and tomatoes have limited storage time, root and tuber vegetables, such as carrots, potatoes, kohlrabi, turnips, red table beets, celery, onions and late cabbage cultivars, can be stored for months. Cold storage at high air humidity is the most appropriate. Table 17.16 lists some common storage conditions. The relative air humidity has to be 80−95%. The weight loss experienced in these storage times is 2−10%. Ascorbic acid and carotene contents generally decrease with storage. Starch and protein degradation also occurs and there can be a rise

Table 17.16. Effect of cold storage temperature on vegetable shelf life

Vegetable	Temperature range (°C)	Shelf life (weeks)
Cauliflower	− 1/0	4−6
Green beans	+ 3/+ 4	1−2
Green peas[a]	− 1/0	4−6
Kale	− 2/− 1	12
Cucumber	+ 1/+ 2	2−3
Head lettuce	+ 0.5/+ 1	2−4
Carrot	− 0.5/+ 0.5	8−10
Green bell pepper	− 1/0	4
Leek	− 1/0	8−12
Brussels sprouts	− 3/− 2	6−10
Red beet	− 0.5/+ 0.5	16−26
Celery	− 0.5/+ 1	26
Asparagus	+ 0.5/+ 1	2−4
Spinach	− 1/0	2−4
Tomato	+ 1/+ 2	2−4
Onion	− 2.5/− 2	40

[a] Kept in pods.

in the free acid content of vegetables such as cauliflower, lettuce and spinach.

17.2 Vegetable Products

A number of processing techniques provide vegetable products which have a substantially higher storage stability compared to fresh vegetables, and are readily converted into a consumable form. As is the case with dairy products, unique vegetable products can be produced by fermentation.

17.2.1 Dehydrated Vegetables

Vegetable dehydration reduces the natural water content of the plant below the level critical for the growth of microorganisms (12−15%) without being detrimental to important nutrients. Also, it is aimed at preserving flavor, aroma and appearance, and the ability to regain the original shape or appearance by swelling when water is added. The dehydration process is accompanied by significant changes. First, there is a concentration of major ingredients such as proteins, carbohydrates and minerals. This occurs along with some chemical changes. Fats are oxidatively degraded and, although present in low amounts in vegetables, this oxidation often diminishes odor and flavor. Amino compounds and carbohydrates interact in a *Maillard* reaction, resulting in a darker color and development of new aroma substances (cf. 4.2.4.4). Vitamin levels may also drop sharply. The original volatile aroma and flavor compounds are lost to a great extent.

In the production of the dehydrated product, the vegetable is first washed, peeled or cleaned, and may be sliced or diced. Blanching for 2–7 min to inactivate the enzymes is then done in hot water or steam. Vegetables may also be treated with SO_2.

Dehydration is performed in a conveyor or tube dryer at 55–60 °C to a residual moisture content of 4–8%. Liquid or paste forms, such as tomato or potato mash, are dried in a spray or drum dryer or, in the case of some special products, in a fluidized bed dryer. Dehydration by freeze-drying provides high quality products (good shape retention) with a spongy and porous structure that is readily rehydrated. Some vegetables used in soup powders, e.g., peas and cauliflower, are prepared in this way. For production of dehydrated potato products (Fig. 17.4), tubers are peeled, cleaned, sliced into strings or chips or diced and, after steam-cooking, dried. For production of dehydrated

mashed potato flakes or potato granulate, the steamed slices are squeezed between rollers into a mash with the least possible damage to cell walls. Cell wall damage allows the gelatinized starch to escape from the ruptured cells and to later impart a gluey-sticky texture to the final product. The mashed potato is dried on rollers for the production of flakes and in a pneumatic dryer for the production of granulate. Since the latter drying process requires a flowable product, the mash is mixed with dried powder containing 12–15% of water in a ratio of 1:2 (add-back process). The mixture obtained is then brought to a final water content of 6–8% in a fluidized bed dryer.

Dehydrated vegetables are light, air and moisture sensitive and therefore require careful packaging. Wax-impregnated paper or cardboard, multilayer foils, metal cans or glass containers are commonly used and, occasionally, the packaging is done under nitrogen or vacuum. Also, the dehydrated product may be pressed prior to packaging.

17.2.2 Canned Vegetables

Canning, which involves heat sterilization, is one of the most important processes in vegetable preservation. The selected and sorted

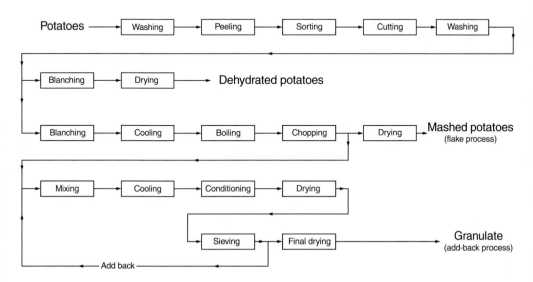

Fig. 17.4. Production of dehydrated potatoes, mashed potato flakes and potato granulate

freshly harvested products are trimmed and blanched as outlined for dehydrated vegetables. Blanching here serves not only to inactivate the enzymes, but to remove both undesirable flavoring compounds (cabbages), and the air present in plant tissue, and to induce shrinkage or softening of the product, thereby increasing packaging density.

Brine (1–2% NaCl solution) often serves as a filling liquid. Sugar (peas, red table beets, tomato, sweet corn), citric acid (up to 0.05%, used for example for celery, cauliflower and horse beans), calcium salts for firming the plant tissue (tomato, cauliflower) or monosodium glutamate (100–150 mg per kg filling) are also added to round-off the flavor.

Sterilization is performed in autoclaves. The autoclaves can be classified according to the heat transfer into water and steam autoclaves and according to the mode of operation into vertical and rotation autoclaves. Rotation autoclaves can be used in a continuous operation only when the cans enter and exit via locks without loss of pressure and steam. The advantage of rotation heating lies in the quicker and more uniform heating of the product. After the required sterilization effect is achieved, the product is quickly cooled to avoid excessive after-heating. As with other foods, vegetable sterilization processes tend toward higher temperatures and shorter times (HTST sterilization) since, in this way, the products retain a better quality (texture, aroma, color).

The nutritional/physiological value of the main constituents of vegetables (proteins and carbohydrates) is not diminished by this common heat sterilization process. Damage due to interaction of amino acids with reducing sugars, which occurs to a small extent, is also negligible. However, there is often a negative effect on vitamins (cf. Tables 6.1 and 6.2). Carotene, a fat-soluble provitamin A, is not affected by the washing and blanching steps, but it is moderately destroyed (5–30%) during actual canning. Vitamin B_1 in carrots and tomatoes does not decrease significantly, while losses are 10–50% for other vegetables (green beans, peas and asparagus). Vitamin B_1 losses are high in spinach (66%) due to the large surface area. Vitamin B_2 is lost (5–25%) by leaching during blanching, but not significantly during further processing. Nicotinic acid losses are similar. Vitamin C losses are due to its water solubility and its enzymatic and chemical degradation, particularly in the presence of traces of heavy metal ions. Vitamin C retention is 55–90% during the canning of asparagus, peas and green beans. Storage of canned vegetables for several years generally results in an additional 20% vitamin loss.

17.2.3 Frozen Vegetables

Beans, peas, paprika peppers, Brussels sprouts, edible mushrooms (*Boletus edulis*), tomato pulp and carrots are particularly suitable for freezing. Radishes, lettuce or whole tomatoes are unsuitable. High quality fresh vegetables are treated with boiling water for 1.5–4 min or steam for 2–5 min for enzyme inactivation. The blanching time is generally shorter than that used in canning, and varies according to type, ripeness and size of vegetable. It is kept as short as possible to prevent leaching. Steam blanching is generally more advantageous than blanching in hot water. The blanching time required for enzyme inactivation is determined by measuring the rate of inactivation of an indicator enzyme (cf. 2.5.4.4).

Immediately after blanching, the vegetable is cooled, frozen at $-40\,°C$ or lower, then stored at -18 to $-20\,°C$. Freezing is mainly conducted using conventional freezing techniques by indirect cold-transfer in plate or air freezers. At present, cryogenic freezing techniques play no appreciable part in vegetable processing.

Freezing preserves vegetable nutrients to a great extent. Vitamin A and its provitamin, carotene, are well preserved in spinach, peas and beans, or are moderately lost (asparagus) after proper blanching, freezing and deep-freeze storage and even after thawing to room temperature. Losses in the Vitamin B group depend mostly on the conditions of the primary processing steps (washing, blanching). The other steps have no effect on B vitamins. Vitamin C leaching by water or steam is detrimental. It is generaly preserved during freezing and thawing. Careful blanching and low temperature storage are critical for vitamin C preservation (Fig. 17.5 and 17.6).

Irreversible textural changes can occur in deep-frozen vegetables. Typical symptoms are soft-

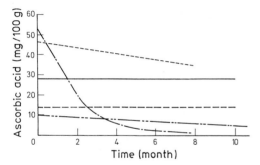

Fig. 17.5. Changes in vitamin C content in frozen vegetables kept at − 21 °C. — Peas precooked, − − − beans precooked, −··−··− beans raw, −·—·— spinach raw, - - - - spinach precooked. (according to *W. Heimann*, 1958)

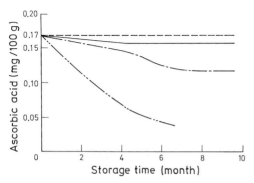

Fig. 17.6. Ascorbic acid losses in frozen peas as influenced by storage temperature. − − − − 40 °C, −−18 °C, −··−··−12 °C, −···−···−9 °C. (according to *Schormueller*, 1966)

ening, ductile stickiness, or looseness or flaccidity (beans, cucumbers, carrots); build-up of a sticky, ductile, gum-like structure (asparagus), or pasty, soggy structure (celery, kohlrabi); or hull hardening (peas).

17.2.4 Pickled Vegetables

Pickled vegetables are produced by spontaneous lactic acid fermentation (white cabbage, green beans, cucumbers, etc.). The fermentation lowers the pH, inhibits the growth of undesirable acid-sensitive microorganisms and, simultaneously, affects the enzymatic softening of cells and their tissues, thus improving digestibility and wholesomeness. The use of salt also has a preservative effect. The acidic pH of the medium stabilizes vitamin C.

While the preservation techniques outlined in earlier sections were aimed at retention of the original odor and flavoring substances of the raw material, including regeneration of lost aroma constituents, this is not important in pickled vegetables since a new typical aroma is developed.

17.2.4.1 Pickled Cucumbers (Salt and Dill Pickles)

Unripe cucumbers, after addition of dill herb and, if necessary, other flavoring spices (vine leaf, garlic or bay leaf), are placed into 4–6% NaCl solution or are sometimes salted dry. Usually, the salt solution is poured on the cucumbers in a barrel and then allowed to ferment and, if necessary, glucose is added. Fermentation takes place at 18–20 °C and yields lactic acid, CO_2, some volatile acids, ethanol and small amounts of various aroma substances. Homo- and heterofermentative lactic acid bacteria like *Lactobacillus plantarum, L. brevis* and *Pediococcus cerevisiae* are involved in the fermentation of pickled cucumbers. In contrast to sauerkraut, *Leuconostoc mesenteroides* does not play a role. The lactic acid (0.5–1%) initially formed is later metabolized partly by film yeast or oxidative yeasts that grow on the surface of the brine. Thus, the original pH value of the fermenting medium (3.4–3.8) is slightly increased.

Apart from spontaneous fermentation, controlled fermentatioin on inoculation with *Lactobacillus plantarum* and *Pediococcus cerevisiae* is also used.

17.2.4.2 Other Vegetables

Green beans, carrots, kohlrabi, celery, asparagus, turnips and others are processed similarly to cucumbers. Sliced green beans, for example, are treated with salt (2.5–3%), subjected to lactic acid fermentation at about 20 °C, and marketed in barrels, cans or glass jars. Some pickled vegetables, mostly those that were not blanched or precooked, will not soften during later cooking.

17.2.4.3 Sauerkraut

Lactic acid fermentation has been used for millenia for the production of sauerkraut (Fig.

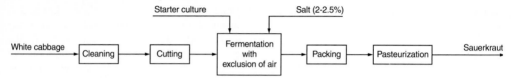

Fig. 17.7. Production of sauerkraut

17.7). It was also customary earlier to place the cabbage into acidified wine or vinegar. White cabbage heads are cut into 0.75–1.5 mm thick shreds, then mixed with salt at 1.8–2.5% by weight. The shreds are then packed into tanks of wood or reinforced concrete, coated with synthetics. After the shreds have been packed in layers, they are tamped and weighted down so that a layer of expressed brine juice covers the surface. The lactic acid fermentation initiated by starter cultures occurs spontaneously at 18–24°C for 3–6 weeks. During the first 48 h of fermentation the pH falls from 6.2 to the range of 3.7 to 4.2. The acid formed inhibits the growth of competing interfering microorganisms. *Leuconostoc mesenteroides* and in addition *Lactobacillus brevis* are the predominating microorganisms during the initial phase of fermentation. Homofermentative bacteria like *Lactobacillus plantarum* and *Pediococcus cerevisiae* appear later. The amount of acid formed depends on the initial sugar content of the cabbage. Hence, sugar is sometimes added (to 1%) to cabbage which does not ferment readily. In addition to *Lactobacillus* ssp., yeasts are also involved in fermentation. The products are lactic and acetic acids (in ratios of 4:1 to 6:1), ethanol (0.2–0.8%), CO_2, mannitol (from fructose) and, most importantly, aroma substances which appear in the prefermentation phase. After fermentation is complete, the sauerkraut pH is about 3.6. Lactic acid values of less than 6 g/l indicate unsatisfactorily fermented cabbage. The end-product is kept in barrels under brine. The sauerkraut is also packaged or canned in retail containers. The cans are filled at 70°C, then exhausted, sealed and sterilized at 95–100°C. In addition, sauerkraut is packed and distributed in plastic foils and containers. Mildly acidic sauerkraut, preferred in South Germany, is produced by stopping the fermentation before all the sugar is degraded. After pasteurization, the product can be stored for a longer time and still retains a clearly sour taste. Sauerkraut is flavored and spiced to some extent by addition of sugar, juniper berries, caraway or dill seeds. For wine sauerkraut at least 1 liter of wine per 50 kg sauerkraut is added after fermentation.

Drained sauerkraut contains on the average 90.7% water, 1.5% nitrogen compounds, 0.3% crude fat, 3.9% carbohydrates, 1.1% crude fiber, 0.6% minerals (excluding NaCl), 0.8–3.3% NaCl, 1.4–1.9% titratable acid (calculated as lactic acid; 0.28–0.42% is acetic acid) and 0.29–0.61% ethanol. There are small amounts of formic, *n*-heptanoic and *n*-octanoic acids, methanol, and compounds important for palatability, i.e., dextran and mannitol. Vitamin C content (10–38 mg/100 g) is not changed when sauerkraut is heated in a pressure cooker. However, after several reheatings about 30% is destroyed.

17.2.4.4 Eating Olives

In 1980, the world production of eating olives amounted to ca. 650,000 t. About 38% of the production were green olives, ca. 31% black lactic-fermented olives, ca. 10% black unfermented olives, and the rest were other products. Table 17.17 shows the composition of the flesh of fresh and green lactic-fermented olives.

For the production of green lactic-fermented olives, the fruit is harvested in a yellow-green to yellow state and placed in 1.3–2.6% NaOH for 6–10 h. During this time, most of the bitter substance oleuropein (Formula 17.22) is hydrolyzed.

$$\text{(17.22)}$$

O – β – D – Glc*p*

Table 17.17. Composition[a] of the flesh of fresh (1) and green lactic-fermented olives (2)

Component	1	2
Water	50–75	61–81
Lipids	6–30	9–28
Reducing sugar	2–6	
Non-reducing sugar	0.1–0.3	
Raw protein	1–3	1–1.5
Raw fiber	1–4	1.4–2.1
Ash	0.6–1	4.2–5.5
Other components	6–10	

[a] Percentage by weight.

The olives are then washed with water and allowed to undergo spontaneous lactic fermentation in a 10–12% NaCl solution. Fermentation is carried out in concrete containers coated with epoxide resin or in polyesters tanks reinforced with glass fibers. In addition to yeasts, *Pediococcus* and *Leuconostoc spp.* are involved in the first fermentation stages and *Lactobacillus spp. (L. plantarum)* in the later stages. After fermentation, the olives are left in the brine or filled into small packs with fresh salt solution and pasteurized. Before packing, the olives are usually stoned and filled (paprika, anchovies, almonds, capers, and onions). The final product has a pH value of 3.8–4.2 and contains 0.8–1.2% of lactic acid. The salt concentration should be at least 7% and at least 8% in products with a longer shelf life.

For the production of black lactic-fermented olives, the ripe, violet to black fruit is washed and directly allowed to undergo spontaneous lactic fermentation in a 8–10% salt solution. Lactobacilli and yeasts are involved, but the yeasts dominate normally. Fermentation proceeds slowly because the olive skin is not as permeable as after alkali treatment. After fermentation, the olives are packed into glass or plastic containers and pasteurized. The final product has a pH value of 4.5–4.8 and contains 0.1–0.6% of lactic acid. The salt concentration is 6–9%.

For the production of black unfermented olives, the ripe fruit is placed 3–5 times in 1–2% NaOH. In between the fruit is washed and well aired to ensure that the flesh is uniformly dyed black by intensive phenol oxidation. Iron gluconate is added to the last wash water to stabilize the color. The olives are then packed in a 3% NaCl solution and sterilized. The product has a pH value of 5.8–7.9 and contains 1–3% of common salt.

17.2.4.5 Faulty Processing of Pickles

Pickled cucumbers are often softened due to the effects of their own or microbial pectolytic enzymes. Brown-to-black discoloration is caused by iron sulfide build-up or by black pigments formed by microorganisms (*Bacillus nigrificans*). Hollowness is caused by gas-forming microorganisms, i.e. gaseous fermentation, and can be prevented readily by pickling in the presence of sorbic acid.

Sauerkraut is darkened by chemical or enzymatic oxidations when the brine does not cover the surface. Reddish color is caused by yeasts. Sauerkraut softening occurs when fermentation takes place at too high a temperature, when the cabbage is exposed to air, too little salt is added; or by faulty fermentation when the lactic acid content remains too low. In addition to faulty fermentation, the kraut can be ruined by infections caused by molds and other flora of the surface film and by rotting (insufficient brine for full protection). Small chain fatty acids like propionic acid and butyric acid cause an aroma defect.

17.2.5 Vinegar-Pickled Vegetables

These products are prepared by pouring preboiled and still hot vinegar onto the vegetables. Vegetables used are cucumbers, red table beets, pearl and silver onions, paprika peppers, mixed vegetables, which also include cauliflower, carrots, onions, peas, mushrooms (in particular the table mushroom, *Boletus edulis*), asparagus, tender corncobs, celery, parsley root, parsnip, kohlrabi, pumpkin and pepperoni peppers.

Only unblemished raw material is used. The vegetable is covered with a solution of 2.5% vinegar. Salt, spices and herbs, herb extracts, sugar and chemical preservatives are usually added. Depending on the vegetable and its preparation method, there are "single pickles" in vinegar (vinegar cucumbers, chili pepper-flavored cucumbers or gherkins, mustard cucumbers, sterilized deli and spiced garlic, dill-flavored cucumbers) and "mixed pickles" in vinegar, which are made partly from fresh and partly from precanned vegetables (unsliced cucumbers, cauliflower, onions, delicate and tender corncobs, paprika peppers). When an infusion of mustard paste is added, they are marketted as "Piccalilly" mixed pickles.

Mixed salads pickled in vinegar are marketed under various trade names. They are made of onions, green or red bell peppers (pimientos), and/or tender corncobs. These products are partially pasteurized.

17.2.6 Stock Brining of Vegetables

Salting is a practical method for preserving some vegetables in bulk until further processing. Usually the vegetable is salted with table salt after being blanched. Brined vegetables are kept for the production of other products. Salted asparagus, for example, is obtained by addition of ~20% by weight of salt and used for the preparation of "Leipzig medley" and mixed fresh vegetables. Stock brining of beans is also important. Blanched or nonblanched beans are soaked in salt brine or are treated with dry salt to 10–20% by weight (added by hand or by machine spreading or dusting) and kept in brine prior to the manufacture of other products. As with other vegetables, the beans are thoroughly drained of brine and rinsed in a stream of hot water before further processing. In the same way, vegetables such as cauliflower, cabbage, carrots, pearly onions and gherkins are stock brined. Mushrooms and morels are also salted; a practice primarily found in Poland and Russia.

12.2.7 Vegetable Juices

The vegetable is cleaned, washed, then blanched and disintegrated in a mill. In some instances, e.g., the tomato, it is first disintegrated and the slurry heated to >70 °C for some time. The juice is then separated in presses or by centrifuging and salt is usually added to 0.25–1%. Nonsour juices are mixed with lactic or citric acid. For storage stability, such products are subjected to pasteurization in plate heat exchangers. Mostly tomatoes and occasionally other vegetables such as cucumbers, carrots, red beets, radishes, sauerkraut, celery or spinach are used for processing into juice.

17.2.8 Vegetable Paste

A vegetable purée or paste is a finely dispersed slurry from which skins and seeds have been removed by passing the slurry through a pulper or finisher. The most important product is tomato purée which, depending on the brand, has a dry matter content of 14–36% and contains 0.8–2% NaCl. Tomato ketchup is made by the intensive premixing of tomato paste (28% or 38%) with vinegar, water, sugar, spices, and stabilizers, followed by fine homogenization via colloid mills, if necessary. Each charge, which is usually made batchwise, is fed via a plate-type heat exchanger (90 °C) and via a degassing device to a hot-filling apparatus with subsequent cooling. If the heat treatment is too long, defects such as caramelization, color change, and bitter taste can be caused. Since the product tends to separate, especially at air bubbles when degassing is inadequate, it is important that the viscosity is sufficient. If the natural pectin content is well preserved (e.g., by hot break tomato puree), the use of thickening agents is unnecessary. The filled bottles are often stored upside down to prevent a relatively frequent defect called "black neck", a browning at the neck of the bottle due to a high proportion of air in the headspace.

Some other vegetable purées are important primarily as baby foods.

17.2.9 Vegetable Powders

Vegetable powders are obtained by drying the corresponding juice with or without addition of a drying enhancer, such as starch or a starch degradation product, to a residual moisture content of about 3%. Drying processes used are spray-drying, vacuum drum drying, and freeze-drying. The most important product is tomato powder. Other powders, such as those of spinach or red beets, are in part used in food colorings.

17.3 Literature

Adler, G.: Kartoffeln und Kartoffelerzeugnisse. Verlag Paul Barey: Berlin. 1971

Bötticher, W.: Technologie der Pilzverwertung. Verlag Eugen Ulmer: Stuttgart. 1974

Buttery, R.G., Teranishi, R., Flath, R.A., Ling, L.C.: Fresh tomato volatiles. Composition and sensory studies. ACS Symposium Series 388, American Chemical Society, Washington, DC 1989, p. 213

Chung, T.-Y., Hayase, F., Kato, H.: Volatile components of ripe tomatoes and their juices, purées and pastes. Agric. Biol. Chem. *47*, 343 (1983)

Elbe, J.H. von: Influence of water activity on pigment stability in food products. In: Water Activity: Theory and Applications to Food (Eds.: Rockland, L.B., Beuchat, L.R.) Marcel Dekker, Inc.: New York. 1987

Fenwick, G.R., Griffiths, N.M.: The identification of the goitrogen, (-)5-vinyloxazolidine-2-thione (goitrin), as a bitter principle of cooked Brussels sprouts (*Brassica oleracea* L. var. *gemmifera*). Z. Lebensm. Unters. Forsch. *172*, 90 (1981)

Fernández Diez, M.J.: Olives. In: Biotechnology (Eds.: Rehm, H.-J., Reed, G.), Vol. 5, p. 379, Verlag Chemie: Weinheim. 1983

Fischer, K.-H., Grosch, W.: Volatile compounds of importance in the aroma of mushrooms (*Psalliota bispora*). Lebensm. Wiss. Technol. *20*, 233 (1987)

Hadar, Y., Dosoretz, C.G.: Mushroom mycelium as a potential source of food flavour. Trends in Food Science & Technology *2*, 214 (1991)

Herrmann, K.: Gemüse und Gemüsedauerwaren. Verlag Paul Parey: Berlin. 1969

Herrmann, K.: Tiefgefrorene Lebensmittel. Verlag Paul Parey: Berlin. 1970

Kaiser, K.-P., Bruhn, L.C., Belitz, H.-D.: Proteaseinhibitors in potatoes. Protein-, trypsin- and chymotrypsininhibitor patterns by isoelectric focusing in polyacrylamidegel. A rapid method for identification of potato varieties. Z. Lebensm. Unters. Forsch. *154*, 339 (1974)

Ross, A.E., Nagel, D.L., Toth, B.: Evidence for the occurrence and formation of diazonium ions in the *Agaricus bisporus* mushroom and its extracts. J. Agric. Food Chem. *30*, 521 (1982)

Salunkhe, D.K., Do, J.Y.: Biogenesis of aroma constituents of fruits and vegetables. Crit. Rev. Food Sci. Nutr. *8*, 161 (1977)

Schobinger, U.: Frucht- und Gemüsesäfte, Verlag Eugen Ulmer: Stuttgart. 1978

Schormüller, J.: Die Erhaltung der Lebensmittel. Ferdinand Enke Verlag: Stuttgart. 1966

Shah, B.M., Salunkhe, D.K., Olson, L.E.: Effects of ripening processes on chemistry of tomato volatiles. J. Am. Soc. Hort. Sci. *94*, 171 (1969)

Stamer, J.R.: Lactic acid fermentation of cabbage and cucumbers. In: Biotechnology (Eds.: Rhem, H.-J., Reed, G.), Vol. 5, p. 365, Verlag Chemie: Weinheim. 1983

Talburt, W.F., Smith, O.R.A.: Potato Processing. AVI Publ., Westport, CT 1975

Tiu, C.S., Purcell, A.E., Collins, W.W.: Contribution of some volatile compounds to sweet potato aroma. J. Agric. Food Chem. *33*, 223 (1985)

Whitaker, J.R.: Development of flavor, odor and pungency in onion and garlic. Adv. Food Res. *22*, 73 (1976)

Whitfield, F.B., Last, J.H.: Vegetables. In: Volatile compounds in foods and beverages (Ed.: Maarse, H.) Marcel Dekker, Inc.: New York. 1991

18 Fruits and Fruit Products

18.1 Fruits

18.1.1 Foreword

Fruits include both true fruits and spurious fruits, as well as seeds of cultivated and wild perennial plants. Fruits are commonly classified as pomaceous fruits, stone fruits, berries, tropical and subtropical fruits, hard-shelled dry fruits and wild fruits. The most important fruits are presented in Table 18.1 with pertinent data on botanical classification and use. Table 18.2 provides data about fruit production.

18.1.2 Composition

Fruit composition can be strongly influenced by the variety and ripeness, thus data given should be used only as a guide. Table 18.3 shows that the dry matter content of fruits (berries and pomme, stone, citrus and tropical fruits) varies between 10–20%. The major constituents are sugars, polysaccharides and organic acids, while N-compounds and lipids are present in lesser amounts. Minor constituents include pigments and aroma substances of importance to organoleptic quality, and vitamins and minerals of nutritional importance. Nuts are highly variable in composition (Table 18.4). Their moisture content is below 10%, N-compounds are about 20% and lipids are as high as 50%.

18.1.2.1 N-Containing Compounds

Fruits contain 0.1–1.5% N-compounds, of which 35–75% is protein. Free amino acids are also widely distributed. Other nitrogen compounds are only minor constituents. The special value of nuts, with their high protein content, has already been outlined.

18.1.2.1.1 Proteins, Enzymes

The protein fraction varies greatly with fruit variety and ripeness. This fraction is primarily enzymes. Besides those involved in carbohydrate metabolism (e.g., pectinolytic enzymes, cellulases, amylases, phosphorylases, saccharases, enzymes of the pentose phosphate cycle, aldolases), there are enzymes involved in lipid metabolism (e.g. lipases, lipoxygenases, enzymes involved in lipid biosynthesis), and in the citric acid and glyoxylate cycles, and many other enzymes such as acid phosphatases, ribonucleases, esterases, catalases, peroxidases, phenoloxidases and O-methyl transferases.

Protein and enzyme patterns, which can be obtained, for example, by electrophoretic separation, are generally highly specific for fruits and can be utilized for analytical differentiation of the species and variety. Figure 18.1 shows protein patterns of various grape species and Fig. 18.2 presents enzyme patterns of various species and cultivars of strawberries.

18.1.2.1.2 Free Amino Acids

Free amino acids are on average 50% of the soluble N-compounds. The amino acid pattern is typical of a fruit and hence can be utilized for the analytical characterization of a fruit product. Table 18.5 provides some relevant data.

In addition to common protein-building amino acids, there are nonprotein amino acids present in fruits, as in other plant tissues. Examples are the toxic 2-(methylene cyclopropyl)-glycine (I) in litchi fruits (*Litchi sinensis*), the toxic hypoglycine A (II) in akee (*Blighia sapida*), 1-aminocyclopropane-1-carboxylic acid (X) in apples and pears, trans-4-methylproline (XXII), 4-hydroxymethylprolines (XXIII–XXV) and 4-methyleneproline (XXVI) in apples and in loquat fruits (*Evio-*

Table 18.1. Eddible fruits: a classification

Number	Common name	Latin name	Family/subfamily	Form of consumption
Pomme fruits				
1	Apple	*Malus sylvestris*	Rosaceae	Fresh, dried, purée, jelly, juice, apple cider, brandy
2	Pear	*Pyrus communis*	Rosaceae	Fresh, dried, compote, brandy
3	Quince apple shaped pear shaped	*Cydonia oblonga var. maliformis var. pyriformis*	Rosaceae	Jelly, ingredient of apple purée
Stone fruits				
4	Apricot	*Prunus armeniaca*	Rosaceae	Fresh, dried, compote, jam, juice, seed for persipan, brandy
5	Peach	*Prunus persica*	Rosaceae	Fresh, compote, juice, brandy
6	Prune/plum	*Prunus domestica*	Rosaceae	Fresh, dried, compote, jam, brandy
7	Sour cherry	*Prunus cerasus*	Rosaceae	Fresh, compote, jam, juice, brandy
8	Sweet cherry	*Prunus avium*	Rosaceae	Fresh, candied, compote
Berry fruits				
9	Blackberry	*Rubus fruticosus*	Rosaceae	Fresh, jam, jelly, juice, wine, liqueur
10	Strawberry	*Fragaria vesca*	Rosaceae	Fresh, compote, jam, brandy
11	Bilberry	*Vaccinium myrtillus*	Ericaceae	Fresh, compote, jam, brandy
12	Raspberry	*Rubus idaeus*	Rosaceae	Fresh, jam, jelly syrup, brandy
13	Red currant	*Ribes rubrum*	Saxifragaceae	Fresh, jelly, juice, brandy
14	Black currant	*Ribes nigrum*	Saxifragaceae	Fresh, juice, liqueur
15	Cranberry	*Vaccinium vitis-idaea*	Ericaeae	Compote
16	Gooseberry	*Ribes uva-crispa*	Saxifragaceae	Unripe: compote; ripe: fresh, jam, juice
17	Grapes	*Vites vinifera ssp. vinifera*	Vitaceae	Fresh, dried (raisins) juice, wine, brandy
Citrus fruits				
18	Orange	*Citrus sinensis*	Rutaceae	Fresh, juice, marmelade
19	Grapefruit	*Citrus paradisi*	Rutaceae	Fresh juice
20	Kumquat	*Fortunella margarita*	Rutaceae	Fresh, compote, jam
21	Mandarine	*Citrus reticulata*	Rutaceae	Fresh compote
22	Pomelo	*Citrus maxima*	Rutaceae	Fresh, juice
23	Seville orange	*Citrus aurantium ssp. aurantium*	Rutaceae	Candied, marmalade
24	Lemon	*Citrus limon*	Rutaceae	Juice
25	Citron	*Citrus medica*	Rutaceae	Peel candied (citronat)
Other tropical/subtropical fruits				
26	Acerola	*Malpighia emarginata*	Malpighiaceae	Fresh, compote, juice
27	Pineapple	*Ananas comosus*	Bromeliaceae	Fresh, compote, jam, juice
28	Avocado	*Persea americana*	Lauraceae	Fresh
29	Banana	*Musa*	Musaceae	Fresh, dried, cooked, baked
30	Cherimoya	*Annona cherimola*	Annonaceae	Fresh
31	Date	*Phoenix dactylifera*	Arecaceae	Fresh, dried

Table 18.1 (continued)

Num-ber	Common name	Latin name	Family/subfamily	Form of consumption
32	Fig	*Ficus carica*	Moraceae	Fresh, dried, jam, dessert wine
33	Indian fig	*Opuntia ficus-indica*	Cactaceae	Fresh
34	Guava	*Psidium guajava*	Myrtaceae	Compote, juice
35	Persimmon	*Diospyros kaki*	Ebenaceae	Fresh, candied, compote
36	Kiwi	*Actinidia chinensis*	Actinidiaceae	Fresh, compote
37	Litchi	*Litchi chinensis*	Sapindaceae	Fresh, dried, compote
38	Mango	*Mangifera indica*	Anacardiaceae	Fresh, compote, juice
39	Melons cantaloups watermelon	*Cucumis melo* *Citrullus lanatus*	Cucurbitaceae Cucurbitaceae	Fresh Fresh
40	Papaya	*Carica papaya*	Caricaceae	Fresh, compote, juice
41	Passion fruit	*Passiflora edulis*	Passifloraceae	Fresh, juice
42	Golden shower	*Cassia fistula*	Caesalpiniaceae	Fresh
Shell(nut) fruits				
43	Cashew nut	*Anacardium occidentale*	Anacardiaceae	Roasted
44	Peanut	*Arachis hypogaea*	Fabaceae	Roasted, salted
45	Hazel-nut (Filbert)	*Corylus avellana*	Betulaceae	Fresh, baked and confectionary products (nougat, crocant)
46	Almond sweet bitter	*Prunus dulcis* *var. dulcis* *var amara*	Rosaceae	Baked and confectionary products (marzipan); flavoring of baked and confectionary products
47	Brazil nut	*Bertholletia excelsa*	Lecythidaceae	Fresh
48	Pistachio	*Pistacia vera*	Anacardiaceae	Fresh, salted, sausage flavoring, decoration of baked products
49	Walnut	*Juglans regia*	Juglandaceae	Fresh, baked and confectionary products, unripe fruits in vinegar and sugar-containing preserves
Wild fruits				
50	Rose hips	*Rosa sp.*	Rosaceae	Jam, wine
51	Elderberry	*Sambucus nigra*	Caprifoliaceae	Juice, jam
52	Seabuckthorn	*Hippophae rhamnoides*	Elaeagnaceae	Jam, juice

botrya japonica), 3,4-dihydroxyglutamic acid (XXXV) in red currants, 4-methyleneglutamic acid (XXXI) and 4-methyleneglutamine (XXXII) in peanuts and 3-amino-3-carboxy-pyrrolidine (LIV) in cashew. The nonprotein amino acids are discussed in more detail in Section 17.1.2.1.2. The Roman numerals given in brackets above correspond to Tables 17.5 and 17.6.

18.1.2.1.3 Amines

A number of aliphatic and aromatic amines are found in various fruits (Tables 18.6 and 18.7). They are formed in part by amino acid decar-

boxylation such as in apples, or by amination (cf. Reaction 18.1) or transamination of aldehydes (cf. Reaction 18.2).

$$R—CHO \xrightarrow[2\,[H]]{NH_3} R—CH_2—NH_2 \qquad (18.1)$$

$$R—CHO + R^1—\underset{\underset{NH_2}{|}}{CH}—COOH$$

$$\longrightarrow R—CH_2—NH_2 + R^1—\underset{\underset{O}{\|}}{C}—COOH \qquad (18.2)$$

Some amines are derived from tyramine (e. g., hordenine, synephrine, octopamine, dopamine and noradrenaline; cf. Formula 18.3).

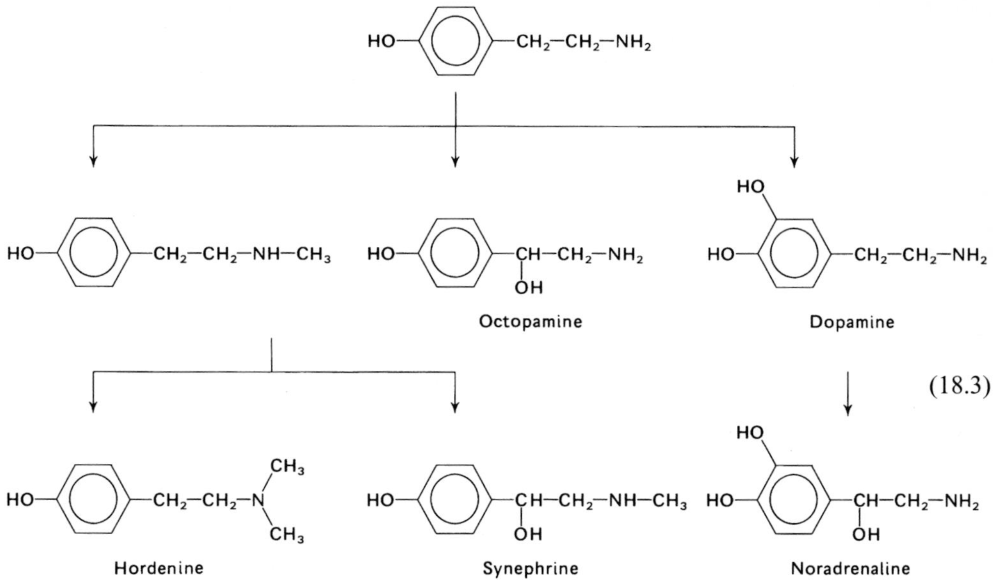

HO—⟨⟩—CH₂—CH₂—NH₂

HO—⟨⟩—CH₂—CH₂—NH—CH₃

HO—⟨⟩—CH—CH₂—NH₂
OH
Octopamine

HO—⟨⟩—CH₂—CH₂—NH₂ (with HO)
Dopamine

HO—⟨⟩—CH₂—CH₂—N(CH₃)₂
Hordenine

HO—⟨⟩—CH—CH₂—NH—CH₃
OH
Synephrine

HO—⟨⟩—CH—CH₂—NH₂
OH
Noradrenaline

(18.3)

Table 18.2. Production of fruits in 1996 (1,000 t)

Continent	Fruits[a], grand total	Nuts, grand total	Grapes	Raisins	Dates
World	413,932	4,996	57,410	1,005	4,492
Africa	54,394	446	2,983	42	1,494
America, North-, Central-	53,884	890	5,625	280	21
America, South-	66,728	270	5,268	11	–
Asia	161,182	2,333	12,405	507	2,969
Europe	72,818	1,022	30,005	81	8
Oceania	4,966	36	1,125	50	–

Continent	Apples	Pears	Peaches + nectarines	Plumes/ prunes	Oranges
World	53,672	13,093	10,409	6,761	59,558
Africa	1,479	443	344	134	4,373
America, North-, Central-	5,967	751	1,123	919	15,152
America, South-	2,978	819	684	200	24,318
Asia	25,906	7,256	3,690	3,019	10,491
Europe	16,520	3,629	4,472	2,455	4,769
Oceania	822	195	95	34	453

Table 18.2 (continued)

Continent	Mandarins[b]	Lemons	Grapefruit	Apricots	Avocado
World	15,954	9,104	5,004	2,387	2,093
Africa	1,018	575	402	262	192
America, North-, Central-	918	2,183	3,129	78	1,273
America, South-	1,524	1,720	385	46	378
Asia	10,384	3,471	1,011	1,147	177
Europe	2,026	1,117	49	813	55
Oceania	85	38	29	42	17

Continent	Mango	Pineapple	Bananas	Mealy bananas	Papaya
World	19,215	11,757	55,787	29,746	5,867
Africa	1,892	1,945	6,803	21,736	775
America, North-, Central-	1,993	1,405	8,508	1,448	549
America, South-	872	1,811	15,052	5,712	2,653
Asia	14,421	6,417	24,082	845	1,873
Europe	–	2	423	–	–
Oceania	37	178	919	5	17

Continent	Strawberries	Raspberries	Currants	Almonds	Pistachio-nuts
World	2,570	343	687	1,210	436
Africa	40	–	–	123	1
America, North-, Central-	858	51	–	392	60
America, South-	57	–	–	3	–
Asia	457	–	–	281	369
Europe	1,145	290	684	403	6
Oceania	13	59	3	9	–

Continent	Hazelnuts	Sweet chestnuts	Cashew nuts	Walnuts
World	590	505	721	1,023
Africa	–	–	217	3
America, North-, Central-	17	–	4	207
America, South-	–	24	165	23
Asia	428	327	335	495
Europe	145	153	–	294
Oceania	–	–	–	–

Table 18.2 (continued)

Country	Fruits, grand total	Country	Nuts, grand total	Country	Grapes
China	45,462	USA	797	Italy	9,000
India	39,197	Turkey	691	France	7,213
Brazil	35,928	China	464	USA	5,030
USA	28,841	Iran	425	Spain	4,486
Italy	17,182	Italy	296	Turkey	3,550
Mexico	12,179	Spain	295	Argentina	2,728
Spain	12,095	Brazil	204	China	2,054
France	11,211	India	174	Iran	1,900
Uganda	10,189	Korea Rep.	104	South Africa	1,670
Iran	9,774	Greece	92	Chile	1,527
Turkey	9,534	Tanzania	88	Germany	1,375
Indonesia	7,430	Vietnam	84	Romania	1,314
Philippines	7,388	Pakistan	78	Greece	1,169
Argentina	6,979				
Thailand	6,577	Σ (%)[c]	76	Σ (%)[c]	75
Colombia	6,428				
Ecuador	6,374				
Pakistan	5,438				
Germany	4,882				
Vietnam	4,355				
Japan	4,121				
South Africa	3,986				
Greece	3,967				
Chile	3,589				
Zaire	3,537				
Russian Fed.	3,386				
Σ (%)[c]	75				

Country	Raisins	Country	Dates	Country	Apples
Turkey	360	Iran	795	China	16,009
USA	273	Egypt	680	USA	4,733
Iran	90	Saudi Arabia	597	France	2,455
Greece	77	Iraq	550	Turkey	2,100
Australia	50	Pakistan	533	Iran	2,000
South Africa	40	Algeria	361	Italy	1,940
Chile	34	United Arab. Emirates	240	Russian Fed.	1,800
Afghanistan	28			Poland	1,700
Syria	12	Sudan	145	Germany	1,594
Argentina	11	Oman	133	India	1,200
		Tunesia	86	Argentina	1,147
Σ (%)[c]	97			Ukraine	1,120
		Σ (%)[c]	92	Japan	963
				Chile	910
				Spain	811
				Σ (%)[c]	75

Table 18.2 (continued)

Country	Pears	Country	Peaches + nectarins	Country	Plums/ prunes
China	5,615	China	2,322	China	2,059
Italy	937	Italy	1,689	USA	849
USA	707	Greece	1,040	Germany	367
Spain	584	USA	927	France	270
Argentina	513	Spain	824	Romania	253
Japan	426	France	464	Yugoslavia FR	240
Turkey	410	Turkey	375	Turkey	205
Germany	370	Chile	270	Ukraine	170
France	350	Argentina	159	Russian Fed.	166
Ukraine	255	Japan	156	Spain	145
				Hungary	140
Σ (%)[c]	78	Σ (%)[c]	79	Iran	137
				Chile	124
				Σ (%)[c]	76

Country	Oranges	Country	Mandarins	Country	Lemons
Brazil	21,811	China	5,922	India	1,700
USA	10,635	Spain	1,414	Mexico	1,001
Mexico	3,556	Japan	1,199	USA	948
China	2,258	Brazil	760	Argentina	713
Spain	2,154	Thailand	650	Iran	655
India	2,000	Iran	630	Italy	545
Egypt	1,608	Korea Rep.	600	Brazil	495
Iran	1,600	Pakistan	540	Spain	436
Italy	1,597	USA	500	Egypt	330
Pakistan	1,380	Egypt	475	Turkey	325
Σ (%)[c]	82	Σ (%)[c]	80	Σ (%)[c]	79

Country	Grapefruit	Country	Apricots	Country	Avocado
USA	2,466	Turkey	460	Mexico	790
Israel	404	Spain	194	USA	173
Cuba	261	France	160	Dominican Rep.	155
Mexico	240	Pakistan	160	Brazil	110
China	217	Iran	121	Indonesia	110
Argentina	203	Italy	105	Colombia	74
South Africa	162	Ukraine	100	Chile	62
India	70	Morocco	90	Peru	52
Sudan	65	Syria	80	Israel	50
Brazil	62	USA	72	Zaire	47
		Russian Fed.	65		
Σ (%)[c]	83	Lebanon	56	Σ (%)[c]	78
		Greece	55		
		Egypt	45		
		Algeria	43		
		China	43		
		Σ (%)[c]	77		

Table 18.2 (continued)

Country	Mango	Country	Pineapple	Country	Bananas
India	10,000	Thailand	2,031	India	9,935
Mexico	1,420	Philippines	1,477	Brazil	5,692
China	1,208	Brazil	1,048	Ecuador	5,309
Pakistan	908	China	836	China	3,298
Thailand	665	India	820	Philippines	3,292
Indonesia	600	Nigeria	800	Indonesia	2,600
Nigeria	500	Indonesia	570	Mexico	2,158
Philippines	480	Colombia	387	Costa Rica	2,100
Brazil	435	USA	315	Colombia	2,100
Egypt	240	Kenia	270	Thailand	1,750
		Honduras	269	Burundi	1,544
Σ (%)^c	86			Venezuela	1,365
		Σ (%)^c	75	Viet Nam	1,282
				Σ (%)^c	76

Country	Mealy bananas	Country	Papaya	Country	Straw-beries
Uganda	9,550	Brazil	2,350	USA	738
Colombia	3,212	Thailand	520	Poland	210
Zaire	2,270	Nigeria	500	Japan	202
Rwanda	2,105	India	490	Spain	191
Nigeria	1,712	Indonesia	460	Italy	190
Ghana	1,642	Mexico	460	Korea Rep.	150
Côte d'Ivoire	1,300	Zaire	210	Russian Fed.	121
Peru	1,060	China	143	France	88
Cameroon	1,000	Peru	140	Mexico	85
Ecuador	681	Philippines	100	Turkey	66
Σ (%)^c	82	Σ (%)^c	92	Σ (%)^c	79

Country	Rasp-berries	Country	Currants	Country	Almonds
Russian Fed.	90	Russian Fed.	190	USA	392
Poland	39	Poland	184	Spain	246
USA	33	Germany	170	Italy	91
Germany	32	UK	23	Iran	69
Canada	18	Czecn Rep.	22	Pakistan	53
Hungary	18	Norway	18	Greece	50
UK	17	Austria	16	Morocco	50
France	8	Ukraine	15	Turkey	47
Ukraine	8	Hungary	12	Tunisia	42
Bulgaria	3	France	7	Syria	35
Moldova Rep.	3	Σ (%)^c	96	Σ (%)^c	89
Romania	3				
Σ (%)^c	79				

Table 18.2 (continued)

Country	Pistachio nuts	Country	Hazelnuts	Country	Sweet chestnuts
Iran	282	Turkey	410	China	112
USA	60	Italy	118	Korea Rep.	93
Turkey	42	USA	17	Turkey	80
China	25	Spain	12	Italy	72
Syria	18	China	9	Japan	35
Greece	4	Greece	5	Bolivia	24
Afghanistan	2	Iran	5	Spain	24
Italy	2	France	4	Portugal	19
Tunisia	1	Russian Fed.	3	Greece	12
		Kyrgyzstan	2	Russian Fed.	12
Σ (%)^c	100				
		Σ (%)^c	99	Σ (%)^c	96

Country	Cashew-nuts	Country	Walnuts
Brazil	165	China	260
India	150	USA	189
Tanzania	82	Turkey	112
Vietnam	80	Ukraine	76
Indonesia	42	Iran	68
Mozambique	32	Russian Fed.	32
Guinea-Bissau	30	France	27
Nigeria	25	India	24
Thailand	20	Greece	21
Kenia	15	Romania	20
Σ (%)^c	89	Σ (%)^c	81

^{a} Without melons and nuts.
^{b} Inclusive of tangerines, clementines and satsumas.
^{c} World production = 100%.

Table 18.3. Average chemical composition of fruits (as % of fresh edible portion)

Fruit	Dry matter	Total sugar	Titratable acidity^a	Insoluble matter	Pectin^b	Ash	pH
Apple	16.0	11.1	0.6 (M)	2.1	0.6	0.3	3.3
Pear	17.5	9.8	0.2 (M)	3.1	0.5	0.4	3.9
Apricot	12.6	6.1	1.6 (M)	1.6	1.0	0.6	3.7
Sour cherry	14.7	9.4	0.7 (M)	1.6	0.3		3.4
Sweet cherry	18.7	12.4	0.7 (M)	2.0	0.3	0.6	4.0
Peach	12.9	8.5	0.6 (M)			0.5	3.7
Plum/prune	14.0	7.8	1.5 (M)	1.3	0.9	0.5	3.3
Blackberry	19.1	5.0	0.6 (C)	9.2	0.7	0.5	3.4
Strawberry	10.2	5.7	0.9 (C)	2.4	0.5	0.5	
Currant, red	16.4	5.1	2.3 (C)	5.9	0.7	0.6	3.0
Currant, black	19.7	6.3	3.2 (C)	5.9	1.1	0.6	3.3

Table 18.3 (continued)

Fruit	Dry matter	Total sugar	Titratable acidity[a]	Insoluble matter	Pectin[b]	Ash	pH
Raspberry	13.9	4.5	1.8 (C)	5.1	0.4	0.5	3.4
Grapes	17.3	14.8	0.4 (T)			0.5	3.3
Orange	13.0	7.0	0.8 (C)			0.5	3.3
Grapefruit	11.4	6.7	1.3 (C)			0.4	3.3
Lemon	11.7	2.2	6.0 (C)			0.5	2.5
Pineapple	15.4	12.3	1.1 (C)	1.5		0.4	3.4
Banana	26.4	18.0	0.4 (M)	4.6	0.9	0.8	4.7
Cherimoya	19	13	0.2			0.9	
Date	80	61				1.8	
Fig	22	16	0.4 (C)		0.6		
Guava	22	4.9			0.7		
Mango	19	14	0.5		0.5		
Papaya	11	9	0.1		0.6		

[a] Calculated as malic (M), citric (C), or tartaric acid (T).
[b] Results are expressed as calcium pectate.

Table 18.4. Proximate composition of shell-nut fruit (as % of fresh edible portion)

Fruit	Moisture	N-Compounds	Lipids	Carbohydrates	Ash	Crude fiber
Cashew nut		16	45.5	21.5		
Peanut	5.0	28.5	47.5	20	2.9	2.8
Hazel nut	7.1	17.4	62.6	9.5	2.5	3.2
Pistachio		21	53.5	16.5		
Almond	4.7	20.5	53.5	16.5	2.3	3.7
Walnut	3.3	15.0	64.4	15.6	1.7	2.1

18.1.2.2 Carbohydrates

18.1.2.2.1 Monosaccharides

In addition to glucose and fructose, the ratios of which vary greatly in various fruits (Table 18.8), other monosaccharides occur only in trace amounts. For example, arabinose and xylose have been found in several fruits. An exceptional case is avocado in which a number of higher sugars are present at 0.2 to 5.0% of the fresh weight (D-manno-heptulose, D-talo-heptulose, D-glycero-D-galacto-heptose, D-glycero-D-manno-octulose, D-glycero-L-ga-

Table 18.5. Free amino acids in fruits (as % of total free amino acids)

Fruit	Asp	Asn	Glu	Gln	Ser	Thr	Pro	Ala	Abu[b]	His	Arg	Pip[c]
Apple (juice)	21	17	15		10	3	2	7	5			
Pear (juice)	10	9	10		11	2	14	9	3			
Grapes	3		13			6	31	9	6		27	
Currant, black		7	17	24	5		8	17	12			
Orange[a]	7–115	20–188	6–71	3–63	4–37		6–295	3–26	4–73		23–150	
Grapefruit[a]	470		280		310	10				76		
Lemon[a]	19–60		6–35		12–28			1–31	4–20		25–106	
Banana	5–10	15		10–15					5–10	10–15		5–10

[a] Values in mg/100 ml juice.
[b] γ-Aminobutyric acid.
[c] Pipecolinic acid.

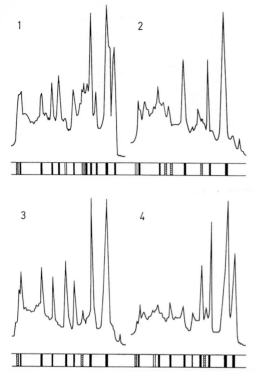

Fig. 18.1. Protein patterns of various wine cultivars obtained by isoelectric focussing (pH 3–10) using Sephadex G-75 as a gel support medium. Staining was done by Coomassie Blue. The figures show the electropherograms and the corresponding densitograms. Cultivation region South Palatinate; 1 Morio Muscat, 2 Mueller-Thurgau, 3 Rulaender, 4 Sylvaner. (according to *Drawert* and *Mueller*, 1973)

Table 18.6. Amines in fruit

Fruit	Amines
Apple	Methylamine, ethylamine, propylamine, butylamine, hexylamine, octylamine, dimethylamine, spermine, spermidine
Plum/prune	Dopamine
Orange	Feruloylputrescine, methyltyramine, synephrine
Grapefruit	Feruloylputrescine
Lemon	Tyramine, synephrine, octopamine
Pineapple	Tyramine, serotonin
Avocado	Tyramine, dopamine
Banana	Methylamine, ethylamine, isobutylamine, isoamylamine, dimethylamine, putrescine, spermidine, ethanolamine, propanolamine, histamine, 2-phenylethylamine, tyramine, dopamine, noradrenaline, serotonin

Table 18.7. Amines in peel and flesh of banana (µg/g fresh weight)

Amine	Peel	Flesh
Serotonin	50–150	28
Tyramine	65	7
Dopamine	700	8
Noradrenaline	122	2

lacto-octulose, D-erythro-L-gluco-nonulose and D-erythro-L-galacto-nonulose). Small amounts of heptuloses have been found in the fruit flesh of apples, peaches and strawberries, and in the peels of grapefruit, peaches and grapes.

18.1.2.2.2 Oligosaccharides

Saccharose (sucrose) is the dominant oligosaccharide. Other disaccharides do not have quantitative importance. Maltose occurs in small amounts in grapes, bananas and guava. Melibiose, raffinose and stachyose have also been detected in grapes. 6-Kestose has been identified in ripe bananas.

Other oligosaccharides occur only in trace amounts. The proportion of reducing sugars to saccharose can vary greatly (Table 18.8). Some fruits have no saccharose (e. g., cherries, grapes and figs), while in some the saccharose content is significantly higher than the reducing sugar content (e. g., apricots, peaches and pineapples).

18.1.2.2.3 Sugar Alcohols

D-Sorbitol is abundant in *Rosaceae* fruits (pomme fruits, stone fruits). For example, its concentration is 300–800 mg/100 ml in apple juice. Since fruits such as berries, citrus fruits, pineapples or bananas do not contain sorbitol, its detection is of analytical importance in the evaluation of wine and other fruit products. Meso-inositol also occurs in fruits; in orange juice it ranges from 130–170 mg/100 ml.

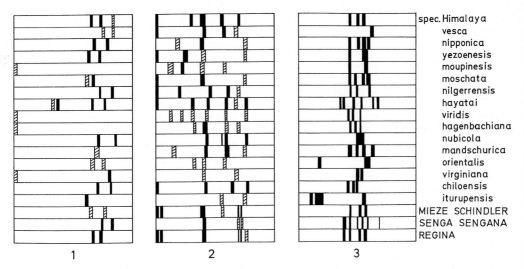

spec. Himalaya
vesca
nipponica
yezoensis
moupinesis
moschata
nilgerrensis
hayatai
viridis
hagenbachiana
nubicola
mandschurica
orientalis
virginiana
chiloensis
iturupensis
MIEZE SCHINDLER
SENGA SENGANA
REGINA

1   2   3

Fig. 18.2. Enzyme patterns of some strawberry species (*Fragaria sp.*) and (*Fragaria ananas*) obtained by PAGE disc gel electrophoresis. Large pore concentrating gel pH 6.7, small pore separating gel, pH 8.9. **1** Peroxidase: incubation with o-toluidine/H_2O_2 at pH 7. **2** Esterase: incubation with α-naphthylacetate at pH 7, the released α-naphthol is diazotized and then coupled with p-chloroaniline. **3** Malate-dehydrogenase: incubation with malate, nitro-blue-tetrazolium chloride and NAD at pH 7.5. (according to *Drawert* et al., 1974)

Table 18.8. Sugar content in various fruits (as % of the edible portion)

Fruit	Glucose	Fructose	Saccharose
Apple	1.8	5.0	2.4
Pear	2.2	6.0	1.1
Apricot	1.9	0.4	4.4
Cherry	5.5	6.1	0.0
Peach	1.5	0.9	6.7
Plum/prune	3.5	1.3	1.5
Blackberry	3.2	2.9	0.2
Strawberry	2.6	2.3	1.3
Currant, red	2.3	1.0	0.2
Currant, black	2.4	3.7	0.6
Raspberry	2.3	2.4	1.0
Grapes	8.2	8.0	0.0
Orange	2.4	2.4	4.7
Grapefruit	2.0	1.2	2.1
Lemon	0.5	0.9	0.2
Pineapple	2.3	1.4	7.9
Banana	5.8	3.8	6.6
Date	32.0	23.7	8.2
Fig	5.5	4.0	0.0

18.1.2.2.4 Polysaccharides

All fruits contain cellulose, hemicellulose (pentosans) and pectins. The building blocks of these polysaccharides are glucose, galactose, mannose, arabinose, xylose, rhamnose, fucose and galacturonic and glucuronic acids. The pectin fractions of fruits are particularly affected by ripening. A decrease in insoluble pectin is accompanied by an increase in the soluble pectin fraction. The total pectin content can also decrease. Starch is present primarily in unripe fruits and its content decreases to a negligible level as ripening proceeds. Exceptions are bananas, in which the starch content can be 3 % or more even in ripe bananas, and various nuts such as cashew and Brazil nuts.

18.1.2.3 Lipids

The lipid content of fruits is generally low, 0.1–0.5 % of the fresh weight. Only fruit seeds and nuts contain significantly higher levels of lipids (cf. Table 18.4). The fruit flesh of avocado is also rich in fat. The lipid fraction of fruits consists of triacylglycerols, glyco- and phospholipids, carotenoids, triterpenoids and waxes.

18.1.2.3.1 Fruit Flesh Lipids (Other than Carotenoids and Triterpenoids)

Table 18.9 presents the lipid fractions of apple flesh. Phospholipids, about 50% of the lipid fraction, are predominant. The most abundant fatty acids are palmitic, oleic and linoleic acids (Table 18.10).

18.1.2.3.2 Carotenoids

Carotenoids are widespread in fruits and, in a number of fruits, such as citrus fruits, peaches and sweet melons, their presence is the main factor determining color. The most important carotenoids found in fruits are compiled in Table 18.11, while Table 18.12 gives the carotenoid composition of some fruits.

Fruits can be divided into various classes according to the content and distribution pattern of carotenoids:

- Fruits with low content of carotenoids (occurring mostly in chloroplasts) such as

β-carotene, lutein, violaxanthin, neoxanthin (e.g., pineapples, bananas, figs and grapes).
- Fruits with relatively high contents of lycopene, phytoene, phytofluene, ζ-carotene and neurosporene, e.g., peaches.
- Fruits with relatively high contents of β-carotene, cryptoxanthin and zeaxanthin. This class includes oranges, pears, peaches and sweet melons.
- Fruits with high amounts of epoxides, e.g.. oranges and pears.
- Fruits which contain unusual carotenoids, e.g., oranges.

The compositional pattern of carotenoids which can be readily analyzed by HPLC is important for analytical characterization of fruit products.

Table 18.9. Lipids of apple flesh (as % of the total lipids)

Triacylglycerols	5	Sterols	15
Glycolipids	17	Sterol esters	2
Phospholipids	47	Sulfolipids	1
		Others	15

Table 18.10. Fatty acid composition of some fruit flesh lipids (as % of the total fatty acids)

Fatty acid	Avocado	Apple	Banana
12 : 0	+ [a]	0.6	+
14 : 0	+	0.6	0.6
16 : 0	15	30	58
16 : 1	4	0.5	8.3
18 : 0	+	6.4	2.5
18 : 1	69	18.5	15
18 : 2	11	42.5	10.6
18 : 3	+	1	3.6

[a] Traces.

Table 18.11. Carotenoids occurring in fruit (Roman numerals refer to their structures shown in 3.8.4)

Number	Carotenoid
1	Phytoene (I)
2	Phytofluene (II)
3	ζ-Carotene (III)
4	Lycopene (IV)
5	α-Carotene (VI)
6	β-Carotene (VII)
7	β-Zeacarotene (Va)
8	Lycoxanthin (16-hydroxylycopene)
9	α-Cryptoxanthin (3-hydroxy-α-carotene)
10	β-Cryptoxanthin (3-hydroxy-β-carotene)
11	β-Carotene-5,6-epoxide
12	Mutatochrome (β-carotene-5,8-epoxide)
13	Lutein (IX)
14	Zeaxanthin (VIII)
15	Cryptoflavin (α-cryptoxanthin-5,8-epoxide)
16	β-Carotene-5,6,5′,6′-diepoxide
17	Antheraxanthin (zeaxanthin-5,6-epoxide)
18	Lutein-5,6-epoxide
19	Mutatoxanthin (XVI)
20	Lutein-5,8-epoxide
21	Cryptoxanthin-5,8,5′,8′-diepoxide
22	Violaxanthin (XIII)
23	Luteoxanthin (XIV)
24	Auroxanthin (zeaxanthin-5,8,5′,8′-diepoxide)
25	Neoxanthin (XX)
26	Capsanthin (X)

Table 18.12. Carotenoid patterns of various fruits

Fruit	Carotenoid	
	Content[a]	Compounds[b]
Pineapple		6, 13
Orange	24	1, 2, 3, 4, 6, 10, 11, 12, 15, 17, 20, 21, 22, 23, 24
Banana		6, 13
Pear	0.3–1.3	2, 3, 6, 7, 8, 11, 12, 13, 14, 16, 18, 20, 24, 25
Fig	8.5	1, 2, 5, 6, 13, 14, 22, 23, 25
Guava		5, 6
Peach	27	1, 2, 3, 5, 6, 9, 10, 13, 14, 17, 18, 19, 22, 23, 24
Plum/ prune		1, 2, 3, 5, 6, 9, 10, 12, 14, 15, 17, 18, 19, 20, 22, 23, 25
Grapes	1.8	1, 2, 4, 5, 13, 14, 22, 23
Canta- loupe	20–30	1, 2, 3, 5, 6, 13, 14, 22, 23

[a] mg/kg fresh weight.
[b] The numerals refer to Arabic numerals in Table 18.11.

18.1.2.3.3 Triterpenoids

This fraction contains bitter compounds of special interest, limonoids and cucurbitacins. Limonoids are found in the flesh and seeds of *Rutaceae* fruits. For example, limonin (II) is present in seeds, juice, and fruit flesh of oranges and grapefruit. The limonin content decreases with fruit ripening in oranges but remains constant in grapefruit. Development of a bitter taste in heated orange juice is a processing problem. Limonin monolactone (I), a nonbitter compound which is stable in the neutral pH range, is present in orange albedo and endocarp. During production of orange juice it is transferred to the juice in which, due to the lower pH, it is transformed into the bitter tasting dilactone, limonin (II; cf. Formula 18.4).
Bitter and nonbitter forms of the many *Cucurbitaceae* are known. The bitter forms contain cucurbitacins (III) in fruits and seeds. For example, *Citrullus lanatus* (watermelon) contains IIIE in glycosidic form; while *Cucumis sativus* (cucumber) contains IIIC and *Cucurbita* spp. (pumpkin) contains IIIB, D, E and I (cf. Formula 18.5).

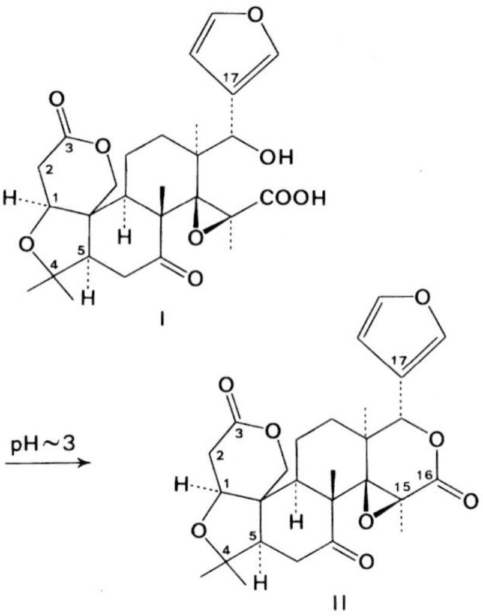

(18.4)

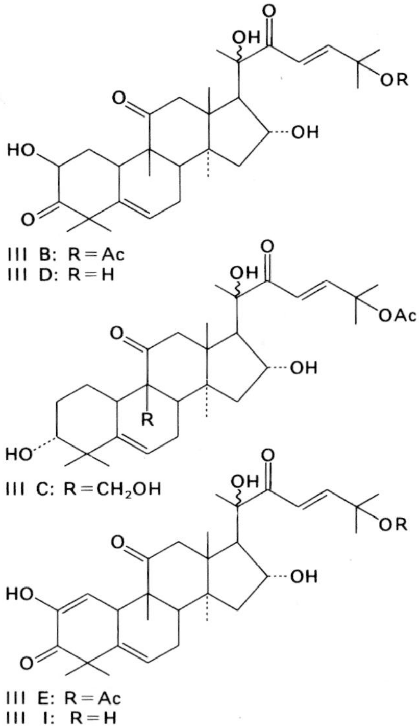

III B: R=Ac
III D: R=H

III C: R=CH$_2$OH

III E: R=Ac
III I: R=H

(18.5)

The common precursor in the biosynthesis of limonoids and cucurbitacins is squalene-2,3-oxide (IV). Based on some identified intermediary compounds, the biosynthetic pathway is probably as postulated in Reaction 18.6.

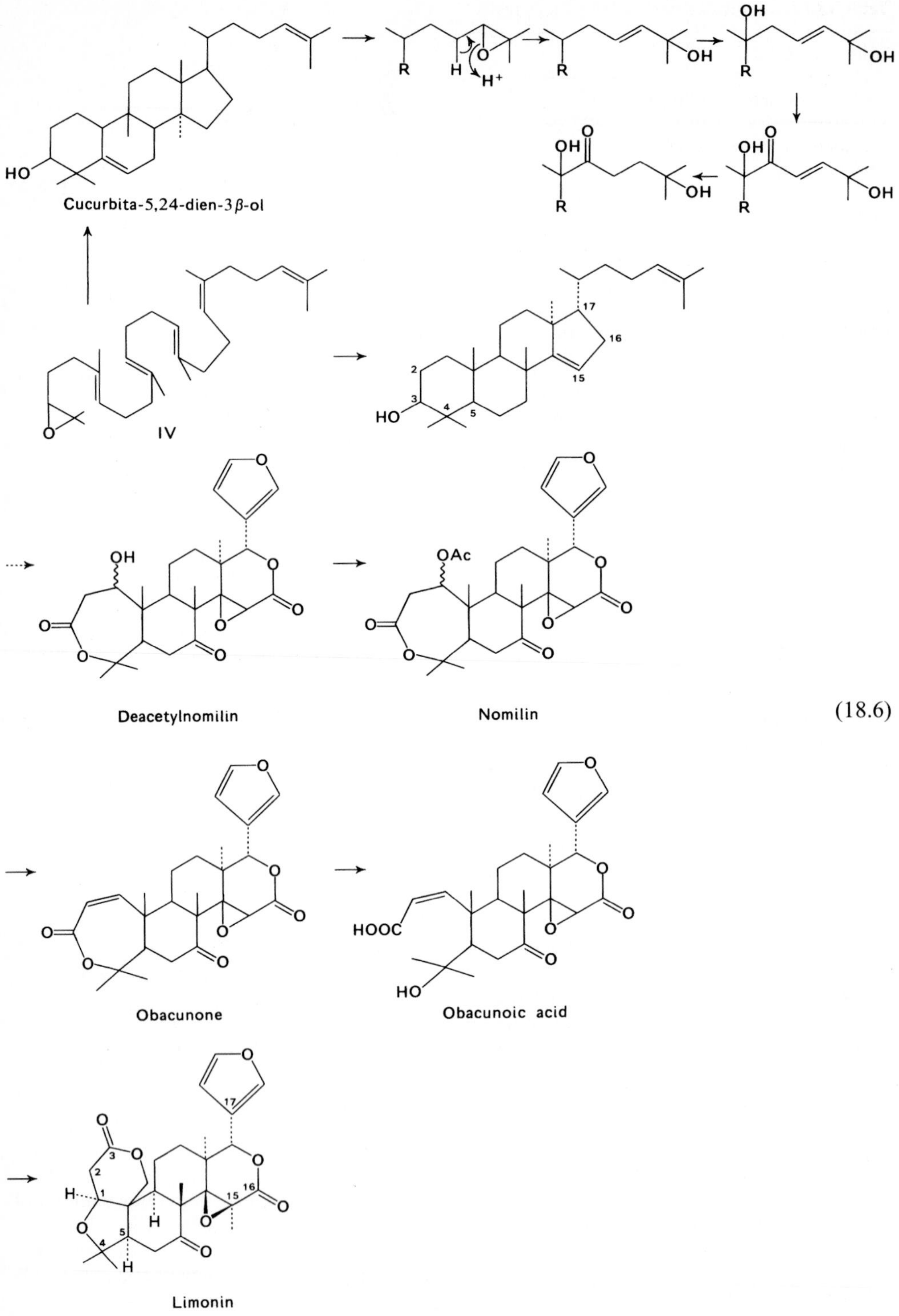

Cucurbita-5,24-dien-3β-ol

IV

Deacetylnomilin

Nomilin

(18.6)

Obacunone

Obacunoic acid

Limonin

18.1.2.3.4 Fruit Waxes

The fruit peel is often coated with a waxy layer. In addition to the esters of higher fatty acids with higher alcohols, these waxes contain hydrocarbons, free fatty acids, free alcohols, ketones and aldehydes. The ester fraction in apples and grapes predominantly consists of alcohols of 24, 26 and 28 carbons, but their fatty acid patterns differ. Apples contain mostly $18:1$, $18:2$, $16:0$ and $18:0$ fatty acids, while grapes contain $20:0$, $18:0$, $22:0$ and $24:0$ fatty acids.

18.1.2.4 Organic Acids

L-Malic and citric acids are the major organic acids of fruits (Table 18.13). Malic acid is predominant in pomme and stone fruits, while citric acid is most abundant in berries, citrus and tropical fruits. (2R:3R)-Tartaric acid occurs only in grapes. Many other acids, including the acids in the citric acid cycle, occur only in low amounts. Examples are cis-aconitic, succinic, pyruvic, citramalic, fumaric, glyceric, glycolic, glyoxylic, isocitric, lactic, oxalacetic, oxalic and 2-oxoglutaric acids. In fruit juices, the ratio of citric acid to isocitric acid (examples in Table 18.14) serves as an indicator of dilution with an aqueous solution of citric acid.

Important phenolic acids, dealt with in Section 18.1.2.5.1, are quinic, caffeic, chlorogenic and shikimic acids. Galacturonic and glucuronic acids are also found.

Tartaric acid biosynthesis in *Vitis* spp. starts from glucose or fructose and probably proceeds through 5-oxogluconic acid or ascorbic acid respectively:

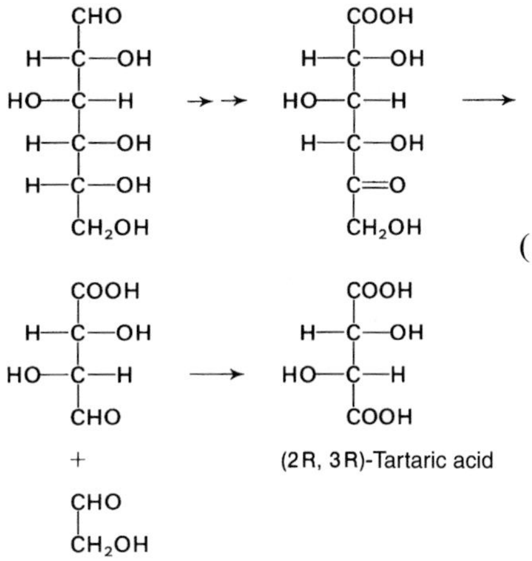

$$(18.7)$$

Table 18.13. Organic acids in various fruits (milliequivalents/100 g fresh weight)

Fruit	Major acid	Other acids
Apple	Malic 3–19	Quinic (in unripe fruits)
Pear	Malic 1–2	Citric
Apricot	Malic 12	Citric 12, quinic 2–3
Cherry	Malic 5–9	Citric, quinic, and shikimic
Peach	Malic 4	Citric 4
Plum/prune	Malic 4–6	Quinic (especially in unripe fruits)
Strawberry	Citric 10–18	Malic 1–3, quinic 0.1, succinic 0.1
Raspberry	Citric 24	Malic 1
Currant, red	Citric 21–28	Malic 2–4, succinic, oxalic
Currant, black	Citric 43	Malic 6
Gooseberry	Citric 11–14	Malic 10–13, shikimic 1–2
Grapes	Tartaric 1.5–2	Malic 1.5–2
Orange	Citric 15	Malic 3, quinic
Lemon	Citric 73	Malic 4, quinic
Pineapple	Citric 6–20	Malic 1.5–7
Banana	Malic 4	
Fig	Citric 6	Malic, acetic
Guava	Citric 10–20	Malic

Table 18.14. Ratio of citric acid (C) to isocitric acid (I) in fruit juices

Fruit juice	C/I
Orange	80–130
Currant	60–140
Grapefruit	50–90
Raspberry	80–200
Lemon	≤ 250
Blackberry	ca. 0.25

$$ \text{(18.9)} \qquad \begin{array}{l} \text{I:} \quad R^1, R^2 = H \\ \text{II:} \quad R^1 = H, R^2 = OCH_3 \\ \text{III:} \quad R^1 = H, R^2 = OH \\ \text{IV:} \quad R^1 = R^2 = OCH_3 \end{array} $$

$$ \text{(18.10)} $$

Human and animal metabolism oxidatively degrade (2R:3R)-tartaric and meso-tartaric acids into glyoxylic and hydroxypyruvic acids respectively:

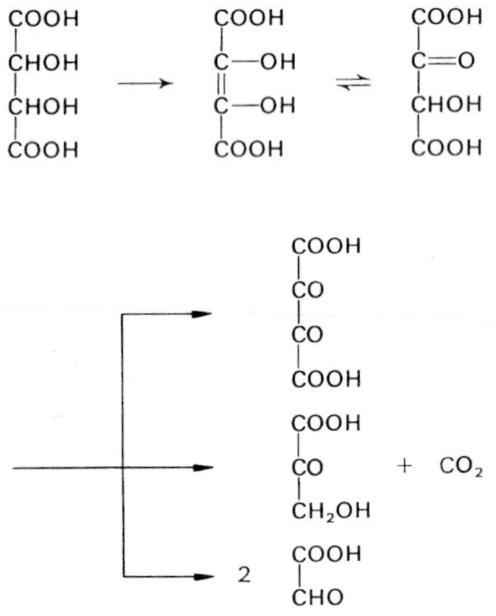

$$ \text{(18.8)} $$

18.1.2.5 Phenolic Compounds

These compounds occur in most fruits and most of them contribute to color and taste. They can form metal complexes during fruit processing, resulting in discoloration of fruit pulp. Table 18.17 provides data for the total phenol content of some selected fruits.

18.1.2.5.1 Hydroxycinammic Acids, Hydroxycoumarins and Hydroxybenzoic Acids

p-Coumaric (I), ferulic (II), caffeic (III) and sinapic (IV) acids are widespread in fruits and vegetables.

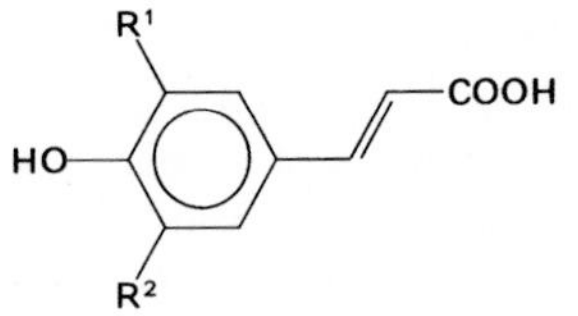

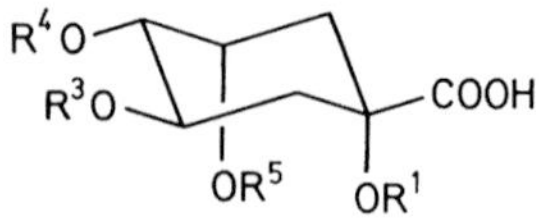

These hydroxycinnamic acids are present mainly as derivatives. The most common are esters of caffeic, coumaric, and ferulic acids with D-quinic and, in addition, with D-glucose (Table 18.15 and 18.16). Since quinic acid has four OH groups, four bonding possibilities exist (R^1, R^3–R^5 in Formula 18.10), the 3- and 5-isomers being preferred. According to IUPAC nomenclature for cyclitols, the 3-, 4-, and 5-caffeoylquinic acids are identical to neo-chlorogenic acid (V), cryptochlorogenic acid (VI), and chlorogenic acid (VII). Isochlorogenic acid is a mixture of di-O-caffeoylquinic acids. Apart from quinic acid and glucose, other alcoholic components are shikimic, malic and tartaric acids and meso-inositol. Sinapine, which is found in mustard seed, is the counter ion of the glucosinolate sinalbin and is the choline ester of sinapic acid, which is as bitter as caffeine. Hydroxycinnamic acid amides are also found in plants.

Scopoletin (VIII, Formula 18.11) in esterified form is the only hydroxycoumarin (Table 18.18) which has been found, in small amounts, in plums and apricots.

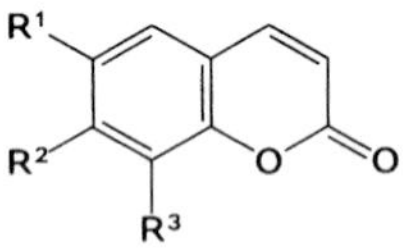

$$ \text{(18.11)} $$

VIII

The hydroxybenzoic acids that are found in various fruits and occur mostly as esters include: salicylic acid (2-hydroxybenzoic acid), 4-hydroxybenzoic acid, gentisic acid (2,4-dihydroxybenzoic acid), protocatechuic acid (3,4-dihydroxybenzoic acid), gallic acid (3,4,5-

Table 18.15. Derivatives of hydroxycinnamic acid in pomme and stone fruits[a]

Compound	Apple	Pear	Sweet cherry	Sour cherry	Plum	Peach	Apricot
5-Caffeoylquinic acid	62–385[b]	64–280	11–40	50–140	15–142	43–282	37–123
4-Caffeoylquinic acid	2	–	+	+	9	–	–
3-Caffeoylquinic acid	–	–	73–628	82–536	88–771	33–142	26–132
3-p-Coumaroylquinic acid	–	.	81–450	40–226	4–40	2	2–9
3-Feruloylquinic acid	–	–	4	1	13	1	7
p-Coumaroylglucose	4	+	–	–	15	–	–
Feruloylglucose	3	–	–	–	5	–	–

[a] mg/kg of fresh weight,   [b] variety "Boskop": 400–500 mg/kg.

Table 18.16. Derivatives of hydroxycinnamic acid in berries[a]

Compound	Straw-berry	Rasp-berry	Black-berry	Red currant	Black currant	Goose-berry	Cultured blueberry
Caffeoylquinic acid	–	1	45–53	1	45–52	3	1860–2080
p-Coumaroylquinic acid	–	1	2–5	+	14–23	1	2–5
Feruloylquinic acid	–	+	2–4	2	4	1	8
Caffeoylglucose	1	3–7	3–6	2–5	19–30	5–13	+
p-Coumaroylglucose	14–17	6–14	4–11	1	10–14	7	+
Feruloylglucose	1	4–7	2–6	+	11–15	1–6	+
Caffeic acid-4-O-glucoside	–	–	–	2	2	2	3
p-Coumaric acid-O-glucoside	+	5–10	2–5	5–16	4–10	6–8	3–15
Ferulic acid-O-glucoside	–	+	–	–	3	2–7	8–10

[a] mg/kg of fresh weight. The hydroxycinnamoylquinic acids are present mostly as the 3-isomers, but in blueberries as the 5-isomer.

trihydroxybenzoic acid), vanillic acid (3-methoxy-4-hydroxybenzoic acid) and ellagic acid (IX, Formula 18.12), the dilactone of hexahydroxydiphenic acid (Table 18.19).

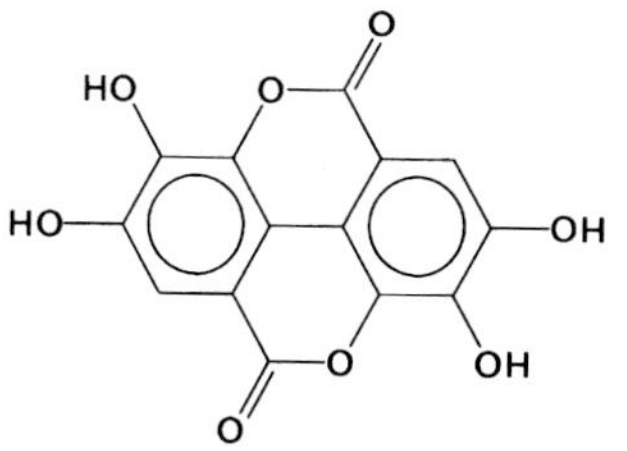

(18.12)

IX

Apart from the proanthocyanidins (cf. 18.1.2.5.2), esters of gallic acid and hexahydroxydiphenic acid form one of the two main classes of plant tanning agents, the "hydrolyzable tanning agents" or tannins. In addition to simple esters with different hydroxy compo-

nents, such as β-D-glucogallin (X in Formula 18.13), theogallin (XI) and the flavan-3-ol-gallates XII and XIII, found, e. g., in tea leaves, complex polyesters with D-glucose are known. They have molecular weights of M_r 500–3000, are generally readily soluble, and contribute their astringent properties to the taste of foods of plant origin.

Table 18.17. Phenolic compounds in fruit

Fruit	Total phenols (g/100 g fresh tissue)
Apple	0.1–1
Pear	0.4
Cherry	0.2
Peach	0.03–0.14
Plum/prune	0.2–1.4
Grapes	0.1–1

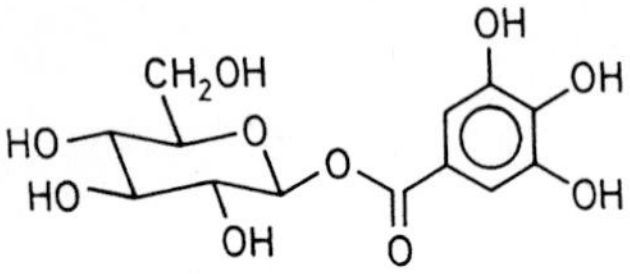

X 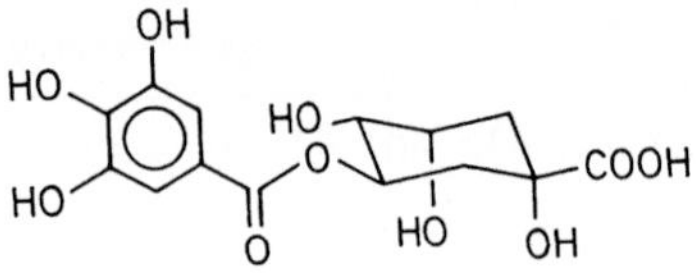   XI 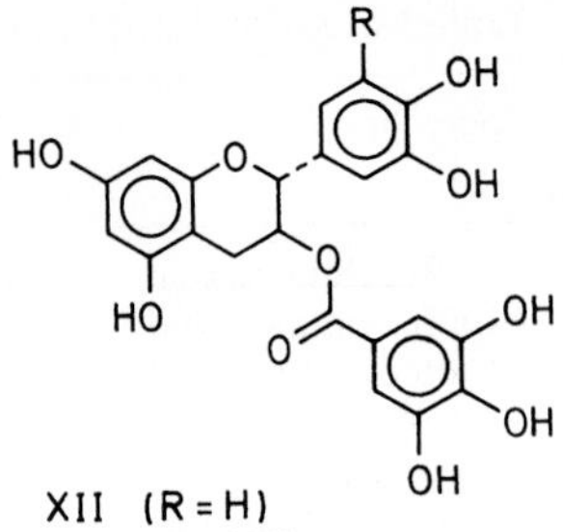   XII (R = H)   XIII (R = OH)   (18.13)

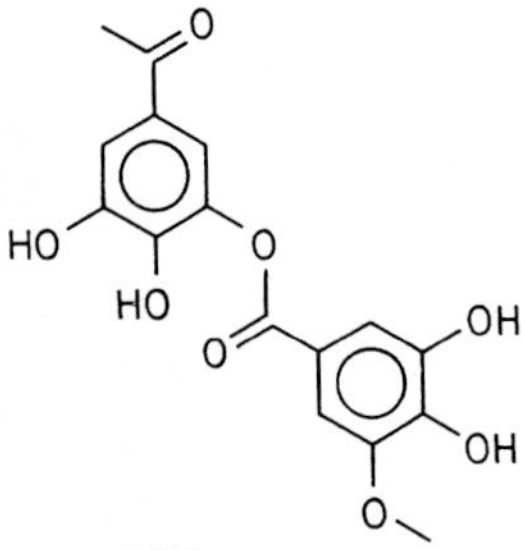

XIV 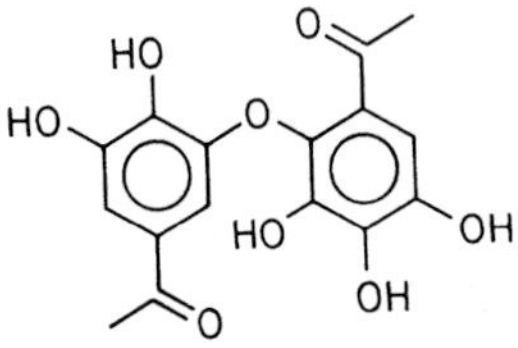   XV 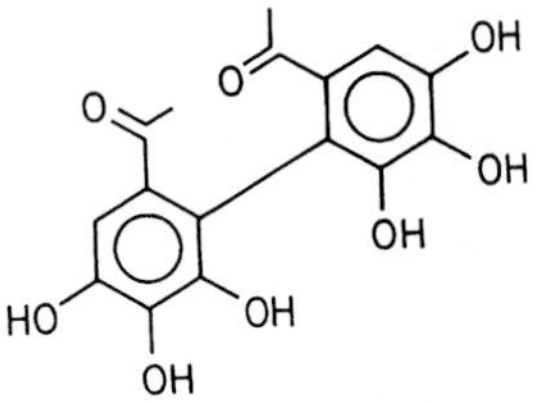   XVI

Apart from gallic acid, most of the tanning agents of this type contain as acyl residues intermolecular gallic acid esters (depsides XIV), their ethers (depsidones, XV), and

hexahydroxydiphenic acid (XVI) formed by oxidative coupling of two gallic acids. Some of the polyphenols derived from β-pentagalloyl-D-glucose are shown in Formula 18.14. They have been found in various *Rosaceae*, e.g., raspberries and blackberries, and contain the structural elements mentioned above.

The changes in the concentrations of phenolic acids in apples and strawberries during ripening and storage are shown in Table 18.20.

Hydroxycinnamic acid biosynthesis starts with phenylalanine [cf. Reaction route 18.15: a) phenylalanine-ammonia lyase; b) cinnamic acid 4-hydroxylase; c) phenolases; d) methyl transferases, R: OH and OCH_3 in various positions].

Table 18.18. Hydroxycoumarins in fruit (VIII)[a]

Compound	Substitution pattern		
	R^1	R^2	R^3
Coumarin	H	H	H
Umbelliferone	H	OH	H
Herniarin	H	OCH_3	H
Aesculetin	OH	OH	H
Scopoletin	OCH_3	OH	H
Fraxetin	OCH_3	OH	OH

[a] See Formula 18.11.

Table 18.19. Occurrence of hydroxybenzoic acids in fruit

Compound	Strawberry	Grapes	Orange	Grapefruit	Lemon
Salicylic acid	+	+	+	+	+
p-Hydroxybenzoic acid	+	+			
Gentisic acid	+		+	+	+
Gallic acid	+				
Ellagic acid	+				
Protocatechuic acid	+				
Vanillic acid	+				

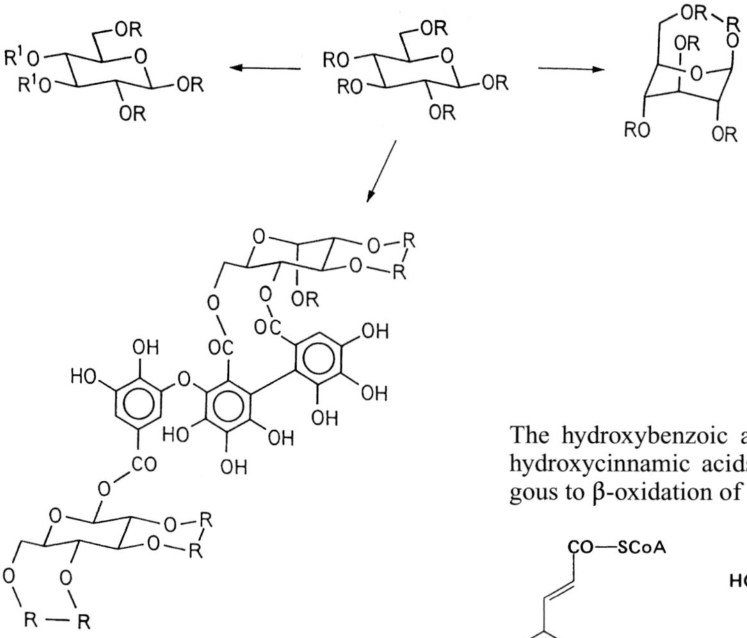

(18.14)

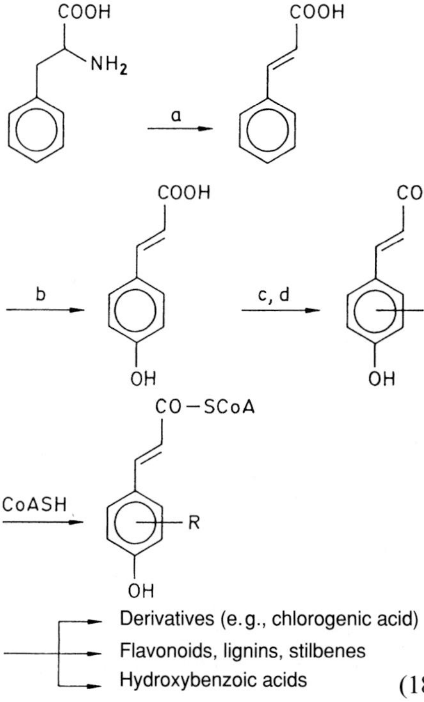

(18.15)

The hydroxybenzoic acids are derived from hydroxycinnamic acids by a pathway analogous to β-oxidation of fatty acids:

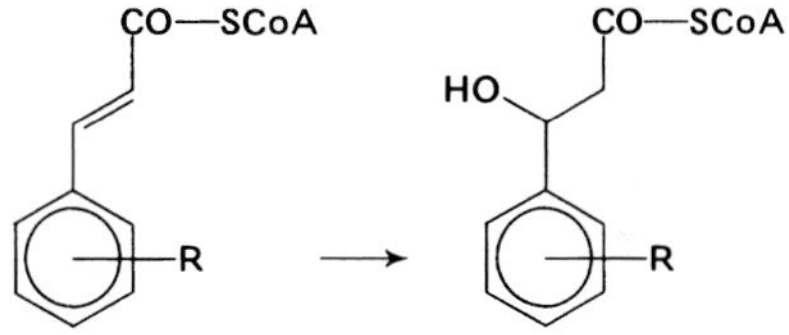

(18.16)

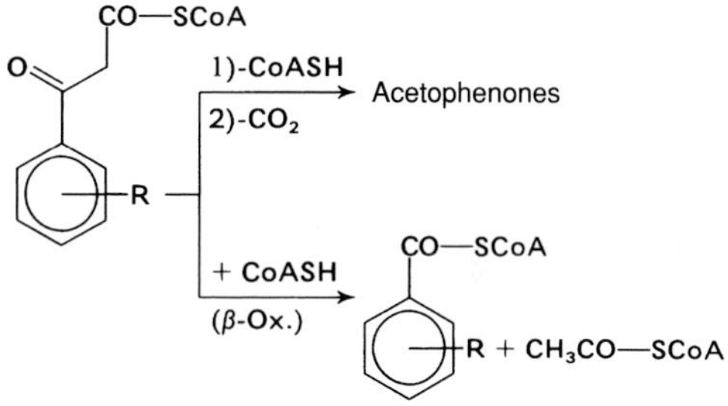

Reduction of the benzoic acid carboxyl groups yields the corresponding aldehydes and alcohols, as for instance vanillin and vanillyl alcohol (XVII and XVIII respectively in Formula 18.17) from 3-methoxy-4-hydroxybenzoic acid and coniferyl alcohol from ferulic acid.

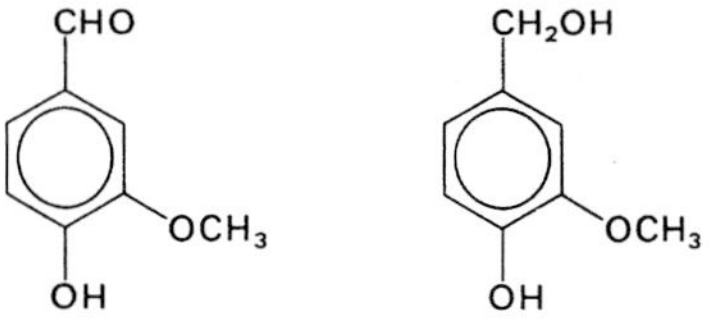

XVII XVIII

(18.17)

Table 18.20. Phenolic acids in apple and strawberry during ripening and storage

	p-Coumaric acid	Caffeic acid	Ferulic acid	Gallic acid	p-Hydroxy-benzoic acid	Vanillic acid
Ripening of apples[a]						
June 01		640				
June 07	250	1,000	60			
June 12	460	1,270	95			
June 23	425	1,010	50			
July 03	250	665	29			
July 14	147	470	22			
August 04	51	214	12			
September 07	28	137	6.4			
October 05	15	85	4.0			
Ripening of strawberries[a]						
June 05	69	15		80		3
June 20	110	34		110	19	23
June 27	119	30		111	87	25
July 01 (ripe)	175	39		121	108	34
Apple storage at 4 °C, picking date: October 19[b]						
October 28	4.4	42	1.0			
December 16	3.0	30.1	0.6			
January 20	2.3	28.5	0.4			
February 01	2.2	19.7	0.4			
February 15	2.2	19.7	0.4			
March 01	1.9	10.4	0.4			

[a] Picking date. [b] Analysis date.

The glucosides of cis-o-coumaric acid are the precursors of coumarins. Disintegration of plant tissue releases the free acids from the glucosides. The acids then close spontaneously to the ring forms (R: OH and OCH₃ in various positions):

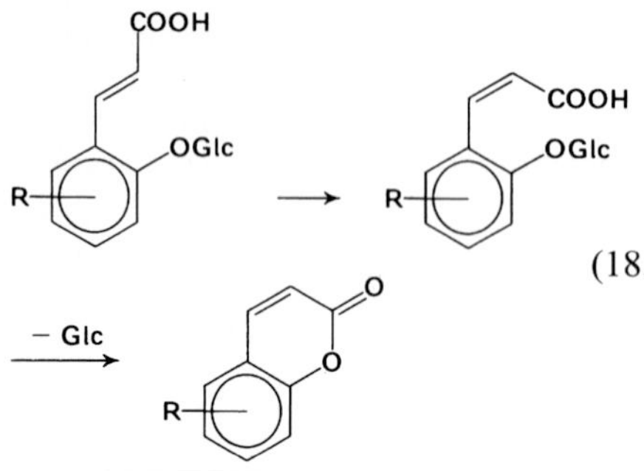

(18.18)

Lignin is formed by the dehydrogenative polymerization of coniferyl, sinapyl, and p-coumaryl alcohol, which is catalyzed by a peroxidase and requires H_2O_2. A section of the structure of lignin formed by the polymerization of coniferyl alcohol is shown in Fig. 18.3. Lignins strengthen the walls of plant cells. They play a role as fiber in foods (cf. 15.2.4.2).

18.1.2.5.2 Flavan-3-ols (Catechins), Flavan-3,4-diols, and Proanthocyanidins (Condensed Tanning Agents)

These colorless compounds (Table 18.26) occur in all commonly grown fruits [R, R¹ = H: a) catechin, b) epicatechin; R = H, R¹ = OH: gallocatechin, epigallocatechin; R = OH: flavan-3,4-diols]:

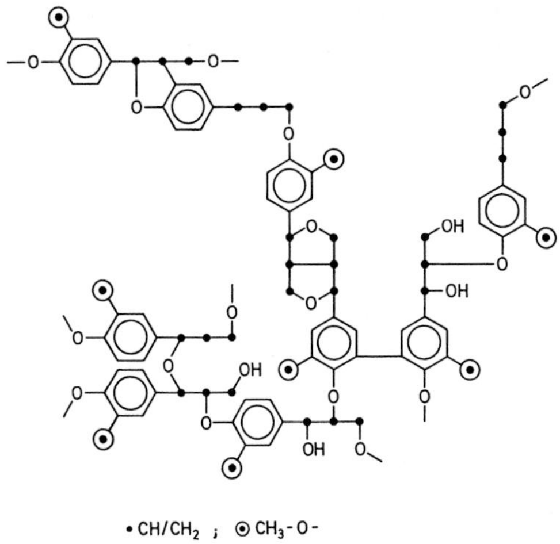

• CH/CH$_2$; ⊙ CH$_3$-O-

Fig. 18.3. Section of the structure of a lignin (according to *H. Kindl* and *G. Wöber*, 1975)

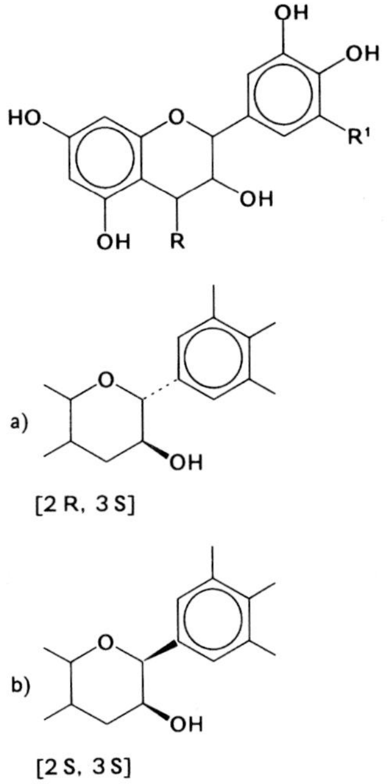

(18.19)

a)

[2 R, 3 S]

b)

[2 S, 3 S]

In flavonoid biosynthesis, flavan-3,4-diols can be converted to flavan-3-ols (cf. 18.1.2.5.6). The intermediate is assumed to be a carbocation which is reduced to flavan-3-ol (Formula 18.20). When the reducing agent, e.g., NADPH, is limited, the cation can react with flavan-3-ol to give dimers and higher oligomers, which are called proanthocyanidins. As

"condensed tanning agents", they contribute to the astringent taste of fruits. The spectrum of oligomers depends on the ratio of the rate of formation to the rate of reduction of the cation. The proanthocyanidins are soluble up to $M_r \sim 7000$, corresponding to ca. 20 flavanol units. Plant tissues also contain insoluble polymeric forms which frequently even predominate and can be covalently bound to the polysaccharide matrix of the cell. Procyanidins (Formula 18.20, R = H) are the most common group of proanthocyanidins; prodelphinidins (R = OH) also occur.

The name proanthocyanidins, previously called leucoanthocyanidins, implies that these are colorless precursors of anthocyanidins. On heating in acidic solution, the C−C bond made during formation is cleaved and terminal flavan units are released from the oligomers as carbocations, which are then oxidized to colored anthocyanidins (cf. 18.1.2.5.3) by atmospheric oxygen (Formula 18.20). Base-catalyzed cleavage via the quinone methide is also possible.

18.1.2.5.3 Anthocyanidins

These red, blue or violet colored benzopyrylium and flavylium salts (Formula 18.21) occur in the form of glycosides, the authocyanins, in most commonly grown fruit varieties and also in some citrus and tropical fruits (Table 18.21).

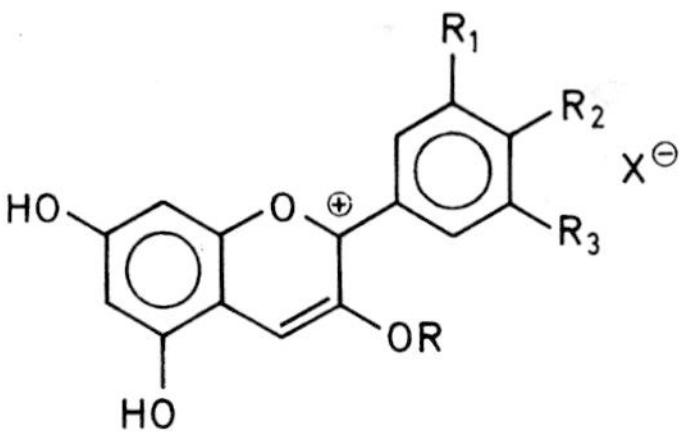

(18.21)

The cation is to be regarded as a resonance hybrid of the following oxonium and carbenium forms:

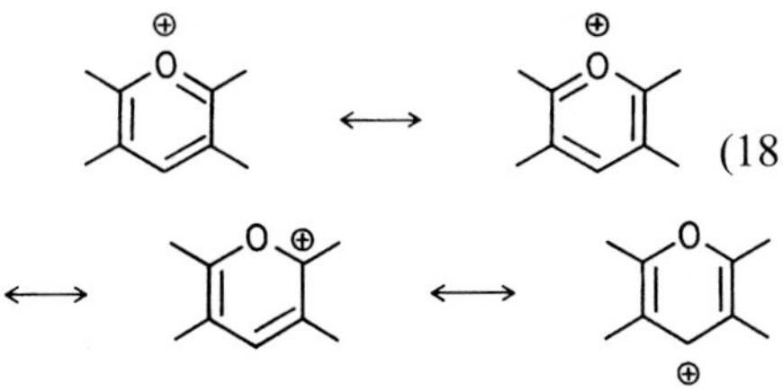

(18.22)

Table 18.22 provides data on the structure and absorption maxima of the most important anthocyanidins. Increasing hydroxylation results in a shift towards blue color (pelargonidin → cyanidin → delphinidin), whereas glycoside formation and methylation results in a shift towards red color (pelargonidin → pelargonidin-3-glucoside; cyanidin → peonidin).

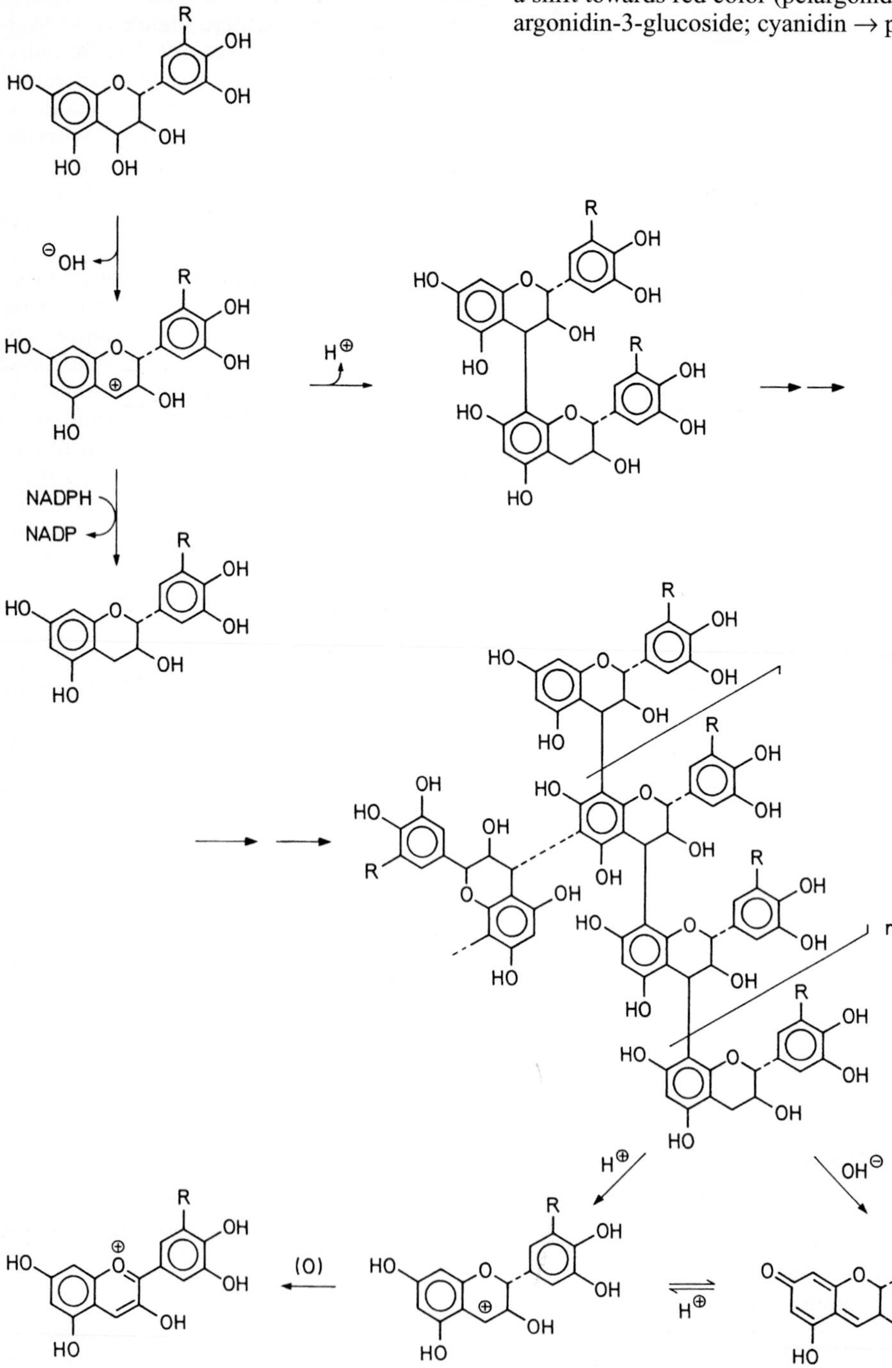

(18.20)

The color of an anthocyanin changes with the pH of the medium (R = sugar moiety; cf. Formula 18.23).

The flavylium cation (I) is stable only at very low pH. As the pH increases it is transformed into colorless chromenol (II). Figure 18.4 shows the decrease in absorption in the visible spectrum at various pH's, reflecting these transformations. Formation of a quinoidal (III) and ionic anhydro base (IV) at pH 6−8 inten-

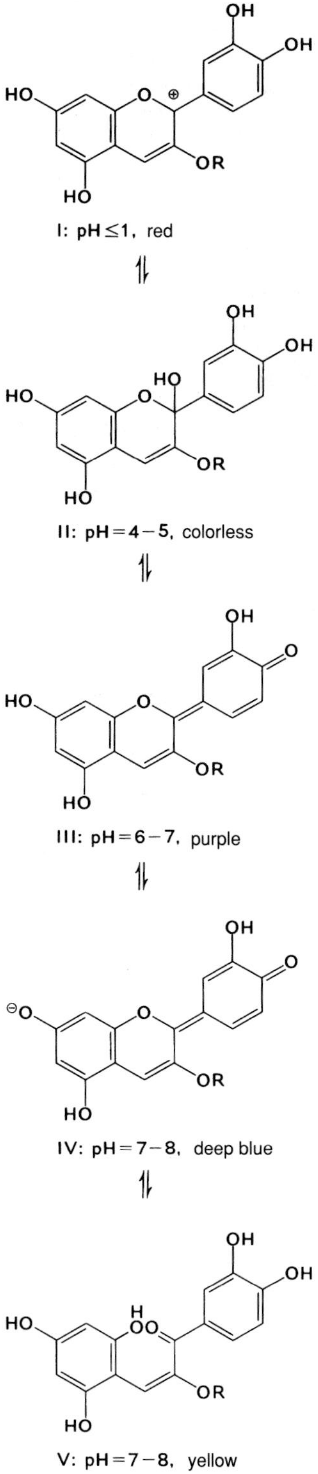

I: pH ≤1, red

II: pH = 4−5, colorless

III: pH = 6−7, purple

IV: pH = 7−8, deep blue

V: pH = 7−8, yellow   (18.23)

Table 18.21. Anthocyanins in various fruits (the major constituents in italics)

Fruit	Anthocyanin[a]
Apple	*Cy-3-gal*, Cy-3-ara, Cy-7-ara
Pears	Cy-3-gal
Peach	Cy-3-glc
Plums/prunes	*Cy-3-glc*, Cy-3-rut, Peo-3-glc, Peo-3-rut
Sour cherry	*Cy-3-sop*, Cy-3-rut, Cy-3-glc-rut, Cy-3-glc
Sweet cherry	Cy-3-glc, Cy-3-rut
Blackberry	*Cy-3-glc*, Cy-3-rut
Strawberry	*Pg-3-glc*, Pg-3-gal, Cy-3-glc
Bilberry	Pet-3-gly, Cy-3-gly, Pet-gly, Del-3-glc, Del-3-gal, Mv-3-glc
Raspberry	*Cy-3-glc*, Cy-3-glc-rut, Cy-3-rut, Cy-3-sop, Cy-3-glc-sop
Currant, red	Del-3-glc, Cy-3-rut, Cy-3-xyl-rut, Cy-3-glc-rut, Cy-3-sop, Cy-3-sam
Currant, black	Cy-3-glc, Cy-3-rut, Del-3-glc
Grapes (*Vitis vinifera*, *V. labrusca*, *V. riparia*, including hybrids)	Cy-, Del-, Peo-, Pet-, Mv-3-glc, Mv-3,5-diglc, Mv-3-p-cumaroylglc-5-glc, Mv-3-p-caffeoylglc-5-glc, Peo-3-p-cumaroylglc-5-glc
Orange	Cy-3-glc, Del-3-glc
Banana	Pet-3-gly
Fig	Cy-3-gly
Passion fruit	Del-3-glc, Del-3-glc-glc

[a] Cy: Cyanidin, Del: delphinidin, Mv: malvidin, Peo: peonidin, Pet: petunidin, Pg: pelargonidin, ara: arabinoside, gal: galactoside, glc: glucoside, gly: glycoside, rut: rutinoside, sam: sambubioside, sop: sophoroside, xyl: xyloside, glc-rut: glucosyl-rutinoside etc. (Sophorose: β-D-Glc$p$(1 → 2)-D-Glc$p$, sambubiose: β-D-Xyl$p$(1 → 2)-D-Glc$p$).

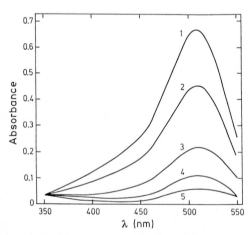

$$(18.24)$$

sifies the color. At pH 7–8 structure IV is transformed through ring opening to yellow chalcone (V). At higher pH's the color can be stabilized by the presence of multivalent metal ions (Al^{3+}, Fe^{3+}). The complexes formed are deep blue (cf. Formula 18.24).

Table 18.22. Anthocyanidins: absorption maxima in the visible spectrum

Compound	R^1	R^2	R^3	λ_{max} (nm)[a]	
				R = H	R = Glc[b]
Pelargonidin	H	OH	H	520	506
Cyanidin	OH	OH	H	535[c]	525[c]
Peonidin	OCH$_3$	OH	H	532	523
Delphinidin	OH	OH	OH	544[c]	535[c]
Petunidin	OCH$_3$	OH	OH	543[c]	535[c]
Malvidin	OCH$_3$	OH	OCH$_3$	542	535

[a] In methanol with 0.01% HCl.
[b] 3-Glucoside.
[c] $AlCl_3$ shifts the absorption towards the blue region of the spectrum by 14 to 23 nm.

Fig. 18.4. Absorption spectra of cyanidin-3-rhamnoglucoside (16 mg/l) in aqueous buffered solution at pH 0.71 (1), pH 2.53 (2), pH 3.31 (3), pH 3.70 (4), and pH 4.02 (5). (According to *Jurd*, 1964)

Figure 18.5 illustrates the shift in absorption maximum from 510 to 558 nm for cyanidin-3-glucoside over the pH range of 1.9–5.4. Readings were taken in the presence of aluminium chloride.

At higher pH's free anthocyanidins (VII) are degraded via chromenols (VIII) and α-diketones (IX) to aldehydes (X) and carboxylic acids (XI):

$$(18.25)$$

Addition of SO_2 bleaches anthocyanins. The flavylium cation reacts to form a carbinol base corresponding to compounds XII or XIII (Formula 18.26). The color is restored by acidification to pH 1 or by addition of a carbonyl compound (e. g. ethanal). Since compounds of type XIV ($R^1 = CH_3$, C_2H_5) are not affected by SO_2, it appears that compound XIII is involved in such bleaching reactions.

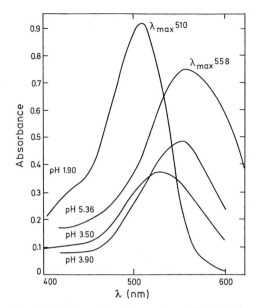

Fig. 18.5. Absorption spectra of cyanidin-3-glucoside (35 µmole/l + 830 µmole/l $AlCl_3$) in aqueous buffered solutions at pH 1.90, pH 3.50, pH 3.90, and pH 5.36. (According to *Jurd* and *Asen*, 1966)

XII

XIII

XIV

(18.26)

18.1.2.5.4 Flavanones

Flavanones ($R^1 = H$, $R^2 = OCH_3$: isosacuranetin; $R^1 = H$, $R^2 = OH$: naringenin; $R^1 = OH$, $R^2 = OCH_3$: hesperitin; R^1, $R^2 = OH$: eriodictyol) occur mostly as glycosides in citrus fruits (Table 18.23):

(18.27)

Table 18.24 shows that flavanone-7-rutinosides are usually nonbitter, whereas flavanone-7-neohesperidosides are generally bitter. The

Table 18.23. Flavanones, flavones and flavonols in citrus fruit

Fruit	Compound
Orange	
flesh	Hes-7-rut
peel	Hes-7-rut, Nob, Nav-7-neo, Isa-7-rha-glc
Bitter orange	Hes-7-neo
Grapefruit	Nar-7-neo, Isa-7-neo, Api-7-rut
Lemon	
peel	Hes-7-rut, Dio-7-rut, Lin, Eri-7-rut, Lol, Api, Lut, Chr, Que, Irh

Api: Apigenin, Chr: chrysoeriol, Dio: diosmetin, Eri: eriodictyol, Hes: hesperitin, Irh: isorhamnetin, Isa: isosacuranetin, Lin: limocitrin, Lol: limocitrol, Nar: naringenin, Nob: nobiletin, Que: quercetin, glc: glucoside, neo: neohesperidoside, rha: rhamnoside, rut: rutinoside.

intensity of the bitter taste is influenced by the substitution pattern. Compounds with $R^1 = H$, $R^2 = OH$, OCH_3 (e.g., naringin, poncirin) are an order of magnitude more bitter than those with $R^1 = OH$, $R^2 = OH$, OCH_3 (e.g., neohesperidin,

Table 18.24. Taste of flavanone glycosides[a]

Compound	R	R^1	R^2	Taste	
				quality	inten-sity[b]
Naringenin-rutinoside	rut[c]	H	OH	neutral	–
Naringin	neo[d]	H	OH	bitter	20
Isosacura-netin-ruti-noside	rut	H	OCH$_3$	neutral	–
Poncirin	neo	H	OCH$_3$	bitter	20
Hesperidin	rut	OH	OCH$_3$	neutral	–
Neohesperi-din	neo	OH	OCH$_3$	bitter	2
Eriocitrin	rut	OH	OH	neutral	–
Neoerio-citrin	neo	OH	OH	bitter	2

[a] Data for R, R^1 and R^2 refer to Formula 18.27.
[b] Relative bitterness refers to quinine hydrochloride = 100.
[c] Rutinosyl.
[d] Neohesperidosyl.

neoeriocitrin). Naringenin-7-neohesperido-side (naringin) is the bitter constituent of grapefruit. Hesperetin-7-neohesperidoside (neohesperidin) is the bitter compound of bitter oranges (*Citrus auranticus*). The non-bitter isomer, hesperitin-7-rutinoside (hesper-idin) occurs in oranges (*Citrus sinensis*). Removal of the bitter taste of citrus juices and citrus fruit pulps is possible by enzymatic cleavage of the sugar moiety using a mixture of α-rhamnosidase and β-glucosidase. These enzymes are isolated from microorganisms such as *Phomopsis citri*, *Cochliobolus miyabe-anus* or *Rhizoctonia solanii*:

Naringin → Naringenin + Rhamnose

+ Glucose

A number of neutral or bitter flavanone glyco-sides can be converted through ring opening to sweet chalcones (II) which, by additional hydrogenation, can be stabilized as sweet dihydrochalcones (III):

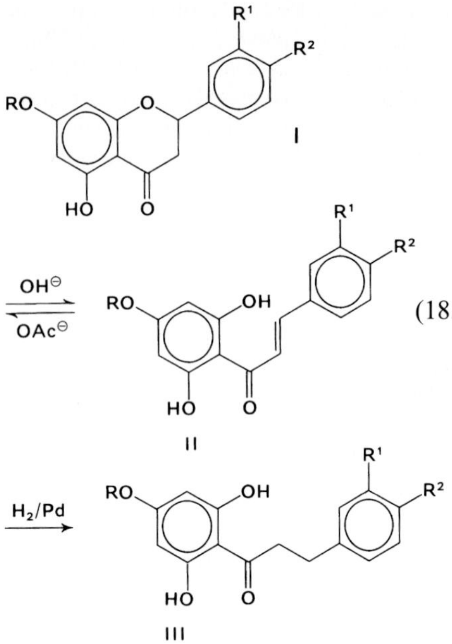

(18.28)

The presence of a free OH-group in position R^1 or R^2 is necessary for a sweet taste. Table 18.25 shows that the dihydrochalcone of naringin corresponds to saccharin in sweet-ness intensity, whereas the dihydrochalcone of neohesperidin is sweeter than saccharin by a factor of 20.

Table 18.25. Taste of dihydrochalcones

Dihydrochalcone from	Taste		
	quality	intensity[a] (µmole/l)	Relative intensity[b]
Naringin	sweet	200	1
Neohesperidin	sweet	10	20
Neoeriocitrin	slightly sweet	–	–
Poncirin	slightly bitter	–	–
Saccharin (Sodium salt)	sweet	200	1

[a] Concentration of iso-sweet solutions.
[b] Related to saccharin.

Conversion of naringin to highly sweet neohesperidin dihydrochalcone (VII) is possible by alkali fragmentation to a methylketone (IV), condensation with isovanillin (V) to the corresponding chalcone (VI), then hydrogenation:

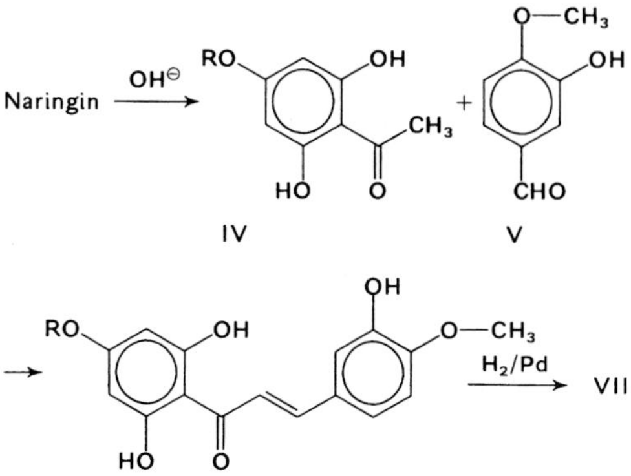

$$(18.29)$$

A sweet compound can be obtained from the neutral-tasting hesperidin of oranges by first converting hesperidin to another neutral-tasting compound, hesperidin dihydrochalcone. The latter can then be hydrolyzed, by acidic or enzymatic catalysis, to remove the rhamnose residue, yielding hesperidin dihydrochalcone glucoside, which is sweet. The use of dihydrochalcones as sweeteners is discussed in Section 8.8.13.

The dihydrochalcone glycoside phloridzin (Formula 18.30) is a suitable indicator of the addition of apples to pear juice (cf. Table 18.37).

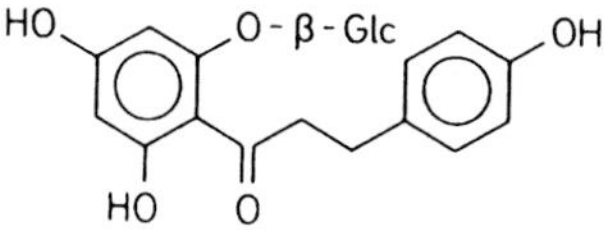

$$(18.30)$$

18.1.2.5.5 Flavones, Flavonols

Flavones (Formula 18.31: I, R = H; R^1, R^2 = H, R^3 = OH: apigenin; R^1, R^2 = OH, R^3 = H: luteolin; R^1 = OH, R^2 = H, R^3 = OCH$_3$: diosmetin; R^1 = OCH$_3$, R^2 = H, R^3 = OH: chrysoeriol; II: nobiletin) and flavonols (Formula 18.31: I, R, R^3 = OH; R^1, R^2 = H: kaempferol; R^1 = OH, R^2 = H: quercetin; R^1, R^2 = OH: myricetin; R^1 = OCH$_3$, R^2 = H: isorhamnetin) occur in all common fruits and citrus and tropical fruits as the 3-glycosides and, less frequently, as the 7-glycosides (Tables 18.23 und 18.26):

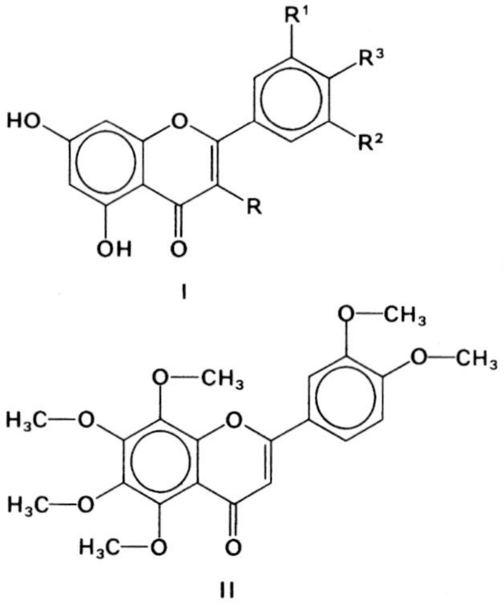

$$(18.31)$$

They are faintly yellow compounds.

Table 18.26. Occurrence of catechins and flavonols in fruit (major constituents in italics)

Fruit	Catechins	Flavonols
Apple	Cat, *Epi*, Gal	Que-3-gal, Que-3-glc, Que-3-rha, Que-3-rha-glc, Que-3-ara, Que-3-xyl
Pear	Cat, *Epi*	Que-3-glc
Peach	Cat, *Epi*	Que-3-glc
Apricot	Cat	Que-3-glc, Kaem-gly
Plum/prune	Cat, *Epi*, Gal	Que-3-glc, Que-3-rha, Que-3-ara
Sour cherry	Cat, *Epi*	
Sweet cherry	Cat, *Epi*	Que-3-glc
Blackberry	Cat, *Epi*	Que-3-glc
Strawberry	*Cat*, Gal	
Currant, black	Cat, Gal	Que-3-glc, Kaem-3-glc, Myr-3-glc, further Que-glc, Kaem-glc
Raspberry	Cat, *Epi*	
Grapes	*Cat*, Epi, Gal	Que-3-rha, Que-3-glc, Que-3-rha-glc

Cat: Catechin, Epi: epicatechin, Gal: gallocatechin. Kaem: kaempferol, Myr: myricetin, Que: quercetin. ara: arabinoside, gal: galactoside, glc: glucoside, gly: glycoside, rha: rhamnoside, and xyl: xyloside.

(18.32)

18.1.2.5.6 Flavonoid Biosynthesis

Flavonoid biosynthesis (cf. Formula 18.32) occurs through the stepwise condensation of activated hydroxycinnamic acid (I) with three activated malonic acid molecules (II). The primary condensation product, a chalcone (III), is in equilibrium with a flavanone (IV) with the equilibrium shifted toward product IV. Recent research suggests that condensation directly yields a flavanone, hence chalcone is not an obligatory intermediary product.

A 2,7-cyclization yields stilbenes (IIIa).

One pathway converts flavanones (IV) to flavones (V) and, through another pathway, flavanones are converted to flavanonols (VI). The latter compounds are converted to flavandiols (VII), flavanols (VIII) and flavonols (IX), as well as anthocyanidins (XII) via endiols (X) and enols (XI).

18.1.2.5.7 Technological Importance of Phenolic Compounds

The taste of fruits is influenced by phenolic compounds. The presence of tannins yields an astringent, harsh taste, similar to an unripe apple (or an apple variety suitable only for processing). Table quality apples are low in phenolic compounds. Flavanones (naringin, neohesperidin) are the bitter compounds of citrus fruits.

Phenolic compounds are substrates for polyphenol oxidases. These enzymes hydroxylate monophenols to o-diphenols and also oxidize o-diphenols to o-quinones (cf. 2.3.3.2).

o-Quinones can enter into a number of other reactions, thus giving the undesired brown discoloration of fruits and fruit products. Protective measures against discoloration include inactivation of enzymes by heat treatment, use of reductive agents such as SO_2 or ascorbic acid, or removal of available oxygen.

Polyvalent phenols form colored complexes with metal ions. For example, at $pH > 4$, Fe^{3+} forms complexes which are bluish-gray or bluish-black in color. Al^{3+} and Sn^{2+} also form intensely colored complexes. Leucoanthocyanins, when heated in the presence of an acid, are converted into anthocyanins. The red color of apples and pears, which is formed during cooking, is derived from leucoanthocyanins.

Phenolic compounds can also form complexes with proteins. These complexes increase the turbidity of fruit juices.

18.1.2.6 Aroma Compounds

Aroma compounds contribute significantly to the importance of fruits in human nutrition. The aroma substances of selected fruits will be outlined below in more detail. The structures and synthesis pathways of common aroma substances are explained in Chapter 5.

The aroma of fruits can change on heating due to the liberation of aroma substances from glycosidic precursors (cf. 5.3.2.4), oxidation, water addition, and cyclization of individual compounds (cf. 5.5.4).

18.1.2.6.1 Bananas

The characteristic aroma compound of bananas is isopentyl acetate. Some esters of pentanol, such as those of acetic, propionic and butyric acids, also contribute to the typical aroma of bananas, while esters of butanol and hexanol with acetic and butyric acids are generally fruity in character. An important contribution to the complete, mild banana aroma is provided by eugenol (I), O-methyleugenol (II) and elemicin (III):

(18.33)

I: R^1, $R^2 = H$
II: $R^1 = H$, $R^2 = CH_3$
III: $R^1 = OCH_3$, $R^2 = CH_3$

18.1.2.6.2 Grapes

Methyl anthranilate is the character impact compound of grapes. Other aroma compounds are responsible for the specific aromas of some grape varieties.

18.1.2.6.3 Citrus Fruits

Terpenes are of great importance in citrus fruits. Limonene (IX in Table 5.29) at 80–90%, is the major component of citrus oils. (+)-Limonene, with its low odor threshold of 10 ppb, contributes to the aroma of citrus fruits.

The typical aroma note of grapefruit is due to 1-p-menthene-8-thiol (IV) and to nootkatone (V), which is about 5% of the grapefruit oil. Compound IV is an exceptionally powerful aroma constituent (cf. Table 5.31), probably obtained by the addition of hydrogen sulfide to limonene. Traces of hydrogen sulfide and dimethyl sulfide are present in all citrus fruit juices and also contribute to their aromas.

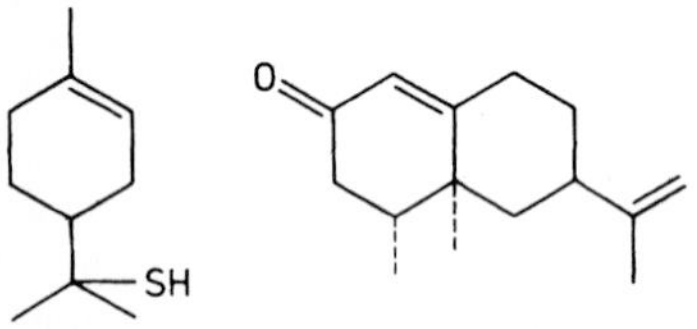

(18.34)

Citral, which is actually a mixture of two stereoisomers, geranial and neral, is the character impact compound of lemon oil (cf. 5.5.1.5):

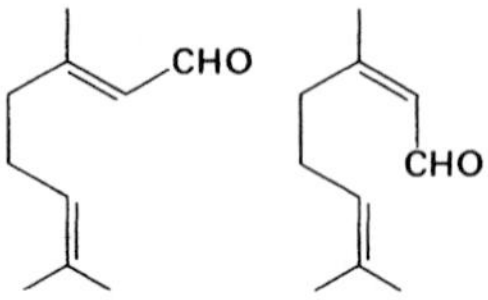

(18.35)

Linalool, myrcene, and limonene also have high aroma values.

The aroma of mandarin oranges is due to: thymol (0.18 weight % of the essential oil), N-methylanthranilic acid methyl ester (0.65), γ-terpinene (14.0) and α-pinene (1.8). Orange aroma is more complicated, but typical odorants are decanal, (Z)-3-hexenal, (R)-limonene, (R)-α-pinene, myrcene, ethyl isobutyrate, ethyl butanoate and ethyl (S)-2-methylbutanoate. Aroma defects, that can appear on storage of citrus juices, are caused by the formation of α-terpineol (from limonene), 4-vinylguaiacol, and furaneol. The taste of orange juice is even

influenced by a mixture of these three compounds in concentrations that correspond to half of the retronasal odor threshold values. The retronasal threshold values (mg/l, orange juice) are α-terpineol (2), 4-vinylguaiacol (0.05), and furaneol (0.05).

18.1.2.6.4 Apples, Pears

The aroma of apples is determined by 2-trans-hexenal, 3-cis-hexenal, the corresponding alcohols, β-damascenone, ethyl butyrate, and 2-methylbutyric acid hexyl ester. The concentration of the last mentioned ester directly correlates with the aroma quality of the variety Golden Delicious. The aniseed-like aroma note of the variety Cox Orange is produced by 1-methoxy-4-(2-propenyl)benzene (estragol). The aroma of the pear Williams Christ is characterized by esters produced by the degradation of unsaturated fatty acids (example in 5.3.2.2): ethyl esters of (E,Z)-2,4-decadienoic acid, (E)-2-octenoic acid, and (Z)-4-decenoic acid, as well as hexyl acetate. In fact, butyl acetate and ethyl butyrate are also involved in the fruity odor note.

18.1.2.6.5 Raspberries

The character impact compound is the "raspberry ketone", i.e. 1-(p-hydroxyphenyl)-3-butanone (VIII). Its concentration is 2 mg/kg and its odor threshold is 5 μg/kg (water). Additional aroma notes are provided by cis-3-hexenol, α- and β-ionone:

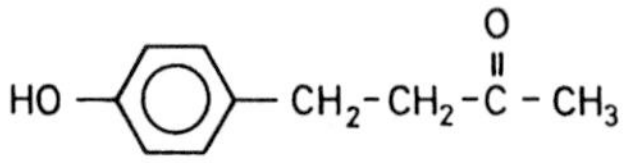

(18.36)

In addition, the ethyl esters of 5-hydroxyoctanoic acid and 5-hydroxydecanoic acid contribute to the aroma. A part of the esters hydrolyzes during cooking and the hydroxy acids released cyclize to the corresponding lactones.

18.1.2.6.6 Apricots

The following compounds are discussed as contributors to the aroma: myrcene, limonene, p-cymene, terpinolene, α-terpineol, geranial, geraniol, linalool, acids (acetic and 2-methylbutyric acids) alcohols, e.g., trans-2-hexenol; and a number of γ- and δ-lactones, e.g., γ-caprolactone, γ-octalactone, γ-decalactone, γ-dodecalactone, δ-octalactone and δ-decalactone.

18.1.2.6.7 Peaches

The aroma of peaches is characterized by γ-lactones (C_6–C_{12}) and δ-lactones (C_{10} and C_{12}). The main compound in the lactone fraction is (R)-1,4-decanolide, which has a creamy, fruity, peach-like odor. Other important compounds are benzaldehyde, benzyl alcohol, ethyl cinnamate, isopentyl acetate, linalool, α-terpineol, α- and β-ionone, 6-pentyl-α-pyrone (Formula 18.37, IX), hexanal, 3-cis-hexenal, and 2-trans-hexenal. Aroma differences in different varieties of peaches are correlated with the different proportions of the esters and monoterpenes.

In the case of nectarines (*Prunus persica* L., Batsch var. nucipersica Schneid), the lactones γ-C_8–C_{12} and δ-C_{10} belong to the compounds with the highest aroma values.

18.1.2.6.8 Passion Fruit

The aroma of the yellow fruit (*Passiflora edulis* var. *flavica*) is supposed to be superior to that of the crimson fruit (*Passiflora edulis* var. *edulis*). Compounds that contribute to the aroma of both varieties are β-ionone and the following esters (% of the volatile fraction): ethyl butyrate (1.4), ethyl hexanoate (9.7), hexyl butyrate (13.9), and hexyl hexanoate (69.6).

Four stereoisomeric megastigmatrienes have been found in crimson passion fruit. A mixture of the isomers Xa and Xb (Formula 18.37) gives a rose-like aroma, that has a hint of strawberries in it (threshold = 100 µg/kg; water).

The following S-containing aroma substances have been isolated from the yellow fruit: 3-methylthiohexane-1-ol which probably gives rise to 2-methyl-4-propyl-1,3-oxathianes (cis/trans isomers in the ratio of 10:1) (XI a, b: Formula 18.37). Of the two cis isomers, only the 2R,4S-isomer (XIb), which has a sulfurous herb-like odor (threshold = 4 µg/kg; water), has been found in the fruit. However, the aroma note more typical of passion fruit is exhibited by the 2S,4R-isomer (XIa).

IX

Xa   Xb   (18.37)

XIa   XIb

18.1.2.6.9 Strawberries

A sensory study which was performed on the basis of analytical data has revealed that 4-hydroxy-2,5-dimethyl-3(2H)-furanone (HDF), (Z)-3-hexenal, methyl butanoate, ethyl 2-methylbutanoate, methyl 2-methylbutanoate, acetic acid and 2,3-butanedione are the character impact odorants of strawberry juice.

The concentration of HDF can vary greatly depending on the variety, degree of ripeness, and processing (Table 18.27).

A heat treatment leads to a change in the overall flavor of strawberries, since HDF, (E)-β-damascenone, (E,E)-2,4-decadienal and guaiacol increase strongly whereas (Z)-3-hexenal and the esters decrease. The flavor changes caused by freezing and thawing were mainly due to a 6-fold increase of HDF and a decrease in the (Z)-3-hexenal concentration from 590 µg/kg to 10 µg/kg.

Table 18.27. 4-Hydroxy-2,5-dimethyl-3(2H)-furanone (HDF) in strawberries and pineapples

Fruit/Product	HDF [a] (mg/kg)
Strawberry	1–34
Strawberry juice[b]	7.1
Strawberry jam (50% fruit content)	7.1
Pineapples	2–25
Pineapple juice (50% fruit content)	1.6
Pineapple jam (45% fruit content)	1.9

[a] Structural formula in Fig. 5.12.
[b] Commercial product.

18.1.2.6.10 Pineapples

The aroma of pineapples is marked by HDF (Table 18.27) and (S)-2-methylbutyric acid ethyl ester, which has a very low odor theshold (6 ng/l, water). In addition, 2-methylbutyric acid methyl ester, ethyl acetate, ethyl hexanoate, ethyl butanoate, 2-methylpropionic acid ethyl ester, and the methyl esters of caproic acid, butyric acid, and heptanoic acid are involved with high aroma values. Contibutions to the aroma are assumed in the case of 1,3(E),5(Z)-undecatriene and 1,3(E),5(Z),8(E)-undecatetraene (cf. 5.3.2.2) because of their very low odor thresholds. However, these hydrocarbons are very quickly enzymatically degraded in chopped fruits.

18.1.2.6.11 Cherries, Plums

The compounds essentially involved in the aroma of cherries are benzaldehyde, linalool, hexanal, 2-trans-hexenal, phenylacetaldehyde, 2-trans,6-cis-nonadienal, and eugenol (Table 18.28).

On heating cherry juice or in the making of jams, the concentration of benzaldehyde increases due to the hydrolysis of amygdalin and prunasin (cf. 16.2.6) and the concentration of linalool increases due to the hydrolysis of the corresponding glycoside (Table 18.28). Since the C_6-aldehydes and nonadienal decrease simultaneously, the fruity-flowery aroma notes are enhanced and the "green" notes diminished.

The important compounds in plums are linalool, benzaldehyde, methyl cinnamate, and γ-decalactone together with the C_6-aldehydes. Benzaldehyde, nonanal, and benzyl acetate contribute to the aroma of canned plums.

Table 18.28. Aroma substances of cherry juice and jams made from it [a]

Aroma substance	Juice (µg/kg)	Jam (µg/kg)
Benzaldehyde	202	1510
Linalool	1.1	13.1
Hexanal	5.6	0.2
(E)-2-Hexenal	8.5	3.8
Eugenol	10.0	4.9

[a] Fruit content: 50 w/w per cent.

18.1.2.7 Vitamins

Many fruits are important sources of vitamin C (Table 18.29). Its biosynthesis in plants starts from hexoses, e. g., glucose, but the biosynthetic pathway has not been fully elucidated. It is postulated that following C-1 oxidation and cyclization to 1,4-lactone (II), the 5-keto compound (III) appears as an intermediary product which is oxidized to the 2,3-endiol (IV) then reduced stereospecifically to L-ascorbic acid (V):

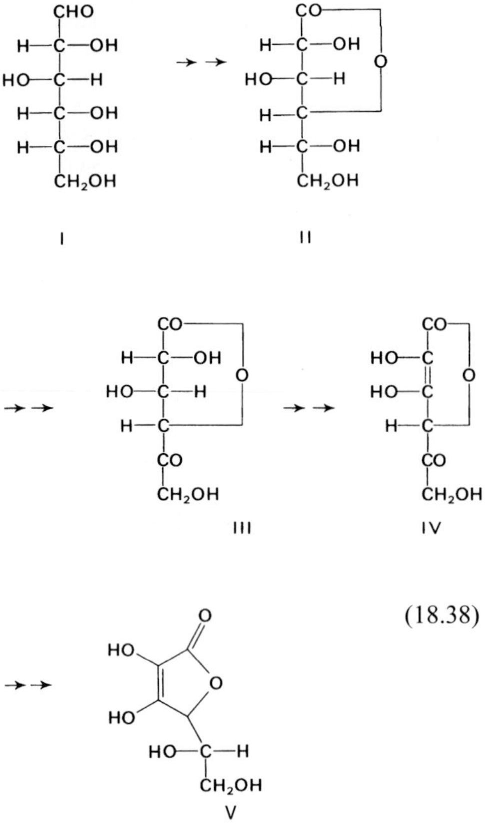

(18.38)

Industrial-scale production of ascorbic acid also starts with glucose. The sugar is first reduced to sorbitol (VI) and then oxidized with *Acetobacter suboxidans* to L-sorbose (VII). The diisopropylidene derivative (VIII) of the latter is oxidized to the corresponding derivative of L-2-ketogulonic acid (IX). After removal of the protecting isopropylidene groups, L-ascorbic acid (vitamin C) is obtained via L-2-ketogulonic acid (X; cf. Reaction 18.39).

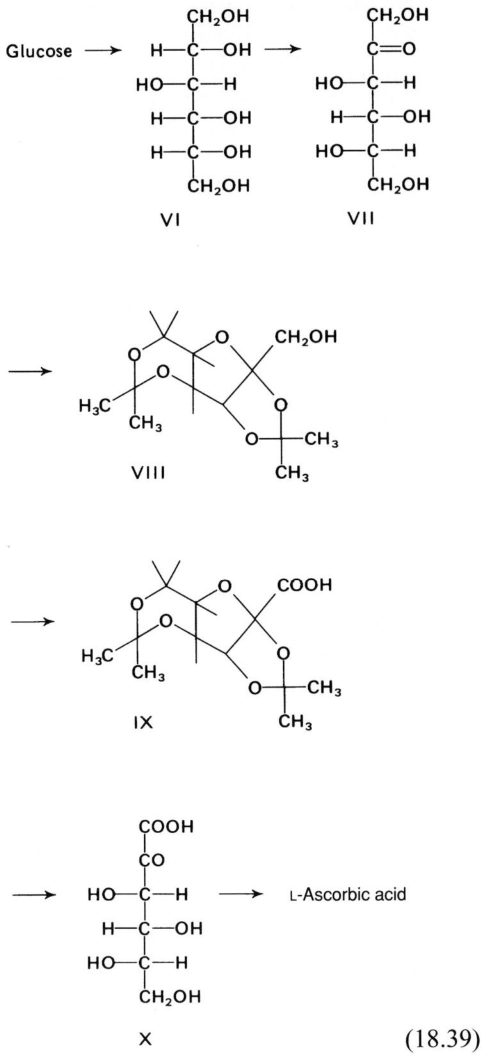

Glucose →

$$
\begin{array}{c}
CH_2OH \\
H-C-OH \\
HO-C-H \\
H-C-OH \\
H-C-OH \\
CH_2OH
\end{array}
\quad\rightarrow\quad
\begin{array}{c}
CH_2OH \\
C=O \\
HO-C-H \\
H-C-OH \\
HO-C-H \\
CH_2OH
\end{array}
$$

VI   VII

→ VIII

→ IX

$$
\begin{array}{c}
COOH \\
CO \\
HO-C-H \\
H-C-OH \\
HO-C-H \\
CH_2OH
\end{array}
\quad\longrightarrow\quad L\text{-Ascorbic acid}
$$

X   (18.39)

Table 18.29. Ascorbic acid in various fruits (mg/100 g edible portion)

Fruit	Ascorbic acid	Fruit	Ascorbic acid
Apple	2–10	Currant, black	210
Pear	4		
Apricot	7–10	Orange	50
Cherry	5–8	Grapefruit	40
Peach	7	Lemon	50
Plum/prune	3		
Blackberry	15	Acerola	1,300
Strawberry	60	Pineapple	25
Raspberry	25	Banana	10–30
Currant, red	40	Guava	300
		Melons	25–35

Table 18.30. Minerals in fruit

Element	Orange juice (% in ash)	Apple (mg/100 g dry matter)
Potassium	40	840
Sodium	1.8	
Calcium	2.8	90
Magnesium	1.7	40
Iron	0.06	2.5
Aluminium	0.12	
Phosphorus	6.8	90
Sulfur	0.8	
Chlorine	0.7	
Zinc, titanium, barium, copper, manganese, tin	≤ 0.03	Zinc 4.8
		Manganese 0.6
Boron	≤ 0.01	Copper 1.2

β-Carotene (provitamin A) occurs in large amounts in apricots, cherries, cantaloups and peaches. B-vitamins present in some fruits (apricots, citrus fruits, figs, black currants and gooseberries) are pantothenic acid and biotin. Other B-group vitamins occur at levels of no nutritional significance. Vitamins B_{12} and D and tocopherols are found in no more than trace amounts.

18.1.2.8 Minerals

Table 18.30 gives the composition of the ash of orange juice and apples. The most important cation is potassium and the most important inorganic anion is phosphate.

18.1.3 Chemical Changes During Ripening of Fruit

Ripening of fruit involves highly complex changes in physical and chemical properties. Softening, increasing sweetness, aroma and color changes are among the most striking phenomena related to ripening. Some changes will be outlined below in more detail.

18.1.3.1 Changes in Respiration Rate

The respiration rate is affected by the development stage of the fruit. A rise in respiration rate occurs with growth. This is followed by a slow decrease in respiration rate until the fruit is fully ripe. In a number of fruits ripening is associated with a renewed rise in respiration rate, which is often denoted as a climacteric rise. Maximal CO_2 production occurs in the climacteric stage. Depending on the fruit, this can occur before or after harvesting. Figures 18.6 and 18.7 show that such a rise occurs a short time after harvest for apples and tomatoes and is accompanied by increased ethylene production.

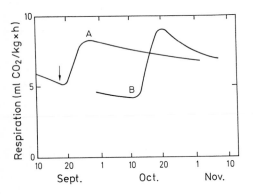

Fig. 18.6. Respiration rise in apples, Bramsley's seedlings. (according to *A.C. Hulme*, 1963.) A, apple picked →, B, left on tree to ripen

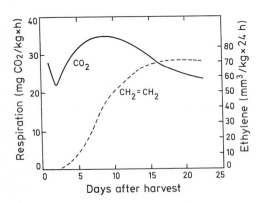

Fig. 18.7. Respiration rise in tomatoes —: CO_2, ---: Ethylene

The climacteric respiration rise is so specific that fruits can be divided into:

- *Climacteric types*, such as apples, apricots, avocados, bananas, pears, mangoes, papaya, passion fruit, peaches, plums/prunes and tomatoes; and
- *Nonclimacteric types*, which include pineapples, oranges, strawberries, figs, grapefruit, cucumbers, cherries, cantaloupes, melons, grapes and lemons.

It should be emphasized that nonclimacteric fruits generally ripen on the plants and contain no starch. The differing effects of ethylene on the two types of fruits are covered in Section 18.1.4.2.

Fruits can also be classified according to respiration behavior after harvesting. Three fruit types are distinguished:

Type 1: A slow drop in CO_2 production during ripening (as illustrated by citrus fruits).

Type 2: A temporary rise in CO_2 production. The fruits are fully ripe after this increase reaches a maximum (e.g., avocados, bananas, mangoes or tomatoes).

Type 3: Maximum CO_2 production in the fully ripe stage, until the fruit is overripe (e.g. strawberries and peaches).

The reason for the increase in CO_2 production is not yet fully elucidated. Physical and chemical factors are involved. For example, a change in permeability for gases occurs in fruit peels. With increasing age the peel cuticle becomes thicker and is more strongly impregnated with fluid waxes and oils. Thus, the total permeability drops, while the CO_2 concentration within the fruit increases. Three possibilities are usually considered for the rise in CO_2 production. The first is related to increased protein biosynthesis coupled with increased ATP consumption thus stimulating enhanced respiration. Secondly, since the respiratory quotient (RQ) increases from 1 to $1.4-1.6$, it is assumed that the additional CO_2 source is not due to respiration but to decarboxylation of malate and pyruvate, i.e. there is a switch from the citric acid cycle to malate degradation. Another possibility is the partial uncoupling of respiration from phosphorylation by an unknown decoupler.

New concepts involving structural factors suggest that fruit flesh possesses marked photosynthetic activity which is then associated with CO_2 uptake. With the onset of ripening, an increased disorganization occurs in chloroplasts and other cell organelles. Photosynthetic activity decreases and finally stops completely. The same is the case for other synthetic activities. Catabolic processes, catalyzed by cytoplasmic enzymes, become dominant. Based on such a perception (*Phan et al.*, 1975) the "climacteric is seen as an indication of the natural end of a period of active synthesis and maintenance, and the beginning of the actual senescence of the fruit".

18.1.3.2 Changes in Metabolic Pathways

Metabolic shifts may occur in several fruits during ripening. For example, during ripening of bananas, there is a marked rise in aldolase and carboxylase activities and thus it appears that at this stage the *Embden-Meyerhoff* pathway becomes dominant and the pentose-phosphate pathway is suppressed.

An increase in malate and pyruvate decarboxylase activities is observed in apples during the climacteric stage. The activities drop as CO_2 production decreases. This provides an explanation for the change in RQ during the climacteric stage. CO_2 production increases more rapidly than O_2 uptake, thus the RQ is greater than 1. The shift from the citric acid cycle to malate degradation in apples is also reflected by the effect of citrate and malate on succinate production. As ripening proceeds, production of succinate from citrate drops to zero. An increase in succinate content after addition of malate in the initial stage of ripening is probably a feedback reaction. In this case, a decrease is also observed later on, suggesting a greater change in metabolic patterns.

18.1.3.3 Changes in Individual Constituents

18.1.3.3.1 Carbohydrates

During ripening of fruits, significant changes occur in the carbohydrate fraction. For example, between picking and onset of decay in apples about 20% of the available carbohydrates have been utilized.

During the growth of apples on trees, the starch content rises and then drops to a negligible level by the time of harvest. This drop appears to be related to the increase in climacteric respiration. Contrary to starch, the sugar content rises. Other sources in addition to starch should be available for conversion to sugars. A decrease in hemicelluloses suggests that they are a possible source. Organic acids may also be an additional source of sugars.

A marked decrease in starch in bananas parallels an increase in the contents of glucose, fructose and saccharose. Biosynthesis of the latter occurs by two pathways:

1) UDPG + Fru-6-P $\longrightarrow$ UDP + Sac-6^F-P
$\longrightarrow$ Sac + P$_{in}$ (18.40)

2) UDPG + Fru $\longrightarrow$ UDP + Sac

The content of hemicelluloses drops from 9% to 1–2% (relative to fresh weight), hence they act as a storage pool in carbohydrate metabolism. There is also a drop in the sugar content in bananas during the post-climacteric stage.

Differences in various fruits can be remarkable. In oranges and grapefruits the acid content drops during ripening while the sugar level rises. In lemons, however, there is an increase in acids.

Decreases in arabinans, cellulose and other polysaccharides are found in pears during ripening. Cellulase enzyme activity has been confirmed in tomatoes.

Remarkable changes occur in the pectin fractions during ripening of many fruits (e.g., bananas, citrus fruits, strawberries, mangoes, cantaloupes and melons). The molecular weight of pectins decreases and there is a decrease in the degree of methylation. Insoluble protopectin is increasingly transformed into soluble forms. Protopectin is tightly associated with cellulose in the cell wall matrix. Its galacturonic acid residues are acetylated at OH-groups in positions 2 and 3 or are bound to polysaccharides as lignin (R^1 = H, CH_3, polysaccharide: arabinan, galactan and possibly cellulose; R^2 = H, CH_3CO, polysaccharide, lignin):

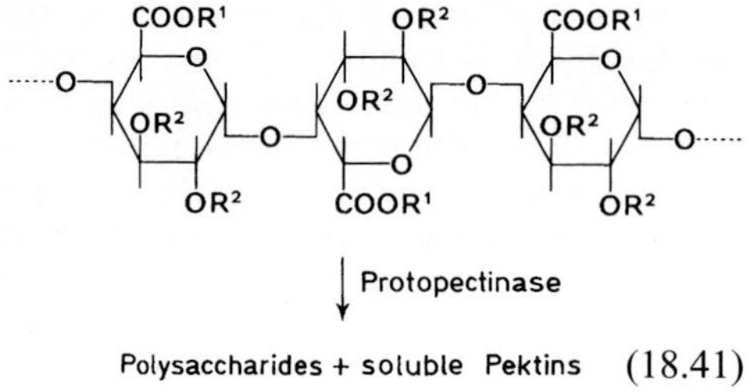

$$\downarrow \text{Protopectinase}$$

Polysaccharides + soluble Pektins   (18.41)

Soluble pectins bind polyphenols, quench their astringent effect and, thus, contribute to the mild taste of ripe fruits.

After prolonged storage there is a decrease in soluble pectins in apples. This drop is associated with a mealy, soft texture. Similar events occur in pears, but much more rapidly and with more extensive demethylation of pectin. Generally, the degree of pectin esterification drops from 85% to about 40% during ripening of pears, peaches and avocados. This drop is due to a remarkable increase in activities of polygalacturonases and pectin esterases. The rise in free galacturonic acid is negligible; therefore it appears that the release of uronic acid is associated with its simultaneous conversion through other reactions.

18.1.3.3.2 Proteins, Enzymes

During ripening of some fruits, although the total content of nitrogen is constant, there is an increase in protein content, an increase assigned primarily to increased biosynthesis of enzymes. For example, during ripening of fruit there is increased activity of hydrolases (amylases, cellulases, pectinolytic enzymes, glycolytic enzymes, enzymes involved in the citric acid cycle, transaminases, peroxidases and catalases). Proteinaceous enzyme inhibitors which inhibit the activities of amylases, peroxidases and catalases are found in unripe bananas and mangoes. The activities of these inhibitors appear to decrease with increasing ripeness.

The ratios of $NADH/NAD^+$ or $NADPH/$ $NADP^+$ pass through a maximum during ripening of fruit. For example, the values for mangoes are $0.32-0.67$ in the unripe stage, $1.44-6.50$ in the semi-ripe stage and $0.57-0.93$ in the ripe stage. During ripening of fruit, shifts also occur in the amino acid and

amine fractions. The shifts are not uniform and are affected by type and ripening stage of fruits.

18.1.3.3.3 Lipids

Little is known about changes in the lipid fraction. Shifts in composition and quantity have been found, especially in the phospholipid fraction.

18.1.3.3.4 Acids

There is a drop in acid content during ripening of fruits. Lemons, as already mentioned, are an exception. The proportion of various acids can change. In ripe apples malic acid is the major acid, while in young, unripe apples, quinic acid is dominant. In the various tissues of any single fruit, various acids can be dominant. For example, apple peels contain citramalic acid (I, cf. Formula 18.42) which is formed from pyruvic acid, and can produce acetone through acetoacetic acid. Acetone is formed abundantly during ripening:

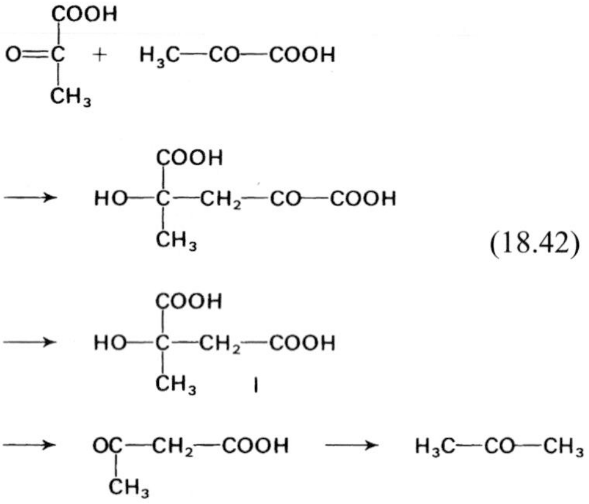

$$(18.42)$$

The synthesis of ascorbic acid is also of importance. It takes place in many fruits during ripening (cf. 18.1.2.7)

18.1.3.3.5 Pigments

The ripening of fruit is usually accompanied by a change in color. The transition of green to another color is due to the degradation of chlorophyll and the appearance of concealed pigments. Furthermore, the synthesis of other pigments plays a big role. For example, the lycopene content of the tomato increases greatly during ripening. The same applies to the carotenoid content of citrus fruits and mangoes. The formation of anthocyanin is frequently enhanced by light.

18.1.3.3.6 Aroma Compounds

The formation of typical aromas takes place during the ripening of fruit. In bananas, for example, noticeable amounts of volatile compounds are formed only 24 h after the climacteric stage has passed. The aroma build-up is affected by external factors such as temperature and day/night variations. Bananas, with a day/night rhythm of 30 °C/20 °C, produce about 60% more volatiles than those kept at a constant temperature of 30 °C.

18.1.4 Ripening as Influenced by Chemical Agents

The regulation of plant growth is achieved with phytohormones. They can be divided into five classes: auxins, gibberellins, cytokinins, abscisic acid and ethylene. There are many synthetic compounds which imitate the activities of phytohormones or act as their antagonists. These compounds are used in practice for various reasons: to induce blossom, to induce parthenocarpy, to enhance yields, to facilitate harvesting, to promote fruit ripening, to prevent sprouting or germination, and to retard aging and thus to increase the shelf life or storage stability of fruits. Some important examples are outlined in the following sections.

18.1.4.1 Compounds with Retarding Effects

A number of compounds retard the ripening and aging of fruits and vegetables and, thus, extend their storage life. Kinetin (I, cf. Formulas 18.43) and N[6]-benzyladenine (II), which are cytokinins, retard chlorophyll degradation and aging of leafy vegetables (e.g., spinach, beans, cucumbers). Compound II influences the storage life of fruits (e.g., sweet cherries and strawberries) and vegetables (e.g., cauliflower, chicory, various cabbage cultivars, radishes, celery and asparagus). It also improves the setting of the fruit and provides

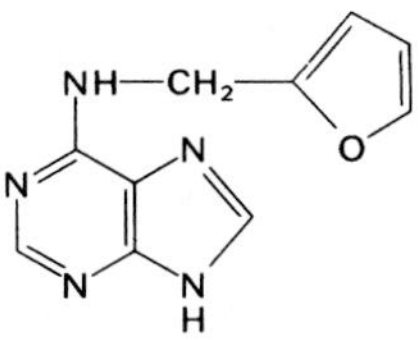

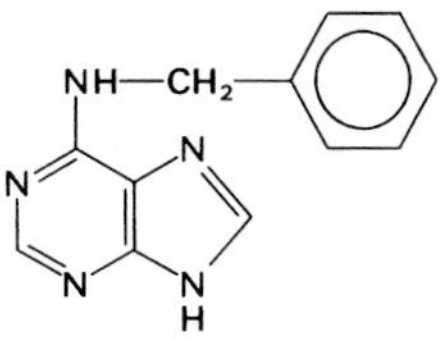

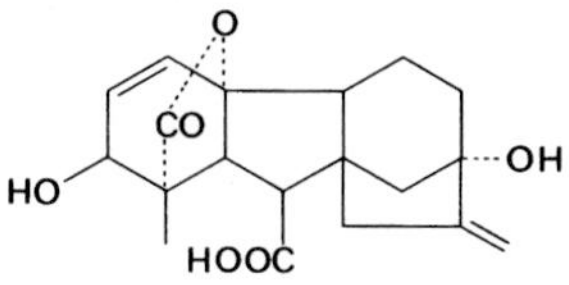

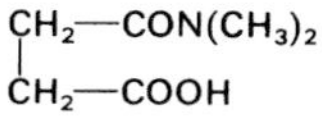

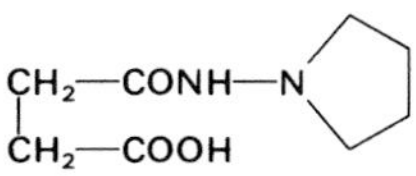

(18.43)

higher yields of grapes, melons and other fruits. Gibberellin A$_3$ (III) can be applied before harvesting (e.g., to oranges or lemons) or after harvesting (e.g., to bananas, guava and tomatoes) to retard fruit ripening. Generally, quality improvements have been reported for many fruits and vegetables. For example, larger grapes are obtained, with a looser arrangement of berries. Parthenocarpy is stimulated in tomatoes and grapes.

Application of gibberellin antagonists, e.g., N,N-dimethylsuccinic acid amide (IV), favorably affects the firmness and color of fruits, accelerates ripening of cherries, peaches and nectarines, and improves formation of grapes. An even more positive influence has been found with N-pyrrolidinyl succinic acid amide (V).

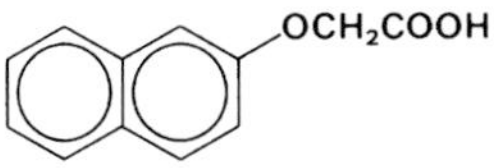

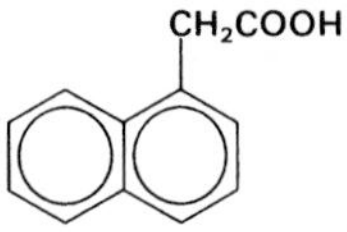

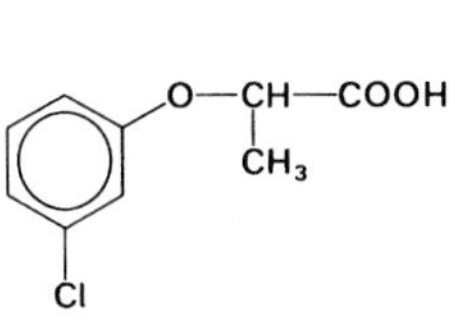

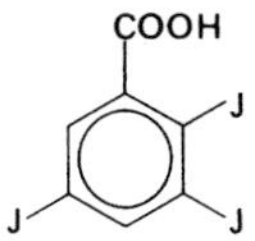

(18.44)

β-Naphthoxyacetic acid (VI, cf. Formula 18.44) is a synthetic compound of the auxin group. Applied before harvesting, it has a favorable affect on the storage properties of mandarin oranges. Fruit weight loss is diminished and the content of ascorbic acid remains high. α-Naphthylacetic acid (VII) is used for the induction of blossoms in pineapples and for the thinning of apples and oranges on the trees. 2-(3-Chlorophenoxy)-propionic acid (VIII) is utilized for peach thinning, while 2,3,5-triiodobenzoic acid (IX) induces blossoming of apple trees and retards tomato growth.

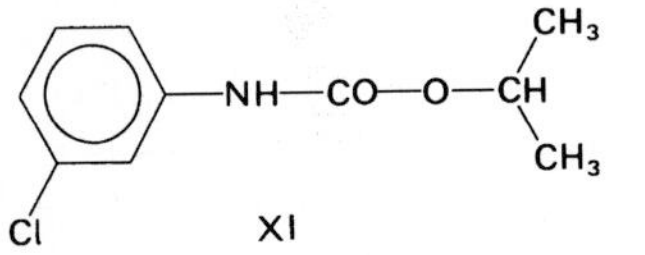

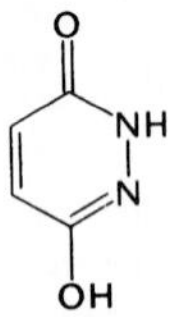

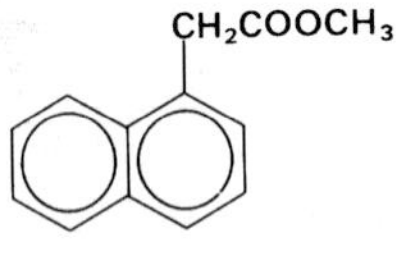

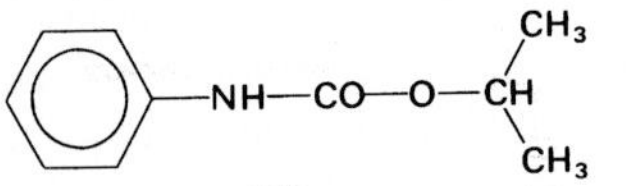

Maleic acid hydrazide (X, cf. Formula 18.45) interferes with cell division and also affects mobilization of storage polysaccharides in favor of mono- and oligosaccharides. It is used as a sprouting inhibitor (e.g., potatoes, carrots or onions). Compound X is also used to retard ripening or various fruits (mangoes and toma-

toes) and to maintain the firm texture of apples during storage.

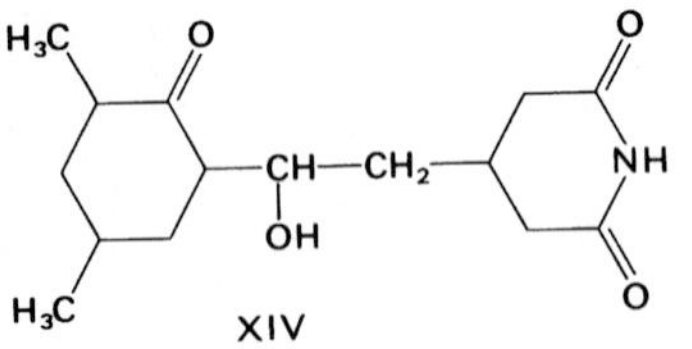

Various carbamates are active sprouting inhibitors. Examples are isopropyl-(3-chlorophenyl)-carbamate (XI), isopropyl-N-phenylcarbamate (XII) and α-naphthylacetic acid methyl ester (XIII).

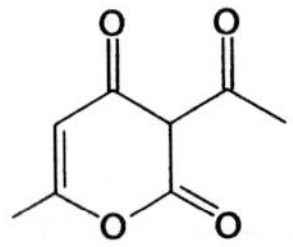

Retardation of ripening of various fruits has also been observed by using cycloheximide (XIV). This compound also facilitates the detachment of citrus and olive fruits from trees.

Actinomycin D, ethylene oxide (e.g., for mango fruits) and dehydroacetic acid (XV; e.g. for strawberries) are also able to retard ripening.

Compounds which bind ethylene (a ripening promoter) also have retarding activities. For example, bananas sealed in polyethylene bags have a prolonged shelf life in the presence of a silica carrier impregnated with alkaline $KMnO_4$.

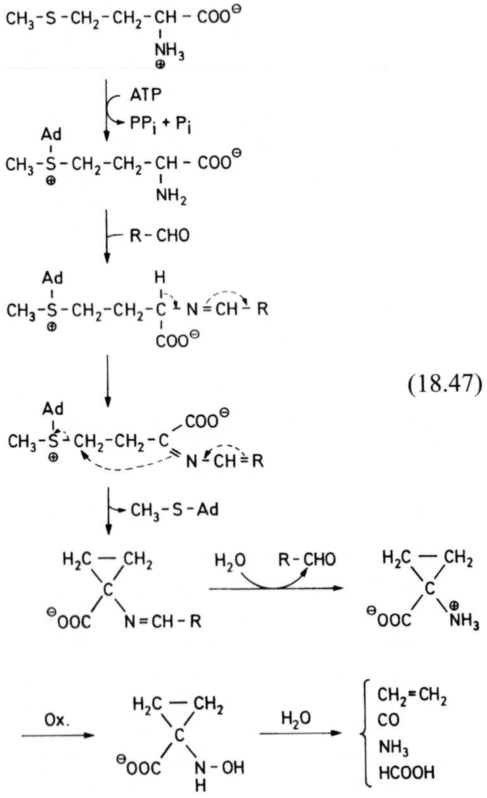

18.1.4.2 Compounds Promoting Fruit Ripening

Fruit ripening is coupled with ethylene biosynthesis. This gaseous compound increases membrane permeability and thereby probably accelerates metabolism and fruit ripening. With mango fruits, for example, it has been demonstrated that before the climacteric stage, ethylene stimulates oxidative and hydrolytic enzymes (catalase, peroxidase and amylase) and inactivates inhibitors of these enzymes.

Climacteric and nonclimacteric fruits respond differently to external ethylene (Fig. 18.8). Depending on the ethylene level, the respiratory increase sets in earlier in unripe climacteric fruits, but its height is not influenced. In contrast, in nonclimacteric fruits there is an increase in respiration rate at each ripening stage which is clearly dependent on ethylene concentration.

The reaction pathway 18.47 is suggested for the biosynthesis of ethylene (R−CHO: pyridoxal phosphate; Ad: adenosine).

Ethylene and compounds capable of releasing ethylene under suitable conditions are utilized commercially for enhancing the ripening process. A number of such compounds are known, e. g., 2-chloroethylphosphonic acid (ethephon; R = H or CH_2-CH_2Cl):

$$Cl-CH_2-CH_2-\overset{\overset{\displaystyle O}{\|}}{\underset{\underset{\displaystyle OR}{|}}{P}}-OH \xrightarrow{pH>4}$$

$$(18.48)$$

$$H_2C{=}CH_2 \;+\; H_2PO_4^{\ominus} \;+\; Cl^{\ominus}$$

The use of ethylene before picking fruit (as with pineapples, figs, mangoes, melons, cantaloups and tomatoes) results in faster and more uniform ripening. Its utilization after harvesting accelerates ripening (e. g., with bananas, citrus fruits and mangoes). Ethylene can induce blossoming in the pineapple plant and facilitate detachment of stone fruits and olives. Vine defoliation can also be achieved.

The activity of propylene is only 1% of that of ethylene. Acetylene also accelerates ripening but only at substantially higher concentrations. Methionine affects the ripening of apples, bananas and mangoes by stimulating ethylene

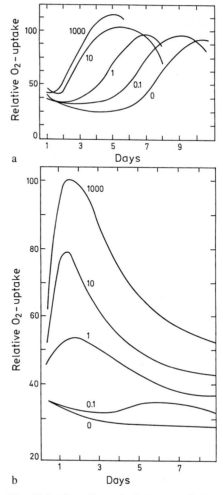

Fig. 18.8. The effect of ethylene on fruit respiration. (a) climacteric, (b) nonclimacteric. Numerals on the curves: ethylene in air, ppm. (according to *Biale*, 1964)

biosynthesis. Other stimulants are auxin, abscisic acid (XVI), ethanedial-dioxime (XVII; used for citrus fruits) and 5-chloro-3-methyl-4-nitropyrazole (XVIII). Application of the latter facilitates harvesting of ripe oranges.

Ethylene biosynthesis from methionine is inhibited by rhizobitoxine (XIX), an amino acid found in the nodules of a symbiotic association of *Rhizobium japonicum* with the root system of soybeans. Carbon dioxide is an ethylene antagonist and, thus, a retardant of fruit ripening.

N,N-bis-phosphonomethyl glycine (XX), which interferes with protein synthesis, accelerates ripening and increases sugar yield in sugar beets and sugar cane. The sucrose yield is increased by up to 10%. A sucrose yield increase of up to 25% is recorded in sugar cane after application of 7-oxabicyclo[2,2,1]- heptane-2,3-dicarboxylic acid (XXI). This compound appears to interfere with lipid biosynthesis and probably causes changes in the permeability of cell wall membranes, and it apparently retards the cleavage of sucrose.

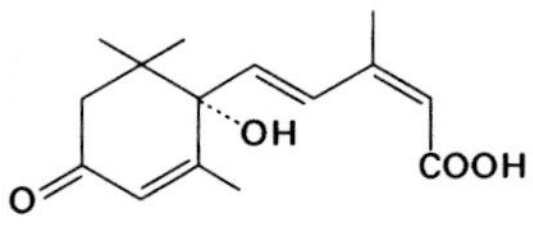

Zeaninic acid (XXII) is a growth promoter with an unknown mechanism of activity. It is recovered from wastes of corn starch production. It facilitates formation of the grape berry.

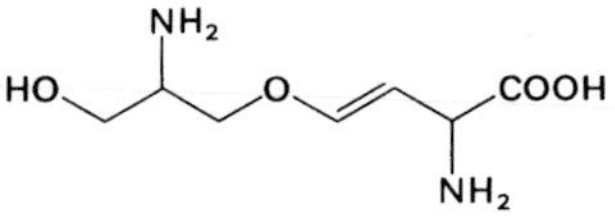

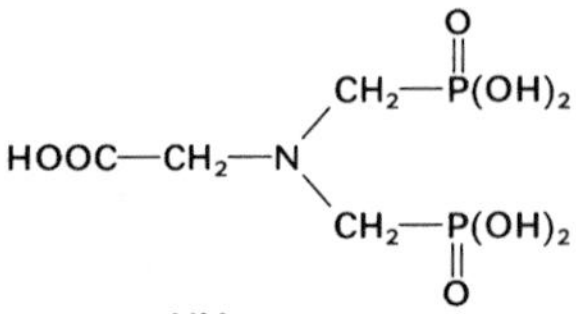

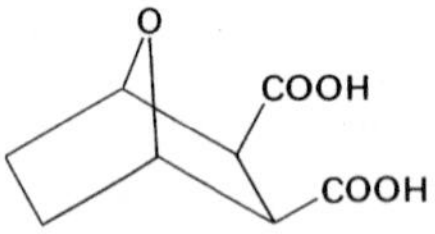

18.1.5 Storage of Fruits

18.1.5.1 Cold Storage

The suitability, duration and required conditions of fruit storage are dependent on variety and quality. Commonly used conditions are $-1°$ to $+2°C$ at $80-90\%$ relative humidity. The storage time varies from $4-8$ months for apples, $2-6$ months for pears, $2-3$ months for grapes, $1-2$ weeks for strawberries and raspberries, and $4-5$ days for cherries. Efficient aeration is required during storage. Air circulation is often combined with purging to remove ethylene, the volatile promoter of fruit ripening. Weight losses occur during fruit storage due to moisture losses of $3-10\%$.

18.1.5.2 Storage in a Controlled (Modified) Atmosphere

This term is applied to an atmosphere which, in comparison to air, has a lowered oxygen concentration and an increased CO_2 concentration. Common conditions for storage of many fruits are shown in Table 18.31. For each fruit variety it is important that optimal condi-

XVI

XVII

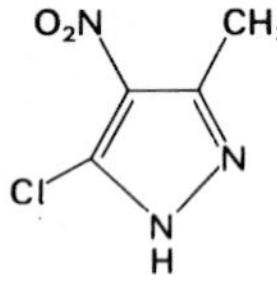

XVIII

XIX

(18.49)

XX

XXI

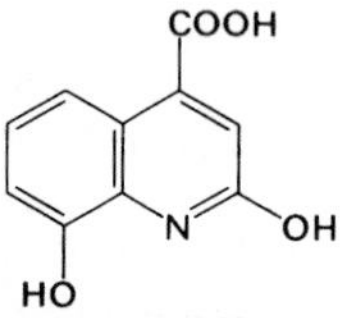

XXII

Table 18.31. Minimal O_2 and maximal CO_2 concentrations in the atmosphere during storage of fruits (temperature $0-5°C$)

Fruit	Minimal O_2 concentration (%)	Maximal CO_2 concentration (%)
Pear	2	2
Apple, kiwi	2	5
Peach, plum		
Pineapples	2	10
Sour cherry	2	15
Citrus fruits	5	10

tions for controlled atmosphere storage be maintained. For example, a high O_2 concentration accelerates ripening, while an overly low O_2 concentration results in high production of CO_2. An overly high concentration of CO_2 promotes glycolysis, which can cause off-flavors due to the formation of acetaldehyde and ethanol. Discoloration can also occur.

18.2 Fruit Products

The short shelf life of most fruits and the frequent need to store and spread out the surplus of a harvest for a prolonged period of time has brought about a number of processes which provide more durable and stable fruit products.

18.2.1 Dried Fruits

Like many other food products, moisture removal from fruits by a suitable drying process results in a product in which microbial growth is retarded and, with a suitable pretreatment, the enzymes present are largely inactivated. Fruit drying is probably the oldest procedure for preservation. It was originally performed in a rather primitive way (spreading the fruit in the hot air of a fireplace or hearth, kitchen stove or oven), thus providing dark "baked products". Solar drying is still a common process in southern and tropical countries for obtaining dried apple slices, apricots, peaches or pears or tropical fruits such as dates, figs or raisins. Predrying is often achieved in sunshine and additional drying by artifical heat in drying installations. The temperature in drying chambers, flat or tunnel dryers is between 75 °C (incoming air) and 65 °C (temperature of the exit air) at a relative humidity of 15–20%. Vacuum drying at about 60 °C is particularly gentle.

Carefully washed and trimmed fruits of suitable varieties are pretreated in various ways: *Pomme fruits* (apples, pears) are initially peeled mechanically and freed from the core and calix (seed compartment). Apples are then cut preferentially into 5–7 mm thick slices, and dried in rings (a yield of 10–20% of the unpeeled fresh weight). Sulfite treatment is used to prevent browning during processing and storage. The sulfurous acid prevents both enzymatic and nonenzymatic browning reactions, stabilizes vitamin C and prevents microbial contamination during storage of the end product. The utilization of dilute solutions of citric acid is also suitable for preventing browning. Whole or sliced pears are heated with steam to achieve a translucent appearance and then are dried at 60–65 °C. The yield is 13–14% of the fresh weight.

The *stone fruits* usually dried are plums/prunes, apricots and peaches. Plums are first dipped for 5–15 s into a hot, diluted solution of sodium hydroxide, or into 0.7% aqueous K-carbonate and then rinsed and dried at 70–75 °C or dried in the sun. Plum peels are often fissured to facilitate drying. In order to clean and to provide a black, glossy surface, dried plums are steamed additionally at 80–85 °C for a short time. The plum yield is 25–30% at a moisture content of not more than 19%. Apricots and peaches are treated alternately with cold and hot water, then are halved, the stone seed is removed and the fruit is dried in the sun or in drying installations at 65–70 °C. The yield, depending on fruit size, is 10–15%. SO_2 (sulfurous acid) treatment is common for apricots and peaches. Cherries play a less important role as dried fruit. To avoid substantial aroma losses, cherries are dried slowly and with a number of precautions.

Grapes are the most commonly dried *berry fruits*. Raisins are dark-colored, dried grapes which contain seeds, whereas sultana raisins are seedless, light-colored, dried grapes. Currants, with or without seeds, are dark and are much smaller in size than the other two raisin products. The surface treatment of raisins, with the exception of currants, involves the use of acetylated monoglycerides to prevent caking or sticking.

The compositions of some dried fruits are presented in Table 18.32. Dried fruits are exceptionally rich in calories and they supply significant amounts of minerals. Of the vitamins found in fruit, β-carotene and the B-vitamins remain intact. Vitamin C is lost to a great extent. Sulfite treatment destroys vitamin B_1. However, fruit color and vitamin C content are retained and stabilized.

Table 18.32. Composition of some dried fruits (g/100 g edible portion)

Fruit	Mois-ture	N-con-taining com-pounds	Fat	Car-bohy-drates	Crude fiber	Miner-als	Vitamin C
Apricots	15–24	5.0	0.4	68	3.2	3.5	0.011
Dates	20.2	1.9	0.5	73.2	2.4	1.8	0.003
Figs	24.6	3.5	1.3	61.5	6.7	2.4	0–0.005
Peaches	24.0	3.0	0.6	65.9	3.5	3.0	0.017
Plums/ prunes	24.0	2.3	0.6	69.4	1.6	2.1	0.004
Raisins	24.2	2.3	0.5	64.2	7.0	1.86	0.001

18.2.2 Canned Fruits

Since the middle of the last century, heat sterilization in cans and glass jars has been the most important process for fruit preservation. Undamaged, aroma-rich and not overripe fruits are suitable for heat sterilization. Aseptic canning is applicable only for fruit purées. Canned fruits used are primarily stone fruits, pears, pineapples and apples (usually apple purée). Strawberries and gooseberries are canned to a lesser extent.

Canned fruits are produced in a large volume by the food industry and also in individual households. Cherries are freed from stone seeds and stems, plums/prunes, apricots and peaches are halved and the stone seeds are removed, strawberry calix is removed, gooseberry and red currant stems are removed, apples and pears are peeled and sliced. Specialized equipment has been developed for these procedures.

With a few exceptions (raspberries and blackberries) all fruits are washed or rinsed. Apricots are readily peeled after alkali treatment at 65 °C. Fruits sterilized unpeeled, e.g., prunes or yellow plums, are first fissured to prevent later bursting. To avoid aroma loss and to prevent floating in the can, fruits which shrink considerably (such as cherries, yellow plums, strawberries and gooseberries) are dipped prior to canning into a hot 30% sugar solution and then covered with a sugar solution, with a sugar concentration approximately twice the desired final concentration in the can. Finally, the can is vacuum sealed at 77–95 °C for 4–6 min and, according to the fruit species, heat sterilized under the required

conditions. For example, a 1 liter can of strawberries is sterilized in a boiling water bath at 100 °C for 18 min, while pears, peaches and apricots are heated at 100 °C for 22 min. Additions of ascorbic and citric acids for stabilization of color and calcium salts for the preservation of firm texture have been accepted as standard procedures for canned fruits consumed as desserts.

Canned fruits used for bakery products, confections or candies are produced like canned dessert fruits, however, the fruits are covered with water instead of sugar solutions.

18.2.3 Deep-Frozen Fruits

Fruits are frozen and stored either as an end product or for further processing. The choice of suitable varieties of fruit at an optimal ripening stage is very important. Pineapples, apples, apricots, grapefruit, strawberries and dark-colored cherries are highly suitable. Light-colored cherries, plums, grapes and many subtropical or tropical fruits are of low suitability.

Rapid chilling is important (air temperature ≤ −30 °C, freezing time about 3 h) to avoid microbial growth, large concentration shifts in fruit tissues, and formation of large ice crystals which damage tissue structure. A blanching step prior to freezing is commonly used only for few fruits, such as pears, and occasionally for apples, apricots and peaches. Some fruits are covered, prior to freezing, with a 30–50% sugar solution or with solid granulated sugar (1 part per 4–10 parts by weight) and are left to stand until the sap separates. In both instances oxygen is eliminated, enzymatic browning is prevented, and the texture and aroma of the fruit are better preserved. Addition of ascorbic acid or citric acid is also common.

Frozen fruit which is stored at −18 to 24 °C is stable for two to four years.

18.2.4 Rum Fruits, Fruits in Sugar Syrup, etc.

Rum fruits are produced by steeping the fruit in dilute spirits in the presence of sufficient sugar.

Fruits preserved in vinegar, mostly pears and plums, are prepared by poaching in wine vinegar sweetened with sugar and spiced with cinnamon and cloves.

Fruits in sugar syrup are prepared by treating raw or precooked fruits or fruit portions (may be precooked under a vacuum) with highly concentrated sucrose solutions which also contain starch syrup. The latter is added to enhance translucency, smoothness and tractability of the product. Candied lemon or orange peels are products of this kind.

Other varieties provide intermediary products processed further into fruit confections: glazed fruits (these are washed fruits treated with a sugar solution containing gum arabic and then subsequently dried at 30–35 °C) or candied fruits in which the dried, glazed fruit is also immersed in a concentrated sugar solution and then dried to form a candied hull. Another product is crystallized fruit in which the dried, glazed fruits are rolled over icing or granulated sugar (sucrose), then dried aditionally and, to achieve a shiny, glossy appearance, are exposed to steam for a short time.

18.2.5 Fruit Pulps and Slurries

Fruit pulp is not suitable for direct consumption. The pulp is in the form of slurried fresh fruit or pieces of fruit either split or whole, and, when necessary, stabilized by chemical preservatives. The minimum dry matter content of various pulps is 7–11 %. For pulp production the fruit, which has been washed in special machines, is lightly steamed in steam conduits or precooking retorts.

The fruit slurry is an intermediary product, also not suitable for direct consumption. The production steps are similar to those for pulp. However, there is an additional step: slurrying and straining, i.e. passing the slurry through sieves. Both the pulp and the slurry can be stored frozen.

18.2.6 Marmalades, Jams and Jellies

Marmalades are thick, spreadable fruit slurries obtained by boiling (thickened by boiling) fresh or fresh stored fruits, or are obtained from fruit pulps or slurries by the addition of sugar. Addition of unpeeled fruits, fruit pectin, starch syrup, and tartaric, citric or lactic acids is common. Marmalades can be made from a single fruit or a mixture of fruits, or can be blends of several marmalades.

For the production of marmalade, the fresh fruits or intermediary products, such as fruit pulps or slurries, are boiled in an open kettle with the addition of sugar (usually added in two batches). Other ingredients (gelling agents, starch syrup and acids) are added before the thickening is completed by boiling. The end of boiling is determined by refractometer readings (the total boiling time is usually 15–30 min). Boiling under vacuum decreases the time and the required temperature. In addition, it enhances color and aroma retention. However, marmalade boiled in a vacuum does not have the characteristic flavor of marmalade boiled in an open kettle since the flavor of the latter is imparted by a limited extent of sucrose inversion and sugar caramelization. The hot marmalade is then poured into appropriate containers.

Jams and marmalades which have not been boiled are pasteurized at temperatures under 85 °C. This assures retention of fruit aroma and color in the product.

Jams are produced similarly to marmalades but usually from one kind of fruit. They are thickened by boiling and constant stirring of the whole or sliced fresh or fresh stored raw material, or of fruit pulp. Unlike marmalades, whole fruits or pieces can be found in ready-made jams. Table 18.33 provides compositional data for some commercial jams.

Jellies are gelatinous, spreadable preparations made from the juice or aqueous extract of fresh fruits by boiling down with sugar. The addition of fruit pectin (0.5 % as calcium pectate) and tartaric acid or lactic acid (0.5 %) is normal. In general, the water content is 42 %, and the sugar content between 50 % and 70 %. The juice is boiled down in open kettles or in vacuum kettles with sugar (about half the weight of the fruit), pectin, if necessary, and the substances mentioned above. The scum is carefully skimmed off and the mixture is boiled further until a moisture content of about 42 % is reached.

Table 18.33. Composition of various jams (average values in %)

Jam from	Moisture	Water soluble extract	Total sugar	Sugar-free extract	Titratable acidity	Ash	Pectin as Ca-pectate
Strawberries	32.2	66.2	57.7	8.5	0.49	0.30	0.34
Apricots	33.1	66.2	51.3	5.0	0.71	0.36	0.50
Cherries	28.6	70.8	61.6	9.3	0.55	0.38	0.42
Blackberries	34.2	64.8	58.0	6.8	0.71	0.32	0.34
Raspberries	30.0	67.2	60.3	6.8	0.90	0.30	0.38
Bilberries	30.1	68.0	60.0	8.0	0.78	0.22	0.37
Plums/prunes	31.1	68.0	59.5	8.3	0.42	0.24	0.43

18.2.7 Plum Sauce (Damson Cheese)

Plum/prune sauce is produced by thickening through boiling of fresh fruit pulps or fruit slurries. The use of dried plums is also common. Normally, the product has no added sugar, but sweetened products or products with other ingredients added are also produced.

18.2.8 Thickened Fruit Syrups

Thickened fruit syrups are made from fresh apples or pears by steaming or boiling, followed by pressing and evaporation of the extract or by evaporation of the juices with or without the addition of sugar. The dry matter content of the end product is at least 65%.

Details of the production are as follows: fresh sweet apples or pears are boiled in water or steamed until a soft, mealy consistency is achieved. The fruit is then subjected to hydraulic pressing. The sap collected is boiled in an open kettle under constant stirring to a thick consistency with a moisture content not exceeding 35%. The same method is used to produce a sweetened product from apples (with a sucrose content of 25% of the weight of the end product), a blend of apples and pears, and thickened sugar beet syrup.

18.2.9 Fruit Juices

Fruit juices are usually obtained by mechanical means, and also from juice concentrates (cf. 18.2.11) by dilution with water. The solid matter content is generally 5–20%. The juices are consumed as such or are used as inter-

mediary products, e.g., for the production of syrups, jellies, lemonades, fruit juice liqueurs or fruit candies. Fruit juice production is regulated in most countries.

Juices from acidic fruits are usually sweetened by adding sucrose, glucose or fructose. Juices used for further processing usually contain chemical preservatives to inhibit fermentation. Some juices from berries and stone fruits, because of their high acid content, are not suitable for direct consumption. Addition of sugar and subsequent dilution with water provides fruit nectars or sweet musts (cf. 18.2.10). Data on the production of fruit juices and nonalcoholic beverages are presented in Table 18.34. Table 18.35 lists data on the composition of some juices and nectars.

Production of fruit juice involves several processing steps: fruit preparation and the extraction, treatment and preservation of the juice.

18.2.9.1 Preparation of the Fruit

Preparation of the fruit involves washing, rinsing and trimming, i.e. the faulty and unripe fruits are removed. The stone seeds and stalks,

Table 18.34. Production of fruit juices and nonalcoholic beverages (W. Germany, 1982)

Type	Amount (10^6 l)
Carbonated drinks (lemonades)	4,130
Fruit juices	910
Fruit nectars	670
Fruit juice beverages	445
Diet fruit juice beverages	125
Others	49

Table 18.35. Composition of some fruit juices (g/l)

	Extract	Total sugar	Volatile acids	Sugar-free extract	Ash	Titratable acidity[a]	Vitamin C
Apple	97–130	72–102	0.15–0.25	14–34	2.2–3.1	4.1–10.4 (M)	0–0.03
Grape	145–195	120–180	0.08–0.25	21.6–35	2.1–3.2	3.6–11.7 (T)	0.017–0.02
Currant, black[b]	120–165	95–145	0.12–0.25	13.3–44.5	2.25–3.2	9.15–12.75 (T)	0.1–0.56
Cherry[b]	126.4–166.4	104.3–138.4	0.08–0.12	17.8–32.6	1.99–3.02	8.0–10.1 (T)	–
Raspberry	45–100	2.7–69.6	–	22.8–64.8	3.5–5.4	13.5–27.8 (T)	0.12–0.49
Orange	87–148	60–110	–	15.2–41.0	2.2–4.0	5–18 (C)	0.28–0.86
Lemon	71–119	7.7–40.8	–	–	1.5–3.5	42–83.3 (C)	0.37–0.63
Grapefruit	76–126	50–83	–	10.3–53	2.5–5.6	5–27 (C)	0.25–0.5

[a] Calculated as malic (M), tartaric (T) or citric acid (C).
[b] Diluted and sweetened.

stems or calyx are then removed. Disintegration is accomplished mechanically in mills, thermally by heating (thermobreak at about 80 °C) or by freezing (less than −5 °C). The yield can be increased by enzymatic pectin degradation ("mash fermentation", particularly of stone fruits and of berries) or by applying procedures such as ultrasound or electroplasmolysis.

18.2.9.2 Juice Extraction

Separation of the juice is achieved using continuous or discontinuous presses or processes such as vacuum filtration or extraction.

Newer approaches involve liquefaction of fruit tissue by pectinolytic and cellulolytic enzymes. Such an approach is very convenient for soft tropical fruits. A schematic production line: fruit preparation → fruit mashing-milling → enzyme treatment → filtration → pasteurization → container filling (omitting dilution with water), provides fruit juices suitable for direct consumption.

18.2.9.3 Juice Treatment

The juice treatment step involves clarification, i.e. removal of turbidity, and stabilization to prevent additional turbidity. This step commonly involves treatment with enzymes, mostly pectinolytic, and, if necessary, removal of starch and polyphenols using gelatin, alone or together with colloidal silicic acid or tannin, or polyvinylpyrrolidone. Finally, proteins are removed by adsorption on bentonite.

Clarification of juice is achieved by filtration through porous pads or layers of cellulose, asbestos or kieselguhr, or by centrifugation.

Since juice production provides juices which are well-saturated with air, oxygen-sensitive products are deaerated. This is achieved by an evacuation step or by purging the juice with an inert gas such as N_2 or CO_2.

Fruit juices (with the exception of citrus juices) are produced as transparent, clarified products, although some turbid juices are available. In the latter case, measures are required to obtain a stable, turbid suspension. This is achieved with stone fruit juices by a short treatment with polygalacturonase preparations which have a low pectin esterase activity and which then partially degrade and, thus, stabilize the ingredients required for turbidity. Citrus juices (lemons, oranges, grapefruits) are heat-treated to inactivate the endogenous pectin esterase, which would otherwise provide pectic acid which can aggregate and flocculate in the presence of Ca^{2+} ions. However, since heat treatment damages fruit aroma, the use of polygalacturonase is preferred. This enzyme degrades the pectic acid to such an extent that flocculation does not occur in the presence of divalent cations.

18.2.9.4 Preservation

Finally, the fruit juice preservation step involves pasteurization, preservation by freezing, storage under an inert atmosphere, or concentration (cf. 18.2.11) and drying (cf. 18.2.13).

Pasteurization kills the microflora and inactivates the enzymes, particularly the phenol oxidases. Since a longer heating time is detrimental to the quality, a short, high-temperature heat treatment is the preferred process, using plate heat exchangers (82–90 °C for 15–150 s) with subsequent rapid cooling. The juice is stored in germ-free tanks. Filling operations for the retail market can lead to reinfection, hence a second pasteurization is required. It is achieved by filling preheated containers with the heated juice, or by heating the filled and sealed containers in chambers or tunnel pasteurizers.

Preservation by freezing generally involves transforming the juice or juice concentrate into an ice slurry (at −2.5 °C to −6.5 °C), then packing and cooling to the retail market storage temperature. The product is stable for 5–10 months in a temperature range of −18 °C to −23 °C.

Storage in an inert atmosphere makes use of the fact that filtered, sterilized juices are microbiologically stable at temperatures below 10 °C and in an atmosphere of more than 14.6 g CO_2/l. To attain such a concentration of CO_2, the filled storage tank has to be at a pressure of 5.9 bar at 10 °C, or 4.7 bar at 5 °C.

Fruit juices are poured into retail containers, i.e. glass bottles, synthetic polyethylene pouches, aluminum cans, or aluminum-lined cardboard containers.

18.2.9.5 Side Products

Pomace is the residue from the production of fruit juices. Citrus fruits and apple pomace are used for the recovery of pectins. Other fruit residues are used as animal feed, as organic fertilizer, or are incinerated.

18.2.10 Fruit Nectars

Fruit nectars are produced from fruit slurries or whole fruits by homogenization in the presence of sugar, water and, when necessary, citric and ascorbic acids. The fruit content (as fresh weight) is 25–50% and is regulated in most countries, as is the minimum total acid content. Apricots, pears, strawberries, peaches and sour cherries are suitable for nectar production. The fruits are washed, rinsed, disintegrated and heated to inactivate the enzymes present. The fruit mash is then treated with a suitable mixture of pectinolytic and cellulolytic enzymes. The treatment degrades protopectin and, thus, separates the tissue into its individual intact cells ("maceration").

High molecular weight and highly esterified pectin formed from protopectin provides the high viscosity and the required turbidity for the nectar. Finally, the disintegrated product is filtered hot, then saturated with the usual additives, homogenized and pasteurized.

Fruit products from citrus fruits (comminuted bases) are obtained by autoclaving (2–3 min at 3 bar) and then straining the fruits through sieves, followed by homogenization.

Fruit nectars also include juices or juice concentrates from berries or stone fruits, adjusted by addition of water and sugar. Such products are commonly denoted as sweet musts.

18.2.11 Fruit Juice Concentrates

Fruit juice concentrates are chemically and microbiologically more stable than fruit juices and their storage and transport costs are lower. The solid content (dry matter) of the concentrates is 60–75%. Intermediary products, less stable concentrates with a dry matter content of 36–48%, are also produced. Fruit juice concentration is achieved by evaporation, freezing, or by a process involving high pressure filtration. Initially, the pectin is degraded to avoid high viscosity and gel setting (undesired properties).

18.2.11.1 Evaporation

Concentration by evaporation is the preferred industrial process. Since the process leads to losses in volatile aroma constituents, it is combined with an aroma recovery step. The aroma of the juice is enriched 100 to 200 times by a counter-current distillation. This aroma is stored and recombined with the juice only at the dilution stage. In order to maintain quality, the residence time in evaporators is as short as possible. In a high-temperature, short-time heating installation, e. g., in a 3- and 4-fold stepwise gradient-type evaporator, the residence time is 3–8 min at an evaporating temperature of 100 °C in the first step and about 40 °C in the fourth step. The concentrate is then cooled to 10 °C. Recovery of the aroma is achieved by rectifying the condensate of the first evaporation stage. A short-time treatment of juices is also possible in thin-layer falling film evaporators. These are particularly suitable for concentrating highly viscous products such as fruit slurries.

18.2.11.2 Freezing

Concentration of juice by freezing is less economical than evaporation. Hence, it is utilized mostly for products containing sensitive aroma constituents, e.g., orange juice. The juice is cooled continuously below its freezing point in a scraper-type cooler. The ice crystals are separated from the resultant ice slurry by pressing or by centrifugation. The obtainable solid content of the end product is 40–50%. This content is a function of freezing temperature, as illustrated with apple juice in Fig. 18.9.

18.2.11.3 Pressure Filtration

Concentration of juice by filtration using semipermeable membranes and high pressure (1–10 bar) is known as ultrafiltration. When the membrane is permeable for water and only to a limited extent for other small molecules ($M_r < 500$, e.g., salts, sugars, aroma compounds), the process is called reverse osmosis.

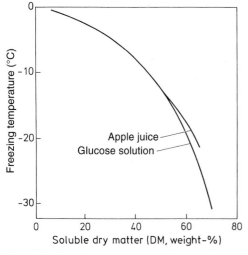

Fig. 18.9. Freezing temperature of apple juice and glucose solution as affected by soluble dry matter (DM). (according to *Schobinger*, 1978)

Concentration of juice is possible only to about 25% dry matter content. The reverse osmosis process is still under development, but may gain in importance as a pretreatment step in combination with other concentrating procedures.

18.2.12 Fruit Syrups

Fruit syrups are thick, fluid preparations made by boiling one kind of fruit with an excess of sugar. They are sometimes prepared without heating by directly treating fresh fruit or fruit juice with sugar, occasionally also using small amounts of tartaric or lactic acids. Fruit syrups from citrus fruits often contain small amounts of peel aromas.

Fruit syrups are rapidly cooled to avoid aroma losses and caramelization of sugar. The boiling process partially inverts sucrose, preventing subsequent sucrose crystallization. Low-acid fruits are treated with tartaric or lactic acid. Boiling in closed kettles permits recovery of vaporized aroma compounds which can be added back to the end product. As in marmalade production, the boiling is occasionally done under vacuum (50 °C starting temperature, 65–70 °C final temperature) in order to retain the aroma. Syrup production by a cold process is particularly gentle. The raw juice flows over the granulated sucrose in the cold until the required sugar concentration has been achieved. Aroma-sensitive syrups which contain turbidity-causing substances, e.g., citrus fruit syrups, are made by adding sugar to the mother liquor with vigorous stirring.

18.2.13 Fruit Powders

Fruit powders are produced by drying juices, juice concentrates or slurries. The hygroscopic powders contain no more than 3–4% moisture. Addition of drying aids (such as glucose, maltose or starch syrup) in amounts greater than 50% of the dry matter can efficiently control clumping or caking due to the presence of fructose in the drying process. Freeze-drying, vacuum foam-drying (1–10 torr, 40–60 °C) and spray-drying are suitable drying processes. The last two are of industrial significance.

18.3 Alcohol-Free Beverages

18.3.1 Fruit Juice Beverages

These drinks are prepared from fruit juices or their mixtures or from fruit concentrates, with or without addition of sucrose or glucose, and are diluted with water or soda or mineral water. Fruit juice refreshments have mainly citrus fruits as their base and contain at least 6% fruit juice. These drinks are occasionally mixed with juices of Seville oranges, mandarin oranges, tangerines, lemons or limes to round-off the base flavor.

18.3.2 Lemonades, Cold and Hot Beverages

These drinks are prepared from natural fruit essences and sugar (sucrose or glucose), fruit acids and soda or mineral water. They are also consumed without added carbon dioxide, either cold or warmed. The drinks are usually colored.

Tonic water is also considered a lemonade. It contains about 80 mg quinine/l to provide the characteristic bitter taste.

18.3.3 Caffeine-Containing Beverages

These are also considered as "lemonades" (particularly in Europe). The most popular are the cola drinks, which contain extracts from the cola nut (*Cola nitida*) or aromatic extracts from ginger, orange blossoms, carob and tonka beans or lime peels. Caffeine is often added (6.5−25 mg/100 ml). Phosphoric acid is sometimes used as an acidulant (70 mg/100 ml). The sugar content of cola drinks averages 10−11%. The deep-brown color of the drink is adjusted with caramel.

18.3.4 Other Pop Beverages

Some effervescent pop drinks are imitations of fruit juices and lemonade-type drinks, however, their sugar content is fully or partially replaced by artificial sweeteners and the natural essence of flavoring ingredients are replaced by artificial or artificially-enhanced essences. Coloring substances are usually added.

18.4 Analysis

As a result of the numerous raw materials and processes involved, the analysis of fruit products is difficult and tedious. Information on the following is important for an evaluation:

- Type, amount, and origin, if necessary, of the fruit and additives used (e.g., acids, sugar).
- Constituents that determine quality (e.g., aroma substances, vitamins).
- Method of processing.

Information of this type is provided by the quantitative analysis of various constituents, determination of species-specific compounds, and by the determination of abundance ratios of isotopes.

18.4.1 Various Constituents

Since the composition of the raw materials varies greatly, deviations from the standard can be recognized only by collective changes in the concentration of as many components as possible. For orange juice, Table 18.36 shows that when the guide values for certain components are exceeded or fallen short of, information is provided on the proportion of fruit, the use of expressed residues, acidification, sweetening, and microbiological spoilage.

18.4.2 Species-Specific Constituents

The occurrence of species-specific constituents is also analytically useful. The composition of the plant phenols of individual fruits can be analyzed quickly and very accurately by using HPLC. These data have shown that certain compounds are suitable indicators of adulteration (Table 18.37).

It must be guaranteed that the selected indicator substance is stable under the production conditions for the particular fruit product. Therefore, anthocyanins are generally not suitable. For fermented products, O-glycosides are not suitable because they are degraded by yeast enzymes. Suitable compounds are C-glycosidically bound flavonoids which are resistant to enzymatic hydrolysis and common chemical hydrolysis, e.g., schaftoside (cf. Table 18.37) can be detected even in wine and

Table 18.36. Guide values for orange juice

Quantity being measured/Component	Mean	Range of variation	Guide value[a]		Indicator[b]
Specific gravity at 20/20°	1.046	1.045−1.055	x	1.0450	1
Extract (°Brix)[c]	11.41	11.18−13.54	x	11.18	1
Soluble solids (g/L)	119.4	116.8−142.9	x	116.8	1
Titratable acids (pH 7.0)					
calc. as tartaric acid (g/L)	9.5	8.0−12	x	8.0	1
Ethanol (g/L)	−	−	o	3.0	3
Volatile acids					
calc. as acetic acid (g/L)	−	−	o	0.4	3
Lactic acid (g/L)	−	−	o	0.5	3
L(+)-Ascorbic acid (mg/L)	350	−	x	200	2
Peel oil (g/L)	−	−	o	0.3	7
Glucose (g/L)	28	20	x	22	1
Fructose (g/L)	30	22	x	24	1
Ratio of glucose to fructose	0.92	0.85−1.0	o	1.0	5
Sucrose (g/L)	33	47	o	45	5
Sucrose (% of total sugar)	−	−	o	50	5
Ash (g/L)	4.0	2.9−4.8	x	3.5	1
Sodium (mg/L)	14	−	o	30	1
Potassium (mg/L)	1900	1400−2300	x	1700	1
Calcium (mg/L)	80	60−120	o	110	1
Magnesium (mg/L)	100	70−150	x	90	1
Chloride (mg/L)	−	−	o	60	1
Nitrate (mg/L)	−	−	o	10	1
Phosphate (mg/L)	460	350−600	x	400	1
Sulfate (mg/L)	−	−	o	150	1
Citric acid (g/L)	9.4	7.6−11.5	x	8.0	4
Isocitric acid (mg/L)	90	65−130	x	70	4
Ratio of citric acid to isocitric acid[d]	105	80−130	o	130	4
L-Malic acid (g/L)	1.7	1.1−2.9	o	2.5	1
Prolin (mg/L)	800	450−1300	x	575	1
Formol value (0.1 mol/L NaOH per 100 mL)[e]	20	15−26	x	18	1
Flavonoid glycosides					
calc. as hesperidin (mg/L)	800	500−1000	o	1000	6
Water soluble pectins calc. as					
galacturonic acid anhydride (mg/L)	300	−	o	500	6

[a] Minimal (x) maximal (o) guide value.
[b] Indicator for: 1 fruit content, 2 heat or oxidation damage, 3 microbiological spoilage, 4 acidification, 5 sweetening, 6 extract of expressed residue, 7 as 6 but aromatized with peel oil.
[c] 1° Brix = 1 g of extract in 100 g of solution.
[d] cf. 18.1.2.4.
[e] Formol titration: after the addition of formaldehyde to a solution of the sample at pH 8−9, the free amino acids are determined by titration with sodium hydroxide solution.

champagne when the must is adulterated with fig juice.

The analytical importance of amino acid (cf. 18.1.2.1.2), protein, enzyme (cf. 18.1.2.1.1), and carotinoid patterns (cf. 18.1.2.3.2) have already been mentioned.

18.4.3 Abundance Ratios of Isotopes

The content of the isotopes ^2H and ^{13}C is a criterion of the origin of the food or of individual constituents, e.g., sugar used to sweeten fruit juice. The method is based on the fact that

Table 18.37. Phenolic compounds as indicator substances for the detection of adulteration of fruit products

Compound	Occurrence	Detection
Quercetin-3-rutinoside	Common, but not in strawberries	Elderberry juice in strawberry juice
Quercetin-3-O-(2''-O-α-L-rhamnosyl-6''-O-α-L-rhamnosyl)-β-D-glucoside	Red currants	Red currants in products from black currants
Naringin or naringenin	Grapefruits	Grapefruit juice in orange juice
Apigenin-6-C-β-D-glucopyranosyl-8-C-α-L-arabinopyranoside (schaftoside)	Figs	Fig juice in grape juice
Phloretin-2'-glucoside (phloridzine) and phloretinxyloglucoside	Apples	Apple juice in pear juice
Isorhamnetinglucoside	Pears	Pear juice in apple juice

isotopomeric molecules, e.g., $^{12}CO_2$ and $^{13}CO_2$, react at different rates in biochemical and chemical reactions (kinetic isotope effect). In general, the molecules with the heavier isotope react slower, so that this isotope is enriched in the products.

The resulting change in the abundance ratio is expressed as the δ-value, based on an international standard (Table 18.38).

$$\delta = \frac{R_{sample} - R_{standard}}{R_{standard}} \times 1000 \ [\text{‰}] \qquad (18.50)$$

$$R = \frac{c_1}{c_2} \qquad (18.51)$$

c_1/c_2: concentrations of heavy/light isotopes.

Table 18.38. Abundance of important isotopes and international standards for their determination

Isotope	Rel. mean natural abundance [atom %]	International standard	
		Name	R^a
^1H	99.9855	V-SMOW[b]	0.00015576
^2H	0.0145		
^{12}C	98.8920	PDB[c]	0.0112372
^{13}C	1.108		

[a] Abundance ratio (Formula 18.51).
[b] Vienna Standard Mean Ocean Water.
[c] Pee Dee Belemnite ($CaCO_3$ from the Pee Dee formation in South Carolina).

The δ(^{13}C) value, which is -8 ± 1 ‰ for atmospheric CO_2, increases during CO_2 fixation as a function of the type of photosynthesis of the plant (Table 18.39). The discrimination in C_3-plants is the greatest and is caused by the kinetic isotope effect in the reaction catalyzed by ribulose-1,5-biphosphate carboxylase. It is considerably less in C_4-plants. CAM plants occupy an intermediate position (Table 18.39) because the C_3- or the C_4-path is taken depending on the growth conditions.

The large differences in the masses of ^1H_2O, $^2H^1HO$, and ^2H_2O result in considerable thermodynamic isotope effects on phase transitions. On evaporation, deuterium (^2H) correspondingly decreases in the volatile phase, so that surface-, ground-, and rain-water contains less ^2H than the oceans. The ^2H enrichment in the oceans is greatest at the equator and decreases with increasing latitude because the amount of water evaporating depends on the temperature.

The hydrogen of plant foods comes from precipitation and from the ground-water in that particular location. Therefore, plants of the same type of photosynthesis, which are cultivated at different places, differ in their δ(^2H) values. Kinetic isotope effects in plant metabolism, which due to the mass difference $^2H/^1H$ are much higher than in the case of $^{13}C/^{12}C$, also have an effect on the δ(^2H) values.

Table 18.39. Isotope discrimination in primary photosynthetic CO_2 binding

Plant group	CO_2 Acceptor	$\delta(^{13}C)$ value ‰	Foods
C_3-Plant	D-Ribulose-1,5-bis-phosphate carboxylase (RuPB-C)	-32 to -24	Wheat, rice, oats, rye, potatoes, barley, batata, soybean, orange, sugar beet, grapes
C_4-Plant	Phosphoenolpyruvate carboxylase (PEP-C)	-16 to -10	Corn, millet, sugar cane
CAM-Plant [a]	RuBP-C/REP-C	-30 to -12	Pineapples, vanilla, cactaceae, agave

[a] CAM: Crassulacean acid metabolism.

For isotope analysis, the sample is subjected to catalytic combustion to give CO_2 and H_2O. After drying, the $^{13}C/^{12}C$ ratio in CO_2 is determined by mass spectrometry. The $^{2}H/^{1}H$ ratio is determined in hydrogen, which is formed by reducing the water obtained from catalytic combustion. The $^{2}H/^{1}H$ ratio can change by $^{2}H/^{1}H$ exchange, e.g., as undergone by OH groups. Therefore, such groups are eliminated before combustion. For example, only the $\delta(^{2}H)$ values of the CH-skeleton in carbohydrates are determined after conversion to the nitrate ester.

Sweetening orange juice with cane sugar or glucose-fructose syrup from corn starch lowers the $\delta(^{13}C)$ value of sugar, which is -25.5 ‰ in the native juice (Table 18.40). On the other hand, the addition of beet sugar (C_3-plant) can be recognized only via the $\delta(^{2}H)$ value. The addition of synthetic products from petrochemicals ($\delta(^{13}C)$: -27 ± 5 ‰) to foods from C_3-plants cannot be detected via the $\delta(^{13}C)$ value, but via the $\delta(^{2}H)$ value in many cases.

Table 18.40. $\delta(^{13}C)$ and $\delta(^{2}H)$ values for orange juice and sugar of different origins

Food	$\delta(^{13}C)$ (‰)	$\delta(^{2}H)$ (‰)
Orange juice, freeze-dried	-25.6 ± 0.8	n.a.
Sucrose isolated from orange juice	-25.5 ± 2.5	-22 ± 10
Beet sugar	-25.6 ± 1.0	-135 ± 25
Cane sugar	-11.5 ± 0.5	-50 ± 20
Glucose-fructose syrup (corn)	-10.8 ± 0.9	-31

n.a.: not analyzed.

Apart from the global ^{13}C and ^{2}H contents of food constituents, the intramolecular distributions of these isotopes are typical of the origin and, therefore, of great analytical importance. They can be measured after chemical decomposition of the substance or with ^{13}C or ^{2}H NMR spectroscopy (example in 5.5.1.5).

18.5 Literature

Bell, E. A., Charlwood, B. Y. (Eds.): Secondary plant products. Springer-Verlag: Berlin. 1980

Berger, R. G., Shaw, P. E., Latrasse, A., Winterhalter, P.: Fruits I–IV. In: Volatile compounds in foods and beverages (Ed.: Maarse, H.) Marcel Dekker, New York, 1991

Biale, J. B.: Growth, maturation, and senescence in fruits. Science *146*, 880 (1964)

Bricout, J., Koziet, J.: Control of authenticity of orange juice by isotopic analysis. J. Agric. Food Chem. *35*, 758 (1987)

Demole, E., Enggist, P., Ohloff, G.: 1-p-Menthene-8-thiol: A powerful flavor impact constituent of grapefruit juice (*Citrus paradisi* MacFayden). Helv. Chim. Acta *65*, 1785 (1982)

Drawert, F., Müller, W.: Über die elektrophoretische Differenzierung und Klassifizierung von Proteinen. II. Dünnschichtisoelektrische Fokussierung von Proteinen aus Trauben verschiedener Rebsorten. Z. Lebensm. Unters. Forsch. *153*, 204 (1973)

Drawert, F., Görg, A., Staudt, G.: Über die elektrophoretische Differenzierung und Klassifizierung von Proteinen. IV. Disk-Elektrophorese und isoelektrische Fokussierung in Poly-Acrylamid-Gelen von Proteinen und Enzymen aus verschiedenen Erdbeerarten, -sorten und Artkreuzungsversuchen. Z. Lebensm. Unters. Forsch. *156*, 129 (1974)

Fowden, L., Lea, P.J., Bell, E.A.: The nonprotein amino acids of plants. Adv. Enzymol. *50*, 117 (1979)

Handwerk, R.L., Coleman, R.L.: Approaches to the citrus browning problem. A review. J. Agric. Food Chem. *36*, 231 (1988)

Haslam, E., Lilley, T.H.: Natural astringency in foodstuffs – a molecular interpretation. Crit. Rev. Food Sci. Nutr. *27*, 1 (1988)

Herrmann, K.: Obst, Obstdauerwaren und Obsterzeugnisse. Verlag Paul Parey: Berlin. 1966

Hermann, K.: Occurrence and content of hydroxycinnamic and hydroxybenzoic acid compounds in foods. Crit. Rev. Food Sci. Nutr. *28*, 315 (1989)

Hulme, A.C.: Problems in the biochemistry of fruits. In: Biochemical principals of the fruit industry (Eds.: Kretovich, V.L., Pijanowski, E.), p. 143, Pergamon Press: Oxford. 1963

Hulme, A.C. (Ed.): The biochemistry of fruits and their products. Vol. 1, Academic Press: London. 1970

Jurd, L.: Reactions involved in sulfite bleaching of anthocyanins. J. Food Sci. *29*, 16 (1964)

Jurd, L., Asen, S.: Formation of metal and copigment complexes of cyanidin 3-D-glucoside. Phytochemistry *5*, 1263 (1966)

Kader, A.A., Zagory, D., Kerbel, E.L.: Modified atmosphere packaging of fruits and vegetables. Crit. Rev. Food Sci. Nutr. *28*, 1 (1989)

Linskens, H.F., Jackson, J.F. (Eds.): Analysis of nonalcoholic beverages. In: Modern methods of plant analysis. Volume 8. Springer-Verlag: Berlin. 1988

Macheix, J.-J., Fleuriet, A., Billot, J.: Fruit Phenolics. CRC Press, Boca Raton, FL. 1990

Maga, J.A.: Amines in foods. Crit. Rev. Food Sci. Nutr. *10*, 373 (1978)

Mosel, H.D., Herrmann, K.: Changes in catechins and hydroxycinnamic acid derivatives during development of apples and pears. J. Sci. Food Agric. *25*, 251 (1974)

Nagy, S., Attaway, J.A. (Eds.): Citrus nutrition and quality. ACS Symposium Series 143, American Chemical Society: Washington, D.C. 1980

Pantastico, E.B. (Ed.): Postharvest physiology, handling and utilization of tropical and subtropical fruits and vegetables. AVI Publ. Co.: Westport, Conn. 1975

Phan, C.T., Pantastico, E.B., Ogata, K., Chachin, K.: Respiration and respiratory climacteric. In: Postharvest physiology, handling and utilization of tropical and subtropical fruits and vegetables (Ed.: Pantastico, E.B.), p. 86, AVI Publ. Co.: Westport, Conn. 1975

Pickenhagen, W., Velluz, A., Passerat, J.-P., Ohloff, G.: Estimation of 2,5-dimethyl-4-hydroxy-3(2H)-furanone in cultivated and wild strawberries, pineapples and mangoes. J. Sci. Food Agric. *32*, 1132 (1981)

Salunkhe, D.K.: Biogenesis of aroma constituents of fruits and vegetables. Crit. Rev. Food Sci. Nutr. *8*, 161 (1977)

Schmid, W., Grosch, W.: Quantitative Analyse flüchtiger Aromastoffe mit hohen Aromawerten in Sauerkirschen, Süßkirschen und Kirschkonfitüren. Z. Lebensm. Unters. Forsch. *183*, 39 (1986)

Schobinger, U. (Ed): Frucht- und Gemüsesäfte. Verlag Eugen Ulmer: Stuttgart. 1978

Schormüller, J.: Die Erhaltung der Lebensmittel. Ferdinand Enke Verlag Stuttgart. 1966

Shaw, P.E., Wilson, C.W., III: Importance of nootkatone to the aroma of grapefruit oil and the flavor of grapefruit juice. J. Agric. Food Chem. *29*, 677 (1981)

Siewek, F., Galensa, R., Herrmann, K.: Nachweis eines Zusatzes von Feigensaft zu Traubensaft und daraus hergestellten alkoholischen Erzeugnissen über die HPLC-Bestimmung von Flavon-C-glykosiden. Z. Lebensm. Unters. Forsch. *181*, 391 (1985)

Stöhr, H., Herrmann, K.: Die phenolischen Inhaltsstoffe des Obstes, VI. Die phenolischen Inhaltsstoffe der Johannisbeeren, Stachelbeeren und Kulturheidelbeeren. Veränderungen der Phenolsäuren und Catechine während Wachstum und Reife von schwarzen Johannisbeeren. Z. Lebensm. Unters. Forsch. *159*, 31 (1975)

Takeoka, G., Buttery, R.G., Flath, R.A., Teranishi, R., Wheeler, E.L., Wieczorek, R.L., Guentert, M.: Volatile constituents of pineapple (*Ananas comosus*). ACS Symp. Ser. *388*, 223 (1989)

Van Straten, S. (Ed.): Volatile compounds in food. 4th edn., with supplements 1–7, Central Institute for Nutrition and Food Research TNO: Zeist, the Netherlands. 1977–1981

Wald, B., Galensa, R.: Nachweis von Fruchtsaftmanipulationen bei Apfel- und Birnensaft. Z. Lebensm. Unters. Forsch. *188*, 107 (1989)

Wegler, R. (Ed.): Chemie der Pflanzenschutz- und Schädlingsbekämpfungsmittel. Bde. 1–8, Springer-Verlag: Berlin. 1970–1982

Williams, A.A.: The flavour of non-citrus fruits and their products. In: International symposium on food flavors (Eds.: Adda, J., Richard, H.), p. 46, Technique et Documentation (Lavoisier): Paris. 1983

Winkler, F.J., Schmidt, H.-L.: Einsatzmöglichkeiten der ^{13}C-Isotopen-Massenspektrometrie in der Lebensmitteluntersuchung. Z. Lebensm. Unters. Forsch. *171*, 85 (1980)

19 Sugars, Sugar Alcohols and Honey

19.1 Sugars, Sugar Alcohols and Sugar Products

9.1.1 Foreword

Only a few of the sugars occurring in nature are used extensively as sweeteners. Besides sucrose (saccharose), other important sugars are: glucose (starch sugar or starch syrup); invert sugar (equimolar mixture of glucose and fructose); maltose; lactose; and fructose. In addition, some other sugars and sugar alcohols (polyhydric alcohols) are used in diets or for some technical purposes. These include sorbitol, xylitol, mannitol, maltulose, isomaltulose, maltitol, isomaltitol, lactulose and lactitol. Some are used commonly in food and pharmaceutical industries, while applications for others are being developed. Table 19.1 reviews production data, while Table 19.2 lists data on relative sweetness, source and means of production, and Table 19.3 gives nutritional and physiological properties. Whether one of the compounds discussed will be successful as

a sweetener depends on nutritional, physiological and processing properties, cariogenicity as compared to sucrose, economic impact, and the quality and intensity of the sweet taste.

19.1.2 Processing Properties

The potential of a compound for use as a sweetener depends upon its physical, processing and sensory properties. Important physical properties are solubility, viscosity of the solutions, and hygroscopicity. Figure 19.1 shows that the solubility of sugars and their alcohols in water is variable and affected to a great extent by temperature.

There are similar temperature and concentration influences on the viscosity of aqueous solutions of many sugars and sugar alcohols. As an example, Fig. 19.2 shows viscosity curves for sucrose as a function of both temperature and concentration.

The viscosity of glucose syrup depends on its composition. It increases as the proportion of the high molecular weight carbohydrates increases (Fig. 19.3).

Table 19.1. Approximate world production of sweeteners of carbohydrate origin (1978)

Sweetener	Amount (1,000 t)
Saccharose	90,000
Glucose syrup (80%)	4,000
Glucose	2,000
Isoglucose (72%)	1,500
Sorbitol (70%) and sorbitol, crystalline	300
Lactose	150
Fructose	20
Mannitol	8
Xylitol	5
Hydrogenated glucose syrup (75%)	4

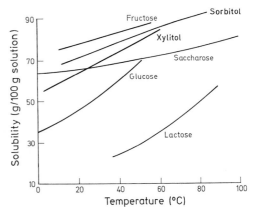

Fig. 19.1. Solubility of sugars and sugar alcohols in water. (according to *Koivistoinen*, 1980)

Table 19.2. Sweeteners of carbohydrate origin

Name	Relative sweetness[a]	Starting material, applied process
Products of economic importance		
Saccharose	1.00	Isolation from sugar beet and sugar cane
Glucose	0.5–0.8	Hydrolysis of starch with acids and/or enzymes (α-amylase + glucoamylase)
Fructose	1.1–1.7	a) Hydrolysis of saccharose, followed by separation of the hydrolysate by chromatography. b) Hydrolysis of starch to glucose, followed by isomerization and separation by chromatography
Lactose	0.2–0.6	Isolation from whey
Mannitol	0.4–0.5	Hydrogenation of fructose
Sorbitol	0.4–0.5	Hydrogenation of glucose
Xylitol	1.0	Hydrogenation of xylose
Glucose syrup (starch syrup)	0.3–0.5[b]	Hydrolysis of starch with acids and/or enzymes; hydrolysate composition is strongly affected by process parameters (percentage of glucose, maltose, maltotriose and higher oligosaccharides)
Maltose syrup		As glucose syrup; process parameters adjusted for higher proportion of maltose in hydrolysate (amylase from *Aspergillus oryzae*)
Glucose/fructose syrup (isoglucose, high fructose syrup)	0.8–0.9	Isomerization of glucose to glucose/fructose mixture with glucose isomerase; conversion degree 45–50%
Invert sugar		Hydrolysis of saccharose
Hydrogenated glucose syrup	0.3–0.8	Hydrogenation of starch hydrolysate (glucose syrup); composition is highly dependent on starting material (content of sorbitol, maltitol and hydrogenated oligosaccharides)
Maltitol syrup		Hydrogenation of maltose syrup
Products of potential economic importance		
Arabinitol	approx. 1.0	Hydrogenation of arabinose
Galactitol		Hydrogenation of galactose
Galactose	0.3–0.5	Hydrolysis of lactose, followed by separation of hydrolysate
Isomaltitol	0.5	Hydrogenation of isomaltulose (palatinose)
Lactitol	0.3	Hydrogenation of lactose
Lactulose	approx. 0.6	Isomerization of lactose
Maltose	0.3–0.6	Hydrolysis of starch
Maltitol	approx. 0.9	Hydrogenation of maltose
L-Sorbose	0.6–0.8	From glucose microbiologically
D-Xylose	approx. 0.5	Hydrolysis of hemicellulose
Palatinit		Isomerization of saccharose to isomaltulose (palatinose) followed by hydrogenation to a mixture of glucopyranosido-sorbitol and glucopyranosido-mannitol

[a] Sweetness is related to saccharose sweetness (= 1); the values are affected by sweetener concentration.
[b] Sweetness value is strongly influenced by syrup composition.

Table 19.3. Nutritional/physiological properties of carbohydrate-derived sweeteners

Sweetener	Resorption	Utilization in metabolism	Effect on blood sugar level and insulin secretion	Other properties
Sucrose	Effective after being hydrolyzed	Hydrolysis to fructose and glucose	Moderately high	Cariogenic
Glucose	Effective	Insulin-dependent in all tissues	High	Less cariogenic than sucrose
Fructose	Faster than by diffusion process	In liver to an extent of 80%	Low	Accelerates alcohol conversion in liver
Lactose	Effective after being hydrolyzed	Hydrolysis to glucose and galactose	High	Intolerance by humans lacking lactase enzyme; laxative effect
Sorbitol	Diffusion	Oxidation to fructose	Low	Slightly cariogenic and laxative
Mannitol	Diffusion	Partially utilized by liver	Low	Slightly cariogenic and laxative
Xylitol	Diffusion	Utilized preferentially by liver and red blood cells	Low	Not cariogenic, available data indicate an anticariogenic effect; mildly laxative
Hydrogenated glucose syrup	After hydrolysis glucose effective; sorbitol by diffusion	Variable depending on composition	Variable, composition dependent	Slightly cariogenic; mildly laxative
Arabinitol	Diffusion	Not metabolized by humans	None	Side effects unknown; probably laxative
Galactose	Effective	Isomerization to glucose	High	Forms cataracts in the eyes in feeding trials with rats; probably laxative
Isomaltitol	None	Probably not metabolized	None	Side effects unknown; strongly laxative
Lactitol	None	Partial hydrolysis to galactose and sorbitol	None	Side effects unknown; strongly laxative
Lactulose	None	No hydrolysis	None	Effects the N-balance; strongly laxative
Maltitol	Effective as glucose after hydrolysis; sorbitol by diffusion	Hydrolysis to glucose and sorbitol	Probably slight	Side effects unknown; laxative
Maltose	Effective after hydrolysis	Hydrolysis to glucose	High	Cariogenic; intravenously given it appears to be utilized directly and, like glucose, it is insulin-dependent
L-Sorbose	Diffusion	Utilized preferentially by liver	Probably slight	Feeding trials with dogs revealed hemolytic anemia at a higher dosage; probably laxative

Table 19.3 (continued)

Sweetener	Resorption	Utilization in metabolism	Effect on blood sugar level and insulin secretion	Other properties
D-Xylose	Diffusion	Not metabolized by humans	None	Forms cataract in the eyes in feeding trials with rats; probably laxative
Palatinit		Partial hydrolysis to glucose, sorbitol, and mannitol	Probably slight	Side effects unknown

Figure 19.4 shows the water absorption characteristics of several sweeteners. Sorbitol and fructose are very hygroscopic, while other sugars absorb water only at higher relative humidities. Chemical reactions of sugars were covered in detail in Chapter 4. Only those reactions important from a technological viewpoint will be emphasized here.

All sugars with free reducing groups are very reactive. In mildly acidic solutions monosaccharides are stable, while disaccharides hydrolyze to yield monosaccharides. Fructose is maximally stable at pH 3.3; glucose at pH 4.0. At lower pH's dehydration reactions prevail, while the *Lobry de Bruyn-van Ekenstein* rearrangement occurs at higher pH's. Reducing sugars are unstable in mildly alkaline solutions, while nonreducing disaccharides, e.g., sucrose, have their stability maxima in this pH region.

The thermal stability of sugars is also quite variable. Sucrose and glucose can be heated in neutral solutions up to 100 °C, but fructose decomposes at temperatures as low at 60 °C. Sugar alcohols are very stable in acidic or alkaline solutions. Relative taste intensity values for various sweeteners are found in Table 19.2. Taste intensity within a food can depend on a series of parameters, e.g., aroma, pH or food texture. Creams and gels with the

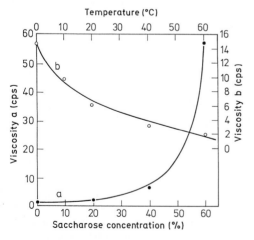

Fig. 19.2. Viscosity of aqueous saccharose solutions as a function of (a) saccharose concentration (20 °C) and temperature (40% saccharose). (according to *Shallenberger* and *Birch*, 1975)

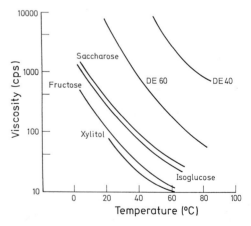

Fig. 19.3. Viscosity of some sugar solutions. Glucose syrup DE40: 78 weight-%; glucose syrup DE60: 77 weight-%; all other sugar solutions: 70 weight-%. (according to *Koivistoinen*, 1980)

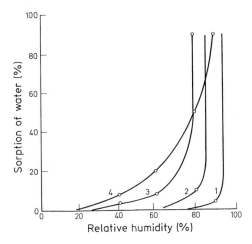

Fig. 19.4. Sorption of water by sugars at room temperature. 1 Saccharose, 2 xylitol, 3 fructose, 4 sorbitol. (according to *Koivistoinen*, 1980)

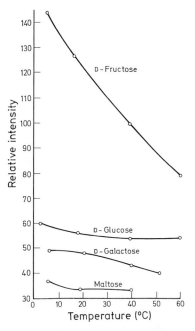

Fig. 19.5. Sugar sweetness intensity versus temperature. At all temperatures the saccharose taste intensity is 100. (according to *Shallenberger*, 1975)

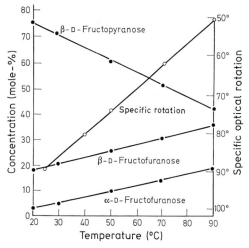

Fig. 19.6. Fructose mutarotation equilibrium as affected by temperature. (according to *Shallenberger*, 1975)

same amounts of sweetener are often less sweet than the corresponding aqueous solutions. The sweet taste intensity may also depend on temperature (Fig. 19.5), an effect which is particularly pronounced with fructose – hot fructose solutions are less sweet than cold ones. The cause of such effects is the mass equilibrium of sugar isomers in solution. At higher temperatures the concentration of the very sweet β-D-fructopyranose drops in favor of both the less sweet α-D-fructofuranose and the β-D-fructofuranose (Fig. 19.6). Such strong shifts in isomer concentrations do not occur with glucose, hence its sweet taste intensity is relatively unchanged in the range of 5–50 °C.

19.1.3 Nutritional/Physiological Properties

The role of carbohydrates in metabolism is primarily determined by the ability of disaccharides to be hydrolyzed in the gastrointestinal tract and by the mechanisms of monosaccharide absorption.

The human organism hydrolyzes sucrose, lactose and oligosaccharides of the maltose and isomaltose type. The enzyme lactase, which is responsible for lactose hydrolysis, is lacking in some adults. Glucose and galactose are actively transported, while all other monosaccharides are transported only by diffusion. Sugar

phosphorylation occurs preferentially in the liver. All monosaccharides which are metabolized can be interconverted. Sugar alcohols are oxidized: sorbitol → fructose, xylitol → xylulose. However, only glucose can enter the insulin-regulated and -dependent energy metabolism and be utilized by all tissues. Galactose is rapidly transformed into glucose and is therefore nutritionally equal to glucose. Oral intake of glucose and galactose causes a rapid increase in blood sugar levels and, as a consequence, insulin secretion. All other monosaccharides are primarily metabolized by the liver and do not directly affect glucose status or insulin release. After fructose intake, insulin secretion is only 50% of that after glucose intake. Sugars to be avoided by diabetics are, therefore, glucose, galactose, lactose and maltose. Fructose, xylitol, sorbitol and mannitol can be well tolerated by diabetics and sucrose, invert sugar and hydrogenated glucose syrup only moderately well.

19.1.4 Individual Sugars and Sugar Alcohols

19.1.4.1 Sucrose (Beet Sugar, Cane Sugar)

19.1.4.1.1 General Outline

Sucrose is widely distributed in nature, particularly in green plants, leaves and stalks (sugar cane 12–26%; sweet corn 12–17%; sugar millet 7–15%; palm sap 3–6%); in fruits and seeds (stone fruits, such as peaches; core fruits, such as sweet apples; pumpkins; carobs or St. John's bread; pineapples, coconuts; walnuts; chestnuts); and in roots and rhizomes (sweet potatoes 2–3%; peanuts 4–12%; onions 10–11%; beet roots and selected breeding forms 3–20%). The two most important sources for sucrose production are sugar cane (*Saccharum officinarum*) and sugar beet (*Beta vulgaris ssp. vulgaris* var. *altissima*). Cane sugar and beet sugar are distinguished by the spectrum of accompanying substances and by the $^{13}C/^{12}C$ ratio, which can be used for identification (cf. 18.4.3, Table 18.40).

Sucrose is the most economically significant sugar and is produced industrially in the largest quantity. Table 19.4 provides an overview of the annual world production of beet and cane sugar. Table 19.5 lists the main producers. Honey is the oldest known sweetener and has relatively recently been displaced by cane sugar. Cane sugar was brought to Europe from

Table 19.4. World production of sugar (beet/cane)

Year	Total production 10^6 t	Cane sugar %
1900/01	11.3	47.0
1920/21	16.4	70.5
1940/41	30.9	62.3
1960/61	61.1	60.3
1965/66	71.1	61.8
1970/71	82.3	64.2
1975/76	92.2	64.6
1980/81	98.4	66.6
1981/82	108.5	66.2
1982/83	98.6	62.9

Persia by the Arabs. After the Crusades, it was imported by Cyprus and Venice and, later, primarily by Holland, from Cuba, Mexico, Peru and Brazil.

In 1747 *A.S. Marggraf* discovered sucrose in beets and in 1802 *F. Achard* was the first to produce sucrose commercially from sugar beets. The new sugar source had great economic impact; the more so when sucrose accumulation in the beets was increased by selection and breeding.

19.1.4.1.2 Production of Beet Sugar

The isolation of beet sugar will be described first because the processes used in material preparation and sugar separation have been developed to perfection. These processes were later transferred to the production of cane sugar from the clear juice concentration stage onwards. In fact, cane sugar was processed fairly primitively for a long time.

Prolonged selection efforts have led to sugar beets which reach their maximum sucrose content of 15–20% in the middle of October. The average for the past 5 years in FR Germany is 16.3%. The early yield achieved by *F. Achard* of 4.5 kg/100 kg beets has been increased to about 14 kg. Currently, beet varieties have a high sugar content and small amounts of nonsugar substances. Anatomically they have a favorable shape, i.e. are small and slim with a smooth surface, and have a firm texture. Since the sugar accumulation in beets peaks in October and since sugar decomposition due to respiration occurs during

Table 19.5. Production of sugar beet, sugar cane and saccharose in 1996 (1,000 t)

Continent	Sugar cane	Sugar beet	Saccharose[a]
World	1,192,555	255,500	124,000
Africa	80,296	3,859	8,357
America, North-, Central-	156,562	25,134	19,668
America, South-	404,864	2,806	20,960
Asia	505,603	38,869	40,930
Europe	169	187,831	28,095
Oceania	45,062	–	5,991

Country	Sugar cane	Country	Sugar beet	Country	Saccharose[a]
Brazil	324,435	France	30,720	India	15,150
India	255,000	Germany	26,433	Brazil	14,850
Thailand	62,422	USA	24,104	China	7,571
China	55,557	Ukraine	22,812	USA	6,383
Mexico	46,980	Russian Fed.	16,132	Thailand	6,300
Pakistan	45,230	Poland	15,200	Australia	5,491
Australia	40,649	Turkey	14,455	France	4,929
Cuba	40,000	Italy	12,125	Mexico	4,660
Colombia	32,500	China	11,400	Germany	4,554
Indonesia	32,053	UK	9,555	Kuba	4,530
		Spain	7,359	Ukraine	2,935
Σ (%)[b]	78	Netherlands	6,416	Pakistan	2,684
				Poland	2,478
		Σ (%)[b]	77	South Africa	2,420
				Indonesia	2,404
				Russian Fed.	2,230
				Colombia	2,043
				Turkey	2,011
				Σ (%)[b]	76

[a] As raw (centrifuged) sugar.
[b] World production = 100 %.

subsequent storage of beets, they are rapidly processed from the end of September to the middle of December.

The beet sugar extract contains about 17% sucrose, 0.5% inorganic and 1.4% organic nonsucrose matter. Invert sugar and raffinose content is 0.1% (in molasses this may be as high as 2%). The trisaccharide kestose (cf. Table 4.14), which is present in the extract, is an artifact generated in the course of beet processing. In addition to pectic substances, beet extract contains saponins which are responsible for foaming of the extract and binding with sugars. N-containing, nonsugar constituents of particular importance are proteins, free amino acids, and their amides, (e. g., glutamine) and glycine betaine ("betaine"). These constituents are 0.3% of beets and about 5% of molasses. Beet ash averages 28% potassium, 4% sodium, 5% calcium and 13% phosphoric acid, and contains numerous trace elements. The nonsugar constituents of the sugar extract also include steam-distillable odorous compounds, phenolic acids, e. g., ferulic acid, and numerous beet enzymes which are extensively inactivated during extract processing. These enzymes, e. g., polyphenol oxidase, can induce darkening through melanin build-up,

Table 19.6. Sugar consumption[a] in selected countries

Continent/ Country	Consumption	Continent/ Country	Consumption
World	19.4	Holland	39.6
Europe	38.1	Switzerland	38.6
North America	40.0	Argentina	37.2
South America	40.1	France	36.1
Africa	13.0	FR Germany	35.1
Asia	9.2	Belgium/ Luxembourg	33.8
Australia	50.9	Italy	30.5
Brazil	46.2	Spain	28.2
Germany DR	45.4	Pakistan	24.0
Hungary	45.0	Japan	23.0
Canada	43.7	India	20.7
USSR	42.4	Egypt	20.2
UK	41.8	Sri Lanka	13.9
Austria	40.8	South Korea	13.4
USA	39.7	China	4.8

[a] kg/year and head.

with the color being transferred during beet extraction into the raw sugar extract.

Beet processing involves the following steps:

- *Flushing and cleaning* in flushing chutes and whirlwashers.
- *Slicing* with machines into thin shreds (cossettes) with the shape of "shoestrings" 2–3 mm thick and 4–7 mm wide.
- *Extraction* by leaching of beet slices. The extraction water is adjusted to pH 5.6–5.8 and to 30–60°dH with $CaCl_2$ or $CaSO_4$ to stabilize the skeletal substances of the slices in the following pressing step. To denature the cells, the slices are first heated to 70–78 °C for ca. 5 minutes (preliminary scalding) and then extracted at 69–73 °C for 70 to 85 minutes. To eliminate thermophilic microorganisms in the extraction system, 30–40% formaldehyde solution is intermittently added to the raw material at intervals of 8–24 hours in amounts of 0.5–1% of the raw juice accumulating hourly.

This was once performed in a so-called diffusion battery of 12–14 bottom sieve-equipped cylindrical containers (diffusers) connected in series and operating discontinuously on a countercurrent principle. Today this battery operation has been replaced to a great extent by a continuous and automatically operated extraction tower into which the shreds are introduced at the bottom while the extraction fluid flows from the top. The extracted shreds (pulp) are discharged at the top. The pulp contains residual sugar of approx. 0.2% of the beet dry weight.

The pulp is pressed, dried on band dryers, and pelleted. It serves as cattle feed. Before drying, 2–3% of molasses and, for nitrogen enrichment, urea is sometimes added.

- *Raw Sugar Extract Purification* (liming and carbonatation). Juice purification results in the removal of 30–40% of the nonsugar substances and has the following objectives:
  - Elimination of fibers and cell residue
  - Precipitation of proteins and polysaccharides (pectins, arabans, galactans)
  - Precipitation of inorganic (phosphate, sulfate) and organic anions (citrate, malate, oxalate) as calcium salts and precipitation of magnesium ions as $Mg(OH)_2$
  - Degradation of reducing sugar (invert sugar, galactose) and, therefore, suppression of the *Maillard* reaction during evaporation
  - Conversion of glutamine to pyrrolidone carboxylic acid and asparagine to aspartic acid. However, these reactions proceed only partially under the usual conditions of juice purification.
  - Adsorption of pigments on the $CaCO_3$ formed.

Moreover, the sludge formed must be easily settleable and filtrable.

The raw juice from the extraction tower is turbid and greyish black in color due to the enzymatic oxidation of phenols, especially of tyrosine, and to the presence of phenol-iron complexes. The raw juice has a pH of 6.2 and contains on an average 15% of solids, of which sucrose accounts for 13.5%. It is first mechanically filtered and then treated with lime milk in two steps (preliming and main liming). Preliming is generally conducted at 60–70 °C up to a pH of 10.8–11.9 with a residence time of at least 20 minutes. Main liming is conducted at 80–85 °C with a residence time of ca. 30 minutes up to a total CaO content of

2–2.5% in the juice. A number of organic acids and phosphate are precipitated as calcium salts and colloids flocculate.

In order to remove excess calcium, decompose the calcium saccharate ($C_{12}H_{22}O_{11}$ × 3CaO) formed, and transform the precipitated turbidity-causing solids into a more filtrable form, the solution is quickly gassed with an amount of carbon dioxide required for the formation of calcium carbonate. Carbonatation is also performed in two steps. In the 1st. carbonatation step at 85 °C, the pH is adjusted to 10.8–11.9. The sludge formed (50–60 g solids/l) is separated at 90–95 °C via decanters and filters and washed on the filters up to a residual sugar content of 0.1–1%. In the 2nd. carbonatation step, a pH of 8.9–9.2 is reached at 94–98 °C. The small amount of sludge (1–3 g solids/l) is filtered off. To lighten and stabilize the color during subsequent evaporation, 50 g/m³ of SO_2 (sulfitation) are frequently added to the thin syrup (juice). Subsequently, the solution is again clarified by filtration, finally producing a clear light-colored thin juice with a solids content of 15 to 18%.

Apart from the classical juice purification processes, different variants are known, which have both advantages and disadvantages. They yield carbonatation juices that can be decanted and filtered more easily. However, these juices are frequently more thermolabile because of incomplete destruction of invert sugar and consequently discolor on evaporation.

Ion exchangers have become important in juice purification. They soften the thin syrup and prevent the formation of hardness scale on the evaporator coils. The substitution of alkaline earth ions (Mg) for the alkali ions is beneficial because it decreases the sugar lost in the molasses by ca. 30% due to the stronger hydration of the alkaline earth ions. Bleaching of thin syrup is possible with activated carbon or with large-pore ion exchangers, which bind the pigments mainly by adsorption.

Extensive elimination (ca. 85%) of non-sugar substances with a corresponding increase in the sugar yield can be achieved by a combination of cation exchangers ($H^{\oplus}$ form) and anion exchangers (OH^- form) (complete desalting). To suppress inversion during the temporary pH drop, the operation must be conducted at low temperatures (14 °C). Higher temperatures (60 °C) can be used if the cations are first replaced by ammonium ions which are then eliminated as ammonia with the help of an anion exchanger or fixed on a mixed-bed exchanger. In comparison with the lime-carbon dioxide treatment, however, complete desalting has not yet gained acceptance.

- *Evaporation of the thin syrup* (15–18% solids) is achieved in multiple-stage evaporators (falling film evaporators, natural or forced circulation evaporators). Mildly alkaline conditions (pH 9) are maintained to prevent sucrose inversion. The boiling temperatures decrease in the range of 130–90 °C. The resultant thick syrup (yield of 25–30 kg/100 kg beet) is once more filtered. The syrup contains 68–72% solids and sucrose content is 61–67%. The raw, thin and thick syrups have purity quotients of approx. 89, 92 and 93, respectively, i.e. the percentage sucrose on a dry matter basis.

  During evaporation, calcium salts precipitate, glutamine still present is converted to pyrrolidone carboxylic acid with lowering of the pH, alkaline degradation of sugar occurs to a small extent, and darkening of the syrup occurs due to *Maillard* reaction and caramelization, depending on the process management (temperature, residence time in the evaporation stages).

- *Crystallization.* Multistage crystallization can be used to isolate 85–90% of the sucrose contained in the thick syrup. The remaining sucrose and practically all the non-sugar substances are found in the last mother liquor called molasses. The crystallization process is predominantly a discontinuous operation. However, efforts are being made to introduce a continuous process (evaporative crystallization and centrifugation).

  The thick syrup is evaporated in a boiling apparatus at 0.2–0.3 bar and 65–80 °C until slight supersaturation is achieved (evaporative crystallization). Crystallization is then initiated by seeding, e.g., by adding a dispersion of sucrose crystals (0.5–30 µm) in isopropanol. The mixture is further boiled until the crystals acquire the required size. In this process, the formation of both new crystals and crystal conglomerates is to be carefully prevented by intensive circulation (steam generation, stirring). The crystal paste (magma) with a crystal content of 50–60% is discharged into mashers for homogenization with constant stirring at a constant temperature.

A further crystallization occurs in part on very slow cooling to 35–40 °C (cooling crystallization). In this process, the viscosity of the mash must be maintained constant by the addition of water or mother syrup. Today, cooling crystallization is generally used only for after-product magma, but it will be of importance for raw sugar and white sugar.

Subsequently, the crystalline sugar from the mashers or massecuite is centrifuged in centrifugal baskets, eliminating the mother liquor called green syrup, which is returned to the process. The sugar (with the exception of raw sugar) is then freed from adhering syrup by washing with hot water and steam in the centrifuge. The resulting sugar solution (wash syrup) is fed back to the crystallization process. The presence of higher concentrations of raffinose in the magma (>1%, based on dry matter) reduces the rate of crystallization of sucrose and produces needle-shaped crystals. For this reason, raffinose is cleaved by α-galactosidase.

In this manner, thick syrup can be processed into raw sugar or consumer sugar (white sugar and refined sugar), depending on the process operation. The different crystallization schemes are simplified in Fig. 19.7. Raw sugar contains 1–1.2% of organic and 0.8–1% of inorganic nonsugar substances and 1–2% of water. It is light yellow to dark brown in color due to the adhering syrup. Like the after-product sugar (3–4% of organic and 1.5–2.5% of inorganic nonsugar substances and 2–3% of water) obtained in the last crystallization stage, raw sugar is generally not suitable for direct use. It is processed to consumer sugar in refineries.

In the refinery, the sugar is mashed into a magma with a suitable syrup, centrifuged, and washed with water and steam (affination). Thus, it directly yields a consumer sugar called affinated sugar. Another possibility is to dissolve the sugar and feed the resulting syrup (liquor) to a crystallization process which then yields refined sugar, a consumer sugar of the highest quality.

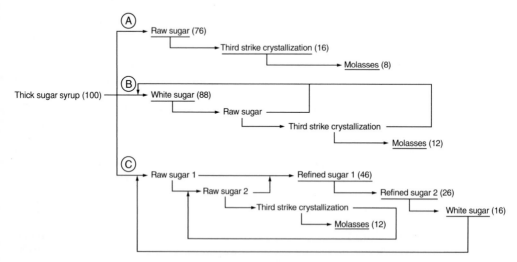

Fig. 19.7. Crystallization scheme for the production of A) raw sugar, B) white sugar, and C) refined sugar. The yields of sucrose (%), based on the amount of sucrose added with the thick syrup, are given in brackets behind the final products (underlined)

A simplified crystallization scheme for the production of white sugar is presented in Fig. 19.8. After affination and dissolving, the raw sugar and after-product sugar accumulating in the course of the process are boiled down together with the thick syrup, and the main part of the sugar finally crystallizes out of the supersaturated solution as white sugar. Centrifugation at 40–45°C yields not only crystals of 2–4 mm (first-product sugar), but also centrifugal syrup (green syrup) which is subjected to two further crystallization steps. The last discharge, a highly viscous brown syrup, is molasses. In the processing of thick syrup to refined sugar, first raw sugar is isolated exclusively. It is then dissolved and fed back into the crystallization process. In this way, the process is independent of variations in the quality of the thick syrup.

Processing losses in sucrose recovery from beets in 1974 were 0.4–0.9% (sugar deter-

Table 19.7. Production losses[a] during saccharose recovery from sugar beet

Processing step	1950	1974
Beet slice extraction	0.4–0.5	0.15–0.25
Sugar extract purification	0.1–0.2	0.02–0.05
Other steps	0.6–0.8	0.25–0.90
Total process	1.1–1.5	0.42–0.60

[a] Sugar amount in % based on the processed beet weight.

mined polarimetrically; and based on processed beet weight) and, when compared to 1950, represent a significant improvement of the sucrose yield (Table 19.7). This technological progress is also reflected in a rise of work productivity (work min/t beets), which was 130–150 in 1950 but only 12–30 in 1974.

19.1.4.1.3 Production of Cane Sugar

Sugar cane processing starts with squeezing out the sweet sap from thoroughly washed cane. For this purpose, the cane moves to a shredding machine where knives shred the stalks and then moves to crushing machines where a series of revolving heavy steel rollers squeeze the cane under high pressure. After the first roller, more than 60% of the cane weight is removed in the form of sap which contains 70% or more of the cane sucrose content. Repeated squeezing provides a sucrose yield of 93–97.5%. The squeezing may be combined with extraction by mixing the "bagasse" (the pressed cane) with hot water or dilute hot cane juice, followed by a final pressing. The experience gained in continuous thin beet syrup production is applied to sugar cane production, with a resultant energy saving and a rise in sugar yields.

Clarification and neutralization of the mildly acidic, raw extract (pH 4.8–5.0) is done by treatment with lime or lime and carbon dioxide. Further processing of the clarified pure syrup parallels that of sugar beet processing. The yield of raw cane sugar is 6–11% of the cane weight.

The "bagasse" is used as fuel, made into wallboard or used as insulation.

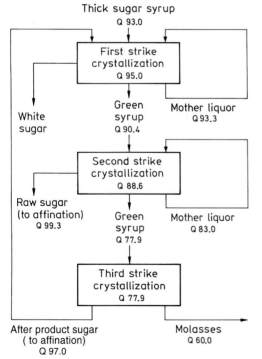

Fig. 19.8. White sugar evaporation and crystallization. Quotient Q: % saccharose in dry matter.

19.1.4.1.4 Other Sources for Sucrose Production

Some plants other than sugar beet or sugar cane can serve as sources of sucrose:

Date Sugar is obtained from the sweet, fleshy fruit of the date palm (Algeria, Iraq), which contains up to 81% sucrose in its solids.

Palm Sugar originates from various palm species, e. g., palmyra, saga or Toddy palm, coconut and Nipa palm grown in India, Sri Lanka, Malaysia and the Philippines, respectively.

Maple Sugar is obtained from the maple tree (*Acer saccharum*), found solely in North America (USA and Canada) and Japan. The sap, which drips from holes drilled in the maple tree trunk, flows down metal spiles into metal pails. This sap contains about 5% sucrose, minute amounts of raffinose and several other oligosaccharides of unknown structures. It is marketed in concentrated form either as maple syrup or as maple sugar. Aroma substances are important constituents of these products. The syrup also contains various acids, e.g., citric, malic, fumaric, glycolic and succinic acids. The main component of maple sugar is sucrose (88–99% of the total solids). Aroma constituents include vanillin, syringic aldehyde, dihydroconiferyl alcohol, vanilloyl methyl ketone and furfural.

Sorghum Sugar. Sugar sorghum (*Sorghum dochna*) stalks contain 12% sucrose. This source was important earlier in the USA. Sugar sorghum is processed into sorghum syrup on a small scale on individual farms in the Midwestern United States.

19.1.4.1.5 Packaging and Storage

Sucrose is packaged in paper, jute or linen sacks, in cardboard boxes, paper bags or cones, in glass containers and in polyethylene foils; the latter serving as lining in paper, jute or wooden containers.

Sugar is stored at a relative humidity of 65–70% in loose form in bins or by stacking the paper or jute sacks. The unbagged, loose or bulk sugar is distributed to industry and wholesalers in bins on trucks or rail freight cars.

19.1.4.1.6 Types of Sugar

Sucrose is known under many trade and popular names. These may be related to its purity grade (raffinade, white, consumer's berry, raw or yellow sugar), to its extent of granulation or crystal size (icing, crystal, berry and candy sugar, and cube and cone sugar) and to its use (canning, confectionery or soft drink sugar). Liquid sugar is a sucrose solution in water with at least 62% solids (of which a maximum of 3% is invert sugar). The invert sugar content is high in liquid invert sugars and invert sugar syrups. Such solutions are easily stored, handled and transported. They are dosed by pumps and are widely used by the beverage industry (soft drinks and spirits), the canning industry and ice cream makers, confectionery and baking industries, and in production of jams, jellies and marmalades. Use of liquid sugar avoids the additional crystallization steps of sugar processing and problems associated with packaging of sugar.

Criteria for the analytical determination of sugars are: (a) color; (b) color extinction coefficient (absorbance) of a 50% sugar solution, expressed in ICUMSA-units; (c) ash content determined from conductivity measurements of a 28% aqueous sugar solutions; (d) moisture content; (e) optical rotation; and (f) criteria based on the content of invert sugar.

19.1.4.1.7 Composition of some Sugar Types

The chemical composition of a given type of sugar depends on the extent of sugar raffination. A raffinade, as mentioned above, consists of practically 100% sucrose. Washed raw beet sugar has about 96% sucrose, <1.4% moisture, 0.9% ash and 1.5% nonsugar organic substances. Berry sugar consists of 98.8% sucrose, 0.70% moisture, 0.20% ash and 0.29% nonsugar organic substances. The presence of raffinose, a trisaccharide, is detected by high optical rotation readings or by the presence of needle- or spearlike crystals.

19.1.4.1.8 Molasses

The molasses obtained after sugar beet processing contains about 60% sucrose and 40% other components (both on dry basis). The nonsucrose substances, expressed as percent weight of molasses, include: 10% inorganic salts, especially those of potassium; raffinose (about 1.2%); the trisaccharide kestose, an artifact of processing; organic acids (formic, acetic, propionic, butyric and valeric); and N-containing compounds (amino acids, betaine, etc.). The main amino acids are glutamic acid and its derivative, pyrrolidone carboxylic acid. Molasses is used in the production of baker's yeast; in fermentation technology for production of ethanol and citric, lactic and gluconic acids, as well as glycerol, butanol and acetone; as an ingredient of mixed feeds; or in the production of amino acids.

The residual molasses after cane sugar processing contains about 4% invert sugar, 30–40% sucrose, 10–25% reducing substances, a very low amount of raffinose and no betaine, but unlike beet molasses, contains about 5% aconitic acid. Cane sugar molasses is fermented to provide arrack and rum.

19.1.4.2 Sugars Produced from Sucrose

Hydrolysis of sucrose with acids, or enzymes (invertase or saccharase) results in invert sugar which, after chromatographic separation, can provide *glucose* and *fructose*. Invert sugar syrup is a commercially available liquid sugar. Invert sugar also serves as a raw material for production of sorbitol and mannitol. Isomerization of sucrose with *Leuconostoc mesenter-*

oides, Protoaminobacter rubrum etc. or with enzymes from these microorganisms gives isomaltulose. Apart from 6-O-α-glucopyranosidofructose (palatinose, Ia, Formula 19.1), 1-O-α-glucopyranosidofructose (Ib) is also formed, the ratio depending on the reaction conditions. The process is operated continuously with immobilized cell systems or enzymes. The fructose component of isomaltulose is present as furanose, the anomer ratio being $\alpha/\beta = 0.25$ (34 °C). The sweetening strength is 0.4, based on a 10% sucrose solution. Isomaltulose is not cleaved by human mouth flora; it undergoes delayed cleavage by the glucosidases in the wall of the small intestines.

Catalytic hydrogenation yields isomaltol (palatinit), a mixture of the disaccharide alcohols 6-O-α-D-glucopyranosidosorbitol (IIa, Formula 19.1), 1-O-α-D-glucopyranosidosorbitol (IIb) (isomaltitol), and 1-O-α-D-glucopyranosidomannitol (III). The sweetening strength is 0.45, based on a 10% sucrose solution. Palatinit is practically non-hygroscopic, non-cariogenic, and is only very slowly cleaved by the glucosidases of the wall of the small intestines of human beings.

This mixture of sugar alcohols can be separated by fractional crystallization. Palatinit is a potential sugar substitute.

Enzymatic isomerization of sucrose with the help of *Leuconostoc mesenteroides* gives an α-

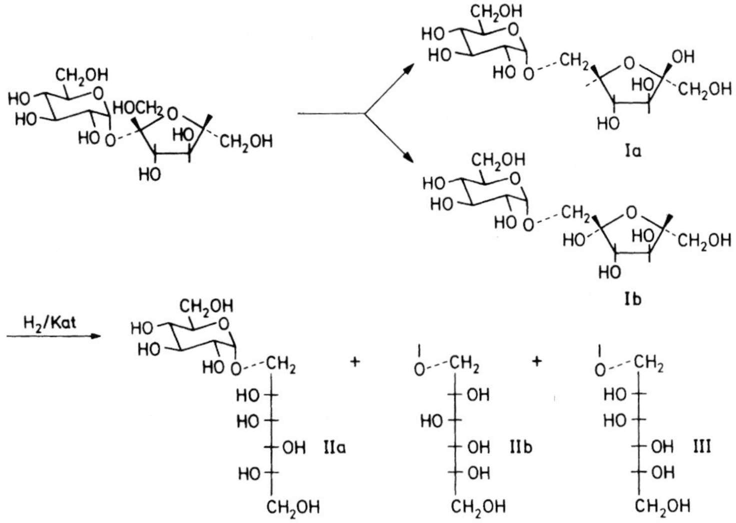

(19.1)

D-glucopyranosido (1 → 5)-D-fructopyranose called *leucrose*. This sugar is fully metabolized but is non-cariogenic.

The transfer of glucose residues from maltose or soluble starch to sucrose with the help of a cyclodextrin glucosyltransferase gives mixtures of oligosaccharides, which are called *coupling sugar*, and are non-cariogenic.

The transfer of fructose residues to sucrose catalyzed by a fructosyltransferase gives an oligosaccharide mixture called *neo-sugar*. It has the general formula ^1F-(β-fructosyl)$_n$-sucrose (n = 1: 1-kestose, n = 2: nystose etc.). Kestose and the higher oligosaccharides can be concentrated to >95% by using chromatographic processes. The caloric value is 7.9 kJ/g.

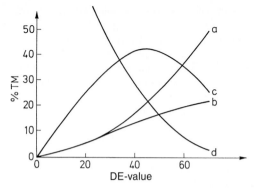

Fig. 19.9. Composition of starch syrups (acid hydrolysis). *a* Glucose, *b* maltose (disaccharide), *c* oligosaccharide (degree of polymerization DP = 3–7), *d* higher saccharides

19.1.4.3 Starch Degradation Products

19.1.4.3.1 General Outline

In principle, either starch or cellulose could be used as a source for saccharification, but only starch hydrolysis is currently of economic importance. Improvements in cellulose saccharification are being sought.

19.1.4.3.2 Starch Syrup (Glucose or Maltose Syrup)

Starch saccharification is achieved by either acidic or enzymatic hydrolysis. Controlled processing conditions yield products of widely different compositions to suit the diversified fields of application. Acid hydrolysis is conducted with hydrochloric acid or sulfuric acid, mainly in a continuous process, and yields glucose syrups with dextrose equivalents (DE value) between 20 and 68. The composition is constant for each DE value (Fig. 19.9).

The raw juice is neutralized and passes through various purification steps. Proteins and lipids from starch flocculate at a suitable pH value and are separated as sludge. Pigments are eliminated with activated carbon and minerals with ion exchangers. The purified juice is evaporated under vacuum (falling-film evaporator) up to a solids content of 70–85%.

During acid hydrolysis, a number of side reactions occur (cf. 4.2.4.3.1). Reversion products are formed in amounts of 5–6% of the glucose used. These are predominantly isomaltose (68–70%) and gentiobiose (17–18%), and, in addition, other di- and trisaccharides. Furthermore, degradation products of glucose are formed, e.g., 5-hydroxymethylfurfural and other compounds typical of caramelization (cf. 4.2.4.3.3) and the *Maillard* reaction (cf. 4.2.4.4).

In enzymatic processes, α-amylases, β-amylases, glucoamylases, and pullulanases are used. First, starch liquefaction is conducted with acid, with α-amylase, or with a combination acid/enzyme process.

The enzyme most commonly used is α-amylase isolated from, for example, *Bacillus subtilis* or *B. licheniformis*. Optimal pH and temperature are 6.5 and 70–90 °C, respectively. The enzyme from *B. licheniformis* is active even at 110 °C. Hydrolysis can be carried out to obtain a product consisting mostly of maltose and, in addition, maltotriose and small amounts of glucose. When, for instance, starch is subjected to combined degradation with bacterial α-amylase and β-amylase or fungal α-amylase, the product obtained has 5% of glucose, 55% of maltose, 15% of maltotriose, 5% of maltotetraose, and 20% of dextrins in the dry matter. Maltose contents of up to 95% (dry matter) can be attained by using pullulanases (cf. 2.7.2.2.4).

A suitable combination of enzymes gives rise to products that cannot be obtained by acid hydrolysis alone.

The extent of starch conversion into sugars is generally expressed as dextrose equivalents (DE value), i.e. the amount of reducing sugars produced, calculated as glucose (DE value: glucose = 100, starch = 0).

The sweet taste intensity of the starch hydrolysates depends on the degree of saccharification and ranges from 25–50% of that of sucrose. Table 19.8 provides data on some hydrolysis products. The wide range of starch syrups starts with those with a DE value of 10–20 (maltodextrins) and ends with those with a DE value of 96.

Starch syrups are used in sweet commodity products. They retard sucrose crystallization (hard caramel candies) and act as softening agents, as in soft caramel candies, fondants and chewing gum. They are also used in ice cream manufacturing, production of alcoholic beverages and soft drinks, canning and processing of fruits and in the baking industry.

19.1.4.3.3 Dried Starch Syrup (Dried Glucose Syrup)

Dried starch syrups with a moisture content of 3–4% are produced by spray drying of starch hydrolysates. The products are readily soluble in water and dilute alcohol and are used, for example, in sausage production as a red color enhancer. The average composition of dried

Table 19.8. Average composition of starch hydrolysates[a]

DE-Value	Glucose	Maltose	Malto-triose	Higher oligo-saccharides
Acid hydrolysis				
30	10	9	9	72
40	17	13	11	59
60	36	20	13	31
Enzymatic hydrolysis[b]				
20	1	5	6	88
45	5	50	20	25
65	39	35	11	15
97	96	2	–	2

[a] All values expressed as % of starch hydrolysate (dry weight basis).
[b] Occasionally it involves a combined acid/enzymatic hydrolysis.

starch syrups is 50% dextrin, 30% maltose and 20% glucose.

19.1.4.3.4 Glucose (Dextrose)

The raw source for glucose production is primarily starch isolated from corn, potatoes or wheat. Saccharification is achieved enzymatically by α-amylase and microbial amyloglucosidase, or by amyloglucosidase after the starch has been subjected to partial acid hydrolysis. The enzyme from *Aspergillus niger*, at pH 4.5 and 60°C, provides a hydrolysate with 94–96% glucose. After a purification step, the hydrolysate is evaporated and crystallized. Glucose crystallizes as α-D-glucose monohydrate. Water-free α-D-glucose is obtainable from the monohydrate by drying in a stream of warm air or by crystallization from ethanol, methanol or glacial acetic acid. Dextrose, due to its great and rapid resorption, is used as an invigorating and strengthening agent in many nourishing formulations and medicines. Like dried glucose syrup, crystalline dextrose is used as a red color enhancer of meat and frying sausages.

19.1.4.3.5 Glucose-Fructose Syrup (High Fructose Syrup)

A glucose-fructose isomerase enzyme occurs in some species of *Bacillus megaterium*, *B. coagulans* and *Lactobacillus brevii*. In a neutral or weakly alkaline reaction medium (pH 8.2) and at 35–60°C, the isomerase converts glucose into fructose. Using immobilized enzymes, large amounts of glucose syrup can be converted into glucose-fructose syrup with a fructose content of 40–50%. When coupled with chromatographic separation techniques, products up to 90% fructose are obtainable. Glucose-fructose syrups are used as invert sugars derived from sucrose.

19.1.4.3.6 Starch Syrup Derivatives

Hydrogenation of glucose syrups results in products which, since they are nonfermentable and are less cariogenic, are used in manufacturing of sweet commodity products. Alkaline isomerization of maltose gives *maltulose*, which is sweeter than maltose, while hydrogenation yields *maltitol* in a mixture with maltotriit. This mixture of sugar alcohols is not crystallizable but, after addition of suitable polysaccharides (alginate, methylcellulose), can be spray-dried into a powder.

19.1.4.3.7 Polydextrose

When D-glucose is melted in the presence of small amounts of sorbitol and citric acid, a cross-linked polymer called polydextrose is formed, which contains primarily 1,6-glucosidic bonds but also other bonds. The caloric value is ≥ 4.2 kJ/g. For this reason, the use of polydextrose as a sweetener for diabetics and for the production of low-calorie baked products and candies is under discussion.

19.1.4.4 Milk Sugar (Lactose) and Derived Products

19.1.4.4.1 Milk Sugar

Lactose is produced from whey and its concentrates. The whey is adjusted to pH 4.7 and then heated directly with steam at $95-98\,^{\circ}\mathrm{C}$ to remove milk albumins. The deproteinated filtered fluid is further concentrated in a multi-stage evaporator and then the separated salts are removed. The desalted concentrate yields a yellow raw sugar with a moisture content of $12-14\%$. The remaining mother liquor still contains an appreciable amount of lactose, so it is recirculated through the process or is used for the production of ethanol or lactic or propionic acids. The raw lactose is raffinated by solubilization, filtration and several crystallizations. The snow-white α-lactose monohydrate is pulverized in a pin mill and separated according to particle size in a centrifugal classifier. Spray drying of lactose is gaining in importance.

To increase lactose digestibility, sweetness and solubility, a 60% lactose solution can be heated to 93.5 °C and the crystallizate discharged to a vacuum drum dryer. β-Lactose is formed. Its moisture content is not more than 1% and it is more soluble than α-lactose. Uses of β-lactose include: a nutrient for children; a filler or diluter in medicinal preparations (tablets); and an ingredient of nutrient solutions used in microbial production of antibiotics.

19.1.4.4.2 Products from Lactose

Enzymatic or acidic hydrolysis of lactose provides a glucose-galactose mixture which is twice as sweet as lactose. A further increase in taste intensity is achieved by enzymatic isomerization of glucose. Such enzyme-treated products contain about 50% galactose, 29% glucose and 21% fructose.

Lactulose is obtained by isomerization of lactose. It is sweeter than lactose. Hydrogenation of lactose yields *lactitol*, while hydrogenation of lactulose yields a mixture of lactilol and β-D-galactopyranosido-1,4-mannitol.

19.1.4.5 Fruit Sugar (Fructose, Levulose)

Fructose is obtainable from its natural polymer, inulin, which occurs in: topinambur tubers (India) or its North American counterpart, Jerusalem artichoke (*Helianthus tuberosus*); chicory; tuberous roots of dahlia plants; and in flowerheads of globe or true artichoke (*Cyanara scolymus*), grown extensively in France. Fructose is obtained by acidic hydrolysis of inulin or by chromatographic separation of a glucose-fructose mixture (invert sugar, isomerized glucose syrup). Only the latter process has commercial significance. Fructose is present in the crystallized state as β-pyranose. Sweeter than sucrose, fructose is used as a sugar substitute for diabetics. It can be partly converted to glucose on boiling for longer periods due to the acid in fruit products.

19.1.4.6 Sorbitol

Sorbitol, a hygroscopic alcohol, is approximately half as sweet as sucrose. It is used as a sweetener for diabetics and in food canning. In baked products and candies, it is usually used in the form of a 70% syrup in amounts between 5 and 10% as a softener and humectant. Sorbitol is also suitable for the production of sugar-free candies (chewing gum). Sorbitol can be produced on a commercial scale by catalytic hydrogenation of glucose. Acid-catalyzed elimination of water yields a mixture of 1,4-sorbitan (85%, I) and 3,6-sorbitan (15%, II). Under more drastic conditions (action of concentrated acids), 1,4 : 3,6-dianhydrosorbitol (isosorbid III) is formed (Formula 19.2).

I

II

III   (19.2)

19.1.4.7 L-Sorbose and Other L-Sugars

L-Sorbose can be formed from glucose via sorbitol. Sorbitol is oxidized by *Acetobacter xylinum* into L-sorbose, an intermediary product for commercial synthesis of ascorbic acid (cf. 18.1.2.7). Sorbose is under discussion as a sucrose substitute for diabetics and as an ingredient with neglible cariogenicity in low calorie foods. It is resorbed only slowly on oral administration.

Until now, other L-sugars have been available only in small amounts. It is assumed that they are metabolized not at all or only to a small extent by human beings and even in low concentrations, they are capable of inhibiting the glycosidases of the small intestine. Therefore, economic methods of synthesis are of interest. A suitable educt is L-arabinose, which yields a L-glucose/L-mannose mixture by chain extension. This mixture can be oxidized directly to L-fructose or after reduction via L-sorbitol/L-mannitol. The isomerization of L-sorbose to L-idose and L-gulose is also under discussion.

19.1.4.8 Xylitol

Xylose is obtained by hydrolysis of hemicelluloses. Catalytic hydrogenation of xylose yields xylitol. Like fructose and sorbitol, xylitol is a sugar substitute for diabetics, has a role in parenteral (intravenous or intramuscular) nutrition and is used as an ingredient of low cariogenicity in the production of "sugar-free" chewing gum.

19.1.4.9 Mannitol

Mannitol can be made by the hydrogenation of invert sugar. As a result of its lower solubility, it is separated from sorbitol by chromatography. Due to its laxative effect, it is predominantly used in chewing gum and in baked products for diabetics.

19.1.5 Candies

19.1.5.1 General Outline

Candies represent a subgroup of sweet commodities generally called confectionery. Products such as long-storage cookies, cocoa and chocolate products, ice cream and artificial "honey" (19.2.2) are also confections.

Candies are manufactured from all forms of sugar and may also incorporate other foods of diverse origin (dairy products, honey, fat, cocoa, chocolate, marmalade, jellies, fruit juices, herbs, spices, malt extract, seed kernels, rigid or elastic gels, liqueurs or spirits, essences, etc.). The essential and characteristic component of all types of candy is sugar, not only sucrose, but also other forms of sugar such as starch sugar, starch syrup, invert sugar, maltose, lactose, etc.

The important product groups include hard and soft caramels (bonbons, toffees), fondant, coconut flakes, foamy candies, gum candies, licorice products, dragees, pastilles, fruit pastes, chewing gum, croquant, effervescent powders, and products made of sugar and almonds, nuts and other protein-rich oil-containing seeds (marzipan, persipan, nougat).

19.1.5.2 Hard Caramel (Bonbons)

For the production of these candies, a sucrose solution is mixed with starch syrup and boiled down to the desired water content either batchwise or continuously (Fig. 19.10). Generally used are vacuum pans ($120-160\,°C$) and film boiling machines in which evaporation takes place in a rotating cylinder ($110\,°C \rightarrow 142\,°C$, 5 s). Volatile labile components (aroma substances) are added after cooling. This applies to acids as well in order to prevent inversion. Air is incorporated into the mass, if necessary. Subsequently, the mass is formed into a cord and processed into bonbons with the help of stamping or casting machines that require a slightly thinner mass. Modern plants have a capacity of $0.6-1.5$ t/h.

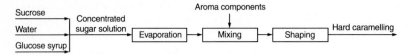

Fig. 19.10. Production of hard caramels

The composition of hard caramels is presented in Table 19.9.

19.1.5.3 Soft Caramel (Toffees)

Milk, starch syrup, and fat are homogenized, mixed with sucrose solution, and boiled down as described above (cf. 19.1.5.2). Labile components are added after cooling. The fat content and, compared to hard caramel, slightly higher water content produce a plastic, partially elastic consistency, which is further improved by the incorporation of air in drawing machines. The mixing of powdered sugar or fondant filler during drawing produces a crumbly consistency due to partial sucrose crystallization. The cooled mass is formed into cords and cut.

The average composition is presented in Table 19.9.

19.1.5.4 Fondant

A sucrose or glucose solution is mixed with starch syrup and boiled down to a water con-
tent of 10–15%. The mass is rapidly cooled while it is subjected to intensive mechanical treatment. With partial crystallization (crystal diameter 3–30 µm), a dispersion of sucrose in a saturated sugar solution is formed. The mass solidifies on further cooling and is meltable and pourable on heating. It is aromatized and processed into various products, e.g., chocolate fillings.

The composition is presented in Table 19.9.

19.1.5.5 Foamy Candies

For the production of these candies, a hot sugar solution (sucrose/starch syrup) is carefully mixed into a stable protein foam (egg white, digested milk protein, gelatin). Apart from conventional beaters, pressure beaters are also used in which all the components are first mixed at 2–9 bar and then foamed by subsequent expansion. The light mass is shaped in a pressing step. Chocolate-coated marshmallows are a typical product.

Table 19.9. Composition[a] of some candies

Component	Hard caramel	Soft caramel	Fondant	Marzipan raw filler	Marzipan
Sucrose	40–70	30–60	65–80[d]	≤ 35	≤ 67.5
Starch syrup[b]	30–60	20–50	10–20	0	3.5
Invert sugar	1–8	1–10		0–10	0–20
Lactose		0–6			
Sorbitol				0	0–5
Fat		2–15		28–33	14–16
Acids[c]	0.5–2				
Milk protein		0–5			
Gelatin		0–0.5			
Aroma	0.1–0.3				
Water	1–3	4–8	10–15	15–17	7–8.5
Minerals	0.1–0.2	0.5–1.5		1.4–1.6	0.7–0.8

[a] Orientation values in %.
[b] Dry matter.
[c] Citric acid or tartaric acid.
[d] Glucose is also used, if necessary.

19.1.5.6 Jellies, Gum and Gelatine Candies

For the production of these products, an aromatized sugar solution is heated with polysaccharides (agar, pectin, gum arabic, thin-boiling starch, amylopectin) and gelatine, poured into starch moulds, and removed with powder after hardening. Typical products are jelly fruit and gum bears.

19.1.5.7 Tablets

Powdered sugar and dextrose are aromatized, granulated with the addition of binding agents (fat, gelatine, gum arabic, tragacanth gum, starch) and lubricants (magnesium stearate), and tabletted under pressure.

19.1.5.8 Dragées

The core (almond, nut, sugar crystal etc.) is moistened with a sugar solution in a rotating boiler and then covered with a layer of sugar by the subsequent addition of powdered sugar. This process is repeated until the desired layer thickness is attained. Chocolate, if necessary, is applied in a corresponding manner.

Sugared or burnt almonds are a well-known product. They consist of raw or roasted almonds which are covered with a hot-saturated and caramelized sugar syrup. The rough crispy surface is formed by the blowing of hot air. Burnt almonds also contain spice and flavoring matter, like vanillin etc. In a larger average sample, the ratio of sugar to almonds should not exceed 4 : 1. The sugar covering of burnt almonds produced in the dragée process can also be colored.

19.1.5.9 Marzipan

In the traditional production of marzipan raw filler, sweet almonds are scalded, peeled on rubber-covered rolls, coarsely chopped, and then ground with the addition of not more than 35% of sucrose. The mixture is then heated in open roasting pans until the remaining water content is not more than 17%. An equal amount of powdered sucrose is worked into this semi-finished product, possibly with the addition of starch syrup and/or sorbitol, to give the actual marzipan.

For reasons of rationalization and bacteriological stability, efforts are being made in modern processes to conduct all the process steps in one hermetically sealed reaction chamber, e.g., in a combined vacuum/boiling/cutting/mixing machine (high speed cooker), to prevent infections. After heating, partial drying occurs here by the application of vacuum. The cooker is aerated with germ-free filtered air. Koenigsberg marzipan is briefly baked on top after shaping. Marzipan potatoes are rolled in cocoa.

19.1.5.10 Persipan

As with marzipan, a raw paste is initially prepared, but instead of almonds, the seed kernels of apricots, peaches or bitter almonds (with bitterness removed) are used. Commercial persipan is a mix of raw persipan filler and sucrose, the latter not more than half of the mix weight. Sucrose can be partially replaced by starch syrup and/or sorbitol.

19.1.5.11 Other Raw Candy Fillers

These are produced from dehulled nuts such as cashews or peanuts. They correspond in composition to raw persipan paste. They are designated according to the oilseed component.

19.1.5.12 Nougat Fillers

Nougat paste serves as a soft or firm candy filling. It contains up to 2% water and roasted dehulled filberts (hazelnuts) or roasted dehulled almonds, finely ground in the presence of sugar and cocoa products. Cocoa products used are cocoa beans; cocoa liquor and butter; pulverized defatted cocoa; chocolate; baking, cream and milk chocolate; chocolate icing; cream and milk chocolate icings; and chocolate powders. The filler may contain a small amount of flavoring and/or lecithin. Also, part of the sugar may be replaced by cream or milk powder. Sweet nougat fillings can also be produced without cocoa ingredients and cream or milk powders. The kneaded nougat paste is often designated just as nougat or noisette.

Recently the trans- and cis-isomers of 5-methyl-4-hepten-2-one have been detected as character impact compounds for the flavor of filberts. The aroma threshold of the trans-isomer (Filbertone) is extremely low: 5 ng/kg (water as solvent).

19.1.5.13 Croquant

Croquant serves generally as a filling for candy. It is made of molten sucrose, which has been at least partly caramelized, and ground and roasted almonds or nuts. It is occasionally mixed with marzipan, nougat, stable dairy products, fruit constituents and/or starch syrup. Croquant can be formulated to a brittle or soft consistency.

19.1.5.14 Licorice and its Products

To manufacture licorice products, flour dough is mixed with sugar, starch syrup, concentrated flavoring of the licorice herb root and gelatin, and the mix is evaporated to a thick consistency. It is then molded into sticks, bands, figurines, etc. and dried further. The characteristic and flavor-determining ingredient derived from the perennial licorice herb is the diglucuronide of β-glycyrrhetinic acid (cf. 8.8.11).

Simple licorice products contain starch (30–45%), sucrose (30–40%) and at least 5% licorice extract. Better quality products have an extract content of at least 30%. The aroma is enhanced, usually, with anise seed oil in conjunction with low amounts of ammonium chloride.

19.1.5.15 Chewing Gum

Chewing gum is made of a natural or a synthetic gum base impregnated with nutrients and flavoring constituents, mostly sugars and aroma substances, which are gradually released by chewing. The gum base is a blend of latex products from rubber trees that grow in tropical forests or plantations. The most important sources are chicle latex from the *Sapodilla* tree of Mexico, Indonesia and Malaysia; jelutongs; and rubber latex. Natural (mastic tree) and synthetic resins and waxes are also used. Synthetic thermoplastic resins are polyvinyl esters and ethers, polyethylene, polyisobutylene, butadiene-styrene copolymerisates, paraffin, microcrystalline waxes etc. The gum base may also contain cellulose as a filler and a break-up agent. The wax portion predominates in normal chewing gum and the gum-like substances predominate in bubble gum.

To be able to process this base into a homogeneous plastic mass, it must be heated to ca. 60 °C before it is kneaded with the sugar components. The mass is made more malleable by the addition of small amounts of glycerol or glycerol triacetate. The mass is cooled to ca. 30 °C before it is rolled out. In fact, very strong kneaders must be used because of the high viscosity of the mass. Recently extruders have been used increasingly in continuous production lines. The production of chewing gum is summarized in Fig. 19.11.

19.1.5.16 Effervescent Lemonade Powders

The powder or compressed tablets (effervescent bonbons) are used for preparation of artificial sparkling lemonades. They contain sodium bicarbonate and an acid component (lactic, tartaric or citric acid). When dissolved in water, they generate carbon dioxide. Other constituents of the product are sucrose or another sweetener, and natural or artificial flavoring substances. Sodium bicarbonate and acids are often packaged and marketed separately in individual capsules or in two separate containers.

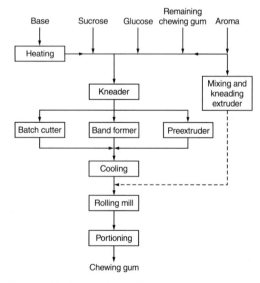

Fig. 19.11. Production of chewing gum

19.2 Honey and Artificial Honey

19.2.1 Honey

19.2.1.1 Foreword

Honey is produced by honeybees. They suck up nectar from flowers or other sweet saps found in living plants, store the nectar in their honey sac, and enrich it with some of their own substances to induce changes. When the bees return to the hive, they deposit the nectar in honeycombs for storage and ripening.

Honey production starts immediately after the flower pollen, nectar and honeydew are collected and deposited in the bee's pouch (honey sac). The mixture of raw materials is then given to worker bees in the hive to deposit it in the six-sided individual cells of the honeycomb. The changing of nectar into honey proceeds in the cell in the following stages: water evaporates from the nectar, which then thickens; the content of invert sugar increases through sucrose hydrolysis by acids and enzymes derived from bees, while an additional isomerization of glucose to fructose occurs in the honey sac; absorption of proteins from plant and bees, and acids from the bee's body; assimilation of forage minerals, vitamins and aroma substances; and absorption of enzymes from the bees' salivary glands and honey sacs. When the water content of the honey drops to 16–19%, the cells are closed with a wax lid and ripening continues, as reflected by a continued hydrolysis of sucrose by the enzyme invertase and by the synthesis of new sugars.

Table 19.10 provides information about honey production in some countries. The honey consumption in Germany in 1996 was 1.0 kg per capita.

19.2.1.2 Production and Types

In the production and processing of honey, it is important to preserve the original composition, particularly the content of aroma substances, and to avoid contamination. The following kinds of honey are differentiated according to recovery techniques:

Comb Honey (honey with waxy cells), i.e. honey present in freshly-built, closed combs devoid of brood combs (young virgin combs).

Table 19.10. Production of honey in 1990 (1,000 t)

Continent		Country	
World	1,210	USSR	270
Africa	108	China	183
America, North-,		USA	84
Central-	211	Mexico	71
America, South-	78	India	51
Asia	305	Turkey	40
Europe	203	Argentina	39
USSR	270	Canada	33
Ozeania	37	Germany	31
		Australia	28
		France	28
		Ethiopia	23
		Spain	21
		Σ (%)[a]	75

[a] World production $\triangleq$ 100%.

Such honey is produced in high amounts, but is not readily found in Germany. In other countries, primarily the USA, Canada and Mexico, it is widely available. Darker colored honey is obtained from covered virgin combs not more than one year old and from combs which include those used as brood combs.

Extracted Honey is obtained with a honey extractor, i.e. by centrifugation at somewhat elevated temperatures of brood-free comb cells. This recovery technique provides the bulk of the honey found on the market. Gentle warming up to 40 °C facilitates the release of honey from the combs.

Pressed Honey is collected by compressing the brood-free honey combs in a hydraulic press at room temperature.

Strained Honey is collected from brood-free, pulped or unpulped honey combs by gentle heating followed by pressing.

Beetle Honey is recovered by pulping honey combs which include brood combs. This type of honey is used only for feeding bees.

Basesd on its use, honey is distinguished as:

Honey for Domestic use. This is the highest quality product, and is consumed and enjoyed in pure form.

Baking Honey. This type of honey is not of high quality and is used in place of sugar in the baking industry. Such honey has spontaneously fermented, to a certain degree has absorbed or acquired other foreign odors and flavors, or was overheated. This category includes caramelized honey.

According to the recovery (harvest) time, honey is characterized as: early (collected until the end of May); main (June and July); and late (August and September).

Honey can be classified according to geographical origin, e.g., German (Black Forest or Allgäu honey), Hungarian, Californian, Canadian, Chilean, Havanan, etc.

The flavor and color of honey are influenced by the kinds of flowers from which the nectar originates. The following kinds of honey are classified on the basis of the type of plant from which they are obtained.

Flower Honey, e.g., from: heather; linden; acacia; alsike, sweet and white clovers; alfalfa; rape; buckwheat and fruit tree blossoms. When freshly manufactured, these are thick, transparent liquids which gradually granulate by developing sugar crystals. Flower honey is white, light-to-dark, greenish-yellow or brownish. Maple tree honey is light amber; alfalfa honey, dark-red; clover honey, light amber-to-reddish; and meadow flower honey, amber-to-brown. Flower honey has a typical sweet and highly aromatic flavor that is dependent on the flavor substances which together with the nectar are collected by the bees; it sometimes has a flavor reminiscent of molasses. This is especially true of honey derived from heather (alfalfa and buckwheat honeys).

Honeydew Honey (pine, spruce or leaf honeydew). This type of honey solidifies with difficulty. It is less sweet, dark colored, and may often have a resinous terpene-like odor and flavor.

19.2.1.3 Processing

Honey is marketed as a liquid or semisolid product.

It is usually oversaturated with glucose, which granulates, i.e. crystallizes, within the thick syrup in the form of glucose hydrate. To stabilize liquid honey, it has to be filtered under pressure to remove the sugar crystals and other crystallization seeds. Heating of honey decreases its viscosity during processing and

filling, and provides complete glucose solubilization and pasteurization. Heating has to be gentle since the low pH of honey and its high fructose content make it sensitive to heat treatment. As with other foods, continuous, high temperature-short time processing (e.g., 65 °C for 30 s followed by rapid cooling) is advantageous.

Processing of honey into a semisolid product involves seeding of liquid honey with fine crystalline honey to 10 % and storing for one week at 14 °C to allow full crystallization. This product is marketed as creamed honey.

19.2.1.4 Physical Properties

Honey density (at 20 °C) depends on the water content and may range from 1.4404 (14% water) to 1.3550 (21% water). Honey is hygroscopic and hence is kept in airtight containers. Viscosity data at various temperatures are given in Table 19.11. Most honeys behave like *Newtonian* fluids. Some, however, such as alfalfa honey, show thixotropic properties which are traceable to the presence of proteins, or dilating properties (as with opuntia cactus honey) due to the presence of trace amounts of dextran.

The specific heat (20 °C; 17.4 % water) is 2.26 J/g/°C. Because of poor heat conductivity, the possibility of heating honey with microwaves

Table 19.11. Viscosity of honey at various temperatures

	Temperature (°C)	Viscosity (Poise)
Honey 1[a]	13.7	600.0
	20.6	189.6
	29.0	68.4
	39.4	21.4
	48.1	10.7
	71.1	2.6
Honey 2[b]	11.7	729.6
	20.2	184.8
	30.7	55.2
	40.9	19.2
	50.7	9.5

[a] Melilot honey (*Melilotus officinalis*; 16.1% moisture).
[b] Sage honey (*Salvia officinalis*; moisture content 18.6%).

is a viable approach. Heating 1 L honey for 1 h from 30 to 55 °C requires 25 kW of energy.

19.2.1.5 Composition

Honey is essentially a concentrated aqueous solution of invert sugar, but it also contains a very complex mixture of other carbohydrates, several enzymes, amino and organic acids, minerals, aroma substances, pigments, waxes, pollen grains, etc. Table 19.12 provides compositional data. The analytical data correspond to honey from the USA, nevertheless, they basically represent the composition of honey from other countries.

19.2.1.5.1 Water

The water content of honey should be less than 20%. Honey with higher water content is readily susceptible to fermentation by osmophilic yeasts. Yeast fermentation is negligible when the water contents is less than 17.1%, while between 17.1 and 20% fermentation depends on the count of osmophilic yeast buds.

19.2.1.5.2 Carbohydrates

Fructose (averaging 38%) and glucose (averaging 31%) are the predominant sugars in honey. Other monosaccharides have not been found. However, more than 20 di- and oligo-saccharides have been identified (Table 19.13), with maltose predominating, followed by kojibiose (Table 19.14). The composition of disaccharides depends largely on the plants from which the honey was derived, while geographical and seasonal effects are negligible.

Table 19.13. Sugars identified in honey

Common name	Systematic name
Glucose	
Fructose	
Saccharose	α-D-glucopyranosyl-β-D-fructo-furanoside
Maltose	O-α-D-glucopyranosyl-$(1 \rightarrow 4)$-D-glucopyranose
Isomaltose	O-α-D-glucopyranosyl-$(1 \rightarrow 6)$-D-glucopyranose
Maltulose	O-α-D-glucopyranosyl-$(1 \rightarrow 4)$-D-fructose
Nigerose	O-α-D-glucopyranosyl-$(1 \rightarrow 3)$-D-glucopyranose
Turanose	O-α-D-glucopyranosyl-$(1 \rightarrow 3)$-D-fructose
Kojibiose	O-α-D-glucopyranosyl-$(1 \rightarrow 2)$-D-glucopyranose
Laminaribiose	O-β-D-glucopyranosyl-$(1 \rightarrow 3)$-D-glucopyranose
α,β-Trehalose	α-D-glucopyranosyl-β-D-glucopyranoside
Gentiobiose	O-β-D-glucopyranosyl-$(1 \rightarrow 6)$-D-glucopyranose
Melezitose	O-α-D-glucopyranosyl-$(1 \rightarrow 3)$-O-β-D-fructofuranosyl-$(2 \rightarrow 1)$-α-D-glucopyranoside
3-α-Isomaltosylglucose	O-α-D-glucopyranosyl-$(1 \rightarrow 6)$-O-α-D-glucopyranosyl-$(1 \rightarrow 3)$-D-glucopyranose
Maltotriose	O-α-D-glucopyranosyl-$(1 \rightarrow 4)$-O-α-D-glucopyranosyl-$(1 \rightarrow 4)$-D-glucopyranose
1-Kestose	O-α-D-glucopyranosyl-$(1 \rightarrow 2)$-β-D-α-fructofuranosyl-$(1 \rightarrow 2)$-β-D-fructofuranoside
Panose	O-α-D-glucopyranosyl-$(1 \rightarrow 6)$-O-α-D-glucopyranosyl-$(1 \rightarrow 4)$-D-glucopyranose
Isomaltotriose	O-α-D-glucopyranosyl-$(1 \rightarrow 6)$-O-α-D-glucopyranosyl-$(1 \rightarrow 6)$-D-glucopyranose
Erlose	O-α-D-glucopyranosyl-$(1 \rightarrow 4)$-α-D-glucopyranosyl-β-D-fructofuranoside
Theanderose	O-α-D-glucopyranosyl-$(1 \rightarrow 6)$-α-D-glucopyranosyl-β-D-fructofuranoside
Centose	O-α-D-glucopyranosyl-$(1 \rightarrow 4)$-O-α-D-glucopyranosyl-$(1 \rightarrow 2)$-D-glucopyranose
Isopanose	O-α-D-glucopyranosyl-$(1 \rightarrow 4)$-O-α-D-glucopyranosyl-$(1 \rightarrow 6)$-D-glucopyranose
Isomaltotetraose	O-α-D-glucopyranosyl-$(1 \rightarrow 6)$-[O-α-D-glucopyranosyl-$(1 \rightarrow 6)]_2$-D-glucopyranose
Isomaltopentaose	O-α-D-glucopyranosyl-$(1 \rightarrow 6)$-[O-α-D-glucopyranosyl-$(1 \rightarrow 6)]_3$-D-glucopyranose

Table 19.12. Composition of honey (%)

Constituent	Average value	Variation range
Moisture	17.2	13.4–22.9
Fructose	38.2	27.3–44.3
Glucose	31.3	22.0–40.8
Saccharose	1.3	0.3–7.6
Maltose	7.3	2.7–16.0
Higher sugars	1.5	0.1–8.5
Others	3.1	0–13.2
Nitrogen	0.04	0–0.13
Minerals (ash)	0.17	0.02–1.03
Free acids [a]	22	6.8–47.2
Lactones [a]	7.1	0–18.8
Total acids [a]	29.1	8.7–59.5
pH value	3.9	3.4–6.1
Diastase value	20.8	2.1–61.2

[a] mequivalents/kg.

Table 19.14. Oligosaccharide composition of honey

Sugar	Content[a] (%)
Disaccharides	
Maltose	29.4
Kojibiose	8.2
Turanose	4.7
Isomaltose	4.4
Saccharose	3.9
Maltulose (and two unidentified ketoses)	3.1
Nigerose	1.7
α-, β-Trehalose	1.1
Gentiobiose	0.4
Laminaribiose	0.09
Trisaccharides	
Erlose	4.5
Theanderose	2.7
Panose	2.5
Maltotriose	1.9
1-Kestose	0.9
Isomaltotriose	0.6
Melezitose	0.3
Isopanose	0.24
Gentose	0.05
3-α-Isomaltosylglucose	+[b]
Higher Oligosaccharides	
Isomaltotetraose	0.33
Isomaltopentaose	0.16
Acidic fraction	6.51

[a] Values are based on oligosaccharide total content (= 100%) which in honey averages 3.65%. Only the most important sugars are presented.
[b] Traces.

The content of sucrose varies appreciably with the honey ripening stage.

19.2.1.5.3 Enzymes

The most prominent enzymes in honey are α-glucosidase (invertase or saccharase), α- and β-amylases (diastase), glucose oxidase, catalase and acid phosphatase. Average enzyme activities are presented in Table 19.15. Invertase and diastase activities, together with the hydroxymethyl furfural content, are of significance for assessing whether or not the honey was heated.

Table 19.15. Average enzyme activity in honey

Number	Enzyme	Activity[a]
1	α-Glucosidase (saccharase)	7.5 –10
2	Diastase (α- and β-amylase)	16 –24
3	Glucose oxidase	80.8 –210
4	Catalase	0–86.8
5	Acid phosphatase	5.07–13.4

[a] **1:** g saccharose hydrolyzed by 100 g honey per hour at 40°C; **2:** g starch degraded by 100 g honey per hour at 40°C; **3:** μg H_2O_2 formed per g honey/h; **4:** catalytic activity/g honey, and **5:** mg P/100 g honey released in 24 h.

For α-glucosidase, 7–18 isoenzymes are known. In a wide pH optimum between 5.8–6.5 the enzyme hydrolyzes maltose and other α-glucosides. The K_M with sucrose as substrate is 0.030 mol/l. It also possesses transglucosylase activity. During the first stage of sucrose hydrolysis the trisaccharide erlose (α-maltosyl-β-D-fructofuranoside) plus other oligosaccharides are formed (E = enzyme, S = sucrose, G = glucose, F = fructose):

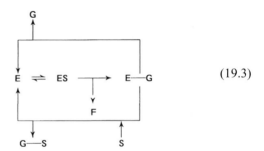

(19.3)

As the hydrolysis proceeds, most of these oligosaccharides are cleaved into monosaccharides.

Thermal inactivation of invertase in honey and its half-life values at various temperatures have been thoroughly investigated. These data are presented in Figs. 19.12 and 19.13. Practically all invertase activity is derived from bees.

Honey α- and β-*amylases (diastase)* also originate from bees. Their pH optimum range is 5.0–5.3. Diastase activity is somewhat more thermally stable than invertase activity (Figs. 19.12 and 19.13).

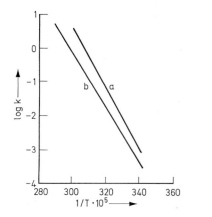

Fig. 19.12. Inactivation rate of (a) invertase and (b) diastase in honey. (according to *White*, 1978)

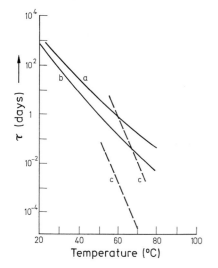

Fig. 19.13. Half-life activity ("τ") of diastase (a), invertase (b), and glucose oxidase (c) in honey at various temperatures (according to *White*, 1978)

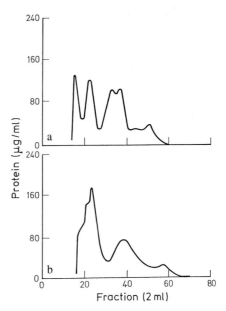

Fig. 19.14. Protein profiles of two honey varieties as revealed by gel filtration on Sephadex G-200. (a) Cottonflower honey (b), honey from sugar-fed bees. (according to *White*, 1978)

Glucose oxidase presence in honey is also derived from bees. Its optimum pH is 6.1. The enzyme oxidizes glucose (100%) and mannose (9%). The enzymatic oxidation by-product, hydrogen peroxide, is partly responsible for a bacteriostatic effect of nonheated honey, an effect earlier ascribed to a so-called "inhibine". The enzymatic oxidation yields gluconic acid, the main acid in honey. Glucose oxidase activity and thermal stability in honey vary

widely (limit values were given in Table 19.13), hence this enzyme is not a suitable indicator of the thermally treatment of honey. *Catalase* in honey most probably originates from pollen which, unlike flower nectar, has a high activity of this enzyme. Similarly, honey *acid phosphatase* originates mainly from pollen, although some activity comes from flower nectars.

19.2.1.5.4 Proteins

Honey proteins are derived partly from plants and partly from honeybees. Figure 19.14 shows that bees fed on sucrose provide proteins with less complex patterns than, for example, cottonflower honey.

19.2.1.5.5 Amino Acids

Honey contains free amino acids at a level of 100 mg/100 g solids. Proline, which might originate from bees, is the prevalent amino acid and is 50–85% of the amino acid fraction (Table 19.16). Based on several amino acid ratios, it is possible to identify the geographical or regional origin of honeys (Fig. 19.15).

Table 19.16. Free amino acids in honey

Amino acid	mg/100 g honey (dry weight basis)	Amino acid	mg/100 g honey (dry weight basis)
Asp	3.44	Tyr	2.58
Asn + Gln	11.64	Phe	14.75
Glu	2.94	β-Ala	1.06
Pro	59.65	γ-Abu	2.15
Gly	0.68	Lys	0.99
Ala	2.07	Orn	0.26
Cys	0.47	His	3.84
Val	2.00	Trp	3.84
Met	0.33	Arg	1.72
Met-O	1.74	Unidentified	
Ile	1.12	AA's (6)	24.53
Leu	1.03		
		Total	118.77

19.2.1.5.6 Acids

The principal organic acid in honey is gluconic acid, which results from glucose oxidase activity. In honey gluconic acid is in equilibrium with its gluconolactone. The acid level is mostly dependent on the time elapsed between nectar collection by bees and achievement of

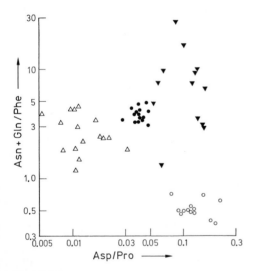

Fig. 19.15. Regional origin of honey as related to its amino acid composition. (according to *White*, 1978)
Honey origin: △ Australia, ● Canada, ▼ United States (clover), ○ Yucatan

the final honey density in honeycomb cells. Glucose oxidase activity drops to a negligible level in thickened honey. Other acids present in honey only in small amounts are: acetic, butyric, lactic, citric, succinic, formic, maleic, malic and oxalic acids.

19.2.1.5.7 Aroma Substances

About 300 volatile compounds are present in honey and more than 200 have been identified. There are esters of aliphatic and aromatic acids, aldehydes, ketones and alcohols. Of importance are especially β-damascenone and phenylacetaldehyde, which have a honey-like odor and taste. Methyl anthranilate is typical of the honey from citrus varieties and lavender and 3,9-epoxy-1,4(8)-p-menthadiene (linden ether) is typical of linden honey.

19.2.1.5.8 Pigments

Relatively little is known about honey color pigments. The amber color appears to originate from phenolic compounds and from products of the nonenzymic browning reactions between amino acids and fructose.

19.2.1.5.9 Toxic Constituents

Poisonous honey (pontius or insane honey) has been known since the time of the Greek historian and general, *Xenophon*, and the Roman writer, *Plinius*. *Xenophon* described a mass poisoning of an expedition by *Kyros* to Asia Minor in 401 B.C. Recent intoxications have been reported in the USSR, the USA and Japan. Toxic honey comes mostly from bees collecting their nectar from: rhododendron species (Asia Minor, Caucasus Mountains); some plants of the family *Ericacea*; insane ("mad") berries; *Kalmia* evergreen shrubs; *Eurphorbiaceae*; and honey collected from other sweet substances, e.g., honeydew exudates of grasshoppers. Rhododendrons contain the poisonous compounds, andromedotoxin (an acetylandromedol) and grayanotoxins I, II and III (a tetracyclic diterpene) used in medicine as a muscle relaxant (I : $R^1 = OH$, $R^2 = CH_3$, $R^3 = COCH_3$; II : R^1, $R^2 = CH_2$, $R^3 = H$; III : $R^1 = OH$, $R^2 = CH_3$, $R^3 = H$) (see Formula 19.4). The poisonous nature of New Zealand honey is a result of tutin and hyenanchin (mellitoxin)

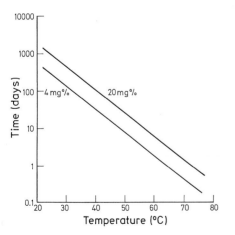

$$(19.4)$$

toxins from the tutu shrub (tanner shrub plant, *Coriaria arbora*). Poisonous flowers of tobacco, oleander, jasmine, henbane *(Datura metel)* and of hemlock *(Conium maculatum)* provide nonpoisonous honeys. The production of these honeys is negligible in Europe.

19.2.1.6 Storage

Honey color generally darkens on storage, the aroma intensity decreases and the content of hydroxymethyl furfural increases, depending on pH, storage time and temperature (Fig. 19.16). The enzymatic inversion of sucrose also continues at a low level even when honey has reached its final density.

Honey should be protected from air moisture and kept at temperatures lower than 10 °C when stored. The desired temperature range for use is 18–24 °C.

19.2.1.7 Utilization

Honey use goes back to prehistoric times. Beeswax and honey played an important role in ancient civilizations. They were placed into tombs as food for deceased spirits, while the Old Testament describes the promised land as "a land flowing with milk and honey". In the Middle Ages honey was used as an excellent energy food and, up to the introduction of cane sugar, served as the only food sweetener. Besides being enjoyed as honey, it is used in baking (honey cookies, etc.) or in the manufacturing of alcoholic beverages by mixing with alcohol (honey liqueur, "beartrag") or by fermentation into honey flavored wine (Met). Preparations containing honey, in combination with milk and cereals, are processed for children. Tobacco products are occasionally flavored with honey. In medicine, honey is used in pure form or prescribed in preparations such as honey milk, fennel honey and ointments for wounds. It is incorporated into cosmetics in glycerol-honey gels and tanning cream products. The importance of honey as a food and as a nutrient is based primarily on its aroma constituents and the high content and fast absorption of its carbohydrates.

19.2.2 Artificial Honey

19.2.2.1 Foreword

Artificial honey is mostly inverted sucrose from beet or cane sugar and is produced with or without starch sugar or starch syrup. It is adjusted in appearance, odor and flavor to imitate true honey. Depending on the production method, such creams contain nonsugar constituents, minerals, sucrose and hydroxymethyl furfural.

19.2.2.2 Production

Sucrose (75% solution) is cleaved into glucose and fructose by acidic hydrolysis using hydrochloric, sulfuric, phosphoric, carbonic, formic, lactic, tartaric or citric acid or, less frequently, enzymatically using invertase. The acid used for inversion is then neutralized with sodium carbonate or bicarbonate, calcium carbonate, etc. The inverted sugar is then aromatized, occasionally with strongly flavored natural honey. To facilitate crystallization, it is seeded with an invert sugar mixture that has already solidified, then packaged with automated

Fig. 19.16. Hydroxymethyl furfural formation in honey versus temperature and time. (according to *White*, 1978)

machines. During inversion, an oligosaccharide (a "reversion dextrin") is also formed, mostly from fructose. Overinversion by prolonged heating results in dark coloring of the product and in some bitter flavor. Moreover, glucose and fructose degradation forms a noticeable level of hydroxymethyl furfural – this could be used for identification of artificial honey.

Liquid artificial honey is made from inverted and neutralized sucrose syrup. To prevent crystallization, up to 20% of a mildly degraded, dextrin-enriched starch syrup is added (the amount added is proportional to the end-product weight).

19.2.2.3 Composition

Artificial honey contains invert sugar ($\geq 50\%$), sucrose ($\leq 38.5\%$) water $\leq 22\%$), ash ($\leq 0.5\%$) and, when necessary, saccharified starch products ($\leq 38.5\%$). The pH of the mixture should be ≥ 2.5. The aroma carrier is primarily phenylacetic acid ethyl ester and, occasionally, diacetyl, etc. Hydroxymethyl furfural content is $0.08 - 0.14\%$. The product is often colored with certified food colors.

19.2.2.4 Utilization

Artificial honey is used as a sweet spread for bread and for making Printen (honey cookies covered with almonds), gingerbread and other baked products.

19.3 Literature[a]

Birch, G.G., Green, L.F. (Eds.): Molecular structure and function of food carbohydrates. Applied Science Publ.: London. 1973

Blank, I., Fischer, K.-H., Grosch, W.: Intensive neutral odorants of linden honey. – Differences from honeys of other botanical origin. Z. Lebensm. Unters. Forsch. *189*, 426 (1989)

Crane, E. (Ed.): Honey. Heinemann. London. 1979

Fincke, A.: Zuckerwaren. In: Ullmanns Encyklopädie der technischen Chemie, 4. edn., Vol. 24, p. 795, Verlag Chemie: Weinheim. 1983

Hoffmann, H., Mauch, W., Untze, W.: Zucker und Zuckerwaren. Verlag Paul Parey: Berlin. 1985

Hough, C.A.M., Parker, K.J.; Vlitos, A.J. (Eds.): Developments in sweeteners-1 ff. Applied Science Publ.: London. 1979 ff

Jeanes, A., Hodge, J. (Eds.): Physiological effects of food carbohydrates. ACS Symposium Series 15, American Chemical Society: Washington, D.C. 1975

Koivistoinen, P., Hyvönen, L. (Eds.): Carbohydrate sweeteners in foods and nutrition. Academic Press: New York. 1980

Lehmann, J., Tegge, G., Huchette, M., Pritzwald-Stegmann, B.F., Reiff, F., Raunhardt, O., Van Velthuysen, J.A., Schiweck, H., Schulz, G.: Zucker, Zuckeralkohole und Gluconsäure. In: Ullmanns Encyklopädie der technischen Chemie, 4. edn., Vol. 24, p. 749, Verlag Chemie: Weinheim. 1983

Pancoast, H.M., Junk, W.R.: Handbook of sugars, 2nd edn., AVI Publ. Co.: Westport, Conn. 1980

Rymon Lipinski, G.-W. von, Schiweck, H.: Handbuch Süßungsmittel. Behr's Verlag: Hamburg. 1991

Schiweck, H.: Disaccharidalkohole. Süßwaren *22* (14), 13 (1978)

Schiweck, H.: Zucker, Rüben- und Rohr-. In: Ullmanns Encyklopädie der technischen Chemie, 4. Aufl., Bd. 24, S. 703, Verlag Chemie: Weinheim. 1983

Shallenberger, R.S., Birch, G.G.: Sugar chemistry. AVI Publ. Co.: Westport, Conn. 1975

Tegge, G., Riehm, T., Sinner, M., Puls, J., Sahm, H.: Verzuckerung von Stärke und cellulosehaltigen Materialien. In: Ullmanns Encyklopädie der technischen Chemie, 4. edn., Vol. 23, p. 555, Verlag Chemie: Weinheim. 1983

White jr., J.W.: Honey. Adv. Food Res. *24*, 287 (1978)

---

[a] cf. 4.5.

20 Alcoholic Beverages

Alcoholic beverages are produced from sugar-containing liquids by alcoholic fermentation. Sugars, fermentable by yeasts, are either present as such or are generated from the raw material by processing, i.e. by hydrolytic cleavage of starches and dextrins, yielding simple sugars. The most important alcoholic beverages are beer, wine and brandy. Beer and wine were known to early civilizations and were produced by a well-developed industry. The distillation process for liquor production was introduced much later.

Figure 20.1 illustrates the *Embden-Meyerhoff-Parnas* scheme of alcoholic fermentation and glycolysis. For related details about the reactions and enzymes involved, the reader is referred to a textbook of biochemistry.

20.1 Beer

20.1.1. Foreword

Beer making or brewing involves the use of germinated barley (malt), hops, yeast and water. In addition to malt from barley, other starch- and/or sugar-containing raw materials have a role, e.g., other kinds of malt such as wheat, unmalted cereals called adjuncts (barley, wheat, corn, rice), starch flour, starch degradation products and fermentable sugars. The use of additional raw materials may necessitate in part the use of microbial enzyme preparations.

Beer owes its invigorating and intoxicating properties to ethanol; its aroma, flavor and bitter taste to hops, kiln-dried products and numerous aroma constituents formed during fermentation; its nutritional value to the content of unfermented solubilized extracts (carbohydrates, protein); and, lastly, its refreshing effect to carbon dioxide, a major constituent. Data on beer production and consumption are given in Table 20.1 and a schematic representation of the production of beer is given in Fig. 20.2.

Table 20.1. World production and consumption of beer (1978)

Continent	Production (10^6 hectoliters)	
World	873.1	
Africa	27.4	
America, North-, Central-	237.2	
America, South-	58.3	
Asia	67.6	
Europe, West-	275.3	
Europe, East- + USSR	152.1	
Australia + South Pacific Islands	23.9	

Country	Production (10^6 hectoliters)	Consumption (l/capita)
USA	190.3	88.6
FR Germany	91.7	145.6
UK	66.4	122.1
USSR	65.0	24.0[a]
Japan	44.3	32.4[b]
Brazil	26.5	
German DR	23.0	
France	22.8	45.9
Czechoslovakia	22.1	130.7
Mexico	22.0	34.0
Canada	20.4	84.9[b]
Australia	19.5	137.7
Spain	18.7	52.1
Holland	14.7	84.9
Belgium	13.4	124.0
	Σ 75.7%[c]	

[a] 1977.
[b] 1976.
[c] World production $\triangleq$ 100%.

20.1.2 Raw Materials

20.1.2.1 Barley

Barley is the most important of the raw materials used for beer production. Different cultivars of the spring barley (*Hordeum vulgare*

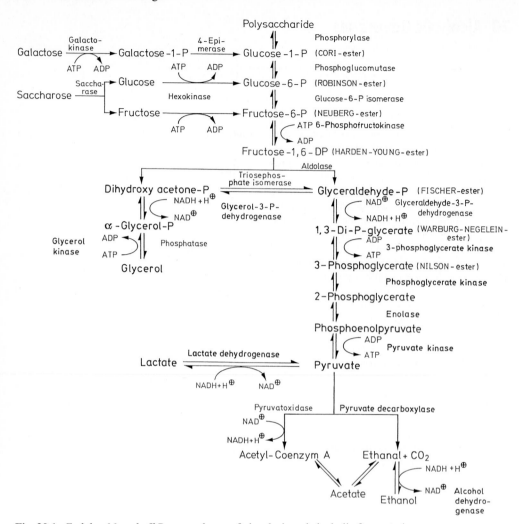

Fig. 20.1. *Embden-Meyerhoff-Parnas*-scheme of glycolysis and alcoholic fermentation

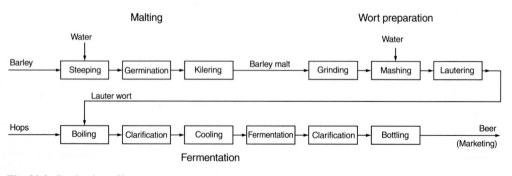

Fig. 20.2. Production of beer

convar. distichon) with exceptionally suitable properties are used as brewing and malting barley in Germany. In addition, six-row winter barley has an increasing role. Barley of high brewing value provides ample quantities of extract from the resultant malt, and has a high starch but moderate protein (9–10%) content, a high degree of germination (at least 95% of kernels), high germination vigor and good swelling ability. Sensory assay (hand appraisal) should also be included in the evaluation of barley.

20.1.2.2 Other Starch- and Sugar-Containing Raw Materials

20.1.2.2.1 Wheat Malt

Wheat malt is mixed with barley malt in a ratio of 40:60 in the production of top fermented beer.

20.1.2.2.2 Adjuncts

In addition to barley malt, supplementary sources of starch are used in the form of un-malted cereals (adjuncts) in order to dilute the mash by 15–50%. The adjuncts are barley, wheat, corn and rice (cracked rice) in the form of whole meal, grits, flakes or flour.

Adjuncts are low in enzyme activity, hence their use may necessitate the addition of microbial enzyme preparations with α-amylase and proteinase activities.

Unmalted barley contains about three times more β-glucans than malted barley. In order to decrease the viscosity of unmalted barley extract to values similar to those of malted barley, β-glucans must be degraded with the enzyme β-glucanase, which is present in microbial enzyme preparations.

20.1.2.2.3 Syrups, Extract Powders

Since adjunct processing may result in undesirable changes, extracts from enzyme- or acid-treated barley, wheat or corn have recently been introduced in the form of syrup or powder. The use of syrup from barley to as much as 45% of the total mash is possible.

20.1.2.2.4 Malt Extracts, Wort Concentrates

For production of hop-free malt extracts or hopped wort concentrates, the usual worts are evaporated in vacuum or concentrated by freeze drying. Such concentrates are diluted prior to use. The content of bitter substances and the tendency to produce cloudiness or turbidity are decreased in such concentrates, since tannins and proteins are removed during the evaporation step.

20.1.2.2.5 Brewing Sugars

Sucrose, invert sugar and starch-sugar are introduced at the stage of hopping or before the beer is bottled.

20.1.2.3 Hops

20.1.2.3.1 General Outline

Hops are a very important and indispensable ingredient in beer production. They act as a clarifier, since they precipitate the proteins in wort, change the wort character to give a specific aroma and bitter taste and, together with ethanol and carbon dioxide, their active antibiotic properties contribute to the stability of beer. Lastly, the pectin content of hops enhances the foam-building ability of beer.

The hop (*Humulus lupulus*) is a tall, hardy, perennial climbing vine. The flowers of the female plants, though lacking pollination, grow well and cluster into a conical blossom which has large thin scales or bracts. This cone, when ripe, is harvested and used commercially. The plant is propagated vegetatively by planting cuttings from fleshy roots. The hop cones are picked in August or September and are dried and pressed into bales. The lupulin gland in the upper and lower portion of bracts contains, in addition to essential oils, bitter constituents. Data on hop production are given in Table 20.2.

20.1.2.3.2 Composition

Table 20.3 presents data on the composition of hops. The constituents of utmost importance are the bitter substances. In fresh hops they occur mostly in the form of α-acids (cf. Formula 20.1): humulon (I), cohumulon (II), adhumulon (III); and in the form of β-acids: lupulon (IV), colupulon (V) and adlupulon (VI). These compounds are susceptible to changes during drying, storage and processing of hops. The changes usually involve isome-

Table 20.2. Production of hops in 1996 (1,000 t)

Continent	Hops	Country	Hops
World	123	Germany	34
		USA	34
Africa	–	China	12
America, North-,	35	Czech Rep.	9
Central-		UK	6
America, South-	–	Slovenia	4
Asia	16	Australia	3
Europe	69	Poland	3
Oceania	3	Ukraine	3
		Korea DP Rep.	2
		Romania	2
		Russia	2
		Spain	2
		Σ (%)[a]	94

[a] World production = 100%.

Table 20.3. Composition of hops

Constituent	Content (%)[a]	Constituent	Content (%)[a]
Bitter compounds	18.3	Crude fiber	15.0
Essential oil	0.5	Ash	8.5
Polyphenols	3.5	N-free extract-	
Crude protein	20.0	able mater	34.0

[a] As % dry matter; moisture content approx. 11%.

rization, oxidation and/or polymerization. As a consequence, a great number of secondary products are found.

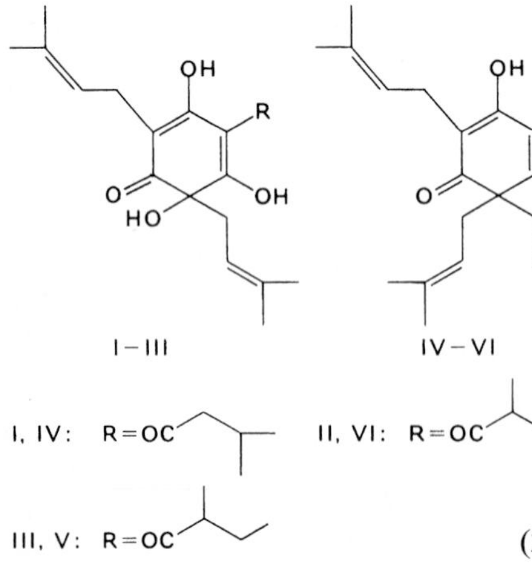

I–III   IV–VI

I, IV: R=OC⟨⟩   II, VI: R=OC⟨⟩

III, V: R=OC⟨⟩   (20.1)

The quality and intensity of the bitter taste derived from these secondary products are different. Evaluation of hops is therefore based on a determination of composition of individual α- and β-acids, rather than of the total content of bitter substances. As seen in Table 20.4, the composition varies greatly with hop origin. During the boiling of hops, humulons isomerize into isohumulons (cis-compounds, VII; trans-compounds, VIII; cf. Formula 20.2), which are more soluble and bitter than the initial compounds. The isohumulons can be further transformed into humulinic acids (IX, X), which have only about 30% of the bitterness of isohumulons.

Hulupons (XI) and luputrions (XII) are the secondary products of the lupulons. They possess an exceptionally pleasant and mild bitter taste which is much less bitter than the compounds from which they are derived. Hence

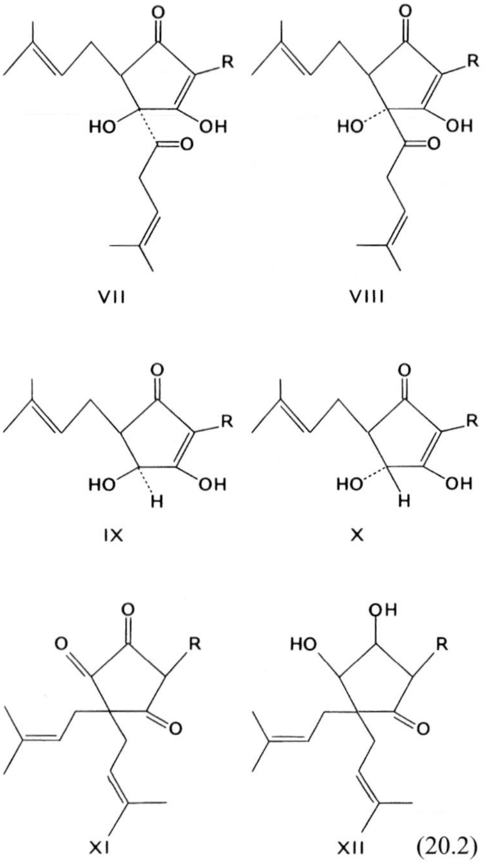

VII   VIII

IX   X

XI   XII   (20.2)

Table 20.4. Content of humulons and lupulons in hops from various sources (values in %)

Hops	α-Acids			β-Acids		
	hu-mulon	cohu-mulon	adhu-mulon	lu-pulon	colu-pulon	adlu-pulon
Japan	46	41	13	21	68	11
America	54	34	12	32	57	11
Hallertau	59	27	14	45	43	12
Northern						
Brewer	64	24	12	46	43	11
Saaz	67	21	12	51	37	12

the bitter taste of beer is primarily due to compounds of the humulon fraction.

More than 150 compounds have been identified in essential oil from hops. Among the most important are the terpenes, such as myrcene, humulene (XIV), which is obtained biosynthetically from farnesyl pyrophosphate (XIII), and caryophyllene (XV), an isomer of humulene:

$$(20.3)$$

Based on the relative proportions of the terpene constituents, hops can be classed as myrcene- and humulene-rich cultivars. The latter (such as cv. Spalt, Hallertauer, Mittelfrueher, Saaz) have a particularly pleasant aroma. The essential oil content decreases during storage and a shift in composition occurs: the hydrocarbons decrease while the oxygen-containing terpenes increase.

20.1.2.3.3 Processing

Freshly harvested hops are dried in a hop kiln in a stream of warm air $(30-65\,°C)$ to $8-10\,\%$

moisture, followed by a readjustment of moisture content to $11-12\,\%$. The dried hops are also fumigated with sulfur to increase their stability.

In addition to hop cones, which are prone to quality loss even under proper storage conditions, processed products from hops are acceptable and utilized.

Hop powder is obtained by grinding the cones, which makes the active aroma ingredients more extractable. Prior to grinding, part of the inert material is separated and thus lupulin-enriched concentrates are obtained.

Hops are extracted with a mixture of water and an organic solvent (e.g., dichloromethane), giving extracts of varying compositions. Recently, a hops extraction process using supercritical carbon dioxide has become important. Such extracts can replace $20-60\,\%$ of the hops.

Isomerized extracts, in which humulon has been converted into isohumulon by heat treatment, are suitable for a cold hopping procedure. In traditional beer hopping this conversion is achieved by boiling the wort for a long time. Isomerized extracts are used in the main fermentation or at a later step in brewing.

Boiling of hops results in the loss of a large portion of oil constituents with the steam. The addition of hops shortly before the end of the boiling process or the use of hop resins or concentrates may greatly enhance the hop aroma of the product. Phenolic constituents in hops contribute to protein coagulation during wort boiling. A part of protein-tannin complexes formed may precipitate at low temperatures after long storage, resulting in turbidity in the beer.

20.1.2.4 Brewing Water

The water used for wort preparation in a brewery has a great influence on beer quality and character. The salt constituents of water can change the pH of the mash and wort. Bicarbonate ions cause a pH increase, while Ca^{2+} and Mg^{2+} ions cause a pH decrease. Heating of water which contains bicarbonates increases the alkalinity according to the equation:

$$HCO_3^{\ominus} + H^{\oplus} \rightleftharpoons CO_2 + H_2O \qquad (20.4)$$

in which the equilibrium is shifted to the left since, during heating, the CO_2 component escapes as a gas. Ca and Mg ions react with secondary phosphates in wort to form insoluble tertiary phosphates, releasing protons which add to the acidity of the water:

$$3\,Ca^{2\oplus} + 2\,HPO_4^{2\ominus} \;\rightleftharpoons\; Ca_3(PO_4)_2 + 2\,H^{\oplus}$$

(20.5)

Magnesium sulfate in high concentrations imparts an unpleasant bitter taste to beer. Manganese and iron salts induce turbidity, discoloration and taste deterioration. Nitrates and silicates interfere with fermentation.

The unique character of different kinds of beer (Pilsen, Dortmund, Munich, Burton-on-Trent), without doubt, can historically be ascribed to the brewing water used in those places, with residual alkalinity playing the major role. Water, low in soluble bicarbonates of calcium, magnesium, sodium or potassium, and soluble carbonates and hydroxides, is suitable for strongly-hopped light beers, such as Pilsener, while alkaline water is suitable for dark beers, such as those from Munich.

Preparation of brewing water is mainly directed to the removal of carbonates. Precipitation by heating with lime is customary. Furthermore, when lime water is used without heating, water softening occurs. Removal of excess salt by ion-exchange resins is also advantageous. Today any water can be treated to match the requirement of a desired type of beer.

20.1.2.5 Brewing Yeasts

Brewing yeasts are exclusively strains of *Saccharomyces*. Two types are recognized: top fermenting yeasts for temperatures >10 °C, and bottom fermenting yeasts used down to 0 °C. The top fermenting yeasts, e.g., *Saccharomyces cerevisiae Hansen*, rise to the surface during fermentation in the form of large budding ("sprouting") associations. They ferment raffinose only partially since they lack the enzyme melibiase. The bottom fermenting yeasts, e.g., *Saccharomyces carlsbergensis Hansen*, settle to the bottom during fermentation and completely ferment all sugars including raffinose. There are yeasts with high

fermentation ability which remain suspended for a long time, giving a high fermentation rate. Yeasts with low fermentation ability flocculate early and settle to the bottom (superflocculent yeasts) and hence are unable to continue active fermentation. Pure cultures of many yeast strains currently in use are derived from a single yeast cell and are used as "starter yeast" in plant operations. After the main fermentation, a part of the yeast is harvested for use in freshly-prepared worts, until the yeast becomes useless due to contamination or degeneration. In this way, it is possible to continuously select suitable yeast strains for a defined goal.

20.1.3 Malt Preparation

The cereals are soaked (steeped) in water and then allowed to germinate. The product, green malt, is dried and mildly roasted into a more or less dark and aroma-rich kiln-dried malt. During processing, the rootlets are removed from the malt. The loss due to malting is 11–13% of the dry weight. Prior to use, the malt is stored for 4–6 weeks.

20.1.3.1 Steeping

Cereal kernels are steeped in water to raise their moisture content to induce germination. The water content is 42–44% for light and 44–46% for dark malt. Usually the steeping water is alternately added and removed. In this way, in contrast to a single-wetting process, the steeping time is shortened and better germination and enzyme build-up are achieved. Good aeration is needed in all phases to remove the CO_2 produced by respiration. The normal steeping temperature is 12–18 °C. Warm (20–30 °C) or hot (35–38 °C) water steeping accelerates germination and thus shortens the process. Alkali treatment (CaO, NaOH) of steeping water serves to reduce microbial contamination and to remove undesirable polyphenols from the hulls.

20.1.3.2 Germination

When the cereals reach the desired moisture content, they are allowed to germinate in germinator compartments, on floors, or in chests or drums. The removal of CO_2 and heat is achieved by turning and mixing in traditional floor malting, whereas a stream of moistened air is used in the newer chest or drum procedures. The optimal germination temperature range is $15-20\,°C$. The process for light malt takes about 7 days; for dark malt, 9 days. Germination can be stimulated by the use of gibberellic acid.

20.1.3.3 Kilning

The germinated cereals, termed green malt, contain $42-45\,\%$ moisture. They are dried in a kiln to give a malt with $2-3\,\%$ moisture. Simultaneously, the color and roasted aroma of the malt are formed by *Maillard* reactions.

Initially, the green malt is dried at $35-40\,°C$ for about 12 h. For light malt production the moisture is decreased rapidly to about 10%. For dark malt, where more extensive hydrolytic degradation of starch is needed, the moisture content is maintained for a longer time at about 20%. The malt is then heated to the roasting temperature within 2 h and is roasted for $4-5$ h at $80\,°C$ for light malt or at $105\,°C$ for dark malt. Finally, the kiln-dried malt is freed of rootlets, cleaned and polished.

20.1.3.4 Continuous Processes

Several kinds of installations have been developed which provide continuous steeping, germination and, occasionally, also kilning, offering substantial savings in time. Steeping in this case is performed as a single washing followed by water spraying and continuous transferring to the germination stage. The process conditions are regulated by means of forced air. Green malt is obtained within $5-7$ days. In some installations the malt is moved for different stages, while in others it remains in the same container from steeping to kilning.

20.1.3.5 Special Malts

Special malts are prepared for many purposes. Dark caramelized malt is held briefly at $60-80\,°C$ to saccharify its starch and is then roasted at $150-180\,°C$ for the desired degree of color. Such color-rich malt is free of diastase enzyme activity, is a good foam builder, and is

mostly used for aromatizing malt beers and strong bock beers. Light caramelized malt is made in a similar way, but is treated at lower temperatures after the saccharification step. This preserves the enzyme activities. It is lightly colored and when used gives beer an increased foaming capacity and full-bodied properties. Colored malt is obtained by roasting the kiln-dried malt at $190-220\,°C$, omitting the prior saccharification step. It can be used to intensify the color of dark beers.

20.1.4 Wort Preparation

The coarsely ground malt is dispersed in water. During this time, the malt enzymes hydrolyze starch and other ingredients. A clear fermentable solution, the so-called wort, is obtained by filtration. When boiled with added hops, the wort takes on the typical beer flavors.

20.1.4.1 Ground Malt

Malt is disintegrated by passing it through several grinding rolls and sifters. The ground products, hull, middlings and flour, are then combined in the desired proportions. By using finely ground meal, the extraction yield increases, but problems arise in wort filtration. Wet milling is commonly preferred for better filtration as it yields a higher proportion of intact hulls. In addition, it provides the desired high extraction yield. For wet milling the water content of the malt is adjusted to $25-30\,\%$. It is then ground by a set of rollers and processed immediately into wort.

Continuous wet meal steepers have been developed to guarantee a defined steeping time and, thus, to prevent the malt grains from becoming slippery and gelatinized by overly long steepage.

20.1.4.2 Mashing

In the mashing step, the malt meal is made into a paste with brewing water (heatable mixing vessel) and partially degraded and solubilized with malt enzymes.

For 100 kg of malt 8 hectoliters of water are needed. This amount of water is divided into a major portion for production of the mash, and into one or several post-mashing rinses used to wash out extract from the hulls. The course of pH and temperature during mashing are of utmost importance for determining wort composition and, hence, the type and quality of beer. The optimum activity of malt α-amylases is from 72–76°C at pH 5.3–5.8, and of malt β-amylases from 60–65°C at pH 4.6, while that of malt proteinases is from 55–65°C at pH 4.6. Hence, wort with a pH near 6 will not, without prior pH adjustment, provide optimal conditions for the action of enzymes. The methods used for temperature control in mashing are of two types: decoction and infusion. In the decoction method, the initial temperature of the total mash is raised by removing an aliquot of mash, heating this to boiling and then returning it to the main mash in the mash tun. In general one-, two- or three-mash return procedures are used commercially. The latter is used exclusively for dark beer brewing; the two-mash return for light beer; and the one-mash return procedure for brewing all types of beer. The three-mash return procedure will be briefly described as an example: The crushed malt is mixed in the mash tun with water at 37°C; the first aliquot is drawn, heated to boiling and returned to the mash tun. In this way the total mash temperature is raised to 52°C. Two repetitions raise the total mash temperature stepwise to 64 and then to 75°C. The mashing process is completed at a terminal mash temperature of 74–78°C.

In the case of poorly "solubilized" malt in which the starch-containing membranes have not ruptured, enzymatic degradation and the extract yield can be improved by stopping briefly at 47–50°C before further temperature increase. This delays enzyme inactivation. On the other hand, when a low alcohol beer is desired, the malt mashed at 37°C is drained into boiling water, increasing the temperature to 70°C and resulting in extensive enzyme inactivation.

In infusion mashing, used mostly in England for brewing top fermented beer, the terminal mashing temperature is achieved not by stepwise increases, but by live steam injection or addition of hot water. As in the decoction method, the temperature program used can vary greatly.

20.1.4.3 Lautering

The separation of wort from hulls and insoluble residues of the grain is done by a classical procedure in a lauter tun, a vessel with a slotted false bottom. The hull and other residues form a ca. 35 cm deep layer in the bottom which acts as a filter through which the extract, or wort, is strained. The initial turbid liquid (turbid wort) with 16–20% extract is pumped back to the tun. Finally, to obtain more wort, the spent grains are rinsed or sparged 3 to 4 times with water.

Modern installations for lautering use strain masters or discontinuous or continuous mash filters. The draff, the lautering residue, is used for animal feed.

20.1.4.4 Wort Boiling and Hopping

Wort boiling with hops or hop products is done in a brew kettle (hop kettle) in which the initial and subsequent worts from the lautering step are collected. Addition of hops is adjusted according to the type and quality of beer desired. The quantity (in hop cones/hectoliter) for light lager beer is 130–150 g; for Dortmund-type beer, 180–220 g; for Pilsener beer, 250–400 g; for dark Munich beer, 130–170 g; and for malt beer and dark bock beer, 50–90 g. The critical factor is the content of bitter substances in the hops selected. Boiling for 70 to 120 min concentrates the wort, coagulates protein ("break forming"), solubilizes hop ingredients and converts the bitter components to their isoforms and, lastly, inactivates enzymes. The hot wort is then chilled, filtered, aerated and, finally, "pitched" with yeast.

In modern processes, the classical brew kettle is replaced by a whirlpool kettle with external cooker. Shorter boiling times and a better quality of beer are achieved with this system. Moreover, separation from the spent hops can be conducted in the same vessel.

Processes that use pressure boiling (high-temperature wort boiling up to 150°C) can produce beer with an unpleasant cooked taste.

20.1.4.5 Continuous Processes

Efforts are being made to introduce continuous processes via heat exchangers and to save energy and make the process environmentally friendly with heat recovery from the exhaust steam.

Wort treatment, i.e., removal of the trub formed during boiling (protein-tannin complexes), is generally conducted in whirlpool vats (possibly combined with wort drying) or via continuous centrifuges. After cooling to the pitching temperature (6–8 °C), the cooling trub obtained is separated by filtration or centrifugation.

20.1.5 Fermentation

20.1.5.1 Bottom Fermentation

Bottom fermentation involves a primary and a secondary step. In the primary fermentation step, the cooled wort with about 6.5–18% dry mass extracted from malt ("stemwort") is pumped into fermenting tanks, located in fermentation cellars cooled to 5–6 °C. In the past the tanks were made of lacquered, microparaffin, wax or asphalt base-treated oakwood, but today the tanks are made of plastic-lined concrete, enamel-coated steel, aluminum or V_2A steel. The wort is inoculated ("pitched") with yeast in the form of a thick yeast slurry (0.5–1 l/hl) and fermented at 6–10 °C until more than 90% of the fermentable extract has been converted. The primary fermentation is completed in 6–10 days, at which point the yeast "breaks", i.e., flocculates and settles to the bottom. The beer is transferred to large clean tanks. The middle layer of yeast, the "core yeast", is removed from the bottom of the fermentation tank and is filtered, washed and reused in the next fermentation.

The young "green" beer is stored for 1–4 months in tanks at 1.5–2 °C for secondary fermentation. During this time, extensive fermentation of residual sugars occurs, accompanied by carbon dioxide enrichment, beer clarification and maturation. The storage time is 6–12 weeks for Munich beer containing 11–14% stemwort, or 14–18 weeks for Dortmund beer with 13–14% stemwort.

20.1.5.2 Top Fermentation

Primary fermentation proceeds in fermentation tanks, but at higher temperatures (18–25 °C) than bottom fermentation, and

requires a total time of 2–7 days. The yeast builds a solid cap at the top of the tank. It is skimmed off into individual fractions (hops flock, yeast flock, post-flock). The secondary fermentation is a very slow process and may continue in tanks or bottles. Top fermentation is used mostly in England and Belgium, while in Germany it is used in the production of Weiss beer, a light tart ale made from wheat.

20.1.5.3 Continuous Processes, Rapid Methods

Several continuous processing methods provide accelerated fermentation. They make use of thermophilic yeasts, higher fermentation temperatures and more intensive wort aeration.

The production of beer with carrier-bound yeast cells has been recently discussed. However, these methods have not yet been introduced.

20.1.6 Bottling

After ageing, beer is filtered through cotton filter pads and some silicates, often having been preclarified through a kieselguhr pad or by centrifugation. Then, with the aid of a special cask/keg filling apparatus, it is foamlessly filled into transportable casks or metal cisterns. In addition to impregnated oakwood casks, specially-lined iron, aluminum or V_2A steel containers are also acceptable. Bottle filling proceeds from a "bottle tank" in a fully automated process. Tin-plated or aluminum cans are also used.

Pasteurization gives the beer biological stability for overseas export. To avoid cloudiness due to protein precipitation and changes in flavor, the beer is heated to 60–70 °C in a water bath or by steam. The beer is often pasteurized at 62 °C for 20 min. For sterile filling the beer is heated to 70 °C for 30 s or is passed through microfilters (with pore size less than the size of bacteria) and then poured into sterilized bottles or cans.

Temperature fluctuations during storage and transport must be avoided if beer quality is to be preserved.

20.1.7 Composition

20.1.7.1 Ethanol

The ethanol content, which has a very important influence on the aroma, is 1.0−1.5% by weight for a low fermented extract-rich beer, 1.5−2.0% for a weak or thin beer, 3.5−4.5% for a full beer, and 4.8−5.5% for a strong beer. Higher alcohols, such as 2-methylbutanol, 3-methylbutanol, methylpropanol and 2-phenylethanol, are also present in very small quantities.

20.1.7.2 Extract

The nonalcoholic constituents of beer vary within a wide range from 2−3% for plain beers to 8−10% for strong beers. These constituents are the beer solids and consist of to 80% carbohydrate, mostly dextrins. It is possible to calculate the solids content of the original wort before fermentation from the solids content (E, weight %) and alcohol content (A, weight %) of the beer product. The calculation is based on the fermentation equation: 2 parts by weight of sugar equal 1 part by weight of alcohol. The initial solids content of wort, which actually represents a measure of malt utilization, is designated as "stemwort" (St) and can be calculated by the formula:

$$St = \frac{100\,(E + 2.0665\,A)}{100 + 1.0665\,A} \qquad (20.6)$$

Thus, for example, if the solids content (E) of a beer is 3% (w/v) and the alcohol content (A) is 5.0% (v/v), then the solids content of the wort before fermentation was 12.6% (w/v). The stemwort content in Germany is 2−5.5% for plain beers, 7−8% for draft beers, 11−14% for full beers and above 16% for strong beers.

20.1.7.3 Acids

Carbon dioxide is responsible to a substantial extent for the refreshing value and stability of beer. CO_2 is 0.36−0.44% in bottom fermented beers, while in Weiss beer the CO_2 content is up to 0.6−0.7%. A CO_2 content below 0.2% gives flat and dull beers. Apart from small amounts of lactic, acetic, formic, and succinic acids, beer contains 9,10,13- and 9,12,13-trihydroxyoctadecenoic acid. In fact, 9.9 ± 2.1 mg/l were found in five types of beer and 9(S), 12(S), 13(S)-trihydroxy-10(E)-octadecenoic acid was the main compound and accounted for 50−55% of the 16 stereoisomers. The pH of beer is between 4.7 (dark, strong beer) and 4.1 (Weiss beer).

20.1.7.4 Nitrogen Compounds

The N-compounds in beer (0.15−0.75%) originate primarily from proteins in the raw materials and from yeast. They consist mainly of proteins plus high molecular weight protein degradation products; both being responsible for cloudiness in beer during cold storage. The free amino acids found in malt are also present in beer. It appears that glutamic acid contributes to beer taste. The presence of volatile amines has also been confirmed.

20.1.7.5 Carbohydrates

The carbohydrate content is approximately 3−5%, while in some strong beers or malt beers it may be considerably higher. Pentosans are also present in addition to dextrins, mono- and oligosaccharides (maltotriose, maltose, etc.). Glycerol normally is 0.2−0.3% of beer.

20.1.7.6 Minerals

Minerals make up 0.3−0.4% of beer and consist mostly of potassium and phosphate. Calcium, magnesium, iron, chloride, sulfate and silicates are also present.

20.1.7.7 Vitamins

Vitamins of the B-group (vitamins B_1 and B_2, nicotinic acid, pyridoxine and pantothenic acid) are present in various beers, often in significant amounts.

20.1.7.8 Aroma Substances

Odor- and taste-bearing substances which contribute to the aroma of beer are presented in Table 20.5, grouped according to their aroma values.

The primary odor- and taste-active substances essentially determine the type of beer. The bitter taste and hops note of Pilsener beers are

Table 20.5. Examples of beer aroma notes and odor- and taste-active substances responsible for them

Attribute[a]	Odorant/taste compound[b]
Primary flavor constituents (A$_x$ > 2)	
1200 Bitter	Hop bitter substances
0110 Alcoholic	Ethanol
1360 Carbonation	Carbon dioxide
0410 Caramel	Furaneol
0130 Estery	Esters (3-methyl-butylacetate)
1000 Sweet	Sugars
0140 Fruity	Ethyl butanoate, ethyl hexanoate
0160 Floral	2-Phenylethanol
Secondary flavor constituents (A$_x$ = 0.5–2)	
0730 Boiled vegetables	Dialkyl sulfides (dimethyl sulfide)
0620 Buttery	Diacetyl
0630 Rancid	Butyric, acid, 2- and 3-methylbutyric acid
0910 Acetic	Acetic and propionic acids
0920 Sour	Lactic, citric and malic acids
1340 Astringent	Polyphenols
1410 Body and others	Amino acids, small peptides, nucleic acid derivatives
Tertiary flavor constituents (A$_x$ = 0.1–0.5)	
0820 Papery	Unsaturated aldehydes (2-trans-nonenal)
0500 Phenolic	Volatile phenols (4-vinylguaiacol)
0721 H$_2$S	Hydrogen sulfide
0724 Light struck	3-Methyl-2-buten-1-thiol plus methanethiol
1100 Salty	Inorganic salts (NaCl)
1330 Metallic	Metals

[a] The number refers to the flavor wheel (cf. Fig. 20.3).
[b] Ranking of the odorants according to decreasing aroma values A$_x$ (cf. 5.1.4). The aroma values based on odor or taste threshold values, which were estimated in a pale lager "null beer".

produced by relatively high concentrations of isohumulons, humulenes (including oxidation products), and myrcene, while larger amounts of furaneol are responsible for the caramel note of dark beers.

The secondary odor- and taste-bearing substances are in the range of the perception threshold (definition cf. 5.1.3). However, some odorants can increase and reach the detection threshold. Appropriate tests of mixtures have shown that similarities in the aroma profile are of greater importance for additive effects than correspondence in functional groups. For example, the odor intensities of butyric acid and diacetyl add up because their aroma profiles contain butter notes. On the other hand, a mixture of acetic acid and caprylic acid separately exhibits the typical notes of both components. The additive behavior of two aroma substances, each present in beer with an aroma value A$_x$ = 0.5 generally results in an aroma with a value A$_x$ = 0.8 ± 0.2. Since the detection threshold for a certain aroma note is exceeded only on addition of several secondary aroma substances, the absence of one component results in only slight changes in the aroma of beer. However, if the concentration of a secondary or tertiary aroma substance is increased to such an extent that an A$_x$ of 1 is reached or exceeded, aroma defects are produced. Examples of critical compounds are diacetyl, acetaldehyde, 2-trans-nonenal, and dimethylsulfide (cf. 20.1.9).

In the production of alcohol-free beer, the concentrations of important aroma substances drop (Table 20.6).

20.1.7.9 Foam Builders

The foam building properties of beer are due to proteins, polysaccharides and bitter constituents. The β-glucans stabilize the foam through their ability to increase viscosity.

Table 20.6. Aroma substances in lager beer and alcohol-free beer

Compound	Lager beer (mg/l)	Alcohol-free beer (mg/l)
3-Methylbutanol	49.6	6.7
2-Phenylethanol	17.5	2.3
Ethyl hexanoate	0.15	0.01
Ethyl butanoate	0.06	0.01
Furaneol	0.35	0.19
4-Vinylguaiacol	0.52	0.13

Addition of semisynthetic polysaccharides, e. g., propyleneglycol alginate (4 g/hectoliter), to beer provides a very stable foam although the addition is judged as unfavorable.

Lysophosphatidyl cholines (LPC), which occur in cereal as amylose inclusion compounds (starch lipids: cf. 15.2.5), reduce the foam stability. The temperature management during the mashing process regulates the LPC concentration because it determines the activity ratio of α-amylase, which contributes to the release of LPC from amylose, to phospholipase B, which catalyzes the degradation of LPC. Temperatures above 65 °C favor the more stable α-amylase, increasing the LPC concentration.

20.1.8 Kinds of Beer

There is a distinction between top and bottom fermented beers.

20.1.8.1 Top Fermented Beers

Selected examples of top fermented beers from Germany are: Berlin weiss beer, brewed from a wort having 7−8% solids from barley and wheat malts and inoculated at fermentation with yeast and lactic acid bacteria; Bavarian weiss beer brewed from weakly-smoked barley malt with a little wheat malt and fermented only with yeast; Graetzer beer made from wheat malt with a smoky flavor and with a stemwort content of 7−8%; malt beer (caramel beer), a dark, sweet and slightly hop-flavored full beer; the bitter beers such as those from Cologne or Duesseldorf (Altbier) which are strongly hop-flavored full beers; top fermented plain beers (Jungbier or Frischbier) with a low stemwort content and often artificially sweetened; Braunschweig's mumme, an unfermented, non-hop flavored malt extract, hence not a true beer or a beer-like beverage. English beers have a stemwort content up to 11−13%. Stout is a very darkly colored and alcohol-rich beer made from concentrated boiled wort (up to 25% stemwort; alcohol content >6.5%). Milder varieties of stout are known as Porter beer.

Pale ale is strongly hopped light beer, whereas mild ale is mildly hopped dark beer. Incorpo-

ration of ginger root essence into these beers yields ginger-flavored ale.

Top fermented beers from Belgium, which are stored for a longer time, are called Lambic and Faro beers.

20.1.8.2 Bottom Fermented Beers

These beers show a significantly increased storage stability and are brewed as light, mildly colored or dark beers.

Pilsener beer, an example of a light colored beer, is typically hop flavored, containing 11.8−12.7% stemwort. In contrast, Dortmunder-type beer is made from a more concentrated wort which is fermented longer and thereby has a higher alcohol content. Lager beer (North German Lager) is similar to Dortmunder in hop flavoring, while the stemwort content is close to a Pilsener beer. Munich beers are dark, lightly hop flavored and contain 0.5−2% colored malt and often a little caramel malt. They taste sweet, have a typical malt aromatic flavor, and are fermented with a stemwort content of 11−14%. Beers with a high content of extract are designated as export beers. Traditional dark beers and currently produced special light beers, are the bock beers (Salvator, Animator, etc.). They are also strong beers with more than 16% stemwort. The dark Nuernberg and Kulmbacher beers are even higher in colored malt extracts and thereby are darker than Munich beers. An example of mildly colored beer is the Maerzen beer (averaging 13.8% stemwort). It is produced from malt of Munich in which the use of colored malt is omitted.

20.1.8.3 Diet Beers

Diet beers exhibit a high degree of fermentation and contain almost no carbohydrates, which are a burden for diabetics. They are produced by special fermentation processes and contain a relatively high alcohol content. Subsequently, the alcohol level is frequently reduced to values typical of normal beer.

20.1.8.4 Low-Alcohol or Alcohol-Free Beers

The production of low-alcohol and alcohol-free beers is possible by the application of suitable technological measures, e. g., throttling of fermentation and use of special yeasts, reduction of the stemwort content, and elimination of the alcohol formed (distillation, ultracentrifugation). The effect of such measures on the aroma has already been discussed (cf. 20.1.7.8, Table 20.6).

20.1.8.5 Export Beers

These originate from widely different kinds of beer. They are mostly pasteurized and additionally treated with flocculating or adsorption agents (tannin, bentonite) or with proteolytic enzyme preparations to remove most of the proteins. The proteolytic enzymes split the large protein molecules into soluble products.

Such beers are free of cloudiness or turbidity (chill-proofed beers) even after prolonged transport and cold storage.

20.1.9 Beer Flavor and Beer Defects

The taste and odor profile of a beer, including possible aroma defects, can be described in detail with the help of 44 terms grouped into 14 general terms, as shown in Fig. 20.3. Apart from a great variety of terms for odor notes, the terms bitter, salty, metallic, and alkaline are used only for taste and the terms sour, sweet, "body" etc. are applied to both taste as well as odor.

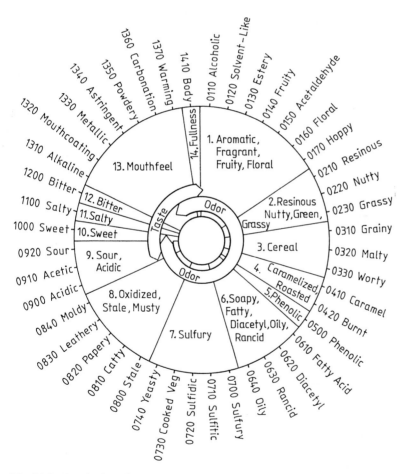

Fig. 20.3. Terminology for the description of odor and taste notes of beer (*American Society of Brewing Chemists*, according to *Meilgaard*, 1982)

Nine of the terms given in Fig. 20.3 describe the most important odor and taste characteristics of a good beer (Table 20.7). They are also suitable for the differentiation of different types of beer (Table 20.7). The odor- and taste-active substances that mark these characteristics are given in 20.1.7.8 and in Table 20.5.

Foaming is an important criterion of the taste of beer. A distinction is made between foam volume (produced by the content of carbon dioxide), foam density, and especially foam stability (caused by protein degradation products, bitter hop compounds, and pentosans). Lower fatty acids that are present in beer bouquet act as defoamers.

Beer defects detract from the odor and taste and are caused by improper production and storage. An example of a taste defect is the harsh, hard, bitter taste produced by the oxidation of polyphenols and some hop constituents. A flat taste, as already mentioned, comes from a low content of carbon dioxide. Diacetyl and ethanal in concentrations greater than 0.13 mg/l and 25 mg/l respectively, produce a taste defect. Acceleration of fermentation caused, e. g., by intensive stirring of the wort, raises the content of diacetyl and higher alcohols in the beer and lowers the content of esters and acids. On the whole, the aroma is negatively influenced. Higher concentrations of ethanal can arise, e.g., at higher fermentation temperatures and higher yeast concentrations.

Beer is very sensitive to light and oxidation. The "light" taste is due to the formation of 3-methyl-2-buten-1-thiol (cf. Table 5.4). Enzymatic peroxidation of lipids contained in the wort and nonenzymatic secondary reactions during wort boiling give rise to the aroma defects listed as No. 8 in Fig. 20.3. A guide substance for defects of this type is 2-trans-nonenal which has a very low threshold value (0.11 µg/l beer) and produces a "papery" aroma defect.

The addition of ascorbic acid or glucose oxidase/catalase (cf. 2.7.2.1.1) is recommended to overcome color and flavor defects caused by oxidation. Therefore, low-oxygen bottling is of great importance. Bottled beer should not contain more than 1 mg O_2/l.

The very potent aroma substance 3-methyl-3-mercaptobutyl formate (cf. Table 5.31) can produce an off-aroma called "catty" (0810 in Fig. 20.3). The concentration of phenylacetaldehyde can also increase to such an extent on the storage of beer that it becomes noticeable in the aroma.

On storage, beer can become cloudy and form a sediment. Proteins and polypeptides make up 40−75% of the turbidity-causing solids. They become insoluble due to the formation of intermolecular disulfide bonds, complex formation with polyphenols, or reactions with heavy metals ions (Cu, Fe, Sn). Other components of the sediment are carbohydrates (2−25%), mainly α- and β-glucans. For measures used to prevent cloudiness, see 20.1.8.5. Undesirable microorganisms, e.g., thermophilic lactic acid bacteria, acetic acid bacteria

Table 20.7. Main characteristics of the odor and taste of various types of beer

Flavor group	Intensity[a]					
	Munich	Pilsner	Pale ale	US lager	Stout[b]	Lambic
Bitterness	3−6	6−10	5−8	2−4	6−10	3−6
Alcoholic flavor	2−4	3−4	3−4	3−5	3−5	3−6
Carbonation	3−4	3−4	1−3	4	3−4	3−5
Hop character	2−6	6−10	5−8	0.5−4	6−10	3−6
Caramel flavor	4−8	0.5−2	3−5	0.5−1	6−100	1−3
Fruity/estery flavor	1−2	1−1.5	1−2	2−3	2−3	3−5
Sweetness	2−3	1−2	1−2	2−3	1−2	1−2
Acidity	1−2	1−2	1−2	1−2	2−3	3−20
Cabbage-like	1−2	1−3	0.2−0.8	1−3	0.2−0.8	1−10

[a] Semiquantitative values on the basis of aroma values.
[b] Top fermented English strong beer with a stemwort content of up to 25%.

(*Acetobacter, Gluconobacter*) and yeasts, can cause disturbances and defects in various process steps (mashing, fermentation, finished product). A well-known aroma defect is, e.g., the diacetyl flavor.

20.2 Wine

20.2.1 Foreword

Wine is a beverage obtained by full or partial alcoholic fermentation of fresh, crushed grapes or grape juice (must). The woody vine grape has thrived in the Mediterranean region since ancient times and Italy, France and Spain are still among the leading wine-producing countries in the world. Other major producers are USA, Argentina, Germany and South Africa. Table 20.8 provides data on wine production and consumption in some countries. An overview of the individual process steps in wine production is presented in Fig. 20.4.

20.2.2 Grape Cultivars

Among the cultivated species of *Vitis*, the most important is the grapevine *Vitis vinifera*, L. ssp. *vinifera* in its many forms; more than 8,000 cultivars are known. The size, shape and color of the grapes vary: there are round, elongated, large or small grape clusters. Grapes are either wine-type grapes, for white or red wine making, or table grapes, which are even grown in greenhouses in some northern countries. The cultivars are different in sugar content and aroma. Table 20.9 provides information about

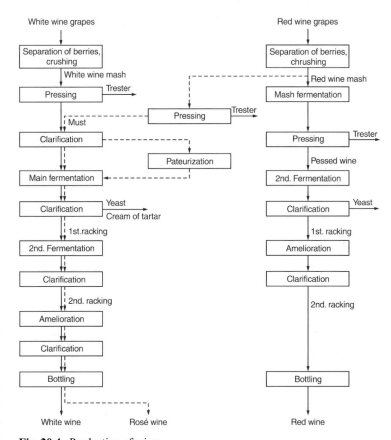

Fig. 20.4. Production of wine

Table 20.8. Wine production 1996 (1,000 t), vineyard area in 10^6 ha (1993) and wine consumption (l/capita)

Continent	Production	Vineyard area (1993)
World	26,842	82.81
Africa	1,071	3.47
America, North-, Central-	2,065	7.93
America, South-	2,442	
Asia	916	13.85
Europe	19,708	56.86
Oceania	640	0.7

Country	Production	Vineyard area (1993)	Consumption	
			(1971)	(1993)
Italy	6,000	9.81	111	61
France	5,897	9.42	107	64
Spain	2,883	13.7	60	39
USA	1,864	3.25	5	6
Argentina	1,710	2.05	85	48
Germany	1,200	1.06	18	23
South Africa	950	–	–	–
Portugal	713	3.7	91	55
Australia	584	0.63	–	–
Romania	550	2.51	23	55
Σ (%)[a]	83			

[a] World production = 100%.

the major grape cultivars of Germany, with some of their characteristics. Table 20.10 shows the share of the major cultivars in German vine growing areas. Table 20.11 gives data on the grape cultivars of some other countries. The European *V. vinifera* and the American vines (*V. labrusca*) have been crossed in order to produce pest-resistant forms (hybrids, "direct producers"), giving plants with pest resistance and good quality must production, although the hybrids still leave much to be desired. The wines are considered rather ordinary, with less character and a more obtrusive flavor than the parent plants.

Grape cultivars providing top quality white wines are:

- *Riesling* – native to Germany; a hardy cultivar grown in the Pfalz (Rhine Palatinate) and along the Mosel (Moselle), Rhine and Nahe rivers.

- *Traminer* – cultivated extensively in Alsace, Baden and Pfalz, and in Austria.
- *Rulaender* (gray Burgundy) – from Alsace and Burgundy regions in the Kaiserstuhl district, and from Hungary.
- *Semillon Blanc* – together with Sauvignon and sometimes with Muscatel, provides Sauternes from the Bordeaux region.
- *Sauvignon* – used for Sauternes, and processed into its own types of wine, such as in the Loire region.
- *White Burgundy* (Pinot blanc) – yields the white wines from Burgundy (Chablis, Meursault, Puligny-Montrachet).
- *Chardonnay* – related to white burgundy, cultivated for example in Champagne.
- *Auxerrois* – also related to white burgundy.

Table 20.9. Important German grape cultivars

Cultivar	Wine type[a]	Acid[b]	Must weight[c]	Maturation characteristics[d]	Yield[e]	Comments about wine[f]
White wine cultivars						
Auxerrois		2	2	4	3	A vivacious wine with distinquished bouquet
Bacchus	M	2	2	3	3	Flowery with a muscat note fragrance
Burgundy, white	S	3	2	4		A full-bodied wine, pleasantly aromatic and more neutral as Rulaender
Ehrenfelser	R	2	2	5	2	Fruity, mildly acidic, a Riesling-like wine
Elbling, white	S	3	1	6	3	A light wine, devoid of rounded body and bouquet
Faber	M		2	2	3	A refined, refreshing and fruity flavored wine
Gutedel, white	M	1	1	3	3	Light wine, pleasing and captivating, mildly aromatic
Huxelrebe	T	2	2	2	3	A mellow wine with muscat-like bouquet
Kerner	R	2	2	4	2	A refreshing wine with a fine Riesling-like bouquet
Morio-Muscat	B	2	1	3	3	A wine with strong captivating muscat aroma
Mueller-Thurgau	M	2	1	2	3	Mild and refreshing wine with fine muscat flavor
Muscatel-yellow	B	2	2	5	1–2	A superior wine with a strong muscat-like aroma
Muscat-Ottonel	B	2	2		2	A pleasing wine with a strong refined muscat bouquet
Nobling	S	2	2	2	2	A full-bodied wine with a fruity flavor and fine bouquet
Optima	A	2	3	2	2	A refined, captivating wine with a fragrant aroma
Ortega	B	2	3	1	1	A wine with refined peach-like aroma
Perle	T	1	2	3	3	A mellow wine with flowery bouquet
Riesling, white	R	3	1	5	2	A superior refreshing and pleasing wine, with a fruity and flowery flavor
Rulaender (gray Burgundy)	T	2	2	3	3	A body-rich wine with burning and passionate perception, and a pleasing bouquet
Scheurebe	T	3	2	4	3	A strong fruity flavored body-rich wine with a bouquet reminiscent of black currants
Siegerrebe	B	1	3	1	1	A wine with highly intensive refined bouquet

846   20 Alcoholic Beverages

Table 20.9 (continued)

Cultivar	Wine type[a]	Acid[b]	Must weight[c]	Maturation characteristics[d]	Yield[e]	Comments about wine[f]
Sylvaner, green	S	2	1	4	3	A mellow pleasing wine with a delicately fruity flavor
Traminer, reddish (Clevner)	T	2	2	4	1	A wine with an exceptionally strong persisting bouquet
Red wine cultivars						
Burgundy, blue, late-		2–3	2	4	2–3	Full-bodied, strongly flavored with a rounded bouquet, dark red mellow wine
Heroldrebe						A superior neutral wine with a tannin-like astringency
Limberger, blue		2	2	5	2	Characteristically fruity, a somewhat herbaceous, tarty and finely astringent bluish-red wine
Muellerrebe (black riesling)		2	2	4	2	Reminiscent of late Burgundy, but of lower quality
Portuguese, blue		2	1–2	1	3	A neutral mellow bluish-red wine with a bouquet deficiency
Trollinger, blue		2	2		3	A mellow refreshing light wine with a pungent flavor and light-red in color

[a] Quality German wines are classified as table wines (Tafelwein, Oechsle degrees less than 60), quality wines (with all the required characteristics of the growing region and an Oechsle degree of at least 60) and the special high quality wines (Oechsle degrees at least 73). The latter are denoted according to increasing quality as Kabinett, Spaetlese, Auslese, Beerenauslese and for the top quality as Trockenbeerenauslese. In addition to the rating, the label might carry a designation as Eiswein (ice-wine, see text).
R: Riesling group of wine (superior, fruity wine with distinct acidity)
S: Sylvaner group (neutral wine devoid of a distinct bouquet)
M: Mueller-Thurgau group (light, flowery with discrete bouquet)
T: Traminer group (wine with a fine bouquet)
B: Bouquet group of wine (wine with strong and aromatic bouquet)
A: Auslese group of wines (fullbodied great wines).
[b] 1: Low (approx. 5 g/l), 2: medium (approx. 5–10 g/l), and 3: high acidity (10–15 g/l).
[c] 1: 60–70 Oechsle degrees, 2: 70–85°, and 3: >85 Oechsle degrees.
[d] 1: Very early maturing (beginning-middle of September), 2: early (middle-end of September, 3: early-medium (end of September, beginning of October), 4: medium late (beginning-middle of October), 5: late (middle-end of October), and 6: very late maturing cultivar (end of October beginning of November).
[e] 1: Low (60 hl/ha), 2: average (60–80 hl/ha), and 3: high yielding cultivar (≥90 hl/ha).
[f] The wine organoleptic quality description has its own wine dictionary. Terms classify and refer to wine (1) aroma or bouquet, (2) body, (3) sweetness and acids, (4) variety or cultivar, (5) age and (6) wine taste harmony (i.e. to which extent are the constituents of wine agreeably blended or related).

Table 20.10. Distribution of major grape cultivars as percent of total vineyard area in Germany (1985)

Vineyard region	Area ha[a] 1[b]	2	3	4	5	6	7	8	9	10	11	12	13	14	15	16	17	18
Rheinhessen	25,082 5	14	25	7	5	10			2	8			95	1	4			5
Rheinpfalz	22,853 14	10	24	10	8	6					1		90	1	8			10
Baden	14,931 7	4	37					9	12		3		79	20	1		1	21
Mosel-Saar-Ruwer	12,747 55		23	6			9						100					0
Württemberg	9,600 24	7	10	8					1				55	4	5	23	12	45
Franken	5,240 3	23	49	2		3				9			98	1	1			2
Nahe	4,607 22	15	28	7	1	7			3		1		97		1			3
Rheingau	2,943 79	4	7	1					1				95		5			5
Mittelrhein	768 74	4	11	6									98	1				2
Ahr	450 16		18	1									38	34	26			62
Hessische Bergstraße	388 52	9	19	1						11			99		1			1
FG Germany	99,609 19	9	26	6	3		4	1	1	3	6	1	2 87	4	3	2	1	13

[a] 1984.
[b] 1: Riesling, 2: Sylvaner, 3: Mueller-Thurgau, 4: Kerner, 5: Morio-Muskat, 6: Scheurebe, 7: Elbling, 8: Gutedel, 9: Rulaender, 10: Bacchus, 11: white Burgunder, 12: Huxelrebe, 13: white grapes grand total, 14: blue late Burgunder, 15: Protugueser, 16: Trollinger, 17: black Riesling, 18: red grapes grand total.

Grape cultivars providing good white wines are:

- *Muscatel* and *Muscat-Ottonel* – cultivars with an exceptionally rich bouquet.
- *Furmint* – the grape cultivar of Hungarian Tokay wines.
- *Sylvaner* – grown in Pfalz, Rheinhessen and Franken regions of Germany.
- *Mueller-Thurgau* – grown widely in east Switzerland and in Germany; it is a cross between Riesling and Sylvaner.
- *Gutedel* (Chasselas, Fendant, Dorin) – often found in Baden, Alsace, West Switzerland, and France.
- *Scheurebe* – a favored cultivar in Germany, obtained by crossing Sylvaner and Riesling.
- *Morio-Muscat*, a cultivar of exceptional bouquet.
- *Veltliner* – of significance in Austria, as is
- *Zierfandler*.

Grape cultivars providing top quality red wines are:

- *Pinot Noir* – the famous red vine cultivated in the Cote d'Or region of Burgundy, and also in Germany along the river Ahr and in Baden.
- *Cabernet-Sauvignon*,
- *Cabernet-Franc*, and
- *Merlot* – are cultivated together and provide the famous red wines of the Bordeaux region.

Other red grape cultivars are:

- *Gamay* – from the southern part of Burgundy and from Beaujolais and Maconnais.
- *Pinot Meunier* – black Riesling; of importance in Champagne, Wuerttemberg and Baden.
- *Portuguese* – found in Pfalz, Rheinhessen, and Wuerttemberg.
- *Trollinger* (Vernatsch) – cultivated in south Tyrol and in Wuerttemberg.
- *Limberger* – found in Wuerttemberg and Austria.
- *Blue Aramon* – the cultivar which provides the wines from Midi, France.
- *Rossary* – widely cultivated in south Tyrol.

Grape vine cultivation requires an average annual temperature of 10–12 °C. The average monthly temperature from April to October should not fall below 15 °C. The northern limit for cultivating the grape vine is close to 50° latitude. The permissible altitude for cultiva-

Table 20.11. Major grape cultivars of selected countries

Country	Grape cultivar	Comments about cultivation area and quality
France	*White wine cultivars*	
	Aligote	Bourgogne, a "vin ordinaire", modest quality wine
	Chardonnay	Cultivated in Champagne and Bourgogne area (Chablis, Montrachet, Pouilly), a very good quality wine
	Chemin blanc	Cultivated in regions of Tourraine, Anjou and Loire
	Folle blanche	Wine used for brandy production in Cognac and Armagnac
	Grenach blanc	Midi
	Melon blanc (Muscadet)	Mellow refreshing wine with a slight muscat bouquet
	Muscadelle	Cultivated in Bordeaux and Charente regions, 5–10% blended into Sauternes and Graves wines
	Pinot blanc	Cultivated in Alsace, Champagne, Loire and Cote d'Or
	Pinot gris	Alsace wine
	Roussane (Rouselle)	Cultivated in Rhone region, a full-bodied, pleasing fragrancy wine
	Sauvignon	Wine of Bordeaux, Loire and Cher regions, a full-bodied fragrant wine, with Semillon used for production of Sauternes wine
	Semillon blanc	As Sauvignon, used for production of Sauternes wine
	Red wine cultivars	
	Cabernet Franc	Spread in Bordeaux and Loire regions, a superior, strong pleasing wine, with Cabernet Sauvignon and Merlot is an ingredient of Bordeaux wines
	Cabernet Sauvignon	As Cabernet Franc, aroma rich, a superior quality wine
	Carignan	Grown in Rhone, Midi and Provence regions
	Cot (Malbec)	Bordeaux, one of the best grape cultivars
	Cinsaut	Grown in Southern France
	Grenach noir	Grown in Southern France
	Gamay noir	Beaujolais, Maconnais; fruity pleasant, refreshing wine
	Merlot	A Bordeaux wine, full-bodied, rich and mellow; as Cabernet Franc and Cabernet Sauvignon is an ingredient of Bordeaux wines
	Petit Verdot	Grown in Bordeaux region, component of Bordeaux wines
	Pinot noir	Bourgogne, and wine of Cote d'Or
	Syrah	Grown in Southern France
Italy	*White wine cultivars*	
	Malvasia, bianca	An important cultivar across Italy
	Mascato, bianco	Grown mostly in Northern Italy, a wine of Asti region
	Trebbiano	Widely grown across Italy
	Vermentino	Grown along Italian riviera, a very good white wine
	Weissterlaner	Wine of South Tyrol
	Red wine cultivars	
	Aleatico	Widely grown in Italy
	Barbera	One of the most important cultivars
	Freisa	Grown in Piemont and Vercelli regions, one of the best Italian cultivars
	Gross-Vernatsch (Trollinger)	The wine of Bolzano, Trento and Como
	Lagrein	Grown in South Tyrol

Table 20.11 (continued)

Country	Grape cultivar	Comments about cultivation area and quality
	Merlot	
	Nebbiolo	A prefered cultivar of Piemont and Lombardy regions
	Pinot Nero	Grown in Northern Italy with Rome as Southern limit
	San Giovese	Spread from Toscana till Latium; major constituent of Chianti wine
Austria	*White wine cultivars*	
	Mueller-Thurgau	
	Muscat-Ottonel	
	Neuburger	A pronounced cultivar bouquet, pleasantly acidic
	Rheinriesling	
	Rotgipfler	Fruity, aroma rich, full-bodied; together with Zierfandler an ingredient of Gumpoldskirchner wines
	Sylvaner	
	Traminer	
	Veltliner, green	A pleasant pleasing bouquet refreshing wine
	Veltliner, red	Fruity wine with a fine bouquet
	Veltliner (early red, Malvasier)	
	Welschriesling	A mellow wine with fine bouquet
	Zierfandler, red	A wine with burning and passionate perception, fragrant aromatic, with a cultivar specific bouquet
	Red wine cultivars	
	Burgundy, blue, late	
	Blaufraenkisch (Limberger)	
	Portuguese, blue	
	Sankt Laurent	A strong wine, dark red colored, with a fine Bordeaux-like bouquet
Switzerland	*White wine cultivars*	
	Gutedel (Chasselas, Fendant, Dorin)	A major Swiss grape cultivar
	Marsanne Blanche (Hermitage)	A mellow wine with a refined bouquet
	Riesling	
	Mueller-Thurgau	Major cultivar of Eastern Switzerland
	Red wine cultivars	
	Burgundy, blue	
	Gamay	Grown in Western Switzerland
	Merlot	The wine of Tessin (Ticino)
Hungary	*White wine cultivars*	
	Furmint, yellow	Used for Tokay wine production
	Red wine cultivars	
	Kadarka	The most important Hungarian red wine cultivar

tion is dependent on the climate (plains in Italy, Spain and Portugal; sunny slopes of Germany; up to 1,300 m on Mt. Aetna in Sicily; up to 2,700 m in the Himalayas). Soil cultivability and quality and weather are of decisive importance.

20.2.3 Grape Must

20.2.3.1 Growth and Harvest

After blooming and fruit formation, the grape berry continues to grow until the middle or the end of August, but remains green and hard. The acid content is high, while the sugar content is low. As ripening proceeds, the berry color changes to yellow-green or blue-red. The sugar content rises abruptly, while both the acid and water contents drop (Fig. 20.5).

The harvest (picking the berry clusters from the vines) is performed as nearly as possible when the grape is fully ripe, about the middle of September until the end of November, or it may be delayed until the grapes are overripe. In USA and Europe, machines are being increasingly used for this very laborious harvesting, e. g., grape harvesters. However, they cannot sort the grapes according to the degree of ripeness. Terms which relate to the time of harvest include "vorlese", early harvest, "normallese", normal harvest, and "spaetlese", late harvest. The latter term, when applied to German wines, identifies excellent, top quality wines. Particularly well-developed grapes of the best cultivars from selected locations are picked separately and processed into a wine called "Auslese". When the grapes are left on the vine stock, they become overripe and dry – this provides the raisins or dried berries for "Beerenauslese", "Trockenbeerenauslese", or "Ausbruch" wine (fortified wine). In some districts, such as Tyrol and Trentino, the grapes are spread on straw or on reed mats to obtain shrivelled berries – this provides the so-called straw wines, Grapes that are botrytised (a state of "dry rot" caused by the mold *Botrytis cynerea*, the noble rot) have a high sugar content and a must of superior quality, consequently producing a superior, fortified wine. Frozen grapes left on the vine stock provide ice-must which, because of freezing, is en-

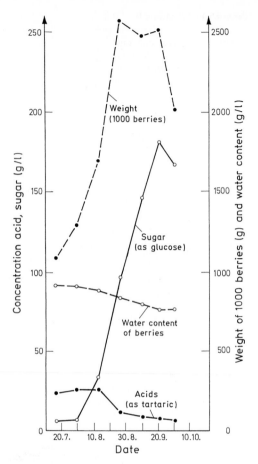

Fig. 20.5. Sylvaner wine grape ripening with measurement of the content of acid (as tartaric), sugar (as glucose), weight of 1,000 berries and water content of the berries

riched in sugar and, as such, is a source of high quality wines (ice-wines).

20.2.3.2 Must Production and Treatment

The grape clusters cut from vine stocks using grape shears are cleaned of rotten and dried berries and then, as fast as possible, separated from the stems. This is done in a roller crusher which consists of two fluted horizontal rolls by which the berries are crushed without breaking the seeds or grinding the stems. The latter are separated out by a stemmer. The crushed grapes are then subjected to pressing to release their juice, the must. The mechanical and partly continuously operated presses are basket-type screw-presses (extruder-like tapered screw), hydraulic or pneumatic presses. The free-running must is collected prior to pressing (first run) and, after mild pressing, the major portion (pressed-must) is produced. The remaining grape skins and seeds (pomace) are loosened or shaken-up and pressed again. This provides the second or post-extract. In red wine making the crushed berries (the mash) are fermented without prior removal of the pomace, i.e. the must is fermented together with the skin. This is done in order to extract the red pigments localized in the skin, which are released only during fermentation. When blue grapes are processed in the same manner as white ones, or blends of blue and white grapes are combined and then processed, pink wines are obtained. They are designated as rosé wines. In red wine making the extraction of red pigments is sometimes facilitated by raising the temperature to 50°C prior to fermentation of the mash, or to 30°C after the main fermentation, followed by a short additional fermentation.

The left over stems, skins and seeds provide the pomace. It is used as feed or fertilizer, or is fermented to provide pomace wine. This is consumed as a homemade drink and is not marketed. Pomace brandy is obtained by distillation of fermented pomace. The average must yield is 75 l/100 kg grapes. Of this, 60% is free juice (must), 30% press-must and 10% must from the second pressing.

The fresh, sweet must is treated with sulfur dioxide (50 mgSO$_2$/l) to suppress oxidative discoloration and the growth of undesirable microorganisms. In order to remove undesirable odors or off-tastes, the must is treated with activated charcoal and, when necessary, is clarified by separators or filters. If required, the must is pasteurized by a short heat treatment (87°C/2 min).

The addition of sugar to and deacidification of must will be discussed in 20.2.5.4.

20.2.3.3 Must Composition

Table 20.12 provides data on the average composition of grape musts. For the quality assessment of grape must, its specific density

Table 20.12. Average composition of grape must

Constituent	Content (g/l)
Water	780–850
Sugar (as glucose)	120–250
Acids (as tartaric acid)	6– 14
N-Compounds	0.5– 1
Minerals	2.5– 3.5

has to be determined at 20°C (must weight M in kg). It is then usually expressed in *Oechsle* degrees (°Oe):

$$°Oe = (M - 1) \times 10^3 \tag{20.7}$$

The sugar content (S in g/l) can be calculated from the following relationship (X is the acid content in g/l; 25 for a good vintage, and 30 for an average year):

$$S = 2.5 \, (°Oe) - X \tag{20.8}$$

In Germany, the must weight determines the later classification of the wine according to grades. This system is presented in Table 20.13.

20.2.3.3.1 Carbohydrates

Ripe grapes contain equal amounts of glucose and fructose, while fructose predominates in overripe or botrytised berries. Saccharose content is much lower (10–12 g/l) and pectins (0.12–0.15%) and small amounts of pentosans are present.

20.2.3.3.2 Acids

The major acids of must are L-tartaric and L-malic acids. Succinic, citric and some other acids (cf. 18.1.2.4 and 20.2.6.5) are minor constituents. In a good vintage, tartaric acid is 65–70% of the titratable acidity, but in years when unripe grapes are fermented, its content is only 35–40% and malic acid predominates. The good vintage year of 1911, for example, yielded grapes with 3.1 g/l malic acid and 6.4 g/l tartaric acid; in the inferior vintage year of 1912, on the other hand, malic acid was 10.7 g/l and tartaric acid 6.0 g/l.

Table 20.13. German grades of wine

Grade	Must weight (°Oe)	Remarks
Table wine		Addition of sucrose or grape juice concentrate permissible
Country wine		Addition of sucrose permissible
Quality wine		Addition of sucrose permissible
Quality wine of a certain area	≥60	
Quality wine with vintage	≥73	No addition of sugar permitted
Cabinet		Fully ripe berries
"Spaetlese"		Late harvest, fully ripe berries
"Auslese"		Like "spaetlese", unripe diseased berries are removed
"Beerenauslese"		Only botrytised (*Botrytis cinerea*) or overripe berries
"Trockenbeerenauslese"	≥150	Like "beerenauslese", largely shriveled berries
Ice wine		Frozen on harvesting and pressing

20.2.3.3.3 Nitrogen Compounds

Proteins, which include various enzymes, peptides and amino acids, are present in low amounts (cf. 18.1.2.1)

20.2.3.3.4 Lipids

The lipid content of must is about 0.01 g/l (cf. 18.1.2.3).

20.2.3.3.5 Phenolic Compounds (cf. 18.1.2.5)

Tannins occur primarily in stems, skin and seeds. In a carefully prepared white must, the tannin content is no more than 0.2 g/l. In contrast, red wines contain high levels of tannin, 1–2.5 g/l or even higher. The pigments in must are anthocyanins.

20.2.3.3.6 Minerals

Must contains predominantly potassium, followed by calcium, magnesium, sodium and iron. Important anions are phosphate, sulfate, silicate and chloride.

20.2.3.3.7 Aroma Substances

The must aroma substances will be discussed together with wine aroma substances (cf. 20.2.6.9).

20.2.4 Fermentation

Wine fermentation may occur spontaneously due to the presence of various desirable wine yeasts and wild yeasts found on the surface of grapes. Fermentation can also be conducted after must pasteurization by inoculation of the must with a pure culture of a selected strain of wine yeast. Wild yeasts include *Saccharomyces apiculatus* and *exiguus*, while the pure selected yeasts are derived from *Saccharomyces cerevisiae var. ellipsoides* or *pastorianus*. The pure wine yeast possesses various desirable fermentation properties. High fermenting strains are used to give high alcohol wines (up to 145 g/l) and those which are resistant to tannin and high alcohol levels are used in red wine fermentation. Other types of yeast are "sulfite yeast" with little sensitivity to sulfurous acid (sulfur dioxide solutions), "cold fermentation yeasts", which are active at low temperatures and, finally, special yeasts for sparkling wines, which are able to form a dense, coarse-grained cloudiness that is readily removed from the wine. The desired yeasts are added to must held in fermenters (vats made of oak, or chromium-nickel steel tanks lined with glass, enamel or plastic). The must is then fermented slowly for up to 21 days at 12–14°C for white wines or 20–24°C for red wines. When must has not been treated extensively with sulfur dioxide, the primary fermentation starts within a day and reaches its maximum after 3–4 days.

As a safeguard against air (discoloration), bacterial spoilage (acetic acid bacteria) and also to retain carbon dioxide, the liquid loss in the fermenter is compensated for by topping up with the same wine. After the end of main (primary) fermentation, which lasts 5–7 days, the sugar has been largely converted to alcohol while the protein, pectin and tannins, along with tartrate and cell debris, settle with the yeast cells at the bottom of the fermenter. This sediment is called bottom mud, dregs or lees. Partial precipitation of tartaric acid in the form of potassium acid tartrate (cream of tartar) is affected by temperature, alcohol content and pH (Fig. 20.6). Tartaric acid also precipitates to a small extent as calcium tartrate. The unfermented residual sugar (residual sweetness) may be retained when necessary, if the secondary fermentation is suppressed by addition of sorbic acid or diethylpyrocarbonate or dimethylpyrocarbonate. Fermentation stops at an ethanol concentration of 12–15% (v/v).

The young wine, which is drunk in some regions of Germany and Austria ("Federweisser" or "Sauser"), is kept in casks for a few weeks up to several months, during which time secondary fermentation occurs. In such wines the residual sugar is degraded, valuable aroma substances (the bouquet) develop and some

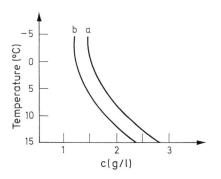

Fig. 20.6. The effect of temperature and ethanol concentration on cream of tartar solubility in wine. (a) 8 vol-%, (b) 12 vol-%. (according to *Vogt*, 1974)

additional dregs, such as yeast cells and cream of tartar, sediment out.

Red wine mash is fermented at somewhat higher temperatures by using various procedures, often in closed double-walled enamel-lined tanks and often by treating extensively with sulfur dioxide. The wine initially drawn off is the better quality free-run wine, followed by the pressed wine, an astringent and dry fraction ("press-wine"). The young wine should not stay on the pomace longer than necessary to extract the pigments, otherwise it will become tannin-enriched and hence harsh and astringent. In industrial production the extraction of the red pigments is not done by fermentation of the mash but by heat treatment of the mash (cf. 20.2.3.2).

The fermentation residue or pomace is processed into yeast-pressed wine or yeast-brandy, into wine oil (for brandy essence) and into tartaric acid. The left-over pomace is used as a feed or fertilizer. Pomace wine, obtained by fermenting a sugar solution containing the dispersed pressed-out pomace, is made only into a household drink and is not marketed.

20.2.5 Cellar Operations After Fermentation; Storage

The following cellar operations develop a particular character in the wine and give it stability and durability.

20.2.5.1 Racking, Storing and Aging

Racking of young wine is required to get rid of the sediment. The wine is drawn-off or decanted into large sulfur-treated vats, with or without aeration. The time for racking is determined by the cellar master's experience. The wine racking is repeated as required. Racking should be carried but as early as possible. When necessary, 5–10% of unfermented sterile grape must is blended with the young wine to round-off and sweeten its flavor.

The objective of wine aging/storage is to further build up the aroma and flavor constituents. Aging requires various lengths of time. In general the wine is removed from vats after 3–9 months and poured into bottles in which aging continues. Duration of aging and storability differ and depend on wine quality. The great Burgundy and Bordeaux wines require at least 4–8 years in order to develop while for an average German wine, maximum development is achieved well within 5–7 years. Only great quality wines endure aging lasting 10–12 years or more without quality loss.

During wine aging or maturation, one of the more important reactions which contribute to flavor is the partial degradation of malic acid to lactic acid, i.e. malo-lactic fermentation under the influence of diverse organisms, mainly of the genera *Pediococcus* (*P. pentosaceus, P. cerevisiae*), *Leuconostoc* (*L. oenos*) and *Lactobacillus*:

$$\text{HOOC—CHOH—CH}_2\text{—COOH}$$
$$\longrightarrow \quad \text{HOOC—CHOH—CH}_3 \; + \; \text{CO}_2 \qquad (20.9)$$

This degradation reduces wine acidity and mellows the wine. Table 20.14 provides information on the time course of this reaction.

Changes induced during wine maturation are not yet well understood. Reactions between wine ingredients, such as ethanol, acids and carbonyl compounds, which form the typical aroma components of wine, are covered in 20.2.6.9.

20.2.5.2 Sulfur Treatment

Crushed grapes (mash) or must are treated with sulfur immediately after grape crushing to preserve the constituents that are sensitive to oxidation, prevent enzymatic browning via phenol oxidation and suppress the growth of undesirable microorganisms (acetic acid bacteria, wild yeast, molds). Sulfur treatment of

Table 20.14. Lactic acid formation during aging of white wine

Date	Total acid[a] (g/l)	Lactic acid (g/l)
June	11.3	0.0
July	11.0	0.4
September	10.5	0.8
February	9.3	2.0
July	8.1	2.8

[a] Calculated as tartaric acid.

wines prior to the first racking serves the same purpose: wine stabilization. Futhermore, a very important effect is the suppression of undesirable aroma notes ("air", "oxidation", "ageing", "sherry" notes) by the binding of carbonyl compounds, especially of ethanal, as hydroxysulfonic acids. Sulfur treatment is achieved by the addition of sulfites in the form of a 6% aqueous solution of sulfurous acid or its salts, or by adding liquid SO_2. Only a part of the added sulfurous acid remains as free acid. A portion is oxidized to sulfate, while another binds to sugars and carbonyl compounds. Use of the right amount of SO_2 is important for fermentation and aging and hence for the quality of the wine. Efforts are made to achieve 30–50 mg of free SO_2/l of finished wine.

20.2.5.3 Clarification and Stabilization

Suitable measures should not only eliminate any turbidity present, but also prevent its formation during storage.

Turbidity-causing solids are mostly proteins as well as oxidized and condensed polyphenols. Furthermore, multivalent metal ions can cause discoloration and sediments. Wine clarification or "fining" is usually achieved by precipitation reactions, filtration or centrifugation. In blue-fining the excess metal ions which are responsible for metal-induced cloudiness (iron, copper and zinc) are precipitated by precisely calculated amounts of potassium ferrocyanide. The blue turbidity formed helps to eliminate the persistent protein turbidity (grayish and black casse). Wine fining with an excess of potassium ferrocyanide causes the formation of hydrocyanic acid. In other fining procedures, gelatin, isinglass (sturgeon bladder gelatin), egg albumen, tannin, iron-free bentonite, kaoline, agar-agar and purified or activated charcoals are added to the wine. This results in adsorption or precipitation of the substances causing cloudiness, the interaction products all being quick-settling coagulums. Iron salts are removed from wine by phytates, phenolic compounds by polyvinylpyrrolidone (detanninizing) and undesirable sulfur compounds by cupric sulfate.

The clarification by filtering involves pads of asbestos, cellulose, infusorial earth, and filter aids such as Hyflo Super Cel and Filter Cel. The filters are built either as leaf filters or as washable filter presses. Sterilizing filters (which remove microorganisms) are suitable for stopping fermentation and thus retaining a desired level of unfermented sugar at a selected stage of fermentation. High efficiency centrifuges (separators) are being increasingly used for rapid clarification in industrial wineries.

Suitable measures to prevent crystalline sediments in the bottle are, e.g., cooling the wine for a few days to $0-4\,^\circ C$, seeding with tartar crystals, and reducing the concentrations of potassium, calcium, and tartaric acid by ion exchange or by electrodialysis. Excessive concentrations of calcium produced by deacidification measures (cf. 20.2.5.4) can also result in additional crystal sedimentation (calcium tartrate, calcium mucate, and calcium oxalate). The elimination of excess calcium with D-tartaric acid is recommended as a countermeasure.

20.2.5.4 Amelioration

Must and wine amelioration is required when unfavorable weather in some years results in grapes with an excess of acids and a low sugar content. Such grapes would provide a must which could not be processed directly into a drinkable, palatable wine. The ameliorated wine should contain neither more alcohol nor less acid than the wine of the same type and origin from a good vintage year. The usual procedures involved are the addition of sugar, deacidification and wine blending.

Addition of sugar, for which regulations exist in most countries, is done before fermentation, though it is usually done at the must or grape crushing stage. Sucrose (dry sweetening) or grape must concentrate are added. The bouquet (aroma) is not improved. Poor or inferior wine is not improved by amelioration.

Deacidification is achieved primarily by adding calcium carbonate, which may give either a precipitate of calcium tartrate or a mixture of calcium tartrate and calcium malate. Ion-exchange resins can also be used for deacidification.

Wine blending is a suitable way of rectifying defects, refreshing old wines, deepening the color of red wines (table wines) and enhancing the bouquet or readjusting the low acid content, thus producing a uniform quality wine for the market.

Other wine treatments involve, for example, the addition of tartaric or citric acid to low acid wines of southern European countries. Addition of gypsum or phosphate treatment, used to increase the color of red wines, is based on color improvement caused by a decrease in pH by $CaSO_4$ or $CaHPO_4$. Saturation with carbon dioxide does not provide substantial amelioration to flat or bland wines.

20.2.6 Composition

The chemical composition of wine varies over a wide range. It is influenced by environmental factors, such as climate, weather and soil, as well as by cultivar and by storage and handling of the grapes, must and wine.

Within the scope of wine analysis, wine extract, alcohol, sugar, acids, ash, tannins, color pigments, nitrogen compounds and bouquet-forming substances are important. Hence, the value and quality of a wine is assessed through the content of ethanol, extract, sugar, glycerol, acids and bouquet substances. With the large number of quality-determining constituents, the evaluation and classification of wine are possible only by a combination of chemical analysis and sensory testing.

20.2.6.1 Extract

The extract includes all the components of wine mentioned above, except the volatile, distillable ones. Many of the extract components are present in must and are described in that section; others are typical fermentation and degradation products. The extract content of 85% of all German white wines is about $20-30$ g/l (average about 22 g/l), while the extract content of red wines is somewhat higher – German "Auslese" wines contain about 60 g/l; other sweet wines, $30-40$ g/l.

Since the sugar content can be manipulated, the "sugar-free extract" (extract in g/l minus reducing sugar in g/l plus 1 g/l for arabinose, which is also detected in the reductometric determination, but is not fermentable) is of greater importance for an evaluation of quality.

20.2.6.2 Carbohydrates

Carbohydrates (0.03–0.5%) present in fully fermented wines are small amounts of the hexoses glucose and fructose and of nonfermentable pentoses. Incompletely fermented wines contain higher concentrations of both hexoses, but substantially more of the slower fermenting fructose. The average ratio of glucose to fructose in the residual sugar of wine is 0.58:1, but it varies to a great extent. Wines with less than 4 g/l of residual sugar are designated as dry. The pentose sugars which are present in fermented wines consist of 0.05–0.13% arabinose, 0.02–0.04% rhamnose, and xylose in trace amounts.

20.2.6.3 Ethanol

The ethanol content of wine varies over a wide range. Normal grape wines, according to vintage and cultivar, have an ethanol content of 55–110 g/l. Light wines contain 55–75 g/l; average wines, 75–90 g/l; and stronger wines 90–110 g/l or more. The ethanol content of strong wines which mainly originate from southern European countries is 110–130 g/l. An alcohol level above 144 g/l indicates addition of ethanol.

20.2.6.4 Other Alcohols

Methanol occurs in wines at a very low level (38–200 mg/l), but much more is present in the fermentation of pomace as a product of pectin hydrolysis. Brandy distilled from pomace often contains 1–2% methanol. Higher alcohols in wine are propyl, butyl and amyl alcohols which, together, constitute 99% of the wine fusel oil. Hexyl, heptyl and nonyl alcohol and other alcohols are present in small amounts. The average butylene glycol (2,3-butanediol) content is 0.4–0.7 g/l and is derived from diacetyl by yeast fermentation. Glycerol, 6–10 g/l, originates from sugars and gives wine its body and round taste. Sorbitol is found in very low amounts. D-Mannitol is not present in healthy wines, but is present in spoiled, bacteria-infected wines at levels up to 35 g/l.

20.2.6.5 Acids

The pH of grape wine is between 2.8 and 3.8. Titratable acidity in German wines is between 4 and 9 g/l (expressed as tartaric acid). Acid degradation and cream of tartar precipitation decrease the acid content of ripe wines. Red wines generally contain less acids than white wines. The wines from Mediterranean countries are low in acid content. Wine acids from grapes are tartaric, malic and citric acids and acids from fermentation and acid degradation are succinic, carbonic (carbon dioxide) and lactic acids and low amounts of some volatile acids. The presence of acetic and propionic acids as well as an anomalous amount of lactic acid is an indication of diseased wine.
Botrytis cinerea can form gluconic acid in concentrations of up to 2 g/l of must. Therefore, this acid is found in the corresponding wines.

20.2.6.6 Phenolic Compounds

The phenolic compounds were described under grape must and in the chapter "fruits" (cf. 18.1.2.5). In the maturation of red wine, tanning agents polymerize (proanthocyanidins, cf. 18.1.2.5.2) in two ways and become insoluble, reducing the astringent taste. Acid-catalyzed polymerization proceeds via the carbocation shown in Formula 18.20. In addition, proanthocyanidins are cross-linked by acetaldehyde, which is formed by the slight oxidation of ethanol that occurs on storage of red wine.

20.2.6.7 Nitrogen Compounds

The nitrogen compounds in must precipitate to a smaller extent by binding to tannins during grape crushing and mashing, while most (70–80%) of them are metabolized by the growing yeast during fermentation. Free amino acids (about 200–800 mg/l) are the major nitogen compounds which remain in wine.

20.2.6.8 Minerals

The mineral content of wine is lower than that of the must since a part of the minerals is removed by precipitation as salts of tartaric acid. The ash content of wines is about 1.8–2.5 g/l, while that of must is 3–5 g/l. The average composition of ash in %, is: K_2O, 40; MgO, 6; CaO, 4; Na_2O, 2; Al_2O_3, 1; CO_2; 18; P_2O, 16; SO_3, 10; Cl, 2; SiO_2, 1.

The iron (as Fe_2O_3) content of wine is 5.7–13.4 mg/l, but it can rise to much higher levels (20–30 mg/l) through improper processing of grapes.

20.2.6.9 Aroma Substances

The volatiles in wine, more than 600 compounds, with a total concentration of 0.8–1.2 g/l, originate in part from the grapes (primary aroma) and in part from fermentation (secondary aroma). Although these compounds have been identified, the actual contribution to aroma has been proved only for a few substances. Most of the grape cultivars used for wine production are neutral in aroma. However, there are aroma-rich grapes, e. g., the Muscatel, Pinot and Sauvignon cultivars.

Terpenes contribute to the aroma of Muscatel wines and, to a smaller extent, other wines. In the must, however, these terpenes are still largely present as odorless glycosides, di- and polyols. As shown in Table 20.15, only linalool is present in a concentration higher than the odor threshold. As the grapes ripen, the increase in the concentration of hydroxylated terpenes and terpene glycosides is at first greater than that of the free terpenes (Fig. 20.7a). It is only at the end of ripening, when the pH value and the sugar level barely change (Fig. 20.7b), that the free terpenes increase at the expense of the bound terpenes (Fig. 20.7, a), enhancing the aroma. Juice processing techniques such as pH adjustment and heating enhances the hydrolysis of the glycosides to yield higher concentrations of linalool, geraniol, nerol and α-terpineol. In addition a broad pattern of aroma-active monoterpenes (e. g., nerol oxide, hotrienol) is formed by cyclization and dehydration reactions of di- and polyhydroxylated monoterpenes (examples, cf. 5.3.2.4). From the limited number of monoterpenes present in the grape juice, numerous and/or larger quantities of aroma compounds are formed during the production of wines.

The monoterpene pattern can be used to differentiate cultivars. For example, a clear distinction can be made between wines from the grape cultivar "White Riesling" and wines from other grape cultivars which are also sold as "Riesling". As shown in Fig. 20.8, the monoterpene concentrations (especially of linalool, hotrienol, α-terpineol, and 3,7-dimethylocta-1,5-*trans*-dien-3,7-diol) in "White Riesling" are considerably higher than in the other "Riesling" wines.

The aroma during fermentation is due to the formation of ethanol, higher alcohols and esters. Assessment of individual compounds for their sensory relevance is very difficult since the aroma is a complex mixture in which the sensory effect of one compound is influenced by the presence of the others.

Moreover other factors, such as the pH value and the concentrations of ethanol, sugars, and other constitutents, are of great importance (Table 20.16). The alcohols methylpropanol

Table 20.15. Free and glycosidically bound monoterpenes in Muscat grapes

Monoterpene	Free (µg/kg fruit)	Bound (µg/kg fruit)	Odor threshold (µg/l; water)
Linalool	100	390	6
Geraniol	< 5	330	130
Nerol	< 5	170	400
α-Terpineol	< 5	40	350
trans-Furan linalool oxide[a]	5	130	7,000
cis-Furan linalool oxide[a]	< 5	< 5	6,000
trans-Pyran linalool oxide	100	20	3,600
cis-Pyran linalool oxide	30	5	5,400
3,7-Dimethylocta-1,*trans*-5-dien-3,7-diol[a]	2,200	300	odorless

[a] Chemical structure: cf. 5.3.2.4.

and 2- and 3-methylbutanol produced in amino acid metabolism are supposed to have a negative effect on the aroma if they exceed the odor theshold. On the other hand, 2-phenylethanol is said to have a positive effect.

The concentration ranges and odor thresholds of esters found in a larger number of white and red wines are presented in Table 20.17. The high variability of the ester fraction has an effect on the intensity of fruity notes in the aroma profile because the ester concentrations can be considerably higher than the odor thresholds, e. g., in the case of ethyl pentanoate.

Table 20.16. Effect of medium on odor threshold values of some wine aroma compounds

Compound	Threshold value (µg/l) in solution[a]			
	1	2	3	4
Ethyl caproate	36	37	56	850
Ethyl acetate	25,000	40,000		160,000
Amyl alcohol[b]	1,900	12,500	60,000	180,000

[a] 1: Water, 2: 12% ethanol with tartaric acid adjusted to pH 3, 3: solution under 2 + 100 g/l saccharose, 4: dry white wine.
[b] Blend (1 : 3) of 2-methyl-1-butanol with 3-methyl-1-butanol.

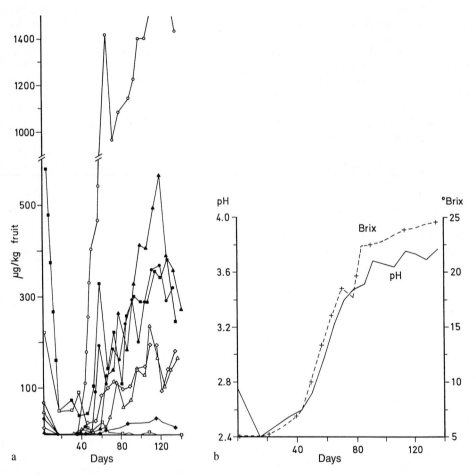

Fig. 20.7 a, b. Changes in concentration of monoterpenes, pH and sugar levels in the juice of ripening Muscat of Alexandria grapes (according to *Wilson* et al., 1984)
a Concentration of free (open symbols) and glycosidically bound (closed symbols) monoterpenes. ○–○, ●–● 3,7-dimethylocta-1, *trans*-5-dien-3,7-diol; □–□, ■–■ geraniol; △–△, ▲–▲ linalool; ◇–◇, ◆–◆ *trans*-pyran-linalool oxide **b** Juice pH and sugar levels (extract) — pH; –+–+ °Brix

Table 20.17. Esters in wine with sensory relevance

Compound	White wine (mg/l)	Red wine (mg/l)	Odor theshold in wine (mg/l)
Ethyl acetate	0.15–150	9–257	12.3
Ethyl propanoate	0–0.9	0–20	1.8
Ethyl pentanoate	1.3	5–10	0.01
Ethyl hexanoate	0.03–1.3	0–3.4	0.08
Ethyl octanoate	0.05–2.3	0.2–3.8	0.58
Ethyl decanoate	0–2.1	0–0.3	0.51
Hexylacetate	0–3.6	0–4.8	0.67–2.4
2-Phenylethyl acetate	0–18.5	0.02–8	1.8
3-Methylbutyl acetate	0.03–0.5	0–23	0.16
Ethyl lactate	0.17–378	12–382	150

Table 20.18. Effect of fermentation conditions on formation of higher alcohols and esters

Temperature (°C)	pH	Higher alcohols total (mg/l)	Fatty acid esters total (mg/l)
20	3.4	201	10.8
20	2.9	180	9.9
30	3.4	188	7.8
30	2.9	148	5.4

The content and composition of the ester fraction is greatly influenced by fermentation conditions. The higher the temperature and the lower the pH during fermentation, the lower the ester concentration (Table 20.18).

During storage and maturation of good wines, changes in the aroma produce the bouquet. Depending on the maturation conditions, the bouquet can be marked, e.g., by oxidative reactions. This is the case with southern wines, like Madeira, which have typically high concentrations of aldehydes and acetals. On the other hand, reductive reactions apparently play a role in the development of the bouquet of fine table wines after the bottles have been stored for long periods. Except for the cultivar-specific aroma substances given in Table 20.19, little is known about the bouquet of wines. The aroma substances formed on storage of bottles include 1,1,6-trimethyl-1,2-dihydronaphthalene. After longer storage, it exceeds the aroma threshold (ca. 20 µg/l water) and contributes a kerosine-like aroma note to the aroma profile in particular of old Riesling wines.

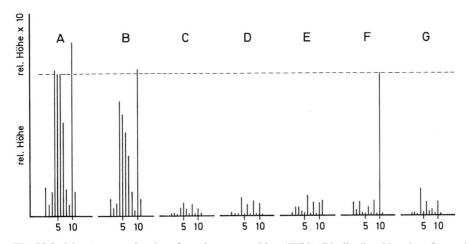

Fig. 20.8. Monoterpenes in wines from the grape cultivar "White Riesling" and in wines from other grape cultivars which are also sold as "Riesling" (according to *A. Rapp* et al., 1985).
A: White Riesling (Rheinpfalz), B: White Riesling (France), C: Welschriesling (Austria), D: Welschriesling (Italy), E: Laski Rizling (Yugoslavia), F: Riesling (Australia), G: Emerald Riesling (USA).
Monoterpenes: **1** *trans*-furan linalool oxide, **2** *cis*-furan linalool oxide, **3** neroloxide, **4** linalool, **5** hotrienol, **6** α-terpineol, **7** not identified, **8** *trans*-pyran linalool oxide, **9** *cis*-pyran linalool oxide, **10** 3,7-dimethylocta-1,5-*trans*-dien-3,7-diol, **11** 3,7-dimethyl-1-octen-3,7-diol

Table 20.19. Cultivar-specific aroma substances of wine

Compound	Cultivar
Ethyl cinnamate	Muscatel wines
β-Ionone	
Linalool	
Geraniol	
Nerol	
β-Damascenone	Riesling, Chardonnay
cis-Rose oxide	Gewürztraminer
3-Isopropyl-2-methoxypyrazine	Sauvignon
3-Isobutyl-2-methoxypyrazine	
4-Mercapto-4-methylpentan-2-one	Scheurebe

20.2.7 Spoilage

As with beer, defects in wine are reflected in appearance, odor and taste and, if not controlled, result in complete spoilage. A full explanation of all defects is beyond the scope of this book; hence only a general outline will be provided.

Of importance is browning due to oxidative reactions of phenolic compounds which, in red wine, may result in complete flocculation of the color pigments. This oxidative darkening process is as much chemical as enzymatic (polyphenoloxidases). Sulfurous acid is the preferred agent to prevent browning. Once the wine is affected by browning, it may be lightened by treatment with activated charcoal. The charcoal treatment can also remove other defects, such as the taste of mash or rotten grapes. Iron-induced turbidity (white or greyish casse) appears as a white, greyish-white or greyish haze or cloudiness and consists mostly of ferric phosphate ($FePO_4$). It is formed by the oxidation of ferrous compounds in wine. Proteins, tannins or pectins can participate in the build-up of such cloudiness (black casse). The so-called copper casse or turbidity is based on the formation of Cu_2S and other compounds with monovalent copper. It originates from the Cu^{2+} ions present in wine and their reduction in the presence of excess SO_2. The so-called "boeckser" of young wine is caused by the smell of hydrogen sulfide. The very unpleasant, rotten and yeasty "boeckser" (= mercaptan) odor is most objectionable and lingers for a long time. It is due to ethylthiol, which can be removed by activated charcoal. The volatile sulfur compounds originate from sulfite which is reduced to H_2S by yeast, later reacting with ethanol to form ethylthiol. Additional wine taste defects are the odd and disagreeable cork tastes which is due to the formation of 2,4,6-trichloroanisole (cf. 5.1.5). Above concentrations of 15–20 ng/l, it contributes to the cork flavor of wine. Geosmin, 1-octen-3-one, 4,5-dichloroguaiacol and chlorovanillin may cause musty off-flavors.

Additional yeast spoilage is induced by species of the genera *Candida* (*Mycoderma*), *Pischia* and *Hansenula* (*Willia*). Other microorganisms are involved in the formation of viscous, moldy and ropy wine flavor defects. Bacterial spoilage may involve acetic acid and lactic acid bacteria. In this case vinegar or lactic acid souring is detectable. It has usually been associated with mannitol fermentation which may result in considerable amounts of mannitol.

Sorbic acid can be converted to 2-ethoxy-3,5-hexadiene by heterofermentative lactic acid bacteria. In concentrations of 0.1 mg/l, this compound produces a "geranium" note.

A "mousy" taint is occasionally detected in fruit and berry wines and, less often, in grape wines. It is thought that 2-acetyl-1,4,5,6-tetrahydropyridine, which has been also identified as an important flavor compound of the bread crust (cf. 15.4.3.2.2), contributes to the "mousy" taint. The compound might be formed by microorganisms in wine.

Likewise, red wines, particularly color-deficient wines, show a microbiologically-induced change reflected in a substantial increase in volatile acids and the degradation of tartaric acid and glycerol. The bitter taste of red wines is caused by bacteria, mold and yeast. The bitter taste is usually a result of glycerol conversion to divinyl glycol. Cloudiness of red wines appears to be due either to bacterial or yeast spoilage or to physical reasons alone, such as the precipitation of cream of tartar. The latter occurs frequently and mostly in bottled wines. Cream of tartar precipitates as a result of oversaturation of the salt solution, as appears to be the case with protein-tannin interaction products. With oversaturation, they sediment as a fine greyish-yellow haze. Cloudiness caused by mucic acid salts also occurs.

20.2.8 Dessert Wines

Dessert wines are those with a high alcohol content, which can not be derived directly from the fermentation of fresh grape must, and/or are wines with a high sugar content. They are made principally by two processes, which can to a certain extent be combined:

- The fermentation of grape juice of exceptionally high sugar concentration, or blending of concentrated grape juices with normal wines (concentrated dessert wine).
- When the must is sufficiently fermented, alcohol or concentrated must mixed with alcohol is added (spirit-enriched dessert wine), stopping further fermentation.

Among the great number of dessert wines coming from southern European countries, several are worthy of mention: Hungarian dessert wines, Szamorodny, fortified Tokay and Tokay essence; and the French Haute Sauternes – they are concentrated dessert wines. Malaga and sherry wines originate from Spain; Port and Madeira from Portugal; and from Italy: Marsalla from Sicily, Calabrese from Syracuse and sweet Muscatel from Syracuse and Catania. Greek dessert wines include Samos and Vino santo (from the island of Santorin). The extract, alcohol and sugar contents of dessert wines are given in Table 20.20.

Table 20.20. Composition of some dessert wines and top quality wines

	Extract (g/l)	Alcohol (g/l)	Sugar (g/l)	Glycerol (g/l)	Tritratable acid (g/l)[a]
Malaga	159.2	143.4	135.8	5.0	5.3
Portwine	67.6	166.5	47.0	2.8	4.5
Madeira	129.0	149.5	107.5		
Marsala	81.0	150.4	52.2	6.2	5.9
Samos	119.0	152.0	82.0	7.5	6.8
Tokay essence	257.5	84.4	225.3	4.1	6.5
Rheingauer top quality	140.6	107.7	99.4	14.3	10.2
Pfaelzer (Palatinate) top quality	171.6	86.7	121.3	10.5	11.6
Sauternes top quality	127.8	101.2	82.7		0.3

[a] Expressed as tartaric acid.

At least 2–5 years are needed to make dessert wines. In the production of sherry the wine is stored in partially filled butts, i.e. in the presence of excess air. Flor yeasts develop on the wine surface in the form of a continuous film or wine cover (sherry yeast). The typical sherry flavor is derived from the aerobic conditions of maturation. During this time the concentrations of the following compounds increase at the expense of alcohol and volatile acids: ethanal, acetals, esters, sotolon (structure in Table 5.15) and 2,3-butylene glycol. In port wine production the wine is drawn off to casks before the end of fermentation and is fortified with wine distillates. The fortifying procedure is repeated several times ("multiple addition") until the desired alcohol content is reached.

20.2.9 Sparkling Wine

Experience has shown that carbon dioxide imparts a refreshing, prickling and lively character to wine (as already mentioned for young wines). Hence, the production of a refined form of wine, enriched with carbon dioxide (sparkling wine) was developed and used in the early 18th Century, originally in the Champagne region of France ("Champagne" wine).

20.2.9.1 Bottle Fermentation ("Méthode Champenoise")

In the production of sparkling wine, young wines from suitable regions are used since fermentation of their grape juice in casks provides the special, fresh, fruity bouquet ("cuvé") desired. Blending of wines ("coupage") from different localities, often with older wines, is aimed at obtaining a uniform end-product. In this way clarified wine is then converted into an effervescent beverage by subjecting it to a second fermentation. Sugar is added (about 20–25 g/l) to wine, together with a pure yeast culture, for the purpose of attaining the desired final alcohol content (85–108 g/l) and carbon dioxide pressure (4.5 bar at 20°C). Special yeasts are selected which, in addition to being good fermenters and insensitive to carbonic acid, sediment as a firm, grainy precipitate ("depot") after fermentation is complete.

The wine is bottled ("tirage") in such a way as to leave a small headspace of air and is then corked with a natural or plastic cork or, very often, with "crown" caps. The cork is finally and firmly secured with an iron clamp ("agrafe"). The bottles are stacked in cellars at normal temperatures (9–12 °C). The fermentation lasts several months while the build-up of carbonation may go on longer, perhaps for up to 3–5 years. During this time, the carbon dioxide pressure within the bottle rises considerably. The sparkling wines are classified in France depending on the pressure: "grand mousseux", high pressure (4.5–5 bar); "mousseux", intermediate pressure (4–4.5 bar) and "cremant", low pressure (below 4 bar).

At this stage the sparkling wine is ready for yeast removal (disgorging). The bottles are restacked upside-down. Then the contents are repeatedly shaken until the yeast is loosened and settles on the cork. After 6–8 weeks the bottles are placed upright and the cork is removed using disgorging pliers. Simultaneously, the yeast is pushed out by the pressure from within the bottle. In order to simplify this production step, which is considered the most difficult step, the neck of the bottle is frozen to about −20 °C and the yeast is forced out as an "ice" plug. Because of this time-consuming, costly procedure, the loss of wine, and other problems inherent in clearing the bottle of its yeast deposit, a "transfer system" has been introduced. The raw wine which has fermented in the bottle is emptied into a tank. The measured wine is filtered from the tank under pressure into a shipping/export bottle. The sparkling, yeast-free wine ("vin brut", dry wine) is then, depending on the market demand, supplemented with "liqueur" (dosage), quite commonly a plain solution of candy in wine. The bottle, with a headspace volume of 15 ml, is then corked and the cork is wired down. For further build-up of CO_2, the finished sparkling wine needs to be stored for an additional 3–6 months. Wine with a small amount of liqueur added is designated as dry ("sec"); with more liqueur added as semi sweet ("demi-sec"); and with more than 12% as sweet ("doux"). Sparkling wine for diabetics is sweetened with sorbitol.

20.2.9.2 Tank Fermentation Process ("Produit en Cuve Close")

With the aim of simplifying the costly and time-consuming classical process, much of the sparkling wine production is now based on fermentation of wine in pressurized steel tanks instead of in bottles. The carbon dioxide-saturated wine is clarified and filtered and then chilled thoroughly and bottled. Fermentation is carried out at a pressure of about 7 bar over a 3 to 4 week period.

20.2.9.3 Carbonation Process

The carbonation of wine ("vin mousseux gacéifié") involves artificial saturation of wine with carbon dioxide, instead of the natural CO_2 developed during fermentation. Thus, the process is identical to the production of carbonated mineral water. The second fermentation, sugar addition and disgorging are omitted. However, sweetening with liqueur, corking and cork wiring are all retained.

20.2.9.4 Various Types of Sparkling Wines

Champagne is obtained by the bottle fermentation of wine from the French grape which grows in the region of Champagne. Sparkling wines produced in this region are the only ones that may be sold under the name of "Champagne". German sparkling wines are called "Schaumwein" and are commonly sold as "Sekt"; such Italian wines are "Spumante"; while in Spain and Portugal they are "Espumante".

Sparkling wines are also made from fruit and berry wines (apple, pear, white and red currant, bilberry). The process is that described above for carbonation.

Perlwine is a prematurely bottled wine which is left to ferment in the bottle.

20.2.10 Wine-Like Beverages

Compositions of some typical wine-like products are given in Table 20.21.

20.2.10.1 Fruit Wines

For the production of fruit wine, pressed juice (fruit must) is made from apples, pears, cher-

Table 20.21. Composition of some wine-like beverages[a]

Beverage	Alcohol	Extract	Acids[b]	Sugar	Minerals
Apple cider	58.4	23.4	3.8[+]	1.7	2.8
Cidre	51.0	29.7	2.8[+]	10.4	2.6
Pear wine	49.3	53.7	6.5[+]	9.0	4.1
Red currant cider	62.1	39.8	18.6[*]	1.8	4.0
Gooseberry cider	96.3	78.6	7.5[*]	55.8	1.8
Sour cherry cider	101.4	62.7	11.7[*]	3.8	3.61
Malt wine	70.6	24.5	4.6[+]	4.9	1.36
Malton sherry	123.0	115.2	8.1[+]	55.9	2.3
Mead	51.4	242.4	3.9[+]	208.0	1.34
Sake	121.2	28.6	5.7[+]	5.5	1.0

[a] Results are given in g/l.
[b] Acids are calculated as malic ([+]) or citric acid ([*]).

ries, plums, peaches, red currants, gooseberries, bilberries, cranberries, raspberries, hip berries and rhubarb. In general, the process used is the same as that for making wine from grapes. Apple and pear mash are first pressed and the pressed juice (must) is fermented, while berry mash is fermented directly in order to extract the color pigments. Natural fermentation is suppressed by inoculation with pure, cultured yeast (cold-fermenting yeast). The vigor of the fermentation of berry musts, which are nitrogen deficient, is increased by addition of small amounts of ammonium salts. Lactic acid (3 g/l) is added to acid-deficient musts, such as that from pears, in order to achieve a clear ferment and, often, sucrose solutions are added to berry and fruit musts to alleviate acidity. The yield and quality of pome fruit must is improved by mixing 9 parts of fruit residue with 1 part of water and adding sucrose to raise the density of the must to 55 Oechsle degrees.

Fruit wines are produced industrially in many countries, e.g., apple wine, which is called cider in France, the UK and the USA, and pear wine, known as "poiré" in France. In Germany fruit wine is made along the Mosel river, around Frankfurt and in the state of Baden-Wuerttemberg. It is a popular beverage and is commonly called "plain must".

20.2.10.2 Malt Wine; Mead

Malt wine is made from fermented malt extract (the hot water extract of whole meal malt). Malton wine is made in the same way,

except that sucrose is added at 1.8-times the amount of malt in order to increase the sugar and alcohol content of the wine. The wort is then soured by the action of lactic acid bacteria (0.6−0.8% lactic acid, final concentration). The acid fermentation is stopped by heating the wort to 78 °C and, after inoculation with a pure yeast culture, the wort is fermented to an alcohol content of 10−13%. The beverage thus formed has the character of a dessert wine, but is different because of its high content of lactic acid and its malt extract flavor. Mead is an alcoholic liquor made of fermented honey, malt and spices, or just of honey and water (not more than 2 l water per kg of honey). Since early times, mead has been widely consumed in Europe and, even today, it is enjoyed the most of all the wine beverages in eastern and northern Europe.

20.2.10.3 Other Products

Other wine-like products include palm and agave wines ("Pulque"), maple and tamarind (Indian date) wines, and sake, the Japanese alcoholic drink made from fermented rice, which resembles sherry and is enjoyed as a warm drink.

20.2.11 Wine-Containing Beverages

Wine-containing beverages are made with wine, liquor wines or sparkling wines and, hence, they are alcoholic beverages.

20.2.11.1 Vermouth

Vermouth was first produced in the late 18th century in Italy (Vermouth di Torino, Vino Vermouth) and later in Hungary, France, Slovenia and Germany. For the production of vermouth, wormwood (*Artemisia absynthium*) is extracted with the fermenting must or with wine, or it is made from a concentrate of plant extracts added to wine. Other herbs or spices are additionally used, such as seeds, bark, leaves or roots, as is the case with thyme, gentian or calamus, the sweet flag plant.

20.2.11.2 Aromatic Wines

These wines are similar to vermouth aperitif wines. They are flavored by different herbs and spices. Ginger-flavored wine is an example of this type of wine.

20.2.11.3 Prescription or Medicinal Wines

Such wines are pharmaceutical preparations. Some are quite bitter, especially those made from Cinchona bark, a source of the alkaloid quinine. Quinine-based wine is used in Europe and South America as a bitter aperitif. Kondurango, camphor and pepsin wines also belong to this class.

20.2.11.4 May Wines and Punches

May wine (a German summer drink) has wine as its base, sweetened with sugar and aromatized with the German Waldmeister plant (*Asperula odorata*, genus Woodruff) or its essence. May punch may also contain sparkling wine and mineral water or fruit wines.
Wine is the basis for other similar drinks, e.g., punches (mixtures of wine, sparkling wine, soda water and fruit such as pineapples, strawberries or peaches), drink mixes with soda water ("Spritzer", "Schorle-Morle"), or cold punch or Cardinal (a mixture of sparkling and red or white wine, mixed with sugar, orange, lemon or pineapple).

20.2.11.5 Wine Punch

Wine punch is a mixed drink consisting of wine and rum or arrack, lemon juice, sugar, water and aromatic substances.

20.3 Spirits

20.3.1 Foreword

Spirits or liquors are alcoholic beverages in which the high alcohol concentration is achieved by distillation of a fermented sugar-containing liquid. Examples are distilled wines (brandies), liqueurs, punch extracts and alcohol-containing mixed drinks. Statistics on the production of liquor are presented in Table 20.22.

20.3.2 Liquor

The term liquor includes all liquids, even pure alcohol, which are obtained by fermentation

Table 20.22. Production of various distilled alcoholic beverages in FR Germany (1983)

Spirits	Amount (10^6 l)	Spirits	Amount (10^6 l)
Grain distillates	84.4	Geneva gin	12.4
Wine brandy	83.4	Whiskey	4.3
Fruit brandy	8.3	Vodka	5.4
Liqueurs	54.2	Other distilled	
Rum and arrack	29.6	beverages	29.5
		Total	311.5

followed by distillation. Some types of liquors contain flavorings.

20.3.2.1 Production

Liquors are produced by removing alcohol from an alcohol-containing liquid by distillation. Such liquids may already contain the alcohol, or alcohol is produced by the fermentation of a sugar-containing mash. The mash may include fermentable forms of sugars (D-glucose, D-fructose, D-mannose and D-galactose), or those forms are prepared by prior hydrolysis of di- and oligosaccharides (sucrose, lactose, raffinose, gentianose, melecitose, etc.) or polysaccharides.
The main raw materials are:

- alcohol-containing liquids (wine, beer, fruit wines, fermented milk);
- sugar-containing sources, such as sugar cane and beet, molasses, fruit and fruit products, fruit pomace, whey, palm extract and sugar-rich parts of tropical plants;
- starch- and inulin-containing raw materials (fruit, cereal, potato, topinambur, sweet potato, cassava, tapioca or chicory);

Saccharification of the starch-containing material is achieved with malt (green malt or kiln-dried malt), or by microbial amylases e.g., from the molds Aspergillus niger and A. oryzae. Fermentation is achieved with Saccharomyces cerevisiae, which converts sucrose and hexoses (glucose, galactose, mannose, fructose). Other substrates can be fermented, e.g., with Saccharomyces uvarum (raffinose), Kluyveromyces fragilis (lactose), and Kluyveromyces marxianus (inulin).

Distillation is performed in various ways, depending on the source and desired end-product. For the distillation of rum, arrack, fruit brandies and cereals, and brandy from wine, the apparatus is often a relatively simple still, used in such a way as to obtain a distillate which contains several other products of fermentation besides ethanol, or which contains the aroma substances of the starting raw material. These aroma substances are alcohols, esters, aldehydes, acids, essential oils and hydrogen cyanide. Repeated distillation is needed to obtain an alcohol-enriched distillate. In the production of pure or absolute alcohol the aim is the opposite: the final product being free from materials other than ethanol.

20.3.2.2 Alcohol Production

Alcohol used for drinks is made primarily from potatoes, cereals and molasses. Distiller's yeast, especially the top fermenting culture (cf. 20.3.2.1), is used for fermentation. Since the fermentation proceeds in an unsterilized mash and at elevated temperatures and since the growth of yeast occurs in mash acidified with lactic or sulfuric acid (pH 2.5–5.5), the yeast must be highly fermentative, tolerant of elevated temperatures ($\leq 43\,°C$) and resistant to acids and alcohol. In addition to saccharification by malt which contains mainly β-amylase, high-activity microbial α-amylases are also used. Molasses does not require saccharification. The saccharified mash is cooled to $30\,°C$ and then inoculated with a yeast starter which has been cultured on a sulfuric or lactic acid medium of the mash or directly with distiller's yeast. After 48 h of fermentation, the ethanol present at 6–10% by volume in the mash is distilled off along with the other volatile constituents. This step and the following rectification of the crude alcohol are achieved by continuous processes.

To facilitate the removal of the fusel oils, the crude alcohol is diluted to 15% by volume prior to rectification. The head product obtained from the rectification column consists nearly of pure ethanol (96.6% by volume) which is used for production of alcohol-fortified beverages. Large amounts of acetaldehyde, methanol and low boiling esters are present in the first runnings of the distillate, while the last runnings contain primarily fusel oil, other high alcohols, furfural and esters. These runnings combined with other intermediate fractions provide technical alcohol. The fusel oil, obtained in amounts of 0.1–0.5 l per 100 l alcohol, is used for technical purposes, while the distillation residue (the wash or stillage) is frequently used as animal feed. The yield of alcohol from 100 kg of mash starch is 62–64 l, i.e. about 89% of the theoretical value.

Technical alcohol is denatured or embittered to prevent its use for other than technical purposes, e.g., for drinking. Burning alcohol is denatured by addition of a mixture of methylethylketone and pyridine and alcohol for industrial use with other solvents, such as petroleum ether, camphor, diethyl ether or dyes.

20.3.2.3 Liquor from Wine, Fruit, Cereals and Sugar Cane

These beverages have a distinct taste and odor and contain at least 38% ethanol by volume. They are called natural, genuine or true liquors. The distillate resulting from a single distillation has a low alcohol content and often contains the specific odor and taste components of the starting material (harsh raw grain or harsh raw juniper liquor-gin).

In the production of liquor, the ultimate aim is to collect most of the desirable, specific fragrance and aroma substances (esters, essential oils) or to develop them (hydrogen cyanide, fermentation products, yeast oil) by using suitable mashing, fermentation and distillation processes. The freshly distilled liquor has a hard, burning taste and unpleasant odor. It is improved by aging, which gives it a new, desirable aroma and flavor. Therefore, aging of liquor is of the utmost importance.

20.3.2.3.1 Wine Liquor (Brandy)

Brandy is distilled wine which contains at least 38% by volume of alcohol. Brandy to which alcohol is added is designated as a brandy blend or adulterated brandy.

The term "cognac" is restricted to brandy made in France in the region of Charente. The brandy produced in southern France, called Armagnac, is close in quality to cognac. Brandy production originated in France. Fermented grape juices (must) are distilled in very simple copper-pot stills on an open fire, often without prior removal of the yeast. The primary distillate (sectionnement) with a harsh, unpleasant odor is refined by repeated distillations ("repasse"). Brandy production soon spread to other countries (Germany, Russia, Spain, Hungary, the USA, Australia) and today brandy is frequently distilled by a continuous process and its production has become a large-scale industry. In Germany imported wines which are fortified with wine distillate serve as starting material. The primary wine distillate contains 52–86% by volume ethanol and is considered as an intermediate product. It is used as the raw ingredient in the production of adulterated brandy by aging from 6 months to several years in wooden casks. Hard oak wood is used predominantly (barrels are made from "limousin" wood, holding about 300 l). Wild chestnut and other woods are also used. During aging, the wine distillate extracts phenolic compounds and colors of the wood, thus acquiring the typical golden-yellow and, occasionally, greenish-yellow color of brandy. Simultaneously, oxidation and esterification reactions mellow and polish the flavor and aroma. In order to improve quality, it is common to add an essence prepared by extraction of oakwood, plums, green walnuts or deshelled almond with a wine distillate and also sugar, burnt sugar ("couleur") and 1% dessert wine to sweeten the brandy. In addition, treatment of brandy with clarifying agents and filtering agents is also common. The desired alcohol content is obtained by dilution of brandy with water. The possibility of shortening the long and costly aging in casks has been repeatedly investigated. Attempts to age brandy artificially have included treatment with ozone or ion-exchange resin or application of ultrasound. None of these methods were successful.

20.3.2.3.2 Fruit Liquor (Fruit Brandy)

Fruit liquors are also called cherry or plum waters or bilberry or raspberry spirits. Production of fruit liquor will be illustrated by cherry and plum liquors. Kirschwasser is made mostly in southern Germany (Black Forest's cherry water), France and Switzerland (Chriesiwas-

ser). Whole fruits of the various sweet cherry cultivars are partly crushed together with the seeds and are pounded into a pulp. The fruit is left to ferment for several weeks, using a pure yeast culture. The fermented mash is then distilled in a copper still on an open fire or is heated with steam. During distillation the first and last fractions are separated. The main distillate contains 60% by volume or more alcohol. It is usually diluted with water to about 40–50% by volume alcohol and is marketed as clear, colorless brandy. The low levels of benzaldehyde and hydrogen cyanide which both contribute to the flavor are derived from the enzymatic cleavage of seed amygdalin. Kirschwasser, as is the case with Marasca from Dalmatia or Italy, is often used as an admixture in liqueur or cordial production, (curacao, cherry brandy, maraschino, etc.).

Plum brandy is produced from fully-ripe plums in a similar way to Kirschwasser, though no seed crushing is involved. It is often marketed under the name "Slivovitz" (from the old Slavic word sliva = plum). Besides Germany and Switzerland (Pfluemli water), major producers are the Balkan states, Czechoslovakia and France. In addition to the common plum, the highly aromatic yellow plum, mirabelle, is also fermented. Mirabelle liquor is a desirable admixture to liqueurs containing fruit extract.

Fruit spirits are obtained from fresh or frozen fruit pulp or juice to which alcohol has been added prior to distillation. Fruits and berries used for this purpose are apricot, peach, bilberry, raspberry, strawberry, red currant, etc. Pome fruit liquor is obtained from freshly fermented apple or other pome fruits, either whole or crushed, or their juices, without prior addition of sugar-containing materials, sucrose or alcohol of some other origin. The alcohol content of liquor from pome fruits is at least 38% by volume. Hydrogen cyanide plays an important role in the chemical composition of fruit liquors of either stone or pome fruit. The cherry liquor sold on the market contains about 0.3–60 mg of hydrogen cyanide per liter of alcohol. In the same range are the concentrations of benzaldehyde (at least 20 mg/l) and the bouquet substances (about 7–15 mg/100 ml). Plum brandy contains less hydrogen cyanide (0.6–21.3 mg/l).

20.3.2.3.3 Gentian Liquor ("Enzian")

Gentian brandy is a product obtained by distilling the fermented mash of gentian roots, or in which gentian distillate is used. The raw materials are the roots of many plants of the gentian family which, in the fresh state, contain substantial amounts of sugars (6–13%) in addition to the bitter glycoside-type compounds, such as gentiopicrin, amarogentin and others. The major production regions are the Alps (Tyrol, Bavaria, Switzerland) as well as the French and Swiss Jura mountains.

20.3.2.3.4 Juniper Liquor (Brandy) and Gin

Juniper brandy is obtained from pure alcohol and/or grain distillate by the addition of juniper distillate or its harsh, raw brandy. The use of juniper oil is uncommon. Juniper spirit is made exclusively from the distillate of whole juniper berries or from a fermented aqueous extract of juniper. The berries of *Juniper communis* are processed into brandy in Germany, Hungary, Austria, France and Switzerland. Pure juniper brandy is also used as an intermediate product for the production of alcoholic beverages with a juniper flavor as, for example, in Geneva gin. The alcohol of this gin is obtained by distillation of a cereal mash prepared from kiln-dried smoked malt. Juniper brandy also flavors the Bommerlunder from the state of Schleswig-Holstein and the Doornkaat of East Friesland in Germany. Common gin is made from juniper distillates and spices, and contains at least 38% by volume alcohol. Dry gin has an alcohol content of at least 40% by volume.

20.3.2.3.5 Rum

Major rum-producing countries are in the West Indies (Jamaica, Cuba, Barbados, Puerto Rico, Guyana and Martinique) and also Brazil and Mauritius.

Rum production in sugar cane-cultivating regions uses the sugar syrup or the freshly pressed extract, often with the addition of such by-products as foam skimmings, molasses, press-skimmings and their extracts, and distiller's wash ("dunder"), the residue leftover from a previous distillation. The sugar-containing solutions are diluted and allowed to ferment spontaneously at a maximum temperature of 36°C and then are usually distilled in simple pot stills. Parts of aromatic plants are occasionally added to increase the aroma of the fermenting mash. This results in rum brands with different aromas. The quality of individual products fluctuates greatly. Especially highly regarded is Jamaican rum, which is marketed in various quality grades. A general classification divides them into drinking and blending types. Export rums have an alcohol content of about 76–80% by volume ("original rum"). Rum has the most intense aroma of all the distilled spirits enjoyed as drinks. This is acquired only after long aerobic aging in casks, by absorption of extracted substances from oakwood, and by formation of esters and other aroma constituents during aging.

Original rum contains about 80–150 mg acids per 100 ml, calculated as acetic acid. A large part occurs in free form as acetic and formic acids, the rest, along with other low molecular weight fatty acids, is esterified. The ester content and composition are of utmost importance for the assessment of aroma quality.

20.3.2.3.6 Arrack

Arrack is made from rice, sugar cane molasses, or sugar-containing plant juices (primarily from sweet coconut palm extract or its bloom spadix) by fermentation and subsequent distillation. Dates are used for the same purpose in the Middle East.

Countries which produce arrack are Indonesia (Java), Sri Lanka, India (Malabar coast) and Thailand. Well-known brands are Batavia and Goa arracks. In comparison to rum, arrack is not available in very many varieties. It is imported as the "original arrack" with an alcohol content of 56–60% by volume, from which "true arrack" is obtained by dilution with water to 38–50% by volume alcohol. At least a tenth of the alcohol in arrack blends must be from genuine arrack. Arrack is used for hot drink preparations, for Swedish punch, as an admixture for liqueurs, and in baking and as a flavoring ingredient in candy manufacture. Batavia brand arrack, with an alcohol content of about 57% by volume, contains on the average 92 mg acids, 189 mg esters, 21 mg aldehydes and 174 mg higher alcohols per 100 ml of ethanol.

20.3.2.3.7 Liquors from Cereals

Typical products are grain alcohol and whiskey (American and Irish brands are usually spelled with an "e", while Scottish and Canadian brands tend to use "whisky"). Different cereals (rye, wheat, buckwheat, oats, barley, corn, millet) are used. The cereals are first ground, mixed with acidified water, and made into an uniform mash by starch gelatinization. Saccharification is then accomplished by incorporating 15% kiln-dried malt in a pre-mashing vat and stirring constantly at 56°C. Saccharification proceeds rapidly through the action of malt diastase enzymes. The enzymes are inactivated by heating the mash to 62°C. This step is followed by rapid cooling of the mash to 19−23°C. The sweet mash is fermented by a special yeast and is then distilled. Grain liquors are obtained by distilling the mash, while malt liquors commonly are produced by distillation of the wort. Simple stills are used for distillation in small plants, while both distillation and rectification are achieved on highly efficient, continuously-run column stills in industrial-scale production. According to the process used, the yield is 30−35 l of alcohol per 100 kg of cereal (e.g., rye), while the quality and character of the spirits vary greatly. Simple stills, with an unsophisticated separation of head and tail fractions, provide characteristic products rich in grain fusel oils. A modern distillery is able to remove the fusel oils to a great extent, yielding a high percentage grain alcohol, from which it is then possible to make a mellow, tasty, pure grain brandy with a subtle aroma. The final flavor of all these products is dependent on well-conducted aging in wooden casks.

Whiskey, depending on the kind, is made by different processes. The raw material for Scotch malt whiskey is barley malt which has been exposed to peat moss smoke during kiln drying. Such smoked malt is mashed at 60°C and filtered. The resulting wort is then fermented at 20−32°C. The distillation is conducted in two steps, sometimes in simple pot stills. The harsh, raw liquor is collected in the first distillation step. The undesirable harsh components are removed in the head and tail fractions in the second distillation.

In the production of Scotch grain whiskey the saccharified starch is distilled in continuous column stills. The character of the distillate is neutral, with less aroma than malt whiskey. In both Scotch whiskey processes, the distillates, with about 63% by volume ethanol, have to be stored/aged in order to develop their full aroma. This is best achieved by aging in old sherry casks or in charred casks. At the end of processing, the alcohol content is reduced to a drinkable level, about 43% by volume. Depending on the desired flavor or current preferences, the malt whiskey might be blended with grain whiskey ("blended whiskey").

American whiskey is made from corn, rye or wheat by saccharification with malt enzymes, fermentation of the wort, followed by double-distillation in column stills and aging, usually in charred oakwood casks. The corn distillate content of bourbon whiskey is at least 51% by volume and that of corn whiskey is at least 80% by volume. Rye whiskey contains at least 51% by volume distillate from rye, while wheat whiskey must contain mostly distillate from wheat.

20.3.2.3.8 Volatile Components of Liquor

In addition to ethanol, distilled spirits contain a great number of volatile constituents which originate from the raw material or arise as byproducts of fermentation. Additional compounds may be formed during aging or maturation of liquors by interactions between the ingredients. Many of these volatile compounds are of great importance to an individual product and may fluctuate greatly in their nature and content, depending on the raw material and the process used. Table 20.23 gives a review of selected volatile compounds and their contents in some liquors, as determined by gas chromatography by direct sample injection, i.e. without a prior enrichment step. Important groups of volatile compounds are:

- Methanol and Higher Alcohols.
  Methanol is found primarily in pectin-rich fruit and pomace wine liquors. Grain spirits or liquors are low in methanol. Higher alcohols are commonly present, though in highly variable amounts. Average values (in g/l) are, for example: cognac, 1.5; whiskey, 1.0; and rum, 0.6. Higher alcohols originate

Table 20.23. Volatile compounds in distilled spirits (average value in mg/100 ml pure ethanol)

Compound	Plum brandy	Cherry liquor	Pear (Williams) brandy	Cognac	German wine brandy	Grain spirit	Blended whisky (Scotland)	Bourbon whiskey (USA)
Methanol	1,137	681	1,408	69	97	30	23	26
1-Propanol	146	806	134	52	41	0.4	44	29
1-Butanol	16	2.5	33	0.6	5	–	0.4	0.7
2-Butanol	23	44	16	2.5	13			
Isobutanol	86	47	56	112	63	+	69	81
2-Methyl-1-butanol	46	28	37	60	53	+	25	129
3-Methyl-1-butanol	143	98	119	218	150	0.8	50	207
1-Pentanol	0.4	+	0.6	–	+	–	+	+
1-Hexanol	2	0.8	11	2	1.5	–	+	0.4
1-Octanol	+	+	+	+	+	–	+	+
Benzyl alcohol	3	4.5	–					
2-Phenylethanol	3	1.5	0.7	3.5	2	+	2.2	7.5
Ethyl formate	0.5	0.5	1.5	3.5	2	+	1.1	2.0
Methyl acetate	9	7	33					
Ethyl acetate	204	295	151	50	55	0.9	26	71
Propyl acetate	–	4.5	–					
Isoamyl acetate	0.7	0.7	0.7	+	0.4	–	1.0	0.5
Hexyl acetate	+	–	–			–	+	+
Benzyl acetate	+	+	–					
2-Phenylethyl acetate	+	–	–	+	+	–	0.4	–
Ethyl propionate	4	4.5	0.6	1	3	–	0.4	0.3
Ethyl lactate	57	100	34	15	9	–	1.1	2.4
Isoamyl lactate	0.4	0.4	+	0.4	+	–	–	–
Diethyl succinate	2	2	+	0.8	0.6	–	+	+
Ethyl butyrate	0.5	+	+	+	1			
Ethyl capronate	0.7	0.4	0.3	0.9	0.8	–	+	0.4
Ethyl caprylate	1.5	0.8	0.5	2	2	–	0.9	1.5
Ethyl pelargonate	+	–	–			–	+	+
Ethyl caprinate	3	2	1	4	4	+	2.9	2.8
Ethyl-*trans*-2-decenoate	–	–	+					
Ethyl laurate	2	0.6	0.4	2.5	2	+	2.5	1.5
Ethyl-*trans*-2-*cis*-4-decadienoate	–	–	11					
Ethyl-*trans*-2-*trans*-4-decadienoate	–	–	5					
Ethyl benzoate	0.8	0.6	–					
Ethanal	18	16	17	21	53	1.5	7.0	8.6
Benzaldehyde	2.5	1	+	+	+	–	+	+
Furfural				+	+	–	0.8	2.0
Acetone	1	1	0.4	1	0.9	+	+	+
Diethylacetal	7	6	11	7	15	0.9	4.4	6.5
cis-Linalool oxide	+	+	+					
trans-Linalool oxide	0.3	+	+					
Terpineol	+	+	–					

either from amino acids which, through oxidation or transamination, give keto acids, which, through decarboxylation, give the corresponding aldehydes; and through reduction give higher alcohols:

$$R\text{—}CH\text{—}COOH \longrightarrow R\text{—}CO\text{—}COOH$$
$$\underset{NH_2}{|}$$
$$\longrightarrow R\text{—}CHO \longrightarrow R\text{—}CH_2OH \qquad (20.10)$$

Higher alcohols are also generated during the biosynthesis of amino acids (cf. 5.3.2.1).

- Carbonyl Compounds, Acetals.
Acetaldehyde is the most important carbonyl compound formed during alcoholic fermentation. Together with diethylacetal, acetaldehyde influences the drinkable quality of liquor even in very low amounts. Other carbonyls present in liquor are: propanal, isobutanal, pentanal, isopentanal, hexanal, diacetyl, 2,3-pentandione, acrolein, furfural, various ketones, vanillin, coniferyl- and p-hydroxybenzaldehyde. Some of these compounds leach from wooden casks during the aging of liquor.

- Organic Acids.
Acetic acid (40−95% of the total acids) is the predominant organic acid found in liquor. In addition, the following acids have been detected: propionic, isobutyric, isovaleric, valeric, caproic, caprylic, capric and lauric acids. A characteristic constituent of rums appears to be 2-ethyl-3-methyl butyric acid.
The total acid content in mg/l is 200 for cognac, 100 for Scotch whisky, 400 for bourbon whiskey and 600 for a rum with a good aroma.

- Esters.
Esters, especially those derived from short chain acids and from aliphatic alcohols ("fruit" esters), play an important role in the odor and taste of distilled sprits. Ethyl acetate predominates, followed by the ethyl, isobutyl and 3-methylbutyl esters of lower fatty acids. Also, there are ethyl esters of higher fatty acids such as caprylic, capric and lauric and, in Scotch whisky, palmitic acid. The effect of the type of process on the composition of the volatile fraction is illustrated by the fact that in a brandy distillate, the amount of higher esters of fatty acids is considerably greater when the distillation is run in the presence of yeast.
The ester content of cherry and apricot distillates is 1.1−4.3 g/l. The various qualities of rum are based on the content of esters. The ester value gives the mg of esters, calculated as acetic acid ethyl ester,

present in 100 ml of pure ethanol. According to the aroma intensity, the ester values of Jamaican rums vary from 80 to 1,600.

- Other Compounds.
This group includes various phenols (p-methyl and p-ethylguaiacol, guaiacol, etc.), terpenes derived from essential oils, the bitter glycosidic compounds of gentian brandy (gentiamarin, etc.) and, finally, the nitrogen compounds (e. g., pyridines, picolines and pyrazines) found in rum and whiskey.
Especially liquors made from stone fruits contain ethyl carbamate in amounts of 0.3−3.2 µg/ml (cf. 8.12.7). This compound is probably formed by the ethanolysis of carbamoyl phosphate from yeast metabolism. It can also be formed from 1,2-dicarbonyl compounds (methylglyoxal, diacetyl, and 2,3-pentandione) and HCN in the presence of ethanol.

20.3.2.4 Miscellaneous Alcoholic Beverages

Many liquors are made "cold" by simply mixing the purified alcohols of various brands with water and are named according to the place of origin: Klarer, Weisser, East-German, etc. Such mixes often contain flavorings (seasonings, spices), e. g., freshly distilled or aged grain liquor, extracts of caraway, anise, fennel, etc., as well as sugar, essence, essential oils or other flavoring substances. These products are designated as aromatized liquors. Some examples are:

Vodka (in Russian = diminutive of water) is made of alcohol and/or grain distillate by a special process. In all cases the characteristic smoothness and flavor must be achieved. The flavor should be neutral. The extract content is 0.3 g/100 ml and the alcohol content is at least 40% by volume.

Aquavit is a liquor flavored primarily with caraway. It is made from a distillate of herbs, spices or drugs and contains at least 35% by volume alcohol (potato alcohol or grain distillate). It is a favorite type of liquor in the Scandinavian countries.

Bitters are made from alcohol and bitter and aromatic plant or fruit extracts and/or their distillates, fruit saps and natural essential oils, with or without sugar, i.e. starch syrup. This group of products includes Boonekamp, bitter drops, English and Spanish bitters, and Angostura. The so-called "Aufgesetzter" is made of black currants and spirit or grain alcohol.

Absinthe is a liqueur flavored with aromatic constituents of wormwood and other aromatic plants. It becomes turbid after dilution with water.

Other Products. Some special liquors of regional importance should be mentioned: tequila and mescal from Mexico and South America, made from fermented sap of the agave cactus; and liquors from the Middle East, made of sultana raisins, figs or dates.

20.3.3 Liqueurs (Cordials)

Liqueurs are alcoholic beverages with 20–35% by volume alcohol and 220–500 g/l sucrose or starch syrup, and flavored with fruit, spices, extracts or essences.

20.3.3.1 Fruit Sap Liqueurs

Fruit liqueurs contain the sap of fruits which give the liqueur its name. The lowest concentration of sap is 20 l per 100 l of end-product (25% by volume alcohol). Addition of natural aroma substances, caramel and some other colors is quite common. Examples of fruit liqueurs are pineapple, strawberry, cherry, blackberry liqueurs, etc. Cherry brandy, a special type of cherry liqueur, consists of cherry sap, cherry-water, sucrose or starch syrup, wine essence and water.

20.3.3.2 Fruit Aroma Liqueurs

These liqueurs are alcoholic beverages made of natural fruit essences, distillates or extracts. Use of synthetic aroma substances (with the exception of vanillin) is uncommon. Liqueurs of this type include apricot, barberry, rose hip, plum, lemon, etc. The designations "triple" or "triple sec" are used only for citrus liqueurs with at least 38% by volume alcohol.

20.3.3.3 Other Liqueurs

Other liqueurs include:
Crystal liqueur, which contains sugar crystals (e.g., "crystal caraway").
Allasch, a special aromatic alcohol- and sugar-rich caraway liqueur with at least 40% by volume alcohol.
Ice liqueur, which is mixed and drunk with ice (e.g., lemon ice liqueur), and has an extract content of at least 30 g/100 ml and a minimum alcohol content of 35% by volume.
Medoc cordial, which contains at least 35% by volume alcohol, at least 20% of which comes from wine distillate or wine liquor.
Gold water, a spice liqueur containing gold leaf as a characteristic ingredient.
Fragrant vanilla liqueur, the aroma of which is derived exclusively from pod-like vanilla capsules (vanilla beans).
Honey liqueur ("Baerenfang", "Petzfang", the "bear traps") has at least 25 kg of honey in 100 l of end-product.
Swedish punch is made of arrack and spices and has an alcohol content of at least 25%.
Cocoa, coffee and tea liqueurs are made from the corresponding extracts of raw materials.
Emulsion liqueurs are chocolate, cream and milk liqueurs, mocca with cream liqueur, egg liqueur (the egg cream, "Advokat"), egg wine brandy, and other liqueurs with eggs added. The widespread and common egg liqueur is made from alcohol, sucrose and egg yolk.
Herb, spice and bitter liqueurs are made from fruit saps and/or plant parts, natural essential oils or essences, and sugar. Examples are anise, caraway, curacao, peppermint, ginger, quince and many other liqueurs.

20.3.4 Punch Extracts

Punch extracts or punch syrups, known simply as punch, are concentrates which are diluted before they are drunk. Rum or arrack punches contain 5% rum or 10% arrack, calculated relative to the total alcohol content. Aromatization with artificial rum or arrack essences, or with fruit ethers or esters, is not commonly done.

20.3.5 Mixed Drinks

Mixed drinks or cocktails are mixtures of liquors, liqueurs, wines, essences, fruit and plant extracts, etc. They are prepared immediately before drinking in restaurants or bars, or are marketed as ready-made cocktail mixes or as their separate constituents.

20.4 Literature

Becker, H., Wucherpfennig, K., Kiefer, W., Stellwaag-Kittler, F., Haubs, H., Dittrich, H.H.: Wein. In: Ullmanns Encyklopädie der technischen Chemie, 4. edn., Vol. 24, p. 381, Verlag Chemie: Weinheim. 1983

Bluhm, L.: Distilled beverages. In: Biotechnology (Eds.: Rehm, H.J., Reed, G.), Vol. 5, p. 447, Verlag Chemie: Weinheim. 1983

Buser, H.-R., Zanier, C., Tanner, H.: Identification of 2,4,6-trichloroanisole as a potent compound causing cork taint in wine. J. Agric. Food Chem. 30, 359 (1982)

Dittrich, H.H.: Mikrobiologie des Weines. Verlag Eugen Ulmer: Stuttgart. 1977

Hamberg, M.: Trihydrooctadecenoic acids in beer: qualitative and quantitative analysis. J. Agric. Food Chem. 39, 1568 (1991)

Hardwick, W.A.: Beer. In: Biotechnology (Eds.: Rehm, H.-J., Reed, G.), Vol. 5, p. 165, Verlag Chemie: Weinheim. 1983

Hillebrand, W.: Taschenbuch der Rebsorten. 5. edn., Zeitschriftenverlag Dr. Bilz and Dr. Fraund KG.: Wiesbaden. 1978

Hoffmann, K.M.: Weinkunde in Stichworten. 3. edn, Verlag Ferdinand Hirt: Unterägeri. 1987

Horak, W., Drawert, F., Schreier, P., Heitmann, W., Lang, H.: Äthanol and Spirituosen. In: Ullmanns Encyklopädie der technischen Chemie. 4. edn., Vol. 8, p. 80, Verlag Chemie: Weinheim. 1974

Jounela-Eriksson, P.: The aroma composition of distilled beverages and the perceived aroma of whisky. In: Flavor of foods and beverages (Eds.: Charalambous, G., Inglett, G.E.), p. 339, Academic Press: New York. 1978

Kreipe, H.: Getreide- und Kartoffelbrennerei. 3. edn., Verlag Eugen Ulmer: Stuttgart. 1981

Lafon-Lafourcade, S.: Wine and brandy. In: Biotechnology (Eds.: Rehm, H.-J., Reed, G.), Vol. 5, p. 81, Verlag Chemie: Weinheim. 1983

Macher, L.: Bier. In: Ullmanns Encyklopädie der technischen Chemie, 4. edn., Vol. 8, p. 462, Verlag Chemie: Weinheim. 1974

Meilgaard, M.C.: Prediction of flavor differences between beers from their chemical composition. J. Agric. Food Chem. 30, 1009 (1982)

Molyneux, R.J., Wong, Yen-i.: High-pressure liquid chromatography in the separation and detection of bitter compounds. J. Agric. Food. Chem. 21, 531 (1973)

Narziß, L.: Abriß der Bierbrauerei. 5. edn., Ferdinand Enke Verlag: Stuttgart. 1986

Nykänen, L., Suomalainen, H.: Aroma of beer, wine and distilled alcoholic beverages. D. Reidel Publ. Co.: Dordrecht. 1983

Palamand, S.R., Aldenhoff, J.M.: Bitter tasting compounds of beer. Chemistry and taste properties of some hop resin compounds. J. Agric. Food Chem. 21, 535 (1973)

Piggott, J.R. (Ed.): Flavour of distilled beverages: origin and development. Ellis Horwood Limited: Chichester. 1983

Pieper, H.J., Bruchmann, E.-E., Kolb, E.: Technologie der Obstbrennerei. Verlag Eugen Ulmer: Stuttgart. 1977

Pollock, J.R.A. (Ed.): Brewing Science, Vol. 1, 2. Academic Press: London. 1979/81

Postel, W., Drawert, F., Adam, L.: Aromastoffe in Branntweinen. In: Geruch- und Geschmacksstoffe (Ed.: Drawert, F.), p. 99, Verlag Hans Carl: Nürnberg. 1975

Rapp, A., Hastrich, H., Engel, L., Knipser, W.: Possibilities of characterizing wine quality and vine varieties by means of capillary chromatography. In: Flavor of foods and beverages (Eds.: Charalambous, G., Inglett, G.E.), p. 391, Academic Press: New York. 1978

Rapp, A., Güntert, M., Heimann, W.: Beitrag zur Sortencharakterisierung der Rebsorte Weißer Riesling. Z. Lebensm. Unters. Forsch. 181, 357 (1985)

Ribereau-Gayon, P.: Wine flavor. In: Flavor of foods and beverages (Eds.: Charalambous, G., Inglett, G.E.), p. 355, Academic Press: New York. 1978

Rinke, W.: Das Bier. Verlag Paul Parey: Berlin-Hamburg. 1967

Schieberle, P.: Primary odorants of pale lager beer. Differences to other beers and changes during storage. Z. Lebensm. Unters. Forsch. 193, 558 (1991)

Strauss, C.R., Heresztyn, T.: 2-Acetyltetrahydropyridines – a cause of the "mousy" taint in wine. Chem. Ind. 1984, 109

Tressel, R., Friese, L., Fendesack, F., Köppler, H.: Gas chromatography – mass spectrometric investigation of hop aroma constituents in beer. J. Agric. Food Chem. 26, 1422 (1978)

Tressl, R., Friese, L., Fendesack, F., Köppler, H.: Studies of the volatile composition of hops during storage. J. Agric. Food Chem. 26, 1426 (1978)

Troost, G.: Technologie des Weines. 5. edn. Verlag Eugen Ulmer: Stuttgart. 1980

Troost, G., Haushofer, H.: Sekt, Schaum- und Perlwein. Verlag Eugen Ulmer: Stuttgart. 1980

Vogt, E., Jakob, L., Lemperle, E., Weiss, E.: Der Wein. 7. edn., Verlag Eugen Ulmer: Stuttgart. 1977

Williams, P.J., Strauss, C.R., Wilson, B., Dimitriadis, E.: Recent studies into grape terpene glycosides. In: Progress in flavour research 1984 (Ed.: Adda, J.) p. 349. Elsevier Science Publ.: Amsterdam. 1985

Wilson, B., Strauss, C.R., Williams, P.J.: Changes in free and glycosidically bound monoterpenes in developing muscat grapes. J. Agric. Food. Chem. *32*, 919 (1984)

Zimmerli, B., Baumann, U., Nägeli, P., Battaglia, R.: Occurrence and formation of ethylcarbamate (urethane) in fermented foods. Some preliminary results. Proc. Euro Food Tox II, Zürich. October 1986, p. 243.

21 Coffee, Tea, Cocoa

21.1 Coffee and Coffee Substitutes

21.1.1 Foreword

Coffee (coffee beans) includes the seeds of crimson fruits from which the outer pericarp is completely removed and the silverskin (spermoderm) is occasionally removed. The seeds may be raw or roasted, whole or ground, and should be from the botanical genus *Coffea*. The drink prepared from such seeds is also called coffee.

Coffee is native to Africa (Ethiopia). From there it reached Arabia, then Constantinople and Venice. Regardless of the prohibition of use and medical warnings, coffee had spread all over Europe by the middle of the 17th century. The coffee tree or shrub belongs to the family *Rubiaceae*. Depending on the species, it can grow from 3−12 m in height. The shrubs are pruned to keep them at 2−2.5 m height and thus facilitate harvesting. The evergreen shrubs have leathery short-stemmed leaves and white, jasmin-like fragrant flowers from which the stone fruit, cherry-like berries, develop with a diameter of about 1.5 cm. The fruit or berry (Fig. 21.1) has a green outer skin which, when ripe, turns red-violet or deep red and encloses the sweet mesocarp or the pulp and the stone-fruit bean. The latter consists of two elliptical hemispheres with flattened adjacent sides. A yellowish transparent spermoderm, or silverskin, covers each hemisphere. Covering both hemispheres and separating them from each other is the strong fibrous endocarp, called the "parchment". Occasionally, 10−15% of the fruit berries consist of only one spherical bean ("peaberry" or "caracol"), which often brings a premium price.

The coffee shrub thrives in high tropical altitudes (600−1,200 m) with an annual average temperature of 15−25°C and moderate moisture and cloudiness. The shrubs start to bloom 3−4 years after planting and after six years of growth they provide a full harvest. The shrubs can bear fruit for 40 years, but the maximum yield is attained after 10−15 years. Fruit ripening occurs within 8−12 months after flowering. Only 3 of the 70 species of coffee are cultivated: *Coffea arabica*, which provides 75% of the world's production; *C. canephora*, about 25%; and *C. liberica* and others, less than 1%. The quantity (in kg) of fresh coffee cherries which yields 1 kg of marketable coffee beans is for *C. arabica* 6.38, *C. canephora* 4.35, and *C. liberica* 11.5. The most important countries providing the world's coffee harvest in 1996 are listed in Table 21.1.

21.1.2 Green Coffee

21.1.2.1 Harvesting and Processing

The coffee harvest occurs from about December until February from the Equator north to the Tropic of Cancer, while south of the Equator to the Tropic of Capricorn harvest occurs from May until August. Harvesting is done by

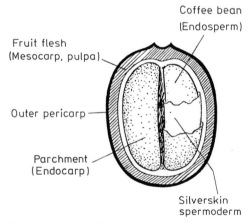

Fig. 21.1. Longitudinal section of a coffee fruit. (according to *Vitzthum*, 1976)

Table 21.1. Production of coffee beans in 1996 (1,000 t)

Continent	Raw coffee	Country	Raw coffee
World	5,931	Brazil	1,290
		Colombia	822
Africa	1,149	Indonesia	431
America, North-, Central-	1,110	Mexico	325
		Uganda	257
America, South-	2,489	Ethiopia	230
Asia	1,123	Guatemala	207
Europe	–	Vietnam	198
Oceania	60	India	180
		Côte d' Ivoire	165
		Ecuador	155
		Philippines	149
		Costa Rica	143
		Σ (%)[a]	77

[a] World production = 100%.

hand-picking of each ripe berry or by strip-picking all of the berries from three branches after most of the berries (often present as clusters) have matured. Harvesting may also be done by sweeping under the tree, i.e. collecting the ripe berries from the ground. Processing commences with removal of the fleshy pulp by using one of the two following processes:

The dry or natural process used in Brazil involves rapid transport of the harvested berries to a central processing plant, where the whole fruit is spread out on sun-drying terraces and dried until the beans separate by shrinking from the surrounding parchment layer.

Dehulling machines – conical screws with a helical pitch increasing toward the discharge end – remove the dried husks and parchment from the dried berries and, as much as possible, the silverskin. The dehulled and cleaned coffee beans are then classified according to size and packed in 60 kg bags. Often, the fresh cherries, instead of being spread on the drying terrace, are piled up, left for 3–4 days under their own heat to ferment the fruity pulp, and are then processed as outlined below. In both cases unwashed beans are obtained.

The wet (washing) process is more sophisticated than the dry process, and by general consent leads to better quality coffee. The method is generally used for Arabica coffee (except in Brazil) in Central America, Colombia and Africa. The freshly harvested berries are brought to a pulper in which the soft fruit is squeezed between a rotating cylinder or disc and a slotted plate, the gap of which is adjustable. The passage of the fruit produces a rubbing action which detaches the skin and the pulp from the beans without damaging the seed. The removed pulp is used as fertilizer. The pulped beans still have the silver-skin, the parchment and a very adhesive mucilaginous layer (mucilage). Hence, such coffee is carried into water stream fermentation tanks made of concrete, the water is drained off and the beans are left to ferment for 12–48 h. During this time, the mucilaginous layer, which consists of 84.2% water, 8.9% protein, 4.1% sugar, 0.91% pectic subtances and 0.7% ash, is hydrolyzed by enzymes of the coffee and by similar enzymes produced by microorganisms found on the fruit skins. The mucilage is degraded to an extent which can be readily dispersed by washing with water. The beans are then collected, sun-dried on concrete floors or dried in mechanical dryers in a stream of hot air (65–85 °C). Beans dried in this way are still covered with the parchment shell ("pergament" coffee or "cafe pergmino") and are further processed by dehulling machines as in the dry process. This yields the green coffee beans. Premium-priced coffee beans are often polished to a smooth, glossy surface and the silverskin, except that retained in the centre-cut of the beans, is removed.

21.1.2.2 Green Coffee Varieties

About 80 varieties of the three coffee bean species mentioned above are known. The most important of the species *Coffea arabica* are *typica*, *bourbon*, *maragogips* and *mocca*; and of *Coffea canephora* are *robusta* (the most common), *typica*, *uganda* and *quillon*. All varieties of *Coffea canephora* are marketed under the common name *"robusta"*.

The names of green coffees may be characteristic of the place of origin; i.e. the country and the port of export. Important washed Arabica coffees are, for example, Kenyan, Tanzanian, Colombian, Salvadorian, Guatemalon or Mexican.

Unwashed Arabica beans are the mild Santos and the hard Rio and Bahia beans. All three are from Brazil. Robusta coffees, mostly unwashed, are, for example, those from Angola, Uganda, the Ivory Coast and Madagascar.

Arabica coffees, particularly those from Kenya, Colombia and Central America, have a soft, rich, clean flavor or "fine acid" and "good body". The Arabica Santos from Brazil is an important ingredient of roasted coffee blends because of its strong but mellow flavor. Robusta coffee, on the other hand, is stronger but harsh and rough in aroma.

The quality assessment of green coffee is based on odor and taste assays, as well as on the size, shape, color, hardness and cross-section of the bean. Major defects or imperfections are primarily due to objectionable off-flavored blemished beans, which are removed by careful hand sorting. Blemished beans consist of: unripe seeds (grassy beans) which stay light colored during roasting; overfermented beans with an off-flavor due to the presence of acetic acid, diacetyl, butanol and isobutanol; frost-bitten and cracked beans; insect and rainfall-damaged beans; and excessively withered beans. Even a single blemished bean can spoil the whole coffee infusion. Additional imperfections are the moldy, musty flavor of insufficiently dried and prematurely sacked coffee and earthy or haylike off-flavors. Coffee varieties grown at high altitudes are generally more valuable than those from the plains or lowlands.

21.1.2.3 Composition of Green Coffee

The composition of green coffee is dependent on variety, origin, processing and climate. A review is provided in Table 21.2. The constituents will be covered in more detail in the section dealing with roasted coffee.

21.1.3 Roasted Coffee

21.1.3.1 Roasting

Green beans smell green-earthy, so they must be heat treated in a process called roasting to bring about their truly delightful aroma. Roasting in the temperature range of 200–250 °C causes profound changes. The beans increase

Table 21.2. Chemical composition of green coffee beans (% dry weight basis)[a]

Constituent	Content	
	Average value	Variation range
Water soluble extract	33	29.0–36.2
Protein		8.7–12.2
Lipids	12.6	8.3–17.0
Reducing sugars[b]		0–0.5
Reducing sugars after inversion[c]		2–9
Saccharose	6–7	
Crude fiber		10–11.7
Citric acid		0.5–1.15
Malic acid		0–0.5
Oxalic acid		< 0.2
Chlorogenic acids		4.5–11.1
Caffeine	1.45	0.9–2.6
Trigonelline	0.63	0.24–1.2
Minerals	4.0	3.0–5.4

[a] Moisture content 9.5% (5.0–12.1).
[b] Calculated as glucose.
[c] Calculated as saccharose.

in volume (50–80%) and change their structure and color. The green is replaced by a brown color, a 13–20% loss in weight occurs, and there is a build-up of the typical roasted flavor of the beans. Simultaneously, the specific gravity falls from 1.126–1.272 to 0.570–0.694, hence the roasted coffee floats on water and the green beans sink. The horny, tough and difficult-to-crack beans become brittle and mellow after roasting.

Four major phases are distinguished during the roasting process: drying, development, decomposition and full roasting. The initial changes occur at or above 50 °C when the protein in the tissue cells denatures and water evaporates. Browning occurs above 100 °C due to pyrolysis of organic compounds, accompanied by swelling and an initial dry distillation; at about 150 °C there is a release of volatile products (water, CO_2, CO) which results in an increase in bean volume. The decomposition phase, which begins at 180–200 °C, is recognizable by the beans being forced to pop and burst (bursting by cracking along the groove or furrow); formation of bluish smoke; and the release of coffee aroma. Lastly, under optimum caramelization, the full roasting phase is achieved, during which the moisture content of the beans drops to its final level of 1.5−3.5%. The roasting process is characterized by a decrease in old and formation of new compounds. This is covered in section 21.1.3.3, which deals with the composition of roasted coffee. The running of a roasting process requires skill and experience to achieve uniform color and optimum aroma development and to minimize the damage through over-roasting, scorching or burning.

During roasting, heat is transferred by contact of the beans with the walls of the roasting apparatus or by hot air or combusted gases (convection). Actual *contact roasting* is no longer of importance because heat transfer is uneven and the roasting times required are long (20−40 min). In the *contact-convection roasting* process (roasting time 6−15 min), efforts are made to increase the convection component as much as possible by suitable process management. Centrifugal roasters (rotating flat pans), revolving tube roasters, fluid-bed roasters (ca. 90% convection) etc. are used either batchwise or continuously. In the new *short-time roasting* process (roasting time 2 to 5 min), the heating-up phase is significantly shortened by improved heat transfer. Water evaporation proceeds by puffing, producing a greater bean volume increase than conventional roasting processes. Therefore, the density in the ground state of coffee roasted by this process is 15−25% lower.

The roasting process is controlled electronically or by sampling roasted beans. The end-product is discharged rapidly to cooling sifters or is sprinkled with water in order to avoid over-roasting or burning and aroma loss. During roasting, vapors formed and cell fragments (silverskin particles) are removed by suction of an exhauster and, in larger plants, incinerated.

There are different roasting grades desired. In the USA and Central Europe, beans are roasted to a light color (200−220°C, 3−10 min, weight loss 14−17%), and in France, Italy and the Balkan states, to a dark color (espresso, 230°C, weight loss 20%).

21.1.3.2 Storing and Packaging

Roasted coffee is freed of faulty beans either by hand picking on a sorting board or, at large plants, automatically by using photo cells. Commercially available roasted coffee is a blend of 4−8 varieties which, because of their different characteristics, are normally roasted separately. Especially strong blends are usually designated as mocca blends.

While green coffee can be stored for 1−3 years, roasted coffee, commercially packaged (can, plastic bags, pouches, bottles), remains fresh for only 8−10 weeks. The roasting aroma decreases, while a stale, rancid taste or aroma appears. Ground coffee packaged in the absence of oxygen (vacuum packaging) keeps for 6−8 months but, as soon as the package is opened, this drops to 1−2 weeks. Little is known of the nature of the changes involved in aroma and flavor damage. The changes are retarded by storing coffee at low temperatures, excluding oxygen and water vapor.

21.1.3.3 Composition of Roasted Coffee

Table 21.3 provides information about the composition of roasted coffee. This varies greatly, depending on variety and extent of roasting.

21.1.3.3.1 Proteins

Protein is subjected to extensive changes when heated in the presence of carbohydrates. There is a shift of the amino acid composition of coffee protein acid hydrolysates before and after bean roasting (Table 21.4). The total amino acid content of the hydrolysate drops by about 30% because of considerable degradation. Arginine, aspartic acid, cystine, histidine, lysine, serine, threonine and methionine, being especially reactive amino acids, are somewhat decreased in roasted coffee, while the stable amino acids, particularly alanine, glutamic acid and leucine, are relatively increased. Free amino acids occur only in traces in roasted coffee.

21.1.3.3.2 Carbohydrates

Most of the carbohydrates present, such as cellulose and polysaccharides consisting of mannose, galactose and arabinose, are insoluble.

Table 21.3. Composition of roasted coffee[a]

Constituent	Content (%)
Moisture	2.5
Protein[b]	9
Polysaccharide, water insoluble	24
Polysaccharide, water soluble	6
Saccharose	0.20
Glucose, fructose, arabinose	0.10
Lipids	13
Formic acid	0.10
Acetic acid	0.25
Nonvolatile acids[c]	0.40
Chlorogenic acids	3.7
Caffeine	1.2
Trigonelline	0.4
Nicotinic acid	0.02
Volatile aroma compounds	0.1
Minerals (ash)	4[d]
Unidentified constituents	35[e]

[a] Arabica-coffee, normal roasting.
[b] Calculated as the sum of amino acids after acid hydrolysis; water soluble fraction amounts to 1.5%.
[c] Lactic, pyruvic, oxalic, tartaric, and citric acids.
[d] Water soluble fraction amounts to 3.5%.
[e] Water soluble fraction amounts to 7.5%.

Table 21.4. Amino acid composition of the acid hydrolysate of Colombia coffee beans prior to and after roasting

Amino acid	Green coffee (%)	Roasted coffee[a] (%)
Alanine	4.75	5.52
Arginine	3.61	0
Aspartic acid	10.63	7.13
Cystine	2.89	0.69
Glutamic acid	19.80	23.22
Glycine	6.40	6.78
Histidine	2.79	1.61
Isoleucine	4.64	4.60
Leucine	8.77	10.34
Lysine	6.81	2.76
Methionine	1.44	1.26
Phenylalanine	5.78	6.32
Proline	6.60	7.01
Serine	5.88	0.80
Threonine	3.82	1.38
Tyrosine	3.61	4.35
Valine	8.05	8.05

[a] A loss due to roasting amounts to 17.6%.

During roasting a proportion of the polysaccharides are degraded into fragments which are soluble. Sucrose present in raw coffee is mostly decomposed in roasted coffee, as are monosaccharides.

21.1.3.3.3 Lipids

The lipid fraction appears to be very stable and survives the roasting process with only minor changes. Its composition is given in Table 21.5. Linoleic acid is the predominant fatty acid, followed by palmitic acid. The raw coffee waxes, together with hydroxytryptamide esters of various fatty acids (arachidic, behenic and lignoceric) originate from the fruit epicarp. These compounds are 0.06–0.1% of normally roasted coffee. The diterpenes present are cafestol (I, R = H), 16-O-methylcafestol (I, R = CH_3), and kahweol (II). Cafestol and kahweol are degraded by the roasting process. Since 16-O-methylcafestol is found only in Robusta coffee (0.6–1.8 g/kg of dry weight, green coffee), it is a suitable indicator for the detection of the blending of Arabica with Robusta coffee, even in instant coffee.

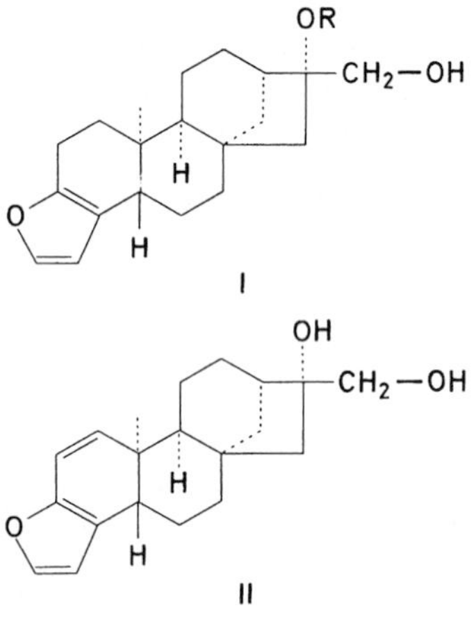

$$(21.1)$$

A diterpene glycoside is atractyloside and its aglycon, atractyligenin:

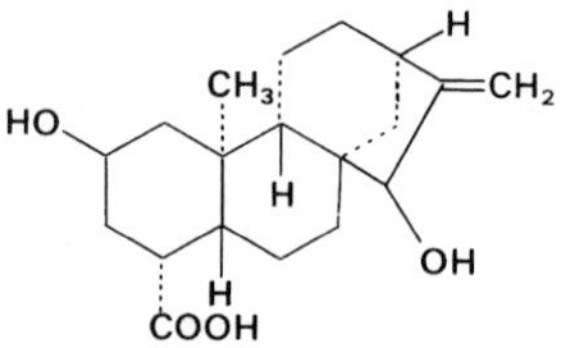

$$(21.2)$$

Table 21.5. Lipid composition of roasted coffee beans (coffee oil)

Constituent	Content (%)	Constituent	Content (%)
Triacylglycerols	78.8	Triterpenes (sterols)	0.34
Diterpene esters	15.0	Unidentified compounds	4.0
Diterpenes	0.12		
Triterpene esters	1.8		

Table 21.6. Chlorogenic acids in roasted coffee beans

Compound	Content (%)
3-Caffeoylquinic acid (neochlorogenic acid)	1.0
4-Caffeoylquinic acid (cryptochlorogenic acid)	0.2
5-Cafeoylquinic acid (chlorogenic acid)	2.0
3,4-Dicaffeoylquinic acid (isochlorogenic acid a)	0.01
3,5-Dicaffeoylquinic acid (isochlorogenic acid b)	0.09
4,5-Dicaffeoylquinic acid (isochlorogenic acid c)	0.01

Sitosterol and stigmasterol are major compounds of the sterol fraction.

21.1.3.3.4 Acids

Formic and acetic acids predominate among the volatile acids, while nonvolatile acids are lactic, tartaric, pyruvic and citric. Higher fatty acids and malonic, succinic, glutaric and malic acids are only minor constituents. Itaconic (I), citraconic (II) and mesaconic acids (III) are degradation products of citric acid, while fumaric and maleic acids are degradation products of malic acid:

(21.3)

Chlorogenic acids (cf. 18.1.2.5.1) are the most abundant acids of coffee. During the normal roasting process, they are decomposed by about 30% and during stronger roasting, by about 70% (Tables 21.2 and 21.3). The percentages of different chlorogenic acids are provided in Table 21.6. Also found are low levels of free quinic, caffeic and ferulic acids and quinic acid esterified with ferulic or coumaric acids.

21.1.3.3.5 Caffeine

The best known N-compound is caffeine (1,3,7-trimethylxanthine) because of its physiological effects (stimulation of the central nervous system, increased blood circulation and respiration). It is mildly bitter in taste (threshold value in water is 0.8−1.2 mmole/l),

crystallizes with one molecule of water into silky, white needles, which melt at 236.5 °C and sublime without decomposition at 178 °C. The caffeine content of raw Arabica coffee is 0.8−2.5%, while in the Robusta variety, it can be as high as 4%. In contrast there are caffeine-free Coffea varieties. Santos, an Arabica coffee, is on the low side, while Robusta from Angola is at the top of the range given for caffeine content. Caffeine forms, in part, a hydrophobic π-complex with chlorogenic acid in a molar ratio of 1:1. In a coffee drink, 10% of the caffeine and about 6% of the chlorogenic acid present occur in this form. The caffeine level in beans is only slightly decreased during roasting. Caffeine obtained by the decaffeination process and synthetic caffeine are used by the pharmaceutical and soft drink industries. Synthetic caffeine is obtained by methylation of xanthine which is synthesized from uric acid and formamide.

21.1.3.3.6 Trigonelline, Nicotinic Acid

Trigonelline (N-methylnicotinic acid) is present in green coffee up to 0.6% and is 50% decomposed during roasting. The degradation products include nicotinic acid, pyridine, 3-methyl pyridine, nicotinic acid methyl ester, and a number of other compounds.

21.1.3.3.7 Aroma Substances

The volatile fraction of roasted coffee has a very complex composition. The 655 substances identified until now belong to a great variety of classes of compounds, as shown in Table 21.7.

The group of aliphatic compounds includes hydrocarbons, alcohols and, above all, carbonyl compounds which are derived during roasting from carbohydrate fragmentation. Also, numerous alicyclic compounds are found, for example, cyclopentanone, cyclopenten-2-one, cyclohexen-2-one, cyclopentanedione-(1,2) and cyclohexanedione-(1,2).

Phenols are predominant among the aromatic compounds, and are most probably derived from thermal decomposition of chlorogenic acids (cf. 5.3.1.9). Phenol ethers, carbonyls, esters and polycyclic compounds are also found.

There is a substantial number of heterocyclic compounds, among which are many 2- and 2,5-substituted furans, probably derived from the pyrolysis of sucrose and other sugars. In addition, many pyrroles, pyrazines, thiophenes, thiazoles and oxazoles are found.

Aroma extract dilution analyses (AEDA) have shown that of the large number of volatile compounds only 28 are so odor-active that they contribute strongly to the coffee flavor.

This fraction consists of the 18 compounds shown for roasted Arabica and Robusta coffee (Table 21.8), and in addition acetaldehyde, propanal, methylpropanal, 2- and 3-methylbutanal, 2-methyl-3-furanthiol, methanethiol, dimethyl trisulfide, 2-ethenyl-3,5-dimethylpyrazine and 2-ethenyl-3-ethyl-5-methyl pyrazine. On the basis of high aroma values (Table 21.8), 2-furfurylthiol, 3-methyl-2-butenthiol and 3-mercapto-3-methylbutylformate are mainly responsible for the roasty/sulfury note in the flavor profile of coffee. The earthy/roasty note might be caused by the pyrazines, the caramel odor impression by the furanones and the phenol/smoky note by guaiacol and 4-vinylguaiacol.

Table 21.8 indicated that the sample of Robusta coffee differed greatly from the Arabica in lower contents of methional, 2-isobutyl-3-methoxypyrazine, 4-hydroxy-2,5-dimethyl-3(2H)-furanone, 5-ethyl-3-hydroxy-4-methyl-2(5H)-furanone and 2,3-pentanedione as well as in much higher contents of the two alkylpyrazines, guaiacol, 4-ethylguaiacol and 4-vinylguaiacol.

The aroma of coffee is not stable, the fresh note is rapidly lost. The highly volatile odorbearing substances methanethiol and 2,3-pentanedione are suitable indicators of freshness. Figure 21.2 shows that methanethiol rapidly decreases even on storage of unground coffee.

Table 21.7. Volatile constituents of roasted coffee (arranged according to the compound classes)

Class of compound	Number
Hydrocarbons	50
Alcohols, ketoalcohols	20
Aldehydes	28
Ketones, diketones	70
Carboxylic acids	20
Esters, ethers	29
N-Compounds (amines, nitriles)	24
S-compounds	16
Phenols	42
Furans	99
Lactones	8
Pyrroles	67
Pyridines	13
Pyrazines	79
Thiophenes	26
Thiazoles	28
Oxazoles	27
Others	9
Grand total	655

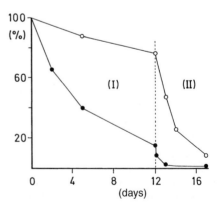

Fig. 21.2. Storage of unground (I) and ground (II) roasted coffee at room temperature (according to *Holscher* et al., 1990) o–o 2,3-pentanedione, •–• methanethiol

Table 21.8. Concentrations and aroma values of potent odorants in roasted Arabica and Robusta coffee samples[a]

Odorant	Concentration (mg/kg)		Aroma value[b]	
	Arabica	Robusta	Arabica	Robusta
3-Methyl-2-butenthiol	0.0082	0.0083	2.7×10^4	2.8×10^4
2-Furfurylthiol	1.7	1.73	1.7×10^5	1.7×10^5
Methional	0.240	0.095	1.2×10^3	5×10^2
3-Mercapto-3-methylbutylformate	0.130	0.115	3.7×10^4	3.3×10^4
2-Ethyl-3,5-dimethylpyrazine	0.330	0.94	2.1×10^3	5.9×10^3
2,3-Diethyl-5-methylpyrazine	0.095	0.31	1.1×10^3	3.4×10^3
3-Isobutyl-2-methoxypyrazine	0.083	0.012	2.1×10^4	3.0×10^3
4-Hydroxy-2,5-dimethyl-3(2H)-furanone	109	57.0	1.1×10^4	5.7×10^3
2(5)-Ethyl-4-hydroxy-5(2)-methyl-3(2H)-furanone	17.3	14.3	1.5×10^4	1.2×10^4
3-Hydroxy-4,5-dimethyl-2(5H)-furanone	1.47	0.63	74	32
5-Ethyl-3-hydroxy-4-methyl-2(5H)-furanone	0.160	0.085	21	11
Guaiacol	4.2	28.2	1.7×10^3	1.1×10^4
4-Ethylguaiacol	1.63	18.2	32	360
4-Vinylguaiacol	64.8	177.7	3.2×10^3	8.9×10^3
Vanillin	4.8	16.1	1.9×10^2	6.4×10^2
(E)-β-Damascenone	0.195	0.205	2.6×10^5	2.7×10^5
2,3-Butanedione	50.8	47.8	3.4×10^3	3.2×10^3
2,3-Pentanedione	39.6	19.8	1.3×10^3	6.6×10^2

[a] Aroma values are calculated by dividing the concentration of the compound by the odor threshold in water.
[b] The Arabica was from Colombia and the Robusta from Indonesia.

21.1.3.3.8 Minerals

As with all plant materials, potassium is predominant in coffee ash (1.1%), followed by calcium (0.2%) and magnesium (0.2%). The predominant anions are phosphate (0.2%) and sulfate (0.1%). Many other elements are present in trace amounts.

21.1.3.3.9 Other Constituents

Brown compounds (melanoidins) are present in the soluble fraction of roasted coffee. They have a molecular weight range of 5–10 kdal and are derived from *Maillard* reactions or from carbohydrate caramelization. The structures of these compounds have not yet been elucidated. Apparently, chlorogenic acid is also involved in such browning reactions since caffeic acid has been identified in alkali hydrolysates of melanoidins.

Secondary products of the thermolysis of mixtures of carbohydrates and proteins are probably involved in the formation of the bitter flavors of roasted coffee. This has been demonstrated in model systems consisting of mixtures of sugars and amino acids. Particularly intensive bitter tastes are obtained by heating sucrose and proline together (Table 21.9). The structures of some of these roasted bitter substances are known (cf. 4.2.4.4.4).

Extracts of roasted coffee have been separated by gel chromatography into fractions with coffee taste. These fractions contain carbohydrates, organic and amino acids and trigonelline.

21.1.3.4 Coffee Beverages

In order to obtain an aromatic brewed coffee with a high content of flavoring and stimulant constituents, a number of prerequisites must be fulfilled. The brewing, leaching and filtration procedures used give rise to a variety of combinations.

While in our society brewed coffee is enjoyed as a transparent, clear drink, in the Orient

Table 21.9. Formation of bitter taste by roasting mixture of amino acids and sugars (molar ratio 1:1, roasted at 190° for 30 min)

Amino acid	Sugar	Yield[a] (g/100 g)	c_{Sbi}[b] (g/100 ml)	Tb[c]	Aroma description
L-Pro	Sac	75	0.01–0.03	3750	Cracker-like, roasted peanuts
L-Pro	Fru	59	0.02–0.03	2360	Cracker-like, roasted peanuts
L-Pro	Glc	39	0.03–0.04	1114	Cracker-like, roasted peanuts
L-Pro	Sac	69	0.01–0.03	3450	Cracker-like, roasted peanuts
L-Lys	Sac	71	0.04–0.05	1578	Alkaline, burnt
L-Lys	Fru	25	0.04–0.06	500	Alkaline, burnt
L-Lys	Glc	22	0.05–0.06	400	Alkaline, burnt
cyclo-Leu	Sac	60	0.07–0.09	750	Mildly burnt aroma
L-Met	Sac	76	0.11–0.13	633	Potato-like, roasty
L-Met	Fru	6	0.05–0.06	109	Potato-like, roasty
L-Met	Glc	19	0.18–0.22	95	Potato-like, roasty
L-Phe	Sac	63	0.12–0.14	485	Floral with cocoa aroma note
L-Phe	Fru	16	0.12–0.14	123	Floral with slight cocoa aroma
L-Phe	Glc	15	0.12–0.13	120	Floral
L-Val	Sac	78	0.18–0.22	390	Cocoa aroma
L-Val	Fru	20	0.13–0.15	143	Mild cocoa aroma, burnt
L-Val	Glc	28	0.20–0.22	133	Mild cocoa aroma, burnt
L-Thr	Sac	61	0.18–0.19	330	Caramel-like, smoky
L-Thr	Fru	18	0.16–0.18	106	Caramel-like, smoky
L-Thr	Glc	25	0.22–0.26	104	Caramel-like, smoky
L-Ala	Sac	80	0.30–0.34	250	Smoky
L-Ala	Fru	7	0.09–0.10	74	Smoky
L-Ala	Glc	7	0.11–0.12	61	Smoky
L-Leu	Sac	81	0.40–0.50	180	Cocoa aroma, mild sweet note
L-Leu	Fru	17	0.13–0.15	121	Mild cocoa aroma, smoky
L-Leu	Glc	16	0.15–0.16	103	Mild cocoa aroma, smoky
D-Leu	Sac	79	0.40–0.50	176	Cocoa aroma
D-Leu	Fru	16	0.15–0.17	100	Mild cocoa aroma, smoky
D-Leu	Glc	22	0.25–0.35	73	Smoky
Gly	Sac	84	0.45–0.55	168	Caramel-like
Gly	Fru	35	0.40–0.50	78	Mildly burnt, scorched
Gly	Glc	41	0.60–0.70	63	Mildly burnt
L-His	Sac	65	0.35–0.45	163	Caramel-like
L-His	Fru	34	0.25–0.35	113	Smoky
L-His	Glc	38	0.35–0.40	101	Smoky
L-Ile	Sac	69	0.40–0.45	162	Cocoa aroma
L-Ile	Fru	16	0.13–0.15	114	Mild cocoa aroma, smoky
L-Ile	Glc	22	0.20–0.24	100	Mild cocoa aroma, smoky

[a] Yield of water soluble substances is based on reaction system.
[b] Detection threshold value for a bitter taste.
[c] Total bitterness Tb corresponds to the solution volume with c_{Sbi}-concentration per gram reaction mixture (yield/c_{Sbi}).

brewed coffee is prepared from pulverized beans (roasted beans ground to a fine powder) and water brought to a boil, and is drunk as a turbid beverage with the sediment (Turkish mocca). Coffee extract is made by boiling the coffee for 10 min in water and then filtering. In the boiling-up procedure the coffee is added to hot water, brought to a boil within a short time and then filtered. The steeping method involves pouring hot water on a bag filled with ground coffee and occasionally swirling the bag in a pot for 10 min. In the filtration-percolation method, ground coffee is placed on a support grid (filter paper, muslin, perforated plastic filter, sintered glass, etc.) and extracted by dripping or spraying with hot water, i. e. by slow gravity percolation. This procedure, in principle, is the method used in most coffee machines. In an expresso machine, which was developed in Italy, coffee is extracted briefly by superheated water (100–110 °C), while filtration is accelerated by steam at a pressure of 4–5 bar. The exceptionally strong drink is usually turbid and is made of freshly ground, darkly roasted coffee. The water temperature should not exceed 85–95°C in order to obtain an aromatic drink with most of the volatile substances retained. Water quality obviously plays a role, especially water with an unusual composition (some mineral spring waters, excessively hard water, and chlorinated water) might reduce the quality of the coffee brew. Brewed coffee allowed to stand for a longer time undergoes a change in flavor.

For regular brewed coffee, 50 g of roasted coffee/l (7.5 g/150 ml cup) is used; for mocca, 100 g/l; and for Italian espresso, 150 g/l. Depending on the particle size and brewing procedure, 18–35% of the roasted coffee is solubilized. The dry matter content of coffee beverages is 1–3%. The composition is presented in Table 21.10.

The taste of coffee depends greatly on the pH of the brew. The pH using 42.5 g/l of mild roasted coffee should be 4.9–5.2. At pH < 4.9 the coffee tastes sour; at pH > 5.2 it is flat and bitter. Coffees of different origins provide extracts with different pH's. Generally, the pH's of Robusta varieties are higher than those of Arabica varieties. Figure 21.3 shows the relationship between pH and extract taste for some coffees of known origin.

21.1.4 Coffee Products

The coffee products which will be discussed are instant coffee, decaffeinated coffee and those containing additives.

21.1.4.1 Instant Coffee

Instant (soluble) coffee is obtained by the extraction of roasted coffee. The first technically sound process was developed by *Mor-*

Table 21.10. Composition of coffee beverages[a]

Constituent	Content (% dry weight basis)
Protein[b]	6
Polysaccharides	24
Saccharose	0.8
Monosaccharides	0.4
Lipids	0.8
Volatile acids	1.4
Nonvolatile acids	1.6
Chlorogenic acids	14.8
Caffeine	4.8
Trigonelline	1.6
Nicotinic acid	0.08
Volatile aroma compounds	0.4
Minerals	14
Unidentified constituents (pigments, bitter compounds etc.)	29.4

[a] Arabica-coffee, medium roast, 50 g/l.
[b] Calculated as sum of the amino acids after acid hydrolysis.

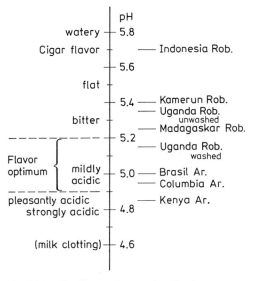

Fig. 21.3. The flavor of roasted coffee brew as related to pH value. (according to *Vitzthum*, 1976)

genthaler in Switzerland in 1938. Ground coffee is batchwise extracted under pressure in percolator batteries or continuously in extractors. The water temperature may be as high as 200 °C while the temperature of the extract leaving the last extraction cell is 40–80 °C.

The extracts exhibit a concentration of ca. 15% and are evaporated in vacuum film evaporators to a solids content of 35–70%. To minimize aroma losses, the extraction can be conducted in two stages. In a gentle stage, the first extract is obtained with a solids content of 25–27% and carries the main portion of the aroma. Without concentration, it is mixed with a second extract which was obtained under stronger conditions and concentrated. In addition, aroma concentrates can be isolated by stripping; they can be added back before or after drying. The technical extraction yields are 36–46%. Further processing involves spray or freeze drying. In the latter method, the liquid extract is foamed and frozen in a stream of cold air or an inert gas (−40°C), then granulated (grain size of 2–3 mm), sifted and dried in vacuum in the frozen state. Spray-dried coffee extract can be agglomerated in vibration fluid beds by steam or spray.

The resultant extract powder is hygroscopic and unstable. It is packaged in glass jars, vacuum packed in cans, aluminum foil-lined bags, flexible polyethylene, laminated pouches or bags, or packaged in air-tight plastic beakers or mugs, often under vacuum or under an inert gas.

Like roasted coffee, instant coffee is marketed in different varieties, e. g., regular roasted or as a dark, strongly-roasted espresso, or caffeine free. Instant coffee contains 1.0–6.0% moisture. The dry matter consists of 7.6–14.6% minerals, 3.2–13.1% reducing sugars (calculated as glucose), 2.4–10.5% galactomannan, 12% low molecular organic acids, 15–28% brown pigments, 2.5–5.4% caffeine and 1.56–2.65% trigonelline.

The products are used not only for the preparation of coffee beverages but also as flavorings for desserts, cakes, sweet cookies and ice cream.

21.1.4.2 Decaffeinated Coffee

The physiological effects of caffeine are not beneficial nor are they tolerated by everyone. Hence, many processes have been developed to remove caffeine from coffee. The first technically usuable process was developed by *Roselius* in Bremen (Germany) in 1908. In this process, which is still used in Europe, the green coffee is treated with superheated steam at high pressure. The swollen beans are then extracted with various organic solvents (dichloromethane, acetic acid ester) with constant stirring for the selective removal of caffeine. The decaffeinated beans are recovered after solvent removal. The moist beans, which, after steaming, acquire a moisture content of about 40%, are then dried under vacuum or in a stream of hot air. In another process, used in the USA, initially all the water-soluble compounds including caffeine are extracted from the green beans. The aqueous extract is decaffeinated with an organic solvent (e. g., dichloroethane), then added back to the green beans and evaporated to dryness with the beans.

A new process uses liquid carbon dioxide as a solvent at 70–90°C and 100–200 bar. Decaffeinated coffee extract can also be prepared from roasted coffee by this process. High-pressure extraction is a process for the separation of mixtures of substances in which compressed gases are used as solvent (e. g., CO_2, N_2O, propane, and toluene). Carbon dioxide (critical temperature: 31°C, critical pressure: 73.8 bar) is of special importance in food chemistry because it is physiologically safe and easy to handle. The high vapor pressure under normal conditions guarantees a product that is free from solvent residues. Apart from the extraction of caffeine, this process can also be applied in the extraction of odor- and taste-active substances from hops and other plant materials.

The market share of decaffeinated coffee products in Germany in 1974 was 12%.

21.1.4.3 Treated Coffee

The "roast" compounds, the phenolic acids and the coffee waxes, are irritating substances in roasted coffee. Various processes have been developed to separate these constituents to make roasted coffee tolerable for sensitive people.

Lendrich (1927) investigated the effect of steaming green beans, without caffeine extraction, on the removal of some substances (e. g., waxes) and hydrolysis of chlorogenic acid. In a process developed by *Bach* (1957), roasted coffee beans are washed with liquid carbon dioxide. In another process, the surface waxes of the raw beans are first removed by a low-boiling organic solvent, followed by steaming, as used by *Lendrich*. The extent of wax removal can be monitored by the analysis of fatty acid tryptamides, which have already been mentioned (cf. 21.1.3.3.3).

21.1.5 Coffee Substitutes and Adjuncts

21.1.5.1 Introduction

Coffee substitutes, or surrogates, are the parts of roasted plants and other sources which are made into a product which, with hot water, provides a coffee-like brew and serves as a coffee substitute or as a coffee blend.

Coffee adjuncts (coffee spices) are roasted parts of plants or material derived from plants, mixed with sugar, or a blend of all three sources and, when other ingredients are added, are used as an additive to coffee or as coffee substitutes. The starting materials for manufacturing such products vary: barley, rye, milo (a sorghum-type grain) and similar starch-rich seeds, barley and rye malts and other malted cereals, chicory, sugar beets, carrots and other roots, figs, dates, locust fruit (St. John's bread) and similar sugar-rich fruits, peanuts, soybeans and other oilseeds, fully or partially defatted acorns and other tannin-free plant parts, and, lastly, various sugars.

Coffee substitutes have been known for a long time, as exemplified by the coffee brew made of chicory roots (*Cichoricum intybus* var. *sativum*) or by clear drinks prepared from roasted cereals. The coffee substitute industry during the 2nd World War processed about 315,000 t of such products. In post-war Germany (1966) the amount was 24,000 t, which is about 27 l per capita per year. For comparison, in the same year coffee consumption was 127 l per capita. About ten years later (1977) the production of coffee surrogates dropped to 10,000 t or 10.3 l per capita in comparison to about 159 l of brewed coffee per capita.

21.1.5.2 Processing of Raw Materials

The raw materials are stored as such (all cereals, figs), or are stored until processing as dried slices (e.g., root crops such as chicory or sugar beet). After careful cleaning, steeping, malting and steaming in steaming vats, pots or pressure vats take place. Roasting follows, with a final temperature of $180-200\,°C$, and then the grains may be polished or coated with sugar.

For the manufacture of substitutes and adjunct essences, liquid sugar juice (cane or beet molasses, syrup or starch-sugar plant extracts) is caramelized in a cooker by heating above $160\,°C$ under atmospheric pressure. The dark, brownblack product solidifies to a glassy, strongly hygroscopic mass which is then ground.

Pulverized coffee substitutes are obtained from the corresponding starting materials, as with true coffee, by a spray, drum, conveyor or other drying process.

The starch present in the raw materials is diastatically degraded to readily-caramelized, water-soluble sugars in the manufacture of coffee substitutes during the steeping, steaming and, particularly, the malting steps. This is especially the case with malt coffee. Caramel substances ("bitter roast") formed in the roasting step, which provide the color and aroma of the brew, are derived from carbohydrate-rich raw materials (starch, inulin or sucrose). Since oilseeds readily develop rancidity, processing of carbohydrate-rich materials is preferred to oil- or protein-rich raw materials.

As aroma carriers, the oils from roasted products have been analyzed in detail, specially for malt and chicory coffees. From the volatiles identified in the coffee aroma, numerous constituents are also found in these oils. However, a basic difference appears to be that important sulfur- and nitrogen-containing aroma compounds present in roasted coffee beans are practically absent in coffee substitutes, or are present in negligible amounts.

21.1.5.3 Individual Products

21.1.5.3.1 Barley Coffee

Barley (or rye, corn or wheat) coffee is obtained by roasting the cleaned cereal grains after steeping or steaming. The products contain up to 12% moisture and have about 4% ash.

21.1.5.3.2 Malt Coffee

Malt coffee is made from barley malt by roasting, with or without an additional steaming step. It contains 4.5% moisture, 2.6% minerals, 74.7% carbohydrates (calculated), 1.8% fat, 10.8% crude protein, 5.6% crude fiber and provides an extract which is 42.4% soluble in water. Polycyclic aromatic hydrocarbons are also detected. Rye and wheat malt coffees are manufactured from their respective malts in the same way.

21.1.5.3.3 Chicory Coffee

Chicory coffee is manufactured by roasting the cleaned roots of the chicory plant possibly with addition of sugar beet, low amounts of edible fats or oils, salt and alkali carbonates. This is followed by grinding of the roasted product, with or without an additional steaming step or treatment with hot water. Chicory contains on the average 13.3% moisture, 4.4% minerals, 68.4% carbohydrates, 1.6% fat, 6.8% crude protein, 5.5% crude fiber andprovides an extract which is 64.6% soluble in water.

21.1.5.3.4 Fig Coffee

Fig coffee is made from figs by roasting and grinding, with or without an additional steaming step or treatment with hot water. It contains 11.4% moisture, 70.2% carbohydrates and 3.0% fat and provides an extract which is 67.9% soluble in water.

21.1.5.3.5 Acorn Coffee

This product is made from acorns, freed from fruit hull and the bulk of the seed coat, by the same process as used for coffee. It contains an average 10.5% moisture, 73.0% carbohydrates and provides an extract which is 28.9% soluble in water.

21.1.5.3.6 Other Products

Coffee substitute blends and similarly designated products are blends of the above-outlined coffee substitutes, coffee adjuncts and coffee beans. Caffeine-containing coffee substitutes or adjuncts are made by incorporating plant caffeine extracts into substitutes before, during or after the roasting step. The content of caffeine never exceeds 0.2% in such products.

21.2 Tea and Tea-Like Products

21.2.1 Foreword

Tea or tea blends are considered to be the young, tender shoots of tea shrubs, consisting of young leaves and the bud, processed in a way traditional to the country of origin. The tea shrub was cultivated in China and Japan well before the time of Christ. Plantations are now also found in India, Pakistan, Sri Lanka, Indonesia, Taiwan, East Africa, South America, etc. Table 21.11 shows some data on the production of tea.

The evergreen tea shrub (*Camellia sinensis*, synonym *Thea sinensis*) has three principal varieties, of which the Chinese (var, *sinensis*, small leaves) and the Assam varieties (var. *assamica*, large leaves) are the more important and widely cultivated. Grown in the wild, the shrub reaches a height of 9 m but, in order to facilitate harvest on plantations and in tea gardens, it is kept pruned as a low spreading shrub of 1–1.5 m in height. The plant is propagated from seeds or by vegetative propagation using leaf cuttings. It thrives in tropical and subtropical climates with high humidity. The first harvest is obtained after 4–5 years. The shrub can be used for 60 to 70 years. The harvesting season depends upon the region and climate and lasts for 8–9 months per year, or leaves can be plucked at intervals of 6–9 days all year round. In China there are 3–4 harvests per year.

Table 21.11. Production of tea in 1996 (1,000 t)

Continent	Tea	Country	Tea
World	2,622	India	715
		China	609
Africa	385	Kenia	255
America, North-,	1	Sri Lanka	246
Central-		Indonesia	169
America, South-	58	Turkey	124
Asia	2,167	Japan	90
Europe	3	Iran	56
Oceania	9	Bangladesh	48
		Argentina	40
		Vietnam	40
		Σ (%)[a]	91

[a] World production = 100%.

The younger the plucked leaves, the better the tea quality. The white-haired bud and the two adjacent youngest leaves (the famous "two leaves and the bud" formula) are plucked, but plucking of longer shoots containing three or even four to six leaves is not uncommon. Further processing of the leaves provides black or green tea.

21.2.2 Black Tea

The bulk of harvested tea leaves is processed into black tea. First, the leaves are withered in trays or drying racks in drying rooms, or are drum dried. This involves dehydration, reducing the moisture content of the fresh leaves from about 75% to about 55−65% so that the leaves become flaccid, a prerequisite for the next stage of processing: rolling without cracking of the leaves. Withering at 20−35 °C lasts about 4−18 h. During this time the thinly spread leaves lose about 50% of their weight in air or in a stream of warm air as in drum drying. In the next stage of processing, the leaves are fed into rollers and are lightly, without pressure, conditioned in order to attain a uniform distribution of polyphenol oxidase enzymes. These enzymes are present in epidermis tissue cells, spatially separated from their substrates. This is followed by a true rolling step in which the tea leaf tissue is completely macerated by conventional crank rollers under pressure. The cell sap is released and subjected to oxidation by oxygen from the air. The rolling process is regarded as fermentation and proceeds at 20−28 °C for 45 min to 4 h for tea leaves spread thinly in layers 5−7.5 cm thick. The fermentation is stopped when the leaves attain the bright, coppery-red color of a minted copper coin and an odor resembling that of sour apples. Then the fermented tea leaves are heated in large ovens or firing machines or desiccators at 87−93 °C for 20−22 min, or, more recently, in a fully automated process. The firing reduces the moisture content to about 3%, the tea aroma is fixed, and the coppery-red color is changed to black (hence "black tea").

India and Sri Lanka tea factories use both rollers and machines of continuous operation − the socalled CTC machines (crushing, tearing and curling). They provide a simultaneous crushing, grinding, and rolling of the tea leaf, thus reducing the rolling and fermentation time to 1 to 2 hours. Earl Grey tea is black tea perfumed with bergamot oil.

21.2.3 Green Tea

In the green tea manufacture, the development of oxidative processes is regarded as an adverse factor. The fresher the tea leaf used in manufacture, the better the tea produced. Since oxidative processes catalyzed by the leaf enzymes are undesirable, the enzymes are inactivated at an early stage and their reactions are replaced by thermochemical processes. In contrast to black tea manufacture, withering and fermentation stages are omitted in green tea processing.

There are two methods of manufacturing green tea: Japanese and Chinese. The Japanese method involves steaming of the freshly plucked leaf at 95 °C, followed by cooling and drying. Then the leaf undergoes high-temperature rolling at 75 to 80 °C. In the Chinese method the fresh leaves are placed into a roaster which is heated by smokeless charcoal, and roasted. After rolling and sifting, firing is the final step in the production of green tea.

During the processing of green tea the content of tannin, chlorophyll, vitamin C and organic acids decreases only slightly as a consequence of enzyme inactivation.

Green tea provides a very light, clear, bitter tasting beverage. In China and Japan it is often aromatized by flowers of orange, rose or jasmin. Yellow tea and red tea (*Oolong*) occupy an intermediate position between the black and green teas, yellow tea being closer to green teas, and red tea to black teas.

Yellow tea production does not include fermentation. Nevertheless, in withering, roasting, and firing, a portion of tannins undergoes oxidation, and, therefore, dry yellow tea is darker than green tea.

Red tea is a partially fermented tea. Its special flavor which is free from the grassy note of green tea is formed during roasting and higher-temperature rolling.

21.2.4 Grades of Tea

The numerous grades of tea found in the trade are defined by origin, climate, age, processing method, and leaf grade. They can be classified somewhat arbitrarily:

- According to leaf grade (tea with full, intact leaves), such as Flowery Orange Pekoe and Orange Pekoe (made from leaf buds and the two youngest, hairy, silver leaves with yellowish tips); Pekoe (the third leaf); Pekoe Souchong (with the coarsest leaves, fourth to sixth, on the young twig).
- Broken-tea, with broken or cut leaves similar to the above grades, in which the fine broken or cut teas with the outermost golden leaf tips are distinguished from coarse, broken leaves. Broken/cut tea (loose tea) is the preferred product in world trade since it provides a finer aroma which, because of increased surface area, produces larger amounts of the beverage.
- Fannings and the fluff from broken/cut leaves, freed from stalks or stems, are used preferentially for manufacturing of tea bags.
- Tea dust, which is not used in Europe.
- Brick tea is also not available on the European market. It is made of tea dust by sifting, steaming and pressing the dust in the presence of a binder into a stiff, compact teabrick.

With regard to the origin, teas of especially high quality are those from the Himalayan region Darjeeling and from the highlands of Sri Lanka.
All over the world there is blending of teas (e.g., Chinese, Russian, East-Friesen blends, household blends) to adjust the quality and flavor of the brewed tea to suit consumer taste, acceptance or trends and to accommodate regional cultural practices for tea-water ratios. Like coffee, tea extracts are dried and marketed in the form of a soluble powder, often called instant tea.

21.2.5 Composition

The chemical composition of tea leaves varies greatly depending on their origin, age and the

Table 21.12. Composition (%, dry weight basis) of fresh and fermented tea leaves and of tea brew

Constituent	Fresh flush	Black tea	Black tea brew[a]
Phenolic compounds[b]	30	5	4.5
Oxidized phenolic compounds[c]	0	25	15
Protein	15	15	+[d]
Amino acids	4	4	3.5
Caffeine	4	4	3.2
Crude fiber	26	26	0
Other carbohydrates	7	7	4
Lipids	7	7	+
Pigments[e]	2	2	+
Volatile compounds	0.1	0.1	0.1
Minerals	5	5	4.5

[a] Brewing time 3 min. [b] Mostly flavanols. [c] Mostly thearubigins. [d] Traces. [e] Chlorophyll and carotenoids.

type of processing. Table 21.12 provides data on the constituents of fresh and fermented tea leaves. In fermented teas 38–41% of the dry matter is soluble in hot water; this is significantly more than for roasted coffee.

21.2.5.1 Phenolic Compounds (cf. 18.1.2.5)

Phenolic compounds make up 25–35% of the dry matter content of young, fresh tea leaves. Flavanol compounds (Table 21.13) are 80% of the phenols, while the remainder is proanthocyanidins, phenolic acids, flavonols and flavones. During fermentation the flavanols are oxidized enzymatically to compounds which are responsible for the color and flavor of black tea. The reddish-yellow color of black tea extract is largely due to theaflavins and thearubigins (cf. 21.2.6), while flavor intensity is correlated with the total content of phenolic compounds and polyphenol oxidase activity. The enzymes are inactivated in green tea, hence flavanol oxidation is prevented. The greenish or yellowish color of green tea is due to the presence of flavonols and flavones. Thus, tea which is processed into green or black tea is chemically readily distinguishable mainly by the composition of phenolic compounds.

Table 21.13. Phenolic compounds in fresh tea leaves (% dry matter)

Compound	Content
(−)-Epicatechin	1−3
(−)-Epicatechin gallate	3−6
(−)-Epicatechin digallate	+[a]
(−)-Epigallocatechin	3−6
(−)-Epigallocatechin gallate	9−13
(−)-Epigallocatechin digallate	+
(+)-Catechin	1−2
(+)-Gallocatechin	3−4
Flavonols and flavonolglycosides	
(quercetin, kaempherol, etc.)	+
Flavones	
(vitexin, etc.)	+
Leucoanthocyanins	2−3
Phenolic acids and esters	
(gallic acid, chlorogenic acids)	
p-Coumaroylquinic acid, theogallin	~5
Phenols, grand total	25−35

[a] Quantitative data are not available.

Changes in the content of the phenols occur during tea leaf growth on the shrub: the concentration decreases and the composition of this fraction is altered. Therefore, good quality tea is obtained only from young leaves.

Among the remaining phenolic compounds theogallin (XI in Formula 18.13) plays a special role, since it is found only in tea and is correlated with tea quality.

The biosynthesis of phenolic compounds occurs via the shikimic acid and the phenylalanine pathway. It is possible that there is also a pathway that proceeds without phenylalanine.

21.2.5.2 Enzymes

A substantial part of the protein fraction in tea consists of enzymes.

The *polyphenol oxidases*, which are located mainly within the cells of leaf epidermis, are of great importance for tea fermentation. Their activity rises during the leaf withering and rolling process and then drops during the fermentation stage, probably as a consequence of reactions of some products (e.g., o-quinones) with the enzyme proteins.

5-Dehydroshikimate reductase which reversibly interconverts dehydroshikimate and shikimate is a key enzyme in the biosynthesis of phenolic compounds via the phenylalanine pathway.

Phenylalanine ammonia-lyase which catalyzes the cleavage of phenylalanine into *trans*-cinnamate and NH_3, is equally important for the biosynthesis of phenols. Its activity in tea leaves parallels the content of catechins and epicatechins.

Proteinases cause protein hydrolysis during withering, resulting in a rise in peptides and free amino acids.

The observed oxidation of linolenic acid to *cis*-3-hexenal, which then isomerizes to *trans*-2-hexenal, is catalyzed by a *lipoxygenase* and a *hydroperoxide lyase* (cf. 3.7.2.3). In particular *cis*-3-hexenal belongs to the character impact flavor compounds of green tea.

Chlorophyllases participate in the degradation of chlorophyll and *transaminases* in the production of precursors for aroma constituents.

Demethylation of pectins by *pectin methyl esterase* (cf. 4.4.5.2) results in the formation of a pectic acid gel, which affects cell membrane permeability, thus resulting in a drop in the rate of oxygen diffusion into leaves during fermentation.

21.2.5.3 Amino Acids

Free amino acids constitute about 1% of the dry matter of the tea leaf. Of this, 50% is theanine (5-N-ethylglutamine) and the rest consists of protein-forming amino acids; β-alanine is also present.

Green tea contains more theanine than black tea. Generally, there is a characteristic difference in amino acid content as well as difference in phenolic compounds between the two types of tea (Table 21.14).

The contribution of theanine to the taste of green tea is discussed. Theanine biosynthesis occurs in the plant roots from glutamic acid and ethylamine, the latter being derived from alanine. The compound is then transported into the leaves. The analogous compounds, 4-N-ethylasparagine and 5-N-methylglutamine, are present at very low levels in tea leaves.

Table 21.14. Amino acids and phenolic compounds in green and black tea (% dry matter)

Tea	Phenolic compounds	Amino acids
Green tea		
Prime quality (Japan)	13.2	4.8
Consumer quality (Japan)	22.9	2.1
Consumer quality (China)	25.8	1.8
Black tea		
Highlands (Sri Lanka)	28.0	1.6
Plains (Sri Lanka)	30.2	1.7

21.2.5.4 Caffeine

Caffeine constitutes 2.5–5.5% of the dry matter of tea leaves. It is of importance for the taste of tea. Theobromine (0.07–0.17%) and theophylline (0.002–0.013%) are also present but in very low amounts. The biosynthesis of these two compounds involves methylation of hypoxanthine or xanthine:

7-Methylxanthine   →

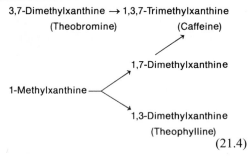

3,7-Dimethylxanthine → 1,3,7-Trimethylxanthine
     (Theobromine)   (Caffeine)

1,7-Dimethylxanthine

1-Methylxanthine

1,3-Dimethylxanthine
     (Theophylline)

$$(21.4)$$

21.2.5.5 Carbohydrates

Glucose (0.72%), fructose, sucrose, arabinose and ribose are among sugars present in tea leaves. Rhamnose and galactose are bound to glycosides. Polysaccharides found include cellulose, hemicelluloses and pectic substances. Inositol occurs also in tea leaves.

21.2.5.6 Lipids

Lipids are present only at low levels. The polar fraction (glycerophospholipids) in young tea leaves is predominant, while glycolipids predominate in older leaves.

Triterpene alcohols, such as β-amyrin, butyrospermol and lupeol are predominant in the unsaponifiable fraction. The sterol fraction contains only Δ^7-sterols, primarily α-spinasterol and Δ^7-stigmasterol.

21.2.5.7 Pigments (Chlorophyll and Carotenoids)

Chlorophyll is degraded during tea processing. Chlorophyllides and pheophorbides (brownish in color) are present in fermented leaves, both being converted to pheophytines (black) during the firing step.

Fourteen carotenoids have been identified in tea leaves. The main carotenoids are xanthophylls, neoxanthin, violaxanthin and β-carotene (cf. 3.8.4.1). The content decreases during the processing of black tea. Degradation of neoxanthin (cf. 3.8.4.4), as an example, yields β-damascenone, a significant contributor to tea aroma (Table 21.15).

21.2.5.8 Aroma Substances

The aroma is greatly affected by the origin and the processing of tea. In black tea the odorants listed in Table 21.15 have been identified on the basis of aroma extract dilution analysis (AEDA, cf. 5.2.5.2). High FD factors indicating high flavor potency were observed for β-damascenone, linalool and two furanones. Headspace analyses showed that diacetyl and the *Strecker* aldehydes methylpropanal, 2- and 3-methylbutanal additionally contribute to the aroma of black tea.

Table 21.15. Odorants of black tea. Results of AEDA

Compound	Odor quality	FD factor
(E)-β-Damascenone	Boiled apple	512
Linalool	Floral	512
3-Hydroxy-4,5-dimethyl-2(5H)-furanone	Seasoning-like	512
4-Hydroxy-2,5-dimethyl-3(2H)-furanone	Caramel	512
3-Methyl-3,4-nonandione	Strawy, hay-like	256
Bis(2-methyl-3-furyl)-disulfide	Boiled meat-like	256
(Z)-4-Heptenal	Biscuit-like	256
1-Octen-3-one	Mushroom-like	128
Vanillin	Vanilla	128
(E,E)-2,4-Decadienal	Deep-fried	64
2-Phenylethanol	Floral	64
(E)-2-Nonenal	Fatty, green	64

Odorants formed by an oxidative degradation of fatty acids occur in black tea (Table 21.15) but they play a much greater role in green tea. In particular (Z)-1,5-octadien-3-one, (Z)-3-hexenal and 3-methyl-2,4-nonandione (concentrations in green tea in Table 21.16) are responsible for the green, hay-like notes in the odor profile of green tea. The first two carbonyl compounds result from linolenic acid while 3-methyl-2,4-nonandione is an oxidation product of furan fatty acids (cf. 3.7.2.1.4). The concentrations of these flavor precursors in green tea are given in Table 3.9. During preparation of a brew the extraction yield of most of the odorants listed in Table 21.16 is higher than 50%. β-Damascenone is an exception, as its yield is only 11% (Table 21.16).

21.2.5.9 Minerals

Tea contains about 5% minerals. The major element is potassium, which is half the total mineral content. Some tea varieties contain fluorine in higher amounts (0.015−0.03%).

21.2.6 Reactions Involved in the Processing of Tea

Changes in tea constituents begin during the *withering* step of processing. Enzymatic pro-

Table 21.16. Concentrations of important odorants in the powder and brew of green tea

Compound	Amount[a]	
	Powder	Brew[b]
(Z)-1,5-Octadien-3-one	1.8	0.012
3-Hydroxy-4,5-dimethyl-2(5H)-furanone	49	0.6
3-Methyl-2,4-nonandione	83	0.56
(Z)-4-Heptenal	112	0.63
(Z)-3-Hexenal	101	0.28
(E,Z)-2,6-Nonadienal	61	0.48
1-Octen-3-one	6	0.03
(E,E)-2,4-Decadienal	127	0.9
(E)-β-Damascenone	9	0.01
4-Hydroxy-2,5-dimethyl-3(2H)-furanone	276	n.a.
2-/3-Methylbutyric acid	5,280	63
2-Phenylethanol	1,140	10.5
Linalool	206	1.0

[a] Values in μg/kg.
[b] Brew (1 kg) prepared from 10 g of the powder.
n.a., not analyzed.

tein hydrolysis yields amino acids of which a part is transaminated to the corresponding keto acids. Both types of acids provide a precursor pool for aroma substances. The induced chlorophyll degradation has significance for the appearance of the end-product. A more extensive conversion of chlorophyll into chlorophyllide, a reaction catalyzed by the enzyme chlorophyllase (cf. 17.1.2.9.1), is undesirable since it gives rise to pheophorbides (brown) and not to the desired olive-black pheophytins. Increased cell permeability during withering favors the fermentation procedure. As already mentioned, a uniform distribution of polyphenol oxidases in tea leaves is achieved during the *conditioning* step of processing.

During *rolling*, the tea leaf is macerated and the substrate and enzymes are brought together; a prerequisite for fermentation. The subsequent enzymatic oxidative reactions are designated as "*fermentation*". This term is a misnomer and originates from the time when the participation of microorganisms was assumed. In this processing step, the pigments are formed primarily as a result of phenolic oxidation by the polyphenol oxidases. In addition, oxidation of amino acids, carotenoids and unsaturated fatty acids, preferentially by oxidized phenols, is of importance for the formation of odorants.

Harler (1963) described tea aroma development during processing: "The aroma of the leaf changes as fermentation proceeds. Withered leaf has the smell of apples. When rolling (or leaf maceration) begins, this changes to one of pears, which then fades and the acrid smell of the green leaf returns. Later, a nutty aroma develops and, finally, a sweet smell, together with a flowery smell if flavor is present."

The enzymatic oxidation of flavanols via the corresponding o-quinones gives theaflavins (Formula 21.5, IX−XII: bright red color, good solubility), bisflavanols (XIII−XV: colorless), and epitheaflavic acids (XVI, XVII: bright red color, excellent solubility). The theaflavins and epitheaflavic acids are important benzotropolone derivatives that impart color to black tea.

A second, obviously heterogenous group of compounds, found in tea after the enzymatic oxidation of flavanols, are the thearubigins

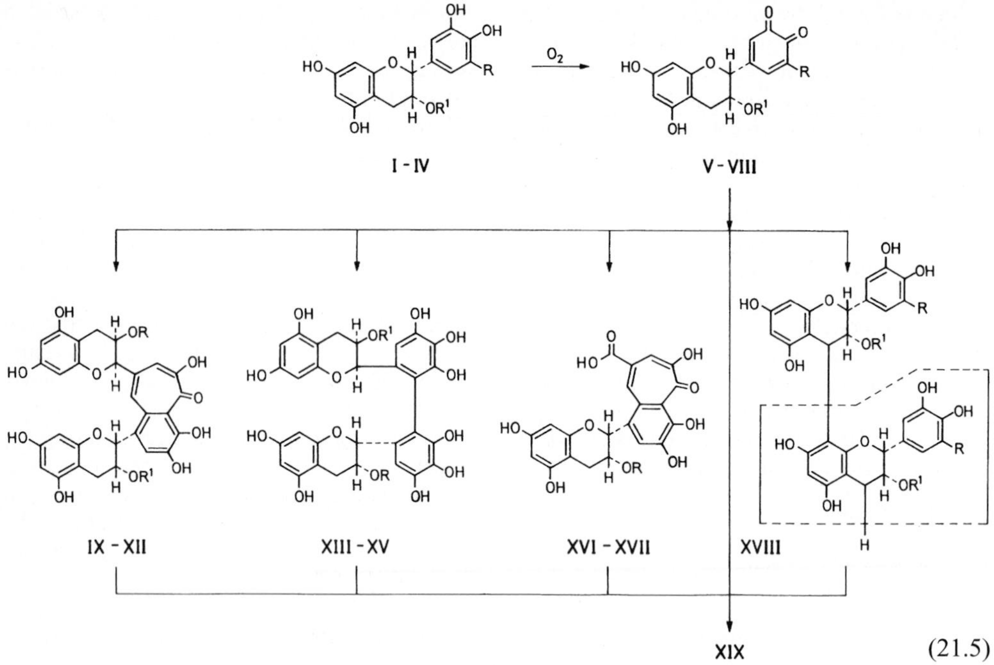

(21.5)

I: (−)-epicatechin, R^1, $R^2 = H$

II: (−)-epicatechin-3-gallate, $R = H$, $R^1 = 3,4,5$-trihydroxybenzoyl

III: (−)-epigallocatechin, $R = OH$, $R^1 = H$

IV: (−)-epigallocatechin-3-gallate, $R = OH$, $R^1 = 3,4,5$-trihydroxybenzoyl

V–VIII: o-quinones of compounds I–IV

IX: theaflavin, R, $R^1 = H$

X: theaflavin gallate A, $R = H$, $R^1 = 3,4,5$-trihydroxybenzoyl

XI: theaflavin gallate B, $R = 3,4,5$-trihydroxybenzoyl, $R^1 = H$

XII: theaflavin digallate, R, $R^1 = 3,4,5$-trihydroxybenzoyl

XIII: bisflavanol A, $R = R^1 = 3,4,5$-trihydroxybenzoyl,

XIV: bisflavanol B, $R = 3,4,5$-trihydroxybenzoyl, $R^1 = H$,

XV: bisflavanol C, $R = R^1 = H$,

XVI: epitheaflavic acid, $R = H$

XVII: 3-galloyl epitheaflavic acid, $R = 3,4,5$-trihydroxybenzoyl.

XVIII: thearubigins (proanthocyanidin-type), $R = H$, OH; $R^1 = H$, $3,4,5$-trihydroxybenzoyl,

XIX: thearubigins (polymeric catechins of unknown structure)

(XVIII, XIX), a group of compounds responsible for the characteristic reddish-yellow color and astringent taste of black tea extracts. Their structures have not yet been fully elucidated (cf. 18.1.2.5.2, Formula 18.20).

Aroma development during fermentation is accompanied by an increase in the volatile compounds typical of black tea. They are produced by *Strecker* degradation reactions of amino acids with oxidized flavanols (Formula 21.6) and by oxidation of unsaturated fatty acids.

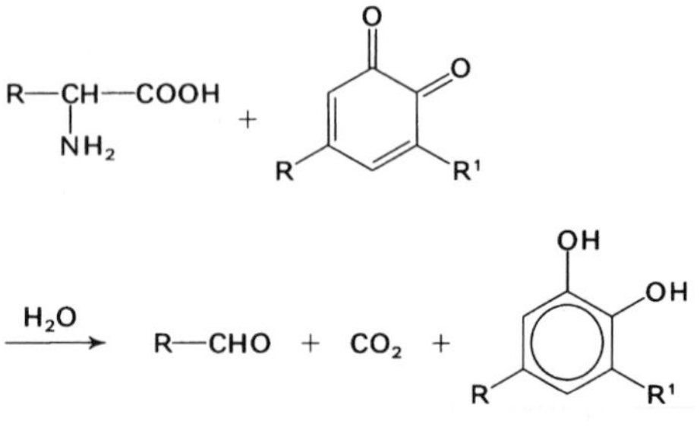

(21.6)

During the *firing* step of tea processing, there is an initial rise in enzyme activity ($10-15\%$ of the theaflavins are formed during the first 10 min), then all the enzymes are inactivated. Conversion of chlorophyll into pheophytin is involved in reactions leading to the black color of tea. A prerequisite for these reactions is high temperature and an acidic environment. The undesired brown color is obtained at higher pH's. The astringent character of teas is decreased by the formation of complexes between phenolic compounds and proteins.

The firing step also affects the balance of volatiles. On the one hand there is a loss of volatile compounds, on the other hand, at high temperatures, an enhancement of the build-up of typical aroma constituents occurs. Furanones (cf. Table 21.15) are most probably also formed in this step as a result of sugar-amino acid interactions.

21.2.7 Packaging, Storage, Brewing

In the country in which it is grown, the tea is cleaned of coarse impurities, graded according to leaf size, and then packed in standard plywood chests of $20-50$ kg lined with aluminum, zinc or plastic foils. To preserve tea quality, the foils are sealed, soldered or welded. China, glass or metal containers are suitable for storing tea. Bags made of pergament or filter papers and filled with metered quantities of tea are also very common.

During storage, the tea is protected from light, heat ($T < 30\,^\circ C$) and moisture, otherwise its aroma becomes flat and light. Other sources of odor should be avoided during storage.

To prepare brewed tea, hot water is usually poured on the leaves and, with occasional swirling, left for $3-5$ min. An initial tea concentrate or extract is often made, which is subsequently diluted with water. Usually $4-6$ g of tea leaves per liter are required, but stronger extracts need about 8 g. The stimulating effect of tea is due primarily to the presence of caffeine.

21.2.8 Maté (Paraguayan Tea)

Maté, or Paraguayan tea, is made from leaves of a South American palm, *Ilex paraguarien-*

sis. The palm grows in Argentina, Brazil, Paraguay and Uruguay, either wild or cultivated, and reaches a height of $8-12$ m. To obtain maté, the palm leaves, petioles, flower stems and young shoot tips are collected and charred slightly on an open fire or in a woven wire drum. During such firing, oxidase enzymes are inactivated, the green color is fixed and a specific aroma is formed. The dried product is then pounded into burlap sacks or is ground to a fine powder (maté pulver, maté en pod). Maté may also be prepared by an alternative process: brief blanching of the leaf in boiling water, followed by drying on warm floors and disintegration of the leaves to rather coarse particles. In the countries in which it is grown, maté is drunk as a hot brew (yerva) from a gourd (maté = bulbshaped pumpkin fruit) using a special metal straw called a bombilla, or it is enjoyed simply in a powdered form. Maté stimulates the appetite and, because of its caffeine content ($0.5-1.5\%$), it has long been the most important alkaloid-containing brewed plant product of South America. It contains on the average 12% crude protein, 4.5% ether-soluble material, 7.4% polyphenols and 6% minerals. About one third of the total dry matter of the leaves is solubilized in a maté brew, except for caffeine, which solubilizes to the extent of only $0.019-0.028\%$, and is 50% bound in leaves.

21.2.9 Products from Cola Nut

Cola (kola) nuts, called guru, goora and bissey nuts by Africans, are not nuts but actually seeds of an evergreen tree of the *Sterculiacea* family, genus *Cola*, species *verticillata, nitida* or *acuminata*, which grows wild in West Africa up to a height of 20 m. The tree is indigenous to Africa, but plantations of Cola are found on Madagascar, in Sri Lanka, Central and South America. Each fruit borne by the tree contains several red or yellow-white cola nuts, shaped like horse chestnuts. The nuts change color to brownish-red when dried, with the typical cola-red color resulting from the action of polyphenol oxidase enzymes. The nuts are on the average 5 cm long and 3 cm wide and have a bitter, astringent taste. The fresh nuts, wrapped in cola leaves and moistened with water, are the most enjoyed plant product of Western and Central Africa. They are consumed mostly in fresh form but are also chewed as dried nuts or ground to a powder and eaten with milk or honey. Cola nuts are used in the making of tinctures, extracts or medical stimulants in tablet or pastille form. They are also used in the liqueur, cocoa and chocolate industries and, especially, in the making of alcohol-free soft drinks, cola-wines, etc. The stimulating effect of cola nuts is due to the presence of caffeine (average content 2.16%), the main portion of which is in bound form. In addition, cola nuts contain on the average 12.2% moisture, 9.2% nitrogen compounds, 0.05% theobromine, 1.35% crude fat (ether extract), 3.4% polyphenols, 1.25% red pigments, 2.8% sugar, 43.8% starch, 15% other N-free extractable substances, 7.9% crude fiber and 3% ash.

21.3 Cocoa and Chocolate

21.3.1 Introduction

Cocoa, as a drink, is different from coffee or tea since it is consumed not in the form of an aqueous extract, i.e. a clear brew, but as a suspension. In addition to stimulating alkaloids, particularly theobromine, cacao products contain substantial amounts of nutrients: fats, carbohydrates and proteins. Unlike coffee and tea, cocoa has to be consumed in large amounts in order to experience a stimulating effect.

Cacao beans were known in Mexico and Central America for more than a thousand years before America was discovered by *Columbus*. They were enjoyed originally in the form of a slurry of roasted cocoa beans and corn which was seasoned with paprika, vanilla or cinnamon. In the first half of the 17th century, cacao beans were introduced into Germany. Cocoa became popular in the Old World only after sugar was added to the chocolate preparation. Initially, cocoa was treated as a luxury item, until the 19th century, when production of pulverized chocolate and defatted cocoa was established and they were distributed extensively as a food commodity.

The world production of cacao was 31,000 t in 1870/80, 103,000 t in 1900 and 1,585 million t in 1979. The main cacao-producing countries are listed in Table 21.17.

The processing of cacao beans into cocoa powder and chocolate is presented schematically in Fig. 21.4.

21.3.2 Cacao

21.3.2.1 General Information

Cacao beans are the seeds of the tropical cacao tree, *Theobroma cacao*, family *Sterculiaceae*. Originating in the northern part of South America and currently grown within 20° latitude of the Equator, the tree flourishes in warm, moist climates with an average annual temperature of 24–28°C and at elevations up to 600 m. The tree, because of its sensitivity to sunshine and wind, is often planted and cultivated under shade trees ("cacao mothers"), such as forest trees, coconut palms and banana trees. The perennial tree grows in the wild to a height of 10–15 m, but on plantations it is kept at 2–4 m by pruning. The tree blooms all year round and the small red or white flowers bear 20–50 ripe fruits per tree. The ripe fruit or pod resembles a cantaloupe, 15–25 cm long and 7–10 cm wide. The pod is surrounded by a strong 10–15 mm thick shell. Embedded within the pod are pulpa, i.e. a sweet, mucilaginous pulp

Table 21.17. Production cacao bean in 1996 (1,000 t)

Continent	Cacao beans	Country	Cacao beans
World	2,954	Côte d'Ivoire	1,254
		Ghana	340
Africa	1,909	Indonesia	274
America, North-,	138	Brazil	256
Central-	454	Nigeria	145
America, South-	417	Cameroon	126
Asia	–	Malaysia	125
Europe	35	Ecuador	88
Oceania		Colombia	65
		Dominican Republik	63
		Σ (%)[a]	93

[a] World production = 100%.

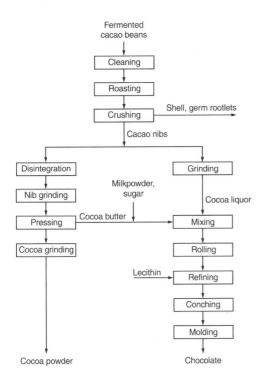

Fermented
cacao beans
↓
Cleaning
↓
Roasting
↓
Crushing → Shell, germ rootlets
↓
Cacao nibs
↓
Disintegration   Grinding
↓   Milkpowder,   ↓
Nib grinding   sugar   Cocoa liquor
↓   ↓
Pressing ← Cocoa butter →   Mixing
↓   ↓
Cocoa grinding   Rolling
↓   Lecithin →   Refining
↓   ↓
                               Conching
                                  ↓
                               Molding
↓   ↓
Cocoa powder   Chocolate

Fig. 21.4. Production of cocoa powder and chocolate

containing 10% glucose and fructose. The pulp surrounds 20−50 almond-shaped seeds (cacao beans). The seed is oval and flattened, about 2 cm long and 1 cm wide, and weighs close to 1 g after drying. The embryo, with two thick cotyledons (nibs) and a germ rootlet, 5 mm long and 1 mm thick, is under a thin, brittle seed coat. The colors in the cross-section of a nib range from white to light brown, to greyish-brown or brownviolet, to deep violet.

The fruit is harvested year round but, preferentially, twice a year. The main harvest time in Mexico is from March through April; in Brazil, February and, in particular, July. The summer harvest is larger and of higher quality. After the tree is planted (progagation by seed or by vegetative methods), it begins to bear pods after five or six years, giving a maximum yield after 20−30 years, while it is nearly exhausted after 40 years of growth. After reaching full beaning capacity, a cacao tree provides only 0.5−2 kg of fermented and dried beans per year. Harvesting at the right time is

of great importance for the aroma of cacao and its products. The fruit is harvested fully ripe but not overripe, avoiding damage to the seed during its removal from the fruit.

The tree species *Theobroma cacao* (the only one of commercial importance) is divided into two major groups. The "Criollo" tree (criollo = native) is sensitive to climatic changes and to attack by diseases and pests. It bears highly aromatic beans, hence their commercial name "flavor beans", but they are relatively low yielding. The second group of trees, "Forastero" (forastero = strange, inferior), is characterized by great vigor and the trees are more resistant to climatic changes and to diseases and are higher yielding. The purple-red Forastero beans are less flavorful than Criollo varieties. Nevertheless, the Forastero bean is by far the most important commercial type of cacao and accounts for the bulk of world cacao production (Bahia and Accra cacaos).

Other varieties worth mentioning are the resistant and productive Calabacillo and the Ecuadorian Amelonado varieties.

Cacao beans are differentiated by their geographical origin, grade of cleanliness and the number of preparation steps to which they are subjected prior to shipment. "Flavor beans" come from Ecuador, Venezuela, Trinidad, Sri Lanka and Indonesia, while "commercial beans" are exported by the leading cacao-growing countries of West Africa (Ghana, Nigeria, Ivory Coast and Cameroon), and by Brazil (the port of Bahia) and the Dominica Republic.

21.3.2.2 Harvesting and Processing

At harvest the fully ripe pods are carefully cut from trees, gathered into heaps, cut open and the beans scooped out with the surrounding pulp. Only rarely are the beans dried in the sun without a prior fermentation step (*Arriba* and *Machala* varieties from South America). The bulk of the harvest is fermented before being dried. In this fermentation step the beans with the adhering pulp are transferred to heaps, ditches or fermentation floors, baskets, boxes or perforated barrels and, depending on the variety, are left to ferment for 2−8 days. From time to time the beans are mixed to make the oxygen in the air accessible to the fermentation process. During this time the temperature of the beans rises rapidly to 45−50°C, the germination ability of the seeds is lost and the pH becomes acidic. Flavor and color formation and partial conversion of astringent phenolic compounds also occur. The adhering pulp is decomposed enzymatically and becomes liquid. It drains away as a fermentation juice. In addition, there are reactions between constituents of the beans and pulp. After fermentation is completed, the beans may be washed (Java, Sri Lanka), and are dried to a moisture content of 6−8%.

Well-fermented cocoa provides uniformly colored, dark-brown beans which are readily separated into their cotyledons. Inadequate or unripe fermented beans are smooth in appearance (violetas) and are of low quality.

The cocoa imported by consuming countries is processed further. The cocoa beans are cleaned by a series of operations and separated according to size in order to facilitate uniform roasting in the next processing step. Roasting is being performed more and more as a two-step process. Roasting reduces the moisture content of the beans to 3%, contributes to further oxidation of phenolic compounds and the removal of acetic acid, volatile esters and other undesirable aroma components. In addition the eggs and larvae of pests are destroyed. The aroma of the beans is enhanced, the color deepens, the seed hardens and becomes more brittle and the shell is loosened and made more readily removable because of enzymatic and thermal reactions. The ripeness, moisture content, variety and size of the beans and preliminary processing steps done in the country of origin determine the extent and other parameters of the bean roasting process. Generally, the roasting temperature should not exceed 150°C; this is substantially lower than in roasting coffee. In the production of cocoa for making chocolate, roasting is more extensive than for beans used for the manufacture of cocoa powder. Losses induced by roasting are 5−8%. As with coffee, roasted beans are immediately cooled to avoid overroasting. The roasters are batch or continuous. Heat transfer occurs either directly through heated surfaces or by a stream of hot air, without burning the shell of the beans. Roasting lasts 10−35 min, depending on the extent desired.

Roasted beans are transferred, after cooling, to winnowing machines to remove the shells and germ rootlets (these have a particularly unpleasant flavor and impart other undesirable properties to cocoa drinks). During winnowing the beans are lightly crushed in order to preserve the nibs and the shells in larger pieces and to avoid dust formation.

The winnowing process provides on the average 78−80% nibs, 10−12% shells, with a small amount of germ and about 4% of fine cocoa particles as waste. All yields are calculated on the basis of the weight of the raw beans.

The whole nibs, dried or roasted, dehulled and degermed or cracked, are still contaminated with 1.5−2% shell, seed coats and germ. The debris fraction, collected by purifying the cocoa waste, consists of fine nib particles and contains up to 10% shell, seed coating and germ. Although the cocoa shell is considered as waste material of little value, it can be used for recovery of theobromine, production of activated charcoal, or as a feed, cork substitute or tea substitute (cocoa shell tea) and, after extraction of fat, as a fertilizer or a fuel. Detection of shells in a cocoa blend is becoming increasingly more difficult due to ever-improving disintegration techniques. Chemical procedures, such as crude fibre analysis, furfural determination or assay of "mucilage", are far from being satisfactory. Microscopic methods, which involve isolation, enrichment and counting of scleroid cells, are more suitable. However, even here many precautions are necessary. Estimation of galacturonic acid, which is present in shells, and also of pectins which are 4.6−7% (dry weight basis) of the shell (but only a negligible 0.2% in nibs), provides another possibility for distinguishing pure, good quality cocoa powder with a maximum shell content of 1.75% from cocoa adulterated with shells.

21.3.2.3 Composition

The compositions of fermented and air-dried cacao nib, cacao shell and germ are presented in Table 21.18.

21.3.2.3.1 Proteins and Amino Acids

About 60% of the total nitrogen content of fermented beans is protein. The nonprotein nitro-

Table 21.18. Composition (%) of fermented and air dried cacao beans (1), cacao shells (2) and cacao germs (3)

Constituent	1	2	3
Moisture	5.0	4.5	8.5
Fat	54.0	1.5	3.5
Caffeine	0.2		
Theobromine	1.2	1.4	
Polyhydroxyphenols	6.0		
Crude protein	11.5	10.9	25.1
Mono- and oligosaccharides	1.0	0.1	2.3
Starch	6.0		
Pentosans	1.5	7.0	
Cellulose	9.0	26.5	4.3
Carboxylic acids	1.5		
Other compounds	0.5		
Ash	2.6	8.0	6.3

gen is found as amino acids, about 0.3% in amide form, and 0.02% as ammonia, which is formed during fermentation of the beans.

Among the various enzymes, α-amylase, β-fructosidase, β-glucosidase, β-galactosidase, pectinesterase, polygalacturonase, proteinase, alkaline and acid phosphatases, lipase, catalase, peroxidase and polyphenol oxidase activities have been detected in fresh cacao beans. These enzymes are inactivated to a great extent during processing.

21.3.2.3.2 Theobromine and Caffeine

Theobromine (3,7-dimethylxanthine), which is 1.2% in cocoa, provides a stimulating effect, which is less than that of caffeine in coffee. Therefore, it is of physiological importance. Caffeine is also present, but in much lower amounts (average 0.2%). A cup of cocoa contains 0.1 g of theobromine and 0.01 g of caffeine. Theobromine crystallizes in the form of small rhombic prisms which sublime at 290 °C without decomposition. In cocoa beans theobromine is often weakly bound to tannins and is released by the acetic acid formed during fermentation of the beans. Part of this theobromine then diffuses into the shell.

21.3.2.3.3 Lipids

Cocoa fat (cocoa butter), because of its abundance and value, is the most significant ingre-

dient of cacao beans, and is dealt with in detail elsewhere (cf. 14.3.2.2.3).

21.3.2.3.4 Carbohydrates

Starch is the predominant carbohydrate. It is present in nibs but not in shells, a fact useful in the microscopic examination of cocoa powders in methods based on the occurrence of starch as a characteristic constituent. Components of the dietary fiber are amongst others pentosans, galactans, mucins containing galacturonic acid, and cellulose. Soluble carbohydrates present include stachyose, raffinose and sucrose (0.08–1.5%), glucose and fructose. Sucrose hydrolysis, which occurs during fermentation of the beans, provides the reducing sugar pool important for aroma formation during the roasting process. Mesoinositol, phytin, verbascotetrose, and some other sugars are found in cocoa nib.

21.3.2.3.5 Phenolic Compounds

The nib cotyledons consist of two types of parenchyma cells (Fig. 21.5). More than 90% of the cells are small and contain protoplasm, starch granules, aleurone grains and fat globules. The larger cells are scattered among them and contain all the phenolic compounds and purines. These polyphenol storage cells (pigment cells) make up 11–13% of the tissue and contain anthocyanins and, depending on their composition, are white to dark purple. Data on the composition of these cells and that of the total tissue are given in Table 21.19.

Three groups of phenols are present: catechins (about 37%), anthocyanins (about 4%) and leucoanthocyanins (about 58%).

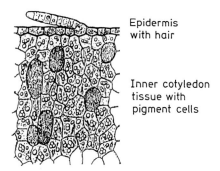

Epidermis with hair

Inner cotyledon tissue with pigment cells

Fig. 21.5. A cross-section of cocoa cotyledon tissue

The main catechin is (−)-epicatechin, besides (+)-catechin, (+)-gallocatechin and (−)-epigallocatechin. The anthocyanin fraction consists mostly of cyanidin-3-arabinoside and cyanidin-3-galactoside.

Pro- or leucoanthocyanins are compounds which, when heated in acidic media, yield anthocyanins and catechins or epicatechins, respectively. The form present in the greatest amount is flavan-3-4-diol (I in Formula 21.7) which, through $4 \rightarrow 8$ (II) or $4 \rightarrow 6$ (III) linkages, condenses to form dimers, trimers or higher oligomers (cf. 18.1.2.5.2, Formula 18.20).

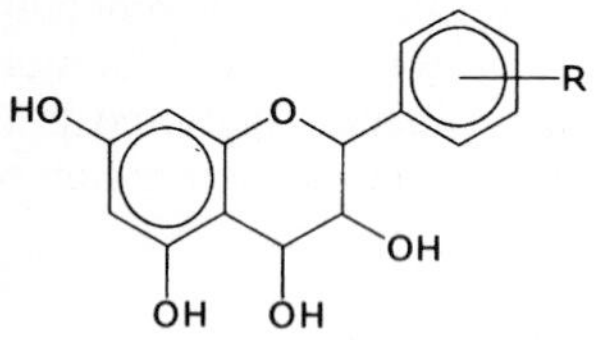

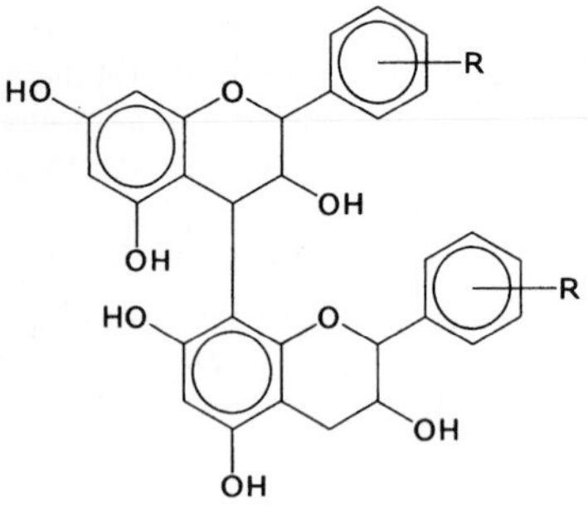

Table 21.19. Composition of polyphenol storage cells of cacao tissue

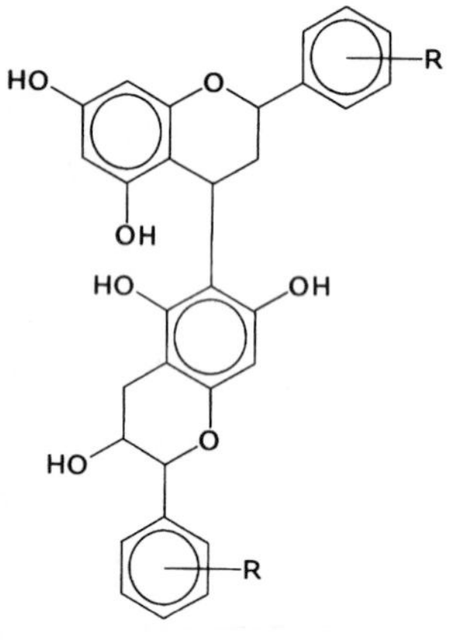

Constituent	Polyphenol storage cell (%)	Cotyledons (%)[a]
Catechins	25.0	3.0
Leucocyanidins	21.0	2.5
Polymeric leucocyanidins	17.5	2.1
Anthocyanins	3.0	0.4
Total phenols	66.5	8.0
Theobromine	14.0	1.7
Caffeine	0.5	0.1
Free sugars	1.6	
Polysaccharides	3.0	
Other compounds	14.4	

[a] As % of dry matter.

Leucoanthocyanins occur in fruits of various plants in addition to cacao; e. g., apples, pears and cola (kola) nuts.

21.3.2.3.6 Organic Acids

Organic acids in cocoa (1.2−1.6%) are formed mainly during cacao fermentation and consist mostly of acetic acid (a flavor constituent), citric acid (0.45−0.75%) and oxalic acid (0.32−0.50%). The amount of acetic acid released by the pulp and partly retained by the bean cotyledons depends on the duration of fermentation and on the drying method used. Eight brands of cocoa were found to contain 1.22−1.64% total acids, 0.79−1.25% volatile acids and 0.19−0.71% acetic acid.

21.3.2.3.7 Volatile Compounds and Flavor Substances

Cocoa aroma is crucially dependent on harvesting, fermentation, drying and roasting. The fresh beans have the odor and taste of vinegar. The characteristic bitter and astringent taste and the residual sweet taste of fermented beans might be impaired by various faults, such as processing of unripe or overripe fruit, insufficient aeration, lack of mixing of the fruit, infection with foreign organisms and/or smoke damage as a result of improper drying.

(21.7)

Table 21.20. Odorants of cocoa mass. Results of AEDA

Compound	Odor quality	FD-factor
3-Methylbutanal	Malty	1,024
Ethyl 2-methylbutanoate	Fruity	1,024
Hexanal	Green	512
Unknown	Fruity, waxy	512
2-Methoxy-3-isopropylpyrazine	Peasy, earthy	512
(E)-2-Octenal	Fatty, waxy	512
Unknown	Tallowy	512
2-Methyl-3(methyldithio)furan	Cooked meat-like	512
2-Ethyl-3,5-dimethylpyrazine	Earthy, roasty	256
2,3-Diethyl-5-methylpyrazine	Earthy, roasty	256
(E)-2-Nonenal	Tallowy, green	256
Unknown	Pungent, grassy	128
Unknown	Sweet, waxy	128
Phenylacetaldehyde	Honey-like	64
(Z)-4-Heptenal	Biscuit-like	64
δ-Octenolactone	Sweet, coconut-like	64
δ-Decalactone	Sweet, peach-like	64

The important odorants of cocoa mass listed in Table 21.20 were evaluated by AEDA (cf. 5.2.5.2). Amino acids which are released during fermentation of the beans are the precursors yielding 3-methylbutanal, phenylacetaldehyde, 2-methyl-3(methyldithio)furan, 2-ethyl-3,5-dimethyl- and 2,3-diethyl-5-methyl-pyrazine during roasting of the beans. The fruity smelling ethyl 2-methylbutanoate is most likely produced in the fermentation step.

The bitter taste is derived from the purines theobromine and caffeine and from dioxopiperazines, which are formed during the thermal degradation of proteins during roasting:

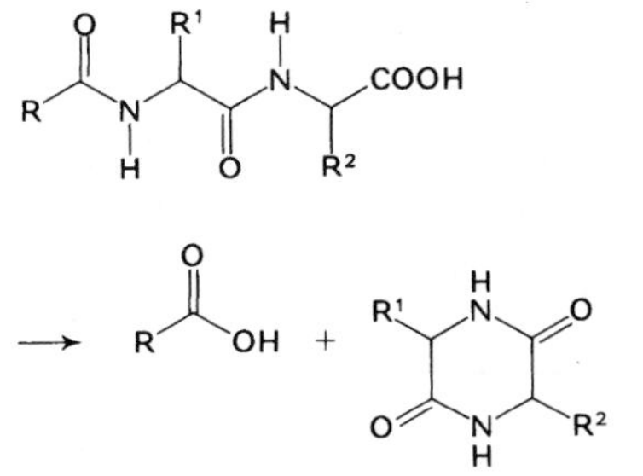

$$(21.8)$$

21.3.2.4 Reactions During Fermentation and Drying

The reactions occurring within the pulp during fermentation of whole cacao fruit can be distinguished from those occurring in the nibs or cotyledons. The pulp sugar is fermented by yeast to alcohol and CO_2 on the first day. Lactic acid fermentation may also occur to a small extent. Pectolytic enzymes and other glycosidases affect the degradation of polysaccharides. This is reflected in the fruit pulp becoming liquid and draining away. This improves aeration, resulting in oxidation of alcohol to acetic acid by acetic acid bacteria during the second to fourth days. The pH drops from about 6.5 to about 4.5 and the temperature increases to $45-50\,°C$. The seed cell walls become permeable, the living cacao seed is killed and an oxidative process takes over the entire mass. From the fifth to the seventh day, the oxidation and condensation reactions of phenolic compounds predominate. Amino acids and peptides react with the oxidation products of the phenolic compounds, giving rise to water-insoluble brown or brown-violet phlobaphenes (cacao-brown and red), which confer the characteristic color to fermented cacao beans. A decrease in the content of soluble phenols mellows the original harsh and astringent cacao flavor. Finally, the oxidation reactions are terminated by drying the seeds to a moisture content of less than 8%.

It is extremely important to properly handle the fermentation process for the formation of cocoa aroma. The growth of detrimental microorganisms, such as molds, butyric acid bacteria and putrefaction-inducing bacteria, must thereby be prevented.

21.3.2.5 Production of Cocoa Liquor

After roasting and drying, the cocoa nib is disintegrated and milled in order to rupture the cell walls of aggregates and expose the cocoa butter. Knife-hammer mills or crushing rolls usually serve for disintegration, while roller-ball, horizontal "stone", steel disc or disc attrition mills are used for fine disintegration of cocoa particles. The resultant product is a homogeneous mobile paste, a flowing cocoa mass or cocoa liquor.

21.3.2.6 Production of Cocoa Liquor with Improved Dispersability

The cocoa nib or the cocoa mass is subjected to an alkalization process in order to mellow the flavor by partial neutralization of free acids, improve the color and enhance the wettability of cocoa powder, improve dispersability and lengthen suspension-holding ability, thus preventing formation of a sediment in the cocoa drink. The process involves the use of solutions or suspensions of magnesium oxide or hydroxide, potassium or sodium carbonate or their hydroxides. It is occasionally performed at elevated temperature and pressure, usually using steam. In this process, introduced by *C.I. van Houten* in 1828 (hence the term "Dutch cocoa process"), the roasted nibs are treated with a dilute 2–2.5% alkali solution at 75–100 °C, then neutralized, if necessary, by tartaric acid, and dried to a moisture content of about 2% in a vacuum dryer or by further kneading of the mass at a temperature above 100 °C. This treatment, in addition to acid neutralization, causes swelling of starch and an overall spongy and porous cell structure of the cocoa mass. Cocoa so treated is often incorrectly designated as "soluble cocoa" – the process does not increase solubility. Finally, the cocoa is disintegrated with fine roller mills. The "alkalized" cocoa generally contains 52–58% cocoa butter, up to 5% ash and up to 7% alkalized mass or liquor.

21.3.2.7 Production of Cocoa Powder by Cocoa Mass Pressing

To convert the cocoa mass/liquor into cocoa powder, the cocoa fat (54% of nib weight on the average) has to be reduced by pressing, usually by means of a hydraulic, mechanical or, preferentially, horizontally-run expeller press at a pressure of 400–500 bar and a temperature of 90–100 °C. To remove the contaminating cell debris, the hot cocoa butter is passed through a filter press, then molded and cooled. The bulk of the cocoa butter produced is used in chocolate manufacturing. The "stone hard" cocoa press cake, with a residual fat content of 10–24%, is disintegrated by a cook breaker, i.e. rollers with intermashing teeth. It is then ground in a peg mill and separated into a fine and a coarse fraction by an air sifter, the coarse fraction being recycled and milled repeatedly. Cocoa powders are divided according to the extent of defatting into lightly defatted powder, with 20–22% residual cocoa butter, and extensively-defatted powder, which contains less than 20% but more than 10% butter. Lightly defatted powder is darker in color and milder in flavor. Cocoa powder is widely used in the manufacture of other products, e.g., cake fillings, icings, pudding powders, ice creams and cocoa (chocolate) beverages.

21.3.3 Chocolate

21.3.3.1 Introduction

Chocolates were originally made directly from cocoa nibs by grinding them in the presence of sugar. Chocolate is now made from nonalkalized cocoa liquor by incorporating sucrose, cocoa butter, aroma or flavoring substances and, occasionally, other constituents (milk ingredients, nuts, coffee paste, etc.). The ingredients are mixed, refined, thoroughly conched and, finally, the chocolate mass is molded. To obtain a highly aromatic, structurally homogeneous and stable form and a product which "melts in the mouth", a set of chocolate processing steps is required, as described below.

21.3.3.2 Chocolate Production

21.3.3.2.1 Mixing

Mixing is a processing step by which ingredients such as cocoa liquor, high grade crystalline sucrose, cocoa butter and, for milk chocolate, milk powder are brought together in a mixer ("melangeur") or paster. A homogeneous, coarse chocolate paste is formed after intense mixing.

21.3.3.2.2 Refining

The refining step is performed by single or multiple refining rollers which disintegrate the chocolate paste into a smooth-textured mass made up of much finer particles. The rollers are hollow and can be adjusted to the desired temperature by water cooling. The refined end-product has a particle size of less than 30 to 40 µm. Its fat content should be 23–28%.

21.3.3.2.3 Conching

The refined chocolate mass is dry and powdery at room temperature and has a harsh, sour flavor. It is ripened before further processing by keeping it in warm chambers at 45–50 °C for about 24 h. Ripening imparts a doughy consistency to the chocolate and it may be used for the production of baking or other commercial chocolates. An additional conching step is required to obtain fine chocolates of extra smoothness. It is performed in oblong or round conche pots with roller or rotary conches. The chocolate mass is mixed, ground and kneaded. This step is usually run in two stages. In the first, the mass is heated at 80 °C for 6–12 h. Loss of moisture occurs during heating, a protion of the volatiles is removed (ethanal, acetone, diacetyl, methanol, ethanol, isopropanol, isobutanol, isopentanol and acetic acid ethyl ester) and the fat becomes uniformly distributed, so that each cocoa particle is covered with a film of fat. The temperature at this stage is not allowed to rise since important aroma substances, e.g., pyrazines (cf. 21.3.2.3.7), may be lost. In the second stage, the mass is liquefied by the addition of residual cocoa butter and homogenized further. Lecithin is then added to reduce the viscosity of chocolate, or rather to give chocolates the required fluidity by the use of less cocoa butter, and homogenization is continued. Conching is a mixing process which produces a fine flavor and the desired texture, which was not attainable in the previous refining step. Chemical processes involved in conching are only partially understood.

Efforts have been made to shorten this time-, energy- and space-consuming final refinement in conche pots. Processes have been developed that are based on the separate pre-refinement of cocoa nibs or cocoa mass. The spray-film technique uses a cocoa mass with its natural water content or, in the case of highly acidic cocoa varieties, with the continuous addition of 0.5–2% of water. In a turbulent film with direct heat transfer, the cocoa mass is continuously dehumidified, deacidified, degassed, and roasted in counterflow with hot air (up to 130 °C). For the final refinement, apart from the time-tested conche pots, newly developed intensive refiners can be used. They reduce the conching time to 8 hours. The development of continuously operated conche pots is also being expedited.

21.3.3.2.4 Tempering and Molding

Before molding, the mass must be tempered to initiate crystallization. For both the structure (hard nibs, filling the mold) and appearance (glossy surface that is not dull), this is an important operation in which crystal nuclei are produced under controlled conditions (precrystallization). Molten chocolate is initially cooled from 50 °C to 18 °C within 10 min with constant stirring. It is kept at this lower temperature for 10 min to form the stable β-modification of cocoa butter (cf. 3.3.1.2). The temperature of the chocolate is then raised within 5 min to 29–31 °C. The process conditions vary according to composition. Regardless of processing variables, tempering serves to provide a great abundance of small fat crystals with high melting points. During the cooling step, the bulk of the molten chocolate develops a solid, homogeneous, finely crystalline, heat-stable fat structure characterized by good melting properties and a nice glossy surface.

Before molding, the chocolate is kept at 30–32 °C and delivered to warmed plastic or metal molds with a metering pump. The filled molds pass over a vibrating shaker to let the trapped air escape. They then pass through a cooling channel where, by slow cooling, the mass hardens and, finally, at 10 °C, the final chocolate product falls out of the mold.

Tempering, metering, filling, cooling, wrapping and packaging machines now provide nearly fully mechanized and automated production of chocolate.

21.3.3.3 Kinds of Chocolate

In a strict sense, chocolate represents a food commodity which may be molded and which consists of cocoa nibs, nib particles, or cocoa liquor and sucrose, with or without added cocoa butter, natural herbs or spices, vanillin or ethyl vanillin. Chocolate contains at least 40% cocoa liquor or a blend of liquor and cocoa butter, and up to 60% sugar. The content of cocoa butter is at least 21% and, when cocoa liquor is blended with cocoa butter, at least 33%.

The composition of the more important kinds of chocolates and confectionery coatings are shown in Table 21.21.

Baking chocolate is made by a special process. Other kinds of chocolates include: cream; full or skim milk; filled; fruit, nut, almond; and those containing coffee or candied orange peels. Cola-chocolate is a caffeine-containing product (maximum of 0.25% caffeine) prepared by mixing with extracts obtained from coffee, cola or other caffeine-containing plants. Diabetic- or diet-chocolates are made by replacing sucrose with fructose, mannitol, sorbitol or xylitol. Information about chocolate coatings is presented in Table 21.21. Choco-lates can also contain nuts and almonds whose oil contents are occasionally reduced by pressing to reach $^2/_3$ of the original amount. This is because the oil has a melting point lower than that of cocoa butter. In filled chocolates, the filler is first placed into a chocolate cup and then closed with a chocolate lid or cover. Fine crumbs of chocolate are made by pressing low-fat chocolate through a plate with orifices. Hollow figures are made in two-part molds, by a hollow press or by gluing together the individually molded parts.

The term "praline" originates from the name of the French Marshal *Duplessis-Praslin*, whose cook covered sweets with chocolate. Only a few of the many processing options will be mentioned. For pralines with a hard core, the hot, supersaturated sugar syrup (fondant) is poured into molds dusted with wheat powder and left to cool. The congealed core (korpus) is dipped into molgen kuverture and, in this way, covered with a chocolate coat (creme-praline). The fondant can be fully or partly replaced by fruit pastes like marzipan, jams, nuts, almonds, etc. (dessert-pralines). Such pralines are prepared with or without a sugar crust. Products with a sugar crust are made from a mixture of thick sugar solution and liqueur by pouring the mixture into mold cavities. The solid crust crystallizes on the outer walls, while the inner portion of the mixture remains liquid. The core so obtained is then dipped into melted chocolate, as described above. For pralines without a sugar crust (brandy or liqueur), the processing involves hollow-body machines in which the chocolate shell is formed, then filled with, e.g., brandy, and covered with a lid in a second machine.

Table 21.21. Composition of some chocolate products

Product	Cocoa mass %	Skim milk powder %	Cocoa butter %	Total fat %	Butter fat (milk) %	Sugar %
Baking chocolate	33–50	–	5–7	22–30	–	50–60
Chocolate for coating	35–60	–	to 15	28–35	–	38–50
Milk cream chocolate	10–20	8–16	10–22	33–36	5.5–10	35–60
Whole milk chocolate	10–30	9.3–23	12–20	28–32	3.2–7.5	32–60
Skim milk chocolate	10–35	12.5–25	15–25	22–30	0–2	30–60
Icings	33–65		5–25	35–46		25–50

The fondant may also contain invertase and, thereby, the praline filling liquefies after several days. Plastic pastes are made by preliminary pulverization of the ingredients in a mill and then refiner by rollers. The oil content of the ingredients (nuts, almonds, peanuts) provides the consistency for a workable paste after grinding. Chocolate for beverages or drinks (chocolate powder or flour) is made from cocoa liquor or cocoa powder and sucrose. It is customary to incorporate seasonings, especially vanillin. The sugar content in chocolate drink powders is at most 65%.

Chocolate syrups are made in the USA by adding bacterial amylase. The enzyme prevents the syrup from thickening or setting by solubilizing and dextrinizing cocoa starch. A fat coating is a glazing like chocolate coatings made from a fat other than cocoa butter (fat from peanuts, coconuts, etc.). It is often used on baked or confectionery products. Tropical chocolates contain high melting fats or are specially prepared to make the chocolate resistant to heat. The melting point of cocoa butter can be raised by a controlled precrystallization procedure. Another option is based on the formation of a coherent sugar skeleton in which the fat is deposited in hollow or void spaces. In this case, in contrast to regular chocolate, there is no continuous fat phase to collapse during heating.

21.3.4 Storage of Cocoa Products

All products, from the raw cacao to chocolate, demand careful storage – dry, cool, well aerated space, protected from light and sources of other odors. A temperature of $10-12\,°C$ and a relative humidity of $55-65\%$ are suitable. Chocolate products are readily attacked by pests, particularly cacao moths (*Ephestia elutella* and *Cadra cauteila*), the flour moth (*Ephestia kuhniella*) and also beetles (*Coleoptera*), cockroaches (*Dictyoptera*) and ants (order *Hymenoptera*).

Chocolates not properly stored are recognized by a greyish matte surface. Sugar bloom is caused by storage of chocolate in moist conditions (relative humidity above $75-80\%$) or by deposition of dew, causing the tiny sugar particles on the surface of the chocolate to solubi-

lize and then, after evaporation, to form larger crystals. A fat bloom arises from chocolate fat at temperatures above $30\,°C$. At these temperatures the liquid fat is separated and, after repeated congealing, forms a white and larger spot. This may also occur as a result of improper precrystallization or tempering during chocolate production. The defect may be prevented or rectified by posttempering at $30\,°C$ for 6 h.

21.4 Literature

Blank, I., Sen, A., Grosch, W.: Aroma impact compounds of Arabica and Robusta coffee. Qualitative and quantitative data. Fourteenth International Conference on Coffee Science, San Francisco, July 14–19, 1991. ASIC 91, p. 117

Bokuchava, M.A., Skobeleva, N.I.: The biochemistry and technology of tea manufacture. Crit. Rev. Food Sci. Nutr. *12*, 303 (1979/80)

Castelein, J., Verachtert, H.: Coffee fermentation. In: Biotechnology (Eds.: Rehm, H.-J., Reed, G.), Vol. 5, p. 587, Verlag Chemie: Weinheim. 1983

Clarke, R.J., Macrae, R. (Eds.): Coffee. Vol. 1: Chemistry. Elsevier Applied Science Publ.: London. 1985

Clifford, M.N., Willson, K.C. (Eds.): Coffee, botany, biochemistry and production of beans and beverage. The AVI Publishing Comp. Inc., Westport, Conn. 1985

Flament, I.: Coffee, cacao and tea flavors: a review on present knowledge. In: International symposium of food flavors (Eds.: Adda, J., Richard, H.), p. 82, Technique et Documentation (Lavoisier): Paris, 1983

Flament, I.: Coffee, cacao, tea: In: Volatile compounds in foods and beverages (Ed.: Maarse, H.), Marcel Dekker, Inc.: New York. 1991

Forsyth, W.G.C., Quesnel, V.C.: The mechanism of cacao curing. Adv. Enyzmol. *25*, 457 (1963)

Harler, C.R.: Tea Manufacture. Oxford University Press: London. 1963

Holscher, W., Vitzthum, O.G., Steinhart, H.: Identification and sensorial evaluation of aroma-impact compounds in roasted Columbia coffee. Cafe, Cacao, The *34*, 205 (1990)

Kleinert-Zollinger, J.: Schokolade. In: Ullmanns Encyklopädie der technischen Chemie. 4. edn., Vol. 20, p. 673, Verlag Chemie: Weinheim. 1981

Lange, H., Fincke, A.: Kakao und Schokolade. In: Handbuch der Lebensmittelchemie, Bd. VI (Ed.: Schormüller, J.), p. 210, Springer-Verlag: Berlin. 1970

Lehrian, D.W., Patterson, G.R.: Cocoa fermentation. In: Biotechnology (Eds.: Rehm, H.-J., Reed, G.), Vol. 5, p. 529, Verlag Chemie: Weinheim. 1983

Maier, H.G.: Kaffee. Verlag Paul Parey: Berlin. 1981

Sanderson, G.W.: Black tea aroma and its formation. In: Geruch- and Geschmackstoffe (Ed.: Drawert, F.), p. 65, Verlag Hans Carl: Nürnberg. 1975

Sanderson, G.W.: Tea manufacture. In: Biotechnology (Eds.: Rehm, H.-J., Reed, G.), Vol. 5, p. 577, Verlag Chemie: Weinheim. 1983

Speer, K., Tewis, R., Montag, A.: 16-O-Metyhlcafestol a quality indicator for coffee. Fourteenth International Conference on Coffee Science, San Francisco, July 14–19, 1991. ASIC 91, p. 237

Vitzthum, O.G.: Chemie und Bearbeitung des Kaffees. In: Kaffee und Coffein (Ed.: Eichler, O.) p. 3, Springer-Verlag: Berlin. 1976

Wickremasinghe, R.L.: Tea. Adv. Food Res. *24*, 229 (1978)

Wurziger, J.: Tee. In: Ullmanns Encyklopädie der technischen Chemie, 4. edn., Vol. 22, p. 405, Verlag Chemie: Weinheim. 1982

Yamanishi, T.: The aroma of various teas. In: Flavor of foods and beverages (Eds.: Charalambous, G., Inglett, G.E.), p. 305, Academic Press: London-New York. 1978

22 Spices, Salt and Vinegar

22.1 Spices

Some plants with intensive and distinctive flavors and aromas are used dried or in fresh form as seasonings or spices. Table 22.1 lists the most important spice plants together with the part of the plant used for seasoning.

22.1.1 Composition

22.1.1.1 Aroma Substances

As shown in Table 22.2, the aroma substances in most spices are present as essential or volatile oils which, as outlined in 5.5.1.1, are obtainable from plants by steam distillation.

Table 22.1. Spices used in food preparation/processing

Number	Common name	Latin name	Class/order family (bot)	Cultivation region
Fruits				
1	Pepper, black	*Piper nigrum*	Piperaceae	Tropical and subtropical regions
2	Vanilla	*Vanilla planifolia* *Vanilla fragans* *Vanilla tahitensis* *Vanilla pompona*	Orchidaceae	Madagascar, Comore Island, Mexico, Uganda
3	Allspice	*Pimenta dioica*	Myrtaceae	Caribbean Islands, Central America
4	Paprika (bell pepper)	*Capsicum annuum, var. annuum*	Solanaceae	Mediterranean and Balkan region
	Chili (Tabasco)	*Capsicum frutescens*		
	Brown pepper	*Capsicum baccatum, var. pendulum*		
5	Bay tree [a]	*Laurus nobilis*	Lauraceae	Mediterranean region
6	Juniper berries	*Juniperus communis*	Cupressaceae	Temperate climate region
7	Aniseed	*Pimpinella anisum*	Apiaceae	Temperate climate region
8	Caraway	*Carum carvi*	Apiaceae	Temperate climate region
9	Coriander	*Coriandrum sativum*	Apiaceae	Temperate climate region
10	Dill [a]	*Anethum graveolens*	Apiaceae	Temperate climate region
Seeds				
11	Fenugreek	*Trigonella foenum greacum*	Leguminosae	Mediterranean region, temperate climate region
12	Mustard	*Sinapsis alba* [b] *Brassica nigra* [c]	Brassicaceae Brassicaceae	Temperate climate region
13	Nutmeg	*Myristica fragrans*	Myristicaceae	Indonesia, Sri Lanka, India
14	Cardamom	*Elettaria cardamomum*	Zingiberaceae	India, Sri Lanka
Flowers				
15	Cloves	*Syzygium aromaticum*	Myrtaceae	Indonesia, Sri Lanka, Madagascar

Table 22.1 (continued)

Number	Common name	Latin name	Class/order family (bot)	Cultivation region
16	Saffron	*Crocus sativus*	Iridaceae	Mediterranean region, India, Australia
17	Caper	*Capparis spinosa*	Capparidaceae	Mediterranean region
Rhizomes				
18	Ginger	*Zingiber officinale*	Zingiberaceae	South China, India, Japan, Caribbean Islands, Africa
19	Turmeric	*Curcuma longa*	Zingiberaceae	India, China, Indonesia
Barks				
20	Cinnamon	*Cinnamomum zeylanicum, C. aromaticum, C. burmanii*	Lauraceae	China, Sri Lanka, Indonesia, Caribbean Islands
Roots				
21	Horseradish	*Armoracia rusticana*	Brassicaceae	Temperate climate region
Leaves				
22	Basil	*Ocimum basilicum*	Labiate	Mediterranean region, India
23	Parsley	*Petroselinum crispum*	Apiaceae	Temperate climate region
24	Savory	*Satureia hortensis*	Labiate	Temperate climate region
25	Tarragon	*Artemisia dracunculus*	Compositae	Temperate climate region, Mediterranean region
26	Marjoram	*Origanum majorana*	Lamiaceae	Temperate climate region
27	Origano	*Origanum heracleoticum, O. onïtes*	Lamiaceae	Temperate climate region
28	Rosemary	*Rosmarinus officinalis*	Lamiaceae	Mediterranean region
29	Sage	*Salvia officinalis*	Lamiaceae	Mediterranean region
30	Chives	*Allium schoenoprasum*	Liliaceae	Temperate climate region
31	Thyme	*Thymus vulgaris*	Lamiaceae	Temperate climate region

[a] Fruits and leaves, [b] white mustard, [c] black mustard.

Table 22.2. Content of essential oils in some spices[a]

Spice	% Vol./Weight
Black pepper	2.0–4.5
White pepper	1.5–2.5
Aniseed	1.5–3.5
Caraway	2.7–7.5
Coriander	0.4–1.0
Dill	2.0–4.0
Nutmeg	6.5–15
Cardamom	4 –10
Ginger	1 –3
Turmeric	4 –5
Marjoram	0.3–0.4
Origano	1.1
Rosemary	0.72
Sage	0.7–2.0

[a] For leaf spices, the values refer to the weight of the fresh material.

The main oil constituents are either mono- and sesquiterpenes or phenols and phenolethers. Examples of the latter two classes of compounds are eugenol (I), carvacrol (II), thymol (III), estragole (IV), anethole (V), safrole (VI) and myristicin (VII):

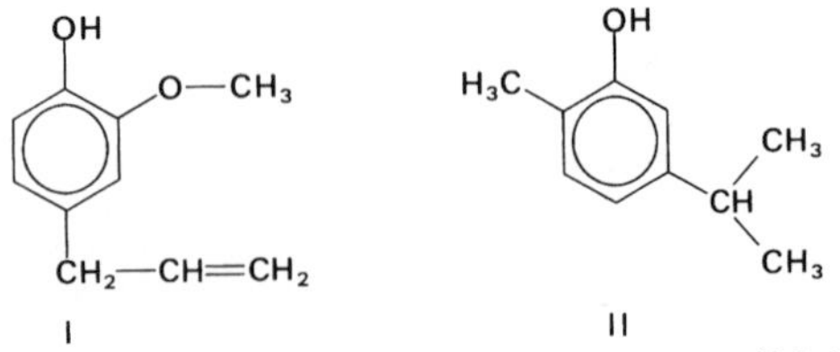

$$(22.1)$$

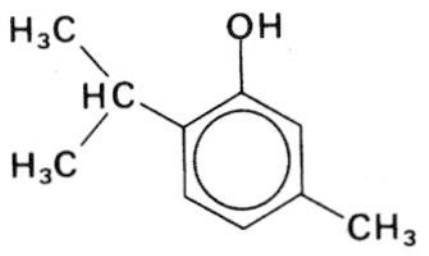

(22.2 cont.)

III

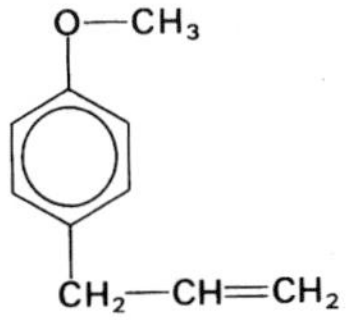

IV

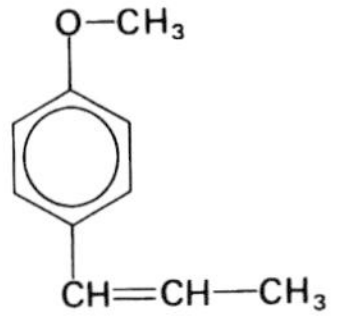

V

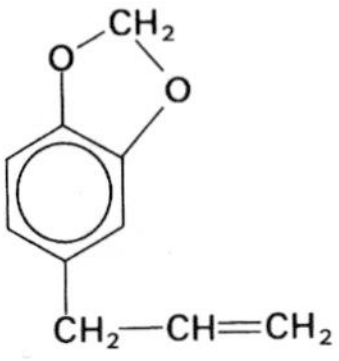

VI

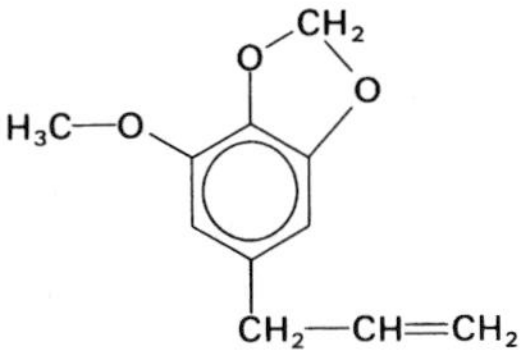

VII

(22.1 cont.)

IX

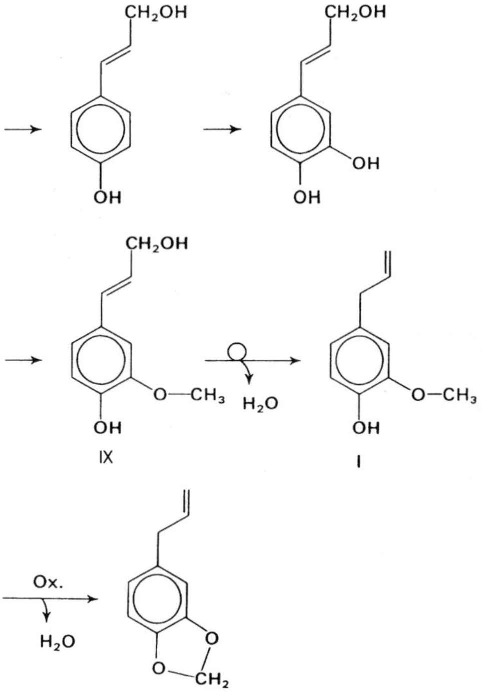

I

Ox.

H₂O → H_2O

VI

Biosynthesis of cinnamaldehyde (VIII), the main constituent of cinnamon bark (Table 22.3), and also of eugenol (I) and safrole (VI) originates from phenylalanine (compare biosynthesis of other plant phenols in 18.1.2.5.1). The following reaction sequence is assumed:

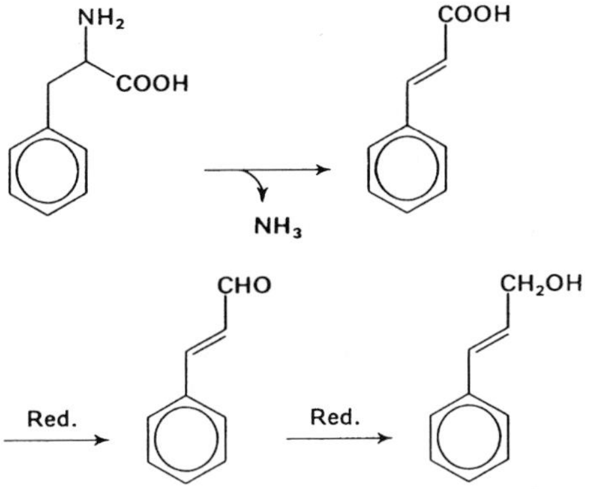

(22.2)

Some aromatic hydrocarbons are probably generated in spices by terpene oxidation. Examples are: 1-methyl-4-isopropenylbenzene (XI, Formula 22.3) derived from p-mentha-1,3,8-triene (X) and (+)-ar-curcumene (XIV) from zingiberene (XII) or β-sesquiphellandrene (XIII) [cf. Formula 22.4].

The formation of (+)-ar-curcumene from the above-mentioned precursor was detected during storage of ginger oil.

Another aromatic hydrocarbon present in significant amounts in essential oils of some spices (Table 22.3) is p-cymene (XV, Formula 22.3).

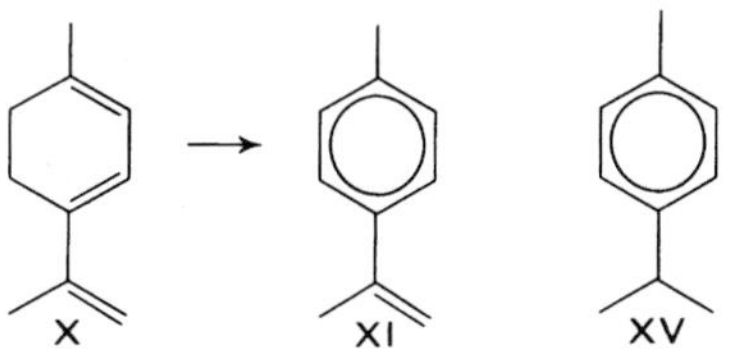

(22.3)

X   XI   XV

Table 22.3. Volatile compounds of spices[a]

Spice[b]	Components[c]
Pepper (1)	1−16% α-Pinene (XXIX*), 0.2−19% sabinene (XXV*), 9−30% β-caryophyllene (XLIX*), 0−20% Δ³-carene (XXXII*), 16−24% limonene (IX*), 5−14% β-pinene (XXX*)
Vanilla (2)	*Vanillin* (1.3−3.8%, dry matter), (R)(+)-*trans-α-ionone, p-hydroxybenzylmethylether* (XX)
Allspice (3)	50−80% *Eugenol* (I), 4−7% β-*caryophyllene* (XLIX*), 3−28% *methyleugenol*, 1,8-cineole (XXIII*), α-phellandrene (XI*)
Bay leaf (5)	50−70%, 1,8-Cineole (XXIII*), α-pinene (XXIX*), β-pinene (XXX*), α-phellandrene (XI*), linalool (IV*)
Juniper berries (6)	36% α-Pinene (XXIX*), 13% myrcene (I*), β-pinene (XXX*), Δ³-carene (XXXII*)
Aniseed (7)	80−95% (E)-*anethole* (V)
Caraway (8)	55% (S)(+)-*Carvone* (XXI*), 44% limonene (IX*)
Coriander (9)	(S)(+)- und (R)(−)-*Linalool* (IV*), linalyl acetate, citral[d], 2-*alkenales* C_{10}−C_{14}
Dill (fruit, 10)	20−40% (S)(+)-*Carvone* (XXI*), 30−50% (R)(+)-*limonene* (IX*)
Dill (herb, 10)	70% (S)(+)-*Phellandrene* (XI*), 17% (3R,4S,8S)(+)-*epoxy-p-menth-1-ene* (XXI), myristicin (VII), (R)-limonene (IX)
Fenugreek (11)	*Sotolon* (cf. 5.3.1.3)
Nutmeg (13)	27% α-Pinene (XXIX*), 21% β-pinene (XXX*), 15% sabinene (XXV*), 9% limonene (IX*), 0.1−3.3% *safrole* (VI), 0.5−14% *myristicin* (VII), 1.5−4.2% 1,8-*cineole* (XXII*)
Cardamom (14)	20−40%, 1,8-Cineole (XXIII*), 28−34% α-terpinyl acetate, 2−14% limonene (IX*), 3−5% sabinene (XXV*)
Clove (15)	73−85% *Eugenol* (I), 7−12% β-*caryophyllene* (XLIX*), 1.5−11% *eugenol acetate*
Saffron (16)	47% *Safranal* (XVIII), 14% 2,6,6-*trimethyl-4-hydroxy-1-cyclohexen-1-formaldehyde* (XVII)
Ginger (18)	30% (−)-Zingiberene (XLII*), 10−15% β-bisabolene (XLI*), 15−20% (−)-sesquiphellandrene (XLIII*), (+)-*ar*-curcumene (XIV), citral[c], citronellyl acetate
Tumeric (19)	30% Turmerone (XIXa), 25% *ar*-turmerone (XIXb), 25% zingiberene (XLII*)
Cinnamon (20)	50−80% *Cinnamaldehyde* (VIII), 10% eugenol (I), 0−11% safrole (VI), 10−15% linalool (IV*), camphor (XXXIII*)
Parsley (23)	*p-Mentha-1,3,8-triene* (X), *myristicin* (VII), *2-sec-butyl-3-methoxypyrazine, 2-isopropyl-3-methoxypyrazine, (Z)-6-decenal, (E,E)-2,4-decadienal, myrcene* (I*)
Marjoram (26)	3−18% *cis-Sabinenehydrate* (XXVII*), 1−7% *trans-sabinenehydrate*, 16−36% 1-*terpinen-4-ol*
Origano (27)	60% *Carvacrol* (II), *thymol* (III)
Rosemary (28)	1,8-Cineole (XXIII*), camphor (XXXIII*), β-pinene (XXX*), camphene (XXXI*)
Sage (29)	1,8-Cineole (XXIII*), camphor (XXXIII*), thujone (XXVI*)
Thyme (31)	*Thymol* (III), p-cymene (XV), *carvacrol* (II), linalool (IV*)

[a] Compounds printed in italics substantially contribute to aroma. With the exception of vanillin and dill (herb), the quantitative values refer to the composition of the essential oil.
[b] The number in brackets refers to Table 22.1.
[c] Roman numerals with an asterisk refer to the chemical structures of the terpenes presented in Table 5.29. Roman numerals without an asterisk refer to chemical structures shown in Chapter 22.
[d] A mixture of neral and geranial (cf. footnote "b" in Table 5.29).

Of the aroma substances of saffron, the terpene aldehydes safranal (XVIII) and 2,6,6-trimethyl-4-hydroxy-1-cyclohexen-1-formaldehyde (XVII) dominate. They are probably formed by the hydrolysis and elimination of water from the bitter substance picrocrocin (XVI, Formula 22.5).

The aroma substances typical of a number of spices have been identified. Their names are given in italics in Table 22.3. As a rule, these are compounds that quantitatively dominate in essential oils. In most of the cases, the importance of very potent odorants that occur in very low concentrations is still unclear. For

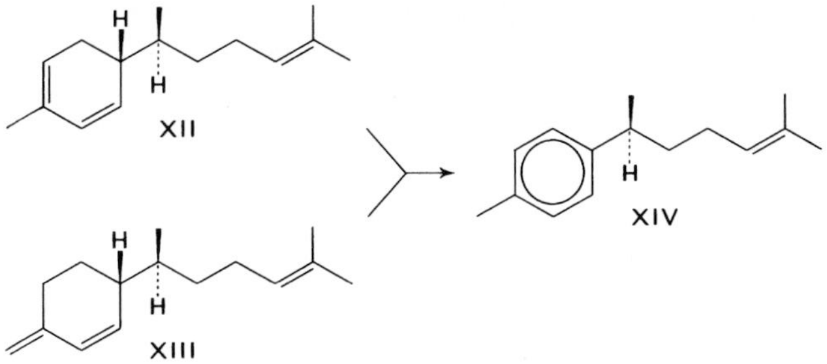

XII

XIV

XIII

(22.4)

XVI

XVII

XVIII

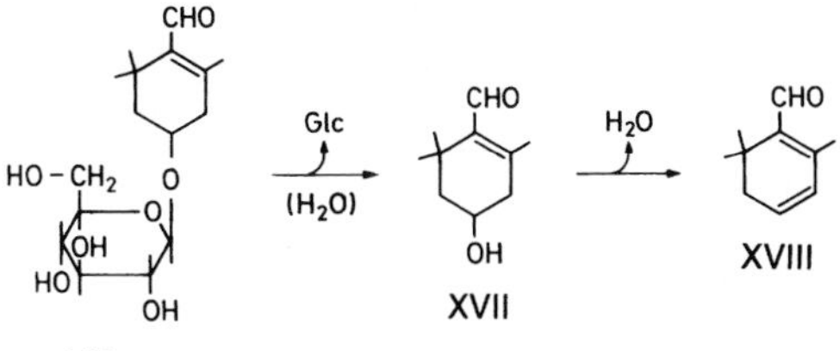

(22.5)

example, it has been found that two methoxypyrazines (Table 22.3), that are present in the μg/kg range, are essential for the aroma of parsley.

The concentrations given in Table 22.3 are guide values which can vary greatly depending on the variety and cultivation conditions.

Mustard and horseradish contain glucosinolates (Table 22.4) which, after cell rupture, are exposed to the action of a *thioglucosidase* enzyme (cf. 17.1.2.6.5), yielding isothiocyanates (mustard oil). Allyl isothiocyanate is obtained from the glucoside sinigrin, a compound responsible for the pungent burning odor and taste of both spices. *p*-Hydroxybenzyl isothiocyanate obtained from sinalbin is only slightly volatile and contributes significantly to the sharp pungent taste of mustard.

The aroma of horseradish is also influenced by methyl, ethyl, isopropyl and 4-pentenyl isothiocyanates which, however, are present only

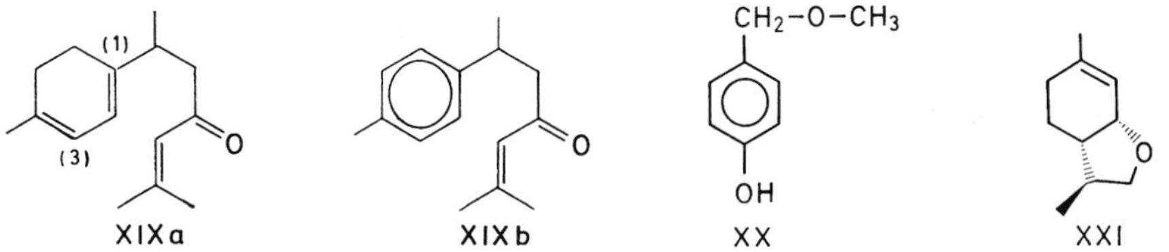

XIXa

XIXb

XX

XXI

(22.6)

Table 22.4. The most important glucosinolates of mustard and horseradish

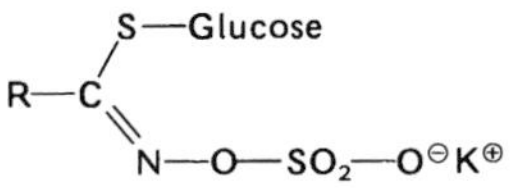

R	Name	Occurrence
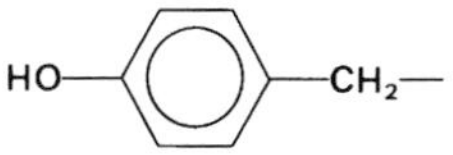 HO—⬡—CH₂—	Sinalbin	Mustard
H₂C=CH—CH₂—	Sinigrin	Mustard, horseradish
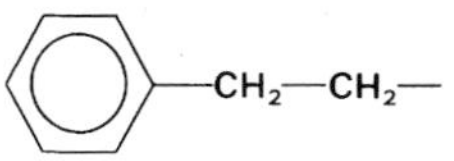 ⬡—CH₂—CH₂—	Gluconasturtiin	Horseradish

in very small amounts in comparison to allyl isothiocyanate.

The aroma of capsicum pepper plants consists of pyrazines, particularly of 2-isobutyl-3-methoxypyrazine (cf. 5.3.2.6).

In the capsular fruit of vanilla, incorrectly called vanilla bean, 170 volatile compounds have been identified. However, the only fact that is certain is that apart from the main aroma substance vanillin, which is released from the glucoside on fermentation of the fruits, and (R)(+)-trans-α-ionone, the p-hydroxybenzyl-methylether (XX) contributes to the aroma since its concentration (115−187 mg/kg) greatly exceeds the odor threshold (0.1 mg/kg, water). A mixture of 99% of sugar and 1% of ground vanilla is sold as vanilla sugar and a mixture of 98% of sugar and 2% of vanillin is sold as vanillin sugar.

22.1.1.2 Substances with Pungent Taste

The hot, burning pungent taste of paprika (red pepper), pepper (black pepper) and ginger is caused by the nonvolatile compounds listed in Table 22.5.

Black pepper contains 3−8% of piperine (XXIV) as the most important pungent substance. Pepper is sensitive to light since the trans,trans-diene system of piperine isomerizes to the cis,trans-diene system of the almost tasteless isochavicin on exposure to light.

In the processing and storage of ginger, gingerol easily dehydrates to shogaol, increasing the pungency (Table 22.5). A retroadol cleavage of shogaol can also occur with the formation of sweet-spicy zingerone and hexanal (Formula 22.7). Above a certain concentration, hexanal causes an aroma defect in ginger oleoresins.

The concentration of the capsaicinoids XXIX, XXX, and XXXI (Table 22.5) in the fruits of capsicum or in various other pepper plants depends on the variety, cultivation, drying and storage conditions, and varies between 0.01 and 1.2%. These compounds are the most pungent spice constituents. Their concentrations are at the upper limit in chillies and tobasco varieties and at the lower limit in sweet varieties.

22.1.1.3 Pigments

Paprika (red pepper) and curcuma pigments are used as food colorants. Paprika pigments are carotenoids, with capsanthin as the main compound (cf. 3.8.4.1.2 and Fig. 3.47). Curcumin (cf. Formula 22.8) is the main pigment of curcuma, a tropical plant of the ginger family.

22.1.1.4 Antioxidants

Extracts of several spices, particularly of sage and rosemary, have the ability to prevent the autoxidation of unsaturated triacylglycerols. Among the most effective antioxidant constituents of both spices, the cyclic diterpene diphenols, carnosolic acid (XXII) and carnosol (XXIII) have been identified:

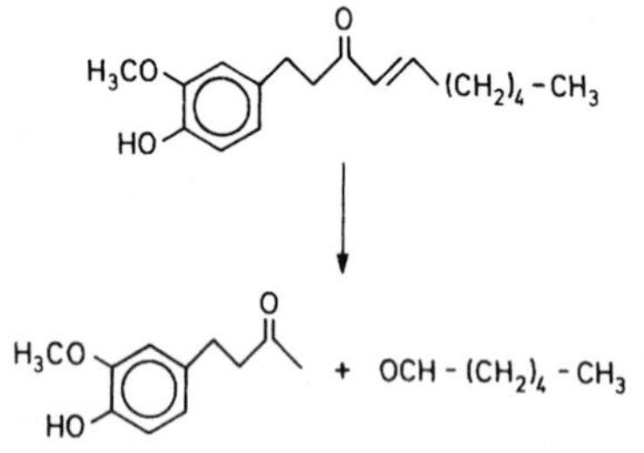

(22.7)

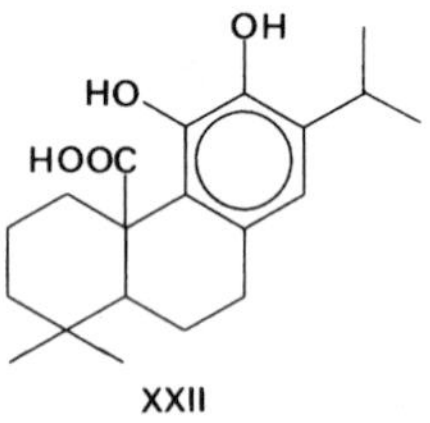

XXII
(22.9)

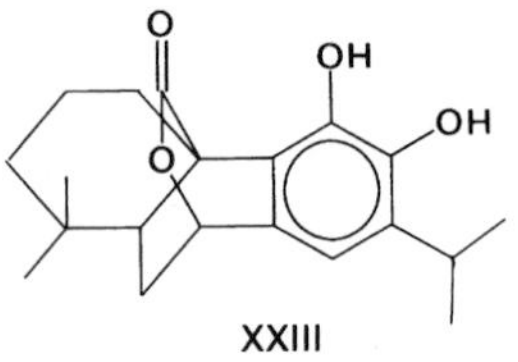

XXIII
(22.10)

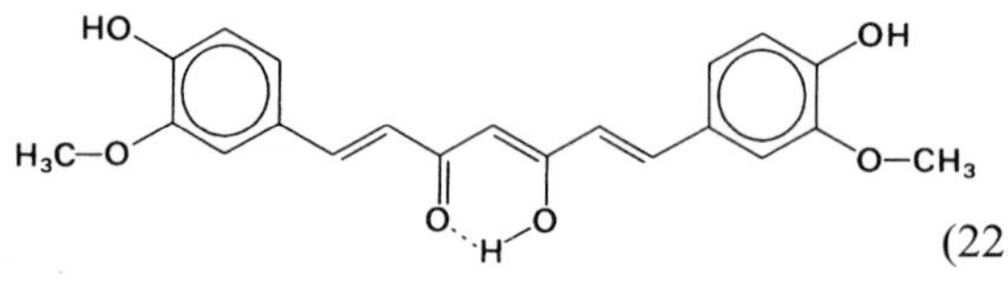

(22.8)

Table 22.5. Compounds present in spices causing a hot burning organoleptic perception

Name	Structure	Occurrence[a]	Relative pungency[b]
Piperine[c]	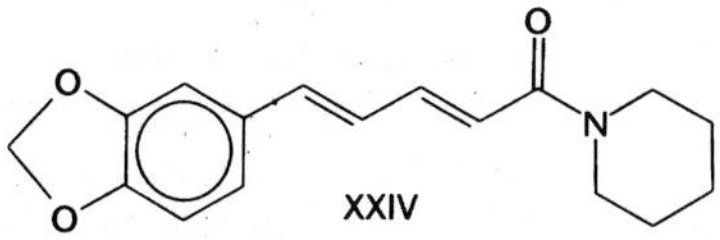 XXIV	Pepper (1)	1.0
Piperanine	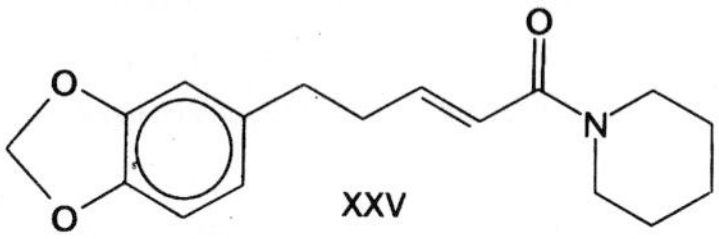 XXV	Pepper (1)	0.5
Piperylin	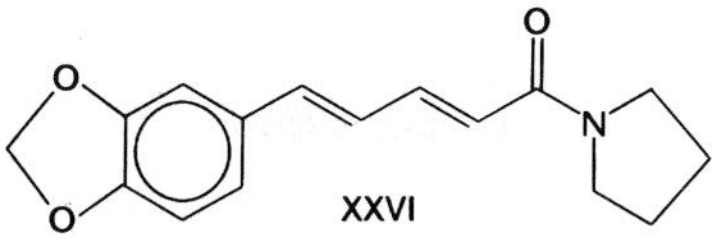 XXVI	Pepper (1)	0–1[d]
Gingerol	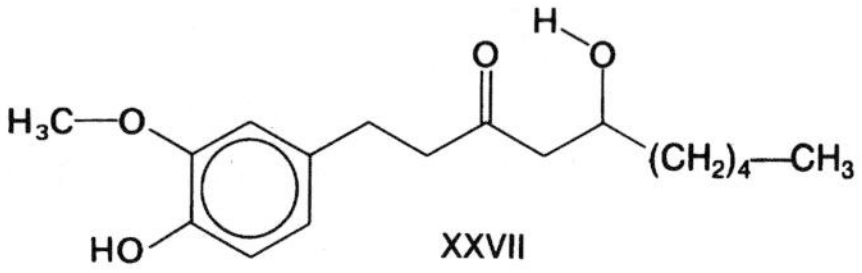 XXVII	Ginger (17)	0.8
Shogaol	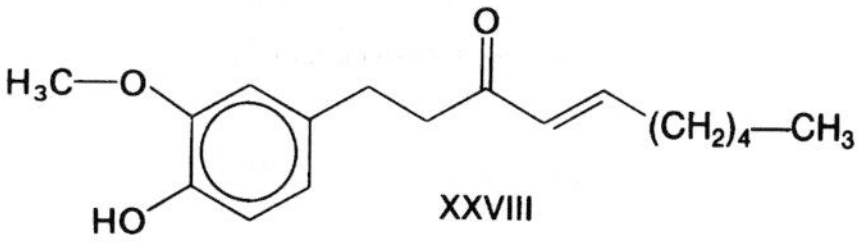 XXVIII	Ginger (17)	1.6
Capsaicin	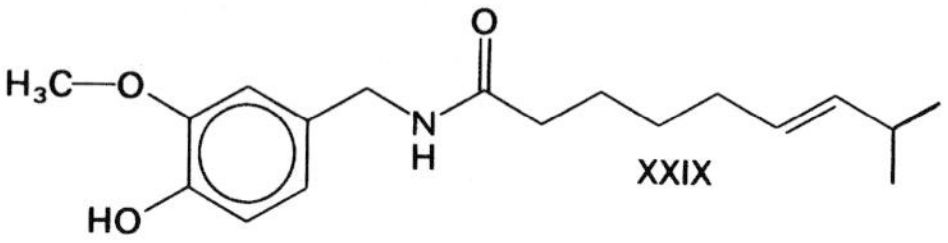 XXIX	Capsicum (4; 7)	150–300[d]
Dihydro-capsaicin	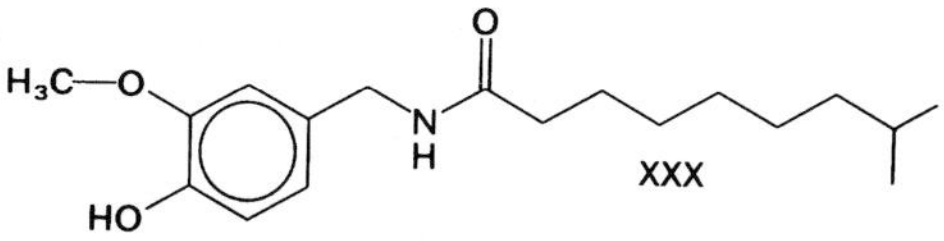 XXX	Capsicum (4; 7)	like Capsaicin
Nordihydro-capsaicin	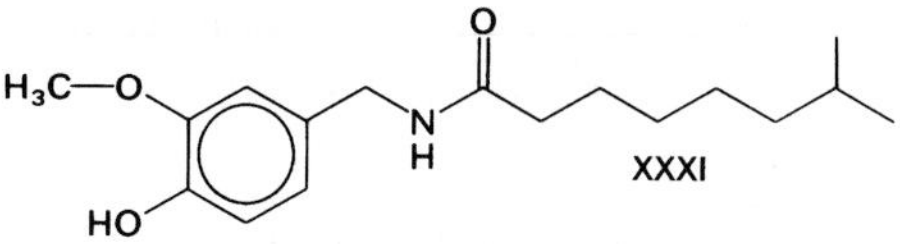 XXXI	Capsicum (4; 7)	75% Capsaicin

[a] The numerals in brackets refer to Table 22.1.
[b] Reference: pungency of piperine = 1.
[c] The corresponding *cis,trans*-compound is devoid of pungent taste.
[d] Literature data are within the range of values presented.

22.1.2 Products

22.1.2.1 Spice Powders

Spices are marketed unground or as coarsely or finely ground powders. The flavor is improved when the spices are ground using a cryogenic mill. After grinding the shelf life of the spices is limited. Favorable storage conditions are the absence of air, a relative humidity less than 60% and a temperature less than 20°C. Crushed spices rapidly lose their aroma and absorb aromas from other sources. Leaf and herb spices are dried before they are crushed. The loss of aroma substances depends on the spice and on the drying conditions. Very high losses are suffered, e.g., by dill (Table 22.6). With regard to aroma preservation, the best results are obtained by freeze drying when the water content is reduced only to 16%. However, gentle drying leads to increased hydrolysis of chlorophylls and to dehydration of the phytol released to phytadienes, e.g., to neophytadiene (7,11,15-trimethyl-3-methylene-1-hexadecene):

Chlorophyll

$$(22.11)$$

Contamination of spice powders with microorganisms is often very high, hence the addition of ground spices to food preparations may accelerate microbial food spoilage.

22.1.2.2 Spice Extracts or Concentrates (Oleoresins)

Spice extracts are being used in increasing amounts in industrial-scale food preparation since they are easier to handle than spice powders and are free of microorganisms. The production of these extracts is outlined in 5.5.1.2. The flavor quality depends on the solvent used and also on the raw material.

22.1.2.3 Blended Spices

Specially blended spices are offered commercially for some food preparations, such as liver sausage which uses a spice blend consisting of sweet marjoram, mace, nutmeg, cardamom, ginger, pepper and a little cinnamon.

Smoked, saveloy sausage spice blend consists of coriander, ginger, mustard kernels, paprika and pepper. Common spices for bread are aniseed, fennel and caraway. Gingerbread spice blend consists of aniseed, clove, coriander, cardamom, allspice and cinnamon.

22.1.2.4 Spice Preparations

Spice preparations are obtained by the addition of spices and blended spices to other substances, such as salt, sugar, glutamate, yeast extract and starch flour.

22.1.2.4.1 Curry Powder

A spice preparation containing a spice blend of turmeric as the main ingredient and paprika, chili, ginger, coriander, cardamom, clove, allspice and cinnamon, mixed together with up to 10% legume meal, starch and glucose, and with less than 5% salt.

22.1.2.4.2 Mustard

A dark yellow paste used as a pungent seasoning for food. It is made from finely ground, often defatted mustard seeds, mixed into a slurry with water, vinegar, salt, oil and some other spices (pepper, clove, coriander, curcuma, ginger, paprika, etc.) and ground repeatedly or refined. During processing, lasting 1–4 h at a temperature not exceeding 60°C, the mustard oil is released from its glucoside, as outlined in 22.1.1.2. "Extra strong" mustard is primarily made from dehulled black mustard seed, while the "medium hot" or "hot" types are made from seeds with hull, using varying proportions of black and white mustards.

22.1.2.4.3 Sambal

A spice preparation from Asia used for seasoning rice dishes. Its base is Sambal oelek, which consists mainly of crushed or pulverized salt-preserved chili.

Table 22.6. Changes in aroma substances in the drying of dill (leaves)

	Fresh	Dried (air)		Freeze dried	
		25 °C/4 h	50 °C/4 h	− 25 °C/59 h	− 25 °C/65 h
Water (w/w %)	90	11	12	16	2
Volatile compounds[a]	326	49	37	188	83
			Volatile compounds[a]		
α-Pinene	5.8	1.2	1.4	3.1	0.6
α-Phellandrene[b]	198.1	13.3	8.1	41.6	14.9
Limonene	10.0	0.7	0.4	2.0	0.7
β-Phellandrene	27.5	2.2	1.1	6.5	1.8
p-Cymene	5.5	1.1	0.4	4.0	0.1
3,9-Epoxy-p-ment-1-ene[b]	39.8	0.5	Traces	8.9	1.4
Myristicin	4.4	0.6	0.3	4.3	1.5
Neophytadiene	1.0	6.3	2.6	38.2	26.0

[a] Values in mg per 100 g of dry weight.
[b] Aroma substances that determine quality (cf. 5.1.2).

22.2 Salt (Cooking Salt)

Common salt occupies a special position among the spices. Salt is used in greater amounts than all other spices to enhance the flavor and taste of food. Also, some foods are preserved when salted with large amounts of NaCl (cf. 0.3.1).
Humans require a certain constant level of intake of sodium and chloride ions to maintain their vital concentrations in plasma and extracellular fluids. The daily requirement is about 5 g of NaCl; an excessive intake is detrimental to health.

22.2.1 Composition

Common (cooking or kitchen) salt is nearly entirely NaCl. Impurities are moisture (up to 3%) and other salts, not exceeding 2.5% (magnesium and calcium chloride; magnesium, calcium and sodium sulfates). Salt also contains trace elements.

22.2.2 Occurrence

Salt is abundant in sea water (2.7–3.7%) and in various landlocked seas (7.9% in the Dead Sea; 15.1% in the Great Salt Lake in Utah) and also in salt springs (Lueneburg, Reichenhall) and, above all, in salt beds formed in various geological periods, e.g., die European Zechstein salt deposits.

22.2.3 Production

In FR Germany salt is mainly mined as rock salt. It is selected, crushed and finely ground. Salt springs are also an important source. Saturated brine is recovered by tapping underground brine springs or by dissolving the salt out of beds with freshwater. For purification, magnesium is first eliminated as the hydroxide with lime milk and then calcium is removed as calcium carbonate with soda. Gypsiferous brine is treated with sodium sulfate containing mother liquor from the evaporation process. Evaporative crystallization occurs in multistage systems at 50–150 °C. The salt is centrifuged and dried. Salt obtained in such a manner is called "boiling" salt.
In warm countries sea water is concentrated in shallow flat basins by the sun, heat and wind until it crystallizes ("solar salt").
The addition of 0.25–2.0% calcium or magnesium carbonate, calcium silicate, or silicic acid improves the flowability. Indeed, 20 ppm of potassium ferrocyanide prevents the formation of lumps in the salt. The latter compound modifies the crystallization process

of NaCl during the evaporation of salt spring water. In the presence of potassium ferrocyanide, the salt builds dendrites, which have strongly reduced volume, density and inclination to agglomerate.

In 1975 the worldwide production of NaCl was 162.2×10^6 t. In 1974 only 5% of the NaCl produced in FR Germany was used for consumption; the remainder, 95%, was used in industry or trade (raw materials, salt for regeneration of ion-exchange resins, etc.).

22.2.4 Special Salt

Iodized salt is produced as a preventive measure against goiter, a disease of the thyroid gland (cf. 17.1.2.9.3). It contains 5 mg/kg of sodium-, potassium- or calcium iodide.

Nitrite salts are used for pickling and dry curing of meat (cf. 12.6.2.4). They consist of common salt and sodium nitrite (0.4–0.5%), with or without additional potassium nitrate.

22.2.5 Salt Substitutes

Some human diseases make it necessary to avoid excessive intake of sodium ions, so attempts have been made to eliminate the use of added salt as a spice or flavoring, without attempting to achieve completely salt-free nutrition. This "low salt" nutrition is actually only related to reduced sodium levels, hence a "low sodium" diet is a more relevant designation.

The compounds listed in Table 22.7 are used as salt substitutes. Their blends are marketed as "diet salts". Peptide hydrochlorides with a salty taste are discussed in Section 1.3.3.

Table 22.7. Substitutes for common salt

Potassium, calcium and magnesium salts of adipic, succinic, glutamic, carbonic, lactic, hydrochloric, tartaric and citric acids;
Monopotassium phosphate, adipic and glutamic acids and potassium sulfate;
Choline salt of acetic, carbonic, lactic, hydrochloric, tartaric and citric acids;
Potassium salt of guanylic and inosinic acids

22.3 Vinegar

Vinegar was known in old Oriental civilizations and was used as a poor man's drink and later as a remedy in ancient Greece and Rome. Vinegar is the most important single flavoring used to provide or enhance the sour, acidic taste of food (cf. 8.12.5).

22.3.1 Production

Vinegar is produced microbiologically from ethanol or by dilution of acetic acid.

$$CH_3CH_2OH + O_2$$
$$\longrightarrow CH_3COOH + H_2O + 494 \text{ kJ} \qquad (22.12)$$

22.3.1.1 Microbiological Production

Acetobacter species are cultivated in aqueous ethanol solution or, to a lesser extent, in wine, fermented apple juice, malt mash or fermented whey. Ethanol, as shown in Fig. 22.1, is dehydrogenated stepwise to acetic acid; the resulting reduced form of the cosubstrate methoxatin ($PQQH_2$) is oxidized via the respiratory chain. Part of the energy formed by oxidation is released as heat which has to be removed by cooling during the processing of vinegar. If there is an insufficient supply of oxygen, the microorganisms disproportionate a proportion of the acetaldehyde, the intermediate compound (cf. Fig. 22.1) in this aerobic reaction pathway:

$$2 CH_3CHO \longrightarrow CH_3COOH + CH_3CH_2OH$$

$$(22.13)$$

Fermentation of ethanol is conducted as a top fermentation and increasingly as a submerged oxidative process. In top fermentation the bacteria are cultivated on spongy, porous laminated carriers (usually beechwood shavings) with the alcoholic solution trickling down over carrier surfaces while a plentiful supply of air is provided from below. The fermentation is stopped at a 0.3% by volume residual ethanol level to avoid overoxidation, i.e., oxidation of acetic acid to CO_2 and water.

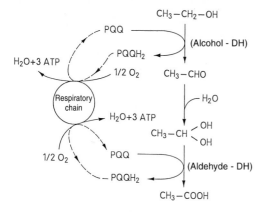

CH₃–CH₂–OH

(Alcohol - DH)

CH₃–CHO

CH₃–CH ⟨ OH / OH

(Aldehyde - DH)

CH₃–COOH

Fig. 22.1. Oxidation of ethanol to acetic acid by *Acetobacter* species (according to *Rehm*, 1980)

22.3.1.2 Chemical Synthesis

Acetic acid is usually synthesized by catalytic oxidation of acetaldehyde:

$$CH_3CHO \; + \; \tfrac{1}{2}O_2 \; \xrightarrow{\;\text{Cat.}\;} \; CH_3COOH$$

$$(22.14)$$

Acetaldehyde is obtained by the catalytic hydration of acetylene or by the catalytic dehydrogenation of ethanol. Formic acid and formaldehyde are by-products of acetic acid synthesis. They are removed by distillation. Chemically pure acetic acid is diluted with water to 60–80% by volume to obtain the vinegar essence. The essence is a strongly corrosive liquid and is sold with special precautions. It is diluted further with water for production of food grade vinegar.

22.3.2 Composition

There are 5–15.5 g acetic acid in 100 g of vinegar. The blending (or adulteration) of fermented vinegar with synthetic acid can be detected by mass spectrometric determination of the $^{13}C/^{12}C$-isotope ratio (cf. 18.4.3); fermented vinegar has 5‰ more ^{13}C isotope than acetic acid synthesized petrochemically. In addition fermented vinegar can be distinguished from synthetic vinegar by analyzing the accompanying compounds. With this method fermented vinegars of different origin can also be distinguished from each other; e.g. spirit vinegar (fermented from aqueous ethanol) from wine, apple, malt and/or whey vinegar. The fermented vinegars contain metabolic by-products of *Acetobacter* strains, such as amino acids, 2,3-butylene glycol and acetyl methyl carbinol, in addition to substances derived from the raw materials used in vinegar production.

22.4 Literature

Boelens, M.H., Richard, H.M.J.: Spieces and condiments I and II. In: Volatile compounds in foods and beverages (Ed.: Maarse, H.), Marcel Dekker, Inc.: New York. 1991

Brockhaus, R., Förster, G.: Essigsäure. In: Ullmanns Encyklopädie der technischen Chemie, 4th edn., Vol. 11, p. 57, Verlag Chemie: Weinheim. 1976

Ebner, H.: Essig. In: Ullmanns Encyklopädie der technischen Chemie, 4th edn., Vol. 11, p. 41, Verlag Chemie: Weinheim. 1976

Ebner, H., Follmann, H.: Acetic acid. In: Biotechnology (Eds.: Rehm, H.-J., Reed, G.), Vol. 3, p. 387, Verlag Chemie: Weinheim. 1983

Gerhardt, U.: Gewürze in der Lebensmittelindustrie. Eigenschaften, Technologien, Verwendung. B. Behr's Verlag: Hamburg. 1990

Gottschalk, G.: Bacterial Metabolism. 2nd edn., Springer-Verlag: Heidelberg. 1985

Herrmann, K.: Übersicht über nichtessentielle Inhaltsstoffe der Gemüsearten. II. Cruciferen (Kohlarten, Radieschen, Rettiche, Speiserüben, Kohlrüben, Meerrettich) sowie Gramineen (Zwiebeln, Porree, Schnittlauch, Knoblauch, Spargel). Z. Lebensm. Unters. Forsch. *165*, 151 (1977)

Huopalahti, R., Kesälahti, E., Linko, R.: Effect of hot air and freeze drying on the volatile compounds of dill (*Anethum graveolens* L.) herb. J. Agric. Sci. Finland *57*, 133 (1985)

Maga, J.A.: Capsicum. Crit. Rev. Food Sci. Nutr. *6*, 177 (1975)

Melchior, H., Kastner, H.: Gewürze. Verlag Paul Parey: Berlin. 1974

Rehm, H.-J.: Industrielle Mikrobiologie. 2nd. edn., Springer-Verlag: Berlin. 1980

Salzer, U.-J.: The analysis of essential oils and extracts (oleoresins) from seasonings – a critical review. Crit. Rev. Food Sci. Nutr. *9*, 345 (1977)

Sampathu, S.R., Shivashankar, S., Lewis, Y.S.: Saffron (*Crocus sativus* Linn.) – Cultivation, processing, chemistry and standardization. Crit. Rev. Food Sci. Nutr. *20*, 123 (1984)

Schmid, E.R., Fogy, I., Schwarz, P.: Beitrag zur Unterscheidung von Gärungsessig und synthetischem Säureessig durch die massenspektrometrische Bestimmung des $^{13}C/^{12}C$-Isotopenverhältnisses. Z. Lebensm. Unters. Forsch. *166*, 89 (1978)

Siewek, F.: Exotische Gewürze. Herkunft, Verwendung, Inhaltsstoffe. Birkhäuser Verlag: Basel. 1990

Wijesekera, R.O.B.: The chemistry and technology of cinnamon. Crit. Rev. Food Sci. Nutr. *10*, 1 (1978)

23 Drinking Water, Mineral and Table Water

23.1 Drinking Water

Drinking water should be clear, cool, colorless and odorless, free from pathogens (low in microorganisms), perfect with regard to taste, cause no materials corrosion, and contain soluble substances only in narrow limits and minerals normally in concentrations of less than 1 g/l. In individual countries, criteria have been defined by law for the quality of drinking water, especially limiting values for microorganisms and contamination. As an example, limiting values stipulated in the German decree on drinking water are presented in Table 23.1.

Drinking water is recovered from springs, groundwater, and surface water. In sparsely populated areas, springs and brooks provide water that can be used without further pretreatment. Frequently, however, the water available does not fulfil the requirements and must be laboriously purified.

In dry areas, drinking water is obtained by desalting brackish or sea water. The usual processes applied are reverse osmosis with the use of semipermeable membranes for slightly saline brackish water and multistage evaporation, mainly as flash evaporation, for sea water.

23.1.1 Treatment

To remove suspended particles, the water is first filtered through gravel and sand layers of different grain size. Humic acids, which may color water yellow to brown, are flocculated with aluminium sulfate. After clarification, the quality of the water is improved still further, if required, by the application of the following processes.

Water should not contain more than 0.2 mg/l of iron, which is present as the bicarbonate, and 0.05 mg/l of manganese (Table 23.1). The iron can be eliminated as iron (III) hydroxide by aeration. In this process, manganese also precipitates as MnO_2 if the pH is higher than 8.5. Biological processes have also been developed for deferrization and demanganization.

Free carbonic acid must be removed because it attacks pipes. The deacidification process applied depends on the hardness of the water and on the concentration of free carbonic acid. The usual process involves aeration and filtration through carbonate rock (e.g., marble or magnesite).

The disinfection of water is mostly achieved by chlorination or ozonation. At a pH of 6–8, the chlorine gas passed into the water forms practically only $HClO$ and ClO^- which, together with the dissolved Cl_2, are expressed as free chlorine. In the case of superchlorination for the killing of very resistant microorganisms, the excess chlorine (> 0.1 mg/l of free chlorine) must be withdrawn with the help of SO_2, Na_2SO_3, $Na_2S_2O_3$ and filtration through calcium sulfite or coal. Disinfection with ozone has the advantage that due to its decomposition into oxygen, no chemicals remain in the water. Interfering odor- and taste-active substances are eliminated by filtration through activated carbon.

Overly high conventrations of nitrate (limiting value in Table 23.1) can be reduced by bacterial denitrification, ion exchange, or reverse osmosis.

The fluoridation of drinking water is discussed in 7.3.2.13.

23.1.2 Hardness

The total water hardness refers to the total concentration of alkaline earths calcium and magnesium in mmol/l. The concentrations of strontium and barium, which are usually very low, are not considered. The following is valid for conversion to German degress of hardness (°d): 1 mmol/l hardness = 5.61 °d. Factors for conversion to the degree of hardness of other countries are given in Table 23.2.

Table 23.1. Chemical and physical analysis of drinking water

Parameter	Limiting value[a]
General values to be measured	
Temperature	25 °C
pH Value	6.5–9.5
Electrical conductivity at 25 °C	2000 µS · cm^{-1}
Oxidizability[b]	5 mg O$_2$/l
Hardness	–[c]
Individual Constituents	mg/l
Sodium	150
Potassium	12
Calcium	–[c]
Magnesium	50
Iron	0.2
Manganese	0.05
Aluminium	0.2
Ammonium	0.5
Silver	0.01
Sulfate	240
Arsenic	0.04
Lead	0.04
Cadmium	0.005
Chromium	0.05
Nickel	0.05
Mercury	0.001
Cyanide	0.05
Fluoride	1.5
Nitrate	50
Nitrite	0.1
Polycyclic aromatic hydrocarbons, calculated as carbon	0.0002
Chlorine-containing solvents, sum of 1,1,1-trichloroethane, trichloroethylene, tetrachloroethylene, dichloromethane	0.025
Carbon tetrachloride	0.003
Pesticides, biphenyls, terphenyls	0.0001[d]
Surfactants	0.2

[a] The limiting values have been taken from the decree on drinking water, Dec. 5, 1990 (BGBL. I. p. 2612)/Jan. 23, 1991 (BGBL. I. p. 277).
[b] Organic substances are detected on the whole by oxidation, e.g., with permanganate.
[c] No limiting value required.
[d] Per individual substance.

The assessment of water involves an evaluation in accordance with the steps of hardness presented in Table 23.3.

Table 23.2. Conversion factors for degrees of hardness

Value	Alkaline earth metal ions (mmol/l)
Hardness[a]	1.00
1 German degree of hardness (°d)	0.18
1 English degree of hardness (°e)	0.14
1 French degree of hardness (°f)	0.10
1 USA degree of hardness (°US)[b]	0.01

[a] Hardness is now expressed as the concentration of the amount of substance (mmol/l). The following correspond: 1 mg/l Ca$^{2\oplus}$ = 0.025 mmol/l; 1 mg/l Mg$^{2\oplus}$ = 0.041 mmol/l.
[b] 1°US = 1 ppm CaCO$_3$.

Table 23.3. Classification in steps of hardness

Step	Range of hardness (mmol/l)	Degree of hardness (°d)	Characteristics
1	<1.3	<7	Soft
2	1.3–2.5	7–14	Medium-hard
3	2.5–3.8	14–21	Hard
4	>3.8	>21	Very hard

On heating, the hydrogen carbonates dissolved in water are converted to carbonates. On boiling, a part of the calcium salts precipitates out as slightly soluble CaCO$_3$. This part of the hardness is called carbonate hardness.

23.1.3 Analysis

The extent and frequency of the analysis of drinking water are regulated by law in many countries. Apart from monitoring the hygienic state of the water resources and of the treated drinking water, maintenance of limiting values is controlled. The data given in Table 23.1 show that extensive analysis of drinking water is a very laborious process.

23.2 Mineral Water

Mineral water comes from a hygienically faultless spring that is protected from conta-

Table 23.4. Classification of mineral water

Description	Requirement
With low mineral content	Solid residue = mineral matter content ≤ 500 mg/l
With very low mineral content	Solid residue ≤ 50 mg/l
With high mineral content	Solid residue > 1500 mg/l
Bicarbonate containing	Hydrogen carbonate > 600 mg/l
Sulfate containing	Sulfate > 200 mg/l
Chloride containing	Chloride > 200 mg/l
Calcium containing	Calcium > 150 mg/l
Magnesium containing	Magnesium > 50 mg/l
Fluoride containing	Fluoride > 1 mg/l
Iron containing	Divalent iron > 1 mg/l
Sodium containing	Sodium > 200 mg/l
Suitable for preparation of infant food	Sodium ≤ 20 mg/l, nitrate ≤ 10 mg/l, nitrite ≤ 0.02 mg/l fluoride ≤ 1.5 mg/l
Suitable for low-sodium nutrition	Sodium < 20 mg/l
"Säuerling"	Carbon dioxide of natural origin > 250 mg/l

mination. It has a nutritional and physiological effect due to its mineral content. In many countries, the recovery and composition of mineral water are controlled by the state and only a few processes for quality improvement are permitted. These are: separation of iron and sulfur compounds, complete or partial removal of free carbonic acid, and addition of carbon dioxide. Mineral water is bottled directly at the place of the spring. With regard to the heavy metal content and possible contamination, limiting values have been stipulated by law. The classification of mineral water is presented in Table 23.4.

In Germany, water used for therapeutic purposes (medicinal waters), because of its chemical composition, is subject to the law governing the manufacture and prescription of drugs.

23.3 Table Water

Table water is made from mineral water, drinking water, and/or sea water by using $NaCl$, $CaCl_2$, Na_2CO_3, $NaHCO_3$, $CaCO_3$, $MgCO_3$, and CO_2. If it contains at least 570 mg/l of $NaHCO_3$ and carbon dioxide, it can be called soda water. Selters is a soda water that comes from Selters on the Lahn.

23.4 Literature

Höll, K.: Wasser, Walter de Gruyter, Berlin, 1979
Quentin, K.-E.: Trinkwasser. Springer-Verlag: Berlin. 1988

Subject Index

Page number in italics: formula
Page number (F): figure
Page number (T): table

Springer
and the
environment

At Springer we firmly believe that an
international science publisher has a
special obligation to the environment,
and our corporate policies consistently
reflect this conviction.
We also expect our business partners –
paper mills, printers, packaging
manufacturers, etc. – to commit
themselves to using materials and
production processes that do not harm
the environment. The paper in this
book is made from low- or no-chlorine
pulp and is acid free, in conformance
with international standards for paper
permanency.

 Springer

Printing: Saladruck, Berlin
Binding: Buchbinderei Saladruck, Berlin